Product Design

Techniques in Reverse Engineering and New Product Development

Kevin N. Otto

Massachusetts Institute of Technology

Kristin L. Wood

University of Texas at Austin

Prentice Hall
Upper Saddle River, NJ 07458

Library of Congress Cataloging-in-Publication Data

```
Otto, Kevin N.
      Product design/ Otto, Kevin N. and Wood, Kristin L.
        p. cm.
      Includes bibliographical references and index.
      ISBN 0-13-021271-7
      1. Design, Industrial. 2. New products. 3. Production Management
      I. Wood, Kristin L. II. Title
```

Acquisitions Editor: Laura Curless
Vice-President and Director of Production and Manufacturing, ESM: David W. Riccardi
Vice-President and Editorial Director of ECS: Marcia Horton
Executive Managing Editor: Vince O'Brien
Managing Editor: David A. George
Art Management: Xiahong Zhu
Manufacturing Buyer: Pat Brown
Marketing Manager: Danny Hoyt

©2001 Prentice Hall
Prentice-Hall, Inc.
Upper Saddle River, New Jersey 07458

Prentice Hall books are widely used by corporations and government agencies for training, marketing, and resale. The publisher offers discounts on this book when ordered in bulk quantities.
For more information, contact Corporate Sales Department, Phone: 800-382-3419;
Fax: 201-236-7141; E-mail: corpsales@prenhall.com
Or write: Prentice Hall, Corp. Sales Dept., One Lake Street, Upper Saddle River, NJ 07458.

ISBN # 0-13-021271-7

Printed in the United States of America
10 9 8 7 6 5 4 3 2

Prentice-Hall International (UK) Limited, London
Prentice-Hall of Australia Pty. Limited, Sydney
Prentice-Hall Canada Inc., Toronto
Prentice-Hall Hispanoamericana, S.A., Mexico
Prentice-Hall of India Private Limited, New Delhi
Prentice-Hall of Japan, Inc., Tokyo
Prentice-Hall Asia Pte., Singapore
Editora Prentice-Hall do Brasil, Ltda., Rio de Janeiro

Dedicated to Ginger—my wonderful wife, partner, companion and friend, whom I dearly love.

Kevin

To those that bring light to my life and spirit: above all, the Lord Jesus Christ; my loving wife Laurie; my precious children Katie, Emily, and Zachariah; their grandparents, Sharon, Del, Bob, and Judith; and in special memory of my grandparents, Bert and Lee Armstrong , who first kindled my inquisitive nature.

Kristin

Table of Contents

Chapter 3: Scoping Product Developments: Technical and Business Concerns 83

Chapter 4: Understanding Customer Needs 111

Chapter 5: Establishing Product Function 147

Chapter 6: Product Teardown and Experimentation 197

Chapter 7: Benchmarking and Establishing Engineering Specifications 259

Chapter 8: Product Portfolios and Portfolio Architecture — 303

Chapter 9: Product Architecture — 357

Chapter written in collaboration with Robert B. Stone, University of Missouri

Chapter 10: Generating Concepts   411

Chapter 11: Concept Selection 477

Chapter 12: Concept Embodiment   535

Chapter 13: Modeling of Product Metrics   603

Chapter 14: Design for Manufacture and Assembly

663

Chapter 15: Design for the Environment

719

Chapter 16: Analytical and Numerical Model Solutions 781

17 Physical Prototypes 833

Chapter 18: Physical Models and Experimentation   891

19 Design For Robustness 979

Appendices 891

Foreword

Many American companies have come to realize that the ability to consistently define and deliver products to the marketplace more rapidly and efficiently than their competitors can become a source of sustainable competitive advantage. These companies have moved to view their product development capability as an end-to-end business process that can be greatly enhanced through re-engineering and can be continuously improved through total quality management. In this environment, the practice of engineering is critical to the success of the process and is as important as the implementation of the engineering sciences. K. N. Otto and K. L. Wood have presented an approach to product design based on "state of the art" engineering methods, tools, and processes that when utilized within the discipline of an overall product development process, depicts a very powerful practice of design. Both students and practitioners will find this approach not only effective and practical, but also immediately applicable to today's design challenges.

Maurice F. Holmes
Vice President, Chief Engineer of Product Design
Xerox Corporation

Preface

Product Design presents an in-depth study of structured design processes and methods. In general, we have found that the exercise of a structured design process has many benefits in education and industry. On the industrial side, a structured design process is mandatory to effectively decide what projects to bring to market, schedule this development pipeline in a changing uncertain world, and effectively create robust delightful products. On the educational side, the benefits of using structured design methods include concrete experiences with hands-on products, applications of contemporary technologies, realistic and fruitful applications of applied mathematics and scientific principles, studies of systematic experimentation, exploration of the boundaries of design methodology, and decision making for real product development. These results have proven true whether at the sophomore introductory level with students of limited practice, or at the advanced graduate student level with students having years of practical design experience.

Based on these observations, this book is intended for undergraduate, graduate, and practicing engineers. Chapter 1 of the book discusses the foundation material of product design, including our philosophy for learning and implementing product design methods. Each subsequent chapter then includes both basic and advanced techniques for particular phases of product development. Depending on the background of the reader, these methods may be understood at a rudimentary level or at a level that pushes the current frontiers of product design.

Historically, this work grew out of a partnership effort between the authors, while we were both teaching product development courses and carrying out research in mechanical design. We both share similar philosophies on design, teaching, and research. Having each developed new methods in design, we were interested in transferring these and others' methods into practice. We also strongly wanted to bring the excitement of the real world, both in physics and the marketplace, to the design classroom.

A fundamental premise of our teaching approach is that reverse engineering and teardowns offer a better paradigm for design instruction, permitting a modern learning cycle of experience, hypothesis, understanding, and then execution. Design instruction is no different than other domains; to learn design one should both follow this learning cycle and DO design. Reverse engineering and teardowns permit us to achieve this combined goal. We begin with a concrete product in our hands, seeing how others have designed products well, rather than rushing straight to the execution stage. With this in mind, we both independently set out to teach and successfully apply advanced methods, such as customer needs analysis, functional modeling, optimization, and designed experiments on real products.

We quickly started sharing experiences, what worked and what did not, and progressively began to string together a series of techniques and that fit naturally together. When one of us had a success, we would brag to the other, or when something failed, we'd lament together. After a bit of systematic testing, we developed the methodology presented in this book, which has proved remarkably robust when applied.

We would like to extend our special thanks to the many persons who directly contributed to this book. These include John Baker, Joseph Beaman, Geoffrey Boothroyd, Ilene Busch-Vishniac, Jim Claypool, Richard Crawford, David Cutherell, Michael Fang, Conger Gable, Javier Gonzales-Zugasti, Matthew Haggerty, Nicholas Hirschi, Maurice Holmes, Jerry Jackson, Jerry Jones, Jennie Kwo, Doug Lefever, Aaron Little, Michael Manente, Robert Matulka, Dan McAdams, David Meeker, Jon Miller, Steve Moore, Jeff Norrell, Caroline Pan, Erick Rios, David Roggenkamp, JoRuetta Roberson, Phil Schmidt, Stephen Shuler, R. S. Srinivasan, Robert Stone, Carlos Tapia, David Wallace, Joe Wysocki, Janet Yu, and Erik Zamirowski. Without their intellectual help, this book would not be.

Many others have sparked our thoughts and inspired us in many ways. These persons include Erik Antonsson, Wolfgang Beitz, Joe Bezdek, Bert Bras, Jonathon Cagan, Uichung Cho, Chin-Seng Chu, Don Clausing, Jim Coles, Ray Corvair, Michael Cusumano, Jack Dixon, John Elder, Steven Eppinger, Rolf Faste, Woodie Flowers, Mark Foohey, Chee-Seng Foong, Douglas Hart, John Hauser, Chester Hearn, Alberto Hernandez, Steve Hoover, Kos Ishii, Gerry Johnson, Nathan Kane, Paul Koeneman, Sridhar Kota, Bill Maddox, Spencer Magleby, David Masser, Ryan Ratliff, David Rosen, Bernard Roth, Warren Seering, Jami Shah, Sheri Sheppard, Alexander Slocum,

George Stiny, David Thompson, Irem Tumer, David Ullman, Bill Weldon, Daniel Whitney, Joseph Wieck, Doug Wilde, and Rick Zayed.

We would like to thank the many persons, companies, and organizations that contributed case studies, important data, and funding that make the examples real world. These include A.T.&T. Corp., W. E. Bassett Co., Design Edge Inc., Desktop Manufacturing Co., Digital Equipment Corporation, Eastman Kodak Co., Ford Motor Co., MIT Bernard Gordon Curriculum Development Fund, June and Gene Gillis, General Electric Inc., International Business Machines Corp., Keurig Inc., Microsoft Corporation, NASA Jet Propulsion Laboratory, National Science Foundation, Robert Noyce, Pré Associates, Product Genesis Inc., Polaroid Corporation, Raychem Corp., Raytheon Corp., Texas Instruments Inc., Verein Deutches Ingineur, and the Xerox Corp.

We would especially like to thank MIT's Bernard Gordon Curriculum Development Fund and to the NSF Center for Innovation in Product Development at MIT, which provided necessary funds to make this book possible. More importantly, the supportive, dynamic and perceptive environment of academic faculty, students, staff and industrial researchers at MIT's Center for Innovation in Product Development cannot be understated, they have made many insights possible. Warren Seering in particular is a great help; he cannot be sufficiently thanked for his vision, insight, advice, and outright help in working in product development.

We would also like to thank the colleagues who reviewed early drafts of the book and provided constructive criticisms. A special group of early reviewers are the faculty of the United States Air Force Academy, Engineering Mechanics Department, including Col. Cary Fisher, Dr. Dan Jensen, Maj. John Wood, Capt. Michael Murphy, and Maj. Mark Nowak. We appreciate their assistance in implementing the material in their courses during Dr. Wood's sabbatical. They truly tested, twisted, shaped, and criticized the material at the most fundamental of levels.

Many others have contributed to the organization and form of the book. In particular, the authors wish to thank Neal Blumhagen, who created the cover artwork and a number of hand drawings in the text. Ann Weeks, artist, Erik Zumalt, digital artist, Michael Young, media coordinator, and Sicily Dickenson, director of the UT Instructional Media Lab, contributed wonderfully to the numerous illustrations and photographs in the book. Finally, Laurie Wood contributed her creativity to a number of the illustrations.

Kevin Otto
Kristin Wood

1 Journeys in Product Development

The **Xerox Corporation's** document systems have multiple systems using hundreds of parts. The design teams number in the dozens, and a project takes years.

The **Microsoft Corporation's** software products have dozens of feature elements, design teams number in the dozen of members, and a project takes months.

The **Raychem Corporation's** products have dozens of constitutive raw materials, research and development teams number in the dozens of members, and a project can take decades.

The **Ford Motor Company's** cars and truck have 20 systems, 166 subsystems, numerous sub-subsystems and thousands of components. There are hundreds of orchestrated design teams that number in the hundreds and a vehicle project takes years.

The design of new products is the key battleground that all companies must master to remain in business—to compete at a basic level. Product designers are the front-line troops who lead and execute the battle. Product design is a set of activities that involves more than engineering. It is fraught with risks and opportunities, and it requires effective judgment over technology, the market, and time. Studying some recent business decisions gives insight to these ideas:

- To avoid losing market share in the 1990s, commercial airplane manufacturers offered contracts to deliver aircraft at prices that were below the current development and production cost (*Wall Street Journal*, 24 April 1995). The companies were betting that they could remain profitable through improvement of their products and development processes.

- In 1985, John Akers, CEO of IBM, was considering eliminating IBM's research laboratory, a popular decision for American CEOs at the time. Instead, the lab director proposed a vision of working to develop a 1 Gbit per square inch storage medium, a vision supposedly proven physically impossible. The lab restructured around this vision and, by 1989, had developed breakthrough technology in thin-film inductive heads. The lab stayed focused on developing ever-increasing storage media, and, into the next decade, IBM maintained a 6–12-month technology lead over its competitors (Tristram 1998).

- In the 1990s, Dell Computer Corporation developed the end-to-end use of digital networks to run their corporation. With information-system intensive marketing, development, and production, Dell made it possible to completely eliminate middlemen in a low-margin industry. They connected directly to the customers via their purchasing Web site, through phone orders, and through its program of offering customized products and services. This "direct customer contact" philosophy enabled them to watch how their customers purchased and rapidly focused on desired features in new product designs. By 1999, such investment in their product delivery process made Dell the #1 computer supplier (*Wall Street Journal*, 10 May 1999).

- In 1996, both Ford and Toyota launched new family sedans. Three years earlier, each had criticized the other's model. Ford decided to increase the options in its Taurus, matching Toyota's earlier Camry, while Toyota decided to decrease the options in its Camry, matching Ford's earlier Taurus (*Business Week*, 24 July 1995). Neither really knew where the market was going or what to do.

To avoid pitfalls, today's design engineers must understand and use many tools of modern product development practice. For almost all products, it is no longer acceptable to develop major enhancements without first consulting customers to forecast the market acceptance of the improvements: The risk is too high to just accept one product manager's belief in their "feel" for the market. Rather, a team must apply statistically sound measurement methods of a product's intended customer population. It is equally important to functionally architect what is required to meet the customer demands, applying rigorous methods for incorporating the best technologies. It is critically important to understand the competition and the time trends of new technology introduction into the market. It is also important to design robust performance into a product, so that the product is as high quality as possible given the price. These goals and associated methods have become the competitive weapons that allow design teams to ensure that their company leads the market. Engineers who wish to harness their drive for success need to understand product development and the application of product development tools and methods. In this book, we develop the underlying theory, methods, and practices to extract an understanding of modern product development.

I. CHAPTER ROADMAP

In this chapter, we first present a general discussion on product development and design. We then offer our view on the advancement of systematic design and a hypothesis on the evolution of design as a practice. We then examine the product development practices of a few representative companies and compare this examination with our philosophies. Finally, we discuss different theories of design that have been developed over time. Figure 1.1 shows the roadmap through the topics of this chapter.

II. AN INTRODUCTION TO PRODUCT DESIGN

By using this book, the reader will learn and practice contemporary theories of effective product design through the adaptive and/or original redesign of consumer products. Engineering design can be challenging and exciting, or it can be taxing, difficult, and unproductive. The objective here is to outline a methodology that highlights the

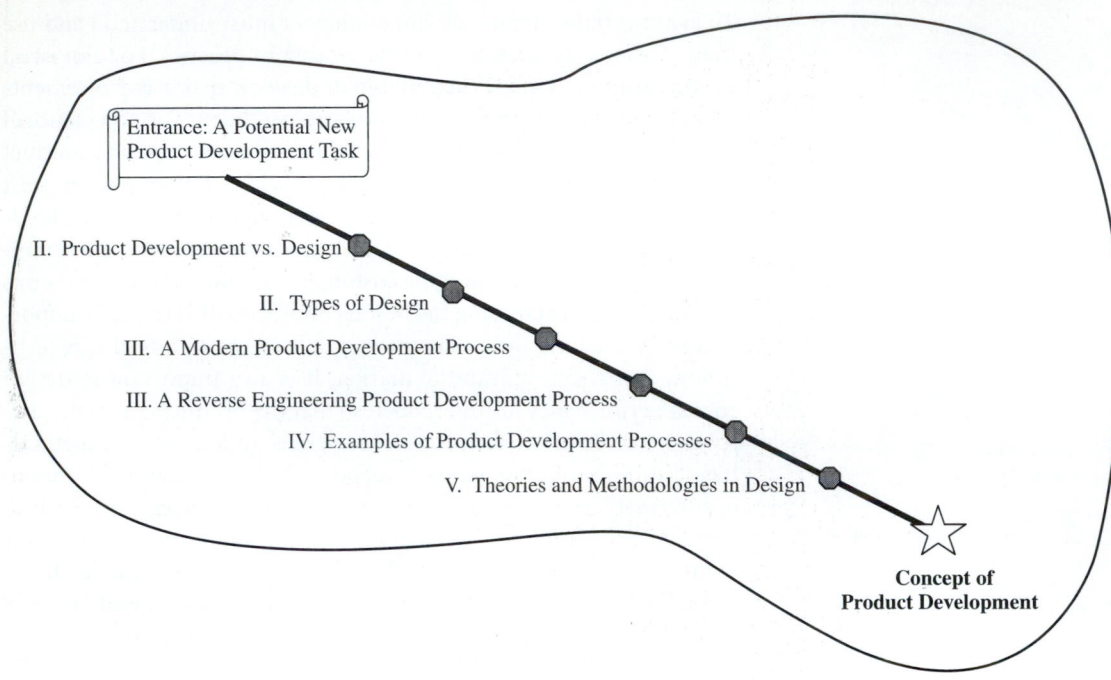

▼ Figure 1.1.
Roadmap of Chapter 1.

exciting challenges of product design and allows the development team to focus on the development of a creative, effective, and profitable solution.

Thoughts for the Reader and Student of Product Design

The particular methodologies explored in this book are selected because they have the potential to satisfy three objectives of engineering design:

▶ to stimulate creative and inventive solutions to problems
▶ to ensure consideration of each of the elements necessary for successful design
▶ to ensure that all consequences of the application of the designed device or process throughout its lifetime are examined

The assignments distributed in design courses are quite different than the relatively well-defined analytical problems delivered in other engineering studies. Because there is a lot less information to start with, we are called on to use our own judgment and make many, many assumptions. Most importantly, there will never be just one right answer.

Assignments of this sort, those that may not be clearly stated and are without a well-defined path to a solution, are, however, much more like those encountered in the "real world" of industry. They are distinct and more challenging than the typical end-of-chapter homework problems. Although the engineering design methodologies presented in this book are, of necessity, more formalized and detailed, there are parallels between them and the problem solution methodologies used in other endeavors. They build on, extend, and strengthen many of the procedures we use in analysis work and courses. Yet, they allow us to move from analysis to the more-complex and much-less-bounded problem of synthesis—design.

The problems encountered in product design are also amenable to many solutions. There will always be more than one way to proceed, and there will certainly never be just one "correct" solution or design. We also can't assume that we have the best answer just because we have *an* answer that appears feasible.

The problem statements we encounter require a good deal more time and effort to understand. We will most likely have to resolve conflicts between the stated functional requirements and external constraints in each of the assignments we are given. We must plan on spending a good deal more time setting up the problem than on comparable analysis tasks. We must also remember that there is a clear progression of material throughout the book. It is cumulative, so we must keep our eyes open!

Product Development Versus Design

A *product development process* is the entire set of activities required to bring a new concept to a state of market readiness. This set includes everything from the initial inspiring new product vision, to business case analysis activities, marketing efforts, technical engineering design activities, development of manufacturing plans, and the validation of the product design to conform to these plans. Often it even includes development of the distribution channels for strategically marketing and introducing the new product.

A *design process* is the set of *technical* activities within a product development process that work to meet the marketing and business case vision. This set includes refinement of the product vision into technical specifications, new concept development, and embodiment engineering of the new product. It does not necessarily include all of the business and financial management activities of product development nor the extensive marketing and distribution development activities.

Neither the product development process nor the design process encompasses the subsequent *manufacturing process,* when the products are physically made. The design of the manufacturing process, however, is generally considered part of the product development process. Often the product design process and the design of its manufacturing system must be carried out simultaneously. Effectively performing this integration is part of the study of *concurrent engineering* (Clausing 1997).

Similar to the manufacturing process that follows the product development process, there are also front-end activities that are required before product design and, based on industry, may or may not be considered part of the product development process. The *Research and Development* (R&D) phase of new product development is when new technology is developed for subsequent incorporation into products. Today, large companies in many industries try to separate the R&D process from the product development process. That is, ideally, R&D efforts create new technology and develop it to the point where the technology is encapsulated into a new system ready for immediate adoption by the product development teams, similar to out-sourced subsystems. Product development then becomes a very rapid process of tailoring technologies into new systems that meet changing market needs. Product development does not get bogged down into researching how to obtain new technology that does not really perform what is required. This paradigm will not work in all industries, but it is effective for many mechanical system industries.

Generally, such R&D to product development transfer is not perfect; the technology passed off to the product development teams may not function well in the new product concept. This mismatch can occur for a variety of reasons. Some causes are social, such as the different cultures between R&D corporate research and product development business units. Other causes include general uncertainty, such as the ability of new technology to be used in ways not foreseen by the R&D group during their development efforts. Often, though, miscommunication of specifications and form is a prime source of error; people understand the same words and descriptions differently.

The entire set of preliminary product development activities that happen before a product is given the go-ahead for development is sometimes called the *fuzzy front-end*. This set includes the decisions on what products to consider for development. Factors include corporate strategy or the determination of what technologies and markets that a company should compete in. Business alliances can therefore impact the decisions as well as forecasted customer markets and business trends. The fuzzy front-end also includes development decisions on what the underlying portfolio architecture should be for a set of products that may be offered by a company. Design engineers play a large contributing role on the teams making these front-end decisions.

The point of this discussion is not to find a classification scheme for development activities that fits all companies and development enterprises; such a classification will always vary by industry and company. Rather, one should be aware of how a design process fits within a product development process of any company and how a product development process fits within a company's business environment.

Types of Design and Redesign

Design tasks may be classified in several different ways. To indicate the extent of the effort required, one approach is to classify a development project as *original design, adaptive design,* or *variant design.* Table 1.1 lists examples of these different types of design for various technologies.

Original design (or *inventing*) involves elaborating original (new/novel) solutions for a given task. The result of original design is

TABLE 1.1.   EXAMPLES OF TYPES OF DESIGN

	Original design	Adaptive design	Variant design
Brakes	Anti-lock brakes	Brake system for a new vehicle	Resized system for a slightly changed vehicle; varying disk diameter or friction material
Steering	Assistive technology for people with disabilities (joystick, foot control)	New column including secondary functions	Resized steering column for resized chassis
Entertainment	CD media	DVD media	Laser disk media
Computing	Microprocessor	Intel 8080 chip	Pentium to Pentium II refinement in chip
Vehicles	Benz's first automobile	Unibody construction (vs. body on frame)	Any new year's model compared to older year
Bearings	Da Vinci's Self-Centering Ball Bearing	First Teflon-coated plain bearings	Different sizes in a family of related bearings

an *invention.* The invention of the transistor, the laser xerography process, and the windowed computer system complete with pointing mouse—these were all original designs. Few successful original designs occur over time, and when they do, they can disrupt the market. Consumers see the new technology, want it (either immediately or eventually), and have to replace not only the old equipment, but also the infrastructure around the old equipment. Automobiles require paved roads, gasoline, repair stations, and insurance; they do not require hay and barns. Computers on the Internet require a microprocessor and memory, a network connection, and a Web browser; they do not require operating systems or application software. Original inventions are often high-risk opportunities for changing a marketplace and then dominating it. Few companies have industrial or engineering designers that manage or are permitted to invent an entirely novel design; even fewer companies exist that can steer such an invention into market domination.

Adaptive design (or *synthesis*) involves adapting a known system to a changed task or evolving a significant subsystem of a current product (such as antilock brakes). Adaptive designs can be very novel, but they do not require a massive restructuring of the system within which the product operates. This type of design dominates the vast majority of design activities. This domination is not due to a mental laziness on the designers' part but is simply a reflection of the demands of the marketplace. Customers generally want new products that fit in their current life-style. Within the boundaries of this life-style, they evolve their system of using the product. Meeting these evolving needs and boundaries can be very profitable with reasonable risk.

Variant design (or *modification*) involves varying the parameters (size, geometry, material properties, control parameters, etc.) of certain aspects of a product to develop a new and more robust design. This type of design usually focuses on modifying the performance of a subsystem without changing its configuration. It is also implemented when creating scaled product variants for a product line. For example, a plain bearing resized to a larger load rating will require a greater surface area, and so a new larger bearing must be developed with a variant design process. Likewise, a standard restaurant-sized food processor may be reduced in power parameters, size of serving bowl, and cutting tool dimensions to create a variant for use by small households.

Redesign is a term that we use to mean *any* one of the above. Redesign does not mean variant design. Rather, redesign only implies that a product already exists that is perceived to fall short in some criteria, and a new solution is needed. The new solution can be developed through any of the above approaches. In fact, it is often difficult to argue against the maxim that "all design is redesign."

What is Engineering Design?

Another way to classify design processes is to consider the discipline within which the process operates. There are many disciplines of design, including mechanical engineering design, electrical engineering design, architectural design, industrial design, food science design, furniture design, materials design, aerospace design, and bridge/roadway design. Name any product, and there will be a design process used in its development. For this disciplinary classification, the question arises over which disciplines would benefit from systematic techniques for product development (the subject of this book). At a high level, they all would benefit, but only certain disciplines will realize the full potential of some of the methods. One approach for investigating this issue further is to explore the modeling detail used in each of the disciplines.

This book applies a model-centric and systems approach to design; it is appropriate for design processes that require and benefit from extensive modeling. A systematic effort, and the contents of this book, is devoted to exploring the activities of a design process: isolating each activity, understanding what is required as input, what is produced as output, and then establishing methods to repeatedly complete the activity. This approach is very appropriate for engineering-design activities, especially in the real-world corporate design environment. It permits many engineers to work in concert and to hand off their results to the team members working on the next or parallel activity. A model-centric approach minimizes the disruptions, explanations, and fixes (engineering changes) of this large development process. Short of removing disruptions, it also permits understanding what must be communicated among activities.

The tasks to be completed for each activity in engineering design are complicated. They require assurance of complex criteria, such as material failure due to stresses, vibrations, thermal characteristics, and dynamic analysis. For example, consider the design of an automotive vehicle. Dynamic analysis of vibration loads is required to distribute mass over the frame, select frame elements, and design joints. These considerations are critical so that the vehicle frame transmits minimal noise to the occupants, is structurally sound, absorbs impacts during a crash, and minimizes cost. Such analysis may be very complex and resource intensive, yet the success and quality of a contemporary product cannot be assured without this analysis of mechanical effects. If mechanical performance issues were not resolved with suitable modeling and analysis, many years of both failures *and* successes would be experienced without any guarantee of future performance. This trial-and-error, or Edisonian, approach would not be acceptable or lucrative. Rather, models of mechanical effects are simply required as part of mechanical product design, and, thus, the methods explored in this book are both applicable and exciting.

On the other hand, as a counterexample, consider the design of furniture. Here, failure analysis is not difficult to visualize or estimate; the craftsman easily understands required material thickness and joint construction without completing rigorous calculations or repeated experimental trials. To support a load, skilled furniture builders empirically understand what material thickness dimensions will fail and what dimensions will work just fine, and it does not take many experiences to develop this understanding. Within this intuitive context, the furniture builders must also design for appearance, fit, and feel; these issues are the crux of furniture design. This "artistic" focus is typically executed with very few customer interviews, no functional models, and no equations of performance. The advanced analytic and systematic methods in this book would be difficult to apply in detail for such furniture design tasks, although the underlying design principles remain the same.

The difference between these two design processes, mechanical systems and furniture concepts, is the level of engineering required to complete the design task. The vehicle design is an engineered product; it requires modeling to attempt the design task, certainly in any reasonably quick, accurate, and cost-effective fashion. *Engineering design* is thus a process that requires modeling to complete the design task. The furniture design problem, on the other hand, is one of *craftsmanship.* Here, one need only estimate effects intuitively, create a mental image or drawing of the concept, and then build the piece. Applying the craftsmanship approach to engineering design tasks is often attempted, but it rarely produces reliable or quality devices for the consumer.

Given this contrast, what constitutes an effective approach when we are confronted with a new design task? We explore this question further in this chapter. To do this, we present a modern product development process for engineered products. Then we show and briefly explain how this approach is applied, or has been applied, at various companies. This discussion provides necessary perspective on differences in product development processes. It also demonstrates the essential qualities that may be derived from a well-developed approach to product design. Satisfying and safe products are not created accidentally.

Context: The Light Side

Product design is much more than a blind search for a successful creation. It is founded on principles we must learn and repeat.

A story (author unknown) provides a further refinement of this idea. The story begins with a company that has designed a machine to fill a market void. The machine has been designed as two primary modules (1 and 2) using a well-orchestrated team split across the two modules. After prototyping the modules, they are tested individually. The tests are resounding successes, and a fully integrated test is implemented. During the testing, the overall product performance is not achieved at all. The modules simply interact to produce erroneous results. After many attempts at debugging, evaluation, and improvement, no advances are made in the product performance. The design team surrenders and becomes distraught as product milestones loom ever so close.

The team decides to hire an outside consultant to review the product and suggest possible failure modes. A well-known product designer is hired, reviews each module with the design team during a single one-day visit, and tests the integrated product. At the end of the day, after some contemplation, the consultant calls together the team, takes a piece of chalk, and marks an "X" on the module interface of Module 1, explaining the failure mode that is suspected. The consultant leaves, and sure enough, the team explores the "X'ed" feature, and corrects the fault. The product now works as expected.

A week later the team receives a bill for the consultant's services. The bill simply reads "Invoice for Consultation, Fee $100,000." Before paying the bill, the team asks the consultant to itemize the bill. A follow-up letter is received with the following itemization: "$1 for piece of chalk to mark "X", $99,999 for knowing where to put it. . ."

The moral of this story is that experience and intuitive insight/estimation are key characteristics of product designers, but also that clear, logical thinking permits insight. We must experience design to be designers. However, our ability to obtain experience is just a "blink of an eye" in the scheme of all possible design tasks and problems we may encounter. It should have been obvious to the separate design teams that interfacing was the problem—both teams had made false assumptions about the other. With structured design methods, this fault would have been clearly recognized and either avoided or overcome. We thus need methods and techniques to help direct our efforts when particular experience cannot be obtained or referenced. We begin to explore such methods in the next section.

III. MODERN PRODUCT DEVELOPMENT

In the subsequent chapters of this book, we present an integrated set of structured methods developed in conjunction with a host of industrial partners. Our approach focuses on the design of products and their constituent assemblies. The process begins with a design task and generates a functional model that culminates in a product specification. Later subprocesses build on the functional model and specification to execute the product development process.

This process is one that we have developed and to which we organize all the material of this book. Each chapter basically takes each of these steps, fully explains them, and then presents current state-of-the-art methods to accomplish the task. Therefore, to effectively explore this book, one should buy into the product development process outlined here or a facsimile of it. Through this exploration, the reader's personal design process will begin to mature and flower. It will be based on conscious forethought and introspection instead of the daunting limits of trial-and-error experience.

The remainder of this chapter initiates the journey into product development. First, to provide an industrial context for the engineering student, a typical *stage-gate* engineered product development process is explained as a comprehensive set of activities. Then, processes from certain companies clarify the process steps and illustrate their application by design teams in industry. Then we introduce companies from other industries with product development processes that are arguably much different and discuss how they are different and how they relate to the one discussed in this book.

Every company has a different development process out of necessity; there is no single "best" development process; *the design process* and *the product development process* are misnomers. The sophistication of the product, the competitive environment, the rate of change of technology, the rate of change of the system within which the product is used: These and many other factors that shape a product development process change for different companies. This change leads to different levels of speed, analysis, and sophistication required for the different tasks of product development as discussed in this book.

Xerox Corporation, producer of office document systems, and the Design Edge Company, a small design-consulting firm, apply a product development process much as developed here. On the other hand, Raychem Corporation, producer of advanced materials, has a longer research-intensive development process. Ford Motor Company applies the methods of this book, but does so hierarchically to the large system development required of a vehicle. We will also look at the very rapid development process of Microsoft Corporation, and the technologically intensive development of Hughes Electronics Corporation.

Some may feel that particular industries are special, and therefore the methods espoused in this book do not really apply to their engineered products. We will show how this attitude is typically presumptuous and false by discussing the varied companies previously mentioned. Our main source of information is our personal experience and observations while working with their products and engineers, working on many product development research interactions, and/or during on-site visits. Characterizing any company as following any particular process is, of course, a dubious task, since design processes always change: Lessons are learned, new processes for various tasks are developed, new design process technology is introduced, and the goals of a company change. Nevertheless, each scenario will paint an industrial snapshot in time, one to consider, ponder, and debate.

Students of engineering design should take away the point that every corporate circumstance is different, and so every company's engineered product development process will be a different sequence and level of application of the methods presented in this book. Further, each product development process changes over time in response to technological and market forces. Every engineer is always obligated to find an appropriate process and to always improve on it.

A Modern Product Development Process

Product development is a process: There are tasks of creating, tasks of understanding, tasks of communication, tasks of testing, and tasks of persuasion. At its highest level, we characterize any product development process with three phases: *understand the opportunity, develop a*

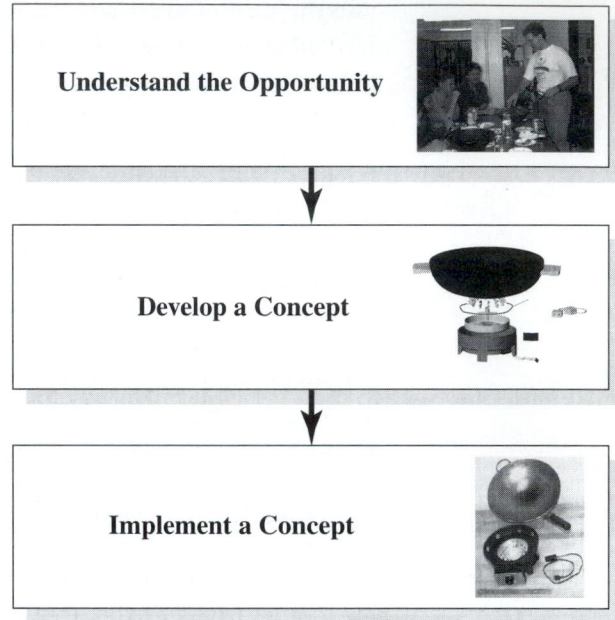

▼ *Figure 1.2.*
Phases in a product development process.

concept, and *implement a concept.* This characterization is outlined in Figure 1.2.

The first phase encompasses all activities needed to make the decision to launch a new product development effort. The second phase encompasses all activities to make the decision on what the product will be. The final phase encompasses all activities to make every product work well all the time. After the final product development phase, the product is ready to be manufactured. In reality, these phases overlap and are complex; yet, they help us to categorize the efforts needed to develop a product, and like all categorizations, there are counterexamples, and the boundaries are fuzzy.

The Stage-Gate Development Process

A product development process can be thought of as a sequence of parallel and serial *activities* or steps to be completed. Within any phase, there are concurrent development activities that occur. Mechanical design proceeds in parallel with electrical design in parallel with software code development. To ensure compatibility of these activities, many companies force the periodic "assembling" of the product as it stands at a given point in time, along with its associated forecasted systems that remain uncertain (such as production, distribution, etc.). This assembly process is executed to obtain a better picture of the design as it is evolving, to evaluate that preliminary system, and, most importantly, to *freeze* parts of the design. Thus, some development decisions are made final at this point, such as general layout, user operation/interface, control, suppliers, etc.

This development process is known as a *stage-gate* process, or *waterfall* development process, where there are *stages*, or phases, or activities in the development work, followed by periodic gates. A *gate* is an evaluation by upper management or within the team structure to ensure the next stage is worth carrying forward. Any product development project must pass through each gate to make it to the point of product launch. Early gates, such as the end of the product definition phase, ensure there is a market for the product, and that it can be developed and manufactured. Later gates ensure "detailed" integration factors, such as ensuring that the software functions with the mechanical hardware. As an example, Figure 1.3 shows the states of completion required at different stages for different development disciplines within Boeing Aircraft Company's product development process.

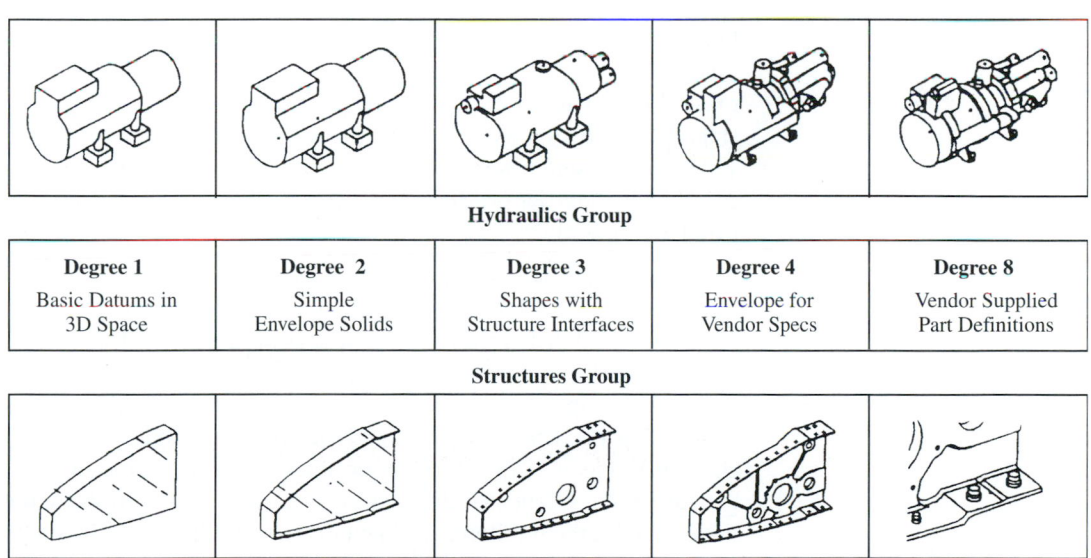

Hydraulics Group

Degree 1	**Degree 2**	**Degree 3**	**Degree 4**	**Degree 8**
Basic Datums in 3D Space	Simple Envelope Solids	Shapes with Structure Interfaces	Envelope for Vendor Specs	Vendor Supplied Part Definitions

Structures Group

▼ *Figure 1.3.*
Development degree-of-completion required for different stages at Boeing (Manente 1995).

As shown in Figure 1.3, stage-gate processes often become very well defined within a company and so permit relatively rapid and reliable execution of the product development. For each gate, the required information that must be assembled is explained beforehand, and typically the product manager (the person in charge of developing the new product within the company) must assemble it and make the case to continue its development. Senior management then evaluates this information and makes a decision.

At each gate, the decision to be made is whether to proceed with the development and its direction or whether to "kill the product" (halt development) or certain features of it. A third option occurs when there is insufficient information to make this pass/kill decision. In such a situation, the decision makers determine what further development or investigation is required to make the decision and what target level the product must attain to either kill or pass it.

In practice, gates operate a bit differently. Typically, few projects are killed outright in the later stages. Instead of halting a project, specifications are revised in light of the development difficulties, and budget allocations are typically expanded. The product managers ask for more funds, and upper management debates the pros and cons. This scenario, of course, requires reanalysis of a revised business case that must be developed before the gate is entered.

Other models beyond a stage-gate process include a so-called *spiral model* of product development. This model is typical in time-compressed industries, such as software products. Here, one repeats the stage-gate process several times before finishing the product to 100% completion, where at the end of any of the stage-gate processes, one has a partial product that "works." It may not be fully featured, but it works.

For example, one could set up a development process for a word processor where the first pass around the spiral is a stage-gate process whose result is a design that simply edits text and saves/opens files. This result is a semicomplete design that works. As a "product," it could then be revised to be capable of different fonts. The revised design is semicomplete and works. This "product" could then be revised to include a spell checker. And so on. This process simply represents a chained set of stage-gate processes, except that during any of the stage-gate processes, the major development criteria also include compatibility considerations with the next product revision. For example, knowing that the next revision must include capability of colored letters may impact how a team designs a data structure to represent letters in the current revision.

Spiral models are effective in time-compressed industries with large uncertainty. Here, an effective strategy is to develop something—anything—to seek user feedback early, before major gates are passed and parts of the design become frozen too soon. Spiral development processes are not as typical in traditional mechanical product design, because mechanical products will not function at all until almost completely designed. For example, a car won't transport any person if it doesn't have a chassis, powerful engine, powertrain, and so forth. Designing only part of the hardware, such as just the steering or just the powertrain, is not designing a complete vehicle that could work. Even though each is developed as a subsystem, its development is not independent of the others. In integrated mechanical/software system development, one often sees hybrid spiral/stage-gate models for the various subsystems being developed.

Given these differences, it should be clear that enumerating a product development process in any more detail than shown in Figure 1.2 would result in discrepancies in its application. That is, if the "Develop a Concept" phase is described as a sequence of activities, then that sequence of activities will depend on the product being developed. Therefore, it is virtually impossible to develop a detailed description of "*the* design process" that will be general to all industries. Nevertheless, it is instructive to consider a typical and effective sequence of activities that one can expect in a product development process. Our view of such a sequence of activities is shown in Figure 1.4. This process is a typical sequence of design activities that will be encountered by any mechanical engineer.

Understand the Opportunity

The first phase, "Understand the Opportunity," we characterize with four activities. The first step in a product development process is often to have a vision for a new product. What product do we wish were out there? What is difficult with the current product we use? Why does it not do something we want it to? The answers to these questions are visions for a new product.

Visions are a dime a dozen. Everyone has an idea for a new product, every user has thoughts on how they wish their devices would work, every CEO has a vision for command of a market, and every research scientist has a vision for how their technology can be applied. The question is whether any vision can be transformed into a successful realization. Can it be developed and implemented into a product at a

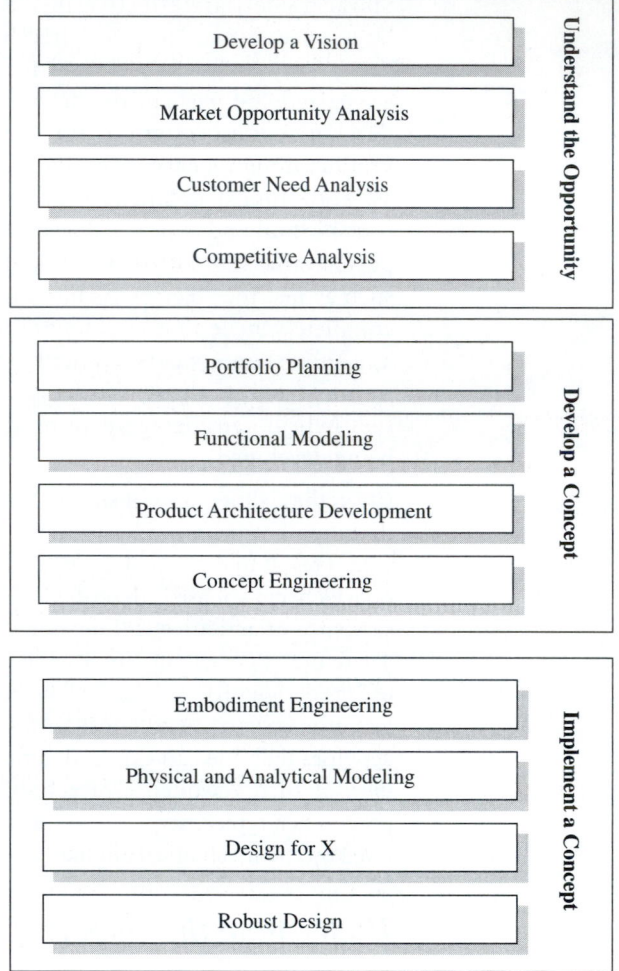

▼ *Figure 1.4.*
Activities in a typical product development process.

worthwhile profit? The first part of this question is the market opportunity. The revenue that might be extracted by a new product from a market has to be estimated. In today's competitive environment, the estimated price and volume that a market will carry is always the starting point. From these estimates, a profit must be subtracted, defining cost constraints that a product must be designed to fit within. Methods for carrying out this analysis are developed in Chapter 3.

Having completed the market analysis and made a decision to proceed to develop a new product, a development team next must analyze and understand what the customer population wants the new product to do. Methods for understanding the customer are developed in Chapter 4.

Once customer needs are understood, the competitive products on the market must be analyzed in terms of how well they satisfy these customer needs. Chapter 7 presents methods for competitive analysis, using the teardown methods presented in Chapter 6.

At the conclusion of these activities, the design team understands the state of the competitive market, the customer population, and any available technologies. A gate evaluation can be made with this information whether to initiate development in the next phase; that is, to proceed in the development of a new concept.

Develop a Concept

Having clarified all available information on what a new product must do, how it must fit into the market, and what cost range constrains the product, the design team can work to meet these expectations. The first activity in this effort is to design a set of general market specifications for the product. This activity must be undertaken with respect to the complementary set of products that the company has to offer. It includes thoughts on product positioning in the marketplace and portfolio planning and development. Chapter 8 develops methods on these topics.

Understanding the portfolio architecture, customer needs, and competition orients a design team toward the generation of a new concept. One of the first tasks in concept generation is to determine what the product must do to supply the customer satisfaction, independent of how it is implemented. This task entails the execution of functional modeling, developed in Chapter 5.

Having a functional model describing the inputs, outputs, and transformations that must happen for a product to work, there are several possible groupings of subsets of these functions into actual subassemblies. Developing the interfaces in the product embodies the product architecting process, developed in Chapter 9.

The functional model and alternative product architectures set the stage for very-effective concept engineering. Here, a product development team generates many concepts for implementing the functional specification. Concept generation methods are discussed in Chapter 10. After synthesizing many concepts, a design team must select one to implement. This selection process is discussed in Chapter 11. The result of this analysis is the concept to develop. Again, a gate can then be established to decide whether to proceed with this project.

Implement a Concept

Having selected a concept, it must be implemented, which is the final phase of product development. Much of this activity is *embodiment engineering,* where a chosen concept is given form through specification of components to purchase, parts to manufacture, and specifications for their assembly into the product. Basic methods of embodiment engineering are given in Chapter 12.

One important aspect of embodiment is *modeling,* the testing of new implementation ideas by physical construction—building it—or by analysis—numerically modeling it. Modeling is typically understood in terms of explaining physical phenomena, in the rich tradition of the physical sciences. In product development, we don't want to explain physical phenomena, we want to develop a novel product that will delight customers, fulfil our dreams, satisfy our ethical responsibilities, and make money. Therefore, we must consider modeling in the real-world context, the product development process, and demonstrate how it supports design decision making. This view of modeling is presented in Chapter 13.

Methods to actually make models of the physical processes in a product are often difficult to initiate and evolve for effective decision making, both for novices and even experienced engineers. Chapter 13 thus develops a methodology for constructing mathematical models of customer needs in terms of variables that a design team can change, using the customer needs, functional model, and teardown results of the earlier chapters.

Based on the developed performance metrics that reflect customer needs, one must use these metrics to select a preferred design configuration. Chapters 17 and 18 develop experimental methods to explore different configuration alternatives and to construct performance models. Chapter 16 develops methods of optimization to study the developed models and to help a design team select values for specific configuration variables.

In addition to the many performance-related metrics of a product, a design team has additional engineering specifications that their product must meet. These specifications are the discussion of the various *Design-for-X* methods, where *X* is one of these requirements. For example, Chapter 14 develops *design for manufacturing and assembly,* where methods are constructed to ensure the ease of manufacturing and assembly of a product. Chapter 15 develops *design for the environment,* to ensure that a product uses minimal-impact materials and operations.

One key factor missing in our discussion is the emphasis that a design team must place on making the product work *well.* Any handy untrained person can make *something,* but it takes analysis and insight to fabricate an *engineered* product, one that can be put together easily, where every product off the assembly line works properly, and where each works well in varying applications and conditions. Ensuring this quality view of a product is the goal of *robust design* discussed in Chapter 19. The philosophy presented in the chapter should be applied to every activity in the product development process.

At the end of this phase, a working prototype exists. Often, production planning and manufacturing process design are also underway. These activities represent the final point in the development to establish the final gate, to determine whether to launch the product. Killing a project at this point is not done without consternation, since the project should not have proceeded to this point if it is to be killed. Though it is rare, if the return on investment looks low, changes are made to reduce costs, and the product is launched. More often, the prototype result performs as expected and particular features that offer strong delight are demonstrated and highlighted for advertisement at launch.

A Reverse Engineering and Redesign Product Development Process

Given the vision of a typical product develop process shown in Figure 1.4, one might be curious why the chapters do not follow this outline. Recall there are many product development processes, each tailored for the many different products and the many different cultures and experiences of different companies. We believe the way to learn product development is to do product development, but this belief presents practical difficulties, as many of the activities overlap. Finding an appropriate sequence through the activities to learn them is difficult. Our approach is to present a *reverse engineering and redesign process,* one that starts with a product in the marketplace and a vision to redesign it for some perceived market defect or envisioned evolution. In this sense, reverse engineering entails the prediction of what a product should do, followed by modeling, analysis, dissection, and experimentation of its actual performance. Redesign follows reverse engineering, where a product is evolved to its next offering in a marketplace.

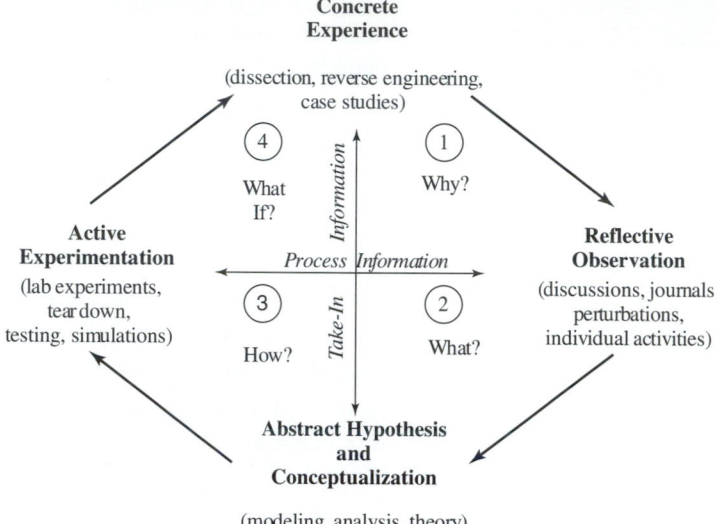

▼ Figure 1.5.
Kolb's model of learning (Stice, 1987).

Figure 1.5 illustrates the motivation for promoting a reverse engineering and redesign process when studying product design. The figure shows Kolb's model of learning, embodied by a cycle that begins with concrete experience, proceeds with reflective observation and conceptualization, and ends, before restarting, with active experimentation. By redesigning a current product, the physical components may be directly experienced with all senses. Design methods may then be used to hypothesize current functions and conceptualize new functions and/or solutions to the current configuration. Observation and active experimentation with the current and refined concepts may then be executed, realizing mental ideas into physical embodiments. The process may then begin again, where further iteration enhances and cements learning as well as actual product improvements.

The Kolb model, as shown in Figure 1.5, swings the pendulum of learning engineering from an emphasis of generalization and theory to a balance with all modes of learning (Stice, 1987). Engineering becomes equally focused on hands-on activities. Without this approach, we have no concrete experience to ground our learning and build a solid understanding. Nowhere is this truth more pronounced than in product design. The grounding in a current product helps nurture our interest for understanding the way things work and for making devices work better.

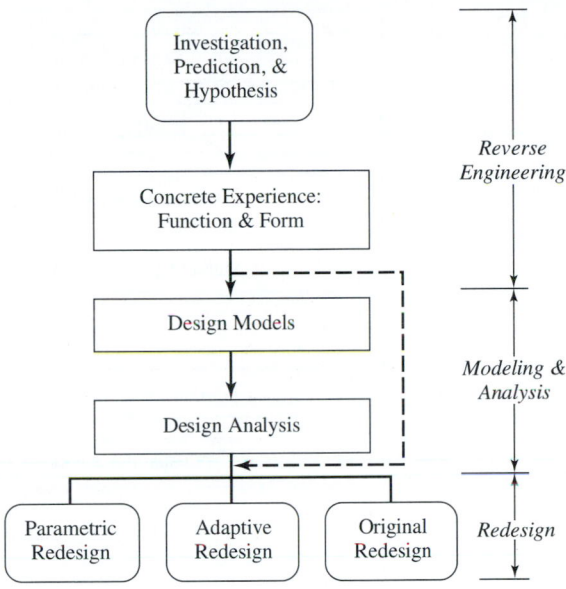

▼ *Figure 1.6.*
General reverse engineering and redesign methodology.

Figure 1.6 shows a general composition of a reverse engineering and redesign methodology. Three distinct phases embody the methodology: *reverse engineering, modeling and analysis,* and *redesign.* This approach allows us to present the necessary material on how to understand the product. For example, it is not likely that a design team will tear down their company's past product to understand how it works—they themselves probably designed it months earlier. Nonetheless, a student of engineering design who has not seen the past product will have to tear down the past design to understand it and will have to do so early. Such considerations lead to the order in which we present the material.

Given the ever-present guideline that activities overlap, Figure 1.7 illustrates an alternative sequence of reverse engineering and redesign activities for studying product development, still within the basic premise of three phases for product development shown in Figure 1.2. For the reverse engineering and redesign product development process, we rename the phases *reverse engineer, develop a redesign,* and *implement a redesign.* Let's consider each of these phases in further detail.

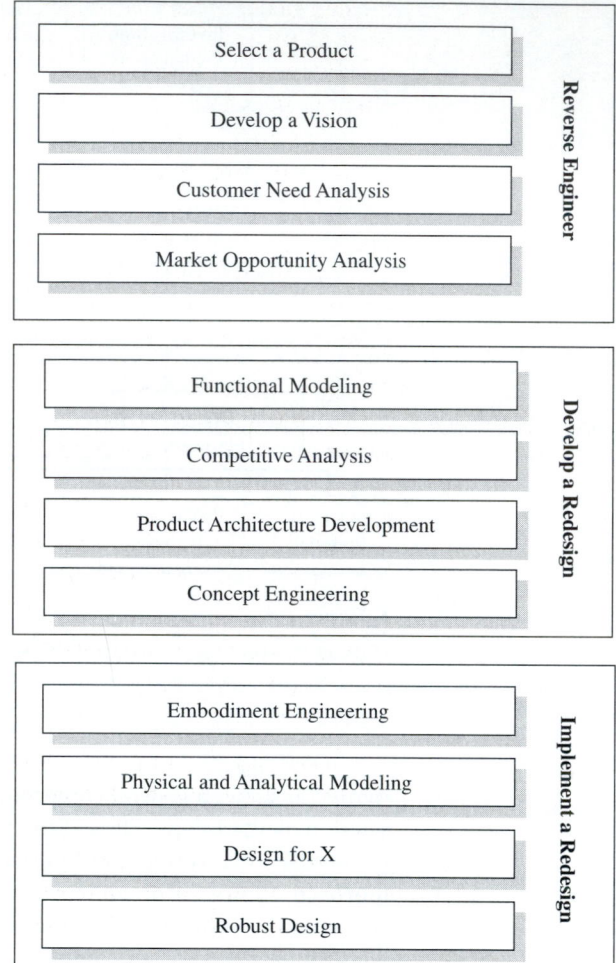

▼ Figure 1.7.
A reverse engineering and redesign product development process.

Reverse Engineering

A reverse engineering and redesign methodology begins with a current product. For something to be redesigned, it obviously must pre-exist. A product that is in the market is well engineered at some level and so makes a good starting point to begin to understand the methods and activities in product development.

One of the first tasks is to understand the market for the current product. Customer needs analysis, as described in Chapter 4, searches for knowledge. This analysis culminates with an understanding of what the customers like and don't like in the product. Based on this understanding, a variety of redesign opportunities will be apparent. One can then complete a business case analysis as in Chapter 3. A business case will define the potential financial gains and risks of pursuing the redesign opportunities.

The next step in our reverse engineering and redesign process is to make intelligent estimates as to what the functional model ought to be, using the modeling in Chapter 5. This step is important to clarify our preconceived notions of how the product ought to function and to adopt a functional view of the design task.

The next step in our reverse engineering development process is to dissect the product and understand how it operates to satisfy or not satisfy the customers. The material in Chapter 6 can be used to help in this process. This reverse engineering activity can also be repeated for many of the competitive products, with additional literature searches on the marketplace to refine the business case. Chapter 7 discusses techniques to support these steps.

Based on the teardown of several products in the marketplace and related products of the manufacturer, the portfolio architecture of the manufacturer's product line can be analyzed, as discussed in Chapter 8. Shared systems among products can be clarified.

Understanding the component and system technology of the company and its competitors, a real/actual function structure for the product can be developed, using the methods of Chapter 5, where the results are compared with the hypothesized structure before teardown to explore preconceptions for superior approaches. The product architecture can be detailed on the function structure, as discussed in Chapter 9.

Develop a Redesign

After reverse engineering the product, new concepts can be explored using one of three redesign strategies. The existing product topology can be maintained and a parametric redesign explored—changes in thickness or geometry of components or changes in materials. Alternatively, the components can be replaced or placed in different topologies of the functional model. Finally, the entire concept can be replaced with a different functional layout or with different core technologies.

There are several tools and methods in the subsequent chapters that can be used to support any of these redesign approaches. First, Chapter 10 presents material on developing new ideas. Chapter 11 presents material on selecting among these ideas. At this point, a redesign strategy and a new concept in that strategy have been developed.

Implement a Redesign

Chapter 12 then begins the study of implementing the new redesign idea. Here, material is presented on how to embody a concept. Chapter 13 presents material on one important aspect of doing this effectively, modeling of performance related to customer needs. Chapters 14 and 15 present particular models that all effective mechanical designs must include beyond performance models of the product.

Chapter 16 then presents material on using models to select highly effective design configurations. Chapters 17 and 18 add to this discussion with experimental methods to accomplish the same goal but with physical models. Finally, Chapter 19 presents robust design material that should be used at every activity in the development process and specific methods that can be used to improve the quality of the embodiment developed up to that point.

Overall, unbiased prediction, customer-driven design, analysis using basic principles, and hands-on experimentation are the philosophical underpinnings of this redesign methodology. The intent of the methodology is to be dynamic, depending on the needed evolution for a product. For some product redesigns, it may be appropriate to perform adaptive or original changes before creating and optimizing a design model. Similarly, the model development of phase two may lead to a better understanding of the product, bypassing parametric redesign and leading directly to adaptive. Alternatively, another product redesign may call for simple parametric modifications to produce dramatic responses in quality and profit margin. This methodology may be used for any of these scenarios or others.

These thoughts illustrate a reverse-engineering-based design process using the material of this book, one that has been tested at several institutions with success. Still, it remains but one sequence through the material; indeed, depending on one's goals, many other sequences could be sensibly applied. One underlying theme of any sequence is to strive to apply the methods *fully* at least once, including wrestling with the details and analysis. Only then can one really understand the extra benefit that one gains through the extra effort, and as a result, one can make intelligent choices whether and how to reduce or even expand the modeling effort for any product development activity.

A natural response to this theme is that we really don't need to complete any number of product development tasks to the level described in this book. "We don't need to complete customer interviews, or we don't need to complete the house of quality, or we don't need to analytically model, or we don't need to trial experiment." Design is full of examples of "*don't need to*'s": It is the most common failure in product development, especially among novices and even among engineers who should know better. Avoid this trap, or learning will not take place and we will fail repeatedly.

At some point, we as engineers will design a product if we have not done so already. All product development processes are different, and it is important to form one's personal and intimate approach. In all likelihood, an engineer's livelihood and the livelihood of others will depend solely on how well one effectively forms that personal product development process. We no longer have any excuse for doing so blindly; we should design a product development process with intelligence and the background understanding and experience of what the *to*'s provide.

IV. EXAMPLES OF PRODUCT DEVELOPMENT PROCESSES

We now consider some industrial product development processes and show how they relate to the generic product development process of Figure 1.4 and to the material in this book.

Systems: Xerox Corporation

The Xerox Corporation is one company that uses a modern product development practice as discussed. Paul Allaire, CEO of Xerox, and Maurice Holmes, the corporate Vice President at Xerox, presented it as the Xerox "Total Time to Market" Product Development Process (1996), where the focus is on rapid time-to-market with a high-quality product. The Xerox Product Development Process is shown in Figure 1.8.

Note the steps in their self-described product development process follow Figure 1.4. The first step is to develop a vision for the market and the product strategy. This process develops a concept for the product family: the platform, how many products are part of the family, key technology components making up the platform, and the

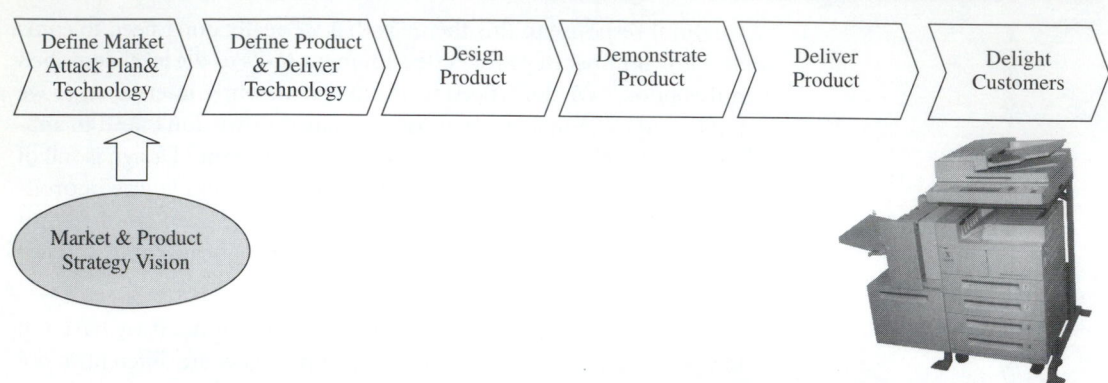

▼ *Figure 1.8.*
Characterization of Xerox's Product Development Process and an example copier.

expected business outcomes. This vision is developed by senior management as part of the yearly corporate planning process. For example, in the early 1990s, Xerox management envisioned a marketplace that was transformed by computers and networks into a digital document environment—information was created from a combination of digital or paper sources. It was distributed over local or enterprise-wide networks and printed at the point of need in monochrome or color. The Xerox copier would therefore be transformed into a digital document machine. This device could copy, print, fax, or scan a document to a file over communication networks, enabling customers to achieve higher levels of office productivity at much lower cost of ownership. To realize this vision, the entire office product line from 12 to 65 pages a minute would be changed from analog copiers to digital document centers through the development of many new product families derived from two new product platforms. Light-lens optical technology would give way to electronic scan bars and raster output scanners. The dominance of Xenographic and electromechanical technology would give way to network controllers, PDL decomposition, raster input scanning, and data compression and decompression technology. This change would in turn be accompanied by changes in manufacturing, service, and distribution as required by the needs of the new digital market.

The next activity is to develop a plan for the particular product in the family. This step defines the market opportunity for the product.

Xerox terms this the *Market Attack Plan;* senior management again develops it. A case is made for a new product line opportunity, typically involving several products that are part of a product family supported by a product platform. Such a business case analysis to determine how to launch a product is developed in Chapter 3.

Individual products are only developed within the context of platforms. The process itself is a sequence of activities that is ostensibly repeated for every new copier development project, but with ever-advancing technological components and ever-advancing design process technology. Such a product platform design is discussed in Chapters 8 and 9.

A product platform is a system of key technological components (platform elements) integrated to form products according to a set of architectural rules. This system includes, for example, the print engine that actually marks ink on the paper, the paper delivery system for feeding paper to the print engine, the user interface, machine controller, network controller, and so forth. Each of these components is a heavily invested-in piece of core technology that Xerox considers a market advantage. For example, the departmental document center document scanner contains a butted silicon scan bar that allowed the scanning of paper documents at far greater speeds and resolutions than typical LCD arrays used in competitive devices. This scanner was used as a platform element for Docucenter 40, Docucenter 55, and Docucenter 65 digital copiers and scan-to-file products.

At Xerox, the product design itself is completed using methods discussed in this book. For example, functional models as discussed in Chapter 5 are applied using a block diagram methodology. The house-of-quality tables presented in Chapter 5 are extensively used. Robust design presented in Chapter 19 is a methodology and philosophy used throughout the design process. The design itself is demonstrated in development with prototypes and mockups and by using performance metric models.

Throughout the development, Xerox also pays strict attention to design process metrics to ensure the development process itself is well designed. This attention to constant design-process improvement is very important to Xerox. It operates in a competitive environment where reduction of development time is critical, since new electronics technology is constantly being developed that impacts Xerox's markets. Methods to ensure good process are discussed in Chapter 2.

In sum, Xerox exercises a well-developed product development process. The activities in the process are developed in this book, and they practice the methods espoused. Further, they are always seeking more rapid, higher performing methods to increase their competitiveness.

Industrial Design: Design EDGE, Austin, TX Product Design Firm

Design EDGE is a product design firm that develops products for a number of industrial clients. Figure 1.9 illustrates a snapshot view of the product development process executed by Design EDGE. The figure lists the major steps in the process in addition to tools that are used to support the development steps. Figure 1.9 also shows recent results (1998) of novel products created from this process and the designers of Design EDGE.

Notice that the process embodied in Figure 1.9 again satisfies the general structure of the process of Figure 1.4. After being approached by a client company, Design EDGE begins the process with understanding the design task, based on customer and client interviews, market analysis/data, and initial clarification of the task through studying functionality, needed attributes, and the feel desired in the product.

The development process then transitions to a focused effort on preliminary concept development. Industrial designers lead this effort with a clear team emphasis, including engineering and industrial design team members. The tangible result of this phase is a product proposal to the client illustrating a number of industrial design concepts and the associated development costs and schedule for executing the designs.

Assuming the client company adopts the proposal, product development now turns to implementing a product concept. Two layout stages comprise this phase: Industrial designers lead the first stage and engineers lead the second stage. During both stages, the goal is to refine the product concept in form, shape, feel, and technical component issues of technological choices and performance. The product team balances physical prototypes and numerical models to make decisions concerning these issues.

To further implement the concept, product tests are performed. These tests are followed by refinements, culminating in tooling design for full-scale product production. Databases and prototypes are transferred to the client as deliverables (after approximately a 6–18-month

Understand Product Task

Market Domain Study
- Interview Clients(s)
- Identify Functionality
- Identify Attributes
- Identify Expected Feel of End Product
- Convince Client of Reasonable Expectations

Tools:
- Semantic Inquiry
- Customer Interviews

Preliminary Conceptual Design

Team: ID/MEE, Led by ID
- Brainstorm Concepts
- Sketch and Record Final Ideas
- Compile Design Notebook
- Repeat Daily/Weekly
- Client Evaluation of Concepts

Tools:
- Whiteboard
- Sketching Equipment:
  – pencils, markers (for accent)
  – large butcher paper
  – scotch tape and masking tape
  – templates
  – large and small rulers, straight edges, and squares
  – large radius tool
  – sissors
- Foamcore
  – exacto knife
  – cutting pad
  – sanding blocks
  – pen
- Conference Roon

Layout Design (Systems I)

Team: ID/MEE, Led by ID
- Identity Components (MEE)
- Gather Component Specs, Pricing, etc.
- Specify Locations and Orientations (ID/MEE)
- Create Spatial Layouts (MEE/ID)
- Construct Component Database (MEE)
- Initiate Concept Refinement (ID/MEE)
- Draw and Fabricate Foamcore Models (ID)
- Create Wire Frame Models (ID/MEE)
- Fabricate Blue and White Foam Models
- Check Spatial Feasibility

Tools:
- Solid Modeler (ProE)
- ID Wire Frame Modeler (Vellum) (for accent)
- Foamcore
- Drawing Equipment
- Blue and White Foam (from hardware store)
  – hot wire cutter
  – exacto knife
  – wire hand saws
  – glue gun
  – tape
  – paint
  – masking tape
  – newspaper

Final Layout Design (Systems II)

Team: ID/MEE, Led by MEE
- Build Component Database with Final Specs
- Integrate ID Geometry
- Embody Mechanisms
- Embody Plastic Parts
- Model Heat Transfer Stresses, Kinematics
- Design EMI Shielding
- Finalize CAD Database
- Prototype Sheet Metal Parts
- Prototype Mechanisms

Tools:
- Sold Modeler (ProE)
- Model Shop
  – machining of plastics
  – wood working
  – sheet metal
  – CNC Mill
- Analysis Codes

Design/Tooling Design

Refine Components

Design Tooling

Prototype Pre-Production Version

⟹ Tooling Design and Database to Client

Legend

ID: Industrial Designer

MEE: Mechanical or Electrical Engineer

▼ *Figure 1.9.*
Characterization of the Design EDGE Product Development Process and example products.

development cycle time), and the client carries the process forward to production, packaging, marketing, and distribution.

Again, the Design EDGE process is very similar to our generic methodology for product development. It includes clear stages of understanding, conceptualizing, and implementing. A number of interesting distinctions do exist, however. For example, Design EDGE stresses a strong emphasis on industrial design. Product feel and expression are critical to consumer products. Design EDGE addresses this issue head on, driving the initial product layouts with the artistic and ergonomic features of industrial design.

Rapid: Microsoft Corporation

Some companies live in a competitive market cycle that is much faster than was just considered. In fact, some exist in technological change cycles that are so fast that the generic development process of Figure 1.4 cannot be entirely completed. Some steps have to be short-changed, or the company will not introduce its product into the marketplace soon enough.

For example, Microsoft Corporation develops software and must do so very quickly. Microsoft Excel version 5.0, for example, was completed in 18 months with as many as 125 people working on the project (Cusumano and Selby, 1995). To meet this schedule, some of the development tasks are abbreviated. For instance, Microsoft does not have an extensive program to interview customers and determine what they like and don't like in a product. Rather, Microsoft relies on a strategy of having a flexible product architecture, understanding the uses of their product, and designing to a delivery schedule.

The product development process of Microsoft can be linearly characterized with three phases: planning, to develop a product specification and development; coding, to complete the coding; and then stabilization, to debug the code. However, Microsoft views its process more of an ever-cycling development process, better described circularly. That is, while finishing one revision of the product, they must necessarily be thinking forward to the next revision. This process is shown in Figure 1.10 (courtesy of Microsoft), with the three key development phases highlighted.

The first phase in Microsoft's development process is as the others': understand the problem and create a specification for the development task. *The Specification Document,* developed by senior management, starts with a vision statement, itemizes it into outline form, and then defines and assigns priorities to new or enhanced product features.

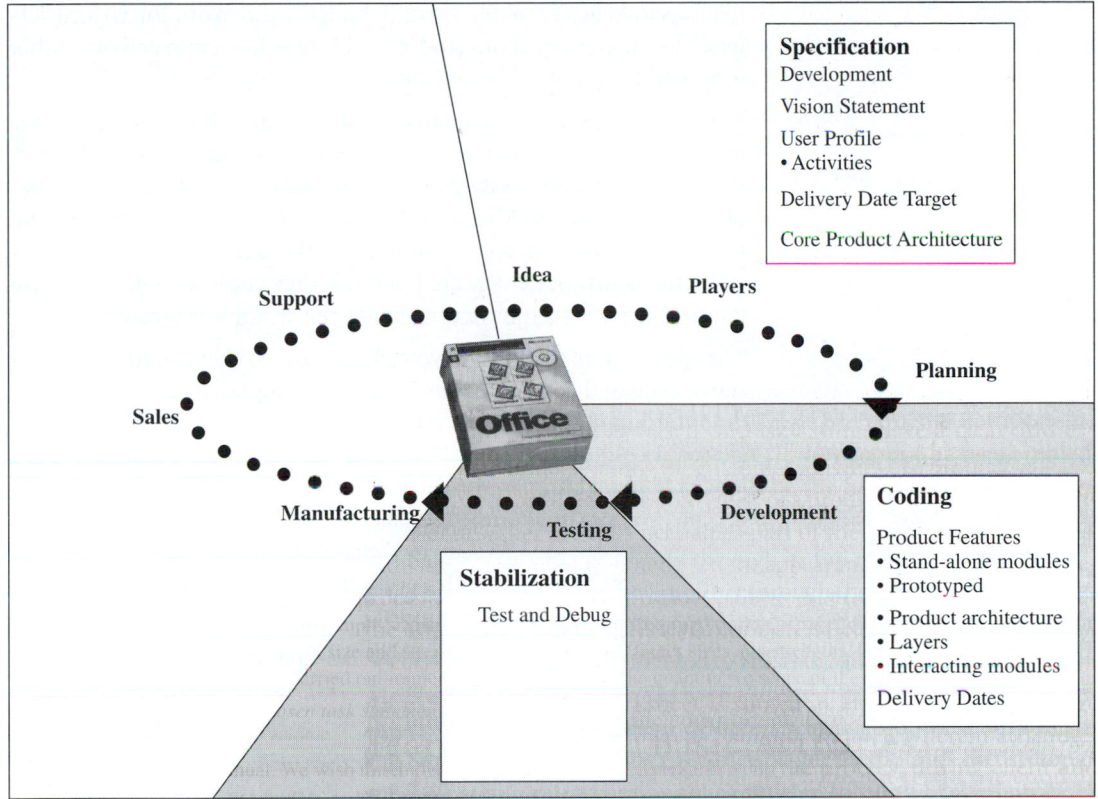

▼ *Figure 1.10.*
Characterization of Microsoft's Product Development Process.

The vision statement is viewed as key. Chris Peters, Vice President of Office Product Unit, states, "A good vision statement tells you what's not in the product, a bad vision statement implies everything is in the product" (Cusumano and Selby, 1995). An example is the MS Office products, where the emphasis of the vision statement was on harmonious use of the multiple desktop applications rather than as a jumble of applications that the user has to determine how to get to work together. Product managers then must refine this choice. With marketing, they develop potential features.

This step is refined based on user activities and data. That is, Microsoft does listen to the customer as espoused here, but in a rapid-response manner. Rather than asking customers for product preferences, Microsoft interviews customers and studies the tasks that people use with the software applications. They uncovered, for example, that many people use Excel for creating budgets and Word for writing letters. They then carried out studies and broke down into activities what is needed to complete these typical tasks.

Features are selected for inclusion in a product based on how they impact these user activities. Thus, Microsoft doesn't carry out interviews on the past product *per se;* rather, they perform market studies of how people work. Microsoft then estimates how new features will impact these use times and estimates market acceptance. If they get it right, the new product feature is a hit. If they guess wrong, the feature is a dud. They do not have time to ask the customers directly.

The specification document typically grows and changes 20–30% from start to end of the project. John Fine, group program manager for Excel, describes how the specification document should be written: "All is from a user's point of view. That's everything—just never worry about anything else except 'What does the user perceive?' When you use a product, what the product does should be understandable enough so that any user can say in a sentence what it does" (Cusumano and Selby, 1995). Clearly, Microsoft understands the need to listen to the customer.

The second phase of Microsoft's development process is to create the code. Here, they develop each module separately as a Visual Basic™ prototype. This prototyping process proves the feature, they then work to integrate it into the product. Key to this modular product philosophy is a flexible product architecture based on modules with standard interfaces, with a flexible common platform core. Every product has a layered platform, which is the inseparable core of functions that a product must perform. They devote considerable time and resources to this step so that the job is completed with high quality.

A key aspect of the rapid development process is to design to a schedule. Here, every activity is given a schedule. They time-delimit each activity, and within that time allotment, every programmer evolves features as much as possible. As expressed by Chris Peters, Vice President of Office Product Unit: "If things don't work out, I'll just cut features I think are less important" (Cusumano and Selby, 1995). They do not extend the schedule and include features as the first development process choice. They can always add the feature into the next revision.

The last phase in the product development process is stabilization; here the focus is on testing and debugging. No new features are added, but rather the product as a system of features is made to work together all the time, that is, to be robust as discussed in Chapter 19.

During this phase, a further key aspect to Microsoft's customer interactions is the use of the customers' feedback during development.

Microsoft has an extensive beta-release network of users, who receive preliminary versions of software to test-run them as a part of the code stabilization phase. This provides excellent feedback and opportunity for adjusting the product before release.

The Microsoft development process does not negate the process of Figure 1.4. Microsoft's activities are the same but are optimized to permit task completion in very short time. The focus of this textbook is to understand what the activities must do, and it presents methods to execute each easily and completely. If one seeks to shorten the cycle at the expense of completeness, one has a basis for making that choice and understanding the ramifications. The rapid result philosophy is appropriate for Microsoft's marketplace—they have to develop products very quickly.

Research Intensive: Raychem Corporation

Some products exist in development cycles much longer than that considered above. The product development cycle may become so long that it becomes difficult to forecast any customer wants and needs. The Raychem Corporation, for example, is a sophisticated manufacturer of products utilizing advanced materials. Historically, they have developed products based on new materials research. Their product portfolio includes heat shrink tubing to easily bundle wires, GelTek™ moldable electrical sealants, arresters that help protect electrical devices from the dangers of surges, and resettable fuses that protect cellular phones, speakers, and batteries from electrical overload.

Some of these products took decades to develop. For example, the PolySwitch™, an electrical overload cutout switch made from a polymer, was first conceived in the 1970s. After several years of R&D, the first prototype was built in the lab in 1981. The first product hit the market a few years later. Yet, the product did not pick up sales until about 1990 and was not really a success until the mid 1990s, when it became a standard component in every laptop computer battery pack. Now the product is widely used and very profitable. How is such long-term product forecasting and technology planning done? What specifications should be considered for a switch, where the product may not be commercially successful for decades? What will the world look like in decades? How could anyone in 1970 forecast the need for a polymer overload switch in laptop computers in the late 1990s?

Obviously there is a bit of luck involved, but not as much as would first appear. With the constant exponential growth in electronics, it was reasonably clear that inexpensive, low-weight, easily produced polymer switches would be in high demand. The issue was whether the polymer chemistry could be reasonably worked out and packaged. Raychem is a high-technology company; its core competency is in working with high-technology materials. This skill is its competitive advantage, which enables Raychem to make materials do most anything: The key is to get a material to do something that will be profitable.

Raychem develops a suite of different products that can require a varied amount of development; some are intensive in determining how to apply a new material, others are rapid applications of existing technology. Even with the high-technology products that incorporate longer cycle new technology development, though, Raychem applies a structured stage-gate product development process; its self-described process is diagrammed in Figure 1.11.

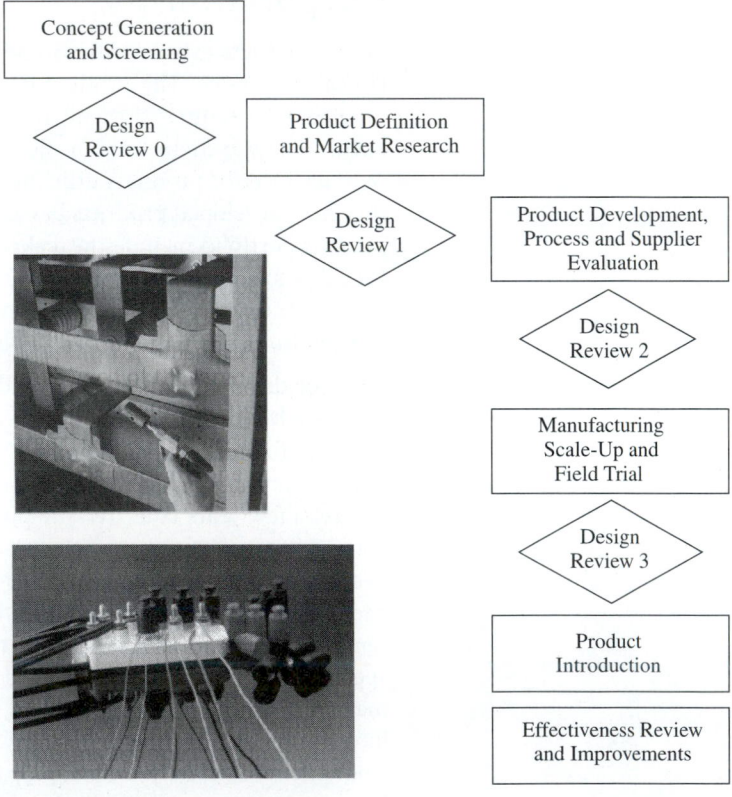

▼ *Figure 1.11.*
Characterization of Raychem's Product Development Process.

The Raychem Product Development Process starts with a vision to fulfil a perceived market need with a new product that Raychem believes it can develop. This is typically initiated through discussions with the customers by the Raychem sales force. Raychem prides itself on having a tight relationship with its customers, always probing for opportunities. They have a saying in the company of "Raychem in Response," meaning the company listens to and responds to its customers.

In the early 1980s, one of Raychem's telecommunications customers asked a salesperson if Raychem could develop a product that would prevent corrosion of connections in outside-mounted terminals. This request was initiated in Raychem's develop process as a document presenting the design and business case, physically instantiated in the "Design Review Book," a documentation of the product to be developed as it is being developed. The decision to consider developing the product is made at the first review meeting, called "Design Review 0," by a design review committee. The committee consists of the Division Manager, Technical Manager, Product Manager, Market Manager, Manufacturing Manager, Quality Manager, and other specialists as needed, representing environmental, legal, and purchasing concerns. The go/no-go decision is made based on whether the product fits within Raychem's product strategy—can Raychem make such a product?

After Design Review 0, the market feasibility and business plan is fully developed and analyzed. The product function is precisely defined complete with technical specifications and customer-need statements. The costs are estimated, including material, manufacturing, distribution, and development costs. Returns on investment analyses are completed. Technical feasibility is agreed to.

For the corrosion-proof outside connector design problem, the customer need was communicated to a design team that evaluated a number of product concepts incorporating standard Raychem technologies. Unfortunately, none of these concepts met the customer's requirements. The design team then came up with entirely new concepts—the "Maybe we can . . ." scenario. The design team decided that a new soft, highly conformable sealant was required. At the next design review, Design Review 1, a technical assessment and business plan was evaluated, and resources were committed to develop such a new soft, highly conformable sealant.

After the Design Review 1, much conceptual and embodiment product design work ensued. The design requirements were flowed down to component specifications, concepts developed, and prototypes constructed and analyzed.

Successful materials development and product design work resulted in the introduction of Raychem's first gel-based product, the Termseal™ terminal cap. This terminal seal was a novel gel inside a plastic housing that could be squeezed conformably around a terminal, thereby sealing it. This concept was shown to many customers, and market acceptance and real-world performance assessed. These results were used in the last design review, Design Review 3, to make the final decision to launch the product full scale. Today, Raychem sells dozens of gel-based products throughout the telecommunication, commercial, and automotive electronics markets.

Raychem has many product families built on core technologies. The gel-based product family is one example. The Termseal™ product was Raychem's original gel-based product. From it, a suite of product variants has been developed into a product portfolio, as discussed in Chapter 8. The Geltek™ family of sealants is the latest generation, including products that provide EMI as well as corrosion sealing.

When Raychem first formed as a company, it operated with a "Cost +" approach: Manufacture the product, count up what it had cost, add a profit, and demand that as a selling price. With the high technology that customers wanted, they could afford to do so. As argued in Chapter 3, this approach is not always the most effective, and so for many years now Raychem has operated on a "value" approach. That is, first Raychem estimates the market price, and from that profits are subtracted to arrive at a design target maximum cost. Against this, feasibility is judged.

The main issue with Raychem has been to develop very advanced research-intensive products. They require significant material science and uncertainty in market acceptance. Nonetheless, at its highest level, the Raychem Product Development Process is a stage-gate process, much the same as other product development processes described in this book.

Complex: Ford Motor Company

Many products are complex assembled systems of many suppliers' components and subsystems that must properly function together. It might be argued that these products form a special class for which the methods in this book do not apply. As an example, consider Ford Motor Company, developer of automobiles and trucks. One can characterize their product development process as in Figure 1.12.

The Ford Product Development Process starts with the "Define Product and Process" phase to define specifications for the new product.

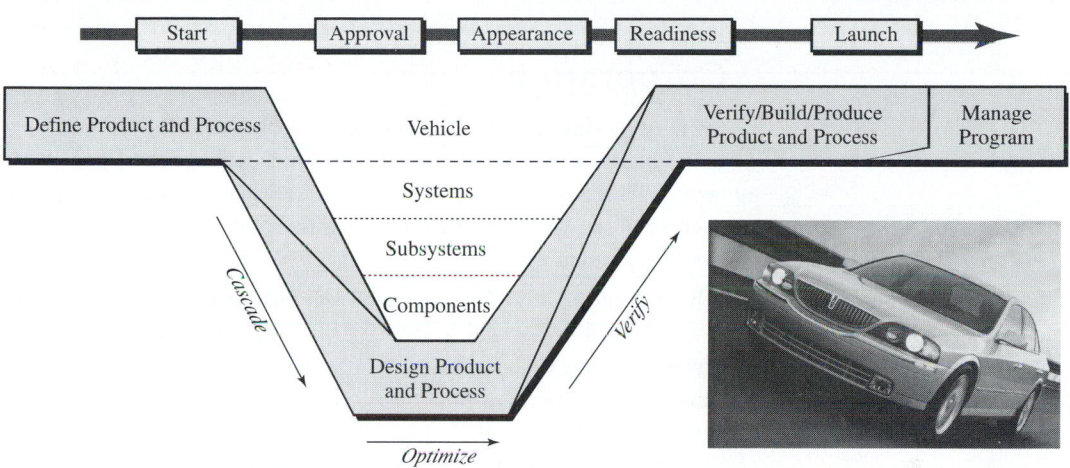

▼ *Figure 1.12.*
Characterization of Ford's Product Development Process.

These are both soft customer criteria and hard functional specifications. These are first gathered at the vehicle level or by using attributes that are dependent on the entire vehicle system. These attributes are then cascaded down to individual systems, subsystems, and eventually components. The added complexity in these large-system products is the interactions between subsystems and the larger product itself and how these interface. The methods of the book can only help make these interactions clear.

In the "Design Product and Process" phase, the actual product components are finalized, including, in particular, how they will be assembled into the vehicle. Significant concurrent activity with manufacturing engineers is required. The "Verify/Build/Produce Product and Process" phase verifies the integrity of the product and process designs and ensures that no negative consequences arise from the interfaces and interactions of the various subsystems. The last phase, "Manage Program," is simply the collection of management tasks that ensure the capability of the program to meet its objectives—managing resources and timing and capturing into the corporate memory lessons that are learned.

The added complexity of this product development process is the subdivision of labor and product functions among the various subsystems. Generally, the tools of this book apply hierarchically, with different detail at the various levels.

Technical: Raytheon Corporation

It might be claimed that if a company's product line is very technical, then customer-, function-, competitor-, or experimental-based concepts do not apply. The communities that design electronics or design precision manufacturing processes often make such claims of "just needing good engineering."

A branch of the Raytheon Corporation, formerly Hughes Electronics, is recognized as a world leader in designing and building electronic systems for defense applications, such as missile guidance systems that can precisely and reliably direct a missile thousands of miles. They are a company that routinely does some of the toughest and most innovative engineering, which has included the development of the world's first laser.

Raytheon characterizes their technology-intensive product development process as shown in Figure 1.13. With Raytheon's defense industry products, there is little apparent market analysis, since there is no statistical population of customers to worry about. Instead, a typical defense product is characterized in terms of scenarios in which it will be used: a *mission profile*. For example, the Cruise missile is a typical product designed to a mission profile. Specifically, a cost-effective missile was described that would have precision strike capabilities which would minimize pilot exposure and minimize population causalities. This was described in a customer-military-approved document that launched the design activity.

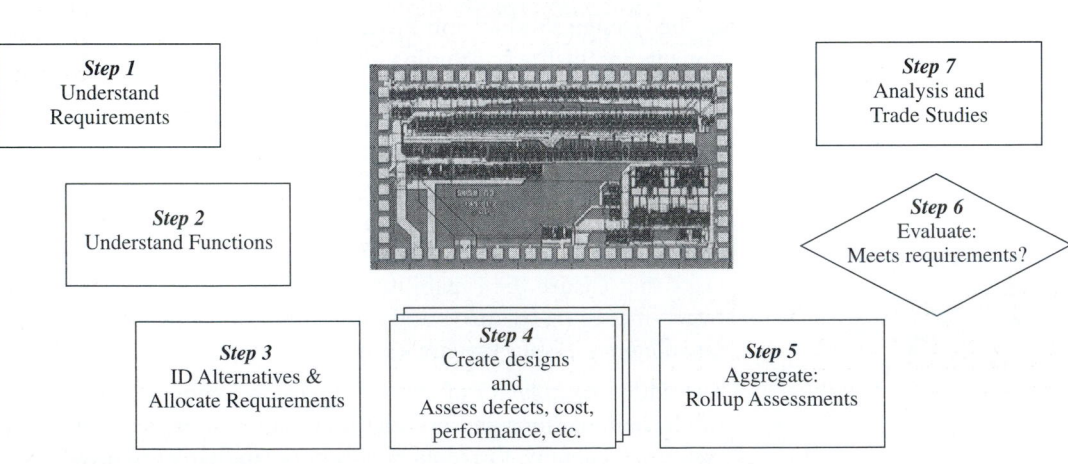

▼ *Figure 1.13.*
Characterization of Raytheon's Product Development Process.

In the second step of the Raytheon Product Development Process, the mission profile is refined into functions for the product to complete. Overall system technical specifications are developed that the product will have to satisfy. In the Cruise missile case, the product required a long-range propulsion system so it could be launched from aboard ship hundreds of miles from a target, with a high-precision targeting capability to ensure target destruction without peripheral casualties.

With the technical specifications, a functional architecture of the product is developed. Then, in Step 3, a system layout and architecture is developed, and further subsystem requirements are allocated. For example, in the case of a fiber-guided air-to-air missile, the missile was divided into a product architecture of seeker, inertial components, ordinance, actuators, and propulsion systems. Each of these modules was then assigned further refined specifications on criteria such as power, mass, and size to ensure the entire missile would perform according to the mission profile.

With the functional and system architecture, subsystem concepts are generated, embodied, and designed as Step 4. During the subsystem designs, periodic system-wide "rollup assessment" meetings are held to ensure compliance with the system specifications, as noted in Step 5. These serve as gates in Raytheon's stage-gate development process. In this entire process, much modeling, experimentation, and optimization are completed.

Raytheon's main emphasis is on developing very technically advanced products. They require significant engineering resources and delicate manufacturing. The design cycle to complete the system design and allocate subsystem specifications is becoming much more rapid with computer-aided concurrent design environments, where multiple designers from different disciplines negotiate specifications while evaluating total system performance. Such systems will be shown in subsequent chapters. However, as shown by Figure 1.11, the process that Raytheon follows is much the same as other product development processes and is as espoused in this book.

V. THEORIES AND METHODOLOGIES IN DESIGN

There is a rich literature on mechanical product design. The methods and approaches presented in this book are a culmination of many persons' efforts and combined wisdom. To understand where these methods and thoughts arose, it is instructive to reflect on the history of mechanical design theory, from its very inception at the start of recorded history to the current state of development. Table 1.2 highlights 50 major design theory developments in world history, in the view of the authors.

As surmised from Table 1.2, design theory is not yet a mature field; new developments are constantly being undertaken. This insight is further understood from the plot in Figure 1.14. Two discernable trends emerge: One is the rich development that occurred up through the time of the Roman Empire, with a plateau around the first millennium. Then, another rapid profusion of design theory development began about the time of the 1700s, which has not let up to this day.

The first development period can be explained as the developments in design theory required of classic civilization, which ended with the fall of the Roman Empire. The advancement stagnated during the Middle Ages. Since craftsmen and architects were the designers of the age, they relied on the handed-down knowledge of ancient Greece and Rome for their understanding. Medieval civilization was, from a design theory point of view, no more complex than ancient civilization. Technology had not advanced significantly, so there was no demand for further developments in design theory. Grossly simplifying: Medieval civilization could develop all required devices with the handed-down design theories of ancient civilization.

It wasn't until after the Middle Ages that this trend changed; technology advanced and the need arose for better design theories. New production systems, new labor-saving technology, new understanding of the physical world, new mathematics—all of these advancements required more than a simple build-and-test design methodology. Complex devices had complex mechanics; this created a need for new design theories incorporating analysis and synthesis of these systems. This complexity trend has shown no signs of any plateau; design theory remains an active field.

It is also interesting to note the contents of Table 1.2, in particular that mechanical design theory is strongly rooted in kinematic theory. The first theories on the design of mechanical devices were basically kinematic. It is only recently that more modern thoughts are being applied from other fields, such as separating function and form when designing a complex mechanical system, applying complex algorithms to routine synthesis tasks, and considering the value that a feature will bring to a device concept.

Where can we expect new developments to occur in mechanical design theory? To answer this question, one must examine what is driving the need for development. As noted in the introduction, there

TABLE 1.2.   DESIGN THEORY DEVELOPMENTS IN WORLD HISTORY

Author	Publication title	Theory or development	Origin	Year
	The cubit	First measurement standard: Pharaoh's forearm plus his palm	Egypt	2500 BC
	Ebers papyrus	First design manual: description of making a soap-like substance	Egypt	1500 BC
Aristotle or Straton	Mêchanika	First known book on engineering: falling bodies accelerate, vacuum is the empty space between atoms, successive gears rotate in opposite directions, the lever and beam balance, compound pulleys, the sling, capstan, windlass, rollers and rolling friction	Greece	350 BC
Archimedes	On floating bodies, On the Equilibrium of Planes, De Mensura Circuli	Calculated pi to ±0.1, 2D center of gravity, hydrostatics and buoyancy; Archimedes law, levers and mechanical advantage, a planetarium, the dolphin; a hoisted missile dropped through a boat	Greece (Syracuse)	240 BC
Hero	Pneumatica (Pneumatics)	Siphons, sprinklers, fountains, steam jets for spinning; first concept of an engine, first coin-operated dispensing machine, check valve, float valve, water force pump with a nozzle, a steam-powered rotating ball, theory of compressible air, first description of a windmill, first discussion of weighing fuel before and after combustion	Alexandria	150 BC
Hero	Mêchanika (Mechanics)	Idea of moment and explicit mechanical advantage, Theory of kinematics using five basic machine elements	Alexandria	150 BC
Vitruvius, Marcus Pollo	Die Architectura libri decem	First design theory—considers design as satisfying human needs, in architecture	Roman	27 BC
Banu Musa brothers	The Book of Ingenious Devices	First description of a feedback control system (to supply an oil lamp at constant flow), description of over 100 devices	Iran	AD 750
Bhaskara		First concept of a perpetual motion machine	India	AD 1159
Walter of Henley	Treatise on Estate Management and Farming	First method of experimental techniques to optimize production output	England	AD 1300
Brunelleschi, Filippo	Patent	First patent, issued by the Republic of Florence, for a canal boat equipped with cranes. More important, in some respects, he is the person who invented perspective drawing	Italy	AD 1421
Unknown	Manuscript of the Hussite Wars	First science fiction (a diving helmet and a deep-sea diver), first depiction of a cannon	Germany	AD 1430
Da Vinci, Leonardo	Notebooks	First professional mechanical engineer; revolving turret mill, notes on armaments; bevel, spiral, differential gears; universal joint; cutting screws, sprockets and roller chains, hydraulics	Italy	AD 1488
Town of Nurmenburg		First industrial exhibition	Germany	AD 1568

TABLE 1.2 (CONT.).   DESIGN THEORY DEVELOPMENTS IN WORLD HISTORY

Author	Publication title	Theory or development	Origin	Year
Stevin, Simon	De Beghinselen der Weeghcons	Triangle of forces: first description of vector math, permitted calculation of static loads on machine elements for the first time	Netherlands	AD 1586
Galilei, Galileo	Discoursi e Dimonstazioni matematiche (Dialogues Concerning Two New Sciences)	First discussion of the responsibility of a designer toward safety; first text on stress analysis in beams; mechanics: formalization of the concept of force and acceleration of bodies, vacuum	Italy	AD 1638
Newton, Isaac	Analysis per aequationes numero terminorum infinitas	Method of infinite series: generality and inverse nature of integration and differentiation	England	AD 1669
Gautier	Text on Bridges	First to complain in writing that the scientists of the day had no interest in practical matters (such as arches)	France	AD 1716
École des Ponts et Chaussées	Charter	First engineering school	France	AD 1747
Hume, David	An Enquiry Concerning Human Understanding	Famous presentation on the philosophical problem of inductive inference, a common thought process in design	England	AD 1748
Smeaton, John		First to call himself a civil engineer: non-military; first use of models to provide quantitative design information: tables of water wheel and windmill data	England	AD 1752
Euler, Leonhard	Recherches sur la veritable courbe que decrivent les corps jete dans l'air, ou dans un autre fluid quelconque	First application of Newtonian mechanics to engineering analysis; solution to differential equations of ballistic flight with air resistance, compared to Robin's experimental data	Germany	AD 1753
Bayes, Thomas	An Essay Toward Solving a Problem in the Doctrine of Chances	Bayes rule for deduction under uncertainty	England	AD 1763
LeBlanc		First to propose interchangeable parts	France	AD 1785
Maudslay, Henry	J. Nasmyth *Autobiography,* 1883.	Precision and accuracy as a machine design theory; Maudslay's Design Maxims: *Get a clear notion of what you desire . . . then you will succeed; Eliminate all material not needed . . . put to yourself the question, "What business has it to be there?" Make everything as simple as possible; Remember the getability of parts;* Standard screw threads; developed compound slide rest lathe; first use of a micrometer in a machine shop	England	AD 1807
Willis	Principle of Mechanism	Definition of a machine as a kinematic train	England	AD 1841

TABLE 1.2 (CONT.).   DESIGN THEORY DEVELOPMENTS IN WORLD HISTORY

Author	Publication title	Theory or development	Origin	Year
Froude, William		Methodology of model testing; similarity theory and scaling from prototypes	England	AD 1869
Reuleaux, Franz	Theoretische Kinematik: Grundzuege einer Theorie des Maschinenwesens	Definition of a machine as a kinematic train that performs work; definition of kinematic degrees of freedom	Germany	AD 1875
Roebling, John	Final Report to the Presidents and Directors of the Niagara Fall Suspension and Niagara Falls International Bridge Companies	Reliable design theory: Reliability comes about by analyzing past failures for failure modes and designing past them; completed the Brooklyn Bridge in 1888	USA	AD 1885
Reynolds, Osborne		Mathematics of testing using scaled prototypes, of the silting of the River Mersey in Liverpool	England	AD 1887
Weber, Max	Wirtschaft und Gesellschaft	Theory of rational behavior (Rationalität)	Germany	AD 1904
Fisher, R. A.	The Design of Experiments	Most influential book on conducting experiments to optimize physical systems	England	AD 1935
Turing, Alan	Can a Machine Think?	Definition of a computer	England	AD 1936
Von Neumann, John & Oskar Morgenstern	Theory of Games and Economic Behavior	Modeling of single-person rational decision making	USA	AD 1944
Miles	Value Analysis	Theory of value as amount of function provided divided by how much it costs to deliver	USA	AD 1947
Zwicky	The Morphological Method of Analysis and Construction	First systematic tool for creativity	USA	AD 1948
Hansen	Konstruktions-systematik	Theory of design as a systematic step-by-step process	Germany	AD 1955
Altshuller	Theory of Inventive Problem Solving	First systematic tool to explore past inventions	Russia	AD 1956
Osborne, Alexander	Applied Imagination	Brainstorming	USA	AD 1963
Simon, Herbert	The Sciences of the Artificial	Design is a study of nonphysical artificial phenomena	USA	AD 1969
	ASME Design Automation Conference	First ASME conference on design automation	USA	AD 1974
Stiny, George	Pictoral and Formal Aspects of Shape and Shape Grammars	Design can be represented as formal geometric rules	USA	AD 1975

TABLE 1.2 (CONT.). **DESIGN THEORY DEVELOPMENTS IN WORLD HISTORY**

Author	Publication title	Theory or development	Origin	Year
Pahl & Beitz	Engineering Design: A Systematic Approach	Systematic design as a complete process	Germany	AD 1977
Taguchi, Genichi	Jikken Keikakuho (System of Experimental Design)	Robust design	Japan	AD 1977
Suh, Nam	Axiomatic Design	Axiomatic design and design information content	USA	AD 1978
Mead, Carver, & Conway, Lynn	Introduction to VLSI Systems	Compiler theory of design synthesis, VLSI design	USA	AD 1979
Ross, Douglas		Structured Analysis and Design Technique (later called IDEF by the U.S. Department of Defense)	USA	AD 1970
Boothroyd and Dewhurst	Design for Assembly	Rules for easy assembly reduced to design principles	USA	AD 1983
Brown, D. & Chandrasekaran, B.	An Approach to Expert Systems for Mechanical Design	First expert system to do mechanical design	USA	AD 1983
Ullman, David	Proceedings, Design Theory and Methodology Conference	First ASME conference on design theory and methodology	USA	AD 1989

are at least two factors that are driving the development of design theory. The first is the need for greater corporate design efficiency. How can a group of engineers, whose members are constantly changing jobs, be made to develop new products of very high value, very quickly, at very low development cost?

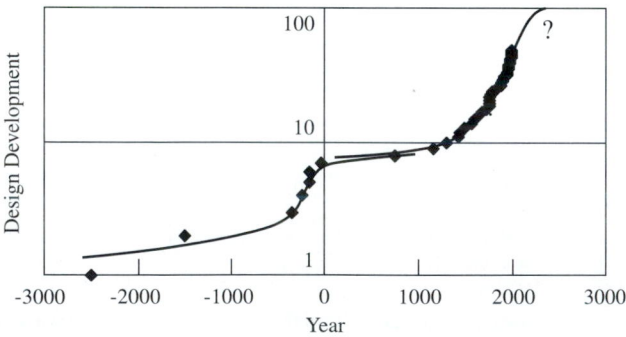

▼ *Figure 1.14.*
Design theory developments over time.

Clearly, design theory can help. If a design theory is developed that is *repeatable,* for example, then increased efficiency could be gained. By repeatability we mean that the results developed are independent of the team members (at least in certain stages and methods), the process is correct when followed, and it produces a consistent result given the problem's input conditions. Some methods presented in this book do achieve this level of rigor, but others only aspire to it. The interview method of developing a list of customer needs in Chapter 4 is repeatable, but the brainstorming methods of Chapter 10 generally are not. It is an aspiration of many design researchers to develop repeatable design methods that are useful in the corporate world. On the other hand, the inventing process will never be repeatable.

Another reason for the development of design theories is the need to understand how to design with our ever-improving new technology that becomes ever more complex. For example, the explosive growth of computing capability has had a tremendous impact on engineering design. Complex static and dynamic analyses of product design concepts are now routinely considered mandatory before any gate decisions are made. Optimization of parametric values to improve the results of these analyses are also being coded and automatically solved. The fraction of design activities that are being reduced to algorithms is growing: Once a task is reduced to an algorithm, its solution can be completed or approximated in comparatively little time, and the result is guaranteed repeatable. The demand for rapid development is pushing for more such advances in algorithmic synthesis of mechanical solutions and is so driving the advancement of design theory.

Such algorithms do not remove the designers from the product development loop. Routine or contentious tasks may become less art and more repeatable. These advancements then free the design team to enjoy and focus on the challenges of design tasks, where creativity, researched knowledge, and hard work can be brought to bear.

At least one other reason for the development of design theories is the need to improve the reliability of any design process. Today, this is enforced through gates, but the typical stage-gate development process is not ideal. A gate is an artificial process constraint: Some activities may not be ready to enter the gate, while other activities may be past due for a gate. Forcing all to be ready at the same level at one point in time forces inefficiencies: Some groups relax and others burn. Similar to what has happened in manufacturing, one can expect product development to become much more instrumented and measured in order to understand where difficulties lie and where to invest.

All of these speculations simply reflect the evolutionary trends of product development as a part of the ever-competing business environment. Product development is the exciting frontline of the continually evolving battle for markets, profits, and business in our capitalist system.

VI. SUMMARY AND "GOLDEN NUGGETS"

Modern product development involves the application of objectively formulated methods that are systematically configured to permit engineers to develop products efficiently:

▶ Every product development process is different, as appropriate for different companys' technological and market environment.

▶ Students of engineering design should examine each subsequent chapter to understand the levels of detail that can be developed in each topic and at least once experience the fully detailed version before making simplifications, shortcuts, and personal versions.

▶ Engineers need to develop their own product development process in any business they work and must continually strive to improve it.

▶ Reverse engineering and redesign is a forum for learning, experimenting, and living product design. After a handful of reverse-engineering experiences, prediction of how products are executed within their "black boxes" is much more apparent. Possible alternatives and ways to execute function are also readily visualized.

▶ The history of methods and the science of product design are rich with tradition and wisdom, but the ink on the historical manuscript is still wet. Many more chapters are yet to be written, contemplated, and analyzed.

References

Allaire, P., and M. Holmes. 1996. *Time to market: A 1996 report to Xerox*. Xerox Corporation.

Clausing, D. 1997. "Concurrent Engineering," *ASM Handbook: Materials Selection and Design*, Volume 20.

Cusumano, M., and R. Selby. 1995. *Microsoft secrets.* New York: The Free Press.

Manente, M. 1995. *The evolution and implementation of electronic prototyping systems,* Master's thesis, Massachusetts Institute of Technology.

Nasmyth, J. 1883. *Autobiography.* Edited by S. Smiles.

Stice, J. E. 1987. Using Kolb's learning cycle to improve student learning. *Engineering Education,* 77 (7): 291–296.

Tristram, Claire. 1998. The big, bad stuffers of IBM. *Technology Review,* July-Aug.: 44–51.

2 Product Development Process Tools

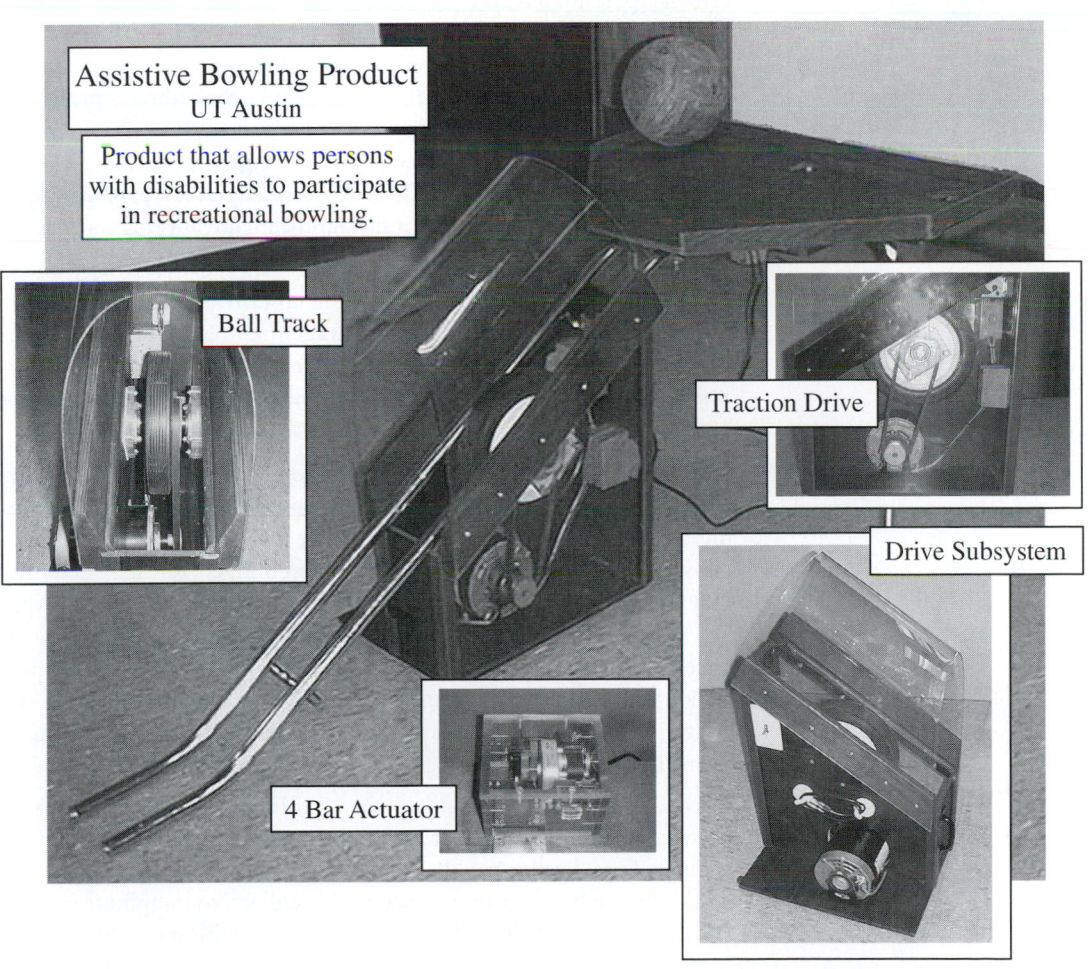

Assistive Bowling Product
UT Austin

Product that allows persons with disabilities to participate in recreational bowling.

Ball Track

Traction Drive

Drive Subsystem

4 Bar Actuator

The cover art for this chapter illustrates a recent assistive technology product (Jackson et al., 1999), a recreational bowling aid for people with disabilities. The product was developed by a graduate student design team, was wonderfully accepted by the clients, and continues to bring the enjoyment of bowling to people with disabilities. It includes a ball platform for either hand or remote activation of ball movement, a traction drive for increasing ball velocity, and a ball track to guide the ball. Without strategic team composition and product planning, this product would not have been successfully developed. It required a four-month development cycle time, and is fabricated as a kit to be assembled by the customer.

Albert Einstein once said, "I am enough of an artist to draw freely upon my imagination. Imagination is more important than knowledge. Knowledge is limited. Imagination encircles the world." (Viereck, 1929). Nowhere is this statement truer than in the field of product development. Product development requires an imagination for new technology, for novel solutions to both common and extraordinary problems, and for unique processes to improve people's lives. Without such imagination, there would be no new products; engineers would be left to routine exercises applying very limited knowledge. New products don't result from routine exercises nor do significant advancements occur in current products.

Product development processes and the task breakdown among team members should enhance and encourage the imagination and effectiveness of the human participants. These processes and their organization should not erect hurdles or roadblocks that must be leaped or bypassed, but must assist in stimulating our natural creativity.

One area of common roadblocks is the operations of a design team. Depending on the scale of a product, from kitchen utensils to commercial airliners, design teams may range in size from a couple of engineers to tens of thousands. Development budgets for these activities may likewise range from hundreds of thousands of dollars to billions. For these ranges of team sizes and budgets, imagination may be squelched simply due to the large bureaucracies, varied human personalities, and an inundating number of tasks. According to the old adage, "It's difficult to separate the trees from the forest"—many people will have many ideas.

In this chapter, we study tools for enabling creative development processes, with emphasis not so much on process mechanics as on organizing for collective creative consensus. These process tools will help us to remove barriers that obscure our collective imagination. Two areas of importance in this study are the composition of design teams and the ability to plan and schedule a product's development. If we can create a vibrant team environment, and if we aptly structure a team's activities, then the horizons of creativity are endless.

I. CHAPTER ROADMAP

Figure 2.1 illustrates the roadmap for this chapter. The first topic is product development teams. Important subtopics include team composition, strategies for teamwork, team-building exercises, and team evaluation. The second topic focuses on planning and scheduling. Basic methods are provided for the planning process and establishing milestones.

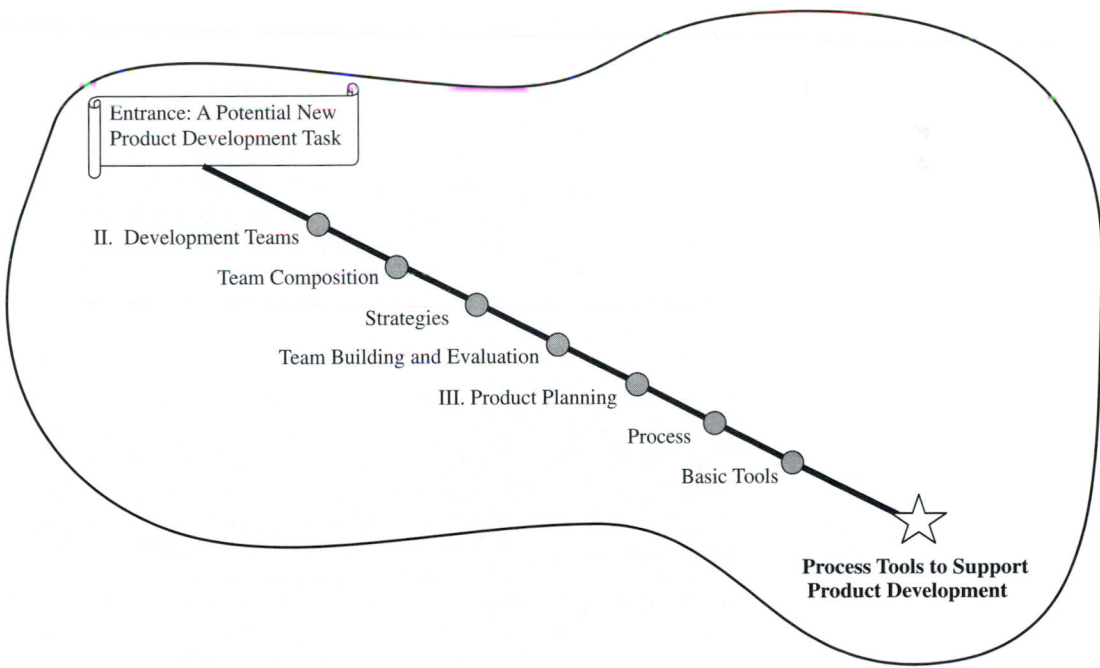

▼ *Figure 2.1.*
Chapter roadmap to product development process tools.

II. PRODUCT DEVELOPMENT TEAMS

Modern product development has provided innumerable artifacts for our use and convenience. While we have made great strides in technology, the trail of progress has been fraught with numerous perils, including technical, social, and political risks. Some of these perils still linger. Perhaps the most prevalent is the possibility of a dysfunctional team, one that does not converge to satisfactory product solutions. The concept of product development teams is thus studied in this section. For context, we begin with a brief history.

At its most basic level, product design has two important aspects: an idea emanating to satisfy some need and a physical embodiment of the idea, a product. The former entails the process of design, whereas the latter is a component of manufacturing. Historically, craftsmen developed both the idea and the product. They had no need to communicate the idea or embodiment to other persons. This craftsman era thus showed a natural unified approach to both design and manufacturing.

Today's approach to product development is much different. As products historically became more complex and the users more diverse and global, a gradual dichotomy between design and manufacturing was inevitable. Specialties and professions emerged in both disciplines, in contrast to their unified genesis. These disciplines greatly enriched our world's economic and technical growth; however, they also created new barriers to the development of products. Knowledge about manufacturing is no longer centered in design. A similar separation occurs for all disciplines. Knowledge about software is not centered in knowledge about mechanical systems; yet today, products worth substantial revenue involve both.

Significant issues emanate due to this role of specialization in product development. Product development requires collective knowledge generation—the key result is a set of control documentation that describes the new product—and so participants must be willing to share intellectually. Effectively organizing a design team is not like effectively organizing a construction project or other endeavors with well-defined actions and outcomes. Participants cannot just show up to work and keep quiet if they cannot stand to be around their coworkers; they have to get along and share, or the product development process will not be effective.

When creating products, the heart of these issues concerns our ability to compose the specialties into a smoothly working process. A number of questions arise about this composition. What defines a team? What team roles are needed to represent and integrate the specialties? Within these roles, how do we deal with the variety of human personalities and traits? What team strategies are effective for progressing from idea to physical product? For these strategies, how do we generate, evaluate, and improve team skills?

The following sections address these questions in succession. As we consider the answers, two important points should be kept in mind. First, successes in product development are shared as a team. Individual accolades must take a subservient role. The words of Mark Twain apply: "Obscurity and a competence. That is the life that is best worth living." Second, teamwork is not merely a mechanism for performing analysis and making decisions. The first requisite to any product's development is a set of novel ideas. "You can't depend on your judgment when your imagination is out of focus" (Mark Twain's Notebook). Fostering collaborative imagination is a fundamental goal in composing and managing a new product development team.

The Basics of Teams

A *team* may be defined as two or more persons engaged in a common goal, who are dependent on one another for results, and who have joint accountability for the outcomes. Think of different groups of people as they interact: people at a restaurant, the employees of a business, a basketball team in the Olympics, and so forth. Are these groups of people teams? If they meet each part of the definition, they are teams.

This definition helps us to identify teams, but it does not help us work as teammates. Beyond the definition, there are a number of necessary characteristics that must exist to be an effective team. The *PRIDE principles* should be followed in any product development project— *Purpose, Respect, Individuals, Discussions,* and *Excellence.*

First, all members must have a clear understanding of a common *purpose* and goal. The team should seek to develop a mission statement (Chapter 3) for this characteristic. Further, all team members should work out any issues with the mission statement as a team. Only then will a team have the common purpose that the mission statement reflects.

Second, all team members should endeavor to act with mutual *respect,* trust, and support. Without these attributes, a team will tend to work either as a set of individuals or as a connected set of bodies with only one head. Repetitive work will occur and progress will be slow (if at all).

Third, a design team should respect and productively utilize *individual* differences. Such differences enhance creativity and the collective imagination of the team.

Fourth, a team should exercise open, honest, and frequent *discussions.* Product development is very dynamic. Without frequent communication, a design team will tend to iterate unnecessarily. Further, team members should exhibit the ability to accept leadership authority and to manage conflict. Ego has no place in a team. Consensus is the goal, with individuals accepting the leadership decisions when they are needed. Conversely, those exercising leadership have the responsibility to ensure that everyone is on board for all key decisions. Nobody can feel outside the team decision making.

Finally, a team must strive for *excellence* in all actions. A team will not have pride in work that is less than the best. Every member should strive for excellence in their work and help the rest of the team also achieve excellence.

While perhaps not explicit, the PRIDE principles are used in some form by every successful product development team; they are a simple necessity. Very successful teams are explicit about it, often visiting each principle in periodic, open team meetings. If they are followed, teams want to work together toward excellence. If they are not followed, teams go astray.

Team Composition: Seeking Synergy, Unity, Competence, and Consensus

Every product development team requires certain roles to be fulfilled. Obvious roles relate to the professional needs of the team. These needs include technical expertise in a variety of areas. Example areas include management, manufacturing, mechanical engineering, electrical engineering, testing and prototyping, materials, industrial design (style, ergonomics, aesthetics, and product feel), solid modeling, suppliers and the supply chain, marketing, quality, and other specialties.

Beyond technical expertise, members of teams must also fulfill roles as general problem solvers and team players. These roles relate to the intraworkings of the team. They also relate to how team members communicate, how schedules are met, how commitments are formed, how members negotiate with customers, how members negotiate among themselves, how decisions are made, and how the team gains energy and momentum.

Team Roles

There exist a variety of ways to describe and present team roles. One approach is to distinguish important roles according to titles and brief distinguishing features. Sixteen team roles are identified below, using this structure. It should be kept in mind that all persons perform all of these roles to some degree or other. The categorized roles are not meant to pigeonhole or otherwise limit a team. They are simply meant to provide guidance and assist in understanding how a design team should function.

Team roles must be carefully studied and considered. Each and every team requires that these roles be fulfilled with strong participation and with high skill. If any of these roles are not adequately fulfilled, a team will be missing an important anatomical component. Problems will then arise that will cost time and resources (perhaps even success).

Roles include (Wilde, 1993; Wilde et al., 1998):

▶ *Administrator/Reviewer:* Monitors project and judges outcomes accurately. Sees the "big picture." Compares results with goals.

▶ *Troubleshooter/Inspector:* Repairs problems and solves difficult impediments to progress.

▶ *Producer/Test Pilot:* Brings tasks to fruition and reality. Treats tasks realistically. Pushes performance envelope. Makes things happen.

▶ *Manager/Coordinator:* Supervises and leads tasks. Encompasses a practical perspective. Focuses effort and saves time.

▶ *Conserver/Critic:* Preserves the team's and project's goals and concerns. Addresses aesthetic and moral issues.

▶ *Expediter/Investigator:* Experiences the team goals, gets facts and know-how.

▶ *Conciliator/Performer:* Detects and fixes interpersonal issues.

▶ *Mockup Maker/Prototyper/Modelmaker:* Builds and tests rough prototypes.

▶ *Visionary:* Imagines various product forms and uses.

▶ *Strategist:* Speculates on and plans project and product future.

▶ *Needfinder:* Evaluates human factors and consumer issues.

▶ *Entrepreneur/Facilitator:* Explores new products and methods, inspires, and motivates.

▶ *Diplomat/Orator:* Harmonizes team, client, and custome·

▶ *Simulator/Theoretician:* Attempts to understand pheno· alyzes performance and efficiency.

> ◗ **Innovator:** Synthesizes new products; improvises solutions.
>
> ◗ **Director/Programmer:** Sets deadlines and breaks bottlenecks.

Psychologist Meredith Belbin describes an alternative nondisciplinary categorization of team roles (Belbin, 1993). Belbin proposes nine different behaviors required for a functional team, regardless of expertise:

◗ **Organizer:** a reliable person concerned about the practical aspects of the design process

◗ **Motivator:** a confident person in charge of the schedule and goals of a design team

◗ **Pusher:** a dynamic person forcing a design team to work faster

◗ **Soldier:** a creative person predominately generating solutions

◗ **Gatherer:** an extroverted person searching for information and communicating with others outside the team

◗ **Listener:** a perceptive person perceiving and combining the ideas and statements of others

◗ **Completer:** a conscientious person eliminating the last flaws of a design

◗ **Specialist:** a dedicated person with extensive knowledge in a special field

◗ **Evaluator:** a strategically thinking person concerned about alternative solutions

To use either category of roles, a team should identify members that will actively and responsibly assume those roles throughout the project. These decisions may be made through personality typing, discussion, managerial decisions, or team-building exercises. The key is that the roles are skillfully performed, with periodic review to rate the team's progress and execution.

Myers-Briggs Type Indicator

> "There is a great deal of human nature in people."—Mark Twain

> "It is easier to go to Mars or to the Moon than it is to penetrate one's own being."—Carl Jung

Building on team roles, let's take a closer look at how people work. The purpose is not to perform the role of an amateur psychologist; rather, we want to understand how human personalities can be made to work effectively together as a team. We all know there are outgoing types of people and there are quiet types of people, each different. There is no worse and no better type, just different.

Different personalities, however, are often better suited to different tasks, ones they naturally prefer. If it makes no difference for other reasons, then it may make sense to have an outgoing team member who enjoys talking to other team members be the one to work on integrating systems rather than on subsystem debugging. Further and more important, it is good to let the other team members know one's tendencies. We may really want to be the systems integrating person, say, for career reasons, even if our personality is not suited to it. Informing our teammates of our tendencies goes a long way toward helping them understand our actions. If they know we tend toward a certain type, they know better how to respond when they are concerned over excessive or lack of action. A brief study of personality types provides the fundamentals for this discussion (Isachsen and Berens, 1988; Wilde, 1993; McCaulley, 1990; Kersey and Bates, 1984; Jensen, Murphy, and Wood, 1998).

The *Myers-Briggs Type Indicator* (MBTI) is a simple measurement indicator of how people behave and contribute in a work environment. The MBTI is based on the work of Carl Jung (1875–1961), who brought psychology to accept the concept of personality. Jung's model of psychological types describes four categories to distinguish personality: how a person is energized (*Extroversion* vs. *Introversion*), what a person pays attention to (*Sensory* vs. *Intuition*), how a person decides (*Thinking* vs. *Feeling*), and what kind of outlook on life a person adopts (*Judgment* vs. *Perception*). For each category of type, a person is assumed to have an intrinsic preference for one of each pair over the other, thereby defining sixteen different *personality types*. Isabel Myers and Katherine Briggs applied Jung's theory of personality types and made it practical through a questionnaire-based measurement system to assess one's personality, the *MBTI personality indicator*.

Taking a closer look at the MBTI types, the first category describes whether a person interacts with his or her environment, especially with people, in an initiating (extroverted) or more reactive (introverted) role. Extroverts tend to gain energy from their surroundings, while introverts usually gain energy by processing information internally.

The second category gives information on how a person processes information. Those who prefer to use their five senses to process the information (sensors) are contrasted with those who view the intake of information in light of either its place in an overarching theory or its future use (intuitors). This sensor versus intuitor category is seen by most researchers to be very important in terms of implications for education and teamwork (Myers, 1985; Lawrence, 1982).

The third category for MBTI preference describes the manner in which a person evaluates information. Those who tend to use a logical "cause and effect" strategy (thinkers) are contrasted with those who use a hierarchy based on values or on the manner in which an idea is communicated (feelers).

The final MBTI type category indicates how a person makes decisions or comes to conclusions. Those who tend to thoroughly consider all of the data (perceivers) are contrasted with those who tend to summarize the situation as it presently stands (facts) and make decisions quickly (judgers). The four letter combination of these indicators ("E" vs. "I" for extrovert and introvert; "S" vs. "N" for sensor and intuitor; "T" vs. "F" for thinker and feeler; "J" vs. "P" for judger and perceiver) constitute a person's MBTI "type". Table 2.1, as adapted from the Myers-Briggs Type Indicator (Myers and McCaulley, 1985; McCaulley, 1990; Jensen, 1998), gives a further overview of the four MBTI categories.

For the purpose of understanding design teams, an MBTI assessment should be taken by all members in a team. Such assessments are administered by trained psychologists or from public-domain sources

TABLE 2.1.   SUMMARY AND OVERVIEW OF THE MBTI TYPES

	Manner in which a person interacts with others		
E	Focuses outwardly on others. Gains energy from others.	Focuses inwardly. Gains energy from ideas and concepts.	I
	EXTROVERSION	INTROVERSION	
	Manner in which a person processes information		
S	Focus is on the five senses and experience.	Focus is on possibilities, future use, big picture, and ideas.	N
	SENSING	INTUITION	
	Manner in which a person evaluates information		
T	Focuses on objective facts and causes & effects.	Focuses on subjective meaning and values.	F
	THINKING	FEELING	
	Manner in which a person comes to conclusions		
J	Focus is on timely, planned conclusions and decisions.	Focus is on adaptive process of decision making.	P
	JUDGMENT	PERCEPTION	

(Kersey, 1984). The results will be a four-letter code, such as "INTJ," with the weightings of the strength for each preference (letter). Summaries will also be provided as to the meaning of each four-letter type.

Based on these summaries, design teams can discuss the inherent differences in the way people with distinct types approach teamwork. For example, judgers (J) and perceivers (P) tend to deal with the world in different ways. J types tend to be goal oriented, responsible, and dependable. P types tend to listen and watch, be curious, and take time in arriving at decisions. Both of these types have very positive contributions to make to a team. However, they will tend to have different timings and processes for decision making. If a team does not understand these differences (up front), conflicts may arise in the team. Teammates may tend to point a finger at one another, even though the only "problem" is that they have different personality types. Through a discussion of personalities, a team can compromise and plan for how the differences will be used as strengths, not as conflicts.

Besides motivating team discussions, the MBTI may also be used to understand team roles and potential weaknesses that may exist in a team. For example, if a team includes introverted-intuitors (IN), these persons will not tend to share their creativity and ideas outwardly, at least as much as extroverted-intuitors (EN). It is then important to include team members with strong extroverted-feeling (EF) types. EF types tend to bring teammates together and facilitate the exchange of information from all other types.

Figure 2.2 shows plots of team roles versus MBTI types. These plots focus on the critical tasks of information-gathering and decision-making in a team. As shown in the figure, roles tend to be defined by pairs of MBTI preferences. The first letter in each pair signifies the greater relative strength between the preference types. While a given team member will typically not gravitate toward only one octant in these plots, they will tend to focus on a subregion, say two or three octants in succession (e.g., strategist through innovator). Based on MBTI test results, team members can use these plots as indicators of their preferred roles. Of course, these preferences may be overridden by member choices regardless of the MBTI test scores.

Overall, the MBTI is a useful model of the way persons interact. However, the user should not attempt to overuse or read too much into the MBTI. It is a model to facilitate effective communication. It is not, however, a tool for us to become amateur psychoanalysts, nor is it ever intended to pigeonhole our teammates with certain labels or certain tasks. We all have the ability to change our behavior depending on the circumstances, and we all have the ability to tackle any job. The MBTI simply points out our preferences for interacting with colleagues and the world around us in any task.

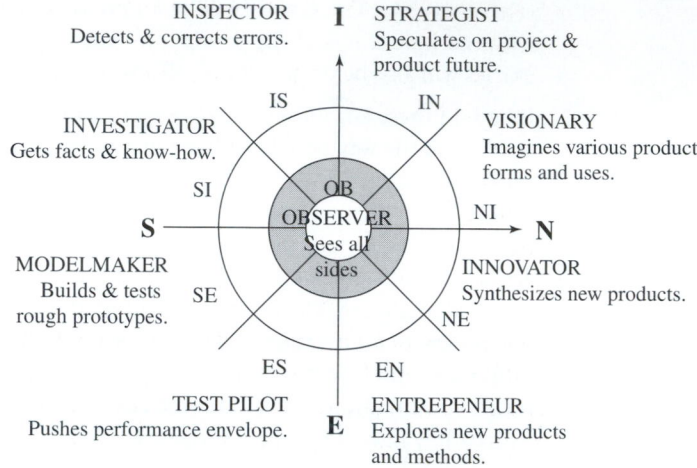

INFORMATION-GATHERING

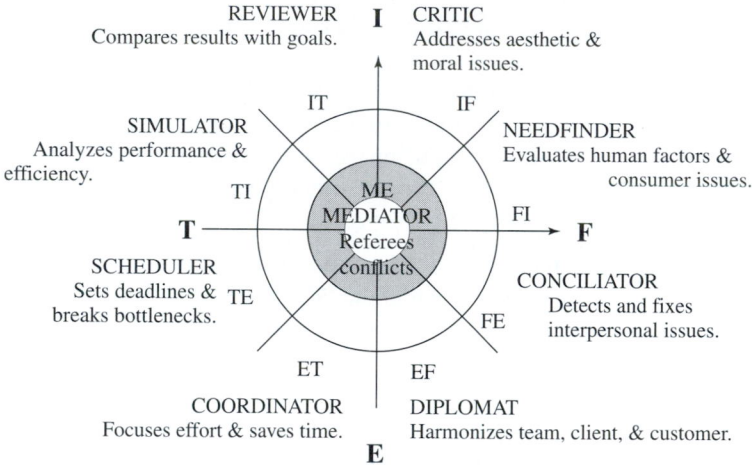

DECISION-MAKING

▼ *Figure 2.2.*
Design team annotated role maps with MBTI types (Wilde et al., 1998).

Strategies: Team Structures

Equally important to team roles is the structure of product develop-ment teams within business organizations. A number of potential or-ganizational structures exist. Examples include sequential functional organizations, project core teams, matrix organizations, and integrated-product teams.

As discussed above, the history of product development led to a dichotomy between design and manufacturing. The craftsman era was transformed into the age of mass production and technical area specialties. This transformation was handled through sequential functional organizations (Figure 2.3) (Funke, 1997). Product development typically began with marketing, followed by product design, engineering, manufacturing, and customer support. At a simplified level, each phase produced outputs for the next stage. Very few communication paths existed among functional groups. Without such paths, walls naturally formed between groups, and engineering changes were continually needed to correct uninformed decisions made in earlier stages.

With increased competition, a focus on customers, and reduced cycle times, this "over-the-wall" approach proved to be inefficient. Skills from all functional groups are needed at all stages of a product's development. This need led to new organizational strategies based on the concept of *simultaneous* or *concurrent engineering*

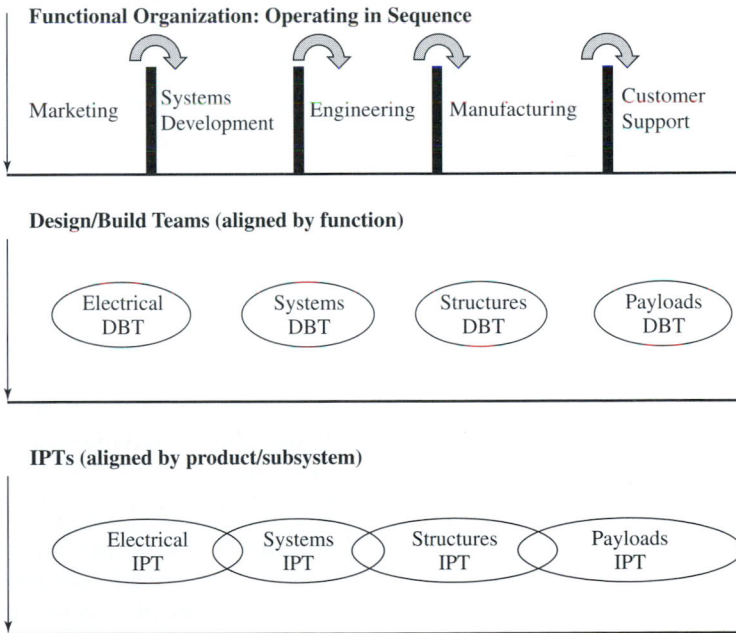

▼ *Figure 2.3.*
Evolution of product development teams at Boeing (adapted from Funke, 1997).

Concurrent engineering is the effective simultaneous development of different disciplinary subsystems required for a product launch. This typically includes product subsystems (mechanical, electrical, software, etc.) and downstream activities, such as manufacturing, procurement, and maintenance disciplines, that all must start gearing up for the new product. The key reason for concurrent engineering is to include the voice of all these disciplines into early product development decisions, before choices are made that adversely affect the downstream team members. Concurrent engineering also works to compress the development cycle by providing the ability for downstream team members to get started on their design activity earlier. If the manufacturing engineers know of and have influence on early key design decisions (injection molded plastic vs. sheet metal for some key parts), they can begin to start their operational decisions, such as whether to buy a new press, with large lead times.

Multifunctional Teams

Concurrent engineering focuses on merging technical and business specialties to produce quality products on short time scales. It also emphasizes the support of design teams and the equal importance of the design, manufacturing, and service processes, not just the product that is produced (Funke, 1997). At the heart of concurrent engineering is the idea of *multifunctional teams,* also known as *integrated product teams.* As shown in Figure 2.3, functional groups are aligned to subsystems or modules of a product. Each of these groups is composed of the required functional skills from marketing through manufacturing and customer support, depending on the subsystem.

Based on this composition of team members, product development tasks are developed in parallel. Figure 2.4 shows an example concurrent engineering process for Boeing (Funke, 1997). Notice in the figure that systems design, analysis, structures, manufacturing engineering, and tooling design all progress in parallel. Decisions do not follow in a sequential process. Instead, the process exploits the natural subsystems of the product, focusing on the early involvement of manufacturing and tooling specialties. For example, this process resulted in a reduction of cycle time (design plus drawing release) for the Boeing 777 from 40 months to 27 months, compared with the Boeing 767.

With this type of process, the question arises: How do we determine the parallel tasks for a given product development scenario? What functional specialties are needed within each of the integrated product teams? The last section of Chapter 9 describes a systematic approach for forming integrated product teams. This approach uses modular design methods to identify parallel subsystems for a product early in the process.

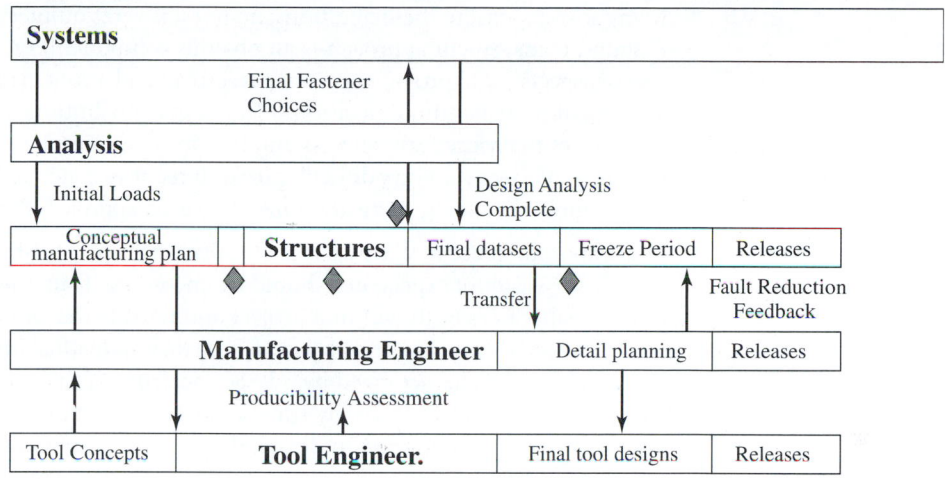

◆ : Denotes milestones.

▼ *Figure 2.4.*
Concurrent engineering process at Boeing.

Team Building (Basic Activities)

A business environment (or university setting) should encourage effective teamwork. There should exist a natural enthusiasm to work together, to support each team member, and to work cooperatively toward a common goal.

Each team member is a dependent in this environment. As dependents, professionalism should be paramount, where the mutual goal is to succeed together. Only through sacrifice of self, before the team, can the full potential of the product development be realized. All members must contribute with mutual trust and to the fullest of their abilities. As spoken by U.S. President J. F. Kennedy, "Ask not what your country can do for you, ask what you can do for your country." In design team parlance, we should ask what we can do for our team and the customers we must support.

Unfortunately, these self-evident truths are difficult to apply. Business environments are not always structured to encourage effective teamwork. Even if they are, human frailties, schedules, and many other factors intrude on the ideal of how a team should function. What can we do to overcome these factors? How do we foster collegial and rewarding work relationships, especially at the initiation of project?

Many answers exist to these questions; no one answer is quintessential. A sound management approach is an obvious component. Other essential aspects are to provide necessary resources and a concurrent environment to make sound decisions. Yet, these contributions are still insufficient to ensure team success and buy-in. The construction and nurturing of a team is a very difficult puzzle. It requires concerted effort and a number of pieces, correctly situated, to even approach the ideal.

One important piece of this puzzle is to actively pursue team-building exercises. We cannot expect, nor should we encourage, team members to be collaborators both within a project and outside the work environment 24 hours per day. Members will have their individual interests. They will also be characterized by a large spectrum of backgrounds, age groups, marital status, family situations, etc. Team-building exercises seek to bring a common understanding and a common communication for the workplace. These exercises are organized activities that are directed by a facilitator. They are usually short in duration and occur outside the context of the actual project. They work by providing a subtle and "risky" task that can only be completed by cooperative team efforts.

Once a task(s) has been attempted, a forum should then be provided for the team to discuss the results of the activity. Successes may be celebrated, imprinting their causes on the team psyche for future reference. Weakness should be similarly analyzed. These weaknesses should be discussed honestly and in the context of member personality types and skills. Through these discussions, team strategies may be adopted for overcoming the weaknesses. As pressures grow in the actual project, where much more is at risk, these strategies may be consciously enacted to transform potential problems into successes.

A variety of team-building exercises may be created and implemented. Common examples include a *blind walk,* where members lead each other blindfolded through sensory tasks. Alternatively, the team may work on outside sports or recreational activities. For instance, the team may travel on a rafting trip, creating circumstances where team members are dependent on each other for guiding and maneuvering a boat. The team may scale a rock face or climb a vertical wall. These activities again have a clear goal that requires a conscious team effort for success.

Besides sports and recreation, analytical and mathematical puzzles may be used as team activities. A seemingly simple task would be described to the team, props provided, and the team is asked to solve the task. This task must again be nonobvious, it should require cooperative effort for solution, and it should be structured to identify team weaknesses and strengths. Example puzzle activities include mystery games or geometric arrangements. In each case, the team members are given clues or hands-on components to help solve the task. Rules are established for sharing and transferring information. However, each member receives different and incomplete information. Only through a positive and thoughtful exchange can the task be solved.

Overall, team-building activities can greatly enhance the ability of a team to perform. Endless lists of possible activities exist. Any one activity, however, should have the following characteristics:

1. a clear goal or set of goals that are nonobvious (i.e., cannot be solved by inspection or previous knowledge)

2. a task the requires team cooperation and leadership for success

3. inherent risk for failure, at least partially

4. a task that is not part of the everyday job or actual project

5. a facilitator to help guide the team when a catalyst is needed

6. an independent observer (maybe a chosen team member) that records the performance and responses of the team, outside the "heat of battle"

7. a forum to discuss the activity, analyzing successes and failures

8. (optional) a competitive environment with other teams, with a prize awarded to the winning team

An Exercise in Tangrams

To illustrate a potential team-building activity, consider the mathematical puzzle known as a *tangram* (Elffers and Schuyt 1997). A tangram is a basic geometric puzzle (game) comprised of seven fundamental geometric shapes, known as *tans*. Figure 2.5 illustrates the seven tans of a tangram. The goal of this puzzle, dating back to its Chinese heritage, is to construct a perfect square from the tans. Beyond the basic square, the tans may be reconfigured into many different layouts, representing recognizable shapes, such as a shark, mountain range, and bridge.

As shown in the figure, the tans consist of one square, one parallelogram, and five triangles. The five triangles, in turn, have three sizes, small (2), medium (1), and large (2). These tans may be arranged in a plane using the operations of translation, rotation, and flipping. Depending on the points of contact between tans, many shapes may be created, as illustrated by the square on the right side of Figure 2.5.

When played as a game, tangrams usually begin with a known shape displayed to the player. The player then attempts to reconst shape by manipulating the tans. The goal is the smallest qu

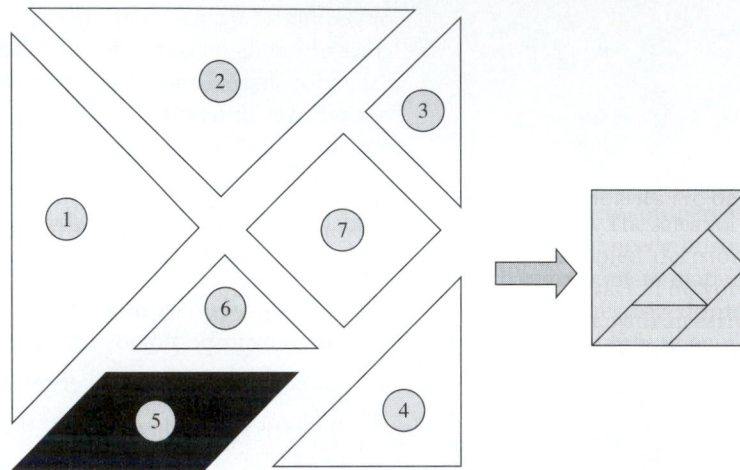

▼ Figure 2.5.
Tangram puzzle layout, showing individual tans.

time to construct the displayed tangram. Alternatively, the goal of the game may be to create a verbally described or interesting shape within a set period of time. The resulting tangrams are then judged for their aesthetics or similarity to the requested shape.

To appreciate these gaming strategies, it may help to understand the geometric relationships between the tans. These relationships are intriguing, and they govern the aggregate tangrams that are produced. Let's assume that the area of the square formed from the tans is 16 units. Each of the large isosceles triangles is then four units of area. Because the square and two small isosceles triangles combine to form the large triangle, and because the two small triangles form the square, the small triangles are one unit of area each, and the square is two units in area. By similar reasoning, the small triangles may be configured to form the medium triangle and the parallelogram. Thus, these shapes are two units in area each. Based on these individual relationships, any tangram configured from the tans will have a total area of 16 units.

Using these relationships, the square shown in Figure 2.5 may be constructed from the tans. This square is unique except for its mirror images. To create any mirror image, the tans need only be rotated and translated, with the exception of the parallelogram. The parallelogram will need to be flipped, as demonstrated by its black shading in the figure.

How might we extend the tangram puzzle for use in team-building exercises? First we provide a set of tans to each member of a team (except for perhaps a chosen observer). Colors may be used for each set of tans. The goal is then stated for the team members to construct a perfect square from the tans, one per team member. A set of communication and interaction rules should be established to create a team learning environment. Likewise, it should be stated that the team is not successful unless all tangrams are completed. Individual success in a short time is not a team success.

When analyzing this proposed exercise, a number of interesting issues arise. First, if each team member is given a distinct seven-piece tangram, the focus will be on individual work. There is no intent to exchange pieces or observe different tasks. To deal with this issue, the pieces from each separate tangram should be mixed up, with each member still receiving seven pieces. Sharing between members will thus be needed to complete the group task.

A second issue concerns the communication rules. If verbal or other types of requests are allowed, both for giving and requesting pieces, then the exercise will be biased to dominant players or good puzzle solvers. To reduce this bias, two rules should be established. Players may only offer pieces to other team members (no grabbing of pieces from other players), and no form of communication between teammates may exist except for hand offerings. No verbal or other body language may be used to assist players in creating the tangram. No grunting approval or swinging arms/pieces to show how to construct the other player's tangram.

Finally, the issue arises concerning the difficulty of the tangram puzzle. Seven geometric shapes must be manipulated using rotations, translations, and flips. This task will take quite awhile for persons unfamiliar with tangrams (and no introduction should be provided). To deal with this issue, an option is to combine individual tans, such as "1" and "2" in Figure 2.5 to create a reduced number. This option can be applied to result in four or five total pieces distributed to individual team members. The combinations of tans have many variations. Likewise, no single team member will have an identical set of tans to create the required square tangram. The result is a more interesting exercise that has reduced complexity in number of tans, but greater teamwork is required in creating tangrams that are not identical in individual tan shapes.

These insights lead to the following tangram team-building exercise (Wilde et al., 1998):

1. Fabricate or purchase tans for use in the exercise. Tans may be constructed from acrylic, laminated construction paper, cardboard, or foam core. These materials will endure many trails

and be easy to manipulate, provided they are large enough. Typically, the legs of the large triangle should be at least 8 in (or 16 cm) in length.

2. Create packets or envelopes of tans, with each packet containing tans that can't be formed into the correct tangram square. As an example, for a team of three persons, four tans may be created for each packet. The three tangram squares may be chosen as {1, 2, 3-7, 4-5-6}, {1-2, 3, 4, 5-6-7}, and {1, 2, 3-5-6-7, 4}. The tans for each tangram are then either the original tans or combinations of originals. To create the combinations, some of the original tan edges are not cut. These tans are then mixed, and a random set of four is entered into each packet. To return the tans to their original packets at the exercise's conclusion, the chosen tans may be colored or numbered.

3. Form groups of 4–5 members.

4. Distribute the packets to each team member.

5. State the goal for the team to create identical squares, one per member.

6. Establish communication rules: (1) tans may only be offered by hand to other members; (2) no tans may be requested; and (3) no communication may exist, verbal or otherwise, except for the offering of tans.

7. The observer should take notes on how the exercise proceeds. How does each team member react? Does frustration occur? What keys and catalysts led to successes? What happened during false starts? Where did team members focus their eyes? How were individual tasks handled versus team tasks? Were team members committed to the exercise? Why or why not? And so on.

8. Hold a post-discussion session where all members state their likes and dislikes regarding the exercise and the team's performance. The results should then be studied in the context of any future team activities.

This exercise will result in a variety of responses by teams. The game may result in squares that are not the correct size, causing false starts or incorrect intermediate states. Likewise, the solution to all tangrams will be difficult to visualize.

These characteristics in tangram game play lead to many observations about a team's responsiveness, communication, leadership, and commitment. Who tried to solve everyone's puzzle? Who worked without looking at any other person? Who overly assisted all others? From these observations, a team will learn many important lessons and discuss topics that are critical for success.

Team Evaluation

As a project progresses, it is important to continually evaluate the psyche of the design team. Any relationship will constantly change. Outside factors and the passing of time simply inhibit a status quo. The goal then is to seek continual growth and maturity in a team setting.

Team surveys are tools to help us in seeking continual growth. These surveys are typically created to ask core questions. The team then answers these questions in an attempt to identify both positive and negative characteristics. Through this process, a team will be able to discuss, at a professional and candid level, any important team issues. If such issues are not identified early and continually revisited, they will become cancers to the team. Unresolved issues even have the possibility to stagnate a team's progress, completely debilitating the team's skills and enthusiasm.

Figures 2.6 and 2.7 illustrate an example evaluation tool for use by a team. It focuses on the important team attributes of unity, self-direction, group climate, communication, distribution of leadership, responsibilities, problem solving, conflict management, decision making, and self-evaluation. These attributes are rated on discrete scales, from poor to best.

A team should periodically complete evaluations of this type. First, individual team members should perform the evaluation, outside the workplace. The team then meets to perform an aggregate group evaluation. During the answering of each question, each member voices his or her honest opinion of performance. To remain at a professional level during such discussions, it should be a team guideline to state one positive comment for every negative criticism.

After completing the evaluation, action items are proposed for improving performance in problem areas. Successes should also be celebrated by some action of the team, such as a reward to the team. The team leadership is then responsible for maintaining the action items and rewards.

It should be noted that the results of team evaluations should not be reported to management, unless extenuating circumstances exist. Upper management should respect this activity. The evaluations are completed by the team, for the team. Team evaluations are not to be used

---

SURVEY OF TEAM DEVELOPMENT (Page 1/2)

TEAM'S NAME: _____ DATE: _____

FOR EACH AREA, CHECK DESCRIPTOR MOST RELEVANT TO THE TEAM

1. UNITY (Degree of unity, cohesion or "we-ness")

_____ Group is just a collection of individuals or sub-groups; little group feeling.

_____ Some group feeling. Unity stems more from external factors than from real friendship.

_____ Group is very close, there is little room or felt need for other contacts and experience.

_____ Strong common purpose and spirit based on real friendships. Group sticks together.

Comments/Clarification: _____
_____
_____

2. SELF-DIRECTION (The group's own motive power)

_____ Little drive from anywhere among members.

_____ Group has some self-propulsion but needs considerable push from leader (or outsider).

_____ Domination from a strong single member, a clique, or the leader.

_____ Initiation, planning, executing, and evaluation comes from total group.

Comments/Clarification: _____
_____
_____

3. GROUP CLIMATE (The extent to which members feel free to be themselves)

_____ Climate inhibits good fun, behavior and expression of desire, fears, and opinions.

_____ Members freely express needs and desires; joke, tease and argue to detriment of the group.

_____ Members express themselves but without observing interest of total group.

_____ Members feel free to express themselves but limit expression to total group welfare.

Comments/Clarification: _____
_____
_____

4. COMMUNICATION (Manner of expressing ideas/views/opinions)

_____ Very little discussion occurs.

_____ Members listen carefully but tend not to express strong views.

_____ There is much expression of "own" opinions but little careful listening or probing.

_____ Communications are open and two-way and discussions are lengthy and deep.

Comments/Clarification: _____
_____
_____

5. DISTRIBUTION OF LEADERSHIP (Extent to which leadership roles are distributed among members)

_____ None of the members take leader roles.

_____ Some of the members take leader roles but many remain passive followers.

_____ Many members take leadership but one or two are continually followers.

_____ All members of the group share leadership responsibilities.

Comments/Clarification: _____
_____
_____

6. DISTRIBUTION OF RESPONSIBILITY (Extent to which responsibility is shared among members)

_____ Everyone tries to get out of jobs.

_____ Responsibilities are carried out by a few members.

_____ Some members accept responsibilities but do not carry them out.

_____ Responsibilities are distributed and carried out by all members.

Comments/Clarification: _____
_____
_____

Composite of rating suggestions from (Dimock, 1987) and (Blake and Mouton, 1985)

▼ **Figure 2.6.**
Example team evaluation form.

SURVEY OF TEAM DEVELOPMENT (Page 2/2)

TEAM'S NAME: _____ **DATE:** _____

FOR EACH AREA, CHECK DESCRIPTOR MOST RELEVANT TO THE TEAM

7. PROBLEM SOLVING (Group's ability to think straight, make use of everyone's ideas, and decide creatively about its problems)

_____ Not much thinking as a group. Decisions made hastily, or group lets leader(s) do most of the thinking.

_____ Some cooperative thinking but group gets tangled up in pet ideas or prejudices of a few concepts. Confused movement toward good solutions.

_____ Some thinking as a group but not yet an orderly process.

_____ Good pooling of ideas and orderly thought. Everyone's ideas are used to reach the final plan.

Comments/Clarification:
_____
_____
_____

8. CONFLICT MANAGEMENT (How group deals with conflict)

_____ Group ignores or avoids disagreement to detriment of task(s).

_____ Group discussion is friendly but little analysis of the problem occurs.

_____ Members argue their own points and try to dominate others and/or always outsider to resolve the disagreements(s).

_____ Differences are presented, thrashed through to sound understanding, and usually acknowledged in decision making.

Comments/Clarification:
_____
_____
_____

9. DECISION MAKING (How group arrives at final decision)

_____ One or two members lead group and dominate decisions.

_____ Decisions are always hard to reach and/or group allows deadlines to dictate course to take.

_____ Decisions are hastily made, without working through options.

_____ Decisions are reached after thorough consideration of options and consequences of each possible option.

Comments/Clarification:
_____
_____
_____

10. GROUP SELF-EVALUATION (How group reflects on it effectiveness)

_____ There is little or no discussion of team performance.

_____ Faultfinding and criticism often dominates self-evaluation by group members.

_____ Self-evaluation by group focuses on how to achieve quick results in the time allowed for task completion.

_____ Group stops periodically to critique its performance as part of members' efforts to develop effective team-working skills.

Comments/Clarification:
_____
_____
_____

Composite of rating suggestions from (Dimock, 1987) and (Blake and Mouton, 1985)

▼ *Figure 2.7 (cont.).*
Example team evaluation form.

for job performance evaluations. They are for the purpose of ensuring a healthy product development activity.

Closing: Product Development Teams

Product development teams produce products, not individuals. When forming a team, while it is important to understand and consciously implement team roles, it is also important to structure team activities according to concurrent engineering practices. The methods in the subsequent sections and chapters of this text support these team goals. These methods seek to form *concensus, commitment,* and *communication.* Example methods include product planning, scoping, customer needs analysis, Quality Function Deployment, Pugh concept selection, concept generation, and robust design.

III. PRODUCT DEVELOPMENT PLANNING

Product development is fraught with risk. Companies are under severe demands to produce successful and high-quality products with the smallest cycle times and investment costs. These demands force us to plan products very carefully. As good product developers, we seek to predict cycle times, costs, and labor within 10% of what is actually needed, yet this prediction is very difficult. Tools are thus needed to assist in this task.

At a very basic level, a product-development product entails a set of activities. This set includes both routine and nonroutine tasks, distinct start and finish dates (at least for distinguished stages), and constraints on resources. Our task is to manage these activities as a product development team. The context, as previously discussed, is always with respect to concurrent engineering and products as families, not just single offerings.

To determine the specific activities for a given product, the following questions should be asked: What is the scope of the product (Chapter 3)? What skills do the team members need, and how will they be effectively coordinated? What are the deliverables, as requested by the customers? What limits exist in terms of time, money, personnel, and equipment? What analogous projects can we reference to aid planning? What standards or regulations exist that must be followed (associated tasks)? What resources do we need? What portfolio and product architecture strategy will we be pursuing (Chapters 8 and 9)?

Once these questions are answered, iteratively and over time, an organized list may be created. This list should be refined and organized, scoping the project as a function of time, resources, and "just-in-time" personnel. With such refinements, a project may be tracked with respect to progress and costs. It may also provide a tool to communicate with the team, customer(s), and management.

With this introduction to product planning, let's consider some very basic tools. These tools will assist in organizing a product, where a rule of thumb states that *10% of our initial time will be spent planning and another 5% spent in process measurements to monitor development.*

Planning Process

A basic tool for product planning is to follow a set of systematic steps. These steps are intended to estimate four basic aspects: the "what"—tasks, the "when"—schedule, the where—equipment and facilities, and the "how"—people, material, facility, and equipment costs.

Steps:

1. ***Identify the tasks and milestones of the project:*** The tasks for a project may be estimated from the questions previously given or from analogous projects. These tasks should be developed in concert with scoping the project (Chapter 3) and customer needs analysis (Chapter 4). For each major task, a milestone should be established to monitor progress. Milestones are important checkpoints. They have clearly defined deliverables and should realistically show a project's success. Example deliverables may be prototypes (proof-of-concept, alpha, or beta), experimental results, analysis data, drawings or sketches, computer-rendered models, data sheets from studies, bill-of-materials, and so on. Major tasks should also be composed into subtasks. Subtasks are specific activities that relate to the *specific* product. They may be assigned to integrated product teams, and, given their specificity, resources may be estimated for their completion. For each task and subtask, clearly defined start and endpoint should be defined in time. These points should not entail a long duration (weeks or days instead of months). They should also be *significant* to the product development. Significant tasks are not stated as generalities, such as "develop concepts." Instead, without overly biasing the project, they are defined with product specific

information, such as "create alternative concepts for power supply module." The techniques in Chapters 5, 8, and 9 will aid in developing tasks of this type, using active verbs followed by nouns or noun phrases.

2. ***Supplement product tasks with team tasks:*** Step 1 identifies tasks that are functional and business oriented. They relate the specialties and skills needed to develop a product. Time must also be devoted to team structure, operation, psyche, and planning. For instance, specific tasks should be planned for team building exercises and team evaluations, as discussed in the previous section. Likewise, product planning should include time to plan (recursive), scheme, and imagine. If such time is not consciously planned, a team will be straitjacketed with artificial pressures.

3. ***Estimate project resources and time (continually updated):*** For each of the tasks, the resources and timing should be estimated. Resources include people and their skills, equipment, materials, and facilities. Time considers days or weeks per task. Optimistic estimates are not desired at this stage, but realistic predictions giving clear bounds on the project. Tasks will typically require a 50% or 100% increase from original estimates. To estimate initial product development costs, break-even analysis should be implemented. Chapter 3 discussed this type of analysis, as part of project scoping. Subsequent estimates of costs will include more information. Chapter 14 discusses cost estimates based on manufacturing and assembly data.

4. ***Assign tasks to a timeline, including parallel and sequential structure:*** This step is completed in concert with Step 3. The goal is to develop a schedule of the tasks for the project. Dependencies should be determined among the project tasks. What tasks can be started with current information and resources? What tasks rely on the results of other tasks? What subsystems have the least dependencies (Chapter 9)? What tasks are the most risky? Answers to these questions will aid in establishing a schedule that divides tasks concurrently, while establishing interfaces where negotiation may occur to resolve conflicts (e.g., Figure 2.4). The next section describes a tool, known as *Gantt charting*, to assist with this step.

5. ***Monitor progress and assign additional tasks:*** The original schedule should be continually monitored for progress. Progress should be graphed as a function of time and resources. In addition, the tasks should be revised, as more information is known. More specific subtasks will be created with this additional information.

Basic Planning and Scheduling Tools

Many tools exist for product planning and scheduling. For example, popular tools include *Gantt charts, critical path methods (CPM), program evaluation,* and *technical review (PERT).* The following discussion focuses on Gantt charts and associated task lists. For smaller scale redesign projects, Gantt charts suffice. For larger development projects, CPM and PERT analysis is helpful for laying out project timelines. For any project, though, a project manager needs to communicate with and keep on top of all parallel current activities and the activities about to start. Charting timelines is not the same as maintaining timelines.

Gantt Charts

Gantt charts are a basic planning and scheduling tool, named after Henry Gantt. They are essentially bar charts that relate product development tasks and activities to time. Activities are typically listed on the vertical axis, and elapsed time is recorded in the horizontal axis. A bar in this axes system denotes the start and finish of each activity.

To develop a Gantt chart, commercial software or spreadsheets may be used. The tasks and activities are derived from the planning process discussed in the previous section. These tasks and activities are then refined and organized into parallel time groups and then ordered into the sequence of the project (from overall start to finish).

An example Gantt chart is shown in Figure 2.8. This Gantt chart illustrates the general tasks required for an entire product development process. Prototype development is also listed as a parallel group of activities to focus on physical milestones common to any project.

The example shown in the figure only provides a general picture of planning for products. It should be advanced to provide a more useful tool. First, time scales with an appropriate resolution should be determined (usually weeks). The time scale should not be too coarse or too fine. Task lists may be added to plan activities on a daily basis. Second, the tasks should not be listed only as broad categories, as exemplified by the figure. Instead, they should be refined to specific product-related tasks. Third, the chart should be augmented with team leader, task responsibility, and resource information. For each task, a member or integrated product team should be identified, with the required functions and skills of the member(s). In addition, equipment and materials should also be identified, with associated costs. Fourth, for each task, the customer and supplier should be identified. Did the customer request this task to be completed, at least indirectly? Fifth, dependencies

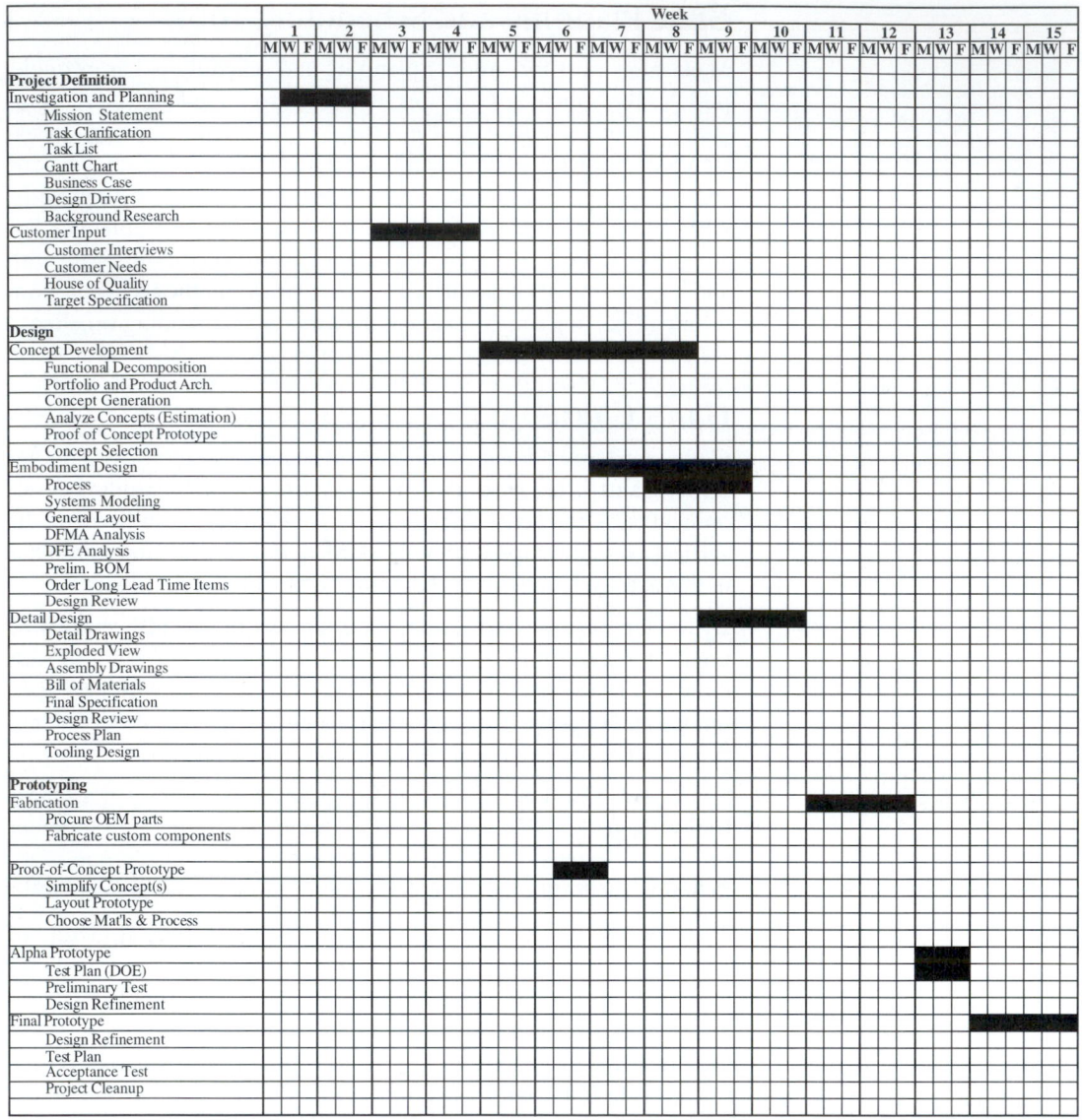

▼ Figure 2.8.
General Gantt chart for any product development.

should be identified in the chart with the use of vertical dashed lines. These lines signify that a given task cannot commence until the results are achieved from previous tasks. The result will indicate potential bottlenecks and critical paths in the product plan. Sixth, milestones should be clearly identified on the chart, for example with diamond symbols. The physical or other nature of the milestone should also be briefly described. Finally, an open-bar should be used to plan tasks, where as a shaded bar should be added to the chart as tasks are completed. This approach will provide a means of monitoring the progress of a project. Actual expenditures per task may also be added to track product development costs.

Task Lists

Project task lists are a useful tool to augment Gantt charts. These lists are simply tables that document the tasks from the Gantt chart, stating the deadline, the team members responsible for the task, and a checkbox to fill in when the task is completed.

Two types of task lists are typically used: an overall project list and weekly lists. The project task list is developed at the beginning of the project. It contains a collection of tasks, where each task is assigned from one day to several months for completion. Task size depends on the project, with usually 20–200 total tasks being listed for a small-to-medium-scale product. This list is used to complete the first two steps of the planning process and is revised as needed.

Weekly task lists are also developed. These lists help a team focus and refine a Gantt chart as a project progresses. Weekly task lists are initially completed at the beginning of a project; however, they are subsequently refined with much more detailed information as a given week approaches. Typically, weekly tasks should be estimated with twice the expected time. They should also be updated weekly to adjust resources of the project. Figure 2.9 shows an example weekly task list for the assistive bowling device pictured on the cover page of this chapter.

IV. SUMMARY AND "GOLDEN NUGGETS"

A number of golden nuggets may be harvested from this chapter:

- ◗ Product development process tools seek to enhance the creativity and imagination of a design team, removing potential barriers.
- ◗ Modern product development results in a dichotomy between design and manufacturing. Concurrent engineering approaches must be used to develop a product effectively and efficiently.

Weekly Task List: Possibowlers

Beginning: October 2

Ending: October 31

Task	Responsibility	Completed?
EMBODIMENT/SYSTEMS DESIGN	Dmel, Amy	
Measure bowling performance of Rosedale customers	Sushant, Amy, Dmel	100%
Build full-scale prototype table (foam)	Sushant, Amy, Dmel	100%
Analyze table strength for impact failure mode	Sushant, Amy, Dmel	100%
Construct flywheel graphical model	Sushant, Amy, Dmel	100%
Build electrically-actuated cam release mechanism	Sushant, Amy, Dmel	100%
Estimate steady-state flywheel rpm	Brad and Murat	100%
Calculate loads on flywheel concept	Brad and Murat	75%
Purchase AC Motor to project spec.	Brad and Murat	100%
Purchase prototype material for flywheel	Damon, Brad, Murat	100%
Construct FLY I (ramp and flywheel) concept	Damon, Brad, Murat	100%
Integrate results	Damon, Brad, Sushant	100%
Test FLY I (following experimental plan, notebook)	Damon, Brad, Sushant	80%
...		

▼ *Figure 2.9.*

Example weekly task list for the assistive bowling product.

- Functional skills are only the first step toward product development. Team members must also assume a number of roles that relate to entirely to the team.

- Personality typing, such as the MBTI, can assist in team development, that is, in recognizing and utilizing personality differences. The MBTI is not to be used to pigeon-hole or label team members, but rather to advance the team.

- A common denominator of effective teams is commitment, consensus, and communication.

- Team building and evaluation exercises are an important component to making progress.

- Although tedious and difficult, product planning must be completed early and evaluated often.

- Tangible milestones should be established early in a development project. Such milestones provide clear measures of success.

References

Belbin, M. R. 1993. A reply to Belbin Team-Role Self-Perception Inventory by Furnharm, Steele, and Pendleton. *Journal of Occupational and Organizational Psychology* 66 (3): 259.

Blake, R. R., and Mouton, J. S. 1985. *The managerial grid.* 3d ed. Houston, TX: Gulf.

Dimock, H. G. 1987. *Leadership and group development.* San Diego, CA: University Associates.

Elffers, J., and Schuyt, M. 1997. *Tangram: The ancient Chinese puzzle.* New York: Stewart, Tabori, & Chang, Inc.

Funke, C. C. 1997. Concurrent engineering in the aircraft industry. *Aerospace Engineering* (Sept.):15–19.

Isachsen, O., and Berens, L. V. 1988. *Working together: A personality-centered approach to management.* Coronado, CA: Neworld Management Press.

Jackson, B., et al. 1999. Modification of a bowling ramp to increase autonomy and normalization. In *Proceedings of the Annual Rehabilitation Engineering Society of North America Conference.*

Jensen, D. D., Murphy, M. D., and Wood, K. L. 1998. Evaluation and refinement of a restructured introduction to engineering design course using student surveys and MBTI data. *ASEE Annual Conference,* Seattle, WA.

Jung, C. G. 1971. *Psychological types.* Vol. 6 of *The collected works of C. G. Jung.* Princeton, NJ: Princeton University Press.

Kersey, D., and Bates, M. 1984. *Please understand me.* New York: Prometheus Press.

Lawrence, G. 1982. *People types and tiger stripes.* 2nd ed. Gainesville, FL: Center for Applications of Psychological Type.

McCaulley, M. H., and Mary, H. 1990. The MBTI and individual pathways in engineering design. *Engineering Education* 80:537–542.

Myers, I. B., and McCaulley, M. H. 1985. *Manual: A guide to the development and use of the Myers-Briggs Type Indicator.* Palo Alto, CA: Consulting Psychologists Press.

Viereck, G. 1929. What life means to Einstein: An interview by George Sylvester Viereck. *The Saturday Evening Post,* 26 October.

Wilde, D., Faste, R., and Roth, B. 1998. *Stanford Workshop on Creativity in Engineering Design Education.*

Wilde, D. 1993. Mathematical resolution of MBTI creativity DATA into personality type components. *ASME Design Theory and Methodology Conference,* Albuquerque, NM, Sept. 1993, 53: 37–43.

3 Scoping Product Developments: Technical and Business Concerns

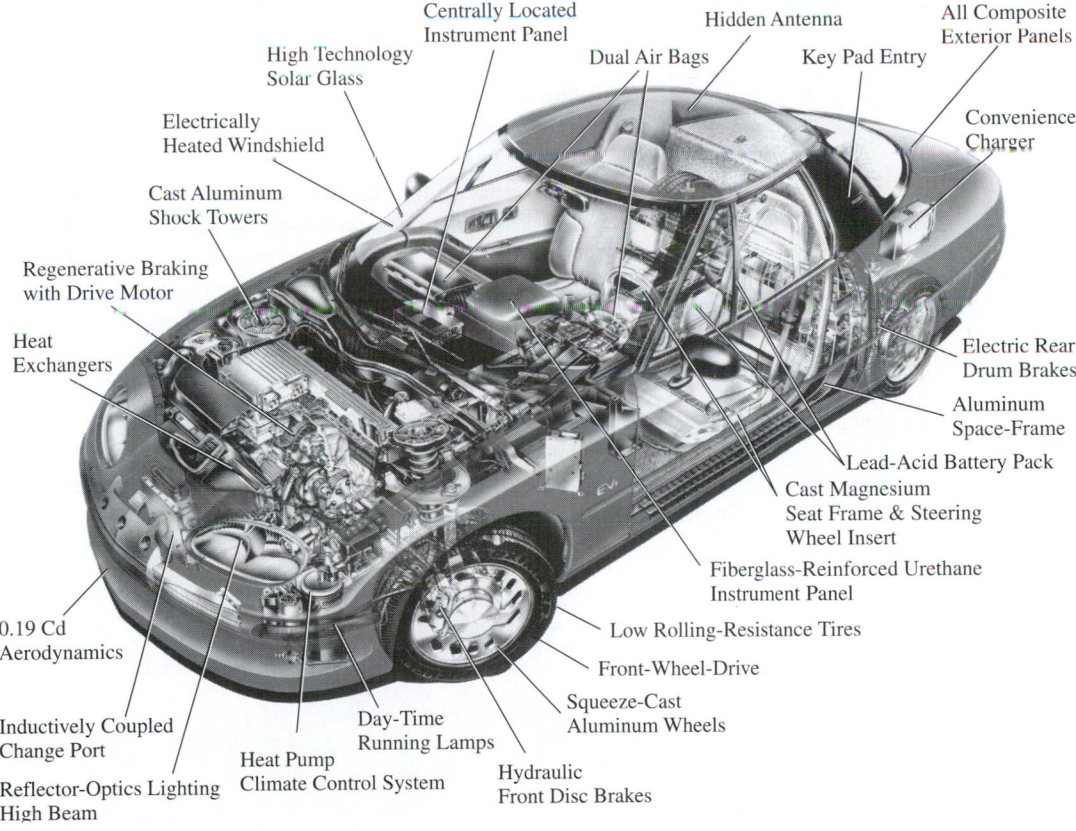

Centrally Located Instrument Panel

Hidden Antenna

All Composite Exterior Panels

High Technology Solar Glass

Dual Air Bags

Key Pad Entry

Electrically Heated Windshield

Convenience Charger

Cast Aluminum Shock Towers

Regenerative Braking with Drive Motor

Electric Rear Drum Brakes

Heat Exchangers

Aluminum Space-Frame

Lead-Acid Battery Pack

Cast Magnesium Seat Frame & Steering Wheel Insert

Fiberglass-Reinforced Urethane Instrument Panel

0.19 Cd Aerodynamics

Low Rolling-Resistance Tires

Front-Wheel-Drive

Squeeze-Cast Aluminum Wheels

Inductively Coupled Change Port

Day-Time Running Lamps

Heat Pump Climate Control System

Hydraulic Front Disc Brakes

Reflector-Optics Lighting High Beam

General Motors' EV-1 includes advanced motor controllers permitting regenerative braking and limited skid around corners, yet also includes simple lead-acid batteries. What new technology can be reasonably incorporated to ensure market success? Technical and business assessment must be completed to answer this question.

Product design, as depicted by the global process in Figure 1.4, begins and ends with the customer, emphasizing quality processes and artifacts throughout. Intertwined with a customer and quality focus are a number of technical and business concerns. Therefore, we initiate the design process with task clarification: understanding the design task and mission, questioning the design efforts and organization, and investigating the business and technological market. Of the many new development tasks, what focus should a product development team adopt? Task clarification sets the foundation for solving a design task, where the foundation is continually revisited to find weak points and to seek structural integrity of a design team approach. In this sense, it should be a *pervasive* activity that does not occur simply at the beginning of the process, but is employed throughout.

In the early 1990s, General Motors Corporation initiated an "out-of-the-box" development approach: They outside-contracted the development of a battery-powered electric vehicle, later to become known as the EV-1. Numerous studies were conducted over the development life of the vehicle, including an initial optimistic market estimate and a business case analysis of the difficult new technologies required. This preliminary analysis was positive, and so the contract development work was initiated. Under the contract, new technology was developed, such as advanced electric motor controllers. Subsequently, with a better understanding of what the vehicle would look like and what its performance would be, a more in-depth business case analysis was completed with the better information. This study included a customer demand analysis under various future environmental scenarios of limited- to greatly-increased air pollution. This analysis helped establish price points and market conditions under which the EV-1 would become a profitable vehicle. Such on-going project scoping and business case analysis is needed to understand the interactions of new developments in a product and the business case for them.

I. CHAPTER ROADMAP

The result of this chapter is a go/no-go decision on a product design activity. It is presented in two approaches (Figure 3.1), where first we present a basic method involving technical questioning and a mission statement. This approach is required for every design activity. In further support of the mission statement, Section 3 goes on to develop a quantitative economic business assessment of an activity, using the Harvard business case analysis approach.

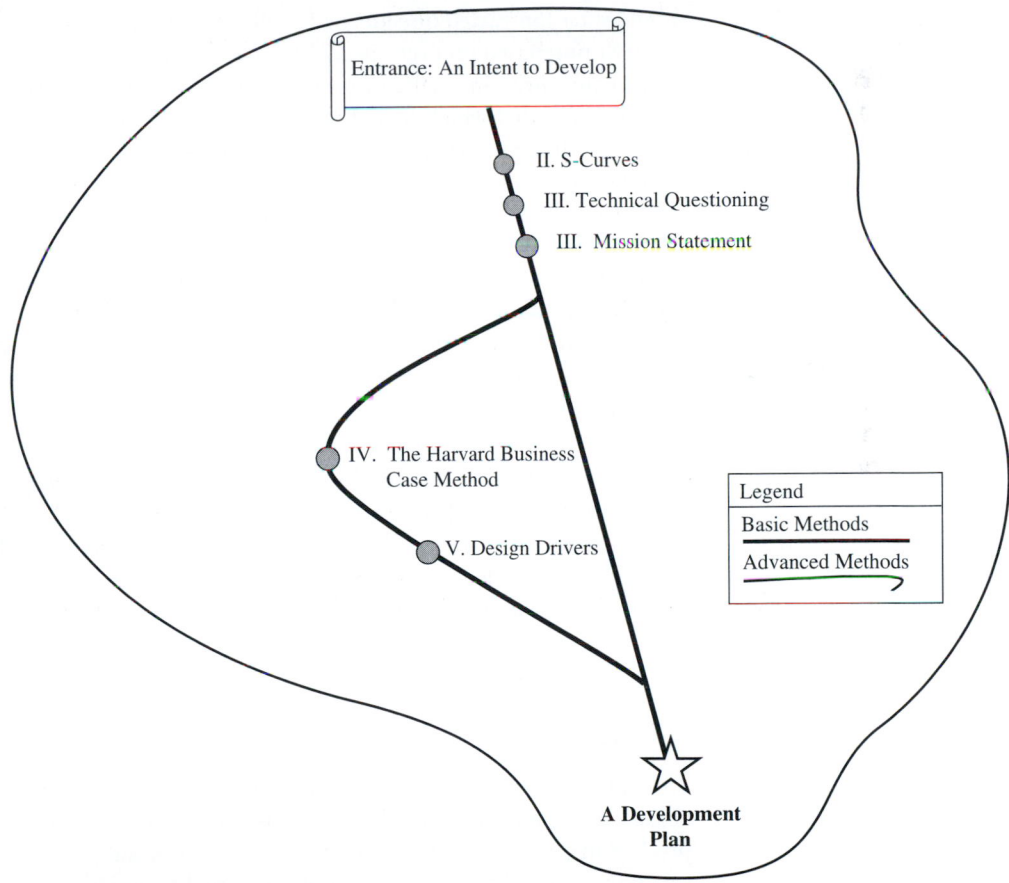

Entrance: An Intent to Develop

II. S-Curves

III. Technical Questioning

III. Mission Statement

IV. The Harvard Business Case Method

V. Design Drivers

Legend

Basic Methods

Advanced Methods

A Development Plan

▼ *Figure 3.1.*
Chapter Roadmap for product scoping.

II. DETERMINING WHAT TO DEVELOP

The first basic issue in new product development is deciding what to develop. Should a completely new technology be introduced? Should the current product be refined and tweaked to better please the customer? Should the product be expanded into variant forms to more comprehensively cover the market?

A very important and related question is one concerning outsourcing: Should a portion of a new product design effort be developed in-house, should it be contracted outside, or should an outside supplied system be OEM purchased on every product? Traditionally, the answer is always to outsource unless the subsystem is the defining characteristic of the product—that which makes it sell or that which is the core competence of the firm. New research (Fine, 1998), however, also highlights the importance of understanding the trajectory of the supplier industries and the system integrating (product) industries, to ensure that one does not outsource subsystems that later may be the defining features of the product. Industry competition among supposedly friendly assembling purchasers and their OEM suppliers can become as fierce or fiercer than competition for customers. The rise of Microsoft as the initial supplier of PC operating systems to IBM represents an example of such competition.

To consider answering, at least in part, questions over levels of new technology to incorporate into a product, we apply *technology forecasting*. We seek to predict what technological developments can occur so as to be prepared to properly introduce technology into a product, whether the technology change is incremental or disruptive (Betz, 1993; Bower and Christensen, 1994; Foster, 1986; Christensen, 1997).

s-Curves

Technological innovations typically manifest themselves into a market along an "s-curve" timeline behavior, as shown in Figure 3.2. For a product in question, consider examining one important product metric. In the case of lighting, for example, one might consider the energy efficiency. For all of the different products (light bulbs) on the market, one can plot each product's metric value (lumens of light output per watt) as a function of the time when each product was introduced. The metric values will naturally fall as an "s-curve" in time. First, the values are low and widely spaced: Not much innovation is occurring in the

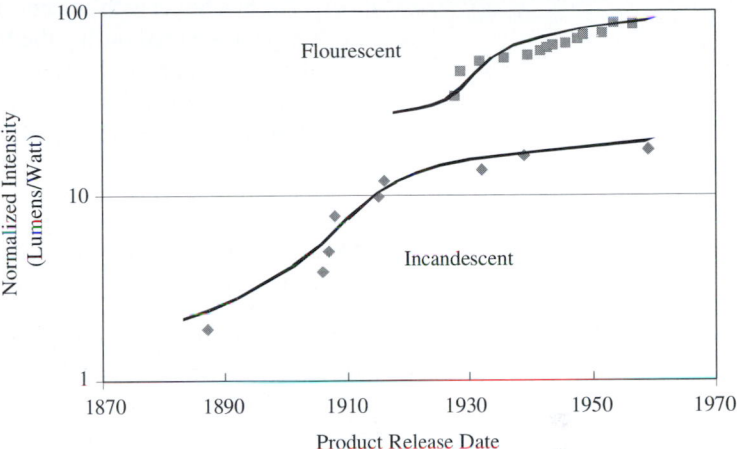

▼ *Figure 3.2.*
Product evolution along s-curves (Betz, 1993).

market, since it is new and difficult to introduce. Next, a rapid profusion of innovation occurs, and many products are launched in time, typically by many competitors who join the market. The lower leg of the "s" is forming. The new technology, however, eventually tops out, physical laws of the process dominate, and engineers cannot extract more performance. The slope of the "s" tops out again, and the curve becomes flatter. Typically, businesses are consolidated into a few mature competitors. This description constitutes the typical technological innovation time cycle, and the market behavior is well characterized by an s-curve.

At the top asymptote of an s-curve, though, all is not ended. Innovators typically conceive of an even newer technology that can provide even better performance. This technology is introduced for the first time into the market and then develops its own s-curve behavior. The result is two different s-curves that can be plotted on the technology introduction graph. The newer s-curve is a disruptive technology that requires changes in the market system for the product to succeed. The improved performance of the technology, though, is sufficient to force these changes. For example, the development of florescent lamps required new fixtures and ballast technology that incandescent bulbs did not require.

A second example of s-curve behavior is shown in Figure 1.18. In this case, two discernable trends are realized in the development of design theory. The first began with ancient civilization through the Roman Empire. A gap then existed during the Middle Ages, followed by the beginning of a new s-curve in the 1700s. This second trend began a serious increase in gradient with the onset of the Industrial Revolution. This upward trend has not reached a plateau through contemporary times.

s-Curves and New Product Development

The s-curve trends of an industry are important for a competitive company to understand. We can, for typical industries, characterize what *technology strategy* a company should employ depending on what regime of the s-curve an industry encounters. A technological strategy is the decision over what level of design to attack: create new technology or improve the current technology.

We should point out this is not a simple design decision that a development team will make. There are many strategic business implications that accompany decisions over incorporation of new technology. For example, when considering whether to adopt new technology that a company has no experience with or that simply does not exist, how should this be done? Senior management must decide whether to develop it through in-house R&D, through collaborations established through buy-outs of start-up firms with the technology or though secrecy agreements with other firms, or finally through development of precompetitive "open systems." Given any decision, what will then be the subsequent technology to adopt? What is the *new technology roadmap* that you forecast for your industry, and what are its implications for the market? Decisions and insight around the development trends of new technology can make or break a company; the decisions are not easy.

To bring the idea to a more concrete level, consider a typical technological s-curve. Consider first when an industry is in a state where the technology is topping out, the top of the "*s*." Here a company should first realize that it is time to invest in exploring new technology, to "jump" to the next s-curve of a newer higher performing technology. A reasonable fraction of investment should be directed toward high-risk advanced research and development. Further, in this state, the marketplace is usually saturated with large market share companies who make incremental improvements. One should then implement two steps: first, really understand how the customers use the product, and second, restrict development activities focused on these uses, each development with a strong business case analysis.

The second step a company needs to implement is to reduce costs to stay competitive. Whereas it is critical to position the product to meet customer needs very well, it is just as critical to be a low-cost competitor. The business case development therefore should *not* be to construct a cost estimate and add a profit to determine a price. Rather, the competitive price should first be determined and then an adequate profit subtracted to determine a target cost for the design team to meet. Any other approach will drive the company out of the competitive market.

On the other hand, if the product is in the rapid change position of a s-curve, then the market is typically shaking out with rapid innovations. In this situation, a profusion of product alternatives are developed and offered by the competition. Some will succeed, others will not. In this situation, a company should pay particular attention to the underlying causes of customer needs and ensure that their technology development is on a rapid development cycle to meet these underlying causes. Then one can be sure to have an evolving winning product in the market. Cost containment, although always important, is not as strong of an issue in this environment, since some innovation failures can be paid for by one success.

Finally, if the product is in the early forming stages of a new s-curve, the company has a technological advantage. In some cases, this can be used to disrupt the current status of the market. This stance can be a risky position to be in, since the new technology may be rejected due to high cost of adoption. Typically, new technologies require some level of new infrastructure to implement the technology. Fluorescent lamps required different light fixtures and ballasts, whereas screw-in light bulbs did not. DVD disks require different players to be purchased by consumers and different recording factory equipment than CD-ROM disks do. Electric- and hybrid-powered vehicles do not require gasoline supply or distribution. Electronic (digital) cameras do not require film. All of these products are examples of new technologies that can disrupt the current support infrastructure of a market.

In addition to high risk, however, there can be high reward. Companies introducing such new products often have the opportunity to replace the companies supplying the older technology in the market and to become the dominant supplier in the new market. In this situation, the company should *not* completely listen to the customers. This statement seems counterintuitive, but it is nonetheless true. Customers are typically calibrated only in the older technology, and so in fact most of the information they provide is relevant to that older technology. Asking customer need information about floppy diskettes has only tangential application to what will become customer needs on DVD technology.

Some needs may apply, such as ease in carrying, but other criteria may have no bearing at all, such as the metal guard having sufficient spring force to keep the media isolated. Further, and more importantly, other new criteria arise that the older technology had no customer consideration of, such as the ability to hot-swap (remove the DVD drive while the computer remains powered on). This situation describes a disruptive technology that can change not only the disk drive technology, but also indeed the computer hardware technology (hot-swap). In this case, the company needs good insight into the future customer environment to determine the product specifications.

Note that a completely new technology is not sufficient to disrupt a current marketplace. That is, it may be the case that the new technology, while new and expensive to develop, was also highly forecasted and understood. Intel sinks vast sums of money into developing new microprocessors, as do its competitors. Yet, the microprocessor market does not get disrupted.

In a great line of research, Christensen (1997) showed that in addition to be being new and improved, if the market is performance saturated, then that technology has the ability to become *disruptive*: to change the market completely. That is, in every case of new technology that eventually changed the marketplace and displaced earlier competitors, the new technology did so by first supplying to a separate niche market and therein improving the technology. This is continued until the mainstream market demand caught up with the oversupply of performance available with the new technology, at which time the mainstream market becomes performance equivalent to the previous niche market. For example, smaller 3.5″ computer disk drives were not initially wanted by desktop computer manufacturers; they instead wanted the larger 5.25″ drives, just at higher capacity. On the other hand, laptop computers did want the smaller drives, providing an initial market. Eventually, though, even the desktop system market caught up and needed to become smaller, and the 5.25″ drives were replaced by 3.5″ drives.

The point is that just because we have a better technology does not necessarily mean our technology will eventually dominate a market. Another condition must also hold: The current old-style technology in the market must oversupply the customers with performance. Eventually they will want higher performance; in fact, performance higher than the old technology can supply, but just not now. Given that situation, we must then find a new market to launch our technology in and improve it there. When the mainstream market then catches up in its demand for performance, our new technology will be debugged and lean, given that it had to exist in a niche group. We can then expand and dominate. This example is the development scenario that disruptive technologies follow.

Comments on s-Curves and Technology Forecasting

s-Curves are useful to understand the typical innovation cycle of new technology, as previously described. Most technology follows this path of market behavior. There are, however, exceptions. Consider, for example, microprocessors. The infamous Moore's law states that transistor density on microprocessors doubles every 18 months (Gwennap, 1996), attributed to Gordon Moore, Intel Corporation cofounder. As shown in Figure 3.3, the s-curve model violates Moore's law; an s-curve model predicts that computing technology must eventually top out. Moore's law states that computing technology *never* tops out. It also states, interestingly, that computing technology innovation always *accelerates*. This law has held true remarkably long, into several decades across several generations of computing hardware research and development engineers.

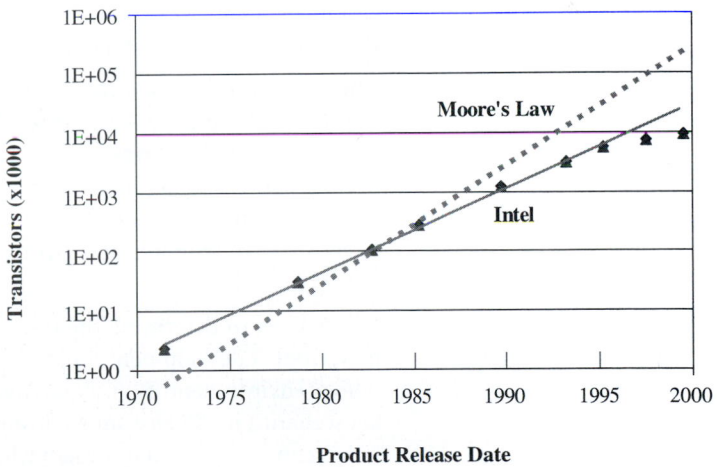

▼ *Figure 3.3.*
Intel microprocessor technology performance.

Moore's law is an example of one caveat that must be understood about s-curves. The reason the curve should top out is that physics starts to dominate in the current technology being used. There is nothing to say that when this happens one cannot switch to a better technology. Indeed, that is the reason Moore's law has held true. Electrical engineers in the computing industry have not been able to rest on and exploit the top end of a technology's s-curve once it is proven. Someone else is constantly working to develop the next greatest technology. Once the competition proves its better technology within the subsequent year or so, any company with older technology will lose out. The industry is constantly jumping s-curves. This competitive cycle has been in play and has kept Moore's law true. Larger wafer sizes have been made possible with ever better mechanical production equipment technology and ever better laser lithography technology, and so on.

Such technology improvements can be understood as an s-curve jumps. The interesting feature in computing technology is that the jumps occur as each technology innovation's s-curve begins to form. Once the s-curve begins to top out, that technology is abandoned. The result is an exponential technology improvement timeline—and Moore's law.

The other interesting aspect of computing technology behavior and Moore's law, however, is the acceleration factor. New technologies must not only be introduced, but either the rate of introduction or the performance improvement per introduction has to always increase. It is not clear that the computing research and development community can continue to keep pace with Moore's law. It is also not clear that it cannot.

Another factor that can cause s-curve heuristics to not be valid is when the market boundary is not crisp. For example, international competitors can suddenly enter a market simply through an instant change in trade law. What was once a mature market dominated by a few manufacturers can overnight become a market saturated by many lean suppliers. Cost containment at high quality becomes even more important in this scenario.

In any case, to determine the underlying conditions for a successful new product development project, a design team must ponder its technology environment. Typically, as discussed above, there are three market scenarios it will face: an environment when they are introducing a new technology, an environment when the technology is rapidly evolving, or an environment when the technology is topping out. A company needs to understand which of these three scenarios it is facing and plan accordingly. Given this search for understanding, we next present a simple approach to ensure that the developments meet these needs.

III. BASIC METHOD: MISSION STATEMENT AND TECHNICAL QUESTIONING

In product design, the ability to frame the problem, to ask the right question, at the right time, and of the right person(s), is essential to success.

> The mere formulation of a problem is far more often essential than [is] its solution, which may be merely a matter of mathematical or experimental skill. To raise new questions, new possibilities, to regard old problems from a new angle requires creative imagination and marks real advances in science.—Albert Einstein

A mission statement and technical clarification of the task are important first steps in the product design process. They are intended to:

▶ focus design efforts

▶ define goals (goals must be stated before they can be met)

▶ translate the business case analysis to the development team

▶ provide a schedule for tasks (define timelines and milestones for task completion)

▶ provide guidelines for the design process that will prevent conflicts within the design team and concurrent engineering organization

Fundamentally, any new product development project has risk from two independent sources. There is *technical risk*—Can we make it? and there is *market risk*—Will they buy it? In all development projects, these two aspects must be weighed, explored, and understood.

In our basic approach, we will explore two simple tools to understand and clarify these two risks: *technical questioning* and *mission statements*.

Technical Questioning

The first step in task clarification is usually to gather additional information. One can explore many developments and gather information on many topics. Focus is always needed. To keep focused, one should first question the current understanding of the development. The questions in Table 3.1 need to be asked and answered, not once, but continually through the life cycle of the design process.

TABLE 3.1.   TECHNICAL QUESTIONS TO CLARIFY STATE OF COMPREHENSION OF A DESIGN

What is the problem really about?

What implicit expectations and desires are involved?

Are the stated customer needs, functional requirements, and constraints truly appropriate?

What avenues are open for creative design and inventive problem solving?

What avenues are limited or not open for creative design? Limitations on scope?

What characteristics/properties must the product have?

What characteristics/properties must the product not have?

What aspects of the design task can and should be quantified now?

Do any biases exist with the chosen task statement or terminology? Has the design task been posed at the appropriate level of abstraction?

What are the technical and technological conflicts inherent in the design task?

The technical questions in Table 3.1 are relatively high level, but they are intended to be that way. These questions force the design team to think critically, to first restate a design task in a more precise way for the project or subset currently under consideration. Immediately answering the questions is not the focus; instead, answers will typically require development exploration. Posing them helps establish what to do next.

Mission Statements

The tangible result of the technical questioning procedure should be a clear statement of the design team's mission, the *mission statement,* sometimes called a market attack plan, a vision statement, or a product plan. Figure 3.4 shows an example template for a mission statement. This template should not be used as a mere statement of intention. Instead, the mission statement is both a calling card and banner stating the design team's intentions. The mission statement is typically first written by a project manager (or a team) and distributed to all development team members at the project kick-off, to explain what they will be doing. An exciting vision of the new product is conveyed to the entire team, to communicate the project goals. The mission statement provides this banner communication.

The mission statement also has another use. When interviewing customers, meeting with potential suppliers, or carrying out design reviews, the mission statement can serve as the calling card to explain what the end goal of the communication supports. The mission should be the lead item of discussion, clarifying, and equalizing the playing field of negotiation, debate, and probing questioning.

Mission Statement: XXXX Product	
Product Description	one concise and focused sentence
Key Business or Humanitarian Goals	• schedule • gross margin/profit or break-even point • market share • advancement of human needs
Primary Market	• brief phrase of market sector/group
Secondary Market	• list of secondary markets, currently or perceived
Assumptions	• key assumptions or uncontrolled factors, to be confirmed by customer(s)
Stakeholders	• 1-5 word statements of customer sets
Avenues for Creative Design	• identify key areas for innovation
Scope Limitations	• list of limitations that will reign back the design team from "solving the world"

▼ *Figure 3.4.*
Mission statement template.

Fingernail Clipper: Clarification and Mission Statement

As an example, consider the common fingernail clipper product. Several nail care products are shown in Figure 3.5. Yet, consider a scenario when there is a perception that the current clippers on the market are difficult to operate, especially for older persons, and could be more compact. A new clipper development project should be explored for possible undertaking (Table 3.2, Figure 3.6).

To develop a mission statement, the technical questions of Table 3.1 are first considered. These questions are shown with brief answers for a fingernail clipper product in Table 3.2. The answers clarify the steps required: first, what is the required business case to make the development profitable, and second, what is the customer interested in that we could provide to meet that business case?

Finally, these results are combined into the mission statement for a fingernail product, as shown in Figure 3.6. Development process and final product goals are listed.

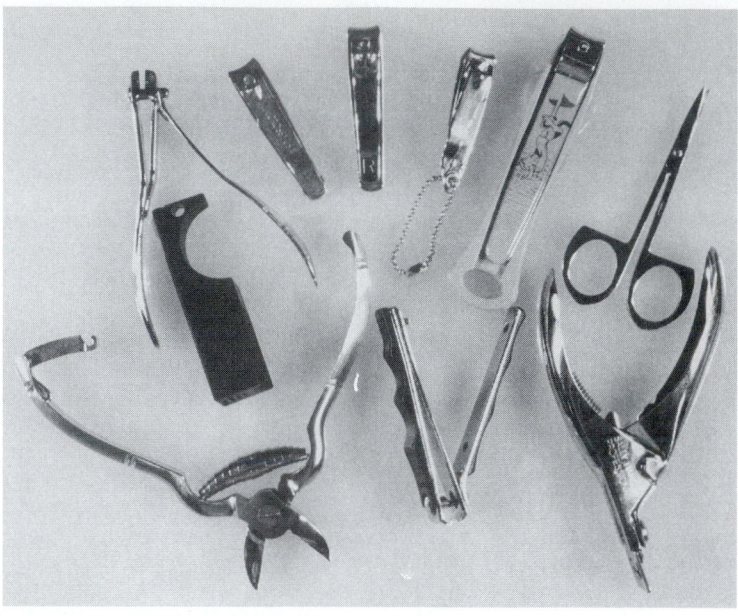

▼ *Figure 3.5.*
Various manicure products on the market.

Mission Statement: Fingernail Clipper Product	
Product Description	Remove and file excess fingernail length.
Key Business or Humanitarian Goals	• 6 month development of beta prototype • 30% profit margin • Initial 5% market share • Supplement fingernail polish business
Primary Market	• Adults of all ages, focusing on fingernail polish users
Secondary Market	• Knife collectors • Business executive
Assumptions	• Small, compact stowage volume • Long life (10-20 years)
Stakeholders	• XXX Corporation; user; salons, retailers
Avenues for Creative Design	• ergonomic shape; store/capture of nails; compact stowage; ease of cutting
Scope Limitations	• Materials: steel processing and moldable plastics.

▼ *Figure 3.6.*
Example mission statement fingernail clipper product.

TABLE 3.2.   TECHNICAL QUESTIONING FOR THE FINGERNAIL CLIPPER EXAMPLE

What is the problem really about?

Clumsy operation of a typical clipper.

What implicit expectations and desires are involved?

Remain a manual clipper that can be operated by oneself.

Are the stated customer needs, functional requirements, and constraints truly appropriate?

Many groups, as documented in trade magazines and consumer studies, have noted clumsy operation. A detailed assessment will be made in post-customer interviews.

What avenues are open for creative design?

Can modify any and all parts. Create further function sharing in a device abundant with components that share functions. Perhaps introduce different materials besides metals as the fundamental fingernail clipper technology. Add functionality, such as the ability to store and dump nail debris.

What avenues are limited or not open for creative design? Limitations on scope?

No electrical power. No service-based solutions.

What characteristics/properties must the product have?

Easy to use. Durable. Safe.

What characteristics/properties must the product not have?

Should not be bulky.

What aspects of the design task can and should be quantified now?

Customer needs analysis: statistical sample size and importance ratings. Required profits to cover development costs. Fingernail characteristics, such as size and strength. Human hand and finger sizes and strengths. Research, estimates, and simple calculations should be carried out immediately to understand the quantitative scope of these issues.

Do any biases exist with the chosen task statement or terminology? Has the design task been posed at the appropriate level of abstraction?

Must be single-person manual. We wish to remove the ends of nails, not necessarily "clip" them.

What are the technical and technological conflicts inherent in the design task?

Compact size vs. large surface area for grasping and large mechanical advantage (small finger force).

IV. ADVANCED METHOD: BUSINESS CASE ANALYSIS

Technical questioning is a qualitative means to make decisions on what to explore next in design developments. Understanding the business market for the results of any activity is a more analytical and detailed approach and can help complete the mission statement of Figure 3.4. During any product design effort, a product's market must be clarified through the development of a business case analysis. A number of financial assessment techniques exist at varying levels of detail. We explore here the *Harvard business case analysis method* (Ronstadt, 1988; McNair, 1954). Typically, this is thought of as a technique in business school education; however, it also serves as an excellent means to form a business plan. By example, this section presents the method to understand

TABLE 3.3. SUMMARY OF THE HARVARD BUSINESS CASE METHOD

Process step	Description
1. Problem Statement	What market problem are you addressing, fixing, improving, making more efficient, etc.? This should be limited to ONE sentence. Only one problem can be addressed. If the problem is complex, with many interrelated sub-problems, the problem should be clarified and refined to the basic (atomic) or kernel problems.
2. Assumptions	Discuss any limiting assumptions made in preparing the business case proposal, such as product costs, direction of the industry/department, etc. This step provides a clear statement of the scope of work.
3. Major Factors	List, briefly, major factors of the environment that affect the decision. This may be the state of the business (capital constraint), critical business needs or directions (strategies), etc.
4. Minor Factors	List, briefly, factors that should be considered, but perceived as not significantly impacting the problem.
5. Alternatives	List concrete or hypothesized alternatives (minimum of three) to address the problem or opportunity defined by the problem statement, assumptions, and major factors. Two or three sentences are sufficient. Under each alternative, list the advantages and disadvantages.
6. Discussion of Alternatives	Review each of the alternatives with respect to the stated problem, assumptions, major and minor factors. Compare alternatives and discuss the relative merits of each (in terms of cost savings/avoidance, cycle time reduction, increase in quality, and head count reduction). From this discussion, a clear leader among the alternatives, i.e., the most feasible alternative, should be identified.
7. Recommendation	State your recommendation. There should be no need to defend it; this should have been covered in the last section. If needed, elaborate on the recommendation to add clarity.
8. Implementation	Describe the implementation plan. Include resource requirements: financial, human, space, etc. Describe the time frame requirements, control, and measurements that will be needed to ensure that goals are met. Measurements should be tied directly to solving the problem, and adequate tracking mechanisms such as ROA, cycle time, or NPV should be used to quantify the success of the project. Contingency plans should be developed that address any high-risk aspects of the solution.

the potential impact of a product development through fabrication. A summary of the methodology is shown in Table 3.3. A simple mechanical product provides the example context: a fingernail clipper.

Harvard Business Case Methodology: Product Evolution

Fingernail clipper devices are widely used consumer products, with markets of the everyday consumer (primary), professional salons, and domestic pet manicurists. For the purposes of this section (and based on our clarification of task), let's assume that our company seeks to improve its current product offering in the primary fingernail clipper market, a complimentary product to our fingernail polish product line. The mission, following Figure 3.4, is to design a fingernail clipper for comfortable use by either the left or right hand. It is assumed that comfort, cost, reliability (consistently remove nails with a simple finger force throughout the product's life), and storage compactness are the driving market needs (to be confirmed or revised through customer interviews). The corporation also seeks only a 30% return on investment, since the goal is to compliment and increase the market share of fingernail polish.

A number of solutions exist for addressing both the technical and process issues associated with a fingernail clipper product development. A business (financial) case may be derived for each of the possible solutions. However, the intent during the early stages of conceptual and configuration design is not to study all possible alternatives in detail, but to determine if a minimal benefit to the business will be realized by improving the clipper problem; that is, comfort, cost, reliability, and compactness. As such, we concentrate here on steps five and six of the Harvard business case method (Table 3.3), where only one generic alternative is considered. A device solution, that is, a new, generic, and hypothetical clipper design, is the alternative considered, emphasizing the possibilities of reduced cost and higher reliability through compactness and fewer components. These possible benefits call for a "break-even" financial analysis for the possibility of developing this new clipper.

Product Development Economic Analysis

When completing an analysis of the market viability of a development project, initially a project manager must weigh the costs of development versus the expected future revenues. Expected future revenues must compensate for development investments made today. Two fundamental considerations, therefore, are apparent: *risk* and the *time value of money.*

First, there is economic risk in pursuing development. The future revenues are not guaranteed. Therefore, to be economically viable for investment today, a project must have a higher projected return on investment than a low-risk alternative does, such as a savings account.

The second consideration is the time value of money. If a savings account is available that pays out 4% interest, then $1000 available today can be placed in the account, and in 5 years the $1000 will accumulate to $1217. The $1000 today is equivalent in value to $1217 in 5 years, or, $1217 of revenues in 5 years is really worth only $1000 today, at a 4% annual rate of inflation. This economic fact places an extra financial burden on all product development projects: Revenues in smaller future dollars must pay for development costs in today's real dollars.

To conduct an economic analysis, several economic variables must be specified or estimated. The amount to be invested—the development and yearly manufacturing costs—are generally fixed and can be estimated. The interest rate is a variable that can be estimated. The sales volume and unit price, however, compose a single coupled decision variable; they are coupled through the economic market. The volume/price will, generally, vary over the life of the product, with an early ramp-up period to a large volume and then typically a market-share decay profile as the product becomes more obsolete.

A development project economic analysis consists of estimating and establishing values for these economic variables and then evaluating the expected economic profit performance of the new development project. Typical economic performance criteria include:

◗ *break-even point:* the time until the investment cost is recovered and the project starts making money

◗ *return on investment:* over the expected sales life in the marketplace, the amount of profit gained compared to the initial investment

◗ *investment risk:* the probability that the market analysis is not correct and there will be less return on investment

To conduct such analysis, a yearly *net-present value* (NPV) *cash flow* model can be used. That is, one can establish the development costs in today's dollars. Then, a timeline is projected into future years, over the life of the product in the marketplace. For each year, below the timeline, the expended dollars by year are indicated, with the initial years expending much and future years expending little. Above the timeline, the revenue generated by each year is indicated, with the initial year showing no revenue, and the revenue from the expected sales indicate for each subsequent year. All amounts are shown in current year dollars. These amounts are then totaled until the point in time where the product becomes profitable. The subsequent revenues then contribute to profits.

Note that more-advanced product development financial analysis can be completed based on the idea of *options* (Faulkner, 1996). NPV calculations assume that once we commit to a project, financially we will maintain that commitment no matter what. Options analysis instead models financial commitments under different outcomes of the early development work and therefore assumes only partial commitment up front. For example, we may commit only half the funds up front, and based on a gate at mid-stream, only then commit the remaining funding for the full project development cost. Such an options-based analysis can be more accurate, presuming a reasonable ability to forecast probabilities of outcomes along the development path (often a tenuous assumption). NPV calculations, on the other hand, are inherently conservative (all fund commitment dates guaranteed at the outset) and so are often overly pessimistic.

Example: Fingernail Clipper

Consider again the fingernail clipper example. An economic analysis here answers the question, "Is a hypothetical clipper concept with less materials (compactness) and fewer components feasible as a business venture?" and begins with a summary of the current costs for fingernail clipper development (as projected from the current product). Because these costs continually change with new technology and market forces, actual-absolute cost values are not shown in this section. The actual costs have been multiplied by a factor. The important issue is the relative cost of the current clipper operations versus a proposed, hypothesized solution.

The current costs for clipper development are listed in Table 3.4. These cost projections are based on 750,000 clippers, with a product distribution of 80% small clippers, for fingernails, and 20% large clippers, for toenails. The average cost for this distribution is $0.31 per product for fabrication, $0.17 for labor, and $0.23 for engineering time.

For the purpose of comparison, the adopted concept for this analysis is a "generic," hypothesized clipper with reduced parts. This device does not exist as a concrete form concept; it is a hypothetical concept. It is assumed that there exists suitable component and fabrication technology for this concept. Such a device would require less materials, piece-parts, assembly, and labor; however, tool costs would potentially increase due to higher precision in the cutter alignment. Based on this new fixture concept, Table 3.5 lists the expected costs for 750,000 products (same distribution of small and large clippers and, as with the current costs, multiplied by a factor). One-time development (engineering) costs account for $187,000 (increase in tooling design), and projected on-going fabrication and engineering costs account for $231,000 ($154,500 fabrication + $76,500 labor), compared with current product on-going costs of $352,500.

The necessary information is developed for a break-even analysis. A comparison between the current and proposed generic clipper costs is carried out to determine the payback period and cost savings. Table 3.6 shows the results of the break-even analysis. The payback period is 6 months, with a potential savings of $121,500 for 750,000 products.

TABLE 3.4.   CURRENT COST SCENARIO—APPLICATION TO A FINGERNAIL CLIPPER DEVELOPMENT

Category	Projected cost ($)	Cost per product ($/clipper)
Labor Costs		
Small Clipper:		
Assembly	$60,000	$0.10
Handling	$36,000	$0.06
Large Clipper:		
Assembly	$16,500	$0.11
Handling	$10,500	$0.07
Total	$123,000	$0.17 (avg. clipper)
Fabrication Costs		
Small Clipper:		
Materials	$96,000	$0.16
Piece-Parts	$72,000	$0.12
Tooling	$6,000	$0.01
Large Clipper:		
Materials	$30,000	$0.20
Piece-Parts	$21,000	$0.14
Tooling	$4,500	$0.03
Total	$229,500	$0.31 (avg. clipper)
Subtotal (on-going costs)	$352,500	
Engineering Costs:		
Avg. 10 weeks per product	$173,600	$0.23
Total Cost	$526,100	$0.71 (avg. clipper)

These results are extremely encouraging. Significant cycle time and cost savings may be achieved for the business if suitable fingernail clipper concepts can be developed. Because of these potential savings, the project should be carried to the next stage of conceptual design and prototype build.

Implications

Even though only a subset of the Harvard business case method is illustrated above, the potential impacts are impressive. A "go/no-go" decision may be made early in the product development process, provided that financial information exists for the current market and that projected costs may be readily assumed for hypothesized concepts. Such decisions should be made in parallel with technical and industrial design clarifications. Also, they should continually be reviewed and updated as new information becomes available, especially as concrete product

TABLE 3.5. PROPOSED COSTS SCENARIO—GENERIC, REDUCED PART-COUNT CLIPPER

Category	Projected cost ($)	Cost per product ($/clipper)
Labor Costs		
Small Clipper:		
Assembly	$30,000	$0.05
Handling	$30,000	$0.05
Large Clipper:		
Assembly	$9,000	$0.06
Handling	$7,500	$0.05
Total	$76,500	
Fabrication Costs		
Small Clipper:		
Materials	$66,000	$0.11
Piece-Parts	$24,000	$0.04
Tooling	$24,000	$0.04
Large Clipper:		
Materials	$21,000	$0.14
Piece-Parts	$9,000	$0.06
Tooling	$10,500	$0.07
Total	$154,500	
Subtotal (on-going costs)	$231,000	
Engineering Costs:		
Avg. 10 weeks per product	$187,000	$0.25
Total Cost	$418,000	

TABLE 3.6. CLIPPER BREAK-EVEN COST ANALYSIS

Issue	Analysis result
Estimated payback period for development costs	6 months
Projected savings for first 100,000 products	$16,200
Projected cost savings for next 650,000 products	$105,300
Expected cycle time savings for each 100,000-product lot	38% of current work days

configurations are derived. Modern research and development investment decision making includes time-stream decision trees with branches that depend on the probability of success at different points in the development cycle and the option of delaying commitments by pursuing multiple technologies or design concepts, depending on expected risk and benefits (Brealey and Myers, 1991; Dixit and Pindyck, 1994).

An important caveat of the Harvard business case approach is the existence of past financial information for a product family or the existence of analogous product data. Whereas a business case should be developed before or during concept generation, cost data are needed to predict a product's potential return on investment. These data must be established considering the marketplace: levels of competition, unit cost and sales volume curves, dependence on other products or systems of the customers, and so forth. These quantities can be notoriously difficult to estimate and can be chosen to make a scenario appear lucrative or poor. Also, if an entirely new product or family of products is under development, cost data may not exist directly. The Harvard business case methodology can still apply, but the data must be extrapolated from a similar or analogous product or rough estimates of preliminary product layouts. Additionally, different scenarios in the financial environments can be explored to ensure one understands all possible scenarios and thus the investment risk.

V. ADVANCED METHOD: DESIGN DRIVERS

Similar to the business case analysis as an advanced analysis for economic factors, there are more-advanced analyses of the technical factors available for the preliminary development phase. An important task in the initial stages of new product development is to get organized and to clarify what is most important and what has to be determined when. One can waste valuable time at the early stages arguing over decisions when it might be that some decisions simply have to drive others.

Design Drivers

It is surprising how often failing to take time at the front end of a project to really "understand" what the problem really is about causes a great deal of time (and money) to be wasted. As an example, consider the development of commercial aircraft.[1] What makes an airplane desirable to airlines and therefore profitable to develop and build? What are the key initial decisions that a design team must make? What are the fundamental

[1]This example is adapted from Dr. Daniel Whitney, Massachusetts Institute of Technology.

design drivers that, through design decisions made by the engineers, essentially distinguish different commercial aircraft in the marketplace? A design driver is an early decision that must be made but that, once made, determines in large part many of the subsequent design decisions.

A development team can expend many resources in time and monetary investment iterating while trying to define project target specifications on the thousands of systems needed in the aircraft. It is often not clear how to manage this specification system or process. Should the engine development team specify available thrust and lift, or should the cabin development team specify required thrust and lift? Who drives whom in the development process, or should it be completed together?

With some thoughtful task clarification analysis of what constrains the various factions in a new airplane design activity, one can distinguish two fundamental equilibrium conditions that must be satisfied. Basically, the plane must first be sound in the business sense: It must carry sufficient passengers to provide a revenue stream to cover operating costs. But, a second view must also be considered, namely the technical sense: the plane must be flyable. It must have enough lift and thrust to fly the number of passengers demanded by the business case. These are two mutually constraining necessary conditions to be a potentially profitable commercial aircraft, and they are coupled. This relationship is shown in Figure 3.7, where the intersection of the two "equations" forms the design drivers (shown in the center of the figure). Different simultaneous solutions to these conditions are demonstrated by the different-sized commercial aircraft that have been successful in the market.

With a diagram such as Figure 3.7, project managers can identify the early specifications that are important to get right, since these specifications ostensibly drive all the other specifications. In the commercial aircraft example, the design drivers are the size of the wings, fuselage, and engines. These form the intersection of the concern that a plane is able to fly and the concern that a plane carries enough passengers to make a profit. Choices for values on these variables must be targeted to a market segment using the economic conditions through an economic analysis and simultaneously through simplified engineering analysis targeted to ensure that the airplane concept is feasible. They drive the design. Once those decisions are made, other decisions such as controls, landing gear, access, and so forth can all be made in accordance with these primary decisions.

To obtain this understanding of design drivers for any arbitrary product, it should be considered under the same two fundamental views:

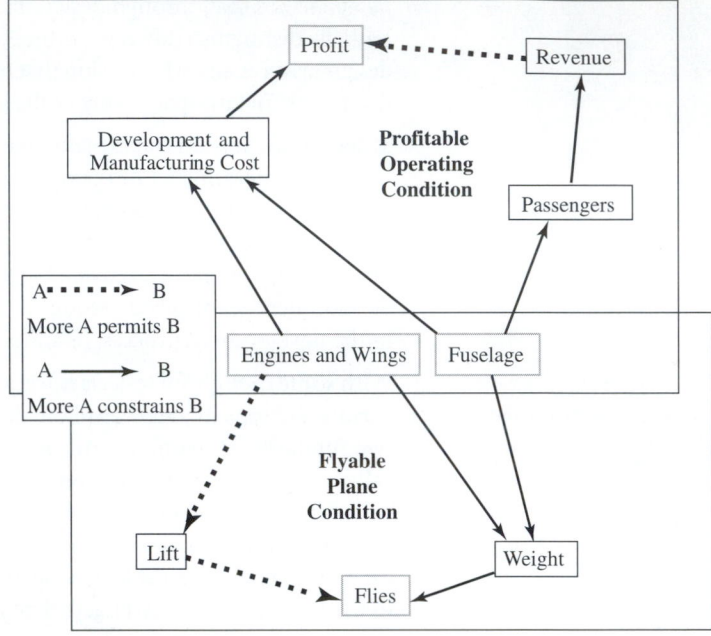

▼ *Figure 3.7.*
Commercial aircraft equilibrium conditions.

first, the view of the business revenue, customer, or *market risk* and second, the view of the physical, engineering, or *technological risk*. Each of these views is an independent source of risk for any new product development project: Can we make it work, and if so, will it sell? These two views form the two product development decision-making efforts that intersect at the key design decisions to be made.

The first loop can be developed with an effective understanding of the customers' needs, as discussed in Chapter 4. The important needs can be related to the preliminary concept design decisions and in particular how the business case is impacted by need value changes and then how the needs are impacted by the design decisions. To start the business case loop, start with an end-node of the business objective: "market share" or "profit." Then ask what can permit an increase in terms of the important customer needs or functional specifications, similar to the "increase in number of passengers" customer need in Figure 3.7. Then ask what can permit this increase of the customer needs in terms of the concept design decisions, similar to the "wings, fuselage, engines" design decisions in Figure 3.7. This forms one half of a business case loop equation, the revenue side. The cost side forms the other half of the loop. Improvements in the design decisions will cause increases in costs, which will force a decrease in the end node of profit. This forms one loop of the design driver equation.

This process should then be repeated for the design's technical constraints after answering the technical questions given in Table 3.1. Here, one should start by listing the major performance constraint that must be met, analogous to the "flyable" node in Figure 3.7. What physically constrains the design from just being low cost and high performance as requested by the business viewpoint? With this end-node, then ask what two competing aspects cause the constraint, such as the "flyable" end node being driven by "lift" versus "weight." One then can continue this process in reverse until reaching the concept design decisions.

Similar to the business end-node, each side of the technical end-node will either be a "constrains" chain or a "permits" chain. In the aircraft example, more lift "permits" more flying capability, and more engine and wing performance "permits" more lift. On the other hand, more weight "constrains" flying capability, and more fuselage "constrains" there to be more weight.

Intersecting the business and technical constraint loops by the common design variables will give one an understanding of the drivers. One should direct the connections out of the design drivers with arrows, with the arrows ending at the "gain market share" node in the business case loop and with arrows ending at the major performance constraint node in the technical constraint loop. Then on each connecting arrow, if the source node constrains more of the sink node, one should highlight the link as dark. If, on the other hand, the source node permits more of the sink node, one should highlight the link as dashed. Basically, for each loop there will be a "constrains" chain and a "permits" chain, starting from the design drivers and ending in the business goal/fundamental constraint node.

Note that the "constrains" causality could also use positive words of "requires more" or negative words of "forces less." The "permits" wording could use word of "allows" or other positive words. The "permits" wording should always be positive, though, where one can change the wording in the nodes to enable the positive link wording. Further note that a "permit" arrow from A to B is the same as a "constrains" arrow from B to A and vice versa.

The student inclined toward thinking in strict legal terms might argue that different design concepts have different design drivers, and so this analysis is limited to postconceptual design. This is not precisely true, because specifications on design drivers serve as overall averages or limits that can be flowed down to specific design variables in different ways for different concepts. The design drivers are meant to be abstract representations of actual design variables so they can take on different forms in different concepts.

The analysis above was presented in terms of the design concept being fundamentally negotiated between business concerns and one primary technical constraint. There are other design concepts that may instead be fundamentally negotiated between two technical constraints. A similar approach can be applied to determine the design drivers that are at the crux of the negotiation and need to be determined early.

With this analysis completed, one has an understanding of how and why a requested increase in business revenue is connected through the design drivers to the fundamental design constraint by "permits" and "constrains" linkages. This analysis should be completed on a continuing basis as the design progresses toward a product. By so doing, vitality of the design will always be questioned and, hopefully, maintained.

Example: Fingernail Clipper

Consider the design drivers of a fingernail clipper. What are they? Thinking as financial analysts and constructing the business loop, we start with a node of "profit." Profit is determined by the two competing factors of revenue and costs. Constructing the "permits" branch, a better perception by the customer would permit an increase in sales or revenue. Suppose, as will be shown in the next chapter, that the generalized customer need of "more comfortable" would accomplish this. To be more comfortable, we might argue that we can change design decisions of generalized "width and length." That forms the "permits" branch of the business loop. We now need the "requires" branch. An increase in the thickness, width, or length requires an increase in material costs. An increase in materials costs requires an increase in revenues. We have completed the business loop.

The next step is to construct the technological constraint loop. Thinking as engineers, the reason we cannot just make the clipper large at the same cost is that we would have to decrease the component thickness until the clipper would bend and fail. So "bending" is the fundamental constraint equation. Two quantities act on the constraint of bending are the applied moment and the component sectional strength. These are the next two nodes. Increasing the length permits an increase in the moment, connecting it to the length design decision. An increase in sectional strength is permitted by an increase in thickness or width.

Combining this into a diagram, we have Figure 3.8. Here, the business case loop result is as expected, a "required" and a "permits" branch exist

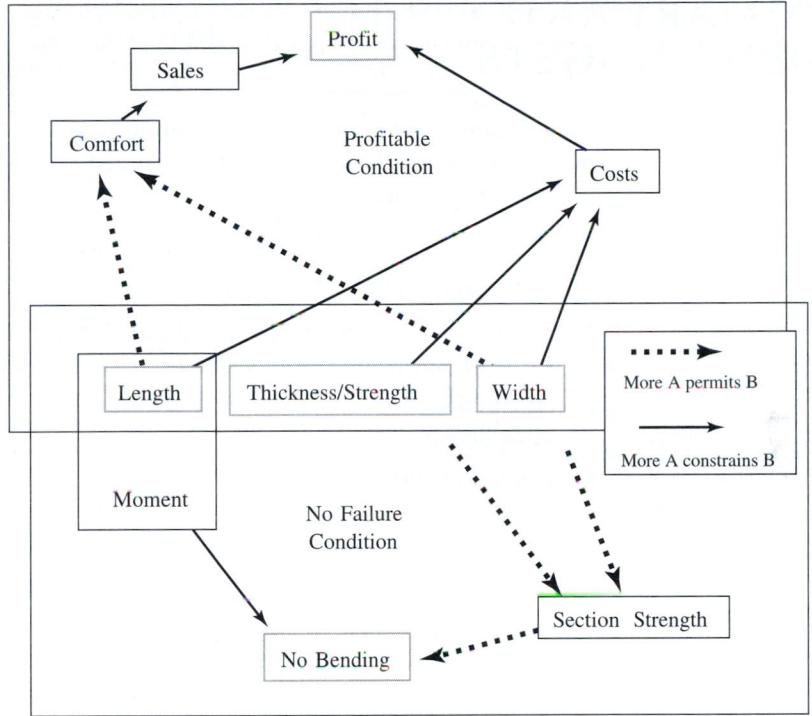

▼ *Figure 3.8.*
Fingernail clipper design driver conditions.

from the design drivers to the end-node of profit. The technical constraint loop is more complex, though. The length-to-moment link ought to be a "required" link to fit the standard format. Because it is not, then as far as decision structure is concerned, "length" and "moment" are equivalent as one node.

The overall analysis result is that an effective target simultaneous specification set for length, width, and thickness type decision variables in any concept must be established early to ensure the product will both not fail and be profitable.

In summary, the student should realize that when entering a design problem, design driver loop "equations" are not obvious and can initially be constructed in different ways that do not have the logical consistency required of the diagram. Different variables and performance requirements typically come to light in the construction of the diagram, which makes the exercise worthwhile.

VI. SUMMARY AND "GOLDEN NUGGETS"

Understanding the economic and technical feasibility of a development project should be done early and reassessed often. Some key ideas in project scoping include:

- understanding the various interests that pull and constrain the development project

- assessing product development risk, which fundamentally arises from two independent sources: market risk and technical risk

- establishing the design driver variables that, when specified, define other decisions directly

- establishing the technical specifications early

- completing an economic analysis to establish project cost limits and marketplace targets

References

Betz, F. 1993. *Strategic technology management.* New York: McGraw-Hill.

Bower, J., and Christensen, C. 1993. Disruptive technologies: Catching the wave. *Harvard Business Review* (Jan.-Feb.): 43–53.

Brealey, R., and Myers, S. 1991. *Principles of corporate finance.* 4th ed. New York: McGraw-Hill.

Christensen, C. 1997. *The innovator's dilemma.* Cambridge, MA: Harvard Business School Press.

Dixit, A., and Pindyck, R. 1994. *Investment under uncertainty.* Princeton, NJ: Princeton University Press.

Faulkner, T. 1996. Applying options thinking to R&D Valuation. *Research Technology Management* 39(3, May-June): pp. 50–56.

Fine, C. 1998. *Clock speed.* Reading, MA: Perseus.

Foster, R. 1986. *Innovation: The attacker's advantage.* New York: Summit.

Gwennap, L. 1996. Birth of a chip. *Byte* 21(12), Dec., p. 77.

McNair, M. P. 1954. *The case method at the Harvard Business School.* New York: McGraw-Hill.

Ronstadt, R. 1988. *The art of case analysis: A guide to the diagnosis of business situations.* 2nd ed. Natick, MA: Lord.

4 Understanding Customer Needs

Interviewing customers of an electric wok to understand their needs.

Many new technology-development initiatives are undertaken with no basis for market acceptance other than management belief. If the developer thinks the technology is amazing and valuable, then everyone else should also. This scenario is known as the *technologist's problem* and is unfortunately very common in the engineering community. Akia Morita, founder of Sony Corporation, has stated, "Our plan is to lead the public to new products rather than ask them what they want. The public does not know what is possible, we do" (Morita, 1986). The result is often products such as the Betamax videotape. Such technology-focused thinking causes market failure of innovative products. They fail to satisfy the customer.

I. CHAPTER ROADMAP

This chapter presents background on modeling customer demands and wishes as a statistical population. Two methods of increasing sophistication are presented for interviewing customers, both with the aim of constructing a statistically representative list of needs. Based on these customer needs lists, the next sections discuss simple methods for weighting the importance of the needs. Finally, customer activity diagrams are described for representing the customer operational system within which a product is used.

Figure 4.1 shows example routes through the chapter. A direct route covers simple methods, with the goal of creating a suitable representation of the customer needs. An alternative route is also shown, augmenting the approach with more sophisticated, alternative methods.

II. CUSTOMER SATISFACTION

Voice of the Customer

In the previous chapter, we clarified what might make a new product a technical and business opportunity. Based on this opportunity, a firm should determine what product features are in actual demand before expending large resources to develop a new or revised product. While the fortunate-technology-push approach can and does work, it is also clear that considering the customers' desires will pull product development into better directions and amplify success. Therefore, we

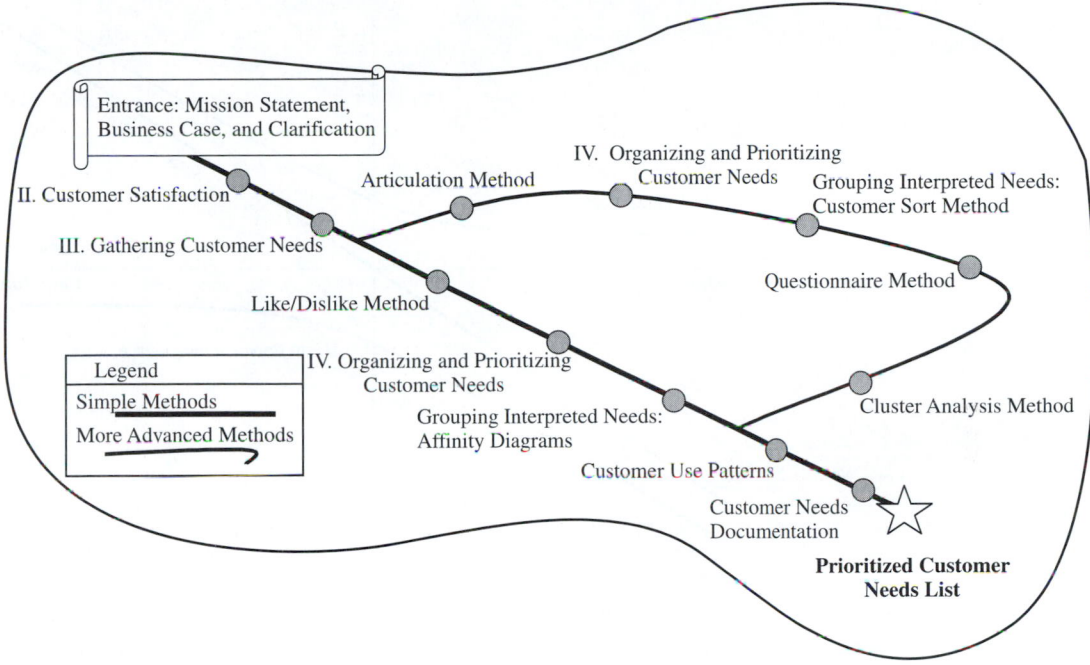

▼ Figure 4.1.
Road map to create a prioritized customer needs list.

explore methods to understand what the customers want, to permit a design team to hear the voice of the customers.

This process is a difficult task, where several issues confound what the designer needs to know. For example, a customer must understand the product that is being designed. Typically, this understanding is conveyed by showing the current product model to a customer and asking for preference information about its features. The product model shown to the customer is not the same as the new product model to be designed. This paradox is a particularly difficult problem for products with rapidly changing technology or for new product domains where technologies are not well established.

Another problem is that customers typically only discuss the failings of the products, what they do not like about their experience in using the product. It requires active prodding by the design team to uncover what is expected by the customer, that is, the latent needs that are not even mentioned directly by a customer. Yet such considerations often drive

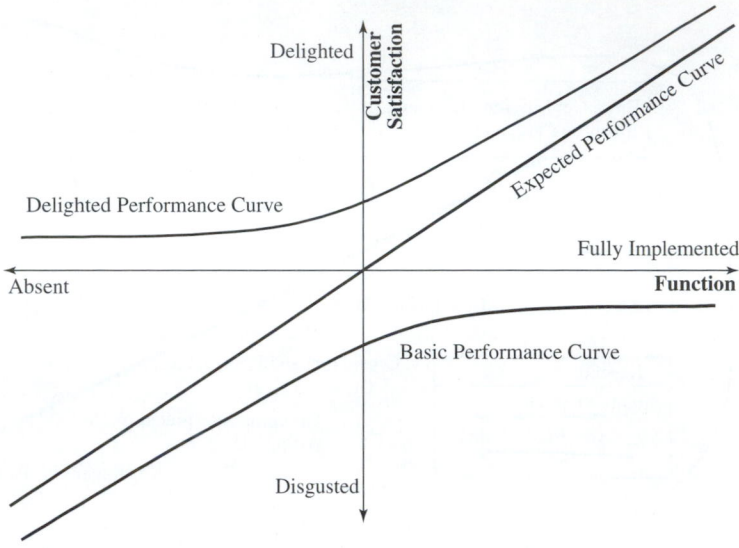

▼ Figure 4.2.
Kano diagram of customer satisfaction.

much of the engineering design. This difficulty is graphically explained by considering customer satisfaction versus any product function, known as the *Kano diagram,* as shown in Figure 4.2.

The Kano diagram (Shiba, Graham, and Walden, 1993) depicts the idea that one might consider customer satisfaction on a scale from disgusted to delighted. Similarly, one might consider product functions on a scale from being absent to being fully implemented. If the product function is a good surrogate for the customer need being considered, one could plot a 45° line, which would indicate the nominal supplied satisfaction for any specification level of the function. This line is known as "one-to-one quality" or "linear quality," the minimum expectation of any *new* product development undertaking. On the other hand, one could also plot a lower curve that would indicate the minimum, basic level of satisfaction that a customer presumes must exist for the function implementation level. This plot, known as "basic performance," represents assumed functionality that must be in the product. It is expected and latent; if it does not exist in the product, the satisfaction of the customer will be greatly deteriorated.

Similarly, one could plot an upper curve that would indicate the *delighted state* that a customer would hope to have for the function implementation level. The delighted state is what a design team should strive for, to provide performance beyond what the customers expect, which delights them. These three states of satisfaction and the spectrum between them forms the background thinking behind customer needs.

There is a spectrum of customer satisfaction from disgust to delight that changes with increased implementation of functions. The more product function implemented, the more the customer expects, and the harder it is to keep delighting the customer. Further, the harder it is to supply basic, expected performance. For example, if every car manufacturer begins including cup holders in the passenger compartments, the added functionality of cup holders no long delights the customer; it becomes expected. Likewise, if we consider car brakes, customers expect reliable disk brakes in passenger automobiles that have a long life (even the pads), that do not make excessive noise, and that reliably stop a vehicle in a short duration of time. When antilock brake systems (ABS) emerged in the market, they represented a significant step above linear quality. Customers were delighted with the improved functionality. As ABS reduces cost, it is placed on a wider array of automobile types and classes, representing linear quality for those models. In the near future, ABS will become expected quality; an automobile function that will simply be expected in virtually all types of road transportation.

The message from the Kano diagram is simple yet profound. Customer expectations increase over time. It is therefore paramount to stay in touch with the customers and to understand their preferences.

Customer Populations

The customer is a statistical concept; there are numerous potential purchasers of a new product being designed. The *customer population* is the set of persons whom we want to be purchasers of our new product. Typically, the customer population is varied; different customers use the product differently, have different objectives for the product, operate the product in different environments, and generally have different expectations. There are different strategies a company might take to such diversity, offering a single product, forcing customers to be happy with that offering; permitting customization of features on the product; or offering a portfolio of different offerings, each tailored to a market niche. To drive this decision making, the customer population must be understood in terms of its criteria for evaluating the product offering.

To obtain this understanding, we must think about the customer population statistically. One obvious way is to determine the average response of the customer population on any set of criteria and call this average the response of the *average customer*. Similarly, we can consider the difficult, hard-to-achieve levels mentioned in the responses of the customer population that occur out at the tails of the customer distribution. We might think in standard deviation terms, and call the 3-sigma level response the *3-sigma customer*. These two customer levels are important for a design team to consider.

Another categorization of the customer population is to consider the different use patterns that the customer population exhibits. For companies operating in global markets, where the design is for customers in many different countries, climates, and economies, then geographical breakdowns are often effective. Socioeconomic categories are also often effective, such as characterizations by income, gender, marital status, and age. The point is to have a clear customer-need-based description (demographics) of the market segment or segments that are to be met by a new product.

Yet another categorization that a design team should consider is to characterize the customer population by lead-lag usage. That is, for products with rapidly changing technology, one can find a fraction of the customer population that is always on the leading-edge use of the product. These persons use the product in a way that the rest of the customer population will also exhibit in a few years or so. These *lead-customers* are important for a design team to identify and open a line of communication with. For example, university students are lead users of computer network technology; they are on-line most often and develop new uses for network technology, such as the first weather reporting sites, the first on-line camera pictures, and the first site to report when a soda machine is empty. Now weather information and traffic pictures are common on the Web. From one perspective, developing product technology that merely satisfies these customers will in turn simultaneously delight the average customer.

Types of Customer Needs

Customer needs can be profitably considered in general categories, based on how easy the customer can express them and how rapidly they change. These are conceptual categories, not distinct objective groupings. It is important, though, to understand the differences among them.

1.1 **Direct Needs:** These are the needs that, when asked about the product, customers have no trouble declaring as something they are concerned about. These are easily uncovered using the methods of this chapter.

1.2 **Latent Needs:** These are needs that typically are not directly expressed by the customer without probing. Customers typically do not think in modes that allow themselves to express these needs directly. Latent needs are better characterized as customer needs, not of the product, but of the system within which the product operates. Other products, services, or actions currently satisfy the needs directly. Yet, these needs might be fulfilled with a developing product, and doing so can provide competitive advantage.

2.1 **Constant Needs:** These needs are intrinsic to the task of the product and always will be. When a product is used, this need will always be there. Such needs are effective to examine with customer needs analysis, since the cost can be spread over time. For example, consider cameras, where the number of exposures is always a customer need, whether implemented as the number that can be recorded on film (24 or 36 exposures) or the number of digital images that can be recorded into digital memory.

2.2 **Variable Needs:** These needs are not necessarily constant; if a foreseeable technological change can happen, these needs go away. For example, digital photography eliminates a customer need of long film storage life. These needs are more difficult to understand through discussions with the customer, since the customer may not understand them yet.

3.1 **General Needs:** These needs apply to every person in the customer population. For instance, every vehicle sold in the continental United States has a customer need of having heat supplied to the passenger cabin.

3.2 **Niche Needs:** These needs apply only to a smaller market segment within the entire buying population. Not every vehicle needs to supply cool, air-conditioned air to the passenger cabin. Some niche customers might also prefer an electric-powered vehicle.

The rate of change of these boundaries in the customer need space is completely dependent on the state of technology and its rate of change. The first category considers observability, and the second considers technology change. The last category arises from an approach to satisfying variance in the customer need space (focus on "niches"), which is different than attempting to satisfy the entire market with a robust technology.

Customer Need Models

In this analysis, a statistical representation of the product customer needs is the desired result. This result can be as simple as a list of needs, as distilled from interviews from the customer. The resulting list might be augmented with importance weightings, determined through questionnaires with the importance of each need determined independently of the others as statistical distributions and an average weight.

At a more advanced level, the customer needs can be uncovered as a list of needs, each need complete with a range of possible target values, any value of which a design team might aim to meet. This representation may be considered a vector space, where the positions of all members in the market population can be represented within this vector space as a probability distribution. Every person in the market has an ideal set of target values on the list of needs, and every person is different. This probability space might be examined for large regions of high probability (cluster analysis), which, if the product is targeted in the center of such clusters, the development effort is offered some assurance of larger success. One might also consider where any competitive offerings are positioned in such a space and respond accordingly through game playing models. Such are the current marketing and product development modeling practices of state-of-the-art companies.

A word should be included on the level of detail desired in the customer need modeling. Often, a customer need list of a couple dozen needs is sufficient for development purposes. This guideline is often true for studies such as understanding market positions, marketing, or early concepts. On the other hand, for detailed embodiment questions, a design team often would like to know what the customer preference is for different selection choices (types of knobs, panel layouts, start-up speed requirements, etc.). For these choices, more detailed postconcept questions, resulting in 200–300 customer needs, are often necessary.

III. GATHERING CUSTOMER NEEDS

Need Gathering Methods

There are several methods available for a design team to understand the customer needs. Different techniques developed and applied to construct a customer needs list include: directly using the product, circulating questionnaires, holding focus group discussions, and conducting interviews.

Interviews: Here a design team member(s) discusses the needs with a single customer, one at a time. Such interviews are usually held in the customer's environment, where the customer uses the product. The design team member records the customer responses. These records can be augmented with video or audiotapes. This process works well for products that have a process associated with their customer use. Interview methods are discussed at length in the next section.

Questionnaires: Here the team develops a list of criteria it thinks is relevant to the customer's concerns. It then ranks the product on these criteria. Alternatively, it forms a list of questions and then organizes the responses provided.

Focus groups: Here a moderator facilitates a session with a group of customers. Usually this session is held in the product developer's environment, typically inside a room with a two-way mirror so that the design team can observe the customers during the session. Effective focus group practice is to have the moderator ask exploratory questions such as "why" to uncover the needs of the customer. Typically, several sequential focus group sessions are held, with customer need analysis of the customer's statements constructed between each session. This analysis provides two essential results: an understanding of what to probe (for the moderator) and development of a needs list (for the design team). The food industry, for example, often implements this approach when developing new food products. Indicators of taste, smell, and appearance involve linguistic, facial, and body expressions, and focus group sessions express these preferences to the design team. Food product needs are also difficult to convert into language syntax and typically do not involve a significant hands-on process when "using" the product.

Be the customer: Design teams will also travel to the locations where their or their competitor's product is used and act as the customer. This activity is typically executed by the design team working hand-in-hand with the customers or, alternatively, when the designer is the customer, such as during the design of equipment for in-house use. This approach can be effective for in-house equipment design at a factory; however, it is impossible for tasks where the design team has no skill, such as might be the case with the development of specialized surgical instruments.

Of these methods, questionnaires and "be the customer" have a number of difficulties. In particular, the questionnaires provide the lowest quality information: The responses only pertain to the questions asked, not necessarily what the customer wants to tell the design team. Being the customer is profitable when it is possible, but often the method places excessive demand on the design team.

Of the remaining methods, conducting interviews is the method that provides the most information per quantity of effort. Griffin and Hauser (1993) report that focus groups are more costly to attain the same amount of information. They also report that interviewing nine customers for one hour each will obtain over 90% of the customer needs that would be uncovered when interviewing 60 customers, as observed on a single function product (industrial lunch pails). Such a rule is not entirely valid for multifunction products.

We next present two methods of forming a list of customer needs through interviews of randomly selected members of the customer population. The first method is a very simple approach, the *Like/Dislike Method.* The subsequent is a more sophisticated approach, referred to as the *Articulated-Use Method.*

Conducting Interviews: Like/Dislike Method

The simplest approach that proves reasonably effective in practice is to interview customers as they use the product. During the interview, a team member asks the customers to describe what they like and do not like about the product at their site of usage. During the process, more detailed questions may be asked to explore different facets of the product. At the same time, a reasonable list of customer needs is developed. It is important when holding interviews to have the customers run through their process of using the product.

While the customers describe and actively use the product, the interviewer should ensure that they describe not only what they do not like but also what they do like. Asking about the likes ensures the design team understands what is expected in the product and what must be provided. Asking about the dislikes in the product establishes what the design team can provide to delight the customers, as it currently is not being expected.

It is important to ask about the needs, such as with "why" questions. Such questions will uncover the latent customer needs of the system within which the product operates. A latent need, in this sense, is an expected function or behavior of a product that is not explicitly stated. Typically, it is not directly visible to the design team without careful inquiring and searching. Such needs might be fulfilled with the current product.

The interviewer might also ask questions about what the customers consider when purchasing the product. Instead of asking this question at the location where the customers use the product, it should be asked in the store where the product is purchased. This addition can lead to added customer needs such as sales information, packaging information, and marketing.

The interviewer might also ask about how the customer would change the product to make it better. The point in this line of questioning is not to have the customer redesign the product, though the design team may wish to include the customer in redesign activities at a later date. Rather, the product improvement question should be posed and immediately probed for why a given change is better. What need does the product change improve on? Again, this follow-up question gets at the latent need inquiry.

Some general hints for effective customer interaction (Ulrich and Eppinger, 1995) are:

- *Go with the flow.* Wherever the customer takes you, follow along, and ask *why* and *how* questions.

- *Use visual stimuli and props.* Bring models of new concepts, competitors' products, products related in the system of operating the product, and analogous products. Ask about all of these.

- *Suppress preconceived notions about the product technology.* Do not bias the interview by presuming any concepts or technology. Uncovering a need representation that is independent of the solution used helps in concept engineering. Thus, visual aids and props should be used at the appropriate times, usually after initial needs are elicited from the customer.

- *Have the customer demonstrate.* Don't just ask about the product; human language is only so expressive. Seeing the need in action will permit much better understanding.

- *Be alert for surprises and latent needs.* Pursue any surprise answers with follow-up questions until we understand the need and follow it beyond questions on the product itself with questions about the system within which the product operates. This additional level of inquiry usually uncovers the latent needs.

- *Watch for nonverbal information.* Again, human language cannot communicate all sensation modes and feelings about a product.

Collecting the Data

Shown in Figure 4.3 is a form that can be used to record the interviews. There is basic header information related to the project and the interview subject. The form itself has three columns. The first column is used to record any particular question that sparks a customer response. The middle column is the actual customer statements. The last column is the conversion of this actual customer statement into a succinct noun-verb-adverb form, using the same words as the subject stated as recorded in the second column.

The Like/Dislike method has general row categories to record the customer statements. If the customer likes the way a need is implemented, it is recorded in "Likes" rows. If the customer does not like how a need is implemented, it is recorded in the "Dislikes" rows. This structure permits immediate understanding of what needs to be focused on when redesigning a product.

Customer Data: Project/Product Name			
Customer: Address: Willing to do follow up? Type of user:		Interviewer(s) Date: Currently:	
Question	Customer Statement	Interpreted Need	Importance
Typical uses			
Likes			
Dislikes			
Suggested improvements			

▼ **Figure 4.3.**
Like/Dislike method data collection form.

The fourth column records linguistic expressions of importance that the customers may have used. Typically, these might include:

must, good, should, nice, poor

any adjective/adverbs used to indicate the customer's preference for the interpreted need statement. A *MUST* is used when a customer absolutely must have this feature, generally when it is a determining criterion in purchasing the product. *MUST* ratings should be used very sparingly; only a few *MUST*s per customer interview is a good rule. A very important customer need should have a *GOOD* importance rating. Needs that are presumed should have at least a *SHOULD* rating. If the customer feels the product should satisfy this requirement, it is important enough for the design team to consider it. The *NICE* category is for customer needs that would be nice if the product satisfied them but are not critical. Typically, added product functions are listed as *NICE;* these functions provide added capability but are required for

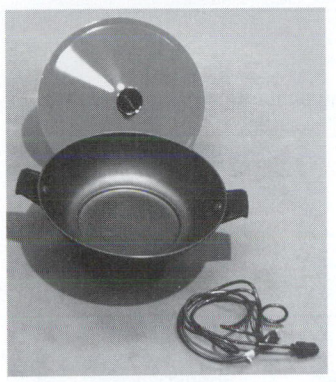

▼ *Figure 4.4.*
Original product—six-quart
electric wok.

normal usage. More rating levels can be used for a more refined resolution, depending on the ability and interest of the subjects.

There are also some other categories worth recording. One such category is "Typical Uses," recorded to ensure that the context of customer statements are understood. Also, the customer may have ideas for improving the product design. These ideas should be recorded as well and considered later during concept generation.

Example: Electric Wok

An electric wok redesign project was initiated through the perception of a need: the inadequacy of current electric woks to satisfy the demands of the young urban dweller desiring to conveniently cook authentic Chinese food. The original product is a six-quart electric wok, shown in Figure 4.4.

A competitive product is a traditional wok, used over a gas flame and heated by convective and radiative modes of heat transfer. Heating occurs uniformly across the entire wok surface, rather than in a concentrated ring. This competitive heating method thus lends itself to uniformly cooking the food placed in it.

Interviews were completed with several persons, as shown in the subsequent interview forms (Figures 4.5–4.6). The "Question" and "Customer Statement" columns were completed at the time of the interview, and the "Interpreted Need" translation afterward. For example, the customer made a statement that the wok "moves around too much when stirring," which needs to be translated into a succinct customer need. This statement was therefore interpreted into "Can grip tabletop," which is a positive phrasing. Similar interpretations were carried out for the other need statements. After completing the customer interviews and translating the results, the next step is to combine the interpreted needs from the various subjects into a single list of aggregate customer needs. This process is discussed in Section IV.

Conducting Interviews: Articulated-Use Method

The previous method of customer interviews is quick and suitable for determining what features of a product are ready for redesign. It is not the best, however, at uncovering latent needs. To add this capability, the design team must understand how the product is used. The *Articulated-Use Method* provides a means of obtaining this understanding.

Customer Data: Electric Wok Customer: John Doe Address: MIT Willing to do follow up? Y Type of user: College student		Interviewer(s): Date: Currently uses: Frying pan	
Question	Customer Statement	Interpreted Need	Importance
Typical uses	Stir-fry Steaming Frying/scrambling eggs Cooking pasta Cooking chili/stews [Everyday cooking]		
Likes	Non-stick surface Size Can stand on its own Temp. response rate high Aesthetically pleasing Depth of dish	Non-stick surface Compact Able to stand on its own Quick temp. response Aesthetically pleasing Deep sides	Good Good Should Good Good Should
Dislikes	Short cord Moves around too much when stirring (doesn't grip surface of table) Entire assembly is too high/tall Handles are hard to grip (esp. if oil splatters) Have to watch constantly Temp. adjustment gets too hot/also too low to read Sides don't get as hot; may overcook on bottom Afraid to get bottom wet	Long extension cord Can grip tabletop Compact (flat) elec. unit Handles are easy to grip Auto shut-off Temp. switch is insulated from heat Temp. switch is in an easily accessible/ readable spot Constant temp. distribution Bottom is watertight	Must Good Good Nice Should Should Good Should Good
Suggested Improvements	Retractable cord Better gripping bottom Hole through handles to grip Make heating element casing flatter Have clip for lid (to flip back & forth) that is also removable Deep-frying accessory (wire mesh shelf) Maybe has ears on both sides in case someone is lefty		

▼ *Figure 4.5.*
Portion of customer interview data for the electric wok redesign: Customer 1A.

Customer Data: Electric Wok			
Customer: Jane Doe		Interviewer(s):	
Address: MIT		Date:	
Willing to do followup? Y		Currently uses: Frying pan/pot	
Type of user: College student			
Question	Customer Statement	Interpreted Need	Importance
Typical uses	Cooking stir-fry Chinese food Frying pan substitute Quick, easy dishes		
Likes	Bottom heats up quickly	Heats quickly	Must
	Little to no smoke emitted when cooking		

	Ears are covered in a type of
	  insulating material
		Ears remain cool (not hot) to
		  touch
			Good

	Non-stick surface	Non-stick surface	Should
	Fairly lightweight	Wok is lightweight	Nice
	Attractive design/color	Aesthetically pleasing	Nice
	Useful for college-age students (w/out easy access to stoves)		
	Easy to clean	Inside can be cleaned easily	Nice
Dislikes	Sides slow to heat up	Sides heat up at same rate	Should
	Need to stir food around a lot to cook well	Smaller bottom area	Must

	Flat bottom: difficult to tilt
	  wok/move food
		Rounded bottom to be able to
		  tilt/shake wok
			Should

	No off button/switch	Include off switch	Should
	Cumbersome to wash	Heating unit is detachable	
	Afraid to get bottom (heating element) wet	wok is watertight	Good
	Needed extension cord to cook on table	Long extension cord	Good
Suggested Improvements	One-sided handle (long) "Flip-out" handle to con- serve shelf space Include off button Make bottom more round Notch on rim for spatula/ stirring utensil to rest on Longer cord (or include extension?)		

▼ *Figure 4.6.*
Portion of customer interview data for the electric wok redesign: Customer 2A.

Bringing an interview sheet with canned questions does not work well for eliciting latent and other types of needs. Canned questions can be asked toward the end of an interview process. Instead, it is much better to bring nothing more than the following single task:

Walk me through a typical session using the product.

Have the customers start by how they approach the product to begin using it. Where is it stored? What must they do to get it out and prepared for use? When they perform any motion at all, ask them what they are doing. Have them also finish with the cleanup and storage of the product. The same interview questioning guidelines apply as in the Like/Dislike method.

The important distinction here is to obtain and understand information on every step in the process of using the product. One should pursue in depth any statements such as "that aspect works well enough" to understand why it works well enough and how the customers use it in their context. One should also pursue in depth any statements such as, "I use a different system than the product to perform that aspect" to understand the environment in which the product is used. These added activities present latent opportunities to redesign the product.

Collecting the Data

An alternative form for collecting customer data is given Figure 4.7. The first column is a prompt to the customers: What did we say to obtain the response? The second column is what the customers said, using

Project Name				
Customer Data				
Customer ID		Interviewer		
Willing to follow up?		Date		
Location				
Type of user (their words)				
Question/Prompt	Customer Statement	Interpreted Need	Weight	Activity

▼ *Figure 4.7.*
Customer interview form.

the actual words they used. These two columns are filled in during the interview. The last three columns should be completed after the interview during a postanalysis/interpretation. The third column will become interpretations of what the customers said. The design team interprets the middle column statements into what it believes the customers want/demand through their statement.

When carrying out this process:

▶ Express the customer statements in terms of what the product must do, not how the product might do it.

▶ Use positive, not negative phrasing.

▶ Express the need as an attribute of the product. This expression ensures consistency and makes subsequent translation into product specifications easier.

▶ Do not use "must" or "should" in the statement. Rather, these modifiers will be incorporated into the importance ratings.

In the fourth column, linguistic expressions of importance are added that the customers may have used. Again, these expressions might include:

must, should, ought, nice, great, poor

any adjective/adverbs used to indicate their preference for the interpreted need statement.

The final column is used to list the activity that the customer is performing during a sequence of customer needs. For example, a customer may be retrieving the product from storage when a set of six customer needed are mentioned. These activities should be recorded for clarification and for later use in the product development process.

After conducting several interviews, the next step is to combine these results into a single list of customer needs for the entire customer population. Usually, tens to hundreds of customer needs will be developed during a customer interview. After an appropriate sample size of interviews, the sheer number of needs may seem overwhelming. Through a combination process, the goals are to develop a manageable set of overall needs with clearly discriminated importance ratings.

Example: Fingernail Clipper

As an example of the Articulated-Use Method, consider the fingernail clipper product. An interview form is shown in Figure 4.8. The first two columns were recorded during the interview and are a record of the conversation. The third column is again an interpretation of what the

Clipper Project Customer Data				
Customer ID:	KNO5	Interviewer:	KNO	
Willing to Follow Up?	No	Location:	Cambridge, MA	
Type of User:	Middle class, white, male, traveling			
Question/Prompt	**Customer Statement**	**Interpreted Need**	**Weight**	**Activity**
When usually use?	In the evening in hotel			
	Keep in my shaving bag	Reasonably Compact	Must	Store
How big is that?	About 3" x 2" x 6", and I have a lot of things in it, it is always full			
Size of things is important?	Very important. I look for the smallest size of everything			
	So I dig it out of my bag, and carry it to the bed, where I usually clip my nails	Striking appearance	Nice	Prepare for filing
		Lightweight	Nice	Prepare for filing
	[Spins handle and rotates simultaneously]	Easy to open file	Should	Prepare for filing
Do you file?	Yes, I file at an angle, with a vertical and an angular motion	File at an angle	Must	Files nails
	[With file between thumb and index finger, and clipper body in fist]			
	The clipper is not the most comfortable in my hand	Comfortable to file with	Should	Files nails
	[Rotates file back in place]	Easy to close file	Should	Return from filing
	[Rotates handle intoposition]	Easy to open clipper	Should	Prepare for clipping
	The handle works reasonably well to open and close			
	[Grabs in hand using thumb and index finger, with tail up against middle finger edge]	Easy to hold clipper	Should	Clip Nails
	[Positions nail to be cut on bottom blade]	Easy to align blade	Nice	Clip Nails
	You need to align the blade			
	[Squeezes finger and thumb to cut]	Low clipping squeeze force	Nice	Clip Nails

▼ *Figure 4.8.*
Customer need collection form, completed for the fingernail clipper example.

customers meant when the need statements were made, using their words. The fourth column is a subjective importance rank, and the last column is the activity the customer was performing when the statement was made. For example, consider the "Dig it out of my bag" customer statement. At the time, this statement referred to the need to see the clipper in the bag, and so "striking appearance" was the interpreted need. This need was not terribly critical, and so it carried a subjective importance of "nice," and the statement was made when the customer was "preparing to file."

Customer Interviews: Product Feel and Industrial Design

In addition to understanding the customer activities of a product, it is also important to capture the desired "feel" of a product. The tools of the industrial designer are thus needed as part of the need-gathering focus. Industrial design is concerned with the appearance, feel, layout, and usefulness of manufactured products. The industrial designer emphasizes the important human qualities of comfort, safety, and aesthetic appeal. Special methods are implemented to satisfy the goals of industrial design.

An example method worksheet is shown in Figure 4.9 (Design Edge 1995). This method is known as *Semantic Inquiry*, wherein attributes and product feel are captured early in the process, during client and

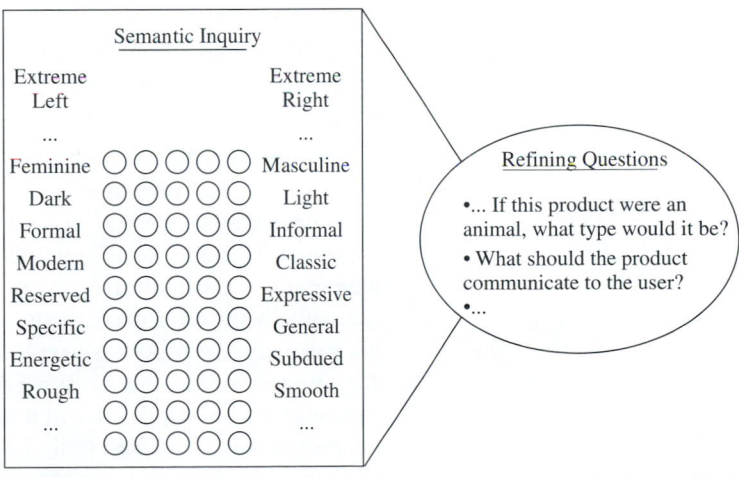

▼ *Figure 4.9.*
Semantic inquiry tool.

customer interviews. This method may be used in conjunction with the previous gathering techniques, augmenting the customer needs with product feel.

IV. ORGANIZING AND PRIORITIZING CUSTOMER NEEDS

Grouping Interpreted Needs

The combination of the varying interpreted needs from the customer population is a sorting process—to combine the slightly different ways that different subjects say ostensibly the same thing. We present two methods for performing this combination. The first simplistic approach, referred to as the *Affinity Diagram Method*, is to have the design team review the interpreted needs list and carry out the sorting themselves. The second, less-biased method is to have the customers complete the sorting. Statistical analysis may then be used to convert the customer results into a single list. This second method is called the *Customer Sort Method*.

Grouping the Needs—Affinity Diagram Method

Having formed interpreted needs lists for sufficient customers (each interview provides no new, different statements), the information must be compiled into a list of interpreted customer needs to which a product will be redesigned. An elementary method is to have the product development team sort the needs into groupings by looking for affinities among the different needs—the *affinity diagram* method.

To group the needs, each interpreted need, with associated customer identification, is copied onto a Post-it® note, as shown in Figure 4.10. Then one takes the Post-it® notes and affixes the first one on a large, white board. The next customer need is compared with the first. If it is different, a new column is constructed on the board. If the statement is essentially the same need, the Post-it® note is placed under the similar statement. The process is repeated for all Post-it® notes (Figure 4.11). This process will create a grouping of customer needs that reflects the customer desires, demands, and importance. As an example, consider the electric wok. The Like/Dislike interviews are combined into the list of Figure 4.12. The result is reasonable with 28 needs uncovered. They are grouped into design-team-oriented categories, such as "Steady-State Temperature Uniform." This categorization is generally not the

CUSTOMER
NEED

Customer ID

▼ *Figure 4.10.*
Customer interpreted need Post-it® format.

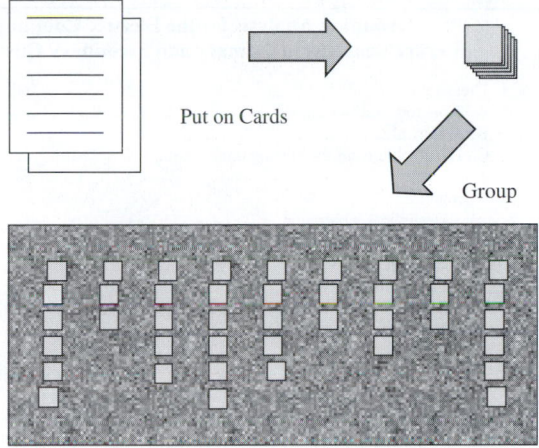

▼ *Figure 4.11.*
Interpreted need sorting.

best, as it is not expressed in the original words of the customers, and the reasoning behind why the customer made the statements can be lost. On the other hand, the method is easy to implement and reasonably effective.

Grouping the Needs—Customer Sort Method

Rather than have the design team conduct the sorting, an alternative approach is to have a few selected customers sort the needs. This process prevents the customer data from being biased by the development team.

To sort the needs, the design team records all of the interpreted needs from every interview onto index cards, where there are N such interpreted need statements. An index number (between $1-N$) is placed on each card as well. These are shuffled and split into n piles, where $n \geq 10$ for reasonable statistics. Each pile is given to a different randomly selected customer. The customers are asked to sort their pile of cards into subpiles, where each card in a subpile says "the same thing" but with slightly different wording.

With this sorting, a *co-occurrence matrix* is constructed. Here, the design team constructs a square matrix of size N, and every column and row represents an interpreted need statement made at an interview. Then in matrix element (i,j), a number is entered corresponding to the sorting customers who placed interpreted need statement i in a subpile with interpreted need statement j. Therefore, every matrix element is between 1 and n and is a measure of their similarity.

Grouping Analysis for the Electric Cooking Wok:
Hierarchical List of Primary and Secondary Customer Needs

I. Cleanable
   A. Non-stick surface ....
   B. Watertight...
   C. Detachable from the heating unit ....

II. Aesthetics
   A. Aesthetically pleasing ....

III. Cooking Shape
   A. Flat bottom for frying..
   B. Small, rounded bottom for stir-fry ....
   C. No ridges on inner surface ....

IV. Size
   A. Compact for easy manipulation/small storage ....
   B. Lightweight ....

V. Stability
   A. Able to stand on own ....
   B. Doesn't slide on tabletop ....
   C. Detachable heating unit to remove heat when cooking............................

VI. Temperature
   A. Heats and cools quickly ....
   B. Temperature uniform across inner surface ....
   C. Steady-state temperature uniform ....
   D. Capable of high temperature..
   E. Heat contained in wok ....

VII. Capacity
   A. Large volume capacity...

VIII. Manipulation
   A. Easy to handle..
   B. Long extension cord..
   C. Handles remain cool/don't get hot ....
   D. Contents can be poured out..

IX. Cost
   A. Cost..

X. Temperature Control
   A. Temperature switch readable ....
   B. Off switch included...
   C. Temperature indicator ....
   D. Temperature controls remain cool/don't get hot..................................

XI. Impact
   A. Impact resistant ....

▼ *Figure 4.12.*
Electric wok: sorted customer needs.

Next, we need to convert the matrix into a tree representing the probability that any two rows are the same. This conversion can be done with the hierarchical clustering procedure. Hierarchical clustering is a procedure that takes co-occurrence values and, if the values are high, places the two matrix elements close in a tree, and if the co-occurrence values are low, places the two matrix elements far apart in the tree. A clustering procedure works by taking any customer need statement and treating it as a cluster, then joining its closest cluster, followed by iterating. The result is a tree with close nodes being similar.

Clustering algorithms are widely available in statistical packages. An example is Ward's method (Norüis, 1988), where the squared distance from the customer need to the cluster centroid is calculated.

The output of the clustering procedure is a tree, from a top-level node down to leaf nodes, where the leaf nodes are the N interpreted need statements. This tree structure needs to be parsed into two levels of interpreted needs and groupings of needs. Low-level branches become different expressions of ostensibly the same thing, so the customer performing the sort can find a single expression for all of these similar needs.

This approach gives very good results. It keeps the entire analysis in the voice of the customer, with very limited opportunity for design team bias.

Determining Need Importance

Having determined a list of customer needs where each line item is important to at least some customers, the next task is to determine just how important each need really is to the whole population. We seek a numerical importance ranking of each need, considered independent of the others. Two simple methods are presented here and more sophisticated methods reviewed. The first method is to use the interview subject ratings data, called the *Interview Data Method*. The second, more-complete approach is to send out survey questionnaires, called the *Questionnaire Method*. Finally, vector space clustering methods are briefly discussed.

Importance—Interview Data Method

An elementary approach to forming a customer importance rating is to first construct a set of normalized weightings by comparing the number of subjects who mention a need versus the total number of subjects, using the original set of customers with no further questionnaire activity.

$$\omega_{CR_i} = \frac{\text{\# times mentioned}}{\text{\# subjects}}$$

where ω_{CR_i} is the interpreted importance rank of the ith customer need. This ranking is flawed, in part, as it includes a measure of the obviousness of the need as opposed to its importance. A need may not be important but may be very obvious, and so every interview subject mentions it. This weighting procedure would rank such a need very important even though it is not.

Because of this concern, typically the design team reviews the different statements in column 2 of the customer response sheets to raise and lower the result from the above. To do this, we first convert the subjective importance ratings into numerical equivalents. A typical transformation used is:

MUST	9		MUST	1.0
GOOD	7		GOOD	0.7
SHOULD	5	or	SHOULD	0.5
NICE	3		NICE	0.3
not mentioned	0		not mentioned	0.0

This mapping is always a subjective interpretation conversion that a design team must agree on to convert linguistic expressions of the customer into numbers. Then the importance assigned to each customer need can be calculated as the average:

$$\omega_{CR_i} = \frac{\omega_{CR_{ij}}}{\# \text{ mentioned}}$$

where $\omega_{CR_{i,j}}$ is the numerical importance rating for the ith need assigned by the jth customer.

This method should be used with great caution. The number of interviews completed is typically just sufficient to attain no more new needs. That is insufficient to attain a random sample on the last few needs that were only mentioned by the last customer. Therefore, the importance assigned to these needs will be very biased, since they were asked of the first subjects. In general, the sampling pool is too small.

As an example, consider the electric wok (Figure 4.13). The number of interview subjects who mentioned the customer need is shown and the weighted value. Note that some needs were mentioned many times yet carry lower importance, as they had low subjective importance as rated by each customer.

Importance—Questionnaire Method

A questionnaire can be sent to more customers, using the uncovered customer needs, asking for importance of each need. This process will

```
┌─────────────────────────────────────────────────────────────────────┐
│ Electric Cooking Wok:                                                 │
│ Customer Needs                                                        │
│ Interviewer:                                          Date:           │
│ Sample Size:  10 customers                                            │
│ Average Customer:   Male/Female, age 20-40                            │
│                     Middle Class                                      │
│                     Wish to cook Chinese food in an authentic manner. │
│                     Has common household kitchen space                │
│                                                      #          WT     │
│ I.  Cleanable                                                         │
│     A. Non-stick surface ...............................(7).......4   │
│     B. Watertight.......................................(3)......2    │
│ C. Detachable from the heating unit ....................(3).......3   │
│                                                                       │
│ II. Aesthetics                                                        │
│     A. Aesthetically pleasing ..........................(4).......4   │
│                                                                       │
│ III. Cooking Shape                                                    │
│      A. Flat bottom for frying..........................(3)......2    │
│      B. Small, rounded bottom for stir-fry .............(7)......2    │
│      C. No ridges on inner surface .....................(5)......3    │
│                                                                       │
│ IV. Size                                                              │
│     A. Compact for easy manipulation/small storage .....(8).......3  │
│     B. Lightweight .....................................(1).......1   │
│                                                                       │
│ V.  Stability                                                         │
│     A. Able to stand on own ............................(7).......4   │
│     B. Doesn't slide on tabletop .......................(2).......3   │
│     C. Detachable heating unit (remove heat when cooking) .(2).......3 │
│                                                                       │
│ VI. Temperature                                                       │
│     A. Heats and cools quickly .........................(6).......4   │
│     B. Temperature uniform across inner surface ........(7).......5   │
│     C. Steady-state temperature uniform ................(4).......5   │
│     D. Capable of high temperature .....................(1).......3   │
│     E. Heat contained in wok ...........................(2).......3   │
│                                                                       │
│ VII. Capacity                                                         │
│      A. Large volume capacity ..........................(4).......3   │
│                                                                       │
│ VIII. Manipulation                                                    │
│       A. Easy to handle.................................(7).......3   │
│       B. Long extension cord ...........................(6).......1   │
│       C. Handles remain cool/don't get hot .............(3).......2   │
│       D. Contents can be poured out ....................(4).......3   │
│                                                                       │
│ IX. Cost                                                              │
│     A. Cost.............................................(1).......4   │
│                                                                       │
│ X.  Temperature Control                                               │
│     A. Temperature switch readable .....................(2).......2   │
│     B. Off switch included .............................(2).......2   │
│     C. Temperature indicator ...........................(3).......3   │
│     D. Temperature controls remain cool/don't get hot...(2).......2   │
│                                                                       │
│ XI. Impact                                                            │
│     A. Impact resistant ................................(1).......1   │
└─────────────────────────────────────────────────────────────────────┘
```

▼ *Figure 4.13.*
Interpreted need sorting: electric wok redesign.

provide a larger statistical sample for ascertaining the importance. Verifying an importance scale requires more samples than identification does, generally at least 30 randomly sampled responses to form a distribution representing the customer population.

The questionnaire should ask for the importance of each need to the customer completing the questionnaire. This inquiry might be implemented by asking the user to respond with a 0–5 rating on each need. This scoring can be difficult, especially with "low price" as a customer need. It is difficult to compare to an independent "importance" scale.

A better approach is to ask for a relative comparison to a decrease in cost. That is, ask, instead of a decrease in 10% off the product cost, how much percentage performance increase is required so that the customer would still want to buy the product. Normalizing these data across all the needs produces an importance scale.

With a questionnaire sent out and responses returned, relative importance ratings can be placed on the needs. The result can be linearly scaled to any other numerical range desired, such as 1–5, where the information content remains unchanged. The variance across the subject pool can also profitably serve as an indication of uncertainty to establish significant figures.

Example: Fingernail Clipper

The interpreted needs must be collated into a single list of customer needs. Table 4.1 illustrates the results of this process as applied to a fingernail product. The interpreted needs in Table 4.1 were developed in the same affinity diagram manner as previously discussed.

Next, the customer importance ratings on these needs must be determined. A questionnaire was sent out to 30 randomly selected customers from the intended market segment for the clipper. A resulting breakdown of importance as reported by the 30 persons is shown in Figure 4.14. Note the different subjective importance and the final calculated overall combined interpretation.

Cluster Analysis Method

The previous methods ignore any inherent model structure that might be associated with the customer needs. They merely identify the needs as labels and weights. This identification assumes the needs have statistically independent, normally distributed importance, and that these importance distributions are independent of the need performance value considered.

TABLE 4.1. INTERPRETED CUSTOMER NEEDS FOR A FINGERNAIL CLIPPER PRODUCT

Inexpensive cost	Act as key-chain
Compact storage	Nonsnag storage
Lightweight	Striking appearance
Easy to open file	Files nails
Picks nails	File rough
File holds filed dust	File has a picking tip
Easy-to-close file	Easy-to-open clipper
Easy-to-align clipper	Easy-to-hold clipper
Body contoured to hold	Curved blade shape
Wide handle	Low squeeze force
Blade can act as pusher	Blade can act as a file
Clips nails	Clips toe nails
Clips hangnails	Sharp blade
Nails fall predictably	Stores cut nails
Easy to clean	Easy-to-close clipper

Other methods can be used to provide greater structure to the customer needs models. The previous methods provide an importance ranking to each customer need. The result is a constant weight, which is generally not true. For example, the importance of the noise generated by a product may fall drastically if the price is cut in half. The two needs have correlated importance over large changes in values. To uncover how the importance ranks change with changes in values of the needs, one must first assign a structure to the need space. This structure represents the range of possible values for each need (noise might have values of decibels, cost might have values in dollars, etc.). Over this space, a relative preference structure can be formed that reflects the customer's perceptions.

Conjoint Analysis is one method to construct such a representation. Here, one constructs a preference structure not over the possible values of the needs but rather directly over different product configurations. It can be difficult to show a customer a set of customer need values (pairs of decibels and dollars, etc.), but it is easy to show a customer a product that the design team believes supplies that set of values in customer need space. With conjoint analysis, a group of customers might be asked to rank product prototypes that are slightly different in configuration features. This ranking can be used to fit a preference structure over these feature attributes. Notice, therefore, that conjoint analysis is really intended as a postconceptual design activity for the new design or as a preconcept analy-

Clipper Project
Customer Requirements

Interviewer(s): KNO

30	Number of Customers
0 5	Weighting Scale

Average Customer:
Male/Females, age 20-60, Middle Class
Not in the hair or nail industry

Customer Requirement	WT	Number of Subjects					
		5	4	3	2	1	0
• Purchase							
- Cost	4	5	21	4			
• Transport in package							
• Unpackage							
• Chain Keys							
- Act as keychain	0				6	2	22
• Store							
- Compact storage	4	6	18		6		
- Non-snag storage	1		6	2			22
- Lightweight	2		12	6			12
- Striking appearance	3		16	4	10		
• Prepare to File/Pick							
- Easy to open file	2		18				12
• Filing/Picking							
- Files nails	2		18				12
- Picks nails	2	10	2			2	16
- File rough	1	6				2	22
- File holds filed dust	0				4	4	22
- File has a picking tip	2	12					18
• Return has a picking tip							
- Easy to close file	2	2	16				12
• Prepare for clipping							
- Easy to open clipper	4	2	18	10			
• Clip nails							
- Easy to align clipper	2		6	2	16	2	4
- Easy to hold clipper	3	2	18	4			6
- Body contoured to hold	3	2	18	4			6
- Curved blade shape	3	2	18	4			6
- Wide handle	2	6	6			6	12
- Low squeeze force	1			2	16	2	14
- Blade can act as pusher	1		6			4	20
- Blade can act as a file	1	4	2			2	22
- clips nails	5	30					
- clips toe nails	1	4	2			2	22
- clips hang nails	1	4	2			2	22
- sharp blade	1	4	2			2	22
- Nails fall predictably	1			2	16	2	10
- Stores cut nails	1	2	4			2	22
- Easy to clean	0				6	2	22
• Return from clipping							
- Easy to close clipper	4	2	26	2			
• Throw away							

▼ **Figure 4.14.**
Clipper customer needs list.

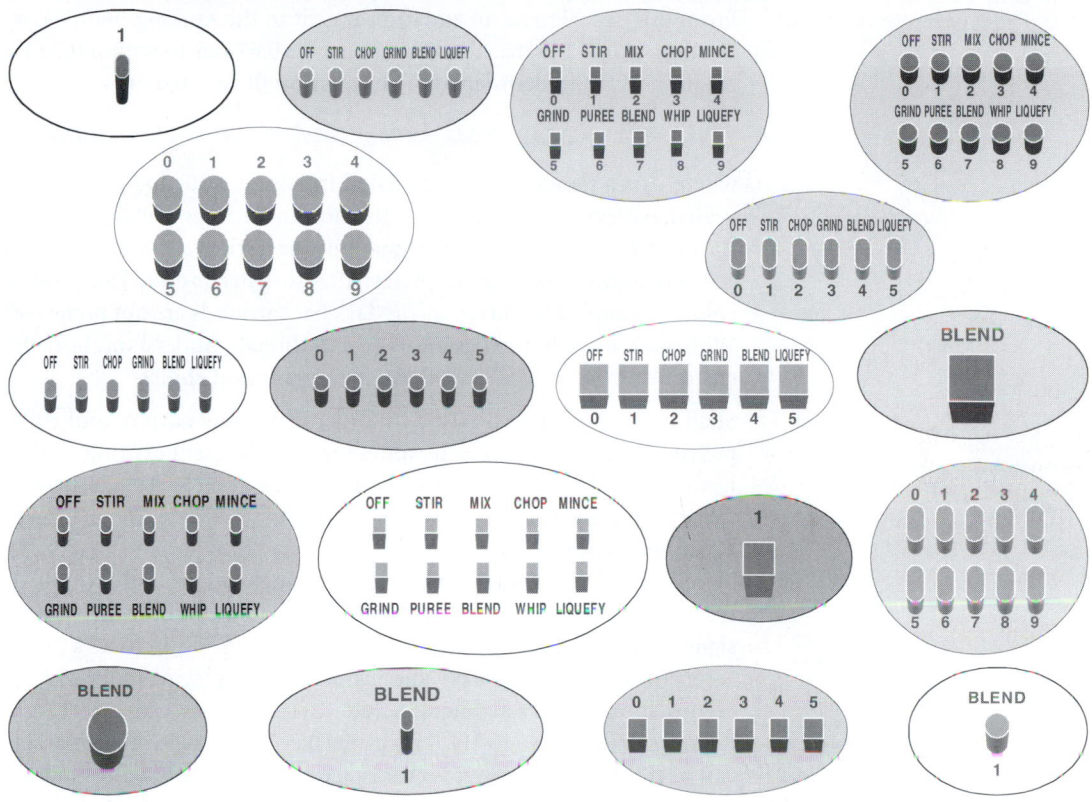

▼ *Figure 4.15.*
Blender panel configurations shown to customers.

sis of a past product's performance decomposed onto different features. A concept with different features is presumed, which is shown to the customers for comment. At this stage, more developed importance rankings become necessary for a design team to represent the customer needs.

As an example, consider an electric blender used to blend ice and other ingredients into cold drinks. The original product had an operating interface panel with 18 different buttons. It was questioned whether some other panel configuration might better serve the customer. To answer this question, eight different design variables, such as label font and size, and button size and shape, were changed in a designed experiment. The different configurations that were shown to the subjects are shown in Figure 4.15. The rank ordering (preferred best to least) of the different configurations resulting from questioning nine different customers are shown in Figure 4.15.

From this experiment, an equation was fit to the rankings with a correlation of 0.971 and a standard error of ± 0.674 or to within 0.7 of a position in the rank ordering. The equation fit resulted in:

$$\mu(\bar{d}) = 12.8 + 3.3d_G + 1.5d_H - 0.7d_B^2 - 1.2d_C^2 + 1.5d_D^2 - 4.5d_G^2 - 0.5d_D d_E,$$

where d_i is a design variable as listed in Figure 4.16, and μ is the revealed average preference. Note the nonlinear importance behavior. Button size, for example, has a quadratic behavior: too-small buttons and too-large buttons are not preferred. Also, letter size and background color are coupled; small letters on dark backgrounds are not preferred. With such models, the design team can directly understand how the customers perceive different design configuration choices.

Such models can also be used to understand how current and competitive products rank. The μ values for such different existing configurations can be evaluated. If μ values can be correlated with market share (typically a very tenuous correlation, however), some (usually weak) indication of market change with respect to design changes can be forecasted. Time variation of these models might also be considered, if one has a reasonably accurate projection of how different design elements will change in time. This process quickly becomes a rather sophisticated market analysis and includes game-playing models of the competition. Some references include Urban and Hauser (1993) and Kinnear and Taylor (1991). The point is to understand that markets can be modeled as random vector spaces. Modern competitive market analysis includes examining positioning of products within such spaces as targets for redesign activity.

	Design Variables								Subject Ranking of the Products								
	A Letter Size	B Button Size	C Button Color	D Letter Color	E Bkgd Color	F Shape	G Number	H Labels	S1	S2	S3	S4	S5	S6	S7	S8	S9
1	-1	-1	-1	-1	-1	-1	-1	-1	3	1	5	5	4	1	1	2	1
2	-1	-1	0	0	0	0	0	0	4	11	17	17	8	15	16	8	12
3	-1	-1	1	1	1	1	1	1	9	16	16	12	13	10	9	14	9
4	-1	0	-1	-1	0	0	1	1	16	18	11	15	17	9	10	15	15
5	-1	0	0	0	1	1	-1	-1	5	2	12	3	3	6	3	3	2
6	-1	0	1	1	-1	-1	0	0	14	12	15	18	9	14	14	16	11
7	-1	1	-1	0	-1	1	0	1	18	13	2	11	14	13	17	12	8
8	-1	1	0	1	0	-1	1	-1	7	9	6	10	15	12	8	18	10
9	-1	1	1	-1	1	0	-1	0	10	5	3	8	2	3	5	6	5
10	1	-1	-1	1	1	0	0	-1	12	8	1	2	16	16	13	7	18
11	1	-1	0	-1	-1	1	1	0	8	15	13	13	7	11	11	13	16
12	1	-1	1	0	0	-1	-1	1	11	3	4	7	5	2	2	1	4
13	1	0	-1	0	1	-1	1	0	1	17	10	14	10	8	12	9	14
14	1	0	0	1	-1	0	-1	1	6	6	18	6	6	5	6	5	6
15	1	0	1	-1	0	1	0	-1	17	7	9	1	11	17	15	11	13
16	1	1	-1	1	0	1	-1	0	13	4	8	4	1	4	4	4	3
17	1	1	0	-1	1	-1	0	1	15	14	14	16	18	18	18	17	17
18	1	1	1	0	-1	0	1	-1	2	10	7	9	12	7	7	10	7

▼ **Figure 4.16.**
Preference rankings from nine subjects on the different display panels.

After identifying and completing an importance ranking of the customer needs, a reasonable model of the customer is complete. However, one also should understand the process within which the product is operated. To represent this process, customer use patterns are represented.

Customer Use Patterns

Any nontrivial product has distinct activities that a user steps through when using the product. A product is purchased, transported, assembled out of packaging, stored away, removed for use, initialized, used in different ways in different environments, perhaps modified by the user, periodically cleaned or maintained, and disposed. For communication to the design team members, these different customer use patterns should be captured and represented, as all can give rise to different product forms (and portfolios). Capturing the customer use patterns helps ensure that each different activity has had customer needs gathered. One can often find complete neglect of some activities in the customer interviewing, in particular the setup of the product for use. We present here a simple network model of the customer activities.

To form the possible use patterns, it is first important to capture the serial sequence or parallel chains of activities for each customer. To carry out this task we complete the last column of the customer data sheets for each customer. Typically, a sub-sequence of customer statements will have an activity associated with it. There is not a one-to-one association of activities-to-customer need statements. This marking of the activity should be completed at the same time as the interpretation of the customer needs, column 3, which is when the interview is still fresh in the mind of the interviewer. After completing column 4 of the customer data sheets, we then transcribe these activities onto a set of cards, one card for each of the activities.

Having transcribed the activities for each customer, we then take the first customer's sequence and form a serial-linked chain of cards across a large board. We then proceed to the next customer and compare the first customer's chain of cards with the first card. If it is the same as any of the cards in the chain, it is overlaid. The second customer's next card is compared with the first customer's subsequent card in the chain. If they are the same, it is again overlaid on the first customer's card. If it is different, then a new use pattern is uncovered. The first and second customer used the product differently. A new serial chain is constructed, starting off in parallel from where the two customers begin to differ.

Now the second customer's next card is compared with all of the first customer's cards beyond where the first and second customer began to differ in use. If there is no match, it is part of the independent parallel chain of activities. If it matches, then the two customers' uses have reconverged, and the second customer's card is overlaid on the first customer's at that point and connected with a new link. This process is continued for the remainder of the second customer's cards and for all other customers.

Figure 4.17 illustrates this process for a fingernail clipper product. An *Activity Diagram* results, beginning with the user activity of purchase through disposal. Note in the figure that three primary parallel activities exist in a fingernail clipper: cutting, filing, and picking (or manicuring). These parallel activities provide a preview of possible modules that naturally can exist in a product. The activities also provide a context for understanding the organized customer needs lists and for choosing the focus user functions that a product must support.

The result is a network of activity cards with the number of cards indicating the frequency of performing the activity by the customer pop-

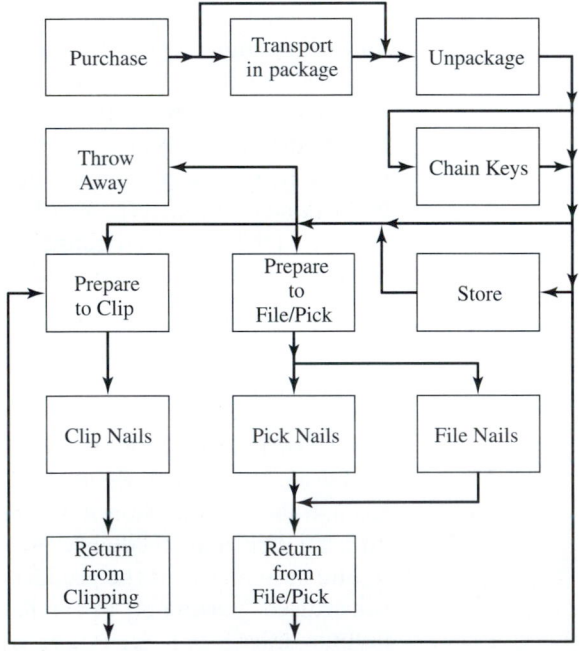

▼ *Figure 4.17.*
Activity network for the fingernail clipper.

ulation. Typically, we also highlight the initial and final activity of the customer with the product. This highlighting can help system-engineer the environment within which the product is used, if it is a topic for a larger design team. The activity diagram differs from the customer needs in that the activities are recorded at a higher level of abstraction; the activities focus on what happens when using the product.

Next, we indicate the *typical nature* of any one of these activities by the line width connecting the activities. For every transition into an activity from another, we examine the number of customers who performed that transition. An accordingly thick line is used: Activity patterns exhibited by many of the subjects are given wide lines; infrequent activity patterns are given thin lines. Figure 4.18 applies this simple representation to the fingernail clipper. Note the typical circular use pattern of clipping and storage. Note the entry and exit from the customer with the purchase and disposal activity.

The important uses of the resulting activity diagram are twofold. First, the activity diagram can communicate to any new and different design

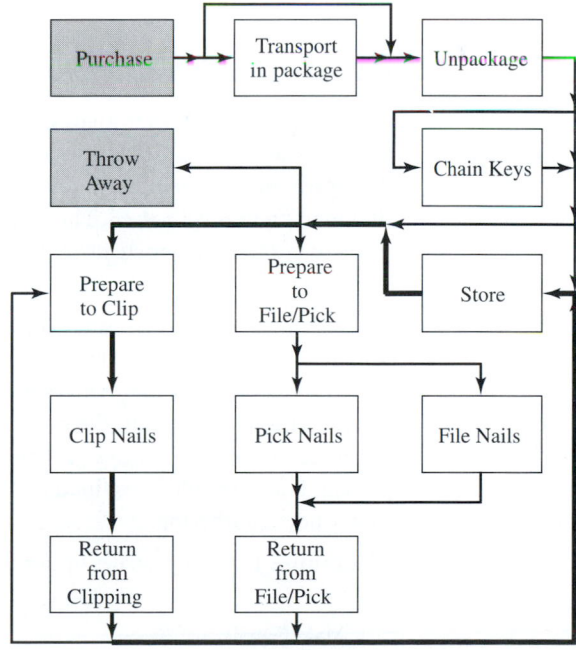

▼ *Figure 4.18.*
Completed activity diagram for the fingernail clipper.

team members what the customer is doing with the product. It helps ensure that a design team is aware of all of the customer lifecycle needs of the product. Note that the activity diagram can also be expanded in the upstream direction to begin to capture activities of the sale, distribution, and manufacture of the product. Similarly, downstream activities such as disposal can be represented.

The second important result of the exercise is that the activity list creates a useful categorization for the list of customer needs. Typically, the customer needs are grouped into more abstract categories based on the design team interpretation. This grouping is usually arbitrary. We propose here that the list of activities provides a more meaningful grouping of the interpreted customer needs. For each activity, the new customer needs introduced in an activity are listed under it. We find this assignment provides a much clearer understanding of the context of the customer needs.

Having completed the activity list, the customer need gathering efforts that are needed to form meaningful specifications are complete. What remains in the customer need gathering effort is to summarize the results into a document as a reference for the design team. The summary document of the customer needs is discussed next.

Customer Needs Documentation

The completed customer analysis must be summarized into a list of interpreted needs for the entire design team to share as a document and agree on. Data templates for presenting the interpreted needs are shown in Figures 4.13–4.15. The resulting customer need list is a milestone document on which all engineers on the team will reference and base their judgments.

Among other items, the header information contains a brief description of a typical customer. The customer description should be a very brief discussion of the target market. This description should have been generated with or given to the product development team from management in the original design brief that initiated the development project. It also should be refined by the interviewers based on the customer needs gathering.

An example for a fingernail clipper is:

Male/Female, age 20–50

Middle-class

Not in the beauty profession

As shown by this example, a core demographic description is developed and then applied when selecting customers to interview. The header includes this information as shown in Figure 4.14, along with any relevant product-particular information.

When tabulating the customer needs in the chart, again we list them by activity. We arrange the activities in serial order, with parallel activities ordered by importance. Within each activity group, the customer needs should be similarly arranged. The numerical importance rating for each need should also be included in the second column. The scale (0–1,1–10, etc.) and normalization (relative to most-important goal, relative to number of goals, etc.) should also be indicated on the top of the form if there is no company standard.

In addition to customer needs, there are also other needs that a product must satisfy, typically legislative or manufacturing-facility related. These needs can be represented as additional items in the customer needs list. All typically have high importance, since indeed they must be met for the product to be legally sold or to be manufactured. Other noncustomer needs can be incorporated in the customer needs list as deemed appropriate.

V. SUMMARY AND "GOLDEN NUGGETS"

Customer needs modeling is required to understand what the customers feel they want from a product.

1. Gathering and interpreting customer needs can be difficult with rapidly changing technology.

2. Interview methods are effective to establish a list of customer needs.

3. Questionnaires are effective to establish importance levels.

4. Advanced statistical methods are used to establish targets in customer needs space with codependent needs and with competition.

5. Activity diagrams are effective to represent the customer environment where a product is used.

References

Clausing, D. 1994. *Total quality development,* New York: ASME Press.
Design Edge. 1995. Product development firm, personal communication. Austin, TX.

Hayes, B. 1992. *Measuring customer satisfaction: Development and use of questionnaires.* Milwaukee, WI: ASQC Quality Press.

Kinnear, T., and Taylor, J. 1991. *Marketing research: An applied approach.* New York: McGraw-Hill.

Morita, A. 1986. Made in Japan. Penguin Books, New York.

Norüis, M. 1988. *SPSS guide to data analysis for SPSS-PC Plus.* Chicago, IL: SPSS.

Shiba S., Graham, A., and Walden, D. 1993. *A new American TQM.* Portland, OR: Productivity Press.

Ulrich, K., and Eppinger, S. 1995. *Product design and development.* New York: McGraw-Hill.

Urban, G., and Hauser, J. 1993. *Design and marketing of new products.* 2nd ed. New York: Prentice-Hall.

5 Establishing Product

Function

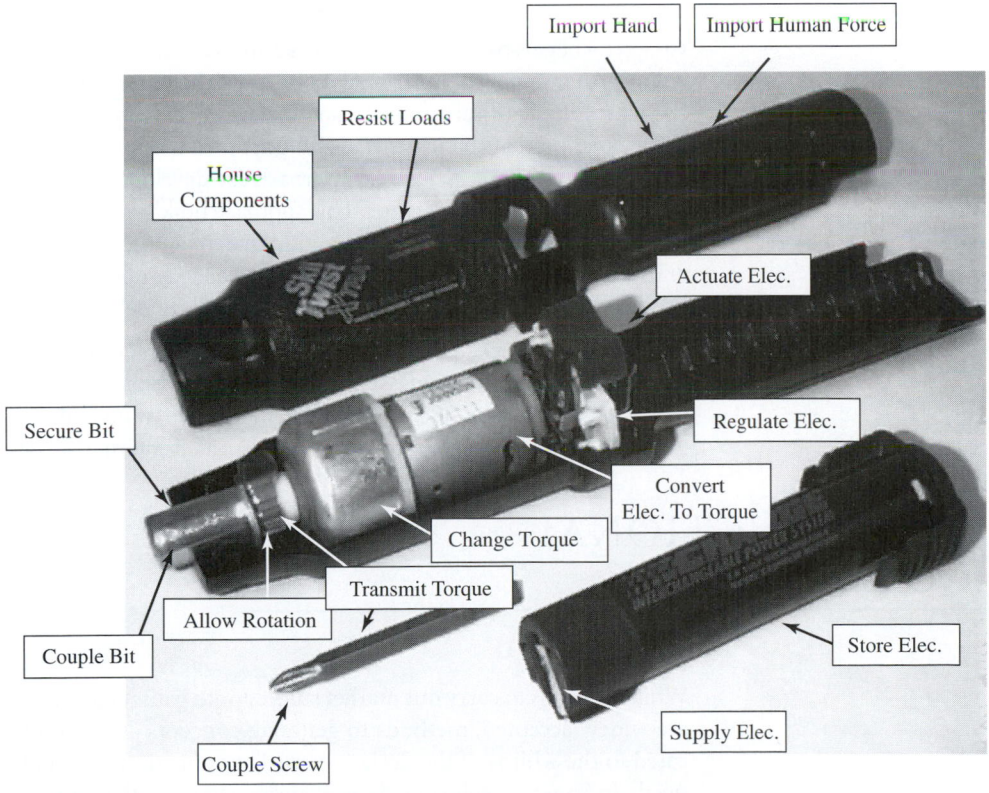

Import Hand | Import Human Force

Resist Loads

House Components

Actuate Elec.

Secure Bit

Regulate Elec.

Convert Elec. To Torque

Change Torque

Transmit Torque

Allow Rotation

Couple Bit

Store Elec.

Couple Screw

Supply Elec.

Example: Functions of a Skil™ Super Twist power screwdriver.

Once we have a representation of what the customer wants from a product, a model of how the product should function is needed to clarify and design the product architecture. Functionally, all products *do something.* Products, therefore, accept "inputs" and operate to produce "outputs," the desired performance. We can model any product, assembly, subassembly, or component as a *system,* with *inputs* and *outputs* that traverse a system boundary. The essence of such a model is the need-function-form definition of engineering design, where our focus is on translating the customer needs for a product to the product functions. This step is needed to engineer how a product will do what it is intended. In the sections below, we construct the necessary machinery for understanding and representing design function, according to a system perspective. This machinery will aid us in synthesizing form solutions, with greater breadth, less bias, and greater technical understanding than ad hoc approaches.

I. CHAPTER ROADMAP

Chapter 5 begins by answering the question, "Why functional decomposition?" An overview of functional modeling is then presented, including basic definitions of functionality and systems. Following this overview, the chapter presents basic methods of functional modeling. The focus is then on a five-phase, systematic approach for developing a more complete model, a function structure. This approach builds on customer needs analysis and ends with a refined function map of a product. The chapter ends with alternative perspectives of functional decomposition, example products, and the "golden nuggets" of functional modeling.

Figure 5.1 shows example routes through the chapter. A direct route covers elementary methods, with the goal of creating a suitable functional decomposition of a product. An alternative route is also shown, augmenting the approach with more sophisticated, alternative methods.

II. WHY FUNCTIONAL DECOMPOSITION?

Motivation

While methods to carry out market studies or to gather customer needs are widely accepted, methods to generate concepts are typically allocated to the whims of the design team. The transition from customer needs to concrete solutions is seen more as an art than a science or method (Dixon and Poli, 1995). In fact, for many consumer products,

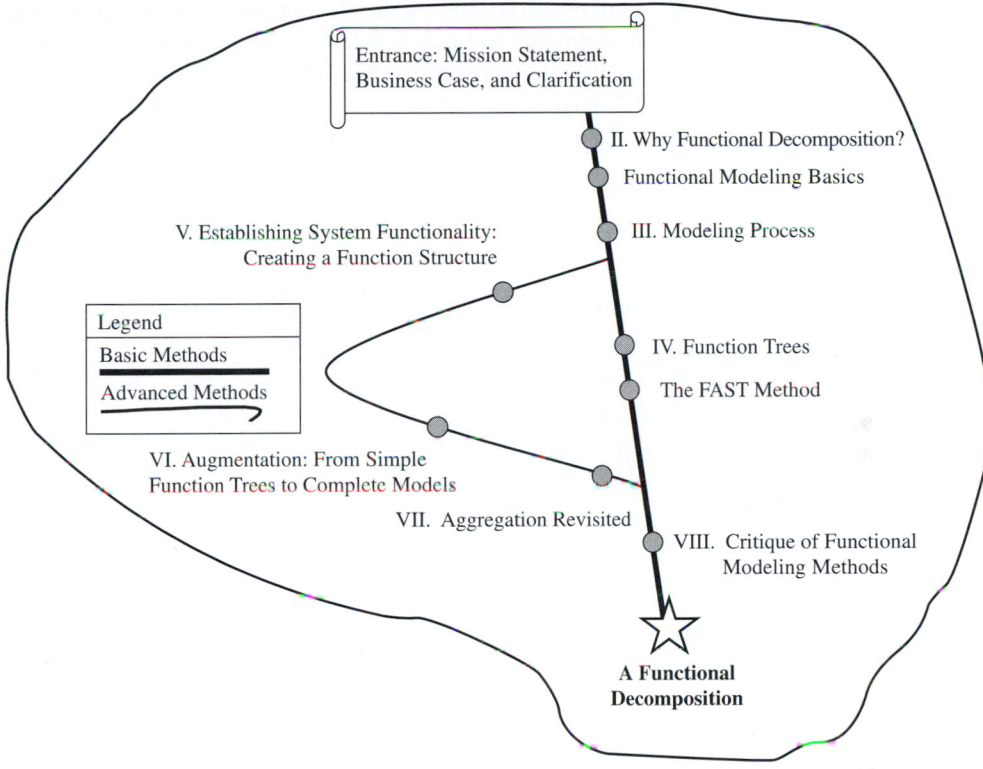

▼ *Figure 5.1.*
Roadmap for alternative routes for developing a product's functional model.

our experience has shown a tendency to seek form solutions directly based on the previous experience of the design team members.

With ever-shrinking cycle times and budgets, and with ever-expanding demands for quality, this approach has a number of limitations. Most notably, the links between customer needs and design concepts are, at best, indirect or implicit. They exist only in the minds of the designers. Customer needs are thus relegated to the criteria for evaluating concepts, not the direct catalysts for generating concepts.

Over the last twenty years, new methods for engineering design have emerged that focus first on mapping customer needs to functional descriptions. These descriptions are then used to generate and select sets of technologies that satisfy the underlying functional requirements (Otto and Wood, 1996; Pahl and Beitz, 1991; Hundal, 1990; Akiyama, 1991; Miles, 1972). We present these methods here, with advancements. When combined, they have a number of intrinsic advantages:

▶ Concentration is on "what" has to be achieved by a new concept or redesign and not *how* it is to be achieved. By so doing, a component- and form-independent expression of the design task may be achieved to comprehensively search for solutions.

▶ Functional modeling provides a basis for organizing the design team, tasks, and process. To the extent that functions of the product are independent, design-process activities may be chosen according to the independent product "chunks." The interactions between the functional elements provide the required key communications needed among the concurrent design activities. Interfacing specifications can be readily developed.

▶ Functions or sets of functions may be derived or generated directly from customer needs. These functions define clear boundaries to associate assemblies or subassemblies of the final design solutions. These boundaries provide a basis for allocating resources to concurrent engineering efforts and for seeking modular concepts.

▶ Creativity is enhanced by the ability to decompose problems and manipulate partial solutions. By first decomposing a design task into its functional elements, solutions to each element are more apparent due to the reduction of complexity and extraneous information.

▶ Functional modeling provides a natural forum for abstracting a design task. Many levels of functional abstractions may be created, from a very high-level single function statement to alternative detailed functional statements for a design's subsystems. Through abstraction, a design team may search for the "real" problem being solved while minimizing biases.

▶ By mapping customer needs first to function and then to form, more solutions may be systematically generated to solve the design problem. "If one generates one idea it will probably be a poor idea; if one generates twenty ideas, one good idea might exist for further development" (Ullman, 1992).

▶ Needs mapped to function and then to form promote set-based concurrent engineering processes (Sobek and Ward, 1996). Feasible regions of technology may be explicitly defined based on functional requirements, not implicitly. Tradeoffs may also be explored in parallel among a wide array of radical and known solutions, since a common functional description is driving the design effort directly from the voice of the customer.

Figure 1.4 in Chapter 1 illustrates the role of functional decomposition in a conceptual design process. The emphasis is on attaining a correct description of what the product is to do as a system of functions. This description will then lead naturally to the solution principles, modularity, and interactions that are effective for the product. The following section describes a simple approach to developing a hierarchical description of product function, but the method has no possibility of checks on its correctness. The next section then develops a systematic approach for establishing functionality of a new design or redesign. A fingernail clipper provides a basic running example to clarify the approach, and further examples are presented throughout the book.

Function Modeling Basics

All products do something. There is some intended reason behind their existence: the product function. Products have secondary functions and constraints, such as "be inexpensive," but these constraints are not the product functions. Rather, the product, when working properly, manipulates matter and energy in the physical world.

Function—What is a "function"? A function of a product is a statement of a clear, reproducible relationship between the available input and the desired output of a product, independent of any particular form.

The *Product function* is the overall intended function of the product—what it is to do.

The product function is the simplest representation of the product, usually just a noun and an active verb. "Chop beans." "Clip nails." "Transport item X." "Produce music (acoustics)." "Make copies." Based on these overall functions, we next need to decompose into subfunctional statements, which, when all are completed, satisfy the overall function. This process is the idea of *functional decomposition*.

Subfunction—A subfunction is a component of a product function. An overall function can, and often must, be divided into identifiable subfunctions corresponding to subtasks. The relationship between some subfunctions and the overall function is often governed by a constraint or input–output relationship. The impact of such constraints on the function must be carefully considered. It is also frequently necessary to examine relationships between various subfunctions, paying particular attention to their logical sequence, any necessary interconnections, consistency of input and output at the system and subsystem interfaces, and physical validity (such as conversation laws).

Abstraction—Abstraction is the process of ignoring what is particular or incidental and emphasizing what is general and essential. Such generalization leads to the crux of the problem. If a product is

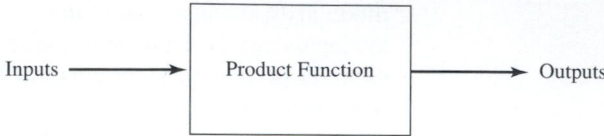

▼ Figure 5.2.
Generic black box model.

viewed in the abstract, one is better able to define overall functional requirements and constraints.

For all functional modeling methods, a basic construct is the *black box,* called "black" because its internal form is deemed unknown. One can think of the product as doing something—the function—by taking in inputs, changing or modifying these inputs, and thereby creating outputs. One can represent this diagrammatically as a simple function statement with a box drawn around it (Figure 5.2). Any black box can be expanded into many interlinked subboxes, just as functions can be decomposed into subfunctions.

Functions and Constraints

Functions, as described above, represent what the product does to satisfy the customer. On the other hand, there can be customer needs that are served not by what the product does, but rather how the product is instantiated in form. For example, airlines are customers that purchase airplanes, and they have demands on dry airplane weight. The need for lightweight aircraft cannot be represented as a function, since the airplane does not do anything to make lightness. There is no identifiable subsystem that generates lightness. Rather, every component on the airplane contributes to weight. It is an intrinsic property of the airplane.

A *constraint* is a statement of a clear criterion that must be satisfied by a product and requires consideration of the entire product to determine the criterion value.

The distinction on whether a customer need can be met with a function or should be considered as a constraint is dependent on the solution principle considered. As another example, consider a customer need for compactness when carrying a fingernail clipper. Here, one might consider features on the clipper that can make it compact: The handle can pivot and fold flat. The handle satisfies the desired compactness functions. On the other hand, some clippers may not fold up at all and attain compactness by simply being small. Here, compactness is a constraint and would have volumetric criteria that the entire clipper must satisfy. Functions are satisfied by subsets of the product through their operation; constraints are satisfied by properties of the entire product.

Typical examples of constraints include criteria of cost, compactness, footprint, mass, and reliability. These examples are very context dependent, however. These criteria can be constraints on one product but functions on another. It completely depends on how the criteria are being satisfied by the product, functionally by subsystems that take care of the criteria, or as system properties where all parts of the product contribute to the criteria.

Using these basic ideas, let's now consider two approaches to functional decomposition. The first is an elementary hierarchical tree approach, and the second is an effective systems approach. A general functional modeling process encapsulates these two approaches.

III. MODELING PROCESS

Results from previous chapters provide us with a clear statement of the customer needs, organized to establish priorities for the design efforts. Functional modeling, as summarized in Figure 5.3, begins the systematic process of transforming these needs to a clear specification of the design task. It also initiates the conceptual design phase, wherein

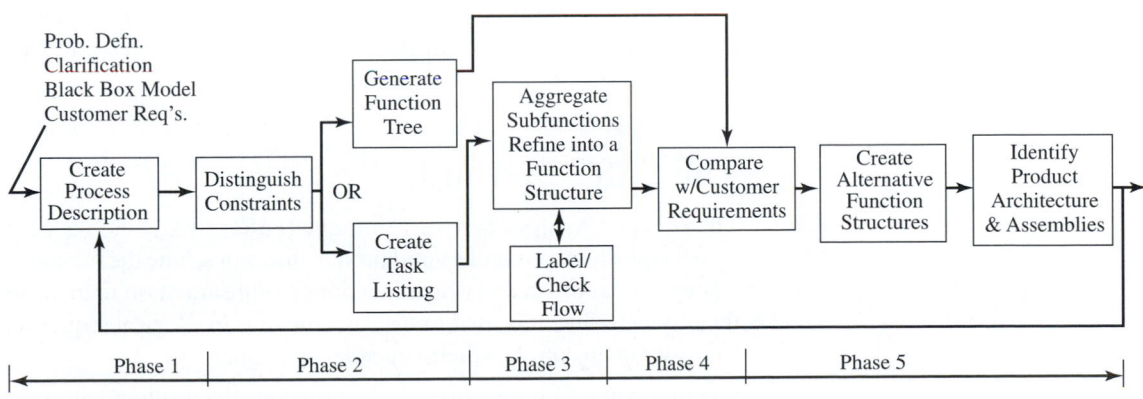

▼ **Figure 5.3.**
Function modeling and analysis process.

we seek a breadth of solutions. Notice, in Figure 5.3, that either function hierarchies (trees) or function structures (task listing) may be used to model a product's functions. These techniques assume that customer needs, clarification, and a mission statement have previously been developed.

IV. A SIMPLE APPROACH: FUNCTION TREES

An elementary approach to developing a functional description of a product is to decompose the prime function hierarchically into subfunctions that, when all are fulfilled, complete the overall product function. Typically, a different subsystem or component of a product satisfies each subfunction. This view can be repeated iteratively down to the functions of simple features of the product.

Function trees are fast and simple to construct, but this ease of construction is gained at the expense of understanding interactions between the expanded subfunctions. Interconnecting links among the subfunction black boxes are not considered. Therefore, the approach is not as effective in helping to establish specifications and structuring of the development process. Nonetheless, it is a rapid first cut.

Developing function trees can be approached in a number of ways, two of which are presented in the subsections below. The first is a top-down approach, using the systematic FAST method (VAI, 1993). The second is a bottom-up approach, using the Subtract and Operate Procedure (Lefever and Wood, 1996), similar to FMEA analysis (Frankel, 1988; O'Connor, 1991). Both work especially well with products that already exist, to reverse-engineer a product's functional description. However, they can work equally well in the embodiment design phase, to embellish a concept into a real form.

The FAST Method

The Function Analysis System Technique (FAST) (VAI, 1993) is used to define, analyze, and understand product functions, how the functions relate to one another, and which functions require attention to increase the product value. It is used to display functions in a logical sequence, prioritize them, and test their dependency (Figure 5.4).

The first step is to brainstorm all the functions the product will serve in the eyes of the customer. One needs to ask "what the product does" rather than "what the product is." To define the functions, a simple

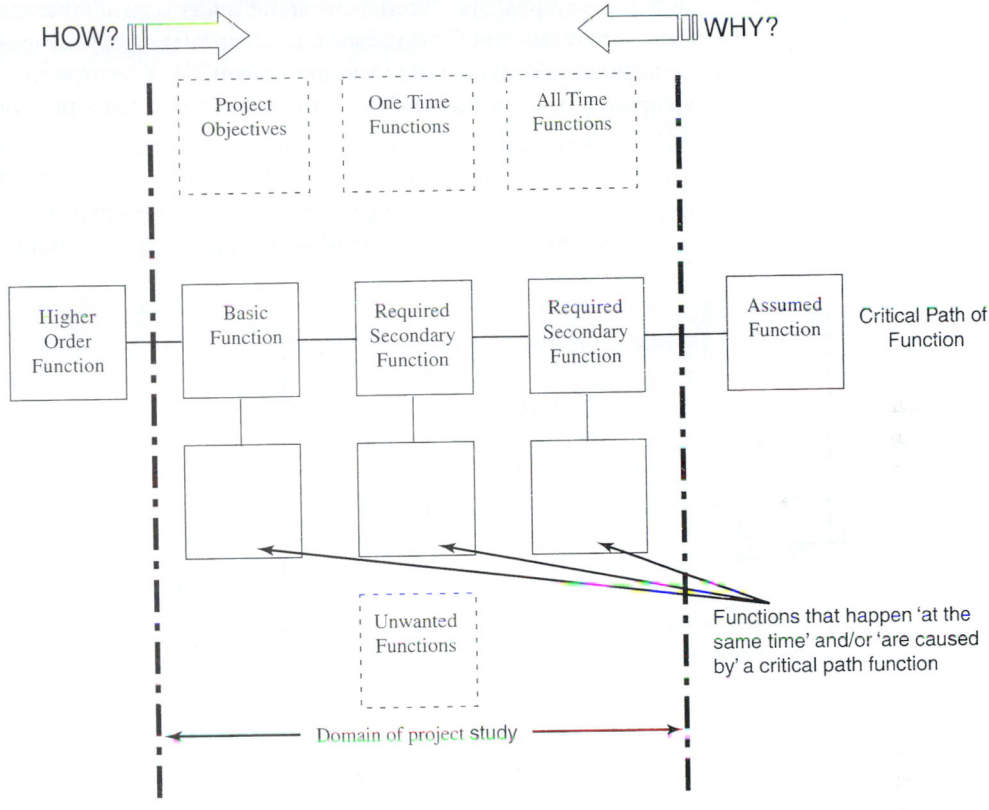

HOW? → ← WHY?

| Project Objectives | One Time Functions | All Time Functions |

| Higher Order Function | Basic Function | Required Secondary Function | Required Secondary Function | Assumed Function | Critical Path of Function |

| Unwanted Functions |

Functions that happen 'at the same time' and/or 'are caused by' a critical path function

Domain of project study

▼ *Figure 5.4.*
The FAST diagram.

verb and noun structure should be used or a verb followed by a noun phrase. When choosing words that define a function, they should be made as broad and generic as possible, such as *produce torque, generate light,* and *shape material.* Appendix A-2 presents a list of function classes and their associated synonyms.

During this process, it becomes obvious that these functions have different levels of importance. Out of all of the functions, one function that is the *overall product function* has to be selected (sometimes called the *basic function*). The product function, again, represents the main reason that the product exists in the eyes of the customer. For example, the basic function of a car seat belt is to restrain a person in a car seat.

A product may have more than one basic function under different conditions. For example, an electric fuse that "conducts electricity under certain conditions" also "breaks the circuit under certain other conditions." These different views initiate our understanding of the complexity inherent to designing time-varying systems. FAST incorporates this complexity by requiring separate analyses for the separate conditions.

As an example, consider establishing a function tree for a coffee grinder. The basic function of a coffee mill is to "chop beans" (Figure 5.5). Once the basic function is identified, all the other functions that the coffee mill performs, either as a whole or within any of its subsystems, are

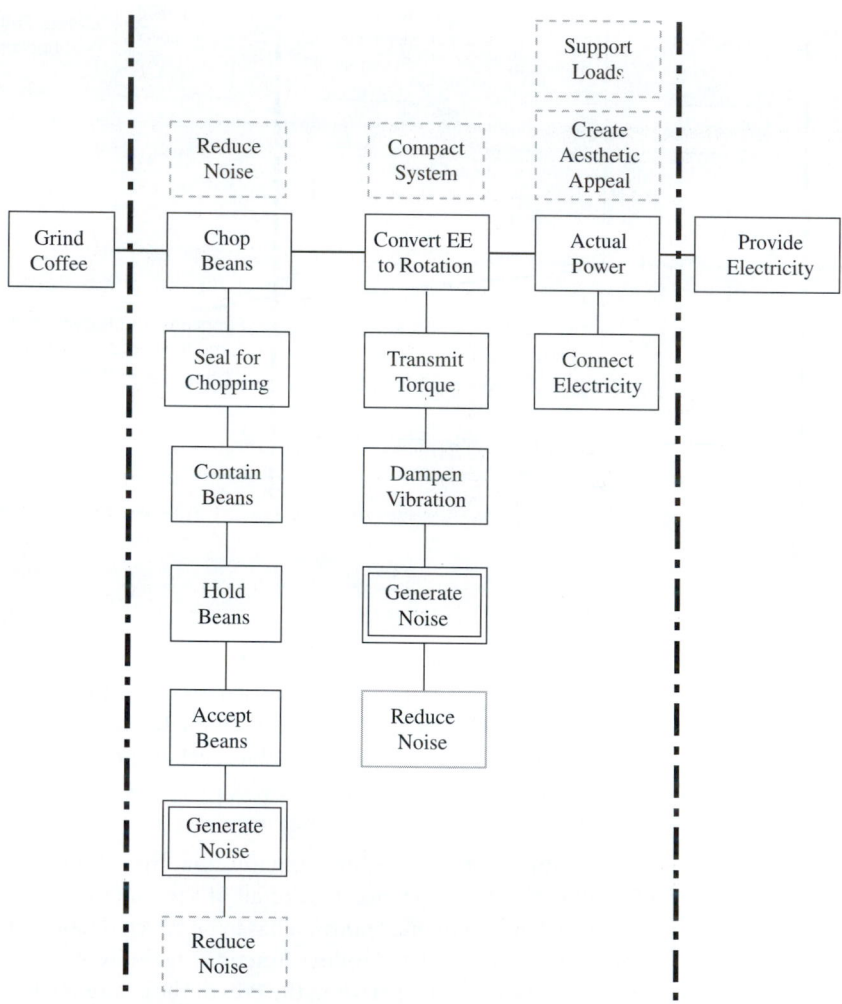

▼ **Figure 5.5.**
FAST diagram for the coffee mill.

subordinate to the basic function. They support the basic function and assist the product to work and sell. These *secondary functions* are essential to the performance of the basic function, and further they are a direct cause of the basic function.

Secondary functions can be categorized into three types: required, aesthetic, and unwanted. For example, "convert to rotary motion" is a necessary function of a coffee mill in order to "chop beans"—the basic function. Similarly, a housing may have the requirements of "look pleasing," an aesthetic requirement. Finally, rubber grommets may have the requirements of "dampen vibration," an unwanted by-product.

With FAST, the function tree is established by asking, "How is this function achieved?" to each function, starting with the basic function. The answer is at least one, usually several, secondary function(s): subfunctions that are essential to, and a direct cause of, the function in question. These ideas are captured in the *FAST diagram* (Figure 5.4). Here, the overall function is placed at the top left of the diagram. The subfunctions resulting from the *how* questioning are listed to the right of the overall function. This structure forms the basic backbone of the function of the product, that is, the string of subfunctions that is critical to achieving the product function. It is called the *critical path*. At the rightmost extreme of the critical path will be a subfunction not performed by the product. For example, for a coffee mill, the mill must be powered, and "supply electricity" is a required subfunction. When we reach a subfunction supplied externally, the critical path is complete, and a system boundary is established. The secondary functions between the product function and the external secondary function forms the critical path of secondary functions.

For each secondary function, typically achieving the subfunction introduces ill side-effects. New functions therefore arise to mitigate these effects. Creating rotary motion introduces vibrations, and a function arises to dampen the vibrations. With FAST, these noncritical path functions are listed under each critical path function that causes them.

Finally, there are other functions that a product may have that are not part of the critical path in any way. These *other* functions are listed at the top of the FAST diagram. Typically, they are listed as "all-time functions" on the right, "one-time functions" in the center, and the objective of the development effort is also highlighted, in the upper left. "All-time functions" are functions that are pervasive to the product. They are supporting functions. For example, a power screwdriver is intended to tighten and loosen screws; however, it also must support loads from the user's hand, from being dropped, and so forth. Thus, the function "support loads" is supportive to the purpose of a power screwdriver and should be listed as an "all-time function."

"One-time functions" are also supportive of a product's main intent. In this case, however, the function is not pervasive to the product's performance, but instead represents a single purpose for "one-time" use. An example is "protect system during transport" for the electric screwdriver. This function, typically solved with packing material, is only needed during the transportation of the screwdriver to a retail store or home. It is not used during the product's principal functionality.

Finally, the FAST diagram documents the development objective. The intent of this step is to ensure that the focus remains on the purpose of product development, not on secondary or auxiliary needs as detailed functions are conceptualized. For the coffee grinder, customer need analysis results in the objective of "reduce noise," as shown in Figure 5.5.

Figure 5.4 illustrates the overall structure exhibited in the FAST method. With FAST, one can check the resulting function tree by applying a *how-why* validation exercise to the critical path. That is, each right node in the critical path must answer *how* the left node is achieved—that is how we constructed the critical path. Similarly, though, every left node should answer *why* the left node is being achieved. For example: How does the coffee mill "chop beans"? By, in part, "converting electrical energy to rotation." Why does the coffee mill "convert electrical energy to rotation"? To rotate the "chopper" and to "chop beans." The function neighbors in the critical path should always be both sufficient and necessary in this way.

Beyond the overall product function, one can expand the functional questioning by asking *why* the overall product function is being performed; the answer will be another function, one that operates in the system of the product operation, not in the product itself. This function is called the *higher order function*. This function is very important to holistically understand the product or project; it is the reason for performing the basic function. In the case of a coffee mill, the higher order function is to "grind coffee." One might think of other ways of achieving the higher order function and thereby develop new product systems and architectures.

Summary: FAST Method

Overall, FAST is a hierarchical approach for modeling the function of a product or system. The steps in the modeling process (and in constructing the FAST diagram) are as follows:

1. Construct two vertical, dashed lines, one to the extreme left and one to the right. These lines define the scope of the product development objective.

2. Place the basic function to the right of the left-hand scope line. Pose the question, "Why is the basic function being performed?" A higher order function will answer this question. Place this function to the left of the basic function and connect with a line, beginning the critical path.

3. Generate functions to the right of the basic function. These functions should always follow a *how* and *why* answering scheme and represent the secondary functions. Connect these functions with lines to define the furtherance of the critical path.

4. The critical path will end with an "assumed function," outside the right scope line. This function is external to the product, such as "supply electricity" in the case of a coffee mill.

5. Generate the remaining secondary functions by placing them under the functions that relate to the basic or critical path secondary functions. These functions either occur at the same time or are caused by the functions on the critical path.

6. State the objective of the development effort above the basic function. In addition, add one-time or all-time functions to the top of the diagram.

The Subtract and Operate Procedure

The FAST approach is a top-down approach which starts with the overall function and then decomposes it. Rather than a top-down approach, one might wish to explore a bottom-up construction. The *Subtract and Operate* procedure is one such bottom-up approach to developing a function tree. The underlying assumption to use this method is that either a form concept or actual product exists. This product or concept will then be reverse engineered using the Subtract and Operate procedure.

We start by considering the smallest isolatable functions of features and components in the product. These functions are not easily decomposed into further subfunctions. For each of these smallest subfunctions, one then removes the feature(s) or component(s) that supplies the function and attempts to operate the product (though not literally, only conceptually, when safety is a concern). By actually subtracting a feature or component from a product and then attempting to operate it, one will thereby establish that feature's or component's critical contribution(s) (or effects) to the overall product.

These lowest functions can be then combined into a function tree structure, following the assembly structure of the product itself. Completing the Subtract and Operate procedure on the subassembly can check each such subsystem. The results of applying the procedure to the coffee mill parts are shown in Table 5.1 and Figure 5.6.

TABLE 5.1.   SUBTRACT AND OPERATE APPLIED TO THE COFFEE MILL

Chamber	Seal	Slicing blade	Shaft	Armature
No defined way of holding content	No protection against contents splattering	Contents will not be chopped	Slicing blade will not be attachable	Shaft does not spin
No measurable volume	No protection against spinning blade	No resistance to torque	Contents will not be chopped	Electricity is not transformed into mechanical energy
No body to measure contents	Safety issues will fail		No resistance to torque	
No body to contain contents	Chamber cannot be closed			
No body to hold the apparatus	Power can't be actuated because electric circuit is not closed			
Aesthetic appearance reduced	Impact noise will not be enclosed			
Difficult to clean undefined body				
Pour out contents				

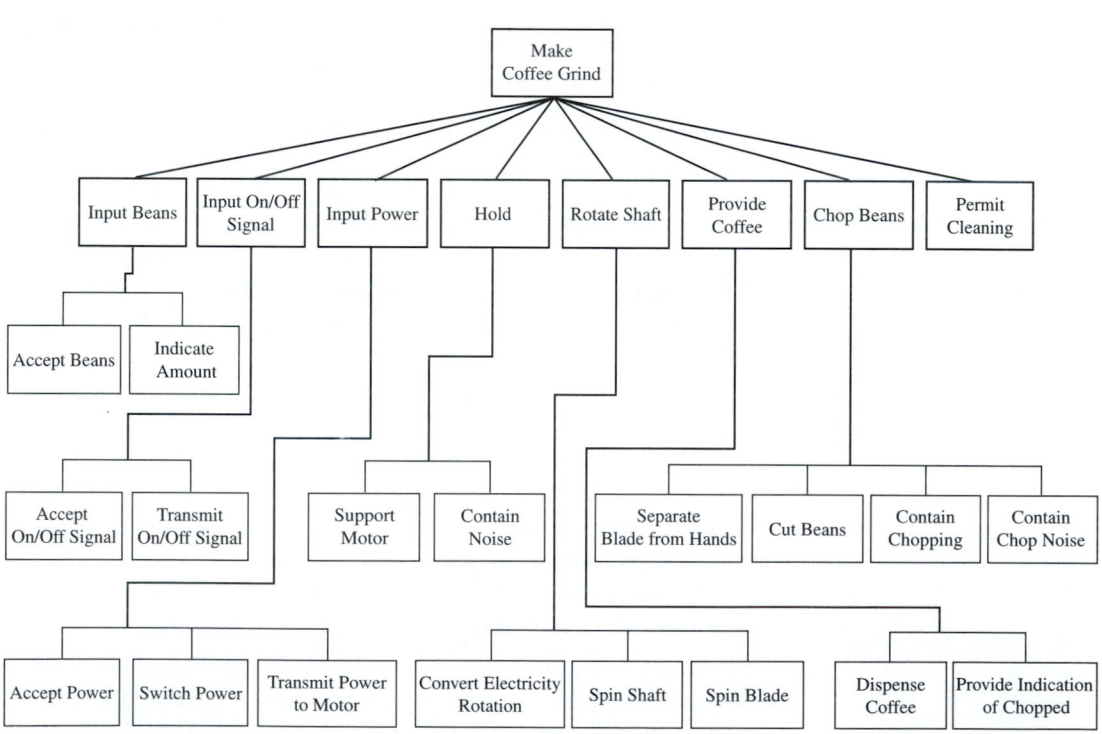

▼ **Figure 5.6.**
Coffee mill function tree generated using the Subtract and Operate procedure.

Summary: Subtract and Operate Procedure

The Subtract and Operate procedure is not difficult; it is a very logical process, based on common sense. What better way of figuring out the function of a component than removing it and operating the system without it, or, alternatively, conceptually removing components from a concept to understand their effects?

The following procedure (as further detailed in Chapter 6 for part count reduction) encapsulates this logical process:

▶ *Step 1:* Disassemble (subtract) one component of the assembly. Removal of components may occur in any order. However, it may be necessary to remove one or several components in order to remove the desired component. These prerequisite components should be reassembled if possible. If they cannot be reassembled, measures should be used to replicate the function(s) of the missing component(s).

▶ *Step 2:* Operate the system through its full range.
This step should test the product through the range of customer requirements. After removing a component, the product should be thoroughly tested. For each customer requirement (structural, ergonomic, kinematic, etc.), the product must be tested to verify the effects of removal of the component.

▶ *Step 3:* Analyze the effect.
This step is most commonly completed through a visual analysis. However, it may be necessary to use a testing device if the effect of removal is not obvious.

▶ *Step 4:* Deduce the subfunction of the missing component.
From Step 3, the subfunction of the missing component may be deduced. A change in any degree of freedom (DOF) during operation should be a major focus. This change is a critical issue in determining component functionality.

▶ *Step 5:* Replace the component and repeat the procedure n times, where n is the number of components in the assembly. Document the results in an effects table (Table 5.1). Step 5 is the reassembly of the removed component. Repeating the process n times allows for the analysis of each component in the assembly. In certain cases, it may be necessary to analyze a product according to subassemblies and components.

▶ *Step 6:* Translate the collection of subfunctions into a function tree.
One does this by grouping the subfunctions of Step 5 into common groups. Each group becomes a higher level functional description node. This process is repeated until the higher level functions collapse into the overall product function as a single node at the root of the tree.

Between the FAST and Subtract and Operate methods, a reasonable function tree can be developed for a product. One should apply both methods independently and then compare the results to merge into an acceptable functional model, one that captures overall high-level intent as well as the functions of important subsystems and components.

V. ESTABLISHING SYSTEM FUNCTIONALITY: CREATING A FUNCTION STRUCTURE

Function trees and the Subtract and Operate procedure are simple, yet effective at revealing subfunctions of a product. While they require practice for effective use, they are not conceptually difficult.

These methods, however, do exhibit a number of limitations, as will be discussed later. Beforehand, though, the following section presents a more robust and complete method for modeling a product's functionality. It follows the process outlined in Figure 5.4 and results in a diagram known as a *function structure*.

The Basics of Function Structures: Black Box and Definitions

Black Box Model: Again, we model a product, abstractly, as a black box, but now with three types of inputs and outputs. A system black box model, as shown generically in Figure 5.7, allows us to focus on the *greatest, overall need* for a product. It also initiates a technical understanding of a product based on its inputs and outputs, known as material, energy, and signal flows. These flow types are sufficient to describe a technical system or product.

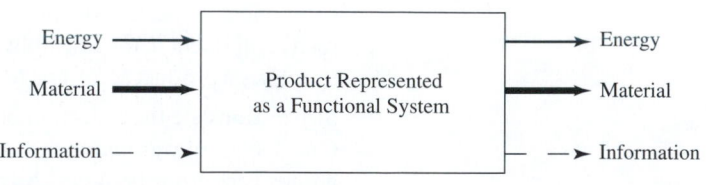

▼ *Figure 5.7.*
Generic black box model.

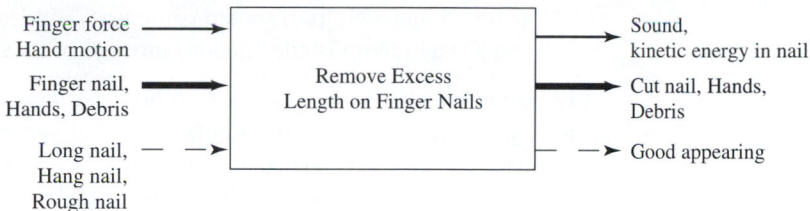

▼ *Figure 5.8.*
Black box model of the fingernail clipper design task.

Figure 5.8 illustrates an example black box model for a fingernail clipper. Notice in the figure that the driving function is to "remove excess fingernail length." It is important to use this model as a definitive statement of the driving function. Quite often, as a design is refined, the overall function may lose emphasis. Details may obscure a product's purpose, resulting perhaps in misdirected and wasted resources. Black box models help maintain the focus on the driving product function.

Black box models also provide a first mapping of customer needs (Chapter 4) to a technical understanding of a design problem. The inputs and outputs of the black box provide this mapping. As shown in Figure 5.8, materials, energies, and signals are transformed by the design system into desired outputs. For a fingernail clipper, the system must convert finger force and attached fingernail material from a variety of people, into an appropriate cut nail. It must also notify the user of a complete cut (sound) and direct the cut nail away or store it.

With this basic notion of a black box model, let's journey into a deeper understanding of how we might create a more refined and useful representation of a product's function. To begin this journey, consider some basic definitions, adapted from Otto and Wood (1996) and Pahl and Beitz (1991) as given below.

Definitions:

System/Subsystems—Technical tasks are usually described in terms of plant, equipment, machines, assemblies, and components. However, these terms do not have identical use in different fields or even for different problems within the same technical field. It is therefore useful to refer to any of the above as a system, with *system* being defined as an entity that is connected to its environment by means of inputs and outputs identified at its boundary. Systems can be defined in terms of mechanical construction (form) or by function.

The functional definition should come first, with each definition being given in terms of the intended purpose of the system.

Any system may be decomposed into *subsystems,* connected to each other by means of inputs and outputs defined at subsystem boundaries. What belongs to a particular system or subsystem is defined by its boundary. Inputs and outputs that cross these boundaries are defined at, and only at, those boundaries.

Boundary interactions (Flows)—For systems that manipulate matter, inputs and outputs are found to be profitably categorized into three types: energy, material, and information. In each flow instance, both the *quantity* and *quality* of the inputs and outputs should be defined.

Information—Information is what we term the provided internal decision-making capability of a device or sensory data provided to or by a device or process. Information is generally given a more concrete expression by means of the term *signal,* that is, the physical form in which the information is conveyed. Signals are prepared, received, compared, combined, separated, transmitted, displayed, or recorded.

Matter/Material—We encounter matter in many forms, either in its natural form or one of the many forms we have impressed on it. Matter without form is incomprehensible. Matter is usually referred to as material with properties of form, mass, color, condition, and so forth. Materials can be mixed, separated, chemically changed, dried, cooked, dyed, coated, finished, packed, machined, shaped, transported, and so on. Mechanical systems always affect material in some way. Material examples are gas, liquid, solid, dust, raw material, samples, work pieces, components, and so forth.

Energy—There can be no transfer or conversion of matter or information without energy, however small. With the development of physics, the concept of force became important. Force was conceived as being the means by which the motion or form of matter is changed. Ultimately the concept of energy was developed to explain the interactions that took place during these processes.

Energy is the ability to make something happen. We speak of energy in terms of its manifest forms, such as electrical, kinetic, magnetic, heat, and optical. Energy must flow in or out of our system for something to happen. It must also be conserved.

Recall that the *function* of a product is a statement of a clear, reproducible relationship between the available input and the desired output of a product, independent of any particular form. As noted above, a system is defined in terms of its physical constituents (material objects, forces, flows, etc.). A function is defined in terms of a description of a process; it is what a system does as opposed to what it is. Functions and subfunctions are generally associated with systems and subsystems but not necessarily on a one-to-one basis. Function and subfunction definitions will, in a general way, parallel those of systems and subsystems. However, it is quite possible that a subsystem (a physical entity) may serve more than one function or that several subsystems might be required to meet a subfunction's requirement.

Careful definitions of the customer needs are essential in design. To solve any technical problem we need to describe, in a clear and reproducible way, the relationship between each of the available (or specified) inputs and each of the desired (or required) outputs. These relationships between the inputs and the outputs are the function of the system. *In this sense, the function is an abstract formulation of the task that is to be accomplished and is independent of any particular solution (physical system) that is employed to achieve the desired result.* Functions are generally stated in terms of physically quantifiable (measurable) effects and/or in terms of mathematical relationships. Textual (or verbal) descriptions of functions usually consist of an active verb and a noun: "increase pressure," "transfer torque," or "reduce speed." There are several broad function categories or classes we will use frequently. They are shown in Table 5.4. Each of these categories can (and will) be decomposed further when it is necessary to be more specific.

Function structure—The overall objective of a design cannot be considered properly defined until it has been clearly stated in terms of its function. Functional relationships must be carefully worked out, that is, designed to accomplish the objective specified. If the objective is material conversion (e.g., stamping out parts), the same input must always produce the same output. In a process (e.g., filling a tank with a fluid), there must be a clearly defined relationship between the beginning and the end.

The meaningful and compatible combination of subfunctions into an overall function produces a "function structure." There are usually a number of function structures that will meet the overall functional requirements and constraints (meet all the customer needs) of the design specifications. Identifying and evaluating these function structures is a very important part of the design process.

Function Structures: Simply Stated

Now we must develop a function structure to abstractly represent a product and its customer needs. How do we think a product should link, connect, or transform its inputs and outputs? We will be discussing a very systematic approach, but for now let's consider a simpler presentation.

To carry out an initial approximation, we can decompose a product's overall function into subfunctions. Again, this approach is a model boundary question, which means there are an infinite number of ways to complete this decomposition. We choose the decomposition that produces the most clarity in our minds as designers (and that we can justify/defend).

Subfunctions should be expressed in terms of measurable effects and/or mathematical relationships. Typically, subfunctions are expressed in

active verb-noun

pairs. For example, "increase pressure" or "transfer torque" or "reduce speed."

The subfunctions are schematically networked together to form an overall function structure (Figure 5.9). This schematic network makes clear the relationship among one possible set of functions that a product must carry out to achieve its overall task in terms of energy, matter, and information.

Having formulated the customer needs and perhaps having observed customers manipulate a current or competitor's product, we should now develop a reasonable first-level function structure for our design problem/product. A five-phase process is now presented to systematically translate customer needs into an equivalent functional description.

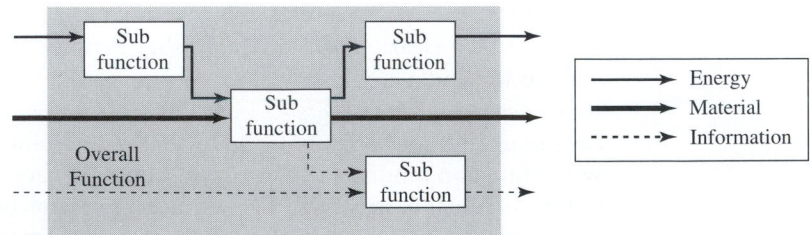

▼ *Figure 5.9.*
A generic function structure format.

The Function Structure Modeling Process: Phase 1— Develop Process Descriptions as Activity Diagrams

Figure 5.3 illustrates the overall functional modeling process. Functional modeling includes developing a process description, represented here with the Activity Diagram and eventually forming a *function structure.*

To start the function modeling process, an important tool is to specify the *process* by which the product being designed will be functionally implemented. A process or process description, in this sense, includes three phases: *preparation, execution,* and *conclusion.* Within each phase, we network high-level user activities to show the full life cycle of a product, from purchase to recycling or disposal. By listing the high-level activities in each phase, a number of product characteristics are chosen, including the product's system boundary, parallel (independent) and sequential paths of function chains, process choices, and interactions between user and device functions. These characteristics are documented with an *Activity Diagram* (introduced in Chapter 4), defined as a network layout of sequential and parallel tasks carried out by the user.

To visualize this process, consider Figure 5.10, which illustrates the Activity Diagram for a fingernail clipper design. To focus on product usage, the system boundary chosen includes all of the customer activities, such as purchasing, transporting, clipping, storage, and picking. Figure 5.10 does not include manufacturing-related activities such as packaging, sales functions such as unpacking, or disposal. Depending on the scope of the design task, it could have. This modeling boundary defines the product system, receiving inputs from and producing outputs to the user and environment. Parallel and sequential activities are given by the Activity Diagram structure. Parallel customer activities will likely lead to parallel product or device functions (since they are needed by the customer as separate or independent entities). These parallel activities provide implicit subsystems or assemblies for each parallel path.

Besides the system boundary and parallel paths, the Activity Diagram clearly shows a number of choices of what will be the customer process. These choices will clearly influence the final design. For example, the activities of "picking" and "filing" are process choices for improving the customer's fingernail appearance through mechanical contact. Chemical "soaking" process choices or others might be chosen as alternatives. They would lead to different activities, functional descriptions, and, ultimately, product architectures and components.

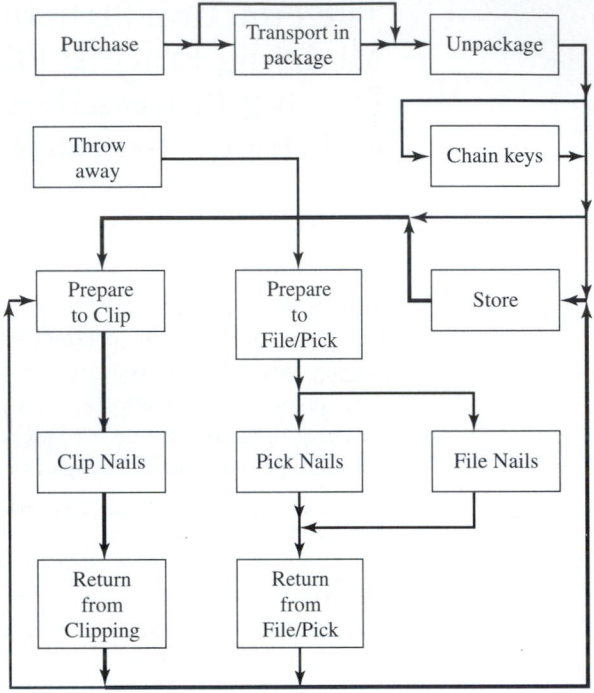

▼ *Figure 5.10.*
Completed activity diagram for the fingernail clipper product.

Phase 2—Formulate Subfunctions Through Task Listing

Using the customer process description (Activity Diagram) and customer needs, a function structure for the product is next formulated (Figure 5.2), where a *function structure* is defined as an input-output model that maps energy, material, and signal flows to a transformed and desired state. Function structure modeling (Otto and Wood, 1996; Little, Wood, and McAdams, 1997; Pahl and Beitz, 1991; Miles, 1972; Hubka, Andreason, and Eder, 1988; Hubka and Eder, 1984; Ullman, 1992; Ulrich and Eppinger, 1995) has historically been used to create a form-independent product expression. We extend common function structure modeling to include a mapping of customer needs to subfunction sequences (called *task listing*), a method for aggregating subfunctions, and a comparison of a functional decomposition with customer needs (Figure 5.3).

The first step is to identify primary flows associated with the customer needs of the product activities. A *flow* is a physical phenomenon, that is, material, energy, or signal (information), intrinsic to a product operation or subfunction. Typically, an energy flow has units of power, material flows might have units of mass per unit time, and information flow might be thought of as having units of information chunks (bits) per unit time.

However, one should not worry excessively about ensuring dimensional consistency of the flows; for example, material flows are best thought of in terms of the material itself. Information flows are best thought of in terms of dimensionless "yes/no" or "high/medium/low" or "zero to one" or whatever the information is. Energy flows are best understood in terms of either of the two components of power (F or v, ma or v, i or V, etc.) with the other component being understood. The important information to convey in the function structure flow is the qualitative phenomena passing through and being changed by the subfunctions. Comprehension of these phenomena by the design team is what must be ensured. As the function structure is refined into a mathematical model, the flows are given dimensional consistency (Chapter 13).

In the context of input–output modeling, a flow enters an operation or subfunction, is manipulated by the subfunction, and exits in a new state. For example, an operation or product task may be to pressurize a fluid. Two critical flows for this operation are an energy to execute pressure change and the fluid material being operated on. A list of some common flows is given in Table 5.2 (Hundal, 1990; Little, Wood, and McAdams, 1997; Stone, 1997; Little, 1997). Appendix A-1 provides definitions of these common flows, including the ideas of compliments, and stand-alone compliments.

Considering the fingernail clipper example, a subset of the customer needs is shown in Table 5.3. We now translate the customer needs to energy, material, or signal flows of the product when effects are exhibited, or are expected to be exhibited, during product use. Primary flows associated with "not compact" are the user's hands, the fingernail dimensions, and storage compartments; for example, pants pockets, wallets, or purses. These flows are material in nature and capture capacity in terms of "volume." Primary flows for the remaining customer needs include:

- "Does not file well"—hand motion (energy), fingernail (material), fingernail roughness (signal)
- "Does not cut well"—generated cutting force (energy), finger force (energy), and fingernail (material)
- "Not easy to open/close"—hand movements (energy) and hand (material)
- "Not easy to hold"—finger force (energy) and hand (material)
- "Not comfortable"—finger force (energy) and hand (material)
- "Not a sharp cutting surface"—generated cutting force (energy) and fingernail (material)

**TABLE 5.2. FLOW CLASSES, BASIC FLOWS, AND COMPLIMENTS
(ADAPTED FROM LITTLE, WOOD, AND MCADAMS, 1997)**

Class	Basic	Compliment		Stand-Alone Compliment	
		Effort Analogy	Flow Analogy	Effort Analogy	Flow Analogy
Energy	Human	Force	Motion		
	Acoustic	Pressure	Particle velocity		
	Biological	Pressure	Volumetric flow		
	Chemical	Pressure	Volumetric flow		
	Electrical	Electromotive force	Current		
	Electromagnetic, Optical, Solar	Intensity	Velocity		
	Hydraulic	Pressure	Volumetric flow		
	Magnetic			Magnetomotive force	Magnetic flux
	Mechanical				
	Rotational			Torque	Angular velocity
	Translational			Force	Linear velocity
	Vibrational	Amplitude	Frequency		
	Pneumatic	Pressure	Mass flow		
	Radioactive	Intensity	Decay rate		
	Thermal			Temperature	Heat flow
Class	Basic	Increasingly specific compliments			
Material	Human				
	Gas				
	Liquid				
	Solid				
Signal	Status	Auditory	Tone, Verbal		
		Olfactory			
		Tactile	Temperature, pressure, roughness		
		Taste			
		Visual	Position, displacement		
	Control				
		Overall increasing degree of specification			

Customer needs for which flows cannot be easily conceived are likely to be satisfied through constraints. For example, the "low cost" customer need cannot be satisfied by the operation of the clipper; it is an intrinsic property of the clipper. Therefore, we do not model the "expensive" effect with a function; rather, we consider it as a constraint. Note that to determine values on a cost constraint, one must consider all parts of the entire product and roll up their individual contributions. Cost is a constraint.

TABLE 5.3.   SUMMARIZED CUSTOMER NEEDS FOR A FINGERNAIL CLIPPER

Need	Effect	Importance
Low cost	Expensive	4
Compact	Not compact	4
Files well	Does not file well	2
Cuts well	Does not cut well	Must
Easy to open/close	Not easy to open/close	4
Easy to hold	Not easy to hold	3
Comfortable	Not comfortable	3
Sharp cutting surface	Not a sharp surface	3

To document the mapping of customer needs to flows, a "black box" model of the product is developed. A black box model lists all input and output flows for the primary, high-level function of the design task, stated in an active verb-noun phrase. Figure 5.8 illustrates a black box model for the fingernail clipper task. This model must now be refined and decomposed to identify the basic product or device functions that will satisfy the overall function and needs.

For each of the flows, the next step (Figure 5.3) is to identify a sequence of subfunctions and specific user operations that, when linked, represent the product when interfacing with the customer during the customer activities. A subfunction, in this case, is an active verb paired with a noun that represents the causal reason behind a product behavior. An operation is a specific action by the user needed to complete the function structure, typically a switch selection or decision-making node. Table 5.4 summarizes typical classes of engineering functions (as defined in Appendix A-2). A useful approach for generating subfunctions is to trace the flow as it is transformed from its initial creation state to its final expected state when it leaves the product's system boundary. This approach may be executed by *playacting* the flow (becoming the flow and asking, "What happens to me?") or brainstorming a hierarchy of functions that must process the flow.

For example, a customer need, expressed in the customers' voice, may exist for "Cuts nail well." A suitable flow for addressing this need is a *force* flow that ultimately acts on the nail material flow. Through playacting these flows, a subfunction sequence may be of the form: capture force, apply force, transform to larger force, transmit force as motion, guide motion, cut material, stop motion, release force, dampen reaction to force, and so forth.

**TABLE 5.4. FUNCTION CLASSES, BASIC FUNCTIONS, AND SYNONYMS
(LITTLE, WOOD, AND MCADAMS, 1997)**

Class	Basic	Flow class restricted	Synonyms
Channel	Import		Input, Receive, *Allow,* Form Entrance, *Capture*
	Export		Discharge, Eject, Dispose, Remove
	Transfer	Transport (M)	Lift, Move
		Transmit (E)	Conduct, Convey
	Guide	Translate	Direct, Straighten, Steer
		Rotate	Turn, Spin
		Allow DOF	Constrain, Unlock
Support	Stop		Insulate, Protect, *Prevent,* Shield, Inhibit
	Stabilize		Steady
	Secure		*Attach,* Mount, Lock, Fasten, Hold
	Position		Orient, Align, Locate
Connect	Couple		Join, Assemble, *Attach*
	Mix		Combine, Blend, Add, Pack, Coalesce
Branch	Separate		Switch, Divide, Release, Detach, Disconnect, Disassemble, Subtract, Valve
		Remove (M)	Cut, Polish, Sand, Drill, Lathe
	Refine		Purify, Strain, Filter, Percolate, Clear
	Distribute		Diverge, Scatter, Disperse, *Diffuse,* Empty
	Dissipate		Absorb, Dampen, Dispel, *Diffuse,* Resist
Provision	Store		Contain, Collect, Reserve, *Capture*
	Supply		Fill, Provide, Replenish, Expose
	Extract		
Control magnitude	Actuate		Start, Initiate
	Regulate		Control, *Allow, Prevent,* Enable/Disable, Limit, Interrupt
	Change		Increase, Decrease, Amplify, Reduce, Magnify, Normalize, Multiply, Scale, Rectify, Adjust
	Form		Compact, Crush, Shape, Compress, Pierce
Convert	Convert		Transform, Liquefy, Solidify, Evaporate, Condense, Integrate, Differentiate, Process
Signal	Sense		Perceive, Recognize, Discern, Check, Locate
	Indicate		Mark
	Display		
	Measure		Calculate

Note: Repeated synonyms are italicized.

Alternatively, for generating a function structure of a product already in hand, one can trace a flow through the product by tracing the flow of parts or subsystems the flow passes through. One can determine a sequence of subfunctions by sequentially removing the part and asking what fails (*Subtract and Operate,* as previously discussed). The opposite of that failure is the subfunction performed by the part in transforming the flow.

Continuing the fingernail design task, Figure 5.11 illustrates the task listing results for a subset of the customer needs and corresponding flows. Each function chain in Figure 5.11(a)–(c) represents a functional decomposition of the functions needed to "Cuts nail well." Customer needs *directly* lead to each of these function chains, a tactical advantage of the method.

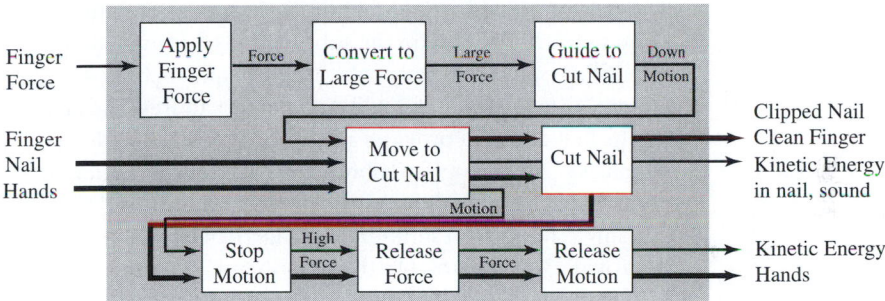

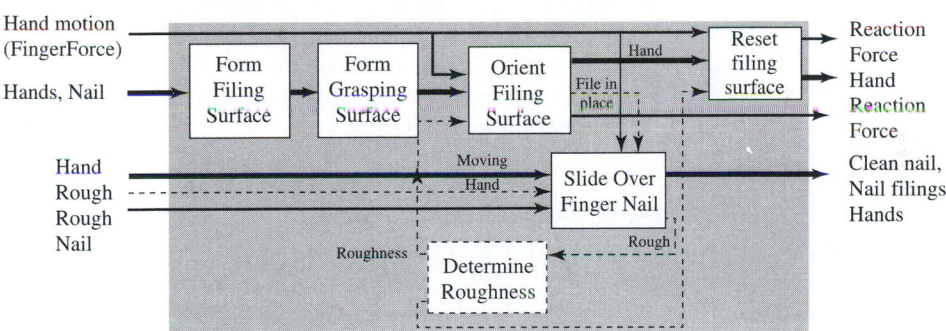

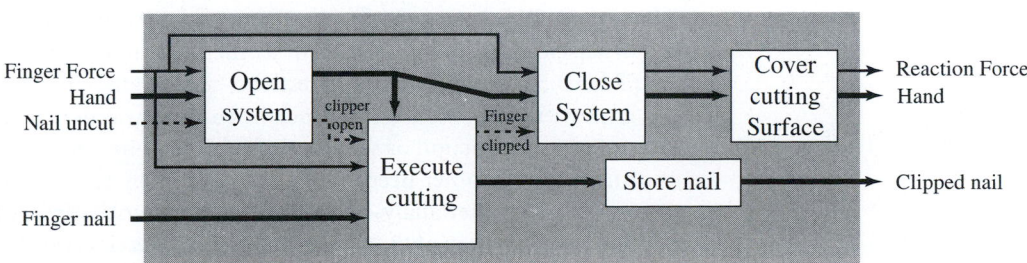

▼ *Figure 5.11.*
Fingernail clipper task listing for each customer need.

Phase 3—Aggregate Subfunctions into a Refined Function Structure

Each sequence of subfunctions for the full set of customer needs is aggregated (combined) to represent the functions of the entire product. This step is accomplished by appropriately connecting flows between each sequence and adding subfunctions that interact or provide control states.

Aggregation and refinement of the function structure ends are based on two criteria: (1) Are the subfunctions "atomic," that is, can they be fulfilled by a *single, basic* solution principle that satisfies the function? and (2) Is the level of detail sufficient to address the customer needs? The first criterion provides a basis for choosing the depth of functional analysis. For example, a subfunction of "control motion in 3-D" should obviously be refined to control in three rotations and translations, since a single, basic form solution in a fingernail clipper technology does not provide 3-D control. On the other hand, the second criterion assures that time is not wasted refining a function structure to the level of miscellaneous and secondary product components, such as fastening.

For the fingernail clipper design effort, an aggregated function structure is shown in Figure 5.12. Notice that subfunctions and flows are combined for overlapping or redundant functionality from Figure 5.11. User functions (as opposed to device functions) are also listed outside of the system boundary for clarity.

Phase 4—Validate the Functional Decomposition

Once the design team completes the subfunction aggregation, functional modeling and analysis comes to a close (at least for the first iteration) through two verification steps. First, all major flows between the subfunctions are labeled and checked according to their state of transformation. By labeling the flows, validity and continuity are ensured, perhaps leading to the addition of further functional representations. Table 5.5 lists the pertinent questions, checks, guidelines, and actions implemented at this stage. Second, the customer needs list is reviewed, and the subfunction or sequence of subfunctions is identified that satisfies each customer need. Needs not covered by the function structure require further analysis, and subfunctions not satisfying a need require confirmation of their incorporation. This verification typically adds more subfunctions to the network, while simplifying or removing others that do not really apply.

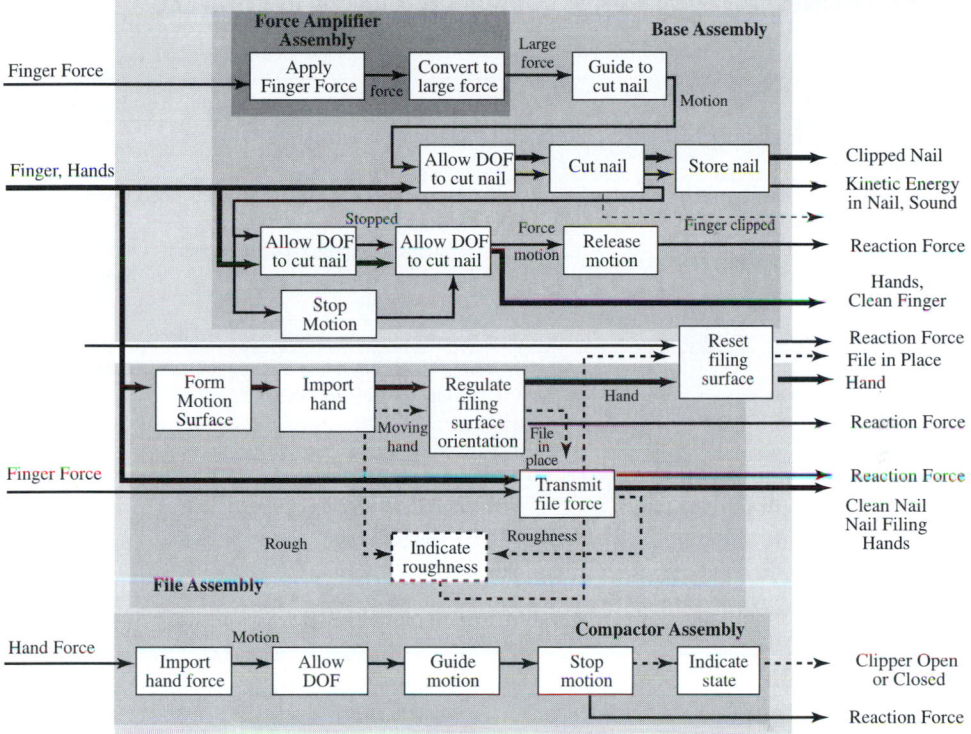

▼ *Figure 5.12.*
Refined function structure for the fingernail clipper design task.

The subfunctions listed for the validation combine to represent the customer need. For example, "compact" will ultimately be governed by the solution principles chosen for the "apply finger force" subfunction, and so forth. The size of these solution principles determines the overall compactness of the final fingernail clipper. This customer need must be balanced with "Cuts well" and the "Cuts nail" subfunction, since a certain size will be needed to cut a nail.

TABLE 5.5.   GUIDELINES FOR FUNCTION STRUCTURE ORGANIZATION AND FLOW CHECKS

Question	Check/Guideline/Action
Are physical laws maintained?	For example, validate conservation of mass and energy.
What is the new state of flow for each subfunction? Is this state valid and correct?	Label input/output states.
Are system and subsystem (assembly) boundaries clearly shown?	Define system boundary from process description or activity diagram. Subsystems (assemblies): starting point—define as a group of functions where the energy flow does not change (e.g., no relative motion)
Are all functions (noun) form independent?	If not, the function should be generalized.
Are the subfunctions atomic? Can a device or structure that only performs that function replace each subfunction?	Refine subfunction into a set of atomic subfunctions without adding unnecessary detail.
Does alternative ordering or placement or number of subfunctions exist?	If so, develop alternative function structures or subfunction structures.
Are all subfunctions product or device functions?	If not, the subfunction(s) is a user function (not performed by the product). Convert the subfunction to the device functions that must support the user function. If leaving the user function in the function structure adds clarity, double box it to show a distinction.
Are certain functions included as "wishes"?	If so, place a dashed box around the function to show that it is auxiliary or secondary.
Are certain functions "prolific," meaning do they affect or are affected by flows of many other subfunctions (e.g., "support impact loads")?	If so, place a ground symbol on the subfunction and only include primary flows through it.
Is the entire function structure sequential, that is, are the subfunctions are connected one after another in a dependent sequence or chain?	If so, product parallelism (assemblies/ subsystems) has not been identified. All of the customer needs are dependent. Parallelism of functions that do not directly depend on one another should be separated.
Do redundant functions exist?	Remove or combine redundant functions.
Do functions exist outside the system boundary?	If so, they should be clearly distinguished and maintained for clarity only, such as a user function.

Phase 5—Establish and Identify Product Architecture and Assemblies

With the functional decomposition verified, it is now important to identify product assemblies that can be addressed by individual designers or cross-functional design teams. The key elements of this phase are to define collections of functions (chunks) that will form assemblies in the product and to clarify the interactions and interfaces between these chunks. By so doing, a product team will have a basis for choosing between *modular* and *integral* architectures (Stone, 1997; Ulrich and Eppinger, 1995; Cutherell, 1996). They will also have a basis for choosing parallel design tasks for the product development, where only the interaction and interface information need be shared continuously among the subteams.

A simple process for establishing the product architecture and assemblies includes the following steps (as refined in Chapter 9):

▶ *Using the functional decomposition of a product, cluster the subfunctions or elements in the function structure.* Dashed boxes around the clusters of subfunctions and a title will serve as an appropriate representation. These clusters are chosen by identifying parallel subfunction chains (each parallel track is a candidate cluster), subfunction chains that have common energy types as flows, and subfunction chains that only have simple interactions between one another.

▶ *Create a rough spatial layout (block diagram with a reference frame) for the product.* This layout is meant to show the relative position of each cluster to understand spatial interactions and clarify interfaces.

▶ *Define interactions, interfaces, and performance characteristics between each cluster.* Given that interfaces are the boundaries between clusters, four types of interactions exist as flows across interfaces: spatial (geometry), energy, information, and material interactions. These interactions represent what must be shared across interfaces within the performance requirements.

Figure 5.12 shows the application of the first of these steps to the fingernail product. Three primary assemblies are identified: the base, force amplifier, and file. Important interactions include the hand flow of the user, attachments to allow relative motion for opening and closing activities, and so forth. These assemblies may now be designed relatively independently, if a modular architecture is chosen. In Chapter 9, Stone (1997) and Cutherell (1996) study further examples, including spatial layouts. In particular, an HP1200C inkjet printer, a power screwdriver, and others are studied with respect to their basic architectures.

VI. AUGMENTATION: FROM SIMPLE FUNCTION TREES TO COMPLETE MODELS

Functional modeling, as presented in the previous section, begins with customer needs and ends in a functional map for a product. It is a bottom-up construction. Depending on background or preferences of the designer, one might wish to take a top-down approach. Rather than starting with subfunctions and chaining, it may instead be desirable to

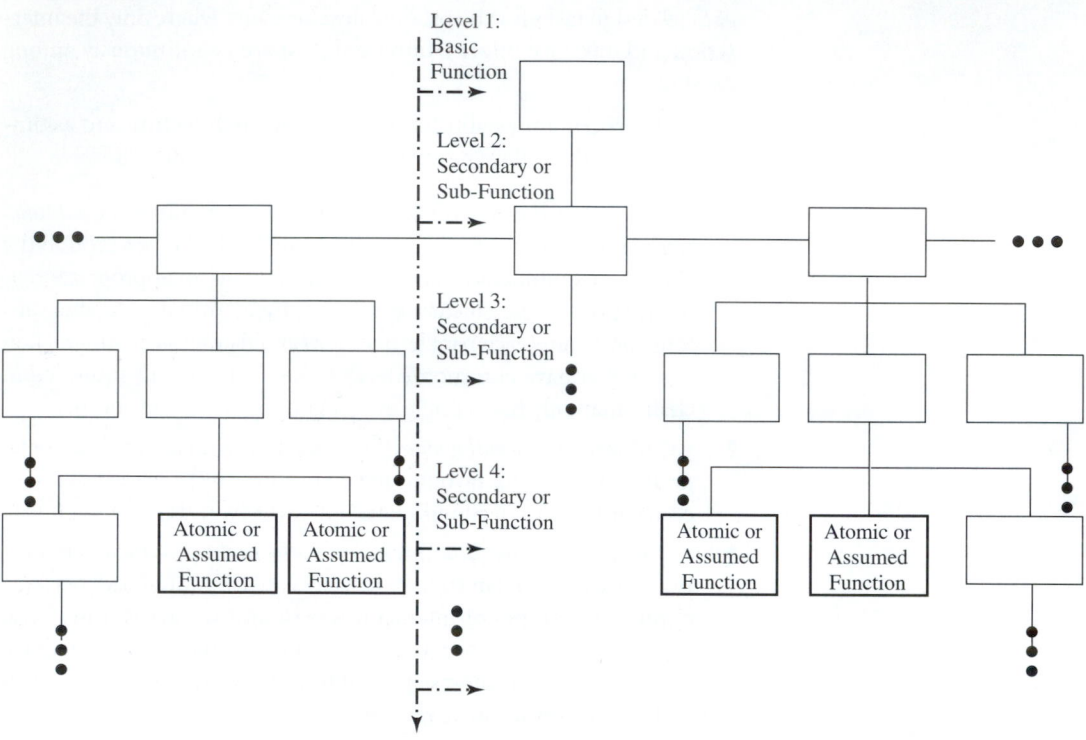

Level 1:
Basic
Function

Level 2:
Secondary or
Sub-Function

Level 3:
Secondary or
Sub-Function

Level 4:
Secondary or
Sub-Function

Atomic or Assumed Function

Atomic or Assumed Function

Atomic or Assumed Function

Atomic or Assumed Function

▼ *Figure 5.13.*
Generic function hierarchy.

generate *function hierarchies* or *trees* (Figure 5.3). Instead of mapping flows directly to function chains, a hierarchy of functions is developed, beginning with the global device function as the root of the hierarchy. We now present a method, entirely parallel to the technique of the previous section, to develop a function structure. As when comparing FAST with Subtract and Operate, neither approach is better or worse. Both bottom-up and top-down approaches can be applied in sequence to check results.

Figure 5.13 illustrates a generic function hierarchy for a product. The leaves (bottom-most level) of the hierarchy ultimately represent the refined functions for satisfying the customer needs. They are derived incrementally by decomposing the global function into its primary function carriers. These might be uncovered as secondary functions with the FAST method, or refined functions with the Subtract and Operate method, or simply brainstormed subfunctions. Then each function carrier is decomposed further using the function vocabulary of Table 5.4.

An Example of Hierarchical Function Structure Decomposition

As an example of hierarchical function structure decomposition, consider a power screwdriver. Many such devices are on the market today, with manufacturers such as Black & Decker and SKIL. The decomposition process begins with a statement of the overall function of a power screwdriver: *loosen/tighten screws*. This beginning point is shown in the black box description of Figure 5.14 along with relevant flows of material, energy, and signals. The black box description also defines the system boundary such that the bit is considered an input to the device.

Next, customer needs are gathered for the problem. Here, interviews were conducted with nine customers, and the stated customer needs are compiled in Table 5.6 and listed in order of importance (on a 0–10 scale).

To continue the decomposition of the power screwdriver, the overall function of *loosen/tighten screws* is decomposed into smaller, simpler subfunctions. Figure 5.15 shows the next level of decomposition performed. Here, five subfunctions are identified to compose the overall function. For each level of decomposition, the customer needs are checked to make sure that some subfunction is meeting each need.

The decomposition process for the power screwdriver continues through two more levels of decomposition before the problem is considered fully decomposed (or the subfunctions are considered refined). This hierarchy is shown in Figure 5.16. The decomposition process ends with Step 4 of the previous process (atomic functions). It is important to note, however, that decomposition is an iterative process that results in a set of subfunctions that, taken together, solve the overall problem. These refined subfunctions are shown in the boxes with darker outlines.

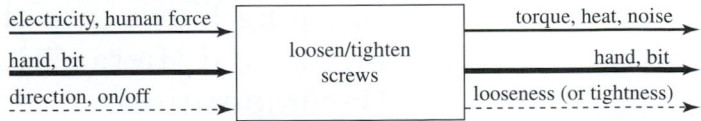

▼ *Figure 5.14.*
A black box representation of a power screwdriver. The overall function is shown in the box as *loosen/tighten screws*. Identified input and output flows are shown on the left and right sides of the box.

TABLE 5.6.   CUSTOMER NEEDS STATEMENTS AND IMPORTANCE RANKING FOR A POWER SCREWDRIVER

Customer need	Importance (1–10 Scale)
Powerful	9
Long-lasting battery charge	9
Fast	8
Reversible, i.e., screw and unscrew capability	8
Lightweight	8
Short charging time	7
Able to use manually	7
Uses different (interchangeable) tips	7
Comfortable handle	6
Automatic shut-off when not in use	6
Small size	5
Variable velocity	5
Maintenance-free	4
Balanced weight	3

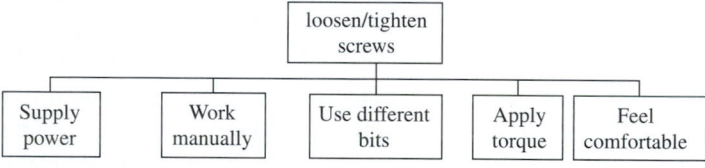

▼ *Figure 5.15.*
The first level of decomposition beneath the parent function (which in this case is the overall function) of *loosen/tighten screws* shows five subfunctions. These subfunctions all contribute to the overall function and meet stated customer needs.

Bringing Flows into the Functional Hierarchical Decomposition

Within the hierarchical category, there are many approaches to decomposition. We may now use the bottom level of the hierarchy to form a function structure. These subfunctions, from the bottom level, trace the flow of energy, materials, and signals through the device as depicted in Figure 5.17. Subfunctions may be further decomposed as needed. In this type of hierarchical decomposition, it is important to

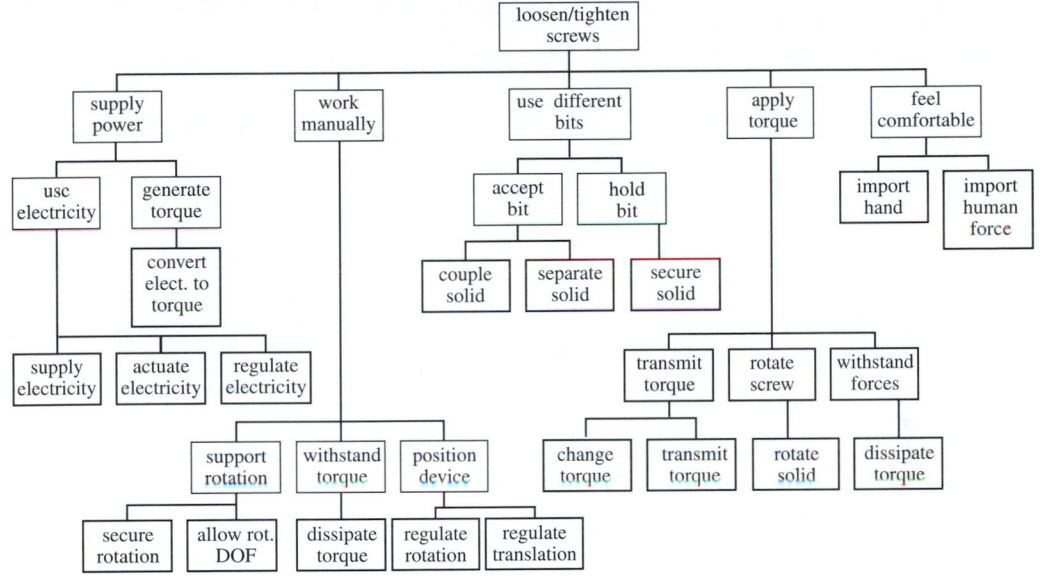

▼ *Figure 5.16.*
An example of a hierarchical decomposition of a power screwdriver. Boxes with heavy outlines indicate the refined functions.

note that subfunctions of the same layer are related to one another through flows.

VII. AGGREGATION REVISITED: SIMPLICITY OF SHOOTING DARTS

The previous sections describe two approaches for generating a function structure. Such structures are functional mappings of the input states of materials, energies, and signals to the output states desired from a product.

The structure captured by a functional mapping has a number of important implications. These implications include: parallel and sequential function chains, denoting possible assemblies, subassemblies, and modules; high-level physical models of the transform of materials, energies, and signals; consistency of flow transformations from function to function; boundary conditions based on the entering flows; and so forth. Because of the robustness of this structure, a functional model has a number of uses as the product design and development process proceeds (as discussed in subsequent chapters).

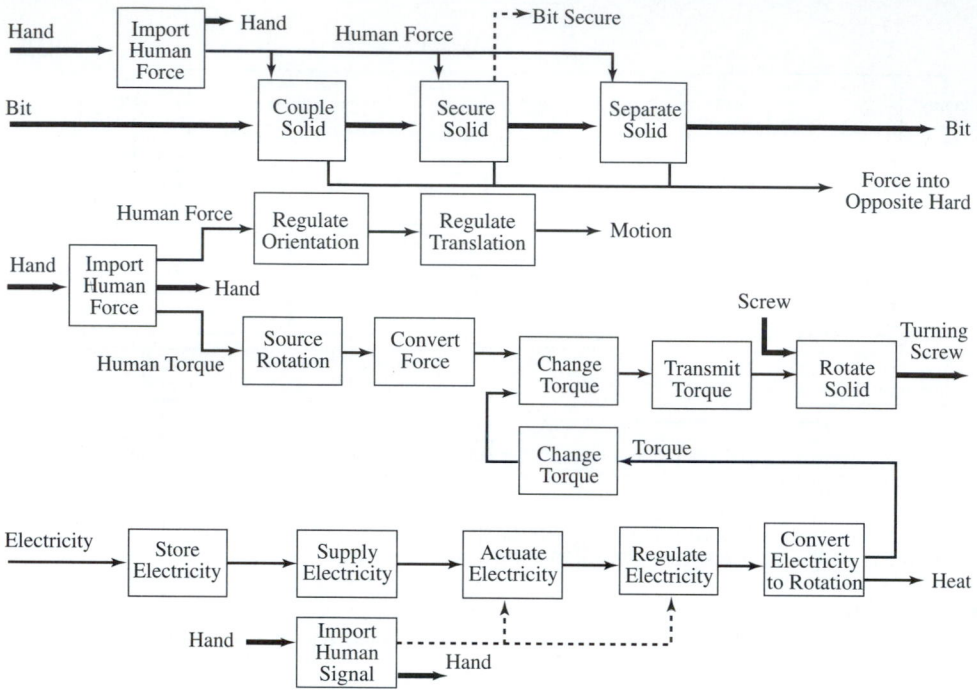

▼ Figure 5.17.
Power screwdriver function structure based on the hierarchical decomposition shown in Figures 5.13–5.16.

However, given some practice with function-structure generation, two problems are typically encountered. First, the natural parallelism that is derived from mapping flows to function chains can be distorted or obscured. Sequences of functions that are really not dependent can result, especially if function trees are generated first.

Another issue concerns the aggregation of function chains into an overall function structure. During aggregation, flows are connected between function chains, some functions are added to combine the chains, and some functions are removed due to redundancies. The logic behind these combinations, however, is undirected and so can be difficult to do. A simple approach is thus sometimes needed to result in a natural aggregation of the function chains.

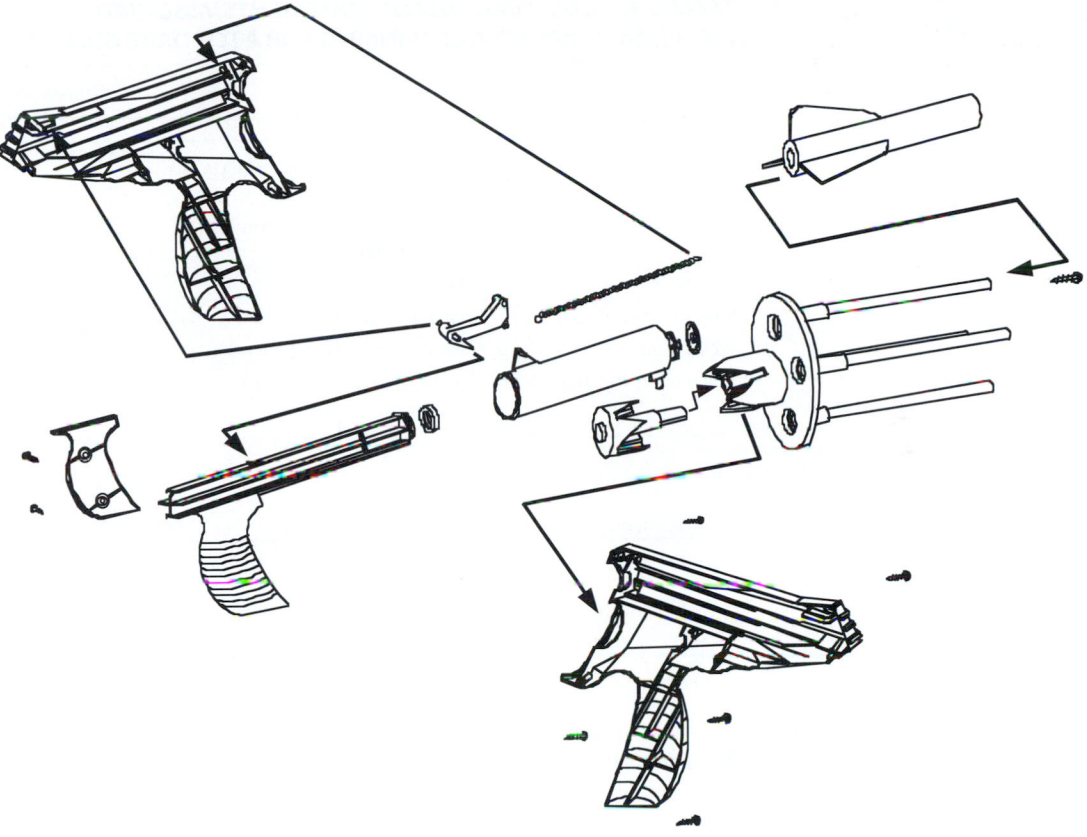

▼ *Figure 5.18.*
Assembly view of a Misslestorm™ Nerf Gun.

To demonstrate a systematic approach for aggregation, consider an example of a toy dart gun product. Figure 5.18 shows an example of such a product on the market. An abbreviated form of the customer needs and associated flows are shown in Table 5.7.

Figure 5.19 shows a partial black box model for the toy dart gun product. It is now our task to develop an *aggregated* functional model of the toy gun based on the black box model. The beginning point is to create a system boundary with an empty or "black" mapping of the input flows to the outputs. Figure 5.20 illustrates this first step. We next consider a flow for a customer; for example, the flow of human energy to "shoot darts a long distance." Playacting what must happen to the human energy, this flow is mapped from the input through to where it exits the

TABLE 5.7. CUSTOMER NEEDS STATEMENTS, ASSOCIATED FLOWS, AND IMPORTANCE RANKING FOR A TOY DART GUN

Customer need	Flows	Importance (1–5 Scale)
Shoots darts a long distance	Dart (M), Human Energy (E) (Cocking), Gravity (E), Air (M)	5
Shoots darts accurately	Pneumatic Energy (E), Aiming (S), Target (M)	5
Lightweight	Hand, Human (E), Gravity (E)	4
Easy to "trigger" gun	Hand, Human (E), Triggering	4
Well-balanced	Hand, Human (E), Gravity (E)	4
Entertaining dart flight	Visual (S) (Colors/Motion)	3
. . .	. . .	. . .

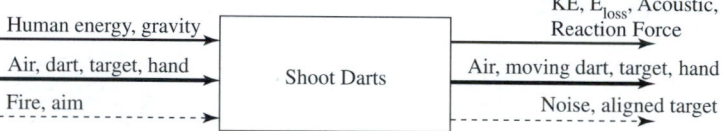

▼ *Figure 5.19.*
Black box of a toy dart gun. Assumes that dart enters the system.

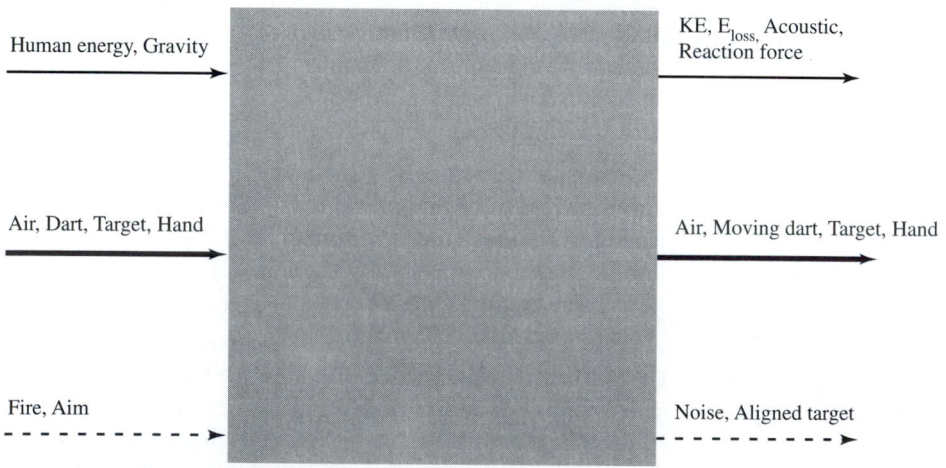

▼ *Figure 5.20.*
Initial step of functional modeling and aggregation: toy dart gun product.

system boundary or where it is needed or must operate on another flow in the system.

Figure 5.21 shows the results of this mapping (task listing). Notice that the human energy must be received, transmitted, stored, and released. The result of this process is energy losses (such as friction) and potential energy that will be needed to compress the incoming air to the system (assuming the process choice of compressing air to shoot a dart).

The next step is to choose the next flow for a given customer need and map it in a similar way in the *same* diagram. To illustrate this step, consider the flow of human energy for the "Easy to trigger" customer need. Figure 5.22 shows its addition as a function chain to the function structure. Human energy for triggering must be received (imported), transmitted, magnified (mechanical advantage), stored, and discharged, where the last function is executed based on a firing signal from the user. This function chain ends with energy crossing the system boundary and with energy available to release air to propel a dart.

The process of adding to the function diagram continues in this way as all relevant customer needs and associated flows are processed. Figure 5.23 shows the result of this process. Notice that a number of flows have been connected to the functions that need the results of their transformations. Examples include the energy from cocking to compress air, energy from triggering to release the air, and the material of air to energize the dart.

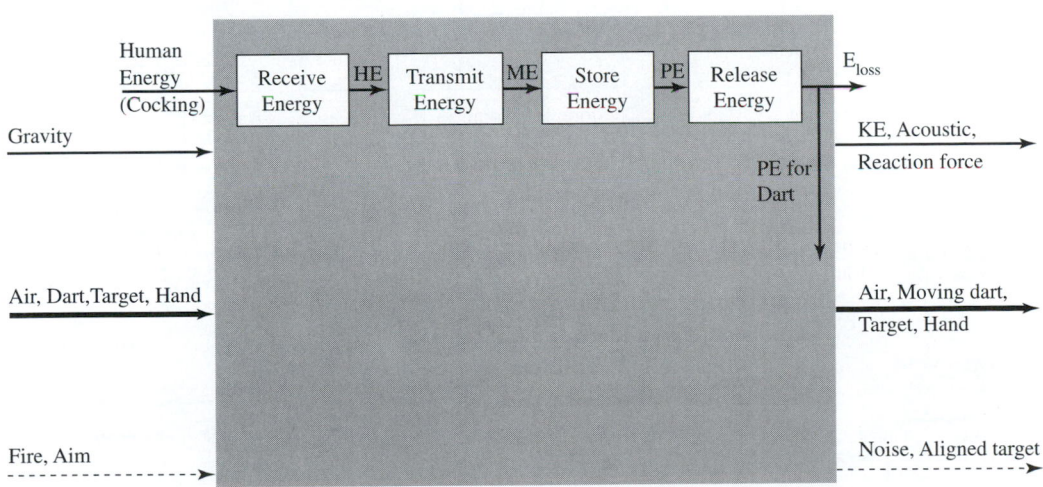

▼ **Figure 5.21.**
Creating first function chain for the customer need of "Shoot darts a long distance."

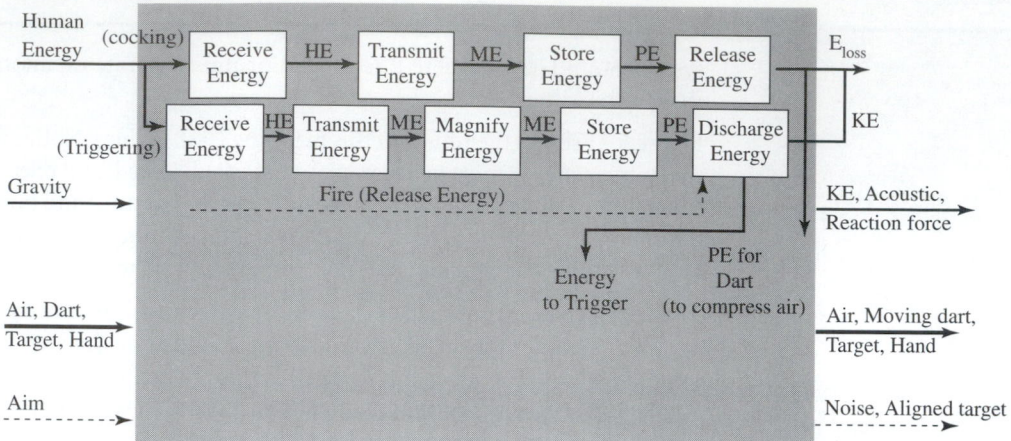

▼ *Figure 5.22.*
Creating next function chain for the customer need of "Easy to trigger."

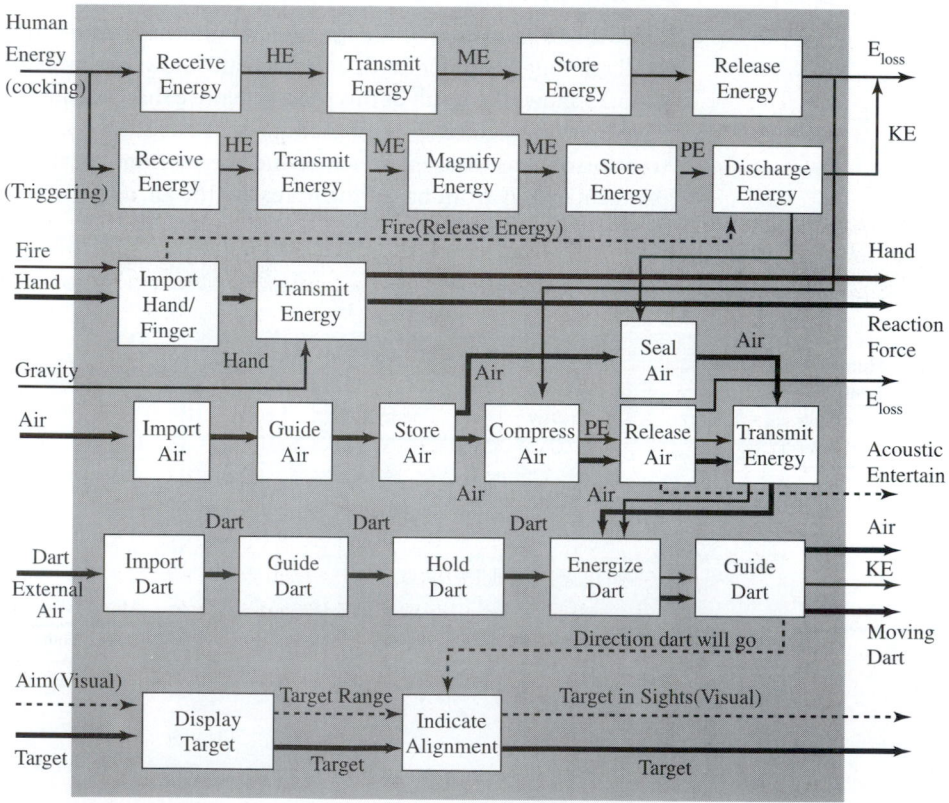

▼ *Figure 5.23.*
Function structure of a toy dart gun. Assumes that dart enters the system.

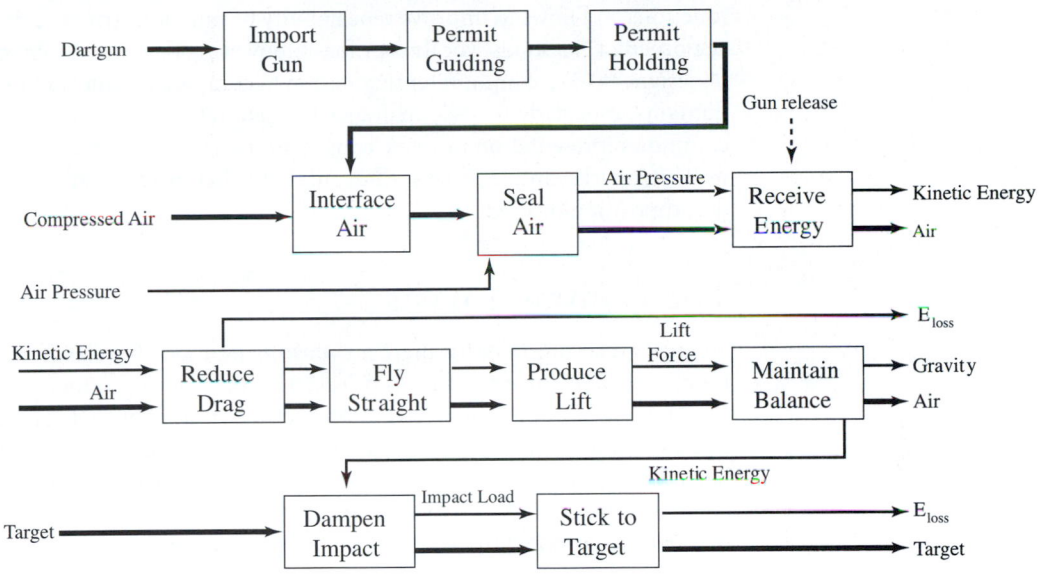

▼ **Figure 5.24.**
Function structure of a dart for a toy dart gun.

To further illustrate the idea of aggregation, a function structure for the dart is also provided (Figure 5.24). The primary customer needs mapped in this process are "Shoot darts accurately" and "Entertaining dart flight" as listed in Table 5.7. The primary connection created from mapping these needs is the energy of the dart as it is transferred to the target.

VIII. A FUNCTIONAL COMMON BASIS

In function-based design methodologies, functional decomposition of a device is a critical step in the design process. By decomposing the overall function of a device into small, easily solved subfunctions, the form of the device follows from the assembly of all subfunction solutions. But how do we define "small, easily solved subfunctions"? Is there a method that will lead to a repeatable functional decomposition between different designers?

One approach for answering these questions is to develop a "common basis" of functions and flows. The goal of such a basis is to form a set of well-defined function verbs and flow nouns that can represent products across product domains. This set will provide a number of potential benefits to product design: (1) a common vocabulary of functions and flows to improve repeatability of function structure development; (2) a robust vocabulary for empowering the expressiveness of designers; (3) a common representation to compare products across domains, especially to seek analogies for generating concepts; (4) a common representation to create consistent metrics for a product and its benchmarks; and (5) a level of detail that, when reached, stops the decomposition process.

The Common Basis

To create a common basis, there are various methods for classifying functions. Based on a careful empirical study of various products, the function classes of our representation are shown in Table 5.4. These classes represent basic or atomic functions that can be solved with a basic device, where the given synonyms are equivalent to the classes, as defined in Appendix A-2. Basic functions, in this sense, uniquely operate on flows in a product to produce a desired performance.

In addition to the function classes, basic flow definitions are needed. The three flow classes of materials, energy, and signals are extended to form the vocabulary of standardized flows. These are shown in Table 5.2. Within each class, flows are separated into basic flows. In practice, a basic flow description is formed from a basic word plus the class it belongs to. For example, a basic flow for a power screwdriver is *human energy*. Basic flows are further specified by adding an effort or flow compliment found in the third or fourth column of Table 5.2. Here the flow description is formed by a basic descriptor plus its compliment. A further specification of the human energy required in a power screwdriver is the *human force* needed to actuate the *electrical energy*. An energy flow may also be specified by using a stand-alone effort or flow compliment from the last two columns. Taking an engine, for example, we may be interested in the *torque* produced by the engine. Note that the material and signal classes do not use the effort and flow descriptions found in the energy class. Instead, compliments are listed where appropriate. Flows are then specified by the basic descriptor alone or compliment plus basic descriptor format.

The degree of specification depends on the type of design and customer needs. Using a more general flow description produces a generic function structure and, thus, a wider range of concept variants. However, if customer needs dictate concreteness in flows, then an increasingly specific compliment is more valuable. Another use of the flow set (and the function set of Table 5.4) is to compare different devices on a functional level. In this case, the flows (and functions) should be expressed in their basic categorization to capture similarities between devices.

Considering the material and energy classes, both have basic flows labeled as "human." The importance in human crossing of device boundaries merits this special inclusion. Often the requirement of human interaction is known at an early stage of design. By its specification, it will guide the design to appropriate solutions faster.

Signals, while in actuality either material or energy, receive their own class because their function is to carry information. Here, signals are treated as two basic flows used for sensing (status) or control purposes.

Careful observation of the energy class of flows reveals the division of basic flows into effort and flow compliments. This consistent categorization of flows eliminates confusion when increasing specification is needed. For instance, in a hand-held power screwdriver, is the relevant flow out of the motor *angular velocity* or *torque*? Of course, both exist, but it may be argued that *torque* is the correct choice to describe the situation, because effort is the more important output of the power screwdriver, based on the customer need of inserting screws easily. That is, for this application, a certain value of torque is necessary, but any value of angular velocity is acceptable that is nonzero and not excessively high, and so torque is a more important variable to keep track of in the function structure than power and angular velocity are. For other applications, the power or the velocity may be the important variable of concern.

Not only is a consistent division of basic flows necessary but also is a clear definition for all flows. Flow definitions are given in Appendix A-1. Combined with the existing definitions of functions in Appendix A-2, the flow definitions provide a complete basis for repeatable functional decompositions. Furthermore, they provide a stopping point for the decomposition.

Transforming Functional Models

These basic flows and functions, as listed in Tables 5.2 and 5.4, compose the formal common basis for our representation. Product function structures may be converted to these functions and flows for consistency, to obtain meaningful results, and to determine a stopping point for the decomposition. When product functions may be mapped to the basic or flow-restricted functions in Table 5.4, the decomposition has reached a stopping point. Note that a product's functional model should not be mapped to just the function classes. If all function structures were converted just to the high-level function classes or to the three high-level flows of material, energy, and signal, significant information would be lost. The results would be too generic a representation.

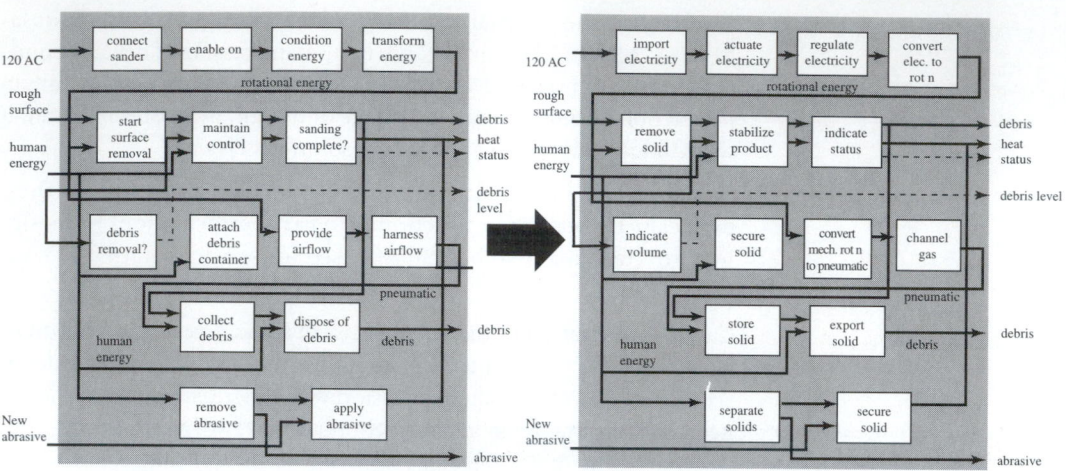

▼ **Figure 5.25.**
Example of transforming a functional model to a common basis.

Using the common basis, the first step in transforming a function structure is to clearly understand the intent of each subfunction. This step may require further study of the product. The next step is to categorize each subfunction into its classification by selecting a function verb from the basic function list and a noun from the list of flows. Each basic function is defined in Appendix A-2, and common synonyms for completing the transform are listed in Table 5.4.

To illustrate the transform process, Figure 5.25 shows function structures for a palm grip sander (cover figure of Chapter 9) before and after transformation. The basic functions listed in the second function structure are obtained directly from Table 5.4, satisfying their definitions. Likewise, the selected flow names are chosen to maintain the information in the function structure, since flows directly relate to customer needs.

Uses of a Common Basis

A number of uses exist for functional models that are converted to a common function and flow form. Chapter 6 uses a common basis to select analogous products for product benchmarking and measurement. Chapter 9 uses functional models, converted to a common form, to systematically determine alternative modular architectures for a product. Chapter 10 uses the common basis to search for analogous solutions to product functions. By so doing, product concepts may be generated from diverse analogies across product domains. Chapter 13 uses functional models as a starting point to develop analytical models of a product concept. A common basis aids in choosing an appropriate metric or performance parameter for evaluating a product with respect to its customer needs. Finally, this chapter advocates the transformation of functional models to a common basis. By performing the transform, a clear stopping point may be determined for the modeling process. In addition, the validity and consistency of the model may be checked as the transform is carried out.

Aggregate Function Study

Based on these uses of a common basis, let's take a closer look at the entire functional analysis process and the data it produces. To understand the potential of a common basis, 60 consumer household products were modeled functionally. These products ranged from kitchen appliances to power tools and children's toys. Each functional model was transformed to its common form, and customer needs, on a scale of 1–5 (importance), were related to each of the functions per product. The analysis resulted in approximately 125 unique basic function-flow pairs for the product, illustrating that the products overlapped significantly in their functional representation.

The results of the 60-product study are shown in the bar chart of Figure 5.26. This chart plots the average importance of a function across the full product set for the most commonly occurring functions (the top 10–20% of the functions). The results reveal many insights into the subfunctions of small household consumer products. For example, one may not have guessed "Import human force" would show up as the number-one subfunction. This result shows ergonomic issues and industrial design must play a crucial role in the development of future products. Other subfunctions in the top 10%, such as "[Convert] torque to pneumatic," may be questioned at first; however, this function becomes important due to the use of fans and impellers, which have many different uses in products, including suction and cooling. In addition, the number-five subfunction, "Remove material," is defined as the separation of part of a material from its prefixed place. This function is used to cut, sand, polish, or chop anything from food to wood.

Besides these single function insights, relationships exist between many functions, for example, "[Convert] torque to pneumatic" and "Dissipate noise." When there is a fan or rotor used in a product, the customers usually want less noise. However, "Dissipate noise" is more important than just "[Convert] torque to pneumatic" because there are other functions that can cause the need to dissipate noise, such as a noisy motor in "[Convert] electricity to torque."

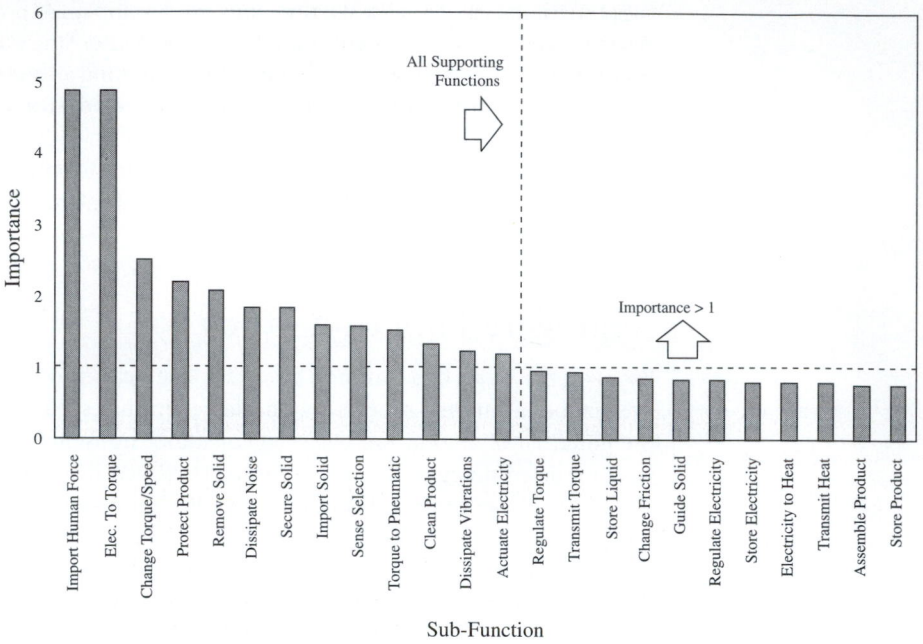

▼ Figure 5.26.
Aggregate functions with respect to average customer need importance for the 60 consumer products.

IX. CRITIQUE OF FUNCTIONAL MODELING METHODS

At its most basic level, this chapter advocates the use of methods to decompose a problem functionally. It also provides an *explicit* tie to customer needs, assuring that they remain a focus of the design effort, as opposed to implicit specifications.

In the process of advocating functional modeling, two categories of techniques are presented, one based on function trees, the other on system maps of input to outputs. Function trees are a simpler method, emphasizing a pure *brainstorming* approach to generating the *what*s of a design problem. What must the product "do" to achieve this goal, not how it will do it. FAST does provide some direction in the brainstorming approach, but generally function trees require an open search for functions at a lower level than the basic function being solved.

A function structure approach, referred to here as task listing, adds additional insights and information through its layout and process. It ties the designer to the technical system and physics of the problem through the basic engineering flows of materials, energies, and signals. It also adds an explicit logic by first mapping customer needs to flows, then flows to dependent chains of functions, followed by chains of functions to aggregated functional models of a product.

Through the additional information and logic, function structures have a number of advantages over a simpler function-tree approach. These advantages have clear benefits as a product reaches later stages of development. Some primary advantages, in brief, may be enumerated as follows:

- Function structures, based on task listing, have a clear and explicit relationship to customer needs. Function trees, alternatively, do not exhibit this characteristic. Since flows are not represented, customer needs are not tied directly to any sequence of functions.

- Function structures illustrate parallel and sequential functional relationships, virtually by default. Such relationships are critical for making product architecture decisions early in the development cycle. These decisions lead to the layout of possible assemblies, subassemblies, and modules in addition to team-resource allocation, including module interface issues. Chapter 9 expounds on this topic.

- Ultimately, a set of virtual (analysis) or physical prototypes will be constructed for any product. To develop such prototypes, high-level models are needed of the product performance. Function structures provide one source of this information in terms of the transform of materials, energies, and signals. These functional transforms, in turn, represent abstractly the underlying physics and physical processes of the product. Even conservation of mass and energy is displayed, provided that a valid functional has been developed.

- Function structures provide a means of "playacting flows." By starting with a given material, energy, or signal, a designer may ask, If I am an input force (for instance), what is the first thing the *product must do with me or to me*? After this action by the product, what state am I in? For this new state, what's the next product action that must happen to me? and so forth. The process ends when the product outputs the desired state of the flow.

- Function structures show a definitive system boundary as chosen by the designer. By defining a system boundary, boundary conditions are available for modeling and understanding a product's interaction with its environment.

▶ The flows represented in function structures also provide a direct means of choosing measurements for a product. The appropriate measurements are simply the output flows of any of the functions along a function chain in the structure. Subsequent chapters describe methods of choosing engineering specifications and measurements.

X. SUMMARY AND "GOLDEN NUGGETS"

Functional decomposition provides an abstract yet direct method for translating customer needs to a functional specification of a design task. Figure 5.4 illustrates a systematic five-phase process for executing a functional decomposition. Using the results of the method, quantitative specifications may be formulated for customer needs, based on the applicable subfunctions. A designer is also tactically situated for generating and then combining concepts for each subfunction, a far simpler task compared with the complexity and potential enormity of an overall design problem without decomposition.

Golden nuggets that should be harvested from this chapter include:

▶ Functional modeling enhances the characteristics of a creative designer. It provides a systematic approach for decomposing a product design problem into simpler subproblems.

▶ Functions represent what parts of the product do in a form independent way; constraints are intrinsic properties of the entire product.

▶ A greater breadth of concepts may be generated in product design using functional modeling.

▶ Functional models provide a basic systems-approach to design, as needed for supporting experimental and analysis methods.

▶ Allocating design-team resources is aided by functional modeling.

▶ Product architecture decisions may be made earlier in the development process through functional modeling.

References

Akiyama, K. 1991. Function analysis: Systematic improvement of quality performance. Portland, OR: Productivity Press.

Cutherell, D. A. 1996. Product architecture. In *The PDMA handbook of new product development,* edited by M. D. Rosenau. New York: Wiley.

Dixon, J., and Poli, C. 1995. *Engineering design and design for manufacturing—a structured approach.* Conway, MA: Field Stone.

Frankel, E. 1988. *System reliability and risk analysis.* Cambridge, MA: Massachusetts Institute of Technology.

Hubka, V., Andreason, M., and Eder, W. 1988. *Practical studies in systematic design.* London: Butterworths.

Hubka, V., and Eder, W. Ernst. 1984. *Theory of technical systems.* Berlin: Springer-Verlag.

Hundal, M. 1990. A systematic design method for developing function structures, solutions, and concept variants. *Mechanism and Machine Theory* 25 (3): 243–256.

Lefever, D., and Wood, K. 1996. Design for assembly techniques in reverse engineering and redesign. *Proceedings of the 1996 ASME Design Theory and Methodology Conference.*

Little, A. 1997. A reverse engineering toolbox for functional product disassembly and measurement. Master's thesis, The University of Texas, Austin.

Little, A., Wood, K., and McAdams, D. 1997. Functional analysis: A fundamental empirical study for reverse engineering, benchmarking, and redesign. *Proceedings of the 1997 ASME Design Theory and Methodology Conference.*

Miles, L. 1972. *Techniques of value analysis and engineering.* New York: McGraw-Hill.

O'Connor, P. 1991. *Practical reliability engineering.* Chichester, England: John Wiley.

Otto, K., and Arhens, G. 1997. Eine Methode zur Definition technischer Produktanforderungen. *Konstruktion* vol. 49 (Nov./Dec.): 19–25.

Otto, K., and Wood, K. 1996. A reverse engineering and redesign methodology for product evolution. *Proceedings of the 1996 ASME Design Theory and Methodology Conference.*

Pahl, G., and Beitz, W. 1991. *Engineering design: A systematic approach.* London: Springer-Verlag.

Sobek, D., and Ward, A. 1996. Principles from Toyota's set-based concurrent engineering process. *Proceedings of the ASME Design Theory and Methodology Conference.*

Stone, R. 1997. *Toward a theory of modular design.* Ph.D. diss., The University of Texas, Austin.

Ullman, D. 1992. *The mechanical design process.* New York: McGraw-Hill.

Ulrich, K., and Eppinger, S. 1995. *Product design and development.* New York: McGraw-Hill.

Value Analysis Incorporated (VAI). 1993. *Value analysis, value engineering, and value management.* Clifton Park16, NY: VAI.

6 Product Teardown and Experimentation

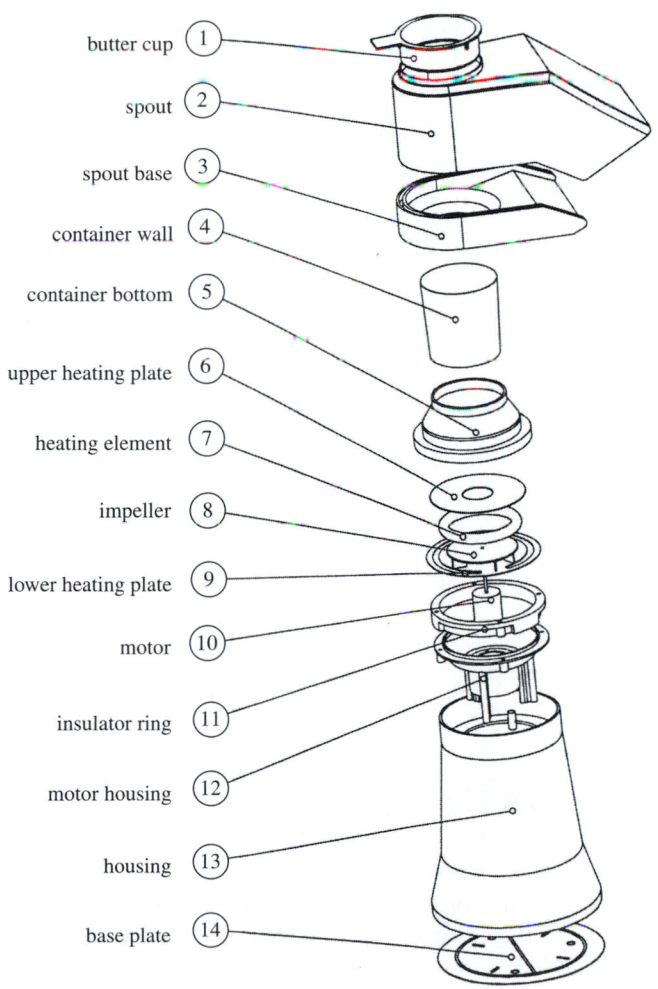

butter cup	①
spout	②
spout base	③
container wall	④
container bottom	⑤
upper heating plate	⑥
heating element	⑦
impeller	⑧
lower heating plate	⑨
motor	⑩
insulator ring	⑪
motor housing	⑫
housing	⑬
base plate	⑭

Exploded view of the Presto™ hot air popcorn popper.

Product teardown is the process of taking apart a product to understand it, and to understand how the company making the product succeeds. A product teardown serves three primary purposes:

1. dissection and analysis during reverse engineering
2. experience and knowledge for an individual's personal database
3. competitive benchmarking

The first of these purposes pertains to the evolution of a product. Before a product is evolved to its next generation, the current version must be analyzed and understood according to its functions (intended and latent), its technology, its strengths, and its weaknesses. Product teardown and associated techniques provides this baseline information, even to an independent design team with no knowledge of how the product was actually developed.

Another purpose for product teardown is simply to understand "how things work." By dissecting products, we gain kernels of information of how to accomplish function. The more we dissect technology, the larger our knowledge base of concepts grows to solve and synthesize solutions to new problems.

Finally, and most important, product teardown pertains to competitive benchmarking. To remain competitive, a design team must compare any proposed concept with the competition. What is the competition doing to succeed? Often a corporate group exists to tear down the competitor's product, estimate costs, plot trends, make predictions on requirements, and work with the design teams. These efforts uncover the clever things that the competition has spent effort on, uncover the principles behind how they work, and uncover predicted costs.

It is important to note that benchmarking can be appropriate at different phases of the product development process. It may be more appropriate to benchmark products before customer needs analysis and functional decomposition. Alternatively, a team may not wish to bias these initial analyses, but update the results with the benchmarking data. Chapter 7 will explore these benchmarking concepts further.

I. CHAPTER ROADMAP

As shown in Figure 6.1, we first present an overview of a step-by-step method for tearing down products. This method works well for both simplified and complex analyses. Any level of analysis will require

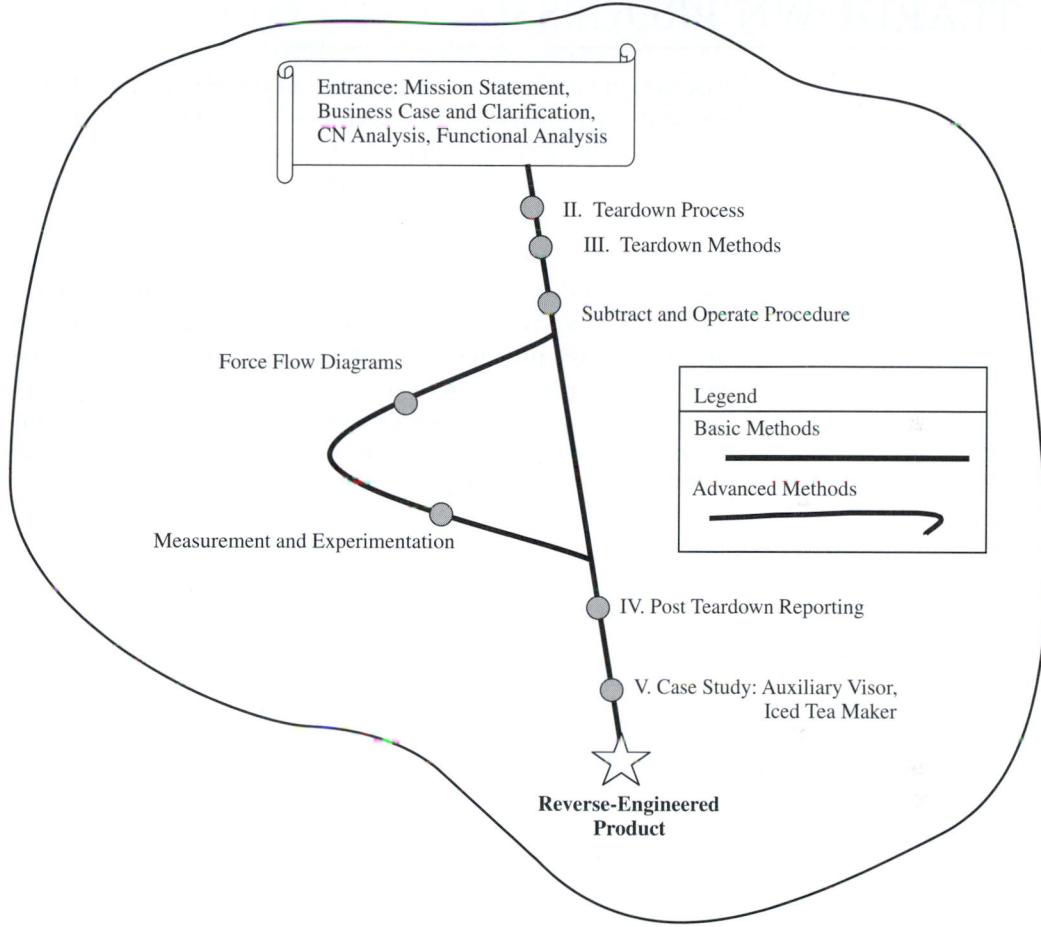

Entrance: Mission Statement, Business Case and Clarification, CN Analysis, Functional Analysis

II. Teardown Process

III. Teardown Methods

Subtract and Operate Procedure

Force Flow Diagrams

Legend

Basic Methods

Advanced Methods

Measurement and Experimentation

IV. Post Teardown Reporting

V. Case Study: Auxiliary Visor, Iced Tea Maker

Reverse-Engineered Product

▼ *Figure 6.1.*
Roadmap for navigating product teardown.

systematic teardowns of products. Following the systematic teardown method, we then present supporting techniques for the steps in the method. The next chapter (Chapter 7) focuses on the particular activity of benchmarking. It makes use of the teardown process presented here, but with particular emphasis on understanding the competition.

II. TEARDOWN PROCESS

The most important aspect of a product teardown is to make clear why we are doing it. What are the objectives of this activity? What are the purposes of the new, evolved product? What is expected to be uncovered through the teardown? It is always fun to tear apart mechanical devices; however, nothing will be accomplished if data are not collected on important factors.

Generally, teardowns are carried out in industry to benchmark work against the competition. They observe the technology and architecture and uncover the principles behind how it works. They also uncover their costs. The competition may use hand-wired power supplies, incorporating many components, wiring and connections, whereby it is realized that the competition must be shipping assembled products from a low-labor-cost country. On the other hand, a different competitor may use expensive multifunctional parts and assemblies that have been designed for automated assembly equipment. This latter situation leads to the realization that the competition fabricates their product in high-labor-cost countries. Corporate development and production strategies can be uncovered.

Many large companies have in-house staffs whose sole job it is to reverse engineer the competition. Provided they break no laws, this is entirely legitimate and ethical. They purchase the competition's product, tear it down, estimate costs, plot trends against previous teardowns of similar models, make predictions on requirements, and work with design teams on understanding competitive capability. Such teams operate by disassembling and analyzing products to evaluate assemblability, potential impact on the environment, recyclability, maintenance, and so forth. As new laws, regulations, and market demands drive lifecycle choices, industry must constantly seek new strategies for evolving products.

List Design Issues

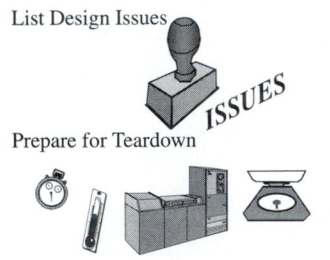

Prepare for Teardown

Examine Distribution and Installation

Disassemble and Measure

Form a Bill of Materials and Summary Report

	Name	QTY
A1	Assembly 1	1
1	Part 1	2
2	Part 2	2
3	Part 3	1
A2	Assembly 2	2
4	Part 4	1
5	Part 5	1
6	Part 6	1

▼ *Figure 6.2.*
Overview of the teardown activity.

Overview

An overview of an effective teardown process is given in Figure 6.2. The teardown process is tightly integrated with the idea of *benchmarking,* comparing one's product with competitive products in the marketplace. It is also tied to the notion of simply understanding the workings of any product, even a company's own product. In this chapter, the focus is on the teardown steps of any benchmarking process. We discuss the documentation of design issues, evaluating distribution and installation of a given product, physically disassembling the product, carrying out needed experimentation and measurement, and analyzing the results. The next chapter considers the remaining steps of identifying competitive and related (analogous) products, information search, plotting industrial trends, and reporting the state of the market, including target-engineering specifications.

A functional product teardown is far more than disassembly of a product to see how it is put together. One must also *analyze* the systems and transform this analysis into information that can be used as a part of the new redesign. As diagrammed in Figure 6.2, steps include: listing functionality to explore, preparing for the teardown, evaluating the distribution and installation, conducting the teardown complete with analysis and labeling of parts, and report generation.

STEP 1: List the Design Issues

First, it must be clear what problems and opportunities the design team is facing on the current project. If it is a new project, the design issues may be unknown, and so any and every facet about the customer market, competitors, and features of competitors products are worth investigating. These issues are the important performance criteria that must be measured. Typically, they are related to the customer needs.

If the project entails a redesign, an investigation can ask of the previous design team:

▶   What was difficult for them?

▶   What design problem did they solve that they are proud of?

▶   What related technologies were they interested in?

To help develop the design issues, a companion study of customer needs (Chapter 4) and predicted product functionality (Chapter 5) can be executed. Customer needs analysis and ranking obviously focuses and augments the list of design issues.

Similarly, predicted functionality of a product helps to focus on the "what" before the "how" and thus can help in forming issues to measure in the teardown. What are all of the products being benchmarked intended to do? What predicted or hypothesized functions do they need to achieve the primary goal or overall function? By answering these questions using the methods of Chapter 5 before the product is dissected, a design team can minimize solution principle bias and have clear goals to direct the product dissection. Without doing this initial generation of a function structure, a team may experience tunnel vision, becoming enamored with the current products and not straying from the current layout. Fundamentally, developing an initial hypothesized function structure helps uncover design issues.

Finally, the last set of design issues that needs recording is basic information on the components in assemblies. This is needed to evaluate the performance criteria in the later analysis. For example, material types and masses are needed to determine part costs. Such basic factors that are typically needed to be known include:

- quantity of parts per product unit
- dimensional measurements
- maximum, minimum, and average material thickness
- weight
- material
- color/finish
- manufacturing process, including sufficient information for a Design for Manufacturing analysis
- geometric, spatial, and parameter tolerances
- primary functions
- cost per part or subassembly
- other notes

STEP 2: Prepare for Product Teardowns

After identifying the design issues, one should identify all tools that will be required to complete teardown. This activity includes identifying all sensors and test equipment required for the measurement process. Camera? Videotape (of product operating)? Multimeter? Hardness tester? Optical sensor? Flow meter? Dynamometer? Calipers? Strobe? and so forth. Typically, one documents this information into a written or electronic report of the teardown.

STEP 3: Examine the Distribution and Installation

Important factors in the product-development decision-making process that must not be overlooked are the means used to acquire parts, contain them, ship, distribute, and market the product. These must also be examined as a part of the teardown process. The distribution packaging of a product should be examined and reported to the design team; often it can be quite expensive. Consumer installation instructions and procedures should be examined for costs, effectiveness, and liability.

STEP 4: Disassemble, Measure, and Analyze Data by Assemblies

Disassembly is the obvious step commonly pictured when thinking of reverse engineering. However, to be effective, this step must be coordinated with measurements and experimentation. First, take pictures and measurements on the whole assembly before disassembly. Then:

- Take apart the assembly.
- Take pictures in an exploded view (and/or produce a solid model assembly diagram—electronic exploded view).
- Take measurements on the parts and assemblies to complete the data sheets.

In this step, it is important to avoid destructive testing during the first iteration. Parts that are fabricated with insert molds, rivets, welds, plastic sonic welds, solder, integral components (such as windings on motors and coil springs in mechanical clocks), and so on should be carefully disassembled so that the product can still function. During subsequent experimentation, further teardown can occur where destructive testing is needed.

STEP 5: Form a Bill of Materials

During disassembly, the team should complete a written form that details the product. A good format for a fingernail clipper product is shown in Figure 6.3. Also, the sequence of assembly photos and an exploded view CAD drawing should be completed. The data collected in each column of the bill of materials (BOM) are that required for subsequent analyses, including cost and performance. Subsequent sections describe methods for systematically collecting data for a BOM. These methods also provide additional data, such as assembly issues.

				Mass		DFM Cost Analysis Data	
Part #	Part Name	QTY	Function	(gm)	Finish	Manufacturing Process	Dimensions
Clipper				14.5		Plated, blade ground, file ground, packaged	
A1	Arm Assembly	1		4.1		Assembly tooling	
	001 Actuator arm	1	Apply finger force	2.9	Chrome	Stamped, deburred	2", shaped 0.25" pivot
	002 Pin	1	Convert to large force	1.3	Chrome	Die Cast, deburred	0.13" round
A2	Cutter Assembly	1		7.9		Spot Welding tooling	
	003 Blade arms	2	Guide to cut nail Cut nail, stop motion Release force	3.9	Chrome	Stamped, deburred	2.00 x 0.44 x 0.13 0.13" blade gap
A3	File Assembly	1		2.5			
	004 File	1	Form filing surface File nail	2.3	Chrome	Stamped, deburred	Scored 1.50 x 0.25 x 0.06
	005 Pivot rivet	1	Orient filing surface	0.3	Chrome	OEM	0.19 rivet

(Table heading, upper-left:)

Materials List
Clipper

▼ *Figure 6.3.*
Bill of materials format, including example of a fingernail clipper.

III. TEARDOWN METHODS

We next present several methods that can be used in support of the teardown process outlined above. These include the Subtract and Operate procedure to uncover the function of subsystems and possible component redundancies, the Force Flow Diagram approach to determine assembly functions and possible component combinations, and experimental methods to obtain needed data.

Subtract and Operate Procedure

As discussed in the teardown process, it is important to gather information regarding the function of components and assemblies in a product. Insights must be obtained regarding "what" the product is intended to do, not just the single means by which it performs its function. To complete this analysis, one can generate function structures for competitive products and a product under study.

It is likewise critical to assess the manufacturability of a product. Often, a product is overdesigned. An overdesigned product is one that uses multiple separate components to solve subfunctions that could more efficiently be solved using fewer components or just a single one. If this situation occurs, there is an opportunity for elimination of components that produce the redundancy. Identifying which components are redundant can be the difficult problem.

The *Subtract and Operate Procedure* (SOP) is the following five-step procedure aimed at exposing redundant components in an assembly or subassembly through the identification of the true functionality of each component (Lefever and Wood 1996):

> *Step 1:* Disassemble (subtract) one component of the assembly. Removal of components may occur in any order. However, it may be necessary to remove one or several components in order to remove the desired component. These prerequisite components should be reassembled if possible. If they cannot be reassembled, measures can be used to replicate the function(s) of the missing component(s). These measures are discussed later in this section.

> *Step 2:* Operate the system through its full range.
> This step should test the product through the range of customer needs. After removing a component, the product should be thoroughly tested. For each customer need (structural, ergonomic, kinematic, etc.), the product must be tested to verify the effect due to the removal of the component.

> *Step 3:* Analyze the effect.
> This step is most commonly completed through a visual analysis. However, it may be necessary to use a testing device if the effect of removal is not obvious.

> *Step 4:* Deduce the subfunction of the missing component.
> From Step 3, the function(s) of the missing component may be deduced. A change in any degree of freedom (DOF) during operation should be a major focus. DOF is a critical issue in determining component redundancies and will be discussed.

> *Step 5:* Replace the component and repeat the procedure n times.
> Step 5 is the reassembly of the removed component. Repeating the process n times, where n is the number of pieces in the assembly, allows for the analysis of each component in the assembly. In certain cases, it may be necessary to analyze a product according to subassemblies and groups of components (n combinations exist in the product). This issue is touched on later in this section.

An example of the SOP is given using a simple rotational constraint mechanism. This mechanism is a simple example used to clearly exhibit the SOP.

SOP Examples

Figure 6.4 is an exploded view of a mechanism used to oscillate an arm through a designated angular range, the arm being the source of power for the rotary motion. The arm is connected to a rotary shaft that oscillates, but the range of rotation is constrained by two pins and top-plate slots.

When applying the SOP five-step procedure, it is useful to construct a table of the components. Part number, part description, and the observed effect of operating the assembly without the part are items that should be documented. In addition, deduced component functions and affected customer needs should be documented. For the mechanism above, the worksheet shown in Table 6.1 is completed.

The SOP exposes five parts that are potential redundant components (Table 6.1). These are denoted as "No effect" parts in the worksheet.

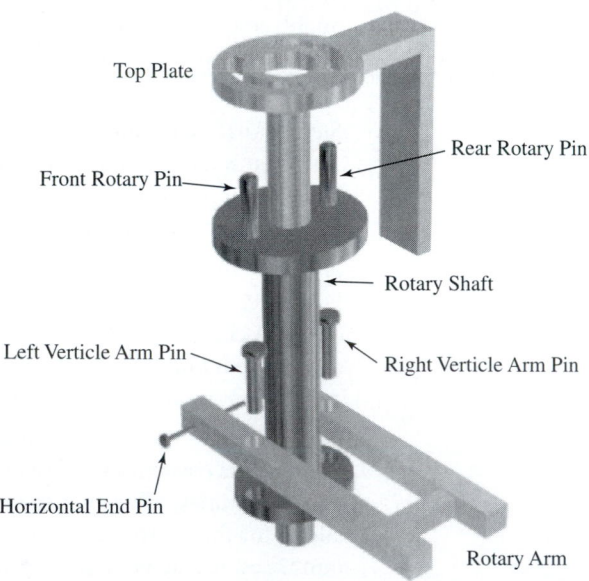

▼ **Figure 6.4.**
SOP example device.

TABLE 6.1.   EXAMPLE SOP DEVICE WORKSHEET

Assembly/Part No.	Part description	Effect of removal	Deduced subfunction(s) & Affected customer needs
A-1	Shaft Assembly		
1	Top Plate	360° rotary freedom	Allow DOF Regulate Motion (Arc)
2	Rotary Shaft	No torque transfer	Transmit Torque
A-2	Arm Assembly		
1	Front Rotary Pin	**No effect**	Allow DOF Support Loads (Durability)
2	Rear Rotary Pin	**No Effect**	Allow DOF Support Loads (Durability)
3	Rotary Arm	No torque transfer	Transmit Torque
4	Horizontal End Pin	**No effect**	Support Loads (Safety)
5	Right Vertical Arm Pin	**No effect**	Support Loads
6	Left Vertical Arm Pin	**No effect**	Support Loads

Examining Figure 6.4, it is observed that only one of the rotary pins is required to constrain the shaft's range of rotation. For the arm pins, only one of the vertical pins (left or right) is required to constrain all DOFs of the arm. This conclusion assumes that large loads are not applied to the arm and that the arm is square, rigid, and tightly toleranced with the shaft. Even if large loads are applied, though, a properly designed single pin should work. Removing the horizontal pin has no effect, as the arm will not slip against the shaft because of the fixed vertical pins. A rule in using the SOP is: If a component has "No effect," then leave it off and continue on to the next component. Hence, with the horizontal pin still removed, one of the vertical pins is removed. When the system is operated, there is still "No effect," as the geometry of the arm and shaft along with the one fixed pin restrict any motion.

Generalizing SOP

A question that must be addressed is: Does the SOP only hold for cases in which the design is overconstrained (as in the example above)? In general, the components that result in "No effect" are normally those that superfluously constrain DOFs that are constrained by another component. However, there exists other types of redundancies. For example, redundant structural components can be found when, if removed, the strength of the system is still within its factor of safety. Aesthetics and ergonomics are examples of other types of redundancies. These are more subjective, but if the customer needs are detailed, they can be discernible.

Overall, classifying components as changing or not changing the DOF of a product when they are removed is the critical determination. This concept leads to the following proposition:

Proposition: Components that when removed cause no change in the DOF or other factors of the design are termed Type 1 redundant components. Components that cause no change in the DOF but do have other effects due to their removal are termed Type 2 redundant components.

1. Type 1 components are always candidates for removal.
2. Type 2 components may be removed if another component can be parametrically redesigned to compensate for the other effects.

Support/Explanation: If a component does not move with respect to another, then the two components can possibly be combined. If no adverse effect is caused, then this is viable and also is a Type 1 redundancy. If adverse effects are caused, then this is viable only if the adverse effects can be compensated and also is then a Type 2 redundancy.

Figure 6.4 exhibits Type 1 redundancies, except for the case of the horizontal end pin (Part A-2-4). In this case, the pin does not affect the DOF when removed. It also does not affect the supporting of loads, assuming that the arm pins have been designed properly. However, its removal does affect the customer need of safety. The horizontal end pin exists to prevent the arm from disengaging if the arm pins fail or are not assembled correctly.

Because of its effect on the customer need of safety, the end pin should be classified as a Type 2 redundancy. A parametric change in the rotary arm is needed if the end pin is to be removed as a component. The affected parameters in this case are the geometric dimensions and shape of the rotary arm. If the arm were cast with a closed end (Figure 6.4, left side of part), and if it were widened only near the brace (Figure 6.4, right side of part), the shaft could be inserted near the brace. It could then be translated to the closed end, and the arm pins could be assembled to join the arm and shaft. This process would remove the redundant component while satisfying the need for safety. Any redesign of this type would have to be judged on the savings of part count and complexity versus the manufacturing and material cost associated with the redesigned arm.

Another example of a Type 2 redundancy is a drawer with a handle that is affixed with a fastener. If the handle is removed, a ledge can be used to pull out the drawer, but the effect is that there is an "Increase in pinch/grip effort." The drawer ledge geometry could parametrically be redesigned to accommodate a hand and act as a handle. The result is an indented or extended geometry of the drawer material. In this sense, the handle is a redundant component, since it does not affect the DOF of the drawer.

Discussion of SOP

The SOP can have limitations depending on the product being disassembled. Many times, the disassembly of a component requires the disassembly of other components. A simple example of a screw joining one block to a fixed block illustrates that to remove one block, the screw will have to be removed as well. Components, such as the screw in this case, can be considered functionally dependent components. This type of component means that the screw's function is dependent on the removed block. Once the block is removed, the function of the screw becomes obsolete, and the operation of the system will not be affected by its absence. In the case where the screw not only holds together one block, but also a second block to the fixed block, the screw would be reinserted before operation, as it is functionally dependent on both blocks.

There are situations in which operation of the product is difficult when a component is removed. For these cases, it is useful to think of a product in a hierarchical manner as shown in Figure 6.5.

Dividing the product into subassemblies and applying the SOP to each subassembly can eliminate the challenging or impossible task of operating the entire system. Components confined to the subassembly can be evaluated without regard to the rest of the assembly. Subassembly components interfacing with components outside of the subassembly should be evaluated as they are integrated. In some cases, this integration may take some insight and creativity. External actuating or mounting tools can be used to resolve operating challenges.

SOP Example II

Consider the electric wok product design discussed in other chapters. The primary market for the product is the young urban dweller desiring to conveniently cook authentic Chinese food. Figure 6.6 shows an exploded view of an electric wok.

Implementing the SOP procedure, each component is removed from the wok, and, if safely possible, the system is operated through its entire range of cooking foods, at a variety of temperature settings. Table 6.2 illustrates the results of this process in the form of a SOP effects table, including DOF information.

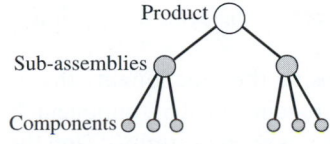

▼ **Figure 6.5.**
SOP hierarchy.

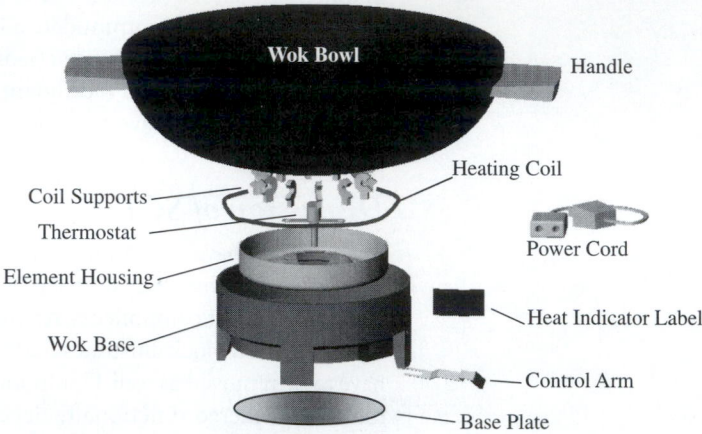

▼ *Figure 6.6.*
Exploded view of an electric wok.

The results in Table 6.2 show opportunities for assembly redesign in parts 002, 004, 005, 006, 007, and 015. The table also identifies the basic functions performed by each component in the wok system. These functions may be connected as a function tree or a function network to represent the actual functionality contained in the wok (Chapter 5). Measurement strategies, new concept generation, specifications, engineering models, and cost may be associated with these functions.

Summary—SOP

SOP is a logical tool in product teardown. It determines component functionality and/or redundancies (or potential redundancies) in an assembly. The process of SOP may be summarized as follows:

▶ *Step 1:* Disassemble (subtract) one component of the assembly.

▶ *Step 2:* Operate the system through its full range.

▶ *Step 3:* Analyze the effect.

▶ *Step 4:* Deduce the subfunction of the missing component.

▶ *Step 5:* Replace the component and repeat the procedure *n* times.

SOP determines redundancies by exposing the functionality that is subtracted from the system resulting from removal of a component. If this functionality is redundant, the design is overconstrained, and the component can be removed. If the DOFs remain unchanged when subtracting a piece, but other functionality is affected, there may exist

TABLE 6.2. ELECTRIC WOK SOP EFFECTS TABLE

Assembly/ Part No.	Part description	Effect of removal	Degree of Freedom type	Deduced subfunction(s) & Affected customer needs
A1	Lid Assembly			
001	Cover	Heat radiates to environment; food splatters		Insulate Energy Flow, Contain Food
002	Cover knob	**No effect** (difficult ergonomics)	2	Import Hand
A2	Bowl Assembly			
003	Wok bowl	Foods falls into base; food directly in contact with coils		Contain Food, Conduct Heat, Direct Food, Heat Food, Enable Cleaning
004	Handles & handle caps	Burn hands	2 (handles) 2 (caps)	Import Hand, Insulate Hand, Transmit Weight
005	Wok stand	Bowl cannot be oriented/ stabilized; heat element is exposed; temp. settings not visible	2 (each redundant foot)	Oriental Cooking, Transmit Loads, Distribute Loads, Insulate Heat, Indicate Temperature Setting, Prevent Skidding
006	Connector prongs	Cannot turn on/off; electricity doesn't flow; exposes electrical current	1 (no connector needed)	Actuate On/Off, Transmit Energy, Protect from Shock
007	PEM studs	Bowl is not connected to base	1 (>1 stud)	Transmit Loads
008	Metal plate	Exposed coils; Safety info. not displayed		Insulate Heat; Indicate Visual Safety Signals
A3	Electrical heating assembly			
009	Electrical cord	No connection to source		Connect to Energy, Transmit Energy
010	Thermostat	Temperature very high, continuous		Regulate Power/Temp., Sense Heat
011	Heating coil	Open circuit; Wok doesn't heat		Convert Electrical Energy to Radiation
012	String	Heating coil is not held in place		Transmit Loads
013	Insulated wires	Temp. rise continuously; no temp. sensing		Transmit Electrical Energy to Sensing
A4	Electrical heating support			
014	Radiation shield	Thermostat heats up; Heat is lower in bowl		Shield Temp. Control, Direct Heat
015	Coil supports	Heating coil misaligned; Housing heats up	2 (>1 support)	Insulate Heat, Direct Heat
016	Element housing	Heating element is free floating; bowl does not heat well		Direct Heat/Radiation (Heat Container)
...	...	...	...	...

opportunities for compensation through parametric variation of remaining components. If these opportunities do not present themselves, SOP is still a useful technique for understanding component functions during a reverse engineering process.

Force Flow (Energy Flow Field) Diagrams

Whereas the SOP is a method for exposing component functionality and opportunities for component elimination, Force Flow Diagrams focus on component combination. This focus, in addition to searching for assembly functions, is accomplished through a modeling of the design followed by an analysis of the model.

Force Flow Diagrams represent the transfer of force through a product's components (Lefever and Wood, 1996). The components are symbolized as nodes using circles (or squares), and the forces are drawn as arrows connecting the components in which the force transfer takes place. Consider the straightforward example of the three-piece paper clip (binder clip) shown in the Figure 6.7.

In this case, the hand transfers the force to each of the lever arms. These lever arms in turn transfer the force to the clip. Representing transfers using a Force Flow Diagram gives Figure 6.8.

The motivation behind constructing a Force Flow Diagram of a product is to map the force flow through a product so that the diagram can be analyzed to help expose opportunities for component/piece combination. Once a model is formulated (Figure 6.8), the first step in analyzing a Force Flow Diagram is to place an "R" on the flows that have relative motion between two components. Once this step is completed, the diagram can be decomposed into groups separated by "R"s. Having performed this separation, the following proposition is applied.

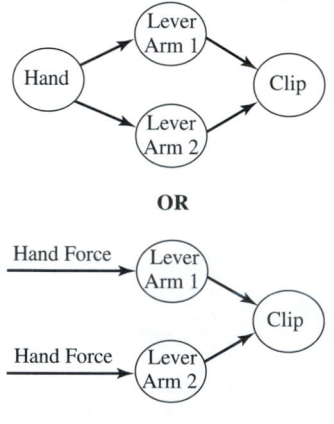

OR

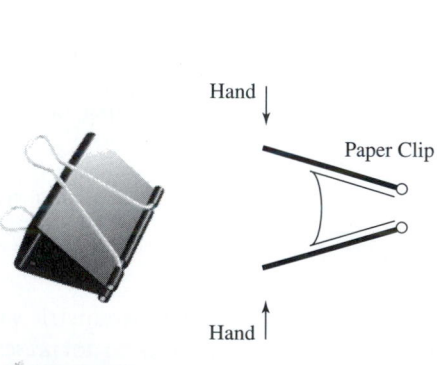

▼ **Figure 6.7.**
Binder clip schematics.

▼ **Figure 6.8.**
Force Flow Diagram of binder clip.

Proposition: The components comprising "R" groups are candidates for combination if not prohibited by material or assembly/disassembly issues. Combination between a member of one group and a member outside of the group requires more complex redesign.

Support/Explanation: Two parts move with respect to each other in a product to provide a function(s). Eliminating a movement means eliminating a function, and redesigning the product functional structure is difficult, requiring an additional subsequent redesign of the concept implementing the new function structure. Therefore, adaptive redesign is required to combine members outside "R" groups that move with respect to each other. Components within "R" groups do not move with respect to each other and so do not require functional redesign unless prohibited by functions carried out by the material or assembly.

Referring to Figures 6.7 and 6.8, let's assume that the arms of the binder clip do not need to rotate. The diagram of Figure 6.8 would thus not include a relative motion, that is, an "R" label on any connect between components. Under this assumption, the lever arm 1 could be grouped and combined with the clip. An identical conclusion exists for lever arm 2. Because no relative motion exists between these two groups, they in turn could be combined. The result is a one-piece binder clip, as shown in Figure 6.9(a). It could be produced as a single mold or sheet metal part, reducing the part count by two thirds. This result is really the novel binder clip design that has existed since 1899, except for

Integral Binder Clip

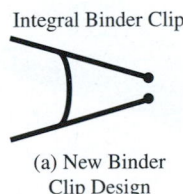

(a) New Binder
Clip Design

(b) Common
paperclip

(c) Variant paper
clip products

▼ *Figure 6.9.*
Binder/paper clip designs.

the wider "mouth" to hold a greater quantity of paper. Figure 6.9(b)–(c) shows examples of the original paper clip geometry in addition to some unique variations.

Of course, the new binder clip design shown in Figure 6.9(a) does not satisfy the full set of customer needs. An important customer need is for the clip to be compact after binding the paper. Because of the long lever arms (to satisfy ease of use), the design principle of "separate in space and time" is implemented. When binding the paper, long lever arms are needed, increasing the size of the clip. After binding the paper, a flat profile is needed to avoid snagging items or taking up excessive space. Thus, joints are added between the lever arms and the elastic clip, resulting in "R"s being added to the Force Flow Diagram (Figure 6.8). Three components are then needed to maintain the relative motion requirement.

Application: Swingline™ Stapler Example

Figure 6.10 illustrates a Swingline™ stapler product. The stapler is composed of twenty-four parts. Figure 6.11 is a schematic diagram showing a side-exploded view of the stapler. Let's apply the force flow approach to this stapler product.

The activity of using any product consists of a set of operations. For instance, the stapler has three sets of operations. Each set has a forward operation and a corresponding reverse operation. The three sets of operations for the stapler are as follows. 1a) When pushed on the top, the stapler extrudes a staple through the front side of a set of papers and bends the staple up into the back of the last page(s). 1b) When released, the stapler springs back into a ready position. 2a) The impact plate can

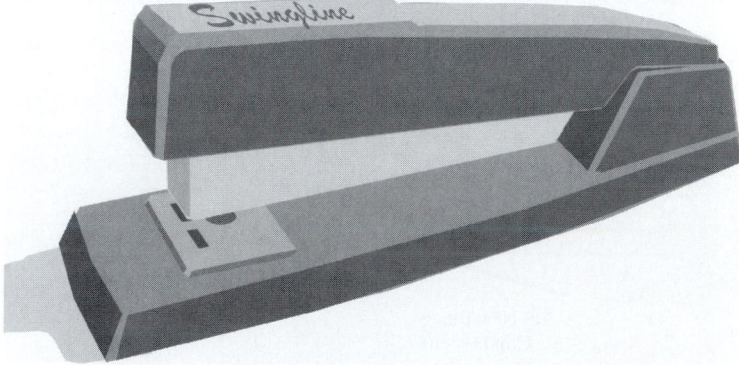

▼ **Figure 6.10.**
Rendering of a Swingline™ stapler product.

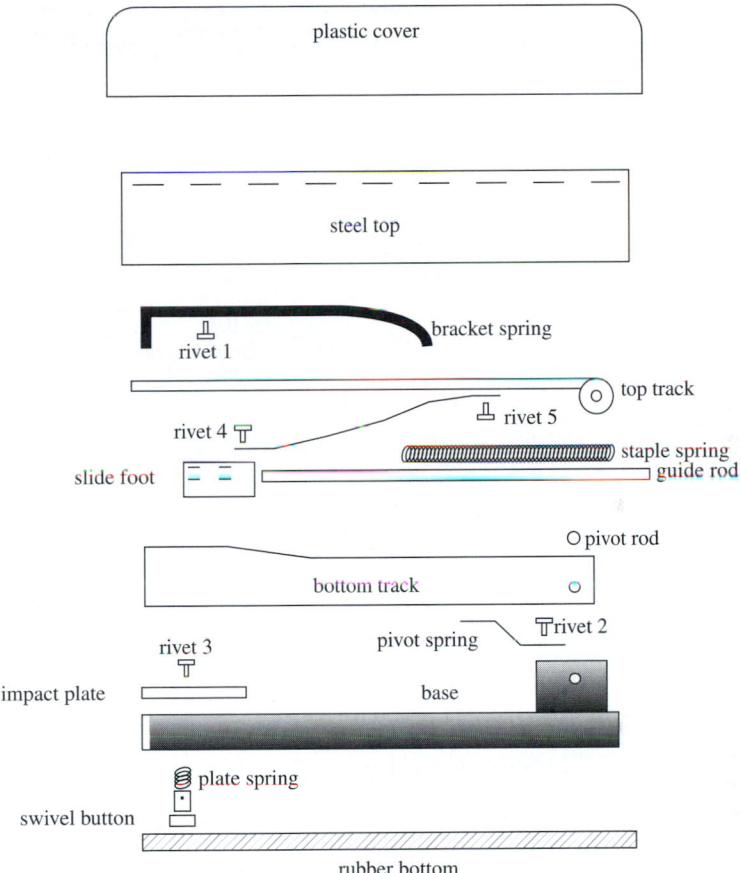

▼ *Figure 6.11.*
Exploded side view of stapler.

be pushed up and rotated 180° using the swivel button and 2b) then released back into the base so staples can be bent inward or outward into the page. 3a) To insert staples or remove failed staples, the top track can be disengaged with the bottom track by lifting up on the top and then 3b) reengaged when completed.

Force Flow Diagrams of each of these operations can be individually constructed. For all three, as should be expected when springs are loaded then unloaded, there will be a great deal of redundancy in the diagrams. The flow direction simply reverses once the spring unloads.

For this example, operation (1) is chosen. The hand provides the external force that pushes down on the top. This force is transferred to the spring-bracket, to the top track, to the bottom track, through the pivot

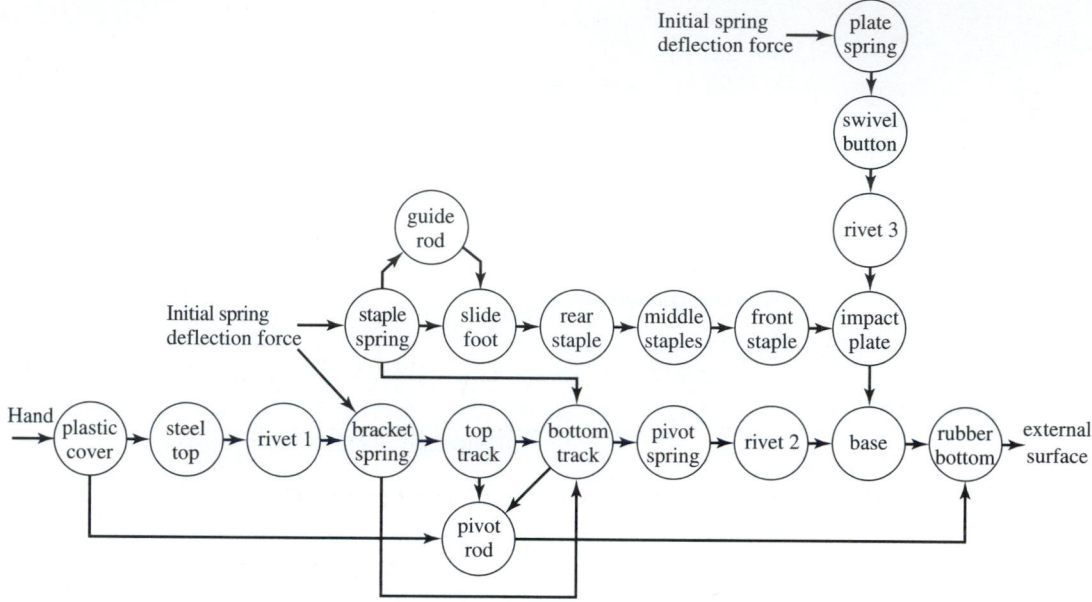

▼ Figure 6.12.
Swingline™ stapler Force Flow Diagram.

spring into the base, and out to an external surface. When the pivot spring is deflected sufficiently, the upward reaction force overcomes the downward spring-bracket force. This force causes relative motion between the spring-bracket and the top and bottom tracks. This relative motion allows the spring-bracket to begin to push a staple out through the pages and onto the impact plate. In the meantime, the staple spring transfers force to the slide foot and onto the staples. This force holds the front staple in place and moves the next staple up (horizontally) after the front staple has been displaced.

Force flow occurs through all components except three. The connectband and rivets 4 and 5 are integral components for operation (3) but are not needed for operation (1). Figure 6.12 shows the Force Flow Diagram generated for the Swingline™ stapler. Placing "R"s between relative moving components and grouping them leads to Figure 6.13.

Considering Group 1, it can be conceived that the plastic cover, the steel top, and the rivet could all be combined as one component. This component would be a plastic top with a snap for the bracket-spring to connect. For Group 2, combination of the staple spring and slide foot could take place by using a spring that could push the staples itself. Moving to Group 3, the top track and bottom track could easily be

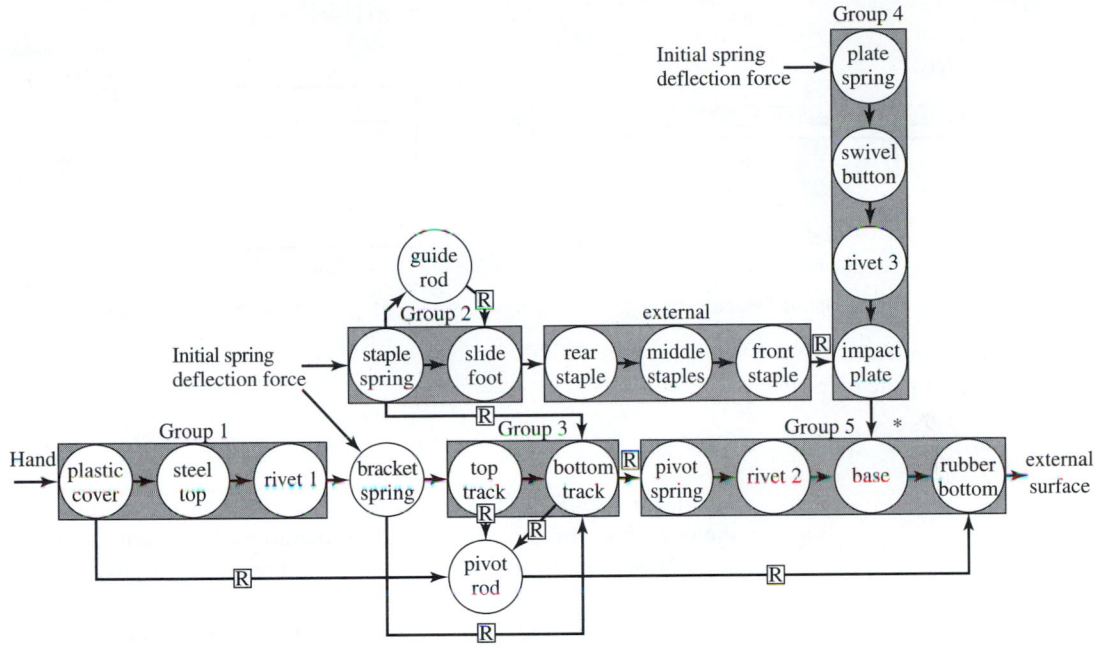

▼ *Figure 6.13.*
Swingline™ stapler evaluated Force Flow Diagram.

combined. The top track provides the cover for the staples and a surface for the spring-bracket to force down on. If the bottom track is enclosed, this combination would eliminate the need for the top track. However, the reason an asterisk is placed between the top and bottom tracks in Figure 6.13 is because these components have relative motion on a Force Flow Diagram for just one of the other operations. In this case, it is operation (3): the input or removal of staples through the disengaging of the two tracks. This analysis does not mean that it is impossible to combine the top and bottom track, as there might exist a design that would eliminate the need for disengaging the tracks.

Examining Group 4, the entire group consisting of the plate spring, swivel button, rivet 3, and impact plate could be visualized as a novel single component, embodied in Figure 6.14. For this redesign, an integral leaf spring (replacing the compression plate spring and rivet) is manufactured into the impact plate stem. As the component is assembled onto the base through a stamped hole, the spring collapses toward the base stem. After the edges of the spring exit the hole, they spring open to their equilibrium state. The impact plate can then be pushed upward and rotated by deflecting the leaf spring.

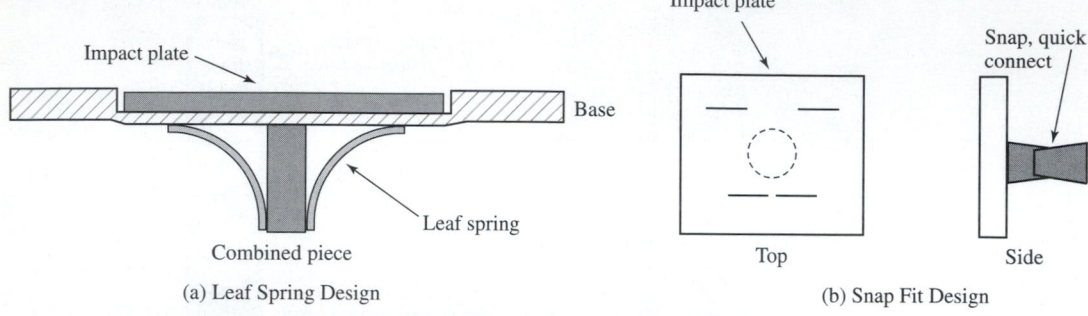

Impact plate

Base

Combined piece

Leaf spring

(a) Leaf Spring Design

Impact plate

Snap, quick connect

Top

Side

(b) Snap Fit Design

▼ **Figure 6.14.**
Impact plate redesigns.

Again, for Group 4, the asterisk between the impact plate and base shows relative motion for a different operation. Disregarding operation (2) would lead to the conclusion that there is no relative motion between any of the parts, and that they could all be integrated into the base, which is not the case. Instead, a novel component, such as the integrated part shown in Figure 6.14(a), is needed, maintaining the relative motion desired in operation (2). Alternatively, the redesign shown in Figure 6.14(b) might be adapted, where the spring and pivot joint functions are replaced by a quick connect that provides for orienting the impact plate for inward or outward bending of the staple prongs.

The final group is Group 5. It is feasible that rivet 2 and the base could be combined, and the pivot spring could be press fit onto a knob. The pivot spring and the base, however, require different materials as do the base and the rubber bottom. These components would be difficult to combine.

Each example combination of components given above is a redesign that could be undertaken without affecting the functionality of other components. Attempting to combine components outside of their groups means combining pieces that have relative motion. This is a higher level of redesign that would likely cause functional alterations in addition to the change in physical form. To visualize this, consider combining the stapler's components that are of different groups. It is not trivial proposition, but the force flow approach points out the opportunities. The ultimate extreme is shown in Figure 6.15, where the stapler has been completely function shared and reduced to a single insert-mold component (Ananthasuresh and Kota, 1995). The return springs are molded as integral plastic components. The only metal parts, the staple actuator and the impact plate, are inserted in the mold during fabrication. No assembly is required.

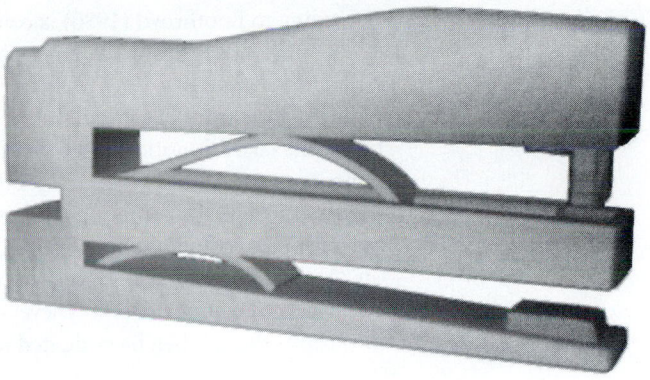

▼ *Figure 6.15.*
Stapler product converted to a single insert mold, with compliant
components (Ananthasuresh and Kota, 1995).

Summary—Force Flow Diagrams

The primary intent of Force Flow Diagrams is threefold: (1) identify the
functions of a sequence of components (subassemblies); (2) identify
potential avenues for component combination; and (3) provide an initial force flow model for subsequent, more-detailed design analysis.
Eight steps comprise the process for executing a force flow analysis:

1. STEP 1: Identify the primary force flows (or energy flows) transmitted through a product.

2. STEP 2: Map the force (energy) flow from the external source
   through each component of the product until the flow exits to
   ground. Special care should be exercised to identify when force
   (energy) flows split (become parallel) through components.

3. STEP 3: Document the result in a Force Flow Diagram, where the
   nodes are components and the connections are forces (energy terms).

4. STEP 4: Analyze the diagram, labeling relative motion between
   components with an "R."

5. STEP 5: Decompose the diagram into groups separated by "R"s.
   Box these component groups.

6. STEP 6: Deduce the subfunctions and affected customer needs for
   each group.

7. STEP 7: Develop creative, conceptual designs to combine the components in each group.

8. STEP 8: Repeat for each force (energy) flow.

According to Boothroyd (1980), a component in a product must exist if: (1) the part moves relative to other parts, (2) the part must be made of a different material than other parts, or (3) assembly or disassembly of other required parts would be impossible if the part is not separate. Force Flow Diagrams take this notion a step further by actually showing which components can be combined for the product as a system. Modeling the force transfer through the components allows the complete product to be analyzed for piece combination. Grouping the products according to criterion (1), relative motion, highlights the sets of components that can potentially be combined for low-level adaptive design. The components of the groups should then be evaluated with criteria (2) and (3).

Measurement and Experimentation

Motivation: The Need for Measurement in Product Design

Measurement in engineering design comes in several forms, notably *specifications* and *benchmarking*. A product's specifications provide a precise and measurable description of what the product must accomplish. Each specification consists of a *metric,* denoting the type of measurement, and a *target value* in the form of a number or range. The product target values are the performance levels that the designer must achieve. The target values are generally determined by examining the uses the product will be put to or by examining performance of other similar products. This second approach is called *benchmarking*. Most design methods initially concentrate on various aspects of acquiring product information, and ultimately the data is converted into a set of quantifiable specifications. These metrics provide a clear goal for evolving or developing a product in the form of target values developed by determining state-of-the-art in product benchmarks.

The importance of measurement in design is illustrated by noticing how design methods are concerned with ensuring that both the customer needs and functionality of the product are quantifiable. Miles (1972) states, in reference to determining the functionality of products in value engineering: "To provide more benefits, functions are named whenever possible using a verb and noun that have measurable parameters." In recent years, researchers have stressed the importance of ensuring that customer needs are integrated into product design, which emphasizes the significance of selecting appropriate and quantifiable specifications using a more systematic method (Hauser and Clausing, 1988; Otto, 1996).

Measurement has evolved quite dramatically over the past 100 years and is a very broad field of study (Little, 1997). Many volumes of analysis could be devoted to the study of measurement theory and techniques; however, we must focus on the application of measurement for the teardown of products. Because the final goal is to develop a set of metrics and measurement devices, the following questions should be answered first: Why/how is measurement important during the disassembly procedure? What are the important criteria of measurements? What are the methods for selecting measurement equipment? How do we determine what to measure? These questions can be answered by investigating the fields of forensics, experimental methods, and current reverse engineering procedures.

Advanced Method: Product Development Forensics

The word *forensic* is defined in part "as pertaining to, connected with, or used in courts of law or public discussion and debate" (*New Webster's Dictionary and Thesaurus*, 1992). Forensics and the product teardown are similar in that both have a common goal of obtaining the pertinent information about the subject using systematic procedures, so it is worth our time to explore forensics and apply the methodology to product development.

Forensics has been applied to a variety of fields, including anthropology, engineering, medicine, and accounting. Each area requires organized gathering and evaluation of data to present a testimony before a court of law. Because of the inherent necessity in forensics for an accurate and unwavering study, measurement has a crucial role in each field's forensic procedure.

Anthropology Forensics: Forensic anthropology is a branch of physical anthropology that deals with the identification of skeletonized remains known to be, or suspected of being, human (Stewart, 1979). Instead of dissecting a consumer product, a forensic anthropologist excavates human remains to identify the human and the possible cause of death. Three issues are discussed in the following sections concerning anthropology forensics that relate to reverse engineering and product teardown. The procedures used in forensic anthropology and the reverse engineering process are very similar and show the importance of the common steps. Secondly, the importance and procedure for removing bias during the extraction of data reemphasizes the significance of reverse engineering. And finally, the concept of taking different sets of measurements to evaluate distinct features of the subject can be adopted in the reverse engineering process.

Procedure for Excavating Human Remains: The objective for the anthropologist is to remove human remains so that lab tests can be completed on each of the pieces to conclude identity and/or cause of death. The procedure is very analogous to disassembling a consumer product:

1. Define and record outline of grave.
2. Slowly remove grave fill.
3. As bones are encountered, they should be left in place and surfaces brushed clean of soil.
4. Photograph w/ markers for depth.
5. As each bone is removed attach identification tags.
6. Add PVA acetone solution to fragile bones.
7. Take sample of soil.
8. Conduct lab tests on bones (Chamberlain, 1994).

The *concrete experience* of the product teardown process follows this procedure similarly:

1. Plan product disassembly.
2. Execute product disassembly.
3. As each component is removed, photograph each component and the remaining product hull.
4. Create BOM, exploded view, and parametric description.
5. Execute SOP.
6. Experiment with product components.

Both procedures start with a preparation step and then execute the disassembly/extraction. Next, both procedures record information about each component and its location, whether by identification tags and photographs or a BOM and exploded view. Finally, both approaches end with experimentation or lab tests. Comparing the forensic procedure to disassembling a product emphasizes the importance of first taking measurements of the product as a whole, identifying the location of each component, listing the components, and finally conducting measurements on each component. Note that measurements play an important role throughout the excavation process: before it starts, during the removal, and after its completion.

Importance of Avoiding Bias: T. Stewart, a highly respected forensic anthropologist, emphasizes the importance of removing bias and how easy it is to become biased. Stewart (1979) states, "I made it a rule in participating in forensic cases always to avoid so far as possible receiving information about them until I had examined the bones and committed my determinations to writing." Similarly, the teardown process must pride itself on providing an unbiased perception of the product through investigation, prediction, and hypothesis before disassembling and experimenting with the product components. Creating biases about consumer products is an equally difficult problem, which stresses the importance for providing systematic methods for quantifying and determining the important aspects of a product.

Measurement Technique: One final idea developed from forensic anthropology is the concept of taking specific measurements related to particular aspects of the subject. For example, one set of measurements is used to determine the age of the human and others for gender, race, stature, and others (Bass, 1987). Similarly, if a product can be decomposed into a complete set of elements such as functions, a set of measurements could be defined for each function and provide a means for measuring consumer products. For example, to measure all aspects of a product, one could first list the functions of the product from the function structures developed before (prediction) and after disassembly (actual). Then one could focus on each function, taking the appropriate measurements as opposed to attempting to determine the measurements from investigating the product or its components. Because the functional analysis of a product ensures a complete representation of the product, including customer needs, operating ranges, and flows (energy, material, or signal), a more accurate set of measurements can be developed by examining each function.

Forensic Engineering

Engineering forensics involves determining whether or not a product failed and why. This field is similar to redesign because the designer is interested in the faults of the product so that it can be improved; however, the legal aspects change the required procedures by greatly prohibiting removal or movement of the subjects parts until extensive measurement and documentation have been finished. The forensic procedure for obtaining data is similar to forensic anthropology in that the subject is investigated as a whole and then examined carefully piece by piece; however, in engineering forensics, the importance of first performing nondestructive tests is highly stressed. "Field testing should be arranged in a hierarchy beginning with techniques that do not contaminate the evidence and ending with those that do" (Brown, 1995).

This concept of nondestructive testing should be applied to the measurement techniques used in redesign methods in two areas. The most important is to ensure that the procedure for measuring the product saves the destructive testing for last. Second, the designer should attempt to remove or limit the amount of destructive testing during the measurement selection process. Another option is to have several of the same products to test, although this course of action can prove to be costly. The removal or limitation of destructive testing is very important; otherwise, future measurements may be impossible or less accurate.

Experimental Methods

The area of forensics has given us insight into overall concepts and procedures for taking measurements, such as removing bias, importance of taking measurements throughout the disassembly, using nondestructive techniques, and developing a formal method of measurement. In addition, we must discuss experimental planning, the criteria of measurements, and current methods for selecting measurement systems to determine the current state of the art.

In planning an experiment, one must consider both what to measure and how to measure it. Holman states:

> The key to success in experimental work is to ask continually: What am I looking for? Why am I measuring this—does the measurement really answer any of my questions? What does the measurement tell me? (1989)

In developing a measurement method, we must focus on *what we are looking for*. This statement can seem trivial; however, great quantities of resources can be spent on measurements that do not reveal any relevant information. Therefore, as mentioned earlier, it is important to develop an approach for determining the crucial measurements of the product.

The second problem is determining how to make the measurement. This problem can be further posed as follows: *Does the measurement really answer any of my questions?* In addition to making sure the measurement provides pertinent information, it is equally important to make sure resources are not wasted on obtaining unnecessary information. For example, if one needed to measure a temperature to within $\pm$ 1°F, it would be unnecessary to obtain a measurement to within $\pm$ 0.001°F. Although the extra accuracy still answers the intended question of the measurement, it is unnecessary. Therefore, it is important to develop a technique that determines the exact specifications of the measurements to ensure that both the measurement answers a needed question about the product and an unnecessary measurement tool is not purchased.

TABLE 6.3.   CRITERIA FOR MEASUREMENT DEVICES

No.	Criteria	Explanation
1	Accuracy	Deviation of the reading from the actual value.
2	Range	The minimum to maximum possible readings
2	Repeatability	Range of variation in repeated measurements.
3	Dynamic Accuracy	Frequency range over which measurements are accurate.
4	Calibration	Comparison of instrument with a known input source.
5	Mass, size, and power	Constraints on available power and desired physical limits.
6	Safety	How safe is the device to the user throughout its operating range?
7	Universality	Ability to interface with and measure a variety of aspects of products.
8	Cost	Purchase price or lifetime operating cost of device.
9	Output Requirements	Visual display, data recorder, etc.
10	Ergonomics	Ease of use.
11	Non destructiveness	How and to what extent does the measurer modify the measurand?

Measurement Criteria: A variety of sources offer information regarding criteria for measurement devices, including literature on designing sensors/transducers, experimental methods, and the study of measurement systems. To insure a thorough list, the criteria may be combined from each of the three areas starting with "general criteria for selection of a transducer" and adding additional criteria found in the other sources (Table 6.3; Norton, 1989; Holman, 1989; Bentley, 1983).

To develop high-quality experiments, each of these criteria must be investigated and used in the selection process; however, universality, range, and accuracy are critical to the application of a product teardown. Creating a toolbox (set of measurement tools) that can measure a wide variety of products requires devices having both a large accurate range and the ability to interface with a variety of measurands.

Current Reverse Engineering Procedures

So far we have determined the important measurement criteria and discussed the methods for selecting measurement equipment by examining experimental methods. We have also answered the question as to how measurement is important during the disassembly process by investigating forensic procedures, which leaves the remaining question: How do we determine what to measure? Once a method is developed for determining what to measure, appropriate measurement equipment can then be selected. Developing a systematic approach for selecting crucial measurements for a given product or product domain is paramount. This process elicits the engineering specifications for a product (as documented and implemented in Chapter 7).

Metrics can be determined by quantifying the input and output of each subfunction in a product (Chapter 13; Otto and Wood, 1996). A valid measurement method should provide a systematic procedure so that any designer would arrive at the same specifications given the same set of background information, such as customer needs and functional analysis. For our purposes, we adopt the view to "contemplate each need" by "quantifying the input and output of each subfunction in the product," because they represent the customer needs.

Measurement Method

A measurement method for product teardown requires four steps, starting with selecting the product domain and ending with actually selecting the measurement tools for the toolbox (Figure 6.16). Before this procedure is initiated, the user should analyze the set of functions needed for the product and those of its competition and product family. The first step is then to select the product domain in which the measurements will apply. The second step involves determining which subfunctions to include in the measurements. The third step is to take the chosen subfunctions and determine how each one is measured. This step includes how to collect the data and organize it for the following step. The last step involves selecting the measurement tools from catalogs using a worksheet or designing a specialized tool or alternative measurement technique. The important issues and development of each step are discussed below with guidelines on execution.

Step 1: Select a Product Domain—The first step in carrying out measurement during teardown is to determine the product domain. Two important issues to consider when selecting the product domain are the similarity of products within the domain and the size of the domain. The similarity of products is achieved by selecting a group of products that have common functional characteristics. This selection can be accomplished in a variety of ways, including selecting a category of products such as yard tools or small electronic devices from product studies (e.g., *Consumer Reports,* 1997). Alternatively, a product hierarchy may be developed that classifies different products according to the functions they solve. Because these products will have analogous solutions, the hierarchy can provide an excellent method for selecting groups of products with similar measurement requirements. An example product hierarchy classifies products by following the primary input flow

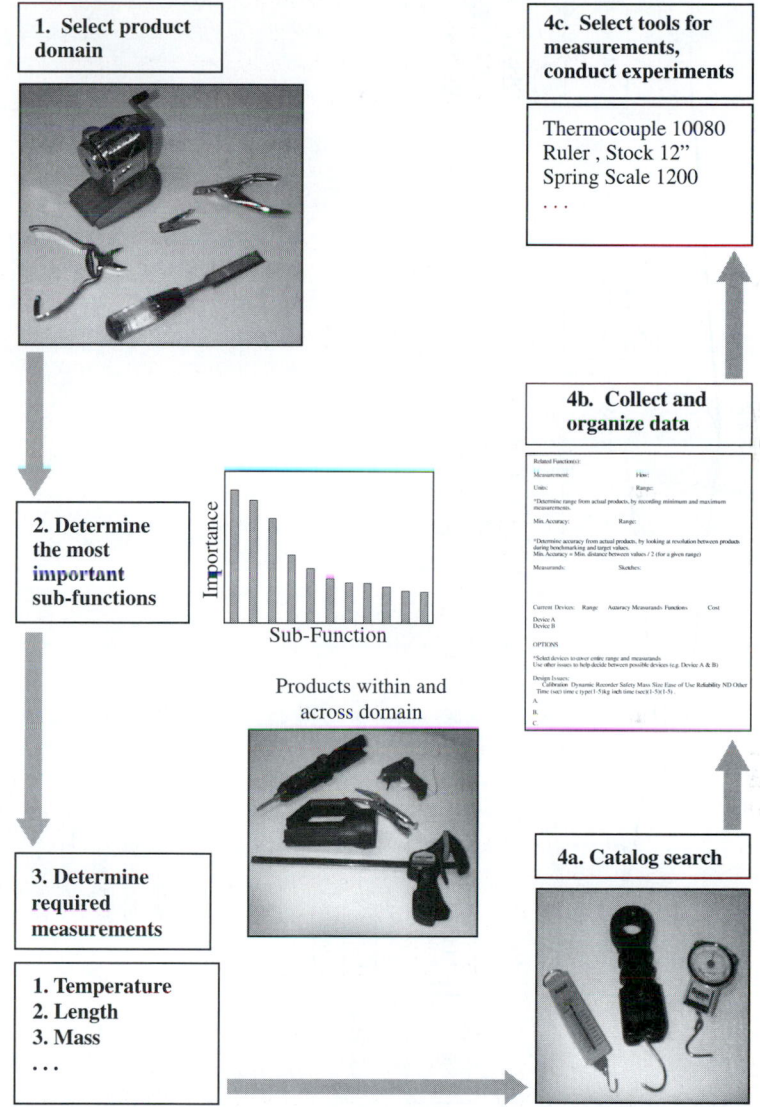

▼ Figure 6.16.
Measurement method for product teardown.

through the product and develops categories such as electricity-heat-material-solid, also known as the "cookers" (refer to Figure 6.17). This class includes products that convert electricity to heat and transfer the flow, *heat*, to a solid material. Example products for this class are an iron, electric wok, or a slow cooker, which have similar primary

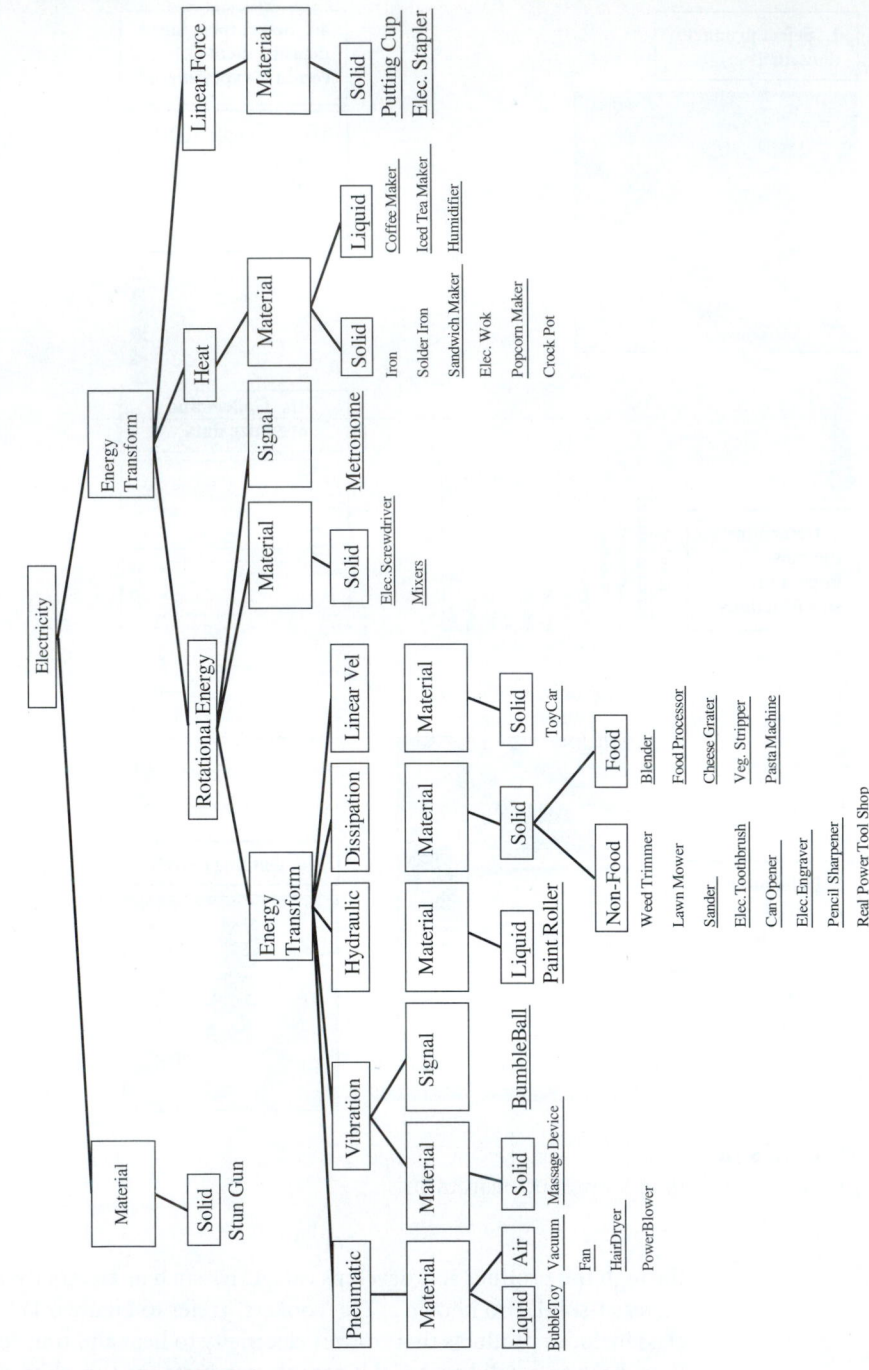

▼ *Figure 6.17.*
Product hierarchy for products with electricity as primary input flow.

TABLE 6.4.   LIST OF POSSIBLE MEASUREMENTS FOR THE REMOVE SOLIDS SUBFUNCTION

Remove solid	Units
Removing Device Geometry	m/deg
Removing Device Material Properties	Hardness
Depth of Cut	cm
Size of Cut	cm
Material Removed per Unit Time	gm/min
Removing Device Energy	rpm, N, N-m
Surface Roughness	μm
Angle of Deviation	deg
Percentage of Removal	%

functions and, therefore, for this approach, will have similar measurement requirements.

The second issue to consider involves the actual breadth of the domain. The smaller and more defined the domain, the more similar the products will be. By having similar products with related primary functions, chosen measurement devices will relate more directly with the products, increasing their effectiveness.

Step 2: Determine the Most Important Subfunctions: The purpose of this step is to set priorities for the measurement effort. Customer needs should first be assigned to important functions for the product(s) being considered, where more than one customer need may be assigned to each function. The importance level of each customer need should then be used to weight each function. The higher-weighted functions become the focus of measurements and experimentation.

Step 3: Determine Necessary Measurements: To facilitate the determination of measurements for a specific subfunction, a database of common measurements related to product functions can be developed. For example, Table 6.4 shows a list of measurements for the "remove solid" function. This database is developed from studying how the function was measured in sixteen products having a remove solid subfunction, such as a power sander.

Given this example, the first phase in determining the crucial measurements for a product domain is to collect measurement information for the selected subfunctions from Step 2:

▶   Each primary subfunction is studied individually to determine how the function was or could be measured in products within and

Customer Needs	Weight	Importance	Categorized Sub-Function
Easy to open	5	1+4+5+5	Import Human Force
Easy to close	4		
Easy to access cooked food	5		

▼ *Figure 6.18.*
Sandwich maker customer needs for the "import human force" subfunction.

outside the selected product domain. Since in most product studies metrics are related to the customer needs, literature and Web sources provide a useful starting point.

▶ The metrics, used to measure each function, are recorded. It is important to record the units measured, the range of measurement values (target, actual, benchmark values, etc.), what was actually measured (e.g., in the case of a transmission, input torque vs. output torque), and the minimum required accuracy for each relevant metric. The accuracy required for each measurement must be sufficient to distinguish between the product and any benchmarks or target values. Therefore, the minimum accuracy is determined by taking the resolution between products during benchmarking (target values for each metric) and dividing it by two.

To illustrate this first phase, let us consider the function "import human force." Starting with an electric sandwich maker, for example, we study how the function is measured by first observing how the customer needs may be correlated with the function. If we determine which customer needs are addressed by the product function, we can add the importance levels of these needs to arrive at a quantified correlation. The product function "import human force" receives a score of 15 for the sandwich maker from three customer needs: easy to clean, easy to close and easy to access cooked food (Figure 6.18). Now we investigate how each customer need was measured for the original product. For each customer need, the metric is recorded, along with the range between target benchmark values, the units, and the minimum accuracy required (Table 6.5). For example, the first metric "human force required" from the customer need "easy to open" was measured in Newtons (N). The force required to lift the upper housing, benchmark, and target values were 6, 3, and 4 N, respectively, which produces a range of 3–6 and a minimum accuracy of $(4 - 3)/2$ or ± 0.5. To complete this phase, data must be collected in a similar fashion for additional products in develop a thorough list of metrics for various products within and outside the product domain.

TABLE 6.5.   SANDWICH MAKER METRICS FOR THE "IMPORT HUMAN FORCE" SUBFUNCTION

Metric	Range	Units	Min. accuracy
Human force required	3.0–6.0	N	±0.5
Magnitude of force amplification	2:1–2.5:1		±0.1:1
Mass of upper housing	.45–.75	kg	±0.025
Human force required	70–118	N	±13.5
Human force required	20–25	N	±2.5
Human force required	20–36	N	±3

The second and final phase of this step is to organize the data to aid the selecting of measurement devices. After collecting measurement information for each product function, the data must be reduced to a more concise form. The final organization will list each required measurement with range and accuracy for the product domain.

> *Procedure:* Each metric (e.g., human force required) is studied individually to locate all other related measurements within the function and across the other primary functions within the product domain. Then, the overall range and the minimum accuracy from all the related measurements are determined.

For example, for the "import human force" subfunction, if there were no other force measurements in the other primary product functions, the final measurement would be a force measurement with a range of 3–118 N and accuracy of ±0.5 N. Because an accuracy of ±0.5 N could be costly to obtain, the accuracy and range information should be segmented into several categories or subsets. For example, from 3–50 N, an accuracy of ±0.5 N could be required, and the range of 50–118 N could require an accuracy of ±13.5 N. This flexibility can decrease the accuracy requirements of the measurement tools; however, it can increase the quantity of tools. With a list of what measurements are required for the product domain, we now proceed to the final step of actually selecting the measurement devices.

Step 4: Selection of the Measurement Devices: The next step of measurement for product teardown is not trivial. Selecting measurement tools from catalogs, which satisfy a specific set of criteria, involves many issues. In addition, if a specific measurement device is unavailable, nonexistent, or too costly, either a new instrument must be designed or an indirect measurement technique must be applied.

There are three primary catalog design issues related to selecting measurement tools: use of different catalogs, uncertainty of data, and range

Related Function(s):				
Measurement:		**Flow:**		
Units:		**Range:**		

* Determine from actual products, by recording minimum and maximum measurements

Min Accuracy:		**Range:**	

* Determine accuracy from actual products by looking at resolution between products during benchmarking and target values

Measurands:		**Sketches:**	

Current Devices:	Range	Accuracy	Measurands	Functions	Cost
Device A					
Device B					
Options:					

* Select devices to cover entire range and measurands. Use other issues to help decide between possible devices

Design Issues:

	Calibration	Dynamic	Recorder	Safety	Mass	Size	Ease of Use	Reliability	ND	Other
	Time (sec)	time C	type	(1-5)	kg	inch	time (sec)	(1-5)	(1-5)	-
A										
B										
C										

▼ *Figure 6.19.*
Worksheet for choosing measurement tools.

and accuracy options. In addition, we have the design parameters, including universality, accuracy, range, calibration, nondestructiveness, and cost. A worksheet is developed that incorporates each of these issues (Figure 6.19). Let us first discuss how the worksheet includes each issue and criterion and then the procedure for its use.

First notice that each of the twelve criteria is included on the worksheet. Universality (measurands), accuracy, and range are considered primary criteria because the measurement tool must meet these basic requirements before it can be considered. The other nine criteria are used for determining which measurement tool to select. Secondly, considering catalog design issues, the worksheet allows the use of different catalogs by providing space for additional design issues/specifications. Therefore, as different catalogs are used, new specifications can be recorded for the measurement devices. In addition, the issue of uncertainty of data is resolved by keeping the uncertain (soft) data such as safety or ease of use in the secondary criteria, which does not affect the selection of other measurement devices. Finally, the range and accuracy issue is captured by the universality criterion that is used as a basic requirement for the measurement tool using hard data from the catalogs.

The procedure for the worksheet is rather simple. A separate worksheet should be filled out for each measurement, decomposing the procedure into three steps: import data, collect data from catalogs, and choose measurement devices.

▶ The data from Step 3 are transferred onto the worksheet, including the range and accuracy required.

▶ The different types of measurands the device must measure are recorded. This information can be obtained from revisiting the initial data recorded in Step 3 discussing what was actually measured. For the example, products were studied to determine how the human force was imported. They were categorized into three types: handles, levers, and buttons. The measurement tool(s) selected should cover all of the measurands listed.

▶ Applicable measurement equipment is then located in catalogs. All the available data for each measurement device are copied onto the worksheet: range, accuracy, measurands it can measure, safety, mass, and so forth must be recorded. If additional information is given, it is recorded in the space allotted for other design issues.

▶ The final step is to actually select the measurement devices using the data on the worksheet. The worksheet provides a means for giving the designer as much information as possible for selecting the tool. The intent is to allow the designer to quickly study the organized data and make a decision. For the human force example, a stock 10 N scale and a fish scale were selected because they cover the range and accuracy requirements and cost less than the other options (Figure 6.16).

After each measurement device has been selected, the tools are purchased. Alternatively, measurement equipment is designed when devices do not exist or are too costly.

Once the measurement equipment is obtained, an experimental plan is developed for the required measurements. This plan is then executed, considering appropriate replicates and control of the experiment. Later chapters discuss methods for planning and executing an experiment.

IV. POST-TEARDOWN REPORTING

We summarize the results of teardown activities into some key documents. These documents include a disassembly plan, an indented BOM, exploded views, and an actual function structure (model) of the product.

Disassembly Plan and BOM

Two documents should be created during the teardown of a product. These documents are the disassembly plan and an indented BOM. A disassembly plan documents when a product was disassembled, who disassembled it, and a step-by-step plan for disassembling the product. The plan includes a number of entries, including the step number, a description of the teardown task for the step, the required tools to perform the step, and the access direction needed for disassembly (with respect to a defined reference frame). The tools and access directions aid in rating the difficulty of disassembling (and, although not 1-to-1, assembling) the product. The overall plan can be used to rate the assemblability of the entire product, and it can be used to provide a guide for reassembling a product (so "no parts are left over").

A template for a disassembly plan is shown in Figure 6.20. This template is incrementally developed as a product is disassembled. A few steps are planned and executed, recording the information as the steps are performed. The process repeats until the product is fully disassembled or until a destructive (irreversible) step is needed that will prevent subsequent reassembly.

Notice in Figure 6.20 that additional information is included in the disassembly plan. This information includes any measurements taken on subsystems and components, observed manufacturing processes of the components, and observed physical principles that are applied by the components. These data aid in understanding the product, especially for later modeling efforts and for comparing to predictions by the design team.

In addition to a disassembly plan, a BOM is developed for a product. A BOM, as shown generally in Figure 6.21, provides a means to record vital structural, physical, and functional information about a product. This information includes the labels chosen for each assembly and the components in the assembly, quantity of each component, descriptions of the components, the components' input and output flows and functions, the component colors, physical data (size, material properties,

Product Dissassembly		
Project Name		
Engineer(s):		
Date:		

Known Desired Information:

Disassembly Plan:

Step	Task	Needed Tools

Measurements:

Comparison with Predictions:

* could be components, features, physics, functions

▼ *Figure 6.20.*
Data template for disassembly and experimentation.

etc.), and the manufacturing process used to created each component. These data will be used to benchmark the product and develop design for manufacturing cost models. Figure 6.3 provides an example of a BOM for a fingernail clipper.

			Functional Analysis			DFM Cost Analysis					
Part #	Name	Qty	Function	Flows In	Flows Out	Manuf. Process	Dimensions	Mass	Material	Finish	Other Variables

Bill of Materials

Project Name

Engineer(s):
Date:

▼ *Figure 6.21.*
Indented BOM Form.

Exploded Views with Highlighted Features

Exploded views, either graphical or photographic, represent a graphical documentation of a product. An example exploded view of a product is shown in Figure 6.22. As shown in the figure, exploded views provide a perspective view of each component, exploded away from the product's center, along the assembly access for the component. All components are appropriately labeled and are detailed with all geometric features.

An exploded view of a product may be used to assist in the development of assembly plans. It may also be used to study important component and assembly interfaces as well as intricate product features.

Actual Product Function Structure

An actual function structure of a product is simply a function network of how the current product transforms energy, materials, and signals to produce the desired outputs. It is not a hypothesized functional model of how a product works as a black box. Instead, it represents the *existent* part functions of a product as they relate to customer needs.

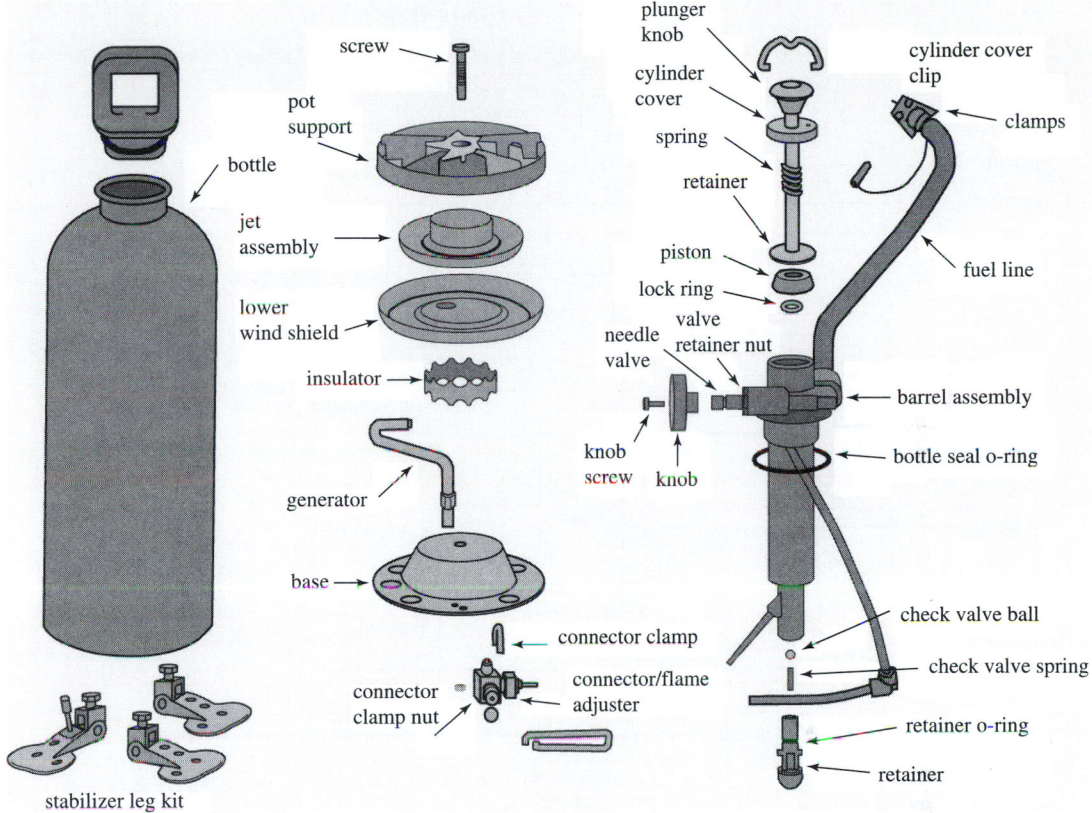

▼ **Figure 6.22.**
Example exploded view of a camping stove product.

Figure 6.23 illustrates an actual function structure for an electric wok product. Note that except for the first function in dashed lines, all functions in the structure represent what the product does, not what the user does or what other products do during a typical use. This function structure provides a qualitative model of the product's physics—we can see what flows where. It also provides a clear decomposition of a product to change a product's architecture or generate new concepts for solving the product functions. By developing an actual product function structure, it may be compared with alternative functional views or be used to predict the value (through value analysis, Chapter 7) of each of the functions. It may also be added to, subtracted from, or integrated to address customer needs not addressed by the current product.

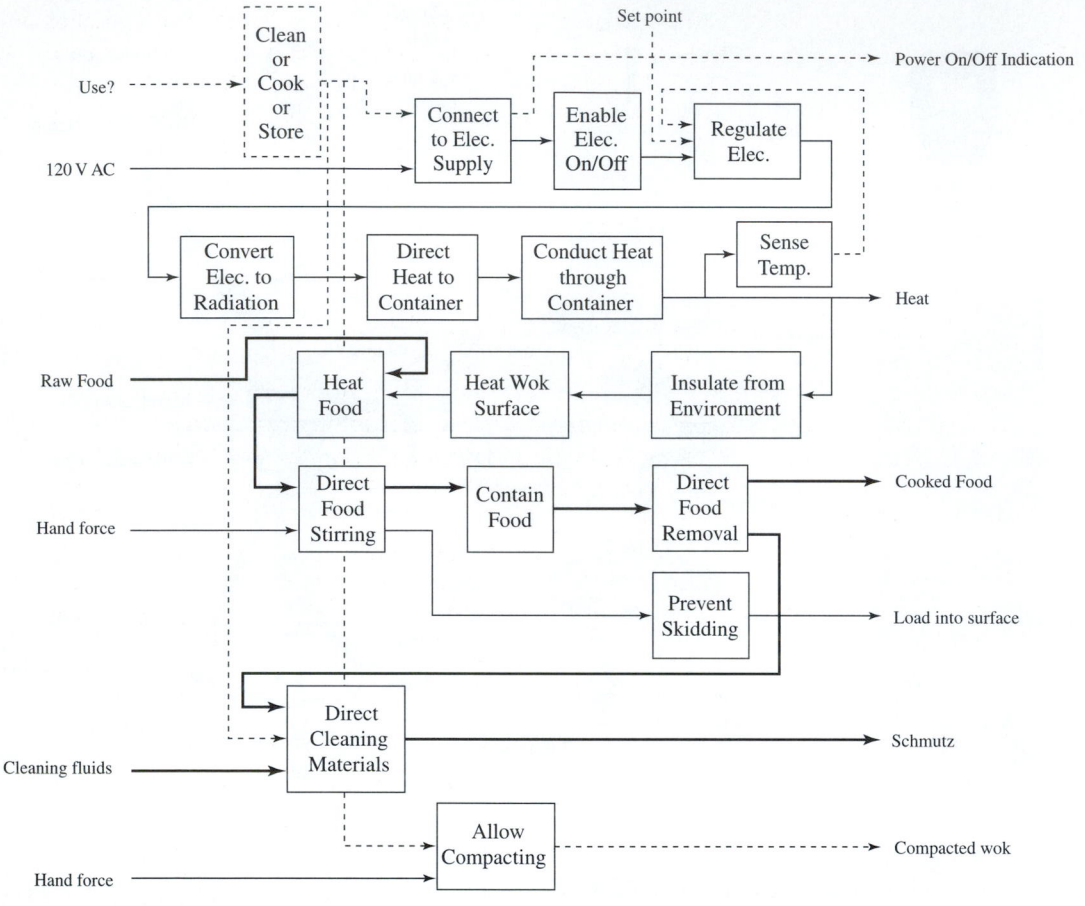

▼ **Figure 6.23.**
Actual product function structure of an electric wok.

V. APPLICATIONS OF PRODUCT TEARDOWN

Product teardown helps us to understand the engineering manifested within a current product. It also exposes creative avenues for product evolution. In this context, this sections studies two product applications. The first, an automobile auxiliary visor subsystem, focuses on the tools for teardown, including SOPs and Force Flow diagrams. Results from these tools, when combined with engineering analysis, lead to an evolved product with a reduced part count. The second application considers the development of metrics and measurement tools for an iced tea brewer. The end result is a set of metrics that address the customer needs of the product and the corresponding transducers for quantifying these metrics from the product and competitive benchmarks.

Application: Slide-Out Auxiliary Visor

An automobile supplier currently manufacturers a slide-out auxiliary visor (SOAV) as a part of a complete overhead ceiling unit. This visor is supplied to a luxury automobile manufacturer. The motivation behind an auxiliary visor is to provide a means of blocking light in the front range for the driver or passenger, while the traditional fold-down visor, in the swiveled position, shields light coming from the side. Most auxiliary visors consist of a second fold down visor. The SOAV, being contained above the headliner, must first translate out, then rotate down to block incoming light. A schematic of the arrangement is shown Figure 6.24. Figure 6.25 is the SOAV in the translated position.

The SOAV assembly consists of forty parts. It is a high-volume product, so reducing the assembly costs for each SOAV is an important issue.

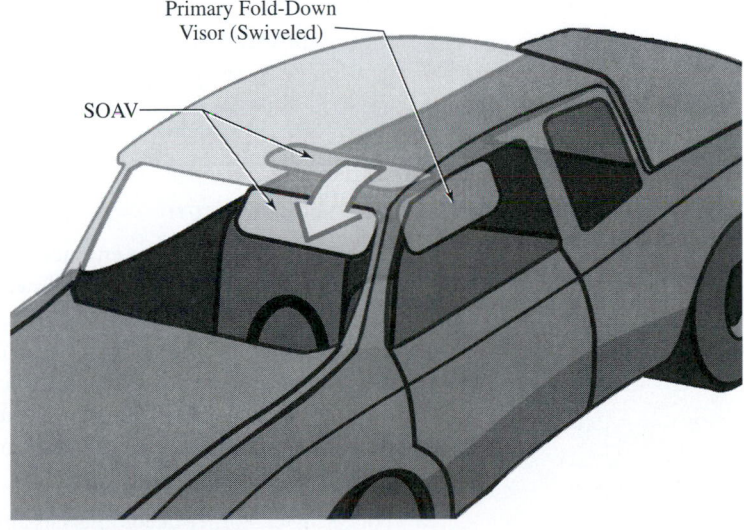

Primary Fold-Down
Visor (Swiveled)

SOAV

▼ *Figure 6.24.*
Simple schematic of SOAV.

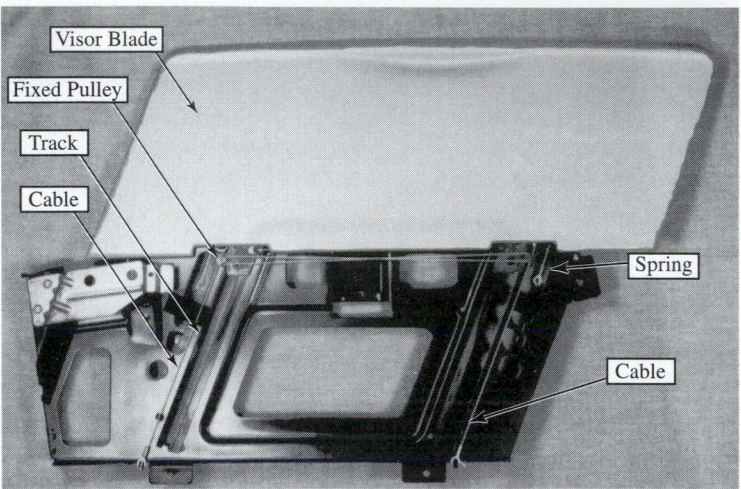

▼ *Figure 6.25.*
SOAV parts.

Assembly time can be reduced through easing the handling and insertion of certain components or by reducing the number of assembly operations. Because forty parts is considered a high quantity for the function that is performed, reducing the number of parts would likely lead to the greatest reduction in assembly times. This reduction in turn reduces the assembly costs.

The established task is to redesign the SOAV for piece count reduction. The following sections present how the SOP and Force Flow Diagrams are used in product teardown and redesign to significantly reduce the piece count of the SOAV.

SOAV Redesign

For the SOAV, our reverse engineering product development process is implemented, emphasizing teardown methods. Each step is not documented completely due to space limitations (Lefever, 1995; Lefever and Wood, 1996). Critical to any product development is the gathering and organizing of customer needs. In developing the customer requirements, all relevant customers must be considered. After carrying out customer interviews, a list of customer needs is assembled and organized as shown in Table 6.6.

Following customer needs analysis and initial functional modeling, the SOAV product is disassembled using the teardown methods. The following procedure is used for this particular product:

TABLE 6.6.   CUSTOMER NEEDS FOR THE SOAV

Functional performance	Interface with overhead unit
Blocks light	Hidden overhead, except handle
Easy to operate	. . .
Comfort to hold	Manufacturing
Easy to rotate and translate	Easy to attach to overhead unit
. . .	Use injection molding and stamping
Isolation	. . .
Does not vibrate	Appearance
Does not generate noise	Aesthetically pleasing: streamlined
. . .	Not bulky
Structural integrity	. . .
Has long life	
Stays closed	
Won't bend	
. . .	

1. In disassembling the SOAV, the strategy is to use two products. One SOAV acts as a reference as the other one is disassembled.

2. Each part is removed one at a time.

3. The component is recorded in a Disassembly Process Table in the order it is removed.

4. Once the device is completely disassembled, it is reassembled using the aid of the Disassembly Process Table. This process helps to fully understand the correct or best order of assembly and validate the access direction, necessary tools, and process.

5. Then, the five-step SOP procedure is implemented.

Table 6.7 illustrates the results of the disassembly process. In particular, important functional insights result from the process. The more generalized subfunctions of the SOAV are decomposed into atomic (finest) level subfunctions. For instance, the "guide motion" function is decomposed into "allow linear DOF," "provide linear resistance," and "constrain DOF during translation." These functions provide a basis for a complete functional statement of the SOAV and motivate the force flow redesign.

Figure 6.26 shows the Force Flow Diagram for the operation of translating the visor, then rotating it down. The diagram shows a great deal of symmetry with the force flow transfer through the SOAV. A line could be drawn horizontally down the middle of the diagram. Those components above the line are components found on the inboard side of the SOAV. Those below the line are components found on the outboard side of the SOAV. The diagram is thus a representation that combines function and form.

TABLE 6.7.   *SOAV DISASSEMBLY PROCESS TABLE, INCLUDING SOP RESULTS*

Step	Part	Task	Necessary tools	Access Direction	Effect of removal	Deduced functions	Change in DOF (Y or N)
1	Inboard rotation club screw (left)	Unscrew	philips screwdriver	z	blade flutter		Y
2	Inboard rotation club screw (right)	Unscrew	philips screwdriver	z	blade flutter		Y
3	Outboard rotation club screw (left)	Unscrew	philips screwdriver	z	blade flutter		Y
4	Outboard rotation club screw (right)	Unscrew	philips screwdriver	z	blade flutter		Y
5	Blade	Now separate	fingers		no blade		Y
6	Inboard cable	Slip cable eye ring over hook, follow through path, unhook from spring	fingers	z	reduced linear resistance		N
7	Outboard cable	same	fingers	z	blade chucks in +phi direction		Y
8	Outboard extension spring	Unhook spring	fingers	x	dependent on cable		–
9	Inboard extension spring	Unhook spring	fingers	x	dependent on cable		–
10	Inboard upper slide block screw (rear)	Unscrew	philips screwdriver	z	gapping		Y
11	Inboard upper slide block screw (front)	Unscrew	philips screwdriver	z	gapping		Y
12	Inboard upper slide block	Not joined now– gather upper slide block	fingers	z	y direction slop		Y
13	Inboard bumper	Remove foam from within upper slide block	fingers	z	noise against end tabs		N
14	Outboard upper slide block screw (rear)	Unscrew	philips screwdriver	z	gapping		Y
15	Outboard upper slide block screw (front)	Unscrew	philips screwdriver	z	gapping		Y
16	Outboard upper slide block	Not joined now– gather upper slide block	fingers	z	y direction slop		Y

TABLE 6.7.   SOAV DISASSEMBLY PROCESS TABLE, INCLUDING SOP RESULTS (CONTINUED)

Step	Part	Task	Necessary tools	Access Direction	Effect of removal	Deduced functions	Change in DOF (Y or N)
17	Outboard bumper	Remove foam from within upper slide block	fingers	z	noise against end tabs		N
	Inboard lower slide block	Cannot disassemble: insert mold onto rod					
	Outboard lower slide block	Cannot disassemble: insert mold onto rod					
18	Rotation rod	Gather sub-assembly	fingers		no lower side block/ rod assembly		–
	Outboard rotation clip	Cannot disassemble: snapped on rod					
	Inboard rotation clip	Cannot disassemble: snapped on rod					
19	Inboard guide rail	Slide off	fingers	y	increase in linear resistance, noise, wear on slide blocks		N
20	Outboard guide rail	Slide off	fingers	y	increase in linear resistance, noise, wear on slide blocks		N
21	Center z-axis clip screw	Unscrew	fingers		z-axis clip not constrained		Y
22	Center z-axis clip	Gather	philips screwdriver	z	headliner will not attach to roof		Y
23	Inboard z-axis clip screw	Unscrew	fingers		z-axis clip not constrained		Y
24	Inboard z-axis clip	Gather	philips screwdriver	z	headliner will not attach to roof		Y
25	Guide tox	Press fitted to pan			reduced linear resistance, SOAV not attach to headliner		Y
26	Guide tox felt	Adhered to guide tox (left in on)			reduced linear resistance, increase wear on blade		N

TABLE 6.7. SOAV DISASSEMBLY PROCESS TABLE, INCLUDING SOP RESULTS (CONTINUED)

Step	Part	Task	Necessary tools	Access Direction	Effect of removal	Deduced functions	Change in DOF (Y or N)
27	Pan	Now separate	fingers		no pan		–
28	Outboard pan foam	Adhered to pan (left it on)	could remove with fingers	z	vibration, noise against roof		Y
29	Outboard foam dampener	Adhered to pan (left it on)	could remove with fingers	z	blade flutter, reduced rotational resistance		Y
30	Inboard foam dampener	Adhered to pan (left it on)	could remove with fingers	z	blade flutter, reduced rotational resistance		Y
31	Inboard top foam strip	Adhered to pan (left it on)	could remove with fingers	z	vibration, noise against roof		Y
32	Outboard top foam strip	Adhered to pan (left it on)	could remove with fingers	z	vibration, noise against roof		Y
33	Outboard spring felt	Adhered to pan (left it on)	could remove with fingers	z	spring twang		Y
34	Inboard spring felt	Adhered to pan (left it on)	could remove with fingers	z	spring twang		Y
35	Inboard diagonal felt	Adhered to pan (left it on)	could remove with fingers	xyz	reduced linear resistance, increase wear on blade		Y
36	Inboard corner felt	Adhered to pan (left it on)	could remove with fingers	xyz	vibration, noise against roof		Y

Analysis of Results from SOP (Table 6.7): The removal of the inboard cable results in no change in the DOFs of the blade. The functions of the cable are identified as constraining a rotational DOF and providing linear resistance. If the DOF is unchanged when the cable is removed, there might be a redundancy. Checking the subfunction "constrain DOF during translation" exposes that there are two solution principles for constraining the negative SOAV rotation: the inboard cable and the angle of the tracks themselves.

Although there is no change in DOF when subtracting the inboard cable, the other function of providing linear resistance is lost, and the effort to pull out the blade is decreased. This means that the cable is not a Type 1 redundant component, but it may be a candidate for a Type 2 redundancy. Type 2 components can have their effect from removal compensated through parametric redesign of other components. The cable's other subfunction, "provide linear resistance," is solved by

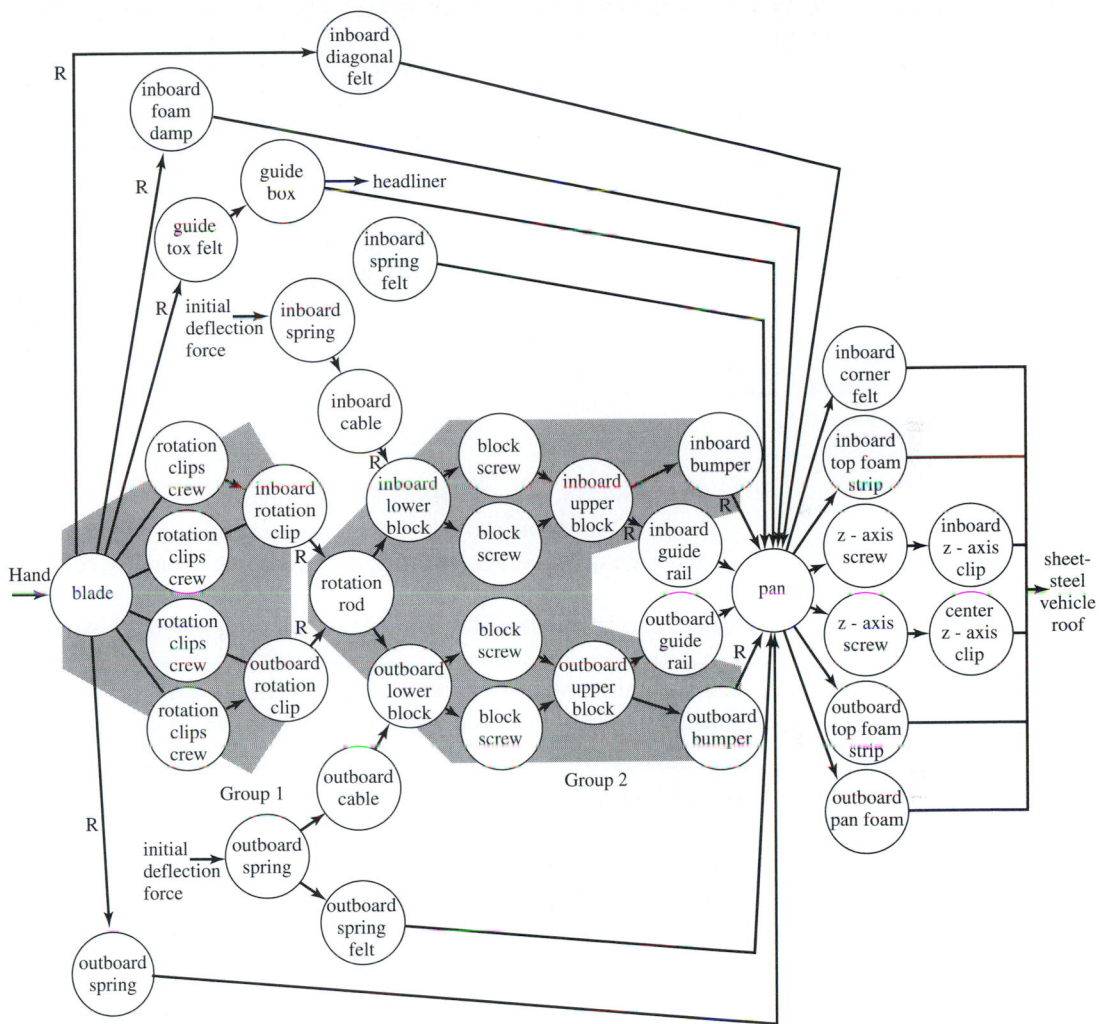

▼ *Figure 6.26.*
Force Flow Diagram of the SOAV.

multiple solution principles. Analyzing these solution principles and the components that embody them, it is apparent that the remaining outboard cable can be used to compensate for the loss in linear resistance. Decreasing the length of the cable or increasing the spring constant of the extension spring will increase the cable's tension and provide more friction across the lower slide blocks.

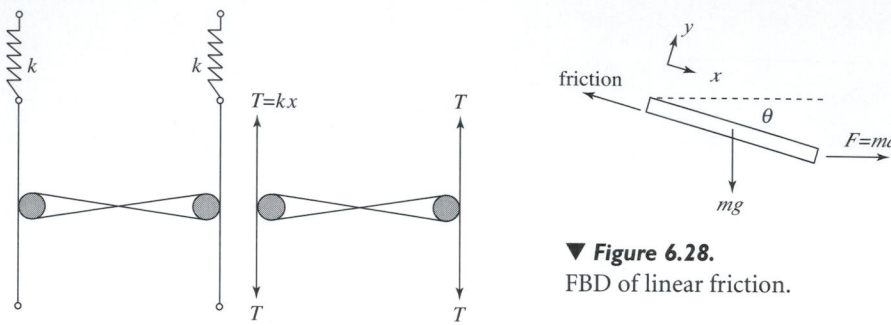

▼ **Figure 6.28.**
FBD of linear friction.

▼ **Figure 6.27.**
Schematic/Free Body Diagram (FBD) of
tension cables.

Engineering Model and Mathematical Equations: An assortment of different components could be the focus of a model for the parametric redesign of the SOAV. For instance, the blade geometry could be analyzed to reduce blade flutter, or the rotation clips could be modeled to check the effect of thickness variations on the rotational resistance. Because the objective of this case study is to reduce the piece count of the SOAV, and a Type 2 redundant component has been identified, the friction provided by the cables is the subject of this model.

Figure 6.27 is a schematic of the cable/spring design. The tension, T, in the cables is provided by the springs. The spring deflection is x, and the spring constant is k. The tension in the cables is the normal force acting across the knobs of the lower slide blocks. The coefficient of friction between the cable and knob is denoted as $\mu_{c,k}$. Thus, the normal force across each knob is equal to $2kx$, and the total friction force is: *friction* $= 4kx\mu_{c,k}$ (the objective function).

This friction provides the necessary resistance so that the visor can maintain any linear position while experiencing a specified deceleration. From Figure 6.28, knowing the mass of the blade and attached components and the deceleration, the required friction can be determined. By summing the forces in the x-direction, the constraint for the frictional force becomes:

$$4kx\mu = ma \cos \theta + mg \sin \theta$$

Therefore, it can be seen that eliminating a cable simply reduces the left side of this equation by half. Parametric compensation can occur by increasing the spring constant, the spring deflection (cable length), and/or the friction coefficient.

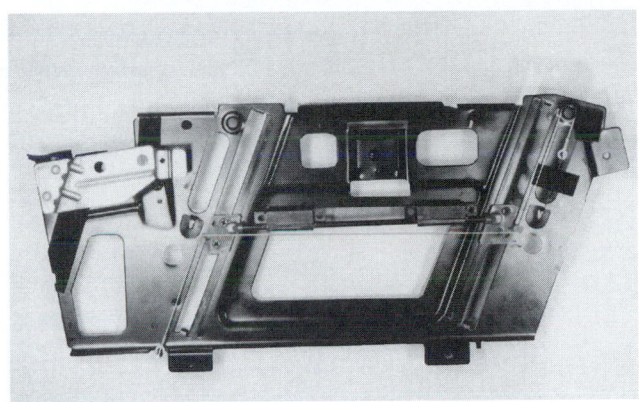

▼ **Figure 6.29.**
SOAV single cable redesign.

Create a Physical Model: Creating a prototype of this parametric alteration is trivial. First, the inboard cable is removed, and the remaining extension spring is replaced with a spring having double the spring constant. Figure 6.29 shows this alteration.

This prototype is then tested for its pull effort. This effort is found to be within specification (4.5 − 16 N), and the redesign is validated. Eliminating the inboard cable allows for the elimination of the inboard extension spring and the inboard extension spring felt. Overall, this is a piece-count reduction of three parts.

Analysis of the Force Flow Results: The Force Flow Diagram for the translation and rotation operation is shown in Figure 6.26. To analyze the diagram, an "R" is placed between components with relative movement. The components are then grouped. Figure 6.26 displays two of the groups found in the Force Flow Diagram. The focus of the adaptive design for the SOAV is on Group 2.

Group 2 is symmetrical. The rotation rod connects a series of inboard components and a series of outboard components. Any redesign to one side can be duplicated on the other side. Combining components across the symmetry is difficult.

Examining Group 2 shows that the block screws that join the upper and lower slide blocks could possibly be eliminated through the use of snaps. However, it would be preferable to simply combine the upper and lower slide blocks. The group shows that there are assembly/

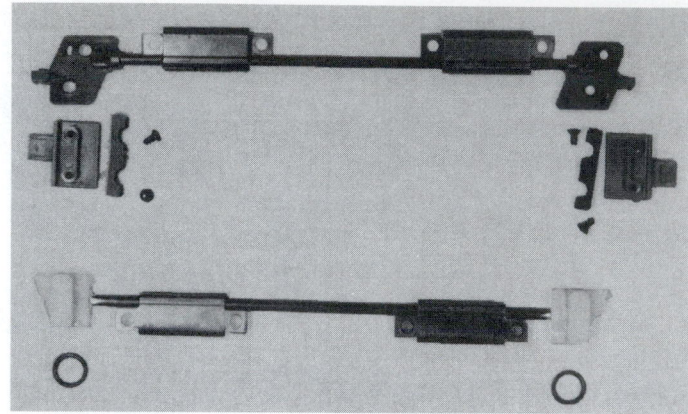

▼ *Figure 6.30.*
SOAV present *(top)* and new single slide block redesign *(bottom).*

disassembly issues involved. The adaptive redesign in Figure 6.30 shows how this constraint is eliminated so that the two components could be combined. The foam bumper that fits between the upper and lower slide blocks is replaced with an O-ring. The assembly issue is handled by swiveling the block subassembly into the tracks. The combination of the upper and lower slide block on the inboard and outboard sides leads to a piece-count reduction of an additional six parts.

Redesigned SOAV: The established task of reducing the piece count of the SOAV is accomplished through the use teardown methods, in the context of customer needs and the product development process. A redesign with a total of nine less components is developed.

Eliminating the cable, extension spring, and spring felt, and purchasing springs with double the spring constant are straightforward adjustments. The supplier implemented these changes. Not only did the piece reduction lower the manufacturing and vending costs, but also it allowed the supplier to move an assembler off of the SOAV line onto another line. Altogether, the change translated into significant annual savings (seven figures in dollars) for the SOAV manufacturer.

The single slide block redesign that eliminates six components was also implemented. Even with the costs of new molds for the single block design, the annual savings from this change was an order of magnitude greater than the cable elimination redesign.

Case Study of an Automatic Iced Tea Maker

This section presents an automatic iced tea maker study, focusing on teardown methods related to measurement and experimentation. The West Bend Iced Tea Maker, shown in Figures 6.31 and 6.32, produces iced tea using a similar method to making automatic drip hot coffee. Tap water is poured into a tank or reservoir and tea is placed into a steeping basket. The device is turned on and the water is "steamed" to the top of the product through a tube, by heating the water using a resistive element at the base of the product. The water is then directed into the steeping basket and dripped over the tea bags. The tea then flows out a hole in the bottom of the steeping basket and into a pitcher full of ice. The design is very similar to automatic coffee drip makers, except the pitcher and machine are taller, and the container is plastic instead of glass.

▼ **Figure 6.31.**
Automatic iced tea maker.

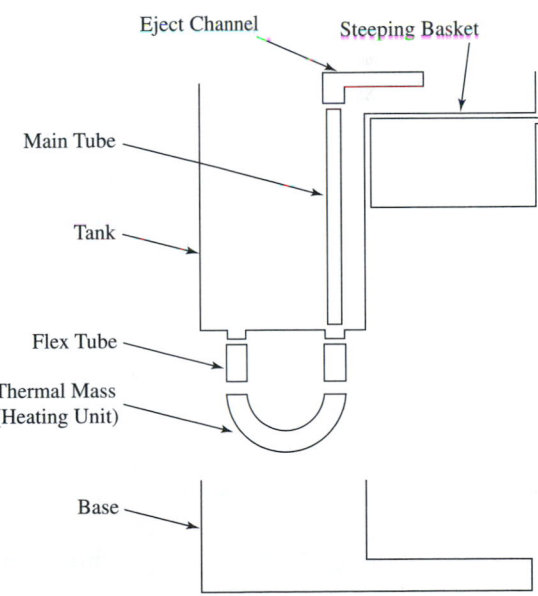

▼ **Figure 6.32.**
Schematic of the iced tea maker.

Set-Up Case Study

Before tearing down the iced tea maker, let us first briefly introduce the background of the iced tea process. Currently there are dozens of automatic coffee makers on the market; however, only a few brands have automatic iced tea makers on the shelf, including, Mr. Coffee® and West Bend®. The Stash Tea Co. reports that now "tea is the most widely consumed beverage in the world next to water and can be found in almost 80% of all U.S. households" (Stash Tea 1997). The traditional approach to making iced tea requires boiling the water, adding the tea, and then cooling it in the refrigerator. There are three primary types of tea: black, oolong, and green. Black tea undergoes several hours of oxidation during its process and constitutes 90% of the U.S. market (Nielsen, 1997). Black tea is the most robust. Because iced tea is most commonly brewed from black tea, it is less complex, having a fixed steeping time of five minutes, and has less of a problem with scorching the tea leaves, whereas green and oolong tea have varying steeping times and need hot water temperatures.

A black box model of the product can be created as shown in Figure 6.33. This model defines the main function of the product and its interactions with the environment as is necessary before customer interviews are conducted.

Nine international customers were interviewed who drink iced tea prepared in various ways: powdered, traditional boiled water, automatic iced tea maker, and ready-to drink. The interviews focused on what they liked and disliked about the product as they brewed the iced tea. The customer base included one Middle Easterner, from whom we learned that iced tea is only an American drink, and three Europeans, some who drink iced coffee daily in Europe. There were no implications that the iced tea maker was not brewing properly to foreign standards, as no additional international standards for iced tea could be found. The overall major dislikes about the product included: "use less ice," "not easy to add tea," and "soak the tea" (Table 6.8). It is important to note that customers without ice machines had less of a problem with

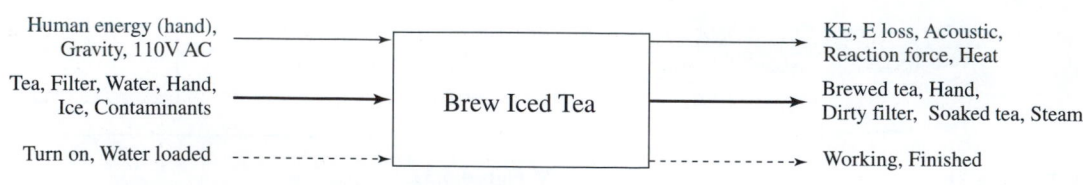

▼ **Figure 6.33.**
Automatic iced tea maker black box model.

TABLE 6.8.   AUTOMATIC ICED TEA MAKER WEIGHTED CUSTOMER NEEDS

Customer needs	Weight
Easy to add ice	5
1. Use less ice	
2. Marks for 1 quart on pitcher	
3. Easy to close cover	
Easy to add tea	5
1. More ergonomic way to attach basket	
2. Non-disposable filter	
Stronger tea	5
Brew tea	
1. Brew large amount	2
2. Good tank lid to contain steam	2
3. Good "on" button	1
Easy to clean	3
1. Access to tank	
2. Dishwasher safe basket	
3. Opaque colored pitcher	
4. Brewer cannot get clogged	
Storage	3
1. Compact	
2. Top for pitcher	

the amount of ice needed; therefore, the scores were either 7 out of 9 for those without ice machines or no response at all from those with ice makers.

Using these results, a function structure is developed as shown in Figure 6.34. This function structure considers each customer need, the flow(s) associated with each need, and the mapping of the flow through a set of atomic functions to satisfy the need. This function structure is then mapped to a common basis (Chapter 5) so that it can be compared with other similar products to choose appropriate measurements. Armed with the customer needs and function structure, along with some insight into the product, we can now determine the most important functions that should be measured during product teardown.

Weight/Correlate Subfunctions to Customer Needs: This step is completed in three short parts. First, the customer needs are scaled to the 1–5 ratings; next we must determine which functions relate to each customer need. Finally, the subfunctions ratings are increased by one to assign the supporting functions an importance of one (Figure 6.35).

There are a few interesting correlations to discuss in the execution of the mapping. First notice that all of the customer needs directly map

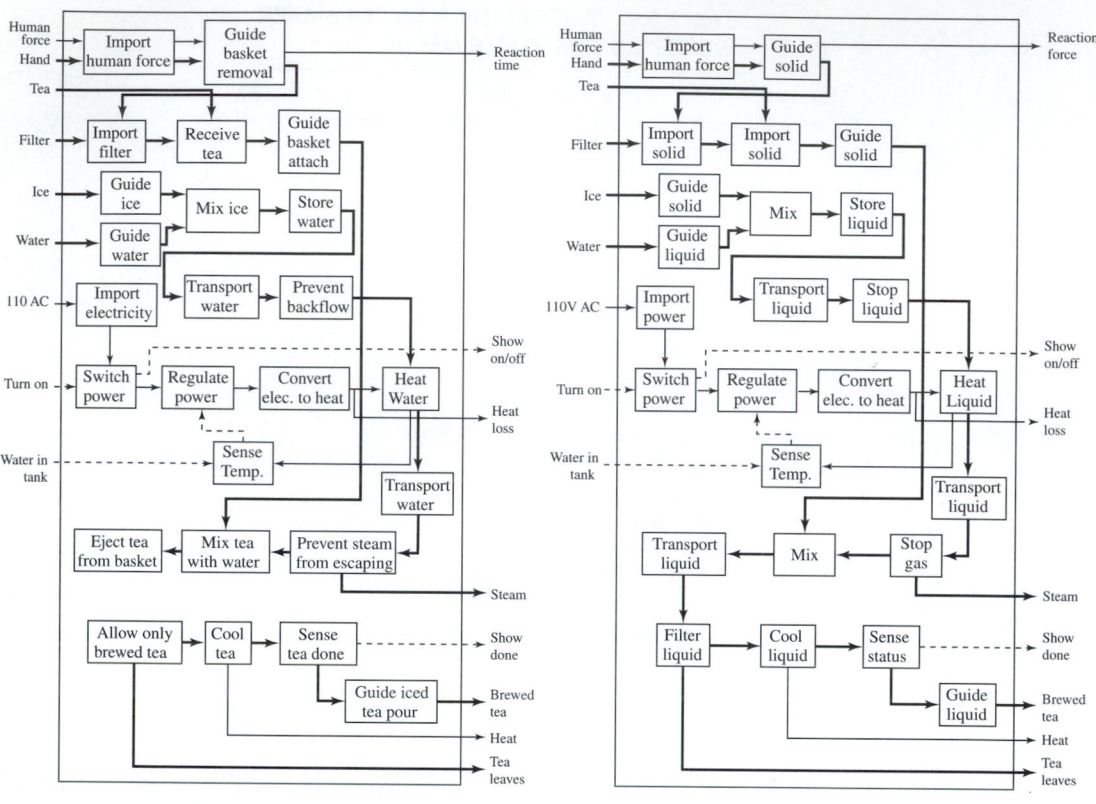

▼ *Figure 6.34.*
Automatic iced tea maker function structure, specific and common.

to between one and three subfunctions. Some are rather direct, such as "contain steam" correlated to "stop gas." Others require more insight, such as "use less ice" that relates to "import human force," because of the ergonomics of adding ice, and "transmit heat," which arises from the initial function of "cool water." The customer need "easy to add tea" relates to four subfunctions, because the process includes both "guiding the hand" to "import the tea" and "adding the actual tea."

Measurement Selection Approach

To begin the measurement selection approach, we first resolve to investigate the "transmit heat" and "mix solid and liquid" subfunctions, since these functions are highly rated in Figure 6.35. Taking in the top five functions could result in a more thorough study; however, such additions are possible after the initial iteration is completed.

Customer Needs	Weight	Scaled Weight		Importance	Sub-Function
Stronger Tea	9	5		9	Mix solid with liquid
Use less ice	5	3		7	Import human force
Easy to add tea	5	3		9	Transmit heat
Easy to clean	3	2		7	Import solid
Easy to store	3	2		6	Convert electricity to heat
Brew largeramounts	2	1		4	Store liquid
Contain steam	2	1		2	Stop gas
				4	Guide hand
				3	Store product
				3	Clean product
				1	Allow DOF of solid
				1	Import liquid
				1	Export liquid
				1	Regulate electricity
				1	Sense temperature
				1	Stop liquid
				1	Filter liquid
				1	Indicate status
				1	Transport liquid
				1	Import electricity
				1	Actuate electricity

▼ *Figure 6.35.*
Weighting of Subfunctions using quantified customer needs for an automatic iced tea maker.

Step 3 of the measurement selection approach involves selecting the measurements in the form of metrics for the iced tea maker. A database of measurements per function reveals the following metrics (Table 6.9). We can now utilize this information to determine metrics and the proper avenues for redesign. Focusing on the first two customer needs, "stronger tea" and "use less ice," which, as shown in the customer needs mapping, both relate directly to the main function contributors of "transmit heat" and "mix solid and liquid." After completing initial research on the steeping process of tea, we decide to base the strength of the tea on the amount of energy causing mixing (temperature of entering water) and the time the water is in contact with the tea. The time of contact is ultimately determined by measuring the geometry of the mixing chamber and the output flow of the mixture.

The second customer need, "use less ice," is primarily related to cool exiting water or transmit heat. After observing the product operation and studying the database, the two metrics "volume of medium" (volume of water in tank) and "final temperature" (temperature of tea entering pitcher) are chosen.

TABLE 6.9.   POSSIBLE MEASUREMENTS FOR AUTOMATIC ICED TEA MAKER

Transmit heat	Units
Initial temperature	(deg) C
Final temperature	(deg) C
Time to medium heat	sec
Volume of medium	m^3
Geometry of transmitter	deg, m
Resistance of transmitter	k

Mix solid and liquid	Units
Input flow of solid	g/sec
Input flow of liquid	g/sec
Geometry of mixing chamber	m, deg
Output flow of mixture	g(L)/sec
Energy causing mixing	rpm, q

Measurement Selection: Selecting a Measurement Device

Since the measurement of temperature can be found in two of these chosen metrics, temperature will be used to illustrate the selection process. There are three parts to choosing a device: import data to worksheet, collect data from catalogs, and select the measurement device. The range and accuracy data are developed for the metrics using, for example, benchmarking. For the iced tea maker, we are measuring hot water; therefore, the range should reach 100°C and at least start at a room temperature of approximately 25°C. We may determine the accuracy of this product based on the overall range and the predicted drop in temperature between the heated water and the exiting water. Since we expect only a slight change in temperature, a delta of 10°–20°C, we need at least a one (±1) degree Celsius accuracy to provide the needed significant digit. Finally, we determine the measurand for this product to be liquid. The range, accuracy, and measurand information is collected on the worksheet shown in Figure 6.36.

The top of the worksheet includes the basic information: measurement, flow, units, overall range, and related functions to the measurement, in this case, "transmit heat" and "mix solid and liquid." Next, the specific minimum accuracy and range is recorded. For this example, ±1°C for the range of 25–100°C is recorded. The measurands follow with optional sketches for reference. For the iced tea maker, the measurement tool must be able to take accurate readings of the water in the steeping

Related Function(s): Transmit Heat, Mix Solid & Liquid

| Measurement: | Temperature | Flow: | Heat |
| Units: | Celsius | Range: | 25 - 100 C |

* Determine from actual products, by recording minimum and maximum measurements

| Min Accuracy: | 1 C | Range: | |

* Determine accuracy from actual products by looking at resolution between products during benchmarking and target values

| Measurands: | | Sketches: |
| A. Hot Water / Tea | | |

Current Devices:	Range	Accuracy	Measurands	Functions	Cost
K Thermocouple	0-2300	± 1 C	Hot Water / Tea	T.H. & M.S.L	$100
T Thermocouple	-300-500	± 0.5 C	Hot Water / Tea	T.H. & M.S.L	$120
Mercury Thermometer	-40-125	± 1 C	Hot Water / Tea	T.H. & M.S.L	$18
Alcohol Thermometer	-20-100 C	± 1 C	Hot Water / Tea	T.H. & M.S.L	$6
RS232 Data Acq. T	-300-500 C	± 0.5 C	Hot Water / Tea	T.H. & M.S.L	$500

Options:

* Select devices to cover entire range and measurands. Use other issues to help decide between possible devices

Design Issues:

	Calibration	Dynamic	Recorder	Safety	Mass	Size	Ease of Use	Reliability	ND	Other
	Time (sec)	time C	type	(1-5)	kg	inch	time (sec)	(1-5)	(1-5)	-
K Type TC	0	0.4	N/A	4	-	-	30	4	5	
T Type TC	0	0.4	N/A	4	-	-	30	4	5	
Mercury	0	2	N/A	4	-	-	10	4	5	
Alcohol	0	3	N/A	4	-	-	10	4	5	
RS232 T	0	0.4	N/A	4	-	-	10	4	5	

▼ *Figure 6.36.*
Worksheet for selecting measurement tools.

basket and of the hot water dripping out. The information, so far, provides the designer an idea of what types of measurement equipment catalogs to search.

The next step involves consulting actual catalogs of measurement equipment. For this example, we select five diverse measurement devices to illustrate all the options: Type K and T thermocouples, mercury and alcohol thermometers, and a data acquisition thermocouple device. The final step of the selection process is to choose the appropriate device. The major design issue here is the time constant of the device. Because we will be measuring the temperature of the water over the duration of the operation of the device, a device is needed that will capture temporal temperature changes. A conventional thermometer, although inexpensive, will not provide an accurate reading shown in Figure 6.36. For this case, a type T thermocouple should be selected to provide better accuracy for the temperature ranges of the water. Finally, a decision must be made as to whether data acquisition equipment will be used. For our redesign, we select the data acquisition equipment shown in Figure 6.36, because it is readily available. Measurements using this and other tools may now be executed, assuming a proper experiment plan is developed.

Overall, the worksheet provides an organized approach to selecting measurement equipment as illustrated in the above example. The worksheet motivates the designer to find the appropriate information about each measurement device. It is simple to complete and provides the designer with all the necessary information to make an informed selection. The worksheet starts out with the important information: accuracy, range, and measurands, and then asks for more detailed characteristics, such as size, calibration, and reliability. The designer can then easily view the different devices, side by side, to make an intelligent decision.

IX. SUMMARY AND "GOLDEN NUGGETS"

Product teardown is a key step in reverse engineering products. It is also essential for benchmarking and understanding competitors. Golden nuggets of product teardown include:

1. Product disassembly must be founded on a knowledge of customer needs and intended function.

2. Simple techniques, such as subtracting components and studying how forces flow through products, can provide profound insights into a product's complexity and operation.

3. An effective search for information will greatly aid product teardown.

4. Measurement and experimentation should seek tools that provide the needed data while minimizing cost and avoiding wasted resources.

References

Ananthasuresh, G. K., and Kota, S. 1995. Designing compliant mechanisms. *Mechanical Engineering* 117 (11): 93.

Bass, W. 1987. *Human osteology: A laboratory and field manual.* 3rd ed. Columbia, MO: Missouri Archaeological Society.

Bentley, J. 1983. *Principles of measurement systems.* New York: Longman.

Boothroyd, G. 1980. Design for Assembly - A Designer's Handbook. University of Massachusetts, Amherst MA. August 1980.

Brown, S., ed. 1995. Forensic engineering. Pt. I of *An introduction to the investigation, analysis, reconstruction, causality, prevention, risk, consequence and legal aspects of the failure of engineered products.* Humble, TX: ISI.

Chamberlain, A. 1994. *Human remains.* London: British Museum Press.

Consumer Reports. 1997. Yonkers, NY: Consumers Union.

Hauser, J., and Clausing, D. 1988. The house of quality. *Harvard Business Review* (May-June): 63–73.

Holman, J. 1989. *Experimental methods for engineers.* New York: McGraw-Hill.

Jensen, C., and Helsel, J. 1992. *Engineering drawing and design.* New York: MacMillan/McGraw-Hill.

Lefever, D. 1995. Integrating design for assemblability techniques and reverse engineering. Master's thesis, The University of Texas, Austin.

Lefever, D., and Wood, K. 1996. Design for assembly techniques in reverse engineering and redesign. *Proceedings of the 1996 ASME Design Theory and Methodology Conference.*

Little, A. 1997. A reverse engineering toolbox for functional product measurement. Master's thesis, The University of Texas, Austin.

Little, A., Wood, K. 1997. Functional analysis: A fundamental empirical study for reverse engineering, benchmarking, and redesign. *Proceedings of the 1997 ASME Design Theory and Methodology Conference.*

Miles, L. 1972. *Techniques of value analysis* and *engineering.* New York: McGraw-Hill.

Nielsen, B. 1997. *Rec.food.drink.tea FAQ version 1.2.* 10 Apr. Available at *http://www.nitehawk.dk/bnielsen/teaminifaq.html.*

Norton, H. *Sensor and analyzer handbook.* Englewood Cliffs, NJ: Prentice-Hall.

Otto, K. 1996. Forming product design specifications. *Proceedings of the 1996 ASME Design Theory and Methodology Conference.* CD-ROM.

Otto, K. and K. Wood. A Reverse Engineering and Redesign Methodology for Product Evolution. *Proceedings of the ASME Design Theory and Methodology Conference,* Irvine, CA.

The Stash Tea Company. 1997. *A world of tea.* 10 Apr. Available at *http://www.stashtea.com.*

Stewart, T. 1979. *Essentials of forensic anthropology.* Springfield, IL: Charles C Thomas.

7 Benchmarking and Establishing Engineering Specifications

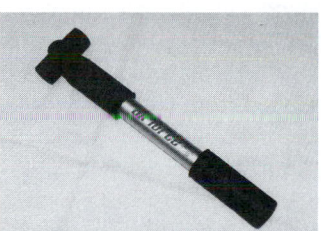
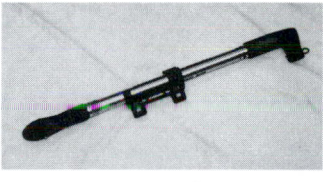

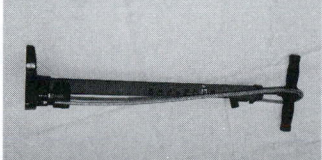
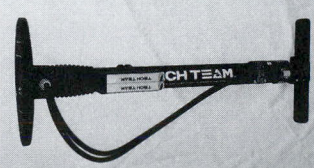
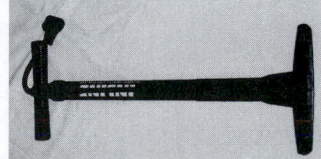

Different bike pumps on the market. Which are successful?

Just as it is essential to interact with customers, product developers must learn from competitors. The alternatives are not palatable. The *Not-Invented-Here* (NIH) syndrome when engineers at a company choose not to use an external system because of the perception that any external system cannot be any good is a cause for many companies to be caught flat-footed when new technology develops in a marketplace. Such an attitude can leave design teams and their companies behind. Companies must understand the importance of newly introduced technology by competitors and be poised to respond as market leaders. Benchmarking is the key product development activity for meeting these goals. It is also an important step in establishing engineering specifications, the kickoff for responding to competition.

I. CHAPTER ROADMAP

Benchmarking as an activity in the product development process overlaps many of the other activities. It generates a wealth of data to understand a product and forecast its future development. In this sense, benchmarking is pervasive to the entire development process. Similarly, engineering specifications are pervasive to the process. They must be developed from customer needs, quantified, and applied to product concepts, refined with more detailed information for chosen product shape and geometries, and continually verified to attain product milestones. To capture these pervasive characteristics, the chapter is organized according to Figure 7.1.

II. BACKGROUND: KNOW YOUR ENEMY TO KNOW YOURSELF

In approximately 500 BC, the Chinese warrior Sun Tzu said, "Know your enemy to know yourself, in a hundred battles you will never peril" (Cleary, 1988). Such advice is as true now in product development as it was then in war. Many development teams understand in great detail how their products work, how they evolved, the reasoning underlying why different features are used and how each was fabricated. They feel they understand the weak points and what could be done to redesign and upgrade their product lines. This understanding is woefully inadequate, however, unless it is grounded in comparisons to other competitive offerings that customers have available. Development teams must understand not only their product, but, more importantly, how

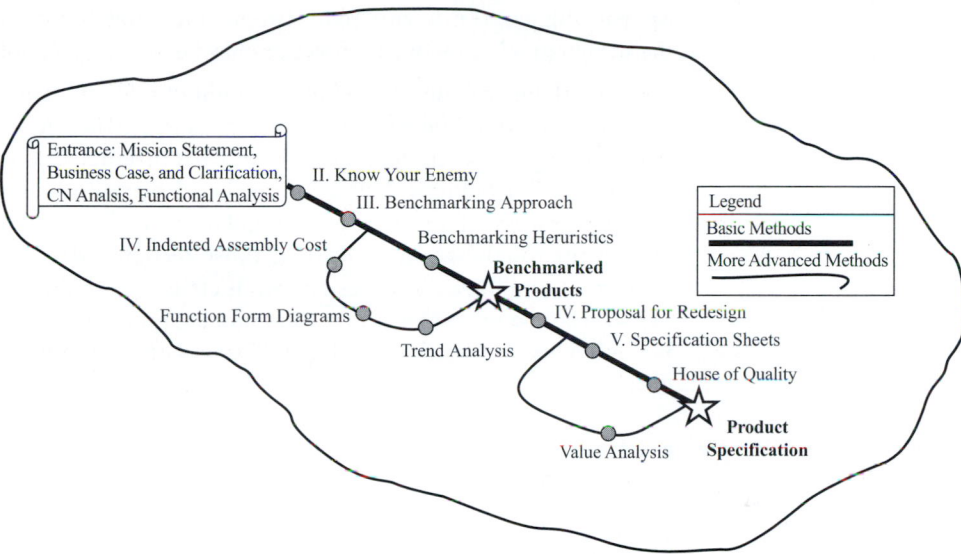

▼ **Figure 7.1.**
Road map for journeying through the world of benchmarking.

other competitors provide similar products. When engineers think they understand their product by mere self-inspection, they are closing doors to a wide array of alternative possibilities.

In 1979, engineers at the Xerox Corporation pondered a question: "How in the world could the Japanese manufacture [a copier] in Japan, ship it over to the United States, land it, sell it to a distributor who sells it to a dealer who marks up the cost to the final customer, and the price the customer pays is about what it would cost us [Xerox] to build the machine in the first place?" (Jacobson and Hillkirk, 1986). At the time, Xerox failed to understand the competition: their products, their production, and their distribution. They now have competitive benchmarking activities, they perform teardowns as a part of their product development process, and they are no longing pondering this question; rather, they ponder how to be successful.

To understand the competition, development teams must tear down and analyze competitive products. Such analyses must not be considered a one-time activity to support a new development effort but rather must be understood as a continuous periodic activity undertaken to understand trends and directions in technology development. Every current teardown activity should compare the competition's new models with past teardown activities of previous models. By doing so, the competition's growth and direction may be fully understood.

Benchmarking provides a standard, or point of reference and range, that can be used to judge quality, value, or performance. A perspective can be gained on what the *best-in-class* product is and what makes it so. For example, Xerox defined benchmarking as "the continuous process of measuring products, services, and practices against the toughest competitors or those recognized as industry leaders" (Camp, 1989). Successful products integrate this detailed best-in-class comparative information into the product development process. This information is vital at all stages of the product development process:

▶ for the front end of the design process for identifying customer needs—to help understand what needs others are satisfying

▶ for improving concept generation—to help start new concept generation off the best-in-class concepts

▶ for embodying a product—to help improve on the state-of-the-art or exciting design features of the best-in-class products that one should also emulate

▶ for establishing product specifications (including cost)—to help ensure the prescription of product goals to outpace all competition

▶ for executing detailed design—to help ensure using the best-in-class components and suppliers

III. A BENCHMARKING APPROACH

This section presents a six-step process for product benchmarking. This process should be executed using a product teardown process as detailed in Chapter 6 (Figure 6.5). We first present the approach and then demonstrate it with an example of coffee mills.

STEP 1: Form a List of Design Issues

A list of design issues must be developed for comparative benchmarking. Further, this list should be continually revised and updated. With a focus for benchmarking efforts, an efficient exploration path may be pursued. The result is a reduction in wasted time and resources. Chapter 6 discusses this issue more thoroughly for product teardown.

STEP 2: Form a List of Competitive or Related Products

Considering the design issues and product function in product development, one must next examine the sales outlets for products that demonstrate these issues. For consumer products, sales outlets are retail stores. For a product, one must list all competitors and their different product models. In addition, all related products in their portfolio should be listed. If the competitors have a family of products under a common platform (they use identical components for some aspects of each product but different components for niche demands), one should detail this information, as it can indicate the competitor's preferred market segments and compromises made for other market segments.

This step should only be an identification of the competitors, basically as company names and product names. With a complete set of different products, vendors, and suppliers to examine, the list should be screened by highlighting the particular competitors that appear most crucial for the design team to fully understand. This approach feeds the next step, conducting a information search.

STEP 3: Conduct an Information Search

The next step is external and active: conducting an information search. The importance of this step cannot be overstated. To benchmark a piece of hardware, the design team must gather as much information about the product as possible. Any printed article that mentions the product, its features, its materials, the company, manufacturing locations or problems, customers, market reception or share, or any other ancillary information will be useful. The first and best choice to begin a search for information is the corporate library. Because of the proliferation of computerized databases and the World Wide Web, a good corporate library is important to provide expertise on which sites have the information needed. The wealth of information available about all business operations across the globe is generous. Before starting any design activity, a team must understand the market demand for product features and what the competition is doing to meet it. A design team should gather information on

▶ the products and related products
▶ the functions they perform
▶ the targeted market segments

All keywords associated with these three categories should be formed and used in informational searches.

Sources of information can be quite varied. Most businesspersons are perfectly happy to discuss the market and noncompetitive business units. Though most will not provide strategic information about their own companies, many people are happy to tell all about their competitors. Suppliers will usually discuss their customers as they can, if it appears that the requester might provide an additional sale. The key is to always be open, honest, and ethical when questioning for information. Once people understand that a design team is designing or redesigning a new product, they naturally want to get involved with new orders and will help the team as far as they legally can. Pursuit of information beyond that point is unethical and not necessary. Most people are happy to share information, and so simple honesty and a friendly attitude can get team members a long way.

Written Sources of Product Information

Public sources of information include:

- ▶ **Libraries:** University libraries are filled with technical engineering modeling references. Also, the librarians have expertise in uncovering obtuse references with limited initial information. A prerequisite for being competitive in today's market is access to a university or corporate library that is capable of accessing the various media, has the infrastructure to process and catalog information from databases, and has the technical staff who can find and filter information.

- ▶ **Thomas Register of Companies:** This set of documents is a "yellow pages" for manufacturing-related businesses. The Thomas Register lists vendors by product name (available at *http://www.thomasregister.com*).

- ▶ **Market Share Reporter:** Published every year by International Thomson Publishers, this book summarizes the previous market research of Gale Research, Inc. It is composed of market research reports from the periodicals literature. It includes corporate market shares, institutional shares (not-for-profits), and brand market shares.

- ▶ **National Bureau of Standards:** This U.S. government branch provides, among other things, national labor rates for all major countries. This information proves very useful for determining competitors manufacturing costs.

▶ **Census of Manufactures:** Taken every 5 years by the U.S. Department of Commerce, this census includes statistics on employment, payroll, inventories, capital expenditures, and selected manufacturing costs. Also, the supplemental *Current Industrial Reports* lists production and shipment data on industries and some products.

▶ **Moody's Industry Review:** Taken every 6 months, this survey provides key financial information, operating data, and ratios on about 3,500 companies. Companies as an industry group may be compared with one another group and against industry average.

▶ ***Consumer Reports* Magazine:** This magazine surveys and tests a number of common consumer products. Useful data are available for customer needs, qualitative benchmarking, engineering specifications, and warranty/maintenance information. If a given product domain is not covered in the magazine, other products can provide analogies as a starting point (available at *http://www.consumerreports.com*).

▶ **Trade Magazines:** Consumer trade magazines such as *Car and Driver, Byte, Consumer Electronics, JD Powers and Associates,* and others provide comparative studies of products within a field. Such studies are very useful to understand how a given product compares with the competition and to understand important customer and technical criteria.

▶ **Patents:** After examining trade journals and uncovering which competitors have bragged about new innovations, gathering the patents on these new innovations explains much. Patent searches based on company names are difficult, however, since companies typically "bury" their patents by filing them under the individual names of designers. Uncovering the individual patents is usually done through refined topical searches. Therefore, absolutely as much information as possible should be supplied to the person searching the patents. Patent information may be obtained from the Classification and Search Support System (CASSIS) or from Web sites, such as *http://www.patents.ibm.com/*.

Market Research Databases

Professional market survey companies now provide their results in on-line databases that can be searched using keywords, for a small fee. These services have electronic text of sources such as newspapers, magazines, wire services, newsletters, journals, company and industry analyst reports, broadcast transcripts, business journals, and academic journals. One can carry out keyword searches on information such as company histories, sales and earnings forecasts, market share projections, R&D expenditures, competitive intelligence, new product development efforts, financial activities, demographics, socioeconomic activities, government regulations, and other events that impact the business environment.

One can use such on-line databases to find information on market and demographic data, financial statistics of competitors, market and technological environments, and others. Intellectual property files can be used to investigate the latest patents issued or to check trademarks or copyrights. Patents can also be compared to create historical technology timelines and be used as a predictor of change. Information may be obtained from a variety of other sources, including sales literature, fax-back services, and organizations. This information helps to provide an understanding of technology, features, market positioning, and competition.

- ▶ **DIALOG** is the world's largest on-line information research service with hundreds of on-line databases of articles, conference proceedings, news, and statistics, all from a diverse spectrum of disciplines. The DIALOG collection contains information from millions of documents in the scientific, technical, and medical literature as well as business, trade, and academic studies, newspapers, and magazines (*http://www.dialog.com/*).

- ▶ **Predicasts,** available through DIALOG, is an annual statistical source that contains tens of thousands of time series data streams for various industries, including economic indicators and industry statistics. The industry statistics usually include production, consumption, exports/imports, wholesale price, plant and equipment expenditures, and wage rate.

- ▶ **American Demographics** is a searchable article database dedicated to analyzing consumer, demographic, and business trends (*http://www.marketingtools.com*).

Other Sources

In addition to the resources above, there are others worth exploring.

- ▶ **The World Wide Web:** Every major manufacturer has a presence on the WWW. Much information can be gathered simply and at no cost, particularly technical information. Material properties of industrial brand plastics and metals, for example, are easily found.

- ▶ **Vendors:** At some point, refinement of specific information is required. Gathering cost quotes from vendors in ordering quantities of the competition's product and components is an easy method to uncover OEM part costs. Usually one can also obtain unsolicited information about the competition.

▶ **Technical Specialists:** University faculty or industry consultants, specializing in a technical area that is exercised by a product, can provide invaluable assistance in redesign or simply as consultants to participate as design reviewers.

▶ **Experts/Friends in the Industry:** Good pointers on nonobvious cost, technical, and market drivers in an industry can be uncovered by asking persons in the industry. Particular overseas labor rates, shipping rates, reliability of supply from geographic areas, and so forth, are all "inside knowledge" that might be uncovered from other public sources only when one knows what to look for.

▶ **NAICS Codes:** Another useful piece of information to track down for a product is the *North American Industry Classification System* (NAICS) code of its manufacturer. The NAICS is the replacement for the former *Standard Industrial Classification* (SIC) system. The U.S. government uses a system of numeric codes to categorize companies by the type of business in which each is engaged (U.S. Economic Classification Policy Committee 1997). Businesses engaged in the same activity, regardless of size or type of ownership, have the same NAICS code. The NAICS code for a product is often used as a key word for database searches. A set of databases to be searched would include general technology, business, patents, and any specific databases that relate directly to the product area of interest. Searching databases for information must often be repeated during the course of the project; it is not a one-time task.

▶ **Industry Associations:** For any NAICS code, another useful source is the association of the manufacturers in that industry. Industry associations typically serve as industrial liaison with government and the media, in addition to providing membership services and meeting and suppliers exhibits. Such organizations can be helpful in providing sales figures and a list of members for a particular product domain.

All of these and other sources must be leveraged to determine as much as possible about what customers want, what competitors are supplying, and what dynamics drive the market. The results of the information search are a well-developed list of design issues, a well-developed list of products to tear down, and a historical and market perspective.

STEP 4: Tear Down Multiple Products in Class

Having conducted the information search, one has a list of products that have been successful in the market and a list of design issues for the teardown. The next step is to complete the teardowns of these products, the heart of a benchmarking activity, using the methods of Chapter 6. The result of this step should be an indented bill of materials for each product, a functional model for each product, an exploded view of each product, and the function-to-form mappings of the functions to the assemblies. Each function should have measured performance levels completed. These measurements should be taken for every product torn down.

STEP 5: Benchmark by Function

After benchmarking several comparable products, the team needs a tangible outcome of the analysis. One approach is to summarize the comparison by product form; for example, with coffee mills, one might list all different versions of containment shells, lids, or power cords that are used. This listing could then be used for comparative purposes.

The problem with this approach is that any component in one product may not be functionally equivalent to the same or similar component in another product. One coffee mill lid may additionally serve as a switch in one product; it may not in another. Additional parts in one product may be additionally required to be "equivalent." To overcome this problem, we will not benchmark products by their equivalent components, but rather we will adopt a functional equivalency approach.

For a new product development effort, one should establish the new product's functional model as in Chapter 5. Then, for each function in the model, find the same function in the other products' function models. For this function, list the various physical forms found among the competition to solve this function, and under each solution list the performance measurements for comparison purposes. Each solution listed is typically a collection of components from each product.

STEP 6: Establish Best-in-Class Competitors by Function

Having listed the various solutions used for every function, comparative analysis can be completed. For each function, the highest performing solution can be called out as *best-in-class*. The least expensive solution can be called out as well. These two bounds are important knowledge for a new development team. Such benchmarking limits can be completed along the hierarchical aggregation of functions (discussed in Chapter 5) up to the overall product function, where one can make the comparison among the products as entire market entities.

STEP 7: Plot Industry Trends

Having uncovered a wealth of information from such sources, the next problem is to arrange and transform the information according to a clear explanation of implications for the design or redesign task. This process should include

- ▶ categorization of the market
- ▶ benchmarking of technical solutions
- ▶ benchmarking of competitors

Market categorization is an approach where one categorizes product solutions by socioeconomic status of the typical customer of the individual product and also by percentage of the market. Here, one identifies the socioeconomic market segments at which different products are aimed and compares them in and across such classes. Such market classifications can be identified using questionnaires (as in Chapter 4), and then sorting these responses into clusters. Urban and Hauser (1993) discuss market clustering techniques.

As an example, one can determine that in the industrialized world there are basically two types of toaster consumers. There are young first-time purchasers of toasters, and there are older recurrent buyers. The first-time buyer has limited income and has one fundamental purchase criterion: minimum cost. They typically fail to consider that the cheapest toasters, by design, only last a year or so. Second-time buyers, being older and typically having higher incomes, have the failed toaster experience and look for higher quality. Summarizing these two market segments for the development team is important so that appropriate design decisions can be made—developing a minimum cost toaster is different than developing a high-quality toaster.

The *benchmarking of technical solutions* is an approach to comparing how products perform. The most effective approach is the development of a thorough technology timeline using s-curves as discussed in Chapter 1. Recall all technological innovations manifest themselves over time into the market along an "s-curve" timeline behavior. These should be plotted as an output of the benchmarking process for the key comparisons made technically in Step 6.

The *benchmarking of competitors* is similar in spirit to the benchmarking of technical solutions and considers the performance over time of the entire portfolio of a company. Corporate strategies may be deciphered from their performance on business criteria such as market share in different regions, price points in the market over time, assets over time, buyouts, inventory costs as a fraction of sales, and labor costs as a fraction of sales.

Benchmarking Example: Coffee Mills

To demonstrate the benchmarking process, consider the coffee mill product class. Coffee mills permit a consumer to convert whole coffee beans into coffee grinds, providing very fresh tasting coffee. Consider benchmarking the Krups 203 Coffee Mill shown in Figure 7.2, a well-performing model in the coffee mill market.

Completing Step 1 for the coffee mills, a design issues list includes price, noise level, size, grind time, and fineness of grind. These issues can be determined, for example, by completing a customer needs analysis for the Krups mill using the methods of Chapter 4 and choosing the customer needs with high importance levels.

Next, for Step 2, a list of competitors is researched and assembled. There are several for Krups in this market. Braun, Molinex, Salton, Bosch,

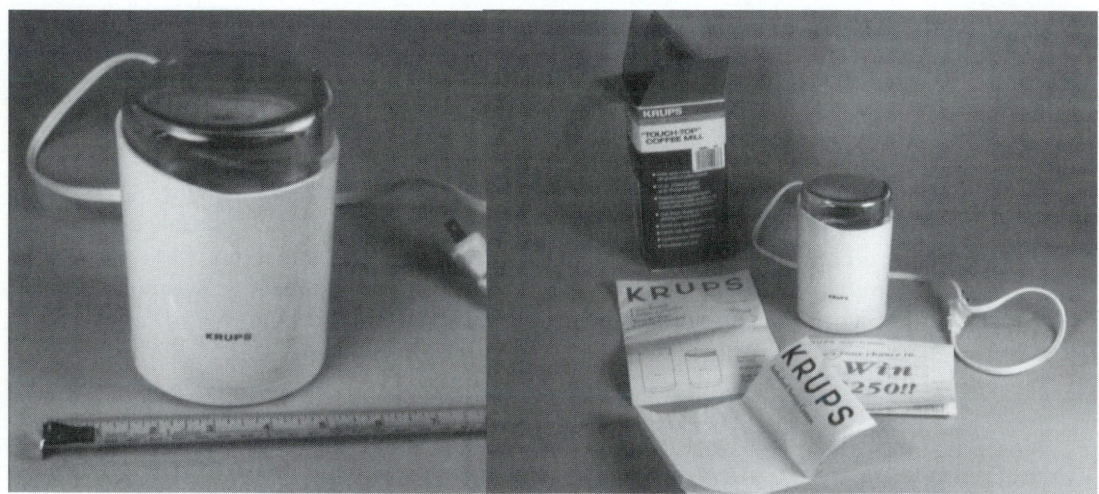

▼ *Figure 7.2.*
Krups Coffee Mill.

Proctor-Silex, and Cuisinart, for example, all produce coffee mills. Coffee-mill-related products might include coffee grinders, coffee brewers, combined coffee mill/brewers, and food processors. These products can be identified, for example, by visiting and documenting the various sales outlets for coffee mills. Searching for products that perform identical primary functions, such as "chop" or "cut a solid" may also identify them.

With this preliminary start at a list of design issues and competitors, Step 3, information search, can be initiated. The issues list and competitor list constitute keywords to use in a search. Using DIALOG, for example, one can find the Dun and Bradstreet credit report for Krups North America Inc. A typical (hypothetical company) credit report is shown in Figure 7.3. Key information includes sales, cost of goods, expenses, income, breakdown of payments, and operation information. The cost of goods and breakdown of payments can be used to help determine the unit manufacturing cost. The operation information can help determine where and how the goods are produced.

As an example, examining a credit report for Krups one would learn that since 1994, Krups has been a wholly owned subsidiary of Molinex S.A. of Paris, France. Krups and Molinex are the same company; thus, one might expect their mills to begin to share features. Also, by examining their breakdown of payments, one sees that Krups North America has made payments only to shipping and distribution. Therefore, Krups performs neither design nor manufacturing in North America. This information can help in determining their costs and can help in understanding why they design features as they have.

Other information that should be gathered includes U.S. patents 336211 for the Krups chopper, 743788 for the Cuisinart chopper, among others. Also, a *Consumer Reports* magazine article (*Consumer Reports* 1991) compares different choppers, as summarized in Table 7.1. The evaluation criteria developed by Consumer Reports should be mapped to the customer needs list to determine their effectiveness. The compared choppers in the *Consumer Reports* list should be candidates for teardowns. Other information gathered includes uses of coffee makers and projections on combined mills and makers that should soon enter the market.

Having gathered this information, the next step is to complete competitive teardowns. The design issues list was formed in Step 1, where the primary issues to be measured and analyzed include mill cost, noise level, size, grind quality, and grind time. Different choppers to tear down included models by Krups, Braun, Cuisinart, and Procter-Silex,

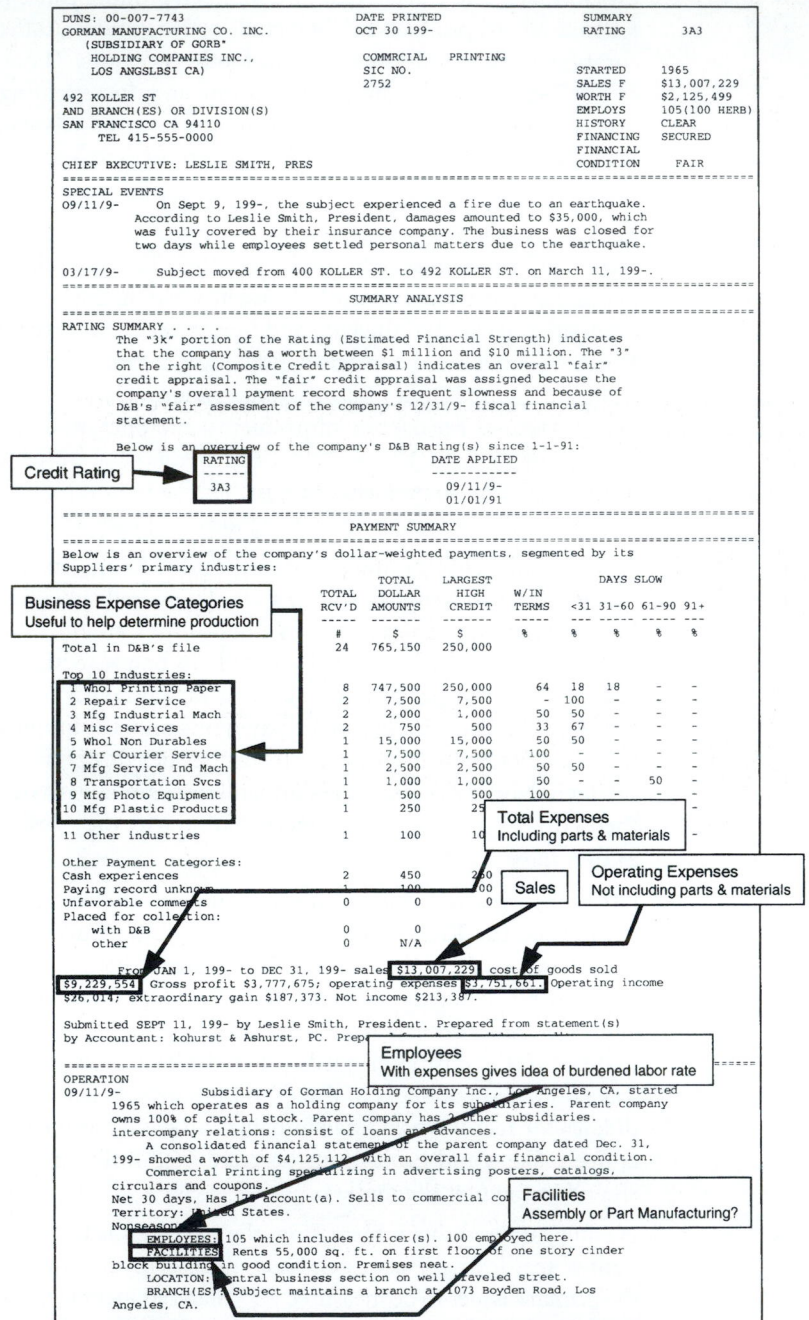

The figure contains the following Dun & Bradstreet credit rating report:

```
DUNS: 00-007-7743                 DATE PRINTED          SUMMARY
GORMAN MANUFACTURING CO. INC.     OCT 30 199-           RATING      3A3
   (SUBSIDIARY OF GORB*
   HOLDING COMPANIES INC.,        COMMRCIAL  PRINTING
   LOS ANGSLBSI CA)               SIC NO.               STARTED     1965
                                  2752                  SALES F     $13,007,229
492 KOLLER ST                                           WORTH F     $2,125,499
AND BRANCH(ES) OR DIVISION(S)                           EMPLOYS     105(100 HERB)
SAN FRANCISCO CA 94110                                  HISTORY     CLEAR
     TEL 415-555-0000                                   FINANCING   SECURED
                                                        FINANCIAL
CHIEF BXECUTIVE: LESLIE SMITH, PRES                     CONDITION   FAIR
======================================================================
SPECIAL EVENTS
09/11/9-      On Sept 9, 199-, the subject experienced a fire due to an earthquake.
              According to Leslie Smith, President, damages amounted to $35,000, which
              was fully covered by their insurance company. The business was closed for
              two days while employees settled personal matters due to the earthquake.

03/17/9-      Subject moved from 400 KOLLER ST. to 492 KOLLER ST. on March 11, 199-.
======================================================================
                         SUMMARY ANALYSIS
======================================================================
RATING SUMMARY . . . .
       The "3k" portion of the Rating (Estimated Financial Strength) indicates
       that the company has a worth between $1 million and $10 million. The "3"
       on the right (Composite Credit Appraisal) indicates an overall "fair"
       credit appraisal. The "fair" credit appraisal was assigned because the
       company's overall payment record shows frequent slowness and because of
       D&B's "fair" assessment of the company's 12/31/9- fiscal financial
       statement.

       Below is an overview of the company's D&B Rating(s) since 1-1-91:

              RATING                      DATE APPLIED
              ------                      ------------
              3A3                          09/11/9-
                                           01/01/91
======================================================================
                         PAYMENT SUMMARY
======================================================================
Below is an overview of the company's dollar-weighted payments, segmented by its
Suppliers' primary industries:

                            TOTAL    LARGEST                DAYS SLOW
                    TOTAL   DOLLAR   HIGH     W/IN
                    RCV'D   AMOUNTS  CREDIT   TERMS  <31 31-60 61-90 91+
                    -----   -------  -------  -----  --- ----- ----- ---
                    #       $        $        %      %   %     %     %
Total in D&B's file  24     765,150  250,000

Top 10 Industries:
 1 Whol Printing Paper  8   747,500  250,000  64    18  18    -     -
 2 Repair Service       2   7,500    7,500    -     100 -     -     -
 3 Mfg Industrial Mach  2   2,000    1,000    50    50  -     -     -
 4 Misc Services        2   750      500      33    67  -     -     -
 5 Whol Non Durables    1   15,000   15,000   50    50  -     -     -
 6 Air Courier Service  1   7,500    7,500    100   -   -     -     -
 7 Mfg Service Ind Mach 1   2,500    2,500    50    50  -     -     -
 8 Transportation Svcs  1   1,000    1,000    50    -   -     50    -
 9 Mfg Photo Equipment  1   500      500      100   -   -     -     -
10 Mfg Plastic Products 1   250      25

11 Other industries    1    100      10                             -

Other Payment Categories:
Cash experiences       2    450      2 0
Paying record unknown  1    100      00
Unfavorable comments   0    0        0
Placed for collection:
   with D&B             0    0
   other               0    N/A

     From JAN 1, 199- to DEC 31, 199- sales $13,007,229 cost of goods sold
$9,229,554; Gross profit $3,777,675; operating expenses $3,751,661. Operating income
$26,014; extraordinary gain $187,373. Not income $213,387.

Submitted SEPT 11, 199- by Leslie Smith, President. Prepared from statement(s)
by Accountant: kohurst & Ashurst, PC. Prep
======================================================================
OPERATION
09/11/9-           Subsidiary of Gorman Holding Company Inc., Los Angeles, CA, started
      1965 which operates as a holding company for its subsidiaries.  Parent company
      owns 100% of capital stock. Parent company has 2 other subsidiaries.
      intercompany relations: consist of loans and advances.
          A consolidated financial statement of the parent company dated Dec. 31,
      199- showed a worth of $4,125,112 with an overall fair financial condition.
          Commercial Printing specializing in advertising posters, catalogs,
      circulars and coupons.
      Net 30 days, Has 120 account(a). Sells to commercial con
      Territory: United States.
      Nonseason
          EMPLOYEES: 105 which includes officer(s). 100 employed here.
          FACILITIES: Rents 55,000 sq. ft. on first floor of one story cinder
      block building in good condition. Premises neat.
          LOCATION: Central business section on well traveled street.
          BRANCH(ES): Subject maintains a branch at 1073 Boyden Road, Los
      Angeles, CA.
```

Annotation callouts on the figure:
- Credit Rating → RATING 3A3
- Business Expense Categories / Useful to help determine production → Top 10 Industries list
- Total Expenses / Including parts & materials
- Operating Expenses / Not including parts & materials
- Sales
- Employees / With expenses gives idea of burdened labor rate
- Facilities / Assembly or Part Manufacturing?

▼ *Figure 7.3.*
Typical credit rating report, summarized (courtesy Dun & Bradsteet).

TABLE 7.1.   COMPARISON OF CHOPPERS

	Bosch MKM6003	Braun KSM2	Krups 203	Regal 505	Salton QC-6	Moulinex 843	Xcell EL1020	Black & Decker CBM122	Melitta CG-2
Price ($)	$25	$20	$28	$19	$15	$30	$20	$22	$22
Performance									
Drip	5	4	4	4	4	3	5	3	3
Espresso	5	5	4	5	4	3	4	3	3
Noise	2	2	3	1	1	3	1	3	3
Max. capacity (oz)	2.75	2.25	2.75	2.25	2.00	3.00	2.25	3.00	3.00

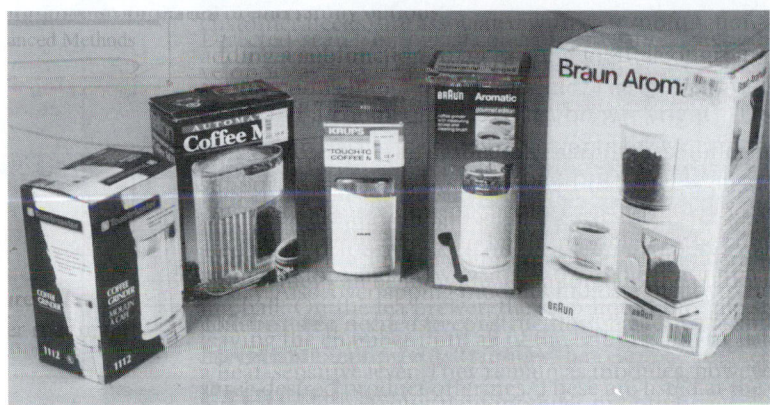

▼ *Figure 7.4.*
Coffee mills benchmarked.

as shown in Figure 7.4, who all compete effectively in the North American market. One does not have to benchmark every possible competing product, but rather the ones that are leaders on some aspect. One will not learn much information from "losing" products, but one will learn much from "winning" products.

The measurements of the mills on these criteria must be completed, a cost analysis broken down by parts and assemblies, and a function structure analysis. In addition, detailed pictures of the mills are taken, including exploded view, assembly, and part pictures. For example, the exploded view of the Krups is shown in Figure 7.5.

After completing the research, inspection, analysis, and teardown of the mills, the information must be summarized and compared. We will discuss a series of teardown tools and use the coffee mill as a context for this presentation.

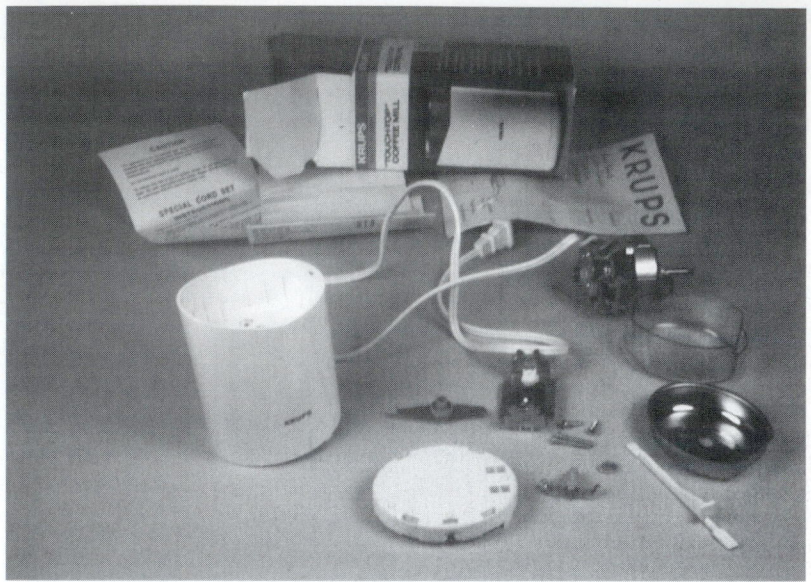

▼ *Figure 7.5.*
Exploded view of the Krups mill.

IV. SUPPORT TOOLS FOR THE BENCHMARKING PROCESS

Indented Assembly Cost Analysis

One key result of a benchmarking activity is a comparative understanding of the cost structures that different competitors face. Part manufacturing in the United States with assembly in China has a different cost structure than part manufacturing in Singapore with assembly in Japan. These different scenarios call for different design configurations, and what may appear a "bad" design may not be, given the production scenario. Assembly in low-labor-cost countries can enable the use of complex assembly processes with many inexpensive parts, whereas assembly in high-labor-rate countries requires designs with complex multi-functional snap-fit parts.

To uncover such understanding, one must analyze the part/assembly structure during the teardown and record the assembly structure—what is attached to what. Then, to each part, one can estimate a part cost, and to each assembly step, one can estimate an assembly cost. Generally, each of these costs is a function of material, equipment, and labor costs. One can estimate these costs and aggregate them into an indented cost estimate. An indented cost estimate organizes the resulting analysis according to the hierarchy of assemblies, subassemblies, parts, and features. Manufacturing cost estimating is covered in Chapter 14. The results of an indented cost estimate are shown in Figure 7.6–7.10 for the Krups mill, indicating manufacturing, assembly, and combined costs, respectively. Such analysis is completed for all of the mills chosen for teardown.

The part cost analysis, shown in Figure 7.6, demonstrates the highest cost component is the motor, which dominates the part costs. This should not be surprising. However, the next most expensive item is the box containing the mill, followed by the lid and housing.

The assembly cost analysis, shown in Figure 7.7, demonstrates how the highest cost assembly is associated with attaching the electric cord. Perhaps a nonsolder attachment system could be developed and used to reduce this cost. The total cost breakdown by subassembly, shown in Figure 7.8, shows where the costs are for the different subsystems. Again, the motor can attribute for much of the costs, but the packaging also constitutes a high percentage as well.

Function-Form Diagrams

After analyzing the cost of several comparable products, as above, one can develop a tangible outcome by comparing costs. As developed earlier, we will benchmark products by their comparative functionally equivalent components. Next, one should offer comments on the various solutions in an outcome document. For example, one should highlight the best, highest performing solution. One should also highlight the least expensive solution. These solutions are benchmarks for the new development effort to surpass.

As an example, consider the coffee mill application. We might consider various ways that one can actuate a mill, turning it on for grinding. This device function is the "Actuate Power" subfunction in the mill function structure, shown in Figure 7.9. Then each benchmarked product is examined for the parts used to satisfy this subfunction. These benchmarks are shown in the Function-Form diagram, as shown in Figure 7.10 for the coffee mill actuate power subfunction. This diagram basically lists the various solutions that are proven on the market for this product function.

The results of the benchmarking analysis should be listed on the Function-Form diagram, such as part count, cost, and any material information. Finally, the best-in-class solutions on the market should be called out, such as the highest quality and the lowest cost models, as shown in the last row of Figure 7.10 for the mill example.

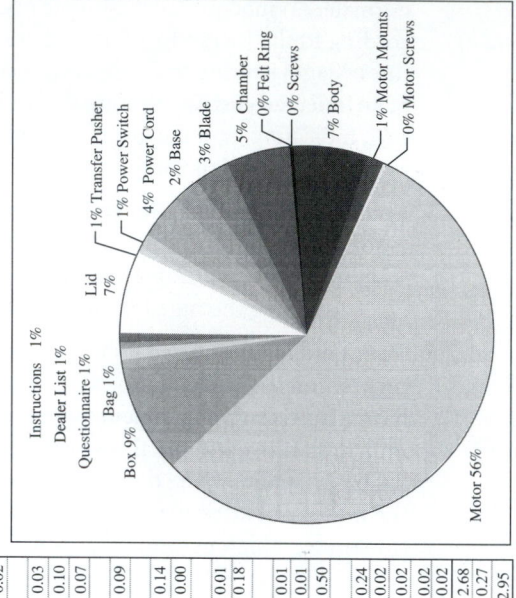

Part #		Part Name	Qty	Labor Time (min)	Labor Cost ($)	Material	Tooling	OEM	Fabrication Totals
A1	001	Lid	1	0.03	$ 0.02	$ 0.16	$ 0.01		$ 0.19
		GRINDER ASSY							
	002	Transfer Pusher	1	0.01	$ 0.01	$ 0.00	$ 0.01		$ 0.02
S1		BASE ASSY							
	003	Power Switch	1					$ 0.03	$ 0.03
	004	Poer Cord	1					$ 0.10	$ 0.10
	005	Base	1	0.03	$ 0.01	$ 0.03	$ 0.02		$ 0.07
S2		BODY ASSY							
	011	Blade	1	0.08	$ 0.04	$ 0.01	$ 0.03		$ 0.09
S21		MECH ASSY							
	012	Chamber	1	0.03	$ 0.01	$ 0.12	$ 0.00		$ 0.14
	013	Felt Ring	1	0.00	$ -	$ 0.00	$ -		$ 0.00
S211		MOTOR BODY							
	014	Screws	2					$ 0.01	$ 0.01
	015	Body	1	0.07	$ 0.04	$ 0.13	$ 0.02		$ 0.18
S212		MOTOR ASSY							
	016	Motor Mounts	2	0.01	$ 0.01	$ 0.00	$ 0.01		$ 0.01
	017	Motor Screws	2					$ 0.01	$ 0.01
	018	Motor	1					$ 1.50	$ 0.50
A2		PACKAGING							
	019	Box	1	0.02	$ 0.01	$ 0.23	$ 0.00		$ 0.24
	020	Bag	1	0.01	$ 0.00	$ 0.02	$ 0.00		$ 0.02
	021	Questionaire	1					$ 0.02	$ 0.02
	022	Dealer List	1					$ 0.02	$ 0.02
		Instructions	1					$ 0.02	$ 0.02
		TOTALS:	22	0.29	$ 0.11	$ 0.71	$ 0.09	$ 1.70	$ 2.68
								10% OH	$ 0.27
							PARTS TOTAL		$ 2.95

▼ Figure 7.6.
Manufacturing part cost for a coffee mill product.

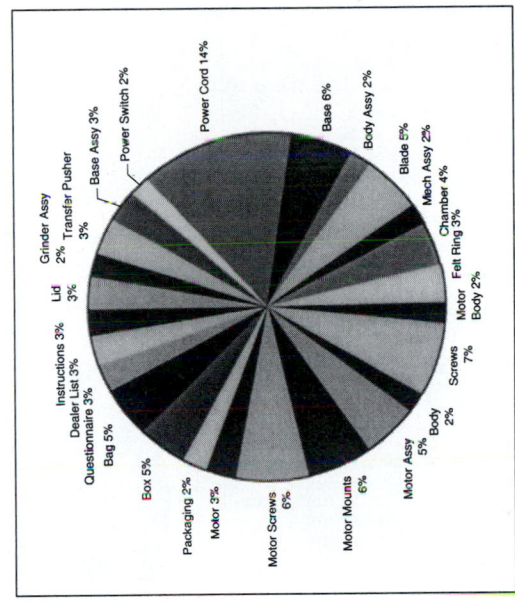

Part #	Part Name	Qty	Acquisition Time (min)	Acquisition Cost ($)	Assembly Time (min)	Assembly Cost ($)	Inspection Time (min)	Inspection Cost ($)	Assembly Totals
001 / A1	GRINDER ASSY	1	-	$ -	0.12	$ 0.06	0.08	$ 0.04	$ 0.06
002	Transfer Pusher	1	-	$ -	0.13	$ 0.07			$ 0.04
S1	BASE ASSY	1	-	$ -	0.12	$ 0.06			$ 0.07
003	Power Switch	1	-	$ -	0.06	$ 0.03			$ 0.06
004	Power Cord	1	0.05	$ 0.02	0.48	$ 0.24			$ 0.03
005	Base	1	-	$ -	0.21	$ 0.11			$ 0.27
S2	BODY ASSY	1	-	$ 0.02	0.08	$ 0.04			$ 0.11
011	Blade	1	0.05	$ 0.02	0.16	$ 0.08			$ 0.04
S21	MECH ASSY	1	-	$ -	0.08	$ 0.04			$ 0.10
012	Chamber	1	-	$ -	0.16	$ 0.08			$ 0.04
013	Felt Ring	1	-	$ -	0.12	$ 0.06			$ 0.08
S211	MOTOR BODY	1	-	$ -	0.08	$ 0.04			$ 0.06
014	Screws	2	0.05	$ 0.02	0.21	$ 0.10			$ 0.04
015	Body	1	-	$ -	0.08	$ 0.04			$ 0.13
S212	MOTOR ASSY	1	-	$ -	0.19	$ 0.10			$ 0.04
016	Motor Mounts	2	-	$ -	0.23	$ 0.12			$ 0.10
017	Motor Screws	2	0.05	$ 0.02	0.20	$ 0.10			$ 0.12
018	Motor	1	-	$ -	0.11	$ 0.06			$ 0.12
A2	PACKAGING	1	-	$ -	0.17	$ 0.09	0.08	$ 0.04	$ 0.06
019	Box	1	-	$ -	0.17	$ 0.09			$ 0.04
020	Bag	1	-	$ -	0.10	$ 0.05			$ 0.09
021	Questionaire	1	-	$ -	0.10	$ 0.05			$ 0.09
022	Dealer List	1	-	$ -	0.10	$ 0.05			$ 0.05
023	Instructions	1	-	$ -	0.10	$ 0.05			$ 0.05
	TOTALS:	22	0.19	$ 0.10	3.45	$ 1.72	0.17	$ 0.08	$ 1.90
							50%	OH	$ 0.95
								ASSEMBLY TOTAL	$ 2.86

▼ Figure 7.7.
Assembly costs for a coffee mill product.

	Part #	Part Name	Qty	Fabrication Totals	Assembly Totals	Sub Totals ($)
	001	Lid	1	$ 0.19	$ 0.06	$ 0.24
A1		GRINDER ASSY	1		$ 0.04	$ 0.12
	002	Transfer Pusher	1	$ 0.02	$ 0.07	
S1		BASE ASSY	1		$ 0.06	$ 0.66
	003	Power Switch	1	$ 0.03	$ 0.03	
	004	Power Cord	1	$ 0.10	$ 0.27	
	005	Base	1	$ 0.07	$ 0.11	
S2		BODY ASSY	1		$ 0.04	$ 0.23
	011	Blade	1	$ 0.09	$ 0.10	
S21		MECH ASSY	1		$ 0.04	$ 0.26
	012	Chamber	1	$ 0.14	$ 0.08	
	011	Felt Ring	1	$ 0.00	$ 0.06	
S211		MOTOR BODY	1		$ 0.04	$ 0.39
	011	Screws	2	$ 0.01	$ 0.13	
	015	Body	1	$ 0.18	$ 0.04	
S212		MOTOR ASSY	1		$ 0.10	$ 1.94
	016	Motor Mounts	2	$ 0.04	$ 0.12	
	017	Motor Screws	2	$ 0.01	$ 0.12	
	018	Motor	1	$ 1.50	$ 0.06	
A2		PACKAGING	1		$ 0.04	$ 0.69
	019	Box	1	$ 0.24	$ 0.09	
	020	Bag	1	$ 0.02	$ 0.09	
	021	Questionaire	1	$ 0.02	$ 0.05	
	022	Dealer List	1	$ 0.02	$ 0.05	
	023	Instructions	1	$ 0.02	$ 0.05	
		DIRECT TOTAL	22	$ 2.68	$ 1.90	$ 4.58
		OH		$ 0.27	$ 0.95	$ 1.22
		TOTAL		$ 2.95	$ 2.86	$ 5.80

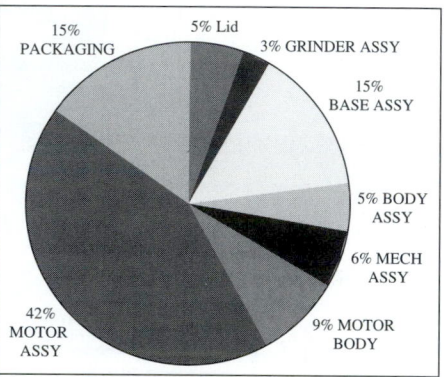

▼ **Figure 7.8.**
Combined manufacturing and assembly costs for a coffee mill product.

Trend Analysis

Final outcomes of the benchmarking analysis are observations that the benchmarking group can make about where the products and industry appear to be headed. These observations can be quantitative, such as plotting s-curves (Figure 3.1), or can also be qualitative, such as uncovered comments by senior executives on where they want to place their company in the market.

In the case of the coffee mill market, the trends over time are better described qualitatively. Ten years ago, most manufacturers had a single

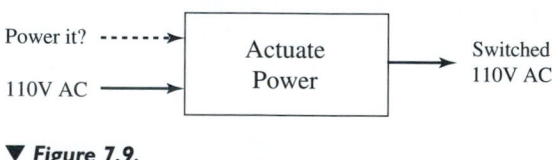

▼ **Figure 7.9.**
"Actuate Power" subfunction.

Concept	Top Switch			Pivot Lid
	Braun	Krups 2	Quisinart	Krups
# Parts	5	5	5	3
Cost	$ 0.25	$ 0.23	$ 0.22	$ 0.20
Materials				
Lid	PS	PS	PS	PS
Pusher	PS	PP	PS	PP
Switch	PVC	PVC	PVC	PVC
Contacts	Brass	Steel	Steel	Steel
Class			Quality	Cost

▼ *Figure 7.10.*
Function-Form diagram for the Actuate Power subfunction.

product on the market. Since that time, most have expanded their product line. Krups and Braun, for example, sell multiple colors of plastic shells. More fundamentally, though, Krups has expanded its single offering into two fundamentally different products, one low-end inexpensive model and a high-quality, more-expensive model. The market is maturing, and so focusing products to particular market segments is appropriate. Compromise designs that try to include features but remain somewhat inexpensive will perform less successfully. A recommendation is to either make a mill that is very inexpensive, or make a mill that works very well, or both separately.

Proposal on Opportunities for Redesign

Having completed this analysis, one can compare the current product to others with respect to customer needs and determine whether each of the customer needs is worth addressing. This is not to say it is feasible to address the need, but rather if there is a strong demand to seek new solutions to the need. For example, it may be feasible to redesign the Krups mill so that it is portable from the electrical outlet without unplugging, but this may not be very important to the customer compared with reducing noise. Reducing noise may be very important to the customer but may be infeasible. The former need might call for a redesign with a base unit that the mill can be decoupled from easily, for example, and the latter need would call for a redesign with noise-dampening features. What direction should be taken?

This is not an easy question to answer definitively, since the answer involves both how important it is to do and how hard it is to do. At this point, we have worked to understand the former but have done no real analysis to understand how hard it will be to improve on any of the needs. For example, with the mill, the importance of portability and noise is understood based on the customer needs surveys, but the difficulty in providing portability versus reducing noise is not at all clear. To determine this difference, we have to carry out design work. Yet, we must now decide what design work to perform. We have a chicken-and-egg problem. The solution to chicken-and-egg problems is just to get going and do something, and so we will simply establish where there are market opportunities for redesign.

To decide where to focus our design efforts, we use a basic method of creating a document delineating where there are opportunities for redesign. More advanced methods include either the *House of Quality* or *value analysis,* discussed later. In the opportunities for redesign document, we simply list each need, its importance, and, based on the comparison just completed, an indication of how well we currently satisfy the need. If we do so well, we do not consider addressing it. If we do not satisfy the need well, we mark it as an opportunity.

As an example, Figure 7.11 shows opportunities for redesigning the Krups Coffee Mill. The opportunities are to focus on reducing the noise of the mill and not, for example, focusing on providing the ability to chop materials other than coffee or on a cordless architecture. The preliminary decisions on realistic opportunities were based on the understanding of the importance of the need and a forecast over how difficult the redesign effort would be.

Since these decisions included a forecast of difficulty, the scoping results (Chapter 3) should be revisited once this better understanding of design difficulty is attained. Perhaps reducing noise is very difficult and so is not worth doing despite the high importance. Perhaps providing portability becomes very easy with the new concept that is eventually settled on. Business case analysis (Chapter 3) can be completed on both scenarios. In this sense, design is iterative.

Thoughts on Benchmarking the Competition

Competitive benchmarking can be very effective for understanding a market, it can help forecast the trends in the industry, and it can identify key innovations and key technologies. Yet, one complaint with benchmarking is that the process will always provide lagging information.

Krups Coffee Mill						
Redesign Scoping						
	Customer Requirement	WT	Low	High	Krups 203	Model/Fix?
¥	Purchase					
1	Low cost ($15)	2	$35	$15	4	Y
2	Namebrand	2	Salton	Krups	4	
3	Sleek shape	3	Box	Quisinart	4	Y
4	Compact adequate box	2	Krups	Braun KSM	5	
	Store on Cupboard					
8	Countertop storage	4	KSM	Quisinart	4	Y
9	Plugged in storage	3	Toaster	Krups	0	Y
10	Small face print	4	KSM	Krups	4	Y
11	Small footprint	4	KSM	Krups	5	
¥	Plug in					
12	Long cord	4	2'	6'	4	Y
13	Standard plug	4	12V Supply	Non-polarized	4	
14	Cordless	2	Anywhere	Short Duration	-	
¥	Pour in Beans					
17	Shape shows how to open	4	Krups	KSM	2	Y
18	Integral measuring system	4	No marks	Tablespoons	4	Y
19	Chamber change point is at one pot amount	3	No marks	Visible Change	3	Y
20	Large chamber capacity	3	8 C	20 C	3	Y
21	Lid cannot be put on backwards	3	Symetric	Non symmetric	2	Y
22	Easy to hold shape	4	Grips	Box	3	Y
23	Size fits in hand	4	2.5"	5"	4	
24	Quiet grinding	4	Mixer	Chopper	1	Y
25	Grinds other than coffee	2	Coffee only	w/o cleaning	2	
26	Switch for power on/off	2	Krups	Side button	2	
27	Can actuate and see chamber simultaneously	4	Krups	Side button	2	Y
28	Durable button	4	Catches	Telephone	4	
29	See-through top to grinding	4	Dark dusty	dust free	4	
30	See-through sides to grinding	3	Opaque	dust free	3	
31	Can set finess-of-grind	4	Krups	KSM	0	Y
32	Sound changes to smooth when ground	4	1X	10X	3	
33	Grinding completeness indicator	3	Krups	KSM	0	Y
34	Fast grinding	3	KSM	Krups	4	
¥	Grinding					
35	Vibration free feel	3	KSM	Krups	4	
36	Doesn't move around when operated	4	Krups	KSM	2	Y
37	Doesn't mark countertop surface	4	Rubber	Plastic	5	Y
38	Motor can run a long time	2	5 sec	30 sec	3	
39	No instructions needed	4	Krups	Braun	3	
40	Uniform grinding	4	Chopper	Grinder	3	Y

▼ *Figure 7.11.*
Proposal on opportunities for redesign for the Krups Coffee Mill.

Generally, it will not uncover the new key innovations being developed by the competition nor what the competition will begin to devote its resources to. These issues do not necessarily negate the need for benchmarking but rather emphasize the importance of the forecasting component within benchmarking.

On the other hand, consider the scenario of a company that is a clear market leader. In this case, the competition may not offer much insight.

Charles Szuluk, Group Vice President of Ford Motor Company and President of Visteon, a supplier of core automotive technologies, such as computer control systems, interior controls, and ABS braking systems to all major foreign and domestic car companies, has said that "benchmarking is a waste of time" (1998). He feels that his company Visteon is well positioned in the market with better technology.

To evolve their products, Visteon has developed a product portfolio roadmap and is devoted to implementing the goals along the roadmap and not worrying what the competition does. A *product portfolio roadmap* is a plan with new technology introduction milestones over time. These are often developed by industry associations for an industry as a whole. They are also developed by individual companies for internal planning, such as Visteon has done. As long as Visteon can remain ahead of the competition on their roadmap, they may remain successful with this approach of not "wasting" development team time on benchmarking.

Before one brushes off competitive benchmarking, one should realize that very few market leaders consistently produce the leading technology in a market. There is always a potential for some unknown competitor to have a great idea that can change the market. Markets rarely have stable, planned behavior, and never do in the long term. Further, in establishing a roadmap, it is often necessary to benchmark technology from industries outside of one's immediate market. For example, Visteon could gain by benchmarking key required sensor, processing, and actuator technologies in markets that already use them, such as aerospace or telecommunications.

Another problem often mentioned with benchmarking is "chasing the competition" type of development practices that one should be careful to avoid. Just because a competitor introduces a new product technology does not necessarily mean that one's company should do so as well. It does mean one should consider a competitive strategy to attack the market, but it does not necessarily mean one should use the same plan as the competition. Doing so will mean that all a company ever develops is products that continually lag the competition. Also, the competition may be taking a gamble on a new technology that may not be accepted. These thoughts, however, again do not negate the need for benchmarking, but instead suggest that one should be careful with the decisions made given the benchmarking information.

Two key points of benchmarking should be reinforced to mitigate this concern. First, one should not benchmark all possible competitive products. Tearing down loser products provides almost no information; it only makes you feel comfortable when you probably

should not. Rather, one should identify competitors who are successful in the market, and then the benchmarking activity should focus on determining what it is the competitors do that makes them successful.

The second point one should realize about benchmarking is one that a student might think would go without saying, except that it is usually *the* key problem with benchmarking. Most often a company will complete or pay for an external benchmarking and then do absolutely nothing with the results. They may read the results, but often companies will "explain away" any shortcomings of their products or processes and not make fundamental change. This is especially true when it will involve a major restructuring of company processes. This *consistently* happens, because people are afraid of necessary change.

The point to take away is that benchmarking is one activity in a product development process, and as discussed in Chapter 1, different companies operate in different markets. As all decisions in product development, the decision over the level and frequency of benchmarking requires wisdom and judgment, and the actions that a benchmarking activity indicate are required should not be lightly ignored.

In any case, the establishment of opportunities and specifications for future products is required whether or not benchmarking is done. We next turn to discuss the activity of establishing product performance specifications.

V. SETTING PRODUCT SPECIFICATIONS

Having benchmarked competitive products on customer and technical criteria, one next step is to use this information to set targets for a new product development effort. Since new product specifications are the purpose behind and culmination of the benchmarking process, we discuss it next. The benchmarking process allows us to understand where there are potential openings in the market and so establish what it would take to take advantage of such opportunities. We now begin to establish these new required levels of performance.

We are therefore leaving behind the first phase of product development—*understand the opportunity*—and are moving on to initiate the second phase of product develop: *develop a concept* (Figure 1.6).

Specification Process

Specifications for a new product are quantitative, measurable criteria that the product should be designed to satisfy. They are the measurable goals for the design team. Specifications, like much design information, should be established early and revisited often.

There are two aspects to a specification that need to be clarified. First, the specification is on a dimension that can support units. That is, there are associated dimensions: meters, degrees Fahrenheit, lumens, horsepower, and so forth. A quantity that has units we will also call an *engineering requirement.* In addition to having units, though, a specification needs a target value. This is a number along the dimensional unit that establishes required performance. A target value can be a specific value or a range: 1, 30–42, $\geq$70, blue, and so on.

Product specifications can occur at many levels at different points in a development process: targets at the preconcept phase are different from refined targets at the embodiment phase. For example, with the coffee mill, knowing or not knowing that one will use a removable chopping chamber in the new concept will obviously make a difference in the appropriateness of a "chamber removal force" specification. Early concept-independent criteria (such as "Opening ease") get refined into performance specifications for a selected concept, which in term get refined into specifications for subsystems, assemblies, parts, features, and so on.

Each specification should be measurable—testable or verifiable—at each stage of the development process, not just at the end of the process when the product is designed and built. In the end, "if it isn't testable and quantifiable, it isn't a specification." The test(s), the means of measuring the performance of the product's system (and subsystems), should always be stated and agreed on up front.

We present here one important milestone of establishing specifications, one that occurs just after a benchmarking activity and just before new concept development. This stage is a point in the development process when overall product specifications should be well considered. Detailed specifications for individual parts and assemblies can wait, but high-level performance targets should be established.

We develop specifications using two approaches, the first from a checklist viewpoint and the second from a viewpoint of the translation of qualitative customer needs. Both are necessary. For the translation of customer needs, we present two methods, a basic approach using the House of Quality and an advanced approach using value analysis. First, however, we will present an important distinction among design spec-

ifications, that of functional performance requirements versus overall product constraints.

Functional Requirements Versus Constraints

When developing the engineering requirements for a product development project, the design team must collect enough information from the customers and other sources to produce a specific set of needs. Engineering requirements fall into two categories, *functional requirements* and *constraints*.

Functional requirements are statements of the specific performance of a design, that is, what the device should *do*. Functional requirements should be stated, initially, in the broadest (most generic) terms. They should focus on performance, be stated in terms of logical relationships, and be stated, initially, in "solution neutral" terms.

A clear definition(s) of the function(s) is essential in design. To solve any technical problem, we need to describe in a clear and reproducible way the relationship between each of the available (or specified) inputs and each of the desired (or required) outputs. These relationships between the inputs and the outputs establish the function of the system (Chapter 5). *In this sense, the function is an abstract formulation of the task that is to be accomplished and is independent of any particular solution (physical system) that is employed to achieve the desired result.* Functions are generally stated in terms of physically quantifiable (measurable) effects and in terms of mathematical relationships. Textual (or verbal) descriptions of functions usually consist of a verb and a noun; "increase pressure," "transfer torque," or "reduce speed." Functional requirements should be stated in these terms, followed by appropriate quantification to measure the specification.

Constraints are external factors that, in some way, limit the selection of system or subsystem characteristics. They are not directly related to the function (or functional objective) of the system, but apply across the set of functions for the system. They are generally imposed by factors outside the designers' control. Cost and schedule are constraints. Size, weight, materials properties, and safety issues such as nontoxic, nonflammable materials are constraints. Specifications relative to surface finish and tolerances may or may not be considered constraints (e.g., in the case of a mirror, a particular surface finish would be considered a functional requirement rather than a constraint).

Constraints can drive the solution of many products, especially large-scale systems. Because of this fact, the constraints should be added with particular care, and no constraint should be added frivolously, but only if it really exists. These guidelines lead to the following guideline:

> Constraints should be established only after critical evaluation.

In addition to identifying functional requirements and constraints, it is useful to guide the specification generation process with a functional decomposition strategy, as in Chapter 5. That is, each specification can be consider as being met when several more-detailed specifications are simultaneously met. By taking this flow-down approach, specifications will be more directly relevant to particular subsystems and components and so have a greater likelihood of attainment. This will be further developed in Chapter 9 in discussions on product architecture and modularity.

Basic Method: Specification Sheets

Customer needs do not necessarily provide a complete picture for a design task. They provide the foundation to focus design efforts, but there also exist other criteria that are important to a design task that the customer may not even perceive, such as standards, ethics, and manufacturing. Therefore, it is important to supplement and complement consumer needs with engineering requirements. One method to supplement consumer needs is to consider a larger "customer" base, including stakeholders, such as the manufacturers, the assemblers, the marketers, and the distributors, and consider them design customers. This approach tends to obscure and diminish the point-of-view of the person who will be buying the product.

Alternatively, we may apply an approach known as *Specification List Generation* that uses decomposition to guide a search for relevant specifications. This approach focuses on specifications that are latent (the customers do need them, but they do not think to express them), such as safety, regulations, and environmental factors. Designating each specification as a required *demand* or a desirable *wish* will communicate its level of importance.

Consider the checklist in Table 7.2 that is quite useful in identifying specifications, (developed by Franke, 1975). Franke studied a number of specification processes in industry to develop this list. It provides a decomposition strategy for developing specifications, listing categories that aid a comprehensiveness and completeness.

Using the Franke breakdown, a convenient procedure for developing general specifications will now be outlined. This procedure represents a basic approach for specification generation, but it must be augmented with the necessary effort and "perspiration."

TABLE 7.2.   CATEGORIES FOR SEARCHING AND DECOMPOSING SPECIFICATIONS (FRANKE, 1975).

Specification category	Description
Geometry	Dimensions, space requirements, . . .
Kinematics	Type and direction of motion, velocity, . . .
Forces	Direction and magnitude, frequency, load imposed by, energy type, efficiency, capacity, conversion, temperature
Material	Properties of final product, flow of materials, design for manufacturing (DFM)
Signals	Input and output, display
Safety	Protection issues
Ergonomics	Comfort issues, human interface issues
Production	Factory limitations, tolerances, wastage
Quality Control	Possibilities for testing
Assembly	Set by DFMA or special regulations or needs
Transport	Packaging needs
Operation	Environmental issues such as noise
Maintenance	Servicing intervals, repair
Costs	Manufacturing costs, materials costs
Schedules	Time constraints

1. Compile specifications. Arrange the functional requirements (FR) and constraints (C) into a clear order. Table 7.3 shows an example specification-sheet template for compiling the specifications. A toy rocket product is shown in the template.

   When compiling the specifications, begin with the functional requirements and then list the constraints. Also, remember that at the preconcept stage, specifications must not be domain or form specific; for example, a specification on "Gear speed" would be inappropriate initially. This guideline on domain specifications only holds true before concepts are developed. Once a preferred concept is selected, the form-independent specifications are expanded into particular form-specific specifications.

2. Determine if each of the functional requirements and constraints is a demand or a wish.

3. Determine if the functional requirements and constraints are logically consistent. Check for obvious conflicts. It is important to make sure that the customer needs (and thus the specifications) can be met and that they are technically and economically feasible. If a system cannot be built to meet the stated specifications or within the stated constraints, the customer should be told immediately.

TABLE 7.3.   SPECIFICATION SHEET TEMPLATE, EXAMPLE OF A TOY ROCKET PRODUCT (PARTIAL)

Date	Demand or wish	Project: Toy rocket design specification sheet functional requirements / constraints	Responsibility	Test/Verification
		Functional Requirements		
1/25	D	Provide thrust for maximum height (velocity > 20 m/s)	DT	Bernoulli and Conservation of Momentum Analysis
1/25	D	Maintain stable vertical flight path (less than than 0.25 m deviation from vertical profile)	JR	Flight tests with prototype, design of experiments
...	...	...	...	...
		Constraints		
		Geometric		
1/25	D	Rocket length $\leq$ 15 cm	WJ	Verify with engr. drawings during concept generation, embodiment, etc.
1/25	W	...	...	...
		Kinematic		
1/25	D	Safe operation ($\leq v_{max}$)	WJ	Verify fluids analysis and prototype testing with impact gauge
...	...	...	...	...
		Safety		
1/26	D	No detachable parts less than 5 cm in diameter (toilet paper roll tube test)	KW	Verify with dimensional check of engineering drawings
...	...	...	...	...

4. Quantify wherever possible. The team may begin with rather qualitative statements, but it is important, in the end, to develop a quantitative statement of the specification—no remaining statements such as "design ease of construction."

5. Determine detailed approaches for ultimately testing and verifying the specifications during the product development process. Examples of tests and verifications include engineering analyses; tests of scaled, full-size, partial, or complete prototypes; checks of engineering drawings; failure modes analyses; or user tests with an appropriate sample size.

6. Circulate specifications for comment and/or amendment. It is helpful to circulate the specifications for comment to all members of the design team, customers, interested colleagues, management, and others.

7. Evaluate comments and amendments. When comments are returned, examine objections and suggested amendments. Resolve the objections and, if necessary, incorporate the amendments in the specifications. It is critical that all specifications be clearly stated and fully justified. If specifications are too restrictive, we may

TABLE 7.4.   EXAMPLE: LOUDSPEAKER DESIGN, QUALITATIVE SPECIFICATIONS VS. QUANTITATIVE

Specification type	Specifications	Quantification
Qualitative:		
	Functional:	
	Broad dynamic range	
	Broad frequency range	
	Very linear	
	Constraints:	
	Use standard box shape	
...	...	
Quantitative:		...
	Functional:	
	Dynamic range	0–100 dB at 1.75 m
	Frequency range	20–20,000 Hz within ± 1 dB
	THD (Total Harmonic Distortion)	less than 0.01%
...	...	...
	Constraints:	
	Geometry	no larger than $X \times Y \times Z$ (m)
...	...	...

miss a better solution. If specifications are not restrictive enough, the goals of a project may not be met. Table 7.4 depicts both quantitative and qualitative specifications for a loudspeaker design before different concepts are developed.

Basic Method: The House of Quality

At this point, from previous work, the design team should understand the customer needs, expressed in their voice. They should also understand the current product (if it exists) and how it satisfies these needs. We now need to determine the priorities for design to achieve the design goals and make the product better. To accomplish this task, we must

▶   find the weakly satisfied customer needs

▶   their dependencies or interrelationships

▶   determine what product changes we can effect to improve these weak points.

This process will define the level of modeling required, both in function and in product components.

2. Determine the customer needs (or WHATs). Customer requirements are the "WHAT IS TO BE DONE" definition of a project. These customer needs may be documented based on the results of Chapter 4.

   ▶ The "what's" can be listed in primary, secondary, and tertiary sequence.

   ▶ List needs in the customer's own voice ("easy," "fast," "lightweight," . . .)

3. Determine the relative importance or priority of the customer needs (scale of 1–5 or 1–10). Importance levels should be determined following the methods in Chapter 4.

4. Translate customer needs into measurable engineering requirements (or HOWs). Determine how the product can be changed in performance to better meet customer needs. The customer domain tells us *what* to do, the engineering domain tells us *how* to do it, at least in terms of measurements. For any customer need, there may be multiple engineering requirements that can be expressed in quantifiable terms. One should document:

   ▶ each *how* in terms of a label and specification value

   ▶ the direction for improvement for each *how*, using a + or − or arrows

5. Determine relationship of engineering design requirements to customer needs. Indicate the relationship and the strength of the relationship between the engineering requirements and the customer needs.

Indicator	Meaning	Strength
⊙	Indicates a strong relationship or much importance	9
○	Indicates some relationship or some importance	5
△	Indicates a small relationship or importance	3
Blank	Indicates no relationship	0

If there are no strong engineering requirements for a given customer need, there is a problem. Possible engineering requirement responses for the customer need should be reconsidered.

6. Perform or execute competitive benchmarking. Here the objective is to determine how the customer perceives the competition's ability to meet each of their needs. Use a simple device to capture customer input, such as a compressed scale such as 1–5, with 1 representing not satisfied and 5 fully satisfied, comparing the benchmark's design attributes with the list of customer needs. This step represents a qualitative benchmarking exercise, capturing the "feelings" of the customer.

7. Rank the technical difficulty of each engineering requirement. Again a pair-wise comparison can be used to determine ranking. The technical difficulty of achieving each customer need in terms of the changes defined by the engineering requirement should also be defined, again using a scale of 1–5 or 1–10.

8. Correlate technical relationships to determine interrelationships of design requirements. This step entails completing the "roof" of the House of Quality. Technical characteristics may be competing rather than complementary. These relationships must be defined and resolved.

Indicator	Meaning
$\oplus$	Indicates high positive correlation
+	Indicates positive correlation
–	Indicates negative correlation
–	Indicates high negative correlation

9. Set engineering requirement targets (specifications) for the product design. One can do this by comparing the requirement measurements of each of the benchmarking products and positioning the new product among these specifications.

Fundamentally, one must consider two factors when setting a target: the cost and the benefit of achieving a value. One might gain some from a very low coffee mill noise specification, but it may be prohibitively costly. One must weigh these qualitatively in the basic House of Quality approach. More quantitative means are discussed in the next section of value analysis.

Setting targets early in the design process is advantageous. Specific values work best for targets. Relatively narrow ranges of values are next best, but if a range is used, be wary of allowing the least satisfactory end of the range to be adopted as a *de facto* target, especially when such an approach is adopted for every range.

10. Select areas for improvement. Similar in spirit to the proposal for redesign above, here we can use analysis of the QFD matrix to define final design targets and to identify areas that need further concentrated effort. To make these decisions, the importance rating of the customer needs must be considered in conjunction with the qualitative benchmarking. This analysis leads to the choice of the most critical engineering requirements through the relationship matrix. The HOWs, technical difficulty, correlation matrix, target values, and quantitative benchmarks should be used to guide further development and product improvements.

Product Example: Automatic Iced Tea Brewer

Figure 7.13 shows a partial House of Quality for an automatic iced tea brewer product. The primary customer needs are listed as the rows of the matrix, ranging from "stronger tea" to "adequately contain steam." These customer needs are converted to measurable engineering spec-

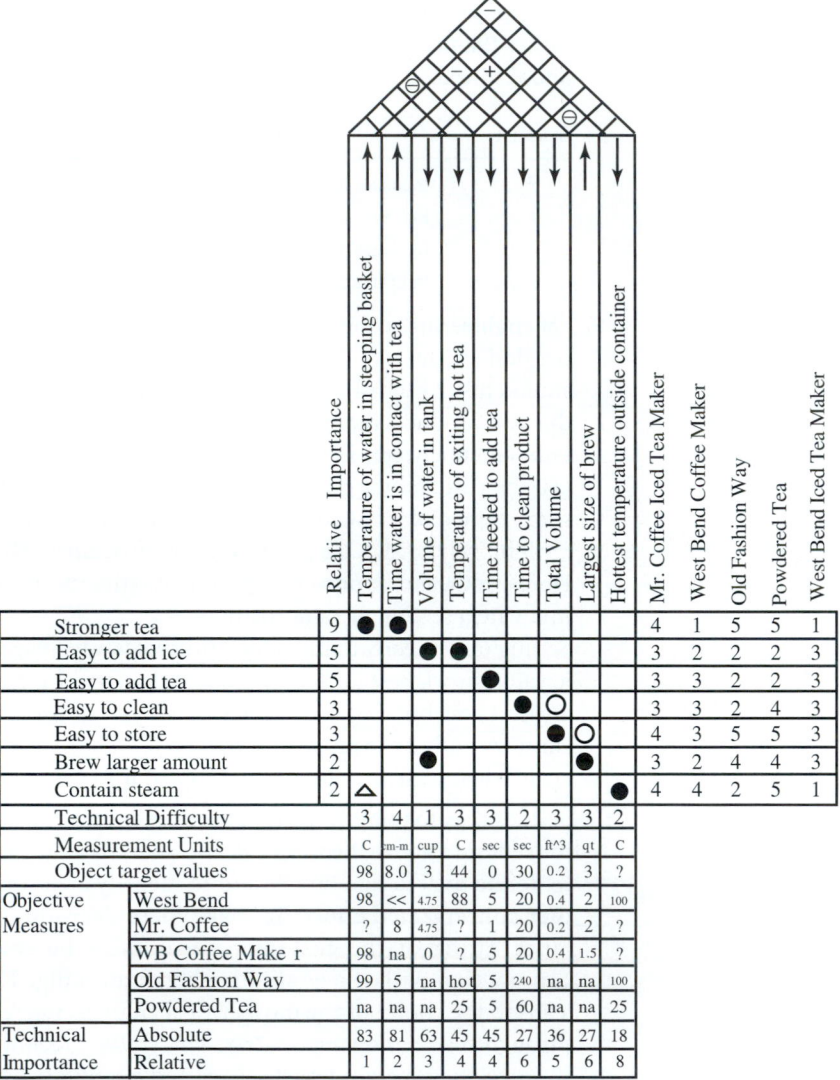

▼ *Figure 7.13.*
Automatic iced tea maker House of Quality (partial).

ifications in the columns of the matrix. In this case, the metrics for stronger tea (not a 1:1 mapping of a customer need to a metric) are temperature of the water and the time that tea is in contact with the water. The metric for easy-to-add ice is the volume of the water in the tank, since this volume, at a high temperature, will define the quantity of ice needed. The greater the volume, the more ice that will be needed. Metrics are listed for the remaining customer needs.

For each of the metrics, units are listed below the customer needs as a row. Arrows are included above the metrics to show the goal of the metric, minimize, maximize, or a target value. In the case of volume of water in the brewer tank, we wish to minimize it, because more hot water will require more ice to cool it. As long as the requisite tea flavor is infused, additional cooler water may be added to the brewed tea to obtain the desired quantity of iced tea.

The matrix cells correlate the customer needs (rows) to the metrics (columns). This correlation is not necessarily 1:1, but is typically 1:many, that is, there will exist more than one metric, on average, for each customer need. The correlation cells are filled with a strong, weak, or no relationship. In the case of the tea brewer, stronger tea, for example, is related strongly to its two metrics, with no (or minimal) correlation to the other metrics.

The roof of the House of Quality, located above the metrics, shows the relationships between the performance metrics. A strong positive relationship indicates that as one metric is significantly improved, the other improves significantly as well (and vice versa). A negative relationship, on the other hand, represents a conflict. If one metric improves, the other will deteriorate. Such conflicts must always be carefully analyzed and monitored. In the case of the tea brewer, the total volume has a strong negative relationship to the largest amount to brew. A large storage container is desired to brew large quantities of tea; however, a small total volume of the brewer is also desired for stowage purposes. These metrics strongly conflict; yet, by separating the storage container from the brewer (separation in time and space), stowage problems are reduced, decreasing the importance of "easy to store." Overall, both of these metrics must be analyzed together to understand the tradeoffs in the conflict.

Qualitative and quantitative benchmark values are also shown on the House of Quality. These values help to understand the market position of a product. They also provide a logical means of setting target values for product evolution. In the case of the tea brewer, ideal goals are given by the benchmark comparisons to the powered tea brewing and old-fashioned method of tea brewing. These goals provide a normalization when comparing products. The tea brewer QFD also shows a comparison of two tea brewers currently on the market. If our product is the West Bend brewer, we need to set aggressive goals in stronger tea containing steam and stowage space to compete in the equivalent value market. Target values of the West Bend show goals for meeting or surpassing the competition.

Finally, the House of Quality shows the relative and absolute ranking of the product metrics, as listed at the bottom of the figure. For the tea brewer, product development should focus on the first three or five metrics to satisfy the voice of the customer. These choices depend on the resources (time and money) available and the technical difficulty expected for improving a metric (as shown in the matrix). The tea brewer shows that the volume of water in the tank is relatively easier to address technically compared to the other high-ranked metrics. This metric should thus be addressed first, in conjunction with temperature of the water due to the strong negative relationship.

In sum, the House of Quality provides a large quantity of information in a very concise and well-organized form. A logical progression through this information leads to the setting of priorities, allocation of resources, and the development of real engineering specifications (metrics) for a product. It also establishes, at a basic level, the current market status of a product and the desired target values for surpassing the competition.

Comments on House of Quality

A number of hints exist for effectively using the House of Quality. For example, one should not let the matrix grow too large; one should keep it under 50 rows and columns. If it gets too large, it becomes unwieldy. To keep it simple, one should operate at different levels in the product.

For example, considering the benchmarking of automobile product, one can develop vehicle-wide specifications with a vehicle-wide House of Quality. Entries might include overall dimensions and weight (measurable) and "ease of unlocking" (not measurable). These specifications can then be flowed down to door-level specifications and a separate House of Quality completed at the door level. Here the "ease of unlocking" specification might flow into a measurable key-turning torque specification. Similarly, the "ease of unlocking" might flow down differently into a different measurable specification on the electronics subsystems House of Quality, where a keyless remote specification might be established for distance that the remote operates. Putting both of these and their counterparts in one large detailed vehicle House of Quality is unreasonable. Separate House of Quality's can be developed along the functional decomposition of the product. This approach will be further developed in Chapter 9 in discussions on product architecture and modularity.

Another hint is to use the function structure to help establish the specifications. Every subfunction has flows in and out. Differences between these flows are readily measurable and so are candidates for specifications. Chapter 6 details an approach for relating product functions to engineering requirements (metrics).

Finally, it should be kept in mind that the intent of the House of Quality is consensus building. It is a tool to ensure that the variety of specifications, typically representing a variety of different disciplines and development subgroups, all converge to a successful product. The matrix does not generate specifications; it documents them.

Advanced Method: Value Analysis

In the approaches discussed so far, target values are established by design team judgement. Basically, for any engineering requirement, a target value is determined by simultaneously judging the cost of attaining that target and the customer desire in delivering that target.

A more quantitative approach is to create models of these two factors. For each specification value, one might create a model of customer preference over the possible target. This model can be developed by using customer questionnaires ("How much more would you pay for twice as much performance") or through conjoint analysis studies, both discussed in Chapter 4.

Similarly, one could estimate the cost of delivering different levels of performance, based on estimation of components required and their cost of manufacture. With these two models of customer desire D and cost to produce C, both measured in dollars, one can determine the foremost target value to use.

We can define *value* or *worth* as the difference in the desire of the customer from the cost of producing it:

$$V = D - C, \tag{7.1}$$

and then we can pick a target value that maximizes this quantity. Other forms include

$$V = \frac{D}{C}, \tag{7.2}$$

a normalized form that can be less sensitive to model errors, such as if all of the cost estimates are made using the same cost analysis tool and are off by the same factor. A normalized form also need not have the desire and the costs expressed in identical units.

This analysis can be completed in detail. For example, the cost function can be expressed over the subassemblies and down to the components of a product. The desired function can be expressed as an overall function of the specifications. The specifications can be functionally flowed down over the systems using the function structure. Using these flow-downs of cost and desire, any subsystem or component can have its relative value determined. Specifications can be established for sub-systems and components by maximizing its value. When moving into the embodiment phase of design, components and subsystem choices can be made based on the value.

Example: Coffee Mill

As an example, we show here value analysis applied to an assessment of the Krups Coffee Mill. The first step to obtaining value figures on each component is to determine the expressed desire for each compo-nent as mapped from the customer needs. We perform this mapping through the functional carriers.

First, the subfunctions of the function structure are listed in a QFD ma-trix versus the customer needs, complete with importance ranks. Then, the fraction of the customer need carried by each subfunction is indicat-ed in the matrix. If the need is a constraint, it is copied over to a column in the matrix. A major fragment of this mapping is shown in Figure 7.14.

Then, at the bottom of the matrix, the importance of each function is determined through a propagation of the need importance ranks in the usual way (Chapter 4). Then, the desire of each function can be deter-mined by taking the selling price and distributing it across these weights:

$$D(\text{function}_i) = \sum_{\text{Needs}j} w_j M_{ij} \tag{7.3}$$

where w_j is the normalized importance of need j, and M_{ij} is the relation-ship matrix value. This process is shown at the bottom of Figure 7.14.

Next, the desires of each function need to be propagated onto the parts and assemblies, either for the previous product or for new embodied concepts as will be developed in later chapters. This propagation can be executed by creating a QFD matrix, similar to the previous section, re-lating the parts and assemblies to the functions they fulfill. This process is shown in Figure 7.15. Note, however, that some functions are not completely fulfilled by the parts or assemblies. The fraction not com-pleted can be listed in a new bottom row called "User Effort," since the user must complete these.

The next step is to determine the cost of each part and assembly. An in-dented cost analysis as discussed in Chapter 14 provides an approach for completing this step and as presented in Figure 7.8 for the coffee mill.

Having developed these data, we can evaluate the value of each part, de-termined in Figure 7.15 using the additive form Equation (7.2). Note that

▼ *Figure 7.14.*
Customer need to function mapping.

					Accept Beans	Measure Beans	Hold Beans	Seal for Chopping	Chop Beans	Contain Chopping	Hold Coffee	Unseal from Chopping	Pour Out	Open Cord	Extend Cord	Plug in Cord	Unplug Cord	Store Cord	Accept Grip	Actuate Power	Chopping Done?	Spin Armature	Seal Shaft	Hold Spinning Armature	Dampen Armature Vibration	Enclose Motor Noise	Support Chopper	Open for Cleaning	Clean Chamber	Close for Reuse	Low cost	Sleek shape	Compact adequate box	Compact for storage	Storable shape	Small faceprint	Small footprint
Part #	Part Name	Value	Cost	Desire	0.35	0.75	0.22	0.82	0.64	0.34	0.43	0.10	1.53	0.13	0.37	0.20	0.20	0.61	0.84	0.55	1.10	0.17	0.06	0.13	0.42	0.14	0.54	0.47	1.62	0.47	0.17	0.22	0.19	0.31	0.28	0.31	0.33
Whole Ginder	MILL	$ 8.03	$ 3.68	$ 11.71																																	
001	Lid	$ 2.05	$ 0.21	$ 2.27	0.1	0.3		0.6		0.6		0.6	0.2							0.3								0.7	0.7		0.3				0.2		
A1	GRINDER ASSY	$ 6.00	$ 3.02	$ 9.02																																	
002	Transfer Pusher	$ 0.08	$ 0.08	$ 0.17																0.3																	
S1	BASE ASSY	$ 2.36	$ 0.43	$ 2.80																																	
003	Power Switch	$ 0.16	$ 0.07	$ 0.22																0.4																	
004	Power Cord	$ 0.07	$ 0.24	$ 0.31											0.4	0.4	0.4																				
005	Base	$ 2.18	$ 0.09	$ 2.27										0.1				1.08						0.5	0.5	0.3	0.9									0.2	0.4
S2	BASE ASSY	$ 3.59	$ 2.47	$ 6.06																																	
011	Blade	$ 0.24	$ 0.14	$ 0.38					0.6																												
S21	MECH ASSY	$ 5.46	$ 0.21	$ 5.67																																	
012	Chamber	$ 1.73	$ 0.18	$ 1.91	0.7	0.4	0.6	0.1	0.2	0.6	0.1																	0.3	0.3	0.3							
S211	MOTOR&BODY	$ 0.76	$ 2.08	$ 2.84																																	
013	Felt Ring	$ 0.03	$ 0.03	$ 0.06																					1												
014	Screws	$ 0.08	$ 0.01	$ 0.10																							0.1	0.1									
015	Body	$ 2.40	$ 0.29	$ 2.69	0.2	0.3	0.4		0.3	0.4	0.4	0.3	0.8																0.7		0.7				0.6	0.4	
S212	MOTOR ASSY	$ (0.79)	$ 1.71	$ 0.93																																	
016	Motor Mounts	$ (0.15)	$ 0.10	$ 0.25																				0.5	0.5												
017	Motor	$ (0.86)	$ 1.54	$ 0.68					0.2													1														0.6	0.6
A2	PACKAGING	$ (0.02)	$ 0.44	$ 0.42																																	
X1	Box	$ (0.15)	$ 0.28	$ 0.13																														0.7			
X2	Foam	$ 0.02	$ 0.03	$ 0.06																														0.3			
X3	Questionaire	$ (0.03)	$ 0.03	$ -																																	
X4	Dealer List	$ (0.03)	$ 0.03	$ -																																	
X5	Instructions	$ 0.20	$ 0.03	$ 0.23	0.1	0.1						0.1							0.2																		
	User Effort		$ 3.29											0.9	0.6	0.6	0.6				1							0.7			1			1			
	TOTAL PRICE		$ 15.00																																	

▼ *Figure 7.15.*
Parts to functions matrix.

some parts carry much value and others do not. The lid carries much value given its function; the user perceives it. The motor, even though it drives the entire mill, does not carry so much perceived value. It is just expected.

Note that the value can be back-propagated onto the functions, to determine which functions carry the most value. These results can then be explored to prioritize redesign efforts.

Comments on Value Analysis

Value analysis is very quantitative and numerical, and so engineers often find it appealing. On the other hand, the numbers are often not so easily quantified. For example, from the company's point of view, the desire function D is the expected revenue that will be drawn off the product. Revenue is notoriously difficult to predict. A new movie may come out where the lead character uses the product, and then its sales and revenues may go through the roof. Or the product could be on the other side of the coin and sales drop flat. The economy, states of war, proximity to Christmas, actions of competitors, price, and many other issues all affect product sales and therefore affect the revenue. It is a noisy, lagging, measurable reflection of market desire D.

Another difficulty is expressing desire D in terms of performance specifications or functions. The weighted sum expression is perhaps valid only locally about some nominal target. Over any range of interest, the desire function D is very nonlinear. In particular, the worst behavior of the formulation is that it enhances desired functions and mitigates expected functions, as viewed on the Kano scale (Chapter 4). In the coffee mill example, when using the weighted sum formulation, the coffee mill lid is deemed more valuable to the customer than the motor. This result is simply mapping down that the customer needs list expressed *desires* (high expressed importance), not *expectations* (low expressed importance). The customer really perceives the lid in many of the product functions and inadequacies. In questionnaires, customers do not comment much on the motor-related functions, which work as expected.

Before dismissing the exercise for this reason, though, it should be pointed out that the linearized result is usually a good thing to know. In the mill example, it is good to know that the lid carries much more of the customer perceived limitations, and so time ought to be spent on it. It is a flag as to the lid's importance, not the motor's unimportance. As with many things in design, one must interpret the analysis results with wisdom. As a final thought on the weighted sum linearization, more-complex models, such as utility theory, attempt to model this complexity as a single decision-maker behaving according to restrictive assumptions, but these are again more-refined approximations that may or may not construct a better representation of market value.

Generally speaking, value analysis is a useful technique for comparing alternatives and alternative specifications in reasonably well-understood domains, where the customers perceive and state well their expected and desired needs, and domains where the technology or markets are reasonably unchanging and can be forecasted.

VI. SUMMARY AND "GOLDEN NUGGETS"

Competitive benchmarking, product teardowns, and establishing product specifications are required practices for effective product development. Doing so systematically offers some assurance in gaining effective understanding from the activity. Some key ideas include:

1. tearing down and documenting the competition on how they solved design problems

2. analyzing the competition on a function-by-function basis, rather than on an equivalent-form basis

3. use of technical and business literature to help understand the competitive forces

4. summarize the results in summary documentation, including QFD, best-in-class values on each function, and technical trends with s-curve plots

References

Clausing, D. 1994. *Quality function deployment.* Cambridge, MA: MIT Press.

Cleary, T. 1988. *The art of war,* by Sun Tzu. New York: Random House.

Consumer Reports. 1991. 56(11):November, pp. 740–746.

Franke, H.-J. 1975. Methodische Schritte beim Klaren konstruktiver Aufgabenstellungen. *Konstruktion* 27: 395–402.

Fuld, L. 1994. *The new competitor intelligence.* New York: Wiley.

Hauser, J., and Clausing, D. 1988. The house of quality. *Harvard Business Review* (May-June): 63–73.

Jacobson, G., and Hillkirk, J. 1986. *Xerox: American samurai.*

U.S. Economic Classification Policy Committee. 1997. *North American industry classification system—United States,* Office of Management and Budget.

Szuluk, C. 1998. Keynote address, *31st CIRP International Workshop on Manufacturing Systems,* 27 May, Berkeley CA. (Reported by K. Otto.)

Urban, G., and Hauser, J. 1993. *Design and marketing of new products.* 2d ed. New York: Prentice-Hall.

8 Product Portfolios and Portfolio Architecture

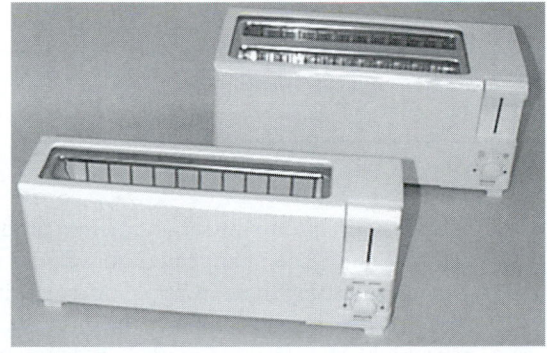

Different product portfolios implemented under different architectures.

Before a product can be developed into a concept or even into a reasonable set of specifications, it is important to develop the larger corporate environment within which the product will exist. Companies typically do not survive based on revenues from a single product but rather offer a variety of closely related products, all of which must be evolved over time. It is critical to make effective configuration choices for the corporate set of products.

In the late 1980s, Sun Microsystems changed their computer hardware architecture from a Motorola 680X0 microprocessor to a SPARC RISC processor. When they made this change, all software compiled for the previous microprocessor would not run on the new systems, making their networked client/server architecture more difficult to use. At the time, Sun was susceptible to market share loss, since customers had to change both their hardware and software. In the early 1990s, on the other hand, Apple Computer made a similar upgrade in processors from the same Motorola 680X0 microprocessor to a PowerPC microprocessor. Apple, however, developed a system where the old 680X0 software programs would run on the new microprocessors. Apple's architectural transition was transparent to the customers and proved to be a success. Understanding how one's products should be compatible among one another and among new generations is critical to profitability.

I. CHAPTER ROADMAP

In this chapter, we review the different types of portfolio architectures that a company can apply. The chapter is divided into three distinct sections, one discussing portfolios, the next providing the underlying customer need analysis to determine how to architect a portfolio, and finally a discussion on product architecture for platformed products. Figure 8.1 provides a chapter roadmap.

II. PRODUCT PORTFOLIO ARCHITECTURE

Background

Product portfolios are the set of different product offerings that a company provides. A company can choose from a variety of different strategies for providing these multiple offerings: it can make each product completely unique, it can make the products share a common system, or it might choose some method in between. There are market and cost advantages to any approach. A *product portfolio architecture* is the

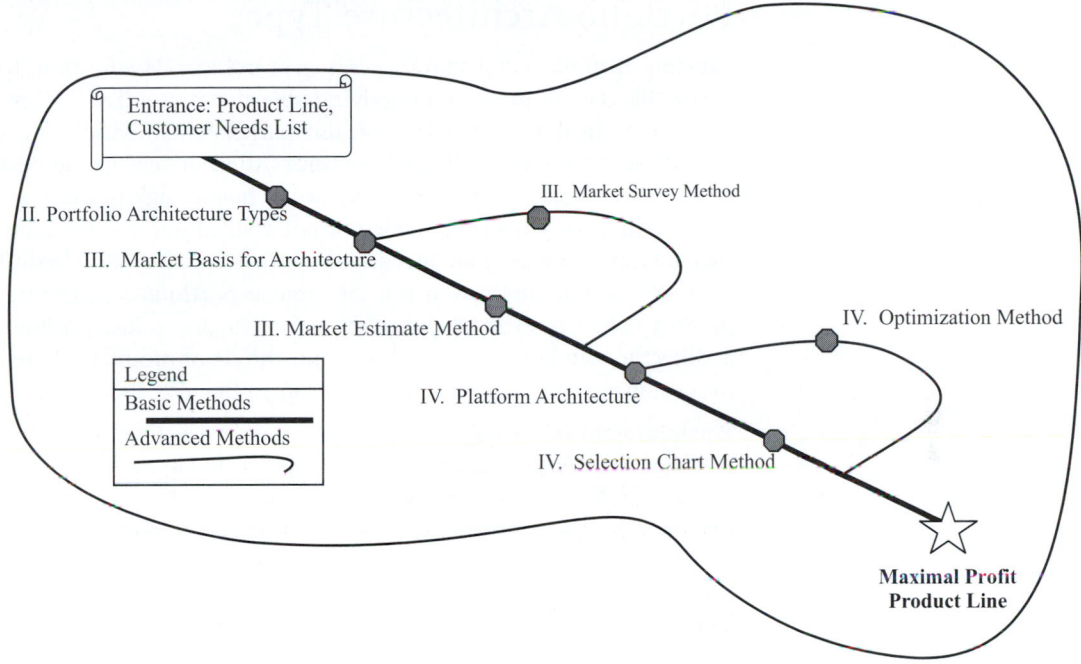

Entrance: Product Line,
Customer Needs List

III. Market Survey Method

II. Portfolio Architecture Types

III. Market Basis for Architecture

IV. Optimization Method

III. Market Estimate Method

Legend

Basic Methods

IV. Platform Architecture

Advanced Methods

IV. Selection Chart Method

Maximal Profit
Product Line

▼ *Figure 8.1.*
Chapter roadmap.

system strategy for laying out components and systems on multiple products to best satisfy current and future market needs.

There are two basic corporate objectives considered in developing a product portfolio architecture: costs and revenue. Revenues increase with expanded offerings in a larger portfolio, as a company can then make products more tailored to each customer in the market. On the other hand, costs go up with the added complexity of developing, supporting, and manufacturing a larger set of different products. Thus, the decisions made about how to support multiple products can have a large impact on the overall profitability of a product portfolio. Choosing a portfolio with a large number of unique products has the potential of high revenue from multiple market segments but at high manufacturing cost due to the required production complexity. A portfolio with only a single product has low production cost but also limited market satisfaction. One portfolio design task is to determine where one might develop subsystems within the products that can be reused across the different products, since this reuse can permit a manufacturer to attain both low cost and large market variety. Yet, this approach is not always appropriate. To explore portfolio design, we first review different portfolio architectures.

Portfolio Architecture Types

Labeling portfolio architecture types is an exercise in classification. To create this classification, an underlying objective is required to form the basis of the distinctions. Here we use market demands as the basis for differentiating portfolio architectures. Alternatively, in the next chapter, we will discuss *product architecture,* or how a single product satisfies its functions by being divided or not divided into modules and thereby classify product architectures based on function and physical form. Product architecture is not the same as portfolio architecture; product architecture can be spoken about for a single product, portfolio architecture can be spoken about only for the set of products offered by a company. This will be elaborated more in the next chapter.

Portfolio architecture falls into three basic categories: *fixed unsharing, platform,* and *massively customizable,* as depicted in Figure 8.2. We now define and develop each of these in detail in separate sections. Note that these categories are not mutually exclusive: One could have a product portfolio architecture that exhibits both fixed unsharing architecture on some customer needs and platform architecture on other needs. This differentiation will be explained in greater detail.

Fixed Unsharing Portfolio Architecture

A *fixed unsharing portfolio architecture* is defined as when each product in a portfolio is unique and shares no components or systems with any other product member in the portfolio. Figure 8.3 displays example product portfolios based on a fixed unsharing portfolio architecture. The portfolio of magnetic media products offered by Sony constitute a fixed unsharing portfolio architecture; each product in the portfolio shares no components with the other member in the portfolio. Similarly,

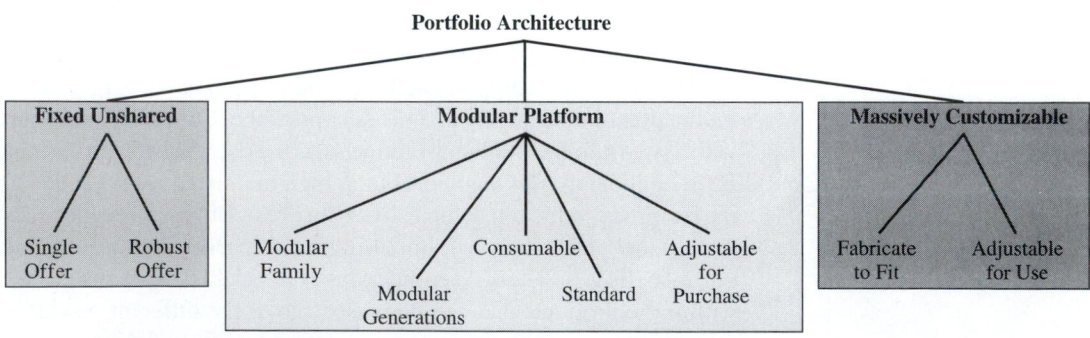

▼ *Figure 8.2.*
Portfolio architecture types.

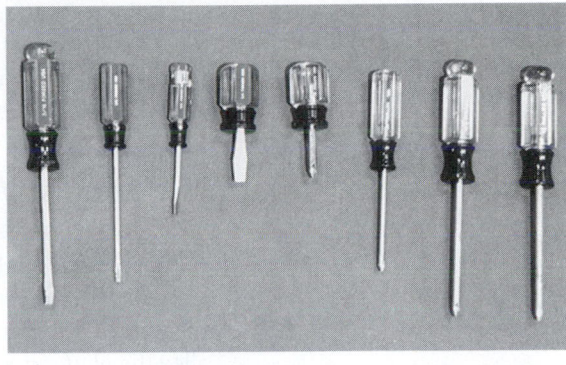

(a)   (b)

▼ *Figure 8.3.*
Examples of integral family architecture. (a) Sony magnetic media portfolio.
(b) Screwdrivers.

the screwdriver portfolios shown in Figure 8.3(b) are fixed unsharing in nature. While similar properties are shared (material properties, etc.), each product in the portfolios includes its own components. These components are produced only for the individual product.

Fixed unsharing architecture is typically applied when the product has very high volumes, implying that economies of scale exist to remain competitive. Magnetic storage media is one such example. Manufacturers do not really care if their VHS tape can share components with their cassette tape; each must be minimal in cost. This choice typically means application of design for assembly principles to the product to reduce the number of parts. This application of assembly principles consequently increases the function sharing of each part (each part serves many functions), implemented using complex injection molded practices, for example. A complex part is less likely to be reused on a different product in the portfolio.

Fixed unsharing architecture can be further classified into two types, those of a single level of performance and those of a robust nature. That is, a market may seek variety. The fixed offering may provide only one option to the entire market and be less than ideal for many customers. The first Model-T Ford was black—and black only. On the other hand, the fixed offering might be naturally robust to the variety in the market and of itself provide for the variety. For example, European, Asian, and American power grids all operate at different voltages and frequencies. A product might be manufactured with a power supply designed for 110 V, 60 Hz input and thereby not be satisfactory to European customers, who might have a power grid system of 240 V, 50 Hz. On the other hand, the single power supply might be capable of being plugged into any voltage from 110–240 V, and any frequency from 50–60 Hz. Such a power supply is robust. It meets all market variety naturally and yet is a fixed unsharing portfolio architecture with other power supplies that a company may offer.

Platform Portfolio Architecture

When a company offers products that share components, modules, or systems to meet market variety, we call the configuration layout, within the set of products and their shared elements, a *platform portfolio architecture*. The common components, modules, and systems are also called the *platform*, and the supported products are called the *variants*. There are at least four types of platform architectures: modular product families, modular product generations, consumable platform architecture, and adjustable-for-purchase architecture. We now detail each of these general types of platform architecture.

Modular product families: A *modular product family* is defined as the set of products supported at any one time by a platform. For example, in Figure 8.4(a), the Krups Toastronic toaster has both a two- and four-slice product model in the Toastronic toaster family. The family has shared internal components such as heating elements, reflectors, and controllers that make up the platform.

One consequential opportunity of a modular platform architecture is the development of *derivative products*. These are family variants that arise historically by taking an original product and making changes to it to offer more products. Because of the established product design of

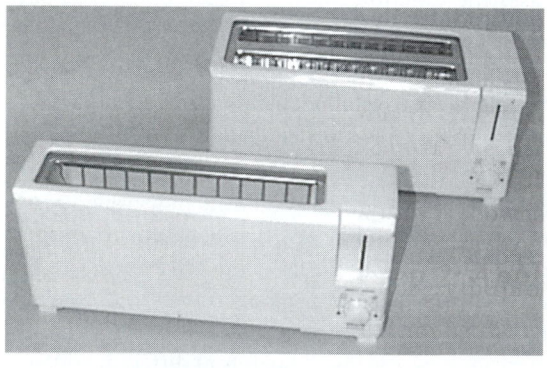

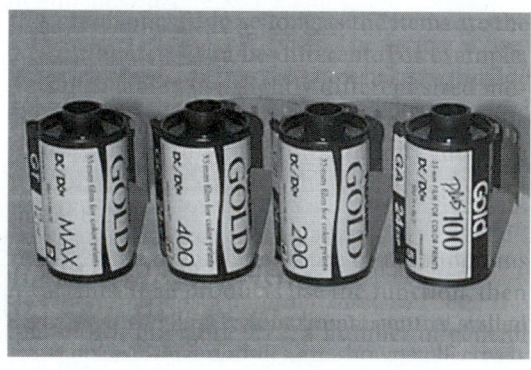

(a)   (b)

▼ **Figure 8.4.**
Examples of modular product families. (a) Krups Toastronic toaster two- and four-slice toast capacity family. (b) 35-mm film speed family.

the original product, the development of derivative products requires less cycle time and project costs.

Cutherell (1996) describes three basic types of derivative products:

▶ *Cost-reduced derivatives:* The cost of an original product in a family is reduced by choosing alternative materials, removing product features, improving production efficiency, or optimizing aspects that were originally overdesigned. The new product is offered separately in the product family, and the majority of the cost savings is passed on to the customer (e.g., Video Cams; plastic vs. metal snow shovels).

▶ *Product-line extensions:* In this case, on a basic product, the features are extended or modified to address more customer needs or a different market or submarket. Alternatively, the current design features of a product are redesigned to improve engineering metrics, such as accuracy, reliability, and durability. This type of derivative product is only an extension to an original product, without significant price increases passed on to the customer. Yet, it naturally forms a platform and a family (e.g., convertible cars, cars with air conditioning).

▶ *Enhanced products:* These products build from a basic model, adding additional features that address alternative markets or more-difficult customer needs. The original product features, however, are redesigned and extended into a platform, where the platform interfaces are planned to support multiple variants. An enhanced product usually requires a significant price increase when it enters the market (e.g., Sony Walkman with and without radio, Swiss Army Knives).

Modular product generations: A portfolio platform can support multiple products not just at any given moment but also over time. A *modular product generation* is defined as the architecture for product offerings that share the same modular components in offerings that succeed each other in time. For example, in Figure 8.5, the Krups Toastronic two-slice toaster is shown for two generations. The internal components are identical and form the platform. The stylish outer shells form the variants and change from generation to generation, permitting easy upgrades in appearance as market tastes change.

Platformed families and generations arise due to requested variety in customer needs and functionality and due to the availability of technology. As new manufacturing processes, technology, and markets become apparent, new platforms are developed and introduced.

Scalable platform Another type of platform architecture is the *scalable platform*. Here, the products in the family share no common components, but are all the same except for size. What is common among the variants is the production or development activities

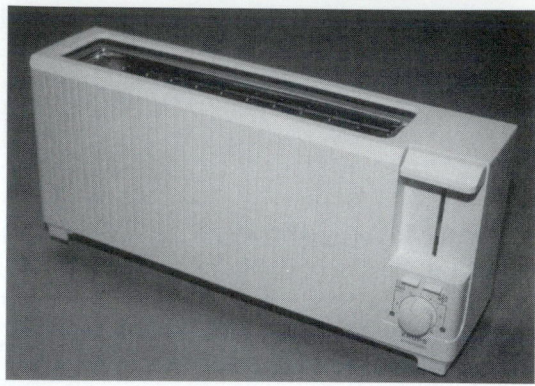

▼ *Figure 8.5.*
Krups Toastronic toaster modular product generations.

used to create the products. With a scalable platform, the functions of the product variants are basically identical. As an example, Pratt and Whitney's jet engine family is a scalable platform. Each jet engine has the same systems that each perform the same functions. Yet, to achieve different thrust levels, the components themselves are all different in size and material.

Consumable platform: Another type of platform architecture is the *consumable platform*. Here, one isolates components that are consumed quickly. For example, 35-mm film is a consumable portfolio platform for photographic cameras, as the film is isolated from any other part of a 35-mm camera. This isolation satisfies the need to take many pictures with a camera before disposal of the camera is required. This distinction should not be confused with the platform architecture also exhibited by 35-mm film on another customer need of film speed. That is, there are several speeds of 35-mm film available, such as 100, 200, and 400 speed (see Figure 8.4(b)), all of which form a platform portfolio architecture for 35-mm film.

Figure 8.6 illustrates two mechanical consumable platforms, 35-mm film and a Canon toner cartridge. The toner cartridge is identical for both the HP Laserjet and the Apple Laserwriter series of printers.

Standard platform: Related to the consumable portfolio platform is the *standard portfolio platform,* where a subset of a product system in a portfolio of products is a platform that conforms to an industry-agreed standard. Standard-sized or standard-quality components that can be supplied by many vendors, such as fasteners and the like, are

 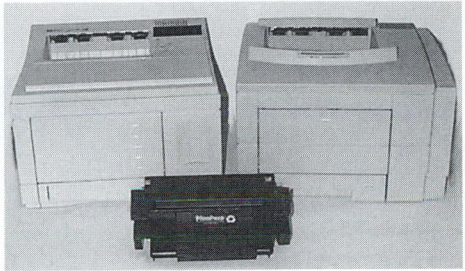

▼ *Figure 8.6.*

Examples of consumable platform portfolio architecture. (a) 35-mm film platform for cameras. (b) Toner cartridge platform for printers.

typical. Also typical, however, are complex products that rely on standardized software systems. Operating systems and file formats both form different standard platforms.

From a corporate management point of view, there are two types of standards, open and proprietary. *Open standards* are those that a company or organization publishes, and anyone can sell a product conforming to the standard and not pay a royalty fee for copyright infringement. Proprietary standards are those that a company or consortium of companies develops, and typically any producer who sells a product making use of the standard must make a royalty payment. Typically, the company developing the standard works to prevent others from selling competing products.

Standards as a platform present interesting issues from this viewpoint of product line business strategy. If we are going to launch a product that requires a standard for people to use it, do we make it an open standard, or do we try and keep proprietary control? Which direction offers greater long-term benefit? Attempts at closed proprietary standards that failed in the sense that the product is no longer offered include Sony Corporation's Betamax video tape and the Digital Corporation's VMS computer operating system. Open systems that survived these are the VHS video tape and the UNIX operating system. Any company can sell a product using these standards and not infringe on patents or copyrights. On the other hand, Microsoft Windows is a very successful proprietary operating system, and newer media formats such as DVD have competing proprietary standards set by consortiums.

Adjustable for purchase: Different market segments may have different requirements for some of the subsystems in a product. For example, computer manufacturers sell to a variety of customers with

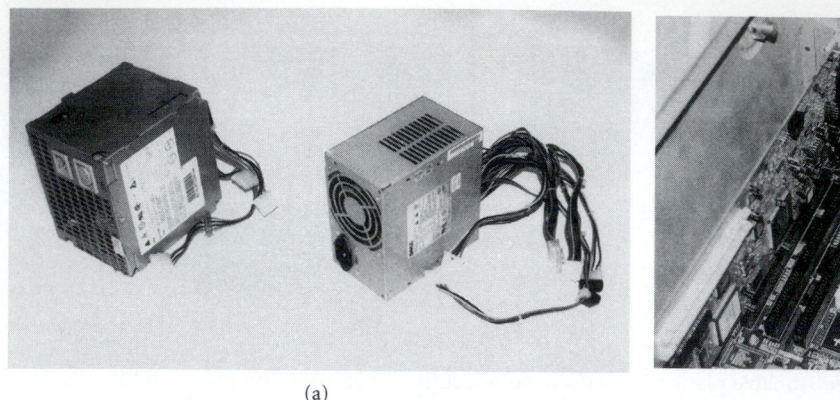

(a)   (b)

▼ Figure 8.7.
Examples of adjustable-for-purchase portfolio architecture. (a) Power supplies. (b) PC
Cards.

different electrical power input requirements. To meet this need, the
power supply function might be isolated from the rest of the product
as a module, such as that employed by Hewlett-Packard on their printers (Feitzinger and Lee, 1997). The different power supplies available
on the basic product form an *isolated adjustable-for-purchase product
family* (Figure 8.7). There is a requirement that is isolated and can be
met by the manufacturer, retailer, or even customer by inserting different subsystem units. During use by the customer, the power supply
setting is fixed and remains so until the customer perhaps purchases a
different power supply and installs it. Once installed, however, it is fixed.
As another example, computer RAM is now typically isolated from the
rest of the computer. At purchase, a customer can specify the amount
of RAM installed. After that time, it is fixed for the customer, unless
an additional purchase of RAM is installed. Once installed, however, the
product configuration is fixed.

Mass Customization

Platforms and product families offer the opportunity for greater variety at lower cost. The variety supplied, however, is fixed at the discrete offerings in the family, which may not be sufficient for some
customers. A *mass-customization portfolio architecture* is defined as a
portfolio architecture with features in the basic platform that can be
varied depending on the desires of an *individual* customer. This permits greater individual tailoring of each product to each customer
(Pine, 1992; Pine, Victor, and Boynton, 1993). The individually requested variations in the platform may require an increase or decrease

in cost of the mass-customized product, or the price may not be affected significantly. There are at least two different types of mass customization architectures: fabrication-to-fit and adjustable-for-use.

Fabricate-to-fit: The adjustable-for-purchase portfolio architecture has each product configuration determined by the modules constructed by the manufacturer; hence performance is set at values determined by the manufacturer. A customer may want different values. A *fabricate-to-fit* mass customization platform is one where the customer can special order the platform at the exact specifications desired. For example, in the early 1990s, Toyota attempted to permit every customer to special order their vehicle from a very large array of options (see Figure 8.8(a)). The color, interior, sound system, wheels, and so forth could all be selected as desired. Ultimately, this available variety introduced such complexity into Toyota's production system that it failed. It proved too expensive to continue and remain competitive, so Toyota dropped the program. Production complexity is a common problem with this portfolio architecture.

Adjustable-for-use: With the adjustable-for-purchase portfolio architecture, the manufacturer sets each product at time of sell, and then the settings remain fixed for the rest of the product life. This approach may not be acceptable to the customer, who may wish to change the settings dynamically at any point in time. For example, a customer may wish to have a power supply architecture whereby the customer can at any time change the input power voltage and frequency, say, by flipping a selection switch. Providing features to the customer to permit the product to become any member in the "portfolio" is an *adjustable-for-use* platform.

As another example, automobile seat positions are adjustable at any point in time by a customer, even after purchase (see Figure 8.9(a)). Cameras that permit adjusting the focus to different distances are an adjustable-for-use architecture; they are not focused by a single lens to one or two fixed distances (see Figure 8.9(b)). These platforms were designed to not have a fixed offering, no multiple fixed offerings, and no offerings fixed at time of sale, but rather to have an adjustment that permits any offering at any point in time.

The differences among the modular family, adjustable-for-purchase, adjustable-for-use, and fixed unshared architectures can be confusing. The power supply example can help clarify. If a power supply is fixed for one and only one input voltage and frequency (say 110 V, 60 Hz), it exhibits an fix unshared architecture on the customer need of different input voltages. If the manufacturer permits selection of two different options (say 110 V, 60 Hz and 220 V, 50 Hz), both of which are fixed in a product once ordered and purchased, it is from a modular

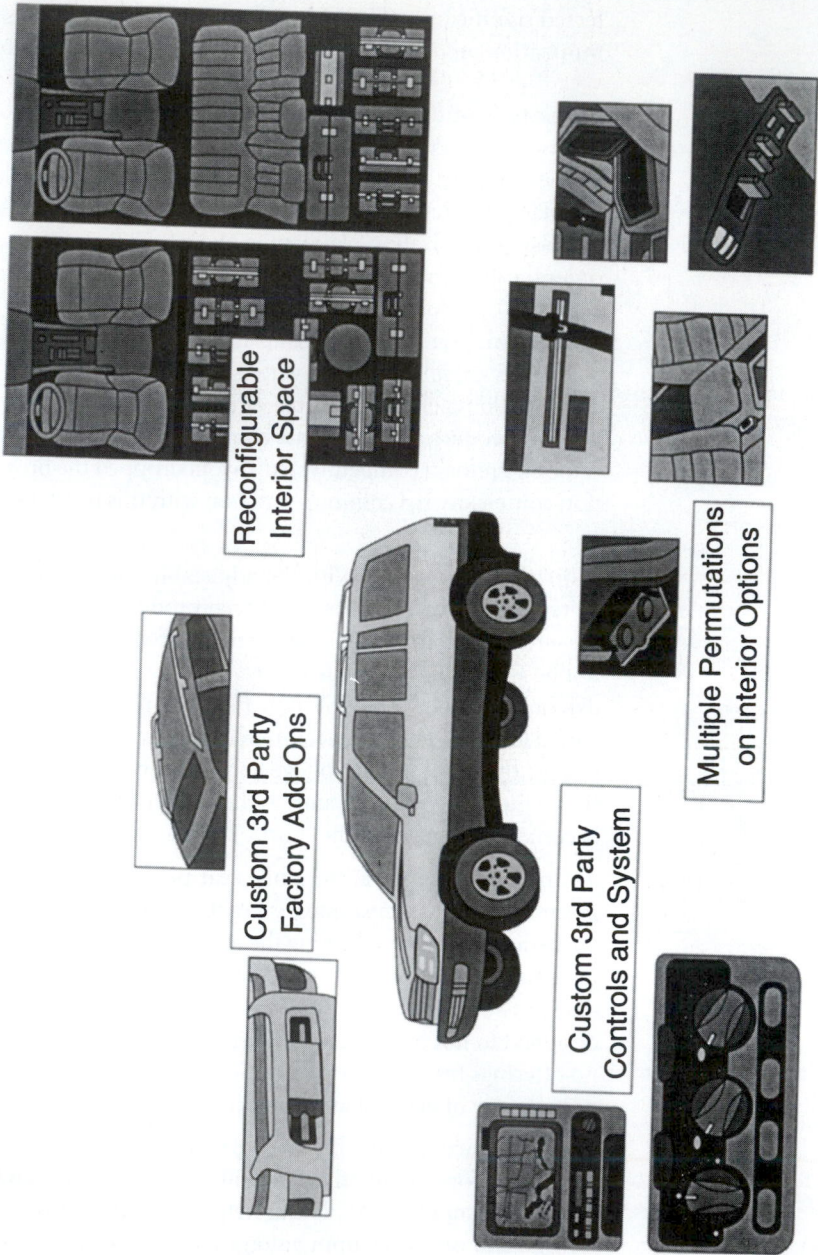

Reconfigurable Interior Space

Multiple Permutations on Interior Options

Custom 3rd Party Factory Add-Ons

Custom 3rd Party Controls and System

▼ *Figure 8.8.*
Examples of fabricate-to-fit massively customizable portfolio architecture.

(b)

(a)

▼ *Figure 8.9.*
Examples of adjustable-for-use portfolio architecture. (a) Fully
adjustable car seat architecture. (b) Fully adjustable focus architecture.

product family. If the two can be changed at a later date through replacement of a module, it is an adjustable-for-purchase architecture. If
the selection between the two input powers is made not by variants
but on the platform by a switch or some other setting change that a
customer can adjust, it is an adjustable-for-use architecture. Finally, if
the power supply is robust and can operate at any input power from
110–240 V and 50–60 Hz (such as on some portable computers), then
the architecture reverts to integral: There are no components on the
product to choose from, select, or adjust for use.

III. CHOOSING AN ARCHITECTURE TYPE

As shown by the previous examples, selecting the portfolio architecture is complex and critical. Much of the content of subsequent product design activity is determined by the decisions made at this stage.
One approach to selecting a portfolio architecture is to base it on customer need variety (Yu et al., 1998).

Theory

The complexity in portfolio architecture stems from a product portfolio exhibiting many types of portfolio architecture simultaneously. For example, 35-mm film is both a modular family (different speed films) and a consumable platform (film is separated from the camera).

To untangle the complexity, it is important to realize that the classification of a product portfolio must be completed on every important customer need. That is, out of all of the customer needs, one should compile the subset that has high importance ranks. For every one of these needs, the portfolio architecture should be questioned and a type selected. Thus, most product portfolios exhibit many portfolio architectures simultaneously.

We now present objective marketing methods to create a portfolio architecture allocation. With a desired architecture type for each important customer need, the development team is faced with creating a combined architecture that exhibits all of the requested portfolio architecture types, one for each need.

Production Cost Assumption

In this chapter, we choose not to include analysis of manufacturing cost but rather look at how the market would point to a desired type of architecture regardless of manufacturing cost. Therefore, we will not consider creating a product architecture that divides along a company's existing organizational structure or its *extended enterprise organizational structure* (which includes outside suppliers and consultants). For example, it may be that a company outsources the design of all electronic controls for a product. In this case, it might make sense to divide the portfolio architecture such that the electronics on all products are isolated and thereby make the electronic controls form a modular family platform. These design-for-organizational-compatibility considerations will not be discussed here, since generally such divisions of the portfolio architecture are more obvious.

We will instead consider what portfolio architecture the customer market desires, independent of how the current organization can deliver it. For example, one customer segment may desire low-performance, low-price control, and another customer segment may desire high-performance, high-price control. Making the electronic controls a platform, thereby forcing all customers to accept the same average, would not necessarily make an ideal design in this case.

Given the best portfolio architecture for a market determined through the subsequent approach, it can then be compared with the product development organizational capability. Perhaps the electronic controls development ought to be coupled with some actuator design to form a complete module. If the current team is not structured to deliver this, then either the organizational structure should change or a compromise portfolio design accepted—one that fits within the current team structure. We will now develop the method to determine a portfolio structure that can best meet a target market, ignoring the production enterprise.

Customer Market Models

Why do different portfolio architectures exist? Fundamentally, the underlying market dynamics that give rise to different portfolio architectures are the interaction between requested market variety from different customers versus how individuals have variety in their uses over time. Some architectures are more adaptable for a design team to meet the variety of different customers; other architectures are more responsive to a customer who wants to use the product under different circumstances and uses.

The first step is to identify a list of customer needs for the market population, as developed in Chapter 4. This should be complete with relative importance of each need to the market.

After this basic market modeling, we poll the population again on desired target values for each important need. Sampled customers respond to questions on what target values they want the product to perform at in order for them to purchase the product and how much more or less they would pay to get that performance.

An important observation is that these desired performance target values are weighted average target values over the different uses that the customer will put the product to. That is, when a customer responds to a survey question on a product, they consider all use circumstances they will put the product to, weight each in some way, and then prescribe a desired target value. With this information, we arrive at a *population distribution* of target values for each need, which can be described by the mean, μ_{pop}, and standard deviation, σ_{pop}. The statistics of the market nominal desired target values are reflected in these distributions, one for each important need.

On the other hand, we also seek to understand how desired target values change according to different uses, since some product architecture types exist to support variety in time. A customer uses a product for many activities that can vary over time. To capture this character, we construct a *time distribution*. This distribution may be constructed by observing the product use through different circumstances by representative customers from each market segment, collecting the target values for each need in the different circumstances. When this distribution is normal, it can also be described by a mean μ_t, and a standard deviation σ_t. Comparing these parameters against μ_{pop} and σ_{pop} forms the basis for making portfolio architecture decisions. The overall process of market-based architecture selection is shown in Figure 8.10.

Customer Need Distributions

To construct the two different types of customer need distributions, data must be collected by surveying customers for target values on a short list of the most important needs. The sampled customers should be from the various market segments in the market. A questionnaire can be formed by showing target values as options labeled 1–5, 1 being the least performing and 5 being the most. Every customer would always ask for the most performing unless a penalty were incurred, so each change in performance must also be given a price difference. In the questionnaire, to get the higher performance, you must pay a bit more and similarly accepting lower performance provides a price break. This type of question is repeated for each of the important needs. Such questionnaire methods are discussed in Chapter 4.

The first use of the questionnaire is to determine the market nominal distribution and its mean target value, $\mu_{pop,i}$, and standard deviation, $\sigma_{pop,i}$, for each customer need i. This information provides the population distribution of desired target values: a representation of the customer population's nominal-over-time desired targets.

Market Basis for Architecture Decisions

The market distributions must be compared with segment usage differences to determine a portfolio architecture. A guideline for this architecture selection is summarized in the flowchart of Figure 8.11. To illustrate this method, consider the example of the Krups Toastronic Toaster. First, determine if the population distribution for a need is fixed with respect to time. If it is not, then the market as a whole is evolving on that need, and one should isolate it by implementing the modular-product-generations architecture. The features that satisfy that need will benefit from being isolated into a module, since it can then be replaced easily without seriously altering the remainder of the design.

For example, because styles change constantly and each manufacturer wishes customers to perceive its product as the newest and thus the most desirable item on the market, appearance is often best disassociated from the rest of the product. The Krups toaster, for example, achieves this goal by isolating the functional components from the stylish plastic shell. Figure 8.5 shows the nearly identical underlying platforms from

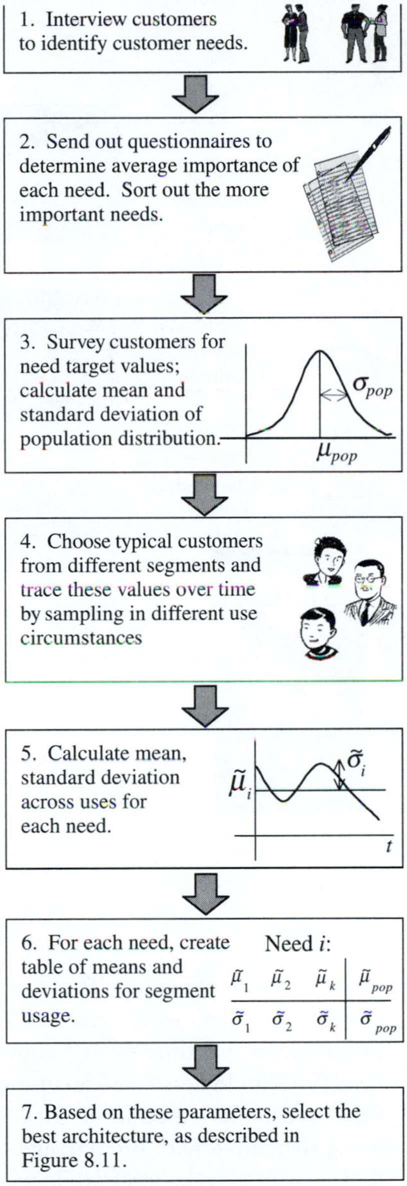

1. Interview customers to identify customer needs.

2. Send out questionnaires to determine average importance of each need. Sort out the more important needs.

3. Survey customers for need target values; calculate mean and standard deviation of population distribution. σ_{pop} μ_{pop}

4. Choose typical customers from different segments and trace these values over time by sampling in different use circumstances

5. Calculate mean, standard deviation across uses for each need. $\tilde{\mu}_i$ $\tilde{\sigma}_i$ t

6. For each need, create table of means and deviations for segment usage.

Need i:

$\tilde{\mu}_1$	$\tilde{\mu}_2$	$\tilde{\mu}_k$	$\tilde{\mu}_{pop}$
$\tilde{\sigma}_1$	$\tilde{\sigma}_2$	$\tilde{\sigma}_k$	$\tilde{\sigma}_{pop}$

7. Based on these parameters, select the best architecture, as described in Figure 8.11.

▼ **Figure 8.10.**
Market-based architecture selection process.

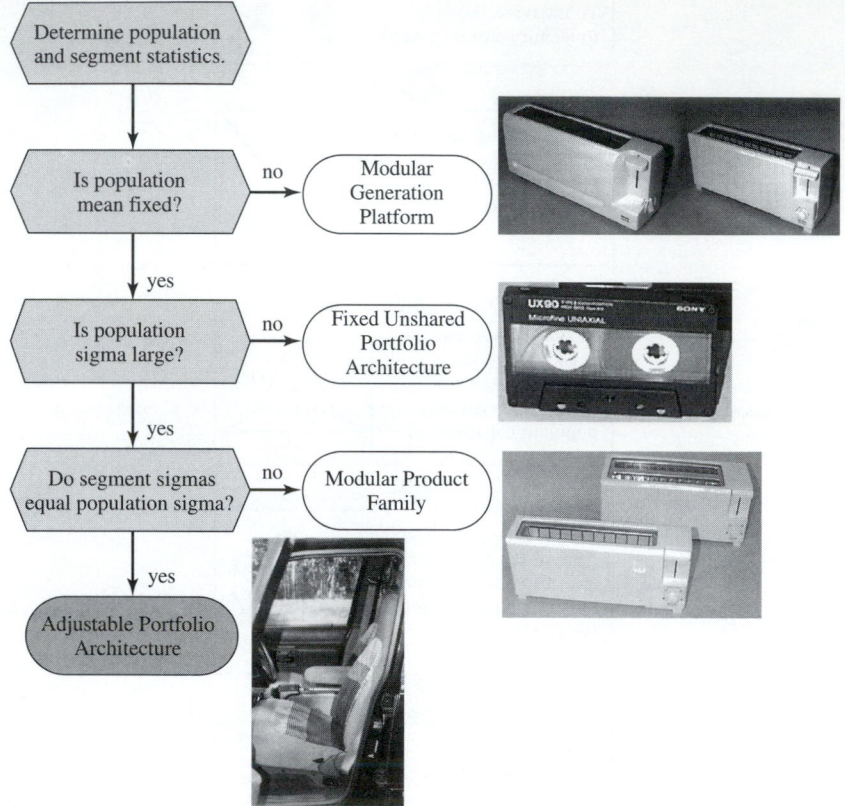

▼ *Figure 8.11.*
Architecture selection flowchart.

two generations of toasters. Thus, the manufacturer is able to sell a toaster with a new look while only having to redesign and retool the shell, toasting lever, and time knob. Another clear example of a time-variable need that should be isolated is computer CPU speed. As technology advances, customer expectations for CPU speed rise. Thus, computers are being structured with interchangeable processors, allowing the basic design to remain the same while providing state-of-the-art processors with increased processing speed. Technology-based and cosmetic concerns typify time-varying customer needs best served by modular generational product portfolios.

Next, we consider the opposite case, when the population distribution is constant with respect to time. Here we must first decide whether the fixed integral product architecture can encompass all the variation in the market for a single customer need. In other words, if the variation is sufficiently small ($\sigma_{\text{pop},i}$ is small), all customers will find a single fixed product satisfactory, because the need value provided by its features does not differ greatly from their target need values. This choice must be made with respect to how much it will cost the company to provide multiple offerings.

If the breadth in the market cannot be answered with a single fixed need value ($\sigma_{\text{pop},i}$ is large), we must endeavor to provide multiple values to capture more of the market. There are two effective methods to provide many values for a given need: modular product families and adjustability. Again, adjustability is often more difficult and expensive to provide, but it allows the product to fulfill a range of need values after purchase.

Adjustable integral product architectures should be constructed around needs whose distributions demonstrate *ergodicity* (Bendat and Piersol, 1986). In our context, ergodicity refers to the condition when the need distribution across the population is equal to usage distributions of customers over time. Hence, ergodicity in a need implies that the entire variation seen in the population distribution is encapsulated in each customer, and so the variation can be observed in his or her usage over time. Thus, a mass customization portfolio architecture is needed: an adjustable feature that captures both usage and population variety should be developed.

For example, the seat adjustment on an automobile must provide the range in leg room required by the entire population. This same amount of variety must also be available from the product at any point in time after purchase, as different drivers may want to adjust the seat to their own needs. Leg room is an ergodic requirement and is best served with an adjustment. In the case of the toaster, ergodicity is demonstrated in the distribution of toasting time target values. Depending on type of bread, whether it starts off fresh or frozen, and personal preferences, a toaster user requires a variety of toasting times for different uses. Similarly, what the toasting population desires at any given time also constitutes a wide toasting time distribution. These distributions are roughly the same. Toasting time is an ergodic requirement. The adjustability of toasting time on most toasters accommodates both of these types of variety.

When the population and time distributions of customer need target values are different, the standard deviation of the time distribution will necessarily be smaller. This analysis is based on the assumption that the population survey is sufficiently comprehensive that it captures all the need variation that exists in all different usage situations. Thus, the standard deviation of a segment k's different uses, σ_{ik}, will not exceed the deviation of the entire population, σ_{pop}. When the time-based deviation is significantly smaller, we see that the product traced in the time distribution only sees a fraction of the need values expressed by the population distribution. Full adjustability is not required. If deviation within the time distribution is sufficiently small, we can cover the population breadth with a family of modular products, with each variant catering to a market segment. For example, with portable stereos, size is often a concern that matters more to some and less to others. Thus, manufacturers provide both bulky and more lightweight versions of stereos with essentially the same features. Customers for whom stereo size is a concern are willing to pay more for a smaller product. Toast capacity is also a need that benefits from offering several fixed options. By and large, there are two-toast users and four-toast users, presumably determined by eating habits, household size, economics, and so forth. Modularization of toast capacity (see Figure 8.4(a)) allows manufacturers to market to different customer segments without being forced to develop and manufacture two completely distinct products. In this case, comparison of the population and time distributions will yield curves with differing means and standard deviations.

Basic Method: Estimated Market Segments

Our first method to execute the actual portfolio architecture allocation is a market segment estimation method. That is, on each important customer need, the flowchart of Figure 8.11 will be answered. This approach will construct the portfolio.

To use this method, we start with a customer needs list as in Chapter 4, complete with importance ranks. From this list, we examine the most important requirements, say those ranked above 80% important. From this list, we create a list of approximately five requirements to ask the customers what performance level is sought. Typically, one can reduce many of the specific needs into a short list. For example, several customer needs in the toaster example basically relate to compactness. Similarly, many relate to toasting uniformity.

After completing this step, one must next assess the standard deviation in the importance ranks, as discussed in the advanced method of Chapter 4. If this information is not available, one must estimate it. For example, consider toasters. As abstracted from the complete list of customer needs, the top customer needs and their approximate weights are:

Abstracted need	WT
Number of slices it can toast	4
Inexpensive	3
Compact	4
Uniformly toasted result	4

The next step is to classify the market population into categories of types of users or *market segments*. These are the different types of persons who purchase the product, as discussed in Chapter 4. A random sample of each of these segments, or more practically one representative from each of these segments, should be selected. That is, one can typically group the entire market population into categories, such as the least cost purchase group, the high-end group, the high-volume group, and so forth. Advanced clustering algorithms applied to the questionnaire results for the entire population are one rigorous means to cluster the market into segments, as mentioned in Chapter 4. Each market segment forms a subset of the market population, and every customer can be effectively placed into one of the different groups.

For example, for toasters, market survey information available via online information services indicates there are three basic toaster segments. First-time purchasers, because they tend to be young professionals, want a toaster that is as cheap as possible. These purchasers then experience the short life span of a cheap toaster and then become more intelligent second-time or later purchasers. At this later time, customers tend to be more affluent and older and willing to pay more for a quality toaster. This group forms a second larger customer segment. Finally, there are niche purchasers, such as cafes, shops, and large families that want large-capacity toasters. These are small in number, but they are willing to pay a premium for the higher capacity.

Having split the market into these three segments, one must establish the target values desired by a typical customer in each segment, one target for each segment for each customer need. These assignments can be done using "high, medium, low, and none" levels of performance.

Next, the time variation of each customer need by each market segment must be determined. For each market segment, considering how much the target value varies as the typical customer uses the product in different use scenarios might create this estimate. Again, a "±high, ±medium, ±low, ±none" scale can be used. For example, in Table 8.1(a), the target values and the time variation on the customer needs are shown for each market segment for the toaster.

Table 8.1(a) provides the estimated data on which the questions in Figure 8.11 can be posed. Answering the questions in Figure 8.11 will provide a suggestion for how a portfolio should be architected. For example, Table 8.1(b) provides the suggested portfolio architecture for the toaster.

With Table 8.1 and understood market segment sizes, a toaster portfolio concept can be developed. Typically, this is best done as a range of portfolios of different sizes. If only one toaster is permitted in the portfolio, then a two-slice, high-quality, not-compact, uniform-toasting

TABLE 8.1. TOASTER MARKET TO ARCHITECTURE MAPPING

Need	(a) Customer Segment Needs								(b) Suggested Architecture
	Population		LowPrice Segment		High-End Segment		High-Capacity Segment		
	μ_{pop}	σ_{pop}	μ_t	σ_t	μ_t	σ_t	μ_t	σ_t	
# Slices can Toast 1 2 4	2	± 1	2	± 0	2	± 0	4	± 0	Modular: Two- and four-slice
Price $50 $30 $10	$30	± 20	$10	$\pm \$5$	$50	$\pm \$10$	$50	$\pm \$10$	Modular: Low cost and quality
Compact 20×20 6×20 6×10	6×20	$\pm 3 \times 10$	20×20	$\pm 3 \times 10$	6×10	$\pm 10 \times 10$	20×20	$\pm 3 \times 10$	Modular: Compact or not
Uniform Result Spots Uniform	M	$\pm M$	M	$\pm M$	H	$\pm L$	H	$\pm L$	Unsharing: Uniform

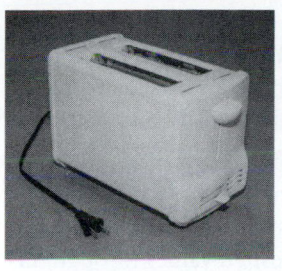

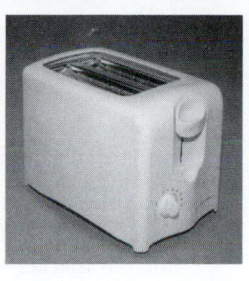

Inexpensive   Two-slice   Compact   Four-slice

▼ *Figure 8.12.*
Suggested toaster portfolio.

product is best. If two toasters are permitted, then both a high-quality and a low-cost two-slice, not-compact, uniform toaster should form the portfolio, since the low-end market segment is larger than the high-capacity segment. Next, if three toasters are permitted, then a low-cost, two-slice, not-compact toaster, a high-quality, two-slice, compact toaster, and a high-quality, four-slice, not-compact toaster should constitute the portfolio, all uniform toasting.

Given there are only three market segments, it ought to be apparent that no more than three toasters are needed to cover the market. But that is true only to the extent that the market segmentation is sufficiently accurate. In the toaster example, one can see that the compactness requirement has large variance in the high-end segment. The nominal toaster requested by the high-end market is compact, which some in the segment are concerned about and some are not. Therefore, two toasters might be developed for this one segment, one compact and one not, but both would be two-slice, high quality, and uniform toasting.

Beyond this, the market requests no further toaster differentiation, as indicated by the market surveys. One of these four toasters in the four-toaster portfolio is very close to the demands of any toaster customer, given the different uses they will put the toaster to and what they are willing to pay. Such a portfolio concept is shown in Figure 8.12. Note the two-slice models have sufficiently wide slots to fit items such as bagels and have sufficiently long slots to fit oversize bread slices.

Advanced Method: Market Surveys

Rather than estimate the time behavior of different market segments, we can evaluate this issue directly through sampling of the customer population. Members from different segments of the population can be randomly sampled and their different uses over time sampled for desired target values on the important needs. Then usage variation can be summarized into means and standard deviations and, using the previous discussion's concepts, compared with the population mean and standard deviation to determine a desired portfolio architecture.

Data collection becomes difficult. To construct a time distribution of desired target values for each need, we must sample customers from different market segments and track their desired values through different circumstances. To do this, we ask each customer, k, to conceive of different circumstances in which they use the product. At each of these different use circumstances, j, we sample the customer's target values, V_{ijk}. We also record for each of these circumstances a rating, P_{jk}, of how important these circumstances are (or how often they apply). This number is representative of the probability that the user will be using the product in such a situation. Therefore, the higher the value of P_{jk}, the more effect the sample taken in situation j, V_{ijk}, should have on the customer need model. We take this issue into account by determining a normalized mean value of criterion i for customer k. Here, the tilde over $\tilde{\mu}_{ik}$ indicates normalization for different P_{jk}.

$$\tilde{\mu}_{ik} = \frac{\sum_j V_{ijk} P_{jk}}{\sum_j P_{jk}} \tag{8.1}$$

Similarly, we must alter the basic equation for standard deviation based on a sample, σ, to take into account the probability of usage in a given scenario. The standard deviation is typically defined as

$$\sigma = \sqrt{\frac{\sum_j (V_j - \mu)^2}{N - 1}} \tag{8.2}$$

where V_j is the value in sample j, μ is the mean value across all samples j, and N is the total number of samples taken. We need a variant equation that takes into account the different P_{jk} values. We create this equation by treating each value V_{ijk} as P_{jk} number of samples with the same need value. Hence, N, the total number of samples in Equation 8.2, is replaced in Equation 8.3 with the sum of P_{jk} for all scenarios j. This process results in a standard deviation normalized across samples of differing importances that approximates the deviation of an actual time-distributed sampling of the customer's target values for need i.

$$\tilde{\sigma}_{ik} = \sqrt{\frac{\sum_j P_{jk}(V_{ijk} - \tilde{\mu}_{ik})^2}{\sum_j P_{jk} - 1}} \tag{8.3}$$

This set of information is completed to determine a portfolio architecture. We compile these data, mean and standard deviation, into a

TABLE 8.2.   POPULATION AND TIME DISTRIBUTIONS OF INSTANT CAMERA NEEDS

	Population		Laboratory Users		High End Users		Household Users	
	mean	std dev	mean	std dev	mean	std dev	mean	std dev
Size	4.333	0.816	3.533	0.516	3.700	0.949	3.800	0.632
Light adjustment	3.667	0.724	2.533	1.187	4.600	0.843	3.300	0.483
Focusing	3.933	0.961	3.000	0.000	4.600	0.843	3.900	0.876
Ruggedness	3.533	0.990	3.333	0.488	3.000	0.000	2.800	1.033
Picture quality	4.267	0.704	3.267	0.458	3.800	0.422	3.700	0.949
Film pack size	3.800	0.862	4.800	0.414	3.000	0.471	3.500	0.527
Style	2.923	1.188	3.000	0.000	2.700	0.675	3.800	1.317

table (Table 8.2) for each segment k traced. In addition, we compile the and mean and standard deviation for the entire population, $\mu_{pop,i}$ and $\sigma_{pop,i}$. We use these results to then compare distributions to determine the best product architecture for the market, using a procedure described in the next section.

Example: Instant Camera

In this section we examine an instant-film camera as an example of comparing population and time distributions to determine product architecture. We first polled potential customers and determined from these data the seven needs that were most important to their evaluation of the product: picture quality, compactness, convenient focusing, ability to adjust to lighting environment, ruggedness, large film pack capacity, and stylish appearance.

Next, we had each customer reply to a survey on these features as shown in Figure 8.13. We asked for desired nominal target values compared with a $15 cost increase or decrease with the associated gain or loss in performance. The means and standard deviations for each need target value, derived from the customer surveys, are shown in the "Population" column of Table 8.2. The values in Table 8.2 do not necessarily represent the entire instant-film camera market population but is an illustration of the concepts described.

The questionnaire was also completed by the customers from different market segments that we tracked in time by questioning them on each of the different scenarios they felt were important to their use and purchase of the camera. The normalized mean and standard deviation values for the customer segments traced in time are also shown in Table 8.2 in the "Segment" columns.

The information in Table 8.2 was used in conjunction with the flow-chart depicted in Figure 8.11 to determine which architecting option

Criteria	$15 less		Current Price		$15 more
Size	Shoebox		Camera		Walkman
	1	2	3	4	5
Light Adjustment	only in daylight		3 Settings		Automatic
	1	2	3	4	5
Focusing	Fixed		3 Settings		Automatic
	1	2	3	4	5
Ruggedness	Survives 1 Drop		5 Drops		10 Drops
	1	2	3	4	5
Picture Quality	Fuzzy		Current		35 mm
	1	2	3	4	5
Film Pack Capacity	5 Exposures		10 Exposures		20 Exposures
	1	2	3	4	5
Style	15 years out of date		Current		Wicked Cool
	1	2	3	4	5

▼ *Figure 8.13.*
Questionnaire on instant-film camera performance.

should be used for that need. The results are shown through graphs of the observed distributions. Figures 8.14 and 8.15 show the normal distributions corresponding to two of the customer needs shown in Table 8.2, "focusing" and "film pack capacity" (or size).

For the focusing feature, we observe that one of the customer segments has measurable difference in desired target value. This segment has content with having a single fixed-focus camera. On the other hand, the other segments sometimes want auto-focus capability, but sometimes not. Their time distribution match the fixed-time population distribution. These segments would like a camera that has auto-focus capability, but they do not need it all the time. Thus, using the guidelines illustrated in Figure 8.11, these two different time distributions would indicate that having two models in the product family would be best for the customer population, one inexpensive fixed-focus and one more expensive auto-focus. Sometimes the persons who purchase the auto-focus would use it, and sometimes not, and so they are not willing to spend excessively for it.

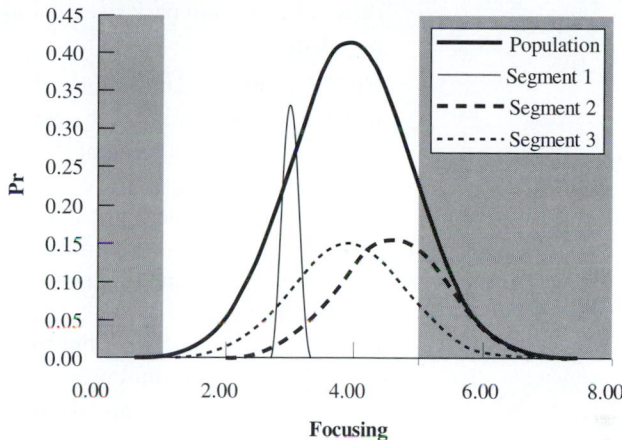

▼ *Figure 8.14.*
Usage distributions for the focusing need for three representative instant-film camera customers showing a need for both a fixed and an adjustable focus feature.

On the other hand, for the film pack capacity feature, we observe that all three of the segments have measurable differences in desired target value, and all three distributions are narrower than the population distribution. Using the method outlined in Figure 8.11, these different time distributions would indicate that a modular architecture with

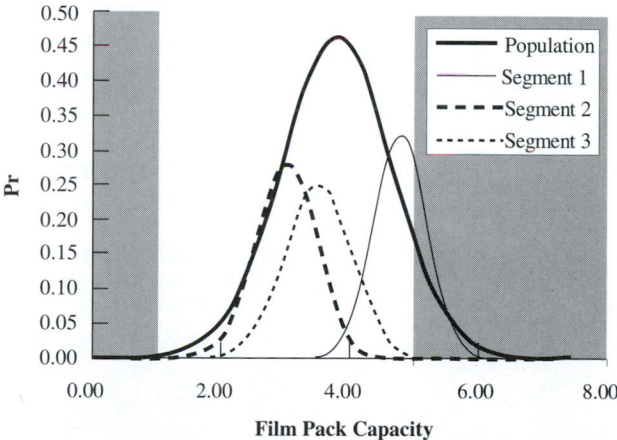

▼ *Figure 8.15.*
Usage distributions for film pack capacity for three representative instant-film camera customers showing a need for different size film packs.

three different film pack sizes would be most pleasing to the customer population.

Similar recommendations for architecting the other features are shown in Table 8.3. The camera size data would indicate that some customers want a small camera some of the time and do not care at other times. But since an adjustable-size camera is not feasible, a single-size camera or possibly two models, one standard and one compact, is best for the market. For the light adjustment feature, users' need values show distinct but large-variance distributions. These values would indicate the need for an architecture with an inexpensive fixed model and a model that can adjust to different lighting conditions. The data for the camera ruggedness show the customers sometimes wanting a rugged camera, but other times not, and some at more times than the others. These data would suggest a modular architecture with a standard and a rugged model. The film picture quality time distribution of each customer segment was approximately equivalent to the population distribution. This result indicates the use of either an adjustable or fixed feature, depending on the amount of variation. For instant film, these data would indicate a fixed integral architecture on this feature, as it is difficult to offer many levels. For other photographic tools, such as digital cameras, providing flexibility in picture resolution is easy and so can be implemented as an adjustment (dots per inch resolution selection). Finally, for the style feature, the data exhibit two distinct time distributions, one for a person who does not care about style and one who does at times. Therefore, a modular architecture supporting two models is indicated, one inexpensive and not concerned about style, and another model with rapidly changing stylish features. Figure 8.16 shows an instant-film camera portfolio with just three cameras that meet all but the film pack size portfolio recommendations in Table 8.3.

TABLE 8.3. ARCHITECTURE RECOMMENDATIONS FOR INSTANT-FILM CAMERA FEATURES

Customer Need	Architecture Choice
Size	Integral
Light adjustment	Fixed and adjustable
Focusing	Fixed- and auto-focus
Ruggedness	2 model modular
Picture quality	Integral
Film pack capacity	3 model modular
Style	2 model modular

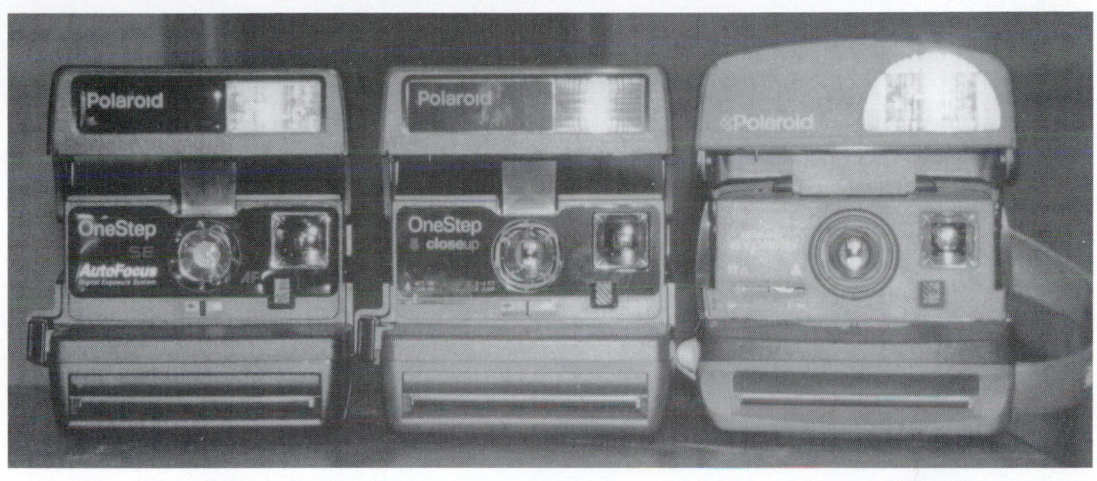

| Auto Focus | Standard | Heavy Duty |

▼ *Figure 8.16.*
Suggested instant-film camera portfolio.

IV. PLATFORM ARCHITECTURE

Of the different product portfolio architectures available, the platform architecture is often a cost-effective approach. A portfolio platform can be reused across the different product variants and thereby reduce design time and gain economies of scale on manufactured components (Meyer and Lehnerd, 1997; Meyer and Utterback, 1993; Sanderson and Uzumeri, 1996). Given the decisions to use a platform architecture, the next problem is to determine, on the physical implementation, what the platform should actually be composed of and how the supported variants should interface the platform.

Following an introduction, three topics are covered in this section. First, an overview of the platformed family concept selection problem is presented. Then, the first method is covered on chart based selection of physical modules, an expanded version of the Pugh selection process of Chapter 11 for single products. Next, a functional architecting method is covered for developing platform family concepts at the functional level, similar to the functional architecting methods of Chapter 9. Finally, a form level numerical modeling approach for developing module designs is developed.

Negotiating a Modular Family Platform

When faced with the task of simultaneously designing several related products, how might a set of engineers determine what should be in a platform, and what should be left unique to different variants? The process we have chosen for designing a product platform can be summarized in the following steps (Gonzales-Zugasti, Baker, and Otto, 1998):

1. requirements and model gathering

2. platform performance design

3. variants design

4. platform evaluation and re-negotiation

We now explore these steps in more detail. A diagram illustrating an overview of the process is shown in Figure 8.17.

First, the engineering team needs to create a system model of the product. The purpose of this model is to accurately describe the performance of the product and to report metrics that the designers can use to judge the design. This system model takes in a set of requirements or specifications (e.g., mass <500 kg) and architectural choices (e.g., engine model X, type Y) for the product and represents them by variables x_i. With these inputs, the model should calculate and display performance metrics to describe the product to the desired level of detail. These metrics would include the expected cost of the product in addition to its technical performance. Cost results, $\text{cost}_i(x_i)$, are obtained through models that utilize the same input variables x_i to calculate development, manufacturing, operations, and other life-cycle costs of the product. We refer to a *system* model because, for most products, the components cannot be designed in isolation from all the other components or subsystems, since there is bound to be coupling among their performance. These dependencies are represented by constraints in performance $g(x_i)$.

Once the system model exists, the engineering team needs to gather the requirements for each product and for the family as a whole. The system model should then be used to create a model of each desired product as if it were to be developed individually, so that it can be compared with the platform-based designs that will result from this whole process. At this point, we envision the beginning of an interactive negotiation process among the members of each individual product design team and the team designing the platform. The platform design team should be composed of members of all the individual teams, since it will be making decisions that will affect them all.

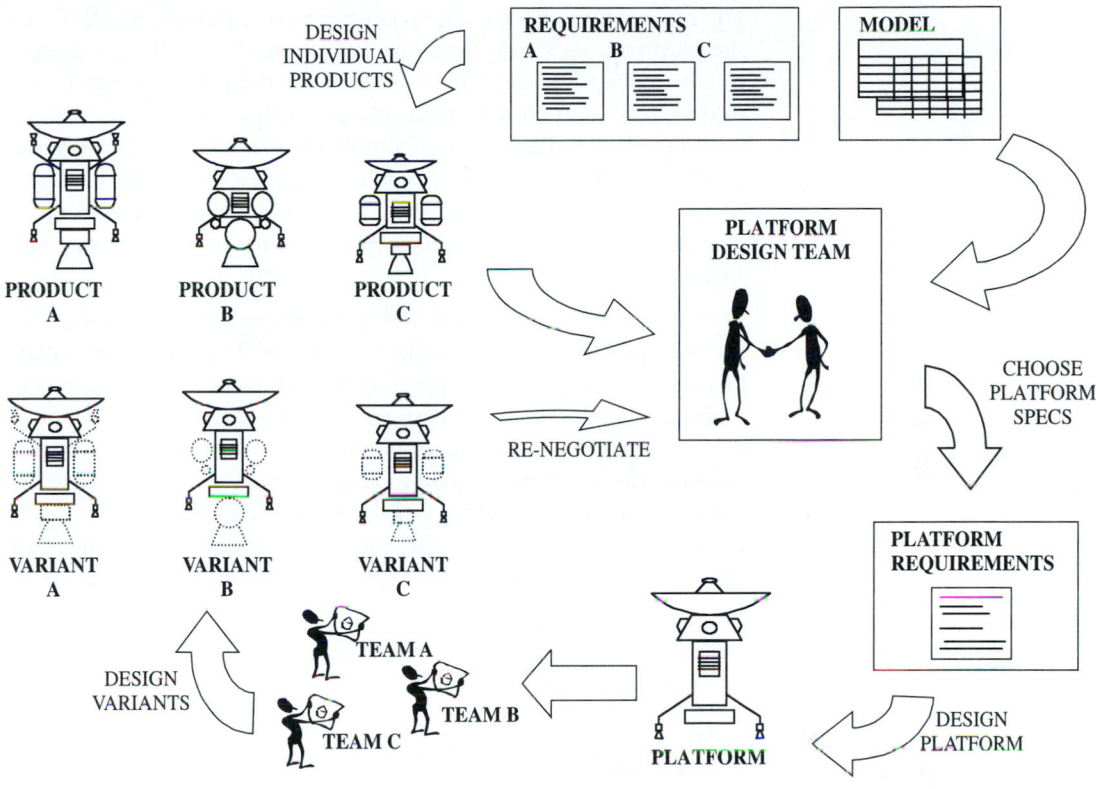

▼ *Figure 8.17.*
Platform design process overview.

As the first step in the negotiation process, the platform team must look at the models for each of the individual products and pick those areas that in their judgment should be made common. These choices can be achieved by compiling all input variables to the system models of each of the products and displaying them side-by-side to the platform designers. Then, those variables that can be standardized across products can be agreed on as well as those to be individually designed for each product.

With the variables to be standardized selected, adequate values for those variables may be chosen, either automatically by the software on which the model is implemented or manually by the group of designers. In either case, the designers must approve the settings for those variables that determine the design of the platform. These tasks form the next step of the negotiation process.

The settings for the common variables are the input specifications for the platform, x_p. Using the same system model for the product, a platform design model can be constructed and platform metrics reported back to the design team. Based on this common base, individual teams can then work on completing the design of each desired product or variants. Those variables that are not standardized can be specified for each variant, and a new set of variant metrics can be obtained. Each product team is then optimizing their particular variant, or x_v.

The finished variant designs can then be compared to those designs for the same products that were designed individually, not based on a platform. If the variants are not appealing due to their cost or any other performance metric, the variant designers can then go back to the platform design team and negotiate again with their peers from the other variant product teams to generate a new platform that may better fit their interests. At the same time, they can also see metrics for the product family, such as the total family cost, and make decisions about the most appropriate platform design, thus optimizing at the product family level.

We now cover two different methods that can be used to accomplish this negotiation. The first is a basic chart-based approach. The second is a more advanced numerical method.

Basic Method: Charts

We present here a simplified version of the optimization exercise described in Gonzales-Zugasti, Baker, and Otto (1998) as an appropriate method to guide platform architecture design decisions. For simple products in the portfolio, ones with subsystems that have simple relationships among them, it is not necessary to build a complex product model and platform design model. For this type of platform and supported family, there is a clear idea of the criteria most important to the success of the product variants. This type of situation allows for simpler ways of evaluating potential portfolio architectures than in more-complex products, where design changes can affect the product performance in ways that are not apparent to all people in the design team.

The simplified platforming method is based on a simple selection chart, which is used to evaluate different concepts for a *single* product, as developed in Chapter 11. Here, though, we evaluate platformed families and take into consideration the different platform and variant options for a whole family. The platform design and evaluation method is outlined in Figure 8.18. The following section illustrates the procedure through an example of the design of a family of toaster products: a two-slice toaster, a four-slice toaster, and a commercial-grade four-slice toaster.

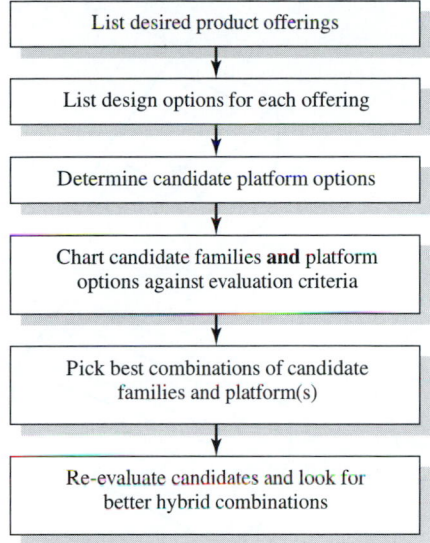

▼ *Figure 8.18.*
Simplified platform design method steps.

Step 1: List Desired Offerings

After exploring the market for household and commercial toaster ap-pliances, a design team decides it could best capture the available mar-ket with a portfolio of three product offerings: a two-slice and a four-slice household model and a four-slice commercial model.

Step 2: List Design Options for Each Offering

The design team must establish some different layouts for each of the members in the product family. There is still uncertainty as to which of the product layouts would be best. For example, with the toasters, there is a design with long slots to fit two slices of bread side-by-side or another design with shorter slots to fit only one piece of bread. These alternatives are shown in Figure 8.19.

Step 3: Determine Platform Options

The next step is to consider different options for components that could be shared among the three products. These are:

1. Electronic control board,

2. Heating elements and reflectors,

3. Sheet metal chassis and toast loading/holding mechanism.

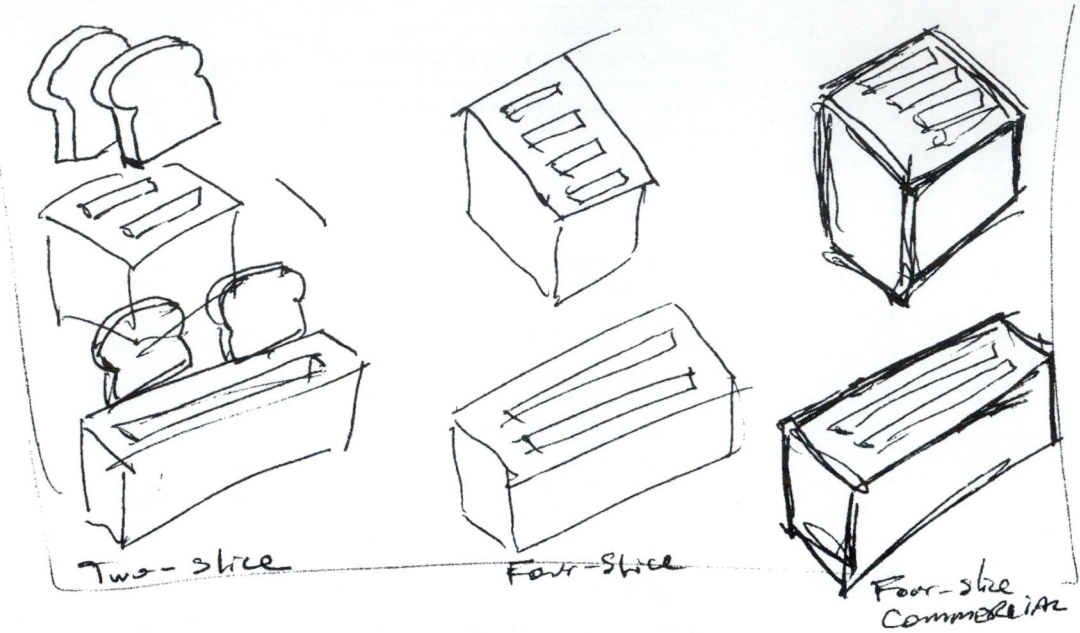

▼ **Figure 8.19.**
Design options for toaster portfolio products.

Sharing these components could result in substantially lower development costs, inventory, and manufacturing equipment. However, a platform of shared components may also cause larger materials cost per unit, since the parts will not be optimized for each of the products on which they are used. What is needed at this point in the process is a method to compare the design and platform possibilities.

Step 4: Chart Platform and Design Options Against Evaluation Criteria

For the purposes of this example, consider four basic toaster evaluation criteria:

1. Materials and manufacturing cost per unit, which will be by far the largest contributor to the final price; due to the large expected number of units produced, development costs are small in comparison,

2. Inventory cost, which decreases as more parts are shared among the models,

3. Visual appeal,

4. Ergonomics.

Product Family	Family 1			Family 2			Family 3			Family 4			Family 5			Family 6			Family 7			Family 8		
Platform Option	A	B	C	A	B	C	A	B	C	A	B	C	A	B	C	A	B	C	A	B	C	A	B	C
Materials Cost	+	+	+	0	0	0	-	-	-	0	0	**0**	-	-	-	0	0	0	-	-	-	-	-	-
Inventory Cost	-	0	+	-	-	-	-	-	-	-	-	**0**	-	0	+	-	-	-	-	-	-	-	-	0
Visual Appeal	+	+	+	0	0	0	-	-	-	0	0	**0**	-	-	-	0	0	0	0	0	0	-	-	-
Ergonomics	-	-	-	-	-	-	+	+	+	0	0	**0**	+	+	+	-	-	-	0	0	0	-	-	-
TOTAL	0	1	2	-2	-2	-2	-2	-2	-2	-1	-1	0	-2	-1	0	-2	-2	-2	-2	-2	-2	-4	-4	-3

Evaluation Criteria (left vertical label)

▼ *Figure 8.20.*
Chart for analysis of platform and family options.

One can construct an evaluation chart of these different platforms and supported family variants. The chart to compare all the options would then look like the one shown in Figure 8.20. Note what is being evaluated is the platform, based on the combined performance of the variants as they can be individually tailored to the platform.

The first step needed to construct the chart such as in Figure 8.20 is to list the different product families that can be designed to provide the three desired product offerings. These are listed at the top of the chart and form the main columns. In this case, the families are represented graphically for easier identification.

Next, each possible platform from which the family could be constructed is listed as subcolumns under each family. For this example, there are three candidate platforms as described in Step 3. Platform option A is the circuit board only, platform option B also includes the wiring and heating elements, and platform option C also includes a sheet metal housing that fits two slices.

Finally, the structure of the chart is completed by filling in the rows of the chart. These are formed by the criteria used to evaluate the possible options. For the toasters, we had four criteria.

Once the chart is constructed, the design team needs to pick one of the options (columns) to be the baseline case used to judge the other options. Figure 8.20 shows the selected baseline as a column with bold text. The baseline is assigned a rating of "0," or neutral, for all evaluation criteria. The team should pick what they think is a reasonable candidate as the baseline. If a poor option is selected, all others will get high ratings in all categories, making the exercise useless as a comparison tool.

The next step is to fill in all other cells in the chart by comparing to the baseline case. Each cell can be determined to be:

"0" if deemed equally good as the baseline design for that criterion,

"+" if better, or

"−" if worse.

If the consensus is that some family options are not acceptable, they can be simply crossed out and not considered. If there is disagreement over the content of a family option, the various versions can be considered as separate new options for the chart.

Step 5: Choose Preferred Combinations of Platform and Family Options

Once all of the cells have been filled out, the plus and minus signs are counted for each column and the total "score" reported at the bottom of the chart. Those columns with high positive counts (lots of "+") are considered to be the best design and platform options. In the toaster example of Figure 8.20, Family #1 (long slots), and in particular Platform Option "C" (sheet metal and electrical module), scores high marks against the baseline, whereas the other families seem to be inferior options according to these evaluation criteria.

Step 6: Reevaluate Candidates and Look for New Options

What needs to be completed next is to choose better options and reevaluate them to ensure that the team members agree with the results of the exercise. The highest scoring column may not always be the one that the team thinks is the best, since there may be criteria that are considered more important than others. Progressing through the exercise of reevaluating the better columns, the team will either reach a consensus on which is the preferred option or find a hybrid solution, one that combines the best qualities of several of the original designs.

The chart of Figure 8.20 reflects a process where the team first considers platforming the two- and four-slice household toaster using the sheet metal and electrical components as a platform module and leaving the commercial toaster as a unique design with more expensive, high-quality components. The process leads to including the commercial toaster in the platform, thereby reducing its cost and the family complexity. This choice, however, requires using more durable components in the two- and four-slice household toasters at added expense. This result is deemed a better portfolio concept.

Advanced Method: Functional Architecting

In this section, we consider architecting a platform portfolio architecture and its associated product family not by possible subsystem options on a physical form level, but rather at the more preliminary and abstract functional level (Zamirowski and Otto, 1999). Platform based product variants are made possible by having the variants share modules. Given that, it should be clear that the single product modularity heuristics of Chapter 9 ought to have platform portfolio architecture extensions. We explore these platform portfolio modularity heuristics in this section. The result is a method to architect the modularity of a complete product family.

As an example, the variants in the Black and Decker VersaPak product family are shown in Figure 8.21. While the complete VersaPak line has many variants, we focus here on the power screwdriver, drill, Wizard, ScumaBuster, and a multi-purpose saw. Across all variants, the product

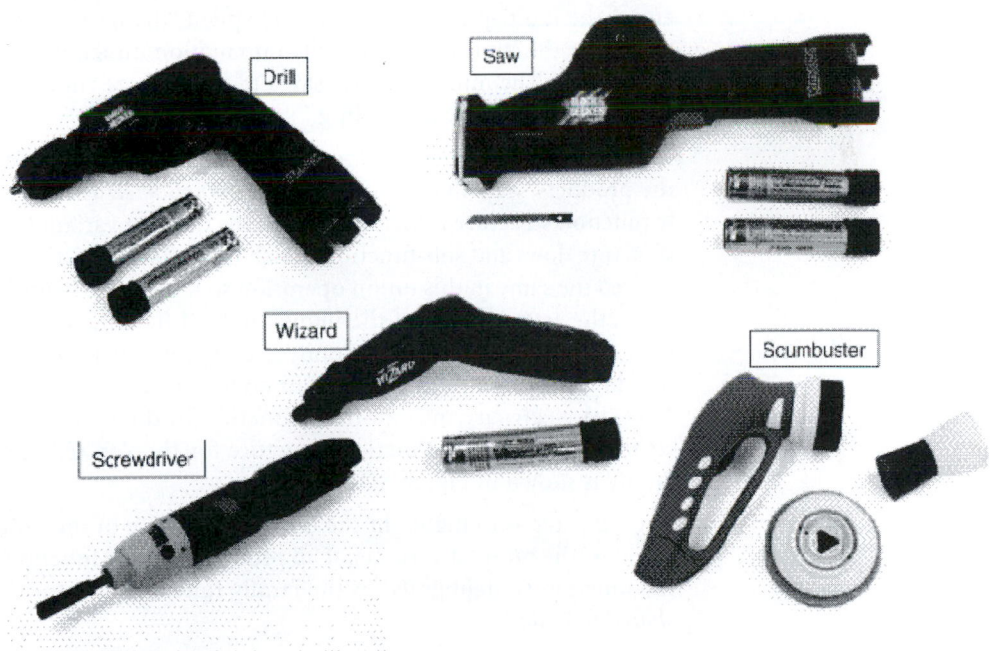

▼ *Figure 8.21.*
Black and Decker VersaPak product family.

family shares the exact same battery pack. Further, within the family, different variants share some modules while others do not. The power screwdriver and ScumBuster share a gearbox and motor, for example.

The first step to functionally architecting the product portfolio is to develop possible function structures for each product variant. Each product variant may have slightly different possible function structures, depending upon concepts considered. The function structures for each product as developed by Black and Decker are shown in Appendix A.

The next step is to transform each product function structure into a common basis, so that each function structure uses the same generic functional terms. The interesting thing about these product variant function structures is their similarity. Each looks remarkably similar, despite their different functions. That is typical of a product family.

The next step is to use the different function structures to determine differences and similarities among the variants in the product family. As an example of similarities, all products have the function "convert electricity into motion." As an example of differences, the multi-purpose saw has an explicit "dissipate heat" function, since the "transform electricity into motion" function generates excess heat. The multi-purpose saw's motor is large enough that it needs a fan to supply air cooling.

To explicitly represent the commonality and the differences among the products, we will construct a *family function structure*. A family function structure is the union of all the product variant function structure flows and sub-functions. Each flow or sub-function is considered the same in this union operation so long as the items are the same, the actual values of the items can be different. For example, the products in the VersaPak family use slightly different sized motors to covert electricity to motion; nonetheless, there is only one "convert electricity into motion" function in the family function structure. The family function structure for the VersaPak product family is shown in Figure 8.22.

The next step is to highlight the differences in use of the functions amongst the product variants. If all products use the function, then the function is highlighted in the family function structure, called *shared functions*.

Since all the shared functions are used in all the products, it might make sense to supply these functions with the same physical hardware across all the product family variants, to create the platform. These shared functions are the ones to consider for such component sharing.

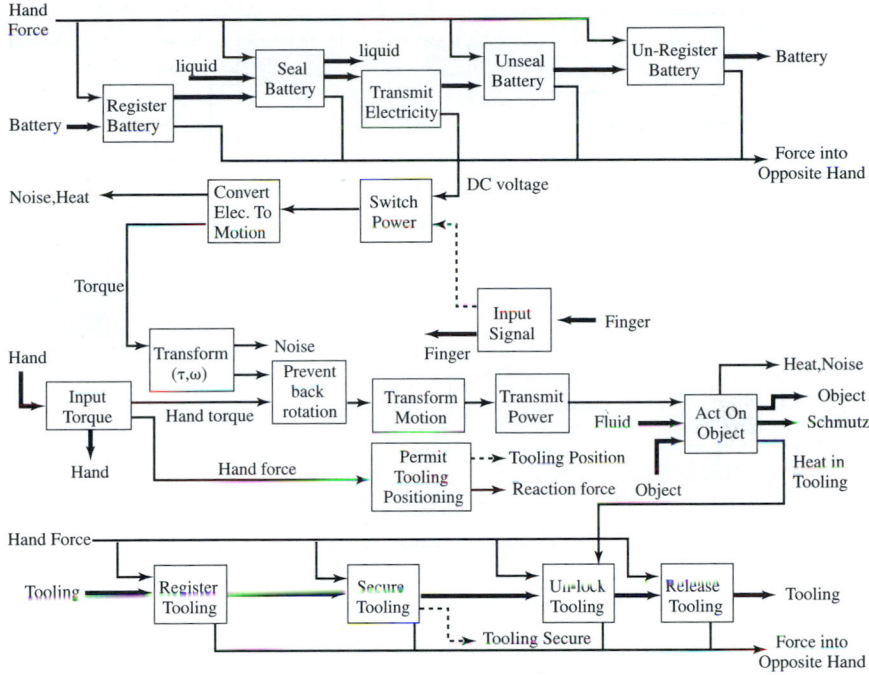

▼ Figure 8.22.
Black and Decker VersaPak product family function structure.

As a further refinement of this idea, it usually makes little sense to incorporate variant functions into the shared components. This would incorporate unused functionality into some of the product variants. For a product family, this leads to the first additional modularity heuristic one must consider beyond the single product modularity heuristics of Chapter 9.

Proposition 1. A module can be defined by the shared functions of a product family. This shared module is called a *platform module.* As with the previous single product heuristics of Chapter 9, the heuristic can be applied to the maximal extent in the function structure (all shared functions) or to substructures within the shared functions (only parts of the shared functions).

Note that a single platform is rarely possible across all products in a product family. An example is the VersaPak battery. More typically, though, a single shared function in a portfolio function structure can be satisfied not by a single shared component, but by a few shared components. Examples are the switches, control circuitry, or electric motors in the VersaPak product family. Here, no single switch, control circuit, or electric motor can serve all product variants. Instead, there are five different switches, four different gearboxes, and four different electric motors shared to various degrees across this subset of the VersaPak product family.

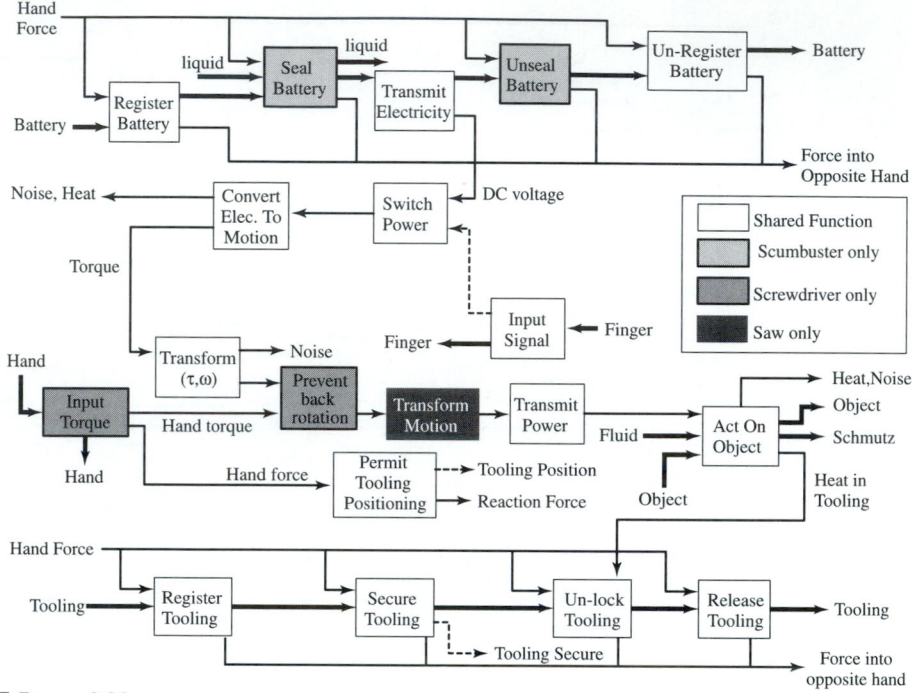

▼ Figure 8.23.
Refined Black and Decker VersaPak product family function structure.

The converse of the function sharing modularity heuristic is also true. With the product family, functions that only a single product variant requires are called *unique functions*.

Proposition 2. A module can be defined by all variant functions that are unique to a single product or subset of products. Since the unique functions appear only in a single product, it often makes sense to bundle these into a module to ease assembly.

Within the family function structure, each block of unique functions can be each differently highlighted. Then, the core of common functions, the unique functions, and the remaining functions that appear in some product variants but not others are all denoted. This refined family function structure for the VersaPak product family is shown in Figure 8.23.

So far this is an incomplete picture of the product family differences. The problem is that each product variant may require different levels of the flows and functions of the family function structure.

For example, consider the commercial jet engine family of Pratt and Whitney, which includes a 52, 64, 73, 84, and a 98-thousand pound thrust engine variant. Figure 8.24 shows the function structure for *any* of the

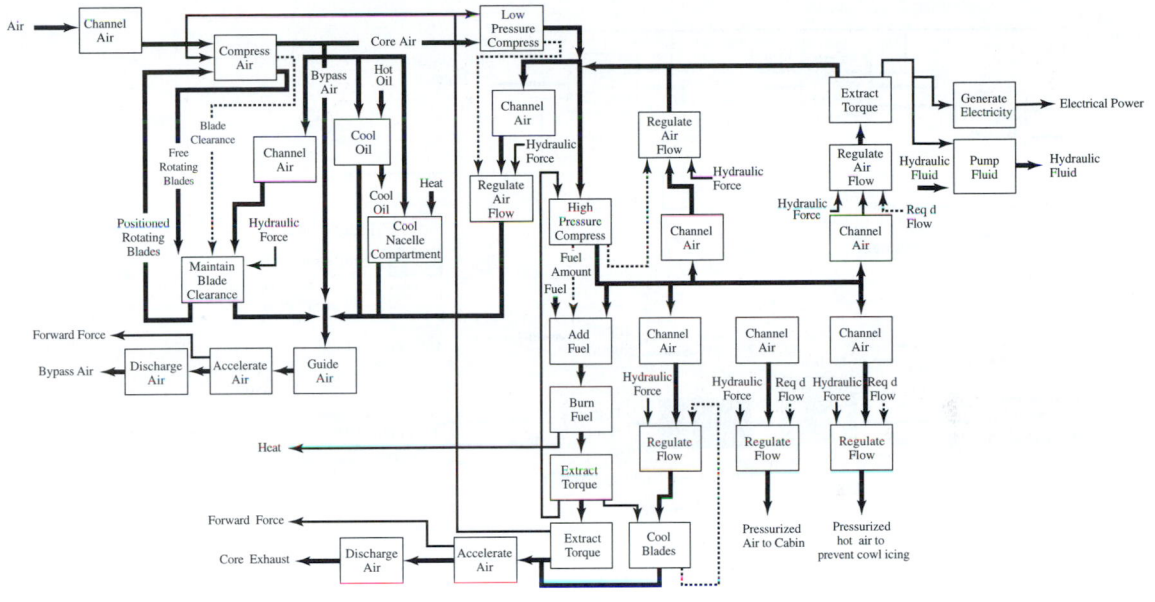

▼ *Figure 8.24.*
Pratt and Whitney commercial jet engine function structure.

Pratt and Whitney commercial jet engines. That is, the function structure is identical for all members of the Pratt and Whitney commercial jet engine family! Each jet engine, no matter the size, has the same sub-functions, since each sub-function is required by any jet engine to operate. Functionally, each jet engine is identical. Further then, the family function structure is then also identical to the function structure for each product variant.

The reason for this identical function structure of the product variants is that Pratt and Whitney's product family is created not by changes to the product functions, but rather by parametric size changes to the systems supplying the basic functions of a modern jet engine. The product family is created by parametric modularity.

To account for these parametric differences, the variety of the function parameters and the flow parameters of the family function structure need to also be highlighted. Different values of function or flow parameters prevent modularization within the family function structure, and so help delimit the family architectural modularity. For example, many of the Pratt and Whitney jet engine flows are of different levels, so the sub-functions are also at different parametric levels, and so the amount of function sharing is limited.

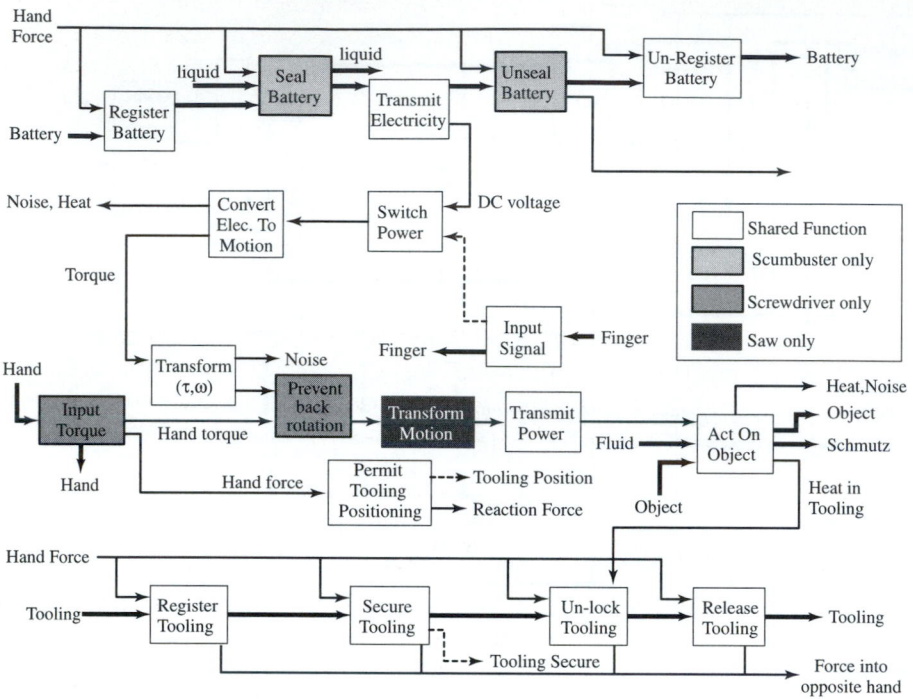

▼ Figure 8.25.
Further refined Black and Decker VersaPak product family function structure.

The VersaPak family function structure also highlighting the family parametric differences is shown in Figure 8.25. Note there are really only two common flows to the product family, all other flows in the family function structure vary from at least two products in the family. Therefore, other than the batteries, currently the VersaPak does not use the same sized flows for its product variants.

It is important to realize that such parametric levels are not fixed in stone. There are options in selecting these values. For example, currently each VersaPak product variant has a different sized electric motor, and so there are different parametric values for the sub-function "convert electricity to motion." Yet this need not be so. Perhaps the product family could be more effective if only three different sized motors were used. Thus, the family architecting task is to layout different permutations of the family function structure, each permutation with unique and common parametric levels. This will thereby develop different architectural modularity concepts. These different concepts can then be evaluated as different family concepts, as was done in the previous section on chart based portfolio selection methods.

Further, however, the family function structure as developed can also have the single product modularity heuristics of Chapter 9 applied to further define the product architecture of all the product variants. The result is, as described in Chapter 9, the generation of different possible product modular architectures. Each of these module concepts can be developed, and a family based concept selection method applied.

In sum, just as there is an opportunity for many different modular product architectures for a single product as demonstrated in Chapter 9, there are many modular platform architectures possible for a product family. This section has laid out the basic function architecting rules for a product family, but the immense set of possible permutations in applying these rules for any family function structure and set of variant function structures makes the task difficult. One can expect continued product development research into this area.

Advanced Method: Optimization Selection

In the previous section, it was noted that parametric values of different flows and sub-functions possibly limit the function sharing that is possible across product variants. Given a proposed set of shared function parametric values, are all the product variant concepts feasible? For example, in the Versapack example, if we limit the motors to three sizes, is it possible to have all the variant products work well with one of these three platform motor sizes? In this section, we develop a design optimization model to answer these questions (Gonzales-Zugasti et al, 1998).

In this section, we develop a design optimization model to develop a platform-based portfolio (Gonzales-Zugasti, Baker, and Otto, 1998). Optimization modeling is fully developed in Chapter 16, but we provide here an application to platform design. The model is demonstrated via an implemented spacecraft design application, where an interactive platform design tool is used to explore platform design tradeoffs when developing a platform that must serve multiple spacecraft missions. First, we present the common design optimization model for a single product and then expand that model into the platform design problem formulation.

Non-Platform-Based Products

The development of any product i that is individually designed, that is, not based on a platform, can be expressed as an optimization problem where one tries to optimize an objective function, f, over a vector of configuration variables, $\bar{x}_i$, subject to a set of technical performance, cost, and other constraints. This problem can be stated as follows:

$$\min_{\bar{x}_i} f(\bar{x}_i) \tag{8.4}$$

subject to performance constraints

$$g(\bar{x}_i) \leq 0$$

and budget constraints

$$\text{cost}_i(\bar{x}_i) \leq B_i$$

where B_i is the product's budget, determined independently by the company's management.

This model is admittedly simplistic; it presumes that there has been much work completed to develop concepts and models so that the design is at a development state amenable with Equation 8.4. Nevertheless, preliminary design equations of the form of Equation 8.4 can be developed for many systems, including spacecraft design, the results of which form specifications lists that flow down to the embodiment design activities.

Platform-Based Products

A platform-based product specification task can also be framed as an optimization problem. One would like to pick a platform by searching over the choice of what aspects of the products should be made common—or form the platform—and which aspects should be individually fulfilled for each of the variants. However, this search is complicated by the observation that once a candidate platform is chosen, each variant has remaining design choices that can be further individually optimized to provide the individual product's unique functionality. All of this is true while also needing to minimize the costs of each variant and the total cost of the whole family. In sum, the design must be optimized at two levels, at the product platform level and at the product variant level, as demonstrated in the general problem definition below.

The performance of a platform-based product will be partly determined by variable settings that describe those areas selected to be shared with the other members of the family, or *platform variables*. It will also be partly determined by performance settings arising from the design of those areas that are unique to that product i, or *variant variables*. Observing the different optimizations that are possible at the platform and variant levels as outlined above, the entire optimization problem becomes:

$$\min_{\bar{x}_p} \sum_{i \text{ products}} \min_{\bar{x}_{v_i}} [f(\bar{x}_p, \bar{x}_{y_i})] \tag{8.5}$$

subject to performance constraints

$$g_i(\vec{x}_p, \vec{x}_{v_i}) \leq 0$$

and budget constraints

$$\sum_{i\ \text{products}} \text{cost}_i(\vec{x}_p, \vec{x}_{v_i}) \leq B$$

where B is the budget for the family of products. This formulation first makes a platform choice of $\vec{x}_p$ in the outer minimization search of Equation (8.5); then each variant chooses a different $\vec{x}_{vi}$ by each minimizing f. The inner searches are repeated for different $\vec{x}_p$ or common platform variables as part of the search for the best platform configuration. The $\vec{x}_p$ are chosen that best meet all variant performances on arithmetic average, as reflected by the sum in Equation (8.5). Rather than simple addition, different norms across the different variant products are possible, such as a weighted sum or worst case norm.

Equation (8.5) is an expression that underlies what must be, in one way or another, determined when developing a product platform. However, from a practical, implementation point of view, setting up and solving such an optimization as a means to design a product platform would be difficult, as establishing target values to use with the constraint functions would be nontrivial, for example.

Nonetheless, the optimization problem can be used as a framework for an interactive tool allowing designers to choose the platform design they feel is the preferred combination of commonality and individual design. This tool would report performance and cost indices for the chosen product family, making it possible for the design team to iterate until a suitable platform/variant combination is achieved. The optimization would thus be implemented interactively by the design teams. The important distinction is that the design tool would collectively freeze some variables ($\vec{x}_p$, or platform), while allowing individual mission changes to others ($\vec{x}_v$, or variants).

Example: Implementation for Interplanetary Mission Design

Performance Models. As mentioned in the previous section, platforming decisions that make subsystems common to a set of products force changes in the performance of the products in question. Also, functions in complex products are coupled. For example, consider spacecraft design. Increasing the data rate may mean designing in a bigger antenna, which would increase the system mass and therefore the propulsion system, attitude control, and other subsystem parameters. Therefore, in order to observe the full effects of changing performance parameters to create a platform, one needs an interlinked system performance model. Developing such models will be fully discussed in Chapter 13.

In this spacecraft design example from the Jet Propulsion Laboratory (JPL) (New Millenium Program 1998), a system model created by the Aerospace Corporation is used to develop early-stage designs for interplanetary missions. This model is named *Concurrent Engineering Methodology* (CEM). It is used by an interdisciplinary team of JPL spacecraft subsystem experts acting as a team to develop system designs based on the requirements of concept missions.

The CEM model consists of several linked components:

- a system model, where spacecraft requirements are entered and architecture choices are made, such as the selection of hardware components such as engines, sensors, and type of propulsion system, and power source

- a database with performance and cost data for hardware as well as planetary information

- a summary display of the key system-level characteristics of the spacecraft

A section of the system model is shown in Figure 8.26. Some of the cells and pick lists in the spreadsheet are inputs to the model; these are used to calculate system performance values through formulas built into the model that link the behavior of the spacecraft subsystems to one another. Such a system performance model is typical of effective product development teams in modern practice.

Requirements Gathering

The CEM model described above was used to create three individual mission models based on the requirements for the Pluto, Europa, and Solar Probe NASA spacecraft concepts, as shown in Figure 8.27. Creating these mission requirements is not a trivial task, requiring day-long interactive design exercises by the design team, where subsystem trade-offs in power, mass, and costs are negotiated and assurances made that the mission hardware can function under various scenarios of the spacecraft mission.

Platform Design

Once the three individual missions had been modeled, the input variables for all three were used as inputs to a *Platform Selection* model developed on top of the CEM model. In the Platform Selection model, the designers are asked to make decisions as to whether a particular

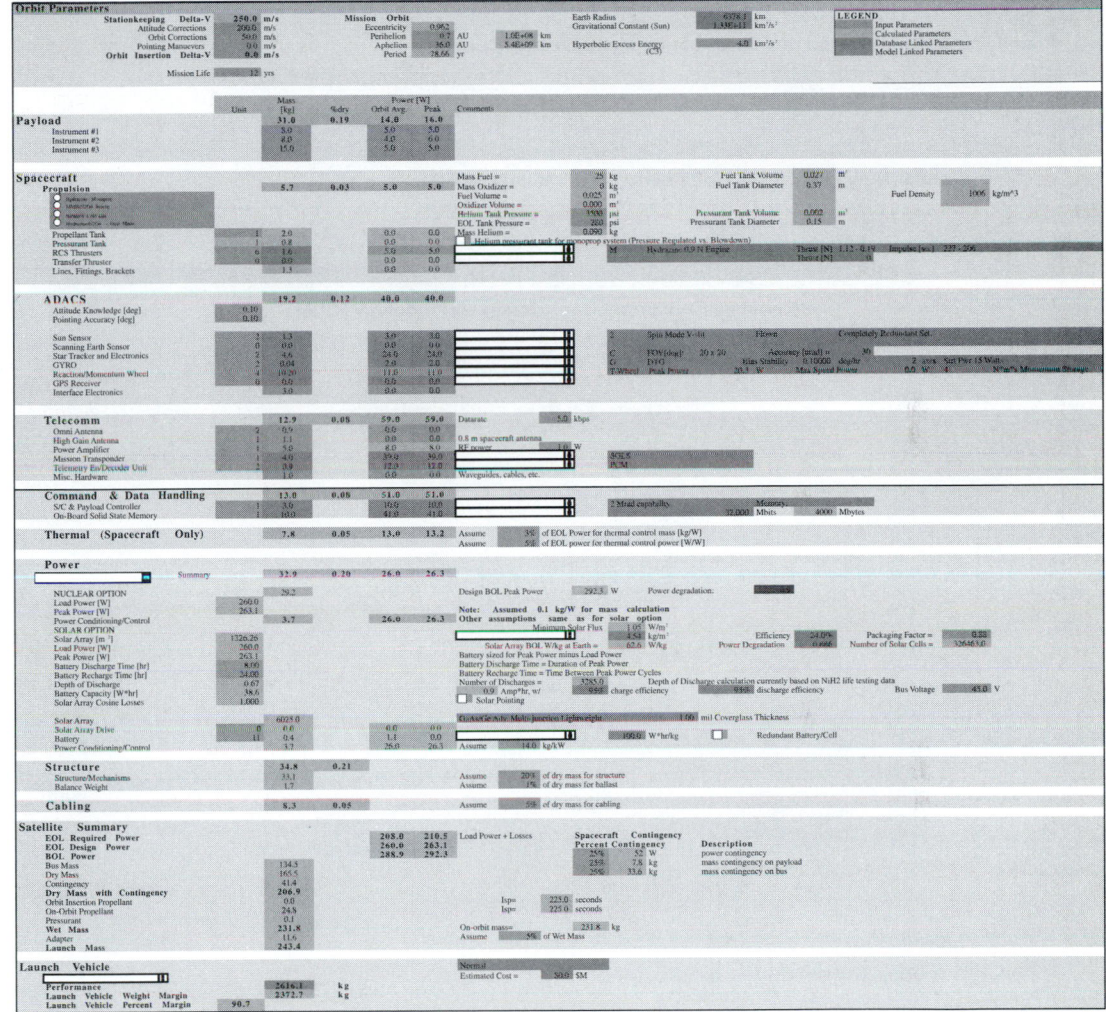

▼ *Figure 8.26.*
System summary portion of CEM spacecraft model.

spacecraft subsystem or performance requirement should be made common to all, some, or none of the missions. This process is diagrammed in Figure 8.28.

The choices are made through checkboxes placed next to each performance variable to be platformed, as shown in Figure 8.29. A setting of "TRUE" indicates that variable will become part of the platform design, whereas a "FALSE" choice indicates that variable setting will be used for that mission only.

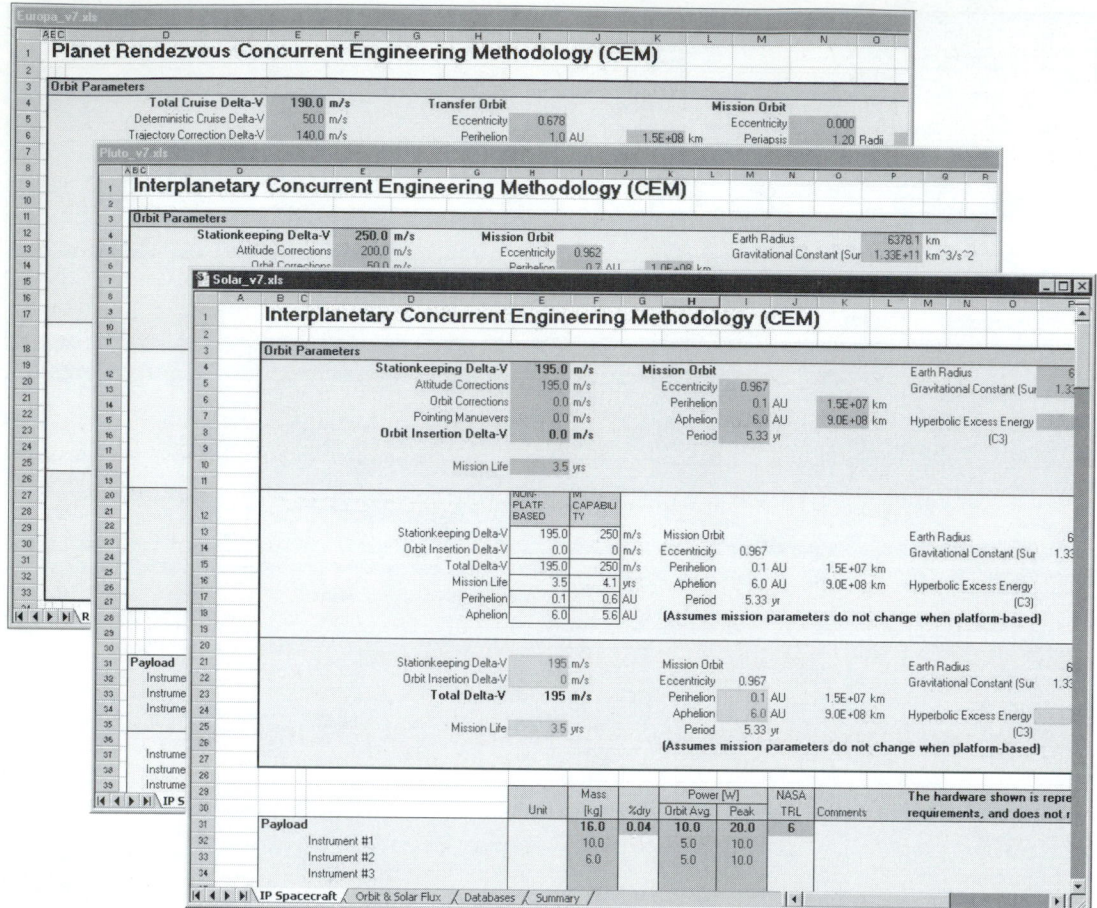

▼ *Figure 8.27.*
CEM models for the three case missions.

These decisions drive the design of the areas shared by the set of spacecraft or the platform. The model automatically calculates the settings for each input variable to be used in the design of the platform, usually by calculating the maximum, minimum, or average values of the individual mission's values that were selected to be platformed. If the calculated platform inputs are not acceptable to the design team, the team has a choice to override the automatically calculated values.

Once the input variable settings for the platform are chosen, the CEM model is used again to design the platform. This platform performance model then becomes the base for the design of each variant

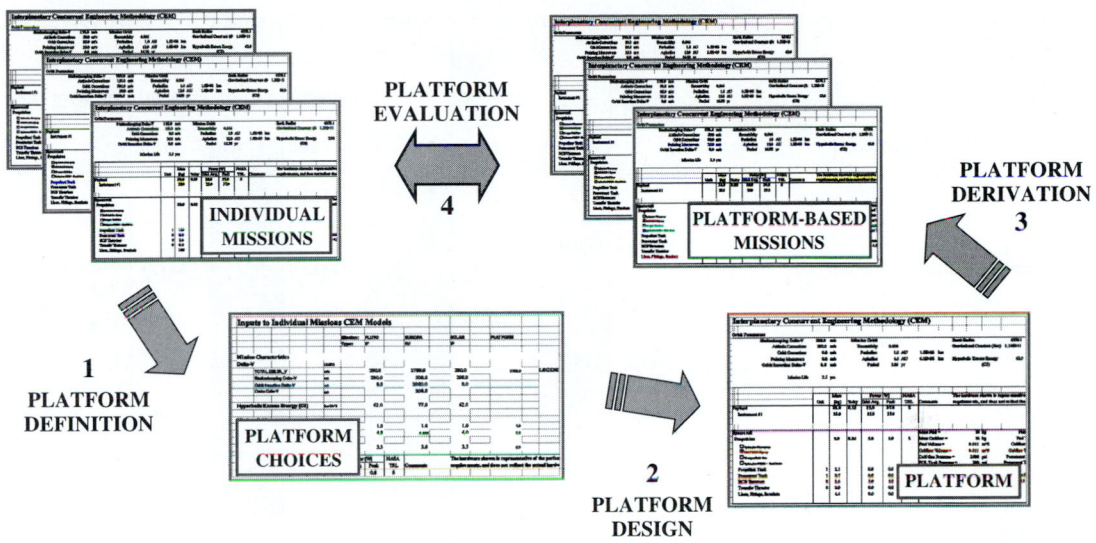

▼ **Figure 8.28.**
Interactive spacecraft platform design process.

mission. In reference to Equation (8.5), this process results in the selection of the variables.

Variants design: The outputs of the platform model are then reported back to the new individual mission design sheets for each mission and placed side by side with the original individual missions for comparison between the platform-based and non-platform-based designs. The user is then asked to modify the design of the platform-based mission, if necessary, to create a viable variant design. In reference to Equation (8.4), this process is the selection of the variables.

A section of the mission spreadsheet showing the individual and platform-based design for one of the missions is shown in Figure 8.30. The figure shows the attitude-control subsystem for the original mission in the top section, the platform design for that subsystem in the middle section, and the final variant design on the bottom section. Note the changes in the variable settings among the three sections, showing slightly different design solutions for the attitude control subsystem, depending on whether or not the design is platform based.

Platform selection and renegotiation: Once the variants for all missions are laid out, system designers can evaluate the chosen platform by comparing the variants to the original, non-platform-based designs.

		PLUTO		EUROPA		SOLAR		PLATFORM
Mission:		**PLUTO**		**EUROPA**		**SOLAR**		**PLATFORM**
Type:		**IP**		**RV**		**IP**		
			Platform?		Platform?		Platform?	Auto Input Value
ADACS								
Attitude Knowledge [deg]	deg	0.10 ☑	TRUE	0.05 ☑	TRUE	0.10 ☑	TRUE	0.05
Pointing Accuracy [deg]	deg	0.10 ☑	TRUE	0.05 ☑	TRUE	0.10 ☑	TRUE	0.05
Sun Sensor	#	2 ☑	TRUE	1 ☑	TRUE	2 ☑	TRUE	2
Type		6		6		6		
Operational Type		Spin Mode V-slit		Spin Mode V-slit		Spin Mode V-slit		
Scanning Earth/Horizon Sensor	#	0 ☐	FALSE	0 ☐	FALSE	0 ☐	FALSE	
Type		19		19		19		
Star Tracker and Electronics	#	2 ☑	TRUE	1 ☑	TRUE	2 ☑	TRUE	2
Type		4		1		4		
FOV	deg	20 x 20		8 x 8		20 x 20		
Accuracy	μrad	30		5		30		
GYRO	#	2 ☑	TRUE	2 ☑	TRUE	2 ☑	TRUE	2
Type		2		5		2		
Operational Type		DTG		DTG		DTG		
Bias Stability	deg/hr	0.1000		0.1000		0.1000		
Axes	#	2		2		2		
Reaction/Momentum Wheel	#	4 ☑	TRUE	3 ☑	TRUE	4 ☑	TRUE	4
Type		20		6		22		
Wheel Rated Peak Power	W	20.3		40		3		
Max Speed Power	W	0		6		6		
Momentum Storage	Nms	4		1.9		5-10		
Peak Power		11		11		11		
Orbit Average Power		11		11		11		
GPS Receiver	#	0 ☐	FALSE	☐	FALSE	0 ☐	FALSE	
Type		4		4	(none)	4		
Interface Electronics		☑	TRUE	☑	TRUE	☑	TRUE	
Mass	kg	3		3		3		3
Peak Power	W	0		0		0		0
Orbit Average Power	W	0		0		0		0
NASA TRL		4 ☑	TRUE	5 ☑	TRUE	6 ☑	TRUE	6

▼ **Figure 8.29.**
Portion of platform selection model.

If the variants are not satisfactory, the designers can optimize the missions by changing the variant settings or by arguing with the other missions for changes in the choices made in the platform selection model, thus optimizing the design at the platform or product family level. The process allows the designers to iterate until a satisfactory platform and variants combination is achieved.

Results

The models described above were used to design a platform for the three case missions, and three variants were derived. For each mission, the variant could be compared to the original design on the performance of each subsystem and on system-level performance metrics

ADACS		24.1	0.06	40.0	40.0	6		
Attitude Knowledge [deg]	0.10							**Non-Platform Design**
Pointing Accuracy [deg]	0.10							
Sun Sensor	2	1.3		3.0	3.0		0.05 deg Sun Sensor	▼
Scanning Earth Sensor	0	0.0		0.0	0.0		none	▼
Star Tracker and Electronics	2	4.6		24.0	24.0		30 arc sec Star Tracker	▼
GYRO	2	0.04		2.0	2.0		Dynamically Tuned Gyro	▼
Reaction/Momentum Wheel	4	15.12		11.0	11.0		3.78 kg wheel	▼
GPS Receiver	0	0.0		0.0	0.0		none	▼
Interface Electronics		3.0		0.0	0.0			
		24.1	0.06	40.0	40.0	6		
Attitude Knowledge [deg]	0.05							**New Variant Design**
Pointing Accuracy [deg]	0.05							
Sun Sensor	2	1.3		3.0	3.0		0.04 deg Sun Sensor	▼
Scanning Earth Sensor	0	0.0		0.0	0.0		none	▼
Star Tracker and Electronics	2	4.6		24.0	24.0		15 arc sec Star Tracker	▼
GYRO	2	0.04		2.0	2.0		Ring Laser Gyro	▼
Reaction/Momentum Wheel	4	9.60		11.0	11.0		2.3 kg wheel	▼
GPS Receiver	0	0.0		0.0	0.0		none	▼
Interface Electronics		3.0		0.0	0.0			

Platform Fixed

Variant Choice

▼ **Figure 8.30.**
Portion of mission spreadsheet showing individual, platform, and variant design for one of the spacecraft's subsystems.

such as launch margin and total power required. These metrics for the different designs for each mission were displayed in a summary sheet within the model as shown in Figure 8.31.

The effect of platforming subsystems, or choosing to share them across missions, could be clearly seen in areas such as the power subsystem. Since two of the missions were designed to travel far away from the sun (Europa and Pluto), whereas the third was meant to head toward the sun, it did not make sense to try to platform on a shared solar power subsystem. This choice could be clearly seen from the models; forcing the missions to share a solar power module caused the Solar Probe mission to carry several hundred extra kilograms of large solar panels. These very large panels were not needed and caused many other subsystems, such as the spacecraft's propulsion module, to grow correspondingly in size.

This issue is an extreme example, since no designer would realistically suggest sharing a solar power module for these three missions. However, it does show that the model accepts platforming decisions and displays their effects on each variant to be derived from the platform, both on the subsystem level and also at the system level through

Spacecraft	Mass (kg)	Average Power (W)
Payload	16.0	10.0
Bus		
Propulsion	8.8	5.0
Attitude Determination and Control	24.1	40.0
Telecomm	12.8	43.0
Command and Data Handling	4.6	17.0
Thermal	6.0	10.0
Power	201.5	26.3
Structure and Mechanisms	78.8	
Cabling	22.5	
Summary		
End of Life Required Power (W)		151.2
End of Life Design Power (W)		199.6
BOL Power (W)		206.8
Bus Mass (kg)	359.0	
Dry Mass (kg)	375.0	
Contingency Mass (kg)	120.0	
Dry Mass with Contingency (kg)	495.0	
Propellant Mass (kg)	45.7	
Wet Mass (kg)	540.8	
Adapter Mass (kg)	27.0	
Launch Mass (kg)	567.8	
Launch Vehicle Performance (kg)	214.3	
Launch Vehicle Performance Margin (kg)	-353.5	
Launch Vehicle Performance Margin (%)	-165%	
Launch Vehicle Cost (M$)	125	

Spacecraft	Mass (kg)	Average Power (W)
Payload	18.0	10.0
Bus		
Propulsion	7.4	5.0
Attitude Determination and Control	19.2	40.0
Telecomm	12.8	43.0
Command and Data Handling	4.6	17.0
Thermal	0.0	0.0
Power	154.6	17.5
Structure and Mechanisms	62.3	0.0
Cabling	17.8	0.0
Summary		
End of Life Required Power (W)		132.5
End of Life Design Power (W)		174.9
BOL Power (W)		181.2
Bus Mass (kg)	278.7	
Dry Mass (kg)	296.7	
Contingency Mass (kg)	94.9	
Dry Mass with Contingency (kg)	391.6	
Propellant Mass (kg)	36.2	
Wet Mass (kg)	427.7	
Adapter Mass (kg)	21.4	
Launch Mass (kg)	449.1	
Launch Vehicle Performance (kg)	364.5	
Launch Vehicle Performance Margin (kg)	-84.6	
Launch Vehicle Performance Margin (%)	-23.2	
Launch Vehicle Cost ($M)	80	

▼ *Figure 8.31.*

Summary of performance metrics for individual and platform-based designs for one of the missions being considered for platforming.

appropriate performance metrics. Choices not so drastic as the one mentioned above, such as changing the type or size of engines or other components, could also be seen in the mass and power totals for each mission. The changes in these metrics could be used by system design engineers to make decisions about what should or should not be shared across the three case missions.

The point of this example is to show that with models, a design team can apply Equations (8.5) to determine a platform design for a variety of support product variants. Modeling and analysis can support the decision making.

V. SUMMARY AND "GOLDEN NUGGETS"

Product portfolios are the set of products offered by a company. The portfolio architecture is a representation of how these products interrelate, both functionally and through sharing common components or standards. There are several types of portfolio architectures, from fixed unshared to modular to adjustable. Customer needs over a market and over time determine how a portfolio should be architected.

Golden nuggets that should be taken from this chapter include:

▶ Each customer need indicates a different type or portfolio architecture. Thus, a portfolio can exhibit several types of portfolio architectures at the same time.

▶ Customer need variation across a population and across time determines how to architect a desired portfolio.

▶ Platform architecture is effective for offering variety at reduced cost.

▶ Determining what constitutes a platform for a family involves considering all of the supported variants and how their remaining functions can best be designed after a platform is selected.

References

Bendat, J., and Piersol, A. 1986. *Random data: Analysis and Measurement Procedures.* New York: Wiley.

Cutherell, D. 1996. Product architecture. Chap. 16 in *The PDMA handbook of new product development,* edited by M. Rosenau, Jr. et al. New York: Wiley.

Feitzinger, E., and Lee, H. 1997. Mass customization at Hewlett Packard: The power of postponement. *Harvard Business Review* (Jan.-Feb.): 116–121.

Gonzales-Zugasti, J., Baker, J., and Otto, K. 1998. A method for selecting optimal product platforms. *Proceedings of the Design Automation Conference,* Atlanta, GA.

Meyer, M., and Lehnerd, A. 1997. *The power of product platforms.* New York: The Free Press.

Meyer, M., and Utterback, J. 1993. The product family and the dynamics of core capability. *Sloan Management Review* 34 (3): 29–47.

New Millenium Program. 1998. NASA Jet Propulsion Laboratory. *http://nmp. jpl.nasa.gov/Program/.*

Pine, B. J., II. 1992. *Mass customization: The new frontier in business competition.* Boston, MA: Harvard Business School Press.

Pine, B. J., II, Victor, B., and Boynton, A. C. 1993. Making mass customization work. *Harvard Business Review* (Sept.-Oct.): 108–119.

Sanderson, S., and Uzumeri, M. 1996. Managing product families. New York: McGraw-Hill.

Yu, J. S., Gonzalez-Zugasti, J. P., and Otto, K. N. 1999. Product architecture definition based upon customer demands. *Journal of Mechanical Design,* 129(3): 329-335.

9 Product Architecture

Chapter written in collaboration with Robert B. Stone,
University of Missouri at Rolla

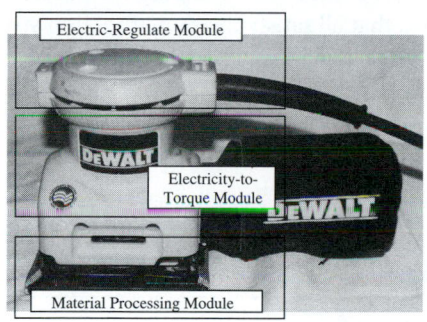

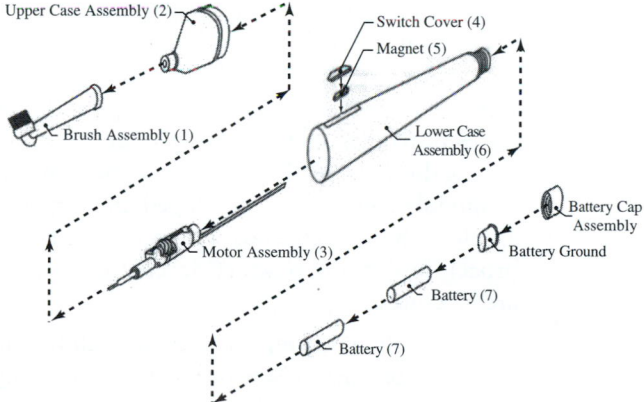

Example product layouts: DeWALT™ power sander and an electric
toothbrush.

In product development, key milestones are apparent throughout the process. Developing a product architecture is one such strategic milestone for any class of products. It is where we begin to make key decisions on how the product will physically operate.

Assuming a successful development of the customer needs list, the business case, the functional model, and so forth, we are transitioning to the "Develop a Concept" stage of product design. The challenge in this stage is to translate the customer needs and business case into a realizable product concept(s). This translation is what we define as the product architecture, which is the mapping from the product function to the product form. It is the division into parts and assemblies of a product and how the functional network matches or cuts across these physical divisions and interfaces. It also provides a sound foundation for organizing and managing the product development effort.

We present this material before the concept generation and selection chapters, since understanding the different possible product architectures that all satisfy the same function structure can help inspire new concepts and ideas. Product architecture is part of the early transition from function to form.

I. CHAPTER ROADMAP

In this chapter, we explore the intricacies of conceptualizing a product architecture, its implications on product performance and service, its impact on product cost and strategic evolution, and its association with forming and managing design teams. Chapter 9 begins by investigating the predominant types of product architectures. It then concentrates on modular design and presents a basic method for identifying modules from a product's functional description. Following this basic method, a more sophisticated approach, referred to as module heuristics, is developed and applied to a number of examples. The chapter ends with an extension of this advanced approach to forming product development teams based on product architectures.

Figure 9.1 shows example routes through the chapter. A direct route covers simple methods, with the goal of creating a suitable architecture of a product. An advanced route is also shown, augmenting the approach with more sophisticated, analytic techniques.

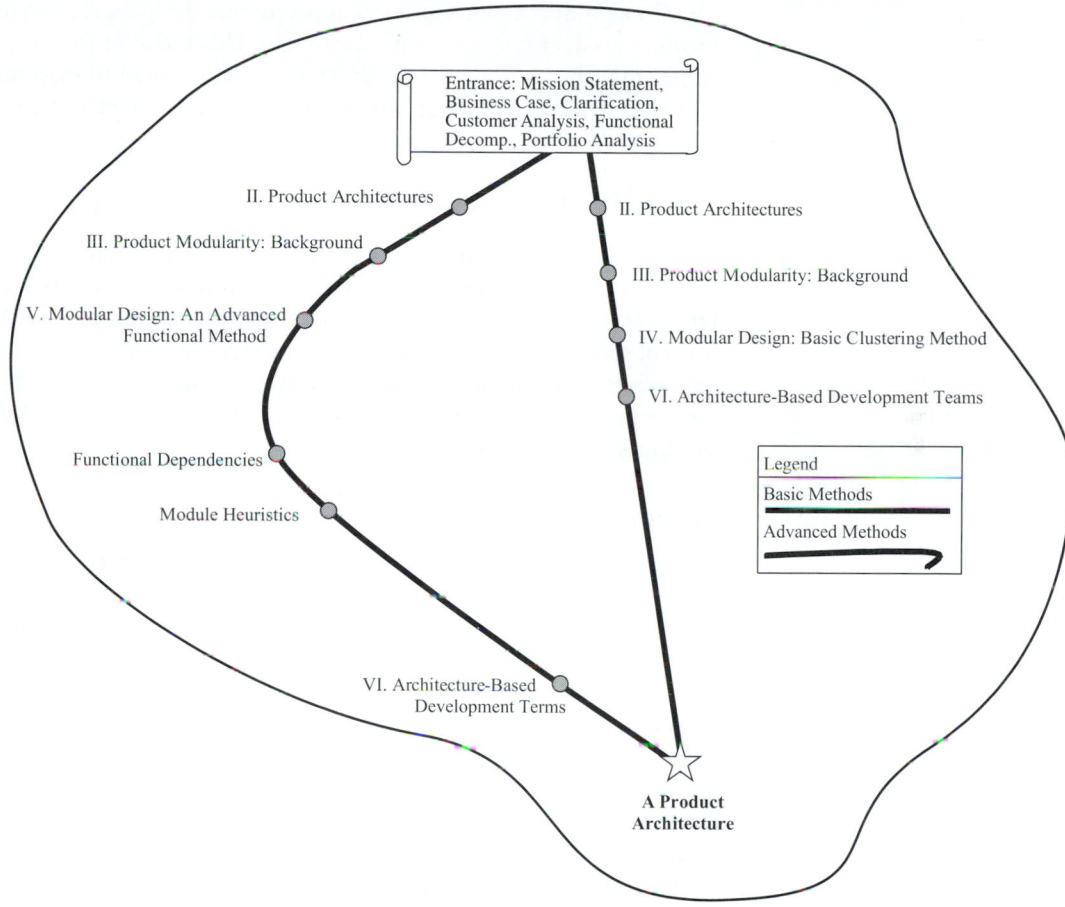

▼ Figure 9.1.
Road map for alternative routes for developing a product's architecture.

II. PRODUCT ARCHITECTURES

Introduction

Product architecting, at a basic level, starts the creation of effective *layouts* of components and subsystems, where different tasks are completed by subsets of the product development team. We seek to answer a number of questions during this stage: what alternative architectures exist, how will the subsystems interact, and how will the subsystems be divided and interfaced. If we look a little closer at the *act* itself, of creating a product architecture, it is apparent that the focus is on transforming product function to product form. This focus, of course, assumes that the proper level of effort has been devoted to mapping customer needs to a functional model of a product (Chapters 4, 5).

Architecture Types

There exist two categories of architecture types: *product* and *portfolio.* Portfolio architectures relate to a group or family of products (Chapter 8). Design strategies revolve around whether to, how to, and the form of sharing components among the products in a portfolio. Product architecture, on the other hand, relates to a specific product layout. In this category, strategies for product design revolve around the product's market and performance.

Integral

Integral product architectures are physical structures where all of the subfunctions map to a single or very small number of physical elements (Cutherell, 1996). This definition implies that the individual physical elements include and perform a large number of functions. There is no attempt made to isolate individual functions into individually isolated components or subsystems; rather, the components function-share. This implies, in the typical case, that the physical elements tend to "blend" together at their interfaces. These interfaces will also have complex interactions and are difficult to define precisely. As a result, changes made to any component in an integral architecture tend to propagate to many, if not all, other physical elements.

Figure 9.2 displays example products based on an integral architecture. Examining each product shown, whether a screwdriver, wrench, or

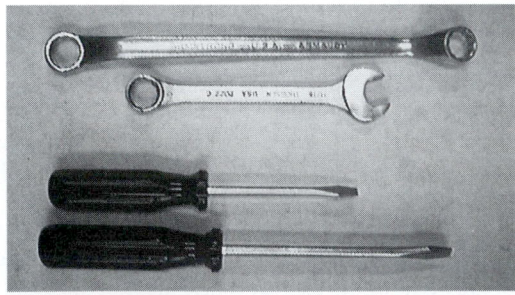 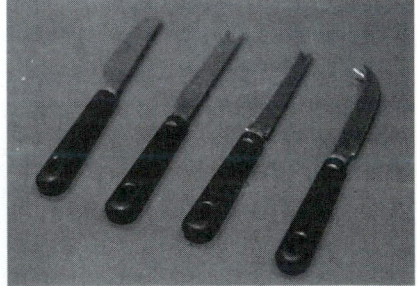

▼ *Figure 9.2.*
Examples of integral product designs.

knife, each component in the product performs many functions. Typically, integral product architectures are applied on very high-volume products where the cost of the product need not be reduced through sharing of subsystems with other products to achieve economies of scale. Rather, the product is so high volume that every amount of cost is taken out through design-for-assembly practices that lead to complex parts that have minimal assembly effort but much function sharing.

Modular

Product modules are defined as integral physical product substructures that have a one-to-one correspondence with a subset of a product's functional model (Ulrich and Tung, 1991; Ulrich, 1995; Cutherell, 1996). Modular products may be defined as machines, assemblies, or components that accomplish an overall function through the combination of distinct building blocks or modules (Pahl and Beitz, 1988).

While modular products have distinct manufacturing advantages in many instances, modularity in mechanical design is often an afterthought. Once a product is designed and developed, components of the product are observed to have other potential uses. This observation, if realized during reverse engineering and redesign, can lead to faster development and reduced costs in future product designs. There is a myth, however, that the initial product design activity will suffer, since the design team will be slowed by having to design the first product to be compatible with subsequent products. If modularity is identified and properly designed for in the initial conceptual effort, then the *immediate* product design will also benefit (U.S. Congress 1992; Pahl and Beitz, 1988; Stone, 1997).

Architecture Examples

Table 9.1 lists examples of the fundamental two types of product architecture—integral and modular. Also listed in the table are some common product derivatives for the listed products. This is done to differentiate the portfolio architecture for the product and its derivatives (Chapter 8) from the product architecture. Note that the portfolio architecture—whether a platform or fixed unsharing portfolio architecture—is not the same as whether the product exhibits integral or modular product architecture.

For example, a platform portfolio architecture can exist with integral product architecture—as it does with a set of different-sized kitchen knifes—or a platform portfolio architecture can exist with modular product architecture, as it does with cordless power tools that all use the

TABLE 9.1. PRODUCT ARCHITECTURE EXAMPLES

Product Architecture		Portfolio Architecture for the Product and Common Derivatives		
Type	Product	Derivatives	Type	Characteristics
Integral	Handled vegetable peeler	Fisted vegetable peeler	Fixed unsharing	Shares no common components
	Wooden pencil	#1, #3 lead pencils	Modular platform	Differing lead hardness
	Kitchen knife	Kitchen knife set	Modular platform	Parametric handle size
Modular	Swiss Army Knife	Complex knives	Modular platform	Expanding width
	PC computer	More RAM, devices	Adjustable for purchase	Standard interfaces
	Black and Decker cordless drill	VersaPak line of power tools	Modular platform	Common battery Common motor
	Tinkertoys	Theme sets	Adjustable for Use	Standard interfaces Component variety

same battery pack modules. This is because product architecture has nothing to do with derivative products, which are a feature of the portfolio architecture. Derivative products can be derived from products exhibiting integral product architecture or modular product architecture. While we may have spent much time elaborating the difference between product architecture and portfolio architecture, it remains a source of confusion both to the academic community and to the industry as a whole. Product architecture is well defined as the mapping of function to form in a product. Product architecture so defined does not cover other architecture decisions, though, such as sharing or not sharing of modules among portfolio elements to reasonably meet market demands for variety, the topic of portfolio architecture.

III. PRODUCT MODULARITY: BACKGROUND

Modular devices make economic sense for a number of important reasons, as listed in Table 9.2. For further specific documented examples, Xerox's use of modular attachment methods and standardized components has made their copiers easier to disassemble, modify, and reassemble (U.S. Congress, 1992). Fiber optic cable splicer units by 3M also incorporate modules to provide sealing around the cables for a variety of enclosures as well as decoupling the seals from the structur-

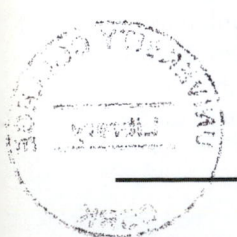

**TABLE 9.2.   COMPARISON OF MODULAR AND INTEGRAL ARCHITECTURES
(ADAPTED FROM CUTHERELL, 1996)**

	Pros	Cons
Modular architecture	Improves device reconfigurability Increases the device variety and speed of introduction for new devices Improves maintainability and serviceability of device Decouples development tasks (and manufacturing, to some extent)	May make devices look too similar Makes imitation of device easier by competitors Reduces device performance Modular design may be more expensive than integral design
Integral architecture	Harder for competitors to copy design Tighter coupling of team with less interface problems Increases system performance Possible reduction in system cost	Hinders change of design in production Reduces the variety of devices that can be produced

al supports that hold the cables together (J. Jackson, personal interview, 1997).

Effective approaches to modular design has been a recognizable goal of industry for some time. Methods are needed to develop modular designs that are intuitive and correct but also generic. The approach we will present here is one based on function structures.

Visits to a number of industries show that systematic function-based design methods are seeing increased use in industry, both from our observations and others (J. Jackson, personal interview, 1997). In the next sections, we take a close look at methods to design modular products, in the context of the larger problem of architecture. Both a simple and an advanced method are described for carrying out a modular design.

Types of Modularity

Two benefits of modular design are standardization of components and reconfigurability of devices. Based on these two attributes, a modular device may be classified as a member of one or more classes (Ulrich and Tung, 1991; Cutherell, 1996). This classification of modularity does not form crisp categories, since any particular device may exhibit qualities of more than one classification. One set of subfunctions may be carried out with an integral architecture; another set of subfunctions may be carried with a modular architecture. However, the classification is useful in understanding the various modular designs that are possible.

We define two macro types of modularity: *function-based* and *manufacturing-based*. The first type is applied to partition the functionality of a product and how these functions are distributed. Manufacturing-based modularity, on the other hand, relates more to the manufacturing techniques and assembly operations associated with a product. A set of components may be bundled into a subassembly for purely manufacturing reasons, such as combining many tangly parts into one subassembly. Consideration of the set of functions performed by the subassembly then defines a module. Modules may be so conceptualized—according to manufacturing and assembly with functions following.

Function-Based Modularity

Four classifications of function-based modularity are defined as follows.

1. *Slot modularity* is where one basic device uses several different components to allow it to perform multiple tasks. It often provides the means to support customizable portfolio architecture: the same module is used across different products in a portfolio, each product has the same interface, and the platform module only performs one function. This class is also associated with the concept of component standardization.

An example of slot modularity is Black and Decker's line of VersaPak power tools shown in Figure 9.3. All of the tools utilize the same battery packs to provide power. Additionally, many use the same motors.

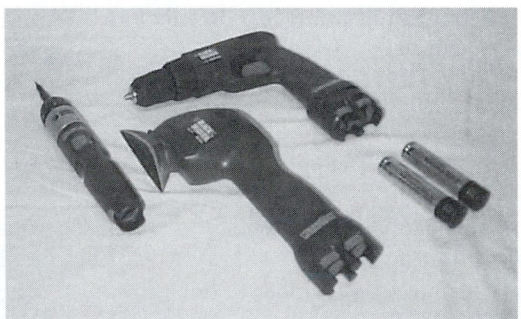 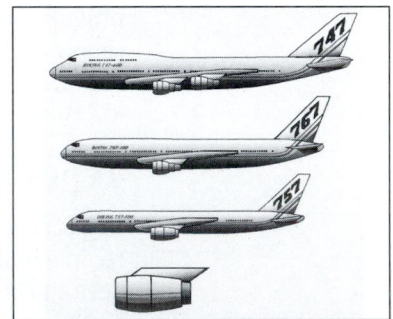

▼ *Figure 9.3.*
The three VersaPak power tools from Black and Decker or the engine mounts on Boeing airplanes represent examples of slot modularity. All of the power tools use the same batteries (two are shown of the right side of the figure) as energy sources; all of the planes can use the same engines.

▼ *Figure 9.4.*
Computer DRAM expansion slots and the computer card cage are examples of bus modularity. Here, two DRAM chips are inserted in the rightmost bus slots with two more bus slots empty. The three empty bus slots on the left are VRAM slots. On the right, the card cage can accept different functioning cards, from disk drives to video cards to I/O boards.

2. *Bus modularity* describes a device (considered the main component of the system) that is equipped with a standard interface that accepts any combinations of different functioning modules. In most cases, the modules have a standard interface that attaches to one of the *slots* on the common bus interface. Memory expansion slots in computers are an example of bus modularity and are shown in Figure 9.4. Track lighting is another example. In this case, the lights are the components and the track is the bus that accepts them.

3. *Sectional modularity* is exhibited by a chained interconnection of modules (sometimes called *sections*), each equipped with an identical interface. The modules can each individually accomplish different product subfunctions, and their recombination on the chain interface then permits different system (product) functions. There is generally no one main module that is called the device; rather, the collection of modules is the product. Each component may have more than one interface. Office furniture is an example of sectional modularity. More mechanical type examples of sectional modularity are shown in Figure 9.5.

4. *Mix modularity* combines several standard components together through webs of modules rather than a simple chain as with sectional modularity. The modules must be equipped with at least two complementary interfaces to create a new device. The beloved Tinkertoy set is an example of mix modularity (Figure 9.6). Its components combine to produce an almost endless number of new devices (in the form of toys).

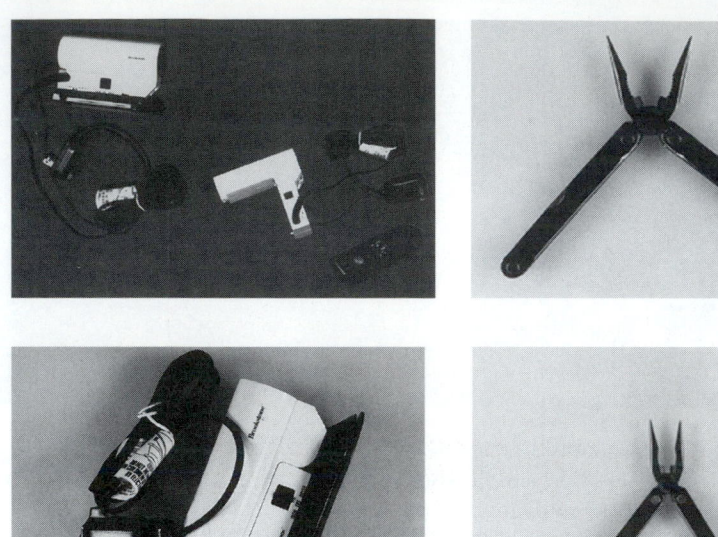

▼ *Figure 9.5.*
The travel iron/hairdryer can be reconfigured either as an iron or as a hairdryer. The Leatherman can be reconfigured into a variety of devices: knife, screwdriver, or pliers.

Generally, very complex systems exhibit mix modularity, such as armies with differing troops, helicopters, planes, guns, tanks, and munitions, each of which is a module. Each module can serve different functions, and each module can be combined in differing ways to accomplish different tasks. Similarly, construction projects make us of standard components such as air handling units, lighting fixtures, I-beams, and pipe fittings, which can be mixed to achieve very different buildings and spaces.

Figure 9.7 summarizes the four types of function-based modularity: slot, bus, sectional, and mix (Ulrich and Tung, 1991). As shown in the figure, iconic descriptions of the modularity types generalize the examples discussed above.

Manufacturing-Based Modularity

In addition to the function-based modularity descriptions above, derived from the basis of function-to-form, four manufacturing-based modularity classes exist that describe other types of product modules (McAdams, Stone, and Wood, 1999). These classes group subassemblies based on manufacturing technique and assembly operations. In

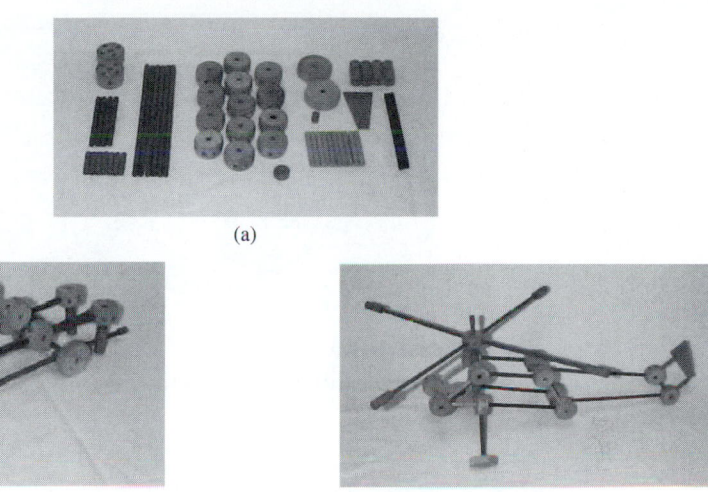

(a)

(b)   (c)

▼ *Figure 9.6.*
Examples of mix modularity. The (b) toy car and (c) toy helicopter are made from the same set of standard components shown in (a).

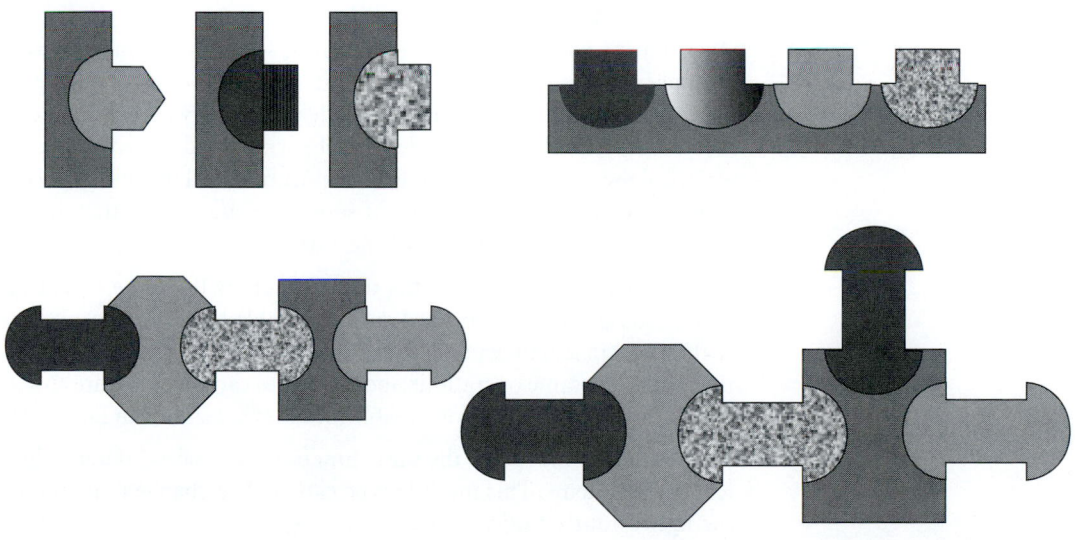

▼ *Figure 9.7.*
Summary of function-based modularity: Iconic descriptions (adapted from Ulrich and Tung, 1991).

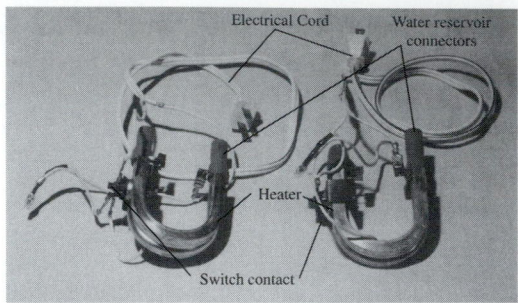

▼ **Figure 9.8.**
Manufacturing-based modularity: Assembly modules.

fact, a device may accurately be classified by both function-based and manufacturing-based modularity descriptions. The manufacturing-based modules are described next.

The most obvious of manufacturing based modules are *OEM modules*—groups of components or functions that are grouped simply because a supplier (Original Equipment Manufacturer—OEM) can provide them at less expense than could be developed and manufactured in-house. Power supplies in computers, for example, are modules that are bought as commodity items.

Assembly modules are components or groups of components that solve related functions but are bundled to increase assembly ease (Figure 9.8). An example is the electricity-to-thermal energy module of a coffee maker. The module includes the electrical cord, switch contacts, electrical resistor heater, the water transport tube and the tube-to-water reservoir connectors. It consists of several separate parts that are assembled before final assembly of the entire device.

Sizable modules are components that are exactly the same except for their physical scale (Figure 9.9). Lawn mower blades are sizable modules as they vary in length depending on the width of cut of the mower cutting deck. The same operations and machines can manufacture sizable modules. Other examples of sizeable modules are shown in Figure 9.9.

Conceptual modules solve the same functions but have different physical embodiments. This module type can apply a change in manufacturing without significant changes to the remainder of the product. Identification of conceptual modules in related devices provides an opportunity for component sharing between the devices.

An example of a conceptual module is shown in Figure 9.10. These photos are of the bottom of the water tanks of a Mr. Coffee coffee maker

Tea brewer   Coffee maker

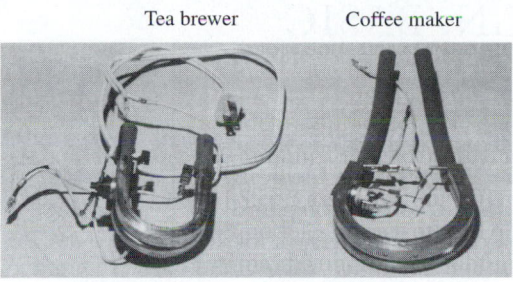 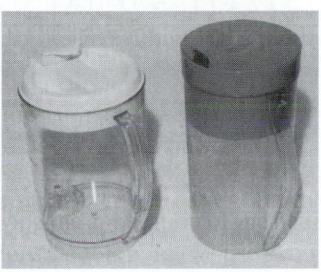

▼ *Figure 9.9.*
Manufacturing-based modularity: sizable modules. (a) Thermal generation modules for
a tea brewer (left) vs. a coffee maker (right). (b) Tea containment modules for two
different models of iced tea brewers.

(left) and a competitor's iced tea brewer (right). The module stops
water from leaving the holding tank until the water in the heating chamber is hot. These modules are conceptually equivalent but have different physical embodiments because of the differences between coffee
and tea brewing. The stop liquid module on the coffee maker incrementally stops and allows water entry to the heating chamber by a floating ball. On the tea brewer, the stop liquid module stops water from
leaving the chamber until all of the liquid is hot by using a stopper on
a heat-sensitive lever. They remain as modules, however, since they interface the other equivalent modules in both brewers.

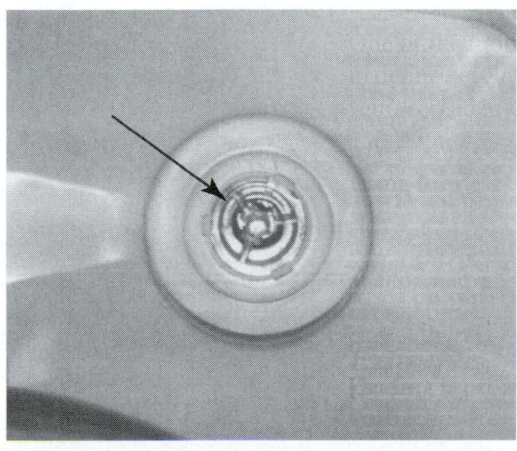 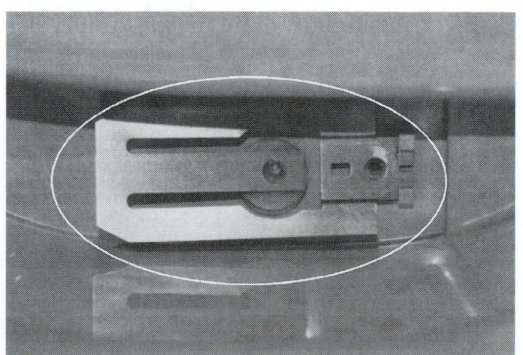

▼ *Figure 9.10.*
Manufacturing-based modularity: conceptual modules.

IV. MODULAR DESIGN: BASIC CLUSTERING METHOD

Based on the types of modularity described above, this section presents a basic technique for establishing a modular product architecture. The overall goal is to translate customer needs into rough layouts of a product. To accomplish this goal, small "chunks" (modules) must be identified in the product. These chunks will provide a means of distributing individual designers or design teams to tractable subsystems in the product design. They will also provide a means of allocating resources to satisfy customer needs and a means of searching for novel and creative layouts of a product.

A four-step process will now be provided for establishing an architecture (Ulrich and Eppinger, 1995; Cutherell, 1996). Two views of a product drive the process: functions and components. The functional view follows the concepts of function hierarchies or function structures (Chapter 5). In this sense, functions are the operations or activities performed by the product. Components, on the other hand, perform the action of the function. For example, a knife cuts food, whereas it is usually made up of two components, the integral shank-blade and the handle.

Using the function-component views, consider the four-step process for laying out modules. A product function structure begins the process, followed by clustering, generating rough geometry, and defining interactions. This process focuses on systematically identifying the viable modules or chunks in a product design. In addition, it seeks to identify the interfaces between modules and the interactions at those interfaces.

Step 1: Create a Function Structure of the Product

The first step entails the creation of a function structure. As defined in Chapter 5, a function structure is an input-output diagram of what a product does. Materials, energies, and signals enter from the environment, are processed by the function structure, and leave the product as new flows (Otto and Wood, 1997).

For the purpose here, subfunctions in the function structure are blocks (in verb-noun form). Lines to show dependencies connect these blocks, just as in Chapter 5. Different line nomenclatures provide information regarding the type of dependency: material, energy, or signal.

A fully connected set of blocks results in a suitable structure for analysis. It is worth noting, however, that full detail should not be included here. We are attempting to identify high-level modules, not generate detailed choices of form, materials, or connectors. It is also worth noting that function structures are not unique for a product. They depend on the physical principles chosen to solve a product design. For example, different functional structures for a screwdriver will result depending on the choice of power: manual operation, power-assisted (e.g., electricity), or a hybrid. At this stage of the development process, alternative structures should be developed and maintained. These alternatives will provide different avenues for product design. Commonality in the structures will also illustrate the core elements that may be used for family architectures.

Chapter 5 provides various methods for creating a function structure for a product. An example result is shown in Figure 9.11. This figure represents the functional structure of a Hewlett-Packard desktop inkjet printer (Bockman, 1994; Cutherell, 1996).

Step 2: Cluster the Elements into Module Chunks

After creating one or more valid structures, the next step is to group the subfunctions into "chunks." These chunks will become the modules or assemblies for the product.

Chunks are chosen based on the "natural" or intuitive groups of subfunctions that depend on each other and/or can be solved together. During the choice of groups, care must be taken in defining the boundaries of each chunk. Simple interactions between modules are preferred. By simplifying the interactions, each chunk will be as independent as possible. This independence will provide greater flexibility and freedom on the part of each design team assigned to a chunk. Simple interactions will also reduce the possibility of problems or faults caused at the interfaces of modules.

The overall process of identifying chunks is known as *clustering*. An example of clusters chosen for the desktop inkjet printer product is shown in Figure 9.12. Notice in the figure that five primary modules are identified for the printer: ink system (print cartridge), chassis, base, paper-handling system, and electronics. Notice also that these modules depend, for the most part, on one or two distinct connections, such as paper in the case of the paper-handling system. Choosing a module based on these dependencies is not part of the method, but it does result in minimal interactions between chunks. The advanced modular design method in the next section formalizes this idea.

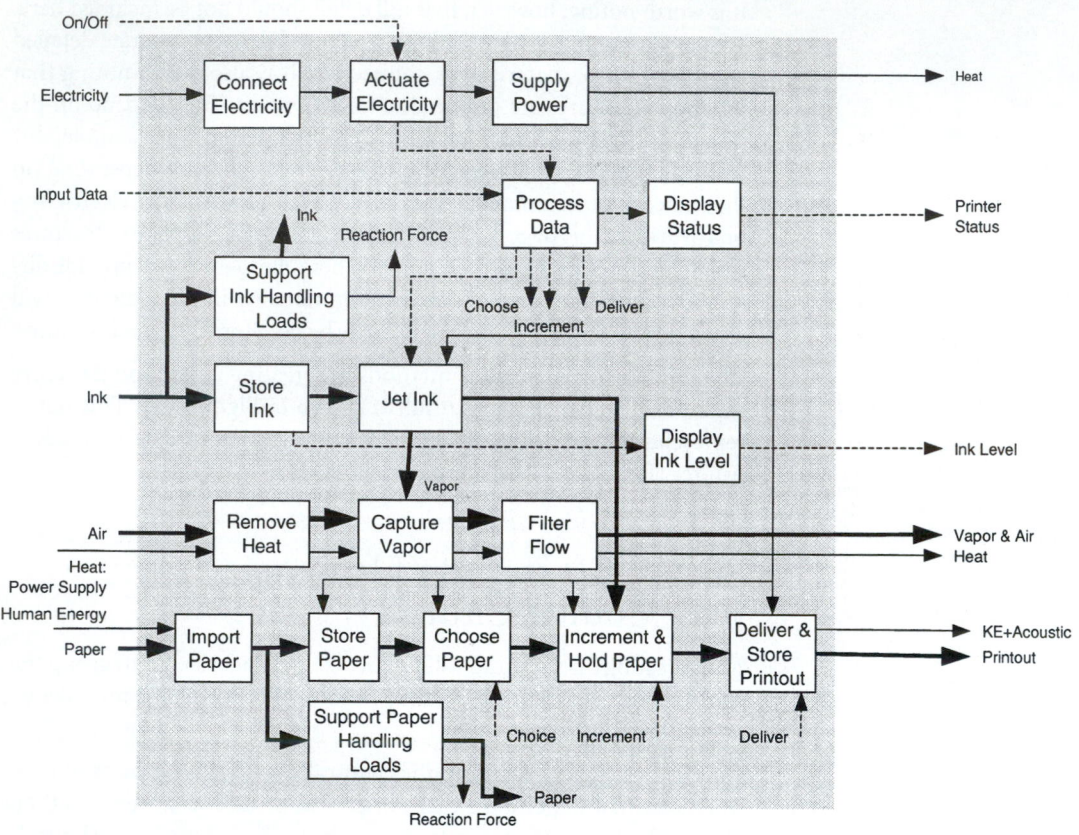

▼ *Figure 9.11.*
Function structure of an HP1200C desktop inkjet printer.

Extending the example shown in Figure 9.12, a number of general guidelines may be recommended for clustering functions into module chunks. These guidelines are neither mutually exclusive nor flawless, but they do address a number of common situations encountered in modular design. These guidelines may be applied, iteratively, throughout the remainder of the product development process:

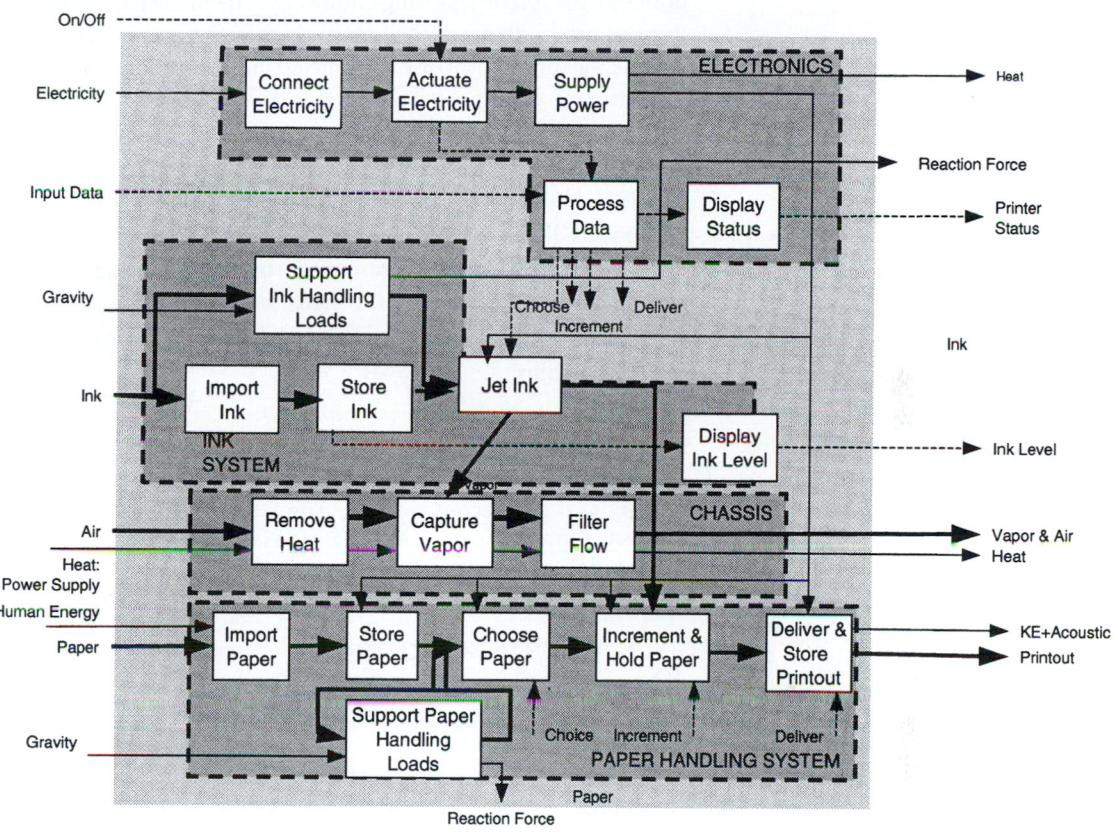

▼ Figure 9.12.
Clustered function structure of a desktop inkjet printer.

> ◗ Combine as many subfunctions into a module as possible, provided that they share a common primary connection (flow of material, energy, or signal). This guideline, if adopted, will provide opportunities for components in the module to share subfunctions. It will also aid in choosing manufacturing processes and reducing costs in production and assembly.

> ◗ If interface standards exist, choose modules such that these interfaces are congruent and not split across modules.

▶ Clearly show all relevant signal flows in the function structure. In doing so, the job of choosing chunks may be made easier. Signals, in most products, carry the ergonomics to the user. They also illustrate the mechanical or electromechanical control needed. Handling these signals properly in modules will greatly increase the likelihood of a successful product.

▶ If it is expected that a product will be designed as a family architecture, chunks should be chosen based on what will be shared by the variety of products. If needed, function structures of each planned product may be developed during the early stages of modular design. Chunks may then be chosen based on the shared functions of the products.

▶ Cluster chunks to take advantage of available manufacturing and tooling capabilities.

▶ Anticipate new technologies and new markets for the product. Based on these predictions, cluster chunks according to expected changes or evolutions. Modules should be arranged, in part, according to the expected changes. These arrangements will minimize effects on the remainder of the product architecture.

Step 3: Create a Rough Geometric Layout(s)

When suitable chunks have been chosen, we transition to the first phase of form design. This stage entails the creation of a rough geometric layout(s) of the product. Typically, a rough geometric layout is known as a *block diagram* of the product modules. It is also known as *configuration design* of product assemblies.

Two substeps embody this activity of modular design: (1) create a hierarchy of the product architecture from the function structure chunks, and (2) map the hierarchy to a 2-D or 3-D sketch(es) of the product layout. In the first substep, a hierarchy is created, delineating major components that are expected within each module of the product. The top of the hierarchy is the product. One level down is a list of the identified chunks from the previous step. The next levels are names of expected components that are needed to fulfill the functions of the modules.

Figure 9.13 contains an example hierarchy for the desktop inkjet printer example. When laying out the modules of the printer, it is important to discuss the 12 major component groups that are expected.

After developing a hierarchy, sketches should be developed of alternative spatial layouts of the product modules. This substep usually entails a cooperative effort between industrial designers and engineers.

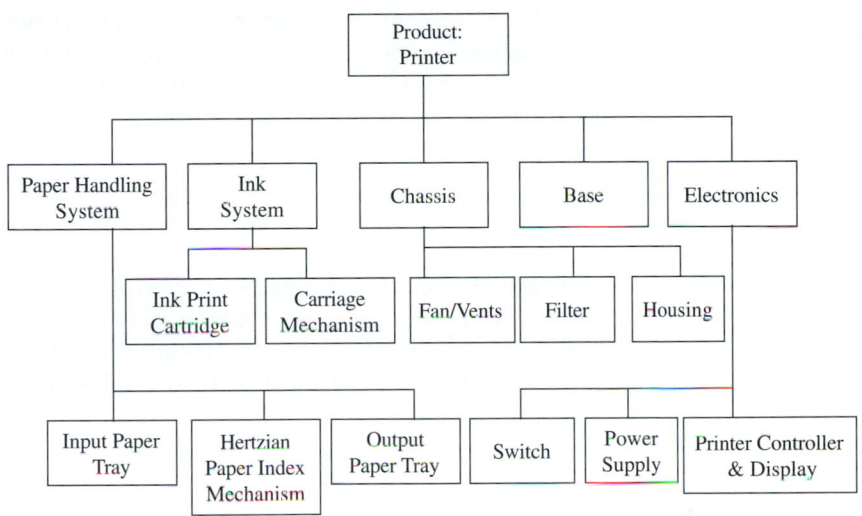

▼ Figure 9.13.
Component hierarchy for a desktop inkjet printer modular design (Cutherell, 1996).

Aesthetics, ergonomics, and spatial configuration decisions are all goals of the layout process. However, it is important to keep the layouts as simple as possible. Quite often a block diagram with prismatic chunks is sufficient to define initial layouts. Figure 9.14 illustrates an example

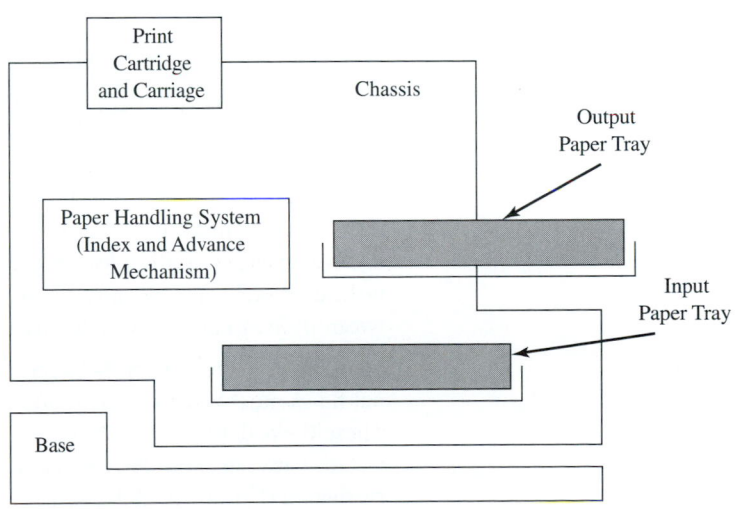

▼ Figure 9.14.
Rough geometric layout of a desktop inkjet printer.

layout for the desktop inkjet printer example. This layout was shown after many configurations were generated (Bockman, 1994). All major modules are shown in the figure. In addition, only a 2-D layout is shown. The depth dimension into the page is not used to distribute modules in this case.

Extending the example presented in Figures 9.13 and 9.14, a number of guidelines are important when creating a rough geometric layout. Notably:

▶ Iteration is important during the creation of rough geometric layouts. The previous step of clustering subfunctions should be revisited as layouts are developed. Boundaries between functional chunks may need to be modified as more is learned about the interactions between modules.

▶ Multiple configurations should be developed. Alternative layouts will provide insights into the geometric form of the product, without the clutter of extraneous details.

▶ Layouts should be refined to include ergonomic and aesthetic information, depending on the customer needs. Applications of industrial design should drive the layout sketches, linking the "feel of the product" to its engineering functions.

▶ Foam mockups, preliminary analysis, and other forms of prototypes (such as rapid prototyping) are applicable at this stage (Chapter 17). Implementing the form ideas into relatively inexpensive physical models will aid in the discussions of module interactions, size, and so forth.

Step 4: Define Interactions and Detail Performance Characteristics

After creating rough geometric layouts of a modular product, interactions between modules must be defined. Two goals drive this step: (1) to develop a specification for generating, embodying, prototyping, and testing of each module, and (2) to enhance the communication between individuals or design teams assigned to each module.

To define the interactions between modules, material, energy, and signal flows must be investigated and refined at each module boundary. These flows define the interactions and the boundary defines the interface between modules. Typically, four types of interactions are investigated (Cutherell, 1996):

1. Material interactions: solid, liquids, or gases that flow from one module to the next,

2. Energy interactions: energies that must be transmitted or shielded between modules,

3. Information interactions: signals (tactile, acoustic, electrical, visual, etc.) that must be processed from one module to the next, and

4. Spatial interactions: geometrical dimensions, degrees-of-freedom, tolerances, and constraints that must be maintained between modules.

After defining these interactions, a specification for each interaction is developed for each module. This specification is used to generate component concepts for the module, carrying these concepts to the next stages of product development.

Summary: Basic Method for Modular Design

Figure 9.15 shows the final product developed for the desktop inkjet printer. As the product transitioned from conceptual design to embodiment and production, changes in the external geometry occurred. However, the primary modules remained intact, with relative spatial position being maintained. Overall, four steps characterize the establishment of this example modular architecture:

1. Step 1: Create a function structure of the product.

2. Step 2: Cluster the subfunctions into module chunks.

3. Step 3: Create a rough geometric layout(s).

4. Step 4: Define interactions and detail performance characteristics.

▼ *Figure 9.15.*
Actual inkjet product.

V. MODULAR DESIGN: AN ADVANCED FUNCTIONAL METHOD

The previous method for establishing a modular architecture uses a function structure to cluster subfunctions into modules. It relies heavily, however, on an undirected search for modules in the structure. Very little direction is provided for systematically assigning modules, especially with respect to simple interactions and product families.

The following section presents a method for addressing this limitation. It arranges a function structure into a proper form for identification of all possible modules. It then provides heuristics for assigning modules directly. This technique relies on the concept of *function dependencies,* or the manner in which one set of subfunctions depends on the results of other subfunctions.

Function Dependencies

In functional models (Chapter 5), a *flow* refers to the energy, material, or signal that travels through the subfunctions of a device. Decomposition techniques trace flows through product functions without regard for the dependence of functions on a specific order. Ulrich and Eppinger (1995), though, note that task dependencies for product development processes are either parallel, sequential, or coupled. Here we extend the concept of parallel and sequential dependencies to product functions and flows of a function structure. In each case, the dependencies are defined with respect to a given flow.

The benefit of classifying and ultimately arranging a function structure based on the dependencies of its flows and functions will become evident in the module identification process, as presented in the next section. For now, suffice it to say that for any design problem, regardless of the concern for modules, any additional information that can be contained in the functional model will expedite decisions in the conceptual phase.

Sequential function chains: In these, the subfunctions must be performed in a specific order to generate the desired result. A flow common to all these functions is termed a *sequential flow.* This concept is directly analogous to a series circuit, which is illustrated in Figure 9.18. This illustration provides a *clarification* of how we define sequential and parallel function chains. It does not illustrate real product functions per se.

In Figure 9.16(a), the resistors R_1 and R_2 and capacitor C_1 are analogous to subfunctions of a device. The current, i, is the flow common to all subfunctions. Because of the physical layout of the circuit, the flow (current

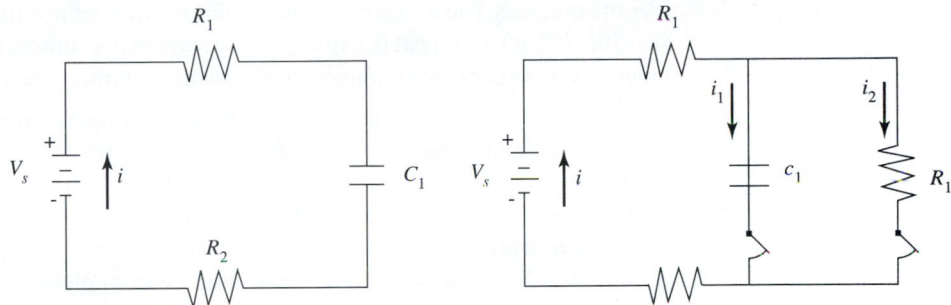

▼ **Figure 9.16.**
Electrical circuit analogy with sequential and parallel subfunction chains. (a) Series
circuit. (b) Parallel circuit.

i) must conceptually "travel" through each subfunction in a specific order
(first R_1, next C_1, and finally R_2), thus the idea of sequential function chains.

Parallel function chains consist of sets of sequential function chains sharing one or more common flows. Graphically, they are represented by
branching flows in a function structure. The chains are called *parallel* because they all depend on a common subfunction and flow but are independent of each other. Independence means that any one of the chains of
the parallel function chain set does not require input from any other chain
within the set. Physically, the parallel function chains represent different
components of a device that may operate all at once or individually. The
circuit in Figure 9.16(b) is analogous to the parallel function chains introduced here. As before, the resistors and capacitor act as subfunctions,
but the voltage is the "flow." Subfunctions C_1 and R_2 are in parallel and thus
experience the same "flow" of voltage potential. Both subfunctions C_1 and
R_2 depend on the flow across R_1, but do not depend on each other. Switches
are included in the sequential legs of the parallel chain to indicate that either one or both of the sequential legs may operate at the same time. This
analogy is exactly the situation described by the parallel function chains.

Overall, parallel and sequential function chains must be identified accurately in a product's function structure. By identifying parallel versus sequential chains, the process of clustering elements into modules
is greatly facilitated.

Module Heuristics

Using the concept of function dependencies, we now move on to the
method by which modules can be identified at a functional level. The
method of module heuristics consists of three separate strategies to

identify modules. The necessary starting point is a well-refined function structure (Chapter 5). That is, a function structure that is sufficiently detailed that it can be readily transformed into the common basis.

Assuming the existence of a refined function structure, we introduce three heuristic methods below as developed from these definitions. These methods offer surprisingly simple clusters of modules from transformed function structures. However, each of the methods may identify overlapping modules or modules that are subsets or supersets of other modules. The choice of which module to implement, in either case, is not always clear. The rule suggested here is to implement the module with the smaller number of subfunctions. This idea keeps with the philosophy that modules should be easily identifiable with a particular function. Ultimately, though, which module to implement requires some engineering judgment.

As the three heuristics are introduced, a function structure of a SKIL Twist power screwdriver is used as a physical example. The power screwdriver is chosen for the example series because it deals with several material flows. Some of the modules identified by the heuristics appear as modules in the current product and others offer areas for future modularity. We will show the identified modules in the power screwdriver for each heuristic, discuss which ones exist and which ones present opportunities for a more modular design.

Dominant Flow

The *dominant flow* heuristic examines each flow of a function structure and groups the subfunctions that the flow travels through until it exits the system or is transformed into another flow. The identified set of subfunctions defines a module that deals with the flow traced through the system. The identified subfunctions form the boundary, or *interface*, of the module. Any other flows, in addition to the traced flow, that cross the boundary are *interactions* between the module and the remaining product. A dominant flow module is shown schematically in Figure 9.17. To implement the module, conduits must be specified to carry the interactions across the interface.

Stated succinctly, the dominant flow proposition is:

Proposition 1: A module defined by a dominant flow is the set of subfunctions, which a flow passes through, from entry or initiation of the flow in the system to exit from the system or conversion of the flow within the system.

Screwdriver: Identified modules: In the SKIL Twist power screwdriver, the highest ranked (from customer needs), nonbranching flow is

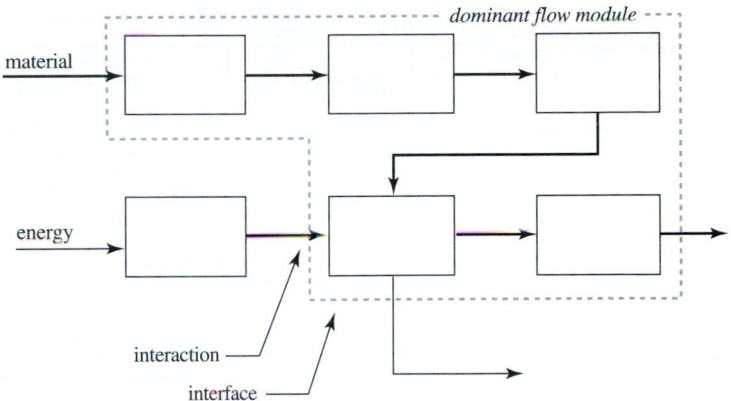

▼ Figure 9.17.
Dominant flow heuristic applied to a generic function structure.

that of electricity. Electricity passes through four subfunctions as outlined by a grey rectangle and shown in Figure 9.18. *Store electricity, supply electricity, actuate electricity,* and *regulate electricity* all have the flow electricity. The dominant flow heuristic identifies that these subfunctions could be combined as one module; here we'll name it the *electrical supply module.*

The next flow to trace is the bit (assuming that it is not designed as part of the screwdriver system). It first enters the system and traces through the two subfunctions *couple solid* and secure solid, with a "separate solid" function to permit decoupling. This set of subfunctions forms a module that we'll call the *coupling module.* The flow torque emerges from two sources, the motor and the human. It travels through the subfunctions *change rotation, change torque,* and *transmit torque.* Here the heuristic identifies five sub-functions that form the *torque transmission module.*

Lastly, the human force flow identifies a *positioning module,* composed of the *regulate orientation* and *regulate translation* functions. In sum, the dominant flow heuristic identifies four modules, one for each purely sequential function chain. Not all function structures have a module associated with every non-branching flow. Flows may enter only one subfunction and then exit the system or be transformed, thus never forming a sequential function chain. Next, we discuss how the identified modules compare with existing modules.

Screwdriver: Actual modules: The identified modules can be compared to the modularity of the actual product. Actual modules will be referenced with respect to the exploded view shown in Figure 9.19. The module associated with the flow of electricity is actually found as two modules in the SKIL Twist power screwdriver. The *store electricity* and

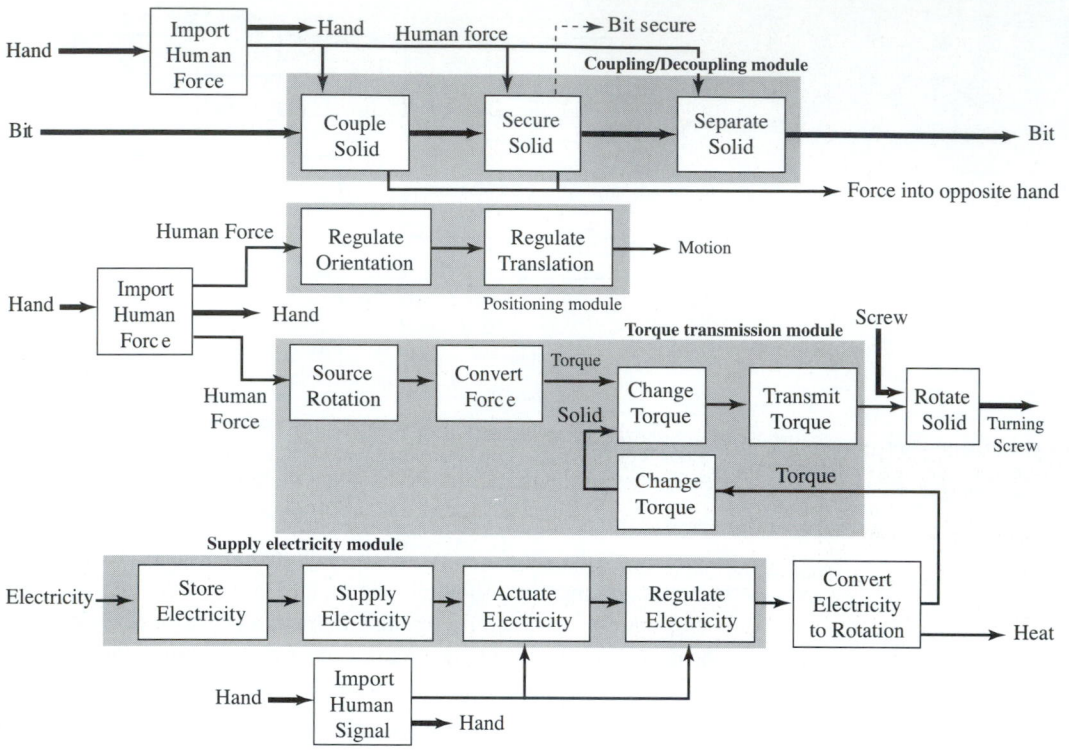

▼ *Figure 9.18.*
Modules identified by the dominant flow heuristic from a function structure of the SKIL Twist power screwdriver.

supply electricity subfunctions are embodied by a rechargeable battery, while the *actuate electricity* and *regulate electricity* functions form a switch module (to turn on as well as change the direction of the screwdriver). These two submodules are shown in Figure 9.19(c) and (f). In this sense, the heuristic method correctly identifies subfunctions that come together as a module. A possible product change is shown in Figure 9.20, which integrates the four subfunctions into a single component that stores, supplies, and actuates electricity as well as interfaces with other drive units besides screwdrivers. The rechargeable battery

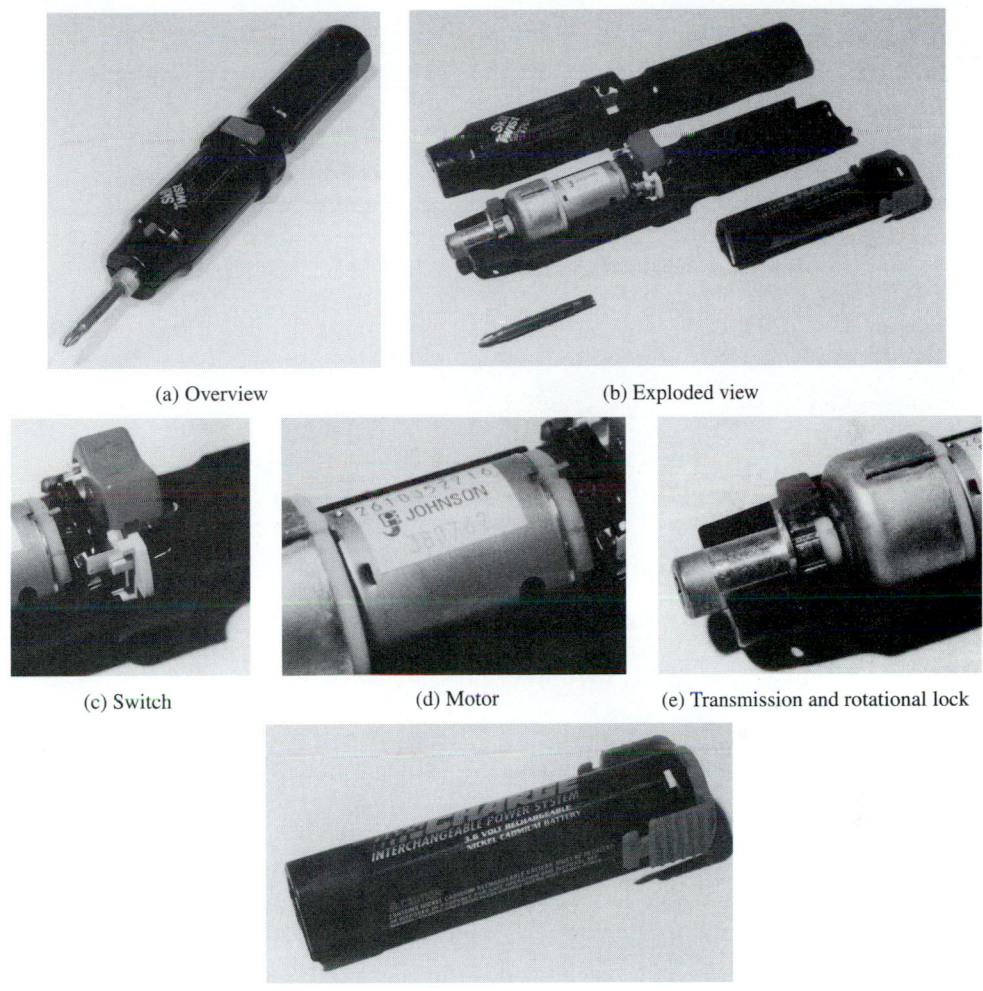

(a) Overview

(b) Exploded view

(c) Switch

(d) Motor

(e) Transmission and rotational lock

(f) Rechargeable battery

▼ **Figure 9.19.**
Views of the SKIL Twist power screwdriver. (a) Overview. (b) Exploded view. (c) Switch. (d) Motor. (e) Transmission and rotational lock. (f) Rechargeable battery.

solves the *store electricity* subfunction. The switch handles the subfunctions *actuate electricity* and *regulate electricity*. The contacts *supply electricity* for associated drive units. With this concept, the *supply electricity module* would interface with different drive modules to create a mix modularity tool. For example, it could attach to a screwdriver, drill, detail sander, or flashlight drive unit. This approach advances the modular battery-powered hand tools available a step further.

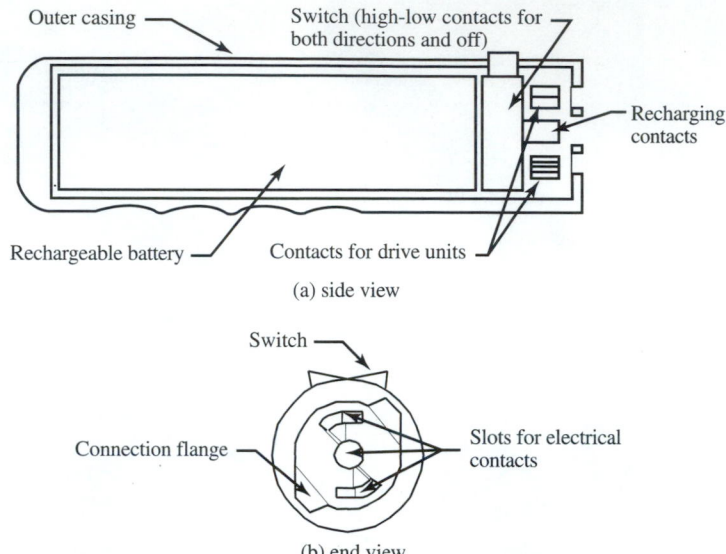

Outer casing

Switch (high-low contacts for both directions and off)

Recharging contacts

Rechargeable battery

Contacts for drive units

(a) side view

Switch

Connection flange

Slots for electrical contacts

(b) end view

▼ *Figure 9.20.*
Schematic of a possible *supply electricity module* for the SKIL Twist power screwdriver. (a) Side view. (b) End view.

Predicted modules associated with the flows "bit" and "torque" exist in the screwdriver as well. The *coupling module* and a subset of the *torque transmission module* (without the *change torque* subfunction) are embodied by the one physical component (Figure 9.19(e)).

Thus, the dominant flow heuristic method predicts modules that exist in the screwdriver. In addition, it provides an innovative way of combining functions into one component that could be shared among products. Another point to note about the dominant flow heuristic is that it defines functions that can be combined into assembly modules, that is, parts that are best connected together before assembling the entire product.

Branching Flows

The second heuristic is referred to as *branching flows*. The method of the branching flow heuristic first requires identification of flows that branch into or out of parallel function chains. Once identified, we will examine them in descending rank order. Each branch of the flow defines a potential module as shown schematically in Figure 9.21. The module is formed by the subfunctions that make up the branch. (Each branch consists of a sequential function chain.) All modules (one per branch) must interface with the product at the last subfunction before the flow

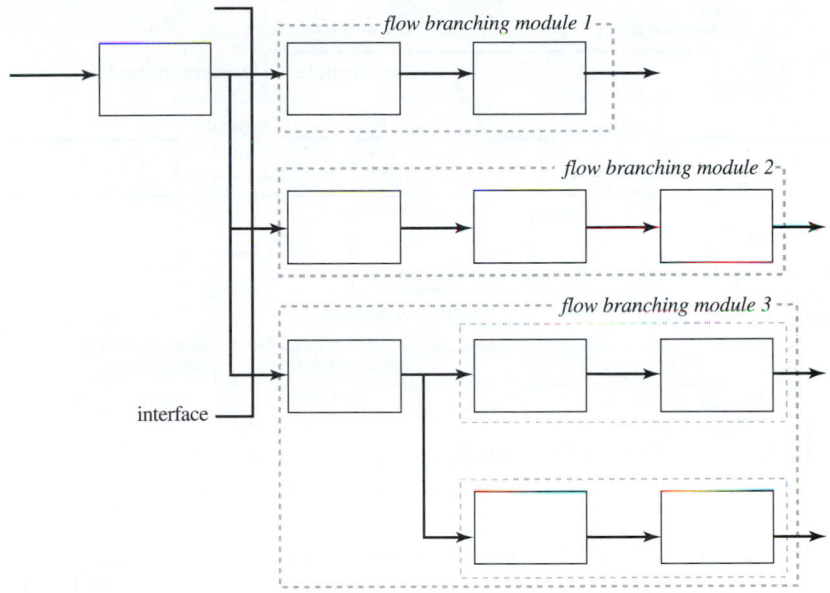

▼ Figure 9.21.
Flow branching heuristic applied to a generic function structure.

branches. All flows that cross this interface are the interactions between the remaining product and the module.

Note that branching flows will identify products capable of slot or bus modularity. The interface boundaries defined are physical connections between module and product. In some cases, they will be well-defined geometric connections, like various end mill attachments on a milling machine. Other times, the interface may be more fuzzy, like the differing interactions between the hand and the SKIL Twist power screwdriver.

The branching flow heuristic is stated formally as:

Proposition 2: Parallel function chains associated with a flow that branches constitute modules. Each of the modules interfaces with the remainder of the product through the flow at the branch location. Similarly, two identical flows that combine define modules back to the points of initiation of the combining flows.

Screwdriver, Identified modules: The power screwdriver has two flows that branch, *torque* and *human force*. Of the two flows, *human force* has the higher rank according to customer needs and will be examined first. Following the subfunction *import human force*, the flow branches into three limbs. Each branch represents a module as shown

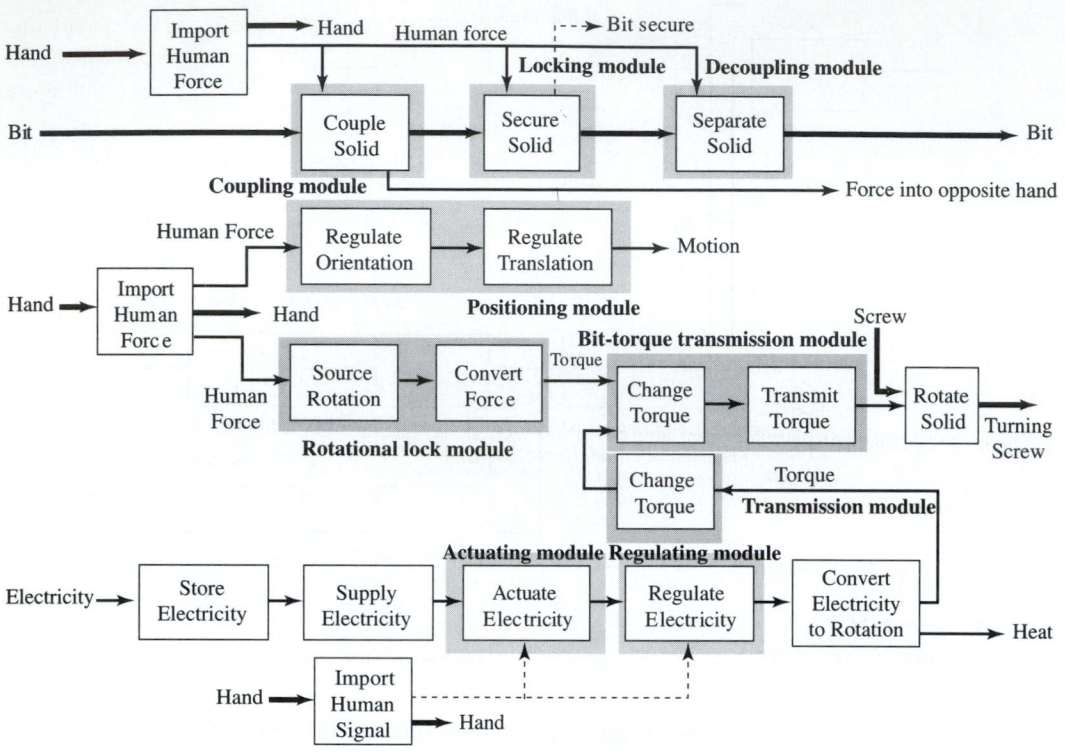

▼ *Figure 9.22.*
Modules identified by the flow branching heuristic from a function structure of the
SKIL Twist power screwdriver.

in Figure 9.22. The three identified modules are the *coupling, locking,*
and *decoupling* modules. Note that subsets of the *coupling/decoupling*
module identified by the dominant flow heuristic are identified here.
The second branching flow of the *hand* splits into two limbs of signal
flows, one for actuating the device, and one for regulating the speed.

The branching flow of *torque* identifies three possible modules: *rotational lock, transmission* and the *bit-transmission* module. This
branching flow heuristic application basically split up the *torque
transmission* module from the dominant flow heuristic into three
sensible, smaller modules.

Screwdriver, Actual modules: The branching flow heuristic identifies several modules that relate hierarchically to the modules from the dominant flow heuristic. The *coupling/decoupling* module is composed of the *coupling, locking,* and *decoupling* modules. The supply electricity module is composed of the actuate and regulate modules, as well as additional functions. This relationship between the dominant flow and branching flow heuristics is natural, as the dominant flow tends to find sets that can be maximally unioned, whereas the branching flow considers a more refined level.

A few remarks about the heuristics usage thus far are warranted. It is evident that they identify overlapping modules in some cases. At this point, we wonder about which module to implement in such a case. For now, we make the observation that the more ways a module is identified (in terms of heuristics and flows), the more important it is to implement (since it must be associated with more customer needs).

Conversion-Transmission Modules

The third heuristic method deals with conversion subfunctions and conversion to transmission chains. Conversion subfunctions accept a flow of material or energy and convert the flow to another form of material or energy. In many cases, these conversion subfunctions are already components or modules themselves. For instance, electrical motors, hydraulic cylinders, and electrical heaters can all be represented by a single conversion subfunction and exist physically as a single component. If, additionally, a conversion subfunction exists in a chain with a transmit subfunction (or transport subfunction for material flow), then the chain presents an opportunity to form a module. This *conversion-transmission module* converts an energy or material to another form and then implements (transmits or transports) that new form of energy or material.

The method of the conversion-transmission heuristic, shown schematically in Figure 9.23, is simple. The essential actions are as follows: identify conversion subfunctions and check for transmit or transport subfunctions downstream of the converted flow. If none exist, then the conversion subfunction is a module by itself. If transmit or transport subfunctions exist without any other subfunctions between them, then the convert-transmit (transport) pair represents a module. If other subfunctions exist between the convert and transmit (transport) subfunctions and those intermediate subfunctions only operate on the

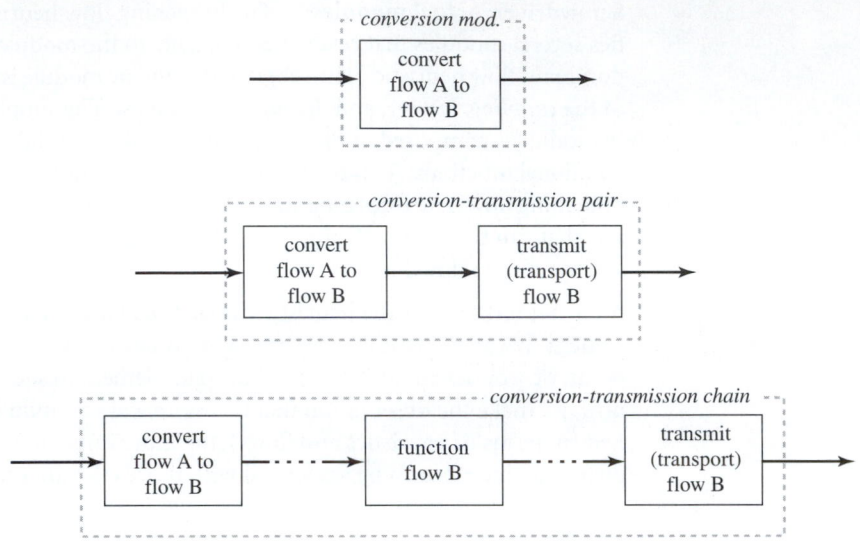

▼ **Figure 9.23.**
Conversion-transmission applied to a generic set of subfunctions.

converted flow (i.e., the object in the subfunction verb-object pair is the converted flow), then the conversion-transmission (transportation) chain represents a module.

Interfaces of a conversion-transmission module are defined in a similar manner as those for a dominant flow module. Two necessary interactions across the interface are the flow to be converted and then the exiting converted flow. Additional flows may also cross the interface.

The conversion-transmission proposition, stated succinctly, is:

Proposition 3: A conversion subfunction or a conversion-transmission pair or proper chain of subfunctions constitutes a module.

Screwdriver: Identified modules: Considering the power screwdriver again, the conversion-transmission heuristic identifies two modules as shown in Figure 9.24. The *electrical to torque* module consists of four subfunctions between the bounding *convert electricity to torque* and *transmit torque*. The *hand to torque* modules consists of four subfunctions between the radially amplifying *convert force* and *transmit torque* modules.

Screwdriver: Actual modules: In the actual product, neither of these modules are implemented as shown. The *convert electricity to torque* subfunction is a distinct component, shown in Figure 9.19(d).

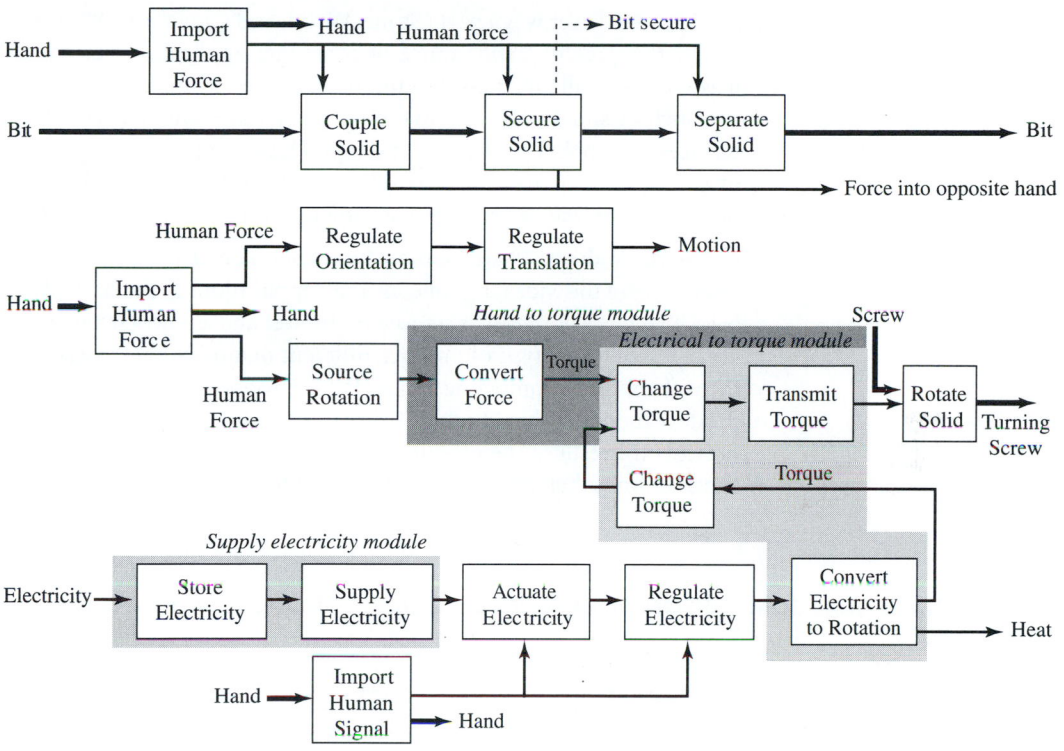

▼ **Figure 9.24.**
Module identified by the conversion-transmission heuristic from a function structure of the SKIL Twist power screwdriver.

The *change torque, resist torque* and *transmit torque* subfunctions, though, are part of the *torque transmission module* in Figure 9.19(e). Similarly, the *convert force* subfunction is a stand-alone, implemented as the radius of the scredriver. In this case, the method provides an innovative approach to the design of a module that incorporates a motor, transmission, and drivetrain. Recall the innovative *supply electricity module* of the screwdriver in Figure 9.20. In concert with the *supply electricity module*, the *electricity to torque module* becomes the drive units in a mix-and-match set of power packs with switches and motor drive units.

Verification of Heuristic Methods

The three heuristic methods just introduced are presented with the SKIL Twist power screwdriver example. The heuristics were developed and verified more rigorously using a database of 70 products, of which the power screwdriver was a part (Stone, Wood, and Crawford, 1998). The products represent a wide range of consumer applications, customer needs, and overall functions. The mechanical or electromechanical products include small construction tools, small kitchen and household appliances, automobile accessories, and toys. This set of products represents over 100 person-years of work in reverse engineering and redesign (Little, Wood, and McAdams, 1997; McAdams, Stone, and Wood, 1999).

Of the 70 products, 18 are examined in detail. The 18 products are representative of the wide range of consumer applications in the entire database. Two types of products, an iced tea brewer and a power screwdriver, are repeated (same type of product, different manufacturer) to examine the differences in competing products. All product function structures produce modules when the three heuristic methods are applied. Table 9.3 presents the results. The products are listed in column one followed by three columns that indicate the number of modules identified by each of

TABLE 9.3. STATISTICAL COMPARISON OF IDENTIFIED MODULES AND ACTUAL MODULES IN 18 PRODUCTS

Product	Identified Modules			Actual Modules			Unique	Actual
	Dom. flow	Branch. flow	Conv.-Trans.	Dom. flow	Branch. flow	Conv.-Trans.	Module poss.	Module total
Mr. Coffee iced tea/coffee brewer	6	4	1	5	3	1	7	6
West Bend iced tea/coffee brewer	6	4	1	4	3	1	7	5
Mr. Coffee coffee maker	5	3	1	3	1	1	6	3
B&D screwdriver	3	6	1	2	3	0	6	4
SKIL screwdriver	3	6	1	3	4	0	6	5
DeWalt sander	4	4	3	4	3	3	9	7
Bissell hand vacuum	5	3	2	4	2	2	6	6
Pencil sharpener	3	2	1	3	2	1	4	4
B&D electric knife	3	4	2	2	2	2	6	4
Presto air popcorn popper	3	2	3	3	2	3	8	6
Krups cafe trio	3	2	1	3	1	1	6	5
B&D sander	5	2	3	3	2	3	7	6
Dazey fruit/veggie peeler	5	3	2	3	2	2	7	4
Dremel engraver	3	4	1	2	2	1	5	3
1974 Chevy tailgate	3	6	1	2	3	1	5	3
B&D VersaPak trimmer	4	2	1	4	2	1	5	5
Cadillac visor	2	0	2	2	0	2	4	2
Super Maxx ball shooter	3	2	2	3	0	2	6	3
Average	4	3	2	3	2	2	6	4

the three heuristic methods: dominant flow, branching flow, and conversion-transmission. The next three columns indicate the actual modules found in the products, associated with the three heuristic methods. Since it is possible for the different heuristic methods to identify overlapping modules, the final two columns give the number of unique modules possible and then the actual number of modules found in the product.

It is important to note from the final two columns of Table 9.3 that the number of unique modules possible is always greater than (or equal to for three of the products) the actual number of modules implemented in current products. This result strongly supports the heuristic methods' validity as a tool to identify possible modules in a product. For this database of consumer, largely hand-held products, four to nine unique modules are possible. The actual products, though, incorporate between two to seven modules, with some exhibiting an impressive degree of modularity. As a comparison, a large heavy-duty maintenance product—a lignite removal system for the power generation industry—was examined (Stone, 1997). It has a unique module possible count of 14 for its two subsystems studied, supporting the heuristics' utility for larger scale design problems. For consumer products with 15–30 subfunctions, we now know that, on average, six unique modules are possible.

Process: Application of the Module Heuristics

Step 1: Develop a Function Structure

The first step in applying the module heuristics is to create a function structure for the product. Chapter 5 and the previous method of modular design discuss approaches for creating a structure. The key issues are to relate the flows of materials, energy, and signals to customer needs and to carefully distinguish between parallel and sequential function chains as the structure is developed. Transforming to a common function basis (Chapter 5) will aid in the verification and completeness of the function structure.

Step 2: Incrementally Apply the Module Heuristics to the Function Structure

With the function structure transformed, we apply the module heuristics. The detailed tasks are listed in Table 9.4. The first three tasks consist of applying the module heuristics one after the other. The precedence of the tasks is important. The dominant flow heuristic is applied first, as it may identify sequential subfunction chains prior to a flow branching. This step can be thought of as identifying the base

TABLE 9.4. STEP 2: APPLY THE MODULE HEURISTICS

Task	Input	Output	Use/Impact
1. Apply dominant flow heuristic	Ranked flows Function structure	Groups of subfunctions that are possible modules Interactions of the module	• Identification of a module based on flows • Reduction of incidental flow interactions between modules
2. Apply branching flow heuristic	Ranked flows Function structure	Groups of subfunctions that are possible modules Interactions of the module	
3. Apply conversion-transmission heuristic	Function structure	Groups of subfunctions that are possible modules Interactions of the module	• Modules that are based on the conversion of material or energy from one form to another and its subsequent transmission
4. Reconcile modules and interactions	Identified modules from the three heuristics	A set of unique modules Module-interaction list	• Removes repeated modules and identifies modules that are subsets of others • Sets the stage for concept variants
5. Reflect	Identified modules	Refined set of modules with interactions	• Checks for thoroughness of interaction identification • Eliminates any clearly infeasible modules

module on which the branching flow modules will connect. Therefore, the branching flow heuristic is applied second. The conversion-transmission heuristic is applied last, as it identifies modules that convert energy or material to another form and typically connect the above modules.

In Task 4, we reconcile the overlapping modules that can occur after application of all the modules. In the case of two or more flows or heuristics identifying the same module, we simply keep one of the modules in our set of possible modules to implement. The other case is where a module is a subset of another module. For this situation, the recommendation is to keep both modules through the concept generation step and use a concept selection process to determine which to implement. For example, consider a hot air popcorn popper, pictured at the beginning of Chapter 6. The dominant flow heuristic identifies a *kernel handling module* and the branching flow heuristic identifies a *kernel heating module*. The latter is a subset of the former. Generating concepts for both increases the variety of design variants and ultimately produces a more creative design.

Also in Task 4, we generate a module-interaction list. This is a list of interactions, classification of the interactions, spatial requirements, and

TABLE 9.5.   A SEGMENT OF THE MODULE-INTERACTION LIST FOR A HOT AIR POPCORN POPPER (COVER PHOTO, CHAPTER 6)

Module	Interactions	M	E	I	Sp	Interfaces In: Modules or (External Systems)	Sp	Interfaces Out: Modules or (External System)
Kernel handling	Kernels	•				(User)		
	Air	•			•	Air handling	•	Butter handling (environment)
	Pneumatic energy		•		•	Air handling	•	Butter handling (environment)
	Thermal energy Popcorn		•		•	Air handling	•	Butter handling (environment)
	Popcorn	•						

Legend: M: Materials
E: Energy
I: Information
Sp: Spatial

interfaces for each module. In addition to identifying interactions as energy, material, or information, we further specify the input and output spatial requirements of the interaction with other modules or parts of the device. The spatial type is checked for the interaction if the module must be placed in close proximity to another module in order to prevent a flow from degrading. The detailed interaction information in the module-interaction list sets the stage for the concept generation step. As an example, a segment of the module-interaction list for a hot air popcorn popper is shown in Table 9.5. Note that the input and output interactions of air, pneumatic energy, and thermal energy require special spatial consideration in order to prevent the interaction from degrading between modules.

Finally, Task 5 reviews the modules. For each module, we recheck the list of interactions to ensure all are documented. At this point, any clearly infeasible modules are eliminated.

Step 3: Generate Concepts for the Modules

In Step 3, we generate concepts (Chapter 10) based on the modules identified from Step 2. These modules are identified based on their flow interactions, an advancement over the basic method of modular design. The tasks of concept generation are listed in Table 9.6.

The first task is to generate rough geometric layouts of the device. We've answered how the flows interact between modules, but this

TABLE 9.6. STEP 3: GENERATE CONCEPTS

Task	Input	Output	Use/Impact
1. Create rough geometric layouts Establish spatial relationships Assign rough dimensions where possible	Identified modules Interactions Ranked customer needs	Rough form layout concepts	• Produces a rough layout of the devices form • Guides the search for module solutions in terms of scale
2. Search for existing components	Vendors Published literature Patents Experts Device database (from quantitative assessment method)	Module concepts	• Incorporating an existing component is cheaper and easier than developing a new one • Allows creative focus on aspects of design that are truly unique
3. Search for creative modules New modules manuf. internally Vendor	Team Open mind	Module concepts	• Generate new forms for the identified modules • Truly novel modules could be used in other devices as well
4. Reflect	Module concepts	Refined module concepts	• Continuous evaluation of concepts

task takes into account the spatial interactions between modules. First, we lay out the modules as they will physically be connected in block form. We roughly scale the blocks to their anticipated physical size. Then, we assign overall dimensions to the layout based on information from customer needs or the environment used for the product. Several rough geometric layouts should be generated for eventual comparison. This task allows varying layouts of the same modules as well as different sets of modules (in the case of overlapping modules from Step 2).

Tasks 2 and 3 search for solutions to the modules of the rough geometric layouts from Task 1. Here, concept search focuses on external and internal components. The ultimate extension of the modular design method is that a device would be built up from a collection of existing modules, and Task 2 would be the only search required. But the reality is that the device will incorporate existing modules and creative modules that are manufactured internally to the company.

At the end of this step, we have a number of modular device concepts. We first should reflect on the concepts to check their feasibility. From the set of feasible concepts, we apply a decision-making process to select one for further development and production.

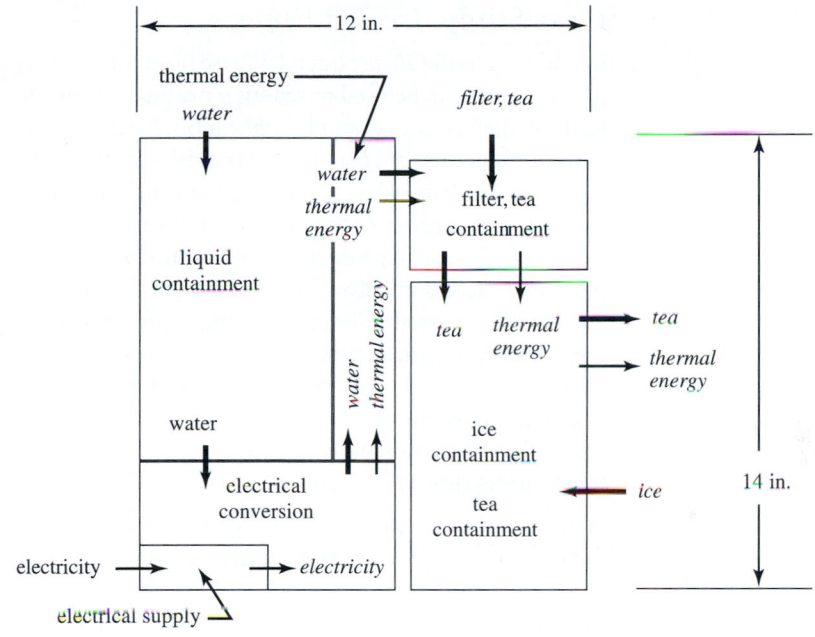

▼ **Figure 9.25.**
A rough geometric layout of an iced tea brewer.

As an example of Step 3, recall the iced tea brewer described in previous chapters. Application of the module heuristics and customer need rankings result in the following prioritized modules: *electrical conversion, filter and tea containment, ice containment, liquid containment, electrical supply,* and *thermal energy.* A rough geometric layout is shown in Figure 9.25. Note that the flows between components are identified and that modules with spatial requirements are placed adjacent to each other. If a module is not butted directly to another module, then the module is a completely separate component not intended to be permanently assembled to the rest of the device. An overall dimension of the product is given. The dimensions are based on similar devices within the iced tea brewer family. Though this specification is not set in stone, it provides an upper bound for the overall dimension of the device. These two actions guide the spatial layout and help to focus our search for module solutions to the right scale.

Case Study: Ice Tea Brewer

In addition to the 70-product database used for verification, the module heuristics can be used to redesign products. Consider a Mr. Coffee Iced Tea Brewer as an example application. Following a modular design methodology, the iced tea brewer is redesigned to meet two important customer needs: take less space on the counter and prevent leaking between the tea brewer base and pitcher (pitcher-to-brewer alignment method). The rough geometric layouts must address these needs. Two possible concepts are shown in Figure 9.26. Concept (a) is similar to a current iced tea brewer model, utilizing a side-by-side design. An alignment foot is shown, and the overall dimensions are smaller to address the customer needs. Concept (b) is a stacked design, departing from the existing iced tea brewer configuration. It has a small footprint and incorporates an alignment notch in the layout. Both layouts show the interactions between the modules.

With the modular concepts in rough layout form, we can now search for existing components and creative solutions to embody the modules. Since this is a redesign, many of the same components from the existing device may be incorporated into the redesigned device. For example, the *electrical supply* and *electrical conversion modules* are solved by a readily available component that is used in virtually all coffee makers and iced tea brewers. This component is shown in Figure 9.27. On the other hand, such modules as the *liquid containment module* are typically brand and model specific, requiring a creative solution.

Completing the redesign: Following the concept generation step, the concepts are then selected through some selection process. These selection processes are well documented and the choice is left to the designer (Chapter 11). Without belaboring the point, a concept screening method selects the stacked concept of Figure 9.26(b) for development. The redesign is completed through the generation of detailed models and prototypes and the eventual production of the next generation iced tea brewer.

The evolved iced tea brewer: After rough geometric layouts were developed for this case study, a new model of the Mr. Coffee Iced Tea Brewer was discovered on local store shelves. The new model, along with the original side-by-side version, is shown in Figure 9.27. Note the similarity to the selected stacked concept from the above steps.

While the stacked iced tea brewer model by Mr. Coffee and our modular design work are two independent events, it does offer strong support for the modular design methodology. The stacked model is a modular device, and it answers the customer needs that drive the redesign: a compact design (footprint) and a method to align the pitcher

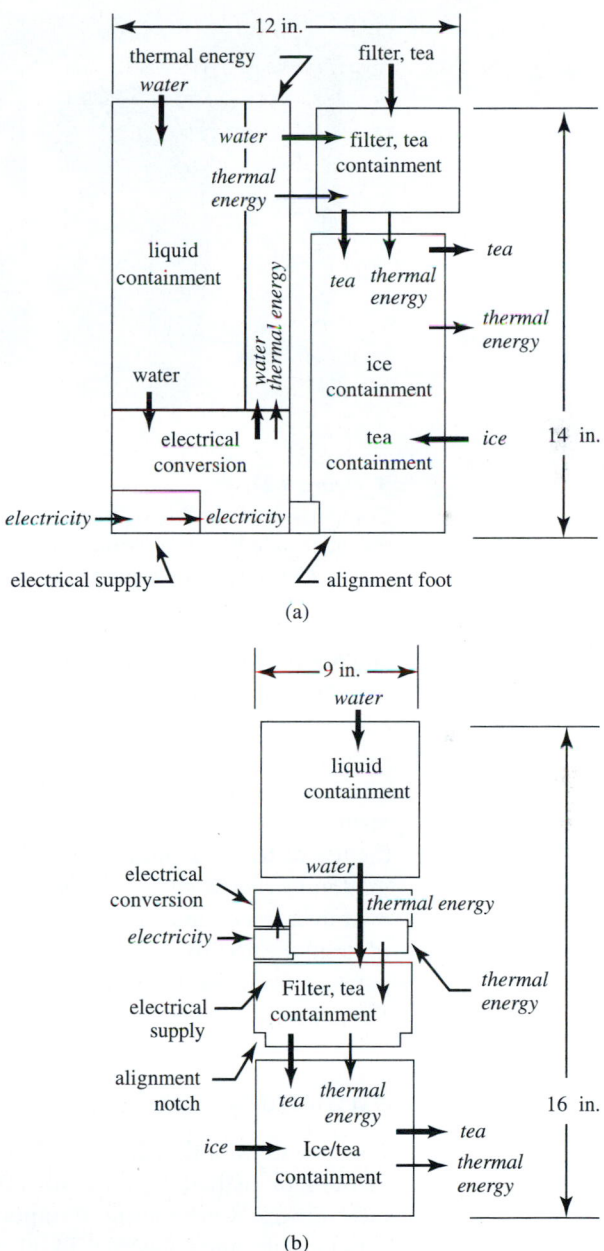

▼ **Figure 9.26.**

Rough geometric layouts for the iced tea brewer redesign. The layout in (a) is the traditional side-by-side design, while the layout of (b) features a stacked concept.

▼ *Figure 9.27.*

Comparison of two Mr. Coffee Iced Tea Brewers. The model on the left is the older, side-by-side version. The model on the right is very similar to the stacked concept presented in Figure 9.26.

and brewer. Many of its modules are contained together in the top assembly module, shown in Figure 9.28. The *liquid containment, thermal energy, electrical conversion,* and *electrical supply modules* are aggregated together, similar to the arrangement shown in Figure 9.26(b). The main difference, though, is the absence of a *filter and tea containment module.* This module's functions are solved instead by requiring the use of tea bags and merging the subfunctions of *mix liquid and solid* and *refine liquid* into the *liquid containment module.* Here, the tea mixture is heated and recirculated through the *liquid containment module* until the tea mixture reaches a set temperature. Then the tea is dispatched from the *liquid containment module* into the *ice* and *tea containment modules.*

Summary

This case study of the Mr. Coffee Iced Tea Brewer shows that the modular design methodology is easily integrated into a product development process. Customer need importance may be directly associated with modules and knowledge of other existing, related devices may be drawn upon for solutions. This case highlights another strength of the modular design method: the ability to begin the search for solutions at the higher modular level rather than the subfunction level. This ability reduces the number of solution principles that must be found but does not diminish the opportunity for creative modular concept solutions.

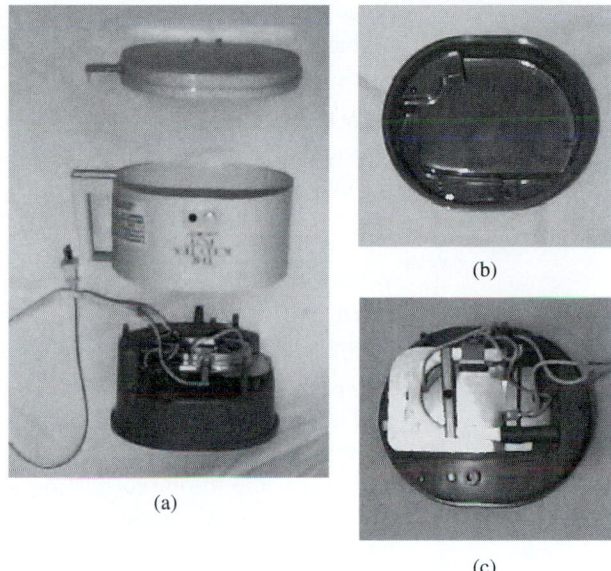

▼ *Figure 9.28.*
Modules from the stacked Mr. Coffee Iced Tea Brewer. (a) An exploded
view of the *liquid containment, thermal energy, electrical supply,* and
electrical conversion modules. (b) The top view of the *liquid containment
module.* (c) The bottom view of the *liquid containment module* and the
interfaces between it and the other modules in (a).

VI. ARCHITECTURE-BASED DEVELOPMENT TEAMS

An important consequence of establishing product architectures is the
formation of development teams based on device modules. In this section, we provide a common-sense approach for forming development
teams. Consider the following scenario to motivate this process.

A large company, the pioneer of the plain-paper copier market, had an
established product architecture and development team structure that
propelled it to the top of its market. In the 1970s, this company faced
competitors that were producing smaller and more reliable copiers
than their own. The competing products implemented very little new
technology and even used the core technologies invented by the pio-
neering company. Despite this, it took the company nearly a decade to
introduce a competitive product into the market. In that time, it lost half
of its market share. That company was Xerox. The lack of a modular

design and the unresponsiveness of the established development team to change are cited as reasons for Xerox's problems (U.S. Congress 1992; Clark 1988).

Since that time, industry has invested in concurrent engineering practices and differing team structures (Chapter 2). One popular practice is *cross-functional product development teams,* an approach that forms teams of members with various areas of expertise to develop, manufacture, market, and support a product (Clausing 1997; Prasad 1997). While there is literature supporting team interaction and duties in such an environment, there are still few formal methods for defining the subsystem design teams of a complex device.

One proposition is to use the module heuristics method to identify device modules and thereby define the responsibilities of the development team. By definition, the identified modules are physical structures that have a one-to-one correlation with functional structures. Thus it makes sense for the modules of a device to define its development teams. Their interactions are also clearly identified, which denote the necessary lines of communications between other teams.

A Method of Forming Module-Based Development Teams

The method is straightforward. In fact, much of it may already be complete after the completion of the modular design methods presented above. Table 9.7 lists the steps and required input and output of the development team method.

TABLE 9.7. STEPS OF THE DEVELOPMENT TEAM METHOD

Step	Input	Output
1. Decompose system on a functional level	Description of the system (e.g., black box)	Function structure
2. Apply module heuristics and rank order the modules according to customer needs	Function structure Device database	Identified modules Module-interaction list Ungrouped subfunctions
3. Associate ungrouped subfunctions with existing modules or each other Use customer weighted subfunction combinations Consider connecting flows	Function structure Identified modules Interactions	Regrouped modules Identified areas which require communication between teams
4. Assign teams to develop modules; maintain a mix of designers and manufacturing personnel on teams	Modules for teaming Identified areas that require communication between teams	Development teams Concurrent design process for remainder of project

Step 1 is the now familiar decomposition process. A functional model, developed in Step 1 of the modular design method, is the result of this step.

In Step 2, we apply the module heuristics to the functional model to identify modules. As an additional part of this step, a customer need ranking to the identified modules is carried out.

Step 3 associates any ungrouped subfunctions from Step 2 with related modules or each other. Following the application of the heuristics, there may be ungrouped subfunctions. At this point we use two pieces of information to group these components: customer-weighted subfunction combinations (based on customer needs: Chapter 6) and the flows connecting these subfunctions. Lone subfunctions that have a high customer weight with another subfunction (both related to highly ranked customer needs) are associated with that subfunction. That may mean adding a subfunction to an existing module or grouping a set of lone subfunctions into a pseudomodule. The alternative is to group lone subfunctions to modules if they have connecting flows (other than the flow that defined the module).

In Step 4, we use the results of Step 3—the list of modules with known interactions—and assign teams to develop the modules. The interactions define the lines of communications between groups that are necessary. Also, overlapping modules indicate that two teams must closely collaborate on the aspect of the design that solves the subfunctions they share.

Application of Module-Based Development Teams

One of the motivations for the theory of modular design is the ability to define development teams based on module groupings. Now, in this vein, consider a climate control system for an automobile (Pimmler and Eppinger, 1994).

Using the development team method, modules of the climate control system are identified. The climate control system is shown schematically in Figure 9.29. This figure shows the components of the system at a form level, that is, the actual forms of the solutions that solve the overall function of *control climate*. The climate control system has two main functions: heat and cool the air in the passenger compartment.

Heating is provided by the circulation of hot engine coolant through heater hoses through the heater core (a heat exchanger). Since the heater core is a secondary system for dissipating thermal energy from the engine, hot engine coolant also circulates through the radiator (another heat exchanger) to dissipate thermal energy to the outside air.

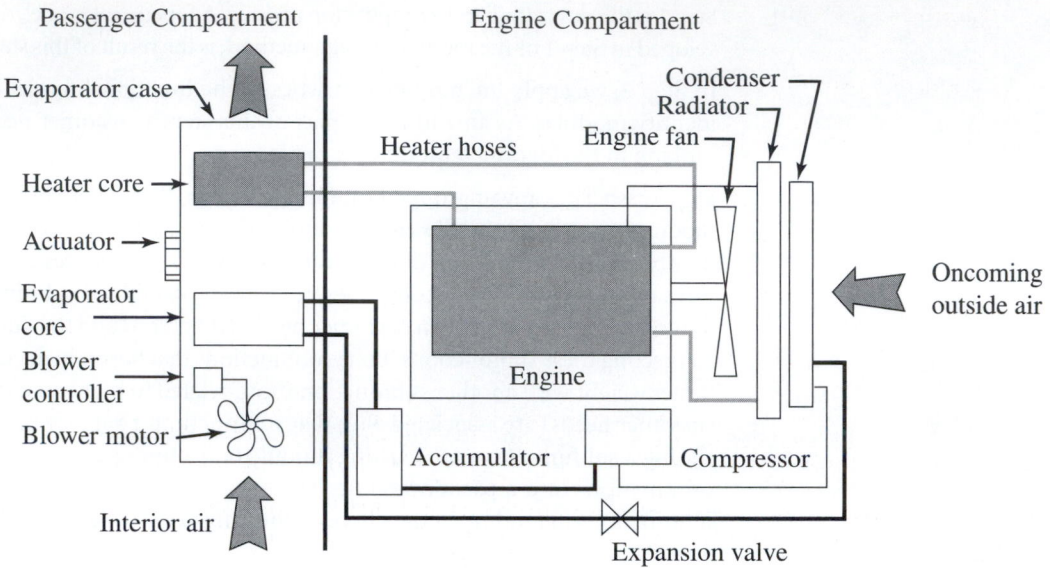

▼ Figure 9.29.
Schematic of an automotive climate control system (adapted from Pimmler and Eppinger, 1994).

Cooling of the passenger compartment air is achieved through the use of a refrigeration loop with five main components: compressor, condenser, evaporator, expansion valve, and accumulator. The compressor accepts low-pressure refrigerant gas and provides high-pressure gas to the condenser. The condenser (a heat exchanger) condenses the gas, dissipating the thermal energy to the outside air. The high-pressure liquid refrigerant then flows through an expansion valve (which maintains the pressure difference in the loop) to the evaporator. The refrigerant expands in the evaporator as the thermal energy of the passenger compartment air is absorbed (the evaporator is another heat exchanger). As the refrigerant exits the evaporator, it is a mixture of gas and liquid. The accumulator stores the liquid, allowing only the low-pressure gas to pass to the compressor, which completes the cycle.

Step 1: Decompose System on a Functional Level

A function structure of the climate control system is given in Figure 9.30. Note that the components are now represented as a subfunction or set of subfunctions that the component solves. If the team formation follows the modular design methodology, then this step is already complete.

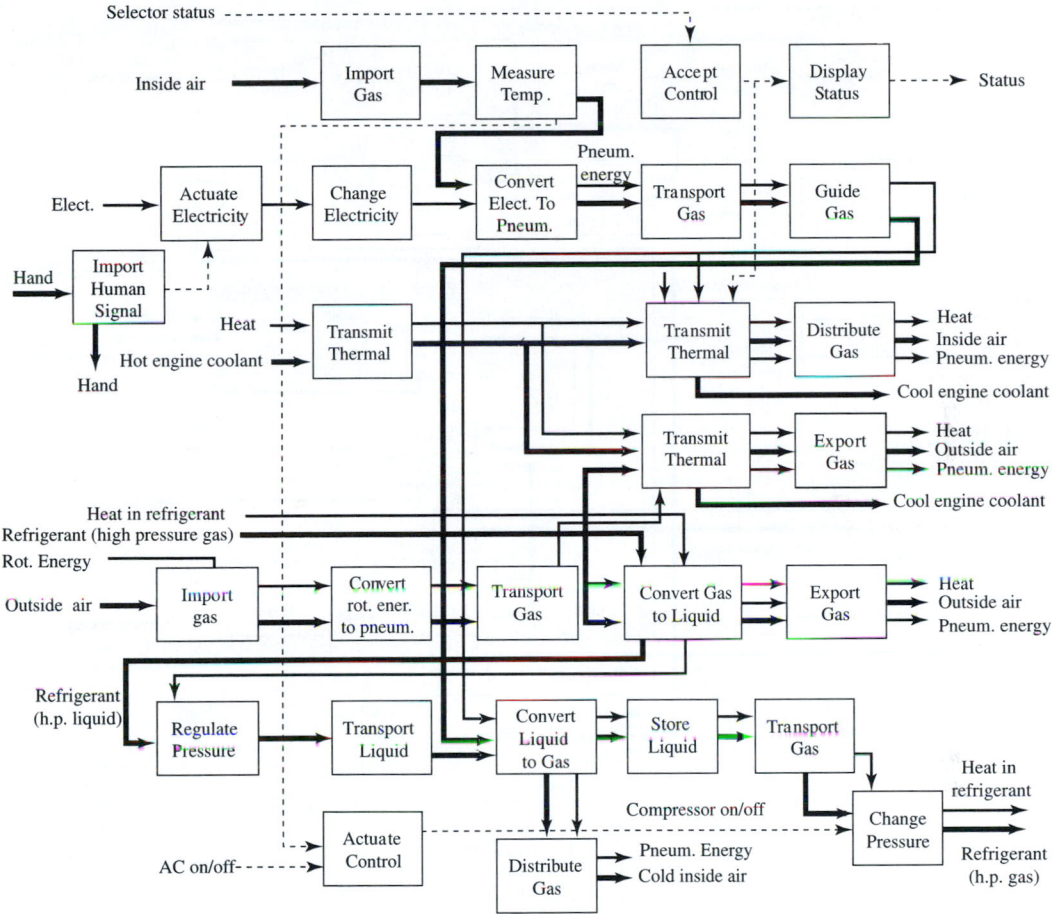

▼ **Figure 9.30.**
Function structure of the climate control system.

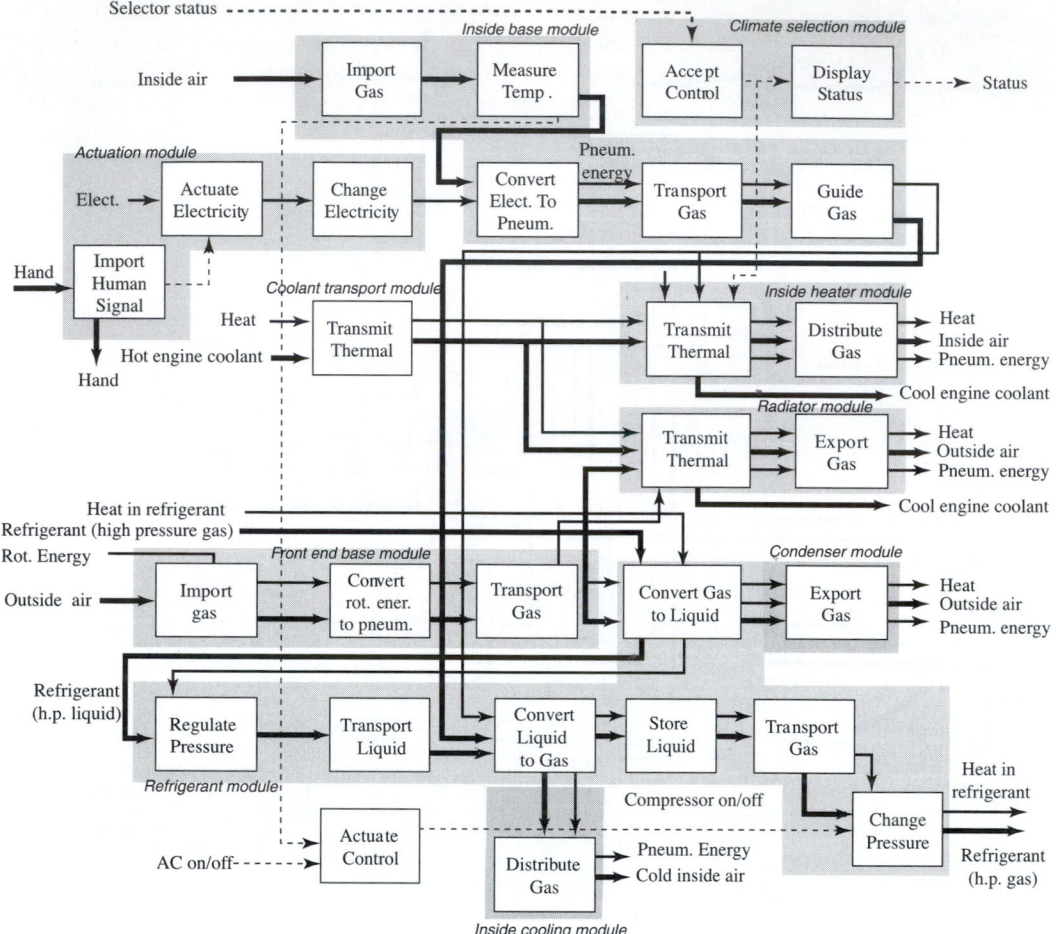

▼ *Figure 9.31.*
Identified modules of the climate control system by the dominant flow and flow branching heuristic.

Step 2: Apply the Module Heuristics

The three module heuristics are applied to the climate control system function structure next. The resulting modules are shown collectively in Figure 9.31. As with Step 1, if the modular design methodology was followed, then this step is already complete.

Considering first the dominant flow heuristic, we follow the flow *inside air* from the entrance of the system until it splits. This group is called the *inside base module.* Next, we follow the flow of refrigerant through the system. It traverses through seven subfunctions labeled the *refrigerant module.* The flow of *outside air* defines the *front-end base module* before it splits. The flows *coolant liquid* and *thermal energy* (exhibiting the material-energy duality of the engine coolant) define the *coolant transport module* before they split. *Electricity* and the signal *selector status* identify the *actuation* and *climate selection modules,* respectively.

For the branching flow heuristic, three main flows branch: *inside air* (and its associated *pneumatic energy*), *outside air* (and its associated *pneumatic energy*), and *coolant liquid* (and its associated *thermal energy*). Note that all three of these flows have modules associated with them based on the dominant flow heuristic as applied before they branch. The flow *inside air* splits and forms two additional modules, denoted by *inside heater* and *inside cooling module.* In fact, these two modules are the heater core and the evaporator. The flow *outside air* splits to form two modules, the *condenser* and *radiator module.* The *coolant liquid* flow also identifies two modules that were previously identified by the flows *outside air* and *inside air.*

For this example, the conversion-transmission heuristic identifies modules that are all subsets of the modules identified by the dominant flow and branching flow heuristic. They are not discussed here, since the development team application is searching for broader modules as opposed to the narrower modules the conversion-transmission heuristic predicts in this case.

The module-interaction list is given in Table 9. 8. Each module and its interactions, the flows crossing the module boundary, are listed in the first two columns. The interactions are classified according to type in columns 3–5. The basic flow classes of **M**aterial, **E**nergy, and **S**ignal are represented. The final four columns list the input and output interfaces associated with the interaction as well as the **Sp**atial requirement between the modules. The spatial interaction, as used in Pimmler and Eppinger (1994), is an indication of whether components must be placed in close proximity to each other in order to successfully transfer material, energy, or signal flows. In this table, the spatial column is simply a binary indication of the need for the module to be located in close proximity to another.

Step 3: Associate Ungrouped Subfunctions

After applying the heuristics, any ungrouped subfunctions must be lumped into an existing module or a new one. This step ensures their assignment to a development team. In this example, we assign the ungrouped subfunctions to the modules that their flows directly affect. For

TABLE 9.8.   LIST OF MODULES IDENTIFIED BY HEURISTICS FOR THE CLIMATE CONTROL EXAMPLE

Module	Interactions	M	E	S	Sp	Interfaces In: Modules or (External Systems)	Sp	Interfaces Out: Modules or (External System)
Actuation	Electricity		•			(Electrical system)		Inside base
	Human force		•		•	(Human hand)		
Inside base	Inside air	•			•	(Passenger compartment)	•	Inside heater, refrigerant, inside cooling
	electricity		•			Actuation		
	Pneumatic energy		•				•	Inside heater, refrigerant, inside cooling
	Temperature			•				(Actuate control)
	Heater/AC			•		Climate selection		
Inside heater	Inside air	•			•	Inside base	•	(Passenger comp.)
	Pneumatic energy		•		•	Inside base	•	(Passenger comp.)
	Thermal energy		•			Coolant transport	•	(Passenger comp.)
	Coolant liquid	•				Coolant transport		Coolant transport
Inside cooling	Inside air	•			•	Inside base	•	(Passenger comp.)
	Pneumatic energy		•		•	Inside base	•	(Passenger comp.)
	Thermal energy		•		•	(Passenger comp.)		
	Refrigerant	•				Refrigerant		Refrigerant
Front-end base	Outside air	•			•	(Environment)	•	Refrigerant, condenser, radiator
	Rotational energy		•		•	(Engine)		
	Pneumatic energy		•				•	Refrigerant, condenser, radiator
Radiator	Coolant liquid	•				Coolant transport		Coolant transport
	Thermal energy		•			Coolant transport	•	(Environment)
	Pneumatic energy		•		•	Front-end base	•	(Environment)
	Outside air	•			•	Front-end base	•	(Environment)
Condenser	Thermal energy		•				•	(Environment)
	Pneumatic energy		•		•	Front-end base, refrigerant	•	(Environment)
	Outside air	•			•	Front-end base	•	(Environment)
	Refrigerant	•				Refrigerant		Refrigerant
Refrigerant	Refrigerant	•						
	Pneumatic energy		•		•	Front-end base, inside	•	Inside cooling, condenser
	Thermal energy		•		•	(Passenger comp.)	•	Condenser
	Outside air	•			•	Front end	•	Condenser
	Inside air	•			•	Inside	•	Inside cooling
	Compressor on/off			•		(Actuate control)		
Coolant	Thermal energy		•			(Engine)		Inside heater, radiator

TABLE 9.8.   LIST OF MODULES IDENTIFIED BY HEURISTICS FOR THE CLIMATE CONTROL EXAMPLE (CONTINUED)

Module	Interactions	Interaction Type				Interfaces In: Modules or (External Systems)	Interfaces Out: Modules or (External System)
		M	E	S	Sp		Sp
transport	Coolant liquid	•				(Engine)	Inside heater, radiator
climate selection	Selector status			•		(Passenger comp.)	
	Heater/AC			•			Inside
Legend:	M: Materials						
	E: Energy						
	I: Information						
	Sp: Spatial						

the subfunction *import human force,* the flow *human force* affects the *actuation module* and is lumped with it. The other ungrouped subfunction *actuate control* is lumped with the *refrigerant module,* the one to which it sends a control signal.

Step 4: Assign Teams to Develop Modules

Now we assign the development teams to the identified modules of Table 9.8. Teams are now defined based on function and interaction as indicated in the table. Instead of traditional, mutually exclusive teams, these teams share responsibility for components that affect more than one module. There are many ways to populate development teams once their responsibilities are defined by the module-based development team method. One such method is to arrange teams based on personality type, as determined by the *Myers-Briggs Type Indicator* (MBTI) tool. Accounting for type differences and, more importantly, making team members aware of other's different strengths, eases tension between the differing member types (Wankat and Oreovicz, 1993).

The module heuristic development team method identifies ten modules for the climate control system (Table 9.9). This method identifies modules similar to the design structure matrix technique of Pimmler and Eppinger, but the modules (first column of Table 9.9) are narrower in their focus (or may be viewed as having greater resolution).

This method represents an addition to standard industry practice. It emphasizes team formation, hence product architecture, based on function and interaction concerns. Other product architecture techniques focus on interactions only *after* an architecture is selected (Ulrich and Eppinger, 1995; Cutherell, 1996).

TABLE 9.9. IDENTIFIED MODULES FOR TEAM ASSIGNMENTS

Development Team Method	Pimmler and Eppinger (1994) Modules	Comment
Inside base Inside heater Inside cooling Actuation	Inside chunk	Development team method provides greater resolution of modules
Front-end base Radiator Condenser	Front-end chunk	Development team method provides greater resolution of modules
Refrigerant	Refrigerant chunk	Same
Coolant transport Climate selection	Controls/Connections chunk	P&E simply lump remaining components together

Summary of the Development Team Method

Engineers are well known for developing the "how to" of problem solving, but the "why" is often not evident. So, why use this method for development team formation? It gives organizations a method to define development teams based on the actual modules of a device. Thus, a team need not be confined to historical team structures. Market leaders have been noted to lose market share when confronted by competing products with different architectures (Henderson and Clark, 1990). This method provides a way to adapt teams and product architecture. It also builds on the information gained from the modular design methods. If the modular design methods have been used, then much of the work to form development teams is complete.

This method translates modular product architecture to the development team formation. Since the modular architecture minimizes incidental interactions between modules, this approach has the potential to reduce the negotiation and conflict between team members.

VII. SUMMARY AND "GOLDEN NUGGETS"

Following a systematic design approach, it is possible to identify modular components at the functional level of a conceptual design activity. The approach is formalized in the modular design theories of this chapter. Direct benefits from identifying modules include greater use of common components in similar devices, a process to search for innovative product concepts, a more logical method of organizing design teams, and a potential reduction in development and manufacturing lead time of a device.

Golden nuggets from this chapter include:

▶ Product architectures may be viewed as integral and modular within the context of a product portfolio.

▶ Architectures translate product function to initial forms.

▶ Valid modules may be identified in products through the systematic application of module heuristics.

▶ Allocating design-team resources may be directly realized after developing a product architecture.

▶ Product architecture decisions may be made earlier in the development process through functional modeling.

References

Bockman, K. M. 1994. HP DeskJet 1200C printer architecture. *Hewlett Packard Journal* 45 (1): 55–66.

Clark, K. 1988. Managing technology in international competition: The case of product development in response to foreign entry. Chap. 2 in *International Competitiveness,* edited by M. Spence and H. Hazard. Cambridge, MA: Ballinger.

Clausing, D. 1997. *Concurrent Engineering.* Vol. 20 of *ASM Handbook, Materials Selection and Design.* Materials Park, OH: ASM International.

Cutherell, D. 1996. Product Architecture. Chap. 16 in *The PDMA handbook of new product development,* edited by M. Rosenau, Jr. et al. New York: Wiley.

Henderson, R., and Clark, K. 1990. Architectural innovation: The reconfiguration of existing product technologies and the failure of established firms. *Administrative Science Quarterly* 35 (1): 9–30.

Little, A., Wood, K., and McAdams, D. 1997. Functional analysis: A fundamental empirical study for reverse engineering, benchmarking and redesign. *Proceedings of the 1997 Design Engineering Technical Conferences,* Sacramento, CA.

McAdams, D., Stone, R., and Wood, K. 1999. Functional interdependence and product similarity based on customer needs. Forthcoming in *Research in Engineering Design.*

Newcomb, P., Bras, B., and Rosen, D. 1996. Implications of modularity on product design for the life cycle. *Proceedings of the 1996 ASME Design Theory and Methodology Conference,* Irvine, CA.

Otto, K., and Wood, K. 1997. Conceptual and configuration design of products and assemblies. Vol. 20 in *ASM Handbook, Materials Selection and Design.* Materials Park, OH: ASM International.

Pahl, G., and Beitz, W. 1988. *Engineering design: A systematic approach* London: Springer-Verlag.

Pimmler, T., and Eppinger, S. 1994. Integration analysis of product decompositions. *Proceedings of the ASME Design Theory and Methodology Conference,* 1994.

Prasad, B. 1997. *Concurrent engineering fundamentals.* Vol. II in *Integrated Product Development.* Upper Saddle River, NJ: Prentice-Hall.

Stone, R. B. 1997. Toward a theory of modular design. Ph.D. diss., The University of Texas at Austin.

Stone, R., Wood, K., and Crawford, R. 1998. A heuristic method to identify modules from a functional description of a product. *Proceedings of DETC'98,* Atlanta, GA.

Ulrich, K. T. 1995. The role of product architecture in the manufacturing firm. *Research Policy* 24 (3): 419–440.

Ulrich, K., and Eppinger, S. 1995. *Product design and development.* New York: McGraw-Hill.

Ulrich, K. T., and Tung, K. 1991. Fundamentals of product modularity. *Proceedings of the 1991 ASME Winter Annual Meeting Symposium,* Atlanta, GA.

U.S. Congress. 1992. Office of Technology Assessment. *Green Products by Design: Choices for a Cleaner Environment.* OTA-E-541.

Wankat, P., and Oreovicz, F. 1993. *Teaching engineering.* New York: McGraw-Hill.

10 Generating Concepts

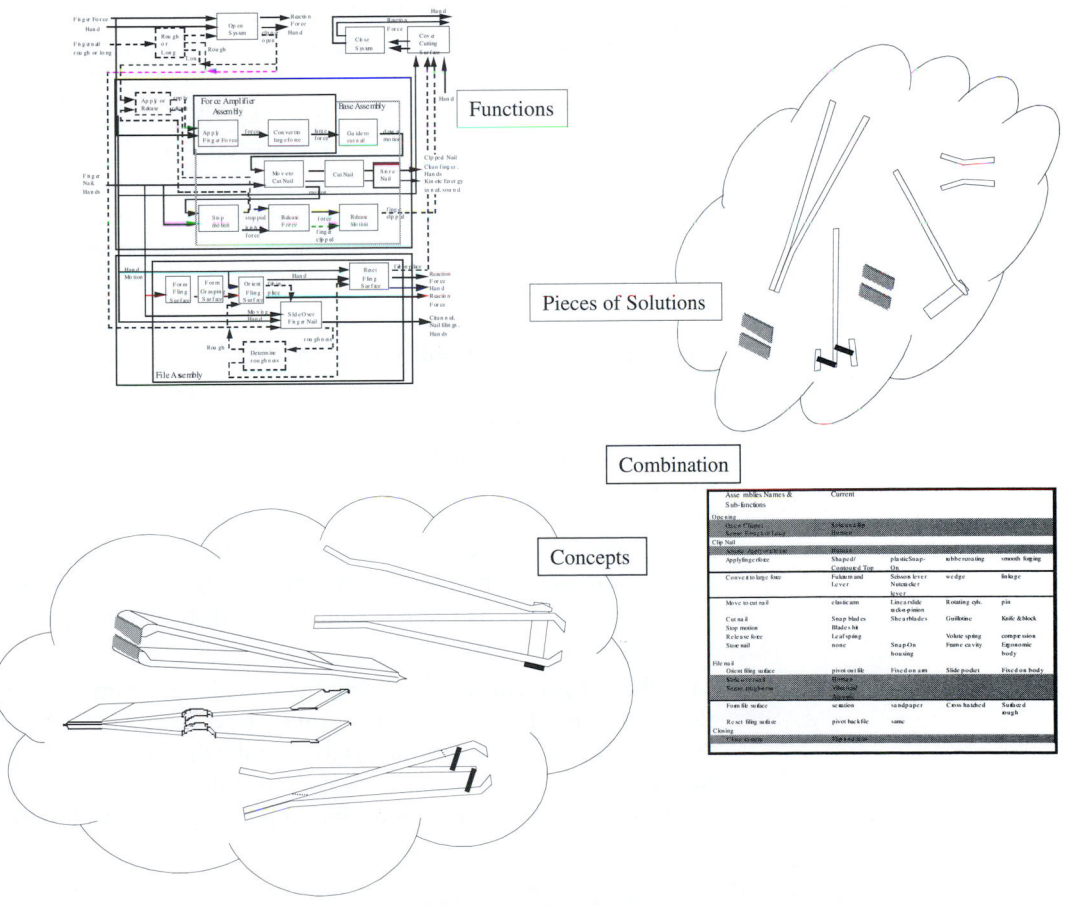

Functions

Pieces of Solutions

Combination

Concepts

Assemblies Names & Sub-functions	Current				
One ring					
Open Clutter	Scissors slip				
Store Energy to Loop	Human				
Clip Nail					
Apply Applying force	Human				
Applying force	Shaped/ Contoured Top	plastic Snap-On	rubber coating	smooth forging	
Convert to large force	Fulcrum and Lever	Scissors lever Nutcracker lever	wedge	linkage	
Move to cut nail	elastic cam	Linear slide index pointer	Rotating cyls.	pin	
Cut nail	Snap blades	Shear blades	Guillotine	Knife & block	
Stop motion	Blades hit				
Release force	Leaf spring		Volute spring	compression	
Store nail	none	Snap-On housing	Frame cavity	Ergonomic body	
File nail					
Orient filing surface	pivot on file	Fixed on arm	Slide pocket	Fixed on body	
Seek over nail	Human				
Setup, grip lever	Abrasive				
Form file surface	serration	sandpaper	Cross hatched	Surface rough	
Reset filing surface	pivot back file	same			
Closing					
Close Energy	Slip and slide				

411

The cover illustration shows snapshots of product ideas generated from abstract specifications and descriptions. Building on functional models and product architecture, this chapter focuses on methods for creating innovative concepts. The activity of concept generation is one of the lampposts of engineering design. It provides a forum for designers to apply creativity and contribute their personal flair. It also represents the time when technology is chosen or developed to fulfill the customer needs.

In this sense, the imaginary clay of product development is molded during concept generation. Artistic skills must be brought to bear that will allow us to shape the clay according to the direction of our inner-eye, experience, and knowledge. Tools are also needed to remove portions of the clay, to wet and reshape it, to spin it into new forms, and to add features that will appeal to customers.

We are all artists. We are all creative. However, we need methods to exercise and direct our skills. We also need methods that enhance and enrich the innate creativity we all possess. This chapter provides some basic methods that will mature our skills and aid us in developing a broad array of concepts for product development.

One danger in any concept creation process is the bias of *preconceived solutions*. Preconceived solutions usually occur early in the process of solving a design task. They are the engineer's idea of what a product should do, not the customer's. Experience shows that the final concept chosen is rarely the first considered. On the contrary, it can often be the twentieth, thirtieth, or more.

Another danger of concept generation is the creation of ideas within the "vacuum" of a design team's experience. With a small number of members, a product development team can generate a large number of solutions from their experience. This number, however, does not compare favorably to the vast array of solutions that are possible from the knowledge contained in the history of humankind. A psychological inertia exists within the design team due to their finite experience. We must avoid the psychology that a much better solution is around the corner, if only we had the extra knowledge and insight.

This chapter presents methods that help the design team overcome these common dangers. *Decomposition,* or breaking down a problem in to smaller parts, is one fundamental principle that helps us overcome these dangers. Many ideas are readily apparent to simple, small-scope problems. Decomposing and recasting a complex product system into such smaller pieces can greatly enhance the number of concepts, in addition to their breadth. Another useful principle is to use our engineering and applied science knowledge to help direct the search for alternative solutions. Concept generation is not a blind search for ideas, where the proverbial "light bulb" is suddenly illuminated. The generation of concepts requires insights that come from a fundamental knowledge of how the world behaves and how we can estimate this behavior. These insights, followed by a great deal of *perspiration* and *hard work,* can greatly advance the breadth of solutions by considering the synthesis of analogous and feasible technologies.

Let's therefore move forward into concept generation with the goal of learning and applying these principles. We must avoid the tendency to rely solely on our innate skills. As engineers, we are well trained in the application of analysis through applied mathematics and science. Very little of our formal training, however, exists in the realm of increasing our skills as innovators. But skills in innovation can be developed and matured; such skills require the understanding of concept generation principles and executing them repeatedly on open-ended problems.

I. CHAPTER ROADMAP

Concept generation, as a journey in product development (Figure 10.1), begins with an overall process and strategy. Alternative products may be created from this strategy through the use of systematic methods. Two types of methods are considered in this chapter. The first is known as *information gathering* and *brainstorming* techniques. These methods focus on the combination of obtaining knowledge of possible technologies with the generation of ideas from the minds of the designers. Usually the creation of ideas from such methods is undirected and "free-wheeling."

The second type of methods are known as *biasing* methods, in that they add "bias" to the search for solutions by using physical insights and documented design principles. By directing the search for solutions, a greater number of ideas may be generated for particular aspects of the design problem. Conflicts in the problem may also be resolved by considering innovative solutions, avoiding unnecessary compromises.

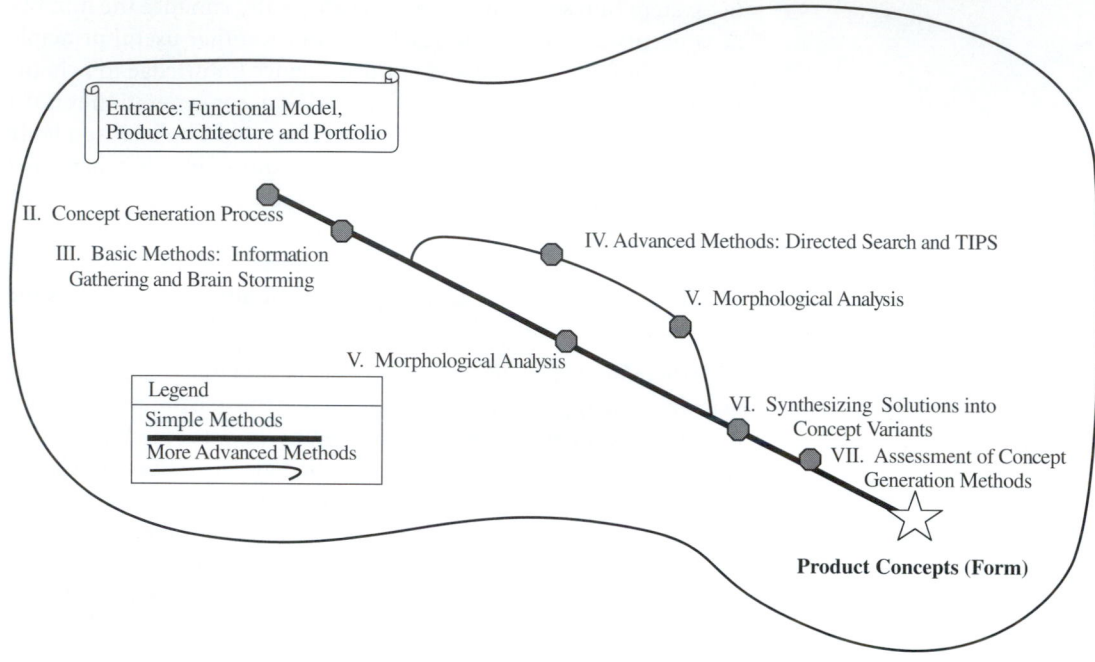

▼ Figure 10.1.
Chapter roadmap of generating concepts, that is, forming solutions.

II. CONCEPT GENERATION PROCESS

The underlying goal of concept generation is to develop as many ideas as possible. One or two alternative concepts are unacceptable. Tens of concepts are acceptable; the more the better.

Figure 10.2 illustrates a process for meeting this goal. The process begins with a review of the customer needs, highlighting the primary needs that are the initial focus. Ultimately, all the needs must be satisfied through concept generation. Yet the process begins by considering the most important needs first. Iteration in the process may then be used to create further concepts for secondary or supporting functions.

Based on the customer-need focus, the design task is decomposed (or divided) into subproblems that may be more easily understood and solved. These subproblems are of three forms: functional models, product architecture, and product portfolio. In each case, the focus is on *what* the product must do, not *how* it will do it. Chapters 5, 8, and 9 describe methods for developing these decompositions of a product design. In terms of

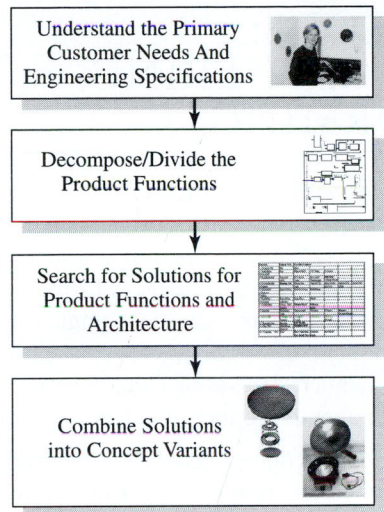

▼ *Figure 10.2.*
Concept generation process.

concept generation, a functional decomposition may be used to generate form solutions to each of the product functions, that is, transforming function to form. Alternative layouts and interfaces may be developed from the product architecture information. Alternative integral or modular concepts may be developed from the portfolio choices.

This approach will lead to a broad number of solution ideas for each of the product functions. These solution principles may be applied to alternative layouts that are created or alternative concept classes defined by the portfolios.

The next step in the concept generation process is to combine the solution ideas per product function into concept variants, that is, alternative designs. This step in the process is challenging and nonlinear. We are essentially "adding" and connecting together solutions to each of the functions by creating the geometry of the solutions, their interfaces to each other, and their interfaces to the environment. An infinite number of solutions are possible in this combining step. It requires sketching and refinement, repeatedly performed to result in a concept that is sufficiently detailed to be evaluated against the engineering specifications.

Based on the process shown in Figure 10.2, methods are needed that help us generate concepts for product functions and combine them into alternative product ideas. The next section discusses basic methods for the first case: product function solution principles.

III. BASIC METHODS: INFORMATION GATHERING AND BRAINSTORMING

Concept generation, as described in this chapter, is the divergent development of many alternatives, where the focus is on innovation, structural layout, and function satisfaction. A convergent strategy is adopted once a breadth of ideas is formed. This ensuing strategy provides a means of converging to a single solution (or finite portfolio of solutions) that will ultimately be the product in the marketplace. The remainder of this book focuses on a convergence toward a single product or a set of products that form a portfolio and product architecture.

Formal concept generation methods may be classified, broadly, into two categories: *intuitive* and *directed* (logical) (see Figure 10.3, Shah, 1998). The intuitive category relates to the methods that focus on idea

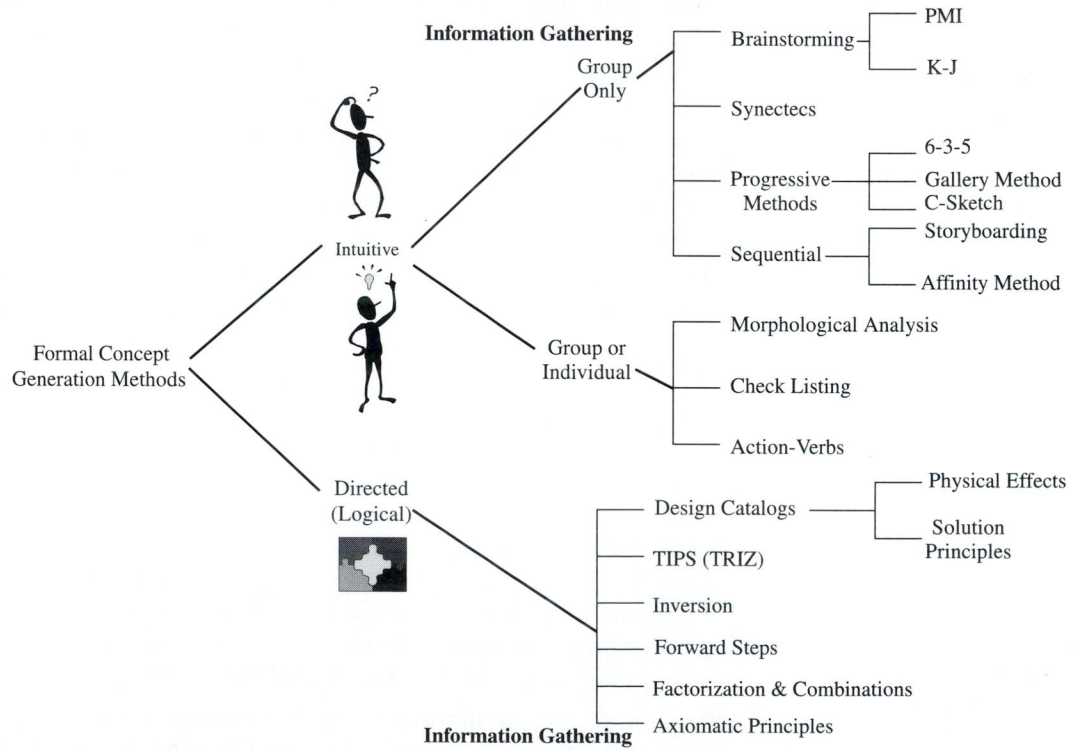

▼ *Figure 10.3.*
Types and names of concept generation methods (adapted from Shah, 1998).

generation from within an individual or group of individuals. The intent of such methods is to remove barriers to divergent thinking so that new connections and features in a product may be visualized. By removing these barriers, the environment of idea generation may be filled with conditions that promote creativity. Example methods include brainstorming and morphological charting.

Directed methods, on the other hand, use a systematic, step-by-step approach to searching for a solution (Shah 1998). These methods rely on technical information, expertise, and guidelines to seek solutions to technical problems. Ideally, they force solutions to be determined along a particular path, although the final solution is not readily apparent at the outset. As such, these methods balance the technical and technological information available with the makeup of the design team or the work environment.

In the remainder of this section, we study the first category of concept generation methods: intuitive. Information gathering begins this study. While information gathering is not an intuitive method per se, the research of information from different sources is critical to the success of any concept generation activity. After describing information gathering, brainstorming, progressive methods, and morphological charting are presented. The subsequent section presents directed methods as advanced approaches to concept development.

Information Gathering: Conventional Aids

As a first method for concept generation, let's consider *information gathering*. This activity entails the dynamic search for data that will contribute to the technology, physical principles, or industrial design of a product. Usually, the search seeks documented ideas on solving a product function with a form solution. It also seeks concepts for producing, analyzing, or testing a product idea once it is conceptualized.

Information gathering, as described above, begins the study of concept generation methods due to its critical stature in the process. As the famous saying goes with regard to brilliant ideas: "I stood on the shoulders of the giants that preceded me." The implication of this saying is that we should never generate concepts in a vacuum. Of course, we should begin generating ideas intuitively to avoid the bias associated with seeing others' ideas first. We do not want to become enamored with a single solution to a product function or subsystem, since it will subsequently be difficult to generate any further ideas. However, we must utilize the innumerable ideas of all persons that have preceded us. Only through such consideration can we expect to make progress. Indeed, a synonym for concept generation is synthesis, the collecting of individual components or ideas into unique and novel combinations. Let's be synthesizers first; very rarely (if ever) are ideas completely spontaneous.

In today's colossal information infrastructure, the task of information gathering can be inundating. A classification of informational sources is thus needed to aid in the process. Figure 10.4 illustrates one such classification. One primary category in this classification is literature. "Published" media represent a large resource for obtaining ideas for

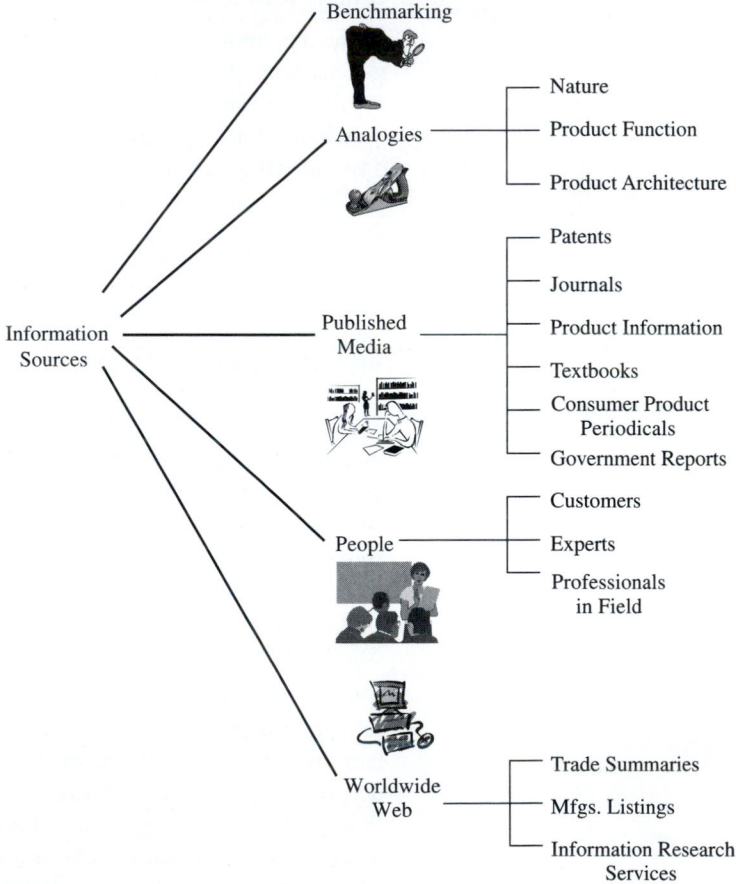

▼ *Figure 10.4.*
Information sources for concept generation.

product functions. Examples of this media include patents, journal or government articles, catalogs, textbooks, consumer product periodicals, and product information, all as discussed in Chapter 7. A patent search is essential in understanding the current technology in a product domain. It is also useful for seeking analogous ideas from products that solve similar functions, although they may be situated in entirely different domains.

In addition to published media, information should be gathered from analogies, the World Wide Web, benchmarking, and people. Analogies consist of a similar product or artifact that operates in a different domain. It is similar because it implements an architecture or function that is in common with the product being designed. For example, a product function for a coffee bean grinder is to "dissipate noise" produced from the milling process. Analogous products include power tools, sound rooms, automobiles, aircraft, and food processors. All of these products also "dissipate noise." By studying their solutions, we can develop analogous solution principles for the coffee grinder. Analogies may also be obtained from artifacts, such as those that exist in nature. If we are considering the function of "support or resist loads," natural analogies may be considered as possible solutions, such as honeycombs.

Other categories of information sources include benchmarking, people, and the World Wide Web. Chapter 7 discusses these sources with respect to benchmarking activities. These sources also provide a wonderful forum for harvesting ideas that may be used to generate further concepts for a particular product development.

In the end, the breadth and novelty of ideas generated during conceptual design are a function of the information available to the product development team. "Knowledge is power," and the power of knowledge leads to innovative ideas. We must devote a significant percentage of our resources to information gathering. In doing so, we can harness the power that is at our fingertips. We can also avoid the "not-invented-here" syndrome, viewing ourselves, instead, as synthesizers of novel concepts from known information.

Traditional Brainstorming

Brainstorming is an intuitive method of generating concepts. It focuses on product function and architecture, where team members communicate ideas verbally during a set time period. All team members are encouraged to be open and uninhibited during the initial sessions of brainstorming. There is no need to strictly adhere to product specifications, but instead to focus on the functional needs of a product.

The overall aim of brainstorming is to obtain several concepts that might work as solution principles to a piece of the design problem. Ideally, the search is comprehensive, where no solution is left unmentioned. Of course, this comprehensive requirement will never be satisfied, but the team does the best they can. A breadth of solutions is desired.

The primary advantage of brainstorming is the ability of a set of individuals to collectively build on each other to generate ideas that would not arise individually. (The group is greater than the sum of the individual parts.) Team members will *piggyback* and *leapfrog* each other. Piggybacking creates building-block ideas to words, body language, statements, and concepts stated by a team member. Responses to team member's statements include relations, inversions, tangential or partial changes, antonyms, synonyms, and so forth. Leapfrogging, on the other hand, results in divergent or discontinuous jumps in the responses. Each team member brings different expertise, skills, and personality to a group effort of piggybacking and leapfrogging. Brainstorming taps into this diversity to create, quickly, a large number of high-level solutions.

Disadvantages of the method include a number of factors. The "right idea" may not come at the "right time." There is no guarantee. Group conventions may sidetrack or inhibit original ideas. The team may be distracted by a misdirected focus, or certain team members may dominate the discussion. Likewise, the team that is assembled for a given session might not be open to new ideas. This situation will tend to condone only known solutions, with a self-prophesizing result.

Although these disadvantages exist, brainstorming is a powerful technique for generating concepts. A committed team creates ideas together (that are *not* judged at the time of the session), which possibly triggers further ideas. Within this context, there are a number of guidelines that should be followed during a brainstorming session:

▶ Designate a group leader/facilitator, to prevent judgments and encourage participation by all. The facilitator should not contribute directly (until another session), but rather direct and record.

▶ Form the group with 5–15 people, usually no more and no less. Less gives inadequate ideas, more can break down the group into multiple conversations or inhibitions toward participation.

▶ Brainstorm for 30–45 minutes. The first 10 minutes are typically devoted to problem orientation and familiarity. The next 20–25 minutes will see a sharp increase of ideas, followed by a flat region, followed by a sharp decrease. During the final 10 minutes, a trickling of ideas will occur, which should be encouraged, but stagnation will also happen quickly.

▶ Don't confine the group to experts in the area. The only way to obtain new ideas is to introduce new knowledge and experiential backgrounds.

▶ Depending on the goals of a brainstorming session, individuals may enter a session with a set of ideas.

▶ Avoid hierarchically structured groups. Bosses, supervisors, and managers should not be included in many of the sessions.

Memory (Mind) Maps

Typically, the facilitator is also the engineer responsible for the particular design problem that the brainstorming session is called to fulfill. This person has a vested interest in obtaining the results. He or she also records the session results by collecting the notepads of all participants after the session but also by recording the ideas as labels during the session.

One effective way to record the results of a brainstorming session as it happens is by *memory (mind) mapping*. Here, the facilitator starts with a clean sheet of paper, writes the problem statement in the center of the paper as two words, and draws a box around it. Then, as ideas are generated to solve the problem, they are recorded quickly, say with two or so words, with circles drawn around them. Each new idea to solve the initial problem is connected to the original problem statement.

As an idea is refined, or sparks another idea, these new ideas are connected to the idea that sparked them. Ideas that are all basically the same concept should branch out of the originally proposed concept. Entirely different concepts should have their own branches emanating from the problem statement.

If the problem statement is refined into a new one, this new form can be recorded by entering a new problem statement and drawing a box around it. Problem statements in boxes distinguish the concepts in circles. The new problem statement should be connected to the concept that sparked the reformulated or refined statement by a directed arrow. Typically, some concepts introduce subproblems that then are brainstormed. Memory maps help a facilitator visualize this process (as the memory map will show a single well-fielded branch) who can then redirect the process back onto the original problem.

The memory map also serves as an effective visual documentation of the brainstorming session. It is called a memory or mind map, as if a single person creates all of the ideas individually. One could imagine this documentation as an indicator of how a person connects or networks ideas in memory.

Product Application of Brainstorming

Consider the design of a product to detect and display the location of a golf ball when it is struck off the tee (or other strokes, in or out of bounds). We are interested in discovering the final resting location. Let's apply the brainstorming method to the "detect golf ball" function of this design problem. Five groups of five individuals generated ideas for this problem following the guidelines of brainstorming, except the session time was shortened. Figure 10.5 shows a partial list of the generated ideas. Notice in the figure that a breadth of ideas across multiple energy domains is represented. This process continued with the generation of further ideas, as memory-mapped by a facilitator (Figure 10.6).

Summary

Based on these descriptions, guidelines, and product example, a procedure for brainstorming is given by the following steps:

▶ Conduct either a free-for-all or an orderly (around the room) process for idea suggestions, directed by the facilitator.

▶ Record all ideas as they are stated, but none are judged at this point (even as to practicality).

▶ Detail suggestions far enough for emergence of a specific solution idea.

▶ Wrap up the session in about 30–45 min.

	Smoke trail
Bright colored ball	Shorter golf course
Electronic Grid with ball emitter	Putt-Putt golf
Sound horn in ball	Spotters paced every 10 m
Exploding ball	Colored golf course
Golf lessons	Trajectory calculation system
GPS System	Robotic arm hits ball
Scent-Human	Mini-camera in ball
Scent-Dog	Light emitting ball
Virtual golf	Ball shoots flare
Pressure sensitive ground	Plexiglass side walls on golf course
String attached to ball	Funnel shaped golf course
	Speaker in ball; use microphone to call yourself
•	•
•	•
•	•

▼ *Figure 10.5.*

Partial brainstorming list generated for the function of "detecting a golf ball."

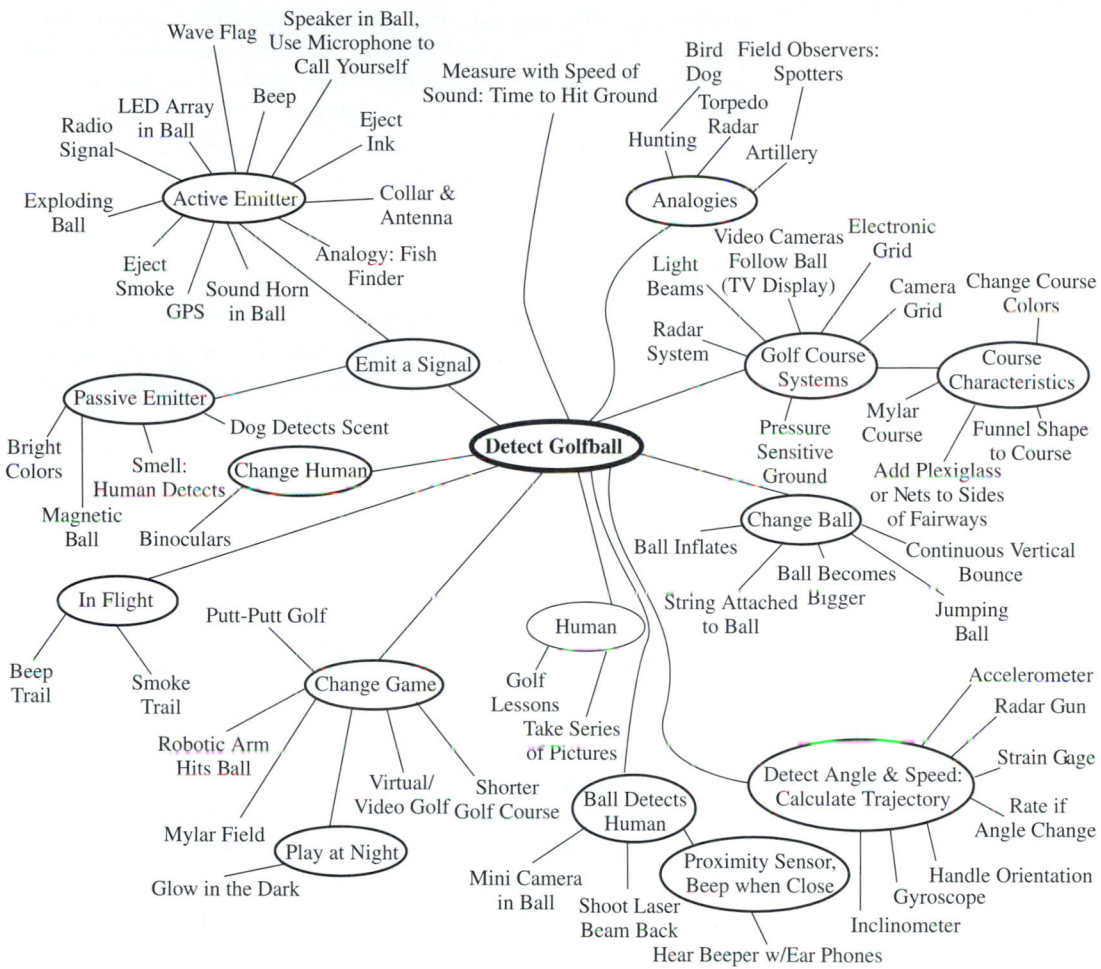

▼ **Figure 10.6.**
Memory (Mind) Map for the function of "detecting a golf ball."

> ▶ After the session is completed, judge the results with experts or the same group. Each person spends one share to vote for a concept. Each person has ten shares with which they can vote.

When executing this procedure, it may be helpful for the facilitator to suggest idea generators. Idea generators are simply catalysts that help a group avoid troughs or low points in a brainstorming session. Momentum should always be developed and maintained during a session. Examples of idea generators include: "Be the problem" (act out the flows of the functions, asking "What happens to me?"); reverse the problem (solve the opposite or work backwards as if a solution exists); generate a list of analogous tasks and products; increase or decrease the scale of the problem (make large things small and vice versa); combine ideas; ask "What if new technologies existed?"; ask "What if a fact doesn't exist?" (such as "no gravity"); think "nasally" (think with one's senses); relax one or more specifications, and so forth.

Brain-Ball

Oral/Verbal methods in brainstorming are wonderful tools to generate concepts; however, there can be a heavy reliance on the "energy" supplied by the facilitator. There may also exist a natural tendency by team members to judge ideas as they are described (left-brain thinking) instead of going with the flow. One approach to create a self-energetic system and minimize negatives is to structure the activity, both the physical environment and roles of the participants. *Brain-ball* (Faste, Roth, and Wilde 1998) is one form of brainstorming that meets this goal.

In brain-ball, the participants form a standing circle where everyone in the circle may see each individual. The process then begins by introducing a hypothetical ball that is passed from individual to another individual, assuring that all participants are included at some point in the passes. A participant may pass the ball to any other member, as quickly and spontaneously as possible without the process becoming unwieldy. In the first round of the game, no speaking is allowed. In addition, throwers and receivers should actively play their roles without introducing ancillary motions or theatrics. The intent is simply to form clear communication of passing the ball.

In the next round, a ball is introduced again by a facilitator and the thrower adds a sound. A receiver restates the sound and passes the ball to some another participant by stating a new sound. Members should not concentrate on thinking of new sounds in advance, or they will be distracted from receiving the ball.

After a few minutes of passing the ball and sound, the facilitator introduces a second ball to an unoccupied member in the circle. Two balls and spoken sounds then pass around the circle simultaneously. Participants must pay close attention to pass or receive one of the balls, maintaining their relaxed attitude and peripheral vision. The facilitator may then introduce a third ball, and the process continues until the team experiences the full activity of responding spontaneously to multiple balls and sounds.

The facilitator ends the second round and then introduces a word or concept to the group, such as "detecting a golf ball downrange." The facilitator throws the hypothetical ball to a member of the group with a solution or response to the concept, such as "golf ball with string." The member receiving the hypothetical ball restates the response and passes the ball to another participant with a new response. The new member repeats the response, and the process continues.

A second and possibly third ball is introduced to the game (by the facilitator) as the time proceeds. Multiple balls with piggyback and leapfrog responses are then simultaneously passed until the brainstorming reaches a repetitive stage. The facilitator can then introduce a new concept, subproblem, solution, or idea generator to create a new round of brainstorming. The participants may also be divided into subcircles to continue new rounds, with redistribution occurring over time.

Variants exist in this process. Different hypothetical items can be passed, circles can grow and shrink in the number of participants, playacting of concepts can occur, and so forth. The idea is simply to create an environment where concepts "flow" between participants. These concepts may be recorded and memory-mapped by scribes. The documented results may then be revised and sorted at later sessions.

C-Sketch/6-3-5 Method

A number of drawbacks exist in the traditional brainstorming method. These drawbacks may be classified according to two primary factors: Idea generation may be dominated by a small number of team members or by an overzealous facilitator, and brainstorming usually relies on an oral means of communication. The first factor creates an atmosphere that squelches participation by members, resulting in fewer ideas being generated. The second factor only provides one medium for expressing concepts. The execution of written descriptions and sketches is usually not employed.

Alternative methods to address these deficiencies are known as the *C-Sketch* and *6-3-5 methods,* also known as *brain-writing.* C-Sketch focuses on sketches as the media for creating concepts. The "6-3-5" method recommends sketches, with a limited use of key words and short descriptions. In the following description, sketching is emphasized, but within the context of the original "6-3-5" method.

The modified "6-3-5" approach uses the following guidelines and process steps. Team members are arranged around a table, usually a circular table to provide continuity. Research shows that a group of six ("6") members is ideal; however, the number of participants can range from 3 to 8 members. Each member writes/describes/lists keywords/sketches for three ("3") ideas for the product functions, architecture, or overall configuration under consideration. (We highly recommend that sketches be the focus of this activity.) Usually the top five product functions with respect to the customer needs are considered. Fifteen to 30 concepts typically result from the process over a period of 25–35 minutes. The ideas may initially be restricted to each individual product function, not combinations of them, or the overall layout drawings from the product architecture analysis (Chapter 9). After developing solution principles for each primary product function, the 6-3-5 method is repeated to aggregate the principles into integrated concept variants. Again, 25–35 concepts typically result.

After each member of the concept generation team spends T minutes of work on the initial concepts, the paper is passed to the right around the table. The team members now have T more minutes to add an addition idea to the paper and to modify and extend the ideas from the previous team members. This timed process continues until all sheets of paper have been added/modified by each team member (a total of 35–60 minutes, typically).

The passing of the papers through one cycle is known as a "round." The method encourages a total of five rounds (spaced in time) to refine and combine ideas. The five ("5") rounds represents the last designation in the "6-3-5" method.

A number of guidelines help to achieve success from the "6-3-5" method. There should be no verbal communication until a cycle (round) is completed. By restricting verbal communication, no one member can dominate the concept generation process. Traditional brainstorming may be implemented after a few "6-3-5" sessions so that team members will be able to interact and spur ideas through fruitful communications.

Besides this guideline, the focus of the modifications during the passing of ideas should be on advancing the ideas, not on negative criticism. As a team member views a pictorial idea, the visual image should be used as a catalyst for imagination. Unintentional product features will be viewed from quick sketches, since team members are prevented, initially, from describing the intent of their sketches (beyond labels and a concise statement). Sketches are the preferred mode for expressing concepts.

Product Application of "6-3-5"

As an illustration of the "6-3-5" method, consider the design of a power screwdriver product. Before concept generation, systematic methods result in significant data to be used for creating concept variants. These data, in part, include organized customer needs, a functional model, an assessment of the product architecture, benchmarking information, and engineering specifications in a House of Quality.

A black box description of the power screwdriver is shown in Figure 5.14 along with relevant flows of material, energy, and signals that identify the primary operating conditions of the product. The black box description also defines the system boundary such that the bit is considered an input to the device (at this point in the development).

Customer needs are gathered for the problem. Here, interviews were conducted with a set of customers, an appropriate sample size for a small consumer product. The stated customer needs are compiled in Table 5.6 and listed in order of importance (on a 1–5 scale).

A functional model is also developed as shown earlier in Figure 5.17. This model focuses on the primary human interface and power train for the screwdriver. Based on this model, the customer needs are associated with the product functions to determine the primary function drivers (Table 10.1). Once the associated functions are identified, each receives the value of the original customer need rating. This approach effectively assigns a customer need rating to those functions that are viewed to directly affect the associated customer need. These customer need ratings are then summed to produce the value of the weighted product functions as shown in Table 10.2.

TABLE 10.1. FUNCTION TO CUSTOMER NEEDS CORRELATION FOR THE POWER SCREWDRIVER EXAMPLE

Customer need	Scaled customer need rating (1–5)	Associated flow(s)	Associated subfunction(s)
Powerful	5	Electricity, torque	Convert electricity to torque, change torque
Fast	4	Electricity, torque	Convert elect. to torque
Long-lasting battery	5	Electricity	Store electricity
Short charge time	4	Electricity	Store electricity
Manual use capability	4	Hand, human force	Import human hand, import human force, secure rotation
Reversible (screw and unscrew)	4	Electricity, torque	Actuate electricity
Lightweight	4	Human force, electricity	Import human force, convert electricity to torque
Weight balance	2	Human force	Regulate rotation, regulate translation
Small size	3	Hand, electricity	Import human hand
Comfortable handle	2	Hand	Import human hand
Automatic shutoff	3	Electricity	Actuate electricity
Interchangeable tips	4	Bit	Secure solid
Maintenance-free	1	Torque, bit, human force	Dissipate torque, store solid
Variable velocity	3	Electricity	Regulate electricity

TABLE 10.2.   WEIGHTED SUBFUNCTION VALUES FOR THE POWER SCREWDRIVER APPLICATION

Subfunction	Associated customer need ratings	Weighted customer need rating
Actuate electricity	4, 3, 1	8
Allow rotational DOF	1	1
Change torque	5, 1	6
Convert electricity to torque	5, 4, 4, 1	14
Couple solid	1	1
Dissipate torque	1, 1	2
Import hand	4, 3, 2, 1	10
Import human force	4, 4, 1	9
Regulate electricity	3, 1	4
Regulate rotation	2, 1	3
Regulate translation	2, 1	3
Rotate solid	1	1
Secure rotation	4, 1	5
Secure solid	4, 1	5
Separate solid	1	1
Store electricity	5, 4, 1	10
Supply electricity	1	1
Transmit torque	1	1

Based on this analysis, the primary function drivers are "convert electricity to torque," "import hand," "store electricity," "actuate electricity," and "change torque." These functions are the emphasis for our concept generation activity. In addition, the product architecture information must be considered to specify the layout configurations for the power screwdriver. From Chapter 9, six unique modules are identified for a power screwdriver. The first is referred to as the coupling/decoupling modules. It provides a means of connecting and removing different bits and attachments. A rotational-lock module focuses on the human input energy. It provides a means of locking the rotational degree-of-freedom of the screwdriver when it is used with only a human force. This module unlocks the rotation when powered by the screwdriver electrical input.

The third module is a positioning module that aids the user in orienting the device to interface with screws. This module focuses on the translation of the screwdriver into and away from a screw. Functions to input rotation form another module. This module, referred to as the actuating module, allows users to tighten *or* loosen a screw depending on their choice.

The final two modules are the supply electricity module and the torque-transmission module. These modules transmit electricity, convert it to a mechanical torque, increase the torque, and then transmit the torque to the screw interface.

This product architecture, combined with the primary product functions, is the focus of the initial concept generation. Let's apply the "6-3-5" method to these issues. Figure 10.7 shows an iteration of the "6-3-5" method. This figure shows the first three concepts generated by one individual of the product development team. This person's concepts focus on the overall architectural layout (coupling/decoupling, positioning, and actuating modules) in addition to the "import hand"

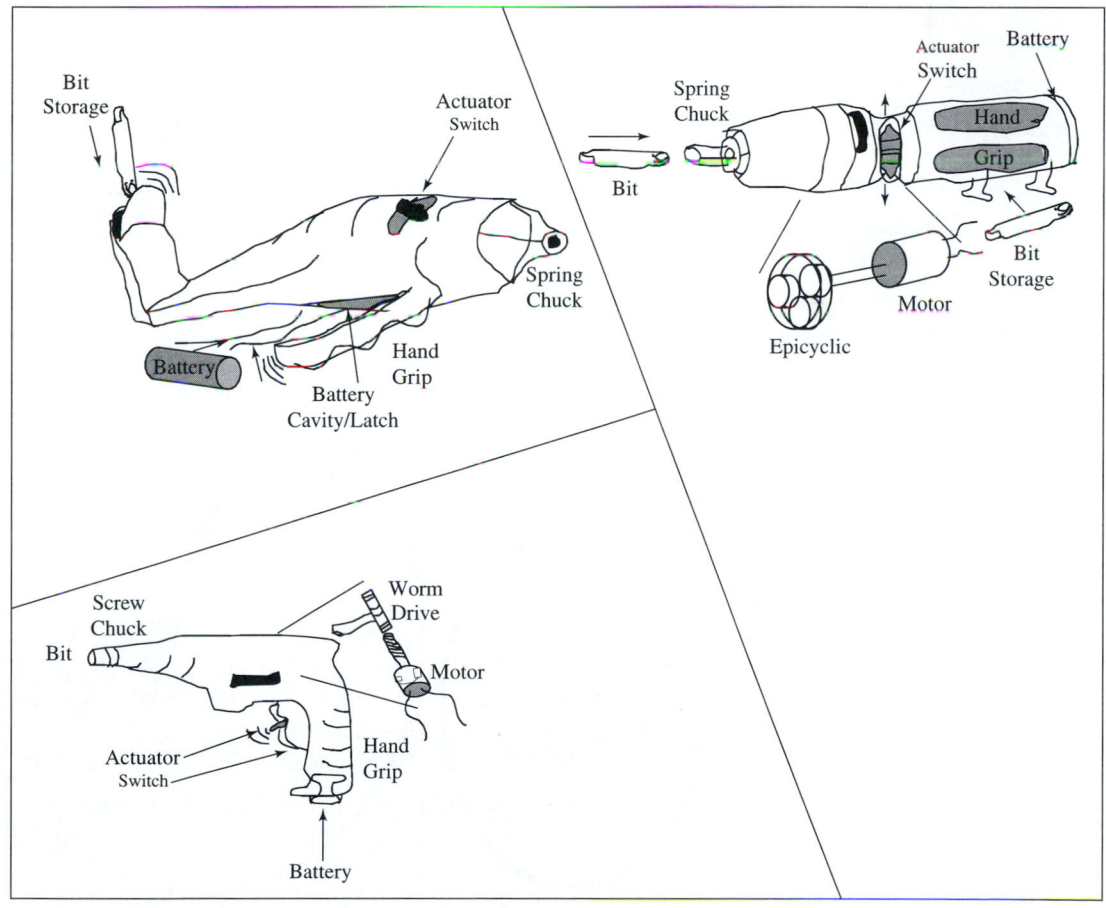

▼ **Figure 10.7.**
First phase of the "6-3-5" method: application to a power screwdriver product.

primary function driver. One concept is an analogy of an electric drill (lower left), the second addresses the positioning modules with a general cylindrical shape (upper right), and the third shows a more contemporary ergonomic form (upper left).

In phase 2, the "6-3-5" concept sheet is passed to the next person on the right. This person spends ten minutes modifying the initial three concepts or adding an additional concept. As shown in Figure 10.8, each concept on the original sheet is modified. The cylindrical screwdriver is transformed into a new concept with the addition of a rotational joint. This joint allows the screwdriver to be varied in time and space between a cylindrical screwdriver and a configuration that emulates a drill shape. In the second concept sketch, the ergonomic screwdriver is evolved with the addition of a new handle shape, a bit holder, and a

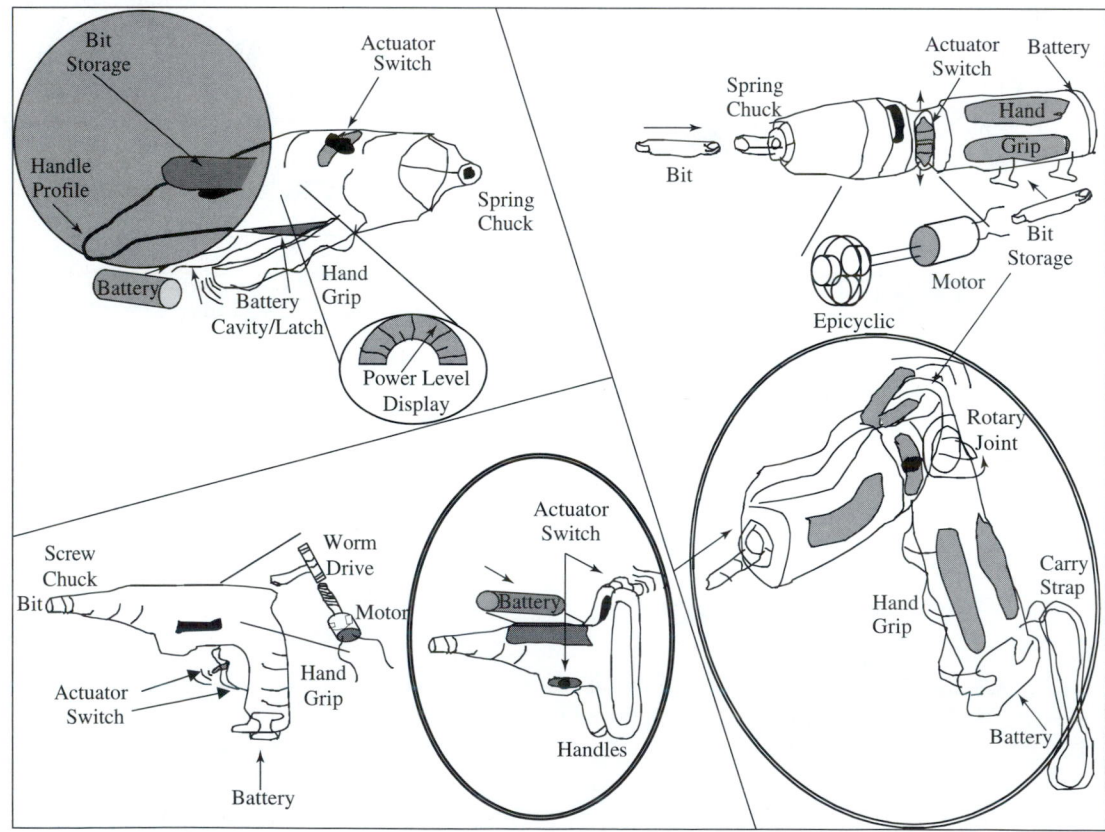

▼ *Figure 10.8.*
Second phase of the "6-3-5" method: application to a power screwdriver product.

display to indicate remaining power level. Likewise, the third concept is modified in the second phase. In this case, the classical drill shape is smoothed to focus on the positioning module architecture and the import hand function.

Summary: "6-3-5" Method

The "6-3-5" method may be summarized by the following process steps:

1. Arrange team members around a table.

2. Each member writes/describes/sketches three ("3") ideas for the primary product functions, usually five or less. The ideas are expressed in clearly distinguished areas (thirds) of the paper, usually on oversized white media such as butcher paper.

3. After T minutes of work on the concepts, members pass their ideas to the person on their right.

4. For the next T minutes, team members modify (without erasing) the ideas on the sheet, with the option of adding an entirely new concept, not contained on their original idea sheet.

5. Passing of the idea sheets continues until a member's original sheet returns, and the round ends. With sufficient time intervals between rounds, five rounds are repeated.

6. After generating ideas for each of the primary product functions, the entire process is repeated to develop alternative layouts and combined concept variants that utilize a summary of the solution principles generated for each function.

7. Postprocess, the ideas are accumulated and summarized.

This process is an effective means to add the sketching dimension to concept generation. Engineers quite often neglect this dimension, since skills are not developed during their formal training. Industrial designers, on the other hand, are quite adept at this dimension and should be used as facilitators of the "6-3-5" method. In any case, this method is intrinsically effective at adding ideas that would typically not be explored.

After a few sessions of the "6-3-5" method, oral brainstorming is typically applied to discuss, refine, and advance the concept variants as a group. The typical brainstorming process is executed at this stage; however, the search is now directed with the concepts recorded by the "6-3-5" method. This directed search for concepts greatly accelerates the refinement of concepts, bringing them closer to ideas that may be evaluated and formalized.

Idea Generators for Intuitive Techniques

Any intuitive method can become stalled by biases in the direction that ideas are being pursued (Adams, 1986; DeBono, 1967; Gordon, 1961). Such methods can also stall due to the working environment or the personality types of the individuals creating the ideas. To overcome any stagnant times during concept generation, idea generators may be applied to redirect the search for ideas. Table 10.3 lists a number of idea

TABLE 10.3. IDEA GENERATORS FOR INTUITIVE CONCEPT GENERATION METHODS

Idea generator	Questions or application
Make analogies	What analogies exist in nature? What analogous products exist in any product domain? How do these products solve the same product functions?
Wish and wonder	What if . . . ?
Sketch/use physical models (e.g., Tinkertoys or LEGOs)	What would an idea look like? How does this model satisfy the function? What can we change?
Eliminate or minimize?	If we remove a feature, how does the device perform? What can we use to replace the feature? What if a feature were smaller? What should I omit? Should I divide it? Split it up? Separate it into different parts? Understate? Streamline? Make miniature? Condense? Compact? Subtract? Delete? Can the rules be eliminated?
Substitute	What can be substituted? Who else? What else? Can the rules be changed? Other ingredient? Other material? Other process or procedure? Other power? Other place? Other approach? What else instead?
Combine	Can we combine purposes? How about an assortment? How about a blend? An alloy? Combine units? What other article or device could be merged with this?
Adapt	What else is like this? What other idea does this suggest? Does the past offer a parallel? What could I copy? Whom could I emulate? What idea could I incorporate? What other process could be adapted?
Modify or magnify	What can be magnified, made larger, or extended? What can be exaggerated? Overstated? What can be added? More time? Stronger? Higher? How about greater frequency? Extra features? What can add extra value? What can be duplicated? How could I carry it to a dramatic extreme? Convert a round action to straight? How can this be altered for the better? What can be modified? Is there a new twist? Change meaning, color, motion, sound, odor, form, or shape? Change name? What changes can be made in the plans? In the process? In the marketing?
Put to other uses (repackage an old idea)	What else can this be used for? Are there new ways to use as is? Other uses if modified? What else could be made from this? Other extensions? Other markets?
Reverse or rearrange	What other arrangements might be better? Interchange components? Other pattern? Other layout? Other sequence? Change the order? Transpose cause and effect? Change pace or schedule? Can I transpose positive and negative? What are the opposites? What are the negatives? Should I turn it around? Up instead of down? Consider it backwards? Reverse roles? Do the unexpected?
What if the product function changes gender?	Feminine and masculine features? What are their opposites?

generators that may be used with any of the intuitive methods. They may also be used by a group or an individual. These idea generators help to break down conceptual blocks by refocusing the group or individual along a different train of thought. To apply the generators, a keyword or activity is stated, followed by a questioning period about the keyword or activity. New ideas are pursued using the same procedure and guidelines for the intuitive method. When ideas become exhausted for a given generator, a new one is stated, and the process repeats.

IV. ADVANCED METHODS: DIRECTED SEARCH

Directed-search or logical concept generation methods are used to develop ideas in a deliberate, step-by-step, comprehensive fashion. Direction is provided by design or physical principles that are previously known. These principles, in conjunction with knowledge of physical effects and technology, drive the process toward particular types of solutions. Let's consider three directed-search approaches: generating ideas from physical principles, using classifying schemes, and implementing the Theory of Inventive Problem Solving (TIPS or TRIZ).

Systematic Search with Physical Principles

For a given functional model of a product, product functions are connected by flows of materials, energies, and signals (information). Each function transforms a flow or set of flows to a new state. This transformation represents a physical process that is governed by physical principles.

Instead of relying on intuitive methods exclusively, we can state possible physical principles that can govern a product function or set of functions. Suppose we have a known physical effect described by a known equation. Then we know the number and type of independent variables. These variables can be studied closely to generate concepts based on variation of each parameter in a methodical way.

Example

Consider the function of "sense rotational speed" for a product application. A possible physical effect that can apply is a capacitive parallel-plate system, governed by the relationship:

$$C = \varepsilon A/d, \text{ where } \Delta C \text{ produces } \Delta e \text{ (voltage)}$$

Figure 10.9 illustrates a systematic search of the variables that can change the capacitance. A change of voltage will result from any of the variable changes. The magnitude of this change represents the detected quantity by the product function.

Three variants are considered. The first varies the distance, d, between the plates. This displacement is inversely proportional to the capacitance, C. By considering this change, we arrive at solutions that are

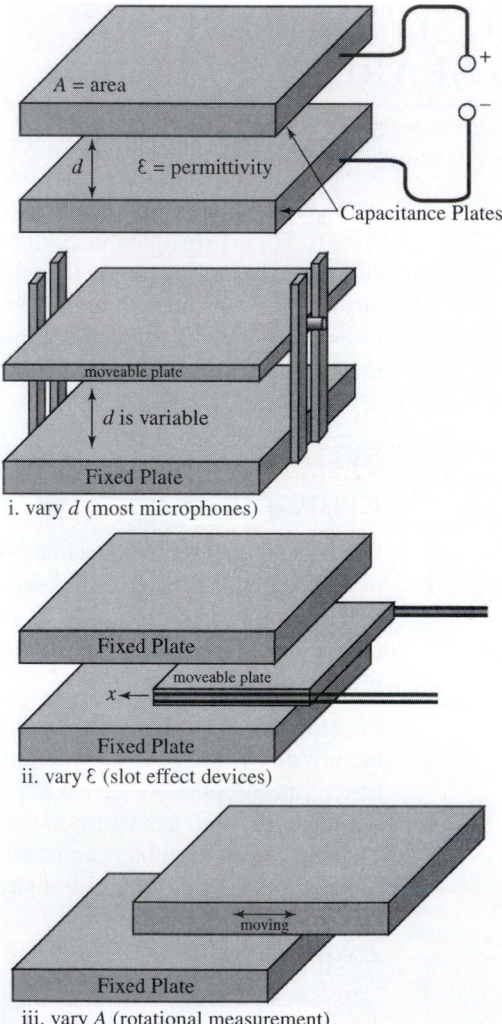

i. vary d (most microphones)

ii. vary ε (slot effect devices)

iii. vary A (rotational measurement)

▼ *Figure 10.9.*
Example application of directed search with physical principles.

equivalent to microphone transducers. The second variable is the material property, ε, between the two plates. If we gradually introduce a solid between the two plates (Figure 10.9), the voltage, e, changes, as detected by a change in capacitance. This change is equivalent to slot effect transducers. The third variable is the incident area, A, of the plates. By moving one plate relative to the other, the area changes, resulting in a voltage change. This change can represent reciprocating or rotational changes of the input on the upper plate (Figure 10.9). We then have an idea for a rotational detection device to satisfy the product function. This device is a capacitive tachometer (when combined with a processor that records and calculates the change as a function of time).

Summary

A procedure for generating concepts from physical principles may be summarized as follows:

1. Model the primary function or subsets of function of a product as a black box, with material, energy, or signal flows.
2. For this black box, determine possible physical principles that can convert the input to the output for the product function(s). This step will require the gathering of information from a variety of sources (Figure 10.3).
3. Write general relationships for the physical principles that relate a measured effect (flow through the product function) to independent (design) variables.
4. Vary each of the design variables to generate a concept for solving the product function(s).
5. Develop a physical realization of the variable changes with sketches. Each of the sketches is a possible concept idea.

This procedure may be applied to any of the function combinations from the product function structure. Intuitive choices are the primary functions of the product, as weighted by the customer needs, and the function chains that form potential modules of the product (Chapter 9).

Systematic Search with Classifying Schemes

Another directed-search method uses classifying schemes to aid in the generation of concepts. *Classifying schemes* are simply categories of high-level physical principles or geometry. They direct the development of concepts by stimulating thoughts in the designers that may not have been considered in a purely intuitive approach. By choosing a category, a product development team can focus on the generation of concepts in a particular technological area or specific line of thought.

Table 10.4 lists example classification schemes for generating concepts. The classifications range from geometry to form properties, and they include subtopics to refine the search. These classifications are chosen to facilitate identification and combination of essential solution characteristics of a product.

To use a classification scheme, the design team focuses on the primary product functions and architecture. A matrix is then formed, listing the product functions as rows and the columns as solutions to these functions according to the categories chosen from Table 10.4. Each

TABLE 10.4. CLASSIFYING SCHEMES FOR GENERATING CONCEPTS (PAHL AND BEITZ, 1996)

Class	Heading/Category	Examples
Working geometry		
	Type	Point, line, surface, body
	Shape	Curve, circle, ellipse, hyperbola, parabola; triangle, square, rectangle, pentagon, hexagon, octagon
		Cylinder, cone, rhombus, cube, sphere
		Symmetrical, asymmetrical
	Position	Axial, radial, tangential, vertical, horizontal; parallel, sequential
	Size	Small, large, narrow, broad, tall, low
	Number	Undivided, divided
		Simple, double, multiple
Working motions		
	Type	Stationary, translational, rotational
	Nature	Uniform, nonuniform, oscillating
		Planar or 3-D
	Direction	x-y-z or about x-y-z
	Magnitude	Velocity, acceleration
	Number	One, several, composite movements
Basic material properties		
	State	Solid, liquid, gaseous
	Behavior	Rigid, elastic, plastic, viscous
	Form	Solid bodies, grains, powder, dust
Basic structural properties		
	Joints	Rigid, rotational, sliding
	Alignment	Horizontal, vertical, angled, truss
	Loading conditions	Tension, compression, bending, torsion, and shear

category is then systematically considered, where solutions are brain-stormed for each category until the ideas are exhausted.

As a simple application of this directed-search method, Figure 10.10 shows the solutions to the product function of "store energy." The classifying scheme is "energy types," and the headings for this scheme are mechanical (motion), hydraulic (fluid), electrical, and thermal. These ideas may be integrated into an alternative design for a product that needs a "store energy" function.

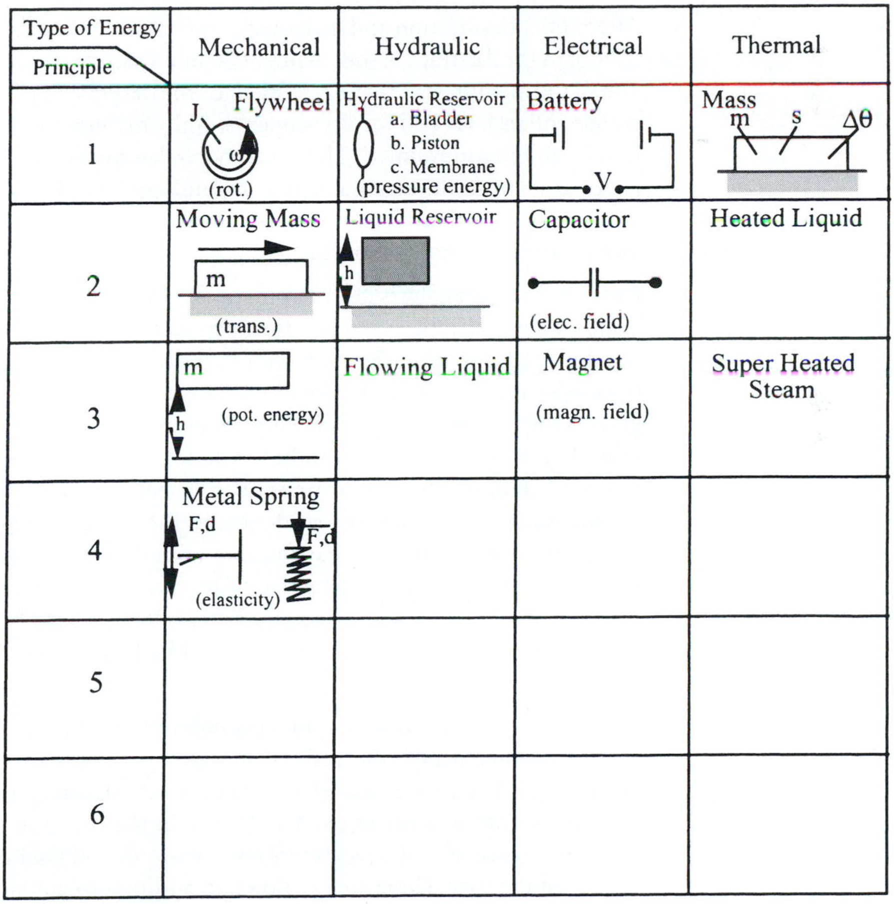

Type of Energy / Principle	Mechanical	Hydraulic	Electrical	Thermal
1	Flywheel J ω (rot.)	Hydraulic Reservoir a. Bladder b. Piston c. Membrane (pressure energy)	Battery V.	Mass m s Δθ
2	Moving Mass m (trans.)	Liquid Reservoir h	Capacitor (elec. field)	Heated Liquid
3	m h (pot. energy)	Flowing Liquid	Magnet (magn. field)	Super Heated Steam
4	Metal Spring F,d F,d (elasticity)			
5				
6				

▼ *Figure 10.10.*
Example of classifying schemes to generate concepts for the function "store energy" (adapted from Pahl and Beitz, 1996).

Device Application of Classifying Schemes

Leisure and recreation are vital for people's physical and mental health. "Evidence indicates that what an individual does during leisure time significantly affects illness, disease, and even longevity" (Wankel, 1994). Primary and secondary students with disabilities need recreational opportunities and skills as part of their educational development so they can carry skills into adulthood and into life in their communities (Auxter et al., 1993). This general need leads to the mission statement "to design a device and/or process that will enhance the recreational activities of students with severe mental and physical disabilities."

After initial clarification and high-level customer interviews at Rosedale School in the Austin, TX Independent School District (Faulkner et al., 1995), the mission statement is refined to "the design of a product that would enhance tee ball for the students, while maintaining their integration in the community." The primary market for the device is educational institutions as an assistive technology. Initial business case analysis demands that the product cost less than $100 as a one-off device produced by the institutions.

Customer needs analysis (10 customers) for this application gives the results shown in Table 10.5. The final product must adapt for changing skill levels for educational development, and it must guide the bat to consistently hit the ball, providing the primary function for integrating the students into community activities. Beyond these "musts," critical needs for a successful product include the ability of a user to adjust the height of the ball for persons in wheelchairs and persons of various sizes, ensure that the student uses the device independently (not hand-over-hand), provide auditory stimulation for cause-effect learning, and reduce the product size for storage and transportation (separation of function in time and space). It must also be "easy to build/duplicate," since the institutions will be fabricating the device themselves.

Through the specification of a process description and task listing of these customer needs, the refined function structure is shown in Figure 10.11. The system boundary for this functional analysis includes only the operational-phase tee ball. Notice in the function structure the convergence of both an external energy source and the human energy from the user. These energy flows, in addition to guidance functions, provide the primary means of satisfying the operational needs of students with disabilities.

TABLE 10.5.   PARTIAL CUSTOMER NEEDS LIST—BATTER UP! APPLICATION

Batter Up Project
Customer Needs

Interviewer(s): Seventh Inning Stretch Team
   10   Number of Customers
    0   5   Weighting Scale
Average Customer:
        Male/Females, age 12–21 yrs. old, students
        Persons with severe mental/physical disabilities

Customer need	WT
• Manufacturing	2
• Easy to build/duplicate	4
• Affordable	3
• Set Up	
• Can be used in several positions	4
• Useful for different grip abilities	1
• Adjustable for different skills	5
• Various social settings	3
• Operation	
• Hits ball consistently	5
• Operates easily	3
• Ball flies differently	3
• User is independent	5
• Resets itself	2
• Stimulates Sensory Modes	
• Visual	1
• Tactile	3
• Auditory	4
• Store/Maintain	
• Compact transport/storage	4
• Simple maintenance	3

A House of Quality for the "Batter Up!" includes measurements of the volume, mass, and number of parts. In addition, the length of bat grips, operating noise, range of bat height, range of handle movement, and range of bat speeds provide a quantified specification of the functions in Figure 10.11(b).

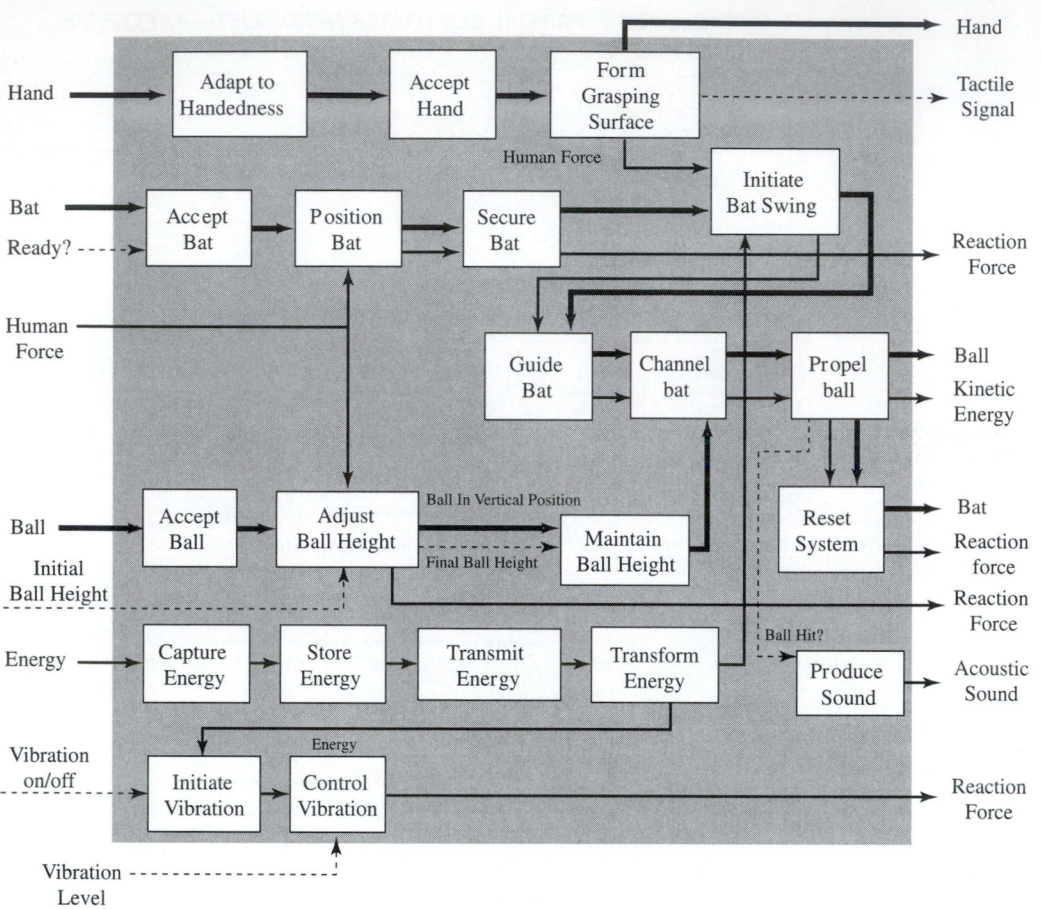

▼ *Figure 10.11.*
Refined function structure—Batter Up! Application.

This specification leads to concept generation of a tee ball assistive technology. Table 10.6 shows a partial analysis matrix of the main functions in the function structure. Because of the critical customer needs of adjustability, hitting the ball consistently, and user independence, the energy flow in the Batter Up! function structure passes through primary functions. Thus, solution principles are listed by directed search, according to

TABLE 10.6.   PARTIAL MORPHOLOGICAL MATRIX—BATTER UP! APPLICATION

Functions	Solution Principles			
	Mechanical	Hydraulic	Electrical	Other
Adapt to Handedness	Gears, Pulleys, Wheels, Pegs, Pins	Water/Air pressure	Solenoid, Servo motor	
Accept Person	Walking, Rolling to product	Riding a water wave, Flying		
Accept Bat	Slot, Groove, Hole, Guide rail, Ridge	Fluid suction		Magnetic attraction
Position Bat	Pegs, Axis movement	Piston cylinder	Solenoid, Servo motor	Magnets
Secure Bat	Clamp, Vice, Belt, Brace, Latch	Fluid suction	Electrical charge (Static cling)	Velcro, Magnets, Adhesion
Initiate Bat Swing	Conveyor, Lift, Mechanical impulse, Catapult	Water jet, Piston cylinder	Electrical impulse	Magnetic propulsion
Guide Bat	Guide rails, Parallel plates, Friction, Loop in bat, String, Hole in bat	Jet Stream		Magnetic restriction/ Attraction
Channel Bat	Human input, Gravity, String	Air flow, Water pressure		Magnets, Explosion
Accept Ball	Mount, Cartridge, Hoop	Tube		
Adjust Ball Height	Jack, Pegs, Pulley system, Lever	Air pressure	Electric current	Magnets, Explosion
Maintain Ball Height	Clamp, Mount	Air pressure, Fluid suction		Magnets, Adhesion
Propel Ball	Loop, Impact	Air suction, Jet stream, Water pressure		Magnetic attraction, Explosion
Accept Energy	Lever, Four bar, Crankshaft, Rope	Tuber, Pipe, Fan, Windmill	Electrical inlets	Metal surfaces, Panels, Fuses
Store Energy	Translational/Rotational Spring, Material Deformation, Rubber band Pendulum, Bat movement	Fluid column, Compressed air (Balloon, Bladder)	Batteries, Capacitor	Magnetic field, Solar panels, Chemical
Transform Energy	Gears, Belt/Sprocket, Lever, Four bar, Cam, Rack and pinion, Universal joint	Piston cylinder	Motor, Generator	
Transmit Energy	Linkages, Bearings	Pipe, Volume deformation	Wires, Volt potential energy	Magnetic field

energy domain. A wide variety of mechanical and electromechanical solutions are shown for the application, especially for the external energy source. These solution principles lead to a number of concept variants.

Figure 10.12 illustrates the selected concept. This concept, based on a pneumatic energy source and a guide wire, was produced from basic hardware supplies to satisfy the fabrication requirements of wood shops in secondary-school systems. The pneumatic air is supplied by a portable air tank, where the air flows through the bat and exits near the end during an assisted swing. A support person provides timing of the air shot to power the swing. After prototype testing, this successful concept was installed into a local school district as a regular recreational activity.

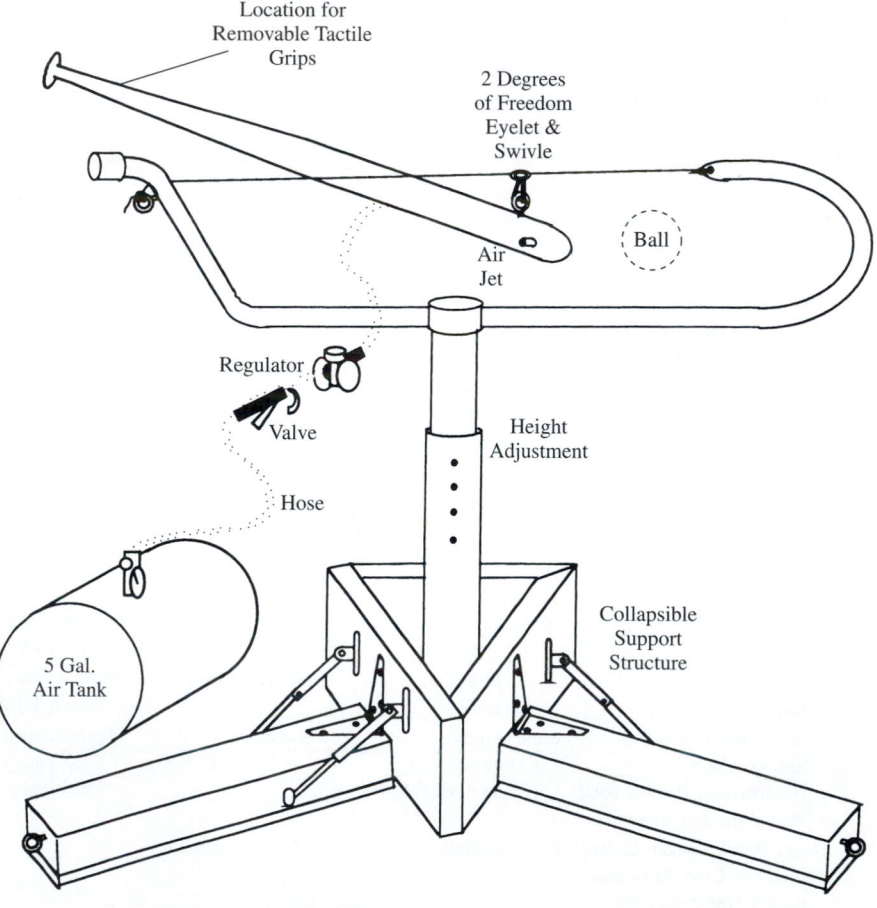

▼ *Figure 10.12.*
Selected design concept for beta functional prototype—Batter Up! application.

Summary

A procedure for executing directed search with classifying schemes may be listed as follows:

1. Model the primary functions or subsets of functions of a product as black boxes, with material, energy, or signal flows.

2. For these black boxes, choose the classification schemes that closely relate to the functions and customer needs (Table 10.4).

3. For one of the classification headings, generate solutions to the function(s).

4. Document the results in a matrix, where the rows are functions and the columns are solutions, organized by classification headings.

5. After ideas are exhausted for a given heading, repeat the process for the next heading.

As stated for direct search with physical principles, this procedure may be applied to any of the function combinations from the product functional model. Intuitive choices are the primary functions of the product as weighted by the customer needs and the function chains that form potential modules of the product (Chapter 9).

Theory of Inventive Problem Solving

The *Theory of Inventive Problem Solving* (TIPS or TRIZ) was developed by Genrikh S. Altshuller in the former U.S.S.R., beginning in the late 1940s (Sushkov, Mars, and Wognum 1995; Altshuller 1984; Domb and Slocum 1998). The basis of this theory is the discovery that patterns exist in patent claims, many of them based on the same working principles. Building on this discovery, Altshuller collaborated with an informal collection of academic and industrial colleagues to study patents and search for the patterns that exist (Sushkov, Mars, and Wognum 1995). Hundreds of person-years were devoted to this effort, and thousands to millions of patents have been studied, resulting in the insight that patents may be classified into five categories. The first two categories were designated as "routine design," meaning that they do not exhibit significant innovations beyond the current technology. These categories are "basic parametric advancement" and "change or rearrangement in a configuration." The last three categories, on the other hand, represent designs that included inventive solutions. These three categories are "identifying conflicts and solving them with known physical principles," "identifying new principles," and "identifying new product functions and solving them with known or new principles."

Based on these categories and patent studies, Altshuller observed a number of trends in historical invention. Some of the key observations, in the context of product design, include the following:

▶ Evolution of engineering systems (products) develops according to the same patterns, independent of the engineering discipline or product domain. These patterns may be used to predict the trends of future evolutions in a product domain. They may also be used to direct the search for new concepts.

▶ Conflicts (or contradictions) are the key drivers for product invention. Principles for eliminating conflicts are universal across product domains. Application of these principles implies that compromise is unacceptable.

▶ The systematic application of physical effects aids invention, since a particular product team does not know all physical knowledge.

These observations lead to the structure of TIPS for solving inventive problems. A number of components comprise this structure. For the purpose of this text, we consider three primary components: (1) laws of engineering system (product) evolution, (2) physical effects, and (3) solution (design) principles. *The laws of product evolution* (nine in number) indicate universal trends of a product's advancement over time. *Physical effects*, on the other hand, document the knowledge of the physical world from many diverse fields. *Design principles*, in turn, are heuristic rules for eliminating conflicts in a design task, creating a high-level concept that is a possible inventive solution.

Using these three TIPS components, a straightforward process may be developed for generating concepts. The process begins with a functional model (Chapter 5). From the functional model (in addition to benchmarks, engineering specifications, product architecture, and other data), conflicts are identified in the design task. These conflicts are then stated as contradictions in *generalized parameters* or *engineering parameters*, where a generalized parameter is a controllable variable or set of variables that embody a physical effect in a product. Design principles are then applied to suggest ways in which the conflict may be resolved. The resulting concepts are refined with known physical effects and analogies to existing solutions. The final step is to refine the concepts, from the principles and effects, into a concrete geometry.

Table 10.8, Table 10.9 and the relationship matrix in Appendix C provide necessary data to execute this process. Table 10.7 lists the 39 generalized parameters for describing product performance metrics. Tables 10.8 and 10.9 list the TIPS design principles with corresponding

TABLE 10.7.   GENERALIZED ENGINEERING PARAMETERS FOR DESCRIBING PRODUCT METRICS

1	Weight of moving object	21	Power
2	Weight of stationary object	22	Energy loss
3	Length of moving object	23	Substance loss
4	Length of stationary object	24	Information loss
5	Area of moving object	25	Waste of time
6	Area of stationary object	26	Quantity of a substance
7	Volume of moving object	27	Reliability
8	Volume of stationary object	28	Accuracy of measurement
9	Velocity	29	Manufacturing precision
10	Force	30	Harmful actions affecting the design object
11	Stress or pressure	31	Harmful actions generated by the design object
12	Shape	32	Manufacturability
13	Stability of object's composition	33	User friendliness
14	Strength	34	Repairability
15	Duration of action generalized by moving object	35	Flexibility
16	Duration of action generalized by stationary object	36	Complexity of design object
17	Temperature	37	Difficulty to control or measure
18	Brightness	38	Level of automation
19	Energy consumed by moving object	39	Productivity
20	Energy consumed by stationary object		

definitions. The generalized parameters and design principles are derived from the large quantity of patents studied as part of TIPS. Appendix C shows a correlation matrix of generalized parameters to design principles. The rows of the matrix represent "What should be improved" versus the columns that represent "What deteriorates." Up to four principles are listed in each cell of the matrix according to the order of applicability. Table 10.10 presents examples of products that utilize design principles to solve engineering conflicts. These example products provide analogies for applying the principles during product development. Finally, Table 10.11 lists a subset of physical effects (Altshuller 1984). These effects may be applied to product conflicts in order to resolve conflicts and seek inventive solutions.

TIPS Example I

Let's consider the evolution of an iron product for smoothing the wrinkles from clothing. An important function of an iron is to transfer force to the clothing to aid in removing wrinkles. It is equally important to import the human hand and reduce the force on the user (comfortable use). The conflict is straightforward; we desire a heavy iron to remove wrinkles, but we do not want a heavy iron due to the impact on ergonomics.

TABLE 10.8. TIPS' DESIGN PRINCIPLES (1–20) TO SOLVE ENGINEERING CONFLICTS

1	Principle of segmentation.	Divide the object into independent parts that are easy to disassemble, increase the degree of segmentation as much as possible.
2	Principle of removal.	Remove either the disturbing part or the necessary part from the object.
3	Principle of local quality.	Change the object's or environment's structure from homogeneous to non-homogeneous. Let different parts of the object carry different functions.
4	Principle of asymmetry.	Make object asymmetrical, or increase asymmetry.
5	Principle of joining.	Merge homogeneous objects or those intended for contiguous operations.
6	Principle of universality.	Let one object perform several different functions. Remove redundant objects.
7	The nesting principle.	Place one object inside another, which in turn is placed in a third, etc., or, let an object pass through a cavity into another.
8	Principle of counterweight.	Attach an object with lifting power or use the interactions with the environment, e.g., aerodynamic lift.
9	Principle of preliminary counteraction.	Perform a counter-action to the desired action before the desired action is performed.
10	Principle of preliminary action.	Perform the required action before it is needed, or set up the objects such that they can perform their action immediately when required.
11	Principle of introducing protection in advance.	Compensate for the low reliability of an object by introducing protections against accidents before the action is performed.
12	Principle of equipotentiality.	Change the conditions such that the object does not need to be moved up or down in the potential field.
13	Principle of opposite solution.	Implement the opposite action of what is specified. Make a moving part fixed and the fixed part mobile. Turn the object upside down.
14	Principle of spheroidality.	Switch from linear to curvilinear paths, from flat to spherical surfaces, etc. Make use of rollers, ball bearings, spirals. Switch from direct to rotating motion. Use centrifugal force.
15	Principle of dynamism.	Make the object or environment able to change to become optimal at any stage of work. Make the object consist of parts that can move relative to each other. If the object is fixed, make it movable.
16	Principle of partial or excessive action.	If 100% is unobtainable, try for slightly less or slightly more.
17	Principle of moving into a new dimension.	Increase the object's degree of freedom. Use a multi-layered assembly instead of a single layer. Incline the object or turn it on its side. Use the other side of an area.
18	Use of mechanical vibrations.	Make the object vibrate. Increase the frequency of vibration. Use resonance, piezovibrations, ultrasonic, or electromagnetic vibrations.
19	Principle of periodic action.	Use periodic or pulsed actions, change periodicity. Use pauses between impulses to change the effect.
20	Principle of uninterrupted useful effect.	Keep all parts of the object constantly operating at full power. Remove test or set-up runs.

TABLE 10.9. TIPS' DESIGN PRINCIPLES (21–40) TO SOLVE ENGINEERING CONFLICTS

21	Principle of rushing through.	Carry out a process or individual stages of a process at high speed.
22	Principle of turning harm into good.	Use harmful factor to obtain a positive effect. Remove a harmful factor by combining it with other harmful factors. Strengthen a harmful factor to the extent where it ceases to be harmful.
23	The feedback principle.	Introduce feedback. If there already is feedback, change it.
24	The go between principle.	Use an intermediary object to transfer or transmit the action. Merge the object temporarily with another object that can be easily taken away.
25	The self service principle.	The object should service and repair itself. Use waste products from the object to produce the desired actions.
26	The copying principle.	Instead of unavailable, complicated or fragile objects, use a simplified cheap copy. Replace an object by its optical copy, make use of scale effects. If visible copies are used, switch to infra-red or ultra-violet copies.
27	Cheap short life instead of expensive longevity.	Replace an expensive object that has long life with many cheap objects having shorter life.
28	Replacement of a mechanical pattern.	Replace a mechanical pattern by an optical, acoustical or odor pattern. Use electrical, magnetic or electromagnetic fields to interact with the object. Switch from fixed to movable fields changing over time. Go from unstructured to structured fields.
29	Use of pneumatic or hydraulic solutions.	Use gaseous or liquid parts of an object instead of solid parts.
30	Using flexible membranes and fine membranes.	
31	Using porous materials.	Make the object porous or use porous elements, e.g., inserts, covers, etc. If the object is already porous, fill the pores in advance with some useful substance.
32	The principle of using color.	Change the color or translucency of an object or its surroundings. Use colored additives to observe certain objects or processes. If such additives are already used, employ luminescence traces.
33	The principle of homogeneity.	Interacting objects should be made of the same material, or material with identical properties.
34	The principle of discarding and regenerating parts.	Once a part has fulfilled its purpose and is no longer necessary, it should automatically be discarded or disappear, e.g., evaporate, or change its shape. Parts that become useful after a while should be automatically generated.
35	Changing the aggregate state of an object.	Change state, e.g., solid to liquid. Use pseudostates and intermediary states, e.g., elastic solid bodies.
36	The use of phase changes.	Use phenomena occurring in phase changes, e.g., use of volume changes, heat dissipation, etc.
37	Application of thermal expansion.	Use expansion or contraction of materials by heat. Use materials with different thermal expansion coefficients.
38	Using strong oxidation agents.	Replace air with enriched air or replace enriched air with oxygen. Treat the air or oxygen with ionizing radiation. Use ionized oxygen. Use ozone.
39	Using an inert atmosphere.	Replace the normal environment with an inert one or a vacuum.
40	Using composite materials.	Switch from homogeneous materials to composites.

TABLE 10.10. DESIGN PRINCIPLES APPLIED WITHIN PRODUCT EXAMPLES/ANALOGIES

Design principles	Examples
1	Papasan Chair; Sectional Garden Hose; Computer Components; Steering Column; Food Processor; Personal Computer
2	Journal Bearing; Mounted Bicycle Pump; Air Cushion Soccer Game; Hover Craft
3	Boeing Fuselage Skin; Bimetallic Skin; Composite Mongol Bow; Stapler
4	Bumble Ball: Eccentric weight on motor creates vibration; Water Buoy: Weight at one end creates orientation; Oval Race Car: Weight shifted to left side of car to aid turning
5	TV/VCR; Cassette Tape Heads; IC Chip
6	Fountain Pen Body; Door Knob; Fingernail Clipper
7	Antenna; Bike Seat Lock; Sleeping Bag Stuff Sack; Boy Scout Glass
8	Hot Air Balloon; Hydro foil; Life Preserver
9	Door Closer; Black-and-White Film
10	Color Coding of Parts; PVC Primer
11	Fuse; Electric Breaker; Shaft Couplers; Slip Clutch
12	Jiffy Lube Pit; Loading Dock; Airport Gate
13	Mill; Lathe; Rock Polisher; Mouse Ball
14	Computer Mouse; Door Jam; Soda Can Lids; Screw Lift
15	Camera Lense; Bicycle Drivetrain and Derailer
16	Rain Parka; Snowboards
17	Book: Open–pages exposed; closed–stored vertically; Computer Mouse: 2-D screen to horizontal mouse pad; Composite Wing: Loads in only one direction per layer
18	Quartz Clock; Reed Pipe; Building Natural Frequency Adjustment
19	Stepper Motor; Hammer Drill
20	Steam Turbine; Mechanical Watch
21	High Speed X-Ray Film; Inkjet Printer Ink; Metal Alloy Quenching
22	Crumble Points on an Automobile; Heat Lamp; Medical Defibrillator
23	Air Conditioning/Thermostat
24	Gear Trains; Bock and Tackle
25	Knife Sharpening Storage Devices
26	Rapid Prototyping; Sand Casting; Crash Test Dummy
27	Paper; Ballpoint Pen; Cardboard Box
28	CD; Microwave; Crane with Electromagnetic Plate
29	Air Shock; Power Steering
30	Astronaut Crew Escape Bubble; High Altitude Balloon; Dome Tent
31	Ivory Soap (floats instead of sinks); Running Shoe Soles; Air Filters
32	Clear Bandage; Roadway Signs; Prescription Sunglasses
33	Shaft and Bushing
34	Multistage Rockets
35	Pipe Freezing Sleeve; Light Stick; Heat Pack
36	Fire Extinguisher; Fuse with Filament
37	Thermometer; Bimetal
38	Metal Forming Ovens; Torch Cutting
39	Heliarc Welding; Aluminum Soda Can; Light Bulb; Goodyear Blimp (vs. Hindenberg)
40	Steel Belted Tires; High Performance Aircraft Wings

TABLE 10.11.   A SUBSET OF TIPS' PHYSICAL EFFECTS FOR CERTAIN SYSTEM (PRODUCT) FUNCTIONS

Product function (required property)	Physical effects (solution principles)
Temperature:	
Lower Temperature	Phase transitions. Jowlie-Tomson effect. Rank effect. Thermoelectric.
Measure Temperature	Heat distribution and change in natural frequency of vibrations; changes in optical, electrical, and magnetic properties. Curie point. Hopkins and Barkhausen effects.
Raise Temperature	Electromagnetic induction, vortical currents, surface effects, dielectric heating, electronic heating, absorption of radiation.
Stabilize Temperature	Phase transitions (including moving through a Curie point).
Objects:	
Change the Dimensions of Objects	Heat distribution, deformation, piezoelectrics, magnetic-electrostriction.
Control Location of Objects	Magnetic, ferromagnetic link, electrical field + charged object, mechanical oscillations, centrifugal forces, heat distribution, pressure.
Control Movement	Capillary action, Osmosis, Toms effect, Bernoulli effect, waves.
Destruct (Destroy) Object	Electrical discharge, resonance, ultrasonics, cavitation, radiation.
Indicate Position/Location of Objects	Marker substances, luminescent traces, reflection of light, Doppler.
Measure Dimensions of Object	Natural frequency of oscillation, apply/read magnetic/electrical markers.
Setup Interaction Mobile/Fixed Objects	Electromagnetic fields.
Stabilize Position of Object	Elec. & magnetic fields, liquids that harden in fields, hydroscopic effect.
Surfaces, Volume, & Structures:	
Check State & Properties of Surfaces	Electrical discharge, reflection of light, electronic emissions, Moire effect, radiation.
Measure Surface Properties	Friction, absorption, diffusion, Bauschinger effect, electrical discharge.
Inspect State & Properties of Volume	Marker substances, change electrical resistance, polarized light, etc.
Change Volume Properties of an Object	Change viscosity by fields, heat action, phase transition, ionization.
Create & stabilize Structure of object	Interference waves, standing waves, Moire effect, magnetic waves, phase transitions, mechanical/acoustical oscillations, cavitation.
Gases & Mixtures:	
Control Aerosol Flows (dust/fog/smoke)	Electrisation, electrical & magnetic fields, light pressure.
Form Mixtures	Ultrasonics, cavitation, diffusion, elec. fields, magnetic fields, and ferromagnetic substance, electrophoresis, solubilization.
Separate Mixtures	Electric and magnetic, change viscosity, centrifugal forces, diffusion.
Forces/Energy:	
Create & Control Forces/High Pressure	Magnetic field + ferromagnetic substance, phase change, centrifugal forces, heat distribution, change hydrostatic forces, conducting liquids.
Change Friction	Johnson-Rabeck effect, radiation, Kragelsky phenomenon, oscillation.
Accumulate Mechanical & Heat Energy	Elastic deformation, hydroscopic effect, phase transitions.
Transfer Energy	Deformations, oscillations, Alexandrov effect, wave movement (& shock waves), radiation, conductivity, convection, induced radiation.
Fields, Light, & Chemicals:	
Indicate Electrical & Magnetic Fields	Osmosis, discharges, Piezo & magneto effects, Hall effect, nuclear magnetic resonance, electronic emissions, gyromagnetic phenomenon.
Indicate/Detect Radiation	Optical acoustic effect, heat distribution, photo effect, luminescence.
Generate Electromagnetic Radiation	Josephson effect, induced radiation, Tunnel effect, Hann effect.
Control Electromagnetic Fields	Screening, increase/decrease electric conductivity, change surface form.
Control/Modulate Light	Refraction/reflection of light, photoelasticity, Kerr/Faraday effects.
Initiate/Intensify Chemical Changes	Ultrasonics, cavitation, ultraviolet, X-ray, shock waves, catalysis.

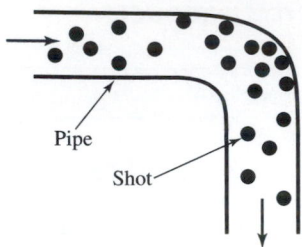

▼ Figure 10.13.
Conflict for a pipe transport system.

Stated in generalized parameters (Table 10.7), the conflict is with regard to the force (#10) versus weight of moving object (#2). Referring to Appendix C, the correlation of force to user friendliness shows that the TIPS principles of "8, 1, 37, 18" apply directly to the problem. Reviewing each of these design principles in Tables 10.8 and 10.9, it is suggested that a counterweight be added, the design be divided into independent parts (mass of iron versus user interface), thermal expansion be added, or mechanical vibration be added to the concept. These suggestions may lead to a levered counterweight in the first case, a foot-operated sandwich iron in the second case, and water spray in the third case.

For the last design principle, mechanical vibration may be added with an eccentric weight that would increase the force into the clothing, while reducing the carrying weight of the iron. This solution creates a conflict, however, since the user, during the operation of the iron, will also feel the vibration forces over the clothing. Adding a vibration absorber between the hand and the vibration source in the clothing may solve this conflict. Alternatively, a different vibration source can be applied that may vibrate the clothing near its resonance frequency using low input amplitudes. The feasibility of this later concept would need to be investigated.

TIPS Example II

As another TIPS example, consider the design of a piping system to transport metal shot (Clausing 1997). Figure 10.13 shows the configuration under consideration. The conflict for this design task arises due to the shot wearing down the turn in the pipe. This conflict states that a coating is desired to avoid wear, whereas a coating is not desired due to the increased expense and short life of the coating. To resolve this conflict, a search through the design principles shows a number of possible inventive resolutions. For example, using the principle of universality (#6) and the replacement of mechanical patterns principles, a magnetic field may be added to the system so that the one of the objects (the shot) performs an additional function (preventing wear). Figure 10.14 illustrates this solution, where, as shot elements are forced off the wall of the pipe by collisions, other shot fill their holes. The system is thus continuously replenishing and does not compromise the needs of the design.

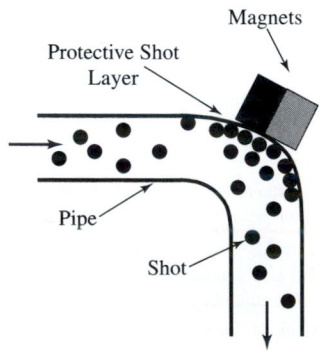

▼ Figure 10.14.
Generated solution to resolve conflict.

Product Application

As another example of TIPS in this section, let's consider the redesign of a fingernail clipper. A generated concept to clip nails is shown in Figure 10.15. As stated in Chapter 3, an intrinsic conflict with fingernail clippers is the need to import the hand (large dimensions and surface area for comfort and nail access) versus the need for compact storage.

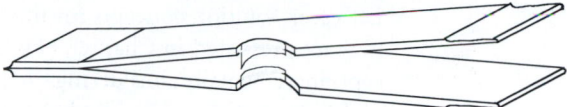

▼ *Figure 10.15.*
Fingernail clipper concept, with the need for compact storage.

To solve this conflict, the principle of dynamism may be applied (changing the nail clipper in space and time). We want the fingernail clipper to be compact during storage and larger during application. An analogous product is a pocket cell phone. It folds to be compact in the product and lengthens to connect the ear and mouth during phone conversations.

Figure 10.16 shows two solutions to the fingernail clipper product, adding a feature (a clip or a quick connect) to lock the height down during storage. This lock can then be released during operation to allow access to the fingernail and blades.

Product Application II

Consider the design of a cat litter box product. The purpose of the product is to house cat litter to permit a cat to produce waste in an indoor environment. Critical needs associated with this product include easy cleaning, adequate odor control, cost, and so forth (see Chapter 11).

An inherent conflict exists in traditional litter box products. While we wish litter to surround and encapsulate cat waste, we also wish to separate the waste easily from the litter. The longer we leave cat waste in the product, the greater the chance to produce smell, germs, and a dissatisfied cat. Thus, a periodical filtration system is needed to separate waste from the litter without discarding the entire litter reservoir. The most prevalent solution currently available is a slotted scoop.

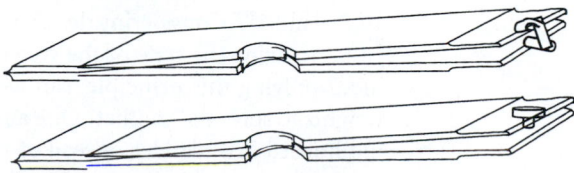

▼ *Figure 10.16.*
Compacted nail clipper product.

Let's seek solution concepts for the filtration problem (further examples are presented in Chapter 11). In this case, we can generate concepts incrementally considering the TIPS design principles (Tables 10.8 and 10.9) and analogies (Table 10.10). For instance, consider the seventh and thirteenth TIPS principles. The ideas of nesting and opposite solution may be applied to a current configuration of a litter box; for example, a hooded system. Can an object in the current system pass through to another part of the system? Can system components have redundant functions? Can the system be turned upside down?

Answers to these questions generate the following insights. The litter may pass through a screen filter, separating out the solid waste. The hood and bottom of the litter box are interchangeable; they can perform the same function, depending on the box's orientation. Since the hood and bottom component can perform interchangeable functions, the litter box may be turned upside down to transfer litter from one container to the other.

Thus, combining these insights, a solution to the litter box problem is to place a filtration screen in the litter box. Turn the box upside down to transfer the litter and separate the solid waste and then remove the screen to dispose of the waste while retaining the litter reservoir. Figure 10.17 shows an example realization of this solution (Lawton 1997).

Product Application III

Consider the general design problem of creating exercise aids for persons that travel extensively as part of their business. Interviews of potential customers show that the need exists to provide exercise weights, such as ankle weights, wrist weights, and small bar bells.

An obvious conflict exists in this problem. Significant weight is needed to aid in exercising; however, no additional weight is desired when traveling with luggage. Using the generalized parameters of Table 10.7, we may consider the first or second parameter (stationary or moving weight) versus the tenth (force). This conflict considers improving the weight of the luggage as it contradicts with the deteriorating effect (carrying force needed) of the exercise weights.

Using Appendix C, the most applicable design principles are "8, 10, 18, 19, 35, and 37." Considering design principle 35 from Table 10.9, it suggests changing the state of the system; for example, from solid to liquid. Applying this principle, can we change the state of the exercise weights to solve the conflict? One answer to this question is to use liquid housed in a bladder instead of metal or solid weights that would need to be carried in luggage. The liquid, such as water, could then be obtained at the travel destination. All that would need to be carried are empty, sealable bladders and lightweight bars to mount the bladders.

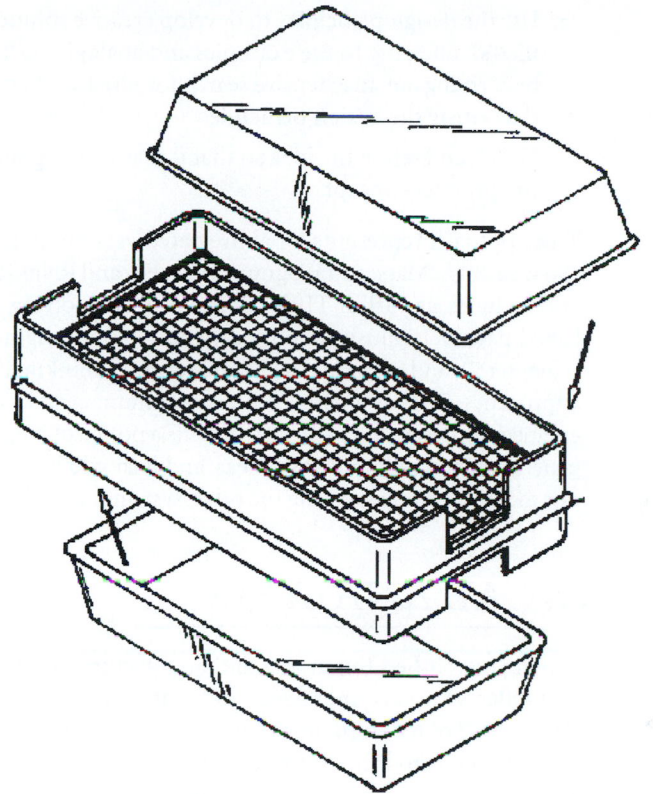

▼ **Figure 10.17.**
Inventive solution to the waste filtration function of a cat litter box product (Lawton, 1997).

This idea appears to be quite innovative and would need to be sketched and embodied.

Summary of TIPS

The central TIPS approach for inventive product design is given by:

1. Determine the conflict(s) in the design problem.

2. Formulate as conflicts in generalized engineering parameters (Table 10.7).

3. Determine the intersections in Appendix C for the numbers of the engineering parameters.

4. Read the principles that apply to help solve the problem (or simply try all 40 design principles), Tables 10.8 and 10.9.

5. Use the design principles to develop creative solutions to the conflict(s), referring to the examples and analogies in Table 10.10 and by carrying out an extensive search for physical effects (Table 10.11) that satisfy the design principle.

6. Sketch and refine the idea so that it may be integrated into the entire product concept.

This approach represents an abbreviated version of TIPS (Altshuller 1984; Sushkov, Mars, and Wognum 1994; Fey and Rivin 1996). Altshuller also includes, as part of TIPS' general problem-solving techniques, selected patents to illustrate physical effects, value engineering analysis (Chapter 7), and methods to enhance creative thinking. Through these approaches, TIPS as a whole addresses a number of basic problems in engineering design. As such, it represents a powerful method for concept generation and refinement of ideas, and it should be wielded as part of our toolbox, complementing the other systematic methods of this book.

V. MORPHOLOGICAL ANALYSIS

Having established functional and architectural models of a product that reflects the customer needs and is at a level of granularity that reflects a level of redesign or evolution, we can apply the above concept-generation methods to explore possible solutions. Any or all of the basic and advanced methods may be applied. The results will be alternative architectural layouts and solutions to the product functions. These results should be assembled so that overall alternatives may be generated in addition to further solutions per function. A systematic method for this assembly and continued idea creation is known as *morphological analysis* or *morphological charting*.

Morphological analysis is a tool that provides a structured search and combination of concepts in product design. The process of executing this analysis is as follows:

1. Consider each product function in the functional model and each module of the product architecture.

2. List the function or module as rows of a matrix.

3. In the first column of the matrix, enter the current solution to the function or module, if the product exists.

4. Apply concept generation methods (as above) and record the concepts in the columns of the matrix for each function.

5. Map the range of solutions per each function to a classification scheme, such as energy domains. Judge if the solutions are too

focused or cover a good breadth. If the solutions are too focused, carry out further sessions of intuitive and directed concept generation.

6. When a good breadth of ideas and technologies are realized in the morphological matrix, combine the ideas into diverse concept variants that seek to satisfy the entire product specification.

Develop Concepts for Each Product Function

We start with the function structure and product architecture of the existing (or original) product. We create a *morphological matrix* and label each row with a product subfunction. Next, we immediately fill in the first column of the matrix with current solutions (if they exist). For each product function (each row), we consider the existing product and identify the component that satisfies the function.

After forming the initial matrix, we may black out functions that are not part of the redesign effort. This action is usually carried out for three reasons. (1) The product function's solution components cannot be changed, usually for regulatory or contractual reasons. (2) The product function's solution components are not part of the redesign effort, and it is not a real concern to the customer or too expensive to invest resources, and so forth. (3) Support functions might not be considered at this stage in the development process. Instead, the focus may be on the primary function drivers for the product. In general, all product functions or only a subset may be considered at this stage, whether or not the product exists. When deciding to disregard certain functions, the development team must clearly understand the risks and tradeoffs.

Example

As an example, consider the fingernail clipper product. A function structure for the product is shown in Figure 5.12, in addition to the listed modules. From this functional model, a morphological matrix is created and initialized with the solution components of the original clipper. The result is shown in Table 10.12.

After creating the initial matrix, ideas are added by applying the various concept generation methods. Table 10.13 shows a range of fingernail clipper solutions per product function. The breadth of this matrix is judged as adequate, since a variety of technologies, geometries, and conflict resolutions exist in the matrix. We now need to seek combined solutions.

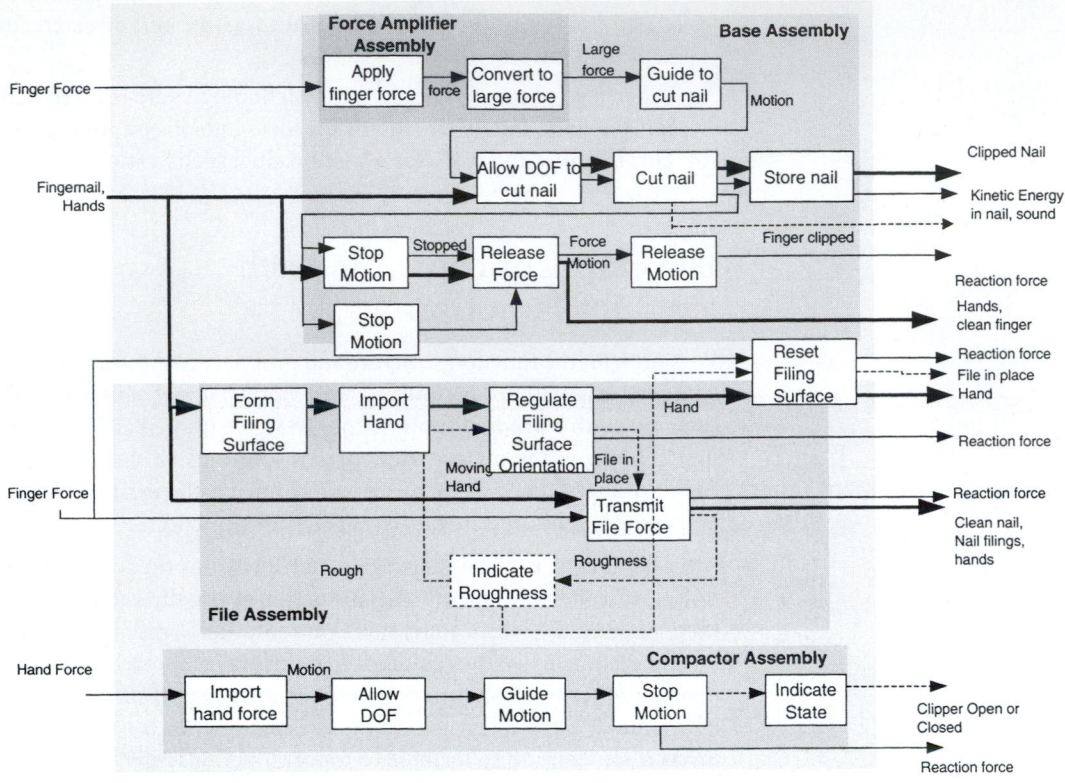

▼ *Figure 10.18.*
Refined function structure for the fingernail clipper product.

VI. COMBINING SOLUTION PRINCIPLES (CONCEPT VARIANTS)

Now that we have a range of possible solutions to each product function, we need to combine solutions (synthesis). A large number of combinations are possible; however, issues exist in geometrical and physical compatibility and function sharing. Our goal, then, is to be compre-

TABLE 10.12.   INITIAL MORPHOLOGICAL MATRIX FOR THE FINGERNAIL CLIPPER PRODUCT

Assemblies Names & Sub-functions	Current			
Opening				
Open Clipper	Spin-and-flip			
Sense Rough or Long	Human			
Clip Nail				
Actuate Apply or release	Human			
Apply finger force	Shaped/ Contoured Top			
Convert to large force	Fulcrum and Lever			
Guide to cut nail	Elastic arm			
Cut nail	Snap blades			
Stop motion	Blades hit			
Release motion	Leaf spring			
Store nail	none			
File nail				
Regulate filing surface	Pivot out file			
Transmit file force	Human			
Indicate roughness	Vibration/ Acoustic Sense			
Form motion surface	Serration			
Reset filing surface	Pivot back file			
Closing				
Close system	Flip and spin			

hensive in creating concept variants while discarding combined solutions that have intrinsic incompatibilities.

It is much easier to be comprehensive than in the case of seeking solution alternatives for product functions, because we know exactly how many solution principles there are for each function. If the first product function has m_1 possibilities, and the second function has m_2 possibilities, and so on, then the total number of possible overall solutions is:

$$m_1 * m_2 * \ldots * m_n$$

We can directly show the possible combinations graphically in the morphological matrix. As discussed in the previous section, the general format of the matrix relates product functions (rows) to solutions (columns) (Table 10.14).

When forming the morphological matrix, a number of guidelines should be followed. First, the product functions should be listed by order of appearance in function structure and product architecture.

TABLE 10.13. FINGERNAIL CLIPPER MORPHOLOGICAL MATRIX WITH ADDITIONAL SOLUTIONS

Assemblies Names & Sub-functions	Current			
Opening				
Open Clipper	Spin-and-flip			
Sense Rough or Long	Human			
Clip Nail				
Actuate Apply or release	Human			
Apply finger force	Shaped/ Contoured Top	Plastic Snap-On	Rubber coating	Smooth forging
Convert to large force	Fulcrum and Lever	Scissors lever Nutcracker lever	Wedge	Linkage
Guide to cut nail	Elastic arm	Linear slide rack-n-pinion	Rotating cycls.	Pin
Cut nail	Snap blades	Shear blades	Guillotine	Knife & block
Stop motion	Blades hit			
Release motion	Leaf spring	Tension spring	Volute spring	Compression
Store nail	None	Snap-On housing	Frame cavity	Ergonomic body
File nail				
Regulate filing surface	Pivot out file	Fixed on arm	Side pocket	Fixed on body
Trasmit file force	Human			
Indicate roughness	Vibration/ Acoustic			
Form motion surface	Serration	sandpaper	Cross hatched	Surfaced rough
Reset filing surface	Pivot back file	same		
Closing				
Close system	Flip and spin			

We should then arrange solution principles so that the columns create logical grouping, for example, for energy type. We should also use brief sketches rather than words only. Sketches provide much more information than do high-level linguistics. Next, we should list the most important characteristics of the solution principles. Why do they solve the functional need? Finally, we should combine compatible product functions. Consideration should only be given to solutions that meet the customer needs and engineering specifications (at least estimated to meet the needs). Through this focus, we can concentrate on the most promising and diverse combinations and establish why these combinations should be preferred.

These suggestions form the global process of creating concept variants. While it is not feasible to consider all combinations of the solution

TABLE 10.14.   GENERAL MORPHOLOGICAL MATRIX FORMAT

Solution / Subfunction	1	2	3	...		n
1						
2						
3						
4						
...						
m						

principles, a large number of divergent alternative designs should be created. This creation process is nonlinear and requires iteration and continued refinement. Concept variants should be sketched as solution principles are "added" together. The product architecture should also be maintained, clearly illustrating the boundaries and interfaces of product modules. As the sketches develop, alternative arrangements and layout should be explored, intuitively judging the direction of preferred choices. Subsequent methods of the product development process will consider more formal methods for concept selection and decisions.

Digression/Caution: Function Sharing

Before considering product applications in combining solution principles, let's digress briefly to study an important issue in functional decomposition methodologies; undue complexity, modularity, and cost. Function sharing must be a conscious act to avoid these problems. Recall the principle of function sharing (the TIPS universality principle, Table 10.8). Let's consider an example that points out the need for function sharing.

Fingernail Clipper

Figure 5.12 shows the function structure for the fingernail clipper product. Two alternative designs for the clipper are shown in Figure 10.19, a concept variant with no function sharing and an actual product.

Example Concept Variant:

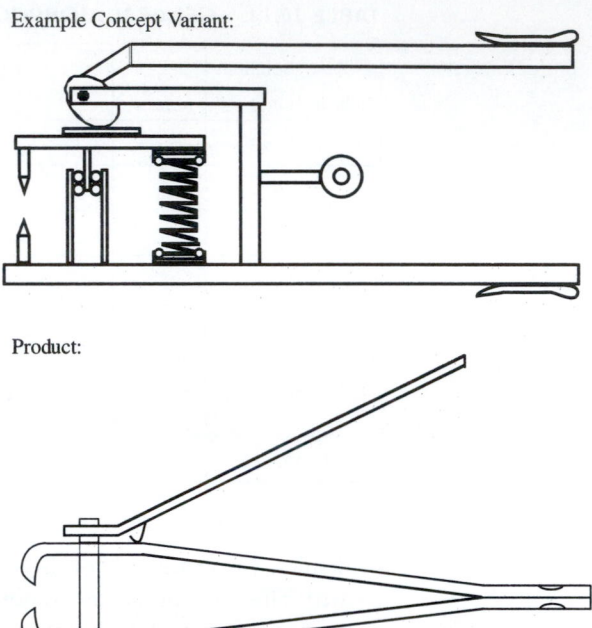

Product:

▼ *Figure 10.19.*
Illustration of function sharing and its antithesis (*Upper:* Concept created by Karl Ulrich, Wharton School of Management, University of Pennsylvania).

Notice that the first of these concepts, while interesting as a mechanical device, is overly complex and costly. Direct application of the product functions' solutions can lead to this situation. To avoid the unnecessary complexity, the principle of function sharing should always be iteratively applied as concepts are being sketched and refined. This process will help to create more innovative solutions but should be tempered with the need to avoid overly complex components and the creation of conflicts within the product.

Product Application: Fingernail Clipper

Let's consider combinations of the solution principles from Figure 10.13. Four concepts are given here, which attempt to evolve the original product.

1. This concept is much more compact and is more comfortable. It places the file on the top arm and has the top and bottom shaped for finger grips.

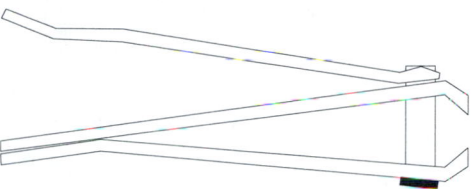

2. This concept has increased mechanical advantage. It uses a linkage to obtain large mechanical advantage at the closed clip position.

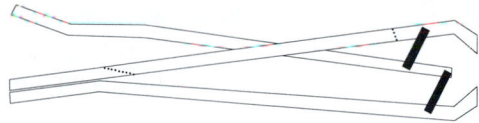

3. This concept uses a shearing scissors to concentrate the force, thereby eliminating the arm. Mechanical disadvantage does occur in the remaining cantilever arms. Figures 10.15 and 10.16 resolve the conflict of removing the arm and adding a scissors-like cutting by applying the principle of "moving in a new dimension" (Table 10.8). The blades are translated and rotated from the end of the clipper to the side.

Product Application: Bilge Water Removal Product

We seek to "design a device to remove water from the bilges of unattended pleasure boats." The customers require natural energy sources. Example energy sources include wind, boat movement relative to mooring post, boat movement relative to water, electricity (battery—not a natural source!!), fuels (not a natural source!!), solar, water temperature differences, differences in concentrations (salt water), reactive compounds (e.g., Alka-Seltzer), falling rain, wave movements, pressure variations (hydrostatic), and water movement relative to a mooring post.

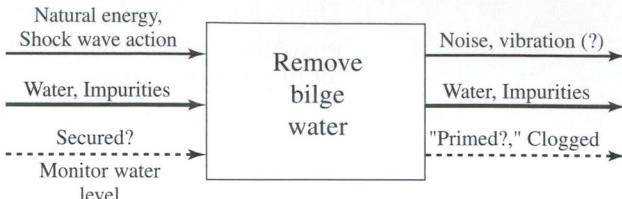

▼ *Figure 10.20.*
Black box for a bilge water removal system.

Based on the overall need of removing water automatically, performance metrics include removal capacity (minimum of 8 L/hr), durability in salt water and exposed to weather, minimize tool usage, stowable in a small volume, cost <$50, life ≥10 years, size <1 cubic meter.

Figure 10.20 shows the black box model for the product (after Hubka, Andreasen, and Eder 1988), and Figure 10.21 illustrates the functional model for the operation phase of the water removal device. Based on

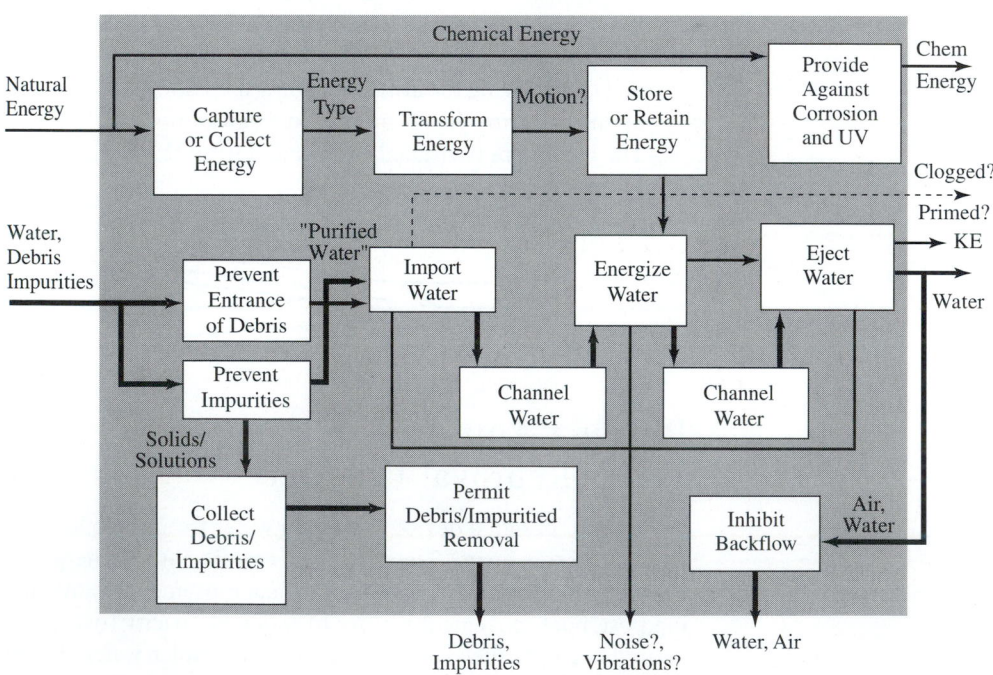

▼ *Figure 10.21.*
Function structure of the water removal system.

the functional model, directed search is used to generate solution principles to the product functions, as documented in the morphological matrix of Table 10.15. Combinations are then developed, focusing on alternative energy sources. Table 10.16 and Figures 10.22 through 10.26 show the results. Notice in the concept figures that mapping ideas from the morphological matrix to actual geometry is nonlinear and filled with design decisions. Again, sketches are needed, which must be continually refined and modified through iteration.

TABLE 10.15. MORPHOLOGICAL MATRIX FOR A BILGE WATER REMOVAL SYSTEM

Energy / Sub-functions	Mechanical		Fluid		Electrical	Misc.
	Principles	Principles	Principles	Principles	Principles	Principles
Capture Energy	Wave - Spring	Moving Column	Wind - Vanes	Reservoir - Rain	Batteries	Solar Panels
	Wave - Pendulum	Mass-Spring	Wind - Cups	Ocean - Pressure	Capacitor	Reactive Compounds
	Wave - Elastic	Salter Duck	Float - Dock	Wave - Bladder		Salt-Water
	Boat Movement		Multiple Floats	Flowing Water		Concentration
	Torsional Spring		Propeller	Wind Mill		Delta Temperature
Transform Energy	Four Bar	Bevel Gear				
	Pendulum	Spur Gears				
	Cam	Belts-Sprokets				
	Universal Joint	Crank-Shaft				
		Rack-n-Pinion				
	Lift	Carousel	Suction			Solidify-Freeze
	Ferris Wheel		Vaporize			Absorb-Sponge
	Archimedes Screw		Atomize			Absorb-Chemical
	Shovel		Syphon			
			Pressure Head			
Transport Water	Ferris Wheel	Spin-Centrifugal	Tube	Water Column		Steam
	Archimedes Screw	Funnel	Pressure	Water Piston		Atomizer
	Piston		Water-Column	Fountain		
			Channel	Squeeze Bladder		
			Weir			
Inhibit Backflow	Actuated Valve	One-Way Resistance	Flapper Valve		Solenoid	
			Butterfly Valve			
			Ball Valve			
Prevent Debris/ Impurities	Screen		Skimmer			Absorb-Sponge
	Permeable Membrane					Chemical Bond
						Oil Eaters

TABLE 10.16. SUBSET OF MORPHOLOGICAL MATRIX, HIGHLIGHTING ONE CONCEPT VARIANT COMBINATION

Energy \ Sub-function	Mechanical	Fluid	Miscellaneous
Capture Energy	**Linear spring,** Torsional spring, Pendulum, Elastic, Mass / spring.	Air: Propeller, Vanes, Cup Water: Hydraulic head, Turbine, Float	Solar cells, Batteries, Fuel, Chemical react., Diffusion
Transform Energy	Crank shaft, Gears, Belt / sprocket, Four bar, Cam, Rack & Pinion	Pneumatic / Hydraulic	Electrical / Mechanical
Import Water	Lift, Wheel (rotary) Archimedes screw, Carousel,	**Suction**, Siphon,	Solidify, Absorb
Channel	Conveyor, Lift, Archimedes screw	**Tube**, Funnel, Jet, V-notch	Atomize, Vaporize
Energize	**Reciprocating**, Screw or Rotary **pump**	Jet pump, Vaporize, Water column,	Heat
Channel	Conveyor, Lift, Archimedes screw	**Tube**, Funnel, Jet, V-notch	Atomize, Vaporize
Eject	Lift,	**Pressure**, Jet	
Inhibit Back flow	Flapper, **Ball**, or Butterfly **valve**,		
Prevent Debris / Impure.	**Screen**, Filter, Permeable membrane	Float, Skim, Vortex	

Product Application: Smart Spoon to Assist Persons with Disabilities

The goal of this product is to design a device to facilitate the handling of kitchen utensils in a single orientation by persons with disabilities. The primary market consists of current and future students in secondary school. The secondary market consists of other persons with similar disabilities who are unable to use a kitchen utensil.

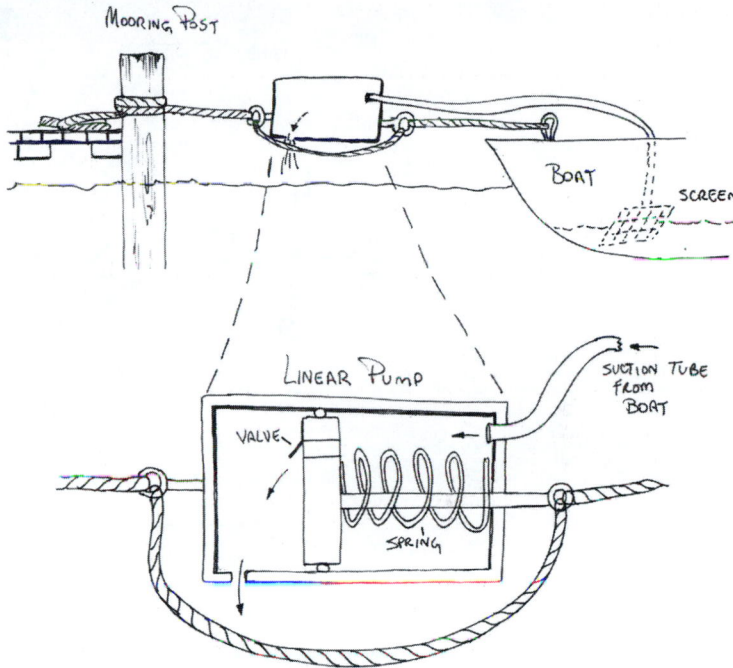

▼ Figure 10.22.
First concept variant based on chosen solution principles in Table 10.16.

The customer needs listed for the smart spoon are shown in Table 10.17. The needs "operable regardless of skill level" and "provides correct orientation" require explanation. The first means that the users are assumed to lack certain cognitive skills and are not able to distinguish between the correct and incorrect ways to hold a spoon. If they are given a normal spoon and hold it wrong, they cannot scoop up food from the plate and are frustrated as a result. The second need indicates that a device is required that will cause the spoon to be oriented in such a way that its bottom surface is parallel to the ground or table (0° to the horizontal), and once this orientation is achieved, any further twist of the wrist must produce a corresponding change in the orientation of the spoon. Thus, when the spoon is turned around by 180° from the horizontal, it drops the food (hopefully into the mouth).

Product specifications provide a quantitative measurement of a subjective need and are used to evaluate the performance of a product against a particular need. They contain a metric and a value that represents the target to be achieved. Product specifications for the more-important needs are tabulated in Table 10.18.

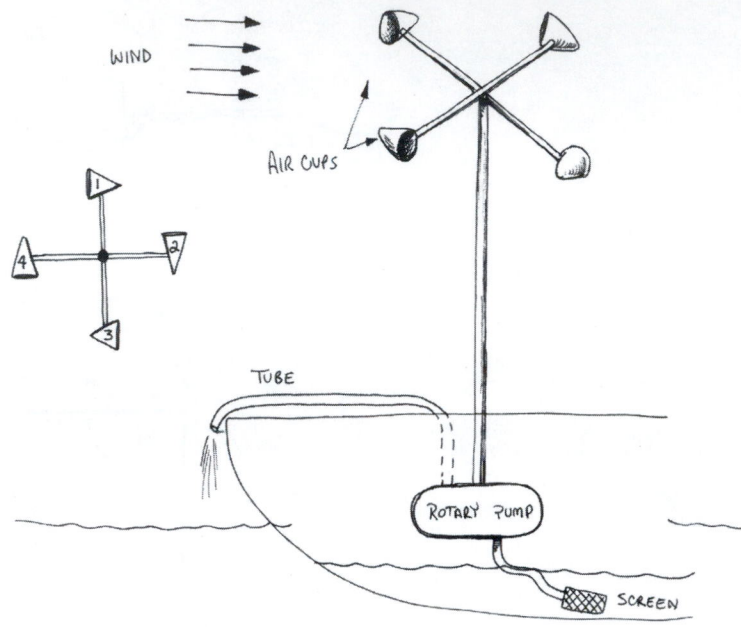

▼ *Figure 10.23.*
Second concept variant, using wind energy.

The black box model in Figure 10.27 graphically defines the main function of the product. In this device, the function is to orient an eating utensil. Decomposing this main function, the function structure for the product is shown in Figure 10.28.

Using the functional decomposition, a morphological matrix is constructed. Each of the product functions is considered individually, and ideas are brainstormed to achieve the necessary function. Figure 10.29 shows a partial morphological matrix for the spoon.

Some of the more promising ideas from the morphological matrix for each of the product functions are listed as follows. The cylindrical shape to "accept" the user has perfect symmetry and will be easy to manufacture. People lacking certain cognitive skills would have no problem identifying a correct side or position. It also has benefits of providing rotation as a means of orienting.

The term *hard lock* implies a situation where there can be no relative rotation between two surfaces, whereas the *soft lock* provides a partial

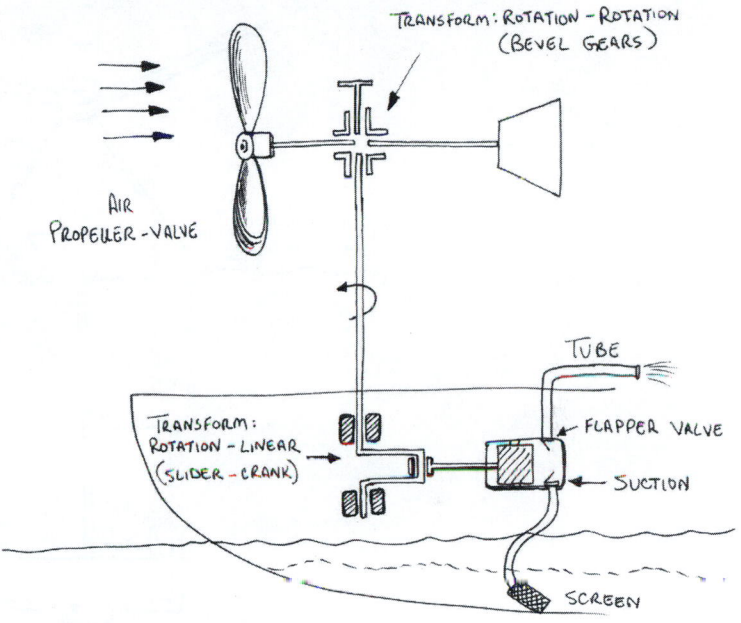

▼ Figure 10.24.
Variant of the wind-energy system.

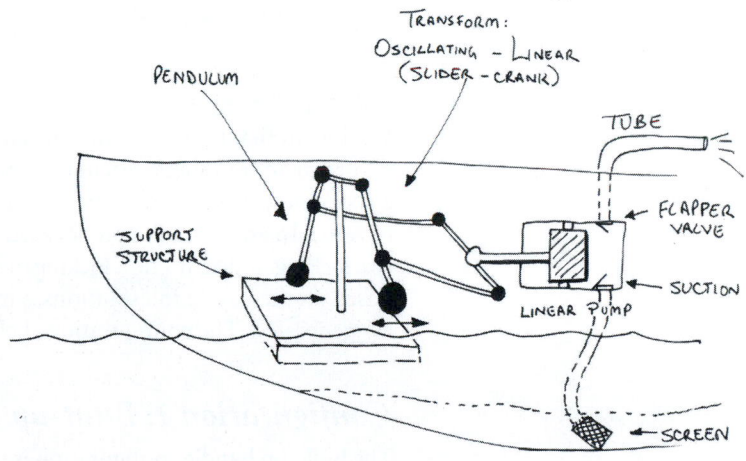

▼ Figure 10.25.
Dual pendulum (third) concept variant.

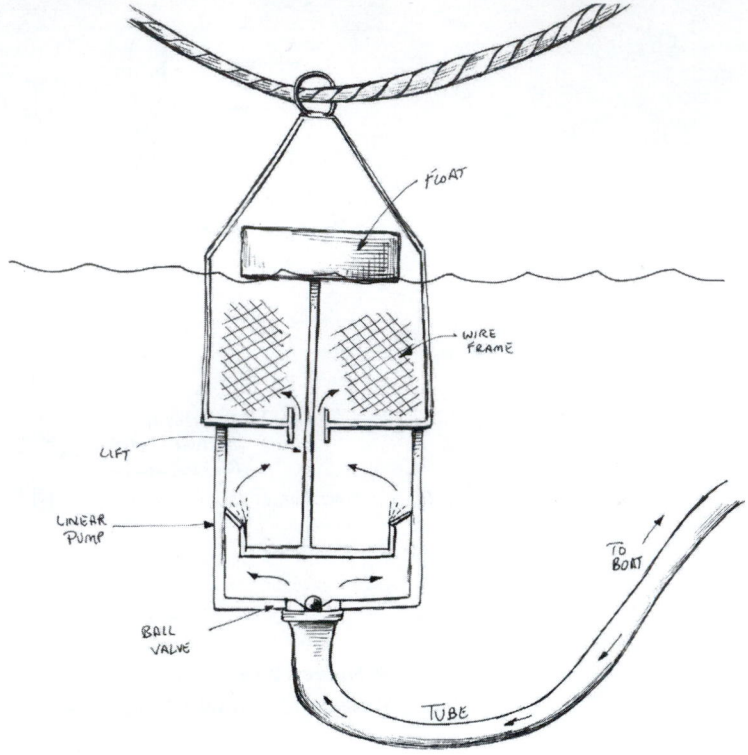

▼ Figure 10.26.
Float (fourth) concept variant.

locking, in that limited relative rotation still exists. The former is useful in providing a single orientation, whereas the latter helps to provide a continuous orientation (angle of spoon to horizontal is always close to zero). In addition, the pin lock, shown in the figure, provides positive locking, unlike friction locking or locking utilizing a magnetic field. Based on these function solutions, a number of concept configurations are generated. The configurations below provide examples.

Configuration 1: Built-up Handle

The built-up handle configuration is the simplest considered. Basically, this configuration is designed to increase the size of the handle on the spoon and provide a better surface for gripping the handle (Figure 10.30). There are three different parts involved in this, as shown in Figure 10.31.

TABLE 10.17. CUSTOMER NEEDS FOR THE SMART SPOON

Customer needs	Importance*
Functionality	
• Provides correct orientation and locks	5
• Needs to hold food during use	5
Operation	
• Operable regardless of skill level (sensory)	5
• Works with all kinds of hand positions	5
• Is easy to hold	3
• Is easy to pick up and set down	3
• Is adaptable to other utensils	2
• Does not interfere with eating action	4
Safety	
• No sharp edges or corners	4
• FDA approved materials to be used	4
Manufacturability & maintenance	
• Device reproducible with equipment at Rosedale (primarily wood-working equipment)	4
• Easy to clean	3
Ergonomics	
• Similar shape to existing spoons	3
• Comfortable grip	3
Low cost	5

*Importance on a scale of one to five, five being the greatest.

TABLE 10.18. PRODUCT SPECIFICATIONS FOR THE SMART SPOON

No.	Need	Primary metric	Target value
1.	Device operable regardless of skill level	Number of steps required for use	1
2.	Provides correct orientation and locks	Angle to horizontal	0°
3.	Works with all kinds of grips	Number of different grips accommodated	3
4.	No sharp edges or corners	Number of sharp edges	0
5.	Easy to make	Availability of manufacturing equipment	Common wood-working equipment
6.	Low cost	Less than $20	$10

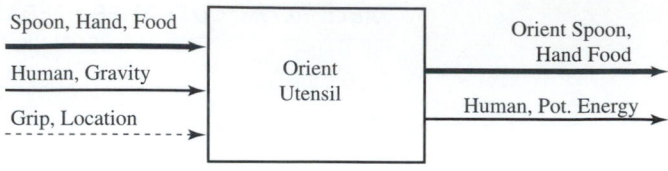

▼ *Figure 10.27.*
Black box model of the smart spoon product.

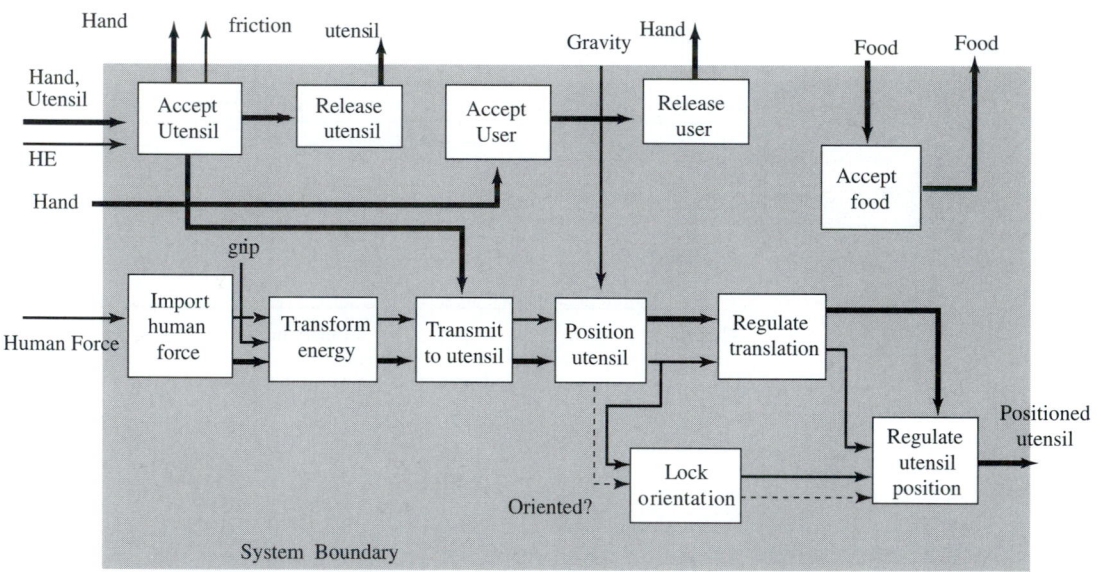

▼ *Figure 10.28.*
Function structure for the smart spoon product.

Functions / Solution Principles	Eccentric Moveable Weight	Direct Translation	Electrical		
Transform Energy					
	Gripteeth	Friction Pad	Direct Attachment	Magnetic Field	
Transmit to Untensil					
	Pre-position	Taper	Eccentric Weight	Track (Fixed)	Relative Rotation
Position - Utensil					
	Clamp	Friction	Magnets		
Lock Position (Hard Locks)					
	Spokes	Rotational Metal	Electro-magnetic	Roller Pin	Screw
Reorient-ation (Soft Locks)					
	Track	User			
Translate Food					
	Cylindrical	Toroidial	Prism	Outer Handle	Deformable
Accept User (shape)					
	Dycem	Ridges	Santoprene		
(surface)					
	Clamping	Permanent Attach	Slot	Magnets	
Accept Utensil					
	Eccentric Weight	Lever	Button	Squeeze Tube	Photocells
Accept Energy					

▼ *Figure 10.29.*
Morphological matrix for a smart spoon product.

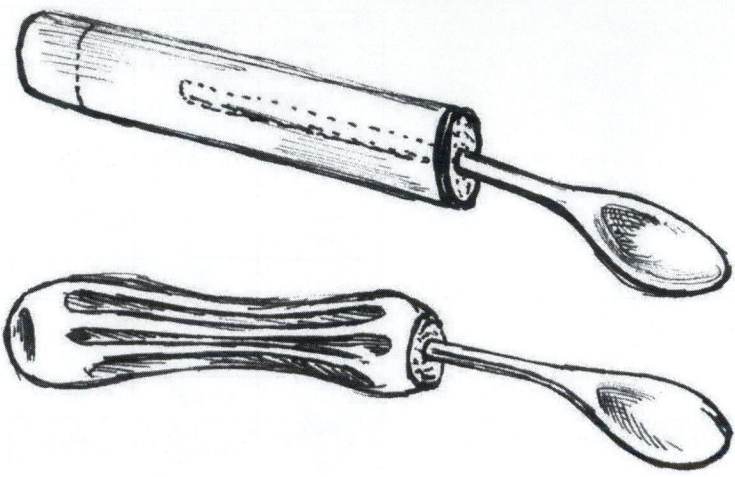

▼ *Figure 10.30.*
Built-up handle configuration (with different "accept" hand solutions-grip).

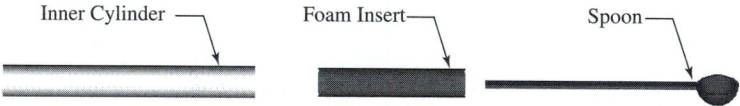

Inner Cylinder ⎯⎯ Foam Insert⎯ Spoon⎯

▼ *Figure 10.31.*
Built-up configuration exploded view.

Configuration 2: Pre-position Configuration

The pre-position configuration, as shown in Figure 10.32, is the first module designed to orient the spoon into the correct position before it is picked up. This concept relies on an eccentric weight attached to the cylinder to rotate the spoon into the correct position when it is set on the table (Figure 10.33).

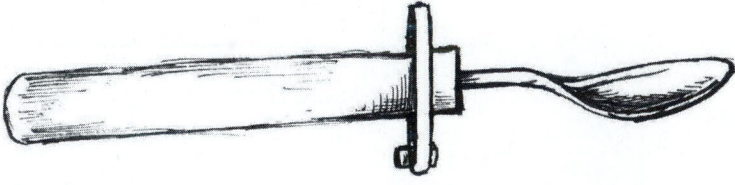

▼ *Figure 10.32.*
Pre-position configuration.

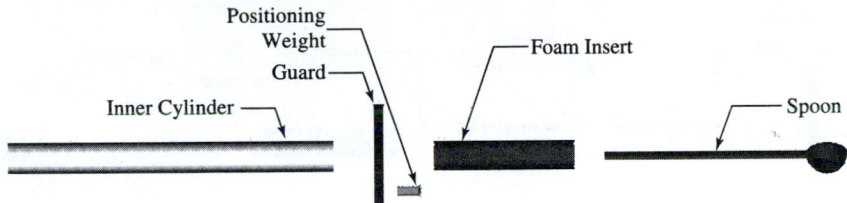

▼ **Figure 10.33.**
Pre-position exploded view.

Configuration 3: Pre-position with Hand-Locking Configuration

The pre-position with hand locking configuration, shown in Figure 10.34, is an architecture that incorporates rotating the spoon into the correct position and then locking it into position when it is picked up. These functions are accomplished by allowing the spoon and inner cylinder to rotate with respect to the table by lifting it off the surface with two brackets (Figure 10.35). These brackets allow the entire inner cylinder to rotate and orient without having to move on the table surface. This behavior is very important when the user sets their food on a towel or other surface where the pre-position module could not rotate into the correct position.

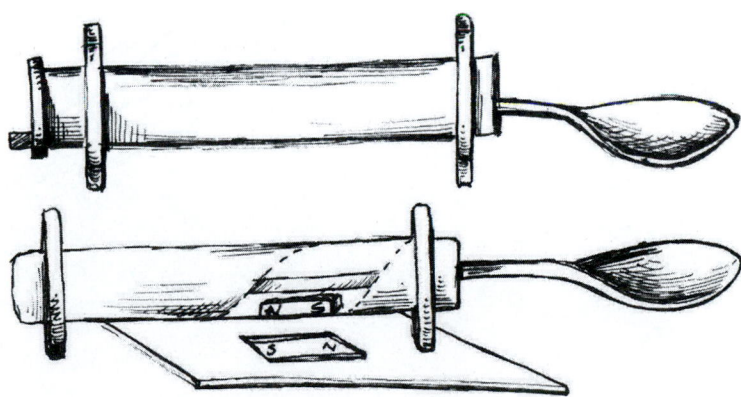

▼ **Figure 10.34.**
Pre-position with hand locking (magnetic and eccentric weight positioning).

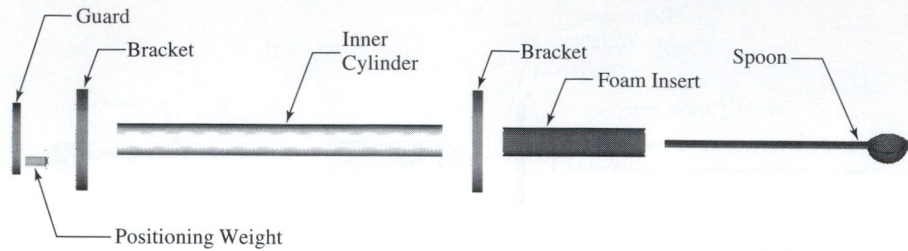

▼ *Figure 10.35.*
Pre-position with hand locking exploded view.

The guard in this configuration has been moved to the rear of the inner cylinder to keep it away from the user's face. While this location is not necessary for the operation of the spoon, it makes the device less obtrusive and helps keep the spoon balanced in the user's hand.

Configuration 4: Pre-position with Weight Locking

Another architecture for the assistive utensil is the pre-position with weight (friction) locking configuration (Figure 10.36). This architecture is the most complex and advanced configuration, because it uses

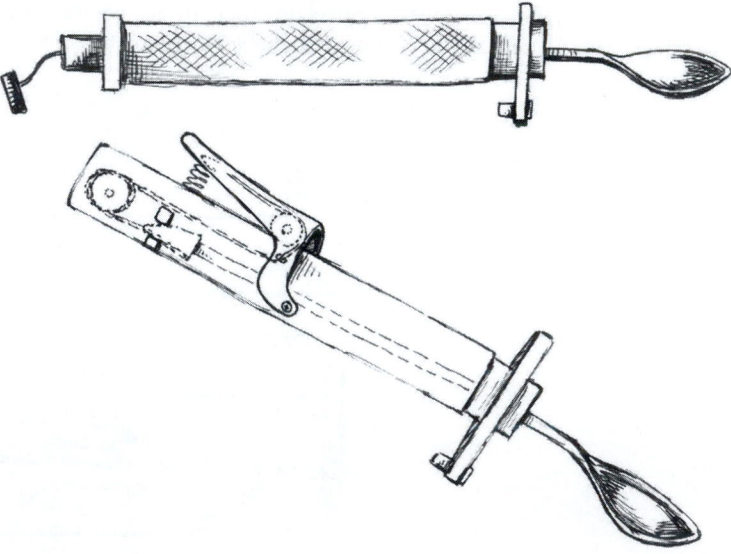

▼ *Figure 10.36.*
Pre-position with locking (weight-friction and lever-friction lock position solutions).

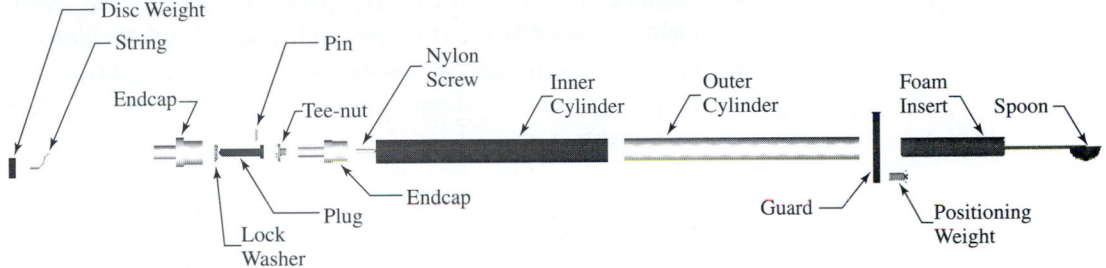

▼ Figure 10.37.
Pre-position with weight locking exploded view.

a mechanical lock to orient the spoon and then hold it into position. The spoon looks similar to the pre-position configuration, except that an outer cylinder is added over the inner cylinder and the locking mechanism is attached to the end of the spoon. The exploded view in Figure 10.37 shows the different parts used in the construction of this spoon.

VII. SUMMARY AND "GOLDEN NUGGETS"

Concept generation represents the time when product function and architecture are transformed to actual geometry. This stage in product development is exciting and challenging. It is the time when creativity and design principles are used to create innovative solutions. It is also the time when the first glimpses of a realized product appears from the design teams' inner thoughts and dreams.

During the concept generation process, solution principles and concept variants should be based on the following:

▶ Preference should be given to main product functions that determine the working principle of the overall solution system.

▶ Drawing and visual thinking should be tapped and encouraged to realize innovative solutions.

▶ Classifying criteria should be derived from identifiable relationships between energy, material, and signal flows; that is, functional models should be used to suggest classifying criteria.

▶ If the physical working principle is unknown, it should be derived from physical effects. If it is known, the form design features (surfaces, motions, and materials) should be chosen and varied.

   ◆ Combinations of intuitive concept generation and directed-search methods should be used to explore a breadth of concept ideas.

   ◆ Final product concepts are not the first ones generated. Many concepts must therefore be explored, seeking continual additions and refinements.

References

Adams, J. L. 1986. *Conceptual blockbusting: A guide to better Ideas.* Reading, MA: Addison-Wesley.

Altshuller, G. S. 1984. *Creativity as an exact science.* Luxembourg: Gordon and Breach.

Auxter, D., et al. 1993. *Principles and methods of adapted physical education and recreation.* 7th ed. St. Louis, MO: Mosby-Year Book.

Clausing, D. 1977. Concurrent engineering. In *ASM handbook: Materials selection and design.* Vol. 20. Materials Park, OH: ASM International.

DeBono, E. 1967. *New think: The use of lateral thinking in the generation of new ideas.* New York: Basic Books.

Domb, E., and Slocum, M., eds. 1998. "The TRIZ Journal." Available at: *http://www.triz-journal.com/.*

Faste, R., Roth, B., and Wilde, D. 1998. *Stanford Workshop on Creativity in Engineering Design Education.*

Faulkner, G., et al. 1995. Designing fun and independence into Tee Ball. ME 392M Project Report, Dept. of Mechanical Engineering, The University of Texas Austin. Contact: Prof. Kristin L. Wood, *http://shimano.me.utexas.edu.*

Fey, V., and Rivin, E. 1996. TRIZ: A new approach to innovative engineering and problem solving. *Target* 12 (Sept.): 7–13.

Gordon, W. 1961. *Synectics.* New York: Harper & Row.

Hubka, V., Andreasen, M., and Eder, W. 1988. *Practical studies in systematic design.* London: Butterworths.

Lawton, G. S. 1997. Litter filtering system. U.S. Patent No. 5,598,810.

Michalko, M. 1991. Thinkertoys: A handbook of business creativity for the 90s. Berkeley, CA: Ten Speed Press.

Pahl, G., and Beitz, W. 1996. *Engineering design: A systematic approach.* 2d ed. New York: Springer-Verlag.

Shah, J. 1998. Experimental investigation of progressive idea generation techniques in engineering design. Paper presented at the *ASME Design Theory and Methodology Conference,* Atlanta, GA.

Sushkov, V., Mars, N., and Wognum, P. 1995. Introduction to TIPS: Theory for creative design. *Journal of AI Engineering.*

Wankel, L. 1994. Health and leisure: Inextricably linked. *Journal of Physical Education, Recreation, and Dance* 64 (4): 28–31.

11 Concept Selection

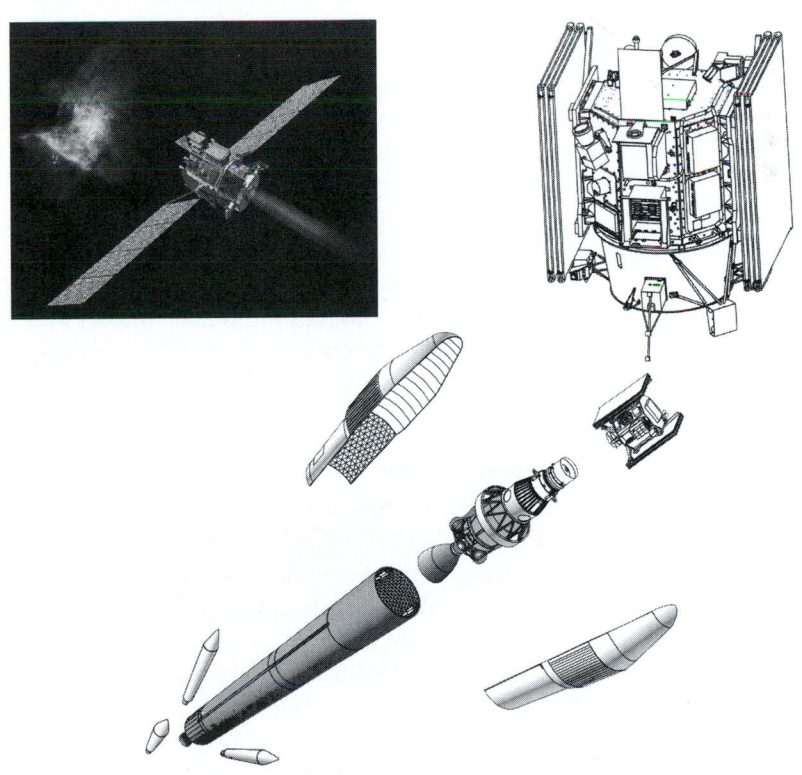

In the late 1990s, the National Air and Space Agency (NASA) developed concepts for new low-cost deep space missions. Various sensing, storage, and transmitting technologies needed to be evaluated and selected for development as well as different power sources and energy storage options. These different technologies manifest into different concepts that must be evaluated when designing each deep space mission. Such concept selection decisions are inherently risky, since the concepts are not complete technologies, a characteristic of all concept selection problems in design. Structured decision-making methods are needed to make such decisions effectively.

I. CHAPTER ROADMAP

Figure 11.1 presents the layout of topics in this chapter. First, basic order-of-magnitude estimation is discussed to determine the technical feasibility of a product concept. Then, to select among concepts that pass such order-of-magnitude analysis, a more refined decision-making analysis must be applied. A basic team-based selection process is presented that focuses on the basic guidelines for effective decision making. A basic method is then presented that can be applied to any design decision. This method is known as Pugh concept selection. After this presentation, the design evaluation process is explored in depth from the viewpoint of measurement theory, which develops the information quality needed for any decision-making method. A more detailed evaluation method is then presented, the common weighted sum approach, as an example of a method making use of more structured information. An error analysis approach of this method is also described and implemented. Finally, a critique of different decision-making methods is given, using measurement theory and design process dynamics as governing criteria.

II. INTRODUCTION

At a very high level, product development can be thought of as having three basic tasks. A development team is either gathering information, making decisions, or disseminating information. From this viewpoint, the decision-making efforts in the development process are critical. They determine how gathered information is transformed and implemented. They also facilitate the forming of team consensus, a critical aspect of an development process.

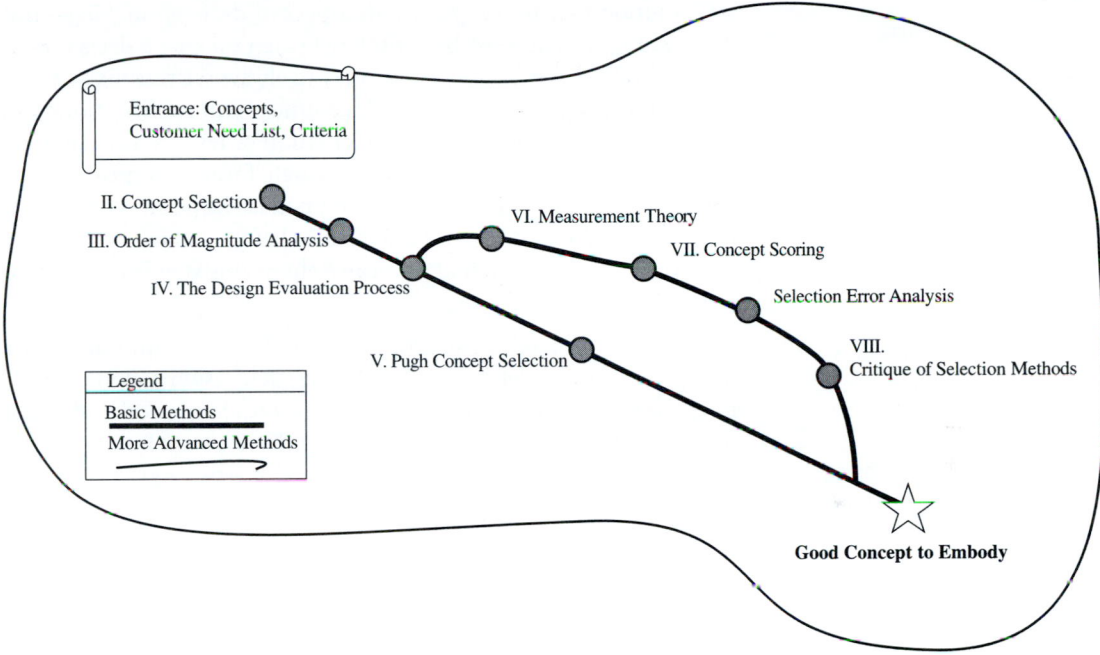

▼ *Figure 11.1.*
Chapter roadmap for concept selection.

Factors that Determine Effective Decision Making

To make decisions effectively, one must basically carry out two steps. The first is to minimize the possibility of misrepresenting a solution principle that may be effective. This oversight often happens when a technology is not in the direct experience of the engineers making the decisions, and so it is not seriously considered. The *not-invented-here* syndrome, of not being willing to consider, adequately, ideas because they are not part of the way a group has considered products in the past, is a cause of market failure.

The second step that one must complete to prevent poor decision making is to fully consider the different ramifications of a decision. For example, if a team does not fully consider what the customer wants in a product, the product can fail in the marketplace. If a product is not analyzed for safety, it can have hazardous failure modes not originally considered (see FMEA in Chapter 12), and so a less safe solution may be adopted. This aspect of decision making—understanding all that must be considered when making a decision—is never easy. With perfect hindsight, one can always see reasons why earlier decisions were less effective, since unforeseen ramifications will make their effects known in implementation or tests. Many ramifications are very difficult to foresee, though foreseeing problems is nonetheless exactly what is asked of engineers. The point is that establishing the different decision ramifications before making a decision is important, and any method that can help in this step is of practical value.

Full consideration of decision options and decision ramifications are the two factors that determine effective design decision making. Therefore, one must be careful to make decisions considering both the widest sets of possibilities and all ramifications of the decision. Structured decision-making methods have been developed to assist engineers in these two factors.

Design Evaluations

Over the course of the product development process, many persons make many decisions. Many of these decisions can be simply made by a single engineer or a small group of engineers, based on information that is readily accessible. The decisions follow from earlier decisions and from the experience of the engineers, all in a logical progression. Clear thinking and perspiration are all that are required.

On the other hand, there are important decisions made by engineers where it is not apparent what the decision outcome should be. There are many possibilities (solution *alternatives*) and many ramifications (*criteria*) that the decision impacts. We call such difficult decisions *design evaluations.*

Design evaluations occur at all phases of the development process, from concept phases through detailed design phases. Design evaluations cannot be effectively made without structured decision-making aids, since the decision alternatives are too many and the simultaneous criteria impacted by the decision too vast to be considered at once by human decision makers.

Information Quality

While design evaluations occur throughout the product development process, the quality of information available during the development process generally changes. Concept selection, for example, is one of the most critical decision-making exercises in a product development process. There are many solution concept alternatives and criteria to consider when making the selection. Yet, how well each alternative would eventually meet each criterion cannot be fully understood, since the decision is happening early. The information quality is low.

On the other hand, consider a designer who might be selecting a bearing. Here, there may still be several alternatives, typically different types of plain bushings made of delrin, sintered bronze, or teflon-coated. There may also be several criteria, such as load capacity, durability, life, and ease of repair. Unlike the concept selection phase, though, each of the criteria might come equipped with know specification levels. Every alternative can be accurately (and reasonably) evaluated on each criterion. Here the information quality is high.

This concept of information quality can be loosely defined. *Information quality* is the level of structure associated with the evaluation criteria as viewed in terms of each decision alternative. This definition will be made precise in the later advanced section of the chapter on measurement theory.

Of the various design evaluations that occur in the development process, the concept selection process is very critical, and so we will focus on it. In any design decision, there are two important aspects. The first is the process used in the decision making, considered as a sequence of steps. The second is the decision-making tools used in the process. We first discuss the concept selection activity as a process, presenting an effective team-based decision-making method to ensure coverage of different concepts and coverage of evaluation criteria. Then we present Pugh selection charts as an effective decision-making support tool that can be used in conjunction with the process.

Design evaluations are time consuming and laborious. Care should be taken that design evaluations are not conducted over decisions that can be made with reasonable estimates. For example, including design concepts that are not feasible in a concept selection process is a waste of time. To eliminate this waste, configuration variable sizes should be first estimated to determine if they are within the range of feasibility. Therefore, before discussing design evaluation, we will first present material on estimating requirements for design concepts.

III. ESTIMATING TECHNICAL FEASIBILITY

Nothing so separates a skilled engineer from a novice as the ability to effectively estimate. After complex analysis is completed, or during the preliminary concept evaluations, the ability to compare the results with known quantities and judge the decision validity is critical. Estimating is not, however, an innate skill; any person can cultivate a very good estimating ability.

Estimating skill of an engineer is dependent on familiarity with dimensional units and familiarity with the different values along the dimensions. For example, when asked to estimate the width of this book, all will have little trouble. When asked to estimate the mass of the same book, most will have little trouble. When asked to estimate the energy released by this book if it were burned, however, many do have trouble, and the estimates can vary by orders of magnitude.

One reason for this disparity is the difference between *perceived* and *derived* dimensional units. Lengths are directly perceived; we can see the difference between 1 cm and 1 m. On the other hand, energy is not directly perceived; we cannot visually detect 1 J and 1000 J. Energy is a derived unit of directly perceived units. We might observe a displacement, feel a force moving through that displacement, and take these two perceptions together to "feel" what a change in energy is, but this process involves the simultaneous sensing of two perceived units and mentally combining the two. Generally this aggregation is difficult to realize.

Thus, we have difficulties in becoming conversant with derived units. They challenge our perceptions and are usually relegated to past experiences, if they exist. For example, what is 1 J/kg K of entropy? Without many varied experiences with entropy calculations and measurements, our ability to estimate this quantity is severely limited. Even if we have experience, a new design situation will not necessarily align with this experience. We therefore need a basic estimation approach that will allow us to make estimates for a wide range of situations.

What permits an engineer to develop familiarity with derived units is to become indirectly familiar with them. Rather than try to feel 2000 W by simultaneous sight and touch, one might realize that 2000 W is 3 hp, and that is what a common lawnmower engine produces. We may not be familiar with 2000 W, but we are familiar with a lawnmower engine and the amount of power it produces. The point is that when one cannot directly perceive a unit, one should directly perceive a concept that exhibits the unit of interest.

Estimation

There are four basic steps to estimating: *imagine, model, compare,* and *judge.* While these steps sound much like any scientific model construction method, the context here is for order of magnitude, simple models, and comparison to known quantities.

The first step is to imagine the concept to estimate. We paint a mental picture of what happens with the concept and imagine points along energy, material, or information flows through the concept where one might measure an input-output change or a capacity buildup in the concept. This mental picture follows directly from the functional model (Chapter 5) and House of Quality (Chapter 7) of the concept.

The second step is to construct a very simple model that relates the capacity or flow (energy, material, or information) through a concept to known quantities. For any dimension, one should memorize, or have readily available, hand-tabulated values of phenomena that one is acquainted. This advice can be understood in two different contexts. The first is that, when completing calculations in an office environment, it is a good idea to have tables that contain examples of different values on various dimensions. These examples can provide comparisons for any calculations made. When making a calculation of power that results in 2.65 hp, it is good to know that 3 hp is typical for a lawnmower engine.

In the second context, let's consider any field or business meeting environment. To hone one's estimating skill, one should also develop an independent personal set of such estimating quantities. For example, one might own an old Volkswagen and thereby intimately know VW engines and so know that 40 hp is the standard for the 1600-cc VW air-cooled engine. This reference point is a personally understood value of power from which no extra effort is required for the value to be memorized; it always exists at ready mental recall. Such personal quantities should be developed and be automatically memorized by every engineer. They should be small in number and dispersed over the range of values on any dimensional unit.

The third step in estimating is to use the concept and the model to provide a comparison with the known quantity. We fill in the model with imagined values on the model variables and calculate a comparison value in the dimension of the known quantity.

The last step is to judge whether the estimated quantity compares with a known quantity. Does the calculated value compare with the known quantity? If so, does it make sense that the concept would have a value on this dimension about the same as the known quantity? If so, the concept is estimated to be viable; if not, then something is incompatible between the model, known quantity, and the concept. When estimating values on preliminary concepts, it should be kept in mind that an estimate is generally accurate only to one order of magnitude. If a conceptual design estimate turns out to be within a factor of two of the final product, as manufactured, the estimate is effectively dead on. This is not to say that one should be sloppy with numbers to add or subtract an order of magnitude. Rather, estimating models is inherently approximate even with precise values; even with the best guesses, estimates are only accurate within one order of magnitude.

Example: Air Conditioning for an Electric Vehicle

As an example, suppose we want to design an air conditioning system for a battery-powered electric vehicle (EV). Current EV air conditioning systems are standard automotive refrigerant systems with the compressor powered by an electric motor drawing energy from the main EV batteries. While not particularly elegant, such a solution eliminates much design work. As an alternative concept, though, consider using prefrozen carbon dioxide blocks that one merely blows outside air over to cool the air as the carbon dioxide blocks sublime. This alternative would use a heat exchanger to prevent the CO_2 from entering the vehicle cabin. The cooling power would be supplied not by drawing down the vehicle batteries but rather by the energy that can be absorbed by the prefrozen blocks, replaced when the vehicle is recharged. Is this concept feasible?

This question hinges on how much frozen CO_2 is required. Current EVs have about a two-hour driving range, so about two hours of cooling are required. The first question becomes, How much heat must be removed over a two-hour period? and the second question becomes, How much dry ice is needed to provide this amount of cooling?

To estimate these quantities, one must examine the cooling load. Suppose the vehicle is driving in Los Angeles, California in 100 °F heat, and it is desired to have a 72 °F cabin. We now paint a mental picture: Where is the heat entering the 72 °F cabin from? Where is the heat absorbed by the air conditioning? What are the heat flows? What are the mass flows?

Imagining the concept, it would seem that heat enters from solar radiation and internal body heat. We could estimate these flows. Alternatively, once in the cabin, the heat flows into the conditioned air mass—that is what keeps the temperature in the cabin low. This heat flow can be estimated. The latter seems a more familiar quantity to estimate, so we will take that approach.

How much air is blown into a vehicle cabin? Consider a steady-state condition with the fan set at a medium setting, blowing out of four vents. The airflow is the cross-sectional area of the vents times the velocity. We might imagine the concept has four vents at 10 cm × 4 cm in cross section. How fast is the air blowing? Probably no faster than running speed, or about 10 m/s. Therefore, we have a volumetric flow rate of:

$$Q = Av = (4 \cdot 0.1 \cdot 0.04)10 = 0.2 \text{ m}^3/\text{s}$$

Now let us test this estimated quantity. Suppose we put an imaginary plastic garbage bag over the air vents and let it fill. In 1 s, we would fill the bag with 0.2 m³ of air. As a reference, consider a cube where the side of the cube is 1 m—a large arm's length. The airflow in 1 s is 1/5 of this reference 1 m³ cube. Judging this estimate, it seems reasonable.

We now must estimate the heat that must be extracted from this heat flow. To perform this assessment, the heat capacity and density of the air must be known.

$$\Delta E = mc_p\Delta T = (Q\rho t)c_p\Delta T$$

For air, c_p is about 1000 J/kg, ρ is about 1 kg/m³, and t is about 2 h. These values, when combined, result in a required energy absorption capability of 24×10^6 J for the dry ice. We now compare this estimate with a known quantity; for example, 5×10^6 J is the amount of heat absorbed by an air conditioner in 1 hr for a typical small room (5000 BTU/hr). Judging this comparison, the 24×10^6 J may be bit high, but it is not unreasonable. We'll continue with 12×10^6 J as a better estimate, guessing the 10 m/s is probably high.

Now we must consider how much heat can be absorbed by dry ice when subliming at −108 °F and rising to, say, 50 °F, a guessed maximum effective temperature for a heat exchanger. A basic energy model, the heat absorbed by the CO_2, will be equal to the heat due to sublimation and the heat absorbed during the temperature rise:

$$\Delta E = m\Delta H + mc_p\Delta T$$

where m is the mass of dry ice required, and ΔH is the energy of sublimation. Chemistry handbooks list c_p and ΔH per mol, so a conversion factor of mols to kg is needed, the inverse M_w of the molecular weight. The result is an equation of the form:

$$\Delta E = m \cdot \Delta H \cdot M_w + m \cdot C_p \cdot M_w \cdot \Delta T$$

where $\Delta H = 25 \times 10^3$ J/mol, $M_w = 22.7$ mol/kg, and $C_p = 37.1$ J/kg C for carbon dioxide. The result is that about 20 kg, or 45 lb, of dry ice is required. One can repeat the calculation with other materials. The calculation with liquid nitrogen, for example, results in 40 kg, or 90 lb, being required.

One can perform a mass estimation for the batteries required to power a motor that drives a compressor of a standard automotive air conditioning system. This results in an estimate of about 25 kg, or 55 lb, of batteries. So, the dry ice air conditioner idea does not save any weight, so it is really not a feasible concept; it should be dropped from consideration.

Estimating Hints

It becomes clear from this example that there are two fundamental aspects to estimating, *model building* and *known quantities*. The first is simple model building—how to model concept phenomena. This is the foundation upon which all engineering is based. Methods of constructing models are developed in Chapter 13, but for simple concept estimation of mechanical systems, it usually reduces to making simple first principle models.

For a mechanical engineer, one key dimension to be adept at modeling is the power flow through any mechanical system. Invariably, estimating mechanical systems reduces to estimating power through it. Therefore, one should pay attention to constructing simple models of power. Generally:

$$P = Fv$$
$$P = M\omega$$
$$P = IV$$

To do this, one should look for where the driving force originates, at what speed the concept operates, and, therefore, what power results. For particular systems, the power formulas can be refined as necessary, such as expanding force F to $F = pA$ for pressure systems. Generally speaking, though, power is a good dimensional unit to be intimately familiar with and adept at modeling.

The other factor of good estimating is to have known quantities at ready recall. For every dimensional unit, such as power, energy, force, pressure, acceleration, and so forth, a skilled engineer will have at least three readily understood reference levels. For example, on the dimension of power, 10 W are consumed by a flashlight bulb, and a human pedaling a bicycle easily feels that pull when powering the 10-W bulb off a bicycle generator. On the other hand, 100 W are consumed by a powerful household light bulb. A typical lawnmower engine provides 3 hp, and 1 hp is 745 W. Table 11.1 lists similar concepts for different dimensional units, which can be referenced.

While the values in Table 11.1 are of use, again, they should generally *not* be the ones that an engineer memorizes. Rather, a skilled engineer should maintain a personal condensed version of such a list for ready recall when synthesizing and analyzing products. Think of the personal interests you have and determine a few handy reference values in these areas. Those are ones you will automatically and easily memorize.

The estimation process should be used to eliminate concepts that are not technically feasible. Power, size, and cost values for the different concepts should be estimated and values outside the bounds of feasibility eliminated from consideration. Generally, this process will retain concepts that the order of magnitude estimation cannot resolve. These concepts require a structured design decision-making process to resolve into a winning (preferred) preliminary concept to expend further resources on.

IV. A CONCEPT SELECTION PROCESS

An industrial concept selection process is a team-based decision-making effort. We present here a process that is very effective at establishing consensus at the product concept selection phase, applying thoughts from Pugh (1990), Ulrich and Eppinger (1995), Ullman (1995), and Otto (1995). Generally, concept selection is a critical point in a product development process. There is no clear solution concept that should be pursued; different members of the team have strong opinions on different solutions, and incorrect decisions can be costly.

The concept selection process presented here is a means to deal with this uncertainty. The process is designed to be a daylong out-of-the-office exercise, where a design team evaluates alternatives and comes to a consensus on the most effective concept to pursue. The process is focused on clearly articulating differences in understanding among team members, forming common definitions, and expanding the options considered that are due to these differences in understanding. When the process is implemented, as described, a design team will have agreement on the concept to pursue, every engineer on the team will understand why they changed their mind to the collective consensus, and finally, every engineer will be supportive of their change in position.

TABLE 11.1. APPROXIMATE REFERENCE VALUES ON DIFFERENT DIMENSIONS ADAPTED, IN PART, FROM (HALLIDAY AND RESNICK 1988).

	Power (W)	Energy (J)	Mass (kg)	Force (N)	Pressure (MPa)	Accel. (m/s^2)	Velocity (m/s)	Length (m)
10^{-6}	Ant crawling up a wall at 1cm/s: 33 μW	Moving 5 g snail: 0.45 μJ kinetic energy	1" × 1" piece of paper: 40 × 10^{-6} kg	Electrostatic attraction between electron and proton in hydrogen atom: 0.08 μN	Moon surface: 0.13 × 10^{-9} MPa	Centripetal acceleration of a regular clock hour hand: 0.3 μm/s^2	Tip speed of a wrist watch hour hand: 20 μm/s 10' snow accumulation rate over 2 winter months: 0.1 μm/s	Human hair thickness: 30 μm
10^{-3}	LED: 40 mW	Bee in flight: 2 mJ kinetic energy	Grape: 10 g Penny: 3 g	Piece of paper, weight: 0.04 N	Blood pressure: 16 × 10^{-3} MPa Mars atmosphere: 0.8 × 10^{-3} MPa	Centripetal acceleration of a regular clock minute hand: 1 mm/s^2	Tip speed of a wrist watch minute hand: 1 mm/s Speed of tide rising from low to high: 0.1 mm/s	Book cover thickness: 2mm
1	Small flashlight: 10 W	A small apple lifted 1 m in gravity: 1 J A small apple falling 1 m: 1 J kinetic energy	Small meal or a large snack: 1 kg	Small apple, weight: 1 N Finger force for appliance buttons: 7 N	25 ft underwater: 1 MPa 1 atm = 1/10 MPa Piston engine compression pressure: 1.3 MPa	Fast car acceleration: 3 m/s^2 Hard braking car: 7 m/s^2 Earth gravity at sea level: 9.8 m/s^2	Falling body after 1/10 s: 1 m/s Walking speed: 1.5 m/s	Person's height: 2 m
100	Bright light bulb: 100 W Typical household appliance: 100–1000W.	90 mi/hr fast ball: 114 J kinetic energy	Average person: 70 kg	Bag of potatoes, weight: 100 N	Piston engine firing pressure: 3.5 MPa	Humans black out: 40 m/s^2 Belly flopping hard in water from a 10-m diving board jump, causing broken bones: 100 m/s^2	Highway speed: 30 m/s Jetliners: 250 m/s	Soccer field length: 100m Height of the Statue of Liberty: 93 m
10^3	Small lawn mower engine: 2000 W	Energy effectively extractable from a AA battery: 1 kJ D-sized battery: 80 kJ	Mid-sized car: 1300 kg Elephant: 5000 kg	Two small people, weight: 1.5 kN	Pressure to create a diamond: 5 GPa Deep ocean trench: 0.11 GPa	Marble dropped from 1m stopping in sand Head-on car collision occupant deceleration: 10 km/s^2 Bullet fired from a rifle: 60 km/s^2	3 times the speed of sound: 1 km/s	Width of a small town: 5 km
10^6	Electrical power to a small town: 1 MW	Car @ 80 mph: 1 MJ kinetic energy Automotive Battery: 5 MJ	A 747 fully loaded: 300,000 kg Ocean liner: 107 × 10^6 kg	Boeing 747: 1 MN thrust.	Center of the Earth: 0.40 × 10^6 MPa	Projectile fired from a rail gun: 800 km/s^2	Voyager 1 traveling in outer space: 17 km/s	Dallas, TX to Denver, CO or Boston MA to Pittsburgh PA: 1000 km
10^9	Electrical power plant: 1 GW	USS Nimitz, 91,400 tons @ 30 knots: 9.9 GJ kinetic energy	Aircraft carrier: 0.5 × 10^9 kg	Saturn V or Space Shuttle: 0.2 GN thrust	Center of the Sun: 20 × 10^9 MPa	Centrifugal acceleration of light trapped in a black hole: 2 × 10^{13} m/s^2	Speed of light in vacuum: 3 × 10^8 m/s	Earth to moon: 3.84 × 10^9 m

Failing this state of consensus, the team will understand exactly why there remains a difference in opinion and be in agreement on what must be further analyzed to resolve the issue. In such a case, the concept selection will then be completely determined by the results of this one analysis.

The concept selection process should be completed in a room with at least three walls that can be written on, paper attached to, or overhead projectors shown on. One wall will have definitions of criteria and alternatives displayed, another wall will be a working wall with the evaluation interaction, and a final wall will be used to keep notes and rejected information.

The selection process is a five-step process plus iterations:

1. Forming consensus on the criteria,
2. Forming consensus on the alternatives,
3. Ranking the alternatives,
4. Evaluating the alternatives,
5. Attacking the negatives.

Earlier steps are revisited during the latter steps as needed for refinement, and the steps themselves are repeated in several sequences. We now detail each step.

Forming Consensus on the Criteria

The first step is to establish evaluation criteria on which the concept selection decision will be based. Generally, most concept selection decisions are based on three evaluation criteria of *cost;* then *development risk, technical difficulty,* or *ability to meet scheduled delivery;* and finally *performance* or *customer satisfaction.* These three general criteria are expressed in a variety of case-specific ways for any specific selection process. The problem is that once the criteria are stated, every design team member will have different perceptions of what these statements mean. An initial task is to resolve these differences into definitions that are common to the team.

To establish criteria definitions, a design team should start with one member articulating a proposed list of evaluation criteria. This list should be developed from the customer needs (Chapter 4) and engineering specifications (Chapter 7) of the product. It should also include the most critical and important specifications, those that will clearly distinguish the differences in the proposed product concepts.

As this list is formed, other team members should chime in with more criteria, until a set of criteria is on the board that everyone agrees could be legitimate criteria. Interpretation of what the criteria mean is not yet argued over. The criteria are instead proposed as labels that can be interpreted slightly differently.

Once this process is complete, each criterion is refined into a common definition. Whoever suggested the criterion provides a definition, and then everyone argues (debates) over its scope. The definitions should be consistent with previous metrics chosen for the criteria (Chapter 7). For example, a cost criterion can include unit manufacturing cost alone or could also include retailer cost, warranty costs, and so forth. The product performance criteria might include criteria of the maintenance personnel or might not. The team must determine what the scope of the criteria does and does not include. As a process, this discussion will lead to further scope of the original criteria label (other team members roll their thoughts into the original label). Alternatively, it will lead to the addition of new criteria (other team members cull out aspects of the criteria into a new and separate criteria), or it may lead to the discarding of some portions of the original criteria. (The person suggesting the criterion is convinced of its lack of importance to the decision because of other factors they were unaware of.)

Note that this process can help with the problem of failure to consider all ramifications of decisions. If each person is not aware of a small fraction of all the decision ramifications, then generally as a group, the team is collectively not aware of a smaller fraction. More persons have more experiences that give rise to understanding more decision ramifications.

For decisions made with high-quality information, each criterion might have understood numerical scales associated with them, such as yield stress in MPa to represent failure, dollars to represent cost criteria, hours to represent operating life criteria, and so forth. Also, the criteria might be sufficiently understood in relative importance, such as product operating life being half as important as the cost. If such scales and/or relative weights can be established, and a design team so wishes, then this step is the time to do so.

A note of caution, however, is appropriate for very early concept selection. Numerical scales may not be appropriate due to low-quality information, and if so, they should not be used. On the other hand, final concept selection is sometimes deferred until extensive analyses are completed, and so information quality is high and numerical scales work fine.

Once complete, the criteria and their definitions (along with measurable scales) should be posted in large, easily readable characters on a side wall during the decision-making process, thereby permitting quick visual reference by any team member. People can only process a few simultaneous thoughts, and expanded memory aids must be made readily accessible.

Forming Consensus on the Alternatives

Once the evaluation criteria are initially established, different alternatives need to be understood on a common basis. Establishing these alternatives will happen in much the same way as the criteria are established. First, team members voice alternatives that are given labels, where only a rough understanding is needed by the members. These alternatives should be those that proceed from concept generation (Chapter 10). They should also be refined to a level where all team members may understand them without ambiguity. Appropriate sketches facilitate this understanding, as shown by the examples in Chapter 10. During the consensus-forming process, team members may make changes and refinements to the concept sketches as appropriate.

Next, each alternative is given a commonly understood definition and, most importantly, articulation of what will be needed to engineer the concept into a final product. It is important that every person voice their understanding of what their specialty will have to do to finish a given alternative in terms of the evaluation criteria previously articulated.

Generally, this definition process will lead to new "alternatives." For example, one engineer will have a perception that a subsystem can be easily delivered by a vendor, and another engineer may not believe this statement and instead suggest that in-house work will be required. If the two engineers cannot agree, then from a decision-making viewpoint, there are two alternatives on the table. They should be labeled as such, even though they are of the same form. The only difference between these "alternatives" is whether a vendor can supply the subsystem or not. If one of the two survives into the preferred or chosen alternative, then resolving that issue becomes the important engineering management task. On the other hand, both may be rated lower than another alternative, so resolving the dilemma is unimportant. For now, all that is required is clear articulation of an alternative, as it is understood with the current incomplete information.

Good mechanical selection practice is to have each alternative drawn in isometric views on at least an 8.5″ × 11″ paper so that the alternative can be visualized as a thumbnail from anywhere in the room. Further, the definition should be written out under the drawing in large, easily read characters. This pair should be stored on a separate side wall for quick visual reference by any team member.

Ranking

The next step is to rank each clearly defined alternative on each clearly defined criterion. There are many different ranking schemes that can be used, depending on the quality of information available. For preliminary concept selection, however, less-structured scales may be used.

The rankings are completed using a decision matrix chart on the main wall and are the focus of the effort. A team should consider each criterion one at a time and rank all the different alternatives on each criterion. A team should not rank each alternative across all criteria before considering another alternative. This process will lead to bias in each criterion's evaluation, since the criteria ranking definitions may drift.

To rank the alternatives, a scale is used, such as $(-,s,+)$, where a $(-)$ is worse than an (s), which is worse than a $(+)$. More-refined scales can be used if the information quality permits it, as discussed later in this chapter. In any case, the transition between levels on any scale, such as from a $(-)$ to an (s) to a $(+)$, must be understood and agreed by the team, and will generally again lead to refinement of the criteria and the alternative definitions.

Assessment

Once the rankings are completed for each criterion, the evaluations should be collected into overall summary rankings on each alternative. The basic reason for creating summaries is that humans cannot process all the rankings simultaneously when comparing two alternatives. The criteria rankings need to be aggregated into one or more overall ranks for simpler ordering of the alternatives into a best-to-worst ranking that can be understood. Generally, this aggregation is a simple exercise in the mathematics of the ranking to see how the alternatives rate overall. Then, an ordering of the alternatives, from overall worst to overall best, is completed, where "overall" implies incorporation of all the different criteria.

Attacking the Negatives

Once the alternatives are ordered by rank, the process is not, by any means, finished. Rather, the alternatives that rate poorly should be taken off the chart, and the alternatives that rate favorably must be more closely examined. In particular, the alternatives that rate high overall but have a few low scores should be closely scrutinized. Pugh (1990) coined this process *attacking the negatives;* it is critical and should not be minimized in importance.

For the alternatives with such negative-ranked criteria, a design team should clearly state what is causing the negative ranking as a chain of physical effects. Then each of these effects should be interrogated by the team to understand how each engineer might apply his or her discipline to eliminate the negative. The Theory of Inventive Problem Solving (Chapter 10) may also be applied at this point to understand and potentially eliminate conflicts in a negatively rated alternative. Likewise, positives from one alternative may be combined with the negatives of another to improve a concept's ranking. This process requires out-of-the-box concept-generation thinking, but as a design team, such new alternatives can be readily embellished and evaluated. Thus, attacking the negatives is not a concept-selection step as much as a new concept-generation step.

In practice, attacking the negatives is very effective at the earlier product concept selection phase and leads to added design effort that need to be further explored. For later embodiment or detailed design evaluations, however, attacking the negatives is less profitable. Typically, in detailed design evaluations, the negatives exist due to physical limitations in materials or physical processes, and, in such circumstances, there simply exists relative trade-offs that the decision is constrained to be within. Attacking the negatives is an out-of-the-box new idea exercise, not an in-the-box trade-off exercise.

V. A BASIC METHOD: PUGH CONCEPT SELECTION CHARTS

The above process can be applied with decision-making tools as developed by Pugh (1990). These tools, known as Pugh charts, use a minimal evaluation scale and three overall ranking metrics. Pugh charts are the most effective known tools for preliminary concept selection when there is minimal information quality available. They are also effective as the information quality increases and the selection scale is refined. The goal of any selection process is to obtain as much information and concept details as cycle time and resources permit.

Establish the Criteria and Alternatives

With Pugh selection charts, the criteria and concept alternatives are established as discussed above. The ranking procedure, however, takes a particular form. The alternatives and criteria are displayed on the main wall labeled *Pugh Selection Chart*, as outlined in Figure 11.2. Here, the concept sketches are displayed at the top of the matrix as columns. Each criterion is listed down each row.

Select a Datum

The next step is to establish the evaluation scale. Pugh (1990) correctly advocated using a minimal scale of only {−, s, +} for evaluating preliminary concepts. Further refinement of the scale, such as with numerical rankings and/or numerical weights, may not be feasible at the preliminary concept selection phase. A team may not be able to supply such detailed ranks for some criteria, and so the evaluation is error prone, where incomplete rough ratings combined with good ratings can numerically alter the result.

Rather, an effective process is to construct a simple evaluation scale (unless more information is known). To do this, the team should select one alternative that will be ranked as (s) (or 0) on every criterion and be called the *datum*. The datum is the alternative to which every other concept will be compared. The other concepts are rated as either worse (−), the same (s), or better (+), on each criterion.

		Concepts					
		Sketch of Datum	Sketch of B	Sketch of C	Sketch of D	Sketch of E	Sketch of F
Criteria	Criterion 1	0					
	Criterion 2	0					
	Criterion 3	0					
	Criterion 4	0					
	Criterion 5	0					
	Criterion 6	0					
	Criterion 7	0					
	$\Sigma + (P_i)$	0+					
	$\Sigma - (N_i)$	0−					
	Σ	0					

▼ **Figure 11.2.**
Template of the Pugh selection chart.

To select a datum, a team can choose one of several approaches. The current product offered by the company might serve as a well understood concept. A competitive product that a team wishes to surpass might serve as a datum.

Pugh (1990) recommends using the alternative that the team votes as best (or preferred) as the datum. That is, the team members vote on each alternative, and the alternative with the majority of the votes serves as the datum. This choice accomplishes two things. First, the distribution of disagreement is understood. Second, since the team-chosen alternative is likely to be carried into the next round of comparisons after attacking the negatives, it ought to be used as the preferred datum. The choice of a datum should not shift from round to round of rankings (e.g., if every alternative advanced to the second round has (+) ranks relative to the existing product).

Ranking and Assessment

Having selected a datum, all the alternatives are evaluated on each criterion, one criterion at a time, relative to the datum. If the team agrees that the alternative offers worse performance than the datum, then the alternative is assigned a (−) rank. If the team agrees that the alternative offers better performance than the datum, then the alternative is assigned a (+) rank. If the team determines the alternative offers the same performance as the datum, it is assigned an (s) rank. These rankings are documented in the Pugh chart, which is on the front wall of the room during the decision-making process.

If the team engages in lengthy debate that repeats the same points without convergence, then the team cannot distinguish between the datum and the alternative. In this case, the alternative should be scrutinized. Rather than debating the rating, the team should ask what must be understood so that either rating is correct. What is the heart of the debate? The team should strive to restate the alternative in question as two *conditional alternatives*.

For example, a coffee mill might need to have a clean chopping container. A team might argue over whether an internal scraping wire mechanism, kept tight against the bowl, that cleans the surface will just end up collecting grinds on its crevices. This single alternative becomes two conditional alternatives, one that is the scrapable bowl alternative, conditional on the wire causing no problems, and one that is the scrapable bowl alternative, conditional on the wire causing a problem.

When a conditional is not readily apparent, then the design team should simply rate the alternative with an (s′) or a (?), and in the overall rating analysis treat this as an (s). Then, the criterion does not enter into the decision in either a "for" or "against" manner with respect to this alternative. If the alternative still has a chance at becoming the preferred choice, then the issue should be debated at length and a plan developed to resolve the debate through more refined engineering.

Alternative Rank Ordering

With the alternatives rated on every criterion, the ratings should be combined into overall scores that can be used to order the alternatives from best to worst. To do this, however, a simple average summation of the ranks, where a $(-)$ cancels a $(+)$, is not adequate. Each criterion is not equally important. A $(-)$ on one alternative may really be $(---)$, and a $(+)$ on another may be only one half of a $(+)$, compared with each other and the datum.

To deal with this discrepancy, one should use three overall scores to compare the alternatives. On each alternative, i, one can determine the average overall summation score:

$$S_i = \sum_{j \, \text{criteria}} R_{ij}.$$

In addition, two more overall scores should be calculated, where each adds up all the $(-)$ scores and all the $(+)$ scores, respectively.

$$P_i = \sum_{j \, \text{criteria}} R_{ij}^+ \quad N_i = \sum_{j \, \text{criteria}} R_{ij}^-,$$

where R_{ij}^+ is a $(+)$ score assigned to alternative i, and R_{ij}^- is a $(-)$ score assigned to alternative i.

These three overall scores should be considered when progressing to the next round of attacking the negatives and reevaluation. The alternatives with high average ranks are good and so should be considered. The alternatives with many $(-)$'s and $(+)$'s also should be considered, as there might be means available to change the alternative.

Attacking the Negatives

Having assessed the alternatives on the criteria, the lower average-rated alternatives should be discarded from further consideration and relegated to a side wall. Then the highly average rated alternatives with $(-)$ scores, as reflected by high N_i and P_i, scores, should be scrutinized, with the hope of out-of-the-box thinking for new alternatives or combinations that eliminate the negative ranks. The precise problem with the alternative has been articulated through the ranking process, and this information should not be lost. The specific concept-generation activity will lead to new alternatives and thus the need for another round of ranking and evaluation.

Iteration and Solution

This process of evaluation, refinement, and attacking the negatives should be repeated until the team converges. One of three outcomes occurs as the process completes:

1. *The team converges to consensus on a winning (or preferred) alternative.* Here, everyone understands the alternatives, why each was rejected, and accepts the judgment.

2. *The team comes to understand a refined choice.* Here, the team could not make a final decision because specific information is lacking. Two or more conditional alternatives cannot be distinguished because their rank on a criterion cannot be determined at this time. It is conditional on the outcome of an engineering exercise. In this case, a plan is developed to obtain the required information, and the concept selection reduces to being based on the outcome of that exercise. Everyone should agree with this determination.

3. *The team comes to understand that they do not understand the criteria.* Here, the team cannot reduce the choice to unclear alternatives that need refining, because they do not agree on the criteria. This situation, while generally rare, can happen. For example, there can be disagreement on what the customer wants from the product. In this case, the team should articulate the varying possible interpretations of the criteria and form a plan on how to establish the criteria for evaluation. In the example of disagreement in customer need interpretation, the team should establish a plan to statistically ask (or re-ask) the customer population for its preference. The selection process becomes deferred until the time of completion of the criteria refinement tasks. Everyone should agree with the plan and agree to be ready to complete the concept evaluation with the refined criteria definitions.

Example: Coffee Mill

As an example, consider the redesign of a coffee mill, restricted to using a chopper concept. Several ideas are developed to improve the mill, focusing on the evolution of the cleaning functions of the mill. These ideas are shown in Table 11.2. A Pugh concept selection process is

TABLE 11.2. COFFEE CHOPPER REDESIGN CONCEPTS

Removable Chamber	Removable Blade	Washable	Scrapered	Removable Unit
• The joint from the unit to the chamber • Noise suppression	• Contaminants in the center hole • Pop-off blade complexity	• Sealing • Heat buildup	• Effective wipe over 3D surface • Flatness during grinding • Dirt along scraper	• The joint from the unit to the chamber • Noise suppression

undertaken to resolve these concepts into one to pursue development. Each of these concepts is depicted on 8.5″ × 11″ sheets of paper and posted on a wall as columns. Discussion is facilitated to elaborate on the positives and difficulties of each concept. These difficulties are recorded as notes under each concept in Table 11.2. Refined sketches are also developed to clarify issues raised during the discussion.

Next, the evaluation criteria are elaborated, as shown in Table 11.3. These criteria arise directly from the customer needs and engineering specifications previously developed for the product. Again, factors elaborated about each criterion are shown in the table. Having completed this process, the next step entails the implementation of the Pugh selection process. The results of the concept ranking are shown in Table 11.4.

Some observations are apparent when completing this first round of the selection process. First, the power set-up criterion did not distinguish or discriminate between the concepts. They were all about the same. Therefore, although it is an important criterion for the product, the different concepts for promoting cleanability do not impact it. Therefore, it is dropped from further discussion.

TABLE II.3.   COFFEE CHOPPER REDESIGN EVALUATION CRITERIA

Cost	Unit manufacturing cost, not development cost, not delivery cost ($)
Store grinder	To put away in a cabinet out of sight (cm^3)
Put in beans	From the time beans are in a bag until the chopper switch can be activated (s)
Take out coffee	Removing all grounds, getting the coffee into a coffee maker (s)
Power setup	From the time the grinder is plugged in until the switch can be activated (s)
Cleanable	After coffee taken out to the point of being spotless (s or number of steps)
Development risk	Difficulty getting a working alpha prototype (number of potential faults or difficulties)

TABLE II.4.   COFFEE CHOPPER REDESIGN CONCEPTS FOR CLEANABILITY

	Removable Chamber	Removable Blade	Washable	Scraper	Removable Unit
Cost	S	+	-	+	S
Store Grinder	S	+	+	+	S
Put in Beans	S	S	-	-	S
Take Out Coffee	S	-	-	-	-
Power Setup	S	S	S	S	S
Cleanable	S	-	S	-	S
Development Risk	S	+	-	+	S
Σ+	0	+3	+1	+3	0
Σ-	0	-2	-4	-3	-1
Σ	0	+1	-3	0	-1

Next, it appears that the decision is reduced to the removable blade concept, and any ideas the team may have to convert any of the other concepts such that their negatives go away. So, each concept to be dropped is scrutinized. A suggestion is made to use a small electric plug at the base of the mill to permit it to be removed from the cord. Then the mill itself is as portable as a removable chamber and would make the scraper concept have a 0 rank on the "pour coffee out" criterion. This modification is then a new concept that competes effectively with the removable blade concept. And so on.

VI. ADVANCED DISCUSSION: MEASUREMENT THEORY

The Pugh chart selection process is effective, but there are many other methods available for design decision making. Numerical rankings, weighted sums, utility theory, probability theory, and fuzzy sets, for example, have all been advocated for different decision-making tasks.

One means to understanding these different methods is to examine the information quality used in the scores applied to each alternative. The basic Pugh process uses a very simple scale: better, worse, or the same. One might instead apply thoughts of "twice as good," or "half as bad," indicating ranks of 2 and $-\frac{1}{2}$, respectively. Measurement theory (Krantz et al. 1971) is an analysis approach to establish, for example, when this refinement in the selection scale can and cannot be done.

There are several levels of information quality to measure alternatives on different criteria. Information quality can be more formally considered as the level of structure associated with the alternatives and the criteria used. That is, alternatives can be *simply ordered* from worst to best, or they can be *comparatively ordered,* where their separation distance has meaning, or they can be *extensively ordered* with a complete and continuous numerical system. Before this discussion, we first review set requirements to form any ordering of the alternatives.

Set Structure of Evaluation

Evaluation methods in product development are utilized to decide among alternatives that a design team has identified: the set of alternatives to be considered. Though evaluation will help spark creative thoughts in a design team, the evaluation itself does not fundamentally generate new options. This viewpoint bounds the study of all strict evaluation: the most general object considered by an evaluation method is a *set*. When a design team needs to evaluate alternatives to make a selection, these alternatives must first be identified. This identification can be as simple as a label. Nonetheless, this requirement is fundamental: Only well identified objects can be evaluated and discussed. Set structure is needed among the alternatives before any evaluation can be made.

For example, the Pugh selection chart represents the set of alternatives with a finite set of labels, each label being listed along the columns of the matrix and each label being understood through the concept sketches and definitions. On the other hand, discussion about "everything in the universe of possibilities," or "anything that could possibly solve my task" is not suitable for evaluation. Only those objects actually described collectively as a set can be evaluated. This formalization requirement is the first axiom of any evaluation method.

Qualitatively, set structure in a collection is identified by two fundamental characteristics. First, given any arbitrary object, one can determine whether or not the object is in the set. Given a new alternative, a design team must be able to determine whether or not the new alternative is already in the collection of alternatives.

The second characteristic of set structure is that any two given objects in the collection can be differentiated. A design team must be able to distinguish the alternatives being evaluated. More complete definitions of sets can be given (Bernays 1968), but this qualitative definition suffices here.

Ordinal Scales

Having a set of product options is not sufficient to make a decision. A "best" or most preferred option must be identified: An evaluation structure must be constructed. The simplest structure is to assign each option a label of "equal or better" or "equal or worse." Such assignments form an ordinal scale over the set: Each option is deemed better or worse than its counterparts and placed in order. No numerical values are used.

To construct this ordering, the design team must be able to judge the alternatives. Out of the set X of all alternatives and in terms of the criterion under consideration, a design team must determine if a given alternative is better or worse than the others. If the design team can do this, then the alternatives can be formed into a weak order across the set X of alternatives:

$$x_1 \leqslant x_2 \leqslant \ldots \leqslant x_n. \qquad (11.1)$$

This ordering is called *weak,* since equivalence can hold. Mathematically, the relation is reflexive, transitive, and comparable. Alternatives can be rated as indistinguishable and so are equivalent to within the resolving power of the design team. The ordering provided by Eq. (11.1) is also called an *ordinal ranking:* The elements are only ordered. An ordinal statement of "$x_1 \leqslant x_2 \leqslant x_3$" says nothing about how far apart x_1 is from x_2 relative to the separation of x_2 from x_3. It says only that x_1 is lower than x_2, which is lower than x_3.

A design team must have a reason for making such statements to form a weak ordering. There must be reasons for the judgment in each evaluation. These reasons might be customer needs such as cost, appearance considerations, or desired colors. The reasons could also be engineering criteria such as stress levels. In addition, availability concerns such as lead times, available sizes, or available off-the-shelf components can be used. Any reasoning on which a product design can be judged forms an *evaluation criterion,* ϕ for rating designs. The label ϕ is not to signify that a criterion is a numerical function; rather ϕ is merely a label. The existence of at least one such reason is the second axiomatic assumption required to form an evaluation.

The Better/Worse Method

One evaluation procedure commonly used is the *better/worse method.* First we select an arbitrary alternative from X to form the beginning of a list $\{x_0\}$ indexed by j. For every other alternative, we start at the bottom of the list and move up through the list, asking

Is design x_i as good as or better than design x_j in ϕ? (Q1)

Once x_i is no longer better than x_j, we then insert x_i into the list at that point. The set S of possible ranks for any evaluation is:

$$S = \{\text{better, same, worse}\}$$

The line of questioning (Q1) interactively produces a bubble sort of the alternatives. The best or most preferred alternative is at the head of the list, and the evaluation among the alternatives is complete.

Effectively, any one of the alternatives serves as a datum for the better/worse method. That is, every alternative is directly compared against all of the others. Therefore, there is a datum in the evaluation; in fact there are many.

Given a single criterion and a set of alternatives, this procedure allows a designer to construct a complete evaluation of the alternatives and thereby make a selection. The alternative at the head of the list is better than the rest for the criteria ϕ and is therefore recommended. For multiple criteria, however, the process must be expanded. Before doing so, we visit another ordinal scale alternative.

Pro/Con Charts

A *Pro/Con chart* is a minimally structured decision-making tool for problems with multiple alternatives x_i and multiple criteria ϕ_j. Here, a design team constructs a chart with the alternatives represented by columns. Two rows are then created. In the top row, criteria are entered that an alternative is rated positively on and are considered as "Pro" for the alternative. In the bottom row, the criteria are entered that an alternative is evaluated negatively on and are considered as "Con" for the alternative. Then the design team considers this chart result as one holistic expression. Typically, such a Pro/Con chart is used to help identify one's thoughts succinctly or to agree as a group on a winning alternative.

When a Pro/Con is used as a basis of evaluation, however, it should be understood what underlies the mechanics of the process. The information required and the elemental steps that must be applied should be analyzed. With a Pro/Con chart, the set of alternatives is ranked into the evaluation levels "Pro," "Con," or "Neither" for each of the different criteria. These evaluation levels can be identified into a simple set of rank values:

$$S = \{\text{Pro, Con, Neither}\}$$

The ordinal question asked on each alternative x_i for each criteria ϕ_j is:

Does alternative x_i get a "Pro" or a "Con" in ϕ_j?   (Q2)

Generally speaking, if ϕ_j is better, then x_i gets a "Pro" rating; if worse, then x_i gets a "Con" rating. If it is neither better nor worse, then the criterion ϕ_j remains blank as a "Neither" on x_i.

Notice that a difficulty arises in this approach: There is no datum for the ranking. For each criterion, the Pro/Con method attempts to rank each configuration in a general reference frame that does not indicate what either "Pro" or "Con" means. "Pro," "Con," or "Neither" as compared to what? Such evaluation without a datum can prove difficult to maintain consistency.

After the ranking is complete for each criterion, the alternatives must still be ordered. Typically, this ordering is completely informal by holistically considering all criteria simultaneously and informally completing the bubble sort using the better/worse method. The original ordinal rank construction question (Q1) is asked not on just one criterion, ϕ, but on all of the criteria $\vec{\phi}$ simultaneously. With a Pro/Con chart, all the criteria must be considered simultaneously when placing each alternative x_i in order.

No numerical calculations are used with a Pro/Con evaluation; rather a design team must mentally make an unsophisticated (though probably very difficult) determination on which option is "on the whole" better than the others. This determination involves simultaneously comparing qualitative measures of incommensurate criteria. The Pro/Con does help in that it provides a two-level ordinal ranking of "Pro" and "Con." The remainder of the evaluation is informal. This informality limits the number of criteria that can be considered with a Pro/Con evaluation, since most humans can only process around seven or eight chunks of information simultaneously. Given this constraint, a Pro/Con chart is limited to around three or four criteria that can be traded off.

On the other hand, very little structure is required to construct the evaluation. The only structure assumed is that the options are well identified and that criteria exist for making the evaluation.

As an example, consider the coffee mill redesign once again. One could list the criteria as "Pro" or "Con" and make a qualitative decision based on these rankings as shown in Table 11.5.

Pugh Charts

The Pugh chart evaluation scheme presented earlier involves sets of alternatives, sets of criteria, and a better/same/worse evaluation against a datum. We can address the mechanics of the Pugh chart in an objective way.

With the set of alternatives and criteria, there are three basic mathematical tasks that are completed:

1. Identify a datum alternative.
2. Rank every other alternative as better/same/worse for each criterion ϕ_j.
3. Sum the ranks for an overall evaluation of each configuration.

Consider the first task (Step 1). When constructing the chart, an effective process demands that a designer first select an alternative to serve as a datum. In measurement theory, an alternative selected as the zero point for a scale is termed the *base point*. Its criterion value effectively serves as a "zero" on the measurement of the criteria over the set of alternatives; all others are rated as higher or lower in the criteria.

Having selected a base point (datum), the next task (Step 2) is to rank each alternative in comparison to the datum, using evaluation levels of

$$S = \{-, S, +\}$$

TABLE 11.5.   PRO/CON CHART FOR THE COFFEE MILL SELECTION

	Removable Chamber	Removable Blade	Washable	Scrapered	Removable Unit
Pro	Cleanable	Cost Store Grinder Development Risk	Store Grinder Cleanable	Cost Store Grinder Development Risk	Cleanable
Con		Take Out Coffee Cleanable	Cost Put in Beans Take Out Coffee Development Risk	Put in Beans Take Out Coffee Cleanable	Take Out Coffee

This set forms a weak ordering (into three groups) of the alternatives in terms of each criterion ϕ_j. Ordinal scales $(X, \preccurlyeq)$ have been constructed over the set of alternatives X, one each in terms of the criterion ϕ_j. Therefore, Steps 1 and 2 are well grounded in measurement theory. The process results in several mathematically correct weak orders among the alternatives.

The next task in the process is to combine these measured results (Step 3). The procedure is to "add" the rankings as if they were numbers. If the overall rank is used as an objective figure of merit to select a preferred concept, then, mathematically, difficulties will arise. To "add" the rankings, then a distance separation has been assigned to the rank set:

$$-1 \cong \text{worse}$$
$$0 \cong \text{same} \qquad (11.2)$$
$$1 \cong \text{better}$$

This mapping, Eq. (11.2), must be closely examined. First, it implies that in the overall rating, S_i, every criterion is assigned an impact of ± 1. This impact must mean that a selection based on the final additive sum is in fact based on the net number of criteria that an alternative offers for improvement. If a final result has an alternative ranked at +3, this means that the alternative is better than the base point by a net of three criteria. This statement is the only interpretation of the final result, whereby the $(-,S,+)$ rank can be evaluated ordinally using the equivalence of Eq. (11.2).

It is questionable if this result is an objective figure of merit for a selection method. Therefore, it should not be used as an overall rank but rather as an indicator of what to do next in the selection process, as discussed earlier. If there are negatives, they should be attacked. The Pugh process is not about making relative trade-offs as a selection technique; it is for improving conceptual designs. In this sense, the ordinal evaluation mathematics and the Pugh analysis relative to a datum are entirely correct.

Interval Scales

Suppose a design team wants information reflecting not just whether one alternative is better or worse than another but how much better or worse. To provide this information, more information quality than that provided by an ordinal scale is needed. The ordering provided by an ordinal scale only states that alternative x_i is better or equal to alternative x_j, $(x_i \leqslant x_j)$, in a criterion. The ordering does not provide information as to how much better. To represent this information, more than order structure must be constructed over the set. An interval scale must be constructed.

Interval scales are constructed for criteria that have no identifiable *units of measure*, but a quantitative scale is desired. For example, consider a typical concept evaluation criterion of "technical risk." A team might have evaluation levels of "low," "medium," and "high." This criterion is identified as an ordinal set and no additional structure is provided by "technical risk," since "technical risk" as identified has no units. What is zero "technical risk?" What is a unit of "technical risk?" Interval scales are used to answer these questions.

To state how much better one alternative is relative to others, additional structure is constructed by first identifying an arbitrary reference alternative, similar to ordinal scales. This alternative will again be called the *base point alternative,* from which all others are referenced relatively. For example, when rating alternatives on technical risk, the least risky alternative can form an acceptable base point.

A base point, however, is not the only structural information needed. A *metric* alternative must additionally be identified to compare the deviation of the criteria from the base point. A metric alternative is an element in the set X of alternatives against which all other alternatives will be compared relative to the base point. For example, with the technical risk criterion, an alternative that exhibits the most technical risk might be used to compare the remaining alternatives in risk compared to the base point alternative with least technical risk. We might assign the metric alternative a risk of 1.0. Each remaining alternative can then be rated between the metric and base point alternatives. A design that is judged midway between the least and most risky alternatives can be rated 0.5 in "technical risk."

The need for identifying both a base point and a metric alternative is axiomatic to forming interval scales. Further, interval scales are required to *quantify* how different alternatives perform. Therefore, without two datum alternatives, quantitative ratings are not possible. This is a fundamental measurement fact that must be incorporated into quantitative design evaluation.

Constructing Measurement Scales

There are various methods available for constructing interval scales over arbitrary sets. Two methods stand out: the basic *lottery method* (Krantz et al., 1971) and a numerical form of the *better/worse method*.

The lottery method: To apply the basic lottery method, a set of alternatives X is required, over a criterion ϕ. This method will allow a designer to construct a real valued scale f that reflects the informal judgement criterion ϕ. The design team must first identify which alternatives in X have the least and most amount of the criterion ϕ, denoted x_0 (the base point) and x_1 (the metric alternative), respectively. These are designated with values of zero and one on the real valued interval scale being constructed. Then. for each other element $x_i \in X$, the design team must answer the lottery question:

> *On a scale of zero to one, what is your belief f that you are indifferent between:*
>
> *1) receiving the criteria performance provided by x_i,*
>
> *or* (Q3)
>
> *2) receiving the criteria performance provided by x_1 with certainty f and receiving the criteria performance provided by x_0 with certainty $(1 - f)$?*

This questions constructs a real valued (measurable) scale $f: \mathbb{R} \rightarrow [0,1]$ that directly preserves the weak ordering of the alternatives X and also the relative separation. A difference of 0.5–0.3 in f means a larger difference in the criterion ϕ.

This basic lottery method also obviously assumes finite sets. On un-countable sets, further continuity assumptions are required for the procedure to be practical. The designer must provide lottery answers on a finite subset that can then be interpolated if an additional piecewise continuity axiom is assumed.

The lottery method is one way of measuring the differences among objects in an arbitrary set. The method starts with a worst and best case object in the set: The worst case object is used as the base point, and the best case object is used as the metric. All other objects are compared with the metric relative to the base point. The measurement scale constructed is always positive; it has an image subset in $\mathbb{R}$ of $[0,1]$. Notice that an interval scale in and of itself does not compare different criteria quantitatively. It rigorously constructs a scale for a single criterion.

The better/worse method: Rather than use the best and worst case product options as in the lottery method, it is also possible to use different, mathematically arbitrary alternatives to construct the measurement. An arbitrary alternative, for example, might be the current product. Doing so, however, requires the designer to measure in both positive and negative directions from the arbitrary base point using the arbitrary metric alternative: Some alternatives may be worse than the base point.

Therefore, if the designer selects arbitrary product options for measurement, questions of "as poor as the metric option is good" must be asked in addition to "as good as the metric option." A method similar to the lottery method might be: Given a base point alternative, b, and a metric alternative, m (that has a different level of performance than b), for all $x_i \in X$:

> Using a real number f, how many times the metric alternative m do you believe alternative x_i is better/worse than the base point b in the criterion ϕ?
>
> (Q4)

Worse ratings are assigned negative numbers. This construction results in a scale whose image subset of $\mathbb{R}$ will have both negative and positive values: $f: \mathbb{R} \to \mathbb{R}$. A designer must be able to rate some alternatives as worse than the base point, given that it is arbitrary and is not necessarily the worst case.

Numerical Concept Selection Charts

A numerical concept selection chart is a simple device used by many design teams to conduct design evaluations. With a numerical concept selection, the alternatives x_i are listed along the columns of a matrix. The criteria ϕ_j that form the basis of the selection are listed along the rows.

Entries in the matrix are the ratings of each alternative on each criterion and are assigned rankings from the set:

$$S = \{--, -, \surd, +, ++\}$$

or perhaps with additional or fewer multiple $(+)$s and $(-)$s. This set has an understood integer scale associated with it in terms of the multiplicity.

In the evaluation process, the rankings are summed across the different criteria for each alternative, and the alternative with the highest sum is nominally determined to be the most promising candidate.

Evaluation Process

When analyzing only the mathematics of the method, we can clarify the need for four separate tasks when constructing the rankings in the chart:

1. Identify a base point and metric configuration.

2. Construct a rating across the configurations for each of the criteria.

3. Normalize each criterion's measurement scale.

4. Summarize the normalized ratings for an overall evaluation of each configuration.

Consider the first task. When constructing the chart, best design practice calls for a design team to first select a reference alternative. In redesign applications, this reference case is usually the existing product or perhaps a competitor's product. This case is assigned $(\surd)$ rankings in every criteria. Every other alternative must be rated on each criterion relative to this reference, assigning ranks from the rank set S.

We can now proceed to the second task. For each criterion, the remaining alternatives are now ranked, scaling relative to the base point in proportion to the deviation of the metric.

We must now combine these measurements. We cannot simply add them, however, since the criteria have different scales. We must "normalize" them. This additional step is a concept that interval scales, in and of themselves, are not capable of handling. The interval scale provides a rigorous numerical construction that can then be quantitatively manipulated.

One approach that is effective in industrial practice is to limit each criterion to a maximum impact in the evaluation, called the *maximum impact normalization*. Here, for each criterion, one establishes maximum values in *S* that may be used during the assigning of rankings. One criterion might be a $(++)$, and others might only be $(+)$s. These values are used to modulate the relative values determined from the better/worse construction.

The better/worse method constructs a relative scale among the alternatives. To combine the rankings into an overall figure of merit, the scale ranges must be modulated so they are appropriate in size. To do this, when using the maximum impact normalization, we multiply each value by the maximum allowed effect and divide by the difference between the highest and lowest score.

Completing the normalization task sets up the final step of concept evaluation. The normalized values are summed. The evaluation process provides a recommendation for the alternative with the highest sum.

Analysis of concept selection mathematics: Analyzing this rigorous form of concept selection using numerical values, several features are clear. First, the concept selection method has a clear base point for each criterion, namely, the reference concept that is assigned a rank of $(\surd)$. Second, any alternative assigned a rating of $(+)$ can be used as a metric option. That is, for numerical evaluation, two alternatives are needed to rank the remaining alternatives. Merely identifying a datum is insufficient. That two alternatives are needed to construct numerical concept selection ranks (and not just one datum alternative) is a measurement fact not commonly recognized.

After identifying the base point and a metric alternative with a $(+)$ rank, each subsequent design can be rated as, for example, "twice as good as the metric from the base point" and assigned a $(++)$, or "as bad as the metric was good from the base point" and assigned a $(-)$.

When restricted to the rank set $S = \{-, \surd, +\}$, questions can arise over whether the evaluation mechanics are ordinal or interval. Such is the case with a Pugh chart. Do the rankings only order the alternatives, or do they provide information reflecting how much better or worse the alternatives are? Pugh originally advocated only the former, but it depends on the interpretation of the final overall figure of merit S_i. If the resultant overall figure of merit is interpreted as the number of criteria that an alternative is better/worse than others, then the individual ranks are ordinal. If, on the other hand, the final figure of merit is interpreted as a numerical value of "overall goodness," then the individual ranks are being manipulated as numerical values. A $(+)$ is better than a $(\surd)$ to the same degree that a $(-)$ could cancel it. Therefore, the assumption when using concept selection charts with only a rank set of $\{-, \sqrt{}, +\}$ is that all of the criteria have equal importance or that the final figure of merit represents the number of criteria. This interpretation is a subtle but significant fact of measurement theory.

Ratio Scales

Interval scales construct a quantitative numerical evaluation over an arbitrary set with arbitrary criteria. The next level of structure that can be considered is criteria that inherently have some reason for anchoring the numerical scale. An interval scale does not define a "zero" value for the criteria. Instead, an arbitrary value (usually zero) is assigned as the base point alternative's criterion value, with the understanding that any arbitrary constant could be "added" to this value and the results of the evaluation remain invariant. The evaluation result is dependent on the numerical differences, not the absolute numerical values.

Rather than assigning the base point alternative a value of zero, perhaps there is a specific value of a criterion that is understood as "zero." Consider evaluation based on the technical risk criterion. Products already developed have no technical risk and could be assigned the value $(\sqrt{})$ rather than an arbitrary base point configuration that may cause some technical risk. Such a "zero" is a *base value* of a criterion rather than the value of the criterion assigned to a base point alternative.

If there is a naturally defined base value for a criterion, there is more structure inherent to the problem. The base value can be used in the evaluation measurements, and the scale is termed a *ratio scale*. There is still insufficient structure to naturally define a metric value (such as 1), and so a complete numerical scale is not yet naturally defined. For example, what is a unit of technical risk? It is not easily defined. There is, however, more structure to operate with than a simple interval construction. "No technical risk" is easily understood.

Constructing Ratio Scales

The procedure for constructing a ratio scale is very similar to constructing an interval scale. The only difference is that the criterion value returned by the base point configuration is replaced by the base value of the criterion.

Given an alternative $m \in X$ chosen as a metric alternative and a base value 0_f of the criterion ϕ, the better/worse method can be used to construct a rating over all other alternatives $x \in X$:

Using a real number f, how many times the metric alternative m do you believe alternative x is better/worse in the criterion ϕ than the base value 0_f? (Q5)

Worse ratings are assigned negative numbers. This construction results in a scale whose image subset of $\mathbb{R}$ will have both negative and positive values: $f: X \rightarrow \mathbb{R}$.

With this scale, no single alternative $x \in X$ may end up getting assigned a 0 value (a $\sqrt{\ }$). Since it is known what 0 means on this scale for ϕ, it may be different from the values of the considered alternatives.

Examples of Ratio Scales in Design

Examples of ratio scales that are commonly used include portions of QFD matrices (Chapter 7), design evaluation charts, and determination of criterion importance. With QFD, ratings are placed inside the matrix, relating customer needs to specification levels. Degrees of association are represented in the matrix with levels from the set $S = \{0, 3, 5, 9\}$, and zero is understood to mean "no relationship between the customer need and the specification."

Another example of using a ratio scale involves the common product development task of establishing the relative importance of multiple real valued criteria. While there are many arguments against using weighting factors in any form (Steuer, 1986), it remains true that they are commonly used in product development practice. Sound methods of construction should be developed. Rating importance must involve a ratio scale. Importance can be considered a "criterion" to be evaluated across the option set Φ of multiple evaluation criteria. Zero importance is a well understood concept as a criterion with no relevance to the decision. A unit of importance, however, is not well understood. Thus, importance falls into the ratio scale domain.

A procedure for determining importance is to apply the ratio scale construction question (Q5). This form is the well known *marginal rate of substitution* question (Steuer 1986). One criterion ϕ_1 must be selected as a metric, either arbitrarily or perhaps by the one that is most understood. Then, for all of the other criteria ϕ_j, we ask:

By how much should ϕ_j be increased to compensate for a loss of one unit in ϕ_1? (Q6)

This question defines relative amounts of importance Δ_j for each ϕ_j. Notice the inherent dependence on the amount $\Delta_i = 1$ used. A problem with weights is that the results can change with smaller or larger

Δ_1. In any case, the results can be scaled by the sum to define absolute importance:

$$\omega_j \approx \frac{\Delta_j}{\Sigma \Delta_i}$$

Extensively Measurable Scales

In many cases, the appropriate scale for rating different alternatives is naturally present and given: The set X of elements in question has additional structure that makes the scale clear. Methods of constructing scales, such as the better/worse method, are not required. For example, criteria that are already represented with real numbers can use the real numbers to order the alternatives on the criteria. Cost, for example, has a base value of no cost and a metric of a dollar (or other denomination), if the alternative is resolved to the nearest dollar. Likewise, length has zero length as a base value and may have a metric of a thousandth of an inch, or a millimeter, depending on the choices of units and resolution.

What these criterion sets have in common is they can be extensively measured: They have a base value, a metric value, and a well defined *concatenation operation* (Krantz et al. 1971). There is a well defined operation over the set, generally called concatenation, that can be used to "add" the metric value (called a *unit of measure* in this case) recursively to the base value to produce any other value in the set. The unit of measure defines the fundamental "units" and "resolution" as commonly understood in engineering practice.

VII. ADVANCED METHOD: NUMERICAL CONCEPT SCORING

The discussion on interval and ratio scales can be applied to complete a numerical concept selection. As an example, consider a material selection problem for a refrigerated food preparation surface material (to be subsequently coated) for use in an ice cream store, where hand mixing of ingredients on the surface is needed.

Scoring with Interval Scales

This material selection example has the benefit of measurable values for the criteria, making it a useful case study. In particular, the required thickness for proper heat transfer, conductivity, thermal mass, diffusivity,

TABLE 11.6. PERFORMANCE VALUES FOR THE MATERIAL SELECTION EXAMPLE

Criteria	(units)	Steel 1020	Stainless 304	Aluminum 5052	Copper	Bronze
Thickness	inches	0.107	0.107	0.407	0.205	0.205
Conductivity	Btu ft/hr °F ft^2	27	9.4	80	200	109
Thermal mass	Btu/°F	2.93E-04	3.43E-04	7.33E-04	5.45E-04	5.26E-04
Diffusivity	ft^2/hr	909	270	3749	6751	3809
Hardness	Brinnell	111	240	47	2	1
Yield	psi	30,000	110,000	13,000	40,000	37,000
Machinability	[0–100]	65	90	30	20	20

hardness, yield strength, and machinability are all metrics with tabulated numerical values for different alternative materials. Suppose the decision has been reduced to considering 1020 steel, 304 stainless steel, 5052 aluminum, copper, and bronze, thus completing the first task of design evaluation. The values of the criteria for each of the alternative materials are shown in Table 11.6.

The level of performance criteria provided at each configuration can be denoted as a mapping given by:

$$p = f(d) \tag{11.3}$$

where p is the level of provided criterion performance, d is an alternative material, and f is the mapping that provides the level of performance, given the configuration (a lookup in a table or handbook for this example).

The next step is to use an interval scale (Q4) or ratio scale (Q5) to transform the performance values of Table 11.6 into relative value scores. We will denote this transformation as:

$$r = g(p), \tag{11.4}$$

where r is a raw *value* score, p is a performance level from the last mapping, and g is a mapping reflecting the designer's judgment of the relative value of p, given the base point and metric alternative values. A linear relationship can be given by:

$$r = \alpha \, \frac{p - p_{\text{base point}}}{p_{\text{metric}} - p_{\text{base point}}}$$

$$\text{where } \alpha = \begin{cases} +1 & \text{metric better than basepoint} \\ -1 & \text{metric worse than basepoint} \end{cases} \tag{11.5}$$

TABLE 11.7. INTERVAL MEASUREMENTS FOR THE MATERIAL SELECTION EXAMPLE

Criterion	Steel 1020	Stainless 304	Aluminum 5052	Copper	Bronze
Thickness	√	√	− − −	−	−
Conductivity	√	−0.5	+	3.5+	1.5+
Thermal mass	√	√	− −	−	−
Diffusivity	√	√	+	+ +	+
Hardness	√	+ +	−	−1.5	−1.5
Yield	√	11.5+	−2.5	1.5+	+
Machinability	√	−	1.5+	+ +	+ +

The results of the raw score mapping are shown in Table 11.7. This mapping is not always necessarily a linear function. For example, stress levels below the ultimate tensile stress (UTS) may provide a monotonic value relationship with stress, but beyond UTS, the value is zero. Asking the interval scale construction question would construct such a nonlinear map.

We must now combine these raw value measurements across the different criteria into a final rating for each option. Although the rated values for any criterion are unitless and proportional, we cannot simply sum across the different criteria, since each criterion has different unitless ranges. Thus, we must "normalize" or "weight" the criteria to reduce/expand their ranges until each can be added with the others. An approach we have often used effectively in industrial practice is to limit each criterion to a maximum impact in the evaluation, which we will refer to as the *maximum impact normalization*. Maximum impact limits for the material example are shown in Table 11.8. These

TABLE 11.8. MAXIMUM ALLOWED EFFECT FOR THE MATERIAL SELECTION EXAMPLE

Criterion	Maximum Effect
Thickness	+
Conductivity	+ +
Thermal mass	+ +
Diffusivity	+
Hardness	+
Yield	0.5+
Machinability	0.5+

limits range from $(+1/2)$ to $(++)$, implying that the criteria are considered to be one to four times as important relative to each other. Ideally, only integer values $(+)$ and $(++)$ would be used. However, sometimes this scale provides an excessively course resolution, and so half-points are required. In theory, any real valued number could be used, determined with a marginal rate of substitution question. That is, choose one criterion and define its impact as $(+)$. Then for each other criterion, we ask:

By how much should ϕ be increased to compensate for a loss of one unit in ϕ_1? (Q7)

This question produces real numbers for the impact on each criterion. Past practical experience, however, suggests starting by rounding to either $(+)$ and $(++)$ and iteratively halving the resolution until satisfactory weightings arise.

These maximum impact values from S, as shown in Table 11.8, are used to modulate the relative values determined from the better/worse construction. The better/worse method constructs a relative scale among the alternatives. To combine the rankings into an overall figure of merit, the scale ranges must be modulated so they are "appropriate," that is, weighted according to the desired impact by the designer or customers. To do this when using the maximum impact normalization, we multiply each value by the maximum allowed effect (impact) and divide by the difference between the highest and lowest score. This calculation introduces another mapping:

$$N = h(r, I) \qquad (11.6)$$

where r is the raw score, and I is the maximum allowed impact. A typical linear transformation is:

$$h(r, I) = I\frac{r - r_{base}}{r_{max} - r_{min}} \qquad (11.7)$$

where I is the impact, $r_{base} = 0$, r_{min} is the minimum raw score, and r_{max} is the maximum raw score value. The results of this mapping are shown in Table 11.9. Note that the linearized value approximation (Eq. (11.5)) can lead to some results that might not be intuitive (Table 11.9). For example, the yield strength of stainless steel is much higher than all other alternatives and so obscures the differences in all other yield-strength normalized ratings in Table 11.9. If this relationship is not acceptable, a different value transformation (Eq. (11.7)) can be applied.

TABLE 11.9. OVERALL RATINGS FOR THE MATERIAL SELECTION PROBLEM

Criteria	Steel 1020	Stainless 304	Aluminum 5052	Copper	Bronze
Thickness	0	0	–	−0.5	−0.5
Conductivity	0	−0.5	0.5+	++	+
Thermal mass	0	0	– –	–	–
Diffusivity	0	0	0.5+	+	0.5+
Hardness	0	0.5+	−0.5	−0.5	−0.5
Yield	0	0.5+	0	0	0
Machinability	0	0	0.5+	0.5+	0.5+
Sum	0.0	0.5	−2.0	1.5	0.0

Summing the results in Table 11.9 produces the overall numerical rating, M_i, for the evaluation:

$$M_i = M(d_i) = \sum_j N_{ij} \qquad (11.8)$$

where i is an index for the number of alternatives. The overall ratings from this calculation are shown in bottom portion of Table 11.9. The analysis would indicate a selection of copper.

Selection Error Analysis

Having clearly represented each step required to complete a numerical concept selection as a mapping, another question might address the uncertainty and risk in the decision. Given the clear definition of the selection process through measurement theory, we can now analyze the evaluation for uncertainties. That is, we can assign and propagate error ranges of uncertainty. The final overall ranking for each alternative can be represented as the composition of each mapping, representing each operation the designer must complete:

$$M_i = M(d_i) = \sum_j h_{ij}(I_j) \circ g_{ij} \circ f_{ij}(d_i)$$

$$= \sum_j h_{ij}(I_j, g_{ij}(f_{ij}(d_i))) \qquad (11.9)$$

where $\circ$ denotes a composition operator. Observing this relationship, one can incorporate uncertainty information within each mapping and propagate the uncertainties into the final rating.

Uncertainty Specification

To model uncertainty for each mapping, one can use tolerances about the nominal selection values. For each value, these tolerances are represented as intervals. There are three such tolerance specifications required for each of the three mappings in the numerical evaluation. These tolerances are:

▶ *Delivered Performance Uncertainty* $p \pm \delta p$. This is the uncertainty in the level of performance that the alternative will deliver, irrespective of how much value the design team interprets in the performance metric value or the importance placed on the performance metric.

▶ *Value Ranking Uncertainty* $r \pm \delta r$. This is the uncertainty in the magnitude of value the design team assigns to the level of each performance metric as compared to the other levels of the same performance metric. It corresponds to the error associated with the rank of an alternative as assigned and is independent with respect to other performance metrics.

▶ *Impact Scaling (Weighting) Uncertainty* $I \pm \delta I$. This is the uncertainty in the level of importance (weight) that the design team prescribes to the performance metric as compared to the other performance metrics.

Consider the first uncertainty in delivered performance. Such uncertainties arise from modeling and measurement errors in the calculations or experiments completed to attain the tabulated values. Table 11.10 lists representative performance uncertainties for the material selection example. As an example, consider the yield strength data in the table. A tolerance of $\pm 15,000$ psi ($\sigma = \pm 5,000$ psi) for steel may be

TABLE 11.10. PERFORMANCE METRIC VALUE UNCERTAINTY FOR THE MATERIAL SELECTION EXAMPLE

Criterion	Units	Steel 1020	Stainless 304	Aluminum 5052	Copper	Bronze
Thickness	inches	±0.01	±0.01	±0.04	±0.02	±0.02
Conductivity	Btu ft/hr °F ft^2	±3	±1	±8	±20	±10
Thermal mass	Btu/°F	±3E-05	±3E-05	±7E-05	±5E-05	±5E-05
Diffusivity	ft^2/hr	±90	±30	±400	±700	±400
Hardness	Brinnell	±10	±10	±5	±0.5	±0.5
Yield	psi	±5000	±5000	±5000	±5000	±5000
Machinability	[0–100]	±5	±5	±5	±5	±5

TABLE 11.11.   INTERVAL MEASUREMENT TOLERANCES FOR THE MATERIAL SELECTION EXAMPLE

Criterion	Steel 1020	Stainless 304	Aluminum 5052	Copper	Bronze
Thickness	±0	±0.5	±0.5	±0.5	±0
Conductivity	±0	±0.5	±0.5	±0	±0.5
Thermal mass	±0	±0.5	±0.5	±0.5	±0
Diffusivity	±0	±0.5	±0.5	±0.5	±0
Hardness	±0	±0.5	±0.5	±0	±0.5
Yield	±0	±0.5	±0.5	±0.5	±0
Machinability	±0	±0	±0.5	±0.5	±0.5

indicative of the measurement error of a tensile test or the variance associated with a replicated design of experiments.

The second uncertainty in value is in subjective design team judgment. It reflects how a design team's certainty or confidence in the value of the performance compares to the other values of the performance. Thus, these error values must be compiled by examining each performance metric one at a time. Note that we define the relative value of the base point to be zero, the comparison point, and the relative value of the metric point to be one. There is no uncertainty in these two assignments. The uncertainty is in comparing all other levels to these two. Table 11.11 contains the uncertainties in value for each metric by each material. Notice that for each performance metric, the base point and metric materials have no uncertainty in their relative value.

To estimate the value ranking uncertainties, a number of approaches might be used. A practical approach for specifying this uncertainty is to ask each member of a design team or to ask a focus group of customers (of appropriate sample size) the better/worse question and use the ensemble statistics to define the uncertainty. Alternatively, the following question might be asked:

> For each interval measurement, what is the uncertainty in the ranking as
> compared to the difference of the metric from the base point?   (Q8)

The results of such questioning can be tabulated, and the value ranking uncertainty can be estimated based on the average. It should be directly apparent that any element in Table 11.7 that is a base point or a metric value will have zero uncertainty with respect to the interview question (Q4). For example, the question, "What is the uncertainty of the difference of steel from steel relative to the difference of bronze from steel?" will obviously result in the trivial result of a zero tolerance. Table

11.11 reflects this approach for the material selection example, where each of the base and metric points has a zero tolerance value. The results tabulate the metric points from each member of a design team or from customers through a semantic inquiry of how much difference in a criterion is significant. In Table 11.11, for each criterion, the error values are the same across all the alternatives. This need not be the case.

The third uncertainty, impact scaling or weighting, is also a subjective designer judgment. It reflects how certain a designer is in the importance (weighting) of each performance metric compared to the others. One approach for specifying impact scaling uncertainties is to choose one of the maximum allowed effects of Table 11.8 and compare all other effects to it (Q7). Importance needs only one of the values defined, since there is a well defined zero of "no importance." For the materials selection example, we choose the thickness criterion, arbitrarily, as a basis for comparing the importance (weighting) of all other criteria. All other metric maximum effects are defined relative to the standard impact of thickness. Table 11.12 tabulates the uncertainty in impact for each metric, where thickness is shown to have a zero tolerance, since it is used for comparison.

The designer may use an alternative approach to specify the impact scaling uncertainty, in a similar manner to the value ranking uncertainty. For example, each member of a design team, or each member of a group of customers, can be asked to independently list the maximum allowed effects (weights) for a concept selection problem. The average of the tabulated results of this inquiry corresponds to the default maximum allowed effect in the form of Table 11.8, and the standard deviation corresponds to one third of the tolerance for the impact scaling uncertainty. As an example, consider the materials concept selection example. Assume that the materials selection problem is part of a

TABLE 11.12.   TOLERANCES IN MAXIMUM ALLOWED EFFECT FOR THE MATERIAL SELECTION EXAMPLE

Criterion	Maximum Effect
Thickness	± 0
Conductivity	± 0.5
Thermal mass	± 0.5
Diffusivity	± 0.5
Hardness	± 0.5
Yield	± 0.5
Machinability	± 0.5

TABLE 11.13.   UNCERTAINTY IN MAXIMUM ALLOWED EFFECT: DESIGN TEAM AVERAGING

Criterion	Designer #1	Designer #2	Designer #3	Designer #4	Designer #5	Designer #6	Average	σ
Thickness	1.00	1.00	1.00	1.00	1.00	1.00	1.0	0.0
Conductivity	1.50	2.00	1.75	2.50	1.75	2.00	1.9	0.3
Thermal mass	1.75	1.75	2.00	2.00	1.50	2.50	1.9	0.3
Diffusivity	1.00	1.50	0.75	1.00	0.75	1.00	1.0	0.3
Hardness	0.50	1.00	0.50	0.50	1.00	1.00	0.8	0.3
Yield	0.25	0.50	0.75	0.50	0.25	0.50	0.5	0.2
Machinability	0.50	0.50	0.75	0.50	0.50	0.25	0.5	0.2

broader product design, where the design team is composed of six members. Each team member assigns the maximum allowed effects (weights) of each criterion, where there is an agreed-on upper bound. Table 11.13 shows the results of this process, with the corresponding average effects and impact scaling uncertainty.

Given these specifications of uncertainties, each must be combined into a final overall error using an appropriate algebra to produce an overall uncertainty in the final scores. The next section develops this overall error estimation.

Linearized Uncertainty Characterization

Given that tolerances are specified for each judgment that composes the numerical ratings, the next task is to combine these into an overall error. There are many possible schemes to perform this combination. To determine which appropriate mathematics to apply, an interpretation must be assigned to the tolerances previously elicited. What does "$\pm \delta X$" mean? If it implies an uncertainty in what choice a designer should prefer, fuzzy set mathematics can be argued as appropriate. If it implies a random uncertainty, probability mathematics can be argued as appropriate.

We choose to apply a probabilistic interpretation to the uncertainty ranges. This probabilistic approach aligns with historical subjective probability theory (Savage 1954). Other applicable methods might include fuzzy set mathematics (Zimmerman 1985; Antonsson and Otto 1995) or interval mathematics (Ward and Seering 1990). Though these other methods might be argued as more appropriate, we choose not to use these approaches due to their current lack of derivable confidence measures in results.

The underlying assumptions of this approach are as follows: (1) The criteria are independent (corresponding to Pugh's method or decision matrices); (2) the error ranges are independent; and (3) the evaluation compositions are linear and follow superposition. Using these assumptions, probability mathematics lead to a tolerance "$\pm \delta X$," implying that the true value of an independent variable x behaves as a normal distribution, centered at x with a standard deviation of σ. The standard deviation relates to the tolerance δX by the following relationship:

$$\delta X = n\sigma, \quad n = 3, \ldots, 6 \tag{11.10}$$

By convention, n is chosen such that $\delta X = 3\sigma$, meaning that a tolerance at any step of the concept selection method is 3σ from the nominal value specified.

With the subjective probability assumption, the resulting uncertainty range on the final evaluation can be readily calculated through application of probability mathematics. Given an invertible mapping $y = f(x)$ and a random variable X, it can be shown (Sveshnikov 1968) that the induced probability density function for Y is given as:

$$pdf_y(y) = \left| \frac{\partial f^{-1}}{\partial y} \right| pdf_x(f^{-1}(y)) \tag{11.11}$$

and for problems of more than one variable:

$$pdf_y(y) = \frac{d}{dy} \int_{-\infty}^{\infty} \cdots \int_{-\infty}^{\infty} pdf_x(x_1, \ldots, f_{x_i}^{-1}(y), \ldots, x_n)$$
$$dx_1, \ldots, dx_{i-1}, dx_{i+1}, \ldots, dx_n \tag{11.12}$$

where $f_{x_i}^{-1}(y)$ returns a value of x_i given y, holding all other values x_j fixed. This forms the basis for propagating the errors into an overall uncertainty range for the final rating.

If we make a first order approximation, it can be shown (Sveshnikov 1968) that the variances are related by:

$$\sigma^2(y) = \left(\frac{\partial f}{\partial x} \right)^2 \sigma^2(x) \tag{11.13}$$

Applying this approximation to Eq. (7), we have our result:

$$\sigma_M^2(d) = \sum_j \sigma_{N_j}^2 \tag{11.14}$$

where

$$\sigma_r^2 = \left(\frac{\partial r}{\partial p}\right)^2\left(\frac{\delta p}{3}\right)^2 + \left(\frac{\partial r}{\partial p_{\text{base}}}\right)^2\left(\frac{\delta p_{\text{base}}}{3}\right)^2$$

$$+ \left(\frac{\partial r}{\partial p_{\text{metric}}}\right)^2\left(\frac{\delta p_{\text{metric}}}{3}\right)^2 + \left(\frac{\delta r}{3}\right)^2$$

$$\sigma_N^2 = \left(\frac{\partial N}{\partial I}\right)^2\left(\frac{\delta I}{3}\right)^2 + \left(\frac{\partial N}{\partial r}\right)^2\sigma_r^2 + \left(\frac{\partial N}{\partial r_{\text{base}}}\right)^2(\sigma_{r_{\text{base}}})^2$$

$$+ \left(\frac{\partial N}{\partial r_{\text{max}}}\right)^2(\sigma_{r_{\text{max}}})^2 + \left(\frac{\partial N}{\partial r_{\text{min}}}\right)^2(\sigma_{r_{\text{min}}})^2 \tag{11.15}$$

where δx is the tolerance the designer has in prescribing a value x.

Equations (11.14) and (11.15) form the basis for a first-order error analysis of a numerical concept selection. If we use the mappings (11.5) and (11.7), we can apply Eqs. (11.14) and (11.15) to derive an overall error formula. Equation (11.11) for the linearized mappings (11.5) and (11.7) becomes

$$\sigma_r^2 = \left(\frac{1}{p_{\text{metric}} - p_{\text{base}}}\right)^2\left(\frac{\delta p}{3}\right)^2 + \left(\frac{p - p_{\text{metric}}}{(p_{\text{metric}} - p_{\text{base}})^2}\right)^2\left(\frac{\delta p_{\text{base}}}{3}\right)^2$$

$$+ \left(\frac{p_{\text{base}} - p}{(p_{\text{metric}} - p_{\text{base}})^2}\right)^2\left(\frac{\delta p_{\text{metric}}}{3}\right)^2 + \left(\frac{\delta r}{3}\right)^2 \tag{11.16}$$

for each raw score r, and

$$\sigma_N^2 = \left(\frac{r - r_{\text{base}}}{r_{\text{max}} - r_{\text{min}}}\right)^2\left(\frac{\delta I}{3}\right)^2 + \left(\frac{I}{r_{\text{max}} - r_{\text{min}}}\right)^2\sigma_r^2$$

$$\left(\frac{-I}{r_{\text{max}} - r_{\text{min}}}\right)^2(\sigma_{r_{\text{base}}})^2 + \left(\frac{-I(r - r_{\text{base}})}{r_{\text{max}} - r_{\text{min}}}\right)^2(\sigma_{r_{\text{max}}})^2$$

$$+ \left(\frac{I(r - r_{\text{base}})}{r_{\text{max}} - r_{\text{min}}}\right)^2(\sigma_{r_{\text{min}}})^2 \tag{11.17}$$

for each normalized score N. This approach is applied to the material selection problem, and the results are shown in Table 11.14. Note that the approach remains valid even without the linearity assumption; however, the calculations become more difficult. Validity of the uncertainty characterization is elaborated in the higher order uncertainty characterization section below.

Examining Table 11.14, it becomes somewhat less clear which alternative should be chosen. The copper rating is 1.5 ± 1.3, and the stainless steel rating is 0.5 ± 1.0. To within three standard deviations, we cannot be assured of the copper recommendation.

TABLE 11.14. UNCERTAINTY IN OVERALL RATINGS FOR THE MATERIAL SELECTION EXAMPLE

Criterion	Steel 1020	Stainless 304	Aluminum 5052	Copper	Bronze
Thickness	±0.0	±0.1	±0.1	±0.1	±0.0
Conductivity	±0.0	±0.1	±0.1	±0.2	±0.1
Thermal mass	±0.1	±0.2	±0.4	±0.3	±0.2
Diffusivity	±0.0	±0.1	±0.1	±0.2	±0.1
Hardness	±0.0	±0.1	±0.1	±0.1	±0.1
Yield	±0.0	±0.2	±0.1	±0.0	±0.0
Machinability	±0.0	±0.1	±0.1	±0.1	±0.1
RMS Error ( )	±0.1	±0.3	±0.5	±0.4	±0.3
Tolerance ( )	±0.3	±1.1	±1.4	±1.3	±0.9

Confidence Levels

The previous decision that seemed clear is now questionable. A quantification of the decision certainty would be helpful. What must be ascertained is if the maximum sum is indeed larger then the other alternatives, the next highest in particular.

In the proposed error analysis methodology, the nominal concept selection ratings (e.g., Table 11.4) represent the average of a normal probability distribution. Denote the maximum rating μ_1 and the next highest rating μ_2. These are the means of two normal distributions X_1 and X_2. The posed question is then whether $\mu_1 > \mu_2$, a well formed statistical hypothesis.

To formulate the solution to this hypothesis, consider the intersection of the two distributions X_1 and X_2. When X_1 and X_2 are identical distributions, our hypothesis fails, $\mu_1 = \mu_2$. We have no confidence that $\mu_1 > \mu_2$. The probability that our hypothesis is false is one, as happens to be the intersection of the two probability distributions X_1 and X_2. Now as μ_1 moves away from μ_2, our confidence in the statement that $\mu_1 > \mu_2$ increases, and the area of intersection of X_1 and X_2 decreases. Thus, the area of intersection of the distributions is a measure of confidence against our hypothesis. This result is shown in Figure 11.3.

To derive the equation for the confidence level, consider a variable c, which uniformly scales both σ_1 and σ_2 until $\mu_1 - c\sigma_1 = \mu_2 + c\sigma_2$. The equation for c is thus:

$$c = \frac{\mu_1 - \mu_2}{\sigma_1 + \sigma_2} \qquad (11.18)$$

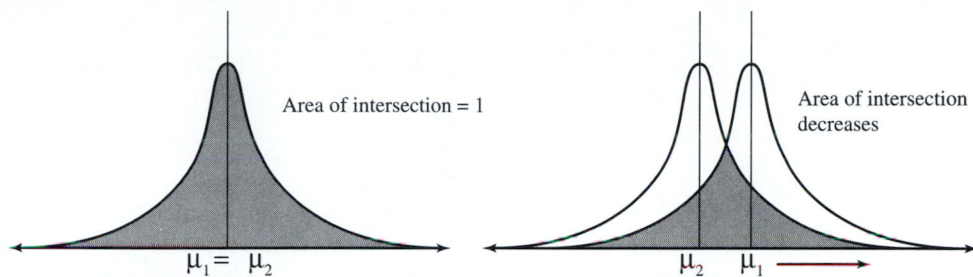

▼ Figure 11.3.
The confidence against is equal to the intersection of the density functions.

This value of c is the value required to stretch the two normal distributions until their areas of intersection are one standard deviation. This definition is shown in Figure 11.4. Note that c is always positive.

Now, when $c = 1$, the distributions are exactly one standard deviation apart each. Then the intersection of the density functions is twice of one minus the cumulative density function. That is,

$$\text{Area of intersection} = 2(1 - \text{Cu}(c)) \qquad (11.19)$$

where

$$\text{Cu}(c) = \frac{1}{\sqrt{2\pi}\sigma} \int_{-\infty}^{c} e^{\frac{(c-\mu)^2}{2\sigma^2}} \, dc$$

as shown in Figure 11.5.

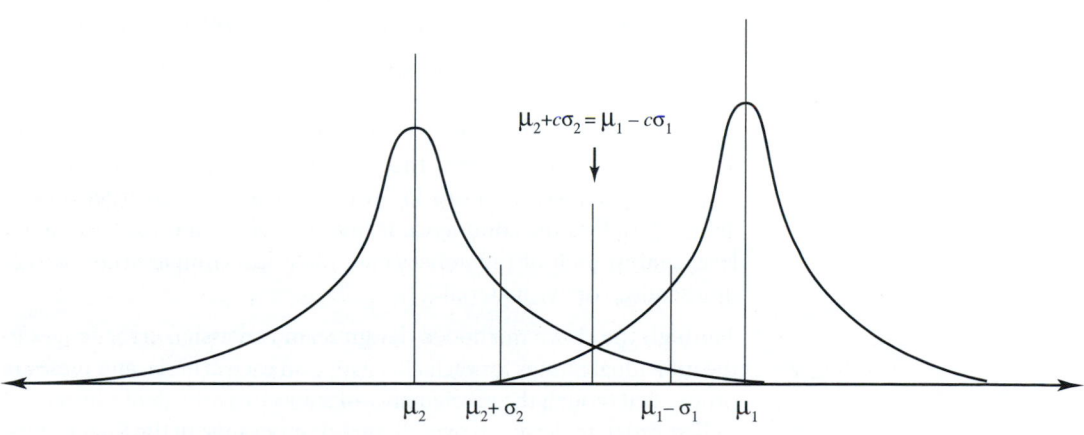

▼ Figure 11.4.
Definition of c.

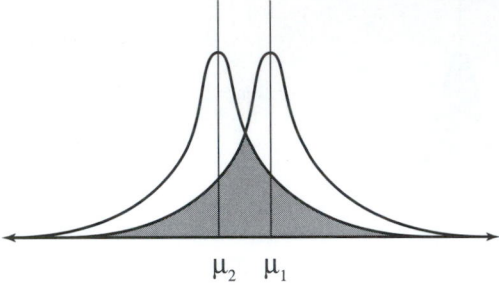

$$\mu_2 \quad \mu_1$$

▼ *Figure 11.5.*
The confidence against is equal to the intersection of the density functions.

Equation (11.19) represents the solution to calculating a degree of belief in the concept selection chart solution. We want the probability that the means are not the same or are one minus the above probability. Therefore,

$$\Pr(\text{decision}) = 2\,\mathrm{Cu}(c) - 1 \tag{11.20}$$

This equation represents a confidence level in the decision, given the specified error estimates. Note that this derivation is analogous to a statistical t-test that gives the confidence that the means of two sample distributions are the same. If the input uncertainties were characterized sampled data and this data were propagated through Eq. (11.9), the output sample distributions would be compared with a t-test, not with Eqs. (11.19–11.20), which are derived with normal distributions. The t-test results approach Eqs. (11.19–11.20) as the sample sizes become large and normal. The c value, therefore, represents the number of standard deviations of error and will be called the *confidence factor.*

For the material selection example, we apply Eq. (11.18) to the copper and stainless steel evaluation values and errors. This results in $c = 1.27$, or 1.27σ error. Then, applying Eq. (11.20) results in $\Pr = 0.80$. That is, we are about 80% confident that the choice of copper over stainless steel is correct, based on our known errors, and a first-order model. Table 11.15 lists the confidence factor (C) values and the confidence level against each of the nonselected materials compared to the recommended copper material.

Through the above method, a design team can assign error ranges to the individual ratings for each alternative on each criteria, and these are propagated though the mathematics of subjective probability, linearized to first order, to derive an overall uncertainty range in the final results. This range can be compared with the differences in overall rating among the alternatives.

TABLE II.I5.   CONFIDENCE AGAINST EACH MATERIAL FOR THE MATERIAL SELECTION EXAMPLE

Criterion	Steel 1020	Stainless 304	Aluminum 5052	Copper	Bronze
Value	2.80	1.27	3.88	N/A	2.03
Confidence	0.99	0.80	1.00	N/A	0.96

Concept Selection with Error Analysis: Design of a Cat Litter Box Product

To illustrate the concept selection process, let's consider the design of a new cat litter box product (Balirian, Fast, and Waldman 1995; Otto and Wood 1995). The critical customer needs for this product are the ability to hold and retain litter, size to comfortably accommodate cats, prevention of odors from permeating the environment, keeping litter off a cat's feet, minimal volume and footprint, easy to clean, portability, and low cost. A number of concepts are developed for these needs, using the techniques discussed in Chapter 10. Because of the difficulty in modeling concepts that remove cat waste from granular material (the litter), proof-of-concept models are created for the most feasible concept variants. These models are created from widely available materials (Chapter 17), such as hardware store components, wood, and foam. Experiments are then performed to estimate the ability of the concepts to remove waste, inhibit smell, and so forth. Actual litter and Play-Doh (to model waste) are used in these experiments in addition to actual cat waste.

The results of these very basic physical models and experiments lead to a subset of concept alternatives. A preferred concept must be chosen from this subset. In this case, the information quality is fairly high to make the decision, based on the resources devoted to the proof-of-concept models.

Table 11.16 displays the remaining concepts, in addition to the chosen evaluation criteria. The criteria listed in the matrix correspond to measurable scales for a subset of the customer needs. These scales relate to the performance of the product, focusing on filtration of cat waste.

The team rates each of the criteria using the proof-of-concept results. A scale of 0-100 is chosen for the ratings, since the information quality is high (ratio scale). For instance, if we consider the criterion of "time to clean," an ideal time exists at 10 s, corresponding to a score of 100 points.

TABLE 11.16.   CAT LITTER BOX FILTRATION CONCEPTS: RANKING CHART

Concepts / Weightings / Criteria		Rotation Filtration	Comb Filtration	No Filtration	Pivot-Gravity Filtration	Rake Filtration
Performance (removal of waste)						
Time to Clean						
Number of Steps to Clean						
Complexity (number of moving parts)						
Portability						
Overall Rating						

Likewise, a worst case time exists at 30 s, rated as a zero score. Every second that a concept requires for cleaning (on average) results in a decrease of five points in its rating. Similar scales exist for the remaining criteria.

In addition to the rating scales for each criterion, the team chooses weightings for the criteria. These weightings are based on the importance levels of the customer needs and associated engineering specifications. Table 11.17 shows the updated concept selection chart with the weightings listed. The updated chart also shows the ratings for each concept.

As shown in the table, the pivot-gravity cat litter concept is the preferred choice. Its performance in reliably removing waste and its time to clean dominate the decision, especially compared with the base point, that is, the no-filtration concept.

Although the information level is high for this case, subjectivity exists in the choices of rankings and criteria weights. Uncertainty analysis, as discussed in the previous section, is needed to determine our confidence in the concept decision. Table 11.18 lists the uncertainty for each of the criteria and chosen weights. For example, a tolerance of ±5 is

TABLE 11.17. TEAM-BASED RATINGS FOR THE CAT LITTER BOX CONCEPTS

Criteria \ Concepts	Weightings	Rotation Filtration		Comb Filtration		No Filtration		Pivot-Gravity Filtration		Rake Filtration	
Performance (removal of waste)	0.45	20	9	40	18	80	36	80	36	40	18
Time to Clean	0.15	50	7.5	85	12.8	0	0	100	15	60	9
Number of Steps to Clean	0.15	100	15	60	9	0	0	60	9	80	12
Complexity (number of moving parts)	0.15	60	9	80	12	100	15	70	10.5	50	7.5
Portability	0.10	20	2	70	7	50	5	90	9	50	5
Overall Rating	1.0	43		59		56		80		52	

used for the "time to clean" criterion. The proof-of-concept measurements only provided accuracy within ± 1 s.

The uncertainty values from Table 11.18 are now used to determine confidence levels for the cat litter box concepts compared with the top-rated choice. Table 11.19 lists the results of this analysis. There exists

TABLE 11.18. UNCERTAINTY VALUES FOR THE FILTRATION CONCEPT SELECTION

Criteria	Rating Uncertainty (tolerance)
Performance (removal of waste)	± 20
Time to clean	± 5
Number of steps to clean	± 10
Complexity (number of moving parts)	± 10
Portability	± 10
Criteria weights	± 0.05

TABLE 11.19. UNCERTAINTY ANALYSIS OF THE FILTRATION CONCEPTS

Filtration Concepts	Ratings	Confidence Level (%)
Rotation filtration	43	99.9
Comb filtration	59	98.7
No filtration	56	99.6
Pivot-gravity filtration	80	—
Rake filtration	52	99.9

at least a 99% confidence level in the choice of the pivot-gravity concept. Although the uncertainty tolerances were liberally chosen, the large disparity in rankings provides confidence in choosing pivot-gravity filtration.

With the filtration system now chosen, the design team must also consider the litter box concepts that focus on odor control. These concepts are considered independent of the filtration subsystem, since only hooded concepts are judged as feasible (Table 11.16). A separate concept selection process may thus be implemented for the odor control concepts.

A large number of concepts are considered for the odor control subsystem. To determine their feasibility, engineering estimates are needed to predict their performance in removing cat waste odors. In this case, analytical models are not readily available. Therefore, simple olfactory tests are needed in a controlled environment to determine the feasibility and relative performance of the concepts.

Based on the results of these tests, six concepts are considered for concept selection. These concepts include vented walls, activated carbon, baking soda, cedar chips, Renuzit™ air freshener, and "do nothing" (the reference). Table 11.20 shows the results of the concept selection. The cedar chips and baking soda alternatives clearly receive the higher ratings. But what does the difference in rating mean between these two alternatives? Can a consensus be formed in choosing one alternative over the other?

Uncertainty analysis may once again be applied to determine the confidence levels. Table 11.21 lists the uncertainty tolerances for each of the evaluation criteria. For example, a ± 10-point rating tolerance exists for the performance criterion (olfactory distance test in detecting an odor). This tolerance corresponds to ± 1 foot in the proof-of-concept tests. Likewise, a ± 10-point tolerance exists for the cost criterion. This tolerance corresponds to a $\pm \$2.50$/year uncertainty in the cost estimates for the replaceable odor control concepts.

TABLE 11.20. CONCEPT SELECTION OF ODOR CONTROL ALTERNATIVES

Criteria / Concepts	Weightings	Do Nothing	Renuzit™ Air Freshner	Baking Soda	Cedar Chips	Vented Walls	Activated Carbon
Performance (olfactory distance-ft.)	0.50	0 / 0	70 / 35	70 / 35	80 / 40	20 / 10	90 / 45
Cost	0.25	100 / 25	52 / 13	76 / 19	84 / 21	100 / 25	20 / 5
Ease of Replacement	0.13	100 / 12.5	70 / 9	90 / 11	40 / 5	100 / 13	20 / 3
Frequency of Replacement	0.12	100 / 12.5	0 / 0	50 / 6	67 / 8	100 / 13	67 / 8
Overall Rating	**1.0**	**50**	**57**	**71**	**74**	**61**	**61**

Using these tolerances, an uncertainty analysis is performed for the odor-control concepts. Table 11.22 lists the results. As shown in the table, a 97% or greater confidence exists in choosing the cedar chips over four of the other concepts. However, only a 35% confidence exists in choosing cedar chips over baking soda. Very little difference exists in these concepts. In fact, the baking soda concept is rated as much easier to replace in the pivot-gravity litter box.

This analysis leads to a number of possibilities. Since the baking soda and cedar chips are shown as virtually indistinguishable, both concepts may be carried forward to embodiment design (Chapter 12). Alternatively, we can seek to combine these concepts into an aggregate solution. This latter choice is adopted as a novel approach for the car litter box product. A pressboard cedar box is designed to load into slots at the roof of the pivot-gravity concept. The small cedar box will absorb significant

TABLE 11.21. UNCERTAINTY VALUES FOR THE ODOR CONTROL CONCEPT SELECTION

Criteria	Rating Uncertainty (tolerance)
Performance: Olfactory distance (ft)	± 10
Cost	± 10
Ease of replacement	± 10
Frequency of replacement	± 10
Criteria weights	± 0.05

TABLE 11.22.   UNCERTAINTY ANALYSIS OF THE ODOR CONTROL CONCEPTS

Odor Control Concepts	Ratings	Confidence Level (%)
Do nothing	50	99.9
Renuzit air freshener	57	98.9
Baking soda	71	35.0
Cedar chips	74	—
Vented walls	61	97.1
Activated carbon	61	98.0

cat odors. In addition, the box is filled with replaceable baking soda (common grocery store variety). As the pivot-gravity concept is actuated during each cleaning, a small quantity of baking soda falls from the box and into the litter. The cat mixes the baking soda into the litter during each usage. The baking soda then neutralizes the acidic odor of the cat urine.

VIII. A CRITIQUE OF DESIGN EVALUATION SCHEMES

From the advanced discussion, it is apparent that there are many possible concept selection approaches. Likewise, there are many proponents for the different approaches. A good criterion to judge these methods is by the level of information quality available at the time of the decision. At a conceptual design phase, information quality may be low, and so ordinal methods such as Pugh charts are appropriate. If more information is available during concept selection due to refinement of the concepts or a redesign situation, the Pugh chart concept may be extended to include importance ratings and interval scale evaluations.

As the product development process progresses, the quality of the information will, by necessity, become better. For example, during an embodiment design phase, more information is available and so more measurable methods can be used, such as the interval scale methods.

There are many other decision-making methods not discussed here. The Analytic Hierarchy Process (Saaty 1980) can be used with decisions where the performance metrics are related hierarchically; goal programming (Osycska 1984) and other approaches are used to convert multiple metrics into a single metric. Utility theory (Raiffa and Keeny 1976), probability methods (Winkler 1972), and fuzzy sets (Antonsson and Otto 1995) all have various advocates for various design decision-making problems.

As mentioned in this chapter, any argument over the inherent mathematics of these analyses largely misses the point of dealing with concept selection. The process used to ensure common consensus on concepts, criteria, and ratings is more important.

On the other hand, as networked computing technology increases, new opportunities are being developed for designers to work on development projects with multiple design decision makers and inherent uncertainty. Application of these advanced methods then becomes important and remains a research area.

IX. CHAPTER SUMMARY AND "GOLDEN NUGGETS"

Concept selection is a very important milestone activity of any product development process. The selection decision will have a large impact on the final success of the new product, and so care must be taken to execute the decisions effectively. Some "golden nuggets" to retain from the chapter include:

- An effective team-based process is needed. This process will inspire and cultivate team consensus.

- Estimation is a critical skill to develop and can help in quickly eliminating poor concepts.

- Based on industrial data, Pugh charts are a very effective technique for making decisions during early concept selection.

- Measurement theory and information quality are rigorous approaches to judging different evaluation schemes.

References

Antonsson, E., and Otto, K. 1995. Imprecision in engineering design. Special combined issue of the *Journal of Mechanical Design* and the *Journal of Vibration and Acoustics,* 117 (B):11–16.

Baliarin, E., Fast, M., and Waldman, J. 1995. Design and prototype of an easy-to-clean cat litter box. Mechanical Engineering Design Projects Program, The University of Texas, Austin.

Bernays, P. 1968. *Axiomatic set theory.* New York: North-Holland.

Krantz, D., Luce, R., Suppes, P., and Tversky, A. 1971. *Foundations of measurement.* Vol. I. New York: Academic.

Halliday, D., and Resnick, R. 1988. *Fundamentals of physics.* 3rd ed. New York: Wiley.

Osycska, O. 1984. *Multiple criterion optimization in engineering with FORTRAN examples.* New York: Halsted.

Otto, K. 1995. Measurement methods for product evaluation. *Research in Engineering Design* 7: 86–101.

Otto, K., and Wood, K. 1995. Estimating errors in concept selection. *Proceedings of the 1995 ASME Design Theory and Methodology Conference,* Boston, MA.

Pugh, S. 1990. *Total design.* New York: Addison-Wesley.

Raiffa, H., and Keeney, R. *Decisions with multiple objectives: Preference and value tradeoffs.* New York: Wiley.

Saaty, T. 1980. *The analytic hierarchy process.* New York: McGraw-Hill.

Savage, L. 1954. *The foundations of statistics.* New York: Wiley.

Steuer, R. 1986. *Multiple criteria optimization: Theory, computation, and application.* New York: Wiley.

Sveshnikov, A. 1968. *Problems in probability theory, mathematical statistics, and the theory of random functions.* New York: Dover.

Ullman. D. 1995. *The mechanical design process.* New York: McGraw-Hill.

Ulrich, K., and Eppinger, S. 1995. *Product design and development.* New York: McGraw-Hill.

Ward A., and Seering, W. 1990. Quantitative inference in a mechanical design compiler. *Artificial Intelligence in Engineering Design and Manufacturing* 4 (1):47–51.

Winkler, R. 1972. *An introduction to Bayesian inference and decision.* New York: Holt, Reinhart and Winston.

Zimmerman, H. 1985. *Fuzzy set theory and its applications.* 2nd ed. Boston: Kluwer.

12 Concept Embodiment

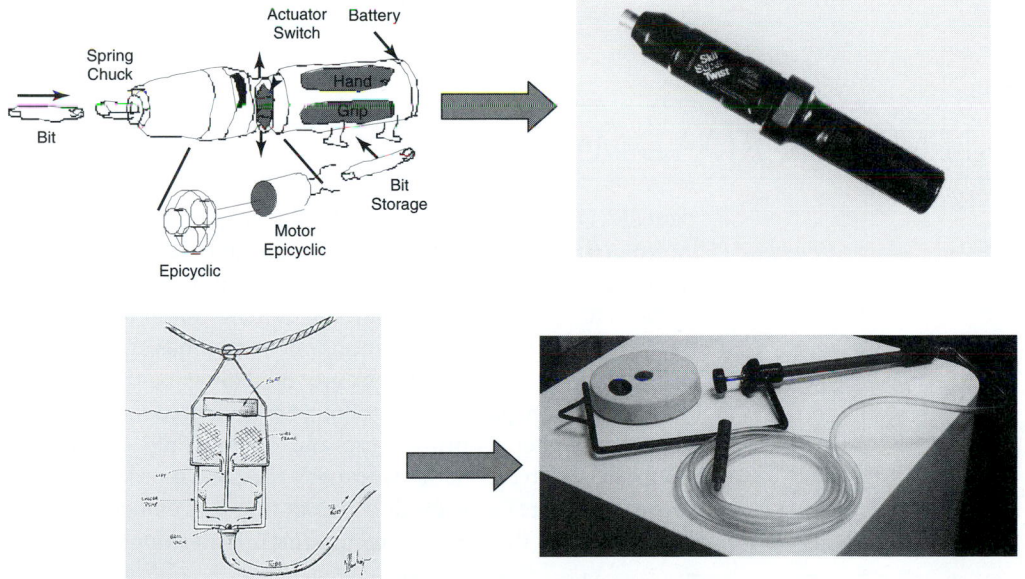

The objective of this chapter, and the remainder of the book, is to transform product concepts and architectures into realized systems. These systems must satisfy the customer needs, be robust with respect to all environmental and user conditions, and be designed to minimize the likelihood of failure. Only through the satisfaction of these goals can we hope to create competitive products that add to the customer's quality of life while minimizing risks to the populace.

The cover figure of this chapter shows snapshots of product concepts and their realizations as physical systems. These concepts begin as mental images of how a product will be architected and configured. They are then transformed systematically and iteratively to processes that allow the fabrication and assembly of the components into a working form. This transformation process depends on the initial description of the product concept. An initial description can range from a hand drawing of simple stick-figure geometry to computer renderings that list initial geometric parameters, texture, and attachments. This chapter discusses the general processes and methods for carrying out this transformation. Subsequent chapters (13–19) present detailed techniques and product examples.

I. CHAPTER ROADMAP

Figure 12.1 illustrates the roadmap for this chapter. Two basic routes are shown in the figure. The first describes the global concept of embodiment design. This description includes a summary of contemporary methods to support concept of embodiment. It also provides a context for understanding when embodiment methods are implemented during the product development process. Following this description, basic methods are presented for refining a product concept into its final geometry, material properties, and supporting manufacturing processes. Two basic methods provide the core approach: a step-by-step process for realizing product function, and an embodiment checklist for ensuring that generalized product specifications are satisfied.

The second route through the chapter includes more advanced techniques in embodiment design. These techniques include the concepts of embodiment principles, mathematical modeling, and failure modes and effects analysis. These techniques complement and supplement the basic methods of embodiment design. They also provide a more thorough means of ensuring the robustness and integrity of a product's performance.

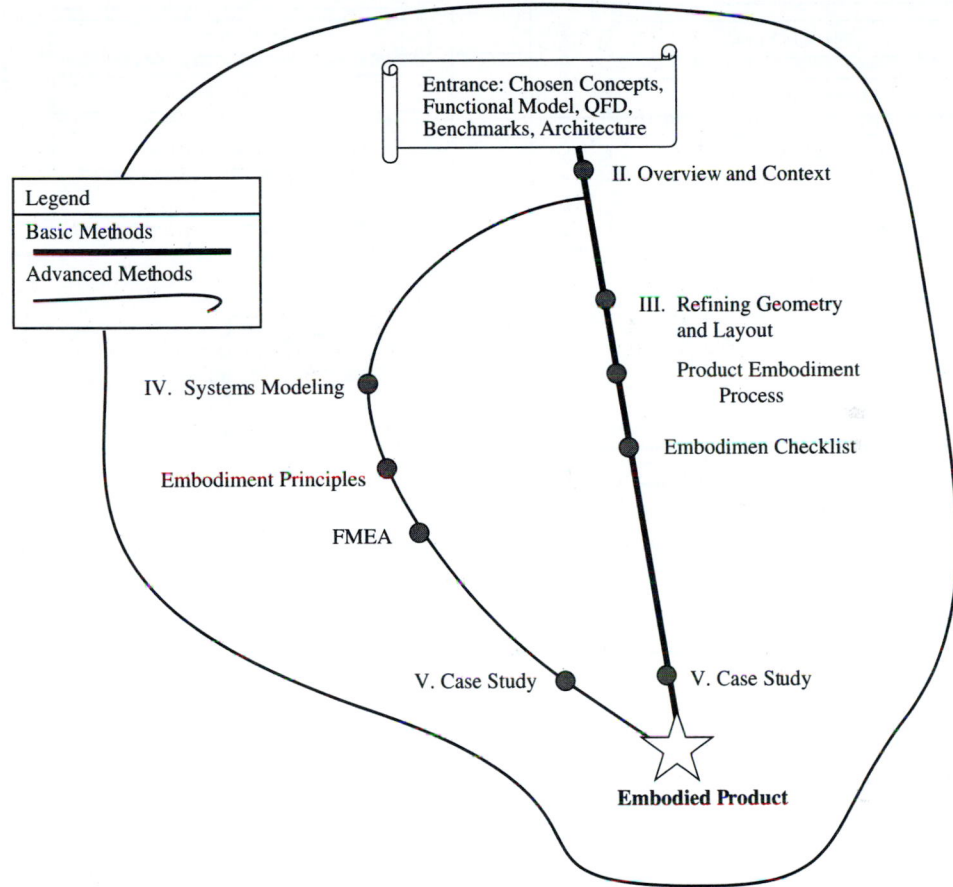

▼ *Figure 12.1.*
Chapter roadmap to concept embodiment.

II. OVERVIEW AND CONTEXT

Concept embodiment is perhaps the task most identified with an engineer in the product development process. During this stage (as shown in Figure 12.2), the engineers (as the product function teams) carry out many activities. Such activities include the choices of components— standard and specialized, component interfaces, materials, geometry (dimensions, shape, and tolerances), surface finish, fasteners and connectors, manufacturing processes, and assembly processes. To make these choices, an engineer must understand a product as a system, one

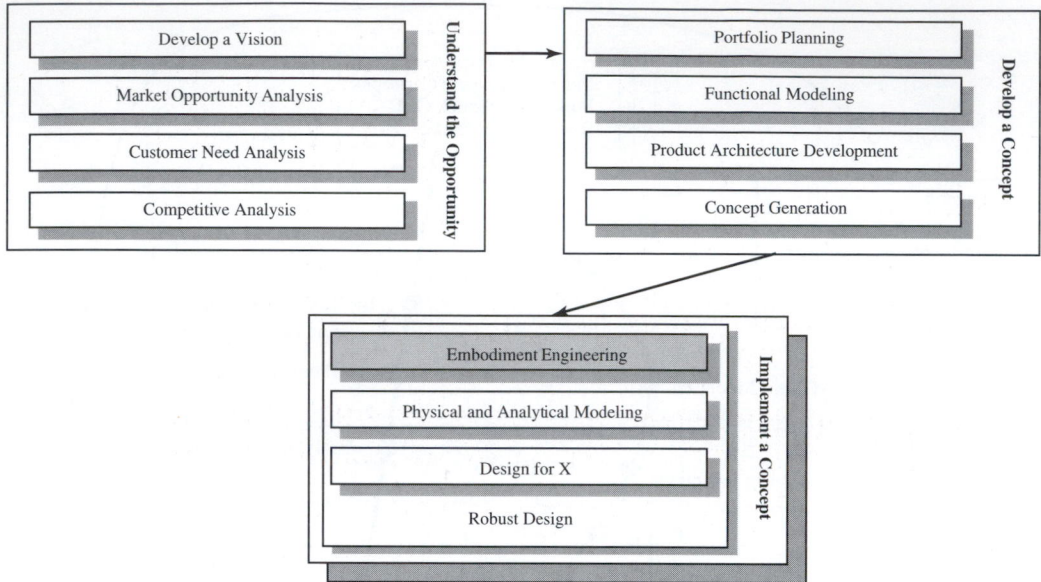

▼ Figure 12.2.
Concept embodiment follows concept generation and selection. It is iterative and seeks to attain a robust design.

that interacts with its environment. This stage is where engineers apply their skills in mathematics and applied science. Products, regardless of scale, are complex systems that must be effectively modeled, tested, and refined. Methods for concept embodiment must aid in this process of system modeling.

Figure 12.3 illustrates various embodiment design methods as a function of a product's life cycle. These methods range from mathematical and empirical modeling of a product's performance to the life cycle issues of service and environmental impact. Modeling methods provide us with a means of representing the performance and potential failure states of a product across its customer needs and engineering specifications. Design for manufacturing and assembly techniques provide a systematic approach for ensuring the producibility, availability, and economics of a product venture. Physical prototyping methods then provide a process for testing the actual performance and manufacturability of a product. The remaining methods of design for service and the environment focus on two particular aspects of a product. These aspects relate to the life cycle of a product, ensuring that it can be maintained and ultimately reused or recycled.

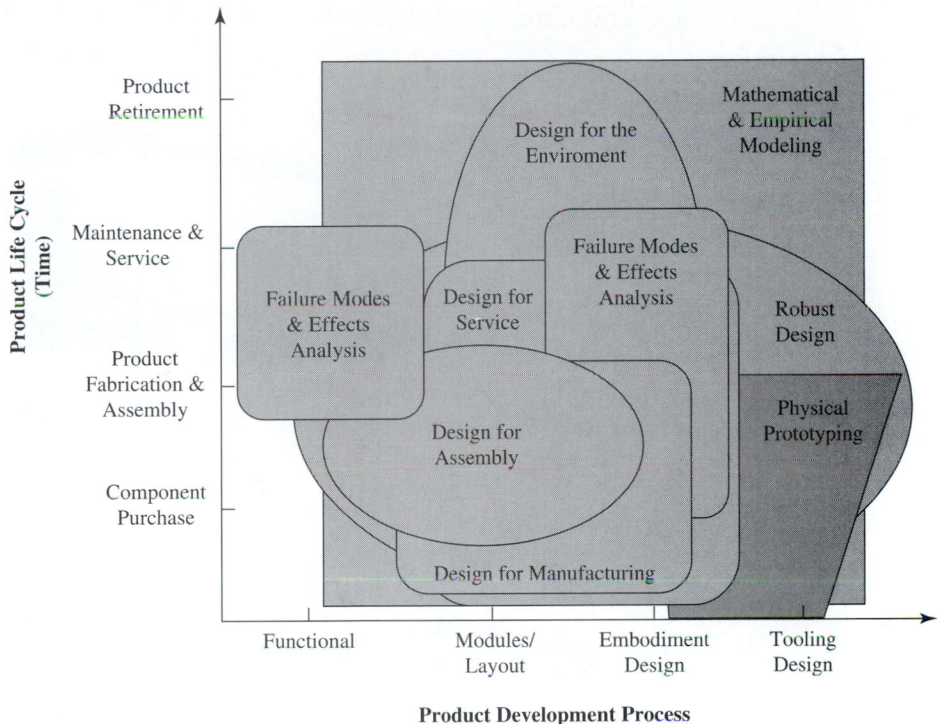

▼ Figure 12.3.
Embodiment design methods: Implementation scope in product development process
and impact on the product life cycle.

Each topic in Figure 12.3 is discussed in great detail in the subsequent
chapters. However, focusing on any one of the topics loses sight of
the overall embodiment design activity. In this chapter, a framework
and approach is provided for these various embodiment design methods. This framework provides a context for implementing concept
embodiment and has proven more effective than others in educational courses. It also illustrates the iterative nature of embodiment
design and the range of tasks that must be completed, balanced, and
negotiated.

When studying this framework, it is important to understand the intrinsic complexity and nonlinearity of the process. Figure 12.4 provides a visual aid for this idea. In the figure, three stages of a
natural-energy bilge pump product (Chapter 10) are captured in time.
The first stage includes customer needs analysis, functional modeling,

UNDERSTAND THE PROBLEM

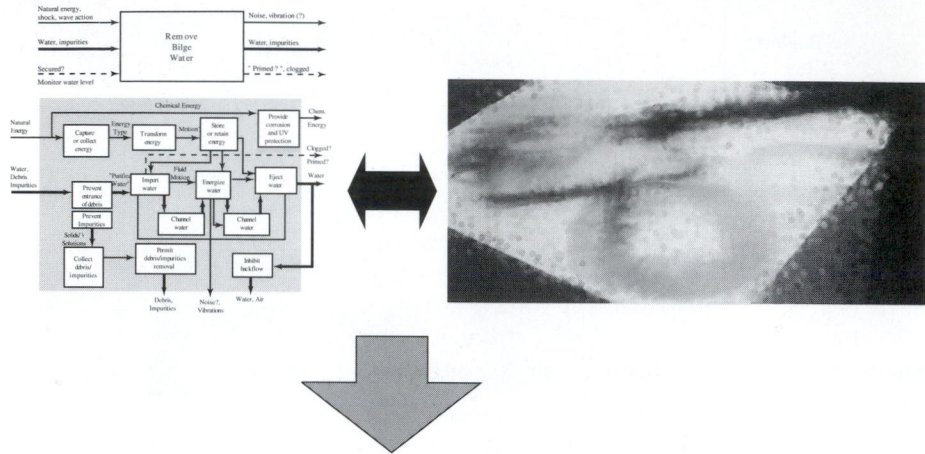

DEVELOP A CONCEPT

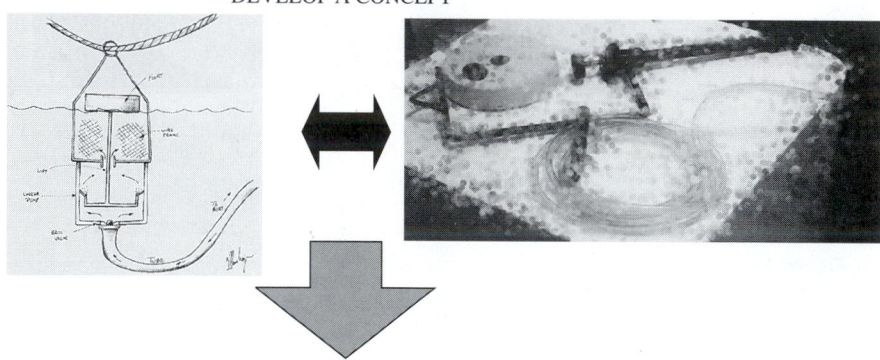

IMPLEMENT A CONCEPT

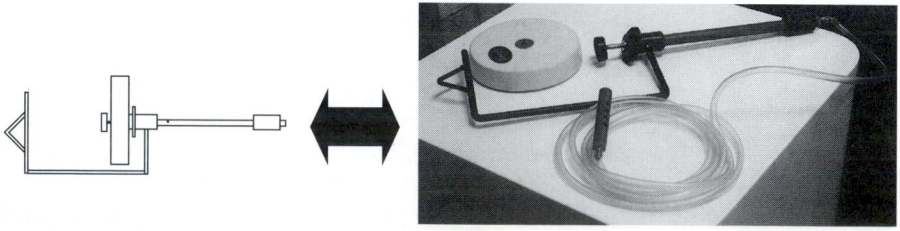

▼ *Figure 12.4.*
Snapshots of three stages of the product development process. As a product is
developed, its realization becomes less fuzzy. Concept embodiment maps product
concepts to crisp physical realizations.

and product-architectural layout. At this stage, the physical embodiment of the product is very fuzzy. The product is understood as an input-output system that converts energy from the environment to water flow from a pleasure craft. The business-case opportunity and engineering specifications are investigated and documented. However, an actual physical concept has not been created in any concrete or tangible vision.

In the next stage of the process, concepts are developed as line drawings and high-level geometric descriptions. The focus is on the operational principles of the product. Detailed geometry, shape, material choices, and so forth are typically unknown. As illustrated in Figure 12.4, the mental picture of the product has become much more focused; yet, the specific elements remain fuzzy and unclear.

The concept embodiment stage, as documented in the figure, seeks to resolve the focus of the product to a singular crisp description. This process involves thousands and thousands of decisions. It also involves from hundreds to tens of thousands of design parameter choices, depending on the scale of the product. Any one choice or decision can potentially affect many subsequent decisions.

So, how do we choose the final layout of the product, such as the weighted cage and rope attachment on the bilge pump (Figure 12.4)? What materials should we use for the various components, such as the ball valves and screen of the bilge pump? What failure modes are possible in the product, such as corrosion and water backflow in the bilge pump? How do we configure the interfaces and fasteners for the product, such as the float attachment and hose interface of the bilge pump? How do we design the product for manufacturing and assembly, such as the processes for coating and reduction of components in the bilge pump?

Embodiment design endeavors to answer these questions and the countless others that arise as we realize a product. The next section provides basic methods toward this end, beginning with a general process for embodying a product. This process applies to original design concepts, where the entire product architecture is being developed from scratch, as illustrated by the bilge pump product in Figure 12.4. The process also applies to redesign situations, where a product's subsystems are being significantly evolved for a new product offering. Figure 12.5 illustrates this later case. As shown in the figure, the original concept of an electric wok product is evolved to a new concept that includes more advanced control of thermal heating, a new configuration to accommodate product cleaning and storage, and a new layout to emulate the historical development of authentic Chinese cooking.

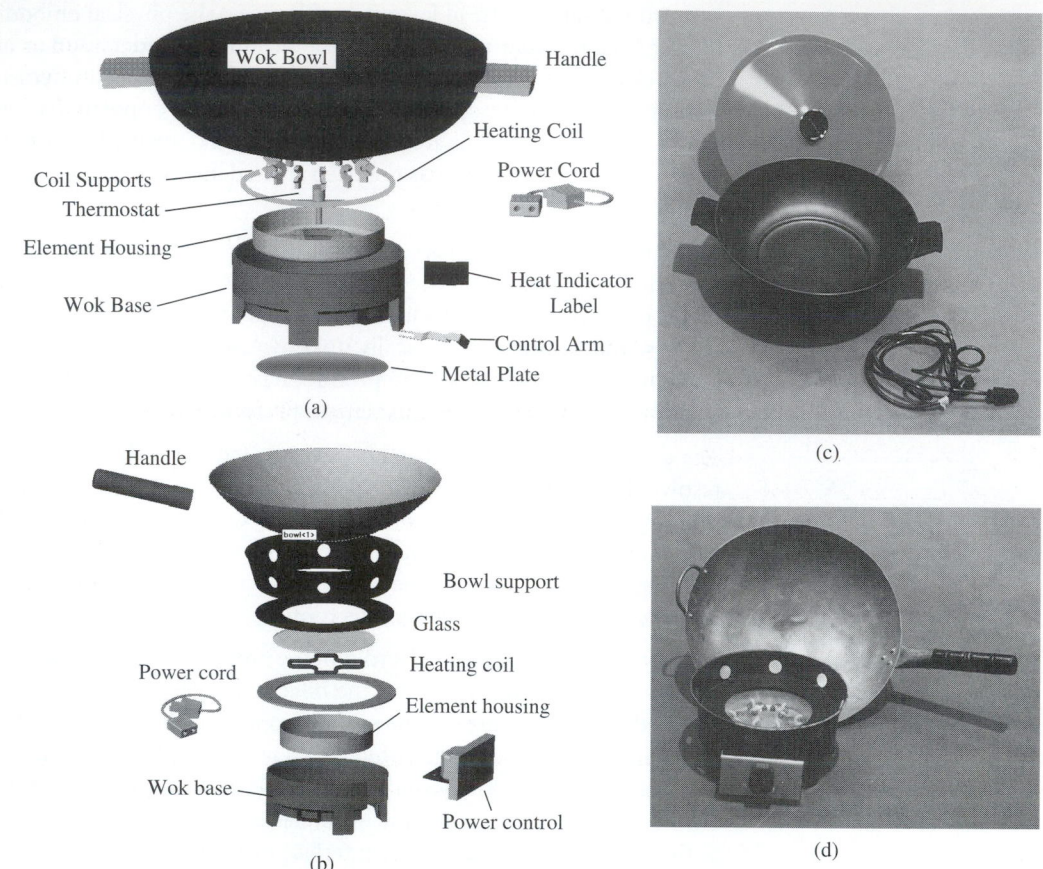

▼ Figure 12.5.
Embodiment example of a new electric wok concept. (a) Original wok concept. (b) Original product realization. (c) Evolved wok concept. (d) Realization of new product concept.

III. BASIC METHODS: REFINING GEOMETRY AND LAYOUT

In the context of creating a robust product or family of products, two issues drive concept embodiment: (1) refining a product's geometry and architecture, and (2) systems modeling toward detail design. These two issues pertain to four design scenarios: original design, adaptive design where a significant new technology is introduced (the bifurcation of the *s*-curve in Figure 3.2), adaptive design where a simple subsystem is modified, or parametric (variant) design. Of these four scenarios, the complexity of developing an embodied product is much greater for the first two due to the number of decisions that are needed. For any of these scenarios, however, the desired results of the embodiment process include the following tangible documentation: detailed drawings, exploded views, assembly diagrams (showing operations and fixturing), tooling design, manufacturing process plans and parameter choices, tolerance design and allocation, packaging, maintenance and warranty information, and user's manuals.

In the following sections, two basic methods are presented for concept embodiment (and associated design documents). The intent of these methods is to guide the transformation of a concept sketch to refined geometry and material choices. These methods focus on the functional performance of a product, including all relevant engineering specifications.

General Process of Product Embodiment

Embodiment design is where we deal with specific parameters and layout, that is, parametric and layout design of the product. At the end of concept development, we have a concept that has been logically chosen from a number of developed solution alternatives.

In embodiment design, we provide form to the selected concept(s) or product family. At this stage, the question arises, "Is the transition between concept development and concept embodiment well defined?" To this point in human technological development, our answer is a resounding "NO."

Ideally, the result of concept development is a single concept, or product family, with one chosen alternative for each subsystem. In reality, one usually ends up with one of three situations:

▶ a single concept with some function choices that are not set because they are not primary

▶ a family or platform of products with single choices for each member of the family, and

▶ two or three alternative concepts that might need further refinement before a choice is made

For the first two cases, embodiment design may progress unhindered, provided the product concept is truly feasible. In the third case, however, the alternative concepts must be developed in parallel until a reasonable decision may be made. The sooner a decision is secured, the more resources that can be devoted to the embodiment of the final product.

As shown in Figure 12.2, the preliminary stages of the product development process transform a problem with infinite possibilities, lots of imprecision, and spotty information into one that is clarified and has a finite set of solutions associated with it. Embodiment design, on the other hand, moves the process iteratively toward a definitive form, including:

- geometric layout
- material composition
- quality and manufacturability issues
- economics

What makes embodiment design challenging is that the parameters in a subsystem can become highly coupled, that is, changes in one parameter affect the others. This scenario means that activities must be performed in parallel. Unlike the considerations of solution alternatives in conceptual design, design changes and choices in embodiment design propagate. One choice may force others in the same subsystem or at the interface of another subsystem. This process is usually iterative. A change is made in a variable or subsystem and the effects are determined. If they are in the right direction, another change in the same direction is attempted. The process stops when the performance becomes acceptable.

In general, how do we deal with these complex characteristics of embodiment design? The general idea is to iteratively refine the geometry and layout of a product from an abstract form to a concrete one. Figure 12.6 illustrates a general process for implementing this idea (Pahl and Beitz 1996).

The process begins by considering the product specifications. Using the customer needs and these specifications (Chapters 4 and 7), the critical needs are identified that will drive the embodiment of the product. Examples include size-determining specifications, arrangement determining specifications, and material-determining specifications. After choosing the driving specifications, an overall layout of the product is drawn to scale based on the concept drawings. At this stage, the drawings (sketches) should not be fully detailed and care should be taken *not* to over constrain the layout. Through these drawings, the following items are illustrated: maximum dimensions of the product, clearances between relative subsystems, installation paths, and the general arrangement of components relative to one another.

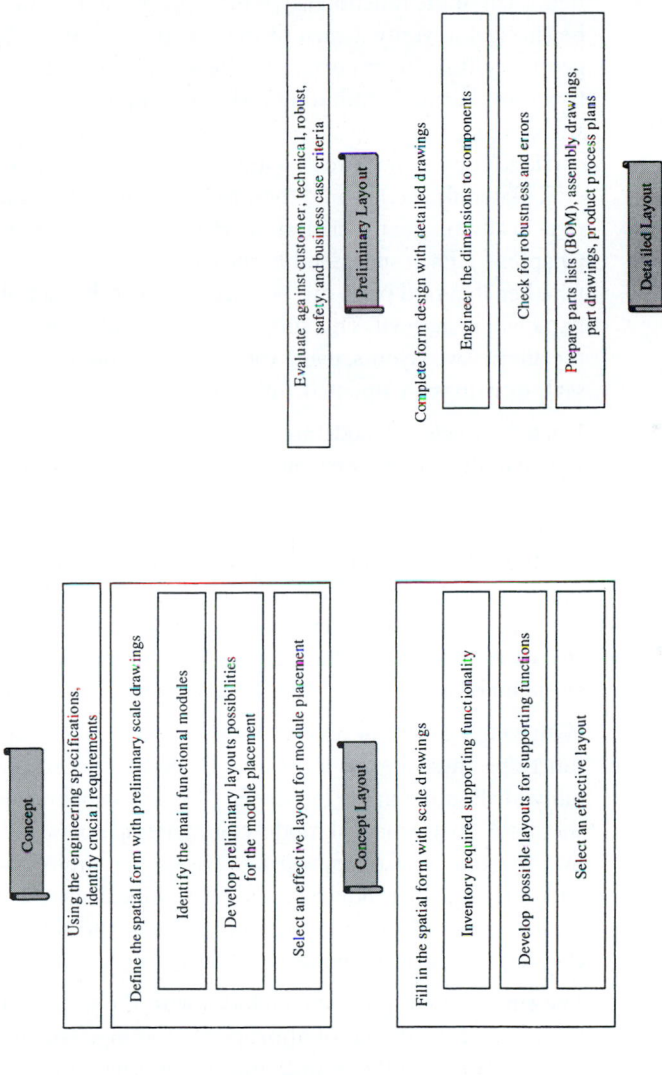

▼ **Figure 12.6.**
A general process for concept embodiment (after Pahl and Beitz 1996).

Once rough scale sketches are completed, the main function carriers of the functional model (Chapter 5) and the product concept are identified. Each of the function carriers (solutions to function chains) must be checked to verify if each of the functions is solved by the sketched geometry. It is also important to check for possible simplifications and function sharing (combining distinct components). Based on the results, preliminary alternative layouts of the remaining main function carriers (subsystems) are developed. Ranges of variable values (geometry, materials, etc.) should then be listed for each of the function carriers in addition to the critical interfaces (complementing the results of Chapter 9). Each subassembly should also be evaluated to determine whether standard parts can be used for the subassembly or if custom parts will be needed. This subprocess concludes by choosing between the alternative layouts, using the primary product specifications. The scale drawings are updated with the choices.

Using the functional model and product-architecture layout, rough layouts should be developed for remaining function carriers. In addition, auxiliary functions should be identified that are needed for the current layout. Auxiliary functions arise due to the choices of form solutions for the original functions. As an example, a battery pack may be chosen to satisfy the function "supply power." Due to the choice of a battery pack, an auxiliary function arises to "route electricity" to the subassemblies of the product. Rough layouts should be added for these auxiliary functions, such as wires, printed circuit boards, or metal conduit in the case of "route electricity."

With rough layouts developed for the main, supporting, and auxiliary functions, these layouts must now be detailed, ensuring the compatibility of all subassembly interfaces. This task usually calls for the use of standards, mathematical models, design guidelines, and experimentation to determine all appropriate parameter choices. In conjunction with this step, the product should be evaluated against customer, technical, robust, safety, and economic criteria, and the layout should be checked to estimate potential faults, including tolerances and clearances.

The embodiment process concludes with the testing of physical prototypes and the design of appropriate tooling. The product layout is updated based on the testing results, and final documentation is prepared, as listed in Figure 12.6. The case study section of this chapter provides a detailed example of this embodiment process.

Embodiment Checklist

To supplement the general embodiment process (Figure 12.6), a second basic method is the application of an embodiment checklist (after Pahl and Beitz 1996). Such a checklist, as illustrated in Table 12.1, provides

TABLE 12.1.   *CHECKLIST FOR EMBODYING A PRODUCT CONCEPT (AFTER PAHL AND BEITZ 1996)*

Embodiment heading	Checklist issue (Partial list)
Function	Are the customer needs satisfied, as measured by the target values? Is the stipulated product architecture and function(s) fulfilled? What auxiliary or supporting functions are needed?
Working principles and form solutions	Do the chosen form solutions (architecture and components per function) produce the desired effects and advantages? What disturbing noise factors may be expected? What byproducts may be expected?

Layout, geometry, and materials	Do the chosen layout, component shapes, materials, and dimensions provide
	    minimal performance variance to noise (robustness),
	    adequate durability (strength),
	    efficient material usage (strength-to-mass ratio),
	    suitable life (fatigue),
	    permissible deformation (stiffness),
	    adequate force flows (interfaces and stress concentrations),
	    adequate stability,
	    impact resistance,
	    freedom from resonance,
	    unimpeded expansion and heat transfer, and
	    acceptable corrosion and wear with the stipulated service life and loads?

Energy and kinematics	Do the chosen layout and components provide
	    efficient transfer of energy (efficiency),
	    adequate transient and steady state behavior (dynamics and control across energy domains), and
	    appropriate motion, velocity, and acceleration profiles?

Safety	Have all of the factors affecting the safety of the user, components, functions, operation, and the environment been taken into account?
Ergonomics	Have the human-machine relationships been fully considered? Have unnecessary human stress or injurious factors been predicted and avoided? Has attention been paid to aesthetics and the intrinsic "feel" of the product?
Production	Has there been a technological and economic analysis of the production processes, capability, and suppliers?
Quality control	Have standard product tolerances been chosen (not too tight)? Have the necessary quality checks been chosen (type, measurements, and time)?
Assembly	Can all internal and external assembly operations be performed simply, repeatedly, and in the correct order (without ambiguity)? Can components be combined (minimize part count) without affecting modular architectures and functional independence of the product?
Transport	Have the internal and external transport conditions and risks been identified and solved? Have the required packaging and dunnage been designed?
Operation	Have all of the factors influencing the product's operation, such as noise, vibration, and handling been considered?
Life Cycle	Can the product, its components, its packaging be reused or recycled? Have the materials been chosen and clumped to aid recycling? Is the product easily disassembled?
Maintenance	Can maintenance, inspection, repair, and overhaul be easily performed and checked? What features have been added to the product to aid in maintenance?
Costs	Have the stipulated cost limits been observed? Will additional operational or subsidiary costs arise?
Schedules	Can the delivery dates be met, including tooling? What design modifications might reduce cycle time and improve delivery?

a systematic approach to apply proven design principles during product development. This checklist is created from generic (and historically proven) design principles of ensuring robustness, clarity, simplicity, and safety in a product.

Robustness is the design principle that seeks to minimize the variability in performance of a product under all expected environmental and user conditions. This principle provides a basis for understanding the impact of noise on a product's performance.

Clarity is the basic principle that all functions should be unambiguously specified, in form, parameters, manufacturing, and assembly. Unintended functions should not be present in a product. It also assumes that product functions (or function chains) will be implemented as independent as possible. In doing so, the performance of each product function (or function chains) can be controlled and modified without deteriorating or compromising the performance of other product functions.

The design principle of *simplicity*, on the other hand, is the minimization of information content within a product design. For this principle, component shapes are simplified to aid in production ease and cycle time. The number of components is also reduced to simplify ease of assembly and increase the reliability of a product. Component actions and motions are also reduced to increase reliability.

The remaining generic design principle is *safety*. Its purpose is to minimize the risks created by the use of a product. As such, this principle seeks to ensure that a product has the desired strength, reliability, environmental impact, ergonomics, and accident prevention measures.

Based on these principles, an embodiment checklist is formulated by considering the range of possible engineering specifications applied to commercial products. These specifications are then arranged in categories, and for each category a set of basic questions are developed. These questions should be exhaustively applied to a product as it is being embodied (during the evaluation steps in Figure 12.6). A mathematical model or physical prototype may be needed to effectively answer each of the questions.

Example: Drive Train Subsystem of a Radio-Controlled (RC) Car

Consider a product found in hobby stores: a RC car. Electric RC cars are entertainment products that emulate full-scale on-road or off-road vehicles. They are controlled by a transmitter (speed and steering) and

Embodied design of a popular
RC car product.

are powered by 6–7 D-cell batteries. An embodied version of a popular RC car is pictured in Figure 12.7.

Applying the checklist to the RC car, one of the questions (Layout,
Geometry, and Materials category) is stated as, "Does the chosen layout ensure freedom from resonance?" Many components are subject to
resonance, including all of the rotating components (bearings, shafts,
etc.) and the helical suspension springs. Of these components, the
driveshafts of the car are the most susceptible due to their nominal
long length and small cross section.

Figure 12.8 illustrates a possible drive shaft layout. One issue with driveshafts is whether the shaft will excessively vibrate and thereby cause
premature failure or annoying dynamics. One mode of the vibration is
when the shaft is not centered as it rotates—when the middle of the
shaft is radially offset outward as the shaft rotates. Based on a downward deflection of the shaft due to its weight, the critical speed of shaft
may be calculated by

$$\omega_{cr} = \sqrt{\frac{5g}{4\delta}},$$

where ω_{cr} is the critical speed for resonance (in rpm), g is the acceleration constant due to gravity, and δ is the lateral deflection of the shaft
(Juvinall and Marshek 1991). Substituting variables from Figure 12.8,
the critical speed for the current shaft layout is approximately 3600
rpm. On the other hand, based on the choices of motor and gear train
variables, the maximum driven speed of the shaft will be 2500 rpm.
Based on this analysis, the shaft will not resonate. If the shaft speed
were close to or greater than the critical speed, the shaft dimensions
would have to be modified (radius increased slightly) to satisfy the resonance checklist item.

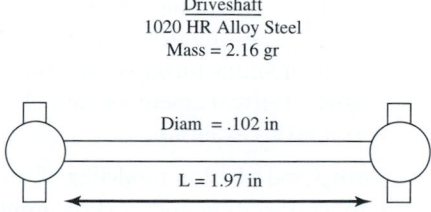

▼ *Figure 12.8.*
Embodiment layout of the driveshaft for a RC car product.

As shown by this example, embodiment checklists such as the one illustrated in Table 12.1 motivate the careful consideration of both performance issues and possible failure modes of a product. Checklists of this type are constructed from historical data and past product developments. They should be used, in conjunction with the general embodiment process, to carefully create detailed subassembly layouts and properly select parameters that will ensure a robust design.

IV. ADVANCED METHODS: SYSTEMS MODELING

In addition to the general embodiment process (Figure 12.6) and embodiment checklist (Table 12.1), systems modeling techniques can be applied when embodying a concept. *Systems models* are representations of a product that predict the product's performance under varying input (environmental or boundary) conditions. This section describes a general approach to systems modeling (where the details and examples are enumerated in Chapters 13–19). Then, more advanced refinements of the embodiment checklist are presented as mechanical design heuristics to help analyze concept layouts and improve them in the face of typical mechanical design problems. Finally, a technique is presented for systematically studying the potential failure modes of a product: *failure modes and effects nalysis.*

Systems Modeling

Functional models, as discussed in Chapter 5, represent high-level systems models of a product. Inputs of materials, energy, and signals are simultaneously converted to desired outputs. During the concept development phase of a product, these models provide a basis for systematically generating product architectures (Chapter 9) and solution concepts. These models need to be extended to facilitate design-parameter and manufacturing decisions during concept embodiment. Mathematical (virtual) methods and physical prototypes are the means to perform this extension.

Virtual and physical modeling of a product provide in-depth insights into its operation and possible improvements. Three tasks are needed to develop a virtual, or mathematical, model. Beginning with each important customer need, engineering specification from the House of Quality (Chapter 7), or possible failure mode, the critical product components should be listed. The governing physical principles and associated modeling assumptions must then be identified for each component or the group of components as a whole. For example, for an electric wok, the customer need of "uniform temperature" is important. Components affecting uniform temperature (spatially) are the heating element, the electrical energy input, the wok bowl, the radiation reflector (housing), supports, and the food. The choices for the governing physical principles might be conduction, convection, and radiation assuming steady-state conditions and lumped masses.

High-level model: After identifying the physical principles and assumptions for each customer need, a balance relationship is created to document a high-level physical model. Control volume methods or cause-and-effect diagrams may be used to document the balance relationship, where the "effect" is the customer need, and the "causes" are the physical principles. The ultimate goal is to refine the "causes" to the level of physical variables.

Balance relationships: The last task in formulating mathematical models for a product is to convert the balance relationships into a set of mathematical equations. Basic engineering principles may be used to choose the appropriate scaling law relationships (Miller 1995). Variable ranges from the product's layout drawings or bill of materials should be used to augment the mathematical relationships with appropriate parameter values. If such variable values are unavailable, appropriate ranges are chosen.

Physical prototype models: In some cases, cycle-time, economic, or product-complexity considerations may prevent the development of a mathematical model. The creation of a physical prototype can be used as an alternative modeling approach. Prototype models should be designed in such circumstances. The intent here is to create a bench-top or other experiment (not an entire product prototype) for a customer need, focusing on the effected product components and variables. For instance, a customer need may exist for an electric wok lid handle to "fit comfortably in the user's hand during operation." A mathematical model of this need is not directly apparent; however, physical prototypes that vary shape, size, and texture of the handle may be designed for analysis and testing.

Example: Bilge Pump Product

Chapter 13 provides complete examples of systems models for various products. However, to illustrate a brief context for modeling in embodiment design, consider the bilge pump product shown in Figure 12.4. Two important customer needs are to "start, automatically, the pumping water" and "remove sufficient quantities of water over time." Let's consider these customer needs.

To automatically initiate pumping, that is, pump priming, we must consider the pressure head due to the change in height from the bilge to the deck of the boat. Applicable assumptions include: (1) hydraulic conditions (incompressible, steady flow, etc.), (2) maximum height difference for a pleasure craft (based on survey and literature) is 2.5 m, (3) a tube with constant diameter is used to channel the water, and (4) the differential fluid velocity from inlet to outlet (control volume) is negligible. Using these assumptions and applying Bernoulli's Equation for the inlet and outlet conditions of the pump, we have:

$$\frac{p_1}{\rho g} + \frac{v_1^2}{2g} + z_1 = \frac{p_2}{\rho g} + \frac{v_2^2}{2g} + z_2$$

$$\Delta p = -\rho g (z_2 - z_1)$$

$$\Delta p_{\text{salt}} = -\left(1030 \frac{\text{kg}}{\text{m}^3}\right)\left(9.81 \frac{\text{m}}{\text{s}^2}\right)(2.5 \text{ m}) = 28 \text{ kPa}$$

$$\Delta p_{\text{fresh}} = -\left(1000 \frac{\text{kg}}{\text{m}^3}\right)\left(9.81 \frac{\text{m}}{\text{s}^2}\right)(2.5 \text{ m}) = 24 \text{ kPa}$$

Thus, for priming purposes, the pump must overcome approximately a quarter atmosphere of pressure for salt or fresh water. This simple pressure model may be varied, depending on the height change of chosen pleasure boats.

To determine if the bilge pump layout of Figure 12.4 can overcome this pressure difference and pump sufficient quantities of water, we need to consider the system as a thermodynamic process. Applicable assumptions include:

1. Thermodynamic conditions: isentropic (reversible, adiabatic),

2. Energy losses are due to friction (mechanical, not thermodynamic),

3. Two cases exist: (a) pump not primed, implying the pumping of air and (b) pump primed, transporting water above bilge (pump outside),

4. Working fluids: air, fresh water, and salt water.

For case (3a), pumping air, and using the isentropic gas law, we consider the pressure differential due to the volumetric change in the pump body. We also assume that due to friction losses, a pressure drop of 36 kPa is needed as opposed to 24–25 kPa. The isentropic gas law, assuming air is the working fluid (with the specific heat ratio, $k = 1.4$), is given by:

$$\left(\frac{p_2}{p_1}\right) = \left(\frac{V_1}{V_2}\right)^k$$

$$\left(\frac{V_1}{V_2}\right) = \left(\frac{p_2}{p_1}\right)^{\frac{1}{k}}$$

$$\left(\frac{V_1}{V_2}\right) = \left(\frac{64 \text{ kPa}}{101 \text{ kPa}}\right)^{\frac{1}{1.4}} = 0.72$$

$$\Delta V = (V_2 - V_1) = 0.28 V_2$$

From this model, the reciprocating pump must accommodate a 28% differential volume. This result gives the minimum volume change to prime the pump.

Considering case (3b), pumping water, we extend an energy approach to the model. Using ideal wave energy equations (Airy), the calculated wave energy per meter of wave crest width ranges from 10–1000 J/m for typical marina conditions. Conservatively, we assume that 10 J is available for use by the pump. The power available is thus a function of this energy and the wave period or frequency (2 sec from research data). This available power must be sufficient to overcome the pressure head and satisfy a customer flow rate specification of at least 8 L/hr. Thus, assuming 50% efficiency of the pump, the available and needed power are given by:

$$P_a = \left(\frac{E_a}{t_{\text{wave}}}\right)\eta = \left(\frac{10 \text{ J}}{2 \text{ sec}}\right)(50\%) = 2.5 \text{ W}$$

$$P_{\text{need}} = \Delta p \cdot Q = 25 \text{ kPa} \cdot 8\frac{1}{\text{hr}} = 25 \text{ kPa} \cdot 0.002 x 10^{-3}\frac{\text{m}^3}{\text{sec}}$$

$$= 0.05 \text{ W}$$

Based on this result, more than enough power is available to pump the minimum flow rate. This approach can be extended to determine a minimum pump size (across all parameters) to accommodate wave forces, manufacturing issues, predicted flow rate/wave velocities, and so forth. Assume that for the reciprocating pumps, a minimum piston radius of 0.5–2 cm must be used, with a 6–20-cm stroke (notice that ranges are used here to model our imprecision in design choice). Is there enough power available for these conditions?

Lower bound:

$$P_{\text{need}} = \Delta p \cdot Q = 25 \text{ kPa} \cdot \frac{\pi r^2 s}{t_{\text{wave}}}$$

$$= 25 \text{ kPa} \cdot \frac{\pi (0.005 \ m)^2 (0.06 \ m)}{2 \text{ sec}} = 0.06 \text{ W}$$

Intermediate:

$$P_{\text{need}} = \Delta p \cdot Q = 25 \ kPa \cdot \frac{V_{\text{pump}}}{t_{\text{wave}}} = 25 \ kPa \cdot \frac{\pi r^2 s}{t_{\text{wave}}}$$

$$= 25 \text{ kPa} \cdot \frac{\pi (0.01 \text{ m})^2 (0.1 \text{ m})}{2 \text{ sec}} = 0.4 \text{ W}$$

Upper bound:

$$P_{\text{need}} = \Delta p \cdot Q = 25 \text{ kPa} \cdot \frac{\pi r^2 s}{t_{\text{wave}}}$$

$$= 25 \text{ kPa} \cdot \frac{\pi (0.02 \text{ m})^2 (0.2 \text{ m})}{2 \text{ sec}} = 3.1 \text{ W}$$

Thus, there exists sufficient energy for the pump to operate with realistic geometries. However, the power need approaches the available power, albeit a conservative estimate, at the upper bound of the geometry intervals. This insight implies that care must be taken in choosing the geometry/materials as concept embodiment progresses. This observation is especially applicable to the float variant due to its direct dependence on wave energy.

This preliminary systems model answers important embodiment questions. It may be developed further by considering appropriate manufacturing and assembly constraints in addition to transient effects and potential failure modes. Chapter 13 describes techniques to formulate such models.

Mechanical Embodiment Principles

Some mechanical assemblies work better than others—we know it when we operate them. What makes a good mechanical design? Beyond proper sizing of components and tolerances, it is often the proper design of the forces through the product. Five design principles are presented in this section. As with any heuristic or principle, they are not guaranteed, and there are times to not follow them. But in most cases, they should be considered.

Alignment of Forces

Forces within an assembly are what make the parts move. The position and orientation of these forces are design choices that the designer makes, and so the reliable and carefree motions of a product are under the control of a designer. They are not the result of arbitrary effects; they are the result of how the designer applies physical principles in a product.

To properly design forces in a mechanical assembly, one should distinguish three forces that act on any moving part. The first is *weight*. The second type is a *frictional force*, which resists motion. The final type is an *applied force (applied load)*, which a designer uses to create the motion. Whether a part moves freely or binds is completely determined by the interactions of these three forces.

To design an exceptional system of movement in a machine, one should consider subassemblies that move relative to one another as distinct moving parts. Then one should analyze and redesign the entire assembly by moving the interfaces so that the motions are carefree. To do this, one should do the following:

1. On an isometric sketch of each moving part, determine and draw on the sketch the center of mass. This location is where the weight acts as a single force down, and the inertial forces act in the opposite direction to the motion.

2. On each moving part, draw each applied force. Then determine the centroid of the applied forces and the vector sum of the applied forces. The vector sum of applied forces acts as a single force at the applied force centroid. The centroid coordinates can be calculated by:

$$x_c = \frac{\sum\limits_{\substack{\text{applied} \\ \text{forces}}} F_{x,i} x_i}{\sum\limits_{\substack{\text{applied} \\ \text{forces}}} F_{x,i}}, \; y_c = \frac{\sum\limits_{\substack{\text{applied} \\ \text{forces}}} F_{y,i} y_i}{\sum\limits_{\substack{\text{applied} \\ \text{forces}}} F_{y,i}}, \; z_c = \frac{\sum\limits_{\substack{\text{applied} \\ \text{forces}}} F_{z,i} z_i}{\sum\limits_{\substack{\text{applied} \\ \text{forces}}} F_{z,i}}$$

3. On each moving part, draw each frictional force. Then determine the centroid of the frictional forces and the vector sum of the frictional forces. The vector sum of frictional forces acts as a single force at the frictional force centroid.

4. Redesign the force locations so that all centers of force—weight/inertia, applied, and frictional—are on top of each other. If they are not, the part will bind as it moves. Be sure to consider the variation of the force values as the device is operated.

Some techniques for redesign of force locations include moving any bearings and bearing surfaces, thereby designing in cantilevers.

As an example, consider a college design competition to deliver hockey pucks to a remote target. One design calls for a rack of pucks (mounted on a wheeled chassis) to be pushed off sequentially by driving a pusher with a worm drive, as shown in Figure 12.9. The left-hand design does not have the center of mass (weight) aligned with either the center of applied forces or with the center of friction. It will bind.

The second design has the center of applied forces aligned with the center of mass but not with the center of friction. It is less likely to bind and will avoid binding only through widening of the space between the linear bearings on the slide rails, creating a cantilever. This design is good for fast-moving applications, where the inertial force is higher than the frictional force.

The third design has the center of applied forces more aligned with the center of the friction forces. It does not have the center of mass aligned with either of the other two centers. This design is good for slower moving applications, where the inertial force is negligible. It is less likely to bind, and again, will only not bind through a wide linear bearing spacing on the slide rails.

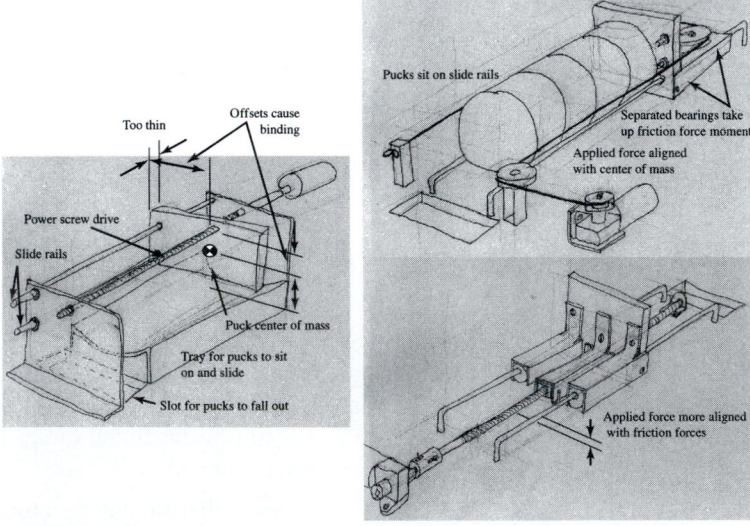

▼ *Figure 12.9.*
Force alignment in drive screw systems.

Note that in the concept of Figure 12.9, it is impossible to align the frictional force centroid with the other two centroids unless one introduces an upper rail to frictionally impede the top surfaces of the pucks. This results in a poor design—again, one should understand the heuristic and use wisdom in applying it.

3-2-1 Alignment

Often subsystems do not properly interface due to poor kinematic layout of the interface. Circuit boards may not fit properly into a bus chassis, casings may not fit properly around internals, supposedly swappable modules may not swap in and out easily. These cases are often very frustrating to repair mechanics and users and are directly the result of poor embodiment design.

To reduce and eliminate these problems, all part and subsystem interfaces should be mechanically designed with sound principles. Any mechanical interface can be broken down into two aspects—how it is *positioned* and how it is *restrained*. These are two separate considerations, where the first is very often ignored.

Any part is a three-dimensional object in three-dimensional space and so has six degrees of freedom—three possible translational directions it can move and three possible rotational directions it can spin. To geometrically position a part, all six degrees of freedom must be properly constrained. Often these restraints are thought of in terms of sufficient strength (Did I design in enough bolts to hold the part down?), but before considering strength, the positioning must be designed.

To constrain any part into an assembly, one should consider how the part will be placed, which determines how it is kinematically positioned in the assembly. Positioning is independent of restraining. The best approach to positioning is to design all parts with the *3-2-1 alignment scheme* (Figure 12.10). Here, every part that must be rigidly attached into an assembly has "3" perpendicular positioning surfaces defined on the part. On one plane, three points are defined as *alignment points*—they contact the greater assembly the part fits within. On a perpendicular plane, "2" more points are selected as alignment points. Finally, on the third perpendicular plane, "1" point is selected as an alignment point. When placing any part into the assembly, the part is first positioned on the three points of the first plane, then slid on that plane into the two points on the second plane, and then slid along all five points until it hits the last point on the third plane. This configuration of three planes defines the position and orientation of the part in the greater assembly, and it is ready to be fastened (restrained) into place.

With a 3-2-1 alignment process, the fasteners chosen for restraints do not define the position and orientation; rather, the six alignment points

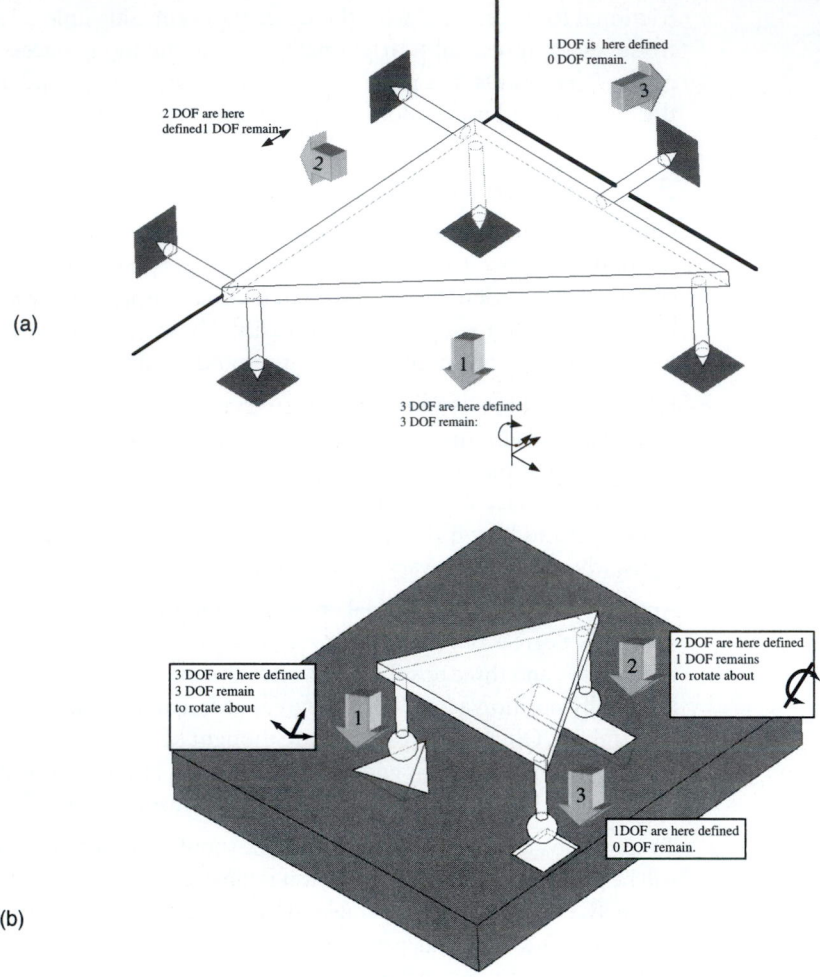

2 DOF are here defined1 DOF remain:

1 DOF is here defined
0 DOF remain.

3

2

(a)

3 DOF are here defined
3 DOF remain:

1

3 DOF are here defined
3 DOF remain
to rotate about

2 DOF are here defined
1 DOF remains
to rotate about

1

2

3

1DOF are here defined
0 DOF remain.

(b)

▼ *Figure 12.10:*
3-2-1 alignment. (a) Translational; (b) rotational.

do. Fasteners should not be used to define position and orientation, since they require adequate tolerance zones around them, and the result would be a sloppy fit.

Note that a 3-2-1 alignment scheme does not mean all parts require six fasteners. One fastener can suffice to restrain any part to a greater assembly. Positioning is not restraining. But, to position the part into the assembly, six points defining three independent face planes are required. Practically, these faces can be on the greater assembly or can be part of an assembly fixture that the greater assembly is first aligned to before the part is added. Also, the first three points need not lie on a plane, but can lie on three parallel planes that the part can slide over during assembly. The same is true of the second two points.

A 3-2-1 alignment scheme is beneficial in that it leaves no uncertainty in how parts or modules interface the greater assembly. Most tolerance problems (where parts do not always fit together properly) are a result of uncertainty in part placement due to poor (or no) design of the part alignment, not as a result of unexpected excessive shape variation.

When a 3-2-1 alignment process is not used, it is for one of two reasons. Either the design process was sloppy or the part is overconstrained.

In the former case, the parts are not well defined with respect to each other, so their positioning accuracy depends not on the design but rather on the assembly process in manufacturing. How careful were the assemblers in positioning the parts before fastening?

The latter case of overconstraining, on the other hand, is often quite proper to use as a designer. That is, consider the first alignment plane that uses three alignment points. Suppose the designer did not restrict the design to three alignment points and rather included multiple points on the plane to contact the greater assembly. Then, if the parts are rigid, only three out of the multiple possible points will contact the greater assembly. One can think of a table with four legs, for example, that rocks back and forth from two different sets of three legs. One can also think of a circuit board that fits into a motherboard, where not all electrical contacts are made. One can also think of two flat surfaces that must be bolted together to make a tight seal between the surfaces.

In all of these cases, the two parts will not mate as desired if the parts are perfectly rigid. Fortunately, they are not perfectly rigid, and both mating surfaces slightly deform and conform to each other. Here, the designer is relying on *elastic averaging*. The table sags under gravity and weights put on the table, and so all four legs touch. Each electrical contact bends to touch its paired contact. Gaskets deform to fill in the space between two rigid surfaces when bolted together. All of these multialignment-point interfaces rely on the deformation of the parts at the interface to define the final alignment surface.

Elastic averaging provides a strong interface, but it is also typically hard to determine precisely, for high-accuracy design, where the interface actually is. Often it is used for no apparent reason, when a 3-2-1 alignment would provide better results. It is necessary, though, for joints that must be pressure tight to prevent escape of liquids or gases.

Deflection Reduction and the Abbe Principle

Despite the fact that most courses on applied mechanics heavily emphasize yield stress/strain and plastic deformation, most mechanical designs encounter excess elastic deflection as a design problem far before actual part failure. These deflections often cause problems in that critical interfaces such as bearing surfaces do not stay aligned as intended by a designer during layout design.

When faced with such a problem or to apply preventative measures, one should design the parts to not elastically deflect. This application is straightforward for most designers when considering stretch. The problem, however, comes when failing to consider two other elastic modes: twists and arm amplification.

Twists are the effects of moments, typically not from applied torque (such as from motors) but rather from two internal forces separated by a distance. To consider this situation, one should always examine the forces (or force flows; see Chapter 6) on a part or assembly and imagine them causing deflections of the fastened assembly as a part. Ask not only how a part stretches, or compresses, but also how does a part twist? As an example, most two-dimensional truss analysis problems have little bearing on practical problems, since usually the truss twists in and out of the plane and does not stretch within the plane. Similarly, most space frames, such as for machines, planes, automobiles, and so forth, do not sag under load; rather, they twist. Airplane wings not only vibrate up and down at the tip, but they torsionally twist along their length about the wing axis. Automobile frames twist up from a wheel shock relative to the center of the frame.

To reduce such twists, one should imagine how to place additional stiffening along the tangent of the twist. At a radius out from the twist axis, we can place a member that absorbs the twist torque. For example, on a commercial jet aircraft, two main ribs are required extending out of the wing structure, since the engines apply a massive twisting torque to the wing. One main rib would not contain the twist.

The second form of deflection, often not fully appreciated, is *arm amplification*. Here, due to the embodiment geometry, a small force is amplified into large moments that cause problems. Similarly and equivalently, small deflections can be amplified into large deflections.

These arm amplifications are shown in Figure 12.11. To eliminate these problems, one must analyze systems to find the arms causing the amplification and eliminate the amplification. This analysis and elimination step is known as the *Abbe Error Principle*.

Methods to eliminate Abbe errors are to extend the other arm that constrains the motion. All Abbe error problems are essentially lever problems, the lever connecting the small force to the fulcrum is too large compared to the lever connecting the fulcrum to the restraining point. One has to shorten the first lever or extend the second. In the extending lever example of Figure 12.11(a), the bearing separation distance D must be made larger, relative to the total reach. As another example, consider the screw drive of Figure 12.12(b). The linear bearing separation down the shafts should be larger than the vertical distance from the slide shaft to the power screw (frictional load binding), and larger

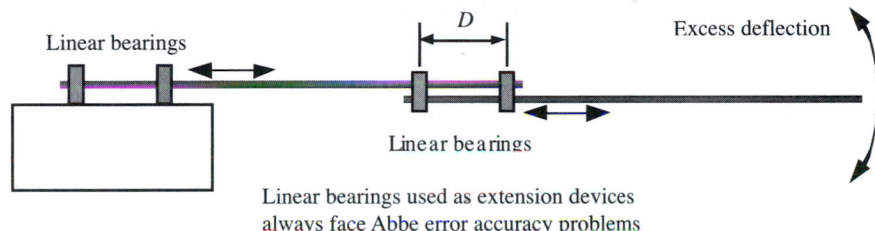

(a)

Linear bearings used as extension devices always face Abbe error accuracy problems

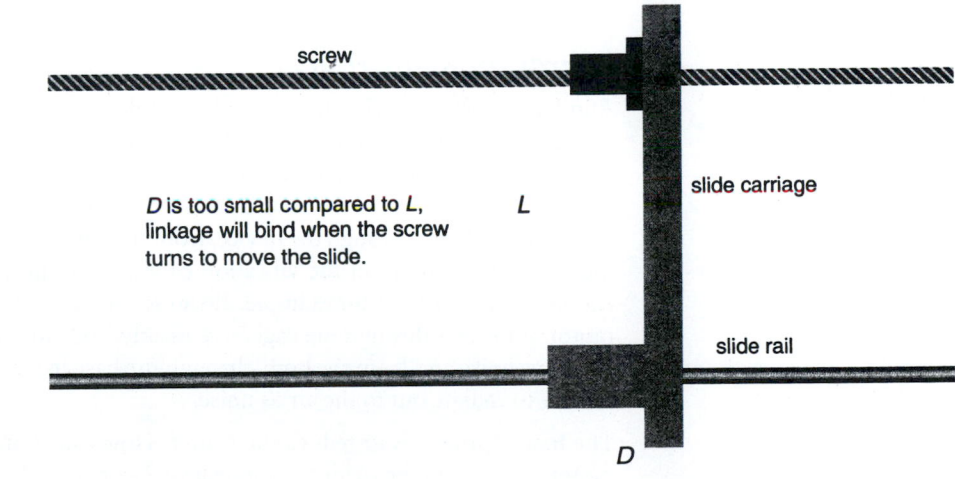

(b)

▼ *Figure 12.11.*
(a) Abbe errors. (b) A binding carriage due to twists.

than the vertical distance from the power screw to the puck center of mass (inertial load binding). .

Forces in Members

Another aspect of mechanical design is to ensure that the parts convey the forces adequately without internal failure. There are times when the part design is as critical as the force interfaces, such as when the material must be minimized (aircraft) or when a part is intended to deform during use (plastic hinges). In such cases, how the forces are transmitted through any part becomes important.

In this case, detailed design activities such as finite element analysis with associated part parameter optimization can improve the design. On the other hand, there are simple techniques that can help.

The first basic principle to reducing internal part loads is to convey moments through triangles. For any part, one should examine the critically stressed regions and imagine how to reduce them through triangular gussetting.

Another approach is to redesign the force application points to reduce the moment arms of the applied forces and thereby the internal moments. One effective approach is to draw the flow-lines of force though a part from application points to absorption points. One can then visualize places of high stress possibly requiring more material to prevent deflection and places of low stress where material can be safely removed.

Vibration Reduction

Another common mechanical embodiment design problem is vibration, either mechanical or acoustic. There are three basic approaches to reducing vibrations: Reduce the source, change the transmission, or reduce the transmission. To execute these approaches, one should map the vibration loads through the device, from the source point, through the parts, to the point of the vibration problem. In the Krups coffee chopper (Figure 12.12), for example, the noise problem originates at the motor, transmits through the cage, screws, and body to radiate out to the air as noise and from the body through the base to any countertop surface to radiate out to the air as noise.

The first approach is to reduce the source. How can sources (usually motors or engines) be made to vibrate less? Again, one should observe the actual source of vibration and its plane of motion and examine where that is being restrained. Is the vibration amplified due to its mounting? Cantilevered motor mounts are a typical problem. For example, the motor in a common coffee chopper is attached on its top

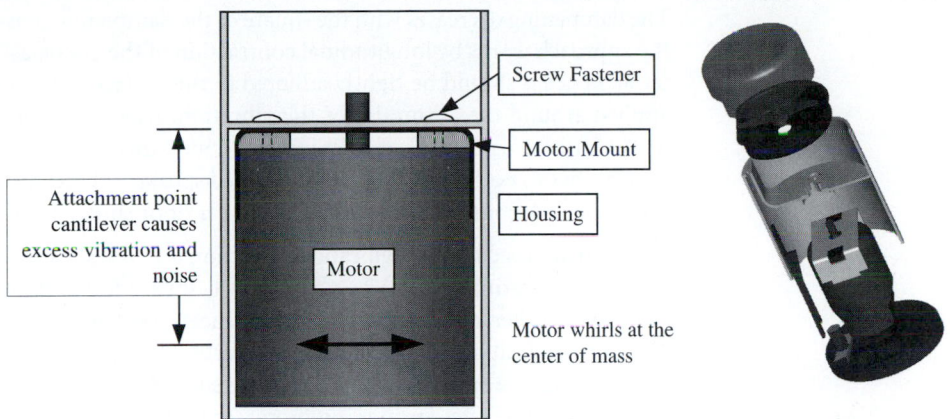

Screw Fastener

Motor Mount

Housing

Attachment point
cantilever causes
excess vibration and
noise

Motor

Motor whirls at the
center of mass

▼ *Figure 12.12.*
Vibration principles: Krups coffee chopper.

surface by two screws (see Figure 12.12), though the vibration center-line plane is lower at the motor midplane. The vibration is amplified as the motor whirls about its top attachment.

The second approach to vibration reduction is to change the natural frequency of the source or transmitting parts. Changes in shape to nonplanar (bowled) and noncylindrical (cone) shapes can greatly stiffen a part, thereby increasing its natural frequency, perhaps to a regime where the vibrations cause no problems.

The last approach is one often applied; it calls for the insertion of dampeners within the travel path of the noise from its source to its area of concern. This approach basically inserts dissipative materials into the path of vibration, permitting the source to vibrate without transmitting to the rest of the system.

One approach is to insert such vibration isolators at the part attachment points. Felt, cork, and fibrous pads work for frequencies above 40 Hz, rubber works in ranges of 5–50 Hz, metal materials work above 1.5 Hz, and air shocks work up to 5 Hz. For high-frequency vibrations, viscoelastic materials can be used as dampers.

Another approach is to dampen the transmitting parts rather than the transmitting joints. For sheet surfaces vibrating in and out of the plane, one approach is to apply a viscous layer to the surface. With viscoelastic rubbers applied to metal surfaces, a general rule is that the viscoelastic material should be three times the thickness of the metal structure, and the mass should be about 20% of the structural mass.

The dampening decreases with the square of the dampening mass. Also, the approach works by longitudinal contraction of the viscoelastic material, hence it should be tightly adhered to the surface and especially applied around the antinodes of the vibration. As an example, some luxury cars have portions of their interior door panels covered with an SBR rubber to both dampen door-mounted speaker vibrations and to provide a nice sound when slamming the car door shut.

A second approach to part vibration reduction is to use shear dampers. Here, the vibrating surface is split into two sheets and a viscoelastic material is adhered to the insides of the sheets, creating a sandwich. This approach, although typically very effective, is expensive. A rule of thumb is that the dampening material mass should be about 10% of the structural mass. The dampening increases linearly with dampening mass. Again, the approach works by longitudinal contraction of the viscoelastic material; hence it should be tightly adhered to both inner surfaces, particularly at the antinodes of the vibration. As an example, space frame construction (vehicles, aircraft, spacecraft, machinery, etc.) can be made far less sensitive to vibration by revising all tube-shaped members to use two thin tubes, one slightly smaller in diameter than the other and injecting a thin layer of visoelastic rubber between.

Other Design Heuristics

These design heuristics are but a few of the many that can be developed to improve design. There are others for heat transfer, fluid mechanics, electrical layout, and any other design discipline. Over a lifetime of design engineering, one should look for elegant designs, compare them against poorer performing ones, and work to formalize the differences into such sound design principles or heuristics.

FMEA Method: Linking Fault States to Systems Modeling

The foundation of robust product design is built on the combined concepts of customer quality and engineering quality. *Customer quality* is to minimize the performance variation of a product for all environmental and user conditions. *Engineering quality*, on the other hand, is to ensure that a product functions as it is intended, without falling short of a customer's implicit expectation. This second type of quality is intended to ensure that a product has adequate strength, reliability, environmental impact prevention, and accident prevention measures. As such, it may be termed as the *expected quality* of a product.

While systems modeling focuses, initially, on the overt customer needs of a product, methods are needed to identify the issues related to the expected quality of a specific product. The embodiment checklist presented in the previous section provides a basic approach for focusing on expected quality. A more advanced and complementary technique is known as *Failure Modes and Effects Analysis* (FMEA) (Stamatis 1996; FMEA 1995; Eubanks, Kmenta, and Ishii 1997). FMEA is an analytical technique used by a product design team as a means to identify, define, and eliminate, to the extent possible, known or potential failure modes of a product system. The technique should be used cooperatively with systems modeling to investigate and determine good choices for variables defining a product.

FMEA focuses on the entire product layout, not just on each sub-assembly, component, and interfacing system of a product. Through this focus, potential problem states of a product are identified as a function of each and every component's operation.

FMEA must also be understood as a process. It entails the continuous application of design team tasks during a product's development. FMEA begins at the initiation of a product's business case and continues through the entire life cycle of the product. It also seeks to identify potential failure modes before a failure can occur in a product, not as a forensic tool for investigating a failure once it has occurred.

According to the automobile industry (FMEA 1995), FMEA supports the product development process in reducing the risk of failure by:

- aiding in the objective evaluation of design requirements and design alternatives
- aiding in the initial design for manufacturing and assembly requirements
- increasing the probability that potential failure modes and their effects on system operation have been considered in the design/development process
- providing additional information to aid in the planning of thorough and efficient design improvements and development testing
- providing an open issue format for recommending and tracking risk reducing action
- providing future references to aid in analyzing field concerns, evaluating design changes, and developing advanced designs

FMEA attempts to satisfy these goals by analyzing each component and associated subassembly in a system to determine its possible failure modes, for example, mechanical, short circuit, and stall. This analysis poses three basic questions in its pursuit of a quality product:

▶ What could fail or go wrong with each component of a product?
▶ To what extent might it fail, and what are the potential hazards produced by the failure?
▶ What steps should be implemented to prevent the failures?

These questions give rise to three basic elements of FMEA: *failure modes, failure effects,* and *failure criticality.* The first element entails identification, the second entails ramifications, and the third measures the relative importance of a given failure state. Through these elements, priorities may be established for product development. In addition, potential hazards may be identified, from no accident potential to a high risk of human/property injury or even death/total destruction.

A systematic process provides the basis for answering the three basic FMEA questions. This process may be summarized with the following steps:

1. List each subassembly and component number, along with the basic functions or function chains of the component. The component numbers may be referenced from a product's bill of materials (Chapter 6). Likewise, the component functions should be consistent with the functional models and architecture developed for a product (Chapters 5 and 9). Any functions listed for a component should concisely represent the design intent. Environmental and operational parameters, such as temperature, humidity, and pressure ranges, should be listed to clarify this intent.

2. Identify and list the potential failures for each product component. Simple prototype models (e.g., fabricated from foam or wood) and brainstorming techniques (Chapter 10) can aid in identifying potential failure modes. Likewise, sketches, storyboards, free-body diagrams, force-flow diagrams, and process-flow diagrams can help in understanding the physics of a failure mode. The checklist of Table 12.1 and the example failure modes of Table 12.2 should be used to check for typical problems with components and product systems. For any listed failure mode, the idea is that the failure *could* occur, but not that it will necessarily occur for the product under study.

TABLE 12.2.   ABBREVIATED LIST OF EXAMPLE FAILURE MODES

List of Example Failure Modes

Corrosion	Ingress	Delamination
Fracture	Vibrations	Erosion
Material Yield	Whirl	Thermal shock
Electrical Short	Sagging	Thermal relaxation
Open Circuit	Cracking	Bonding failure
Buckling	Stall	Starved for lubrication
Resonance	Creep	Staining
Fatigue	Thermal expansion	Inefficient
Deflections or deformations	Oxidation	Fretting
Seizure	UV deterioration	Thermal fatigue
Burning	Acoustic noise	Sticking
Misalignment	Scratching and hardness	Intermittent system operation
Stripping	Unstable	Egress
Wear	Loose fittings	Surge
Binding	Unbalanced	
Overshooting (Control)	Embrittlement	
Ringing	Loosening	
Loose	Scoring	
Leaking	Radiation damage	

3. List possible potential causes or mechanisms of the failure modes. Example causes include tolerance stack-up, assembly errors, poor maintenance, impact loading, overstressing, and so forth. These causes will provide insights into modeling of the failure mode. They will also indicate appropriate preventive measures that might be adopted.

4. List the potential effects of the failure, including impact on the environment, property, or hazards to human users. Example effects include noise, poor appearance, flying debris, unpleasant odor, inoperative, erratic operation, and so forth.

5. Rate the likelihood of occurrence (O) of the failure. The ratings should be on a scale of 1–10, as given by:

1	No effect
2/3	Low (relatively few failures)
4/5/6	Moderate (occasional failures)
7/8	High (repeated failures)
9/10	Very high (failure is almost inevitable)

6. Estimate the potential severity (S) of the failure and its effect. Again, a 1–10 scale should be used. The following meanings are associated with this scale:

1	No effect
2	Very minor (only noticed by discriminating customer)
3	Minor (affects very little of the system; noticed by average customer)
4/5/6	Moderate (most customers are annoyed)
7/8	High (causes a loss of a primary function; customers are dissatisfied)
9/10	Very high and hazardous (product becomes inoperative; customers are angered; the failure may result unsafe operation and possible injury)

7. List current or expected design controls/tests for detecting (D) the failure before the product is released for production. A 1–10 scale is used to assess detection:

1	Almost certain
2	High
3	Moderate
4/5/6	Moderate—most customers are annoyed
7/8	Low
9/10	Very remote to absolute uncertainty

8. Calculate the Risk Priority Number (RPN). An RPN prioritizes the relative importance of each failure mode and effect on a scale of 1–1000. It can be calculated with the following relation:

$$\text{RPN} = (S) \times (O) \times (D).$$

A "1000" rating implies a certain failure that is hazardous and harmful, whereas "1" rating is a failure that is highly unlikely and unimportant. Ratings above "100" will occur, whereas ratings below "30" become reasonable for typical applications. So notice that the RPN scale, while easy to calculate, is nonlinear in risk.

9. Develop recommended actions for the failure modes, assign responsibilities to appropriate parties and team members, and set a schedule for implementing the actions. Corrective actions should be first developed for the highest ranked failure modes based on the RPN. Example actions include revised component or subassembly design (perhaps through modeling or testing), revised test plan or material specification, design of experiments and prototypes, "none," and so forth. These actions should not be stated in their general form for a particular product component and failure state. Instead, specific actions should be listed within these general types (or others).

10. Implement the corrective actions, update the S-O-D ratings, and recalculate the RPN for the updated design.

TABLE 12.3. FMEA TEMPLATE FOR PRODUCT DESIGN AND DEVELOPMENT

Product Name: _____ _____ System _____ Subsystem Name: _____ _____ Component	Devel. Team: _____ _____	Page No. _____ of _____ FMEA Number _____ Date: _____

Part # & Functions	Potential Failure Mode	Potential Effect(s) of Failure	Severity (S)	Potential Causes/ Mechanism(s) of Failure	Occurrence (O)	Current Design Controls/ Tests	Detection (D)	Recommended Actions	RPN

The results of this FMEA process may be documented with the template provided in Table 12.3. This template provides an inventory of a product's parts, following a parallel structure to a bill of materials (Chapter 6). Information may be entered into the template early in the product development process and iteratively updated through the product's final development and ultimate life cycle.

Example: Automotive Product

An abbreviated FMEA example is shown in Table 12.4 (FMEA 1995). This example concentrates on a front door subassembly of a four-door wagon vehicle. Subsequent FMEA's must consider the detailed components of the subassembly, along with the parallel subassemblies in the vehicle.

One failure mode is studied in this example. A number of potential effects and causes are also listed. Notice that the failure mode (corroded panel) is considered of high severity, implying that design actions must be taken. Of the five causes listed, four have a likelihood of occurrence of at least a moderate rating. Each of these failure causes result in an RPN of greater than 100 in magnitude. The recommended actions should thus be implemented for these failure causes, and they should be rerated to determine their updated RPN numbers.

Summary of FMEA

FMEA entails a deliberate, thoughtful, and sincere process of satisfying the engineering quality of a product. As shown by the example, FMEA requires significant resources and follow-up by all interested parties.

TABLE 12.4. PARTIAL FMEA EXAMPLE OF AN AUTOMOBILE FRONT DOOR (AFTER FMEA 1995)

Product Name: 200X/Lion 4dr/Wagon	Devel. Team: _____	Page No. __1__ of __N__
___ System __X__ Subsystem Name: Body Closure ___ Component	_____	FMEA Number __1234__ Date: __0X-03-02__

Part # & Functions	Potential Failure Mode	Potential Effect(s) of Failure	S E V E R (S)	Potential Cause(s)/ Mechanism(s) of Failure	O C C U R (O)	Current Design Controls/Tests	D E T E C T (D)	Recommended Actions	RPN
Front Door LH, H8HX-0000-A Ingress to and egress from vehicle Protect occupant from weather, noise, and side impact Support door hardware including mirror, hinges, etc. Provide proper surface for appearance items (display surface) Adhere paint and soft trim	Corroded interior lower door panels	Deteriorated life of door leading to: Unsatisfactory appearance due to rust through paint over time and impaired function of interior door hardware	7	Upper edge of wax for inner door panel is too low	6	Vehicle general durability test vah. T-118, T-109, T-301	7	Add laboratory accelerated corrosion testing	294
				Insufficient wax thickness specified	4	Vehicle general durability testing as above	7	Add laboratory accelerated corrosion testing; Conduct design of experiments of wax thickness	196
				Inappropriate wax formulation specified	2	Physical and Chem Lab test—Report No. 1265	2	None	28
				Entrapped air prevents wax from entering corner/edge access	5	Design aid investigation with nonfunctioning spray head	8	Add team evaluation using production spray equipment and specified wax	280
				. . .	. . .	. . .	. . .	. . .	. . .
				Insufficient room between panels for spray head access	4	Drawing evaluation of spray head access	4	Add team evaluation using design aid buck and spray head	112

It also focuses not only on the "end user" customer, but on all customers of the product development process, including suppliers, manufacturing, and others.

It is important to note that FMEA produces a living document. This document reflects the very thoughts, feelings, and day-to-day concerns and assessments of a product development team. As such, it snapshots the mental pictures of a product as it is developed (Figure 12.4), formalizing the expectations of the design team. This formalization has a number of legal implications. It also systematically provides a means to satisfy our ethical responsibilities that a product will be safe and effective in all reasonable operational modes.

It should be noted here that this section's focus is on the concept of a "Design" FMEA. This type of FMEA is used for product development and life cycle analysis. Other types of FMEA exist. For example, a "Process" FMEA is used to focus on manufacturing and assembly processes (FMEA 1995). The template shown in Table 12.3 remains essentially the same, with the exception that the first column focuses on process functions as opposed to product functions. This type of FMEA redirects the focus to failure modes caused by or within manufacturing and assembly operations.

V. CASE STUDY: COMPUTER MONITOR STAND FOR A DOCKING STATION

To provide a context for the entire phase of concept embodiment, let's consider the development of a computer monitor stand for docking stations (Burhan 1998; Design Edge 1998). This product represents a new offering as the technology of the notebook computer closes in on the market share of the desktop computer (Burhan 1998). When a notebook is used as a desktop replacement at home or the office, the user sometimes uses a desktop computer monitor instead of the notebook screen itself. In this case, the notebook screen is closed, and the notebook is docked on a docking station. A desktop computer monitor and the docked notebook need to be positioned, as illustrated in Figure 12.13.

The computer stand, shown in Figure 12.13, seeks to save desktop space by locating the notebook below the monitor. One height position would be ideal for this product task; however, there are various types of notebooks, and docking stations have different dimensions. The

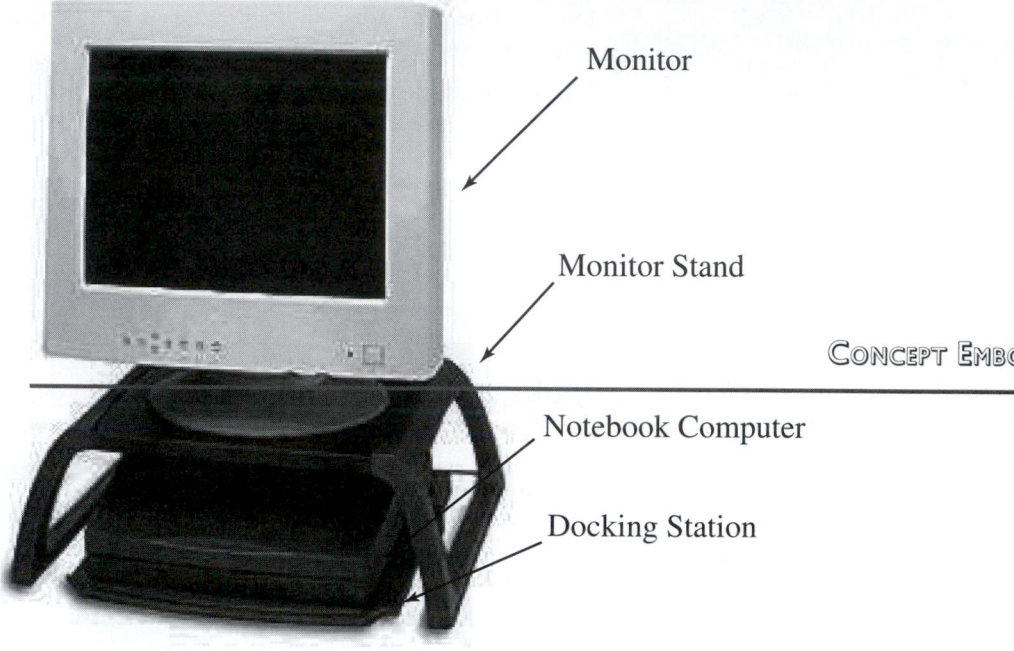

Monitor

Monitor Stand

Notebook Computer

Docking Station

▼ **Figure 12.13.**
Schematic of a notebook computer as a desktop replacement (Burhan 1998).

position of the computer monitor is expected to be as low as possible in order to give the user a natural view as well as to save space. Therefore, a monitor stand has to have different height adjustments to fulfill these requirements.

Within this product context, customer needs are gathered (abbreviated list in Table 12.5), and a business case is secured to judge the potential market. Once these tasks are completed, the project moves forward into the concept development stage. Figures 12.14 and 12.15 illustrate results from the functional modeling of a docking station computer stand. Key functions for the product include adjusting height, locking height position, supporting the monitor forces, and receiving the monitor (ergonomics).

After completing functional analysis and laying out the product architecture, concepts are generated for the monitor standing, implementing the methods of Chapter 10. Figures 12.16–12.18 illustrate partial results of solutions generated for the product functions.

These solutions to the product functions are systematically combined to create a number of concept variants (Chapter 10). Four of these alternative concepts are illustrated in Figure 12.19. These alternative con-

TABLE 12.5.   CUSTOMER NEEDS

Customer statement	Interpreted need
Product must be able to support up to 21″ monitor	• Product supports weight up to 80 lbs. • Product accommodates a 12″ × 12″ base footprint of a 21″ monitor. • Product passes a monitor drop test from a height of three inches.
Must span existing base (freestanding)	• Product allows access to peripheral drive bay. • Product allows access to eject lever. • Product does not rest on existing base for support.
Able to adjust for two types of docking stations	• Product accommodates at least two types of monitor stands.
Cannot use metal over electronic components	• Product made of materials that do not cause electronic and electromagnetic problems.

cepts illustrate the main function carriers of the product in addition to their general operating principles.

Concept selection methods are now applied to these concept variants to choose the preferred concept (if possible) and/or refine and combine the concepts into a hybrid version that most fully satisfies the customer needs. In this case, the concept selection process reveals the telescoping concept as the product development team's preferred choice. Com-

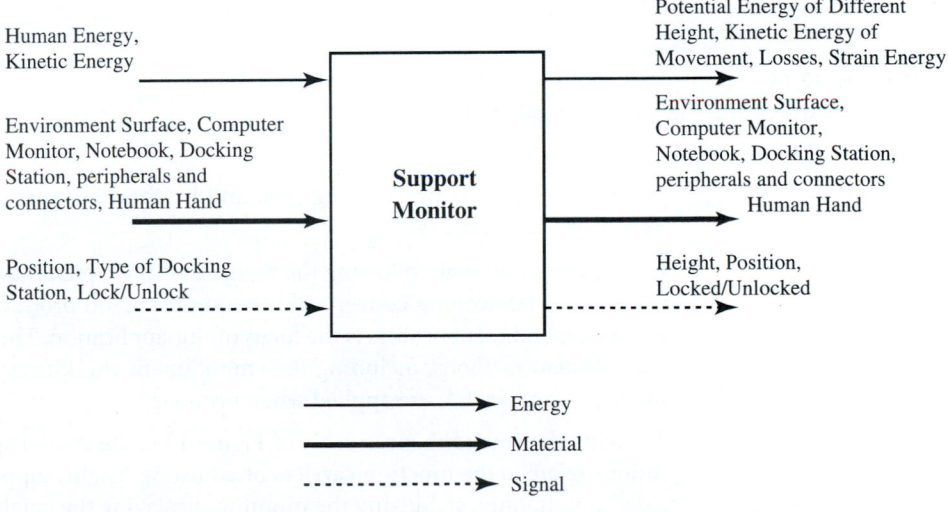

▼ *Figure 12.14*
Black box model of the computer monitor support product.

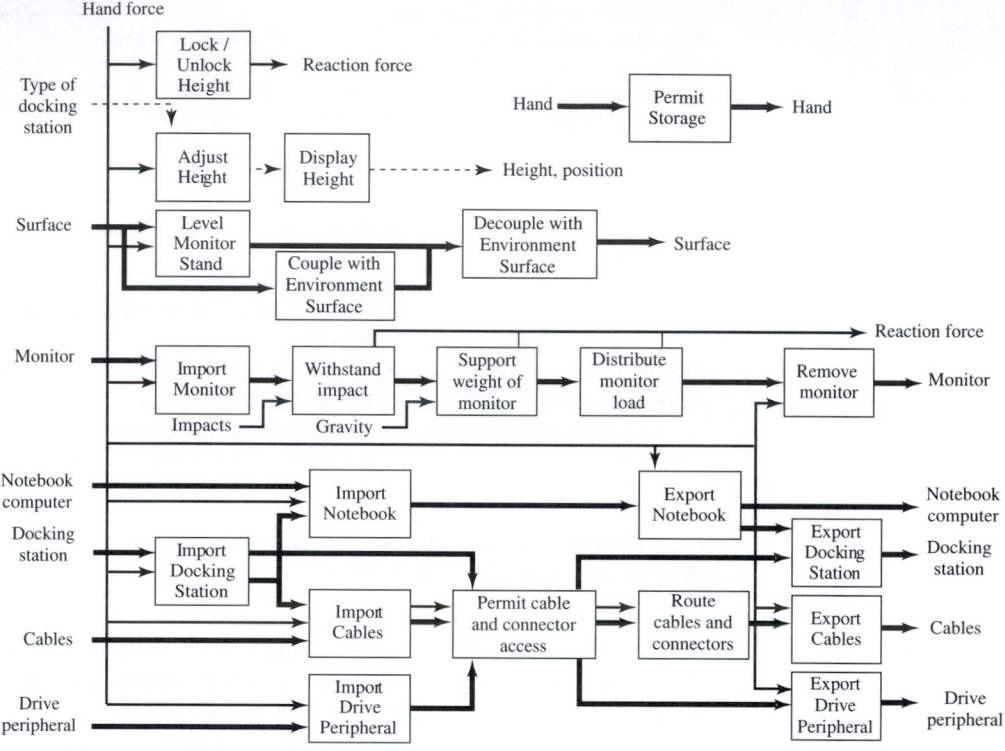

▼ *Figure 12.15.*
Function structure of the monitor stand product.

plexity, estimated manufacturing cost, and height-adjustment capability drive this decision.

Concept embodiment, following the theme of this chapter, is now applied to the telescoping concept of the monitor stand product. The general embodiment process is the focus of this application. The other embodiment methods, including the embodiment checklist, systems modeling, and FMEA, are applied when necessary.

Applying the embodiment process of Figure 12.6, the critical specifications relate to the function carriers of adjusting height, supporting a desktop monitor, stabilizing the monitor, displaying the height, and locking/unlocking the height. A rough layout for the main function carriers is shown in Figure 12.19.

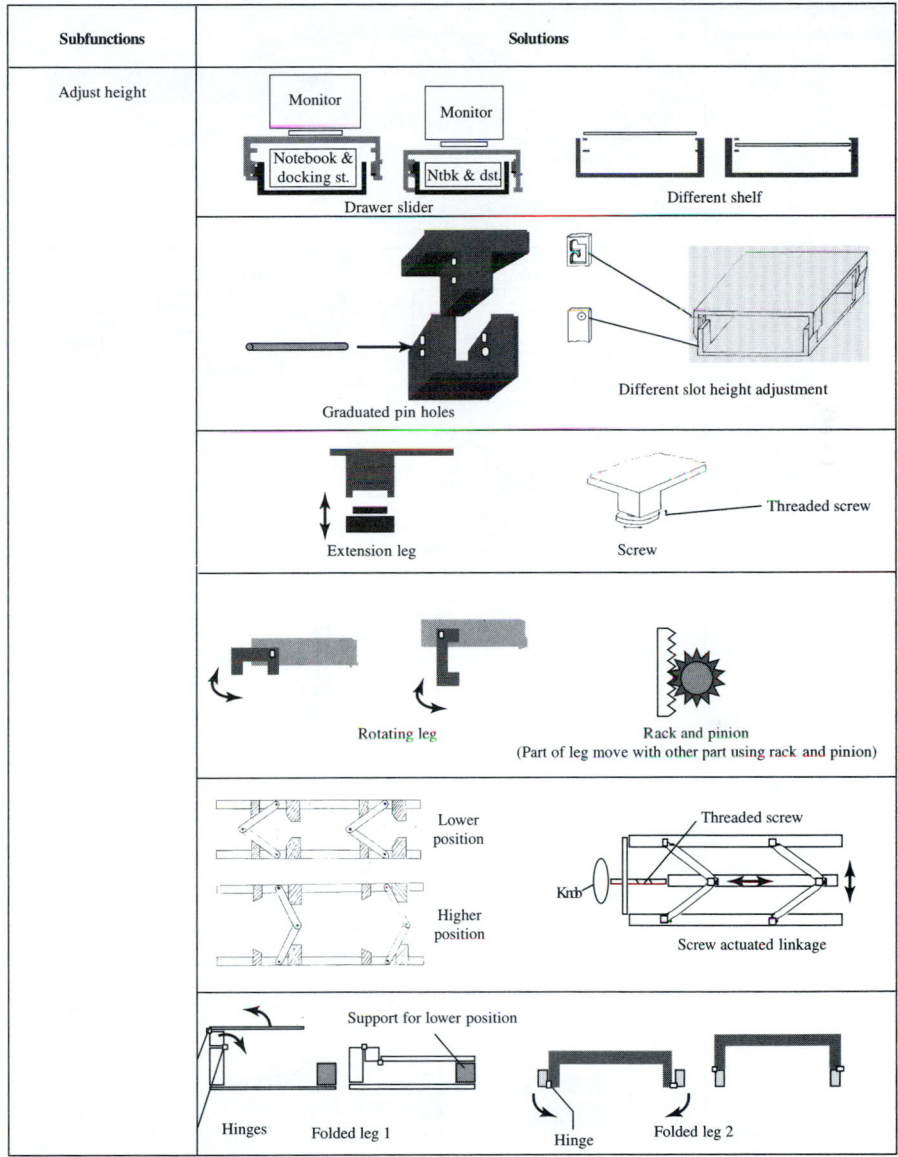

▼ *Figure 12.16.*
Partial morphological matrix for the monitor stand product (Part 1).

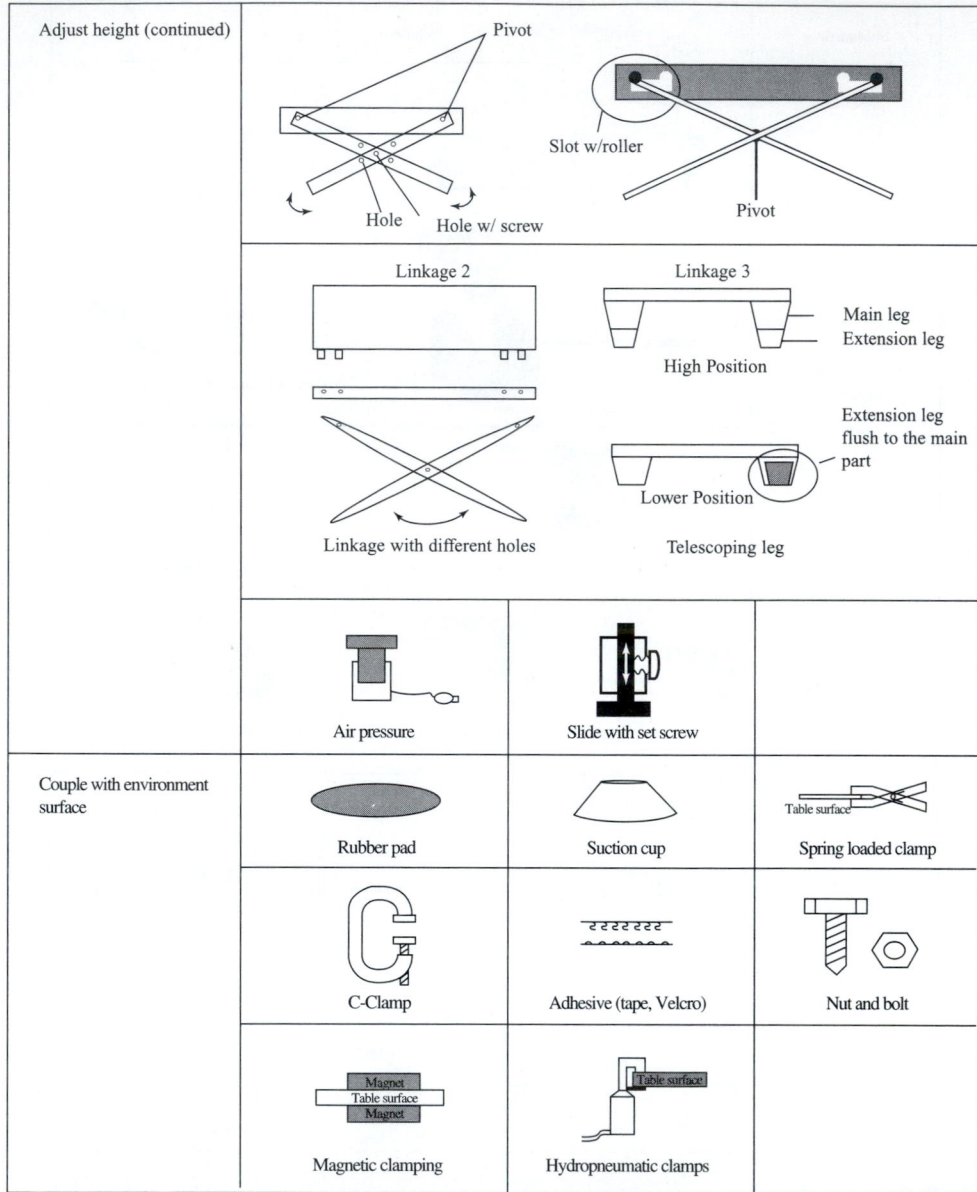

▼ *Figure 12.17.*
Partial morphological matrix for the monitor stand product (Part 2).

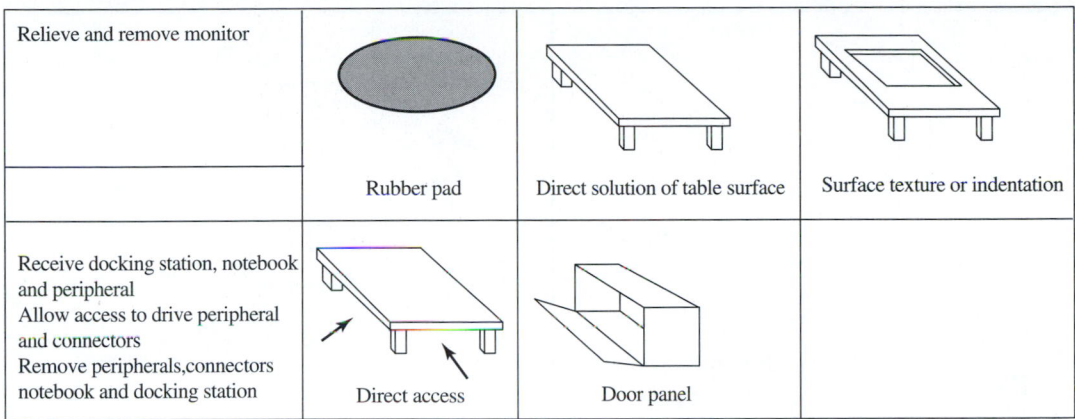

Relieve and remove monitor			
	Rubber pad	Direct solution of table surface	Surface texture or indentation
Receive docking station, notebook and peripheral Allow access to drive peripheral and connectors Remove peripherals,connectors notebook and docking station	Direct access	Door panel	

▼ *Figure 12.18.*
Partial morphological matrix for the monitor stand product (Part 3).

Because height adjustment is an embodiment-determining function carrier, alternative layouts are considered. Applying the embodiment checklist (Table 12.1) to the rough layout, concerns exist about stress concentrations caused by the slots. Figure 12.20 shows a refined concept for the telescoping design. The basic idea behind this concept is that the lower part has four external splines that mate with four internal splines in the upper part. When the splines of the two parts are lined up in the same position, they will support each other, thus placing the monitor at a high-position state. When the splines are in the position where they can slide, the telescoping parts will become flush with the other parts, thus placing the monitor stand at a lower position.

This layout is now added to the original layout of the monitor stand product. The checklist is also applied to the other main function carriers, and their layouts are modified appropriately.

Let's now consider the embodiment of the main function carrier: support monitor. Manufacturability drives the embodiment of this function carrier. To satisfy cost, volume of production, and material constraints, plastic is chosen as the primary material for the product (load-carrying components). Initial design principles for this choice, focusing on injection molding, include the following items (Chapter 14 provides a more complete discussion of design for manufacturing):

▶ All wall sections should have a uniform wall thickness.

▶ Wall sections should be designed with a suitable draft angle to allow for ease of removal from a mold.

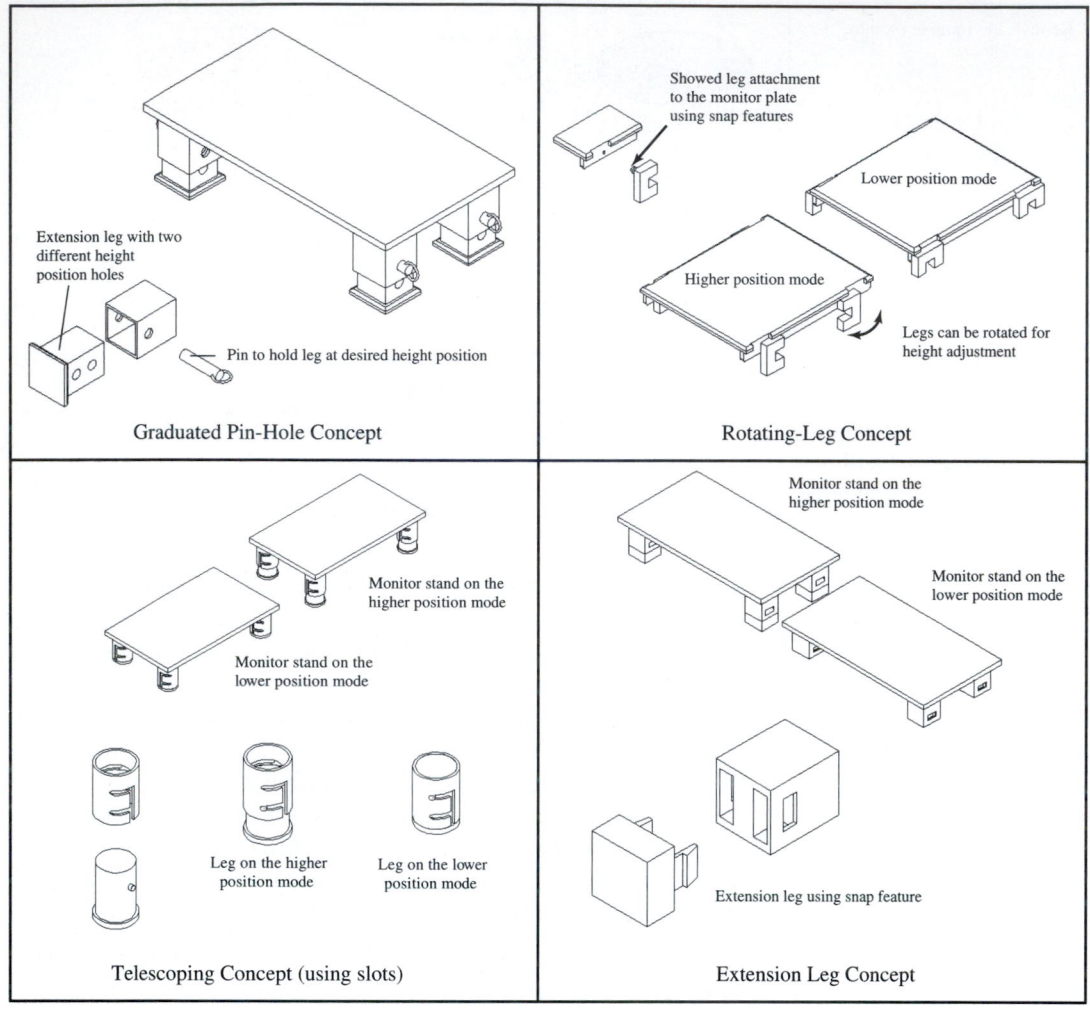

Extension leg with two
different height
position holes

Pin to hold leg at desired height position

Graduated Pin-Hole Concept

Showed leg attachment
to the monitor plate
using snap features

Lower position mode

Higher position mode

Legs can be rotated for
height adjustment

Rotating-Leg Concept

Monitor stand on the
higher position mode

Monitor stand on the
lower position mode

Leg on the higher
position mode

Leg on the lower
position mode

Telescoping Concept (using slots)

Monitor stand on the
higher position mode

Monitor stand on the
lower position mode

Extension leg using snap feature

Extension Leg Concept

▼ **Figure 12.19.**
Example concept variants for the monitor stand product. (a) Graduated pin-hole
concept. (b) Rotating-leg concept. (c) Telescoping concept (using slots). (d) Extension
leg concept.

▶ Round all inside corners and outside edges. Sharp corners will inhibit material flow.

▶ Thick wall sections will cause sink marks on outside walls. Thick
sections should be redesigned or cored to remove mass.

▶ The part should be designed to be free of undercuts.

Internal and external splines in   Internal and external splines
position to slide by each other   position to support each other

▼ *Figure 12.20*
Basic working principle of height adjustment using splines.

▶ Reinforcing ribs should be used to support loads and minimize material usage. The rib thickness should also be less than two-thirds the nominal wall thickness. If the ribs are any thicker, they will create sink marks.

▶ The distance between two ribs should not be less than two times the nominal wall thickness.

▶ Typical wall thicknesses range from 0.060″ to 0.125″.

Figure 12.21 illustrates three of the critical principles from this list. As shown in the figure, rib dimensions, draft angles, and plastic feature sizes are important considerations.

Based on these principles, an initial layout wall thickness of 0.125″ is chosen for the product. Ribs are added to the primary base platform of the product at an initial separation distance of 2/3″. Based on the manufacturing process, tolerances are initially assumed to be the standard 0.010″ for dimensions less than 10″. For any dimensions longer than 10″, we add an extra 0.010″ tolerance for each additional 10″. For example, the dimension of the long side of the monitor stand is about 25″; therefore, the tolerance of the initial layout is 0.030″.

The main function-carrier layout of the product now consists of the following parts: telescoping legs (four pieces), a ribbed base, and a cosmetic top. Let's consider detailing the layout of each of these components with respect to the manufacturing principles, customer needs, and engineering quality. Beginning with the telescoping legs, there are two different slots that mate with the internal splines (Figure 12.20) of the base to provide the two different height settings. The bottom surface of the leg has a handle for users to rotate, pull, and push this part to the desired position. When designing this part, one of the

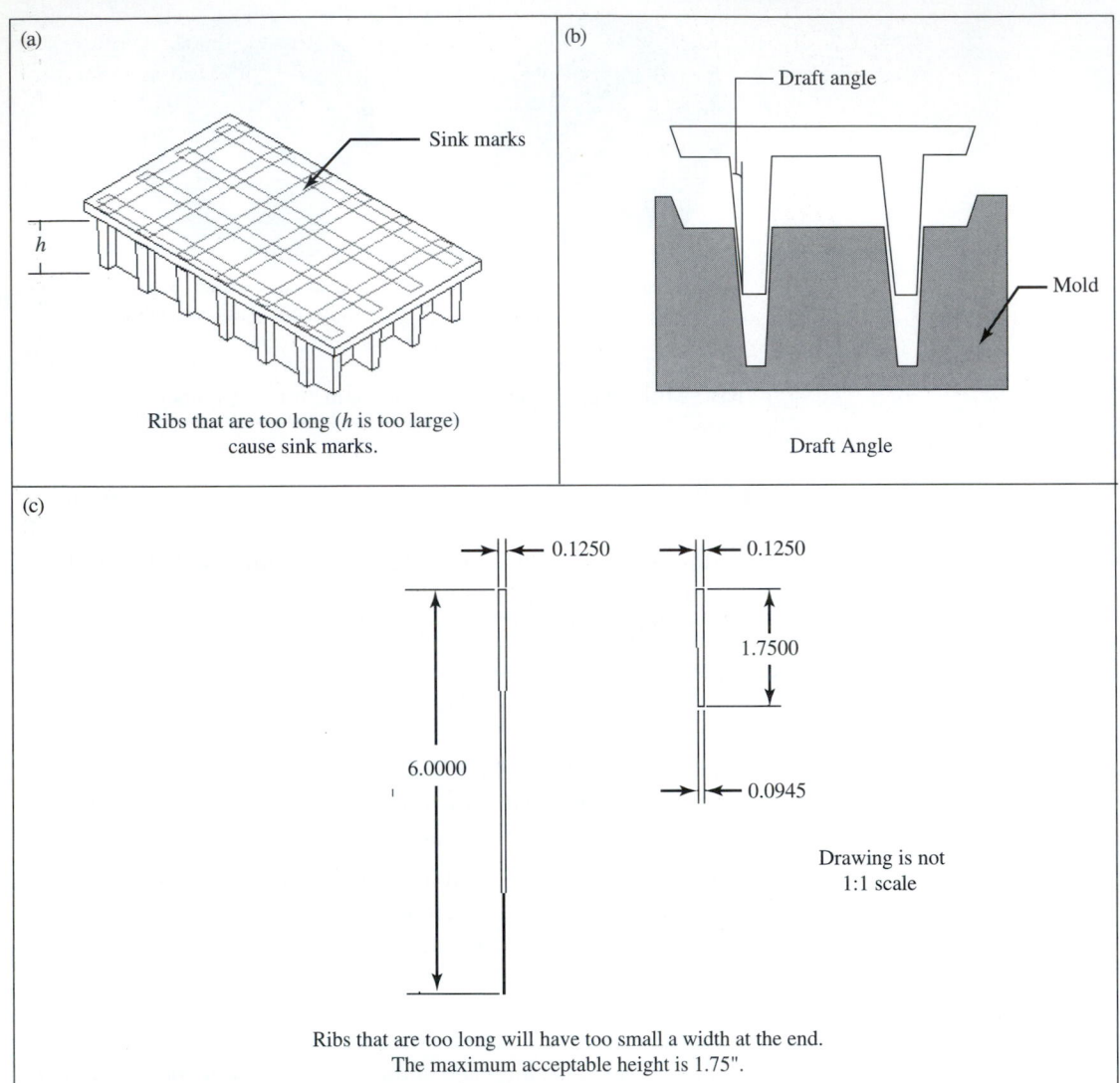

(a)

Sink marks

h

Ribs that are too long (h is too large)
cause sink marks.

(b)

Draft angle

Mold

Draft Angle

(c)

0.1250

0.1250

1.7500

6.0000

0.0945

Drawing is not
1:1 scale

Ribs that are too long will have too small a width at the end.
The maximum acceptable height is 1.75".

▼ *Figure 12.21.*
Example manufacturing principles for the monitor stand product. (a) Ribs that are too
long (h is too large) cause sink marks. (b) Draft angle. (c) Ribs that are too long will
have too small a width at the end. The maximum acceptable height is 1.75 inches.

considerations is for the handle to fit the thumb of the user. This consideration is a starting point for dimensioning the bottom leg. The edges where the user's fingers interface with the leg are rounded for comfort and safety and to improve the aesthetics of the product.

Based on these considerations, the legs are detailed in layout and geometry, as shown in Figure 12.22. Iterative sketching and application of the manufacturing principles lead to this detailed result.

Checking manufacturability from the embodiment checklist of Table 12.1, there are two options for injecting material into a leg mold, either from the bottom or from the top. Based on manufacturing principles, injecting material from the bottom will make the process easier and cheaper. If the material is injected from the bottom, the gate runner of the injecting material may cause appearance problems. There will be a center round gate at the bottom surface of the bottom leg, as shown in Figure 12.22. Based on the angled geometry of the leg, this added geometry from manufacturing will not be a problem.

Besides the manufacturability of the leg, let's consider the operational issues from the embodiment checklist. On one side of the top external spline, there is a tab to snap fit into a hole on the leg of the base to prevent rotation of the bottom leg when the stand is in the higher position.

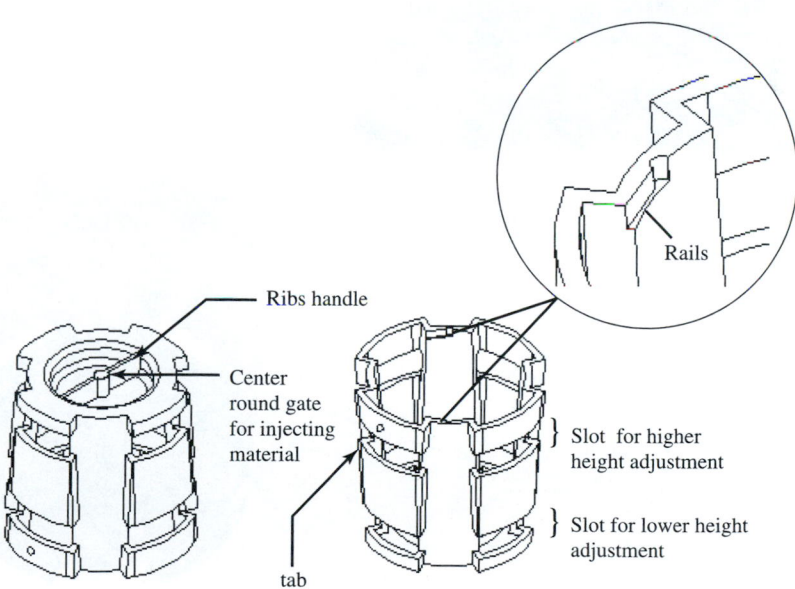

Rails

Ribs handle

Center round gate for injecting material

Slot for higher height adjustment

Slot for lower height adjustment

tab

▼ **Figure 12.22.**
Bottom leg layout.

As indicated in Figure 12.22, the top surface of the bottom leg also has two rails that are positioned opposite each other. The purpose of the rails is to allow the bottom leg to mate with the bead on each rib of the cosmetic top, as shown in Figure 12.23. These components stop the leg motion and support the monitor when the monitor stand is in the lower position. For both locking positions, the leg is turned counterclockwise to lock the position and clockwise to unlock the position.

Building on these embodiment checks (and others), systems modeling is now needed to refine and determine the choice of a specific material, material thickness, and choices of other dimensions. Based on the critical customer need of reliably supporting a desktop computer

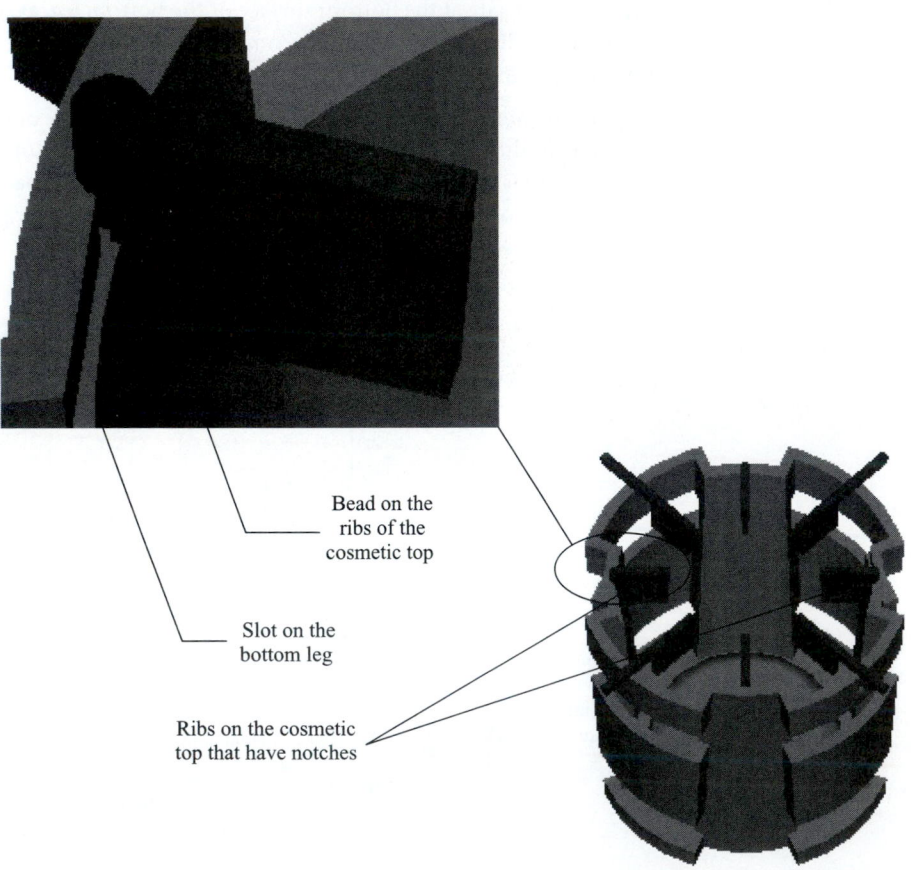

Bead on the ribs of the cosmetic top

Slot on the bottom leg

Ribs on the cosmetic top that have notches

▼ *Figure 12.23.*
Mechanism locking of the bottom leg at the lower position.

monitor, a stress model, using a balance of forces, is needed. Let's consider an analytical model for this case of the monitor stand legs.

It is assumed that the static or dynamic force caused by the weight of the monitor is equally distributed to the four legs and represented with a force R (see Figure 12.24). Each leg consists of four splines and eight stress walls. A stress wall is the surface contact between a spline and the cylinder of bottom leg. Let F be the force that acts on the stress wall. On each spline, there are two stress walls, and hence, $2F$ forces. Therefore, F is $1/8 * R$. The value of R is 30 lbs. for static condition and 75 lbs. for dynamic (impact) conditions.

Based on this high-level model, analytical relations may be derived for the shear stress in the spline and bending stress on the walls. Using the nominal layout thickness of 0.125", the stress states are calculated as

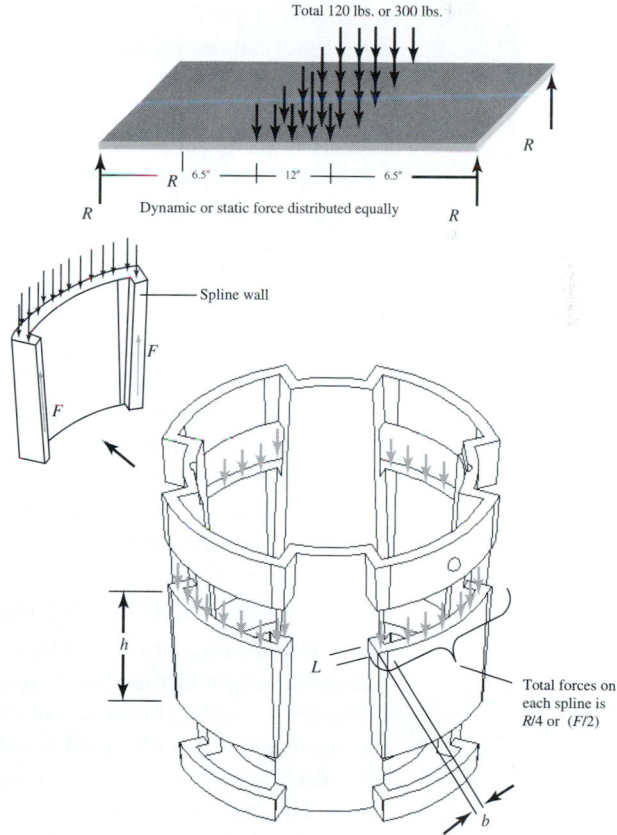

▼ *Figure 12.24.*
Systems model (partial) for a bottom leg.

TABLE 12.6. PROPERTIES OF RELEVANT MATERIAL CHOICES

Materials	$\sigma_{tensile}$ (lb/in^2)	σ_{comp} (lb/in^2)
ABS	6,000	8,000–12,000
Acetal	9,000	5,000
Acrylic	8,800–11,300	13,000–17,000
Polycarbonate (PC)	9,000	12,000
PC/ABS Blend	8,500	
PC Glass Fiber 20%	17,000	17,000
PC Glass Fiber 30%	20,000	20,000
PC Glass Fiber 40%	21,000	21,000

60 psi and 6 psi, respectively. Comparing these results to the material properties of possible plastics (Table 12.6), the modeled stresses do not approach the yield limits of the full range of materials. Thus, these thicknesses can be optimized for cost, or, if 0.125″ is chosen (minimal return in material costs savings), the calculated stresses imply that fatigue and stress concentrations will not be critical issues. Of course, FMEA should be applied to the leg components to extend the systems modeling and verify this assessment.

Let's now continue the embodiment process by considering the layout of the base component (Figure 12.25). The base is the largest component of the monitor stand. It is also the part to which all the other parts assemble. The shape of this part is similar to a tabletop. The legs of the base are cylindrical and hollow and have internal splines that rest on the bottom leg components to allow adjustment of the height of the monitor stand (refer to Figure 12.26). Ribs are added to the splines on the base legs to provide required support without adding excessive material. In each leg cylinder of the base, there is a hole that mates with the notch on the bottom leg component. As shown in Figure 12.26, the leg of the base has a draft angle of 3°, which is the minimum draft angle needed to allow the part to release from a mold and have a textured surface on the part. The angle of the leg on the base also prevents the bottom leg component from dropping down.

The ribs of the base provide most of the support for the monitor. The acceptable length of the ribs is approximately 1.75″. Longer ribs will be too thin according to the manufacturing principles presented above. A systems model may be used to estimate the number of ribs, based on deflection specifications and checking of the stress states. (Deflection will be the governing constraint here, since greater stiffness will be needed for deflection specifications compared to strength specifications.) This model assumes that Hooke's law applies to the base component. The stiffness may then be estimated as a function of the

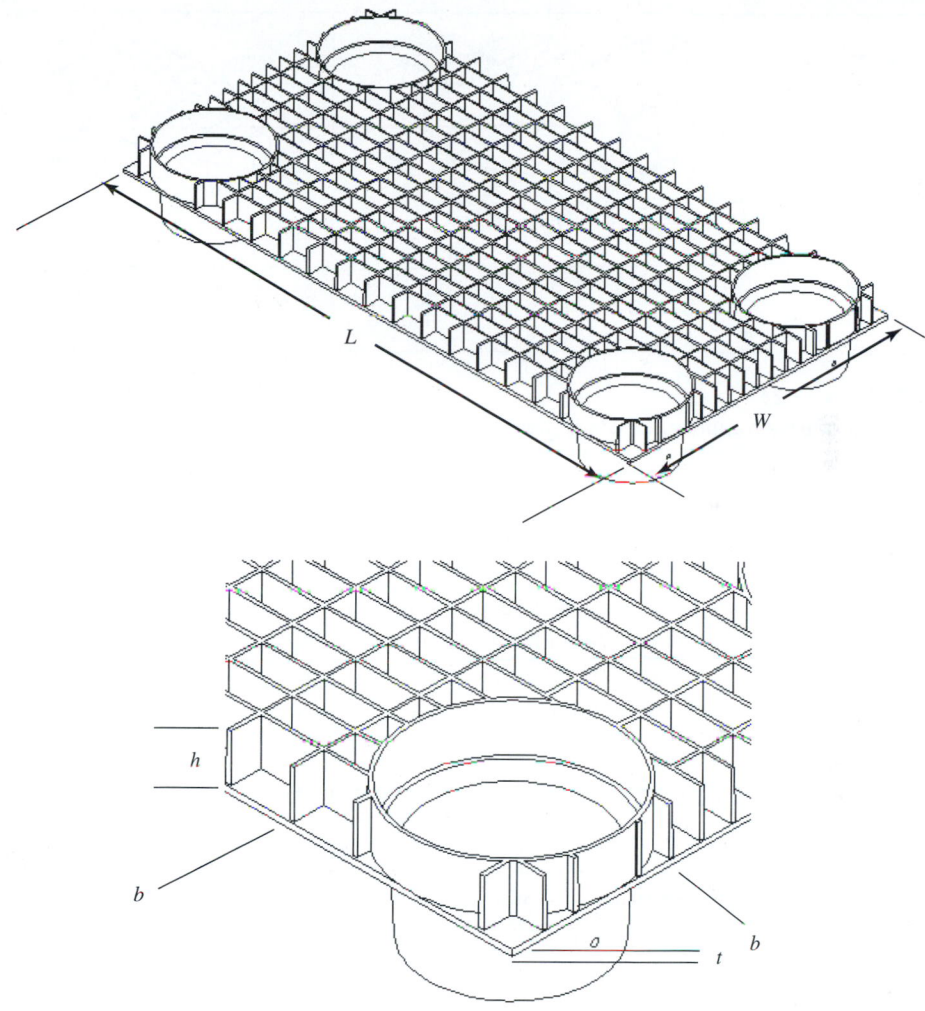

▼ *Figure 12.25.*
Important geometric design parameters for the base.

applied monitor load (worst case impact load), Young's modulus, moment of inertia (which is rather complex), number of ribs, and base dimensions (as shown in Figure 12.25).

The results of this analysis for different materials are shown in Table 12.7, where the minimum number of ribs is 17 long ribs and 15 short ribs. Because of the complex geometry of the base, this systems model

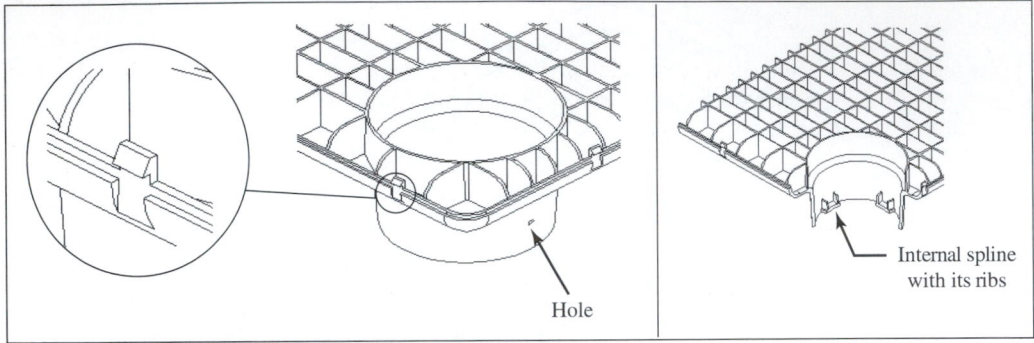

▼ **Figure 12.26.**
Layout of base component. Details show the snap, the slot cut under it, and the internal splines.

may be extended to a more complete model using the finite element method. The base component, shown in Figure 12.25, may be modeled as a quarter-section due to symmetry conditions. Constraints must then be added to the model, as illustrated in Figure 12.27.

Initial choices of nominal layout parameter are given as $L = 24$ in., $W = 12$ in., $b = 0.125$, $h = 1$, and $t = 0.125$. Using these initial

TABLE 12.7. SUMMARY OF DEFLECTIONS USING DIFFERENT MATERIALS CHOICES

Materials	E (lb/in2)	Long Side		Short Side		Average	
		Dynamic Deflection (in)	Static Deflection (in)	Dynamic Deflection (in)	Static Deflection (in)	Dynamic Deflection (in)	Static Deflection (in)
ABS	3.63E+05	0.992	0.265	0.122	0.049	0.557	0.157
Acetal	6.24E+05	0.577	0.231	0.071	0.028	0.324	0.130
Acrylic	5.08E+05	0.709	0.283	0.087	0.035	0.398	0.159
Polycarbonate (PC)	4.35E+05	0.827	0.331	0.102	0.041	0.464	0.186
PC ABS Blend	3.70E+05	0.989	0.396	0.120	0.048	0.555	0.222
PC ABS Glass Fiber 20%	1.20E+06	0.305	0.122	0.037	0.015	0.171	0.068
ABS Glass Fiber 20%	8.80E+05	0.416	0.166	0.051	0.020	0.233	0.093
PC ABS Glass Fiber 30%	1.40E+06	0.323	0.129	0.032	0.013	0.177	0.071
PC ABS Glass Fiber 40%	1.60E+06	0.284	0.113	0.028	0.011	0.156	0.062

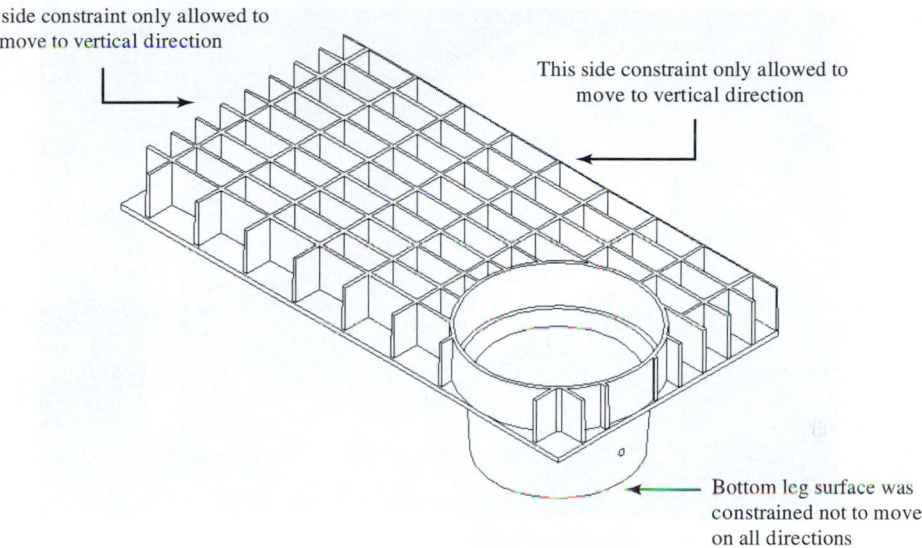

▼ *Figure 12.27.*
Constraint on the quarter finite element analysis.

parameter values, loading conditions are applied to the base model, assuming the a monitor is dropped onto the stand from a short height above the base. A solid model of the quarter-base portion is then entered into an FEA software system, in this case, COSMOS/M. Results are calculated for various materials and dimension variations, where example deflection and stress plots are shown in Figure 12.28.

Based on the FEA analysis, detailed layout parameters for the base are chosen. In this case, the nominal dimensions and tolerances are chosen, as presented above. In addition, the material choice is PC ABS glass fill 30%. For this material, maximum stress under impact load is 5.74×10^2 psi, while the PC ABS glass fill 30% material has a strength of 1.4×10^6 psi. Likewise, the maximum deflection for the base is 0.105″ for this material. This deflection result is acceptable and leads to the choice of PC ABS glass fill 30%.

A similar embodiment approach may now be applied to the cosmetic top of the product. This analysis should focus on the main function carriers of support monitor, level monitor, and so forth, as above. Figure 12.29 shows a detailed layout of the cosmetic top.

Continuing with the embodiment process, let's now consider auxiliary and supporting function carriers for the monitor stand product. One auxiliary function is to "couple with an environmental surface," that

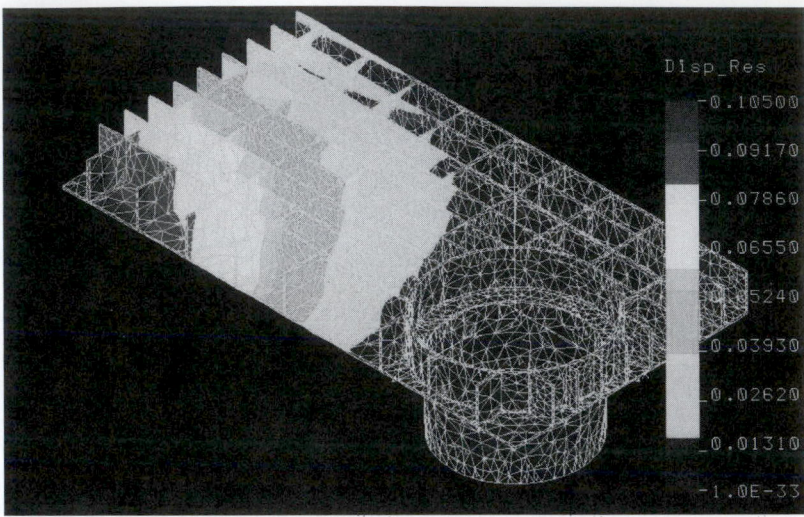

Monitor Stand Deflections Finite Element Analysis Under an Impact Load (in.)

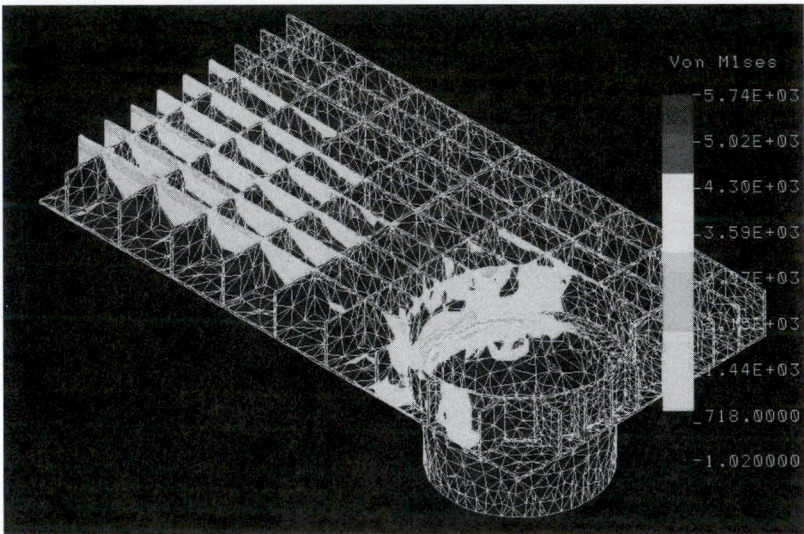

Monitor Stand Stresses Finite Element Analysis Under an Impact Load (lb/in.^{2})

▼ *Figure 12.28.*
Example FEA results for the monitor stand base component. (a) Monitor stand deflections finite element analysis under an impact load (in.). (b) Monitor stand stresses finite element analysis under an impact load (lb/in^2).

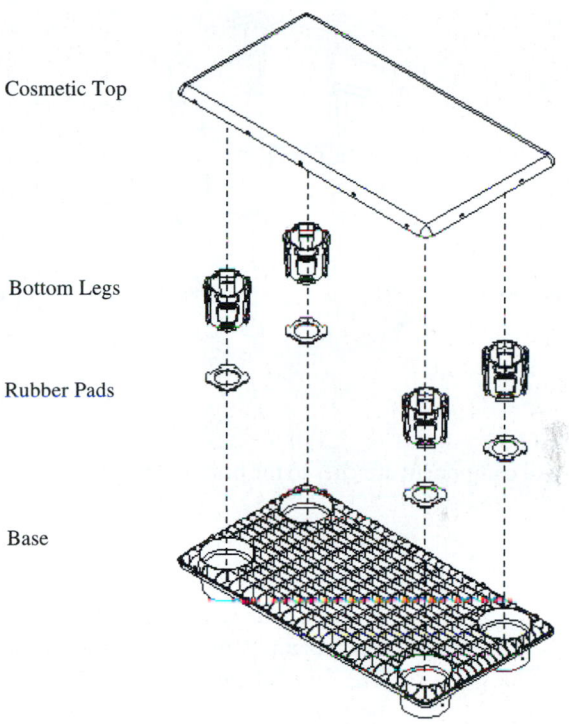

Cosmetic Top

Bottom Legs

Rubber Pads

Base

▼ *Figure 12.29.*
Exploded view of the monitor stand product.

is, to prevent skidding and marking of a desktop or table. A rubber pad is chosen as the solution to this function.

To embody this concept, analogies to other products are consulted to develop an initial layout. Based on analogies, the rubber pad component is designed as a cylinder with external splines and a hole in the middle (Figure 12.30). The external splines on the pad match the splines on the bottom leg component. Through this choice, the rubber pad will support the lowest bottom spline on the bottom leg. This configuration will provide extra support when the monitor stand is positioned in the lower mode. The rubber pad is attached to the bottom leg by an adhesive (see Figure 12.30).

Using the component designs from the product function embodiment, the final assembly layout is considered. In this case, the final assembly is rather simple. The four bottom legs are dropped through the holes in the base. The side slots on the bottom legs are positioned on the internal splines of the base. The legs are then rotated to the desired

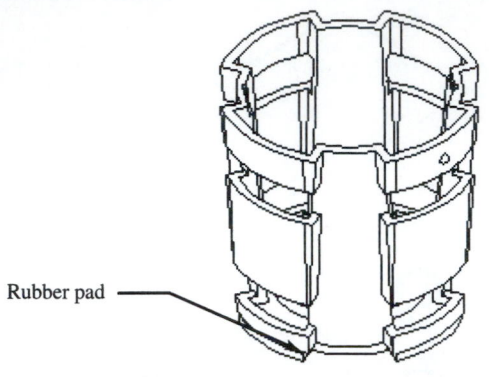
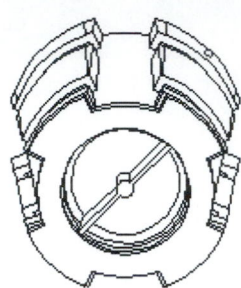

Rubber pad

▼ Figure 12.30.
Rubber pad component attaches to the bottom leg.

position. The cosmetic plate snaps to the top of the base (adhering to assemblability principles of minimum part count—integral fasteners). The four rubber pads are then attached to the bottom surfaces of the bottom legs. An exploded view of the monitor stand assembly is shown in Figure 12.29. In addition, Figure 12.31 shows how the monitor stand operates for both the high- and low-mode height adjustments.

This assembly may now be evaluated against system criteria, such as the economics of the product. Using data from plastic component and toolmakers, Tables 12.8 and 12.9 summarize the cost analysis for the monitor stand product. Table 12.8 shows the mold cost, tool life, cycle time, and cost for each material. Tool life is given in terms of the number of parts that can be fabricated before the tool wears significantly. Cycle time is simply the time to produce each part. Table 12.9 is a summary of the overall number of tools needed and the total tool cost for each part. This table presents lower and upper bounds for each category. The time needed to produce the desired quantity of parts is obtained by multiplying the cycle time and the number of parts needed. The number of tools needed is obtained by dividing the time needed to produce the desired number of parts in days by the number of days in one month (30 days).

The cost estimates of Tables 12.8 and 12.9 do not include all cost factors; they only focus on the tooling costs. Based on this focus, the initial costs of manufacturing this product range from $750,000 to $1,367,500 (for 70,000 units). The monitor stand will thus cost between $10.57 and $19.29 for the minimum of 70,000 units. This cost

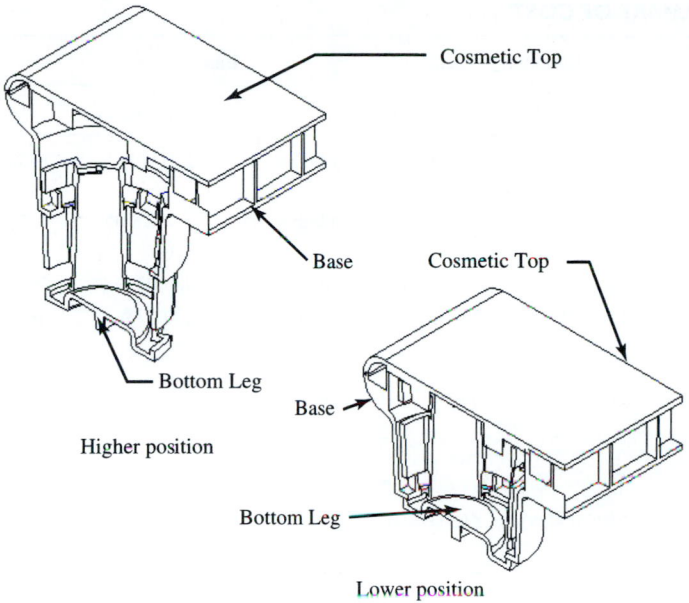

▼ Figure 12.31.
Basic operation of the monitor stand product.

is significantly reduced as we approach 500,000 units (between $1.50 and $2.75).

Assuming that we add the material cost per unit, labor costs, overhead, and markup, this product should be competitive with the similar products on the market. These products range from $24 to $70 retail. Some of these benchmarks do not include a height adjustment capability; those that do have height adjustment use more complicated mechanisms, and all have more parts.

TABLE 12.8.　COST, CYCLE TIME, AND TOOL LIFE OF EACH PART OF A MONITOR STAND PRODUCT

	Bottom leg (each)	Base	Cosmetic top
Tooling cost each mold	$35–$40k	$135k–$170k	$85k–$115k
Tool life (# of parts)	500–1,000k	500–1,000k	500–1,000k
Cycle time	25–30 sec	60–120 sec	50–60 sec
Parts needed /month	280,000	70,000	70,000
Estimated cost per part	$0.25	$2.00	$1.00

TABLE 12.9   SUMMARY OF COST

	Bottom leg (each)		Base		Cosmetic top	
	Lower bound	Upper bound	Lower bound	Upper bound	Lower bound	Upper bound
Tooling cost ($) –(each)	35,000	40,000	135,000	170,000	85,000	115,000
Tool life (# of parts)	5.00E+105	1.00E+06	5.00E+105	1.00E+06	5.00E+105	1.00E+06
Cycle time (seconds)	25	30	60	120	35	60
Parts needed /month	280,000	280,000	70,000	70,000	70,000	70,000
Time needed to produce desired quantity (hours/month)	1944.44	2333.33	1166.67	2333.33	680.56	1166.67
Time needed to produce desired amount (days)	81.02	97.22	48.61	97.22	28.36	48.61
Number tools needed	3	4	2	4	1	2
Total tooling cost perpart	$105,000.00	$160,000.00	$270,000.00	$680,000.00	$85,000.00	$230,000.00

After completing this evaluation of the product economics, the embodied layout must be checked for errors and disturbing factors. This analysis can begin with failure modes and effects analysis (which actually should be developed continuously as the product is developed). A partial FMEA is shown in Table 12.10. A single failure mode is listed in this figure, the potential binding of the bottom leg with the cosmetic top. Two causes are listed for this failure mode, both of which have high RPNs.

At this point, the recommended design actions could be implemented to reduce the likelihood of these failures occurring. In this case, however, let's consider the fabrication of physical prototypes (Chapter 17) to test the operation of the monitor stand.

Figure 12.32 shows such prototypes fabricated from the rapid prototyping process known as Selective Laser Sintering (SLS). Only 1/4 of the product is shown due to the symmetry of the product, size of the SLS build chamber, and interests in testing a single leg's operation.

Tests of the prototypes reveal that the predicted failure mode does exist in the product. As shown in Figure 12.33, the relatively large angle of the ribs on the cosmetic top can cause the leg to bind when it is rotated to the locking position. One of the recommended actions to correct this problem is to add more ribs. This action would create a smaller distance between the ribs and increase the smoothness of the rotation. Another option is to create a single circular rib. This later recommended action is shown in Figure 12.33.

TABLE 12.10.   PARTIAL FMEA OF A MONITOR STAND PRODUCT

Product Name: Burhan Monitor Stand	Devel. Team: ___XXX___	Page No. __1__ of __N__
__X__ System		FMEA Number __KLW-1001__
____ Subsystem Name: Monitor Stand		Date: ___0X-12-27___
____ Component		

Part # & Functions	Potential Failure Mode	Potential Effect(s) of Failure	S E V E R (S)	Potential Cause(s)/ Mechanism(s) of Failure	O C C U R (O)	Current Design Controls/Tests	D E T E C T (D)	Recommended Actions	RPN
Bottom Leg MS-03 Coupled with Cosmetic Top	Binding of leg in lower height setting	Deteriorated operation of product, perhaps leading to excessive force to adjust, inoperable condition of entire leg, or excessive wear on	9	Tolerances are too loose on cosmetic top ribs and the leg's interfacing ring	5	Alpha and beta prototype test of monitor stand operation	7	Add more ribs to the cosmetic top Control process tolerance to 0.0008	315
Support monitor weight				Ribs do not align with the leg in all angular positions	4	Alpha and beta prototype test of monitor stand operation	3	Add more ribs to the cosmetic top Convert discrete ribs to a single circular tab	108
Adjust monitor height									
Stabilize monitor			. . .	. . .	. . .	. . .	. . .	. . .	. . .
Display height level									
Lock height setting . . .									

Many other failure modes can be investigated with FMEA in conjunction with physical prototypes. Figure 12.34 shows two corrective actions for other possible failure modes encountered on the embodied layout of the monitor stand product. The first case shows the addition of a slot to the base's leg cylinder so that the locking tab on the bottom leg component will not bind with the insert. The second case, on the other hand,

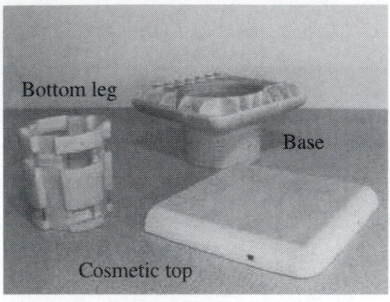 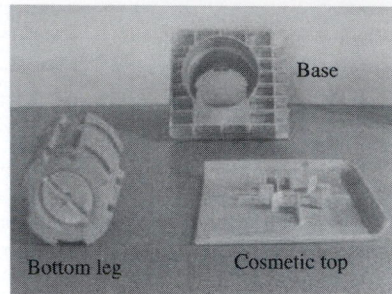

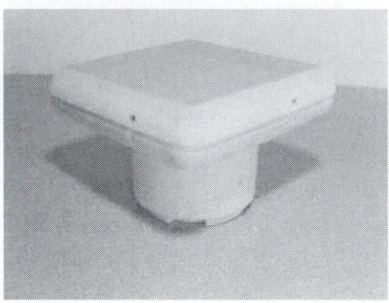 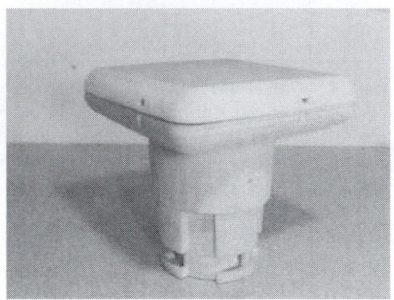

▼ *Figure 12.32.*
Selective laser sintering prototypes of the monitor stand product.

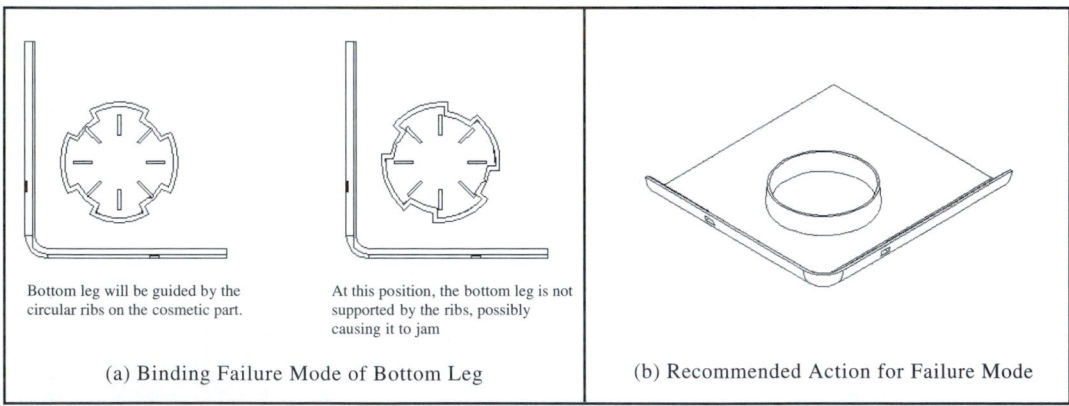

Bottom leg will be guided by the circular ribs on the cosmetic part.

At this position, the bottom leg is not supported by the ribs, possibly causing it to jam

(a) Binding Failure Mode of Bottom Leg

(b) Recommended Action for Failure Mode

▼ *Figure 12.33.*
Example failure mode from FMEA. (a) Binding failure mode of bottom leg. (b) Recommended action for failure mode.

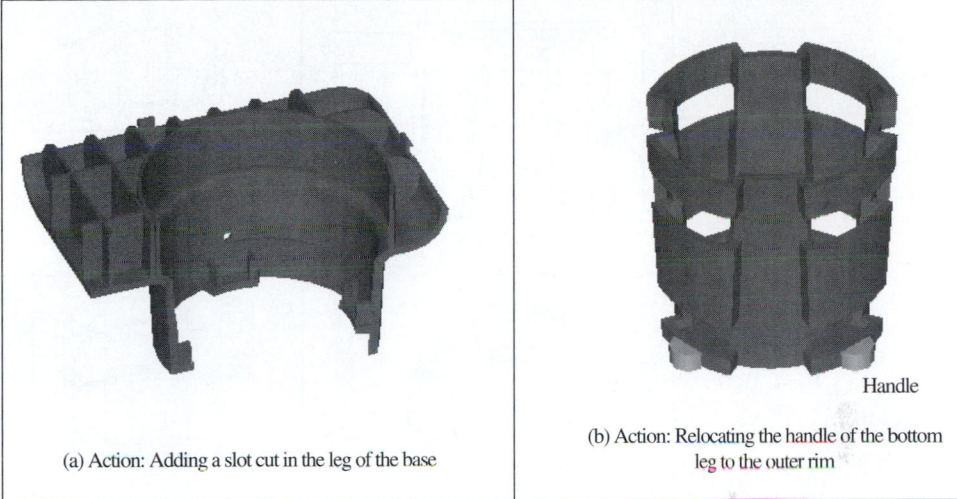

(a) Action: Adding a slot cut in the leg of the base

Handle

(b) Action: Relocating the handle of the bottom leg to the outer rim

▼ *Figure 12.34.*
Examples of corrective actions to other potential failure modes in the monitor stand product. (a) Action: Adding a slot cut in the leg of the base. (b) Action: Relocating the handle of the bottom leg to the outer rim.

focuses on the failure mode of possibly fracturing the handle on the bottom leg. The corrective action relocates the handle to the outer rim and the lowest point on the leg. This action will provide more cross-sectional area, in addition to a larger lever arm for the user to rotate the leg.

Summary

While every step, detail, and decision are not shown in this case study, the monitor stand product illustrates the complexity typically encountered during concept embodiment. It also demonstrates how embodiment methods are applied, explored, and iterated.

As presented by this case study, embodiment design progresses toward a detailed layout of a product through the simultaneous consideration of customer needs, product functionality, engineering specifications, and engineering quality/robustness. This process is iterative. It is also consumed with modeling, analysis, careful thought, and decision-making.

All in all, concept embodiment must come to a closure due to cycle time and resource pressures. The deliverables that are created from this stage of product development include a complete bill of materials, detailed engineering drawings, assembly diagrams, process plans, tooling

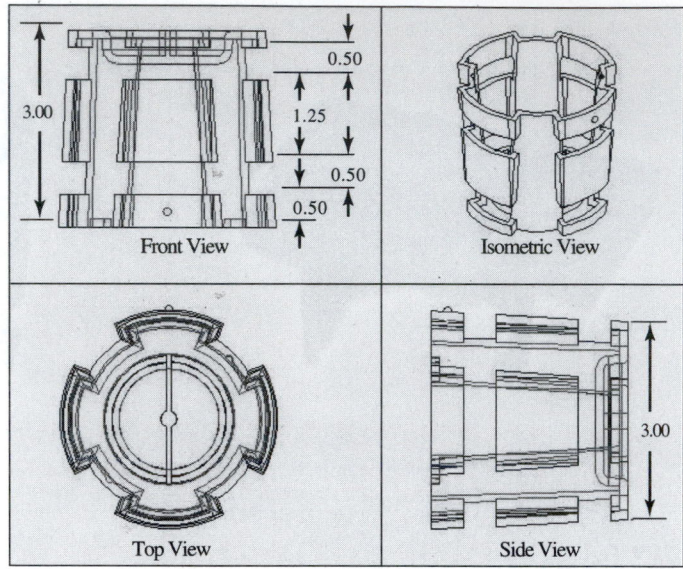

▼ *Figure 12.35.*
Detailed drawings of the bottom leg component.

design, user's manuals, and an FMEA with low RPNs for all expected failure modes. As an example of these results, Figures 12.35–12.39 show detailed drawings of the monitor stand's components.

VI. SUMMARY AND "GOLDEN NUGGETS"

The focus of this chapter is concept embodiment, the stage of product development where concepts are transformed into physical realizations. A number of basic and advanced methods are available to aid in this transformation process. Yet, even with these methods, concept embodiment is a highly nonlinear, iterative, and complex process. It requires sound engineering skills, ranging from modeling and experimentation to manufacturing, assembly, and tooling design. Because of these characteristics, the remainder of this book is devoted to the topic of concept embodiment. Techniques and applications are developed in these chapters to help us understand and exercise the skills that are required.

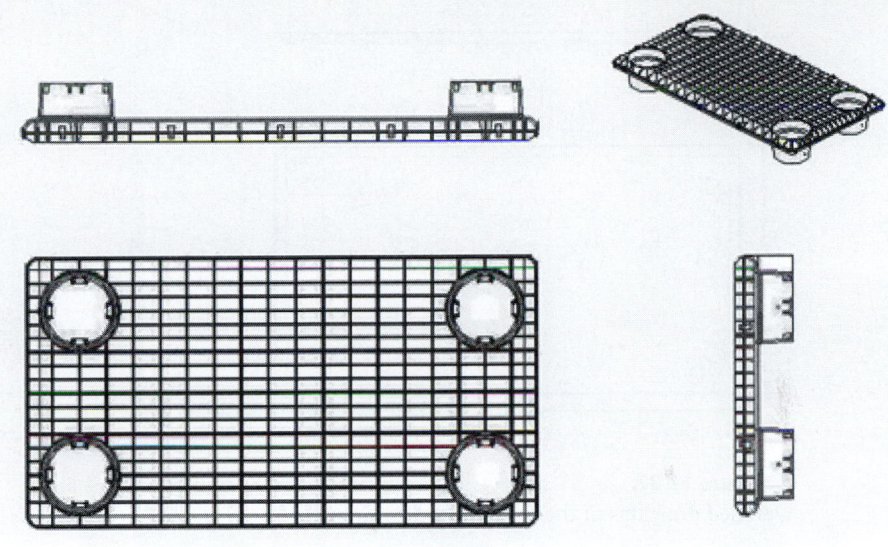

▼ *Figure 12.36.*
Detailed drawings of the base component.

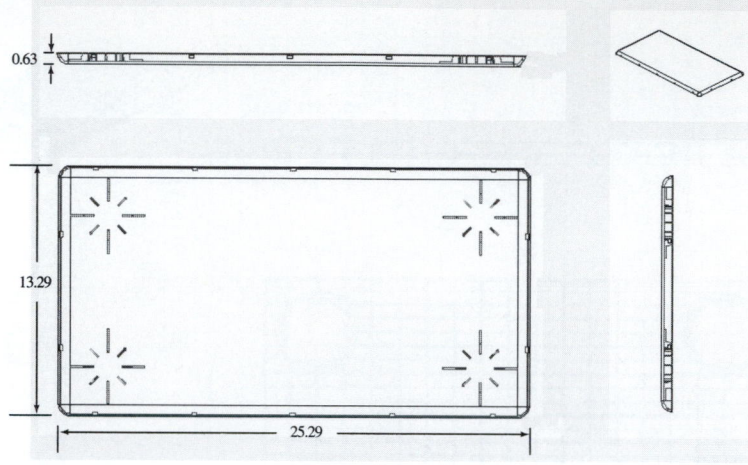

▼ Figure 12.37.
Detailed drawings of the cosmetic top component.

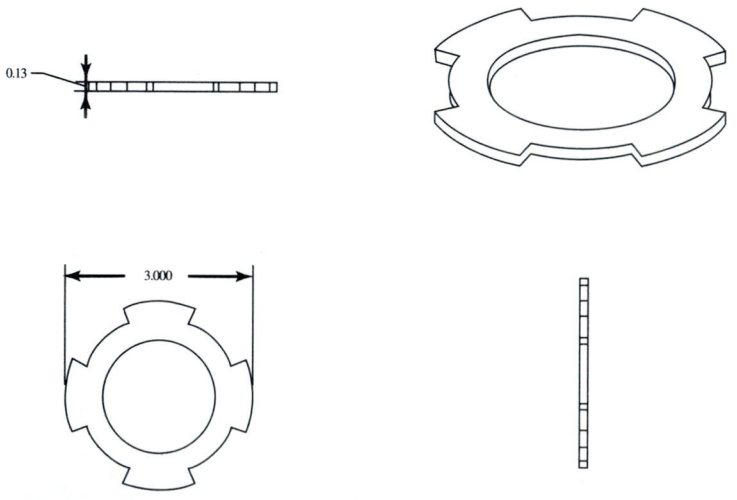

▼ Figure 12.38.
Detailed drawings of the rubber pads components.

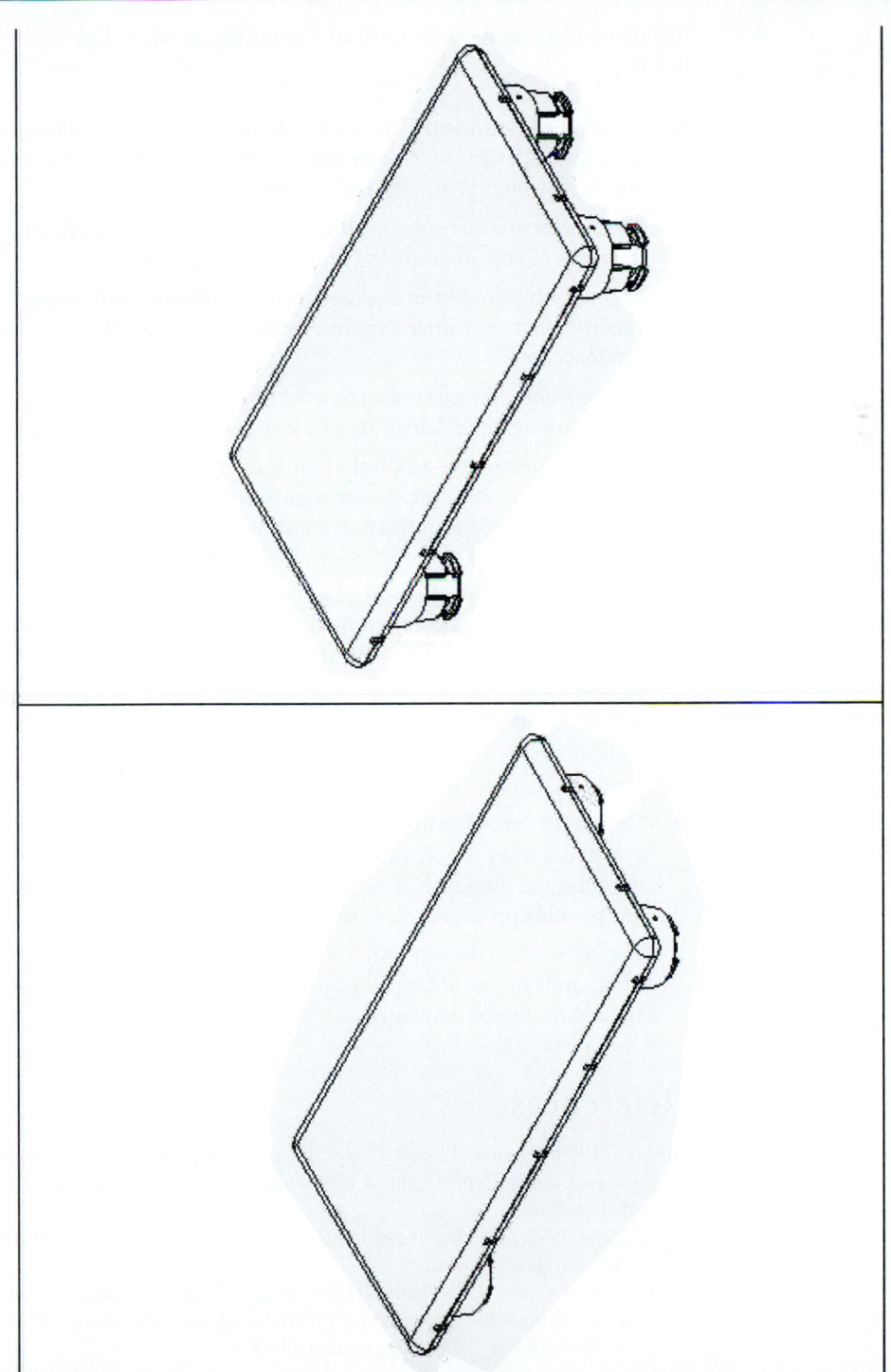

▼ **Figure 12.39.**
Final layout of the monitor stand product, illustrating the lower and higher position settings.

Golden nuggets that can be derived from this chapter include the following:

▶ Concept embodiment requires the input of a creative and feasible conceptual design. A great embodiment of a poor concept will result in a product that fails in the market.

▶ To realize a product concept, the focus should be on the combined concepts of customer quality and engineering quality.

▶ Engineering quality is equivalent to the implicit or "expected quality" that customers assume will exist in a product (at low relative cost).

▶ Concept embodiment must consider the entire life cycle of a product, not merely the delivering of a product to the retail market.

▶ Systems modeling is a skill that must be honed and applied during concept embodiment. Through systems modeling, products may be studied and refined using physical insights into a product's operation.

▶ Design principles, based on past designs and physical principles, should be consciously applied during embodiment design. Through conscious application, a product's functions may be added to or modified, and its layout will change. The result will be a more robust offering.

▶ Failure Modes and Effects Analysis (FMEA) is essential to create products that are robust and ethically responsible.

▶ Design for manufacturing and assembly must be considered before and during concept embodiment. It is thus important to understand the manufacturing and assembly principles that govern any possible process choices for a product.

▶ The nature of concept embodiment is complex and nonlinear. Thousands upon thousands of parameters and decisions must be systematically contemplated.

References

Burhan, D. 1998. Design of a portable computer docking station monitor stand. Master's thesis, Department of Mechanical Engineering, The University of Texas, Austin.

Design Edge. 1998. A Product Development Company, Austin, TX. Contact: Mr. Chris Cavello.

Eubanks, C., Kmenta, S., and Ishii, K. 1997. Advanced failure modes and effects analysis using behavior modeling. *Proceedings of the ASME Design Theory and Methodology Conference,* Sacramento, CA.

FMEA. 1995. *Potential failure modes and effects analysis.* Reference Manual, 2nd ed. SAEJ-1739 Equivalent. Automobile Industry Action Group.

Juvinall, R., and Marshek, K. 1991. *Fundamentals of machine component and design.* New York: Wiley.

Pahl, G., and Beitz, W. 1996. *Engineering design.* 2nd ed. Berlin: Springer-Verlag.

Stamatis, D. 1996. *Failure mode and effect analysis.* Milwaukee, WI: ASQC Quality Press.

Stamatis, D. 1995. *FMEA from theory to execution.* Milwaukee, WI: ASQC Quality Press.

13 Modeling of Product Metrics

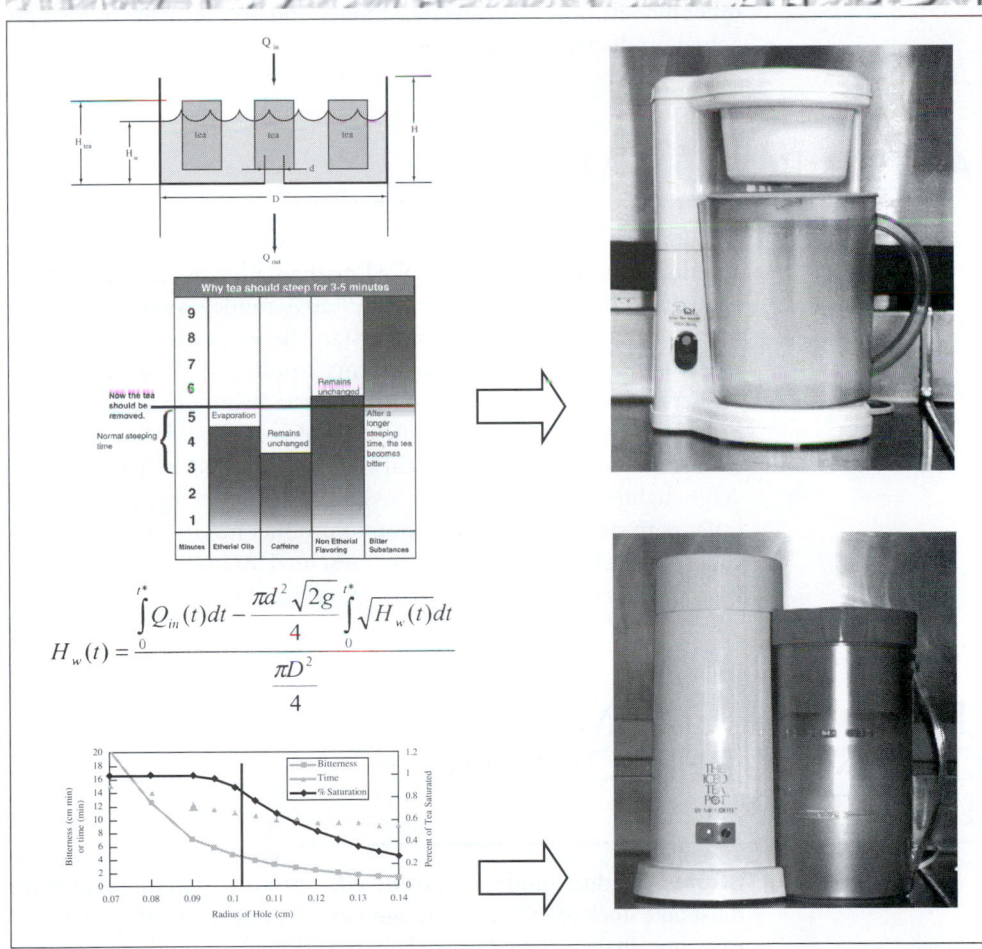

Mathematical modeling of products leads to physical insights and the ability to quickly search for innumerable variations that satisfy the customer needs, support design functions, fit within a product portfolio, and satisfy product architecture constraints.

A *model* of a product metric is simply a representation, simplification, or estimation of a product's realization to aid in making product decisions. Modeling of product metrics separates the engineer from other professions. We seek to answer the How, Why, Where, and When of all aspects of a product's performance. Trial-and-error approaches are too costly and risk the required quality demanded by customers. Effective models must therefore be created during the product development process, so that a team may quickly move forward. Such models may be based on applied mathematics and science or on physical prototypes. In this chapter, we focus on analytical (applied mathematics and science) types of models, including trade-offs with other models, how to formulate a model, and how to create models for actual products.

I. CHAPTER ROADMAP

Developing product models allow us to wield special tools as engineers and designers. Insights into applied mathematics and science may be brought to the table. Whether quantitative models are developed or merely estimations, we seek to understand how a product will perform under all circumstances and operating conditions. This chapter describes two paths toward models of products (Figure 13.1). Both paths begin and end at the preparation stage of modeling, followed by the distinction between mathematical and physical models. They then diverge slightly, with one path discussing basic modeling methods. The other path studies the construction of product models with more advanced tools. Product examples are presented for both paths, from fingernail clipper products to iced tea makers and electric woks.

II. INTRODUCTION: MODEL SELECTION BY PERFORMANCE SPECIFICATIONS

Before a product model is considered, a product development team must take stock of its current design status. A wealth of data usually exists when analytical models are being developed for product concepts. These data include organized customer needs lists, activity diagrams,

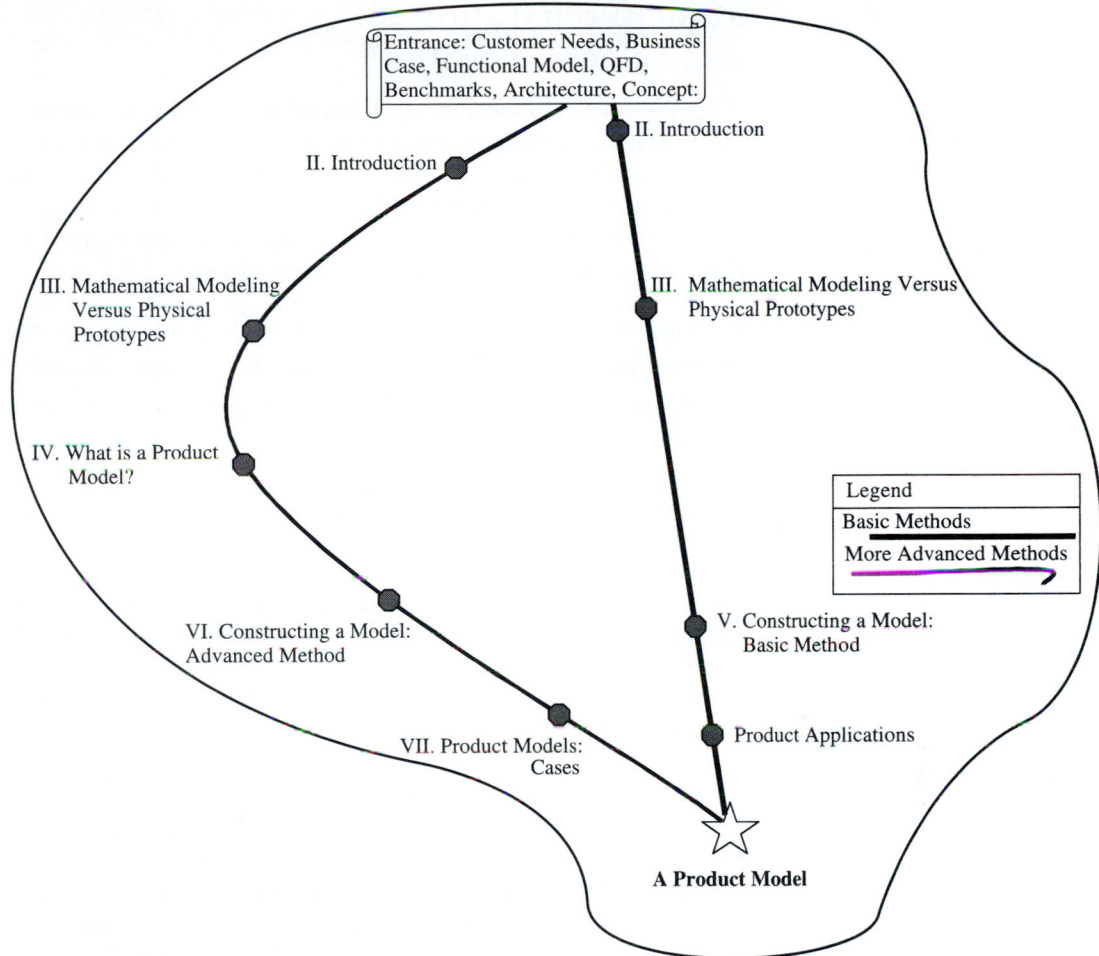

▼ Figure 13.1.
Roadmap of alternative routes for developing a mathematical model of product metrics.

business cases, functional models, engineering specifications within a house of quality, teardown data of past and/or competitive products, s-curve trend data for forecasted improvements, product architecture layouts, and chosen product concept(s). It is now time for the design team to use these data to begin the model creation process. The created models will permit embodiment decisions to be made with understanding of their impact on performance.

Model Preparation
and Selection Method

When creating a product model, the overriding goal is to formulate a representation that may be tested to measure a customer need over a range of possible choices (Chapter 6). We are thus modeling customer needs as metrics, creating simplified versions of the product for testing, and measuring the product against clear target values. Four steps initiate the creation of a product model for this purpose:

1. Map or relate the customer need weights to the product functions.
2. Identify the functions that relate most strongly to the customer needs.
3. Choose the metrics (engineering specifications, Chapter 7) that may be used to quantify the material, energy, or signal flows for these functions.
4. Identify target values for these metrics based on benchmarking results (Chapter 7).

The first step provides a means of relating the customer need weights to the physics of the product. Recall when constructing product-function models, we ensure the functions themselves are a technical reflection of the customer needs. Step 2 then simplifies the modeling process by focusing on the primary function carriers of the product. Not all aspects of a product may be modeled mathematically. Practical limitations exist in the complexity of the model. It may require excessive computational resources, or take too much time to develop accurately, and there will be limitations of the model in predicting performance. By identifying the small subset of functions that relate most strongly to the customer needs, the scope of a model may be realistically reduced, while having the greatest potential for measuring real needs. Of course, as the product is developed, *all* functions of the product must be refined and tested fully. We just choose an initial subset to focus the development efforts on the most important and tractable issues. While all product functions affect the design, it is better not to be concerned with detailed secondary functions at this stage.

The third step of the process is implemented to choose a means of measuring the performance of a product. The choice of metrics must be tied to the flows of the primary product functions, since a transformation of flow models the physics of the product. Such choices may be made directly from benchmarking and engineering specifications (House of Quality, Chapter 7).

In preparing to model a product, the final step is to bound the metrics with target values. These target values provide goals for the final product performance, and may be chosen from benchmarking results.

Choices of the target values are not trivial, nor should they be taken lightly. Targets will ultimately define the competitiveness of the product, its challenges for success, and its future marketing value. Chapter 7 discusses forecasting, the House of Quality, and other methods for choosing target values.

Product Application: Model Preparation and Selection

In selecting a model metric for a product, we are clarifying and honing a product concept so that engineering principles may be applied. Many possibilities exist for the choice of metrics. Many possibilities also exist for formulating a mathematical model, for example, static versus quasi-static versus dynamic; and analytical versus numerical. Model preparation provides a means of reducing the possibilities by allowing us to systematically choose a metric and modeling approach that will instill confidence in the design team.

Consider the model preparation of a household iced tea brewing product. This product development case is introduced in Chapters 6 and 9, where the concept under consideration is currently on the market (Figure 6.31), and where a new product architecture is shown in Figure 9.26(b). Applying the preparation process, we first consider the type of model that should be developed: mathematical or physical. Should we develop mathematical equations based on analysis of the product physics, or should we carry out many experiments?

In this case, a functional model of the product, shown in Figure 6.34, represents the high-level physics of the product. It is governed by fluid mechanics, heat transfer, thermodynamics, and the infusion of a solid in water. Based on the functional model, research, and benchmarking, a mathematical model is feasible for the product. Tea infusion times are known as a function of tea taste. The product geometry and mechanics are not new technologies as individual subsystems, nor are they expected to behave with a high level of nonlinearity.

In preparing to develop a model, we first consider the most important customer needs. These customer needs are mapped to the functions (Figure 6.34) in the product to identify the critical function carriers. Figure 13.2 shows the mapping for the iced tea maker.

Notice in the figure that the customer needs are associated with the functions that will address the needs. When more than one customer need relates to a function, the customer-need weights are summarized

Customer Needs	Weight	Scaled Weight		Importance	Sub-Function
Stronger Tea	9	5		9	Mix solid with liquid
Use less ice	5	3		7	Import human force
Easy to add tea	5	3		9	Transmit heat
Easy to clean	3	2		7	Import solid
Easy to store	3	2		6	Convert electricity to heat
Brew larger amounts	2	1		4	Store liquid
Contain steam	2	1		2	Stop gas
				4	Guide hand
				3	Store product
				3	Clean product
				1	Allow DOF of solid
				1	Import liquid
				1	Export liquid
				1	Regulate electricity
				1	Sense temperature
				1	Stop liquid
				1	Filter liquid
				1	Indicate status
				1	Transport liquid
				1	Import electricity
				1	Actuate electricity

▼ *Figure 13.2.*
Weighting of product functions using quantified customer needs for an automatic iced tea maker.

to assign importance to the function. Functions that are not directly associated with the high-weighted customer needs are assigned an importance of "1" to identify them as supporting functions in the product. Such secondary functions are important to achieve the entire technical process carried out by the product. However, functions with higher importance weights are more useful for measuring the performance of a product model.

In the case of the iced tea maker, two functions are very highly rated: the "transmit heat" and "mix solid and liquid" subfunctions (Figure 13.2). These functions are chosen through propagation of importance. If other customer needs are considered for modeling, then other functions can be chosen from the importance correlation mapping.

The next preparation step involves selecting metrics for the iced tea maker. From benchmarking, a list of measurements taken of each function reveals the metrics shown in Table 13.1. This information provides alternative metrics and avenues for modeling the iced tea maker. Let us focus on the first two customer needs, "stronger tea" and "use less

TABLE 13.1: POSSIBLE METRICS FOR AN AUTOMATIC ICED-TEA MAKER

Transmit Heat	Units
Initial temperature	Celcius (C)
Final temperature	Celcius (C)
Time to heat medium	sec
Volume of medium	m^3
Geometry of transmitter	deg, m
Resistance of transmitter	k

Mix Solid and Liquid	Units
Input flow of solid	g/sec
Input flow of liquid	g/sec
Geometry of mixing chamber	m, deg
Output flow of mixture	g(L)/sec
Energy causing mixing	rpm, q, C

ice," which, as shown in the customer needs mapping, both relate directly to the main product functions of "transmit heat" and "mix solid and liquid." Using the metric list, we decide to base the strength of the tea on the amount of energy causing mixing (temperature of entering water) and the time the water is in contact with the tea. The time of contact is ultimately determined by measuring the geometry of the mixing chamber and the output flow of the mixture. These metric choices are supported and justified by background research into the steeping process for tea.

The second customer need, "use less ice," is primarily related to "cool exiting water" or "transmit heat." After observing the product operation and studying the measurements list, the two metrics "volume of medium" (volume of water in tank) and "final temperature" (temperature of tea entering storage pitcher) are chosen. Mathematical models are possible for these metrics based on the fluid mechanics and thermodynamics of the brewing process. These metrics are also identical to the metrics measured during benchmarking exercises for the domain of iced tea makers (Chapters 6 and 7).

The last stage of preparation involves the identification of target values for the chosen metrics. From background research and benchmarking results (Chapter 7), the target values may be determined (Figure 7.13). For iced tea makers, the target values include: "temperature of entering water" (slightly below boiling temperature of water), "time in contact" (5 minutes), "volume of water in tank" (3 cups and minimize to ensure tea saturation but reduce thermal mass), and "temperature of exiting tea" (minimize to reduce quantity of ice added to the brewed tea).

III. MATHEMATICAL MODELING VERSUS PHYSICAL PROTOTYPING

Assuming that model preparation is complete, metrics are then identified for measuring a product's performance. We must now further the development process by developing a model to test our concepts. At this stage, an issue when considering a product model is *descriptiveness*, also known as *detail* or *fidelity*. How much work should we as engineers put into constructing a model that reflects the product design problem? The product development team does not have infinite resources.

This decision comes down to the comparison of model descriptiveness versus model construction difficulty:

▶ Should we use a lumped mass model or a finite element model?

▶ Should we model an idealized steady-state condition, or use a transient model?

▶ Should we make detailed numerical analyses of all the product options?

▶ Should we construct a physical model or a virtual (mathematical) model?

Determining the level of a model is a trade-off decision between model descriptiveness versus model construction and solution determination time.

Example

For a fingernail clipper product, consider a scenario where we may ultimately derive a model that represents finger force, such as:

$$f = F\frac{L_F}{L_f}$$

$$F = \tau\frac{W_b}{h_0}t^2$$

where f is the finger force, F is cutting force at the blade, L_f is the length of lever arm, L_F is the distance to the blades, t is the nail thickness, τ is the shear strength of the nail material, W is the width of the blade, and h_0 is the blade height. Alternatively, for a iced tea maker, we might derive a representation of the bitterness flavor of tea by the expression:

$$B = \int_5^{t^*} H_w(t-5)dt + \int_{t^*}^{t_{final}} H_w(t)dt$$

where H_w is the water level in the steeping chamber.

To use these models, we have to calibrate them with a number of physical trials. We have to make at least one experimental prototype: Cut a nail of measured dimensions and measure the force required for a measured set of product dimensions and, similarly, brew tea in varying container sizes and times and measure bitterness. These tests will provide us with a shear strength for the nail and bitterness as a function of time so we can use the models. We can then find a set of product dimensions, which gives us a sufficiently low f or an adequate bitterness level.

Let's say we execute this "optimization" procedure. Then what? We still have to build the final product to test it at this new optimal configuration. Even with analytic models, we have to build hardware and carry out experiments.

On the other hand, why not just build approximately 10 different blade and geometric configurations as prototypes and compare the hardware, selecting a better performing configuration? Similarly, could we build different steeping chambers in the tea brewer? This physical prototyping approach is also possible (Chapter 18).

The point of this discussion is that with engineering models, at some point, sets of experiments have to be carried out. Sooner or later, we have to measure the inputs to a product definition, and we have to verify the output performance of a product. Because of this necessity, we should consider the trade-offs between analytical and physical models.

To contemplate this trade-off, let's consider two axes. We may classify prototypes as either *analytical—physical* or *focused—comprehensive* (Ulrich and Eppinger 1995).

We must choose between an analytical or physical model (Table 13.2). Combinations of analytical and physical prototypes are also used. An example is to embed sensors in physical prototypes to provide feedback to computer simulations to enhance the accuracy of the simulations. The other continuum determines whether the prototype has some, most, or all the attributes of the final product (Table 13.3).

TABLE 13.2.   ANALYTICAL VERSUS PHYSICAL

Analytical	Physical
Simulations	Hardware
"Virtual" prototyping	Material and physical property correlation
Computer animations	Prototyping of manufacturing techniques
Optimization	Experimental setups
	Fully functional mock-ups (alpha and beta prototypes)

TABLE 3.3. FOCUSED VERSUS COMPREHENSIVE

Focused	Comprehensive
Testing limited performance	Full-scale, fully functional dimensions version of product
Just representative enough to answer the question, and no more	As representative as possible
As cheap as possible	As true to real product as possible

The other continuum determines whether the prototype has some, most, or all of the attributes of the final product.

One way to consider the decision from these continua is depicted in Figure 13.3. As we initiate a project involving the construction of a formal model, typically we will have a project resource depletion constraint. We do not have infinite time or money to build a "perfect" description of our product, infinite in detail. Thus, we have a project management decision to make, namely, "How detailed should we model?"

To answer this question, we have data from past history or design process prediction studies. We must use our engineering judgment to decide how complex the performance metric is in terms of the design variables and how much safety margin we desire using our resources. Some guidelines in this decision are:

▶ The comparison is in predicted model accuracy *over the entire development process,* including the prototyping stage versus modeling cost in terms of our resources. The reason for making a detailed model is to provide a clear understanding of the design concept with a relation, as opposed to providing the understanding purely by prototyping. More analytic modeling is used to produce less experimental hardware.

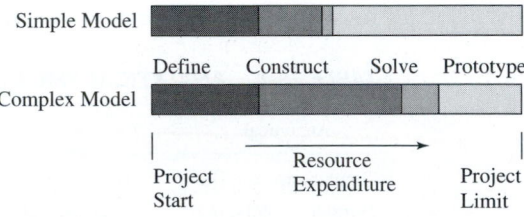

▼ *Figure 13.3.*
Axes of product model decisions.

▶ We will always use the available resources in time or money. It has been allocated for the design team to spend, and a design team gains very few points by saving, but gains many points by a successful result. We therefore will always use up all time and money available, if for no other reason than carrying out repeated analyses for different environments for lower technical risk. We will also make more prototypes if we have more time or money.

▶ It is *not* clear that making a more detailed engineering model will produce better results than simply completing many experiments. It depends on the difficulty of the prototype construction and experimentation process. If we make one experimental prototype with each design variable being adjustable, we can experimentally construct a first-order statistical model of the performance metric in terms of the design variables through structured variations of the prototype (Chapter 18). This statistical model may produce the needed results more quickly than constructing a detailed calibrated engineering (analytical) model.

▶ It is *not* clear that completing many experiments will produce better results than a more detailed engineering model. Analysis will often provide insight through the derived structure of the problem. Some variables drop out of the analysis, and experimenting over them would only waste time and money. Other variables can be uncovered as very sensitive. The analytical model may produce the needed results more quickly than building many pieces of hardware and fitting a statistical model.

▶ In general, it is smart to make detailed engineering models when prototyping is expensive and when we have reasonable expectations in obtaining an accurate model. This idea is depicted in Table 13.4, adapted from Ulrich and Eppinger (1995).

TABLE 13.4. DECISION TRADE-OFF BETWEEN ANALYTICAL AND PHYSICAL MODELS

High	Model It	Model It	Difficult Problem
Model accuracy / **Prototype accuracy**	Model It	Doesn't Matter	Prototype It
	Difficult Problem	Prototype It	Prototype It
Low		**Model expense** / **Prototype expense**	**High**

IV. ADVANCED TOPIC: WHAT IS A PRODUCT MODEL?

As introduced above, we never analyze a real system but only an abstraction of it. The abstraction is a *model*; it abstracts the real world, giving approximations of complex physical phenomena as part of physical systems. In this section, we consider the range of models, from informal descriptions to formal analytical representations. The latter types of models are now our goal, as they help us to explore alternative product configurations with speed, confidence, physical insights, and (potentially) reduced costs.

Informal Models

When a typical design concept is developed, it is first conceived in informal terms. Customer needs lists use informal (English) descriptions. A function structure uses English terms with graphical structure. Morphological charts use rough pictorial sketches. Even quantities that eventually become precise start informally. Requirements that a design must satisfy, for example, are usually first described in a natural language, complete with personal connotations. The concept selection process methodology is a means to develop consensus among team decision makers when operating with informal descriptions (Chapter 11). Such descriptions are not necessarily linguistic, but could be visual, acoustic, tactile, or aesthetic:

▶ Automobiles must have the "right" sound.

▶ Rechargeable battery operated designs must recharge "quickly."

▶ Vehicles must have a "comfortable" ride.

▶ Food production plants must produce the "right" taste.

To enable analytical treatment, these informal requirements must be converted into formal model requirements, as we discussed in Chapter 7 with the House of Quality.

Similar to customer needs, a typical designer performs the initial concept modeling in informal terms. "Back of the envelope" concepts are described informally, perhaps with vague drawings. They are never formal models on which precise calculations or computation can occur.

For example, the first description of an electric wok radiant heater may be "a halogen bulb." At this stage, computation (for layout, sizing, etc.) cannot occur: The description is not *formal*. It must be detailed and transformed to a state where a consistent understanding may be perceived from the one description. Different people have different perceptions of what is meant by "a halogen bulb" until it is made formal.

Such initial concepts, consisting of desires along with informal descriptions of possible solutions (visual, acoustic, linguistic, or whatever the form), are termed an *informal model* of the design.

Definition. An *informal model* is a designer's interpretation of a description of the customers' needs, engineering requirements, manufacturing requirements, and any other product requirement, along with the designer's interpretation of the conceived solutions.

Informal models are not precise, are without measure, and have the characteristics of intentions, interpretations, and connotative meanings. Descriptions can be acoustic, visual, or aesthetic, for example, which are (generally) unmeasurable concepts. All design projects originate in an informal model.

Each desire in the informal model will be called an *objective*, maintaining the terminology of decision theory. We have also called some of them "customer needs," and "effects." Similarly, on the configuration side, we will define a *concept* as the informal comprehension of a design configuration. We have also called concepts by informal descriptors such as "configurations," "causes," and "physical principles." Objectives and concepts are informal and unmeasurable. This formalism only discusses the designer's intentions, not any other person's intentions. Thus engineering embodiment design typically entails the translation of a design from an informal model to a formal model.

Example

Let's consider a fingernail clipper product. We have a number of concepts for the product, as depicted in Figure 13.4. An interpretation of these concepts is our informal model of each. We have a list of criteria for evaluating the concepts (Table 13.5). Our interpretation of what these criteria mean is our informal model of the objectives.

Formal Models

What is Formal?

We are seeking to construct a computable (or analytical) model of the design problem. The fundamental characteristics assumed to define a formal model are twofold. The first characteristic is that we can elucidate the alternatives to choose among them; the alternatives are assumed to have the structure of a set.

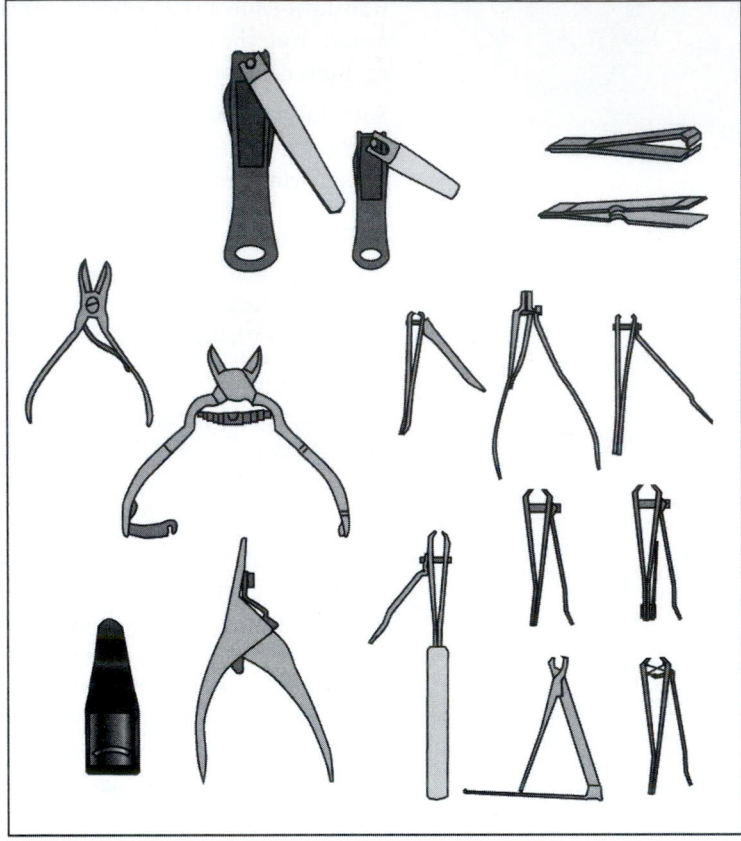

▼ *Figure 13.4.*
Fingernail clipper concepts.

A *set* is a formal concept. A set of objects has the properties that we can:

▶ identify whether objects are inside or outside the set

▶ distinguish among the individual elements within the set

Assuming the formal structure of a set means that when we are presented with an object, we can determine whether the object is in the collection of possible alternatives. We will be able to determine whether any alternative can function as a solution to the problem. When presented with the problem of designing a doorstop, we can, with enough analysis, state whether an object can function as a doorstop—whether it "is" a doorstop.

The second characteristic of defining a formal model is that when shown two different objects in the set, we can determine whether or not they are

TABLE 13.5.   CRITERIA FOR FINGERNAIL CLIPPER CONCEPT MODELS

Evaluation criteria	Subcriteria
Cost	Manufacturing cost
	Transport cost
	Advertising cost
	Retail costs
Store clipper	Compact
	Nonsnag
	Lightweight
	Striking appearance
Open/close	File easy open
	Clip easy open
File/pick	Files
	Picks
Clip	Align easy
	Hold easy
	Contoured body
	Low squeeze force
	Wide handle
Manufacturing	# Piece parts
	Forming
	Material
	Assembly
	Finish
Risk	Experience
	Number of prototypes
	Modeling

distinct. We can, with enough analysis, distinguish between the different alternatives. If our collection of concepts, performance ratings, noise values, or any other development category has these two properties, then the category has set structure, and we can hope to complete expressive analysis. If not, then we have to carry out some formalization work first.

Advanced Aside: A Philosophical Thought Exercise

Consider, though, this "set":

$$R = \{A \mid A \text{ is a set, and } A \notin A\}$$

This relationship is the set of all sets, in effect. Is R a set? Let's claim that it is.

proof:

Assume it is.

Assume R ∈ R. But by the definition of R, R ∉ R.

So assume the converse, R ∉ R. But by our assumption, R is a set. Since R is a set and R ∉ R, we must have that R is equivalent to A ∈ R, so R ∈ R.

This result is a contradiction. •

R is arbitrary; it could contain anything. R could be the set of everything in the universe. We just proved that it cannot exist, or that

There is no universe.

What we really mean is the "universe of discourse," meaning anything of importance to the discussion. Assuming that this universe R is also a set leads to a contradiction. R is and is not simultaneously in R; we could not and never can tell. Hence R is not a set, it is what mathematicians call a "proper class," which is something that cannot be formally defined. We cannot determine, no matter how much computation or anything we do, whether one particular object (namely R) is in the class R.

This example demonstrates a fact about all formal systems that we use in mathematics, science, engineering, and computer science. All models are fundamentally based on set structure, but the definition of a set is not and cannot be well defined. This idea means everything we ever calculate is based on axioms that are not completely defined—they are axioms. We can never completely define the universe of objects from which our set of objects come from. This example R is known as *Russell's Paradox.*

Here is another related example.

$$D = \{X | X \text{ is a fingernail clipper}\}$$

Is D a set? Can we, given any two objects X, determine whether they can function as a fingernail clippers and whether they are distinct? (Yes). Can we enumerate or identify all such X? (No.) This example is the philosophical question of Russell's paradox, applied to engineering design. Can we as engineers ever determine "all possible solutions" to a design problem? Russell's paradox argues that we can do so only in informal terms, not in any computational sense. Given a set of solutions, we can never deduce the universe of all possibilities that the set is from.

We can never, given a collection of designs that satisfy a requirement, compute the universe of all possible designs that satisfy the same requirement. We can only consider it informally. All the conceptual design computational systems that exist or ever will exist will not overcome this fact; all they will do is synthesize solutions in predefined ways from predefined components. While synthesis (combining current knowledge into novel solutions) can and will do much for design practice in the years to come, in the end, all true innovation is provably an exercise requiring creativity and cannot be reduced to an algorithm. Do you agree? How much of product development is formal (has mathematical set structure)? How much can be made formal?

Formal Product Models as Sets

Given this understanding of any formal system, we still want to use engineering models to make our informally perceived product better. This desire means we want to parameterize a set of possible configurations using engineering expertise and solve for the most preferred result.

Definition. A *design space*, $\mathbb{D}$, is the set of considered possible alternative configurations, described using *design variables*, over which we have a direct choice.

Thus, $\mathbb{D}$ is a formal concept, having a set structure. This structure is weak (only requiring the two restrictions as discussed above) but nonetheless fundamental, having identified the $\mathbb{D}$ means that it circumscribes *all* of the considered alternatives. Only the alternatives formalized within the $\mathbb{D}$ are possible outcomes.

The development of a formal set of alternatives is not sufficient to enable the use of a formal solution procedure to "solve" the design model. Informal interpretations continue to enter a design process in the analysis used to differentiate the elements in the design space.

Evaluations are made on alternatives in the $\mathbb{D}$. Such evaluations can be informal. To allow computation, such informal characterizations must be identified so that we know whether an alternative has an informal property or not. That is, to allow computation, the evaluations must also have set structure.

So, to formally evaluate different alternatives in the design space, performance metrics must be developed corresponding to the informal objectives placed on the design problem. As discussed above, performance metrics are evaluations placed on each alternative.

Definition. A *performance space*, $\mathbb{P}$, is a dependent set of evaluated performances determined at each point in the $\mathbb{D}$, described using *performance variables*.

The word "dependent" in the definition means that the $\mathbb{P}$ is related to the $\mathbb{D}$ by a set map. Performance metrics could be simple equations relating the design variables. For example, equations of stress (performance metrics) relate forces, dimensions, and material constants (all of which are design variables). The performance metrics can also be results of differential equations, integration, computer programs, physical experiments, or simple "IF/THEN" logic. A means of determining the identified performance given an identified design configuration is what is required.

Formal models as above could now be formally "solved"; that is, one must determine the element in the $\mathbb{D}$ that has "maximum" performance in the $\mathbb{P}$.

Many formal models, however, include more complex effects. For example, there are uncertainty effects that make the map from the $\mathbb{D}$ to the $\mathbb{P}$ inexpressible. Uncertainty influences can result in possible errors in measurement or in manufacturing. They could be variations in environment. They could as well be differences in the decisions of agents other than the designers. To incorporate such influences and still allow computation, we must determine what the influences are.

We must determine whether the manufacturing or measurement noise is large enough to be of concern. We must determine the range of possible environments in which the design must function. These are all examples of an identification requirement: We must identify the confounding influences; that is, the uncertainty must have set structure.

Definition. A *noise space*, $\mathbb{N}$, is a set of possible configurations, described using *noise variables*, required to evaluate any point in the $\mathbb{P}$, which we do not have direct choice over.

Note that not all modeling error is represented by noise variables. Only the noise that is sufficiently large and predictable is represented with distinct noise variables. There is always, however, a default noise variable ε representing residual error between what actually happens in the physical world and what a representative model predicts. Sometimes it is ignored, but it is always there.

Given this characterization of engineering design, the complete formal engineering model is now defined.

Definition. An *engineering product model* consists of a $\mathbb{D}$, a $\mathbb{P}$, and a (possibly trivial) $\mathbb{N}$.

Inherent in this definition is a set map $f{:}\mathbb{D} x\, \mathbb{N} \to \mathbb{P}$, implied by the dependency of the $\mathbb{P}$. This mapping simply means that given a well specified configuration, we must be able to determine the performance under the influence of uncertainties.

Model Completeness

Given that a formal model hopefully reflects a design's informal description, the validity of the formal model comes into question. We may have a perfect model of engine vibration of an automobile, but this model may not reflect the real concern of steering wheel vibration, for example.

Performance metrics: Recall that customer needs are typically expressed as informal *objectives*, where each objective expresses a particular desire. For the $\mathbb{P}$ to be a valid reflection of our objectives, it must be *comprehensive* and *measurable*:

▶ There must be a one-to-one correspondence between a performance metric and an informal objective.

▶ A performance metric is comprehensive if, by knowing the value of the performance metric, the associated objective is achieved at a target value. Thus, at the end of a modeling process, knowing a performance metric value will allow an informal interpretation of the achievement level of the associated objective.

▶ A performance metric must also be measurable. Formal definitions of measurability exist (Chapter 11), but essentially, the performance metric must be representable on a quantitative scale (a partial order); otherwise, the performance metric is no more formal than the objective itself.

Design space: The accuracy of a $\mathbb{D}$ model as a reflection of informal concept descriptions must also be considered. The same restrictions apply to $\mathbb{P}$. Any informal description of something that must be designed will require formalization, and each formal representation must be one-to-one, comprehensive, and measurable:

▶ One-to-one ensures the design model accurately reflects our conceptualization of a product.

▶ Comprehensiveness ensures that the formal model is true to the informal interpretation. For example, if a design is, in informal terms, a "halogen bulb," this can be formalized into a bulb with specific dimensions. The formal model is comprehensive, in that a formal result of 2–4 inches allows an informal interpretation of the bulb length.

- Measurability ensures we can distinguish among the configuration options. When we ask if two elements in the $\mathbb{D}$ are different, we must be able to tell. We must have a quantifiable structure (ordering) over the set or a separation metric to distinguish it.

Noise space: Finally, variations in the product usage, environment, or fabrication can be considered. The noise variable space is, in the ideal case, formal. Noise variable modeling should be as objective as possible and not based on varying interpretations. Noise variables values should be based on measured observations of the environment, which are then directly applied to the formal model. This situation is rare (some say impossible); however, subjective probabilities of noise may be used, and again the $\mathbb{N}$ is derived from subjective interpretations:

- For a perfectly accurate formal model, it must be one-to-one with the actual environmental disturbances that occur in the real world.
- It must be comprehensive in that its modeled values represent the intensity of the real world disturbance.
- It must be measurable, so that a value in the model can be distinguished.

The point is that *all* variables within a formal model must be comprehensive and measurable. The intent is to transform a model from intuition into something that is comprehensive and measurable. The identification of variables and structure is a fundamental feature of any engineering design process.

V. CONSTRUCTING PRODUCT MODELS: BASIC METHOD

Early stages of product development provide us with the necessary informal description of what we need to model a design problem. Model preparation and selection (Section II) establishes the first link of this informal description to a quantified metric. Having a complete functional model and architectural layout provides us with additional structure to construct a formal model.

A functional decomposition is the first step to constructing a full engineering model. A function structure contains a simple identification of informal information. The entries are not necessarily model variables, but simply flows of energy, materials, and signals through component concepts that solve the functions. More structure is needed to maximize product performance. This structure is simply obtained by relating performance values to the design configuration options through design variables.

A Basic Modeling Approach

To transform informal models into formal ones, we can consider a structured approach to modeling:

1. Identify a flow for the informal effect.
2. Identify a balance relationship for the flow.
3. Identify a boundary for the balance relationship.
4. Formulate an equation (or set of simultaneous equations) for the balance relationship in the system.
5. Use the resulting model to explore design configuration options.

This approach requires rigor and diligence, even if the designer or design team has experience modeling the technical system for the product. It also provides a great "crutch" in times when we don't know how to model the problem. A structured approach such as this one can go a long way toward helping to model difficult concepts.

STEP 1: Identify a Flow

We first identify a material, energy, or information flow associated with each effect of the product concept. This identification is a direct choice from the functional model of the product and overlaps with the preparation phase of product modeling.

Example: Let's examine our customer-needs-to-flow diagram used to construct the function structure for a fingernail clipper product, as shown in Table 13.6.

Now let's choose a customer need (informal objective) to model, say "Easy to squeeze ≡ Low finger force." The energy flow is "finger force," based on the function chain for the clipper, and the metric is "low finger force" as identified from benchmarking. Tracing finger force through the clipper, we have the representation shown in Figure 13.5. This representation helps us to develop a new fingernail clipper concept, understand its physics, and choose appropriate design parameter values to satisfy the customer needs.

The cutting motion acts like a scissors across the fingernail. A valid force flow is the force from the fingers, through the clipper, through the fingernail.

TABLE 13.6. MATERIAL, ENERGY, AND INFORMATION FLOWS ASSOCIATED WITH FINGERNAIL CLIPPER CUSTOMER NEEDS

	Customer Need	FLOW			Customer Need	FLOW
•	Purchase			•	Return from Filing/Picking	
-	Cost	-		-	Easy to close file	**Fingers**
•	Transport in package			•	Prepare for clipping	
				-	Easy to open clipper	**Fingers**
•	Unpackage					
				•	Clip nails	
•	Chain Keys			-	Easy to align clipper	*Line of sight*
-	Act as keychain	**Keys**		-	Easy to hold clipper	**Fingers**
				-	Body contoured to hold	**Fingers**
•	Store			-	Curved blade shape	**Fingers**
-	Compact storage	-		-	Wide handle	**Fingers**
-	Non-snag storage	**Cloth**		-	Low squeeze force	Force
-	Lightweight	-		-	Blade can act as a pusher	Force
-	Striking appearance	*Light*		-	Blade can act as a file	**Nail**
				-	Clips nails	Force
•	Prepare to File/Pick			-	Clips toe nails	Force
-	Easy to open file	**Fingers**		-	Clips hang nails	Force
				-	Sharp blade	Force
•	Filing/Picking			-	Nails fall predictably	**Nail**
-	Files nails	**Nail**		-	Stores cut nails	**Nail**
-	Picks nails	**Nail**		-	Easy to clean	-
-	File rough	**Nail**				
-	File holds filed dust	**Nail dust**		•	Return from clipping	
-	File has a picking tip	**Nail**		-	Easy to close clipper	**Fingers**
					Material Flow	
					Energy Flow	
					Information Flow	

STEP 2: *Identify a Balance Relationship*

Having identified a material, energy, or information flow, we now must identify a balance relationship for the flow.

Example: In the fingernail clipper, the forces above the cutting teeth must equal the forces below. The moments about the fulcrum must

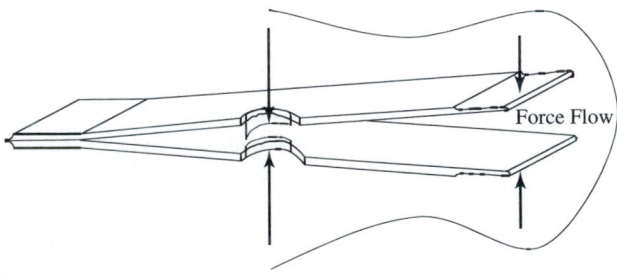

▼ **Figure 13.5.**
Force flow representation for a new fingernail clipper concept.

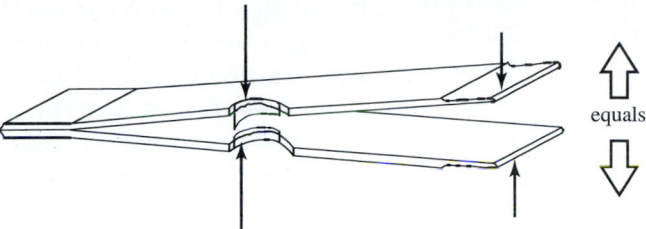

▼ **Figure 13.6.**
Force balance for the fingernail concept.

also sum to zero (assuming quasi-static analysis). The concept balance relationship is shown in Figure 13.6.

STEP 3: Identify the Boundaries

This step entails the recognition of the boundary conditions of the product concept. How is it loaded, and how does it interact as a system with its environment (including the human interface)? What are the inputs and outputs across the boundaries, and what are the limits or ranges of these inputs and outputs?

Example: In the fingernail clipper, it is profitable to model the forces by starting with the finger forces acting across the system boundary (Figure 13.7).

For this model, we could include the finger forces that are internal to the system and create a model of the forces through the finger, hand, and thumb. Such an approach is not necessary. We could zoom in on the cutting blades and not model the finger forces. Then we only have forces acting on the blades, which are not directly related to the forces required of the fingers. This approach is insufficient, since it does not capture the chosen product metric of finger force.

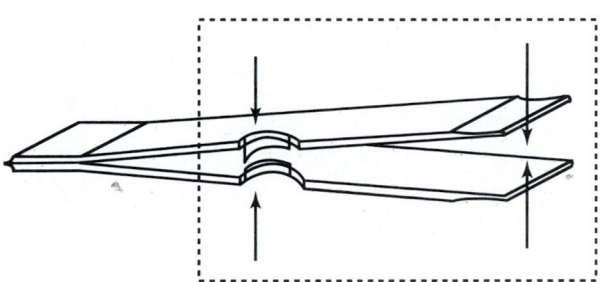

▼ **Figure 13.7.**
Finger force boundaries.

This example hopefully makes it clear how to select a proper level of abstraction for a model. The issue is pervasive to all engineering problems. The effect-to-flow-to-balance-to-system method is a reasonable approach—and generalizes. For example, fluid dynamics, heat transfer, solid mechanics (material stress and strain), and manufacturing process modeling typically use a control volume, and flows in and out are balanced.

STEP 4: Derive an Equation (or Set of Simultaneous Equations)

The next step involves converting the balance relationship to a mathematical form. This step is one of assigning geometric variables, material property constants, and others to formulate an equation (or set of simultaneous equations) that can be solved. The engineer's toolbox of applied mathematics and science is required here. In addition, a number of assumptions and simplifications will be required to develop a suitable model.

Example: Assigning dimension variables in the first control volume, the labeled fingernail clipper concept is shown in Figure 13.8.

Summing moments about the fulcrum, the finger force required to actuate the clipper is:

$$f = F\frac{L_F}{L_f} \tag{13.1}$$

The distances from the fulcrum to the force applications are design parameters (components of the $\mathbb{D}$). The force F from the nail is unknown,

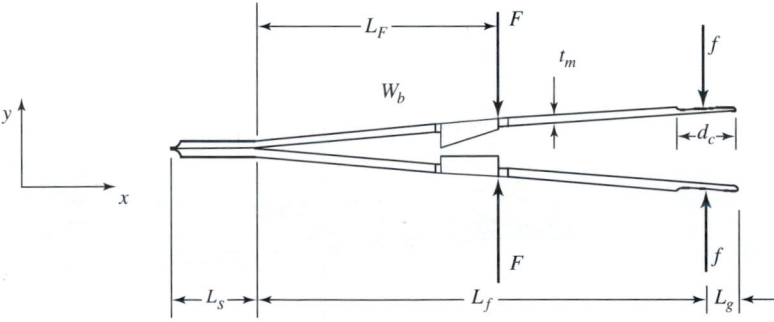

▼ **Figure 13.8.**
Parameterized fingernail clipper concept.

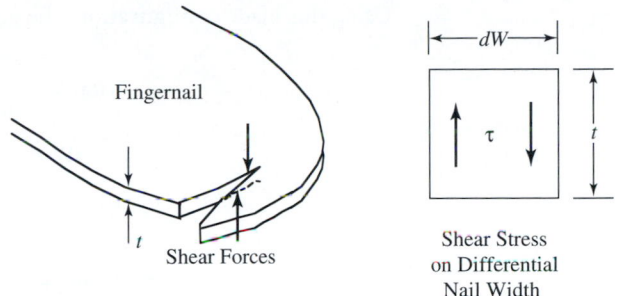

▼ Figure 13.9.
Shear forces as a fingernail is being clipped.

however. We need to repeat the modeling process for the fingernail force, where we have a system of forces within a system of forces. To derive the force F, we model the fingernail being clipped. Figure 13.9 illustrates the fingernail force system.

We may model the nail force as the shear stress over the nail cross-sectional area. Therefore:

$$F = \tau \, dWt \tag{13.2}$$

Now, how do we represent dW? Another system and model is needed. We can visualize that dW depends on the angle of the blade, as shown in Figure 13.10.

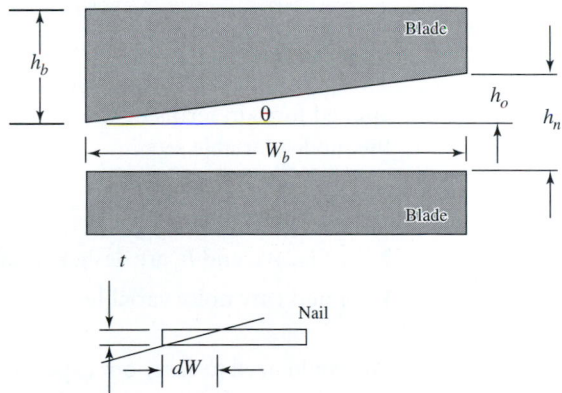

▼ Figure 13.10.
Blade geometry with respect to the nail.

Using this blade configuration, the geometry defines dW as:

$$\tan \theta = \frac{t}{dW}$$

$$dW = t \cot \theta \tag{13.3}$$

where

$$\tan \theta = \frac{h_o}{W_b}$$

$$\cot \theta = \frac{W_b}{h_o} \tag{13.4}$$

Combining Eqs. (13.3) and (13.4), we obtain a relationship for dW:

$$dW = \frac{W_b}{h_o} t \tag{13.5}$$

Substituting this result into Eq. (13.2), we have an expression for the nail force:

$$F = \tau \frac{W_b}{h_o} t^2 \tag{13.6}$$

This expression serves as a first-order model of the fingernail force. Combining this result with Eq. (13.1), we obtain a relationship for the finger force, our product metric:

$$f = \frac{L_F}{L_f} \tau \frac{W_b}{h_o} t^2 \tag{13.7}$$

This performance relationship is a simple engineering model of the applied force in terms of the model variables. Note that, in terms of the model-variable sets:

- f is a performance metric
- L_f, L_F, W_b, and h_o are design variables
- τ and t are noise variables

We would need to carry out experiments to calibrate what τ and t values are appropriate for this model or consult medical or human factors literature to obtain published empirical data. Ninety-fifth percentile users would serve as a starting point for modeling the shear strength and thickness of fingernails.

At this point, a model exists for one of the fingernail clipper's customer needs. More models may be created for any of the remaining customer needs, based on the material, energy, and signal flows through the product concept's functions. Decisions exist on whether to choose analytical or physical models for the remaining needs. For the fingernail clipper, let's consider further analytical models.

Continuing the modeling of the fingernail clipper, let's derive a model of the "compact" customer need; that is, customers desire the fingernail clipper to be compact when not in use. "Flows" for this customer need are the user's hand, pants pocket, wallet or purse compartment, or other storage locations. In fact, we are concerned with volume as the product metric. This metric may be used to determine a singular cubic length (volume) analysis or three separate one-dimensional length analyses, one in each direction of an orthogonal coordinate system.

A balance relationship for the orthogonal coordinate system is the summation of individual feature distances along the fingernail clipper that must be added to a value that is less than a target value for the dimension. Considering this metric, a model of the clipper along the length (x) direction is given by Figure 13.8. Figure 13.11 provides additional geometry for modeling the height and width of the fingernail clipper.

From these parameterized geometric models, we can convert the informal balance relationship into equations that model the "compact" customer need. Adding the boundaries for the manufacturing process (such as tearout and bearing stress requirements on the blade rivets),

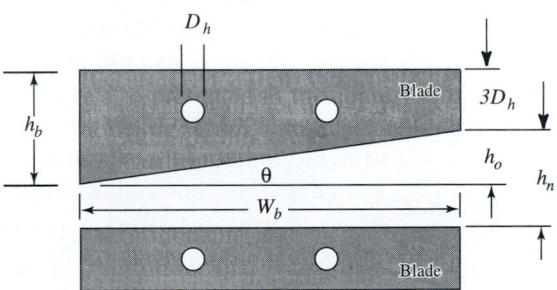

▼ *Figure 13.11.*
Parameterized fingernail clipper blade model.

and including the balance relationship for the finger force, we arrive at the following simultaneous equations:

$$f = \frac{L_F}{L_f} \tau \frac{W_b}{h_o} t^2$$

$$W_b \geq 5D_h$$

$$L_F \geq W_b$$

$$L_{\text{overall}} = L_s + L_f + L_g$$

$$L_s \geq 5t_m$$

$$L_f \geq L_F + \tfrac{1}{2}d_c \tag{13.8}$$

$$L_g = \tfrac{1}{2}d_c$$

$$H_{\text{overall}} = 2h_b + h_n + 2t_m$$

$$h_b \geq h_o + 3D_h$$

$$h_n \geq h_o + 1.1t$$

$$t_m \geq 1.25D_h$$

$$W_{\text{overall}} \geq W_f$$

Examining the variables of these balance relationships, we have the following sets:

▶ f, W_{overall}, L_{overall}, H_{overall} are performance metrics
▶ t_m, D_h, h_o, L_f, L_F, are design variables
▶ d_c, W_f, t, τ are specified or are noise variables

Repeating the analysis for the "long life" customer need, we can derive balance relationships for the elements in the fingernail clipper concept that are most susceptible to failure. Applying failure modes and effects analysis, the elements most likely to fail over the number of cycles of clipper are the lever arm and the blade rivets. Let's first consider the lever arm. Figure 13.12 shows the simplified geometry of the top lever arm, including the force balance. Figure 13.13 then shows the corresponding shear and bending diagrams.

Based on the models in Figures 13.12 and 13.13, the maximum normal stress in the arm (disregarding transverse shear and considering fatigue) is given by:

$$\sigma_{\max} = \frac{My}{I} = K_{\text{cycles}} \frac{6f\left(\dfrac{L_f}{L_F} - 1\right)L_F}{W_b t_m^2} \tag{13.9}$$

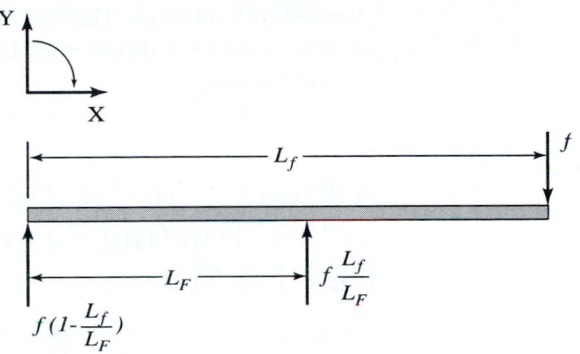

▼ **Figure 13.12.**
Lever arm model of the fingernail clipper.

Likewise, the maximum shear stress in the blade rivets is given by:

$$\tau_{\text{rivet}} = \frac{1}{2}\frac{4F}{\pi D_h^2} = \frac{2\tau W_b t^2}{\pi h_o D_h^2} \leq 1.5 S_{\text{sy,rivet}} \quad (13.10)$$

We may now continue the analysis and model other customer needs for the fingernail clipper concept. The separate models are assembled into a set of simultaneous equations, including objectives (balance relationships) and boundaries. Once a model has been assembled, we apply the fifth step of the modeling process, wherein we solve the model and explore different configurations based on parameter changes. Spreadsheet solutions or optimization codes (Chapter 16) may be applied to

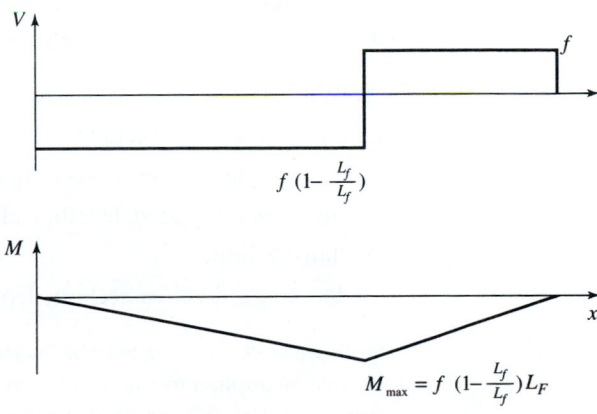

▼ **Figure 13.13.**
Shear and moment diagrams for the fingernail clipper arm.

carry out such solutions. The final choices will have to be verified with physical models. For this chapter, however, we stop with the development of the metric models.

A Product Application in Constructing Basic Models: Iced Tea Maker

This chapter begins with illustrations of product models for household iced tea brewing machines and introduces model preparation with iced tea makers as the exemplifying application. From the product concept earlier, two customer needs are chosen to focus the modeling efforts: "increase tea strength" and "use less ice." The following models of the "steeping process" (the tea leaves losing their particle flavors to the water) and "cooling iced tea" provide insights into possible improvements that may be achieved parametrically by varying the product configuration.

Steeping Model

The main objective for modeling the iced tea maker is to improve the primary customer need of stronger tea (Figure 6.31–6.32). The subfunctions that satisfy this need are the "mix tea and water" and "transmit heat" subfunctions. These subfunctions are completed around the tank and heating elements. Therefore, the heating elements, tank, water and tea bags constitute the physical system to be modeled.

Considering the subfunctions and flows in and out of them reveals that many variables affect the strength of the tea:

1. Water: temperature, flow rates, and soak time
2. Environment: ambient temperature
3. Heating element: cross-section, maximum temperature
4. Tea: how it is placed, length, width, height, type
5. Tank: volume
6. Steeping area: cross-section, depth, and hole size

Since we are saturating tea, the "water" is an important material flow. It must be applied to the tea at some rate and then leave with the assumption that little or no water remains in the system. Therefore, it seems to be a prime flow to examine. Fluid mechanics will be used to balance the water flow.

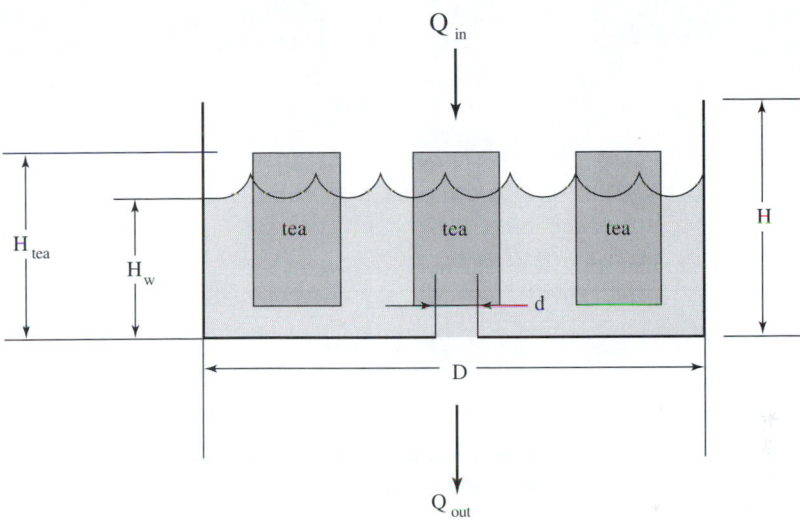

▼ *Figure 13.14.*
Diagram of steeping process.

The primary energy flow, "heat," represented by the temperature of the water, is equally important during the steeping process of tea; however, the current temperature of the water, at 99°C, can obviously not be improved much further.

The boundaries of the control volume are the boundaries of the components fulfilling the subfunctions as above and contain the steeping area (Figure 13.14). The tea brewer operates by providing a small steady stream of hot water to the steeping volume, and a steady stream of brewed tea exits the steeping volume. The input flow is slightly larger than the output flow, and so the steeping volume dynamically fills for a period of time and then empties when the input flow shuts off.

Assuming steady flow, we first use Bernoulli's equation (fluid flow out the bottom of a body of water) to solve for the exiting volume flow rate (Q_{out}), through the duration of the steeping process.

$$Q_{out}(t) = \left(\frac{\pi d^2}{4}\right)\sqrt{2gH_w(t)} \quad \text{Bernoull's Equation} \quad (13.11)$$

As time progresses, the height of the water is determined using a flow balance of the fluid streams entering and exiting the steeping volume.

$$H_w(t^*) = \frac{\displaystyle\int_0^{t^*} Q_{in}(t)dt - \int_0^{t^*} Q_{out}(t)dt}{\dfrac{\pi D^2}{4}} \qquad \text{Volume Flow Balance} \quad (13.12)$$

Here, t^* is some particular time, varying how we determine H_w as a function of time. Substituting Eq. (13.11) into Eq. (13.12) yields:

$$H_w(t^*) = \frac{\displaystyle\int_0^{t^*} Q_{in}(t)dt - \frac{\pi d^2 \sqrt{2g}}{4}\int_0^{t^*} \sqrt{H_w(t)}dt}{\dfrac{\pi D^2}{4}} \qquad (13.13)$$

This result is an integral equation for H_w with initial condition $H_w(t) = 0$, similar to an ordinary differential equation. The form of this equation is $x + a(\dot{x})^2 + b = 0$ which we will need to solve numerically.

Steeping process: We now have a model of the fluid height, from which we can derive how long the water sits in the steeping volume before moving on. We also want to know how long the water and tea need to mix in the steeping volume. To do this, we need to study the steeping process in more detail.

To construct this model of required steeping time, we could construct a model of tea/water mixing using the same approach. Alternatively, we could explore the tea brewing literature, using an approach discussed in Chapter 7 on benchmarking. There, we would find a required tea steeping time model (Tillberg 1997).

Figure 13.15 illustrates what is occurring during the brewing of black tea. The most important column is "bitter substances." According to this figure, as the tea steeps beyond 5 minutes, no more flavor is drawn, and the bitterness begins to rapidly increase. Other types of tea require more precise steeping times; however, they are currently not preferred for iced tea. Therefore, we need a model to determine how much of the tea is saturated too long, producing a bitter taste, or too short, yielding weak and underdeveloped flavor. Figure 13.15 presumes an optimal steeping temperature of just under boiling water temperature.

Modeling saturation and bitterness: The first step is to determine a dimensional unit to measure both the saturation time and bitterness, to form the steeping model. Consider Figure 13.15: To achieve the best saturation process, the steeping chamber must instantaneously fill to the height of the tea in the bag (H_{tea}) and then instantaneously leave after 5 minutes. Therefore, the optimum fluid filling perfor-

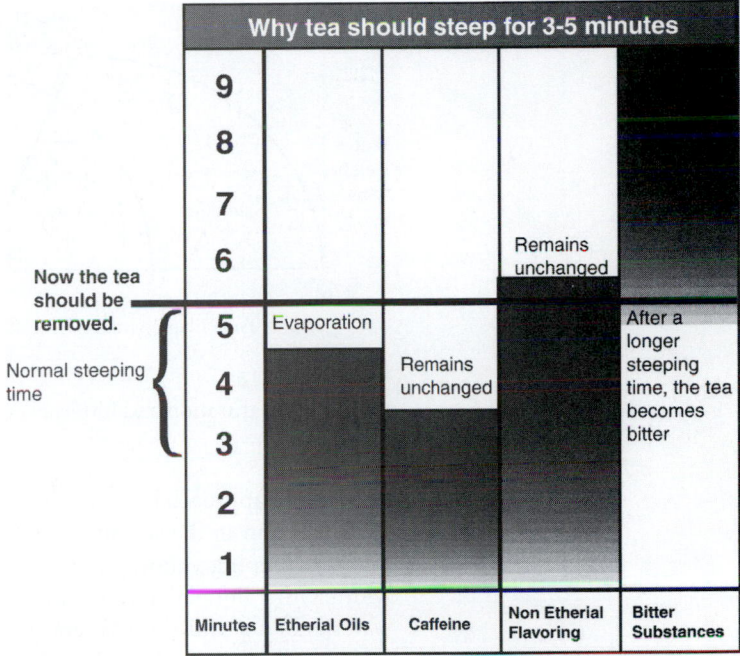

▼ *Figure 13.15.*
Model of steeping process (Tillberg 1997).

mance, or *saturation*, would be H_{tea} cm of fluid for 5 minutes, or $5*H_{tea}$ cm per minutes. Since we cannot achieve instant filling, complete containment, and instant drain of the water, what we would like instead is a filling process that gives the ideal saturation of $5*H_{tea}$ cm per minutes. Saturation, measured in cm per minutes, is the unit metric for us to model the steeping performance (Figure 13.16). Noting that the nominal tea within a bag height is about 2.2 cm (height containing tea), then anything less than 11 cm per minutes will be a percentage of the optimum steeping performance point. A filling/draining rate profile that will provide 11 cm per minutes of saturation would be ideal.

The saturation performance is, however, not as simple as the area under the height-time curve of filling. In particular, according to Figure 13.15, any bit of fluid that itself remains in the steeping volume after 5 minutes turns bitter. It does not meet brewing specifications, and so we need to cancel out all water that has been in the steeping volume for longer than 5 minutes from the saturation area calculation. The question is how to accomplish this task.

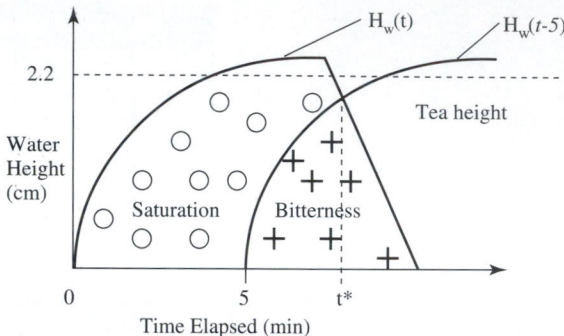

▼ *Figure 13.16.*
Diagram of saturation and bitterness during the steeping process.

One simple approach is to imagine there is no fluid mixing in the vertical direction in the steeping volume. Hot water enters from above, hits the steeping volume floor, and widens to the stepping volume width. Other liquid that flows into the steeping chamber then hits this water below it and simultaneously widens to the steeping chamber width, not mixing with the liquid below it. The entire width lowers to the bottom, then instantly contracts to the exit diameter and drains at the lower exit flow rate.

This is not a good model of tea mixing, but it is a satisfactory model of flow *rates* and the *time* the average liquid element spends in the steeping chamber. With this model, the saturation of good tea is the integral of the steeping volume height minus the excess steeping time (beyond $5*H_{tea}$ cm per minutes) at which point the tea element becomes bitter. Therefore, the saturation performance (S) is:

$$S(t^*) = \int_0^{t^*} H_w(t)dt - \int_0^{t^*} H_w(t - 5)dt \qquad (13.14)$$

where t^* now is the time where good tea, and not bitter tea, is brewed. This is the point in time where $H_w(t) = H_w(t - 5)$.

In addition to this model, we should ensure a constraint that $H_w(t) \leq H_{teabag}$, because the water height above the tea bag has no effect on saturation. Since Q_{out}, Q_{in}, and H_w are dependent, we will select Q_{in} as the design variable. Therefore, the procedure so far for this model is as follows:

▶ For a given flow rate in, $Q_{in}(t)$, the height of water, $H_w(t)$, is calculated numerically using Eq. (13.13).

▶ Then, by substituting $H_w(t)$ into Eq. (13.11), $Q_{out}(t)$ can be solved to determine t_{final}.

▶ Note that when $Q_{out}(t_{final}) = 0, (t_{final})$, a performance metric, equals the time to steep the tea.

▶ Finally, the two functions $H_w(t)$ and $H_w(t - 5)$ are substituted into Eq. (13.14) to solve for S, the saturation of the tea.

Using similar justification, the bitterness of the tea can be calculated as:

$$B = \int_5^{t*} H_w(t - 5)dt + \int_{t*}^{t_{final}} H_w(t)dt \qquad (13.15)$$

This equation calculates the amount of tea that is saturated longer than 5 minutes in the same units as S (cm-minute). Because we want to eliminate the tea's bitterness altogether, the target for B is zero. Note that this is impossible unless $t_{final} \leq 5$ minutes.

Heat Transfer Model

The iced tea maker model, to this point, focuses on the saturation of the tea; however, as $Q_{in}(t)$ changes, the amount of ice needed varies. For example, if less hot water is ejected into the steeping chamber, less ice is needed to cool the hot water. This issue affects the second customer need for using less ice, which is initially linked to the metric of exiting temperature of the hot tea. Therefore, a correlation between the amount of ice and $Q_{in}(t)$ must be established.

Following our process for establishing a model, the ice enters the function structure in the "transmit heat" subfunction to cool the tea. The important flow is "heat," which occurs in the pitcher, the system volume to model. Therefore, heat transfer between the ice, water, and hot tea in the pitcher forms the system, and the heat flow must be balanced (Figure 13.17).

Using heat capacity and phase change equations:

$$q_{ice} + q_{water} = q_{tea} - q_{loss} \qquad (13.16)$$

$$V_{ice}\rho_{ice}C_{p_{ice}}(T_{ice}) + V_{ice}\rho_{ice}h_{ice} + V_{water}\rho_{water}C_{p_{H_2O}}(T_{tap})$$
$$= V_{tea}\rho_{tea}C_{p_{tea}}(T_{tea}) - q_{loss} \qquad (13.17)$$

The total volume of the brew leads to

$$V_{water} = V_{total} - V_{ice} - V_{tea} \qquad (13.18)$$

In order to calibrate this model, the following assumptions must be made:

1. The densities of ice and tea are equal to water. Note that V_{ice} refers to the volume of the *melted* ice.

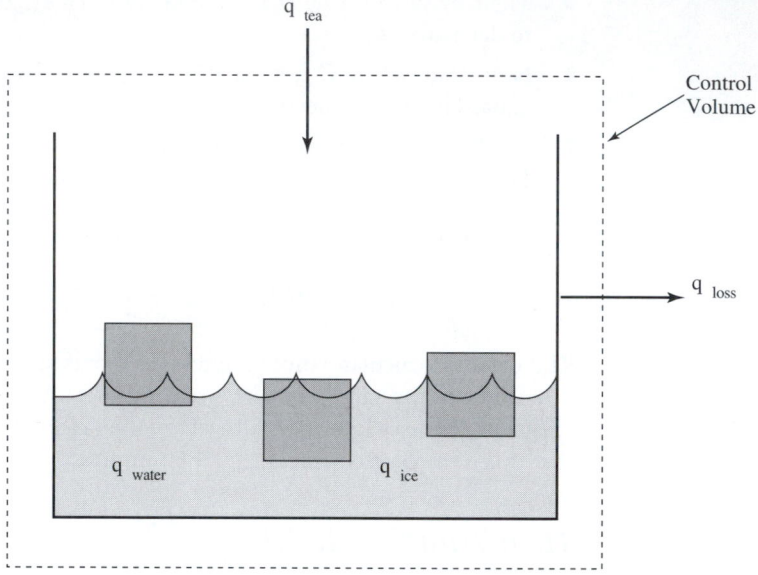

▼ **Figure 13.17.**
Heat transfer model of cooling the tea.

2. C_p of tea is equal to C_p of water.

3. Assume all of the ice is melted, and the temperature of the tea is 0°C at the end.

Several variables in Eqs. (13.17) and (13.18) must be determined experimentally to solve for the heat loss out of the pitcher. We need the initial volume of ice and the volume of tea not including the ice produced during the brewing process. In addition, we must determine the temperate of the exiting tea.

Measuring the volume of ice and tea in a desktop experimental setup of a tea brewer is rather straightforward. Using a measuring cup, the amount of tea without ice is measured. Then the tea is added to the ice, and once melted, the total volume is measured. The difference between the two measurements reveals the volume of the ice. To measure the temperature of the exiting hot tea of an experimental setup, a data acquisition thermocouple device is selected. The temperature is measured over a 10-minute brewing cycle to yield an average temperature of 85°C.

Using the results of the experiments, q_{loss} can now be determined.

$$
\left.\begin{array}{ll}
V_{\text{total}} = 2 \text{ quarts} & T_{\text{ice}} = -10° \text{ C (Assumption)} \\
V_{\text{ice}} = 3.25 \text{ cups} & T_{\text{tap}} = 25° \text{ C} \\
V_{\text{tea}} = 4.75 \text{ cups} & T_{\text{tea}} = 85° \text{ C} \\
V_{\text{water}} = 0 \text{ cups}
\end{array}\right\} \begin{array}{l} \text{Experimental} \\ \text{results} \end{array}
$$

From Eq. (13.17), $\dfrac{q_{\text{loss}}}{\rho} = 537.2 \dfrac{\text{cups*kJ}}{\text{kg}}$.

Calibration of heat transfer model: Although we have determined the heat loss for the current brewing parameters, q_{loss} changes with the volume of hot tea. Therefore, we need to determine the actual percentage of heat that is lost, P, from the total energy in the tea. Assuming that the heat loss is primarily from the hot tea, we can calibrate the model:

$$q = mC_p\Delta T \tag{13.19}$$

$$q_{\text{tea}} = (V_{\text{tea}}\rho_{\text{water}}*C_{p_{\text{water}}}*(T_{\text{tea}} - 25°C)) \tag{13.20}$$

$$q_{\text{loss}} = P\%(V_{\text{tea}}\rho_{\text{water}}(C_{p_{\text{water}}}*(T_{\text{tea}} - 25°C))) \tag{13.21}$$

From the experimental results above, $P = 45\%$. This result states that 45% of the heat from the ejected hot tea is lost to the environment.

Substituting Eqs. (13.21) and (13.18) into Eq. (13.17), and then solving for V_{ice}, we have:

$$
V_{\text{ice}} = \frac{V_{\text{total}}C_{p_{\text{water}}}T_{\text{tap}} + V_{\text{tea}}(C_{p_{\text{water}}}T_{\text{tea}} - C_{p_{\text{water}}}T_{\text{tap}} - PC_{p_{\text{water}}}(T_{\text{tea}} - 25))}{C_{p_{\text{ice}}}T_{\text{ice}} + h_{\text{ice}} + C_{p_{\text{water}}}T_{\text{tap}}} \tag{13.22}
$$

where the design variable $V_{\text{tea}} = \displaystyle\int_0^{t_{\text{final}}} Q_{\text{in}}(t)dt$. Note that:

S, B, t_{final}, and V_{ice} are the performance metrics.

d, D, V_{total}, T_{tea}, and Q_{in} are design variables.

H_{tea}, T_{tap}, T_{ice}, $C_{p_{\text{water}}}$, $C_{p_{\text{tea}}}$, $C_{p_{\text{ice}}}$, ρ_{water}, P are noise variables.

The performance metric values are the variables in the model that determine how well the product performs. The design variables can be changed to improve the performance metrics. The noise variables are the uncertainties or uncontrollable variables in the model.

Optimization

We have now reached the stage where the iced tea maker models may be used to explore design configurations. As a preview to such exploration, consider solutions to the iced tea steeping model and heat transfer model. Later chapters develop methods for solving such models.

We have developed two models related to the customer needs "stronger tea" and "easy to add ice." We can search these models to determine if any changes in the design variables will improve the overall performance of the product.

There are four performance metrics:

Maximize $S = f(Q_{in}, D, d)$ (Saturation time Eq. (13.14))
Minimize $B = f(Q_{in}, D, d)$ (Bitterness time Eq. (13.15))
Minimize $t_{final} = f(Q_{in})$ (Time to brew iced tea $Q(t_{final}) = 0$)
Minimize $V_{ice} = f(Q_{in}, V_{total})$ (Amount of ice needed to cool tea: Eq. (13.22))

The first step in parametrically optimizing the iced tea maker is to investigate the design variables and then form constraints to the problem.

Equality constrained design variables: The diameter of the steeping basket, D, cannot be changed unless a new filter design is added or provided. The current diameter is set by the common disposable filter size. Next, the maximum brew size of 2 quarts will be held constant, because varying V_{total} would create complexities in the brewing containers and because the 2 quart size is generally acceptable to the customer. In addition, a complexity arises in trying to relate V_{total} to Q_{in}, which would add another level to the model.

Regional constrained design variables: The diameter of the steeping basket hole (d) and the Q_{in} will be varied to optimize the redesign. The only constraint on the diameter is that it must be positive; however, Q_{in} is constrained to changes in the length of the steady-state portion of the curve (Figure 13.18). This constraint is achieved by changing the amount of water in the tank. As more water is added, the transient responses at the beginning and end will not significantly change, because the unit turns on, reaches a steady state, and then turns off when the water is almost gone. This figure is developed experimentally by measuring the volume flow rate of the hot tea into the steeping chamber for 1 quart of tank water yielding a time of 10 minutes. The tank water is decreased by approximately one fifth, twice, to prove the transient responses are the same as shown in Figure 13.18. This rate is measured by recording the height of water in the steeping basket over time using a ruler. This approach results in a natural measurement for the "mix solid and liquid" product function that relates directly to the customer need, "stronger tea."

An additional constraint must be placed on the lower limit of Q_{in} to prevent supersaturation of the tea. To determine the saturation level of the tea, an alternative measurement technique is used. Basically, we

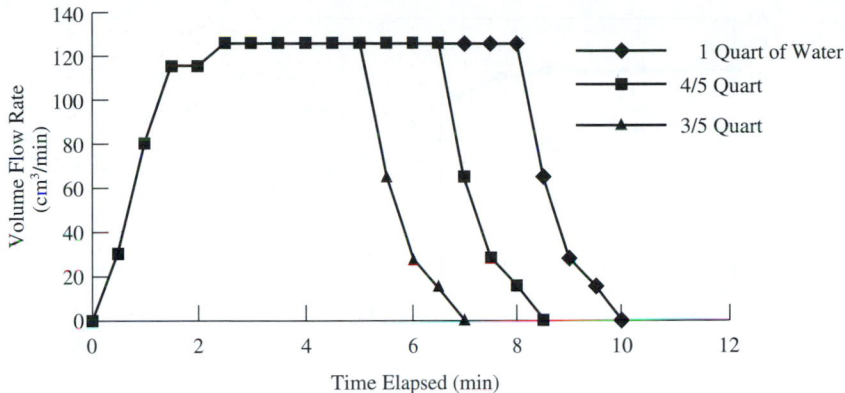

▼ **Figure 13.18.**
Hot water volume flow rate vs. time elapsed for various volumes of water.

need to determine the point at which adding more tea to a set volume of hot water would not change the strength of the tea. Measuring the strength of tea becomes the primary issue. It is too costly, both in time and money, to develop a chemical test to determine the concentration level beyond an unsuccessful litmus test. Therefore, the strength is determined by visual color using a clear glass against white paper. Tea is brewed with various quantities of tea bags until the resulting color of the tea does not change. The experiment shows that the amount of tank water can be decreased up to 50%.

Due to complexity, no effort is placed into changing the transient or steady-state temperatures of Q_{in}. Changing the temperature of the resistive element, tube sizes, or the height the water that is pumped could complete these changes.

Varying discharge hole diameter: To optimize the four-performance metrics, the equations are coded into an Excel spreadsheet. The diameter of the hole is varied from 1.4 mm to 2.8 mm, while plotting three of the performance metrics at each size (Figure 13.19). The amount of ice needed is not affected by the hole size.

The results show that as the diameter is decreased, both the saturation time and bitterness increase as well as the time to brew. The optimum point is hard to distinguish, because it is difficult to determine how much bitterness the customer will allow; however, the optimum diameter can be selected by comparing to a benchmarked product. The vertical line in Figure 13.19 shows the choice, meeting the stiffest competition in the market (for tea taste only).

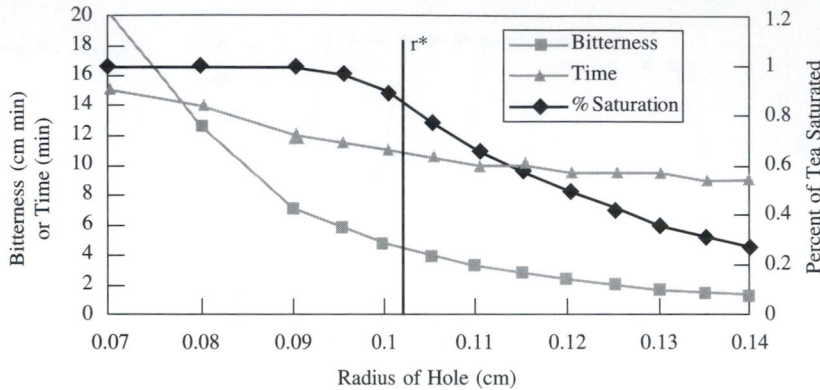

▼ *Figure 13.19.*
Performance of iced tea maker for various drainage hole diameters.

Varying volume flow rate: The next step is to vary the amount of water in the tank. This parameter change is accomplished by decreasing the steady-state portion of Q_{in} in the same spreadsheet, thereby decreasing the duration and total volume of the hot water (Figure 13.20). Notice that the time and bitterness drop off at a much faster rate than

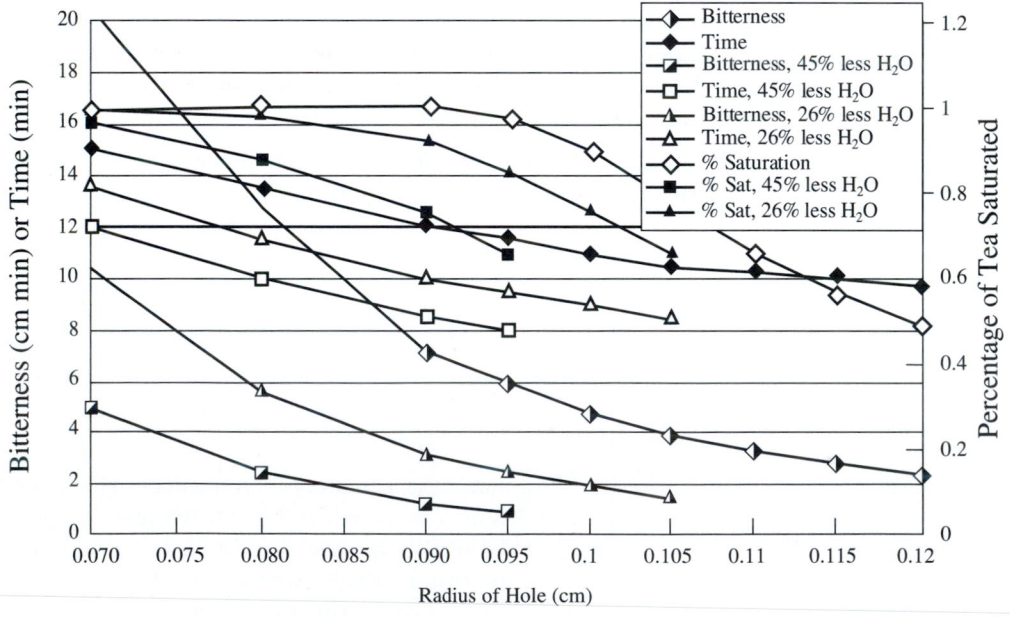

▼ *Figure 13.20.*
Iced tea maker performance vs. hole radius.

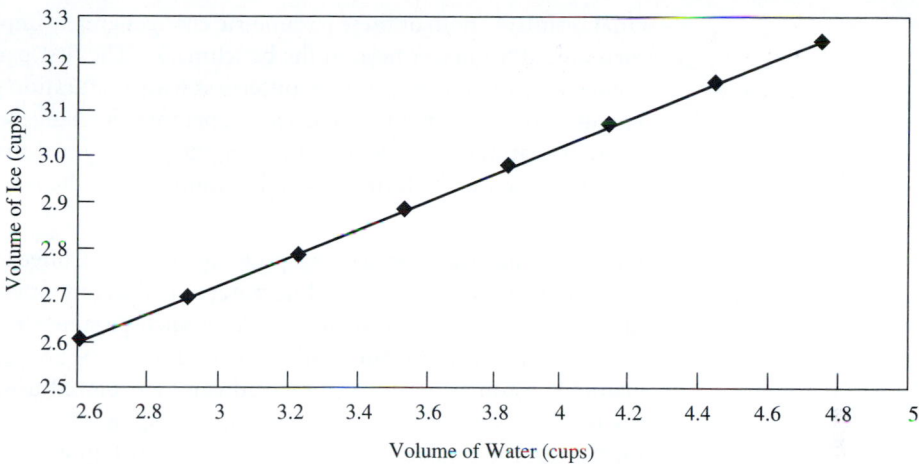

▼ *Figure 13.21.*

Amount of ice needed vs. volume of water in tank.

the saturation percentage does. In addition, a separate graph is created to display the fourth performance metric, amount of ice needed (Figure 13.21).

There are several possible optimization points at which the iced tea maker can be operated. All the results are relative to the benchmarks.

▶ Set target value of saturation percentage at 75%, where Mr. Coffee currently operates, and then go to the boundary of 45% decrease in steeping water.

RESULTS: with a 1.8-mm-diameter hole and 45% decrease in tank water

1. 71% reduction in bitterness

2. 20% reduction in time to brew from 10.5 minutes to 8.5 minutes

3. 17% reduction in amount of ice needed

▶ Keep time at 10.5 minutes, Mr. Coffee operating point, and then go to the boundary of 45% decrease in steeping water.

RESULTS: with a 1.55-mm-diameter hole and 45% decrease in tank water

1. 20% reduction in bitterness

2. 19% increase in saturation

3. 17% reduction in amount of ice needed

Final results: Both of these parametric changes greatly improve the current iced tea maker beyond the benchmarks. The first parametric change yields a 71% reduction in bitterness with a saturation percentage equal to the benchmarks. The second parametric change would increase the saturation percentage by 19%, the primary customer need, and also decrease the bitterness and amount of ice, the second customer need.

The above analysis is based on keeping the same Mr. Coffee tea brewer components and product architecture; only the component design variables were altered in the model. After such parametric redesign exploration of the current product using models as above, the remaining customer needs not modeled can be explored using adaptive redesign, where actual subsystems and components are altered. In such cases, the models of the primary functions above remain the same in objective, though their design variables may change with component or architecture changes. On the other hand, if new functions are added to a new concept, then new measurements and specifications need to be developed, and also new models. For example, the initial adaptive redesign goal for the iced tea maker is to reduce greatly the amount of ice needed by decreasing the exit temperature of the hot tea. This goal would involve adding a function to cool the hot water. This new function can be transformed to "transmit heat," which indicates the need for concept ideas to remove heat from the exiting tea liquid. A new model of this function would have to be developed for each new concept.

VI. CONSTRUCTING PRODUCT MODELS: ADVANCED METHOD

The previous section presents a basic model construction process as containing five fundamental steps. The focus of this process is to transform an informal description of product desires, obtained from earlier analysis, to a formal quantified model. Model preparation begins the process by establishing links of the informal description with product metrics. A functional model of the product concept, as well as the established architectural layout, provides additional links.

Functional models of a product concept relate material, energy, and information flows from the environment to the outputs produced by the product. These relationships, while informal, provide a basic structure for representing the underlying physics of the product's performance. This structure provides a solid foundation for incrementally building a formal analytical model. The chosen product metrics, in fact, are simply measures of the flows exiting functions in the product.

Approach

To build on the functional model foundation, transforming the informal models to formal ones, a systematic development approach is considered. Figure 13.22 shows an advanced modeling approach for product metrics, after Ashby (1992; 1987) and Beaumont (1998) and the previous discussion. It begins by considering the important effects (customer needs and associated engineering specifications), advances the model through physical mechanisms and interrogation, and ends with implementation of the model to explore alternative product configurations. Let's briefly examine the steps involved in this modeling approach.

Method

STEP 1: Identify the Effect (Customer Need/Engineering Specification)

We first contemplate what the model is for. Customer needs mapped to the main function carriers and then mapped to engineering specifications (metrics) is the essence of this step.

STEP 2: Identify a "Flow"

We next identify a material, energy, or signal flow associated with each customer need/metric of the product concept. This identification corresponds to the flows in the product's functional model. It may be used to understand the input-output transformation of the product in the sense of control volume analysis.

STEP 3: Identify the Physical Mechanisms

But how is the identified flow converted to a desired state by the product? How will the environment affect this transformation? We begin to answer these questions in this step. Physical principles must be listed that govern the product functions.

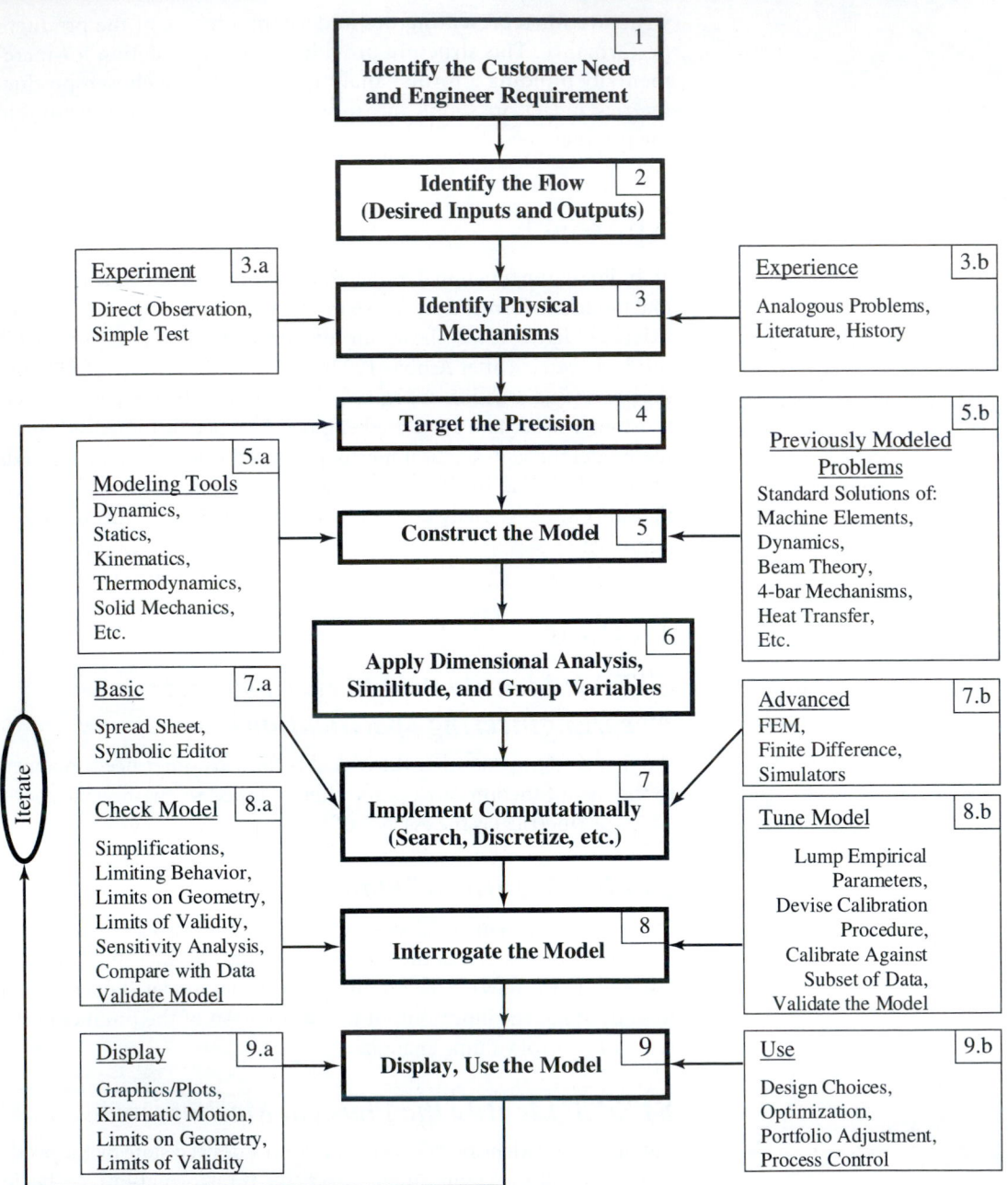

▼ *Figure 13.22.*
Advanced modeling approach for product metrics.

STEP 4: Target the Precision

With the metrics, flows, and physical mechanisms identified, the required precision of the model must be established. The precision must be chosen so that the model will produce effective results to make design decisions. Usually a percentage of the target values for the product metrics is used to set the precision of the product model.

STEP 5: Construct the Boundary and Balance Relationship

This step entails the establishment of system boundaries and the derivation of balance relationships that represent governing physical mechanisms. Basic mechanics from a variety of fields may be called on to create these balance relationships, equations, and boundary representations. It is also expected that the targeted precision is continuously referenced so that simplifications and assumptions may be made and justified (or not). This step concludes with an integration of all balance relationships and boundary conditions as a set of simultaneous equations or equivalent representation.

STEP 6: Apply Similitude/Dimensional Analysis

It is always a good idea to manipulate analytical product models into appropriate groups of dimensionless parameters whenever possible. Dimensionless parameters provide "scale lengths" for understanding the models and obtaining physical insights and trends. They also minimize problems encountered with unit conversions.

STEP 7: Embody the Model Computationally

The model should be converted to a computational form. This form may entail detailed finite element or difference models or other simulations. Alternatively, it may entail the simple coding of formulas in a spreadsheet or symbolic analysis. In either case, a computational form will provide a useful model for interrogation and design application.

STEP 8: Interrogate the Model

Once a model is constructed and integrated, it must be simplified, verified, and tested for its limits. Physical tests and known first-order solutions (such as beam theory) provide a sound basis for model interrogation. A subset of physical data (more than one point) must be collected to calibrate and validate the model. Unknown constants and model order must be adjusted during the calibration to improve the predictive capability of the model.

STEP 9: *Display and Use the Model*

Following interrogation, calibration, and validation, the model is ready for use. The design parameters must be clearly identified and bounded (if not already done so). The range of possible design configurations is then explored, followed by focused experiments/physical models to verify design choices and to uncover physical phenomena not included in the model.

This nine-step process augments and supplements the basic method of modeling product metrics. It adds keys steps such as interrogation and calibration of a model. The next section develops examples using this methodology.

VII. PRODUCT MODELS: CASES

A number of product cases are studied in this section, beyond the fingernail clipper and iced tea maker. These cases provide modeling examples for product metrics. They follow, generally, the advanced approach for model construction.

Electric Wok Product

This case considers the redesign of an electric wok. The wok product concept is shown in Figure 6.6. Having identified the customer needs for this redesign problem (Chapter 4) and the functional model (Chapter 6), we can identify measurable performance metrics and design variables in addition to the functional expressions relating these modeling parameters. We perform this identification using engineering analysis, as described by the model construction methods.

Based on the engineering specifications, the required modeling includes metrics as follows:

- steady-state temperature uniform in space
- steady-state temperature uniform in time
- weight
- volume
- handle temperature
- fast heat-up
- fast cool-down

Some of these metrics require no modeling at this stage, based on the new product architecture and component concepts. For example,

▶ We have selected an open loop controller that restricts the power to the wok. Therefore, by definition, the steady-state temperature uniformity (in time) is completely solved and is no longer a problem.

▶ The volume capacity is ensured by the wok bowl used, that is, a traditional wok.

▶ The handle temperature is a problem that can be best considered through analysis, even though a new handle design exists.

These analyses result in the following required modeling:

▶ steady-state temperature uniform in space
▶ weight
▶ handle temperature
▶ fast heat-up and cool-down

We now explore the development of models for these metrics.

Steady-State Temperature Uniform in Space

To develop a model of the temperature uniformity over the surface of the wok, we apply the model construction approach. We have our metric, the steady-state temperature. A "flow" for this metric (from our function structure and/or physical principle analysis) is heat flow (energy).

We need to define boundaries on the "system." The basic start-to-finish heat flow is from the heater, through space to the wok bowl, through the bowl, into the food, and out to the environment. Now what we must ensure is the uniformity across the wok surface. Therefore, we reduce the modeling scope to the wok bowl itself. We can model the heat flow from the heater as q entering the bowl. By changing where and how much q enters the system, we have a representation of changing the heaters. Figure 13.23 depicts a cross section of the bowl under these conditions.

There are three primary modes of heat flow (heat transfer) that we must consider:

▶ conduction, q_k, the heat flow through the bowl metal
▶ convection, q_h, the heat loss into the air
▶ radiation, q_e, the heat loss out to the environment by glowing

At this point, there are several options in modeling. One could develop a complete finite element heat transfer model of the wok and all its parts, splitting the wok up into many little bits and establishing heat transfer relationships between these tiny finite element bits. One could

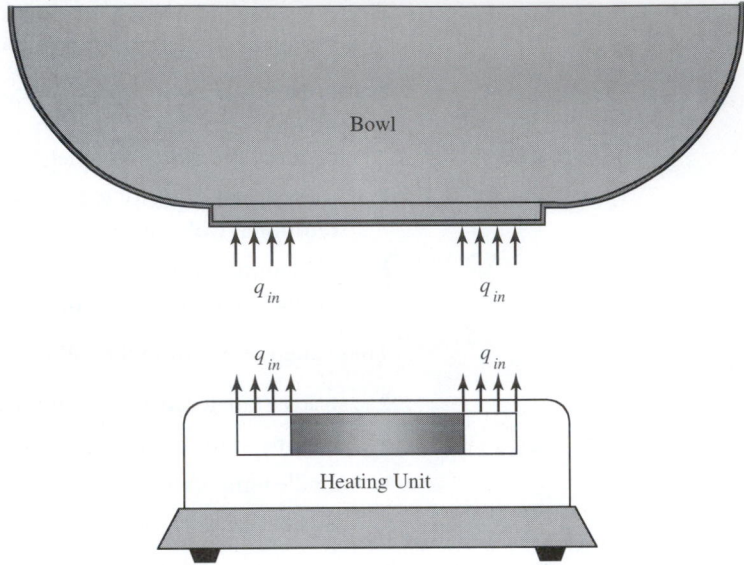

▼ *Figure 13.23*
Model boundaries for the wok bowl.

also develop a simple lumped-mass model, treating the wok as a single bit of material. Consider a development cycle where a quick model is needed but is minimally sufficient to cover the difference in temperature across the wok bowl face. In this case, we need to split the wok surface into different regions, either with a parameterizing variable such as radius and angle or with a finite discretization.

With our goal of a quick model, we need to create a simple representation for a first-order model. The first clear simplification is to model the wok bowl as a disk; this approach reduces a spatial component (Figure 13.24). This means we ignore heat rising effects.

From here, the next simplification is to observe that the disk is radially symmetric, and we can model the system one dimensionally along

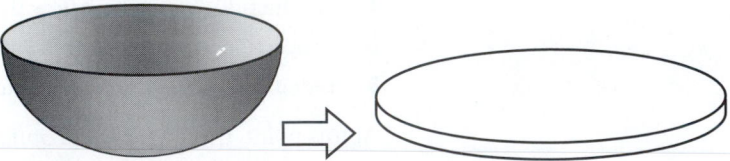

▼ *Figure 13.24.*
Simplified bowl geometry for heat transfer analysis.

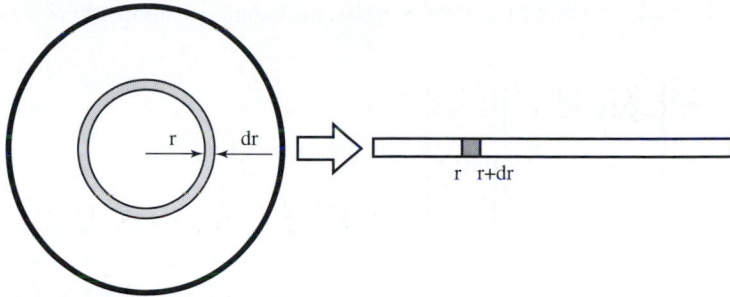

▼ *Figure 13.25.*
Radial symmetry of the wok bowl.

the radius (Figure 13.25). This means we ignore uneven angular mass distribution, such as due to the controls, and so forth.

Now we recognize that the heat fluxes define three different regions along the radial direction, as shown in Figure 13.26. Heat enters in region 2. In the others, heat only escapes the wok bowl metal. As a minimum, we must model each of these regions as a single lumped mass. For each of these regions, we can create a differential "sub" model to construct a heat balance across for steady-state conditions (Figure 13.27). Note that we also can refine each into a chain of finite-sized lumped masses.

Considering a classical heat transfer approach, we know that the following equations hold:

$$q_k = kA\frac{dT}{dx}$$

$$q_h = hA(T - T_\infty) \qquad (13.23)$$

$$q_\varepsilon = \varepsilon\sigma A(T^4 - T_\infty^4)$$

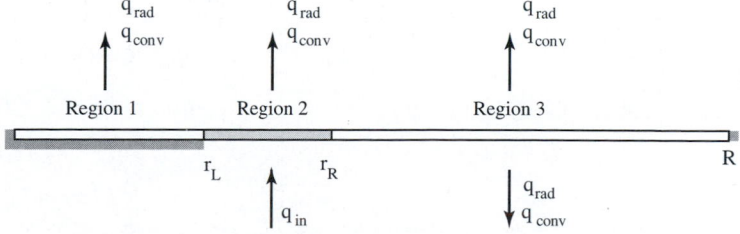

▼ *Figure 13.26.*
Heat transfer model for the electric wok product.

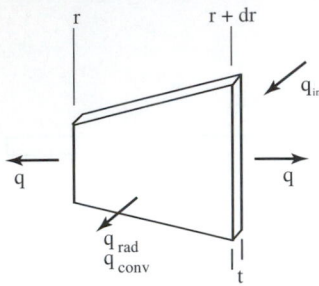

▼ Figure 13.27.
Heat transfer submodel of a
piece of the wok bowl.

Now, the balance relationship is (based on conservation):

$$\sum_{\text{region}} q = 0 \qquad (13.24)$$

at any position.

To apply this balance relation, it is necessary to model the wok geometry. We can satisfy this need (using, say, the set of variables in the bill of materials) by measuring distances to the top and outer edge from the center of the wok and approximating the wok by a cone:

$$r_{\max} = \sqrt{(OD/2^2) + H^2}$$

The outer diameter is 18 cm, and the height is 10 cm, so $r_{\max} = 20.59$ cm, an approximation.

Lumped Mass Analysis

A straightforward method to analyze the model depicted in Figure 13.26 is to decompose the geometry into three regions:

▶ center region

▶ middle region

▶ outer region

We now model each of these regions as lumped thermal masses.

The center region:

$$q_{\text{rad}} = \varepsilon\sigma\pi r_L^2(T_c^4 - T_\infty^4)$$
$$q_{\text{conv}} = h\pi r_L^2(T_c - T_\infty) \qquad (13.25)$$
$$q_L = k2\pi r_L t(T_M - T_c)$$

and

$$q_L - q_{\text{rad}} - q_{\text{conv}} = 0 \qquad (13.26)$$

The middle region:

$$q_{\text{rad}} = \varepsilon\sigma\pi(r_R^2 - r_L^2)(T_M^4 - T_\infty^4)$$
$$q_{\text{conv}} = h\pi(r_R^2 - r_L^2)(T_M - T_\infty)$$
$$q_L = k2\pi r_L t(T_M - T_c) \qquad (13.27)$$
$$q_R = k2\pi r_R t(T_M - T_c)$$

and

$$q_{in} - q_R - q_L - q_{\text{rad}} - q_{\text{conv}} = 0 \qquad (13.28)$$

The outer region:

$$q_{\text{rad}} = 2\varepsilon\sigma\pi(r_R^2 - r_L^2)(T_o^4 - T_\infty^4)$$

$$q_{\text{conv}} = 2h\pi(r_R^2 - r_L^2)(T_M - T_\infty) \qquad (13.29)$$

$$q_R = k2\pi r_R t(T_M - T_o)$$

and

$$q_R - q_{\text{rad}} - q_{\text{conv}} = 0 \qquad (13.30)$$

This decomposition creates three equations for three unknowns, (T_c, T_M, T_o), which we can solve simultaneously for the steady-state temperatures.

Transient Heat Transfer Equations

The minimum heat-up time requires a transient heat transfer analysis. We can again use a three-region lumped-mass model. The difference is that the left side of Eqs. (13.25)–(13.30) do not equate to zero, but to:

$$\sum q_i = \rho V c_p \frac{dT}{dt} \qquad (13.31)$$

where ρ is the material density, V is the region volume, and c_p is the heat capacity. We now need to solve this equation as a nonlinear ordinary differential equation in time. We can do this numerically, using a discretized variable p to represent time t. This equation can be solved iteratively for a time step $dt \cong \Delta p$:

$$T_{p+1} = T_p + \frac{dT}{dt}\Delta p \qquad (13.32)$$

for each region (for each lumped thermal mass). At some discrete time p, the T_p will become indistinguishable from the steady-state temperature.

Weight

The weight can be modeled as the mass, and summing the component masses:

$$W = \sum \rho V \qquad (13.33)$$

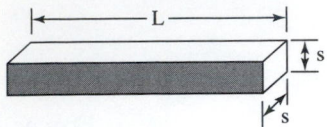

▼ **Figure 13.28.**
Simplified geometry of the wok handle.

Handle Temperature

The handle temperature can be modeled from the steady-state heat transfer. Figure 13.28 depicts a simplified geometry for the wok handle.

The heat transfer out of the system is due to convection, and the heat transfer into the system is due to conduction:

$$q_{\text{out}} = h4Ls(T_k - T_\infty)$$
$$q_{\text{in}} = ks^2(T_o - T_k) \tag{13.36}$$

After first solving for T_o above, equating these expressions is a simple approximation for the handle temperature.

Model Calibration

We must calibrate the engineering (mathematical) model with real data. Based on classical heat transfer literature, we see that for aluminum:

k	200 W/m°C
ε	0.8
h	4.4 W/m^2
ρ	2800 kg/m^3
c_p	870 J/kg°C

$$(13.37)$$

Steady-State Temperature

The steady-state model reflects the average behavior with respect to time. We must determine the q_{in} from the heater. Considering the measured steady-state temperature profiles above the heating coil (example shown in Figure 13.29), we see that in steady state, about 1/3 of the time is devoted to increasing the temperature (heating/power required). Thus, for the recorded data and temperatures, an equivalent 33% duty cycle on the 1000 W rating applies. The 1000 W is surely high; more likely, the power consumption is 700 W, or again 70%. Further, since conduction also occurs through the nonmodeled bottom material, a further reduction will occur, probably 30%. These approximations result in an expected equivalent duty cycle of 16%.

At this condition, the measured center temperature is about 210°F, the mid-temperature is about 220°F, and the outer temperature is about 160°F. We must validate these results with our model and adjust the constants until the model is reasonably calibrated with the known data. Then, as a validation of the model, we need to change the conditions, measure the results, and compare this with the prediction.

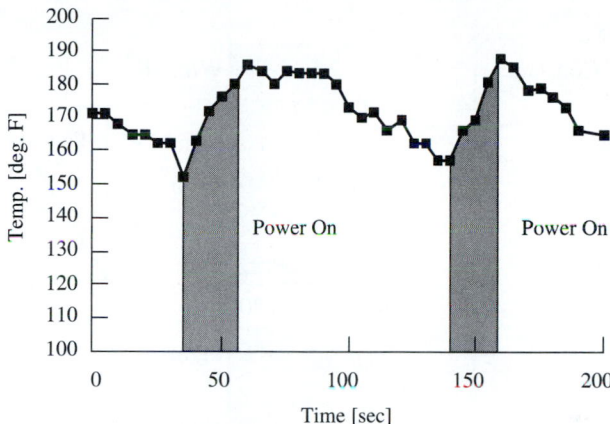

▼ *Figure 13.29.*
Steady-state temperature profile for the wok surface above the heater.

The problem in entering these data is to determine the temperatures that make the equations hold. We can complete this task with a spreadsheet solver, by creating a variable ‖ sum q ‖ that we can minimize to zero, using an unconstrained search (using the solver) across the temperatures. We can then adjust k, h, and the duty cycle (or in fact any of the constants we feel we are unsure of, within their domain of reasonable uncertainty) until these temperatures reflect the measured ones.

Carrying out this approach for the steady-state heat transfer model (Figure 13.30), we observe that reasonable agreement is obtained with the original estimates of the constants. However, the predicted temperatures are a bit high compared to the measured ones. We can lower the duty cycle slightly until the temperatures agree. A duty cycle of 12% makes the model temperatures agree very well with the measured data.

Transient Heat Transfer

We now construct a transient heat transfer model in a similar fashion. This model is slightly different, however, in that we are numerically simulating the time domain differential equation. Unless care is taken, this simulation will cause some implementation difficulties; differential equations can be difficult in spreadsheets. They need not be.

Our basic equation is

$$\sum q_i = \rho V c_p \frac{dT}{dt},$$

(13.38)

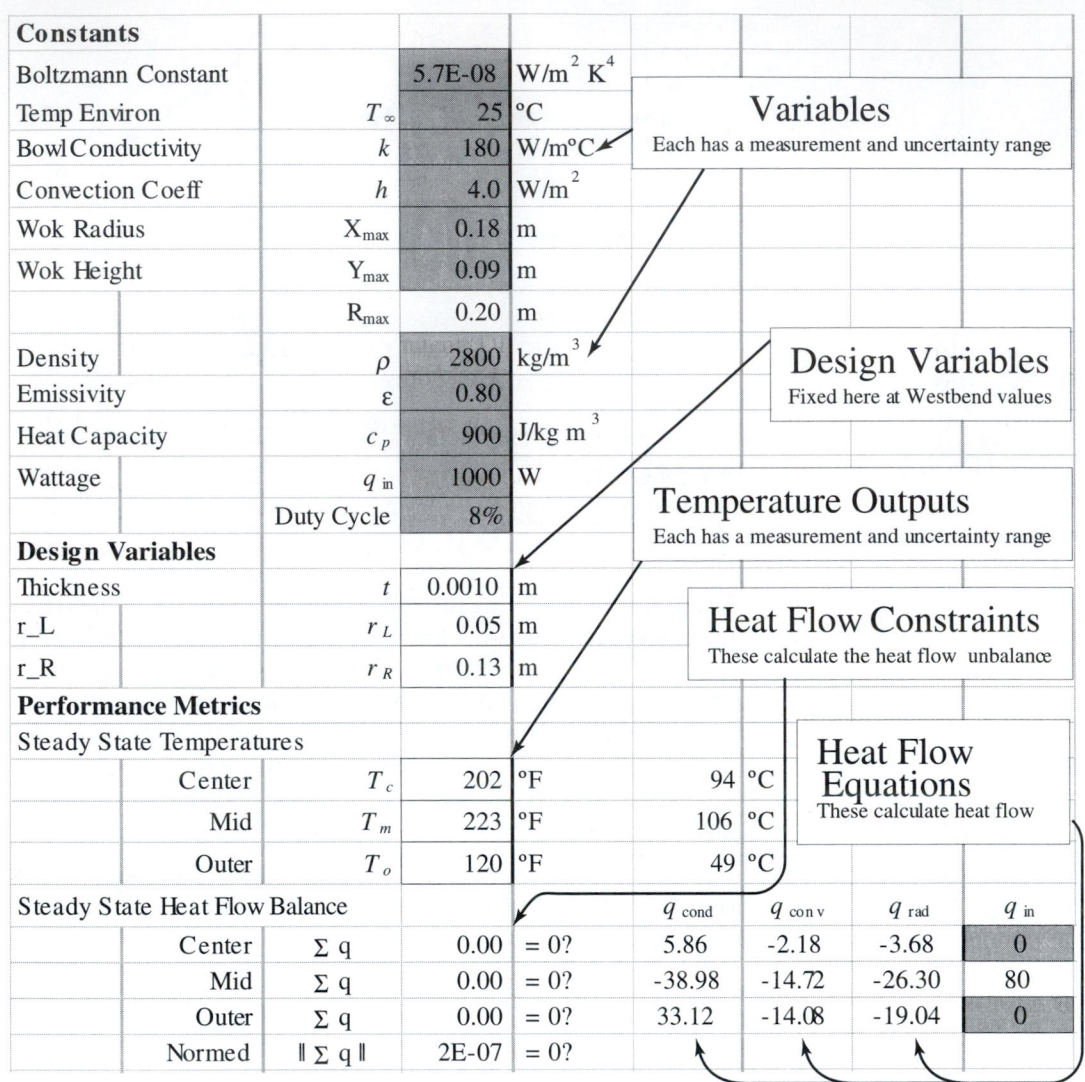

Constants			
Boltzmann Constant		5.7E-08	W/m^2 K^4
Temp Environ	T_∞	25	°C
Bowl Conductivity	k	180	W/m°C
Convection Coeff	h	4.0	W/m^2
Wok Radius	X_{max}	0.18	m
Wok Height	Y_{max}	0.09	m
	R_{max}	0.20	m
Density	ρ	2800	kg/m^3
Emissivity	ε	0.80	
Heat Capacity	c_p	900	J/kg m^3
Wattage	q_{in}	1000	W
	Duty Cycle	8%	

Variables
Each has a measurement and uncertainty range

Design Variables
Fixed here at Westbend values

Temperature Outputs
Each has a measurement and uncertainty range

Heat Flow Constraints
These calculate the heat flow unbalance

Heat Flow Equations
These calculate heat flow

Design Variables			
Thickness	t	0.0010	m
r_L	r_L	0.05	m
r_R	r_R	0.13	m

Performance Metrics							
Steady State Temperatures							
Center	T_c	202	°F		94	°C	
Mid	T_m	223	°F		106	°C	
Outer	T_o	120	°F		49	°C	

Steady State Heat Flow Balance				q_{cond}	q_{conv}	q_{rad}	q_{in}
Center	Σ q	0.00	= 0?	5.86	-2.18	-3.68	0
Mid	Σ q	0.00	= 0?	-38.98	-14.72	-26.30	80
Outer	Σ q	0.00	= 0?	33.12	-14.08	-19.04	0
Normed	$\| \Sigma$ q $\|$	2E-07	= 0?				

▼ *Figure 13.30.*
Spreadsheet to calibrate and validate the wok heat transfer model.

with $T(0) = T_\infty$. We can restate the differential equation as a first-order system:

$$\dot{T} + f(T) = 0 \tag{13.39}$$

We numerically approximate this relationship with a time step $\Delta p \sim dt$, using Euler's method. Thereby,

$$T_{p+1} = T_p - f(T_p)\Delta p \tag{13.40}$$

We have introduced three new constants compared to the steady-state model. These are ρ, V, and c_p. Of these, we don't have much confidence in c_p. While we do have tabulated values for aluminum, we are not quite sure the grade of this aluminum nor the impact of the surface materials compared to the thin material. We can adjust the c_p value within this domain of uncertainty until the model agrees reasonably with our measurements or revisit the varying of both the transient and steady-state models together.

The next difficulty is the time step size Δp. This time step is a concern due to the model accuracy that we can control. We want Δp to be as large as possible, where the results are still meaningful. If Δp is too large, the differential equation behavior will oscillate due to numerical instability. If Δp is too small, we are performing excess calculations and wasting time. The largest possible Δp, however, changes with the differential equation values.

The next difficulty is choosing the stopping condition for the calculations and implementing this in a spreadsheet. We seek the time it takes for the wok to attain a desired temperature. This objective requires that we determine the time until the predicted temperatures are "close" to the final steady-state temperatures. In a spreadsheet, the basic equation (Eq. 13.40) is implemented on a separate row for each time step. Then, for each other time increment, the row can be copied down to lower rows. Each subsequent row is a new time increment. Then time is a question of how many rows of the spreadsheet we use. We can simply add a column that flags when the condition of "close enough" to steady state is met, such as the change in temperatures being within a tolerance. When the flag changes, that time point t_p is the time to steady state. Rows beyond the flip from zero to one in the flag column are extra computations not really needed but are required for a spreadsheet implementation.

We must again calibrate the model with the actual measured data. We know the wok reaches the final temperature in about 3–5 minutes, based on measurements. (An example is shown in Figure 13.31.) A sample calibration is shown in Figure 13.32.

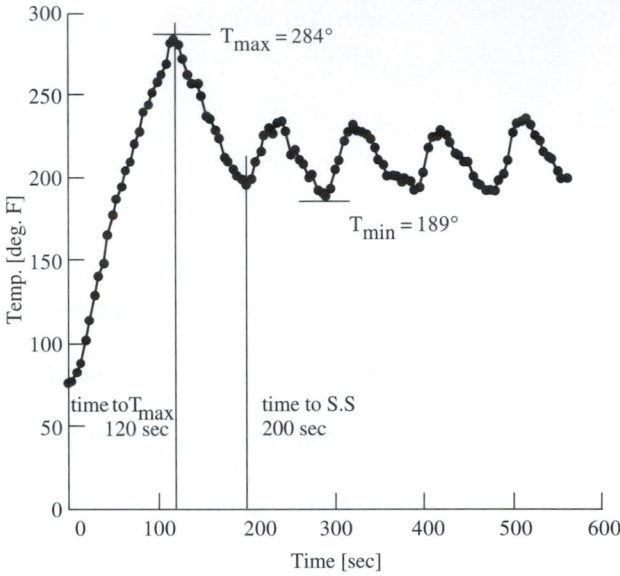

▼ *Figure 13.31.*
Measured wok temperature as a function of time.

Other Metrics to Integrate a Complete Model

We can combine these submodels into a complete engineering model of the wok. The wok bowl weight is very simple to include. The handle temperature can be derived from the previous modeling as:

$$T_h = \frac{k_h s T_o + 2h_h L^2}{k_h s + 2h_h L^2} \tag{13.41}$$

We can calibrate h_h and k_h until T_h agrees with the physical measurements. Note how this model ignores radiation effects, which would make the equation impossible to separate, and further presumes the heat draw off the outer ring into the handle is negligible with respect to the outer ring.

Integrated Model

Based on these results, we can combine the models into a representation of the product metrics, that is, a single integrated design model. Here we no longer keep the design variables fixed in Figures 13.30 and 13.32, but we change them to better performing values.

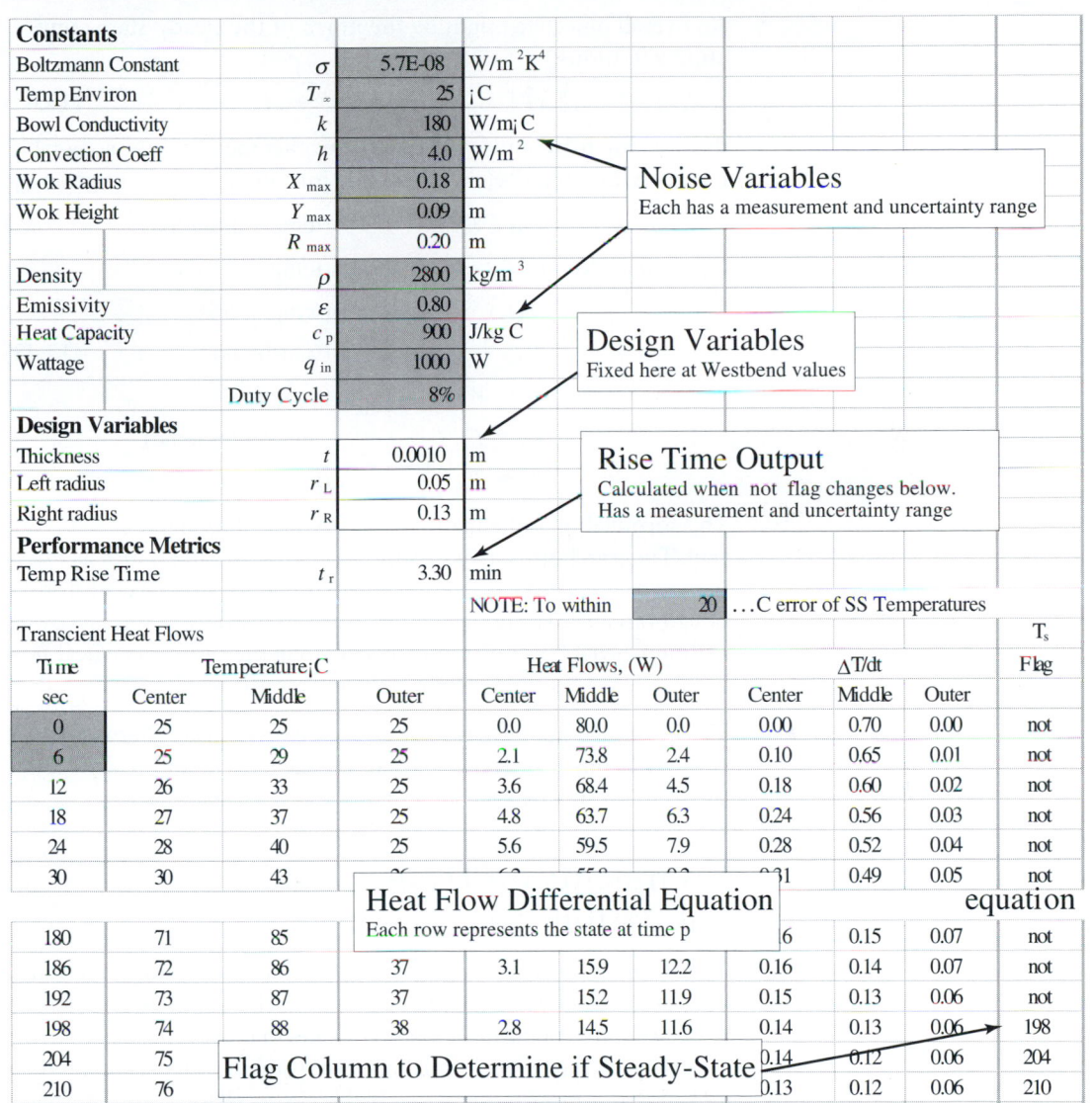

▼ **Figure 13.32.**
Calibration spreadsheet for the wok's transient heat transfer.

An overall objective might be the norm of the steady-state temperature variations:

$$\|\Delta \vec{T}\| = \sqrt{(T_m - T_c)^2 + (T_m - T_o)^2} \qquad (13.42)$$

that we want to minimize, subject to the other metrics being acceptable. Based on this objective, the integrated formulation might be given by:

$$
\begin{array}{lll}
\text{minimize} & \|\Delta \vec{T}\| & \text{(temp variation)} \\
\text{subject to} & t_r \leq t_\tau & \text{(temp rise time)} \\
& |T(t_r) - T_{ss}| \leq \varepsilon & \text{(temp at SS)} \\
& \dot{T}(t) + f(t) = 0 & \text{(transients)} \\
& W \leq W_\tau & \text{(weight)} \\
& T_c \geq T_{\tau c} & \text{(min center temp)} \\
& T_h \leq T_{\tau h} & \text{(max handle temp)}
\end{array} \qquad (13.43)
$$

This formulation is the mathematical representation of our wok concept. The search space is across the geometric variables, such as thickness and radii. In fact, due to our formulation, we have additional design variables. The nodal temperatures are also design variables, with additional equality constraints representing the steady-state heat flow balances. We seek these steady-state temperatures.

The integrated model may now be used to investigate alternative wok configurations. Chapter 16 discusses theoretical and practical approaches for solving such a model.

Comments on Design Model Validation

The astute reader should note that the wok model above is calibrated but not validated. That is, within a range of believability, we selected the preferred values of the noise variable values that fit the data on both the input and output model variables. From sources such as handbooks and/or direct measurement, we had an expectation on a reasonable value for each input and a range we could comfortably adjust the value, if by doing so we could better fit the output data.

This approach is acceptable, provided we then subsequently test the model as a *validation* step. That is, we change a different variable in the model and make a prediction. Then we test the prediction by measuring the physical system in this new configuration. In the wok example, we should now change the design variables and make a prediction of the output temperature behavior. If the prediction agrees, we have a validated model.

Notice the problem, however. To validate the model, we have to build a new configuration. If we have to construct a detailed prototype, it might as well be the optimal one. That is, the validation of the model is whether it provides a better performing product. If it fails, then we need to investigate further modeling, just like any other scientific model validation procedure.

The point is that in the limited resource environment of product development, building the optimized result usually validates a model. We only know the model is valid when the result is a success. Perfectly validated models are a luxury that design teams rarely have.

Other validations might be pursued at less cost. For example, with the wok, one could raise the power-input level and validate the increase in wok surface temperature. However, this test does not say what will happen when the design variables are changed, which is what we are interested in. The model may still be inaccurate. We might test a reduced cost prototype, if time is available for construction, and the differences in the prototype from the product do not impact the physics to be validated. Rapid prototyping and similitude techniques aid in the development of such prototypes, as discussed in Chapter 17.

Model validation is important but often expensive. As with all design process activities, a design team will have to make a cost/benefit consideration and choose how much validation it makes sense to complete.

VIII. SUMMARY AND "GOLDEN NUGGETS"

In this chapter, basic and advanced product modeling techniques are developed. These methods are cast in the light of model preparation, formal and informal models, and analytical versus physical models. They are also presented in the context of customer needs and functional modeling. This context ensures a level of consistency for all model construction activities.

After studying this chapter, it is hoped that the reader gains an appreciation for modeling and its power in product development. Modeling is the fodder that transforms concepts to the beginnings of reality, the glue that infuses our insights into the product reality, and the engineer's toolbox that we bring to bear on the challenges of product evolution. Product modeling thus should be our passion. Without modeling, our innovative ideas may be lost in the swamp of implementation and product fabrication. Other novel ideas may never be discovered due to a lack of knowledge and understanding of our product system. We should not model for the mere sake of modeling, but rather abstract a product concept to fulfill its potential.

"Golden nuggets" from this chapter include the following:

▶ Product models are abstractions of physical reality. As abstractions, they require creativity and perseverance, but they also simplify critical product decisions.

▶ Model preparation wields the product data generated in earlier phases of the development process. These data reduce modeling cycle times and help ensure valid metrics and physical understanding.

▶ When considering a model, there always exists a trade-off between analytical and physical models (prototypes). This trade-off, at the most basic level, comes down to model accuracy versus prototype expense, that is, "$$"!

▶ Informal models are abundant as product descriptions. From customer statements to graphical sketches, informal models describe the essence of what a product should be. It is the design team's challenge to formalize the informal descriptions so that repeatable and defendable product decisions may be made.

▶ Metrics, boundaries, and balance relationships form the core modeling concepts. These concepts transform a product concept to a parameterized form. Calibration and focused experiments extend this form to a useable framework.

▶ Product modeling applications add to our experience and modeling toolbox. We must file away this experience for future development of product models.

References

Ashby, M. F. 1992. Physical modeling of material problems. *Materials Science and Technology* 8: 102–111.

Ashby, M. F. 1987. Technology of the 1990's: Advanced materials and predictive design. *Philosophical Transactions of the Royal Society of London* A322: 393–407.

Beaumont, P. 1998. Predictive design and certification procedure: Modeling fracture stress, fatigue strength, . . . of composites and structures. Tech. Rep. No. ISSN 0309-6505, Dept. of Engineering, University of Cambridge, Cambridge, U.K.

Tillberg, M. 1997. Why shall black tea draw 3-5 minutes? *Internet*, 10 April. Available at *http://www.vtek.chalmers.se:80/~v92tilma/tea/prep/draw.html*.

Ulrich, K., and Eppinger, S. 1995. *Product design and development*. New York: McGraw-Hill.

14 Design for Manufacture and Assembly

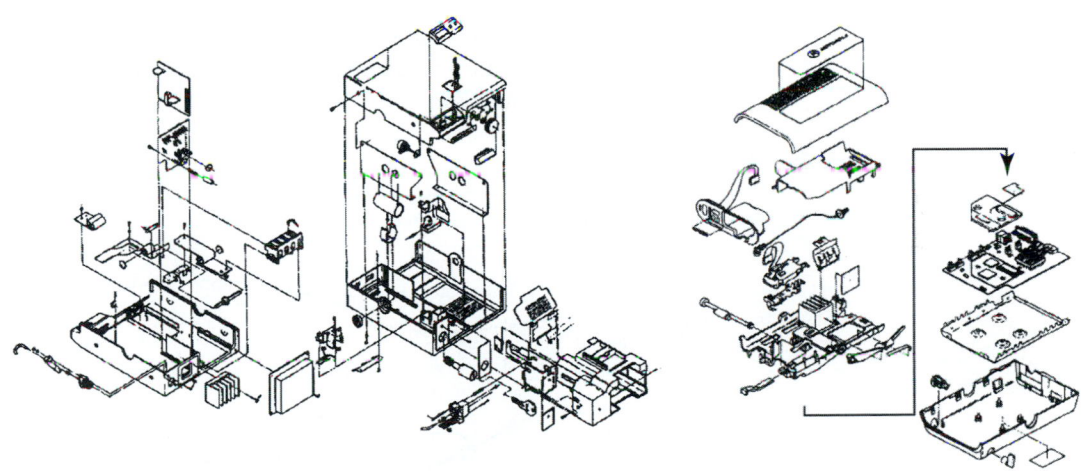

In the late 1980s, Motorola introduced a new vehicle-mounted remote radio for base-remote communication. Their product at the time, the MX Converta Com™, had 217 distinct parts requiring over 2700 seconds of assembly time. Motorola redesigned the product using design-for-manufacturing and assembly methods and reduced the part count to 97 with 1350 seconds of required assembly time, an 87% reduction in direct cost. The number of screw fasteners was reduced from 72 to zero. (Figures from Branan 1991.)

Acommon failure in product development is making products that work but that are also very difficult to build. Difficulty in manufacture makes a product expensive—it is hard to fabricate, takes extra time and is unreliable—the requested geometry is hard to do and requires extra care in production that is hard to maintain. *Design for manufacture and assembly* is the analysis and redesign of a product or concept to make it easier to produce.

I. CHAPTER ROADMAP

In this chapter, we first discuss why design for manufacture is important (Figure 14.1). Then we present basic design guidelines that one can apply to simplify a design in assembly and in piece-part production. Then we discuss manufacturing cost analysis to determine what portion of a product best serves as an area on which to focus. We present a basic part costing method based on a comparison with benchmark parts and a basic assembly cost method using the Xerox producibility analysis. Then more advanced methods are discussed, including Boothroyd and Dewhurst's method. Finally, a comparison among different design for manufacture and assembly methods is presented.

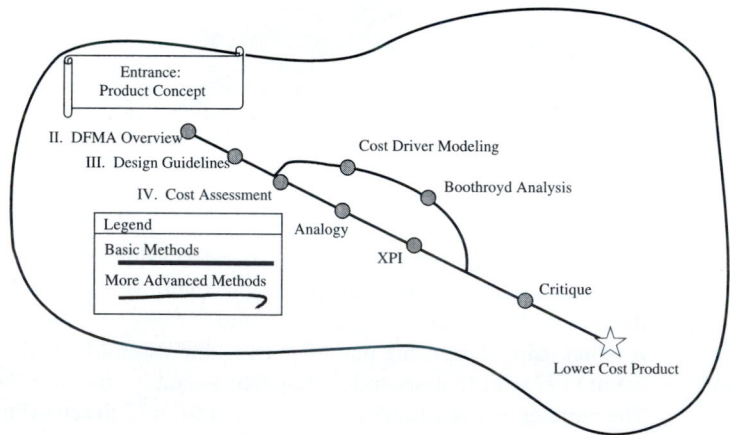

▼ *Figure 14.1.*
Chapter roadmap.

II. OVERVIEW AND MOTIVATION

Design for manufacture and assembly analysis and synthesis is, as with many design process methods, applicable during many phases of a product design process. It can be used in benchmarking analysis as in Chapters 6 and 7, in simplifying new concepts not yet built, and in simplifying fully embodied designs.

There are two components, design for manufacture and design for assembly. *Design for manufacture* (DFM) entails making piece parts easier to produce from raw stock. For example, one can make a plastic part easier to injection mold by changing its draft angle—the angle formed by the difference in wall thickness from the part at the inside of a mold compared with the wall thickness at the end of the mold. Small draft angles make for difficult part ejection. Design for manufacture involves application of part-forming models, whether they are basic rules, analytic formulas, or complex finite element process simulations.

Design for assembly entails making attachment directions and methods simpler, for example, making a part easy to attach by using snap fits instead of machine screws. Design for assembly involves application of attachment time and complexity models, whether they are basic rules, tables based on simplified time studies, or full-time and motion industrial engineering studies.

Design for manufacture and assembly is important for design because it has three beneficial impacts. First and foremost, it reduces part count. From that, it thereby reduces cost. If a design is easier to produce and assemble, it can be done in less time, and so it is cheaper. Design for manufacture and assembly should be used for that reason if no other.

On the other hand, as a second beneficial impact, consider products that are used in extremely critical applications and where cost is basically not an issue. Satellites used in exploratory NASA interplanetary missions, for example, have systems that must work; saving even thousands of dollars on assembly or part cost is meaningless. What is important, however, is reliability. Any design activity that can increase reliability will be applied, and this again is a benefit of design for manufacture and assembly. For example, Motorola conducted a study showing that application of design for manufacture and assembly reduced their failure rates as shown in Figure 14.2. The reason behind this increase in reliability is basically that if the production process is simplified, then there is less opportunity for outright errors.

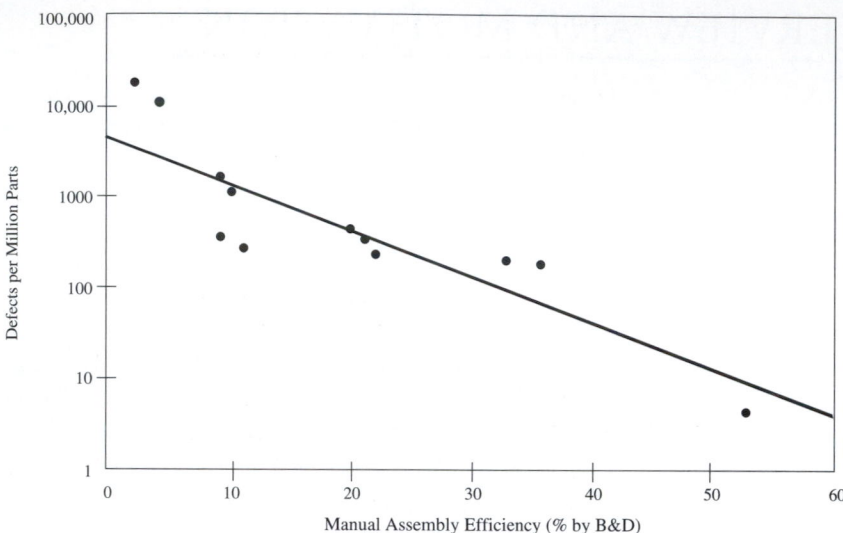

▼ *Figure 14.2.*
Increase in reliability with application of DFA at Motorola (from Branan 1991).

Finally, design for manufacture and assembly also generally increases the quality of a design, basically for the same reason as why it increases reliability (Branan 1991; Barkan and Hinckley 1993). If a part is easier to produce, less machine capability is required to achieve the same tolerances. A part that is easy to injection mold does not require extra tight process control of pressure, time, and temperature to achieve dimensional tolerances. Conversely, for the same machine capability, the easier to produce design will have tighter tolerances.

III. BASIC METHOD: DESIGN GUIDELINES

The most basic approach to design for manufacture and assembly is to apply design guidelines. After developing a design concept, one should examine it on each of the design guidelines and change the design to make it satisfy the guideline.

A note on design guidelines: Like all guidelines, they are heuristics that generally hold true. To every rule there are exceptions, and that holds true here as well. One should use design guidelines with an understanding of the design goals, and ensure that application of the guideline improves the design concept on those goals.

Design for Assembly

Table 14.1 lists DFA guidelines, adapted from several sources such as Andreasen (1983), Baldwin (1966), Digital (1993), Huthwaite (1990), Iredale (1964), and Xerox (1986). If a concept is compatible with these guidelines, one can be reasonably assured that the design will fare well in the subsequent more detailed analysis.

Each of these guidelines will be further explored by breaking them down into categories. First, system design guidelines that reduce assembly can be considered. Next, for the parts that are required, one can consider guidelines to ease handling. Then guidelines for easing the assembly insertion of the parts can be considered. Finally, guidelines for easing the actual attachment can be considered. We now explore the design for assembly guidelines in that order.

TABLE 14.1. DFA GUIDELINES

 1. Minimize part count by incorporating multiple functions into single parts. (Iredale 1964)
 2. Modularize multiple parts into single subassemblies. (Crow 1988)
 3. Assemble in open space, not in confined spaces. Never bury important components. (Tipping 1965)
 4. Make parts to identify how to orient them for insertion. (Tipping 1965)
 5. Standardize to reduce part variety. (Tipping 1965)
 6. Maximize part symmetry. (Iredale 1964; Paterson 1965)
 7. Design in geometric or weight polar properties if nonsymmetric. (Tipping 1965)
 8. Eliminate tangly parts. (Iredale 1964; Tipping 1965)
 9. Color code parts that are different but shaped similarly.
10. Prevent nesting of parts. (Iredale 1964; Tipping 1965)
11. Provide orienting features on nonsymmetries. (Iredale 1964; Tipping 1965)
12. Design the mating features for easy insertion. (Iredale 1964; Tipping 1965; Baldwin 1966)
13. Provide alignment features. (Baldwin 1966)
14. Insert new parts into an assembly from above. (Tipping 1965)
15. Insert from the same direction or very few. Never require the assembly to be turned over. (Tipping 1965)
16. Eliminate fasteners. (Iredale 1964)
17. Place fasteners away from obstructions.
18. Deep channels should be sufficiently wide to provide access to fastening tools. No channel is best.
19. Providing flats for uniform fastening and fastening ease.
20. Proper spacing ensures allowance for a fastening tool.

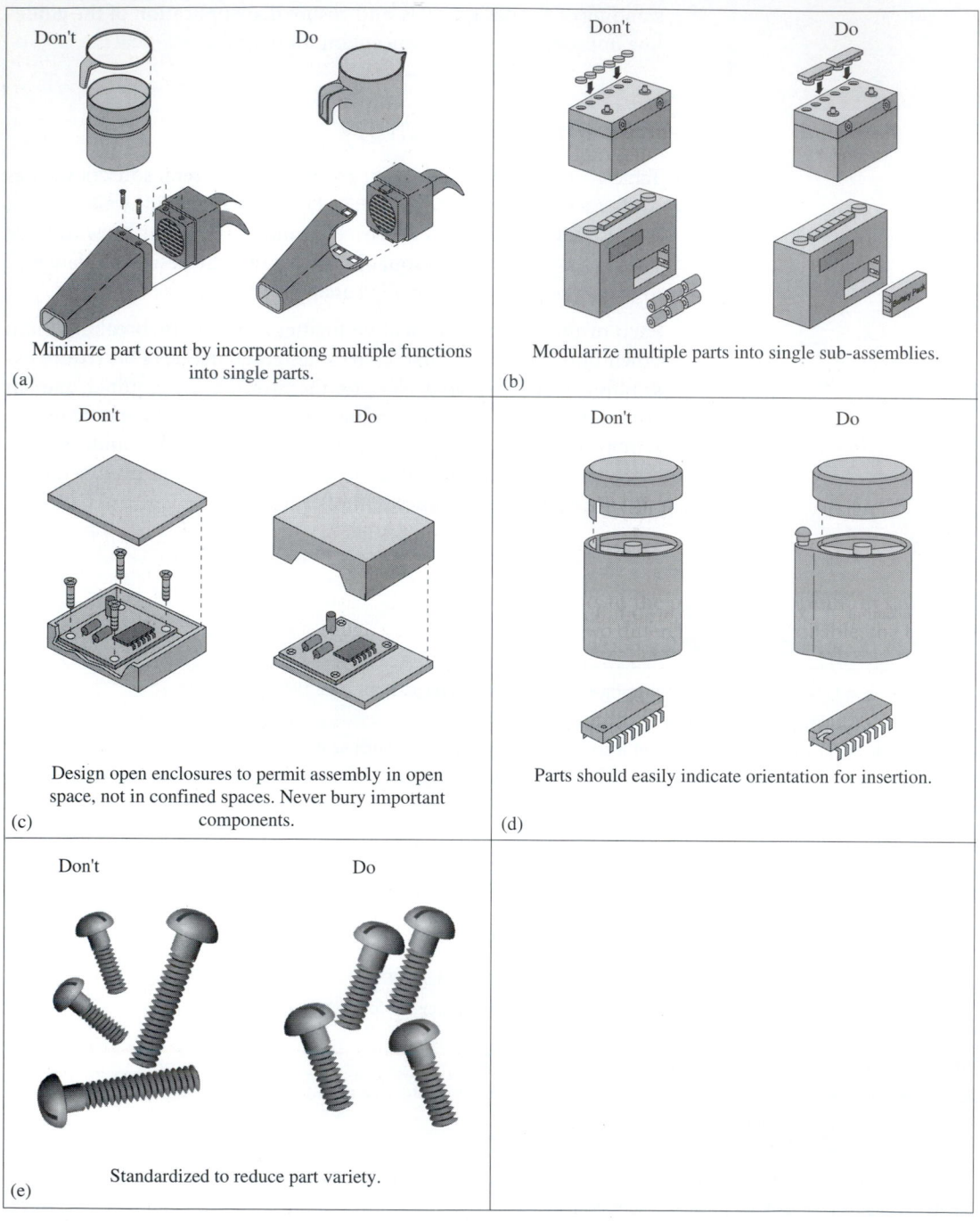

▼ *Figure 14.3.*
Design for assembly system guidelines.

System Guidelines

The first guidelines reduce the number of parts through functional modularity (Figure 14.3) as discussed in Chapter 9. Examine each part, and ask how the part function can instead be completed by a neighboring part. One might also be able to fabricate several parts as one part by using other fabrication processes, such as sheet-metal forming or plastic injection molding (force-flow analysis, Chapter 6).

The second guideline is to reduce the number of parts through assembly modularity as discussed in Chapter 9. Here, several difficult to manipulate parts are bundled together onto a feature such as a board that is easy to manipulate and assemble. Several side benefits are also gained through subsystem modularity. For example, assembly modularity can help reduce defects, as it makes quality problem identification easier when one can test subassemblies rather than having to diagnose the whole product.

The third guideline suggests that one should design a product so that it is assembled outwardly. Do not design a product that requires parts to be fastened on the inside of an enclosure. This makes assembly possible with no reorientations and without having to cram one's hands or tools into tight spaces.

The fourth guideline suggests designing parts so that they are easily oriented. Parts should have self-locating features so that precise alignment by the assembly process is not required. If not that, colored tick marks or indents make orientation easier. This is especially true on electrical components, where one way pin patterns or pin identification labels should be used.

The fifth guideline suggests reducing the variety of parts. In particular, using the same commodity items such as fasteners can avoid errors. It also increases the economies of scale for the part.

Handling Guidelines

All parts must be handled to be assembled. Handling includes picking the part up from a feed location, conveying it to a location for assembly insertion, and orienting the part for the assembly insertion. Some parts are difficult to handle; for example, springs and wires are hard to handle. Guidelines can be developed to help simplify this handling aspect of the assembly process as shown in Figure 14.4.

The first guideline suggests maximizing part symmetry to make orientation unnecessary. If one cannot make parts symmetric, then force the asymmetry to have an obvious asymmetry—plainly mark the orienting features.

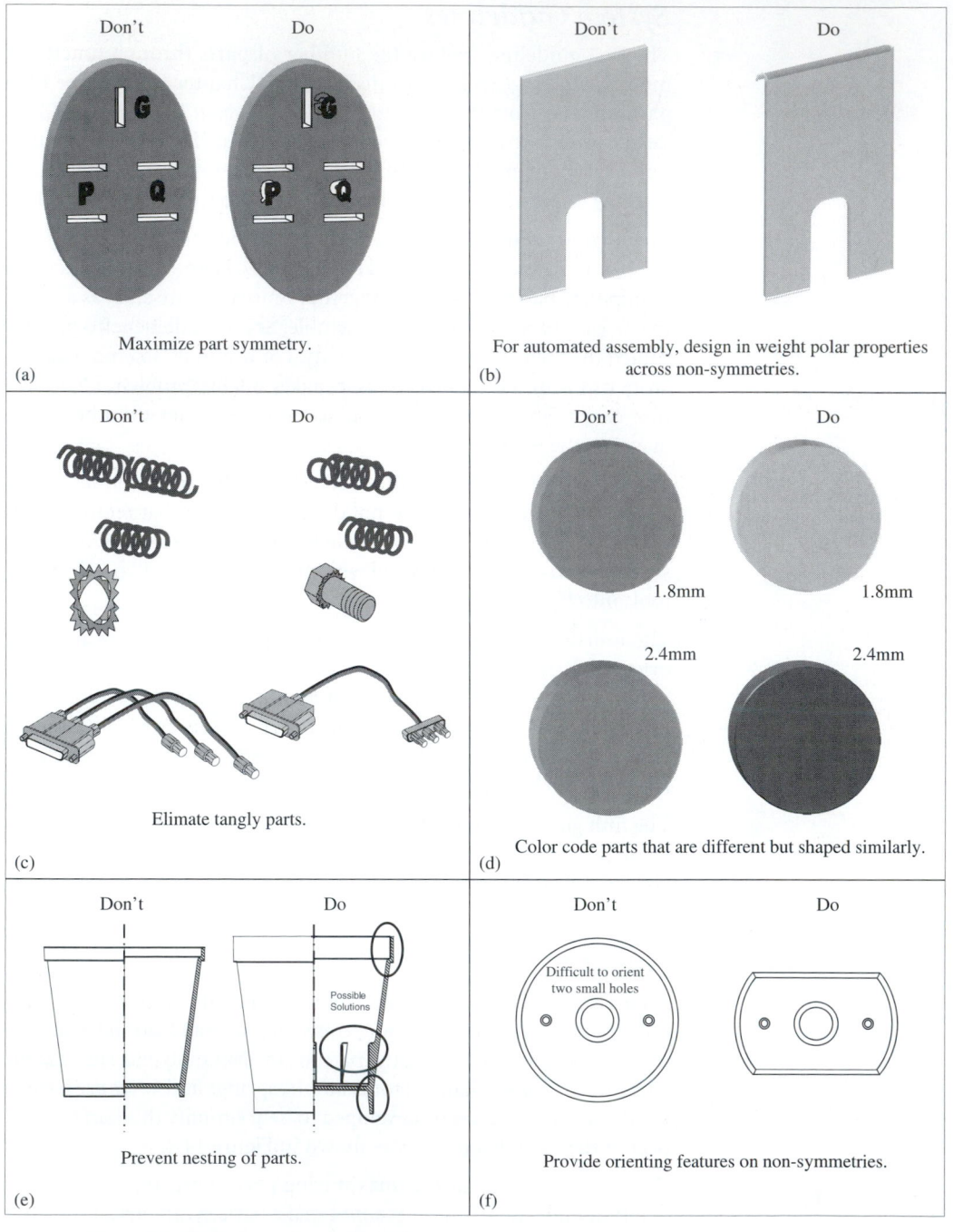

▼ Figure 14.4.
Design for assembly handling guidelines.

The second guideline suggests augmenting asymmetries such that they can be easily oriented. Specifically for automated assembly equipment, often this can be done with the aid of gravity, though gravity does not help orienting parts manually.

The third guideline suggests designing parts so they do not tangle or stick together. Scrutinize the tangle-prone feature, and change it so that it does not tangle. Similarly, parts that are difficult to handle should be changed to make them easier to handle. Slippery or messy parts should have handling features designed in.

The fourth guideline suggests distinguishing different parts that are shaped similarly through nongeometric means. One should color code or mark different thickness shims, for example, to ease identification.

The fifth guideline suggests avoiding parts that nest. *Nesting* is when parts that are stacked on top of one another clamp to one another, such as with vacuum formed plastic coffee lids or cups found in convenience stores. One can design in features to prevent the nesting.

The sixth guideline suggests designing parts with orienting features so that one can align the asymmetries inherent in the part. Often the asymmetric aspects of a part do not necessarily provide a means for alignment.

Insertion Guidelines

After being handled, the part must be inserted into the partially assembled product. Guidelines can be developed for this process in particular as shown in Figure 14.5.

The first guideline suggests adding chamfers to make parts easier to insert. One should design in allowances on all fits, so that if there is variation from part to part, it does not prevent assembly.

The second guideline suggests providing alignment features on the assembly so new parts are easily oriented without measurement. One should do this using a kinematic attachment scheme, such as the *3-2-1 alignment process* as discussed in Chapter 12. Here, first we provide "3" points on the assembly that a new part is placed against. The part is slid along the three points up against "2" more points that are in a perpendicular plane on the assembly. Then the part is slid along the five points up against a final sixth ("1") point, thereby kinematically constraining the new part into the assembly in a predictable way. Also, the geometry defining these six points is candidate geometry for tighter tolerance control compared to other points on the part and assembly.

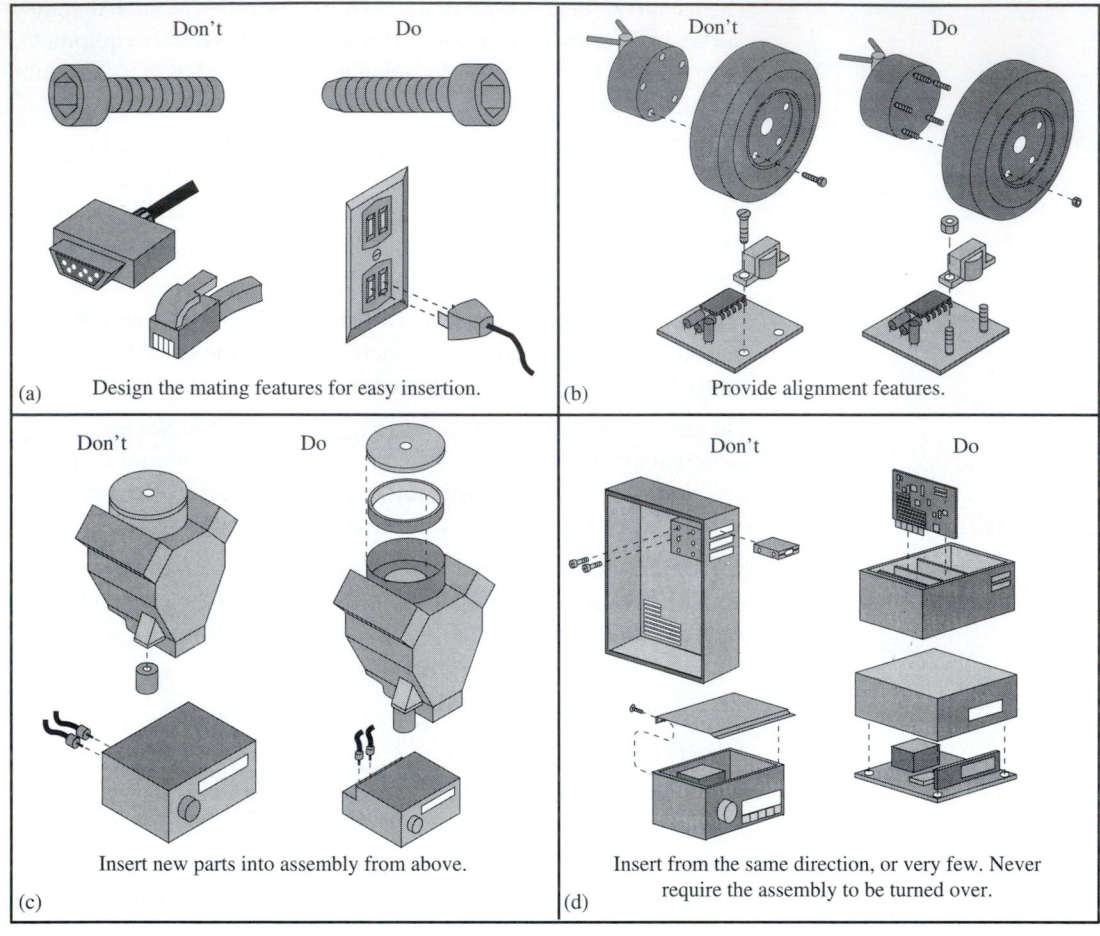

▼ Figure 14.5.
Design for assembly insertion design guidelines.

The third guideline basically suggests that we not fight gravity when placing and maintaining parts for fastening. Make the first part large and wide to be stable and then assemble smaller parts on top of it sequentially. Having to grasp and hold parts from below or from the side while they are being fastened is poor design.

The fourth guideline refines this to suggest that if we cannot assemble parts from the top down exclusively, then apply as few insertion directions as possible. Assemble only from the top and have fasteners come in from only one side, for example. This eliminates reorientation of the product during assembly. One should also consider the assembly sequence and ensure that all initial assembly work is completed on one surface, and then only one reorientation is required to finish the assembly work on the other surface. Do not make the assembly system constantly reorient the product. The worst case is when the subassembly requires turning over to continue assembling parts. It is difficult to keep parts precisely located when they are partially fastened and turned over.

Joining Guidelines

After a part is inserted onto the assembly, it must be firmly attached through some joining process. This can be with fasteners, snap fits, welds, or adhesives. Again, guidelines can be developed to simplify this aspect of the assembly process as shown in Figure 14.6.

The first guideline suggests reducing the number of fasteners. This does not mean to reduce the factor of safety by reducing the attachment strength but rather to change a portion of the fasteners to be of a quick insert type, such as one side of a part being held down by a tongue-in-slot joint on the assembly.

The second guideline suggests that it is better to locate fasteners in places where one has access to the fastener. The third guideline is similar with respect to enclosed spaces.

The fourth guideline suggests designing the assembly of parts per the instructions of the fastener. Typically this means to not fasten against angled surfaces. The fifth guideline is similar: One should space fasteners sufficiently to permit socket or rivet tool heads access to the fastener.

Theoretical Minimum Number of Parts

The thought behind the first two guidelines in Table 14.1—to modularize multiple parts into a single part—is so important it requires special emphasis. There is no better way to simplify an assembly step than to eliminate it. Fundamentally stated, the most effective DFA guideline is to "Simplify your design by eliminating all unnecessary separate parts." This idea goes all the way back to Henry Maudsley, the father of precision engineering, who said, "Eliminate all material not needed . . . put to yourself the question, 'What business has it to be there?' Make everything as simple as possible." Kelly Johnson, Chief engineer of Lockheed's famed Skunk Works development site, whose products include the F-80 Shooting Star fighter aircraft, the U-2 spy plane, the SR-71 Blackbird spy plane, and the F-117A Nighthawk stealth fighter, was the first to express this succinctly as "KISS—Keep It Simple Stupid" (Garmon 1999). The question then becomes whether a part needs to be a separate part

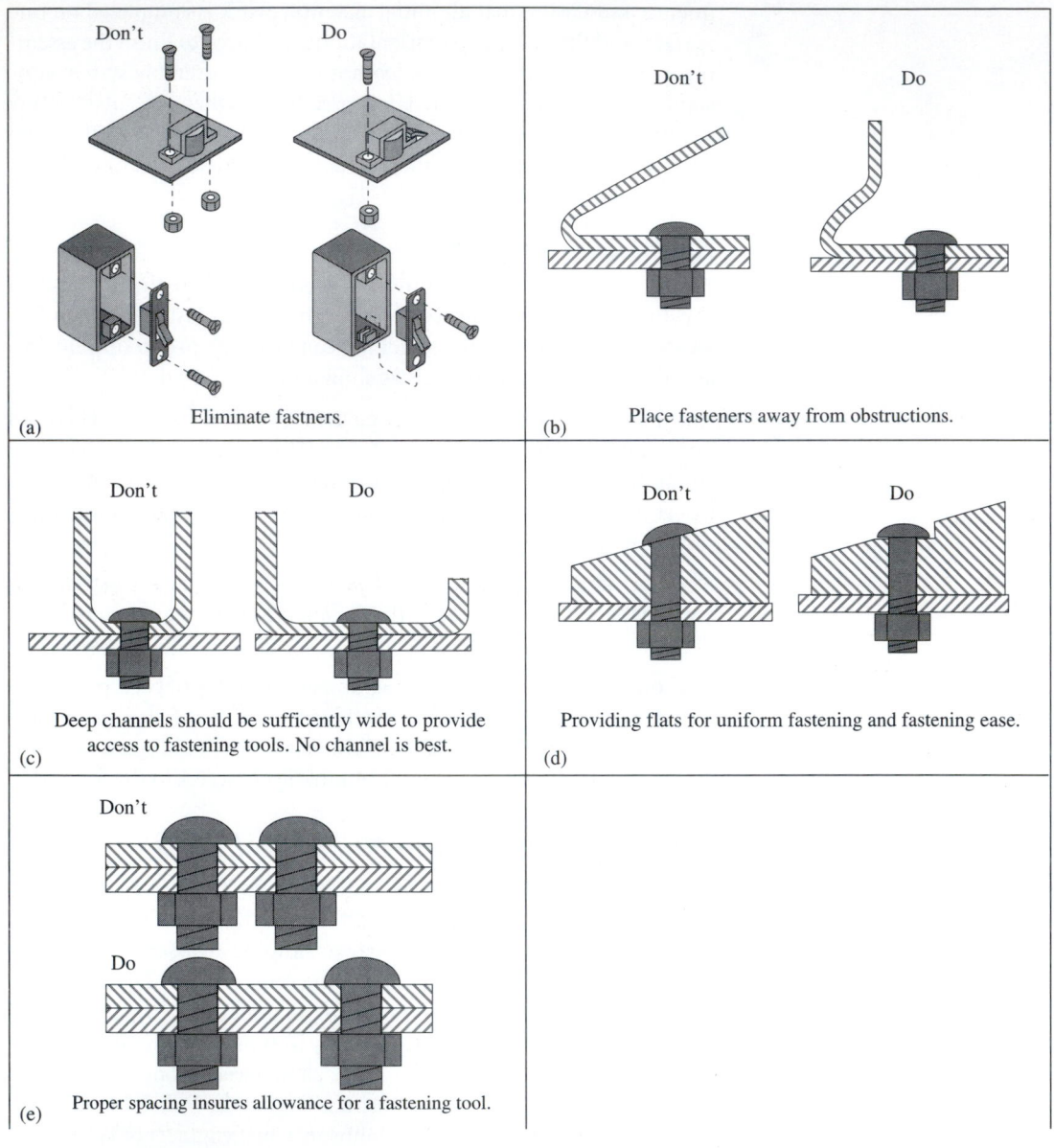

(a) Eliminate fastners.

(b) Place fasteners away from obstructions.

(c) Deep channels should be sufficently wide to provide access to fastening tools. No channel is best.

(d) Providing flats for uniform fastening and fastening ease.

(e) Proper spacing insures allowance for a fastening tool.

▼ *Figure 14.6.*
Joining design guidelines.

or whether it can be modularized into other parts. Chapter 9 provides means to combine parts and functions into modules.

Assembly modules (Chapter 9) can always be defined; the module is simply a subassembly. The real question is how to define assembly modules, how to determine whether it is possible to combine parts into a larger, more-complex part. At a basic level, some tests of neighboring parts can be applied. These include:

◗ Must the parts move relative to one another?
◗ Must the parts be electrically isolated?
◗ Must the parts be thermally isolated?
◗ Must the parts be of different materials?
◗ Does combining the parts prevent assembly of other parts?
◗ Will servicing be adversely affected?

If one can make the answer to these questions "No," then one should find a way to combine the two parts.

The part simplification concept may seem obvious, but at first it was not. The first to analyze it in detail and highlight its importance were Boothroyd (1980, 1982), who argued against other design guidelines in the literature that suggested using more parts that are individually simpler to fabricate. Whereas others such as Iredale (1964) and Tipping (1965) had previously noted the importance of part reduction as a design heuristic, Boothroyd developed a systematic methodology and comparison standard. Generally speaking, it is better to make more complex parts that are individually more expensive but with that added individual part cost more than made up for in less assembly cost and also typically less total part cost due to the less number of parts. Every part requires documentation, control, and inventory.

These thoughts lead to the concept of the *theoretical minimum number of parts* for a product, originally proposed by Boothroyd (1980, 1982). During assembly of the product, generally a part is required only when a kinematic motion of the part is required, when a different material is required, or when assembly of other *required* parts would otherwise be prevented. If none of these statements are true for a part in question, then the part is not needed to be a separate entity and design-for-assembly suggests combining it with another part.

Design for Piece Part Production

DFA is only one aspect of easing manufacturing; the other aspect is to make each part easy to produce. Here, general design guidelines are more difficult to formulate; rather, guidelines align with the processes used to make each part. Sheet metal DFM guidelines are not related to plastic part DFM guidelines. The best approach is to consult production engineers in the process to be used, and ask for advice on the design of the part.

For many processes, though, guidelines and design analysis tools have been established. For plastic parts, mold flow analysis can help indicate better part shapes for mold filling and cooling solidification. Guidelines also exist for nominal wall thickness. Sheet metal parts have guidelines for minimum bend radii and feature separation. Cast parts have recommendations for feature locations to provide material flow paths during mold fill. Machining operations have guidelines on dimensioning and tolerancing.

It is impossible to provide guidelines for every production process; new refined processes are constantly under development. However, Figure 14.7 provides guidelines for injection-molded parts (Xerox, 1984), Figure 14.8 provides guidelines for sheet metal parts (Dayton Rogers, 1998), Figure 14.8 provides guidelines for cast parts adapted from Bradney (1994), and Figure 14.9 provides guidelines for machined parts. Malloy (1994), Wakil (1998), Rosato and Rosato (1995), Bradney (1994), and Boothroyd and Knight (1989) provide additional reading. Each should be considered when designing a part requiring these processes.

Example: Krups Toastronic Toaster

As an example of applying these design guidelines, a product design can be analyzed in terms of Figures 14.3–14.10 and suggestions for simplification made. There may be reasons for not adopting any one of the suggestions, such as aesthetic considerations, but nonetheless they should be considered. Generally, failure to adopt the guidelines results in higher costs and/or reduced part quality.

An exploded view of the Krups toaster is shown in Figure 14.11. Generally speaking, each part in the toaster is designed well, adequate draft angles have been applied, uniform material thickness and adequate corner radii are all used. When considering each guideline, however, there are redesign suggestions that can be made.

For example, the toaster has over 25 separate parts, complicating its assembly. A theoretical minimum of six parts are needed: one as the housing, one for the thermally isolated heating element, one to connect these two, one for the toast pop-up-and-down motion, one for the knob adjustment motion, and one as the electrically isolated circuit board.

Examining further, the sheet metal housing is made of seven parts. It could be made with just one part as shown in Figure 14.12. Another ex-

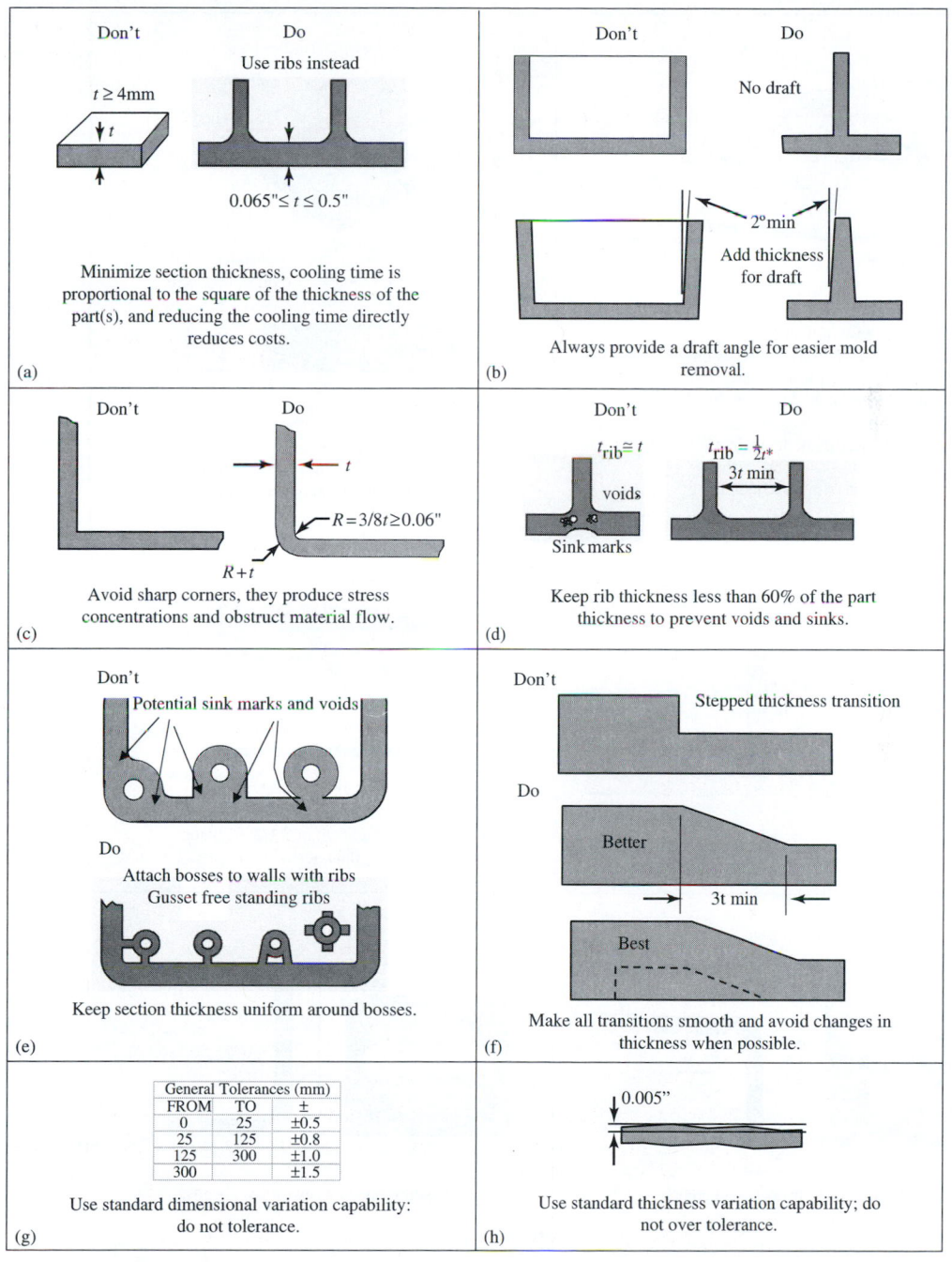

▼ *Figure 14.7.*
Injection-molded part design guidelines.

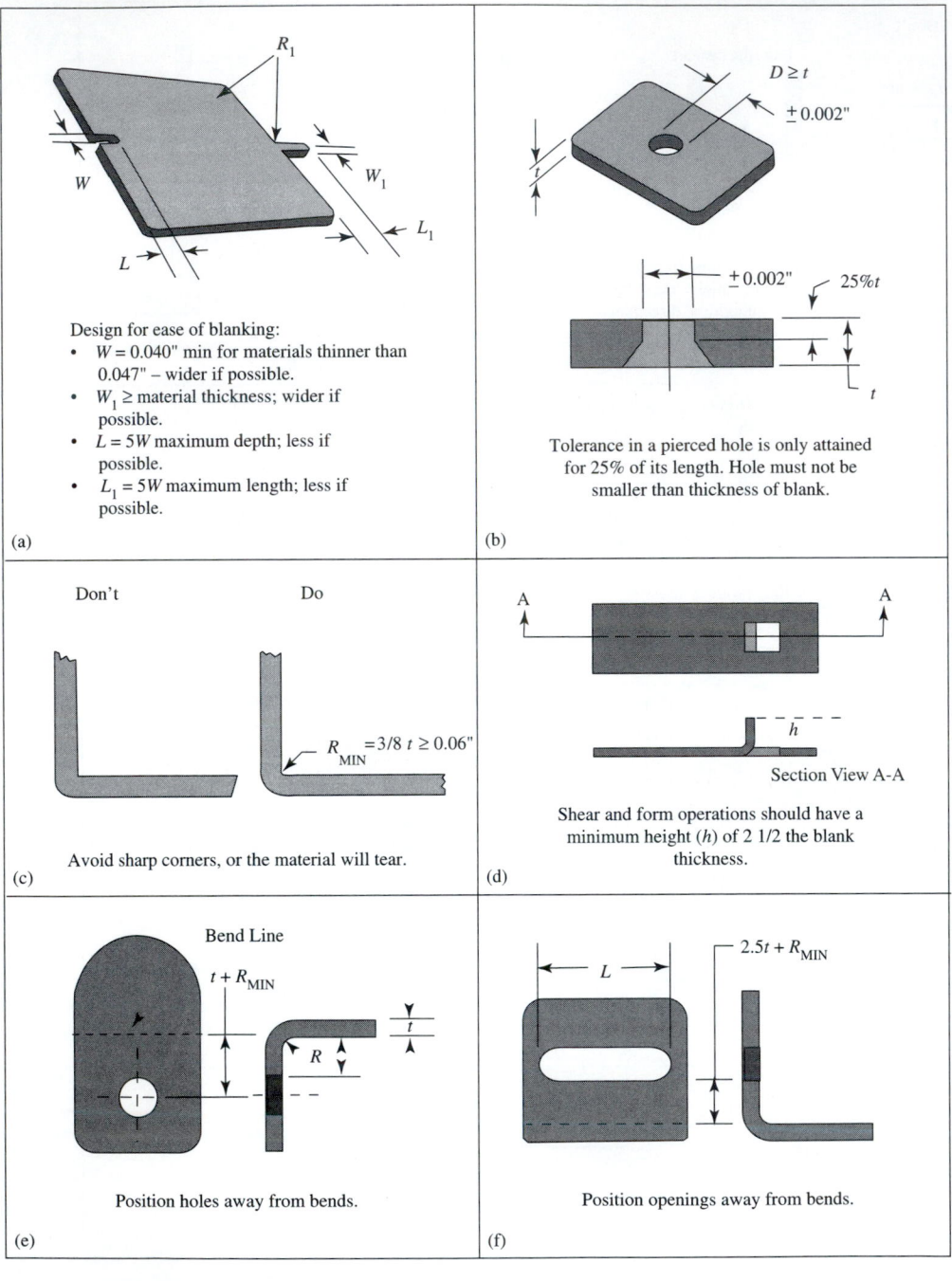

(a)

Design for ease of blanking:
- $W = 0.040"$ min for materials thinner than 0.047" – wider if possible.
- $W_1 \geq$ material thickness; wider if possible.
- $L = 5W$ maximum depth; less if possible.
- $L_1 = 5W$ maximum length; less if possible.

(b)

$D \geq t$

$\pm 0.002"$

$\pm 0.002"$ 25%t

Tolerance in a pierced hole is only attained for 25% of its length. Hole must not be smaller than thickness of blank.

(c)

Don't   Do

$R_{MIN} = 3/8\ t \geq 0.06"$

Avoid sharp corners, or the material will tear.

(d)

Section View A-A

Shear and form operations should have a minimum height (h) of 2 1/2 the blank thickness.

(e)

Bend Line

$t + R_{MIN}$

R

t

Position holes away from bends.

(f)

$2.5t + R_{MIN}$

L

Position openings away from bends.

▼ **Figure 14.8.**
Sheet-formed part design guidelines.

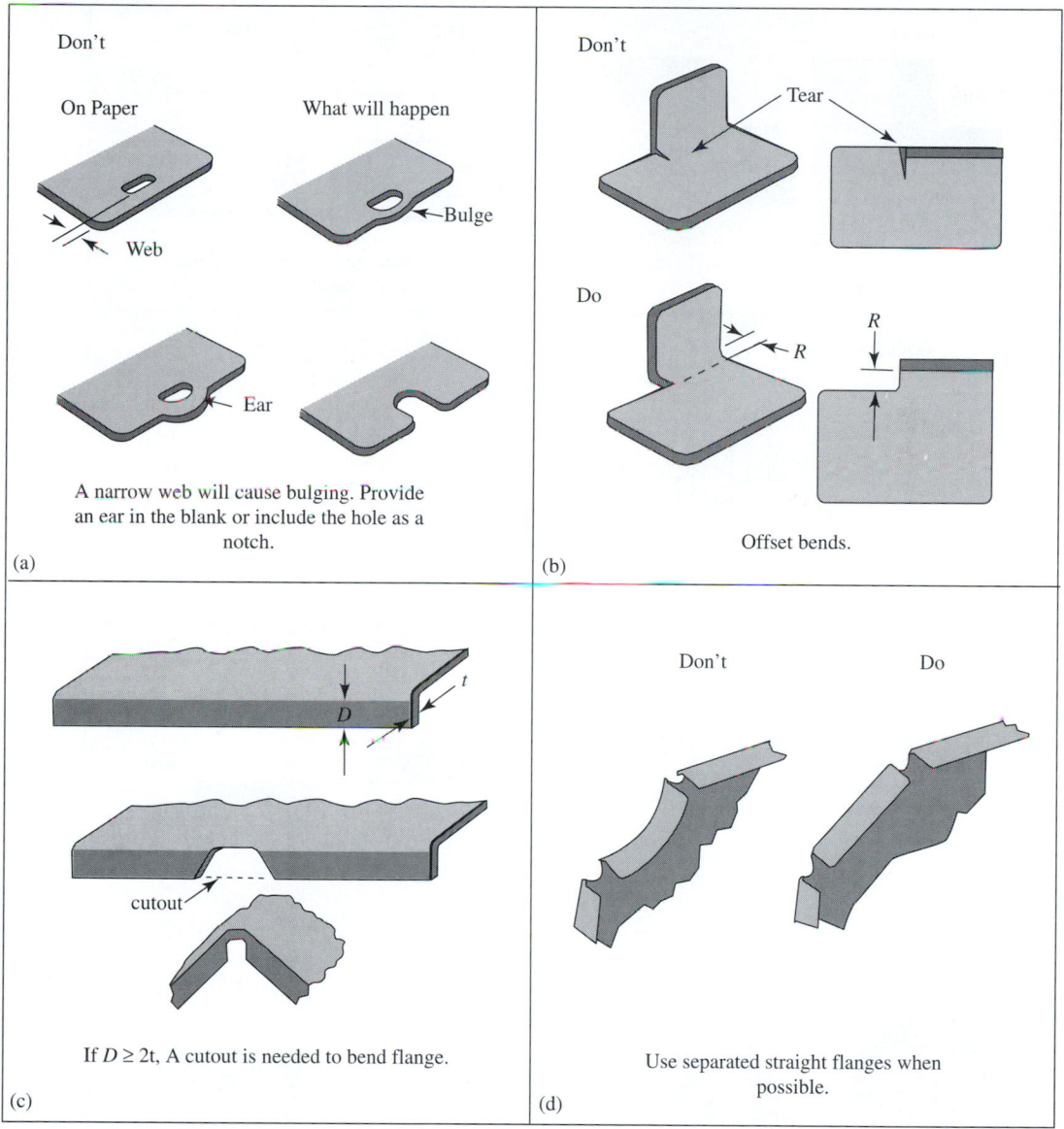

Don't

On Paper   What will happen

Web

Bulge

Ear

A narrow web will cause bulging. Provide
an ear in the blank or include the hole as a
notch.

(a)

Don't

Tear

Do

R

R

Offset bends.

(b)

D

t

cutout

If $D \geq 2t$, A cutout is needed to bend flange.

(c)

Don't   Do

Use separated straight flanges when
possible.

(d)

▼ *Figure 14.8.*
Sheet-formed part design guidelines (continued).

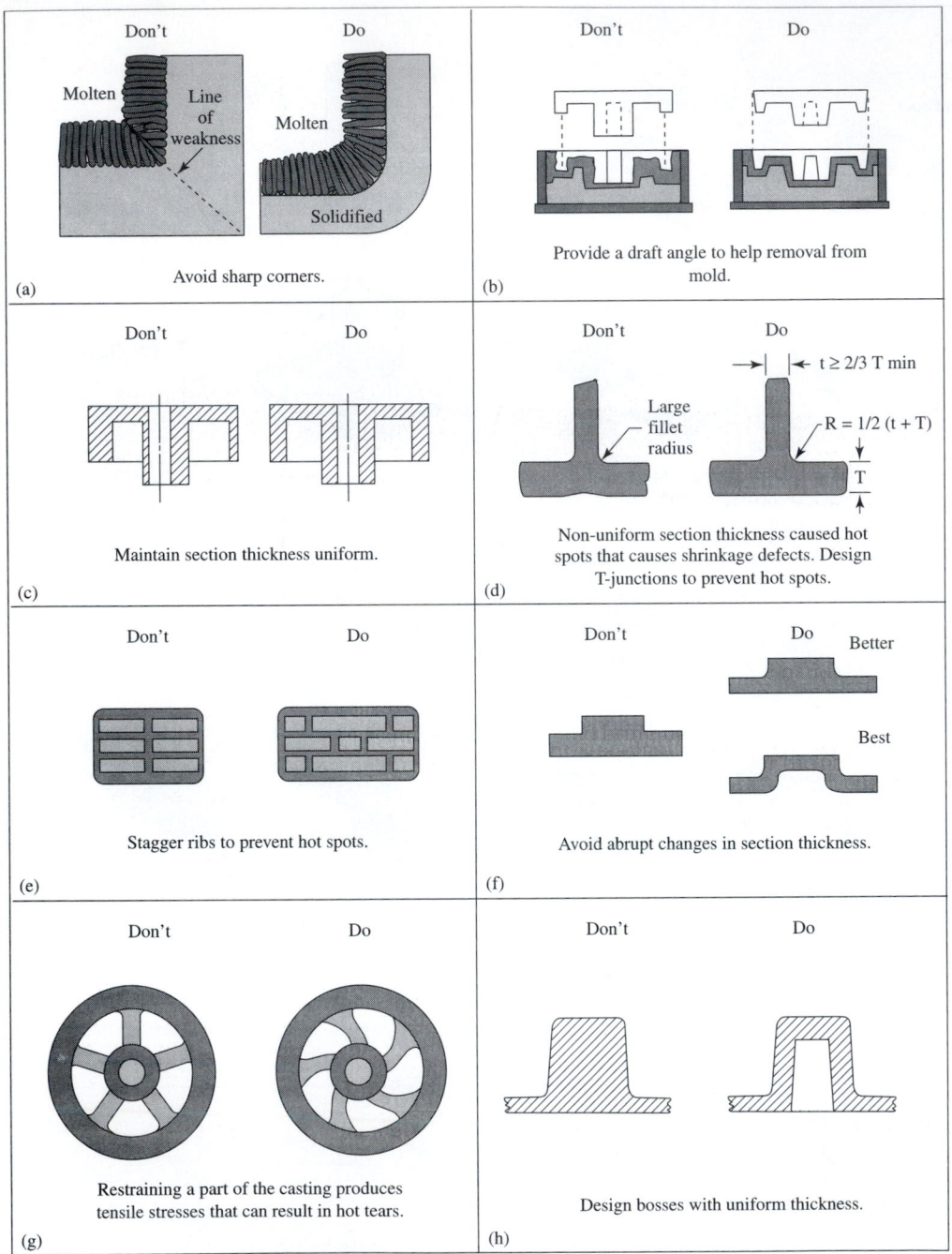

▼ *Figure 14.9.*
Cast part design guidelines.

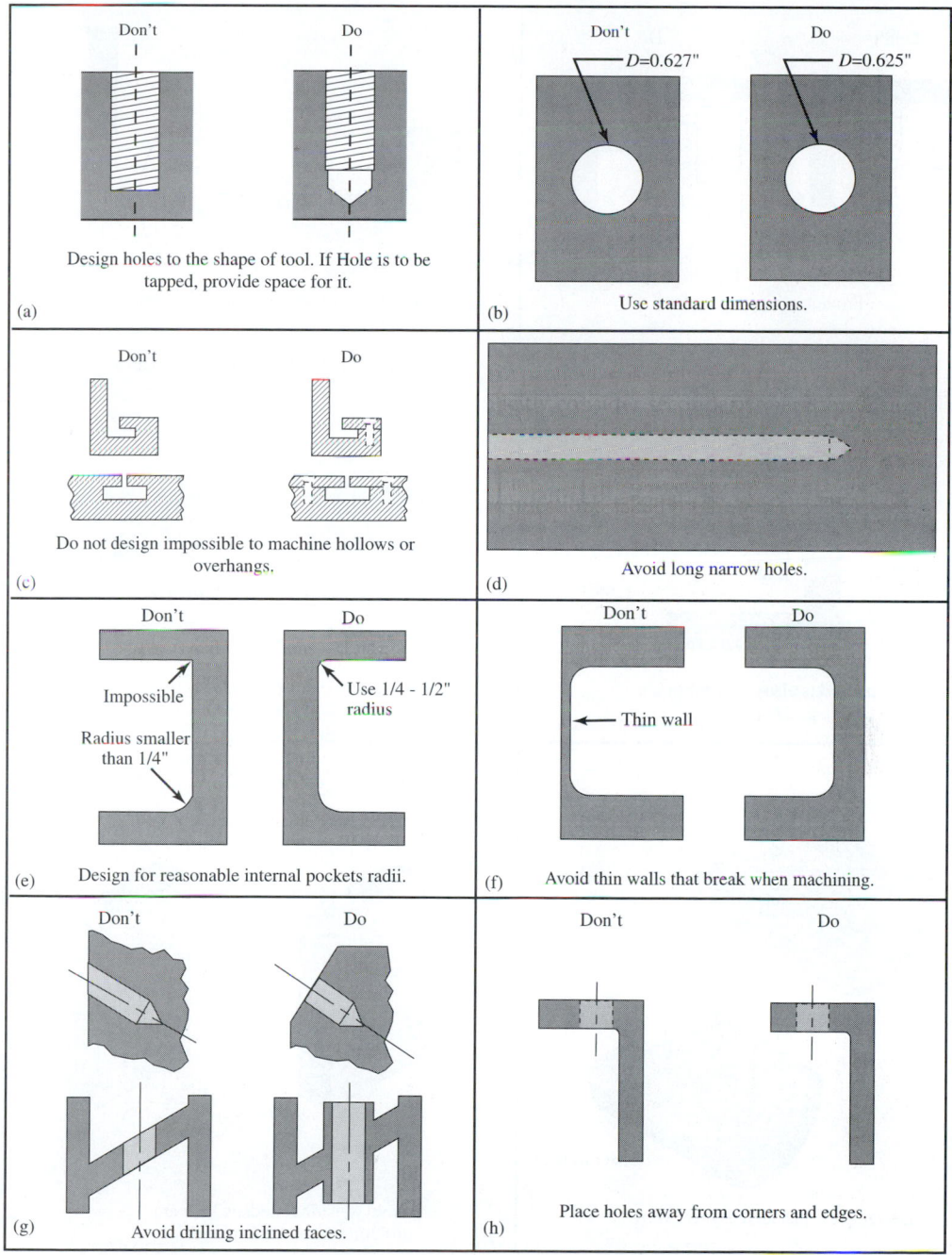

▼ **Figure 14.10.**
Machined part design guidelines.

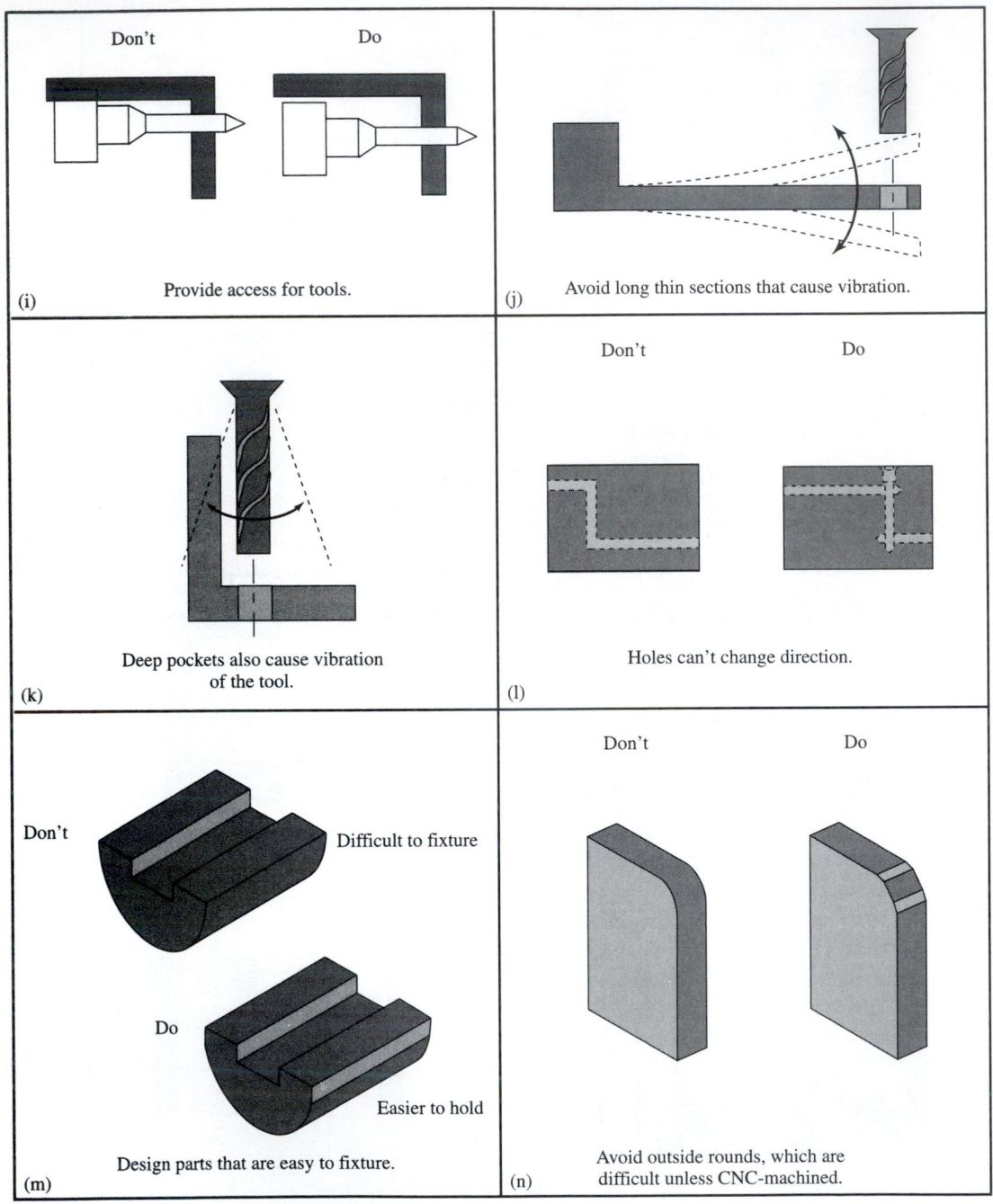

Don't

Do

(i) Provide access for tools.

(j) Avoid long thin sections that cause vibration.

(k) Deep pockets also cause vibration of the tool.

Don't

Do

(l) Holes can't change direction.

Don't

Do

Difficult to fixture

Easier to hold

(m) Design parts that are easy to fixture.

(n) Avoid outside rounds, which are difficult unless CNC-machined.

▼ **Figure 14.10.**
Machined part design guidelines (continued).

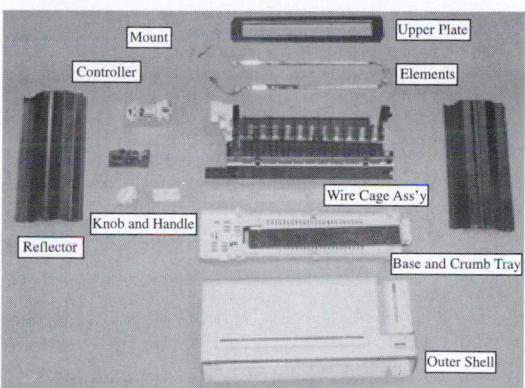

▼ *Figure 14.11.*
Exploded view of the Krups Toastronic Toaster.

ample is that the toaster is held together with four screw fasteners that screw through the base up into internal standoffs that are a part of the outer shell. This requires four internal cams to be designed into the mold, so that the part can be removed from the mold. Each additional cam adds complexity, and internal cams are particularly difficult. The design could be simplified by eliminating the standoffs. Using snap fits, for example, could replace the entire screw fastening system, including the standoffs and screws as shown in Figure 14.12. The snap fits use external cams, simplifying the part production, and also assemble the toaster without fasteners, simplifying the assembly.

As with many design decisions, there are trade-offs with these suggestions. The snap-fit solution can impact the appearance, for example. Nonetheless, it is always worth examining a design concept for ways to decrease manufacturing complexity. This is especially true for assembly complexity—one should always first determine the theoretical minimum number of parts and then question every part beyond this minimum and how the design could be simplified to this minimum.

Potential Conflicts between DFA and DFM

Application of DFA and DFM guidelines can reduce manufacturing costs and increase quality. However, the guidelines do not necessarily always do this. For example, one can increase part functionality to the point where the parts become so complex they cannot be easily produced. For example, in Figure 14.13, the part shown was used in the Kodak single-use camera and was very complex, supplying three separate lens functions for aiming light, also a switch actuator, and also snap fits for attachment functions. This part was sufficiently complex

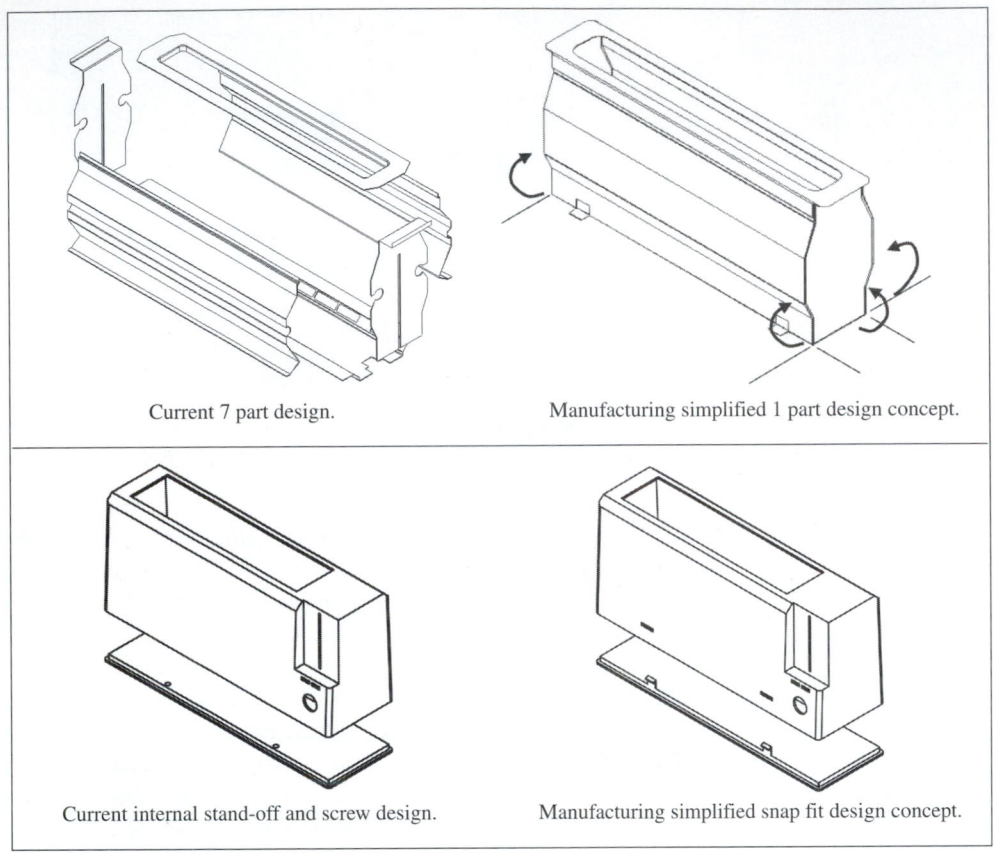

Current 7 part design.

Manufacturing simplified 1 part design concept.

Current internal stand-off and screw design.

Manufacturing simplified snap fit design concept.

▼ *Figure 14.12.*
Simplified toaster concepts.

that in the next generation of Kodak single-use camera, the part was separated into two parts: one providing the lens function and a separate part providing the switch actuation function. This violates the DFA guideline of minimizing parts but was nonetheless a more cost-effective design, since the original part was sufficiently complex that it was excessively difficult to produce reliably.

Typically, this situation does not occur; it is almost always better to simplify a design by eliminating parts. To address such trade-off decisions, we must explore actual concept costing methods. Then, numerical comparisons can be made between designs with more complex assembly versus designs with more complex piece parts.

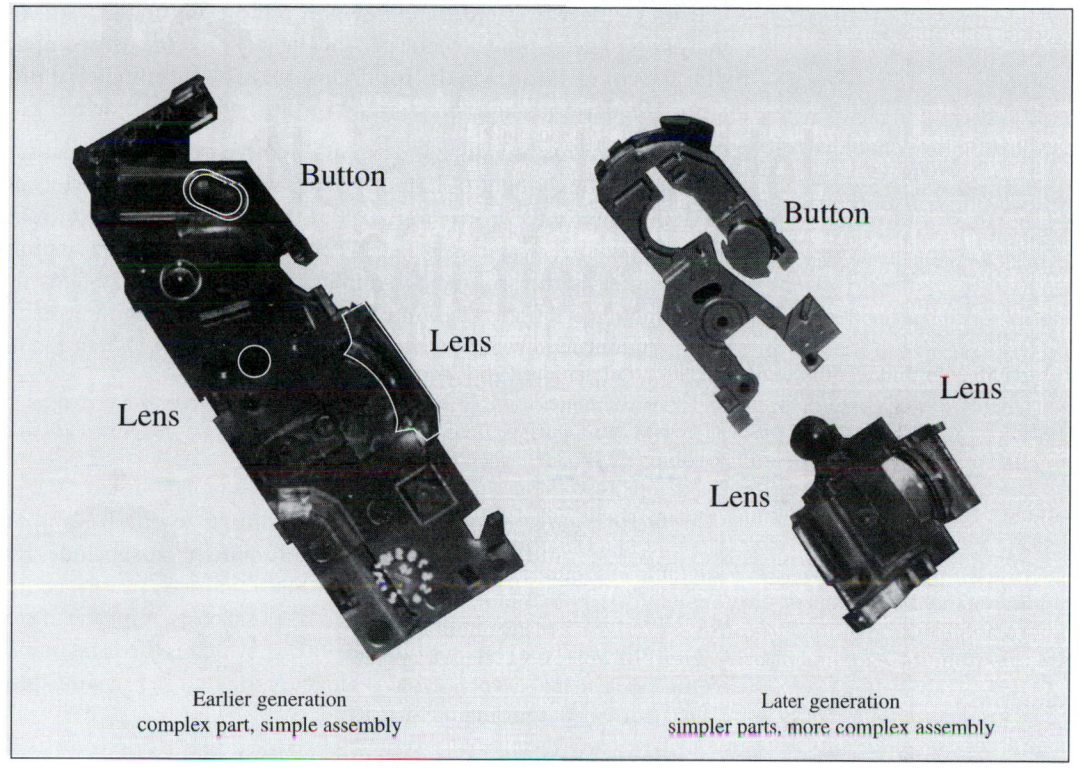

▼ Figure 14.13.
Kodak single-use camera parts (a) Earlier generation: complex part, simple assembly.
(b) Later generation: simpler parts, more complex assembly.

IV. ADVANCED METHOD: MANUFACTURING COST ANALYSIS

Understanding the cost structure designed into a product is important for deciding what portion of a design is more appropriate for detailed cost-reducing design activity and for comparing different design concepts. To do this, basic product cost accounting will be covered, followed by cost models for the various portions of the cost model, including part costs and assembly costs.

The design for manufacture and assembly philosophy underneath cost analysis is to determine how the product delivery major costs compare with the competition. Major cost drains can arise from material procurement, part production, assembly, or finished product delivery.

The design for manufacture and assembly cost-based redesign activity itself should be thought of in terms of an 80-20 rule: 80% of the cost reduction can occur in the top 20% of the relative high cost drains within a product. One can attain 80% of the possible cost reductions through the design-for-assembly simplification of a few key systems, design-for-manufacture redesign of a few expensive components, and adequate bidding on expensive purchased *original equipment manufacturer* (OEM) components.

Cost Driver Modeling

Manufacturing cost is the sum of all expenditures to purchase inputs and dispose of outputs to the manufacturing system. This includes raw materials, purchased components, employee's efforts, energy, equipment, maintenance materials, and plant disposal costs. A metric of cost commonly used is a *unit manufacturing cost,* which is the total manufacturing expenses over a period divided by the number of units produced during that period.

Several system modeling issues complicate this approach:

▶ What are the boundaries of the "manufacturing system"? Do we include product development costs, for example?

▶ How do we charge a product for equipment that is only partially used for this product?

▶ How do we charge a product for equipment inherited from past products?

Nonetheless, one can consider an average situation and develop a reasonable model for comparing product design alternatives.

Manufacturing Cost Accounting

Typical managerial cost modeling separates an activity's cost into *fixed costs* and *variable costs*. Fixed costs are those that do not change in direct proportion to production volume, such as the cost of equipment. A mold for a plastic injection molding machine is a fixed cost; it is needed whether a thousand or a hundred thousand are produced. Variable costs are those that grow in direct proportion to production volume. Material cost is a variable cost. This fixed/variable cost breakdown

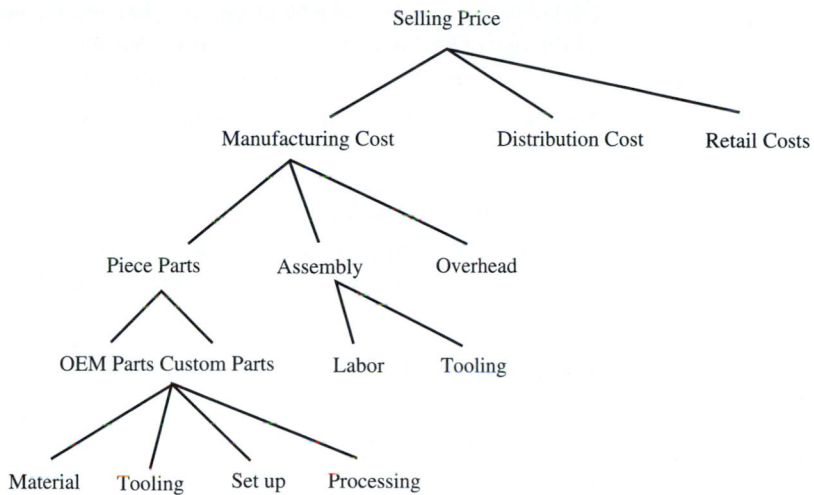

▼ *Figure 14.14.*
Production cost analysis breakdown.

is fuzzy. Sometimes labor is considered fixed: You must hire a person to do the work. Sometimes labor is considered variable: You can hire a number of persons proportional to your production.

For product development purposes, a more effective breakdown is to consider costs according to physical manufacturing processes. Each of these can then be costed and different processes considered. As shown in Figure 14.14, we will form a contributing manufacturing cost breakdown consisting of:

▶ *Piece part costs.* This covers the costs of both parts made and bought from suppliers.

▶ *Assembly costs.* This covers the costs of assembling the parts into the product.

▶ *Overhead rate.* This covers the costs of supporting direct production of parts and assembly.

A product typically contains many subassemblies that are purchased as custom assemblies, or OEM parts. In our costing approach, we model these as individual component costs, much like a material cost.

Note that this includes only the costs of producing the units. Beyond these manufacturing costs, there are remaining costs that complete the retail price. The product must be put in a box for shipping. It must be shipped. These costs might need to be considered, for example, if different labor markets in different geographic locations are used. They will have different levels of production expertise, and that impacts the product design. The design of a product fabricated in the United States is typically far different from the design of a product where labor rates are much cheaper.

Even beyond the distribution system, the product must be placed on a retail shelf, advertised, and sold. These also add to the final retail price. In general, such costs beyond the manufacturer's selling price generally double the cost of many products. This depends on the retailer. Such concerns are not typically part of a design process, but they are very much in the consideration of the entire product development process.

Cost Modeling

The basic approach we will take to cost modeling different manufacturing processes is to consider cost drivers. A *cost driver* is a configuration or process variable that largely determines the cost of the process. A cost model as a performance metric can be constructed in terms of the cost drivers. This will be true no matter the level of analysis completed and no matter which processes are examined.

For example, consider piece part production costs. If the piece parts are OEM parts, the unit price the manufacturer pays is the obvious cost. This can be determined by obtaining quotes from vendors. A crude model is to look up the retail price in a catalog and apply a discount factor for volume purchase. Note also that often OEM parts can be checked for reasonable quoted prices through cost analysis, an important activity for many companies to ensure fair pricing.

On the other hand, parts fabricated by the manufacturer require a better model. We will consider part costs in three categories: material cost, tooling cost, and production costs. *Material costs* are the costs of the raw materials for the part, generally expressed in price per pound. For plastic parts, this is the cost of the plastic pellets. A basic formula for material costs is

Material cost per unit
$$= \text{Part weight} \times \text{scrap} \times \text{material cost/lb} \qquad (14.1)$$

Tooling costs are the fixed costs to buy the equipment to convert the material into the finished part. As an example, for injection-molded parts, this is typically the cost of the molds and their replacement. Other examples are dies and fixtures. The cost includes the cost to design and fabricate the tooling. A basic formula for tooling cost is

Tooling cost per unit
$$= \text{Equipment costs/parts per equipment} \qquad (14.2)$$

Tooling costs are fixed costs. Note that the costs of *capital equipment* are usually included in the overhead. Capital equipment is the large equipment that is used over many products, such as the factory floor, the machinery that molds or dies are fitted in, and so forth.

Set-up costs are the costs to set the machine up to begin a batch of production. The mold must be attached to the injection-molding machine, the runner system must be purged, and so forth. Set-up costs are fixed costs.

Processing costs are the labor and maintenance costs to keep the equipment running. The maintenance component is often estimated by either increasing the labor rate or by incorporating it into the final overhead. A basic formula for processing cost is

Processing cost per unit
$$= \text{Labor rate per hour} \times \text{hours per unit} \qquad (14.3)$$

Piece part costs can be reduced if features that require difficult tooling can be eliminated. If such changes do not affect the product performance, the economic success of the product can be increased by such design changes.

Next, consider assembly costs. Mechanical assembly is almost always done manually for quantities of less than several hundred thousand. A basic formula for manual assembly cost is

Assembly cost per unit
$$= \text{Labor rate per hour} \times \text{hours per unit} \qquad (14.4)$$

Mechanized assembly is similar, with the addition of the fixed tooling costs. For manual assembly, assembly costs are often dominant. Further, with manual assembly, quality is in large part determined by the assembly of the piece parts, not by the individual piece part quality. For these two reasons, designing a product to minimize assembly difficulty has become an important factor for design teams.

All of these costs can be readily measured and accounted for. These costs are therefore often called *direct costs;* they can be directly identified. All other expenses a manufacturer has are indirect costs, or *overhead.* These include costs such as administration, insurance, very large equipment, and so on. The means to determine overhead costs are fuzzy; by definition, overhead is used to cover the quantities that assigning a direct cost to is too difficult to bother with.

Usually, overhead is determined by an *overhead rate,* applied to direct *overhead cost drivers* of a product. In this context, an overhead cost driver is a variable in the direct cost formulas to which the overhead costs are known to be highly related; for example, labor rates. Often, labor rates have an overhead rate applied to them to cover costs of keeping the factory livable and the work force happy. Another overhead cost driver might be material costs. It is reasonable to presume that required machine maintenance will be determined by the amount of material being processed, and so an overhead rate is assigned to augment the direct material costs.

Using overhead permits different overhead rates for the different aspects of manufacturing costs. An overhead rate of 50–300% might be used for labor overhead. Depending on the equipment, an overhead rate of 10–20% might be used for material overhead. The total labor rate including direct and overhead costs is called *fully burdened*.

Even with the breakdown among overhead cost drivers, however, the overhead rate method is not always appropriate; it implies that overhead is directly proportional to the direct costs. *Activity-based accounting* partially addresses this problem and attempts to directly assign overhead costs to the activities that drive the overhead; that is, model costs as direct costs only.

To estimate the manufacturing cost of a product, the previous basic formulas are applied to each part, subassembly, and assembly of the product. A simple hierarchical spreadsheet of costs is useful to represent the subassembly and piece part constitutive costs. Ultimately, this is how all forms of manufacturing cost analysis methods operate.

Manufacturing Cost Analysis

Cost analysis can be completed at different levels of detail. We will consider three categories of cost analysis (Meeker and Thornton 1996). A *Level 1 Estimate* is a first impression by a knowledgeable engineer of what a part, assembly, or system would cost, based on prior experience. This would involve less than 10 minutes of work for a system of 50 parts and is generally accurate to within 20%.

A *Level 2 Analysis* is an estimate based on a breakdown of costs and itemized cost data based on prior experience with similar products, budgetary estimates, vendor quotes, and expert opinion and experience. This would involve a day of work for a system of 50 parts and is generally accurate to within 5%. The methods in this chapter are generally Level 2 analyses.

A *Level 3 Cost Accounting* applies a detailed costing of every part, accomplished by using material cost estimation data bases and time/motion studies. A high degree of accuracy is achieved by comparisons to industry standards and requests for vendor quotes. This would involve a week of work for a system of 50 parts and is generally accurate to within 1%. Many major manufacturers have cost analysis teams that provide their product development teams with Level 3 cost analyses of their competitors' products (Tsiantar, 1988).

We now present cost models of different production processes. First, we provide a very simplistic model of part costs by providing tables of different parts. Then, we provide a simple assembly cost model. These can be used to cost account in a crude way as a Level 1 cost estimate.

Basic DFM Part Cost Method: Analogy Approach

For piece part manufacturing, we now present different cost models for different piece part production processes. Table 14.2 lists typical material costs, Table 14.3 lists typical processing labor rates, Table 14.4 lists typical processing times, and Figures 14.15–14.19 list typical tooling costs, all in 1998 dollars. The data in these tables can be used in as-

TABLE 14.2   TYPICAL BULK MATERIAL COSTS

Material	Typical cost ($/lb)
Plastics (virgin pellets)	
Polyethylene terephthalate (PET)	0.40–1.00
Polypropylene (PP)	0.30–0.50
Polycarbonate (PC)	0.80–2.00
Polystyrene (PS)	0.40–0.60
Polyvinyl chloride (PVC)	0.30–0.40
Acrylonitrile butadiene styrene (ABS)	0.50–1.40
Phenolic	0.60–0.80
Nylon	1.10–2.00
Rubber	
Butyl	1.50–3.00
Latex	1.00–3.00
SBS	1.00–2.00
Silicone	4.00–8.00
Metals (sheet or bulk)	
1014 carbon steel	0.40–0.80
304 stainless steel	1.00–2.00
1100 Aluminum	1.00–2.50
Copper	1.00–2.00
Glass (sheet)	
Fiberglas^{TM}	2.00–5.00
Woods (board)	
Douglas fir	6.00–9.00
Oak	10.00–15.00

Note: 1998 dollars.

TABLE 14.3. DIRECT LABOR COSTS, UNITED STATES

Process	Processing cost ($/hr)
Injection molding	$30 + 0.06T$ (where T is the tonnage of machine)
Die casting	30.00
Machining	50.00
Powder metal	40.00
Sheet metal	30.00
Manual assembly	30.00
Electrical work (assembly, PCB)	30.00
Finishing (paint, labeling)	30.00

Note: 1998 dollars.

TABLE 14.4. PROCESSING TIMES, ROUGH ESTIMATES

Injection molding

$$\text{Processing Time} = \frac{5. + th^2/\alpha}{\# \text{ cavities}}$$

where
- 5 is the mold open and pack time (seconds)
- α is the material thermal diffusivity
- th is the nominal part wall thickness

and the result is in seconds.

Setup time per lot is 1 hour for 25–125-ton press, 2 hours for up 500-ton press, and 4 hours for over 600-ton press.

Die casting

$$\text{Processing Time} = \frac{10. + 200th}{\# \text{ cores}}$$

where
- 10 is the die open and fill time (seconds)
- 200 is a conduction rate factor
- th is the nominal part wall thickness (mm)

and the result is in seconds.

Setup time per lot is 1–3 hours.

Powder Metal

$$t = t_{\text{pack}} + 0.9T_{\text{m}} \frac{L_{\text{oven}} W_{\text{oven}}}{L_{\text{part}} W_{\text{part}}} + t_{\text{finish}}$$

where
- t_{pack} is the time to pack the fill into a green compact (typically about 15 seconds per cycle, but factor by the number of cavities).
- T_{m} is the required time to fire a material as given in Table 14.7.
- L_{pack} and W_{part} are the length and width of the part.
- L_{oven} and W_{oven} are the length and width of the continuous belt-driven firing oven. Typically ovens are 75 feet $\times$ 3 feet
- t_{finish} is for any deburring, machining, and coining operations.

and the result is in seconds.

Setup time per lot is 2–6 hours.

CNC Operations

Operation	Time	Setup
Turning down diameter	0.3 min/cm	30 min
Facing off end	0.3 min/cm	30 min
Drilling	0.5 min/hole	30 min
Tapping	0.5 min/hole	30 min
Sawing	0.3 min/cm²	15 min
Grinding	0.3 min/cm²	30 min
EDM	2 min/cm²	1 hr

Source: adapted from Santandreu (1993).

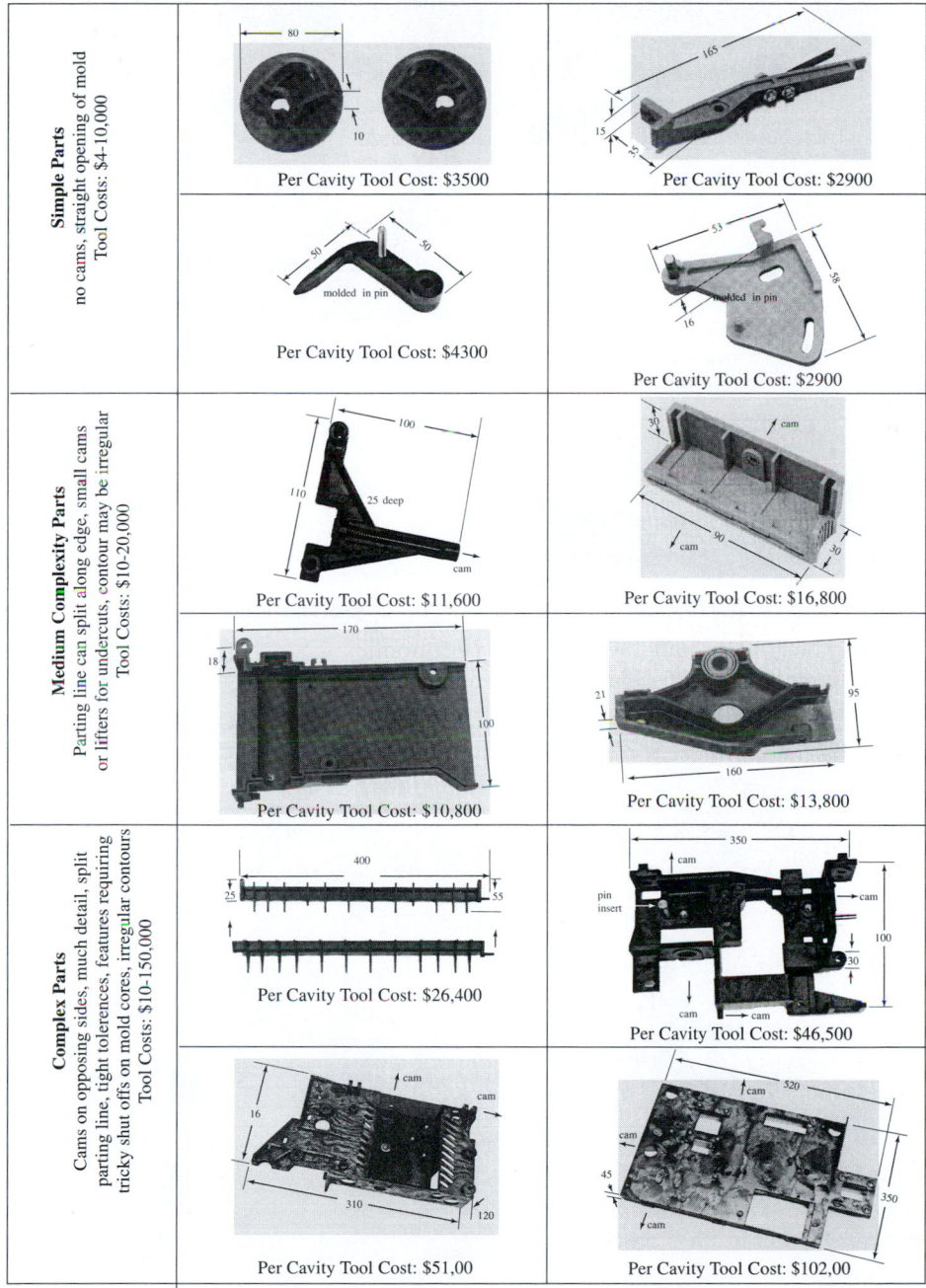

Simple Parts no cams, straight opening of mold Tool Costs: $4-10,000	Per Cavity Tool Cost: $3500
	Per Cavity Tool Cost: $2900
	Per Cavity Tool Cost: $4300
	Per Cavity Tool Cost: $2900
Medium Complexity Parts Parting line can split along edge, small cams or lifters for undercuts, contour may be irregular Tool Costs: $10-20,000	Per Cavity Tool Cost: $11,600
	Per Cavity Tool Cost: $16,800
	Per Cavity Tool Cost: $10,800
	Per Cavity Tool Cost: $13,800
Complex Parts Cams on opposing sides, much detail, split parting line, tight tolerences, features requiring tricky shut offs on mold cores, irregular contours Tool Costs: $10-150,000	Per Cavity Tool Cost: $26,400
	Per Cavity Tool Cost: $46,500
	Per Cavity Tool Cost: $51,00
	Per Cavity Tool Cost: $102,00

▼ *Figure 14.15.*
Typical plastic injection-molded per cavity tooling costs (dimensions in mm). Adapted from Xerox (1986).

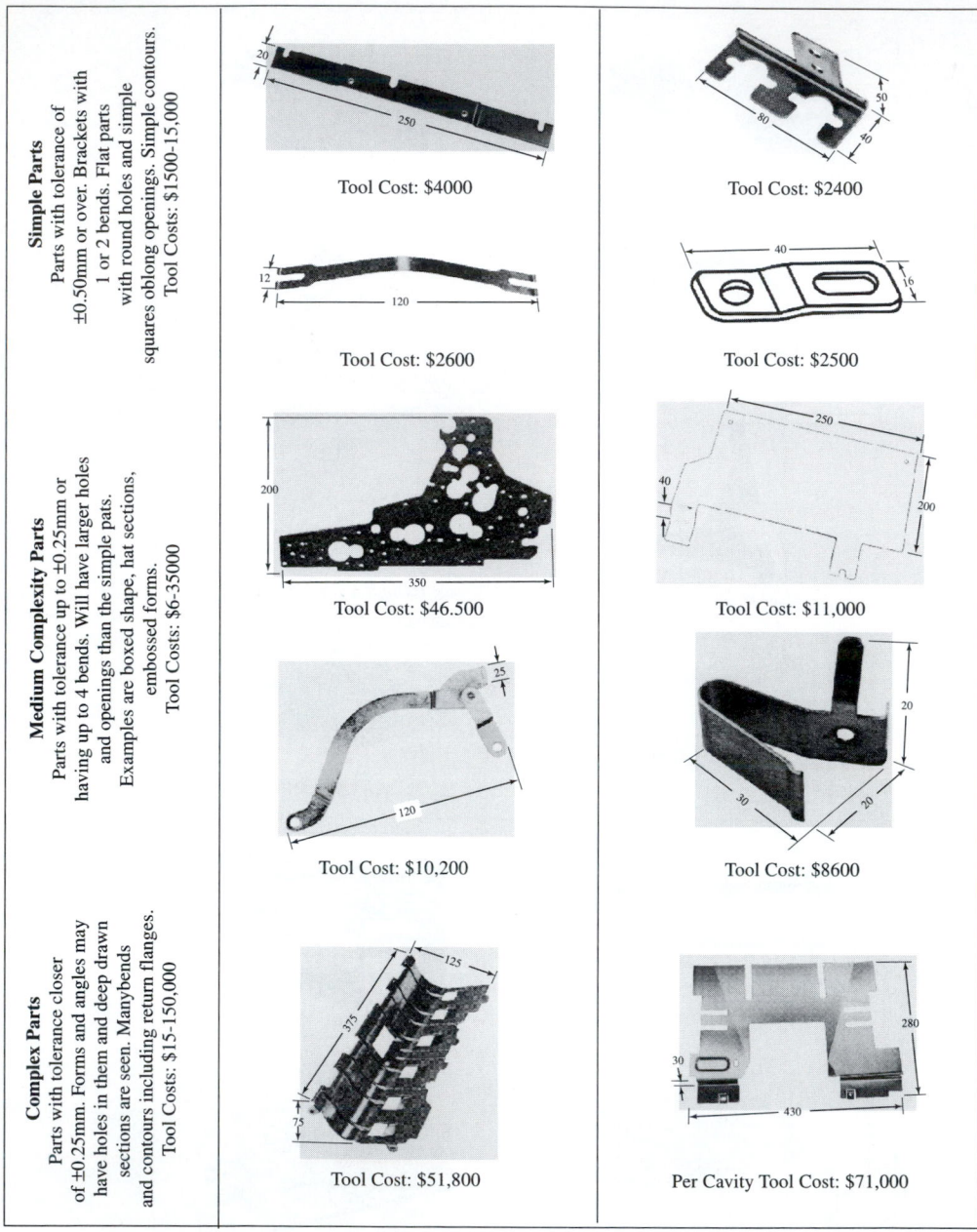

Simple Parts
Parts with tolerance of ±0.50mm or over. Brackets with 1 or 2 bends. Flat parts with round holes and simple squares oblong openings. Simple contours.
Tool Costs: $1500-15,000

Tool Cost: $4000

Tool Cost: $2400

Tool Cost: $2600

Tool Cost: $2500

Medium Complexity Parts
Parts with tolerance up to ±0.25mm or having up to 4 bends. Will have larger holes and openings than the simple pats. Examples are boxed shape, hat sections, embossed forms.
Tool Costs: $6-35000

Tool Cost: $46.500

Tool Cost: $11,000

Tool Cost: $10,200

Tool Cost: $8600

Complex Parts
Parts with tolerance closer of ±0.25mm. Forms and angles may have holes in them and deep drawn sections are seen. Manybends and contours including return flanges.
Tool Costs: $15-150,000

Tool Cost: $51,800

Per Cavity Tool Cost: $71,000

▼ *Figure 14.16.*
Typical stamped sheet-metal tool costs (dimensions in mm). Adapted from (Xerox 1986).

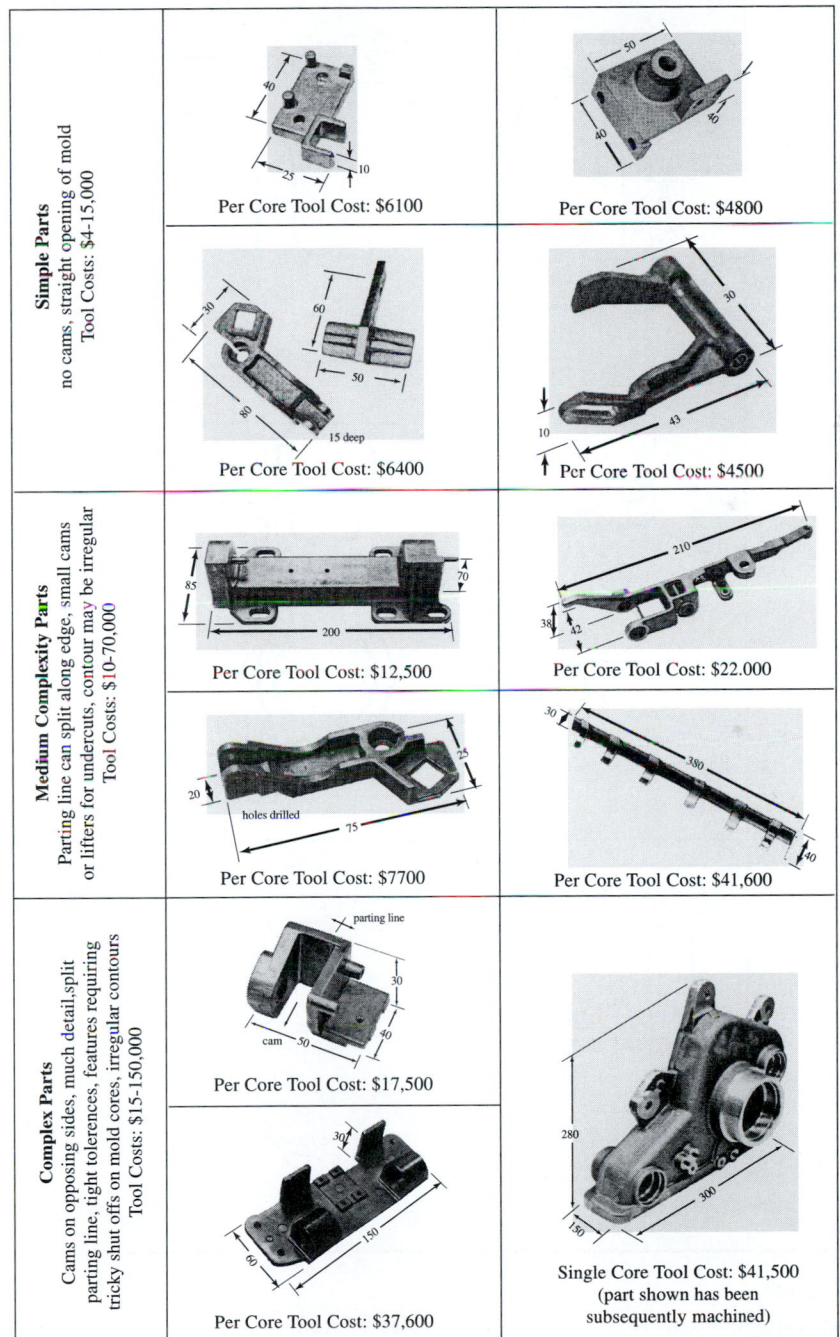

▼ *Figure 14.17.*

Typical die cast per core tool costs (dimensions in mm). Adapted from Xerox (1986).

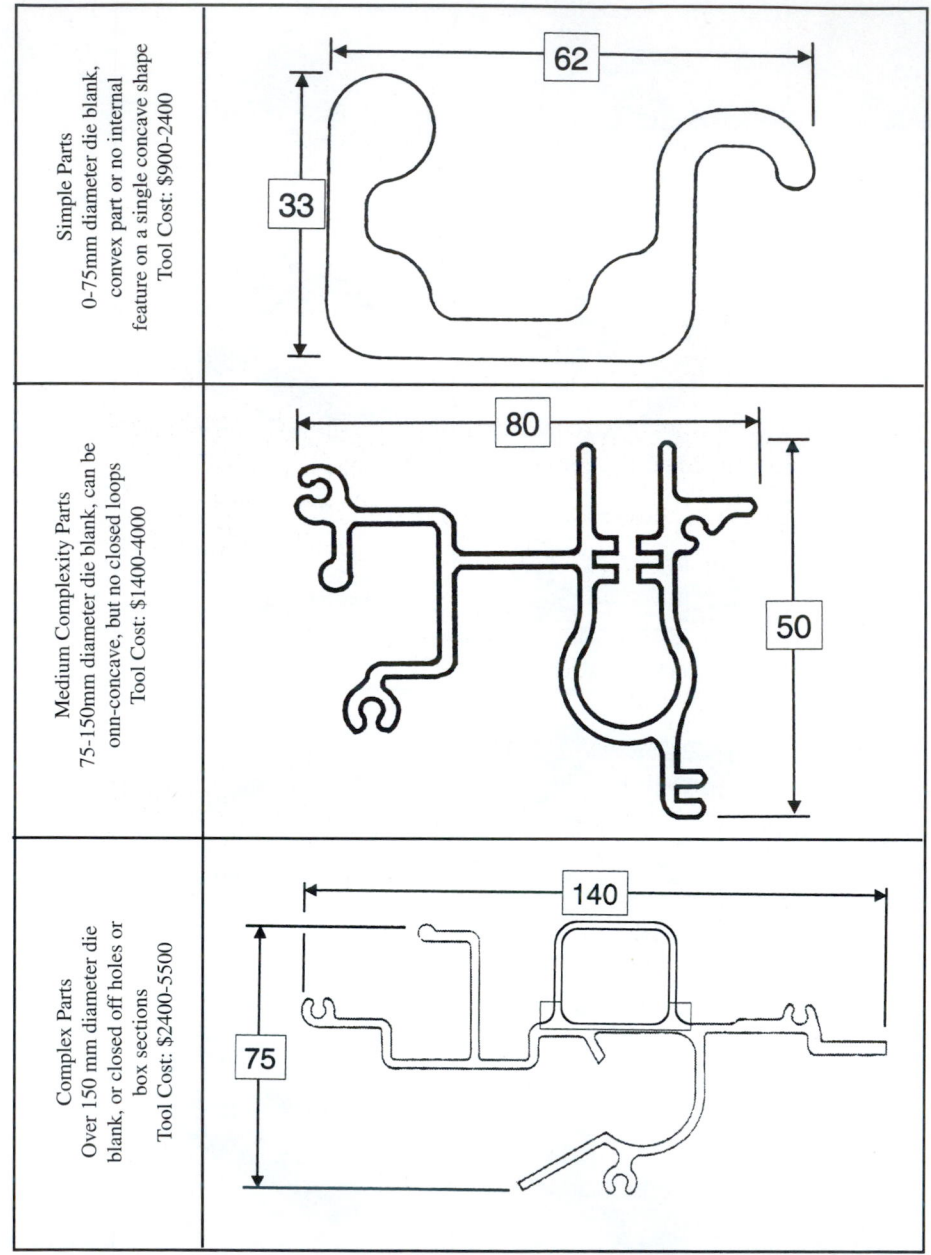

Simple Parts
0-75mm diameter die blank, convex part or no internal feature on a single concave shape
Tool Cost: $900-2400

Medium Complexity Parts
75-150mm diameter die blank, can be onn-concave, but no closed loops
Tool Cost: $1400-4000

Complex Parts
Over 150 mm diameter die blank, or closed off holes or box sections
Tool Cost: $2400-5500

▼ *Figure 14.18.*
Typical extrusion tool costs (dimensions in mm). Adapted from Xerox (1986).

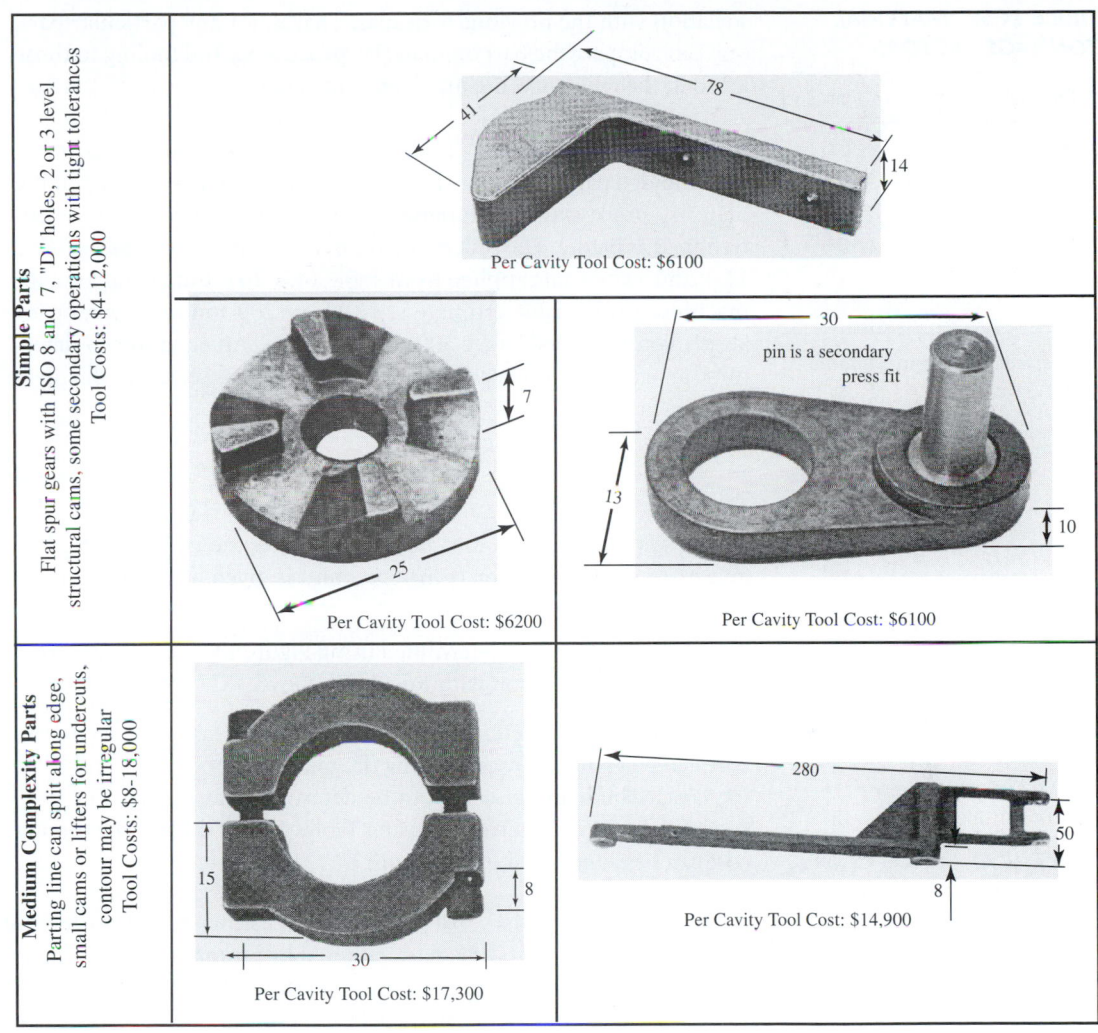

Simple Parts

Flat spur gears with ISO 8 and 7, "D" holes, 2 or 3 level structural cams, some secondary operations with tight tolerances
Tool Costs: $4–12,000

Per Cavity Tool Cost: $6100

Per Cavity Tool Cost: $6200

pin is a secondary press fit

Per Cavity Tool Cost: $6100

Medium Complexity Parts

Parting line can split along edge, small cams or lifters for undercuts, contour may be irregular
Tool Costs: $8–18,000

Per Cavity Tool Cost: $17,300

Per Cavity Tool Cost: $14,900

▼ *Figure 14.19.*
Typical powder-metal part tool costs (dimensions in mm). Adapted from Xerox (1986).

TABLE 14.5. MATERIAL TONNAGE FACTORS

Material	σ (tons/cm^2)
ABS	0.5
Styrene	0.5
Polycarbonate	0.8
Nylon	0.8
Sulfone	1.1
PEI	1.1

Source: Adapted from Santandreu (1993).

sociation with the previous formulas. That is, for any particular part, one can compare the part on material, processing, and tooling to those listed in the tables and use an equivalent value.

Injection-molded parts: To determine an estimate for plastic injection-molded parts, a cost per part can be gathered by estimating of the material, processing, and tooling costs. The material costs can be determined as before. The processing time can be determined using Table 14.4, and a labor rate applied from Table 14.3. To estimate the tonnage machine, a reasonable estimate is a 100- or 200-ton machine, or for large manufacturers, larger 500-ton machines. An equation to determine an effective number of cavities for a mold compared to the press tonnage is:

$$N = \frac{T}{\sigma A} \tag{14.5}$$

where A is the projected area of the part along the mold parting line and σ is a material factor (tons/cm^2) and is given approximately by Table 14.5.

The tooling cost can be determined using Figure 14.15 and Tables 14.5 and 14.6, in conjunction with Eq. (14.5) above.

Die cast parts: To determine an estimate for die cast parts, a cost per part can be gathered by estimating the material, processing, and tooling costs. The material costs can be determined as before. The processing cost can be determined using Tables 14.3 and 14.4. The tooling cost can be determined using Figure 14.17.

TABLE 14.6. TOOL COST INCREASE WITH NUMBER OF CAVITIES

Part complexity	Cavities	Mold cost increase factor
Simple	4	2
	12	3
	24	4
	48	5
Medium	2	2
	4	2.5
	8	3
Complex	2	1.5
	4	3

Powdered-metal parts: To determine an estimate for powdered-metal parts, a cost per part can be gathered by estimating of the material, processing, and tooling costs. The material costs can be determined as before. The tooling cost can be determined using Figure 14.19. The processing cost is a combination of the green mold filling, an oven-sintering operation, and then subsequent finishing operations. Generally, the process times are dominated by the firing times, given by Table 14.7. The total processing time is given by Table 14.4.

Cardboard packaging material costs: Before a product ships, it must be put into retail packaging. This can become surprisingly expensive, especially when considering the assembly costs of inserting the product into the packaging. A formula for the cardboard shipping container packaging itself, with no printing on the packaging, is

$$C = \frac{0.6 + 0.3V}{N} \tag{14.6}$$

TABLE 14.7.   REQUIRED FIRING TIMES FOR DIFFERENT POWDER-METAL MATERIALS

Material	Firing time (min)
Copper, brass, and bronze	10–45
Iron and iron-graphite	8–45
Nickel	30–45
Stainless steels	30–60
Alnico alloys (for permanent magnets)	120–150
Ferrites	10–600
Tungsten carbide	20–30
Molybdenum	120
Tungsten	480
Tantalum	480

where V is the envelope volume of the product, N is the number of products per container, and C is in 1998 dollars (adapted from Santandreu 1993).

More accurate cost accounting: This chapter has presented some simple formulas for estimating costs of some manufactured components. These are very approximate; more detailed models are available to obtain a more accurate analysis, such as Boothroyd and Dewhurst's DFM suite of analysis tools, Galorath's SEER-DFM, (Galorath 1996), Cognition Systems' cost estimating tools, and a wide range of university developed software tools.

Beyond the selected materials and processes, there are a large set of other manufacturing processes and materials. Each can be cost-estimated in a similar way, by first determining the cost drivers of the process and then establishing cost coefficients against these drivers. Cost estimating is a well established profession. Every bid submitted on any job has a professional cost estimate developed with cost analysis tools, and professionals strive to improve these tools to submit good cost estimates. Many job shops and industry groups are willing to share these tools.

We now turn to estimating the assembly cost of a product design.

Basic Assembly Method: Adapted Xerox Producibility Index

We present here a basic method for determining an indented manufacturing assembly cost model. That is, an account of the costs built-up in the product hierarchically by subassembly down to individual parts. The method is a combination of *assembly trees* developed by Ishii (1994), and an adapted version of the *Xerox Producibility Analysis* (XPI), originally developed at the Xerox Corporation (Lewis 1985; Waterbury 1986), and an experimentally determined time equivalent labor factor against the XPI score.

Assembly trees: The first step in establishing an assembly cost model is to establish the assembly sequence hierarchy. Using a tree diagram with the final product as the trunk and each attached part as a leaf node, one should diagram what is attached to what. Exploded view diagrams are very helpful in constructing the assembly tree.

On the assembly tree, one can characterize each assembly step by indicating fixturing needs (the symbol "F"), reorientation (the symbol "R"), and insertion directions (straight and rotational arrows). This information feeds directly into assembly analysis worksheets. As an ex-

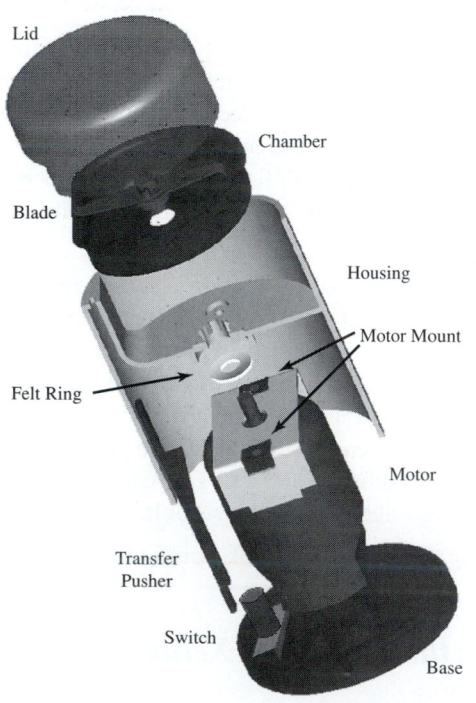

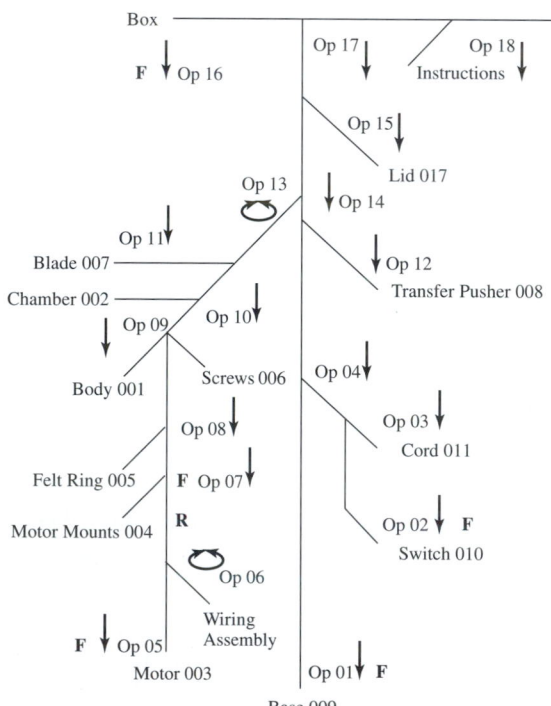

▼ *Figure 14.20.*
Coffee mill and assembly tree.

TABLE 14.8.   ADAPTED XPI ANALYSIS WORKSHEET

Subassembly/Part	Operation	Approach	Held?	Tightening	Repetitions	XPI Score
					SUM:	

ample, Figure 14.20 depicts an exploded view and assembly tree for the Krups Coffee Mill (Chapter 7).

Xerox producability analysis: To estimate assembly difficulty, one very simple method is to use a variation on the *Xerox Producability Index* (XPI) as an assembly difficulty indicator. The subsequent material is adapted from the original method developed at the Xerox Corporation (Lewis 1985; Waterbury 1986). It is a very simplified design for assembly method requiring simple tabulation of efficiency values. Application of the method is straightforward:

Step 1:   Draw the assembly sequence diagram (tree), as discussed above.

Step 2:   For each subassembly as shown in Table 14.8, create a table consisting of:

1. Each part down each row.
2. A column of assembly approach direction (top, side, rotated, bottom).
3. A column of whether a part must be held during assembly (Y/N).
4. A column of tightening required (weld, solder, stake, adhesive, pin, nut, tape, screw, ring, snap).
5. A column of number of repetitions.
6. If there is more than one operation for each part, enter multiple rows.

Step 3:   Using the data of Step 2 and the adapted XPI scoring of Table 14.9, determine the XPI score for each part.

TABLE 14.9. ADAPTED XPI SCORE SHEET

					Assembly Tightening and Tooling						
		Special tool			Small tool						No tool
Assembly approach	Hold?	Weld	Solder	Stake	Adhesive	Pin	Nut	Tape	Screw	Ring	Snap insert
Assemble from top	N	1	10	20	30	40	70	60	70	90	100
	Y	0	1	10	20	30	40	50	40	70	90
Assemble from side	N	0	0	5	10	20	50	40	50	75	80
	Y	0	0	1	5	15	25	35	25	55	75
Assemble from bias	N	0	0	0	1	10	40	30	40	65	70
	Y	0	0	0	0	5	15	25	15	45	85
Rotated parts	N	0	0	0	1	10	40	30	40	65	70
	Y	0	0	0	0	5	15	25	15	45	65
Assemble from bottom	N	0	0	0	0	0	30	0	30	20	60
	Y	0	0	0	0	0	0	0	0	0	55

Source: Permission of Xerox Corporation.

$$XPI_{part} = \frac{\sum\limits_{\substack{part \\ operations}} XPI_{operation}}{\text{number operations}} \qquad (14.7)$$

As a part of Step 2.6, subtract 10 from the score for each multiple approach and multiple tightening of the XPI Scoring Table required for the insertion that must be done simultaneously.

$$XPI_{part} = \frac{\sum\limits_{\substack{part \\ operations}} XPI_{operation} - 10n_{repeated}}{\text{number operations}} \qquad (14.8)$$

Step 4: Repeat Steps 1–4 for each subassembly.

Step 5: Average the XPI subassembly scores for an XPI of the whole product.

This method is very easily implemented and provides good hints on what to change in the design concept. The XPI score is useful for comparing two concepts. To use as an estimate of assembly cost, the XPI score must be converted to an equivalent assembly time so that a labor rate can be applied as above. This is beyond the intent of the original development and is not particularly accurate, but one model is:

$$t = T_{\text{nominal}}e^{K(XPI_{\text{nominal}}-XPI)} \tag{14.9}$$

where t is the new time estimate, K is a scaling factor, XPI is the new XPI rating, XPI_{nominal} is the old XPI rating of the original design before any modifications, and T_{nominal} is the assembly time for that original design, measured during a teardown, for example. K is a means to bound the new estimate result—a reasonable value is $K = 0.69 = \ln(2.0)$, meaning the new time estimate t will never be smaller than $1/2\,T_{\text{nominal}}$ and never larger than $2\,T_{\text{nominal}}$. This model captures only the time trends as increased or decreased from the nominal design. It does not necessarily predict a realistic new design assembly time. Therefore, this model only suffices for a simple comparative estimate of redesign options.

Example: Coffee Mill

As an example, consider the Krups coffee mill, shown in Figure 14.20. The assembly process is broken down into an assembly tree diagram as shown in the figure. Then, for each of these operations, the XPI analysis is completed. This is shown in Table 14.10. As a cost model for design changes to the coffee mill, an assembly time equation of

$$t = 3.00e^{0.69(0.82-XPI)} \tag{14.10}$$

can be used (in minutes), where 3 minutes is used as the nominal assembly time, using $\ln(2.0)$ for a value for K. Note that at 3 minutes, this equates to about \$3 of assembly labor cost at \$60 per hour burdened labor.

In terms of assembly difficulty and candidate alternatives to improve the assembly, one should examine the components for low XPI ratings. Basically, the analysis indicates opportunities for assembly ease in the cord attachment, the wiring assembly (for high temperature motor cut out), the body attachment to the motor, the blade attachment, and the assembly of the body assembly and pusher to the base assembly.

Next, the individual parts are costed using the tooling costs of Figures 14.15–14.19 and estimated material and OEM part costs, all at an approximate 100,000 units per year. The results are shown in Table 14.11. Note the majority of the part costs appears in the electric motor.

TABLE 14.10. XPI ASSEMBLY ANALYSIS OF THE KRUPS COFFEE MILL

Subassembly/Part	Operation	Approach	Held?	Tightening	Repetitions	XPI Score	XPI Totals
Base 009	01	Top	No	No	1	100	100
Switch 010	02	Top	No	No	1	100	100
Cord 011	03a	Rotated	Yes	No	1	65	33
Cord solder	03b	Top	Yes	Solder	1	1	
Motor 003	04	Top	No	No	1	100	100
Wiring Assembly	05	Top	No	No	1	100	50
Wire solder	05	Side	Yes	Solder	1	0	
Motor Mounts 004	06	Top	No	No	2	100	80
Felt Ring 005	07	Top	No	No	1	100	100
Body 001	08	Top	No	Screw	1	70	50
Screws 006					2		
Chamber 002	09	Top	No	No	1	100	100
Blade 007	10	Top	Yes	Screw	1	40	40
Wire solder	11a	Top	Yes	Solder	1	1	60
Pusher 008	11b	Top	Yes	No	1	90	
Body Assembly	12	Top	Yes	No	1	90	
Lid	13	Top	No	No	1	100	100
Box	14a	Top	No	No	1	100	100
Mill Assembly	14b	Top	No	No	1	100	100
Foam	14c	Top	Yes	No	1	90	90
Instructions	15a	Top	No	No	1	100	100
Cord	15b	Top	Yes	No	1	90	90
Close up	15c	Top	Yes	No	1	90	90
						XPI TOTAL	82

This analysis would indicate the best way to reduce the cost of the mill is to find a cheaper motor, which directly impinges on the product quality and so is not really an option. The remainder of the parts are not particularly expensive, and so reducing component costs is not a high benefit proposition. Rather, if cost reduction must be done, then the best approach to reducing costs in this concept is to ease the assembly, whose cost at around $3 is about half of the total manufacturing cost.

Boothroyd DFA Analysis

Boothroyd has developed a design for assembly method that makes us of operation times. For the assembly of any two parts, the analysis is broken down into two categories, *handling time* and *insertion time*. Handling time is the time it takes to pick up a part, including the time it takes to look at it and identify how it is oriented so it can be grasped,

TABLE 14.11.   KRUPS COFFEE MILL PART COST ANALYSIS

Part	Tooling	Material	Processing	Total
Body 001	$ 0.01	$ 0.02	$ 0.02	$ 0.05
Chamber 002	—	$ 0.12	$ 0.01	$ 0.13
Motor 003	—	$ 1.50	—	$ 1.50
Wiring Assembly	—	$ 0.10	—	$ 0.10
Motor Mount 004	—	$ 0.01	$ 0.01	$ 0.02
Felt Ring 005	—	—	—	—
Screws 006	—	$ 0.01	—	$ 0.02
Blade 007	$ 0.04	$ 0.01	$ 0.03	$ 0.08
Transfer Pusher 008	$ 0.01	—	$ 0.01	$ 0.02
Base 009	$ 0.02	$ 0.03	$ 0.01	$ 0.06
Switch 010	—	$ 0.03	—	$ 0.03
Cord 011	—	$ 0.10	—	$ 0.10
Lid 012	$ 0.01	$ 0.16	$ 0.02	$ 0.19
				$ 2.30

grasping it, and carrying it. Insertion time is the time it takes to orient the part, insert it, and fasten it down.

The basic manual calculation method (Boothroyd 1992) has been extended to a suite of very effective and more accurate computer analysis tools, including manual assembly analysis and individual part manufacturing analysis for such parts as injection-molded, cast, machined, and electronic parts.

Application of the manual method is straightforward using the subassembly worksheet and two pages of manual handling and manual insertion times. That is, the worksheet of Table 14.12 should be completed for each subassembly and for the final assembly.

Manual handling time is determined using Figure 14.21. For each part, one must know:

▶ the number of hands required

▶ the symmetry of parts in terms of α, the angle of symmetry about an axis perpendicular to the insertion direction, and β, the angle of symmetry about the insertion direction

▶ the difficulty in grasping

These can be represented using a "manual handling code" in the handling time table.

TABLE 14.12. DFA ANALYSIS WORKSHEET (BOOTHROYD 1980)

1	2	3	4	5	6	7	8	9	10
Part	Part number	Number of times the operation is repeated	two-digit BD manual handling code	manual handling time per part (sec)	two-digit BD manual insertion code	manual insertion time per part (sec)	Part theoretically needed?	Operation Time (sec) (3) x [(5) + (7)]	Operation cost ($) [$/hr] x (9) / 3600
	Theoretical Minimum Number of Parts =								$
	Total Assembly Time (sec) =								
	Boothroyd Dewhurst DFA Efficiency = $\dfrac{3 \times \text{Min \# Parts}}{\text{Total Time}}$ =								

Manual insertion time is determined using Figure 14.22. For each part, one must know:

- how the part is secured
- the resistance and alignment during insertion
- the view during insertion

These can be represented using a "manual insertion code" as in the insertion time table.

		No handling difficulties			Part nests or tangles			Part can be lifted with 1 hand but requires 2 because of tangle, nest, flex, or forming		
		thickness > 2mm		2mm	thickness > 2mm		2mm	180		= 360
		size > 15mm	6 mm < size 15 mm	size 6mm	size > 15mm	6 mm < size 15 mm	size 6mm	size > 15mm	6 mm < size 15 mm	size 6mm
symmetry (deg)		0	1	2	3	4	5	0	1	2
sym < 360	0	1.13	1.43	1.69	1.84	2.17	2.45	4 4.1	4.5	5.6
360 sym < 540	1	1.5	1.8	2.06	2.25	2.57	3.0			
540 sym < 720	2	1.8	2.1	2.36	2.57	2.9	3.18			
sym = 720	3	1.95	2.25	2.51	2.73	3.06	3.34			

▼ *Figure 14.21.*
Selected handling codes for DFA analysis (Boothroyd and Dewhurst 1999).

The analysis is basically completed in five steps.

Step 1: Draw the assembly sequence diagram (tree) as discussed above.

Step 2: For each subassembly, complete the manual assembly time table consisting of:

▶ each part down each row

▶ number of repetitions for the part

▶ handling code

▶ handling time, with a basic handling time of 1.13 seconds

▶ insertion code

▶ insertion time, with a basic insertion time of 1.5 second

▶ calculated operation time in seconds

▶ theoretical determination of whether the part is required to be unique

Step 3: Calculate the summary statistics of:

1. TM, the total assembly time

2. CM, the total assembly cost at $0.011 per second, or $40 per hour burdened labor rate

3. NM, the theoretical minimum number of unique parts required (determined as discussed earlier in the chapter)

4. DE, the design efficiency. Boothroyd was the first to propose such an assembly efficiency metric. It is based on comparing to a 3 second time per part to assemble only the minimally necessary number of parts.

		For part inserted but not secured immediately or secured by snap fit							Part inserted and secured immediately by screw fastening with power tool		
		secured by separate operation or part				secured on insertion by snap fit					
		no holding down required		holding down required							
		easy to align	not easy to align	easy to align	not easy to align	easy to align	not easy to align			easy to align	not easy to align
		0	1	2	3	4	5			0	1
no access or vision difficulties	0	1.5	3.0	2.6	5.2	1.8	3.3	No access or vision difficulties	3	3.6	5.3
obstructed access or restricted vision	1	3.7	5.2	4.8	7.4	4.0	5.5	restricted vision only	4	6.3	8.0
obstructed access and restricted vision	2	5.9	7.4	7.0	9.6	7.7	7.7	obstructed access only	5	9.0	10.7

▼ **Figure 14.22.**
Selected insertion codes for DFA analysis (Boothroyd and Dewhurst 1999).

$$DE = \frac{3 \times NM}{TM} \qquad (14.11)$$

Step 4: Repeat Steps 2 and 3 for each subassembly.

Step 5: Repeat Steps 2 and 3 for the final assembly, putting together the sub–assemblies.

Example: Krups Coffee Mill: As an example, let's revisit the Krups coffee mill and complete this DFA analysis. The assembly analysis cost table is shown in Table 14.13. Each part is analyzed in terms of the manual and insertion difficulty, and a time required to assembly the mill of about 2 minutes is thereby derived using the charts. The fractional breakdown of time for each assembly operation is what is important to examine in order to understand where the assembly is difficult. One should examine all parts with zero in column (9), that is, the parts that are not theoretically required, and work to eliminate the parts with high assembly times.

The DFA analysis indicates which parts are most difficult to assemble, as cost pie charted in Figure 14.23. This proves interesting; for example, the pusher has a disproportionate fraction of the assembly cost given its meager function. Also, the motor mounts are rather expensive to assemble; changing their design to ease insertion would be beneficial. The other interesting aspect is that completing the packaging is also a rather large fraction of the product assembly cost given its function. All of these various costs should be examined for their impact and their importance and opportunities for redesigning the product considered.

Even though the manual method above is useful for simple educational examples, the reader should not believe that it can serve as a replacement for modern design for manufacture and assembly software analysis tools. For an in-depth analysis, the more extensive Boothroyd-Dewhurst design for manufacture and assembly software should be used (Boothroyd Dewhurst Inc. 1999). This software suite of analysis tools includes more accurate assembly times with further explanatory variables than the rough estimates of Figures 14.21 and 14.22 and more extensive databases of different kinds of assembly operations. Much more important to a practical product development process, the software can be integrated into other CAD and software analyses. Further, the software has very extensive piece part DFM analysis tools also integrated into the system. The material in this section is merely intended to give a sense of the power of design for manufacture, provide a tool for very simple analyses, and show what theory an effective tool is built on.

V. CRITIQUE OF DESIGN FOR ASSEMBLY METHODS

There are many tools available for making a preliminary estimate of assembly difficulty of a product. It is worth comparing the results of different systems and the purposes for which they are intended. To do this, a comparison was conducted on a case study of the Kodak Funsaver camera shown in Figure 14.24 (Gonzales-Zugasti et al. 1997).

Each of the DFA methods compared are listed in Table 14.14, from Gonzales-Zugasti, Meeker, and Otto, (1997). Each method uses slightly different variables to estimate some index (or indices) of the assemblability. This section explores the variables used, classifying them into four major categories, each in turn divided into subcategories. Also, different methods ask different questions for any given variable. We present and classify the type of user input required by each method. Finally, the methods are also classified by the amount of time required to analyze an assembly. The required times to analyze the Kodak disposable camera are listed.

We have restricted our analysis to the *manual* mechanical-assembly-related sections of these DFA methods. Some of the tools also have sections for analyzing automated assembly lines and electronics (PCB) assembly, but we have concentrated our analysis to just the manual assembly of mechanical components. Second, some of the methods also include other questions not listed here because they are not strictly related to assembly, but to manufacturing costs. For example, SEER DFM has a section devoted to assembly, which is analyzed here, but is mainly a design-for-

TABLE 14.13. DFA ANALYSIS FOR THE KRUPS COFFEE MILL

1 Part	2 Part number	3 Number of times the operation is repeated	4 two-digit BD manual handling code	5 manual handling time per part (sec)	6 two-digit BD manual insertion code	7 manual insertion time per part (sec)	8 Part theoretically needed?	9 Operation Time (sec) (3) x [(5) + (7)]	10 Operation cost ($) [$/hr] x (9) / 3600
Base	009	1	30	1.95	00	1.5	1	3.45	0.03
Switch	010	1	30	1.95	30	2	1	3.95	0.03
Cord	011	1	30	1.95	95	8	1	9.95	0.08
Motor	003	1	30	1.95	00	1.5	1	3.45	0.03
Wiring Assembly	WA	1	30	1.95	95	8	0	9.95	0.08
Motor Mount	004	2	32	2.7	31	5	0	15.40	0.13
Felt Ring	005	1	00	1.13	00	1.5	0	2.63	0.02
Body	001	1	33	2.51	49	10.5	1	13.01	0.11
Chamber	002	1	33	2.51	30	2	0	4.51	0.04
Blade	007	1	38	3.34	58	10	1	13.34	0.11
Pusher	008	1	30	1.95	06	5.5	0	7.45	0.06
Motor Assembly	MA	1	30	1.95	51	9	0	10.95	0.09
Lid	017	1	30	1.95	02	2.5	1	4.45	0.04
Box	–	1	30	1.95	00	1.5	0	3.45	0.03
Instructions	–	1	03	1.69	06	5.5	0	7.19	0.06
Final Assembly	FA	1	30	1.95	12	5	0	6.95	0.06
Foam	–	1	03	1.69	00	1.5	0	3.19	0.03
Close up the package	–	1	–	1	–	0	0	1.00	0.01
	Theoretical Minimum Number of Parts =						7		$ 1.04
	Total Assembly Time (sec) =							124.27	
	Boothroyd Dewhurst DFA Efficiency $= \dfrac{3 \times \text{Min\# Parts}}{\text{Total Time}} =$							17%	

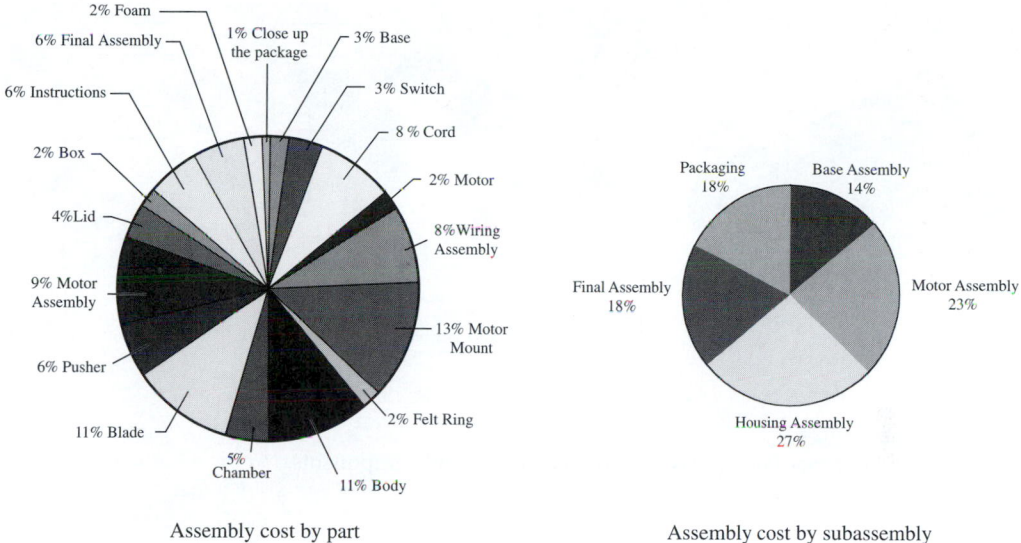

Assembly cost by part

Assembly cost by subassembly

▼ *Figure 14.23.*
Krups Coffee Mill assembly cost breakdown. (a) Assembly cost by part. (b) Assembly cost by subassembly.

manufacturing tool (Galorath 1996). Assembly View and Boothroyd-Dewhurst contain questions about tooling costs and part costs, but again those were considered DFM questions and are not strictly assembly related. Finally, many of these methods can be modified by users to account for variables that are not explicitly covered in their analysis. For example, the contribution to assembly costs from secondary operations could be accounted for by calculating difficulty indices or time penalties for particular operations and merging them with the other analysis results in methods that do not cover them. However, if a variable is not explicitly analyzed by the DFA tool, it is listed as not considered.

Functional analysis questions are present in a few of the DFA methods. It is not used to estimate the asesmbly time for a product, but as a tool for use during the design phase in order to improve the assembly of a product. The goal of this analysis is to point out all the parts that are not necessary for the function of the product, so they can be eliminated. All the DFA tools that perform a functional analysis ask similar questions regarding a part's movement in relation with other parts, material, service, or disassembly requirements to determine whether the part is theoretically necessary or could be combined with existing parts to simplify the assembly. The result of this analysis is generally a ratio com-

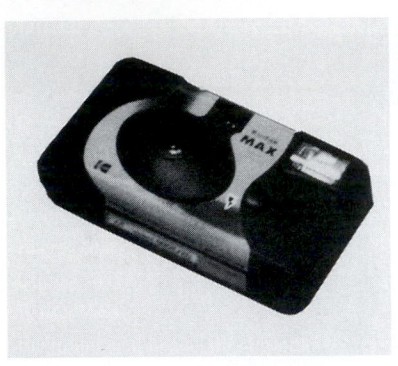
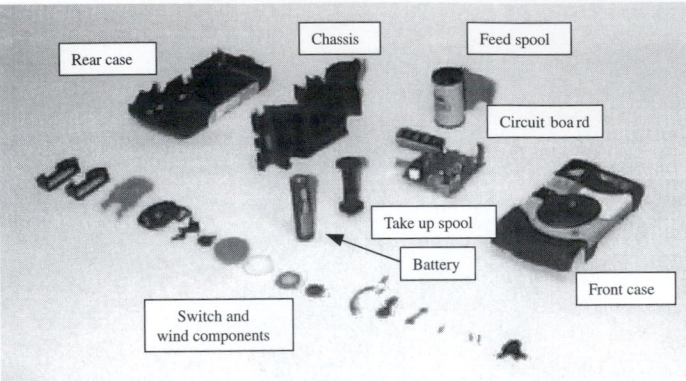

▼ *Figure 14.24.*
Kodak FunSaver Pocket Flash disposable camera and components.

paring the theoretical minimum part count to the actual part count for an assembly. If this ratio or percentage figure is low or below a recommended threshold value, the methods suggest simplifying the assembly, and some of them provide rules or suggestions for the redesign.

Most DFA tools require information on the handling of the parts for the product under analysis. The data are used to determine the difficulty of acquiring each part and bringing it to the point of insertion. In general, the questions related to the handling of the parts can be grouped into four categories:

1. Part attributes such as dimensions, weight, and shape.

2. Handling distance that the assembler has to reach to get to the part.

3. Handling method, or what is required in terms of operators, tools, and aids to handle the part.

TABLE 14.14.   DFA METHODS COMPARED

Name	Format
Boothroyd & Dewhurst DFA	Software (v. 7.1)
Lucas Engineering and Systems DFA	Handbook
Xerox Producibility Index	Handbook
SEER DFM	Software (v. 3.1)
LASeR	Software (v. 1.0)

4. Handling difficulties or conditions that make it difficult to get the part to the assembly point.

On the other hand, not all methods consider secondary operation questions or those centered around extra assembly difficulty due to secondary operations such as testing, measuring, and inspection. These operations can add up to significant percentages of the assembly cost for an assembly, depending on the type of product, quality requirements, degree of automation, and numerous other factors. The DFA tools that do explicitly consider these extra operations provide a few predefined, standard operations but also allow the user to define his/her own set of operations to account for the wide range of possible operations present in a particular assembly. Note that some of the DFA tools that do not explicitly consider secondary operations can be modified by the user to add penalties or indices to account for these extra assembly steps.

Even though the questions asked by the various DFA tools are in general very similar, the level of detail of required user input varies widely. Some of the methods consider only small subsets of the variables required by the more complex tools. Also, even when two methods consider the same variable, one might ask a simple yes or no question to check for a certain property of a part or assembly task, while another may ask for a numerical measurement of that property. For example, some methods ask for the dimensions of each part to determine if it is difficult to handle, and the user enters numbers for the length, width, and depth of the part. Other methods only ask whether the part is large or heavy and only require a yes or no answer, while others do not even consider the size of the parts at all.

The user inputs to the DFA questions were classified into five types:

- *Yes/No.* User specifies whether a certain condition exists in an assembly operation or part (e.g. Are two hands required?)
- *Options.* User chooses one or more of a few options to describe an assembly or part property (e.g., fastening procedure: snap, weld, screw, etc.)
- *Levels.* User picks one out of a few discrete ranges or options of a certain variable (e.g., handling distance: <500 mm, 500–1000 mm, 1000–2000 mm, >2000 mm)
- *Number.* User enters a figure for a certain assembly or part attribute (e.g., part length in mm)
- *Formula.* User enters a formula or index to specify a custom-defined operation.

Most of the analysis performed by the different DFA tools studied use the same set of variables but to different levels of detail. This affects (among other things) the time needed to analyze an assembly and the accuracy of the estimates. In the case of the camera, assembly analysis times ranged from 20 minutes to almost 3 hours. Assembly time estimates (for those methods that report estimates in seconds) ranged from 155 to 229 seconds for the disposable camera, compared to an actual manual assembly time of about 200 seconds ± 30 seconds. This was measured by persons putting the camera together in the laboratory in the same sequence as analyzed with the tool. Therefore, all of the tools provide reasonable overall assembly time estimates.

The simpler DFA tools, such as XPI, which only consider a few variables in the assembly analysis, may not provide enough information about certain types of products. For example, in analyzing the camera, XPI returned a high score for the assembly, since almost all the parts are simply placed or snapped into place and very few pose fitting difficulties. Similarly, in analyzing the Krups Coffee mill in the previous section, the XPI design efficiency was 82%, whereas the more detailed Boothroyd-Dewhurst resulted in a 17% design efficiency. The higher XPI rating is due to the XPI analysis being very course, not recognizing subtle assembly difficulties, whereas Boothroyd-Dewhurst could. With respect to the Kodak Funsaver camera, the XPI method could not differentiate the camera parts enough to be able to point out possible product improvements. It simply helps to eliminate design choices that should not have been considered. On the other hand, it is a quick method to apply and may be useful as a tool to compare alternative designs that do show significant differences in part fitting.

Boothroyd-Dewhurst, Lucas, and LASeR are tools meant to not only estimate assembly time but also reduce part count. They all use the same criteria to determine the minimum possible part count. All of these methods reported large possible part-reduction improvements for the camera product analyzedDesign for manufacture and assembly is the most effective methodology to reduce product cost. Golden nuggets that should be retained from this chapter include:

- ◗ The basic techniques to improve a design are mostly a collection of common-sense rules.

- ◗ One can determine the most effective approach for redesign through stacked-up cost analysis.

- ◗ Modularize to minimize part count, design for top-down insertion with alignment features.

TABLE 14.15.   SUMMARY OF DFA VARIABLES, USER INPUTS, AND ANALYSIS TIME

	Detailed methods		Quick methods		
	BDI	Lucas	XPI	SEER	LASeR
Analysis time (camera)	2–3 hours		$<\frac{1}{2}$ hr	<1 hr	<1 hr
Estimated assembly time	220 sec	195 units	N/A	155 sec	35.2 Td
Accuracy (to nearest 10%)	110%	100%	—	80%	—
Functional analysis					
Theoretical part count	Yes/No	Yes/No			Yes/No
Handling analysis					
Part attributes					
Standard part					
Part size	Levels + Number	Levels			
Part weight (heavy?)	Yes/No	Yes/No			
Part shape/symmetry	Options +Levels	Options			
Handling distance	Levels				
Handling method					
More than one hand	Yes/No	Yes/No			
More than one person	Yes/No	Yes/No			
Tool requirements	Options	Yes/No			
Aids requirements	Options	Yes/No			
Handling difficulties					
Handling conditions	Options	Options			
Orienting ease		Levels			
Insertion analysis					
Wrong insertion possibility		Yes/No			
Insertion method					
Holding requirement	Yes/No	Yes/No	Yes/No		Options
Fastening method	Options	Options	Option		Options
Insertion direction		Options	Option		Options
Multiple/simultaneous processes	Options + Number	Options	Yes/No		Options + Number
Tool acquisition	Levels				
Number of fasteners	1	1	1	Number	1
Mechanization	2	2		Levels	
Insertion difficulties:					
Access/vision restrictions	Yes/No	Yes/No			
Alignment ease	Yes/No	Yes/No			
Resistance to	Yes/No	Yes/No			
Large-part difficulties	Options				
Installation difficulty	3	3	3	Levels	3
Whole-assembly difficulty	4	4	4	Levels	4
Special problems (assembly environment requirements)					
Secondary assembly					
Standard operations	Options	Options	Option		Options
Explicit user-defined	Formula				

TABLE 14.15. SUMMARY OF DFA VARIABLES, USER INPUTS, AND ANALYSIS TIME (CONTINUED)

	Detailed methods		Quick methods		
	BDI	Lucas	XPI	SEER	LASeR
Secondary operations Nonassembly operations Predefined (Library) Explicit user-definable Functional test requirement*	Options Formula 5	Options Formulas 5		Yes/No	

Notes: [1]Fasteners are considered in the same manner as regular parts by all methods except SEER.

[2]Boothroyd-Dewhurst and Lucas have separate analysis modules to model assembly automation.

[3]All methods but SEER consider installation under separate part orientation, acquisition, and insertion questions.

[4]All methods but SEER consider this variable separately under size, weight, and tolerances questions.

[5]SEER only considers functional testing of the assembly; Boothroyd-Dewhurst and Lucas explicitly analyze other nonassembly operations.

VI. CHAPTER SUMMARY AND "GOLDEN NUGGETS"

Design for manufacture and assembly is the most effective methodology to reduce product cost. Golden nuggets that should be retained from this chapter include:

▶ The basic techniques to improve a design are mostly a collection of common-sense rules.

▶ One can determine the most effective approach for redesign through stacked-up cost analysis.

▶ Modularize to minimize part count, design for top-down insertion with alignment features.

▶ Think throughly and simplify the fabrication difficulty of each feature on every part.

References

Adler, R., and Schwager, F. 1992. Software makes DFMA child's play. *Machine Design* 9 April.

Alting, L. 1982. Manufacturing engineering processes. New York: Marcel Dekker.

Andreasen, M. 1983. *Design for Assembly*, IFS Publications, Ltd., U.K., Springer Verlag.

Baldwin, S. 1966. How to make sure of easy assembly. *Tool Manufacturing Engineering. May 67.*

Barkan, P., and Hinckley, C. 1993. The benefits and limitations of structured design methodologies. *Manufacturing Review* 6 (3):211–220.

Boothroyd, G. 1992. Assembly automation and product design. New York: Marcel Dekker.

Boothroyd, G. 1980. Design for Assembly – A Designer's Handbook. University of Massachusetts, Amherst, MA.

Boothroyd, G. 1982. Design for Producability–The Road to Higher Productivity. *Assembly Engineering.*

Boothroyd and Dewhurst, Inc. 1999. http://www.dfma.com/.

Boothroyd, G., and Knight, W. 1989. Fundamentals of machining and machine tools. 2nd ed. New York: Marcel Dekker.

Bradney, D. 1994. The NFFS guide to aluminum casting design: Sand and permanent mold. Des Plaines, Il. Non-Ferrous Founder's Society.

Brannan, B. 1991. Six sigma quality and DFA. *Boothroyd and Dewhurst DFMA*^{TM} *Insight* 2 (1): 1–3.

Crow, K. 1988. Ten steps to competitive design for manufacturability. The Second International Conference on Design for Manufacturability. Chestnut Hill, MA.

Dayton Rogers. 1997. How to Design Metal Stampings. Minneapolis, MN: Dayton Rogers Manufacturing Co.

Digital Equipment Corporation (DEC). 1993. Design for assembly: General principles applied to power design/manufacture. Power Systems Business Unit, Digital Equipment Corp.

El-Wakil, S. 1998. Processes and design for manufacture. 2nd ed. Boston: PWS Publishing.

Galorath Associates. 1996. SEER-DFM user's manual.

Garmon, L. 1999. The American experience: Spy in the sky. Available at PBS On-line, *http://cgi.pbs.org/wgbh/amex/u2/u2.html,* 18 May.

Gonzales-Zugasti, J., Meeker, D., and Otto, K. 1997. Assembly analysis comparison of various DFA methods. Tech. Rep. Engineering Design Research Laboratory, MIT, Cambridge.

Huthwaite, B. 1990. How to reduce part counts. *Design News* 154.

Iredale, R. 1964. Automatic assembly—components and products. *Metalworking Production* 8 April.

Ishii, K. 1994. Life-cycle engineering design. *ASME Journal of Mechanical Design.*

Kalpakjian, S. 1995. Manufacturing engineering and technology, 3rd ed. Reading, MA: Addison-Wesley.

Keyes, K. 1983. Stamping design thru maintenance. Dearborn, MI: Society of Manufacturing Engineers.

Lewis, G. 1985. Design for assembly & automation. Webster, NY: Xerox Automation Institute.

Malloy, R. 1994. Plastic part design for injection molding. Cincinnati, OH: Hanser Publishers.

Miles, B., and Swift, K. 1992. Working together. *Manufacturing Breakthrough* 69–73.

Meeker, D. and A. Thornton. 1995. Benchmarking within the Product Development Process. ASME Design Theory and Methodology Conference, Boston, MA.

Munro, S. 1994. Design for manufacture can work. *OEM Off-Highway.*

Paterson, R. 1965. Recent developments in feeding and orienting. Assembly Fastening and Joining, A PERA Conference and Exhibition, Production Engineering Research Association of Great Britain, Sept.

Rosato D., and Rosato, D. 1995. Injection molding handbook. 2nd ed. New York: Chapman and Hall.

Sachs, G. 1966. Principles and methods of sheet-metal fabricating. New York: Reinhold.

Sturges, R. 1989. A quantification of manual dexterity: The design of an assembly calculator. *Robotics and Computer-Integrated Manufacturing* (3): 237–252.

Tipping, W. 1965. Component and product design for mechanized assembly. Assembly Fastening and Joining, A PERA Conference and Exhibition, Production Engineering Research Association of Great Britain, Sept.

Tsiantar, D. "George Smiley Joins the Firm," *Newsweek,* May 2, 1988, pp. 46–47.

Waterbury, R. 1986. Applying design for assembly principles. *Assembly Engineering* 42–45.

Wakil, S. 1998. Processes and Design for Manufacturing, PWS Publishing, Boston.

Xerox Corporation. 1986. Generic tool scoping manual, RMO/QA/PQE/AME. 2nd ed. Xerox Corp.

15 Design for the Environment

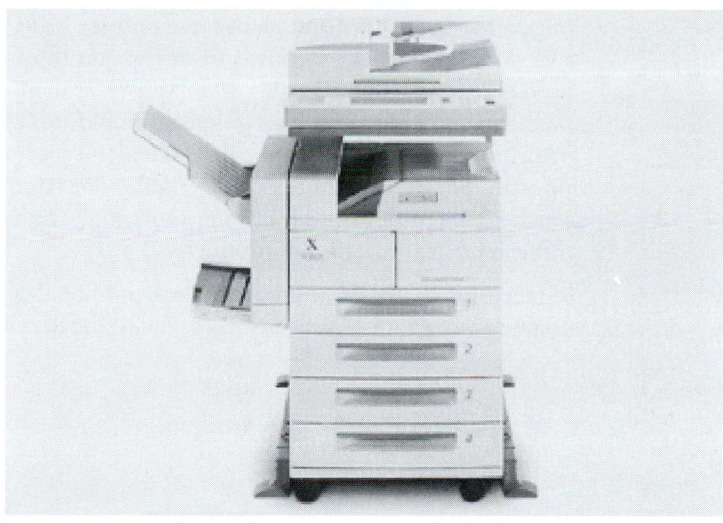

In the late 1990s, the engineers at the Xerox Corporation set about to do a complete clean sheet design of a new mid-sized networked copier/printer/scanner/FAX. Demonstrating that design for the environment is possible now and not just off in some distant future, the new DocuCenter product line is entirely 100% zero-to-landfill, all parts and systems (including electronics) are remanufactured or recyclable as commonly recycled materials. The result was an efficient machine—the total part count was reduced from thousands in the previous generation down to 235 unique part numbers. The focus on a modular product family architecture, design for manufacturability, for repairability, and for environmental efficiency permitted this revolutionary design.

Agrowing concern in the latter part of the twentieth century has been the tremendous impact humans now have on the environment. We can no longer think of our technical and industrial society as an independent subentity of a much larger system that we can extract materials from and dump waste into. Society generates and consumes such a large fraction of the Earth's resources that we must consider our impact on the environment in our technical decision making. A growth area for society, engineering, and design is to simply maintain the standard of technological living we now enjoy into the next centuries, but at a sustainable level of low environmental impact. Design for the Environment (DFE) is a product design approach for reducing the impact of products on the environment.

Products can have adverse impact on the environment during their manufacture through the use of highly polluting processes and the consumption of large quantities of raw materials. They can also have adverse impact through the consumption of large amounts of energy and long half-lives during disposal. Because of these issues, one must consider a product's entire *life cycle*, from creation through use through disposal, as shown in Figure 15.1. In this life cycle, there are many events of creating pollution and many opportunities for recycling, remanufacturing, reuse, and reducing environmental impact. Merely designing a product to use nontoxic materials is not enough. As product designers, we must bring all our ingenuity to bear on the challenging problem of creating efficient products.

To meet this challenge, we must understand the idea of life cycle assessment that adopts a holistic view by analyzing the entire life cycle of

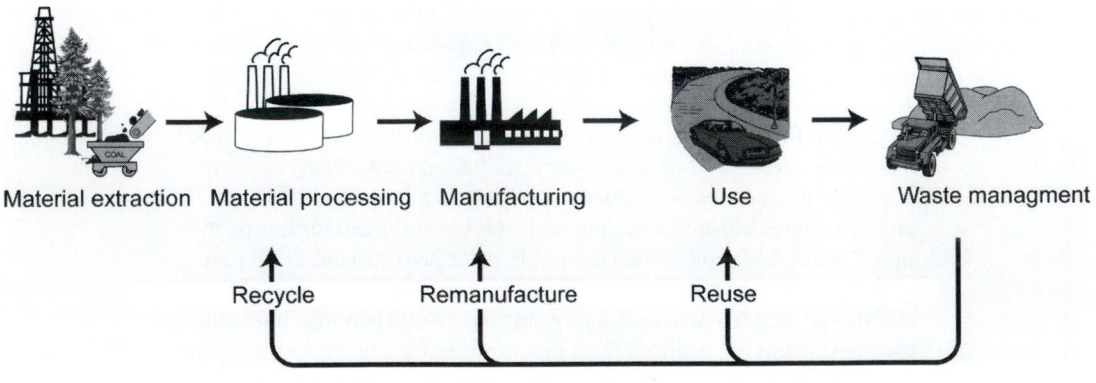

▼ *Figure 15.1.*
Stages of a product life cycle.

a product, process, package, material, or activity. Life cycle stages encompass extraction and processing of raw materials; manufacturing, transportation, and distribution; use/reuse/maintenance; recycling and composting; and final disposition.

I. CHAPTER ROADMAP

In this chapter, we first review the environmental-impact causes—mitigating these causes form the underlying objectives behind any environmental assessment. Then, in Section II, we present a basic approach to assessing the environmental impact of a product concept through guidelines on material selection and energy usage. In the subsequent section, we present more-advanced quantitative methods for making this assessment. We then review detailed life cycle assessment methods. Finally, we present design-for-X techniques on environmental design topics such as recycling and disassembly analysis. Figure 15.2 diagrams this chapter roadmap.

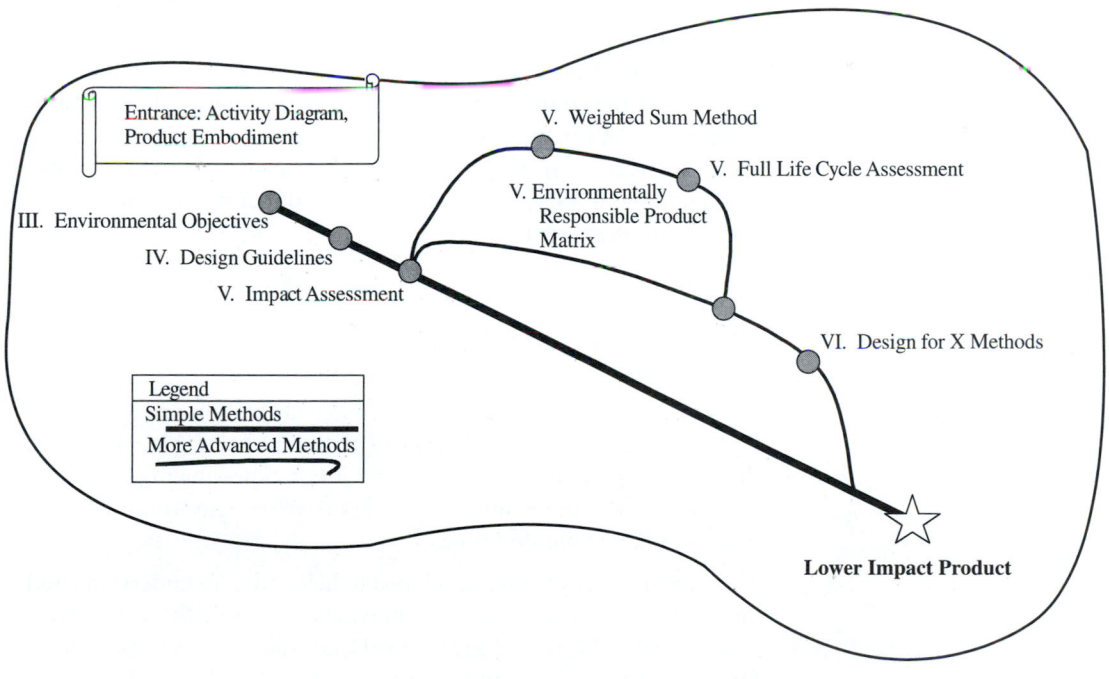

▼ **Figure 15.2.**
Chapter roadmap.

II. WHY DFE?

Design for the environment is an important activity for a design team because environmental damage is, as are most things, greatly influenced in the early design phases. Just as with production cost, a reasonable heuristic is that 80% of the environmental damage of a product is established after 20% of the design activity is complete.

From a business case analysis viewpoint, the reasons a design team should choose to complete design for the environment studies is that modern customers are now demanding products with less environmental impact. As the world's population is becoming aware of society's influence, society is attempting to minimize its impact. Creating a product that impacts the environment less becomes a market advantage. As an example, Xerox has set a commitment to be a waste-free company; as a part of its operations and its products, it will generate no waste material that cannot be recycled or remanufactured (Xerox, 2000).

Governmental agencies are also being asked by their constituents to develop and enforce reduced environmental impact standards for products. Such regulatory pressure will only grow. Many countries now have mandatory postuse buy-back requirements for products such as packaging, computers, and transportation vehicles, complete with required recycling of the components. Standards are also being developed to support design for the environment as a practice. Underlying all of these activities are the market forces that demand design for the environment as a necessary part of modern product development. Customers now demand it.

III. ENVIRONMENTAL OBJECTIVES

The reason customers demand environmentally friendly products stems from a concern over several types of pollution being generated. It is thus important to understand these pollution types, their range of impact, and what can be done in product development to mitigate the impact of these pollution types.

There are many guidelines developed to help industry understand and deal with its impact on the environment. For example, the *Valdez Principles* (*Financial Times*, 27 March 1991), later adopted by the Coalition for Environmentally Responsible Economies (CERES) as the *CERES Principles*, establishes objectives and guidelines to help companies maintain good environmental performance. These principles were adopted by the Exxon Corporation after the Exxon Valdez oil spill (March 24, 1989) and later adopted by hundreds of companies, including ITT Industries, Polaroid Corporation, General Motors Corporation, and many others. The content includes:

▶ *Protect the biosphere*: Companies will minimize the release of pollutants that endanger the earth.

▶ *Sustainable use of resources*: Companies will use raw materials at a level where they can be sustained.

▶ *Reduction and disposal of waste*: Companies will minimize waste wherever possible. When waste cannot be avoided, recycling will be adopted.

▶ *Wise use of energy*: Companies will use environmentally safe energy and invest in energy conservation.

▶ *Risk reduction*: Companies will minimize health risk to employees and the community.

▶ *Marketing of safe products and services*: Companies will sell products that minimize environmental impact and are safe for consumers to use.

▶ *Damage compensation*: Companies will take responsibility through cleanup and compensation for environmental harm.

▶ *Disclosure*: Companies will disclose to employees and the community incidents that cause environmental harm or pose health and safety hazards.

▶ *Environmental Directors*: At least one member of a company's board will be qualified to represent environmental interests, and a senior executive for environmental affairs will be appointed.

▶ *Annual audit*: Companies will conduct annual self-evaluations of progress in implementing these principles and make results of independent environmental audits available to the public.

To understand how to live up to these principles through design of products that are more environmentally friendly, a review of the various types of pollution is necessary. We organize pollution types by the scope of their environmental impact, from global to regional to local impacts.

Global Issues

There are pollution problems whose manifestations exist on a global scale. These include concerns over climate change, ozone depletion, and biodiversity loss.

Climate change is a concern over the probable consequences of possible large changes in the Earth's climate due to increases in greenhouse gases. This issue mainly becomes a concern in energy usage, which, through burning of fossil fuels, increases the carbon dioxide levels in the atmosphere. Increased carbon dioxide levels in the atmosphere may cause increases in global surface, ocean, and atmospheric temperatures, which in turn may cause drastic climate changes. From a product design point of view, developing products that use less energy will help mitigate this problem.

Another global pollution concern is the *depletion of the ozone layer*. The ozone layer is a thin layer in the upper atmosphere that blocks most ultraviolet radiation from reaching the Earth's surface. Fluorocarbon gases, shown to have come from our industrial society, may also react with and reduce the ozone gas in this layer, and efforts are underway to reduce our usage of these gases. From a product design viewpoint, developing products that do not make use of or release these harmful chemicals, either in use, manufacture, or disposal, will help mitigate this problem.

Minimizing *biodiversity loss* is another global environmental impact objective. Loss of habitat for different plant and animal species due to our expanding society can cause loss of ecosystems and extinction of species. Developing products that use less new raw material will help mitigate this problem.

Regional and Local Issues

Other environmental problems exist on a more-regional, societal level. These include problems of acid rain, where pollution byproducts in one region can cause acid rain in another region. Air pollution and smog also are regional problems. Water pollution, either in the ground water, river, bay, or ocean, is also a regional problem, often caused by herbicides and pesticides, in addition to suburban and urban street water run-off.

Acid rain is a regional pollution problem caused by excessive fossil fuel air emissions for a regional area. The fuel combustion products are released into the air and cause rain in the surrounding environment to have a lower acidic pH level, which can cause regional plant and aquatic life to suffer. From a product design viewpoint, developing products that use less energy will help reduce this problem.

Air pollution is a similar problem caused by excessive fossil fuel emissions for a regional area. Nitrogen compounds, ozone, carbon monoxide, and sulfur dioxide are common problem fuel combustion products that are released into the air. Developing products that use less energy will help mitigate this problem.

Other contaminants can enter the environment through water flow streams and landfills as *water pollution*. Herbicides and pesticides are typical problem compounds whose amounts introduced to a regional area must be controlled. Many other chemical compounds that are used in products must similarly be understood and controlled.

This list forms the basic summary of environmental issues. Given their related environmental objectives, we next discuss design methods to understand the environmental impact of design decisions. We then present some methods for designing products with less environmental impact.

IV. BASIC DFE METHODS: DESIGN GUIDELINES

As the most basic approach to implementing design for the environment, design guidelines can be used that, when followed, will result in products with less damage to the environment. The German VDI (1993), British ICER (1993), the University of Manchester (Dowie and Simon 1994), Fiksel (1996), Bras (1998) and General Electric Plastics (1995) all provide guidelines from a design point of view: product structure, material selection, labeling, and fastening. The guidelines from these different sources are summarized in Tables 15.1–15.4.

These guidelines are simple and effective when implemented. After developing a concept, the guidelines should be consulted, every guideline questioned, and the underlying concepts modified to increase the guideline performance. During the embodiment and detail design phase, a guideline should again be so consulted, to ensure the product developed is compatible with these guidelines.

Application: Paper Carrier Design

Consider the design of an electronic desktop organizer (Miller and Wood 1996; Miller 1995). This product is based on the conceptual design of an actual product in the market, a kitchen assistant. The purpose of this device is to store and retrieve electronic recipes. It is also intended to create a shopping list from selected recipes, producing a hardcopy printout of the list.

The organizer uses a thermal printer similar to those found on many adding machines. Manufacturers sell the printer paper in rolls, and the organizer must safely store this paper. Schematics of the organizer's general layout and physical form are shown in Figure 15.3.

TABLE 15.1. PRODUCT STRUCTURE GUIDELINES

Guideline	Reason
Design a product to be multifunctional.	More ecoefficient than many unique-function products.
Minimize the number of parts. Create multifunctional parts.	Reduces disassembly time and resources.
Avoid separate springs, pulleys, or harnesses. Instead, embed these functions into parts.	Reduces disassembly time and resources.
Make designs as modular as possible, with separation of functions.	Allows options of service, upgrade, or recycling.
Design a reusable platform and reusable modules.	Allows options of service, upgrade, or recycling.
Locate unrecyclable parts in one subsystem that can be quickly removed.	Speeds disassembly.
Locate parts with the highest value in easily accessible places, with an optimized removal direction.	Enables partial disassembly for optimum return.
Design parts for stability during disassembly.	Manual disassembly is faster with a firm working base.
In plastic parts, avoid embedded metal inserts or reinforcements.	Creates the need for shredding and separation.
Access and break points should be made obvious.	Logical structure speeds disassembly and training.
Specify remanufactured parts.	Stimulate demand for remanufacturing, reducing raw material consumption.
Specify reusable containers for shipping or consumables within the product.	Reduces raw material consumption.
Design power-down features for different subsystems in a product when they are not in use.	Eliminate unnecessary power consumption for idle components.
Lump individual parts with the same material.	Eliminates the need for disassembly during recycling. Neighbor parts may be ground or melted as a group.

Given this product description, we now apply design-for-the-environment guidelines to the desktop organizer. For the purposes of illustration, only one guideline is considered here: minimization of material mass and type (Table 15.2). We will also only apply this principle to one subsystem of the product.

To apply this guideline, let's first study the product's intended functionality. A portion of the function tree (Chapter 5) for the electronic organizer is shown in Figure 15.4. The highlighted functions, denoted by bold borders, are those functions performed by the paper carrier components, the focus of the study. Notice that some of the highlighted functions are secondary in nature; that is, they contribute less perceivable quality to the user than the primary functions for printing and ergonomic interface. For instance, the "hide components" function originates from the customer need for aesthetics. The "affix to back housing" (or enable degree of freedom [DOF]) function results from the architecture solution choice that utilizes a separate paper carrier for printing.

TABLE 15.2.   MATERIAL SELECTION GUIDELINES

Guideline	Reason
Avoid regulated and restricted materials.	They are high impact.
Minimize the number of different types of material.	Simplifies the recycling process.
For attached parts, standardize on the same or a compatible material. Eliminate incompatible materials.	Reduces the need for disassembly and sorting.
Mark the material on all parts.	Many materials' value is increased by accurate identification and sorting.
Use recycled materials.	Stimulate the market for material that has been recycled.
Use materials that can be recycled, typically ones as pure as possible (no additives).	Minimize waste; increase the end-of-life value of the product.
Avoid composite materials.	Composites are inherently not pure materials, and so not amenable to recycling.
Use high strength-to-weight materials on moving parts.	Reduce moving mass and therefore energy consumption.
Use low-alloy metals that are more recyclable than high-alloy ones.	More pure metals can be recycled into more-varied applications.
If the same base metal can be used, different metals can be fastened.	Aluminum, steel, and magnesium alloys are readily separated from shredder output and recycled.
Hazardous parts should be clearly marked and easily removed.	Rapidly eliminate parts of negative value.

The original paper carrier design is shown in Figure 15.5(a). A removable aesthetic and protective cover encompasses the paper carrier. The paper is then intended to drop into the carrier well, where it is fed into the printer. Figure 15.5(a) shows the organizer's geometric features given the printing functions listed in Figure 15.4.

For this example, again the metric chosen is the mass of the injection-molded plastic paper carrier, a subsystem of the organizer's housing. Mass is a controllable design parameter in the injection molding process, as with many other manufacturing processes. As the mass of the paper carrier decreases, the energy of production will be reduced. Likewise, the quantity of virgin resin/oil resources is likely to decrease.

TABLE 15.3.   LABELING AND FINISH GUIDELINES

Guideline	Reason
Ensure compatibility of ink where printing is required on parts.	Maintain maximum value of recovered material.
Eliminate incompatible paints on parts—use label imprints or even inserts.	Many label-removal operations for paints cause part deterioration.
Use unplated metals that are more recyclable than plated.	Some plating can eliminate recyclability.
Use electronic part documentation.	These parts can be reused.

TABLE 15.4.   FASTENING GUIDELINES

Guideline	Reason
Minimize the number of fasteners.	Most disassembly time is fastener removal.
Minimize the number of fastener removal tools needed.	Tool changing costs time.
Fasteners should be easy to remove.	Save time in disassembly.
Fastening points should be easy to access.	Awkward movements slow down manual disassembly.
Snap fits should be obviously located and able to be torn apart using standard tools.	Special tools may not be identified or available.
Try to use fasteners of material compatible with the parts connected.	Enables disassembly operations to be avoided.
If two parts cannot be compatible, make them easy to separate.	They must be separated to recycle.
Eliminate adhesives unless compatible with both parts joined.	Many adhesives cause complete contamination of parts for material recycling.
Minimize the number and length of interconnecting wires or cables used.	Flexible elements slow to remove; copper contaminates steel, etc.
Connections can be designed to break as an alternative to removing fasteners.	Fracture is a fast disassembly operation.

Using this perspective, we seek to reduce the mass of the original carrier design, shown in Figure 15.5(a). Structural, spatial, paper-handling, and aesthetic requirements do not allow us to remove the carrier completely. It would thus be desirable to generate alternative form solutions to the carrier. As part of this synthesis process, analogous products may be considered to provide new form ideas. In this case, let's consider the analogy of packaging tape dispensers, a product that handles and feeds rolls of material (similar to paper rolls). A survey of tape dispensers shows that they utilize much less housing material for the same primary functions of the organizer's paper carrier. A new carrier design, based on tape dispenser configurations, is thus developed for the desktop organizer. Figure 15.5(b) illustrates this new concept.

Comparing the carrier designs, the original design is approximately 120 gm of material. This mass further decomposes into three sets. The first is the mass of all the features that contribute to the set of functions: store paper, relinquish paper, accept new paper, and release used paper. The second and third allocations are for other function sets (protect paper, shed liquid contaminants) and the set of all secondary functions for the paper carrier. The mass or environmental cost of a function should be indicative of the importance of that function.

The mass allocation for tape dispensers applies primarily to the first set of functions. The new carrier design thus realizes a mass reduction of approximately 10 gm to a new mass of 110 gm, using the volume

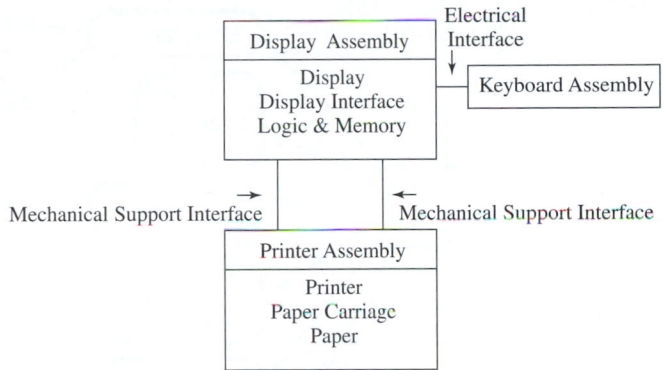

(a) General product layout of a kitchen assistant

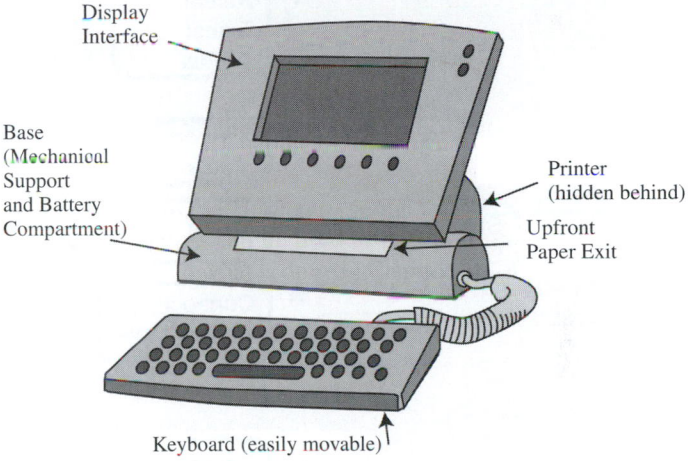

(b) Physical embodiments of the assistant

▼ *Figure 15.3.*
General product layout and physical form of a desktop organizer product, i.e., a kitchen assistant (Brother Co.). (a) General product architecture of a kitchen assistant. (b) Physical embodiments of the assistant.

from a solid model and the density of a polycarbonate/acrylonitrile-bu-tadiene-styrene (PC/ABS) blend. In the new design, the dispenser can rotate backward for access to the paper. Changes to the paper carrier design have not formed any new primary functions. A new secondary function of forming an axis of rotation or pivot for the paper dowels is created. This function conveniently substitutes into the third set of paper carrier functions, replacing the secure paper cover function, which is no longer necessary.

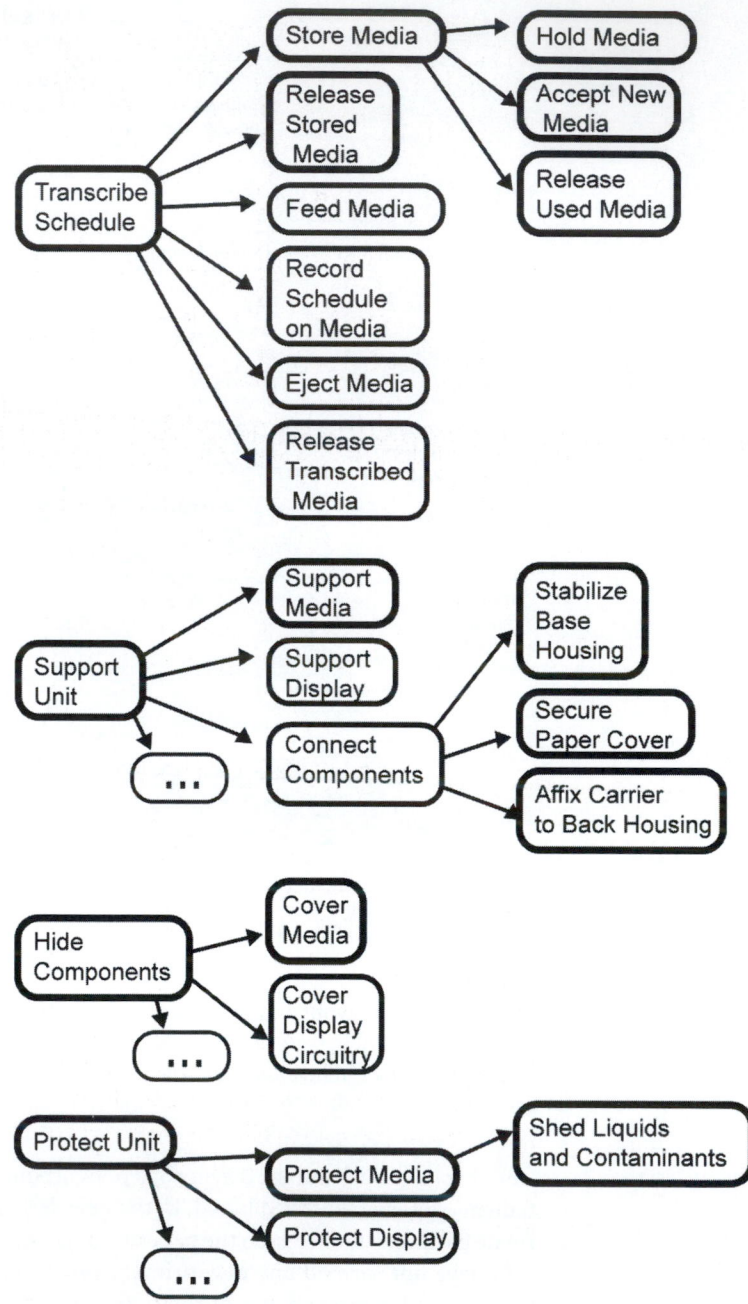

▼ *Figure 15.4.*
Simplified function tree of the desktop organizer product.

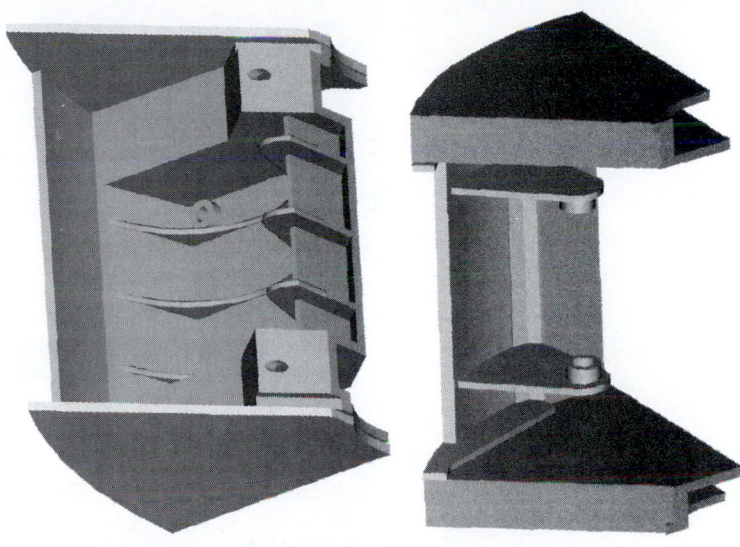

(a) Original paper-carrier design   (b) Alternative paper-carrier concept

▼ *Figure 15.5.*
Paper-carrier design for a desktop organizer product. (a) Original paper carrier design. (b) Alternative paper carrier concept.

Estimating, conservatively, a production quantity of 50,000 units, the projected mass savings is 500 kg of injection-molded plastic material. Given that many devices implement similar functions of the paper carrier, these savings are very significant in terms of a design-for-the-environment metric.

Extensions: Preview to Value Analysis for DFE

While the application of material DFE guidelines might appear to be justified, it is difficult to measure the actual benefit to the customer, the environment, or the company. One might argue for a means to quantify the contributions of the guidelines. One approach is to consider costs to discriminate the perceived benefits using the project economic assessment approach of Chapters 3 and 14.

Assuming a production quantity of 50,000 units, and using activity-based power generation models (Miller 1995), it is possible to estimate the monetized environmental cost of air emissions avoided by the paper carrier design change. These costs are estimated using the material intensities provided by the *IJM Handbook* (1995). Figure 15.6 contains these estimates. The emissions have been amortized over the expected life cycle of the desktop organizer product. The monetized costs of the

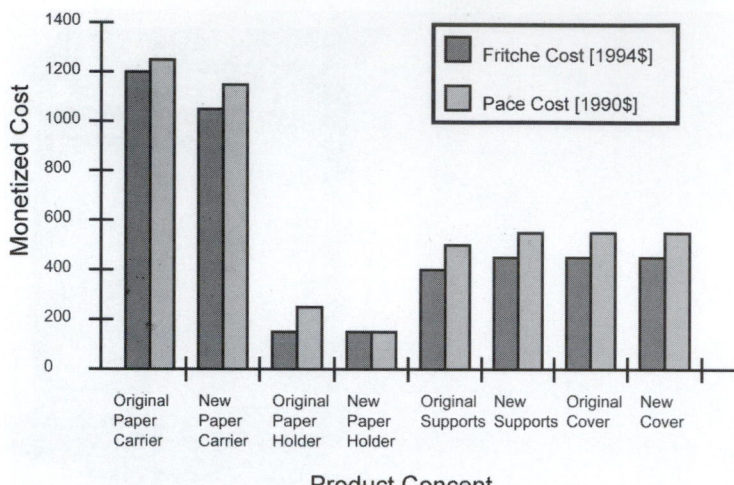

▼ *Figure 15.6.*
Estimated air pollution costs per year for paper carrier design
modifications.

relative simple improvements to the paper carrier result in savings of
roughly $50 using either of two different models available, the Fristche
or Pace monetizations.

The data in Figure 15.6 provides a cost and benefits in terms of dollars.
Yet the question remains: Is this a significant environmental cost savings? A complete life cycle assessment (discussed below) would answer
this question. In this section, however, we choose not to analyze the
issue at a very detailed level. Yet, designers still need a ruler by which
to gauge their achievements. One approach is to compare the dollar
costs and benefits to known environmental impacts that a design team
can interpret. For example, Figure 15.7 displays a graph of estimated environmental emissions costs for product housings, all for completely
different products (Miller 1995). Looking at this chart can help a design team understand their housing relative to other housings. In this
case, the functions of housings are similar across the products.

Comparing Figure 15.6 with Figure 15.7, the design team notes that
the environmental impact from the mass of the paper carrier is more
than the impact of a VCR bezel and close to the impact of a predomi-
nantly plastic Braun shaver housing. Considering the more-detailed
functionality required of a electric shaver housing, the design team
might conclude that the new paper carrier design is a wonderful step
forward, but further refinements are possible to be equitably comparable with a shaver housing.

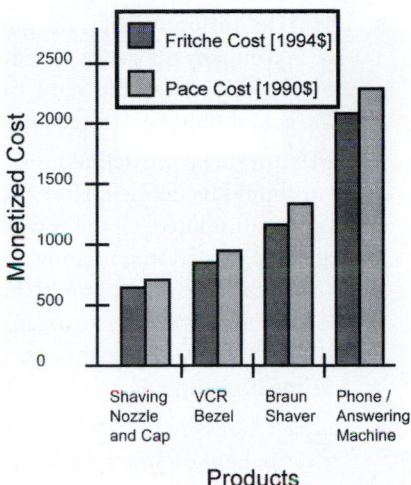

▼ *Figure 15.7.*
Environment quality ruler: Air emission costs for different products.

Such an approach can be used for products of similar complexity levels. Complex systems may be decomposed into subsystems to reduce the complexity of the system being measured by the environmental quality ruler.

V. LIFE CYCLE ASSESSMENT

The quality ruler presented in the previous section provides a basic method of assessing the environmental impact of a product and its components. It also generates a rough estimate for assessing possible product improvements. To develop a more complete analysis, a full life cycle assessment of a product is needed. This analysis is the topic of this section.

Overview

The SETAC Approach

To design a product for the environment, an assessment of environmental impact must be completed. The Society of Toxicology and Chemistry (SETAC) has developed a four-step process for completing a Life Cycle Assessment (LCA). This process allows us to understand the environmental impact of a product, from manufacture through use through disposal. This process include the following steps:

> **Goal Definition:** What is the purpose of the environmental analysis? A company may wish to reduce CO_2 emissions, meet certification, reduce energy use, reduce component toxicity, and so forth. The goal should be defined and revisited.

> **Inventory:** Having defined the goal, the scope of the product system should be defined. How much of the manufacturing process will be considered, all the way back to energy consumed and pollution generated when removing raw materials? For what expected customer use patterns should the typical energy utilization pattern be analyzed? Does there exist many possible alternative utilization patterns? What disposal routes should be considered? Should recycling be assumed?

> **Impact Assessment:** Having scoped the system for study, an impact assessment of each step in the system can be made. We present two methods below.

> **Interpretation:** The assessment must be interpreted for areas of high-leverage impact reduction and for comparisons against other alternatives.

We now review each of these steps in greater detail.

Goal Definition

The obvious first step in considering environmental assessment in product design is to establish clear objectives. Reducing energy usage is different from reducing toxic material content or redesigning a product for disassembly. What are the objectives? Some common objectives include meeting eco-labeling requirements, meeting regulations, or working to meet standards certification. Within these objectives, clearly defined engineering specifications (metrics), such as minimum mass and lumped materials, are established to evaluate a product.

Overall product function: Once a goal is defined, the next step for a design team is to establish the boundary of the system to analyze. Typically, this task includes all activities and operations in the product life cycle. Having defined this boundary, one important item in the system definition is the *overall product function*, often called the *system function* in the LCA literature. The overall product function, as defined in Chapters 4 and 5, is the intended purpose of the product under environmental analysis. What is the purpose of the product? This function is important to state clearly and definitely, as it then permits environmental impact comparisons with other alternative technologies for the same function.

For example, comparing the environmental impact of coffee grinders with telephones is not particularly insightful; they have completely different purposes. It can make sense to compare similar function subsystems such as the housings of a coffee grinder with the housing of a telephone, such as we did with different housings in the previous section (Figure 15.7). But basically one should compare similarly functioning systems. To environmentally compare the effects of a coffee grinder, it makes sense to compare the environmental impact of different technologies that convert coffee beans to coffee. For example, home electric choppers, home electric grinders, home hand-powered grinders, and prepackaged ground coffee are all reasonably equivalent alternatives that one could compare in environmental impact studies, since they all perform the same function: provide ground coffee from beans.

The functional unit: With the overall product function defined, the design team must then establish a *functional unit*. This concept is the basic unit of output representing the product function. A functional unit for a coffee grinder might be one day's worth of ground coffee, or one cup of grounds. A functional unit for an automobile might be a mile of travel or a gallon of gasoline. A functional unit for a lawnmower might be the size of an average lawn in square meters, square yards, or acres.

Emissions and environmental impact are calculated on a per-functional-unit basis to compare alternatives. To understand whether chopping beans at home or buying preground coffee has more or less environmental impact, one must compare these alternatives on a per-cup of-ground-coffee basis.

In summary, the outcomes of the goal definition activity include an objective for the assessment, a system to examine, and a clear overall product function statement, complete with a functional unit.

Environmental Impact Inventory

After establishing the system boundary and functional unit, the system itself needs to be described as a sequence of activities, each called a *life cycle stage*. Each life cycle stage takes in materials and energy and produces the desired activity outcome along with waste material and energy. We represent this process using material and energy flows, as presented in Chapter 5 when considering functional analysis. The basic life cycle stage is shown in Figure 15.8, where the inputs and outputs of concern are material and energy flows.

In a generic product life cycle stream, each stage can be connected to others in complex ways, as shown for a typical analysis in Figure 15.9.

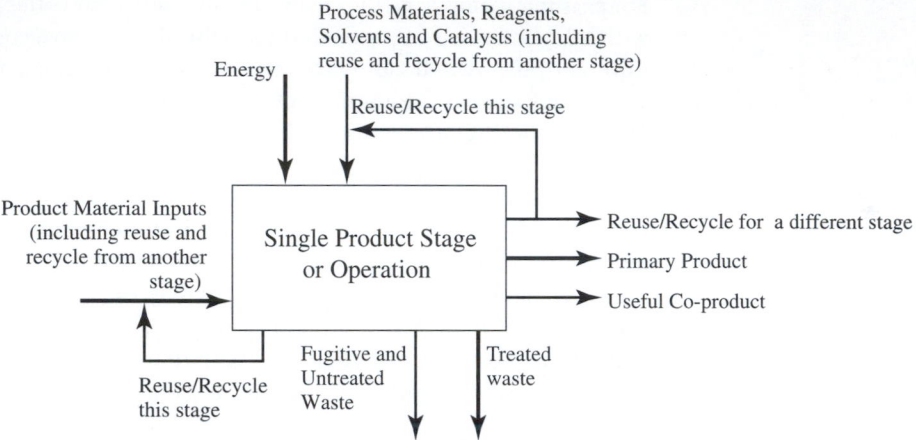

▼ Figure 15.8.
Life cycle stage.

If materials are recycled, or if components are returned to the manufacturer for reuse *(remanufacturing)*, or if the product can be used for many cycles by the customer directly, these connections lead to different impacts on the environment. Understanding the quantity on each flow path of the material and energy is the data collection analysis portion of a life cycle assessment.

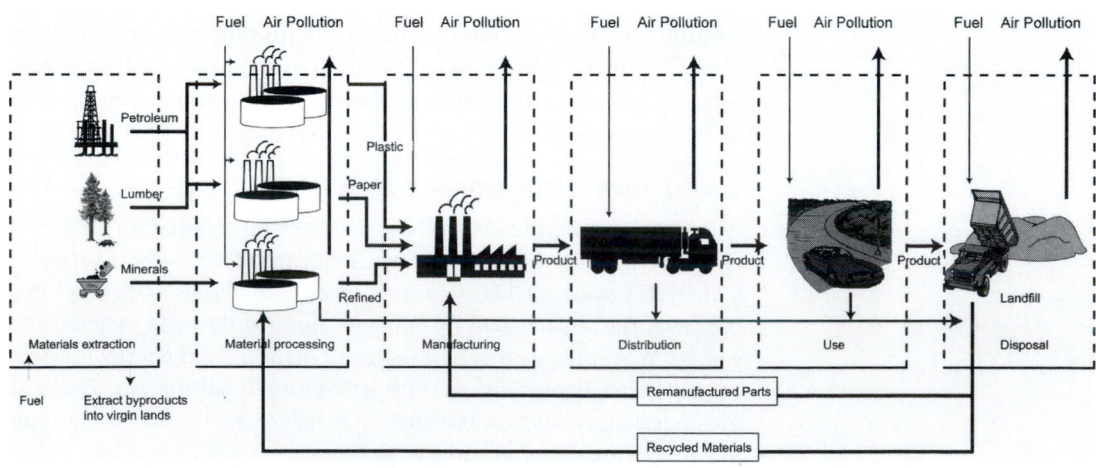

▼ Figure 15.9.
Typical material and energy flows over a product life cycle.

Impact Assessment

Having mapped the system and identified the flows in and out of each life cycle stage, the next step is to quantify these flows in terms of environmental impact. This step represents the assessment activity. We present two basic methods below for carrying out this analysis.

Assessment methods: The basic method, AT&T's *Environmentally Responsible Product Assessment Method*, does not complete a numerical LCA. Rather, estimates of environmental impact are applied. Other similar methods include Motorola's *Product Lifecycle Matrix* (Hoffman, 1995) or the *Environmental Impact Factors Analysis (EIFA) Method* from Stanford University (Ishii and Lee 1996). With the second more-advanced method, *The Weighted Sum Method* from Pré Associates (Goedkoop et al, 1996), we assign numerical impact weightings against the materials we use and the process we use. Here, though, the weightings used are average numbers for a region, not specific values for a particular production line at a particular factory. A completely accurate LCA assessment would calculate weightings based on measurements of the actual production equipment used. Such an assessment would consider the actual recycling, energy efficiency of plants, and so forth that the product in question utilizes. We will only briefly highlight such analysis.

Environmental objective assessment combination: Whether using actual assessments for a particular product life cycle or using average impact weightings, all environmental impact assessments provide impact on the different environmental objectives presented at the start of the chapter. A design team generally wants a single environmental impact rating. To create such a rating, the different objectives must be combined in some way. Therefore, all impact assessment methods that provide a single rating rely on subjective relative rankings of the different types of environmental impacts. That is, one might reduce landfill impact by replacing a styrofoam container with a recyclable paper container. However, the paper container might use more raw materials (lumber) and require more energy to manufacture it, resulting in more air pollution. Determining which solution strategy is better in such trade-off decision making requires some form of relative assessment of the different environmental damage.

Typically, these relative assessments of, for example, how much worse air pollution is compared with a landfill, are regionally dependent. Southern California has a greater problem with air pollution than windswept New England does. New England has a greater problem with acid rain than the Midwest does. Therefore, the relative rankings among different types of pollution are generally valid only regionally.

In any LCA system, these relative rankings are manifested in the rankings of each material, process, use, and disposal method. The result is that the scores of any LCA assessment are valid in absolute terms only regionally. Given the exact same life cycle, an LCA assessment final number calculated for Southern California will be different than a final number calculated for New England. The product life cycle, energy, and waste streams are the exactly same, but that activity's impact on the environment is different due to different local environmental sensitivity.

Given this caveat, LCA assessments become difficult. This difficulty is especially true with the lack of established government- or organization-approved LCA methods for the different engineering disciplines at different levels of abstraction, as is clearly needed for product development. Nonetheless, there are LCA assessment methods available, two of which are provided in the next section. While any LCA method may not be accurate in predicting the exact impact for any particular region, all is not lost. One is always guaranteed that any improvement in an LCA assessment ranking will be better for any region's environment on some weighting of the criteria.

Interpretation

After an assessment is completed, the results can be interrogated for opportunities to change the design to improve the environmental objective and to compare amongst alternatives. Design choices that have a high impact on the environment can then be explored and refined.

We now present two methods for completing the environmental assessment: a basic matrix approach based on qualitative ratings and a more advanced approach of quantitatively assessing impact.

Basic Method—AT&T's Environmentally Responsible Product Assessment

Graedel and others at AT&T have developed a basic guideline and matrix approach to life cycle assessment (Graedel and Allenby 1995). Here, the product life cycle is broken down into five stages: resource extraction, product manufacture, product packaging and transport, product use, and refurbishment/recycling/disposal. For each of these five stages, five environmental criteria are ranked, thereby resulting in 25 environmental metrics for the product.

TABLE 15.5.   AT&T'S ENVIRONMENTALLY RESPONSIBLE PRODUCT MATRIX

Life cycle stage	Materials choice	Energy use	Solid residues	Liquid residues	Gaseous residues
	Environmental concern				
Resource extraction					
Product manufacture					
Product packaging and transport					
Product use					
Refurbishment, recycling, disposal					

The 25 criteria are evaluated and presented in matrix form, as shown in Table 15.5. Each matrix entry is filled with a ranking from 0 to 4, where 0 is a high-impact negative evaluation, and a 4 is low-impact positive evaluation. To complete each rating, guidelines for the matrix elements must be consulted, as shown in Table 15.6.

After completing each of the 25 ratings, an overall rating is determined by combining the scores. The product is rated as:

$$R_{ERPT} = \sum_{i,j} M_{ij}$$

where R_{EPRT} is called the *Environmentally Responsible Product Rating*, and M_{ij} is the matrix element (i,j). Note the maximum R_{EPRT} rating is 100. While the overall rating is important, the real use of the method is in the breakdown of the scores, to indicate where the product DFE activity should focus. An effective visualization tool associated with the matrix is to polar plot each of the 25 scores. This plot provides a visual breakdown of where the product might be improved.

Example: Krups Coffee Mill

As an example, we will again consider the Krups Coffee Mill shown in Figure 15.10. The first step is to complete the matrix by filling in the elements, based on an analysis of the product. Utilizing the guidelines from Table 15.6, the matrix is filled as in Table 15.7.

The matrix requires careful thought and analysis to complete, as the answers to the guideline questions are not always clear. For example, an understanding of the environmental impacts of various materials is

TABLE 15.6. GUIDELINES FOR COMPLETING THE ENVIRONMENTALLY RESPONSIBLE PRODUCT MATRIX

Matrix element	Questions to pose when rating: "Does the design . . ."
1,1	• use materials that are least toxic and most environmentally preferable for the performed function? • minimize the use of materials in restricted supply? • utilize recycled materials wherever possible?
1,2	• minimize the use of materials whose extraction is energy intensive? • avoid materials whose transport will require significant energy use? • avoid producing residues whose recycling will be energy intensive?
1,3	• minimize materials whose extraction or purification involves production of much solid residues? • avoid materials whose transport will result in signification solid residues?
1,4	• minimize materials whose extraction or purification involves production of much liquid residues? • avoid materials whose transport will result in signification liquid residues?
1,5	• minimize materials whose extraction or purification involves production of much gaseous residues? • avoid materials whose transport will result in signification gaseous residues?
2,1	• avoid or minimize incorporating materials that are in restricted supply? • avoid or minimize toxic materials? • avoid or minimize radioactive materials?
2,2	• minimize energy-intensive production steps such as high heat differentials, heavy motors, cooling, etc? • minimize energy-intensive evaluation steps such as testing in a heated chamber? • use production that uses cogeneration, heat exchange, and other techniques to utilize wasted energy?
2,3	• minimize and reuse as much as possible manufacturing solid residues (mold scrap, cutting scrap, etc.)? • minimize packaging material and use the fewest possible different materials? • have suppliers take back packaging material in which the products enter the facility?
2,4	• minimize and use substitutes for solvents or oils that are used in any production process? • minimize toxicity and optimally of liquid product residues? • utilize the maximum amount of recycled liquid species from outside suppliers?
2,5	• apply alternatives to CFCs or HCFCs? • minimize greenhouse gases used or generated in any manufacturing process? • minimize any odorants used or generated in any manufacturing process?
3,1	• minimize the product packaging mass and the number of different materials? • avoid toxic materials in the product packaging? • use recyclable packaging materials?
3,2	• minimize the use of packaging materials whose extraction or processing energy intensive? • avoid energy intensive packaging procedures? • minimize energy use in the product distribution plan?
3,3	• minimize packaging mass at all three levels of distribution (primary, secondary, tertiary)? • use packaging designed to make it easy to separate the constituent materials (foam, cardboard, etc.)? • use product packaging with take-back arrangements made for recycling and reuse?
3,4	• use refillable or reusable containers for liquid products? • avoid packaging with toxic or hazardous substances that might leach if improperly disposed of? • ensure hazardous products prone to spilling or venting are transported on safe routes by trained drivers?
3,5	• apply product distribution plans designed to minimize gaseous emissions from transport vehicles? • with pressurized gases have installation procedures designed to avoid their release? • with incinerated packaging result in any toxic materials?

TABLE 15.6.   GUIDELINES FOR COMPLETING THE ENVIRONMENTALLY RESPONSIBLE PRODUCT MATRIX (CONTINUED)

Matrix element	Questions to pose when rating: "Does the design . . ."
4,1	• minimize single use and dispose consumable components? • minimize restricted supply materials in consumable components? • minimize toxic or otherwise undesirable materials in consumable components?
4,2	• minimize energy use while in service? • have enhanced insulation or other energy conserving features? • monitor and display its energy use while in service?
4,3	• avoid use of periodic disposal of solid materials such as cartridges, containers, or batteries? • have all alternatives to consumables thoroughly investigated? • avoid intentional dissipative emission to land?
4,4	• have all alternatives to liquid consumables thoroughly investigated? • avoid intentional dissipative emission to water? • avoid liquids materials that have the potential to be unintentionally dissipated during use?
4,5	• have all alternatives to gaseous consumables thoroughly investigated? • avoid intentional dissipative emission to air? • avoid gaseous materials that have the potential to be unintentionally dissipated during use?
5,1	• minimize the number of different materials that are used in its manufacture? • minimize use of toxic materials? • ensure that different materials are easy to identify and separate?
5,2	• minimize the use of energy-intensive process steps in disassembly? • reuse materials while retaining their embedded energy? • minimize transport of recycled materials to the recycling facilities?
5,3	• assemble with fasteners such as clips rather than chemical bonds or welds? • avoid joining dissimilar materials in ways difficult to reverse? • apply ISO markings as to their content?
5,4	• contain liquids in the product such that they can be recovered rather than lost at disassembly? • recover and reuse generated liquid residues generated during subassembly disassembly? • recover and reuse generated liquid residues generated during material extraction?
5,5	• contain gases in the product such that they can be recovered rather than lost at disassembly? • recover and reuse generated gases generated during material extraction? • use plastics that can be incinerated without requiring sophisticated scrubbing devices?

necessary to understand if a particular material is in fact the least impact alternative. The analysis requires the designer to make a number of value judgments regarding the different impact of materials or processes that, to be objective, must be substantiated by actual research. The energy use during the resource extraction phase is a good example of one such decision. Natural copper is more difficult to refine than iron, because it is found in lower concentrations, so the assumption is made that copper requires more energy during the resource extraction phase. Although this assumption may be valid, it is unsubstantiated until analysis of the coffee production operations are understood.

▼ *Figure 15.10.*
Krups Coffee Mill.

Explanation of Matrix Elements

The values entered into Table 15.7 contain a high level of uncertainty, which is typical for most design teams. For example, a high level of uncertainty can occur on questions of the resource extraction phase, because this stage is often not translucent to the designer. These numbers are wholly dependent on the materials that are chosen for the product.

TABLE 15.7. AT&T PRODUCT MATRIX FOR THE KRUPS COFFEE MILL

Life cycle stage	Environmental concern					
	Materials choice	Energy use	Solid residues	Liquid residues	Gaseous residues	Total score
Resource extraction	2	2	3	3	3	13
Product manufacture	2	3	2	3	3	13
Product packaging & transport	2	3	3	4	2	14
Product use	4	2	4	4	4	17
Refurbishment, recycling, disposal	3	3	3	4	3	16
Total score	13	12	15	18	15	73 /100

Material choice and energy use are rated low because of the difficulty in extracting copper. Residues are assumed to be present but relatively minimized during the extraction phase for all of the materials.

To complete the matrix, knowledge of the recycling efforts must be understood. For example, the plastics that are used in the body of the mill should be available from a recycled source. Many of the metal parts that comprise the motor are stamped, hence there is material left over that becomes waste and available for recycling. Additionally, the injection molded plastic pieces will always have scrap available for recycling.

The packaging and transport of the product have relatively little impact on the overall environmental impact of the product. The choice of foam as a portion of the packaging is questionable because folded cardboard could also be used in place of the more environmentally damaging foam. Very elaborate printing is used over the package. Energy use during transport and packaging appears to be kept to a minimum.

During the normal operation of the coffee mill, only energy is consumed or left as residue. There is some residual coffee grinds left in the mill after operation; however, this residue is considered to be negligible, because it is organic and decomposed easily (basically equivalent to dirt). The energy consumed by the product could potentially be optimized through a combination of grind time and motor power and consequently obtains a lower rating.

There appears to be little environmental impact during the disposal phase of the product. Nearly all of the materials used in the manufacturing of the product are recyclable, and the product is easily disassembled for recycling. Refurbishment of the product is not likely, because significant financial savings will not result from this activity. (The mill is not worth a lot.) Residues will occur during the decomposition of the product; however, most are not toxic and appear to be minimized.

Interpret the Results

The matrix suggests that the majority of the amendable environmental damage results, using the product life cycle stages' view, from resource extraction and product manufacturing. Alternatively, using a product view, waste occurs from materials and energy use. A review of the guidelines and matrix elements suggests several potential improvements. The use of recycled materials in the mill, especially in the plastics, would help reduce the impact. Additionally, materials that are more common than copper could be used in the product, such as a steel wire electrical cord. An effort to reduce the foam and printing in the packaging could also result in environmental gains for the product.

Weighted Sum Assessment Method

The previous methods are basic heuristics to reduce environmental impact. For comparisons between alternatives, however, such an approach may not provide a clear set of changes to investigate. A more quantitative approach of relative importances may thus be desired. A more-advanced approach is to complete a full life cycle assessment—assess the impact of the material usage and waste generation of each stage in the product life cycle.

Rather than a full life cycle assessment, however, one might instead inventory the parts used in a product and weight them by "average" impact weightings. That is, one might break the product down into quantities of materials in the product by weight. Then one might establish environmental impact of different materials by weight and sum this score. This process approximates the life cycle stages of the product as an "average" typical use for each material.

One such approach was developed in Europe on the initiative of the Ministry of Housing, Spatial Planning and the Environment in the Netherlands, called the *Eco-indicator 95* (Goedkoop, Demmers, and Collignon 1996). The Eco-indicator system provides weightings by mass for materials, treatment processes, transport processes, energy generation processes, and disposal scenarios. As shown in Appendix D, Figure D.1, and Figure 15.11, the weightings themselves are based on a valuation of damage to public health and to the ecosystem through contribution to several effects, such as ozone layer depletion, smog, and so forth, as discussed in Section I. The actual effects considered, and from what impact sources, are shown in Appendix D, Table D.1, and Figure 15.11.

Because the Eco-indicator was developed in Europe, it is based on average European values (shown in Appendix D, Table D.1, and Figure 15.11) for the processes that describe material production, treatments, transportation, and energy generation. Therefore, the application of the Eco-indicator to non-European regions will not be entirely appropriate. For example, product factors that increase acid rain are overweighted for many regions in the United States. On the other hand, the tool is reasonable and provides indications of relative environmental friendliness of different design scenarios. Software versions are available.

The Eco-indicator system operates by having the analyst first establish the mass of component materials in the product, their means of production, and the means of disposal. A worksheet to evaluate the product is then used, as shown in Appendix D, Table D.2, and Figure 15.12.

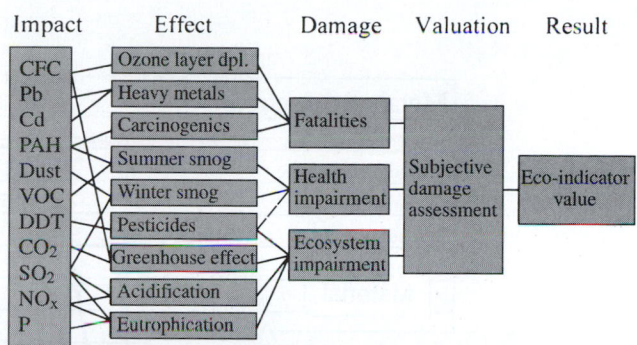

Environmental effect	Weighting Factor	Criteria
Greenhouse effect	2.5	0.1YC rise every 10 years. 5% ecosystem degredation
Ozone Layer Depletion	100	Probability of 1 fatality per year per million inhabitants
Acidification	10	5% ecosystem degredation
Eutrophication	5	Rivers and lakes degradation of an unknown number of aquatic ecosystems (5% degradation)
Summer smog	2.5	Occurrence of smog periods health complaints, particularly amongst asthma patients and the elderly prevention of agricultural damage
Winter smog	5	Occurrence of smog periods, health complaints, particularly amongst asthma patients and the elderly
Pesticides	25	5% ecosystem degredation
Airborne heavy metals	5	Lead content in children's blood, reduced life expectancy and learning performance in an unknown number of people
Waterborne heavy metals	5	Cadmium content in rivers ultimately also impacts on people (see airborne)
Carcinogenic substances	10	Probability of 1 fatality per year per million people

▼ *Figure 15.11.*
Development of weightings for the Eco-indicator 95.

For impact weighting values, the tables in Appendix D, Table D.3, Figure 15.13 are used.

Other life cycle data must be estimated to complete the numerical analysis. In particular, how long the product is used, how it is delivered to the consumer, and how it is disposed must all be estimated. Table 15.8 provides a list of typical product life spans that can be used as a first approximation of the actual useful life for different products (Cheney 1971; Chapman 1983). Delivery can be estimated based on the source, domestic or international. Disposal should simply be approximated as landfill, unless either the material is very hazardous and laws apply to its disposal or the material is very valuable.

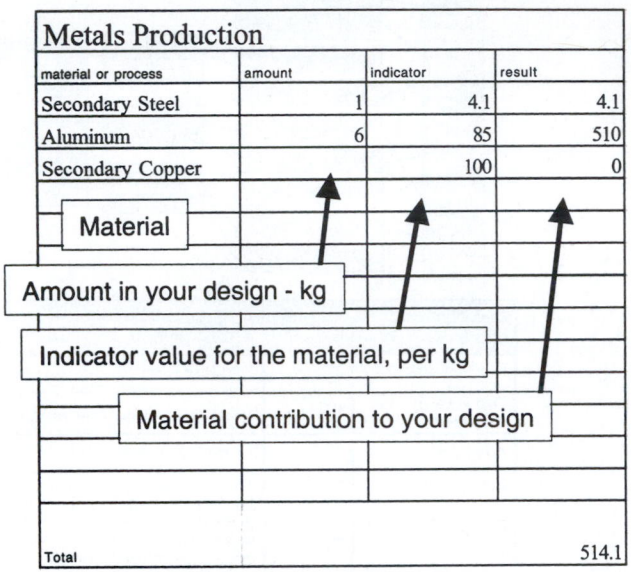

Metals Production

material or process	amount	indicator	result
Secondary Steel	1	4.1	4.1
Aluminum	6	85	510
Secondary Copper		100	0
Total			514.1

Material

Amount in your design - kg

Indicator value for the material, per kg

Material contribution to your design

▼ *Figure 15.12;*
Eco-indicator 95 worksheet.

Example: Coffee Mill

Define the life cycle: We first define the product life cycle for the Krups Coffee Mill. That is, we determine exactly what inputs and outputs during a product's life cycle should be considered in the analysis of the product. Figure 15.14 shows the life cycle determined for the Krups Coffee Mill. The shaded areas are all sections that are not being considered in the analysis. The preparation of coffee beans is not considered because the designer of the coffee mill has no impact on this process. Likewise, the assumption is made that the coffee mill is disposed of in regular municipal waste. The Eco-indicator 95 model assumes a certain fraction of municipal waste to be recycled, another fraction to be incinerated, and the remainder to end up in a landfill. Although recycling every unit could reduce the product's impact on the environment, the assumption is made that because the product is a consumer product, it would be disposed of with the rest of the household trash. Several inputs (such as felt) are not considered in the process tree because they are not mentioned in the Eco-indicator 95 model.

Transport (in millipoints)		
	Indicator	Description and explanation of score
Truck (28 ton)	0.340	per ton kilometer, 60% loading, European average
Truck (75m^3)	0.130	per m^3 km, 60% loading, European average
Train	0.043	per ton kilometer, European average for diesel and electric traction
Container Ship	0.056	per ton kilometer, fast ship, with relatively high fuel consumption
Aircraft	10.000	*per kg!*, with continental flights the distance is not relevant.

Production of energy (in millipoints)		
	Indicator	Description and explanation of score
Electricity high voltage	0.57	per kWh, for industrial use
Electricity low voltage	0.67	per kWh, for consumer use (230V)
Heat from gas (MJ)	0.063	per MJ heat
Heat from oil (MJ)	0.15	per MJ heat
Mechanical (diesel, MJ)	0.17	per MJ mechanical energy from a diesel engine

Production of metals	(in millipoints)	
	Indicator	Description
Secondary aluminum	0	made completely of secondary material
Aluminum	18	containing average 20% secondary material
Copper, primary	85	primary electrolytic copper from relatively American factories
Copper, 60% primary	60	normal proportion secondary and primary copper
Secondary copper	23	100% secondary copper, ralatively high score through heavy metal emissions
Other non-ferrous metals	100	estimate for zinc, brass, chromium, nickel etc.; lack of data
Stainless steel	17	sheet mateial, grade 18-8
Secondary steel	1.3	block material made of 100% scrap
Steel	4.1	block material with average 20% scrap
Sheet steel	4.3	cold-rolled sheet with average 20% scrap

Material

Indicator value for the material, per kg

Helpful description

▼ **Figure 15.13.**
Eco-indicator 95 impact weightings.

TABLE 15.8. COMMON PRODUCT LIFE CYCLES WITHOUT RECYCLING

Product type	Useful life (years)
Novelties	1
Photographic film	1
Disposable dinnerware and hospital goods	1
Packaging	1–2
Construction film	2
Footwear	2
Apparel	4
Household goods	5
Toys	5
Jewelry	5
Saucepans	5
Sporting goods	7
Domestic appliances	7–10
Luggage	10
Cameras	10
Furniture	10
Motor vehicles	10
Electrical goods	10–15
Hardware	15
Aircraft	15
Wire and cable	15–30
Construction	25–40
Machinery	30

Comparisons to other materials of the excluded inputs are made to determine if their absence would affect the overall result of the model. Felt, for example, is compared with paper, cardboard, and PET plastic. The results confirm the assumption that the excluded inputs are negligible.

The next step in the analysis is to quantify the materials and processes used in the production, use, and disposal of the coffee mill. The application of Eco-indicator 95 to the coffee mill is quite simple once the bill of materials is obtained. The materials and processes are summed, and then their impact is calculated by using the weights from the Eco-indicator 95 list of Appendix D and Figure 15.13.

Production & processing: The production matrix for the Krups Coffee Mill is the most complex of the three matrices to expand. The motor for the coffee mill is examined separately because the motor comprises the majority of the weight of the coffee mill, and a large quantity of the

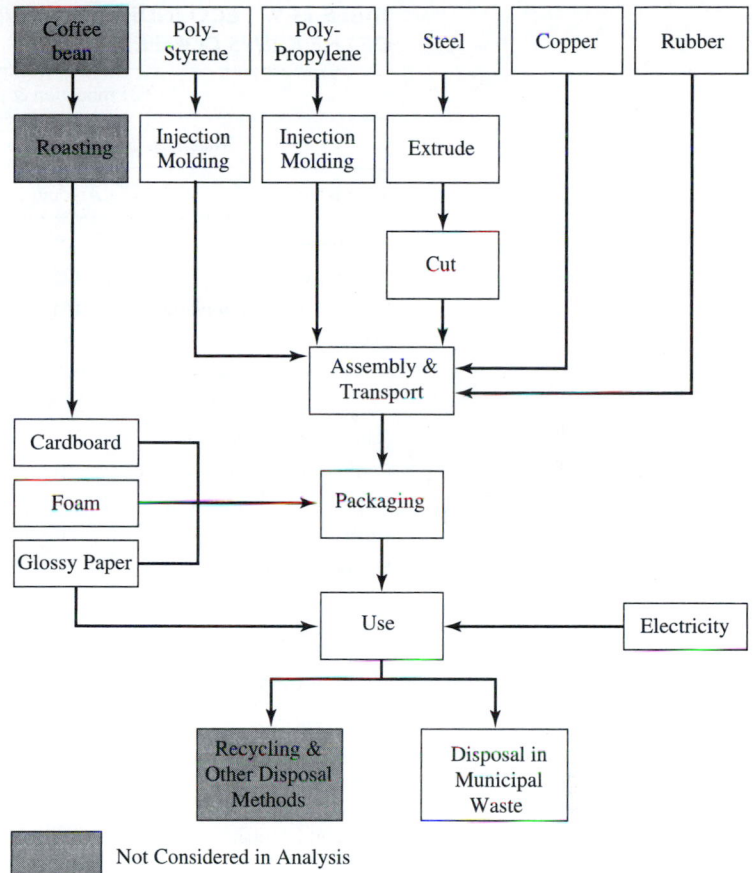

▼ Figure 15.14.
Process tree for Krups Coffee Mill.

environmental impact of the coffee mill results from the motor. A number of assumptions are needed to allow the actual materials and processes to fit within the categories of the impact weightings that are part of Eco-indicator 95, as shown in Table 15.9. For the purposes of the study, brass is assumed to be equivalent to 60% primary copper, the foam used in packaging is assumed to be PUR Rigid Foam, the cord is assumed to contain 50% natural rubber product and 50% copper. The assumption is also made that the coffee mill is shipped an average of 2,000 miles by a 28-ton truck, although the environmental impact of the shipping appears to be inconsequential. The assumptions with the most potential impact, and therefore uncertainty, are the material percentages present in the motor.

TABLE 15.9.   ECO-INDICATOR 95 RESULTS FOR PRODUCTION OF THE KRUPS COFFEE MILL

	Production & processes		
	Materials, processing, transport and extra energy		
Material or process	Amount	Indicator	Result in micropoints
Polystyrene (g)	35	5.8	203
Polypropylene	126	3.3	416
Injection molding of plastic	161	0.53	85
Steel	26	4.1	107
Brass	0.01	60	1
Copper	27.5	60	1650
Natural rubber product	27.5	4.3	118
Cardboard	70	1.4	98
Foam	1	8.4	8
Paper	20	3.3	66
Subtotal			2752
Motor			
Paper (1%)	3.85	3.3	13
Steel/Iron (65%)	250.25	4.1	1026
Copper (15%)	57.75	60	3465
Ceramics (19%)	73.15	0.47	34
Motor total	385		4538
Transport by truck	2.88	0.34	1
Grand Total			7291

Use: An effective determination of the environmental impact of the use of the coffee mill is very dependent on the frequency and duration of use. These data points must be determined by a survey of the customers. The assumptions used are an average product life of 10 yr, a 10-sec daily usage, and power consumption of the motor around 125 W, as shown in Table 15.10. This estimated impact would probably be greater than the actual impact of the coffee mill due to the generous usage assumptions.

The usage indicator provides a good opportunity to highlight several differences that result from the use of a European indicator. The indicator is written for the European standard 230 V electricity. Additionally, the emissions per unit of electricity are determined from the European power grid, supplied by, on average, older, and therefore more harmful, power generation facilities than those found in North America.

TABLE 15.10.   ECO-INDICATOR 95 RESULTS FOR USE OF THE KRUPS COFFEE MILL

	Use		
	Transport, energy, and any auxiliary materials		
Process	Amount	Indicator	Result in micropoints
Electricity, low-voltage (kWh)	1520	0.67	1018

Disposal

There are a number of methods by which a product can be disposed. However, as mentioned above, consumers utilize their household trash-cans, and most consumer waste ends up being processed as municipal waste. A certain fraction of the municipal waste is incinerated, while the rest is disposed of in a landfill. Certain metals are reclaimed and recycled from the incinerator slag. Consequently, the indicator for the disposal of these metals is negative. The negative number represents an environmental gain because the recovered metals are used in place of newly processed metals and because electricity is also generated. The results for the Krups mill are shown in Table 15.11

Interpretation of the results: The Eco-indicator model provides some significant insight into the fundamental environmental stress of the Krups Coffee Mill. Table 15.12 provides a summary of some of the relevant data from the above analysis. One of the most notable observations that result from the analysis is that 85% of the environmental damage occurs during the production of the coffee mill. And, even more notable, all but 1% of that damage occurs during the production of the materials before they ever arrive on the factory floor. The design team will thus have to assign the greatest priority to examining the production phase of the product.

The Eco-indicator 95 model of the Krups Coffee Mill leads to the suggestion that improved material selection could greatly reduce the damage that the product impacts on the environment. The most damaging of the materials in the product, copper, accounts for 57% of the total environmental damage of the product. Potential improvements could involve using aluminum wiring instead of the copper wiring that is used in both the motor and cord. Although the reduced conductivity of the aluminum wiring will result in significantly hotter wires, the short operating time of the motor may prevent the unit from heating up noticeably. Even if the replacement of all copper wiring is not feasible, portions of the wiring, such as the power cord, could integrate both copper and aluminum or be all aluminum. A significant effort would be required to fully examine this option.

TABLE 15.11. ECO-INDICATOR 95 RESULTS FOR DISPOSAL

	Disposal		
	Disposal process per type of material		
Material and type of processing	Amount	Indicator	Result in micropoints
Plastics and rubbers	188.5	0.69	130
Steel	276.25	1.2	332
Paper and cardboard	90	0.33	30
Copper	85.25	−2.6	−222
Total			270

Life Cycle Assessment Method

A life cycle assessment is a more-complete numerical analysis of environmental impact. Rather than use average values for regional impact, actual environmental impact is inventoried, considering the actual production system, operation, and disposal system used. An inventory of

TABLE 15.12. SUMMARY OF RELEVANT ECO-INDICATOR DATA

	Category	Score in micropoints	Percentage of total damage
Life cycle phase	Production & processes	7291	85%
	Use	1018	12%
	Disposal	270	3%
	Total score	8579	100%
Impact by material	Copper (whole product)	4893	57%
	Plastics	834	10%
	Non-copper metals	1464	17%
	Other materials	320	4%
	Subtotal	7512	88%
Impact by phase	Motor	4538	53%
	Power: use of motor	1018	12%
	Cord	1768	21%
	Base & housing	779	9%
	Packaging	202	2%
	Other	250	3%
	Total Impact	8579	100%

all emissions that occur from all processes during the life cycle of a product is summarized into a list of all emissions released, all raw materials used, and all waste generated.

For each of these steps, the energy and raw material requirements are determined for the actual product as well as emission levels produced. The values for each stage of the product's life cycle are then summed with appropriate units.

There are several LCA tools available on the market. For example, *TEAM* from Ecobalance/Ecobilan Inc., *SimaPro* from Pré Associates, LCA Advantage from Batelle Inc., *EcoSys* from Sandia National Laboratories, *the Boulstead Model* from Boustead Consulting Ltd., and *Ecobalance DFE* from Ecobalance Inc.

A life cycle assessment result is complete, comprehensive, accurate, and expensive. The analysis is currently reserved for special products that have high impact materials, such as cleansers, chemical film products, and medical products. Nonetheless, LCA analysis technology is developing, and costs of analysis are decreasing. In the near future, it will be a standard tool for product development.

VI. TECHNIQUES TO REDUCE ENVIRONMENTAL IMPACT

In support of the analysis methods above, redesign techniques are needed to improve the assessment scores. Tools have been developed to help redesign a product to improve its environmental impact. There are several basic approaches:

1. design to minimize material usage
2. design for disassembly
3. design for recycling
4. design for remanufacturing
5. design to minimize hazardous materials
6. design for energy efficiency
7. design to regulations and standards

Design to minimize material usage is where one designs systems so they use minimal amounts of materials. *Design for disassembly* is where one designs a product's subsystems such that they can be taken apart to the point where the components can be reused or recycled. *Design for recyclability* is an approach where one designs a product so that it can be easily recycled. *Design for remanufacturing* is where one designs a product so that its components, once disassembled, can be easily cleaned, inspected, and reused. *Design to minimize hazardous materials* is where one designs a product to use few hazardous materials. *Design for energy efficiency* is often one of the most important considerations in a product's development to minimize energy consumed. It is where one designs a product to reduce losses and effectively use its energy source. Finally, *design to regulations* is where one ensures a product is compatible with regulations or a labeling program.

Design to Minimize Material Usage

One of the more-effective methods to alter environmental impact is to reduce the amount of material used in the product over the life of the product. This action is especially true of high-impact materials. Generally, material reduction can be achieved in three areas: in packaging and distribution factors, in production system factors, and finally in the product itself.

Many products are delivered in single-use packaging with artistic colored labeling, such as shoe boxes, kitchen appliances, and household items. Are there ways that one can eliminate or reuse the packaging in which the product is delivered? For example, many computer companies have programs to accept back a computer's shipping boxes and internal foam packaging. Landfill of the packaging is then reduced.

The production system should also be kept in mind for high emissions and product design changes that eliminate these impacts explored. For example, in 1997, the Chrysler Corporation introduced a new car with body panels made of plastic that required no paint. Rather than redesigning the paint booth system in the factory or examining low-volatility paints, they entirely eliminated this high-environmental-impact process, while still offering a vehicle with shiny body panels.

In addition to the consideration of the packaging or the production process, the product itself can be examined for high-impact components that can be changed. Each subassembly should be examined for possible elimination of parts. Chapter 6 discusses the Subtract and Operate Procedure and force flow analysis to aid in part reduction.

Of particular concern is the compatibility of the components in the assembly. Two materials are *compatible* if they can be processed together as mixed materials. If they are compatible, they may not need disassembly, since the two parts can be reground. Tables 15.13–15.15 show compatibility among different materials.

TABLE 15.13. PLASTIC MATERIAL COMPATIBILITY, ADAPTED FROM (VDI, 1991).

● Compatible	Additive											
◐ Compatible with limitations												
◉ Compatible in small amounts												
○ Not compatible												
Base Material PE	●	○	○	○	●	○	○	○	○	○	○	○
PVC	○	●	○	○	○	○	○	●	◐	○	○	●
PS	○	○	●	○	○	○	○	○	○	○	○	○
PC	○	◉	○	●	○	○	○	●	●	●	●	●
PP	◉	○	○	○	●	○	○	○	○	○	○	○
PA	○	○	◉	○	○	●	○	○	○	◉	◉	○
POM	○	○	○	○	○	○	●	○	○	◉	○	○
SAN	○	●	○	●	○	○	○	●	●	○	○	●
ABS	○	◐	○	●	○	○	◉	○	●	◉	◉	●
PBT	○	○	○	●	○	◉	○	○	◉	●	○	○
PET	○	○	◉	●	○	◉	○	○	◉	○	●	○
PMMA	○	●	◉	●	○	○	◉	●	●	○	○	●

TABLE 15.14. GLASS COMPATIBILITY, ADAPTED FROM (UNITED NATIONS, 1997).

Legend	Additive							
● Compatible	Bottle	Window	Drinking	Crystal	TV screen	TV cone	TV neck	LCD screen
◐ Compatible with limitations								
○ Compatible in small amounts								
Base Material Bottle	●	○	○	○	○	○	○	○
Window	●	●	●	○	○	○	○	○
Drinking	●	◐	●	○	○	○	○	○
Crystal	○	○	○	●	○	◐	◐	○
TV screen	◐	◐	○	○	●	◐	○	○
TV cone	○	○	○	◐	○	●	●	○
TV neck	○	○	○	◐	○	○	●	○
LCD screen	◐	◐	○	○	◐	○	○	●

TABLE 15.15.   METAL ADDITIVE COMPATIBILITY, ADAPTED FROM (UNITED NATIONS, 1997).

Base metal	Incompatible elements	Value-reducing elements
Copper (Cu)	Hg, Be, PCB	As, Sb, Ni, Bi, Al
Aluminum (Al)	Cu, Fe, polymers	Si
Iron (Fe)	Cu	Sn, Zn

Generally speaking, keeping materials in an assembly as pure, unmixed, and unglued as possible increases the recyclability. Less additives and mixing is better. Figure 15.15 clearly shows how the value per pound of a material increases with increased concentration, no matter the material. Pure materials are worth more, and mixed materials reduce the recycling value.

The information in Tables 15.13–15.15 is an indication as to whether materials are physically compatible. For example, can they mix when melted and be reprocessed through an injection molding operation? On the other hand, mixing materials reduces their value, often to the point where the recycled materials become ostensibly worthless. High-density polyethylene (HDPE) and low-density polyethylene (LDPE) are compatible and can be recycled together if attached; however, the HPDE will be recycled down into LDPE and its higher value lost. By design, many recycled plastics end up in such a dynamic value downgrade cycle, reducing the effectiveness and cash flow of neighborhood recycling programs. Design teams should take care to minimize this problem by keeping materials as pure as possible in specification and attachment.

Ensuring material compatibility is a rather straightforward idea. It requires basic design modifications, although ingenious solutions can be developed to satisfy a product's full functionality. On the other hand, making design changes that eliminate packaging, processes, or product attributes requires more out-of-the-box thinking. This requirement makes it difficult to implement and has high risk but also high reward if successful. Eliminating materials or steps invariably leads to cost savings, and so companies will always have an ulterior motive to advance this approach to design for the environment.

Design for Disassembly

Design for disassembly is very similar to design for assembly as discussed in Chapter 14. The basic guidelines are given in Table 15.4, such as minimize the number of parts. Generally, a product that is designed to be easily assembled can also be easily disassembled. However, there are some major differences. For example, in snap-fit design (integral

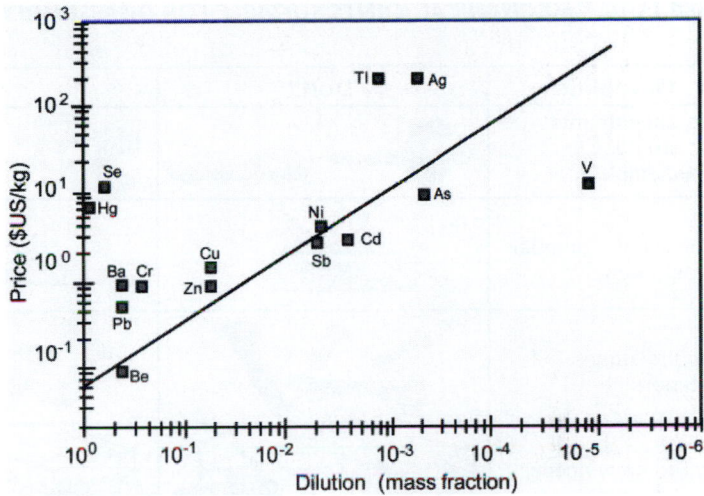

▼ *Figure 15.15.*
Price per unit mass of recycled steel as a function of impurities for steel, adapted from (NRC, 1997).

fasteners), extra care is needed to permit the snap's action during removal as well as insertion, should the part be needed for remanufacturing. On the other hand, if a part is intended for material recycling only, then the disassembly methods typically require tearing apart the assembled connections, and so the disassembly times are completely different from the assembly times. Further, what must be disassembled is not the same as what must be assembled, since some subassemblies may not be worth disassembling, or some subassemblies might be made of compatible materials.

Basic Method: Design Guidelines

Some basic guidelines for mechanical connections are given in the guideline VDI 2243 (1991) as shown in Table 15.16. These are the basic DFA guidelines of minimizing the joint number and difficulty of removal. General Electric Plastics (1995) also provides guidelines specifically for plastic-to-plastic joints, shown in Table 15.17, and for plastic-to-metal joints, shown in Table 15.18.

Advanced Method: Disassembly Time Analysis

A more-detailed analysis is possible by developing a *disassembly tree* (Ishii and Lee 1996), also known as a disassembly diagram. A disassembly tree is a tree structure describing which parts are removed in

TABLE 15.16.   MECHANICAL JOINTS SUITABLE FOR DISASSEMBLY

Guideline	Don't	Do
Use attachments that are easy to disassemble		
Minimize the number of fasteners		
Use the same fasteners		
Ensure easy access for disassembly		
Use simple standard tools		
Avoid long disassembly paths		
Design for damage free dissassembly		
Use the same tools for assembly and disassembly		
Use one disassembly direction to avoid reorientations		
Design for multiple detachments with one operation		

TABLE 15.17.   PLASTIC-TO-PLASTIC JOINT DESIGN GUIDELINE (GE, 1995).

Type		Disassembly Method	Rating
Mechanical Joints			
Hook		Slipped Loose	◉
Snap fit		Snapped Out	◉
Press fit		Ripped Out Pressed Out	◎
Screw		Unscrewed	◎
Screw Insert		Unscrewed Boss Chiseled Off	◎
Welded Joints			
Welded – compatible materials		No separation needed	●
Solvent Bonded – compatible materials		No separation needed	●
Welded (with separate welding material)		Cut off welded area	◎
Stud welded		Chiseled off Milled away	◎
Molded in (insert)		Ripped out Pressed out Drilled out	◎
Glue Bonded		Economically not feasible	○

Rating	Meaning
◉	Excellent
●	Good
◎	OK
○	Bad

TABLE 15.18.   PLASTIC-TO-METAL JOINT DESIGN GUIDELINE (GE, 1995)

Type		Disassembly Method	Rating
Side Hook		Slipped loose	⦿
Snap fit		Snapped out	⦿
Hook press fit		Ripped out Pressed out	◎
Screw		Unscrewed	◎
Screw insert		Unscrewed Chiseled off	◎
Rolled in		Cut off at arrow area	◎
Press fit		Ripped out Pressed out Drilled out	◎
Stud weld		Chiseled off Milled away	◎
Mold in (outsert)		Economically not feasible	○
Glue bond		Economically not feasible	○
Tape weld		Apply electric control	◎

Rating	Meaning
⦿	Excellent
●	Good
◎	OK
○	Bad

TABLE 15.19.   DISASSEMBLY PART REMOVAL TIMES

Degrees of freedom	Horizontal removal		Vertical removal	
	1 hand	2 hands	1 hand	2 hands
2	0.3 sec	0.5 sec	0.6 sec	1 sec
1	0.5 sec	2 sec	1 sec	2.5 sec

Time to Move Parts/Tools	
Pick up	0.7 sec
Put Down	0.7 sec

Source: Dowie and Kelly (1994).

what order from the product and its subassemblies. A disassembly tree is similar to the assembly tree (or diagram) discussed in Chapter 14. The difference is that some subassemblies may not need to be separated on removal. Likewise, fixturing may not be necessary for disassembly.

With a disassembly tree, a disassembly time analysis can be completed, again as discussed in Chapter 14 for assembly. What are needed are estimates of removal times for each disassembly operation. Dowie and Kelly (1994) provide estimated disassembly times for various operations. Their results can be used as part of a design-for-disassembly analysis, similar to what is done in Chapter 14 when discussing design for assembly. Their data are shown in Tables 15.19–15.21. The fastening, which is most critical to separate (in terms of different or hazardous materials) as compared to the difficulty, can be examined.

TABLE 15.20.   SEPARATION TIMES OF TWO FASTENED PARTS

Fastener	Removal method	Time
Screws	Manual	0.6 sec/rev
	Power screwdriver	0.15 sec/rev
Snap fits	Manual breaking	1.5 sec/snap
	Breaking with tool	3 sec/snap
Clips	Manual	1 sec/clip
	Tool	2 sec/clip
Glues, etc.	Manual breaking, 1 hand	3 sec
	Manual breaking, 2 hands	1 sec
	Breaking with tool	2 sec
Cutting cords	Tool	0.5 sec
Cutting wire	Tool	0.25 sec
Disconnect wire	Manual	1.5 sec

Source: Dowie and Kelly (1994).

TABLE 15.21.   MODIFIERS TO FASTENER REMOVAL TIMES (REMOVABLE DIFFICULTIES)

	Motion obstructions				
	Easy access (sec)	More than one direction, around an obstruction (sec)	More than one direction around an obstruction, with restricted vision (sec)	Extended reach (sec)	Severely obstructed access (sec)
No resistance	+0	+3	+9	+12	+17
Holding down part	+6	+9	+15	+18	+23
Corroded	+9	+12	+18	+21	+26

Source: Dowie and Kelly (1994).

Beyond these charts, it is usually straightforward to collect disassembly data for a product by simply taking it apart and measuring the time required for disassembly. In conjunction with this approach, there are many design for disassembly tools available on the market, such as *BDI Design for Environment* from Boothroyd and Dewhurst Inc., *Ametide*, from the Green Manufacturing Program at the University of California, Berkeley, *DFR-Recy*, from the Helsinki University of Technology, *EUROMAT*, from the Technical University Berlin, *LASeR*, from the Manufacturing Modeling Laboratory at Stanford University, *MoTech* from the Technion University in Israel, and *ReStar*, from the Green Engineering Corporation. Also, there are other chart-based worksheet methods, such as the method in McGlothlin and Kroll (1995). All of these tools provide estimates of disassembly difficulty and provide indications of where it is most effective to improve a product.

Example: Coffee Mill

Performing the actual disassembly of the product greatly aids in understanding the difficulties in disassembly and determining actual disassembly times, as shown in Table 15.22. For example, step 2 in Table 15.22, removing the base from the body, is very difficult to figure out without having had experience disassembling the product. The product is snapped together during assembly, and the snap fit positions around the circumference of the mill are very well hidden from the customer.

TABLE 15.22.   DISASSEMBLY TIME ANALYSIS FOR KRUPS COFFEE MILL

Step	Fasteners	Removal method	Time (sec/fastener)	# of fasteners	Total time (sec)
1. Pick up base and tool					1.4
2. Remove base from body	Snap fits	Break with tool	3	4	12
3. Put down tool and pick up cord cutters					1.4
4. Cut cord		Tool	0.5	2	1
5. Pick up slotted screw driver and pliers					2.1
6. Unscrew blade	Screw		10 rev	1	6
7. Put down pliers					
8. Remove chamber	Snap fits	Break with screwdriver	1.5	6	9
9. Pick up Phillips screwdriver					1.4
10. Remove motor from body	Hammer	Chisel out	2 impacts	2	4
11. Remove motor mounts	Hammer	Chisel out	2 impacts	2	4
Total Time					58.3

To disassemble the mill, the operator must use a cutting tool to break the four snap fits holding the base to the shell. If the location were more obvious to the worker disassembling the product, the disassembly time would be improved. An indent on the bottom of the mill indicating the snap-fit center would make for improved disassembly.

Another long disassembly time item is to remove the blade from the motor shaft. It is currently fastened with a screw. A snap fit on a key-shaped shaft would make for easier removal as well as for easier assembly.

The grinding chamber pops down into the mill shell, and removing it is a long disassembly time item. There is no simple point of tool insertion to pop out the metal chamber. Providing this insertion point would reduce disassembly time.

Removing the motor from the shell also requires time and leaves a small plug with a screw and plastic material. Designing a snap fit into the shell might be possible to reduce this waste; however, this modification would require snap fits inside the shell. These additional fasteners would make for a very complex mold and would excessively increase the mold cost.

Design for Recyclability

Another avenue to decrease the environmental impact of a product is to reduce the waste stream of the product. When a product fails and needs to be disposed of, the waste stream can be greatly impacted through recycling. General heuristics as presented in previous sections are effective.

Background

The first requirement for recycling a material in a product is that there must be a demand for the material, regardless of how easy the material is to remove. Therefore, one should use commonly recycled materials, such as metals and plastics. Table 15.23 lists commonly recycled materials for the United States. In general, however, many curbside recycling programs only take PET and HDPE plastics.

Design for Recycling Method

Bras (1996), in association with the USCAR initiative, has developed a very simple yet effective approach to reuse and recycling assessment. Given an indented bill of materials as in Chapter 6, each part can be rated for recyclability of the material and for separability of the component from other product components.

Each product component should be rated on recyclability and separability. Ratings for recyclability are give in Table 15.24, and ratings for separability in Table 15.25. With these ratings, each part can be scored for each criterion. The guideline is that the component is recyclable if both ratings are lower than 3. The individual ratings can then be combined on a mass or value basis into ratings for each subassembly and for the product as a whole. For example, a rating $r_{assembly}$ for a subsystem

TABLE 15.23. COMMONLY RECYCLED PLASTICS (BILATOS AND BASALY, 1997)

| | 1993 | | |
Plastic	Sales (million lbs)	Recycled (million lbs)	Recycling rate
High-density polyethylene (HDPE)	4243	450.2	10.6%
Polyethylene terephthalate (PET)	1598	447.8	28%
Low-density polyethylene (LDPE)	4593	88.3	1.9%
Polystyrene (PS)		35.6	
Polypropylene (PP)	1639	13.6	1.5%
Polyvinyl chloride (PVC)	717	5.5	0.8%

TABLE 15.24   RECYCLABILITY RATINGS (BRAS, 1996).

Rating	Description	Examples
1	Part is remanufacturable	Starter motor, alternator
2	Material in a part is recyclable with a clearly defined technology and infrastructure	Most metals, PETE, HDPE
3	Material is technically feasible to recycle— infrastructure to support recycling is not available	Most thermoplastics, glass, thermosets
4	Material is technically feasible to recycle with further process or material development required	Armrest, airbag modules, single metal with single thermoset
5	Material is organic—can be used for energy recovery but cannot be recycled	Multithermoplastics, wood products
6	Material is inorganic with no known technology for recycling	Heated glass, fiberglass

might be developed by combining the scores of each constituent component, each score in proportion to its mass, such as:

$$r_{\text{assembly}} = \sum_{\text{components}} r_i m_i \qquad (15.1)$$

where r_i is a rating for component i, and m_i is the mass of the component. When completed hierarchically for each subsystem, system, and product, this procedure results in two indented DFE ratings for the product (recycling and separability). These ratings can be then used to focus efforts on the subsystems with a higher percentage of environmental impact.

TABLE 15.25.   SEPARABILITY RATINGS

Rating	Description	Examples
1	May be disassembled easily manually, less than 1 minute	Pull-apart plastics
2	May be disassembled with effort manually, less than 3 minutes	Instrument cluster, radio
3	May be disassembled with effort and some mechanical separation or shredding to separate. The process has been fully proven.	Engines, sheet metal, uncorroded screws
4	May be disassembled with effort and some mechanical separation or shredding to separate. The process is under development.	Instrument panels, corroded screws, adhesives
5	Cannot be disassembled. There is no known effective process for separation.	Heated backlights

Example: Krups Coffee Mill

Table 15.26 shows the recyclability and separability ratings for each of the components listed in the bill of materials. Note that the wiring assembly is not included in the recyclability analysis, because it only represents 0.04% of the total mass of the product and is easily separable from the rest of the product.

Explanation of rankings: The recyclability of the individual components is determined based on the materials or material combination. All steel, copper and cardboard are given ratings of 2 because of the existing network of recycling centers for these metals. Polypropylene (PP) and polystyrene (PS) are given a rank of 3 because although these materials are recyclable, their consumer recycling network is not

TABLE 15.26. RECYCLABILITY RATINGS BY COMPONENT AND ASSEMBLY

Part name	Qty	Material	Mass (gm)	Component		Assembly	
				r_{rec}	r_{sep}	r_{rec}	r_{sep}
Base assembly			81			3.0	2.4
Power switch	1	—	1	3	3		
Power cord	1	—	55	3	3		
Base	1	PP	25	3	1		
Body assembly			40			3.0	1.2
Lid	1	PS	35	3	1		
Blade	1	Steel + PP	3	3	4		
Transfer pusher	1	PP	2	3	1		
Mechanical assembly			25			2.0	1.0
Chamber	1	Steel	25	2	1		
Motor & Body			95.9			3.0	1.0
Felt ring	1	Felt	0.1	5	1		
Screws, self-tapping	2	Steel	0.8	2	1		
body	1	PP	95	3	1		
Motor Assembly			386			2.0	3.0
Motor mounts	2	PP	1	3	1		
Motor	1	—	385	2	3		
Packaging			91			2.7	1.0
Box	1	Cardboard	60	2	1		
Foam	1	Blow foam	1	6	1		
Questionnaire	1	Cardboard	10	2	1		
Dealer list	1	Glossy paper	15	5	1		
Instructions	1	Glossy paper	5	5	1		
Product total			718.9			2.4	2.2

clearly defined. The glossy paper used in the packaging cannot be recycled because of the coatings on the paper. However, energy could possibly be recovered from the combustion of the paper in an incinerator, so it receives a 5. The same logic applies to felt. The foam that is used in the packaging is assumed to be unrecyclable.

Many of the parts are very easily separated from the assembly. The ease of disassembly of the product is probably a testimony to the effort placed into optimizing the assembly of the mill. Several parts that are obtained from outside vendors are much more difficult to disassemble into their component materials. The components with the largest impact are the motor and the cord. With a slight amount of effort, the cord can be broken down to remove the copper. The motor has a significant amount of copper and steel within it. With a few strategic cuts by a band saw, the copper can be easily separated from the steel.

Interpretation of results: Several useful design improvements can be derived for the Krups Coffee Mill. First, the mill is completely mass dominated by the electric motor. The motor is a supplied component, and so avenues for recycling and/or remanufacturing the motor should be explored with the vendor. The second suggestion from the analysis would be to replace PS or PP with a plastic that is more readily recyclable. For example, HDPE costs about as much as PP and PS, but is not as stiff as the other two materials. To use HDPE, an increase in thickness would probably be required. This suggestion would have to be explored in terms of change in product performance and cost. The analysis also considers the packaging material. If the glossy paper used in the packaging could be replaced with plain recyclable paper, the environmental impact of the product would be slightly reduced.

Design for Remanufacturing

Remanufacturing is an end-of-life strategy where the product is taken to a central facility and disassembled. Rather than recycling the parts for their material content only, the parts are instead sorted, cleaned to some degree, and then inspected. Upon certification of quality and performance, the parts are reentered into the manufacturing part supply stream. In 1997, Xerox remanufactured equipment from more than 30,000 tons of returned machines and kept more than four million cubic feet of materials—the equivalent of 250 single-family homes—out of landfills. The automotive industry has sold remanufactured parts such

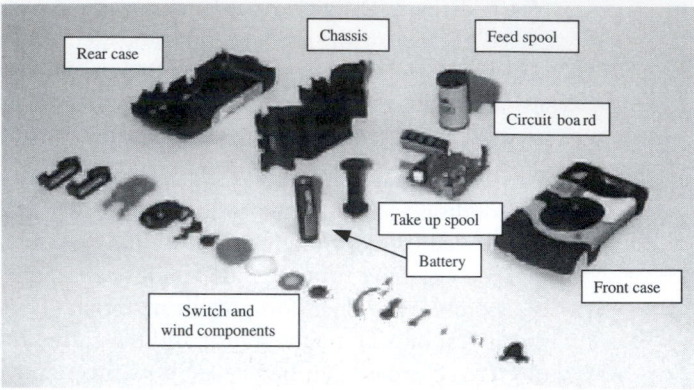

▼ Figure 15.16.
Parts of the Kodak Funsaver single-use camera are remanufactured. Parts must be removed, cleaned, inspected, and returned to the factory for reuse.

as starter motors, alternators, and engines for many years. As a more recent development, consider the Kodak single-use camera, as discussed by Munz (1995), shown in Figure 14.24.

After use with one roll of film, the product line is returned to Kodak where it is disassembled and remanufactured.

The first step in remanufacturing is a visual inspection for disposition. If the product is in bad shape, it is rejected. This step is usually a quick go/no-go visual judgment, erring on the conservative side. For the Kodak camera, this step involves taking pallets of used cameras and unloading them onto conveyer belts. A worker then watches the cameras sequence forward and picks out ones that are so damaged they cannot be reused.

The second step is disassembly of the product into subsets that can be potentially reused or recycled. This step involves the most labor and therefore the most cost, and so design decisions made to simplify this process, as discussed previously, can make or break the effectiveness of this step. The Kodak camera, for example, has been designed for simple separation. Workers remove items in a disassembly line, where the separated parts are placed into bins.

The next step in the remanufacturing process is cleaning and inspection. Parts that are to be remanufactured must be returned to a new condition and their quality assured. Designing parts for cleaning requires careful choices regarding the surfaces of parts, in particular, any labels used. Andreau (1995) gives recommendations as in Table 15.27.

**TABLE 15.27.   DESIGN FOR CLEANING GUIDELINES
(ANDREU, 1995)**

Guideline	Reason
Avoid using labels with glue. Try pop-outs, mold writing.	Preferable to disassemble than to contaminate.
Avoid printing writing on the components.	Printing material is incompatible.
Avoid having closed angles in the components.	Adds difficulty in cleaning.

For the Kodak camera, the only part that is highly cost effective for re-manufacturing is the printed circuit board that powers the flash on the flash camera models. All other plastic parts have such intrinsically low value that it is very expensive to clean and inspect them for return to the plant as opposed to regrinding them into recycled material. Nonetheless, Kodak does inspect and reuse many of the plastic parts, such as the polycarbonate lenses, despite the fact they undoubtedly lose money in doing so.

To design a product for inspection, all that exist as tools and methods are basic guidelines as discussed earlier. Assemblies should be complete with a solid base for easy removal of parts, and parts should be complete with simple test features that indicate quality.

The point in design for remanufacturing is to ensure both design for disassembly and design for inspection. Both are required to support remanufacturing.

Design for High-Impact
Material Reduction

Different functionally equivalent materials can have drastically different impact on the environment. Different functionally equivalent parts made from different production processes can have different impact on the environment. A design team can examine the material and production process for a component and replace them with less-impacting alternatives.

In its 33/50 Program (EPA, 1999), the EPA has developed a list of 17 chemicals targeted for reductions in usage, as shown in Table 15.28. Further, the European Union has restricted many of these chemicals (Hermann and Urbach 1995) as well as biphenyl ethers and poly-brominated biphenyls. Cadmium, chromium, lead, mercury, and nickel are used in the electronics industry, in batteries, electrical components, or electrical joints. Ketones and ethylenes are used in different plastics.

TABLE 15.28. LIST OF CHEMICALS TO AVOID

Benzene	Cadmium
Carbon tetrachloride	Chloroform
Chromium	Cyanides
Dichloromethane	Lead
Mercury	Methyl ethyl ketone
Methy isobutyl ketone	Nickel
Tetrachloroethylene	Toluene
Trichloroethane	Trichloroethylene
Xylenes	

Bipheynls are used as additives in flame retardant materials. Generally, it is not clear if these materials are constituents in the plastics or compounds selected by a designer, and so one must ask. Nonetheless, the chemicals listed in Table 15.28 should be avoided.

From a mechanical design viewpoint, a list of materials and the relative impact on the environment would be more useful than the above lists of chemicals, since often it is not clear what the constituent chemicals are in many materials. Table 15.29 lists a comparison of different materials from best to worst on "average" environmental impact. Using such tables and changing materials in a product can permit reduction in environmental impact. For example, switching from using polystyrene to a less-impacting plastic such as high-density polyethylene can have no impact on design performance, but it can have a reduction in environmental impact.

TABLE 15.29. MATERIAL IMPACT COMPARISON (MICROPOINTS) ADAPTED FROM GOEDKOOP (1995)

Plastics		Metals		Other	
High-density polyethylene (HDPE)	2.9	Aluminum (100% recycled)	1.8	Ceramics	0.5
Polypropylene (PP)	3.3	Steel	4.1	Wood	0.7
Low-density polyethylene (LDPE)	3.8	Sheet steel	4.3	Cardboard	1.4
Polyvinyl chloride (PVC)	4.2	Stainless steel	17	Paper (100% recycled)	1.5
Polyethylene Terephthalate (PET)	7.1	Aluminum (0% recycled)	18	Glass	2.1
Polystyrene (PS)	8.3	Copper (100% recycled)	23	Paper (0% recycled)	3.3
Acrylonitrile butadiene styrene (ABS)	9.3	Copper (60% recycled)	60	Cellulose	3.4
Nylon (PA)	13	Copper (0% recycled)	85	Rubber (NR)	15
		Other nonferrous	50–200		
			0		

Design for Energy Efficiency

A major source of environmental impact is the energy consumed by a product during its use. One should examine the product for opportunities to reduce the energy consumption of the product. Thermal insulation and higher electrical efficiency are often drivers of energy usage, and often energy use is one of the most important factors in a product's environmental impact. Means to increase energy efficiency of components is a good opportunity. There are guidelines to promote energy efficiency that one can follow, as shown in Table 15.30.

Design to Regulations and Standards

There are two basic types of regulations on the environment, mandatory regulations and voluntary compliance standards. Both are important to understand and to comply with.

Environmental Law

In the United States, there are no actual laws on the content of products per se; rather, there are several federal laws that manufacturers of products must conform to. For example, it is up to the Environmental Protection Agency (EPA) to regulate companies' operations and discharge into the environment, and it is up the Department of Occupational Safety and Health Administration (OSHA) to control toxic levels in the workplace. We discuss each of these next.

U.S. environmental law: U.S. Environmental law is enforced by the Environmental Protection Agency, covering air pollution, water pollution, and land contamination. The enforcement is imposed under

TABLE 15.30.   ENERGY EFFICIENCY GUIDELINES

Specify best-in-class energy efficiency component.	Reduces energy usage and societal fossil fuel consumption
Have subsystems power down when not in use.	Reduces energy usage and societal fossil fuel consumption
Permit users to turn off systems in part or whole.	Reduces energy usage and societal fossil fuel consumption
Make parts whose movement is powered as light as possible.	Less mass to move requires less energy
Insulate heated systems.	Less heat loss requires less energy
Solar-powered electronics are better.	Does not create harmful by-products
Choose the least harmful source of energy.	Reduce harmful by-products
Avoid nonrechargable batteries.	Reduce waste in streams
Encourage use of clean energy sources.	Reduce harmful by-products

Source: Bras lecture notes, 1998.

various permitting programs used to implement the Federal law. U.S. environmental laws themselves are listed in Table 15.31.

For air pollution, the EPA enforces the Clean Air Act through the *Code of Federal Regulations* (CFR). The enforcement is therefore a permitting program—manufacturers are permitted to introduce a certain quantity of pollutants into the environment. The act does require waste reduction and product review to achieve clean air emission control. The review comes in the form of reporting and tracking of pollution emissions.

For water pollution, the EPA enforces the Clean Water Act (CWA) through the *National Pollutant Discharge Elimination System* (NPDES), as directed in Section 305 of the CWA. This section calls for a permit

TABLE 15.31. MAJOR FEDERAL EPA-ADMINISTERED ENVIRONMENTAL LAWS

Law	Focus	Content
Clean Air Act (1970)	Air pollution	National Emissions Standards for Hazardous Air Pollution and National Ambient Air Quality Standards. Includes emission limits on lead, hydrocarbons, nitrogen oxides, suspended particulates, sulfur oxides, asbestos, beryllium, mercury, vinyl chloride, benzene, arsenic, radionuclides, copper, nickel, and phenol.
Water Pollution Control Act (1972); later Clean Water Act (1977)	Water pollution	Makes discharging pollutants from a point source to navigable waters illegal without a permit. Includes emission limits on 126 individual chemicals, including volatile organic compounds, acids, pesticides, and heavy metals.
Insecticide, Fungicide, and Rodenticide Act (1972)	Pesticides	Gives federal government control over pesticide sale, use, and distribution. Gives EPA authority to study pesticide use consequences, to require pesticide registration by farmers and businesses, and to require pesticide users to take certification exams.
Resource Conservation and Recovery Act (1976)	Hazardous waste	Cradle-to-grave control of hazardous substances. Use of alternative disposal technology and reduction in waste streams. 400 discarded commercial chemical products and specific chemical constituents of industrial waste streams that are to be destined for treatment or disposal on land.
Comprehensive Environmental Response, Compensation, and Liability Act (1980)	Hazardous waste	Clarified responsibility and liability of parties in hazardous material management. Established Superfund to pay for remediation.
Hazardous and Solid Waste Amendments (1984)	Hazardous waste	Established national efforts to improve hazardous waste management.
Emergency Planning and Community Right-to-Know Act, also known as Superfund Amendments and Reauthorization Act (1986)	Toxic substances	Toxic release inventories (TRI): Companies must report any hazardous materials released. Over 320 toxic chemicals and chemical categories. Over 360 chemicals for which facilities are required to prepare emergency action plans if these chemicals are present above certain threshold quantities.
Pollution Prevent Act (1990)	Life cycle	Life cycle approach to pollution prevention planning and flexibility. No real content.

system to control the *Total Maximum Daily Loads* (TDML) on any given body of water, such as lakes, rivers, or bays. The TDML is used to understand and establish the relationship between pollution sources and in-stream water quality conditions and to then set maximum loading levels for *point sources* of water pollution. A point source is any clearly identifiable source of pollution, such as a treatment plant or a manufacturing facility.

Workplace environmental safety law: With respect to workplace toxic limits, there are two primary laws that cover the health and safety of workers, federal and state OSHA. The federal OSHA is administered by the federal government; state OSHA agencies are administered individually by each state, where state OSHA agencies can impose more stringent regulations as they see fit. Other environmental safety laws are listed in Table 15.32.

To provide employees with their right to know about possibly hazardous materials, OSHA enforces a manufacturer's development, maintenance, and management of the *Material Safety Data Sheets* (MSDS). MSDS forms provide a material's toxicity and safety information. An MSDS sheet must include a product identity, reactivity hazards, hazardous ingredients, spill cleanup, physical/chemical properties, protective equipment needed, fire and explosion hazards, special precautions, and any health hazards. Any employer must manage and provide MSDS sheets for the hazardous materials used. Further, hazardous material products must be sold with information on how to attain MSDS sheets, typically by providing a telephone number on the product to call the manufacturer for the MSDS sheets.

TABLE 15.32.   *MAJOR FEDERAL OSHA ADMINISTERED ENVIRONMENTAL SAFETY LAWS*

Law	Focus	Content
Occupational Safety and Health Act (1970)	Workplace safety	Ensure safe and healthful working conditions for employees; enforcement of the standards developed under the Act; assisting and encouraging states in their efforts to ensure safe and healthful working conditions by providing for research, information, education, and training in the field of occupational safety and health.
Toxic Substances Control Act (1976)	Toxic substance	Regulate, test, and screen all new substances imported or produced.
Food, Drug, and Cosmetic Act (1938)	Food, drugs, cosmetics	Prohibits the manufacture, commerce, receipt for delivery, the delivery or attempted delivery, and the receipt of adulterated food, drugs, or cosmetics.
Resource Conservation and Recovery Act (1976)	Hazardous waste	Cradle-to-grave control of hazardous substances. Use of alternative disposal technology and reduction in waste streams. 400 discarded commercial chemical products and specific chemical constituents of industrial waste streams that are to be destined for treatment or disposal on land.

Labeling Programs

Environmental purchasing awareness is also being spurred by several private, nonprofit, and governmental initiatives. As an example, consider eco-labeling programs. Here, provided that a product meets a set of environmental specifications established by the program, one can advertise the environmental advantages of a product. Some examples are shown in Figure 15.17, and the federal eco-labeling voluntary programs administered by the EPA are listed in Table 15.33. Customers see such labels in product advertising, and then potentially choose to buy the product because of a company's environmental practices.

An example of an effective eco-labeling programs is Germany's *Duales System Deutschland* (DSD) recycling program focused on product packaging. Here, one can advertise with a green dot symbol when their product packaging meets the program requirements. The green dot may be applied to the product for a licensing fee. This dot-labeling system is the oldest eco-labeling program, in existence since 1991.

The program requirements include a waste collection system to collect, separate, and recycle the used packaging. The specific targets include an 80% collection rate of the packaging from consumers and then a 90% content recycling rate of the collected material.

In practice, the program has not met with this target success. As of 1995, the recycling rate was 75% for bottles, 70% for glass, 70% for tinplate, 70% for aluminum, 65% for laminates, 65% for paper, and 65% for plastics.

On the other hand, there was a 95% awareness rate of the program in Germany. Yet the program was expensive. It cost $6 billion to establish and $1 billion per year to run. This equates to $160/ton of disposed material. Also, there was a supply-and-demand problem: More recycled material was generated than Germany could effectively introduce back into production systems. Nonetheless, such programs are in demand by members of society and will only grow. Good product design will make them successful.

ISO 14000

Another related program is the International Standard's Organization ISO 14000 standards on environmental impact management. The *International Organization for Standardization* (ISO) is a private sector,

▼ *Figure 15.17*
Eco-labeling programs.

international standards body based in Geneva, Switzerland. Founded in 1947, the ISO promotes the international harmonization and development of manufacturing, product, and communications standards. The ISO has developed more than 8,000 internationally accepted standards for everything from paper to screws. More than 120 countries belong

TABLE 15.33.   MAJOR EPA-ADMINISTERED AND VOLUNTARY ECO-LABELING PROGRAMS

Program	Focus	Content
Green Lights	Lighting	Companies commit to installing energy efficient lighting
Green Buildings	HVAC	Companies commit to installing energy-efficient heating and air conditioning and insulation
Golden Carrot	Refrigeration	Competition to design an energy-efficient refrigerator
Energy Star	Electronics	Maximum power levels and power down requirements
Climate Wi$e	Greenhouse gases	Voluntary program to reduce greenhouse gas emissions
Waste Wi$e	Solid waste	Voluntary program to reduce landfill solid waste
Environmental Leadership Program	Manufacturers	Program to recognize activities taken by manufacturers
Industrial Toxics Project (33/50)	Toxic substances	Encourage companies to reduce 17 toxic substance releases by 33% by 1993 and 50% by 1995.
Environmentally Preferable Products	Federal Agencies	EPA guideline on a 1993 Presidential directive for federal agencies to purchase low-impact products.

to the ISO as full voting members, while several other countries serve as observer members. The United States is a full voting member and is officially represented by the *American National Standards Institute* (ANSI). The purpose of the ISO is to promote worldwide standards in industry to facilitate international exchange, improving operating efficiency and unify foreign markets.

The ISO 14000 standards are evolving; they seek to ensure that organizations follow practices so that they understand their environmental impact. This goal is achieved through LCA assessments, documentation, and organizational practices. The standard is basically a series of environmental management systems to address the impact that its products, services, and operations have on the environment.

Complying with ISO 14000 does not, however, guarantee that a company is environmentally friendly. Rather, to become certified under ISO 14000, a company must establish and maintain a procedure to identify and access all environmental regulations. They must be applicable to the facilities site, activities, products, and service. Therefore, ISO 14000 certification ensures that you know how environmentally friendly you are, not that you are environmentally friendly. Being ISO 14000 certified is not a legal responsibility; it is a voluntary action.

The ISO 1400 standard, as all standards, is a living document and in a state of change as analysis and tools rapidly evolve. The different sections of the standard are shown in Table 15.34. From a product design viewpoint, the tools in this chapter are of use to ensure the product development process applied can qualify for ISO 14000 certification.

TABLE 15.34.   SECTIONS OF THE ISO 14000 STANDARD

Section	Title
ISO 14001	Environmental Management Systems—Specification and Guidance for Use
ISO 14004	Environmental Management Systems—General Guidelines on Principles, Systems and Supporting Techniques
ISO 14010	Guidelines for Environmental Auditing—General Principles
ISO 14011/1	Guidelines for Environmental Auditing—Audit Procedures—Auditing of Environmental Management Systems
ISO 14012	Guidelines for Environmental Auditing—Qualification Criteria for Environmental Auditors
ISO 14015	Environmental Site Assessments
ISO 14020	Goals and Principles of all Environmental Labeling
ISO 14021	Environmental Labeling—Self Declaration Environmental Claims—Terms and Definitions
ISO 14022	Environmental Labeling—Symbols
ISO 14023	Environmental Labeling—Testing and Verification Methodologies
ISO 14024	Environmental Labeling—Guiding Principles, Practices and Criteria for Multiple Criteria-Based Practitioner Programs (type I)—Guide to Certification Procedures
ISO 1402X	Type III Labeling
ISO 14031	Evaluation of Environmental Performance
ISO 14040	Life Cycle Assessment—Principles and Guidelines
ISO 14041	Life Cycle Assessment—Life Cycle Inventory Analysis
ISO 14042	Life Cycle Assessment—Impact Assessment
ISO 14043	Life Cycle Assessment—Interpretation
ISO 14050	Terms and Definitions—Guide on the Principles for ISO/TC 207/SC Terminology Work
ISO Guide 64	Guide for the Inclusion of Environmental Aspects in Product Standards

VII. CHAPTER SUMMARY AND "GOLDEN NUGGETS"

Understanding the environmental impact of a product and intervening as a design team to mitigate effects is a responsibility that must now be adopted. Legislative and consumer demands will only increase in this area. Some key ideas in the design for the environment include:

▶ Assessment must be completed considering the entire product life cycle, from the time material is extracted from the earth until it is returned to the earth.

▶ Design-for-the-environment techniques and their associated metrics represent one component of a product's systems model.

▶ Basic methods include smart material selection, minimizing energy usage, and increasing recycled content.

▶ Numerical scoring is possible.

▶ Full life cycle assessments provide quantification of environmental impact.

References

Allen, D., and Behmanesh, N. 1994. Wastes as raw materials. In *The greening of industrial ecosystems*, edited by B. Allenby and D. Richards. Washington: National Academy Press.

Andreau, J. 1995. The remanufacturing process. Tech. Rep. DDR/TR24, Manchester Metropolitan University, Manchester, UK.

Billatos, S. and N. Basaly. 1997. *Green technology and design for the environment.* Washington: Taylor and Francis.

Bras, B. 1998. Lecture Notes from *Environmentally conscious design & manufacturing.* Georgia Institute of Technology, Atlanta.

Cheney, R. 1971. Design trends in glass containers. *Proceedings of the Solid Waste Resources Conference on Design of Consumer Containers for Re-use or Disposal.*

Chapman, P. 1983. *Metal Resources and Energy.* Butterworth, London, UK.

EPA 33/50 Program, (EPA, 1999)

Dowie, T., and Kelly, P. 1994. Estimation of dissasembly times. Tech. Rep. DDR/TR15, Design for Environment Research Group, Manchester Metropolitan University, Manchester, UK.

Dowie, T., and Simon, M. Guidelines for designing for disassembly and recycling. Tech. Rep. DDR/TR18, Design for Environment Research Group, Manchester Metropolitan University, Manchester, UK.

Environmental Protection Agency, *33/500 Program, The Final Record.* EPA Report EPA -745-R-004, March 1999.

Fava, J., Denison, R., Jones, B., Curran, M. A., Vigon, B., Selke, S., and J. Barnum, eds. 1994. *A technical framework for life-cycle assessment.* Society of Environmental Toxicology and Chemistry (SETAC).

Fiksel, J., ed. 1996. *Design for the environment.* New York: McGraw-Hill.

General Electric Plastics, Inc. 1995. Design for recycling. Tech. Rep., Pittsfield, MA.

Goedkoop M. J. 1995. The Eco-indicator 95, Final Report; NOH report 9523; Pré Associates; Amersfoot Netherlands.

Graedel, T., and Allenby, B. 1996. *Design for the environment.* Englewood Cliffs, NJ: Prentice Hall.

Graedel, T., and Allenby, B. 1995. *Industrial ecology.* Englewood Cliffs, NJ: Prentice Hall.

Hermann, F., and Urbach, H.-P. 1995. Restricted substances in products in the European Union. *Proceedings of the 1995 IEEE International Symposium on Electronics and the Environment*, Orlando, FL, May 1–3.

Hoffman, W., 1995. A tiered approach to design for environment. *Clean electronics products and concepts (CONCEPT)*, Edinburgh: Institution of Electrical Engineers, London, UK.

IBM Inc. *IBM environmental packaging design guide.* Part #GA23-2201-00.

Industry Council for Electronic Equipment Recycling (ICER). 1993. *Guidelines: Design for recycling: General principles.*

Ishii, K., and Lee, B. 1996. Reverse fishbone diagram: A tool in aid of design for product retirement. *Proceedings of the Design for Manufacturability Conference (ASME)*, Irvine, CA, Sept.

Keoleian, G. and D. Menerey, Life Cycle Design Guidance Manual: Environmental Requirements and the Product System (Washington: DC: Risk Reduction Engineering Laboratory, Office of Research and Development, U.S. Environmental Protection Agency) EPA 600/R-92/226, 191 pages.

McGlothlin, S. and Kroll, E. 1995. Systematic estimation of disassembly difficulties: Application to computer monitors. *Proceedings of the 1995 IEEE International Symposium on Electronics and the Environment*, Orlando, FL.

Munz, B. 1995. Cost analysis of product recovery process in single use camera life cycle. Master's thesis, Massachusetts Institute of Technology.

Miller, J. 1995. New shades of green. Master's thesis, The University of Texas at Austin.

Miller, J. and K. Wood. 1996. Green Product Design. Technical Report, Manufacturing and Design Laboratory, The University of Texas at Austin.

Society of Environmentally Toxicology and Chemistry. 1993. *Guidelines for life cycle assessment: A code of practice*. Brussels: SETAC.

Verein Deutches Ingineur (VDI). 1991. Design of technical products for ease of recycling. *VDI* 2243 (May).

United Nations Environment Program/Industry and Environment (UNEP/IE). 1997. Ecodesign: A Promising Approach to Sustainable Production and Consumption. Tech. Rep. CP18.

Xerox Corporation, http://www.xerox.com/downloads/envcall.pdf, January 2000.

16 Analytical and Numerical Model Solutions

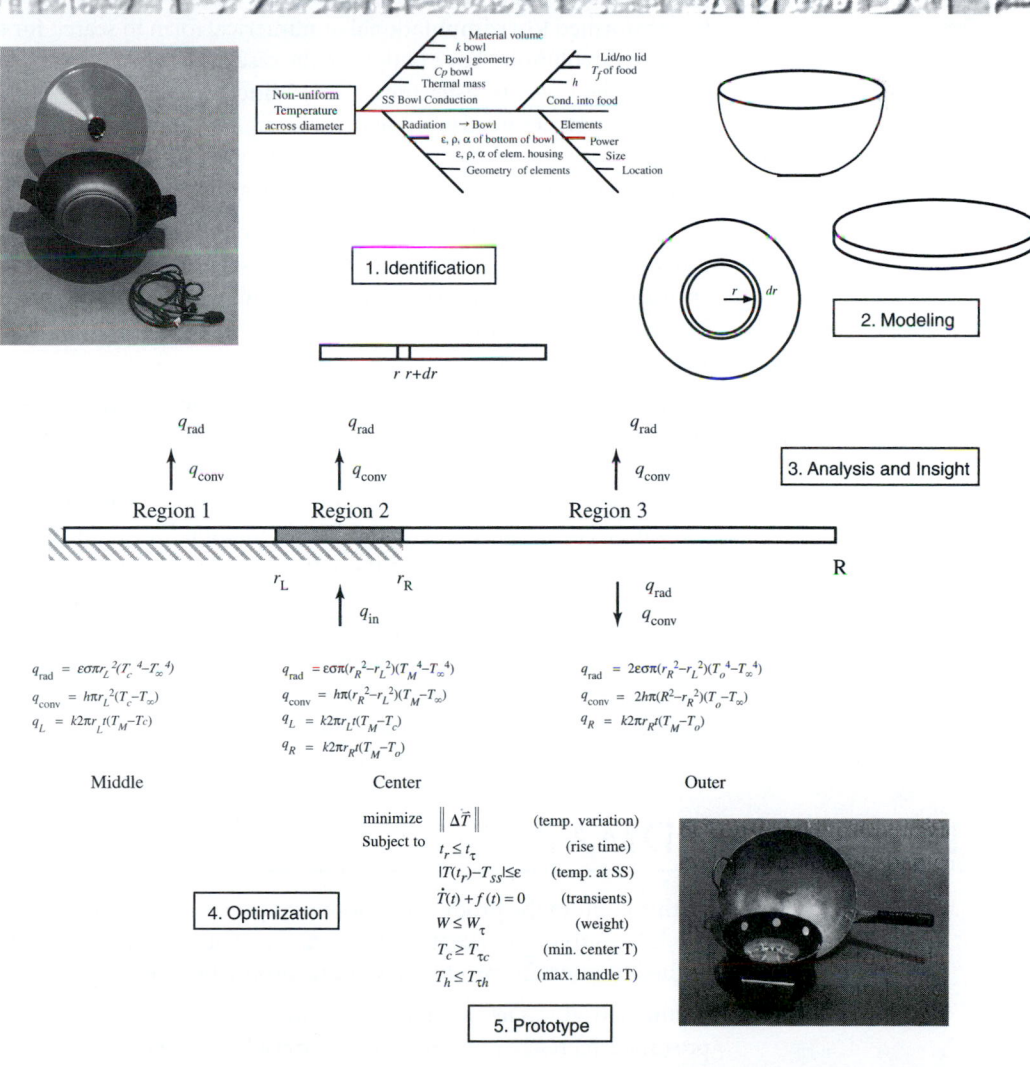

Material volume
k bowl
Bowl geometry
Cp bowl
Thermal mass
Lid/no lid
T_f of food
h
SS Bowl Conduction
Cond. into food
Non-uniform
Temperature
across diameter
Radiation → Bowl
ε, ρ, α of bottom of bowl
ε, ρ, α of elem. housing
Geometry of elements
Elements
Power
Size
Location

1. Identification

2. Modeling

r $r+dr$

q_{rad} q_{rad} q_{rad}

↑ q_{conv} ↑ q_{conv} ↑ q_{conv}

3. Analysis and Insight

Region 1   Region 2   Region 3

r_L r_R R

↑ q_{in} q_{rad}
                                    ↓ q_{conv}

$q_{rad} = \varepsilon\sigma\pi r_L^2(T_c^4 - T_\infty^4)$

$q_{conv} = h\pi r_L^2(T_c - T_\infty)$

$q_L = k2\pi r_L t(T_M - T_c)$

$q_{rad} = \varepsilon\sigma\pi(r_R^2 - r_L^2)(T_M^4 - T_\infty^4)$

$q_{conv} = h\pi(r_R^2 - r_L^2)(T_M - T_\infty)$

$q_L = k2\pi r_L t(T_M - T_c)$

$q_R = k2\pi r_R t(T_M - T_o)$

$q_{rad} = 2\varepsilon\sigma\pi(r_R^2 - r_L^2)(T_o^4 - T_\infty^4)$

$q_{conv} = 2h\pi(R^2 - r_R^2)(T_o - T_\infty)$

$q_R = k2\pi r_R t(T_M - T_o)$

Middle   Center   Outer

minimize $\|\Delta\bar{T}\|$ (temp. variation)

Subject to $t_r \leq t_\tau$ (rise time)

$|T(t_r) - T_{SS}| \leq \varepsilon$ (temp. at SS)

$\dot{T}(t) + f(t) = 0$ (transients)

$W \leq W_\tau$ (weight)

$T_c \geq T_{\tau c}$ (min. center T)

$T_h \leq T_{\tau h}$ (max. handle T)

4. Optimization

5. Prototype

The cover illustration for this chapter paints a picture of product evolution using mathematical models. The original product, in this case an electric wok, is transformed to a new concept that has the potential to raise the product to a new performance level. This concept is modeled, in this case analytically, so that multiple configurations may be searched to realize preferred solutions. It is this search process that is the focus of this chapter.

The remainder of the illustration shows snapshots of the mathematical model being converted to an appropriate formulation. This formulation may be solved for parametric values, and realized in an alpha prototype, to test the parametric decisions. A product model is usually transformed to a computational or numerical form to search for the parametric solutions. For example, in the case of the wok, a steady state and transient heat transfer model, in addition to weight and size, are transformed to a spreadsheet model. This transformed model is solved using optimization schemes, converging to a set of solutions that are feasible (i.e., satisfy the engineering specifications).

A number of computational solution methods exist to solve product models. These methods include finite element analysis, finite difference analysis, modal analysis, dynamic simulations, optimization, etc. While each of these methods has its merits, only optimization is considered in this chapter.

Optimization, as a topic of study for product design, represents a general search method for determining preferred product design variables. It focuses on a well-formulated model, and is one of the many tools that we can wield to embody and improve a product. In this light, optimization should not be overstated as *the* method for solving product models. If we start with a poor, inaccurate, or invalid model of a product, optimization, no matter how effectively applied, will result in poor solutions. However, assuming a well-developed and robust model (Chapters 12 and 13), optimization can be a very powerful technique. It can save time in setting appropriate design variable values and instill confidence in the validity of our solutions.

I. CHAPTER ROADMAP

At this point in the product development process, it is assumed that a candidate product concept has been chosen. This concept must be embodied to move the product toward its production version.

Mathematical models are tools for refining a concept through an approximate representation. Such models must be solved to make decisions

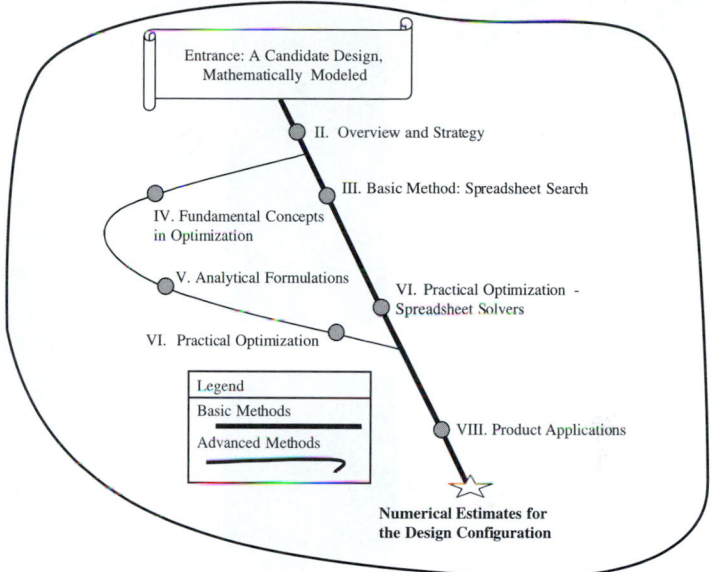

▼ *Figure 16.1.*
Roadmap for creating and solving numerical product models.

about the size, shape, thickness, material properties, etc. of a concept. Figure 16.1 illustrates a roadmap for studying solution methods. One route on this roadmap takes us through basic optimization principles through practical solutions and product applications. The alternative route adds more theoretical understanding to these principles and describes more advanced optimization techniques. Let's begin by considering an overview and strategy for numerical modeling.

II. OVERVIEW AND STRATEGY

As we transition from conceptual design, or early embodiment, the goal is to refine a model of our design concept(s). This model should represent a physical understanding of the product. As an example, consider the simple retail can food container as a product concept. For equal amounts of material, pressure loading, and all other conditions, why are can geometries different? Some cans are short; some cans are tall (long).

Figure 16.2 shows a canned food container (soup or soda) and a graphical representation of its basic parametric model. Assuming a focus metric for the design is the cost of material to produce the can, we seek

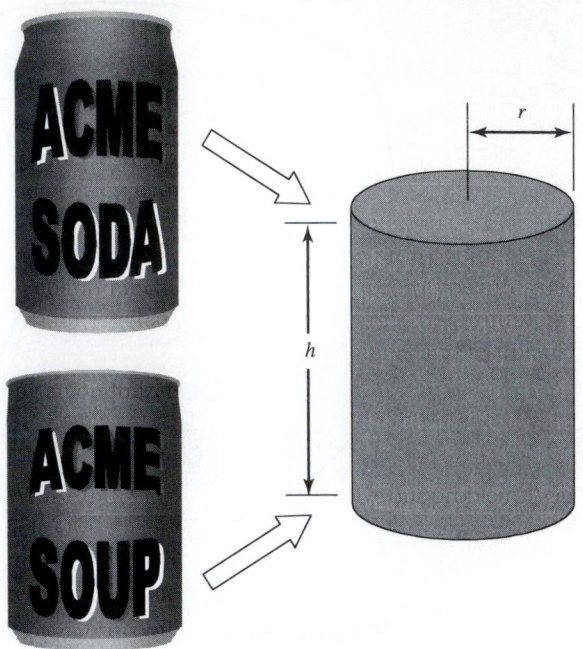

▼ *Figure 16.2.*
Soup or soda can product.

to specify the can's height, radius, and thickness. These variables can form a simple representative model of the product, when a relationship is developed to estimate cost. Let's assume that a minimum thickness is predetermined for the manufacturing processes of the can. Let's also consider minimization of surface area as an equivalent metric for cost. What are appropriate choices of the can height and radius to minimize surface area?

When creating a food product, we are constrained to include a certain volume in the can. Such a constraint will limit the minimum surface area we can obtain. Thus, we might state a model for the can product as a set of two simultaneous equations:

Minimize

$$A_S = 2\pi r^2 + 2\pi rh = 2\pi r(r + h)$$

(16.1)

Subject to

$$V = \pi r^2 h = C$$

While Equation (16.1) represents a simple model of the can product, we are still faced with the task of solving it: choosing the can radius and height given the volume of food product, C. How can we systematically represent and solve such a decision-making problem in product design, especially as the number of variables and equations becomes large?

In this chapter, we will consider a *design optimization approach*, wherein our goal is to improve or refine a design concept so as to achieve the most preferred or robust product, within the available means. To implement this approach, we must answer the following questions: How do we describe the design concept functionally? What is our criterion (or criteria) for a robust design? What are the available means or boundaries (constraints)?

These questions are answered as part of the modeling of product metrics (Chapter 13). But we need to transform the model into a valid optimization statement. This statement involves the following steps:

1. The selection of a set of variables to describe the design concept(s).
2. The identification of noise variables that will affect the robustness of the modeling result.
3. The selection of an objective function (metric) from the model, expressed in terms of the design and noise variables, that we seek to minimize or maximize (or to achieve a target value).
4. The determination of a set of constraints, expressed in terms of the design variables and noise variables, that must be satisfied for an acceptable design.
5. A check of the constraints to make sure that the problem is well bounded.
6. The determination of a set of values that minimize or maximize the objective, while satisfying the constraints (a feasible design that satisfies the product metrics to the best degree).
7. A determination of the sensitivity of the solution. This step involves the comparison of how much the design variables change when optimizing different metrics as the objective function. It may also involve an examination of the solution of the objective function with respect to unit changes in the design parameters.

The data for these steps come primarily from the modeling of the product metrics. But how do we solve such mathematical models/statements, such as Equation (16.1)? One alternative is to set up a spreadsheet and iteratively or combinatorially search the design space defined by the range of choices of design variables. If we have 10 design variables in our product model, and if we consider 10 possible valid values per variable, how many combinations must be calculated? The answer is 10^{10}; this task is difficult, if not impossible!

While such an approach (with fewer tests) is a useful tool for understanding the model, a more efficient technique is needed to solve the model and provide repeatable results. The techniques presented in the following sections satisfy this need. As a first step toward such techniques, let's consider further what optimality means with respect to product design.

Solution Definition

Consider a full engineering model of a product. As a first step, we may want to consider typical values for every noise variable (Chapter 19 considers the full impact of noise). This approach simplifies the noise space to a single point. Each noise variable is then called a "constant" of the formal design problem. (Note: *From now on, every time we hear the word "constant" in an analysis technique, it should set off a warning bell. Physically, how constant is it? What is the constant's range of variation? How is it set? How is it measured?*) We may then consider one of the performance metrics, and determine the configuration in the design space that maximizes the performance.

Typically, classical optimization methods represent performance metrics that seek the smallest values, i.e., minima. Given this description, a performance metric

$$p = f(d) \tag{16.2}$$

can be minimized by finding $d*$ such that

$$f(d*) = \inf\{f(d) | d \in \mathbb{D}\}. \tag{16.3}$$

This result is the minimum solution $(f(d*) = \min(f(d)))$ to the analytically posed and formulated design problem. Similarly for performance metrics that have a target value, we can define a minimization by the variation from a target

$$p = |f(d) - \tau| \tag{16.4}$$

This approach demonstrates the problem formulation for all optimization problems. Basically, we set up a search. This formulation has two aspects:

▸ What do we search for? What is "best" in a formal way? Here, it is the minimum of the product performance metric.

▸ What do we search across? What bounds $\mathbb{D}$? What is feasible, and what do we know is available?

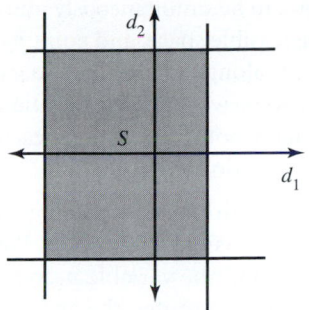

▼ Figure 16.3.
The feasible space S of the design space $\mathbb{D}$.

These are, at the core, problem definition issues. They are not computational complexity problems or numerical problems. What do we want to achieve from the product?

Pareto Optimality

One answer to the previous question is to improve all of the performance criteria, subject only to basic availability constraints. We might then try to minimize the cost and also maximize the performance criteria. This approach is not always possible. For example, when booking a flight, we cannot acquire a ticket that provides the best service at minimum cost. Two objectives exist in this case: greatest comfort (first class) and, at the same time, lowest cost (coach seating). We can either buy a ticket that makes us most comfortable—the first class ticket solution that compromises the cost objective, or we can buy a ticket that is inexpensive—the coach ticket solution that compromises the in-flight service objective.

We might think about this situation in terms of the model spaces and $\mathbb{P}$ (Chapter 13). The constraints define a feasible sub-region S of $\mathbb{D}$ that are available and, consequently, are possible, as shown in Figure 16.3. We might think of the mappings f_i together as a vector mapping

$$\vec{p} = \vec{f}(\vec{d}) \qquad (16.5)$$

and work to minimize this vector $\vec{p}$. Consider mapping every point in S through Equation (16.5), thereby defining a feasible space Z in $\mathbb{P}$. This result is shown in Figure 16.4.

In the performance space, we know that we want to minimize the performance metrics p_i. Yet notice an interesting feature of Figure 16.4.

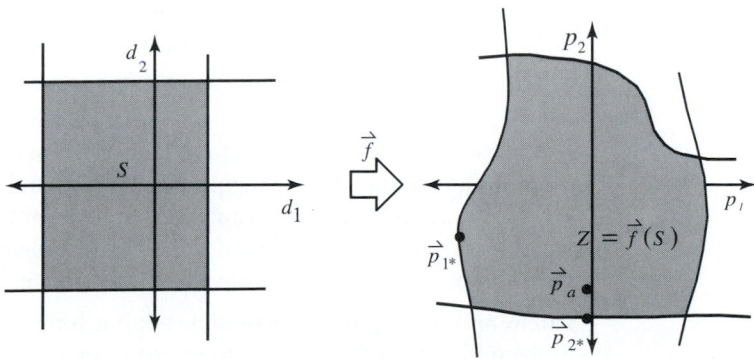

▼ Figure 16.4.
The feasible space Z of the performance space $\mathbb{P}$.

The space Z is not shaped so that p_1 and p_2 can be simultaneously minimized. The point $\bar{p}_{1*}$ minimizes p_1 in the feasible space, and point $\bar{p}_{2*}$ minimizes p_2 in the feasible space. All points along the short boundary of Z, connecting and $\bar{p}_{1*}$ and $\bar{p}_{2*}$, minimize some weighted combination of $\bar{p}_{1*}$ and $\bar{p}_{2*}$. In other words, there are many optimal solutions, each which optimizes a different weighted combination of the criteria.

Put this in the example context of booking an airline flight. All flights that reach the destinations are the space of feasible alternatives. We might consider the variable p_1 to be price, and the variable p_2 to be flight time. Across all airlines and their provided routes, the one airline ticket that provides the shortest flight time (has a direct flight, for example) is the point $\bar{p}_{1*}$. The airline ticket that has least price is the point $\bar{p}_{2*}$, probably by an airline that requires a layover in a hub airport and uses a small propeller plane for part of the journey. Further, there are tickets that perhaps are not least cost nor shortest flights, but are an optimal compromise between these two concerns— such as a moderately priced flight with a layover using all jet service.

Consider, though, a point $\bar{p}_a$ as shown in Figure 16.4. It has the same level of performance on p_1, but worse performance on p_2. Under these circumstances, it would not be rationally selected as a solution. One could always instead choose $\bar{p}_{2*}$ instead. For example, if two airlines offer flights at the same cost (p_1), the flight that has fewer layovers and so a shorter flight time (p_2) is a better solution. Similarly, if two concepts provide the same performance (p_1), the one that uses less costly components (p_2) is a better solution.

This basic thinking is the principle behind what is known as Pareto optimality. A point $\bar{p}$ is *Pareto optimal* if the only way to improve any of its components p_i is by worsening other components p_j. All such points basically lie in the lower quadrant boundary of the space Z, as shown in Figure 16.5. This region is also called the *Pareto Frontier*, denoted by F. Any point in F can be a reasonable optimal solution.

This idea has implications when determining a solution to a modeled design problem. One must search across the design variable space for a useful solution as above. Once found, however, one should then explore around this point by changing the variables, and obtain a feel for how much improvement one can attain in some of the performance variables at how much sacrifice of the others. A different point on the Pareto frontier may offer better overall performance.

Pareto optimality provides a sensible approach for making decisions in the midst of conflicting objectives. However, this concept does not obviate the responsibility of a designer to seek solutions to design problems that remove conflicts altogether. If conflicts may be removed

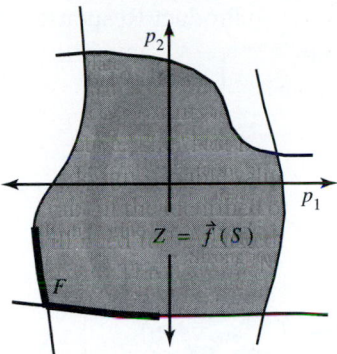

▼ *Figure 16.5.*
The Pareto Frontier.

through a drastic change or metamorphosis of a design configuration, a compromise will not be needed, potentially improving all performance variables. While conflicts will exist in all practical product development, removing as many conflicts as possible will result, by necessity, in novel product solutions. Novel solutions should be our primary mission. Therefore any design problem solution, model optimal or otherwise, is permitted as final only after resources have been expended. A design team should accept an optimal solution only when they are to the point where they cannot go back to develop new concepts that remove trade-offs (Chapter 10).

III. BASIC METHOD: SPREADSHEET SEARCH

A basic method to explore and solve product models is what we call a *spreadsheet search.* Spreadsheets are powerful analysis and data representation tools. They offer a simple means to study a range of product scenarios resulting from the variation of a number of variables.

Spreadsheet programs originated out of the need for tools to perform accounting and finance analysis. With personal computers on desks in industry and in the home, all types of professionals now use spreadsheets. Engineers in particular are heavy users of spreadsheet applications.

While more powerful and specialized analysis tools usually exist for the engineer, spreadsheets are prevalent due to their flexibility and relatively benign learning curve. The flexibility of spreadsheets comes from their architectural features. Spreadsheets represent data in "cells." These cells hold

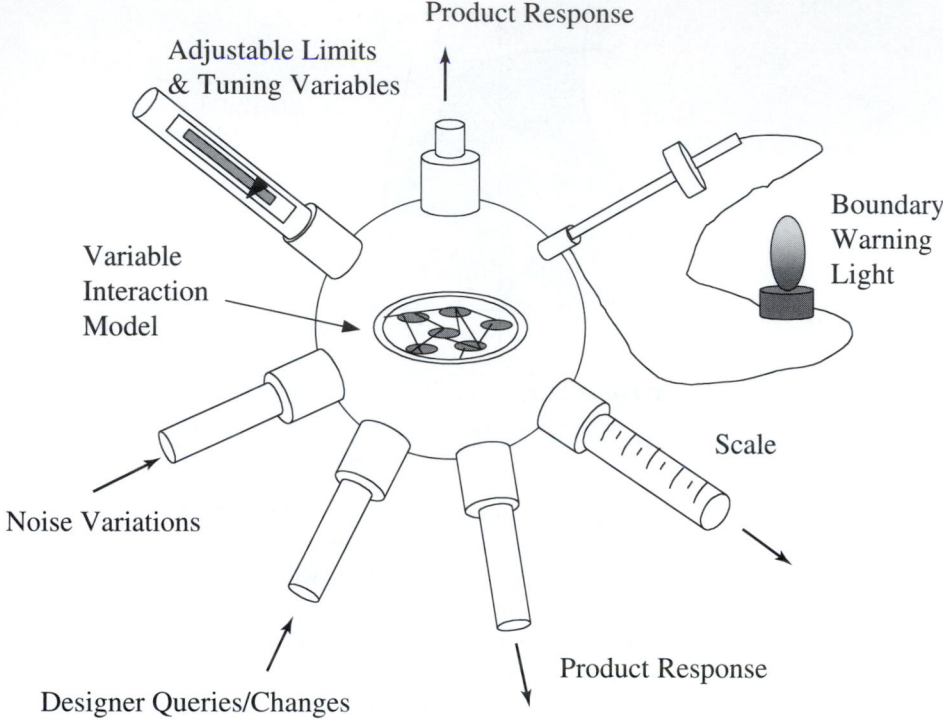

Product Response

Adjustable Limits
& Tuning Variables

Boundary
Warning
Light

Variable
Interaction
Model

Noise Variations

Scale

Designer Queries/Changes

Product Response

▼ Figure 16.6.
A hypothetical machine that solves product models and shows causal dependencies
(after Serrano, 1987).

values or formulas that may be directly linked to other operators or logic
for analysis. The results of an analysis may then be easily plotted or ex-
ported for use in other applications, or for display. Spreadsheets are not
the ultimate flexible numerical tool, however. There are no data types
such as integers, vectors, matrices, etc., instead every number is a floating
point number. Further, algorithms with loops are difficult to implement.

In terms of product design, we wish to solve a model to determine preferred design variable values and gain insight into the physics and sensitivity of how a product will perform. This concept of what an ideal
design model would be is shown in Figure 16.6 (after Serrano, 1987).
In this figure, design variables are input through designer queries. Adjustments are inputs to a product model representing product responses
to the end users' queries. Noise variations from sources such as manufacturing, material, and wearout may also be explored. When limits

or constraints are violated with the query, a signal is produced notifying the designer or end user.

Spreadsheets can satisfy these model solution goals, at least at a basic level, through a series of steps executed by the designer. These steps may be classified, broadly, into two categories: (1) planning and design of a product model worksheet, and (2) developing and executing a solution procedure.

Let's consider these categories for a product model worksheet. The first category involves the implementation of four basic steps. The first step is to identify the purpose of the worksheet. This step entails the title of the worksheet, writing a one-line concise statement of its purpose. Information for this step comes directly from the product model we are attempting to solve.

The next step is to determine the outputs that are desired from worksheet analysis. The outputs, for product development, are the metric values that represent the customer needs. Again these are determined from the product modeling results. However, it is important to also identify intermediate results that should be displayed as outputs. Intermediate results are necessary to validate a model (i.e., compare it to known or simple solutions) and to obtain particular physical insights that may be used to improve a product.

Once the outputs are chosen, the inputs should be identified. The inputs are of two forms, from the product model: noise variables and design variables. Noise variables may be initially viewed as "constants" and then later varied during model analysis to understand their impact on performance. (Chapter 19 studies more complete methods for handling noise variables as part of robust design.)

Design variables, on the other hand, represent the decisions we are seeking to make as part of the model solutions. These variables should be clearly displayed in the worksheet, in addition to their boundaries or ranges of choice. This information, again, may be extracted from the product model.

After identifying the inputs and outputs, the calculations and their corresponding solution strategy, must be chosen. For closed-form equations, the calculations will involve a set of simultaneous expressions that must be solved over the choices of design variables. A "brute force" or divide-and-conquer scheme may be adopted, where each of the design variables is discretized and combinations or subsets of the variable values are substituted. The resulting calculations are then plotted and compared with the product target values, in addition to the boundaries to determine feasibility and the level of the objective(s). Alternatively, this set may be solved with matrix/linear algebra techniques or with adjacency matrices to determine a plan for computing the equations in a dependent order (Ramaswamy et al., 1993).

Other strategies exist for solving a product model in a spreadsheet, depending on the type of product model. If it includes differential equations, a first-order solution strategy may be to implement Euler's method, representing each differential as the first term (linear) from a Taylor series expansion. For example, a differential velocity stated as dv/dt would be represented as $\Delta v/\Delta t = (v_{(i+1)} - v_i)/h$, where h is a small time increment Δt, and all velocities are calculated using a function. That is, updated values of the velocity may be calculated, $v_{(i+1)}$, by choosing Δt and solving iteratively from a known initial value of v_i. For other types of models, similar strategies would be adopted.

Once a strategy is chosen, a worksheet layout must be developed. This development entails the arrangement of the input, calculation, output, and summary blocks for the worksheet, in addition to comment blocks to aid subsequent uses of the worksheet. Many spreadsheet programs provide different color palettes, fonts, borders, shading, etc. to aid in the worksheet layout. These facilities should be used to aid in the utility of the spreadsheet.

When keying in cell formulas, one should never key in numerical model constants in the spreadsheet formulas. If there is a constant such as emissivity ε or a duty cycle K, such factors should be explicitly entered into separate cells, and the formulas linked to such cells. We should not bury model constants within cell formulas, for we may forget about them later. Designs may turn out to be very sensitive to such a variable, especially as a design model is explored and calibrated early in the process.

It is also good practice to color code cells so we can understand them by inspection. All cells containing input data, for example, should be colored to distinguish them from cells containing formulas. A good practice is to color all constants with red fonts and all design variable values with green fonts. Another guideline is to use simple blocks to distinguish between different worksheet elements. The product application below demonstrates these principles at a basic level.

Product Application: Spreadsheet Search for a Toy Rocket Product

To illustrate the use of a spreadsheet for solving product models, let's consider the design of a toy rocket. As illustrated in Figure 16.7, this product uses water and compressed air to launch a lightweight, single-stage rocket. The first introduction of this product occurred during the mid-1950's by Park Plastics Company, based on a German Patent (Dixon, 1966).

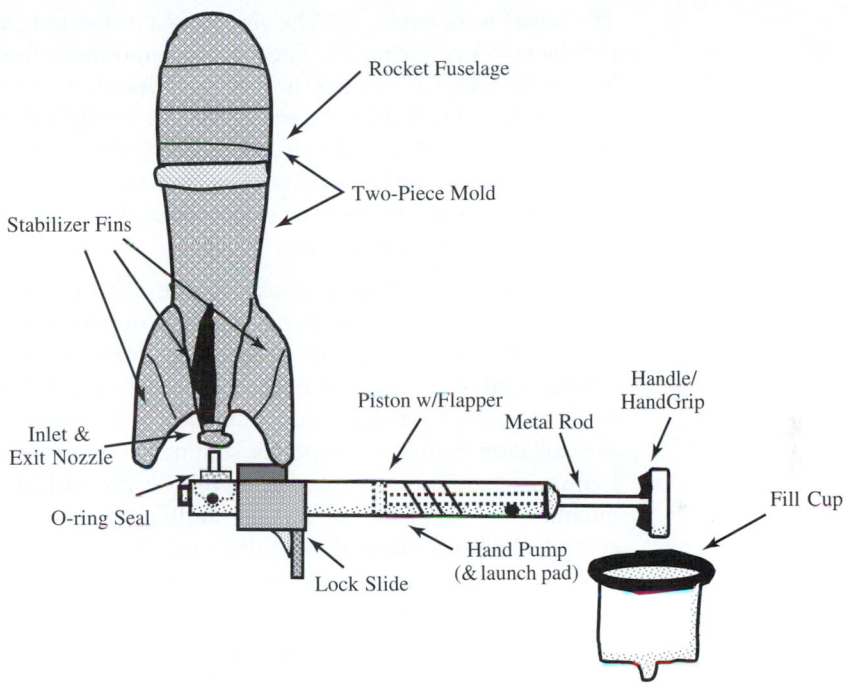

▼ Figure 16.7.
Product concept of a single-stage compressed-air-water toy rocket.

Single-stage compressed-air-water rockets are still popular toys today. We wish to design a portfolio of plastic rockets that will satisfy a number of customer needs, ranging from power and high flight to safe operation and inexpensive production. For the purposes of this example, let's focus on the performance of the rocket, i.e., high-vertical flight.

The basic operation of the rocket is to fill the hollow fuselage of the rocket with water, using a fill cup. Only a percentage of the total rocket is used to allow the remaining air volume to be pressurized. After filling the rocket with water, a hand pump is attached to the rocket, and the rocket is rotated to a vertical-up position for firing. The lock forms a seal between the rocket and the pump. It also provides a release mechanism for the rocket off of the launching pad (i.e., the pump—function sharing).

After the rocket is rotated to a vertical position, the hand pump is used to pressurize the rocket (using 20–30 strokes). The design is based on the ability of a young child to pump a large bicycle tire; thus, 20–30 strokes result in 50 to 65 psi (340 to 440 kPa) gauge pressure.

The rocket is then released. The air acts as a piston to push the water out the nozzle of the rocket. The momentum transfer from the water leaving the rocket accelerates the fuselage vertically (assuming that the rocket is aimed up). After the water exits the nozzle, only gravity and drag forces are acting on the rocket, implying that it will reach a final height, depending on its initial velocity from the transfer of water momentum, at the instant the water exits or when the pressure in the rocket reaches atmospheric pressure.

Based on this description, and based on the customer need of soaring as high as possible, an engineering model is developed for the rocket travel. This model assumes that the rocket can range across a number of sizes (initially a 5.5 in. [14.0 cm] height and 1 in. [2.54 cm] diameter fuselage due to molding constraints). It also utilizes the momentum equation from fluid mechanics, continuity (conservation of mass), isentropic thermodynamic process of initially pressurizing and subsequent expansion of the air, and Bernoulli's equation (assuming incompressible and steady-flow of the water flux through the nozzle). Based on these assumptions, the model may be described by a set of simultaneous equations, applied in the accelerating reference frame (control volume) of the rocket. These equations will predict the height of the rocket as a function of the design variables (such as rocket volume, rocket shape, nozzle diameter, etc.). This model is represented as

$$\frac{dv_r}{dt} = \frac{\rho_w A_n}{m} v_n^2 - g - \frac{C_d \rho_a A_r}{2\,m} v_r^2 \quad \text{(momentum)}$$

$$v_n^2 = \frac{2C_n(P_a - P_{atm})}{\rho_w} \quad \text{(Bernoulli)}$$

$$P_a = \frac{C_{isentropic}}{\left(V_r - \dfrac{m_w}{\rho_w}\right)^{k=1.4}} \quad \text{(isentropic process)} \quad (16.6)$$

$$\frac{dm}{dt} = -\rho_w A_n v_n \quad \text{(continuity)}$$

where m is the total rocket mass (with water) as a function of time, v_r is the rocket velocity, g is the gravitational constant, C_d is the drag coefficient, ρ_a is the density of air, A_r is the cross-sectional area of the fuselage, ρ_w is the density of water, v_n is the jet velocity across the nozzle as a function of time, A_n is the area of the jet (nozzle), m_w is the mass of the water (as a function of time), C_n is the nozzle coefficient, P_a is the pressure of the air (as a function of time), P_{atm} is the ambient pressure, C is the isentropic constant for the air expansion, V_r is the volume of the rocket, and k is the ratio of specific heats.

The model given by Equation (16.6) is valid to the time where the water is ejected from the rocket. After this point, the momentum equation simplifies to

$$\frac{dv_r}{dt} = -g - \frac{C_d \rho_a A_r}{2\,m} v_r^2 \qquad (16.7)$$

where m is the mass of the rocket fuselage (constant).

Based on the product model in Equations (16.6–7), we may apply the spreadsheet method to design a worksheet to solve this model. First, the purpose of the spreadsheet is to calculate rocket heights for varying operating conditions (initial pressure and percent volume of water) and for the choices of design variables (rocket volume, etc.). The purpose is also to investigate the validity of the rocket concept and its extremes, such as determining the time required for the water to eject from the rocket (inertial effects). In conjunction with this worksheet purpose, the inputs to the model are the variables described by Equations (16.6–7), except for the velocity of the rocket and jet. These two remaining variables are the output and an intermediate variable, respectively. The output of the model is velocity of the rocket as a function of time, integrated to calculate the vertical height.

With the inputs and outputs defined (as bounded by plastics' manufacturing and ergonomics), a strategy must be developed for solving the model and creating a worksheet layout. Assuming a relatively well-behaved system, Equation (16.6) is converted to a discrete form using Euler's method. Based on this form, a small time step is chosen and checked for convergence, and the equations are solved iteratively until the mass of the water is zero or until the pressure in the water reaches atmospheric conditions. To solve the equations iteratively, initial values of rocket velocity, water mass, air pressure, and jet velocity are given. A new jet velocity is calculated at the instant the rocket is released, followed by a new rocket velocity. The exiting water mass and expanding air pressure are updated for these new values, and the cycle repeats.

Figure 16.8 shows a spreadsheet layout that corresponds to this strategy. The worksheet shown in Figure 16.8 arranges the inputs, initial conditions, and calculations in a blocked format. The resulting outputs, given the iterations shown at the bottom of the worksheet, are the velocity of the rocket, initial height, and remaining water in the rocket after the acceleration stage of travel. Notice in the worksheet that only the operating conditions are under examination at this point. That is, it is important to understand the preferred values of percent volume of water and initial air pressure before studying the ranges of design variable choices. The worksheet need only vary the design variable values to study their effects.

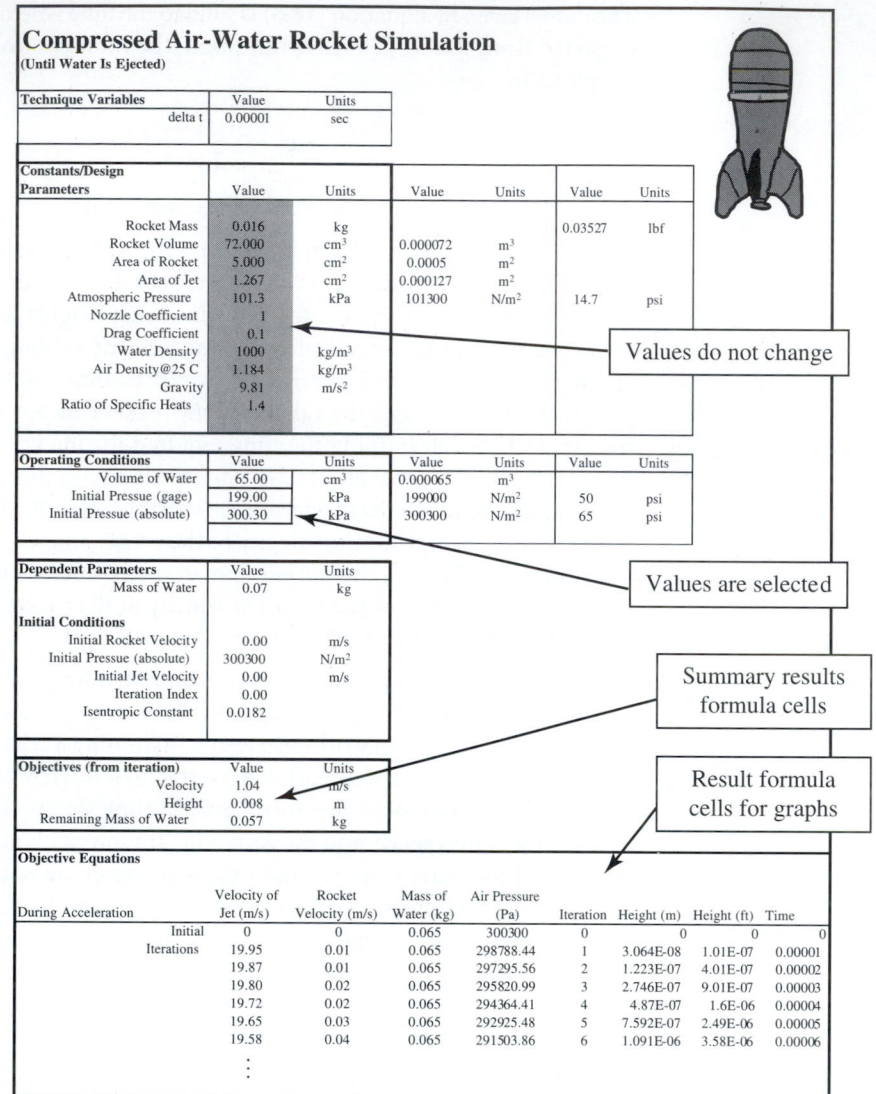

Compressed Air-Water Rocket Simulation
(Until Water Is Ejected)

Technique Variables	Value	Units
delta t	0.00001	sec

Constants/Design Parameters	Value	Units	Value	Units	Value	Units
Rocket Mass	0.016	kg			0.03527	lbf
Rocket Volume	72.000	cm^3	0.000072	m^3		
Area of Rocket	5.000	cm^2	0.0005	m^2		
Area of Jet	1.267	cm^2	0.000127	m^2		
Atmospheric Pressure	101.3	kPa	101300	N/m^2	14.7	psi
Nozzle Coefficient	1					
Drag Coefficient	0.1					
Water Density	1000	kg/m^3				
Air Density@25 C	1.184	kg/m^3				
Gravity	9.81	m/s^2				
Ratio of Specific Heats	1.4					

Values do not change

Operating Conditions	Value	Units	Value	Units	Value	Units
Volume of Water	65.00	cm^3	0.000065	m^3		
Initial Pressue (gage)	199.00	kPa	199000	N/m^2	50	psi
Initial Pressue (absolute)	300.30	kPa	300300	N/m^2	65	psi

Values are selected

Dependent Parameters	Value	Units
Mass of Water	0.07	kg

Initial Conditions		
Initial Rocket Velocity	0.00	m/s
Initial Pressue (absolute)	300300	N/m^2
Initial Jet Velocity	0.00	m/s
Iteration Index	0.00	
Isentropic Constant	0.0182	

Summary results formula cells

Objectives (from iteration)	Value	Units
Velocity	1.04	m/s
Height	0.008	m
Remaining Mass of Water	0.057	kg

Result formula cells for graphs

Objective Equations

During Acceleration		Velocity of Jet (m/s)	Rocket Velocity (m/s)	Mass of Water (kg)	Air Pressure (Pa)	Iteration	Height (m)	Height (ft)	Time
	Initial	0	0	0.065	300300	0	0	0	0
	Iterations	19.95	0.01	0.065	298788.44	1	3.064E-08	1.01E-07	0.00001
		19.87	0.01	0.065	297295.56	2	1.223E-07	4.01E-07	0.00002
		19.80	0.02	0.065	295820.99	3	2.746E-07	9.01E-07	0.00003
		19.72	0.02	0.065	294364.41	4	4.87E-07	1.6E-06	0.00004
		19.65	0.03	0.065	292925.48	5	7.592E-07	2.49E-06	0.00005
		19.58	0.04	0.065	291503.86	6	1.091E-06	3.58E-06	0.00006
		⋮							

▼ **Figure 16.8.**
Worksheet for the toy rocket product model.

A second worksheet is created to calculate the final height of the rocket based on Equation (16.7). This worksheet, shown in Figure 16.9, references the outputs of the first worksheet and iteratively solves for the rocket height, defined by when the velocity of the rocket becomes zero. This height value is shown in the output block of the worksheet.

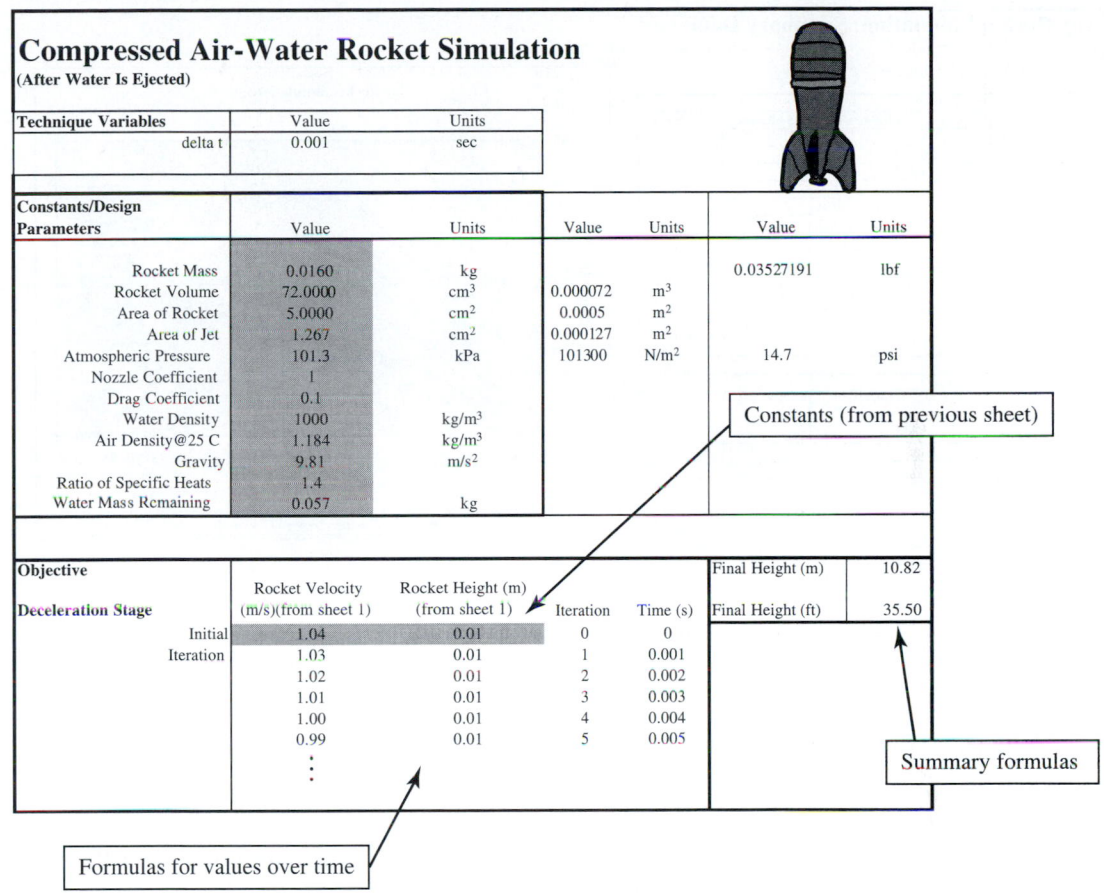

Compressed Air-Water Rocket Simulation
(After Water Is Ejected)

Technique Variables	Value	Units
delta t	0.001	sec

Constants/Design Parameters	Value	Units	Value	Units	Value	Units
Rocket Mass	0.0160	kg			0.03527191	lbf
Rocket Volume	72.0000	cm^3	0.000072	m^3		
Area of Rocket	5.0000	cm^2	0.0005	m^2		
Area of Jet	1.267	cm^2	0.000127	m^2		
Atmospheric Pressure	101.3	kPa	101300	N/m^2	14.7	psi
Nozzle Coefficient	1					
Drag Coefficient	0.1					
Water Density	1000	kg/m^3				
Air Density@25 C	1.184	kg/m^3				
Gravity	9.81	m/s^2				
Ratio of Specific Heats	1.4					
Water Mass Remaining	0.057	kg				

Constants (from previous sheet)

Objective					Final Height (m)	10.82
	Rocket Velocity (m/s)(from sheet 1)	Rocket Height (m) (from sheet 1)	Iteration	Time (s)	Final Height (ft)	35.50
Deceleration Stage						
Initial	1.04	0.01	0	0		
Iteration	1.03	0.01	1	0.001		
	1.02	0.01	2	0.002		
	1.01	0.01	3	0.003		
	1.00	0.01	4	0.004		
	0.99	0.01	5	0.005		

Summary formulas

Formulas for values over time

▼ *Figure 16.9.*
Second toy-rocket worksheet, calculating final rocket height.

A third worksheet is also created to aggregate the results of the simulations for ranges of operating conditions. Figure 16.10 shows the summary worksheet, in addition to an output plot of the rocket height versus the operating variables.

For the sake of calibration and validation, the original Park Plastics' rocket may be experimentally tested, recording rocket heights as a function of initial air pressure and water volume. To provide a valid comparison, the simulation in the first three worksheets must be run based on the Park Plastics' geometry. Figure 16.11 schematically presents the experimental setup for the tests. Notice that a two-point sight system with triangularization is used to measure the rocket height. The angle

Toy Rocket Simulation: Summary Data

Series: Rocket Height vs. %Vol. of Water and Pressures:

	Height (m.)			
%Vol. Of Water	570 kPa-abs.	465 kPa-abs	375 kPa-abs	300 kPa-abs
6.94	4.02	3.11	2.35	1.69
13.89	10.66	8.72	6.2	4.42
20.83	17.41	13.43	9.96	7.01
27.78	22.94	17.56	12.95	8.92
34.72	26.78	20.36	14.78	9.91
41.67	28.75	21.58	15.31	9.78
48.61	28.72	21.07	14.39	8.41
55.56	26.64	18.77	11.83	5.11
62.50	22.19	14.36	6.78	2.49
69.44	14.88	6.97	2.87	1.22
76.39	5.72	2.86	1.26	0.57
83.33	2.01	1.06	0.52	0.23
90.28	0.52	0.28	0.14	0.06

	Height (ft.)			
%Vol. Of Water	570 kPa-abs. (68 psi)	465 kPa-abs (53 psi)	375 kPa-abs (40 psi)	300 kPa-abs (29 psi)
6.94	13.19	10.20	7.71	5.54
13.89	34.97	28.61	20.34	14.50
20.83	57.12	44.06	32.68	23.00
27.78	75.26	57.61	42.49	29.27
34.72	87.86	66.80	48.49	32.51
41.67	94.32	70.80	50.23	32.09
48.61	94.23	69.13	47.21	27.59
55.56	87.40	61.58	38.81	16.77
62.50	72.80	47.11	22.24	8.17
69.44	48.82	22.87	9.42	4.00
76.39	18.77	9.38	4.13	1.87
83.33	6.59	3.48	1.71	0.75
90.28	1.71	0.92	0.46	0.20

Toy Rocket Simulation

▼ *Figure 16.10.*
Summary worksheet for the toy rocket product model (operating conditions).

measured with respect to the horizon is recorded, in addition to the horizontal distance from the sight system to the point where the rocket reaches its apex.

Given this experimental setup, tests on the Parks Plastics' rocket are conducted by first mounting a pressure gage in line with the pump (to measure initial air pressure). The hand pump shown in Figure 16.7 is modified for this purpose. A launch rod is also fabricated to guide the release of the rocket.

Water is then introduced into the rocket with a syringe; the rocket is pressurized, attached to the launch pole, and released. Measurements are taken by two sighting systems at varying distances from the launch location, and all data are recorded. The height for each test run is calculated based on the sighting angle, initial sight height, and distance from the launch location.

Implementing this test procedure, test data are shown in Figure 16.12. In addition, the simulation data are plotted on the same axes. Notice from the figure that some variation occurs between the model results and the test data. However, the overall results are very consistent in trends and

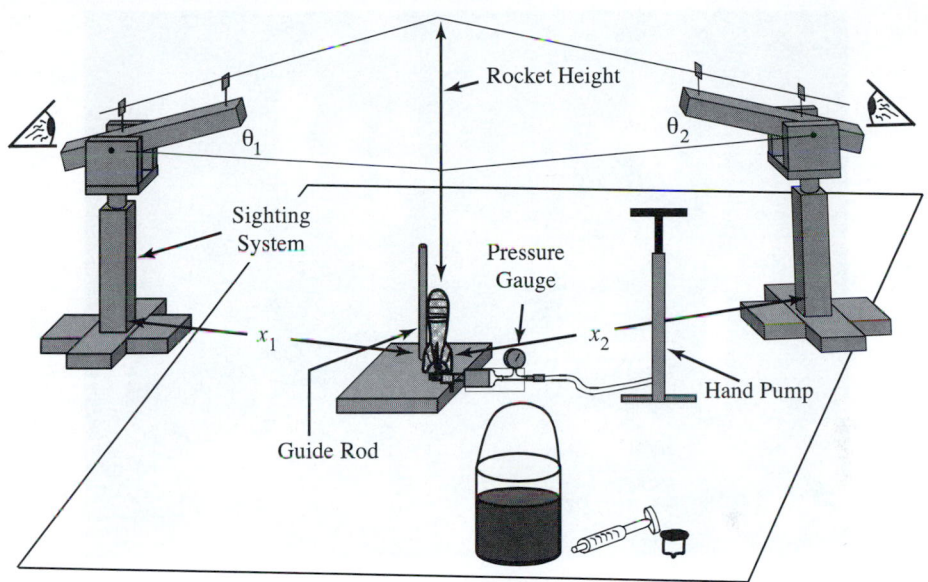

▼ *Figure 16.11.*
Experimental setup of the toy rocket height test.

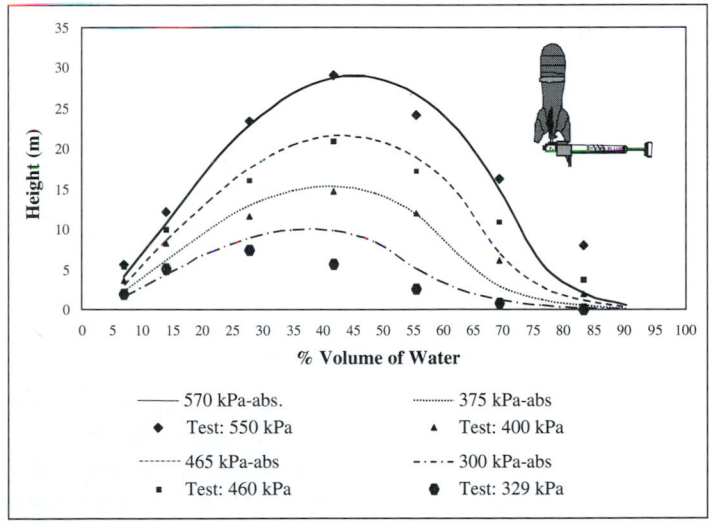

▼ *Figure 16.12.*
Validation of toy rocket model.

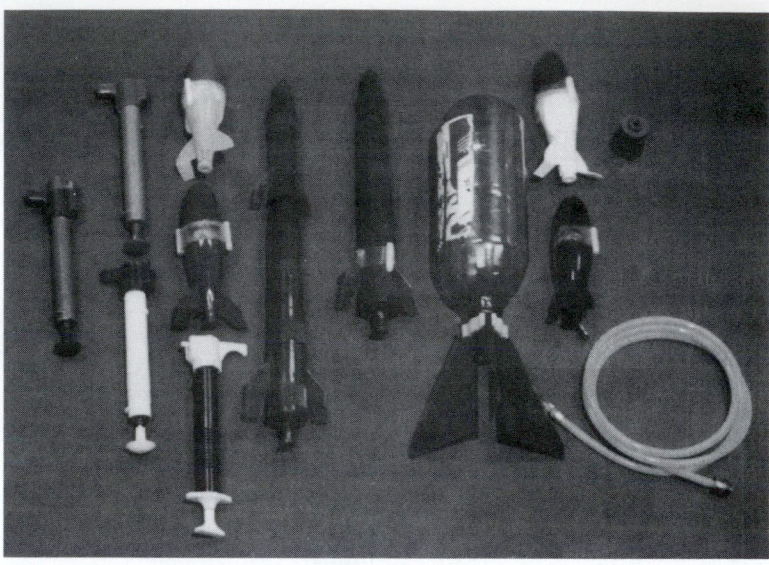

▼ *Figure 16.13.*
Example product portfolio of toy rocket products.

magnitudes! The model appears to need very little adjustment in physical constants or other model variables. The limits of the model must now be explored, in addition to the varying of design variables. The results of the design variable study may be used to establish the rocket geometry and operating specifications for the customer needs under consideration. Further customer needs, such as stability, low cost, durability, etc. may also be modeled, either numerically or physically. Design decisions from these models, with appropriate refinements and prototype testing, can lead to a product portfolio, as exemplified by the rockets shown in Figure 16.13. There are four different pumps, four different single stage water rockets, two different multistage water rockets, and one 2-liter bottle rocket kit.

Summary

The toy rocket product provides an example of using spreadsheet applications to solve engineering models. The model is a powerful predictive tool, as illustrated by its comparison to experimental results. A spreadsheet, then, is a tool for searching the model to determine design solutions.

Based on this example, we may summarize the basic numerical model of spreadsheet search according to the following steps:

1. State the purpose of the product-model worksheet(s).

2. Identify the outputs desired from the worksheet.

3. Identify the inputs, constants and design variables.

4. Determine the calculations needed to map inputs to outputs.

5. Create a layout for the worksheet that adds utility in solving for design choices.

6. Develop a strategy for solving the calculations within the worksheet(s).

7. Explore the Pareto frontier and vary the constants (potential noise parameters).

This method, as embodied by seven high-level steps, is very effective for solving product models. Constraints imposed on the product boundaries may be searched with all combinations (assuming a small number of variables). The design variables may be perturbed to understand the relative importance of the variables and to determine the sensitivity of the outputs to noise. Physical insights may also be obtained by studying the limits and variations in the numerical results.

While spreadsheet search is powerful, it also has distinct limitations. As a product model grows in size and complexity, brute-force search is not feasible or desired. It is not feasible to search through all combinations of design variables. An alternative method is thus needed to direct the search and to instill confidence in the designer that feasible and preferred design choices are being computed with a high level of efficiency. The remainder of this chapter studies optimization approaches as an advanced method to satisfy this need.

IV. FUNDAMENTAL CONCEPTS IN OPTIMIZATION

In previous chapters, our engineering product model concept considered arbitrary sets. Now let's consider numerical models, ones where the set representing the physical design consists of numbers. This step places us in the predominately embodiment design phase.

Constraints

When we need to determine the "best" point in a set $\mathbb{D}$ of design configurations, we cannot usually do so by simply minimizing a performance metric. We usually have some restrictions on the values we can choose among, typically arising from the specification of target values on our performance metrics. Informally, the other performance metrics we use for this purpose we will call *constraints*.

Definition. A *constraint* is a performance evaluation metric and target value pair that is used to constrain the final solution to exhibit additional properties beyond belonging to $\mathbb{D}$.

Regional Constraints

Regional constraints are constraints that are one sided:

- Stress must be less than a yield stress value,
- Cost must be less than a quoted price,
- Geometry must fit within a volume.

We will also call such constraints *inequality constraints*, and denote them by g_j. For mathematical ease, we will always convert such expressions to be bounded above by zero.

For example, if a stress $\sigma(x)$ must be less then a yield stress s_y, our constraint is

$$\sigma(x) \leq s_y \tag{16.8}$$

We will convert this constraint to

$$\sigma(x) - s_y \leq 0, \tag{16.9}$$

or

$$g_j(x) \leq 0. \tag{16.10}$$

Note that constraints of the form

$$g_j(x) < 0 \tag{16.11}$$

in theory cannot be satisfied, we will never be able to reach such an open boundary arbitrarily close.

Equality Constraints

The other general type of constraint we will use is an *equality constraint*. This type of constraint restricts certain design variables to be simultaneously related. We will denote equality constraints by h_j. We will also perform similar transformations on equality constraints h_j so they equal zero:

$$h_j(x) = 0. \tag{16.12}$$

Note the reliance of all numerical constraints on the existence of a "zero." The algebraic structure of real valued numbers are intrinsically used to convert our previous "$\in$" of Eqn (16.3) to a numerical "$\leq$" statement.

Objective Functions

The performance metric feature we need, beyond constraints, is to identify an *objective function*. This objective will be the function we seek to minimize.

Definition. An objective function is a performance metric f we use to define to determine which configuration is better than all others, by minimizing f over $\mathbb{D}$.

Standard Null Form

When we convert a product's performance metrics into constraints, g_j and h_j in the form as above, and select a single metric to serve as an objective function, we have cast the problem in a *negative null form*, the "null" signifying the zeros. If the $\leq$ on $\vec{g}$ were instead $\geq$, the problem is in *positive null form*. In this development, we will always use the negative null form.

This approach leads to the formulation of an optimization problem:

$$\text{minimize} \qquad f(\vec{d})$$

$$\text{subject to} \qquad \vec{g}(\vec{d}) \leq \vec{0} \qquad\qquad (16.13)$$

$$\vec{h}(\vec{d}) = \vec{0}$$

where $\vec{d} \in \mathbb{D} = \mathbb{R}^n$, and all performance metrics are real valued. Any point $\vec{s} \in \mathbb{D}$ satisfying the constraints in Equation (16.13) is a *feasible point*, and all such points compose the *feasible space S*. This formulation of an optimization problem is called the *standard null form* of a *nonlinear programming problem*. The term "programming" arises historically. Whenever possible, we shall adopt this formulation. It is the clearest expression of the problem, one that can be readily encoded for evaluation by various algorithms and computer software.

Example

Let's consider a product design example, a fingernail clipper, as shown in Figure 16.14 (Chapters 10 and 13).

Based on a subset of customer needs analysis, we have five performance metrics:

- the finger force,
- the material stress, and
- the 3 coordinate dimensions (overall lengths).

We can choose the finger force as the objective function, where the stress and dimensions are constraints. From these choices, a model may

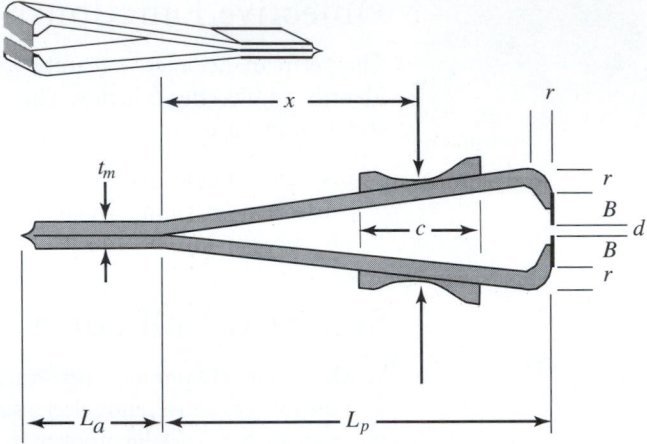

▼ Figure 16.14.
Parameterized fingernail clipper concept.

be developed, similar to the approach used for the side-blade clipper in Chapter 13. This model is given by

$$f = \frac{L_p}{x}\tau\frac{W}{h_b}t^2 \qquad\qquad d \geq h_b + 1.1t$$

$$r = 2t_m$$

$$L_p = x + \frac{c}{2} + r \qquad\qquad L_a \geq 5t_m$$

$$W \geq 5D_h \qquad\qquad (16.14)$$

$$L_{\text{overall}} = L_a + L_p$$

$$H_{\text{overall}} = 2r + 2B + d \qquad \sigma_{\max} = K_{\text{cycles}}\frac{6xf}{Wt_m^2}\left(1 - \frac{x}{L_p}\right)$$

$$B = h_b + x3D_h$$

where the variables are defined in Chapter 13. This model can be rewritten into standard null form

$$\text{minimize}\quad \frac{x + \frac{c}{2} + (2t_m)}{x}\tau\frac{W}{h_b}t^2$$

$$\text{subject to}\quad 2(2t_m) + 2(h_b + 3D_h) + d - H \leq 0 \qquad (16.15)$$

$$5D_h - W \leq 0$$

$$L_a + x + \frac{c}{2} + (2t_m) - L \leq 0$$

$$5t_m - L_a \leq 0$$

$$K_{\text{cycles}}\frac{6xf}{Wt_m^2}\left(1 - \frac{x}{L_p}\right) - \sigma_{\max} \leq 0$$

where the search is across

$$(t_m, D_h, h_b, x, d) \in \mathbb{R}^5 \qquad (16.16)$$

with

$$(c, \tau, t, L, W, H, \sigma_{\max}, K_{\text{cycles}}) \in \mathbb{R}^8 \qquad (16.17)$$

as constants or noise variables.

V. ADVANCED TOPIC: A DISCUSSION OF ANALYTICAL FORMULATIONS

The previous section presents optimization problems as formulated by the standard null form. We now explore how we might solve this formulation analytically using methods of calculus.

Unconstrained Problems

Consider optimizing a function f over a design configuration space of $\mathbb{R}^n$. We seek the point where f is a minimum. From elementary vector calculus, we know that a function attains it minimum over $\mathbb{R}^n$ where its derivative is zero. That is, an optimum point $\vec{d}_*$ satisfies the relations

$$\frac{\partial f}{\partial d_i}(\vec{d}_*) = 0 \qquad (16.18)$$

This result gives us n equations for our n unknowns $\vec{d}_*$. Equation (16.18) is a necessary but not sufficient condition for finding all possible $\vec{d}_* \in \mathbb{R}^n$. This equation only ensures that an optimum point is a minimum, maximum, or inflection. Likewise, it does not ensure the uniqueness of the solution. f may have multiple solutions $\vec{d}_*$. For example,

$$f(x) = \sin(x)$$

has solutions $-\pi/2 \pm \pi$.

For equation (16.18) to define a $\vec{d}_*$ as a possible minimum, f must have positive curvature at this candidate optimum point. Further, for $\vec{d}_*$ to be a unique minimum (or "the only solution"), f must be greater than its value at $\vec{d}_*$. One way to ensure this is for f to have positive curvature everywhere. This condition is satisfied if

$$\frac{\partial^2 f}{\partial d_i^2} > 0 \qquad (16.19)$$

over *all* $\mathbb{R}^n$. This condition is a restriction on $(f, \mathbb{R}^n)$, or f and the domain.

Consider the basic approach, though, of using calculus to solve minimization problems. It has converted a minimization problem (Equation 16.13) into solving simultaneous equations (Equation 16.18). If this conversion can be carried out symbolically, and then solved numerically, the result may be evaluated rather easily. If the simultaneous equations must be solved entirely numerically, the solution will be much more difficult to obtain.

Example

Minimize

$$f(x) = (1 - e^{x^2})\sin(x) \qquad (16.20)$$

We can solve this problem using our calculus machinery by solving

$$\frac{\partial f}{\partial x}(x_*) = 0 = (1 - e^{x_*^2})\sin(x_*) + (1 - e^{-x_*^2})\cos(x_*) \quad (16.21)$$

which we must now solve for x_*. Which is easier, solving this equation or iteratively searching for the minimum? Both, in the end, will require a numerical method.

Lagrangians

Consider a standard null form over $\mathbb{R}^n$ with differentiable functions and no regional type constraints, rather equality constraints only:

$$\begin{aligned} \text{minimize} & \quad f(\vec{d}) \\ \text{subject to} & \quad \vec{h}(\vec{d}) = \vec{0} \end{aligned} \qquad (16.22)$$

We have a fully developed set of calculus tools; thus, we should be able to solve this problem. Such a solution will help us understand and appreciate optimization strategies, in general. But, we will also develop a simple argument as to why it is not always easy to solve constrained optimization problems with calculus.

For this problem, we augment the previous unconstrained problem to consider certain constraints. That is, consider a problem where f is to be minimized, but always along a boundary for which $\vec{h}(\vec{d}) = \vec{0}$. To consider how to do this analytically, it is helpful to first consider the problem graphically, such as in Figure 16.15. Here, $\mathbb{D} = (d_1, d_2)$ is a two dimensional design space. Consider f a linear function with positive coefficients, and so is minimized by choosing $\vec{d}$ values always more toward the lower right quadrant. With f a linear function, the gradient of f is constant, its length and direction is the same everywhere. Next con-

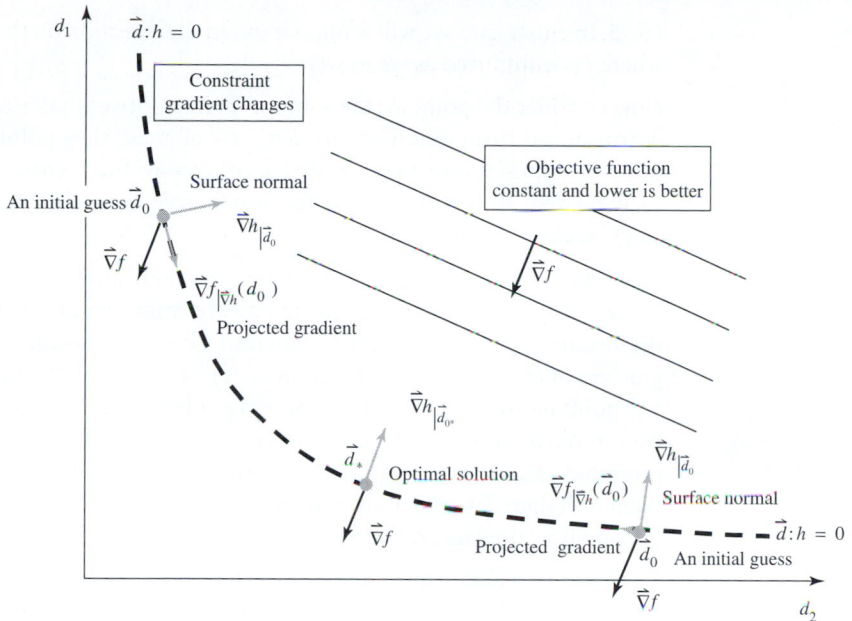

▼ Figure 16.15.
Two dimensional example numerical optimization problem.

sider a single constraint $h(\vec{d}) = d_1 d_2 - \tau$, or $d_1 d_2$ must always equal some positive number τ.

Considering making an initial valid guess for the minimum, say $\vec{d}_0$, as shown as a couple of possible guesses in Figure 16.15. An initial guess must be one that satisfies all constraints, or is on the $h = 0$ curve. To minimize f, we would like to move in the direction that minimizes f, or

in the direction $\vec{\nabla} f(\vec{d}_0) = \left(\dfrac{\partial f}{\partial d_1}, \dfrac{\partial f}{\partial d_2} \right)_{|\vec{d}_0}$. When we consider this

while looking at Figure 16.15, imagine that we do not know by inspection the entire curve $\vec{d} : h = 0$, we only know the tangent vectors at some point $\vec{d}_0$. When searching, we cannot move in the direction $\vec{\nabla} f$, as then we would stray from the constraint equation $h(\vec{d}) = 0$. Instead, we can only move in one of the two tangent directions that is

perpendicular to the path gradient $\vec{\nabla} h(\vec{d}) = \left(\dfrac{\partial h}{\partial d_1}, \dfrac{\partial h}{\partial d_2} \right)$. These two

tangent directions are the only two that will keep to the curve $h(\vec{d}) = 0$. In general $\mathbb{R}^n$, we have a larger choice of moving in any direction defined by the entire tangent plane of $\vec{\nabla} h$, not just two directions as in Figure

16.15. In either case, we will want to move in the direction on this plane where *f* is minimized as we move.

Now consider the point $\vec{d}_*$, the solution point. Notice what is uniquely true about this point: $\vec{\nabla}f_{|\vec{d}_*}$ and $\vec{\nabla}h_{|\vec{d}_*}$ are aligned, they point in the same or opposite directions. Stated another way, the vectors are linearly related, all vector components are linearly related by the same single scaling number.

Now lets reconsider any point $\vec{d}_0$ on the constraint surface. Examining Figure 16.15, one can see that the quantity to minimize in constrained optimization is not the objective function *f*, but rather is the *projected gradient* of *f* onto *h*, or the projection of $\vec{\nabla}f$ onto the tangent of $\vec{\nabla}h$. At any point on the surface *h*, the projected gradient points are the direction to move in order to reduce its length. At the solution point, the projected gradient has zero length, and so the gradient of *f* and the gradient of *h* align. Since they align, their vector components are each proportional by the same quantity.

This gives us additional conditions to solve for $\vec{d}_*$ as a set of simultaneous equations. First, the ratio of the lengths $\vec{\nabla}f$ to $\vec{\nabla}h$ we will denote by a single unknown quantity λ_h. Then, our vector alignment condition becomes $\vec{\nabla}f + \lambda_h\vec{\nabla}h = \vec{0}$, which in this example gives two equations to solve. Also applying the constraint equation $h = 0$, we then have three equations in three unknowns (λ_h, d_1, d_2), which can be solved analytically. That is how to solve constrained optimization problems using calculus.

Analytically, finding the projected gradient minimum is equivalent to solving the derivative equality equation $\vec{\nabla}f + \lambda_h\vec{\nabla}h = \vec{0}$. Each component of $\vec{\nabla}f$ and $\vec{\nabla}h$ is proportional by the same factor λ_h. This is key to understanding constrained optimization, and so the reader should examine Figure 16.15 until it is understood. The projected gradient is minimized to zero length at the solution point, and there the gradient components of *f* and *h* are identically proportioned.

We can extend the derivation to multiple dimensions by expressing the projected gradient derivative in *n* dimensions with *m* constraints as

$$\vec{\nabla}f(\vec{d}_*) + \vec{\lambda}^T[\vec{\nabla}h(\vec{d}_*)] = \vec{0} \qquad (16.23)$$

where $\vec{\lambda}$ is the vector of *Lagrange multipliers*. A useful interpretation of λ is that it is a first order indication of the change in the objective function that would result from a positive unit step change in *h*.

This entire qualitative derivation can be made more concrete by defining a *Lagrangian* (Lagrange function):

$$L(\vec{d}, \vec{\lambda}) = f(\vec{d}) + \sum \lambda_i h_i(\vec{d}) \qquad (16.24)$$

We then find $(\vec{d}_*, \vec{\lambda}_*)$ such that

$$\frac{\partial L}{\partial d_i}(\vec{d}_*, \vec{\lambda}_*) = 0$$

$$\frac{\partial L}{\partial \lambda_j}(\vec{d}_*, \vec{\lambda}_*) = 0$$

(16.25)

This formulation results in a $(n + m)$ set of simultaneous equations that must be solved for $(\vec{d}_*, \vec{\lambda}_*)$:

$$\frac{\partial L}{\partial d_i}(\vec{d}_*, \vec{\lambda}_*) = \vec{\nabla}f(\vec{d}_*) + \vec{\lambda}_*^T[\vec{\nabla}\vec{h}(\vec{d}_*)] = 0$$

$$\frac{\partial L}{\partial \lambda_j}(\vec{d}_*, \vec{\lambda}_*) = \vec{h}(\vec{d}_*) = \vec{0}$$

(16.26)

These expressions define the equations to solve for $\vec{d}_*$ using calculus. As such, it is directly observed that a constrained optimization solution is analogous to an unconstrained result. A candidate minimum exists where the first derivative of the function equals zero.

To expand the calculus formulation to consider regional constraints $g(\vec{d}) \leq 0$, one can simply add a slack variable s to the formulation to convert the regional constraint into an equality constraint:

$$h_g(\vec{d},s) = g(\vec{d}) + s^2$$

(16.27)

where s has interpretation of the distance (in units of g) that any $\vec{d}$ is from hitting the $g(\vec{d}) = 0$ boundary. If $g(\vec{d})$ ever becomes positive, then h_g cannot remain zero, and so the point $\vec{d}$ is invalid.

This completes the calculus solution approach to a standard null form optimization problem. Using this Lagrangian approach, any optimization problem can be recast into a set of simultaneous equations. Except for certain types of analytical models, however, it is doubtful this calculus formulation result is any easier to solve than the original constrained minimization problem. A Newton-Raphson iterative method would need to be applied to solve this set of simultaneous nonlinear equations after symbolically performing the differentiation. On the other hand, the formulation does provide the insight necessary to work with numerical methods that we will use to search for solutions to product models.

Example

As an example, let's return to the simple product model proposed from Figure 16.2. This model seeks to determine the relationship between geometric variables of a can product, where surface area (proportional to cost) is minimized. The Lagrangian for the product model may be represented as

$$L(r, l, \lambda) = (2\pi r^2 + 2\pi rl) + \lambda(\pi r^2 l - C) \qquad (16.28)$$

Differentiating this result with respect to the design parameters and Lagrange multiplier, and setting the results equal to zero, we have

$$4\pi r + 2\pi l + \lambda 2\pi lr = 0$$
$$2\pi r + \lambda \pi r^2 = 0 \qquad (16.29)$$
$$\pi r^2 l - C = 0$$

These equations may be solved simultaneously to produce

$$(l, r, \lambda) = \left(2r, \left(\frac{C}{2\pi} \right)^{\frac{1}{3}}, \frac{-2}{r} \right) \qquad (16.30)$$

This result is interesting. Notice the relationship between r and l. The cylindrical soup or soda can is ideally short and squat, versus a long cylinder. Soda cans don't satisfy this idea strictly due to the ergonomic constraint of fitting in the user's hand. If we extend this concept to oil storage tanks, adding a stress constraint on the cylinder walls due to the fluid head, we find that the product becomes even more short and squat. Refined oil storage products on the market exhibit this result.

This example illustrates the use of analytical approaches to optimization problems. Lagrangians and calculus provide insights into the solution approach and underlying model results; however, the approach is simply not tractable for problems of any real world complexity.

Discussion

We have shown that every optimization problem involving real-valued numbers can, in theory, be solved using tools of modern calculus. Calculus of variations has been known since the time of Lagrange. Its difficulty in implementation, however, stems from the results. It is usually more difficult to solve simultaneous sets of nonlinear equations than to solve nonlinear minimizations with search algorithms. Further, we have to be able to differentiate accurately. These difficulties prevent calculus methods from being used in practice.

Of course, the calculus of variations form the reasoning behind all numerical solution methods used in practice. It can give us insight into the numerical approaches. For example, consider constraint boundaries. For monotonic objective functions, the solution point of an optimization problem $\vec{d}_*$ *always* exists on one of the inequality constraints, g_j. This insight has implications for a design problem formulation. If we say

$$g(\vec{d}) - \tau \leq 0 \qquad\qquad (16.31)$$

then we must realize at $\vec{d}_*$, $g(\vec{d}_*) = \tau$. Thus, we must specify the constraint boundary values with the understanding that this value will be used at the final solution.

Calculus of variations underlies all numerical solution methods used, which is why we discuss the approach above. Numerical methods build upon the concept of Lagrangians to solve constrained optimization problems. Let's therefore continue our development to the point of representing a fully constrained optimization problem.

VI. PRACTICAL OPTIMIZATION

In the previous sections, we have discussed both analytical and simple numerical approaches for solving constrained objective functions (product models) with optimization strategies. These approaches provide the needed background to use commercially available optimization tools, to understand the integrity of results. We will also be able to implement troubleshooting approaches when problems occur, and use optimization output results to seek even more improved solutions.

In this section, we build on the optimization background to develop practical solutions to product models. These solutions are based on commercial programs for desktop computers. Many other packages exist that may be used to create equivalent results.

To begin the discussion, we need to consider four concepts of numerical methods: numerical search, stopping criteria, sensitivity analysis, and global optimality. We will need these concepts to understand the results from practical optimization packages.

Numerical Search

Numerical search procedures underlie numerical optimization methods. There is voluminous material to discuss in this field (Gill, et al., 1981); we overview the topic by considering the basic technique of gradient search, and by touching briefly on the technique of sequential quadratic programming.

Steepest Descent

Given no other information about a function other than its continuity and its expressive form $f{:}\mathbb{R}^n \to \mathbb{R}$, how can one find points $\vec{d}_*$ where f is minimum? We need to determine where the f's derivative is zero, but we do not have an expression of its derivative.

In a design space of one dimension $\mathbb{R}$, one method for finding the minimum of a function is *sub-division search*. In such methods, a function $f{:}\mathbb{R} \to \mathbb{R}$, is minimized over an initial "wide" interval $[a,b] \subset \mathbb{R}$. The interval $[a,b]$ is iteratively subdivided, the function is evaluated at a point in each subdivision, and the half with a higher returned value is eliminated from further exploration (divide and conquer). Functions that are bounded and do not have many local minima within $[a,b]$ work well for this solution strategy; more complex functions present numerical difficulties.

In a design space of n dimensions, the search is more complicated than sub-division search due to the multiple dimensions. We can search along many different unit vectors. An effective method is to sub-division search along the direction of the function that is decreasing most quickly, or over the one-dimensional line along the gradient ∇f. There are many forms of such *gradient search* methods (Gill et al, 1981), such as conjugate gradient methods, that ensure the search path remains focused on the overall steepest path.

Linear Programming

Constrained optimization presents further difficulties than simply searching anywhere in $\mathbb{R}^n$. Consider a simplified subset of all possible constrained optimization problems, the set of problems when all of the constraints are linear equations. In this case, one must determine the solution point that minimizes f subject to linear h and g constraints.

In this situation, different solution methods are available than steepest descent methods. One can think of linear equations as hyper-planes in $\mathbb{R}^n$. If f is linear, or quadratic, then it is monotonic. In this special case, the minimum of f must occur at a constraint boundary. Constraint boundary extrema occur where the constraint hyperplanes intersect. This means that one can reduce the search from all of $\mathbb{R}^n$ to those points where the constraint boundaries intersect. Conceptually, one can list the finite number of hyperplane intersection points and minimize f only over these few points. This form of search, called *simplex search*, can be carried out very rapidly (Gill, et al., 1981), with relatively reduced computational effort.

Sequential Quadratic Programming

Most engineering optimization problems are not linear. To solve non-linear problems, one approach is to repeatedly numerically approximate the problem as a locally linear one.

To do this, one can start a numerical search at a feasible point in $\vec{d}_0$ in $\mathbb{R}^n$. At this point, one can numerically approximate the non-linear constraints by evaluating them at points around $\vec{d}_0$, and fitting a linear surface to f, g, and h. Then, one can solve this linear problem as above with a fast simplex method for a new point $\vec{d}_1$. Then one can re-linearize the problem at $\vec{d}_1$, and so on, until it converges to a solution. This approach is called *sequential linearization*. In particular, *sequential quadratic programming* is a method whereby the constraints are linearized and the objective function is reduced to the nearest quadratic approximation.

Sequential quadratic programming is a very fast and accurate method to solve a general nonlinear optimization problem. It works best with smooth functions and will only converge when there are sufficient constraints to bound the search, yet not overly constrain it (a solution must exist to the problem—there must be a point that satisfies all the constraints). The function should not have multiple local minima (or the final local optima converged to will depend upon the starting point $\vec{d}_0$), and the constraints should define a single smooth convex constrained region. Problems that do not exhibit these qualities can be expected to be more difficult to numerically solve. Sequential quadratic programming is implemented as a tool in many software systems, such as spreadsheets and Matlab™.

Stopping Criteria

Given some numerical quantity to minimize over a search space, how do we know when we can stop the search? There are three answers to this question:

▶ When the separation in successive iterations becomes small:

$$\| I_k \| \leq \varepsilon \tag{16.32}$$

This criterion is valid, but sometimes the function changes quickly, and so the prediction has $f(\vec{d}_{k-1}) \gg f(\vec{d}_k)$.

▶ When the difference in functional evaluations becomes small:

$$\frac{|f(\vec{d}_{k-1}) - f(\vec{d}_k)|}{|f(\vec{d}_k)|} \leq \varepsilon \tag{16.33}$$

This criterion will define $\vec{d}_*$ as any point that is "close enough," but sometimes the function might be very flat near $\vec{d}_*$, and so we might be placed very far away from $\vec{d}_*$ as the agorithm slowly converges. Equation (16.33) must be checked to ensure f itself is not approaching zero (becomes less than ε). If so, one can simply use the numerator. Generally, one should check both the relative and the absolute difference.

▶ When the difference in some arbitrary scalar function becomes small:

$$|L(\vec{d}_{k-1}) - L(\vec{d}_k)| \leq \varepsilon \tag{16.34}$$

This criterion has use when we discuss constraints, since we might want to define a stopping criterion as a function of both the function we are minimizing and the constraint functions.

These definitions imply that we need an order of magnitude estimate of f, and we then choose ε to be a few orders of magnitude less than this estimate. This approach will assure that we stop within some proximity of a minimum for our model. In commercial packages, the stopping criterion concept should be understood so that one can vary ε as needed from a typical default value of 10^{-6} which may be too conservative or not conservative enough, depending on the units of f. Also, by varying ε, we can confirm if a numerically valid solution has been reached.

Sensitivity Analysis

When solving product models, the boundaries on the design variables will typically define the solutions to the model. We should consider what effect changes on the boundaries have on the objective function of the model. By considering such changes, we may be willing to relax certain constraints slightly, to obtain a significant improvement in the objective function. To seek such improvements, we are determining the sensitivity of our model solutions to the design constraints and to the design variable values. We are exploring the Pareto frontier.

Based on the Lagrangian formulation of optimization problems, we may derive the relationship

$$\frac{\partial f}{\partial \zeta} = -\lambda \tag{16.35}$$

where ζ is an f or g constraint. This relationship states that for a change in a constraint by approximately $\Delta\zeta$, the objective function will decrease (for negative null form) by $-\lambda$ multiplied by the change. This result means that a unit change in a constraint will locally result in a change in the objective function equal to the value of the Lagrange multiplier.

Many commercial packages will automatically scale the constraints and output Lagrange multipliers. From this information, we may direct changes to the model that will result in the greatest impact on our product metric. It should be noted that Equation (16.35) is a linear approximation (Taylor series expansion) to the objective function change. Thus, any time we vary a constraint, the optimization model should be re-solved to verify an improvement.

Global Optimality

The optimization methods presented in this chapter only determine a minimum over a restricted set. No guarantee is provided to determine a global optimum when there may exist many local optima. In general, global optimality, or numerically finding the best point over a domain, remains a research topic. Methods such as *simulated annealing* take a probabilistic approach, *genetic algorithms* apply domain splitting heuristics and searching heuristics.

For general nonlinear equations, we cannot guarantee that a global optimum will be found with a numerical search, since it is provably an exhaustively hard problem. However, we can use multiple starting points in our space, until we are reasonably sure we have found the global optimum based on the different local optima we uncover. This approach of using multiple starting points is known as *shotgunning*, i.e., we are shooting a shotgun of starting points to cover the possible optimal solutions.

Solution Method: Matlab

In this section, we discuss solutions of constrained product models using a numerical manipulation package, *Matlab*™, with the *Optimization Toolbox*. While there are many contributed sets of optimization code available, the commercially provided Matlab Optimization Toolbox can manipulate optimization problems of the form

$$\begin{aligned}
\text{minimize} \quad & f(\vec{x}) \\
\text{subject to} \quad & \vec{g}(\vec{x}) \leq \vec{0} \\
& \vec{h}(\vec{x}) = \vec{0}
\end{aligned} \qquad (16.36)$$

where $\vec{x} \in \mathbb{R}^n$.

```
function   [f,g] = fun(x)         % We are defining a vector valued return
                                  % function that depends upon a vector x.

f =        {the function f expressed in terms of x(1), ..., x(n) };
g(1) =     {the function h(1)expressed in terms of x(1), ..., x(n) };

g(p) =     {the function h(p)expressed in terms of x(1), ... , x(n) };
g(p+1) = {the function g(1)expressed in terms of x(1), ... , x(n) };

g(m) =     {the function g(m) expresses in terms of x(1), ..., x(n) };
end                               % Done defining the function.
```

▼ *Figure 16.16.*
Matlab function for defining an optimization model.

We use the `constr` function to solve our problem. Again, it is only available with the optimization toolbox. We start by defining a return function as a Matlab m-file, call it `fun.m`, that passes in the vector $\vec{x}$ and returns the value of f, $\vec{h}$, and $\vec{g}$. The Matlab vector g is not the same as $\vec{g}$; it composes both $\vec{h}$ and $\vec{g}$. Figure 16.16 shows a general Matlab function.

Before Matlab can solve this function, we must start with a feasible point. That is, we must initialize `x(1)` through `x(n)` with a point $\vec{x}$ such that `g(1)` through `g(m)` are all zero or negative, depending upon their being equality or regional constraints.

After defining an initial feasible point, we call the SQP algorithm using the `constr` Matlab routine. We also enter the number of equations in g, representing the equality constraints $\vec{h}$, using the 13th value in the vector `options`. Figure 16.17 shows an example for implementing the Matlab routine. Matlab returns with a result after a number of iterations.

Matlab uses a quadratic estimate of the derivative, and a conjugate gradient search. The initial position in $\vec{x}_0$ in $\mathbb{R}^n$ defines the given initial values for the Matlab vector x. These initial values should be shotgunned to find any multiple local minima, if they exist.

```
>> x = [  x̄₀  ];                  % Enter a feasible point.
>> options (13) = p ;             % The number of equality constraints
>> x = contr('fun',x,options)     % solve it, with the file 'fun.m'
```

▼ *Figure 16.17.*
Matlab function calls.

Solution Method: Spreadsheet Solvers

In this section, we apply the background development of optimization methods to product models using a commercial spreadsheet package, Microsoft Excel™. Excel can manipulate optimization problems of the form

$$
\begin{aligned}
&\text{minimize} && f(\vec{x}) \\
&\text{subject to} && \vec{g}(\vec{x}) \le \vec{0} \\
& && \vec{h}(\vec{x}) = \vec{0} \\
& && \vec{i}(\vec{x}) \in \text{integers}
\end{aligned}
\tag{16.37}
$$

where $\vec{x} \in \mathbb{R}^n$ or integers.

We use the `Formula:Solver` menu option to solve optimization problems. We start by setting a column of cells to be the design variables, $\vec{x}$, say cells $X1:Xn$. Then, we enter the formulas for $f(x)$ in another cell, say $F1$, and we enter constraint formulas in other cells, say $G1:Gm$, $H1:Hp$, and $I1:Iq$, respectively.

Before one can use Excel to solve this model, the model must start with a feasible point. That is, $X1:Xn$ must be initialized with a point $\vec{x}$ such that $H1:Hp$ are all zero, and $G1:Gm$ are all negative. The `Formula:Goal Seek` helps to find an $\vec{x}$ which satisfies an equality constraint. Then we choose the `Formula:Solver` menu option.

The `Solver Parameters` dialog box will be displayed. Then we `Set Cell` $F1$ equal to `Min` to minimize f, and direct Excel to perform this minimization over $\vec{x}$ by `Changing Cells: X1:Xn`. We subsequently identify the constraints in Excel by minimizing `Subject to the Constraints:` $G1:Gm$, $H1:Hp$, and $I1:Iq$ by `Add...`ing them in. We then choose `Add...` and then enter `Cell Reference $G1$` to be $\le$ 0, and so on to Gm. For $H1:Hp$, we use the =, and for $I1:Iq$ we use the `int`. When finished `Add...`ing constraints, the `Solver Parameters` dialog box reappears, where one can direct Excel to `Solve`.

Excel uses either a tangent or a quadratic estimate of the derivative, which can be set in the `Estimates` option box. One can also choose between a quasi—Newton or a conjugate gradient search in the option box. Note the initial position in $\mathbb{R}^n$ is set by entering initial values for $X1:Xn$. These initial values should be shotgunned to find multiple local minima, if they exist.

Example: Fingernail Clipper Product—Shear Blades at End

Let's consider the fingernail product concept shown in Figure 16.14. This concept uses a shearing cutting mode at the end of the clipper, without the use of an additional lever arm. Similar to the fingernail clipper model developed in Chapter 13, a product model for this concept is given by

$$\text{minimize} \quad \frac{x + \frac{c}{2} + (2t_m)}{x} \tau \frac{W}{h_b} t^2$$

$$\text{subject to} \quad 2(2t_m) + 2(h_b + 3D_h) + d - H \leq 0$$

$$5D_h - W \leq 0$$

$$L_a + x + \frac{c}{2} + (2t_m) - L \leq 0 \qquad (16.38)$$

$$5t_m - L_a \leq 0$$

$$K_{\text{cycles}} \frac{fx}{Wt_m^2} \left(1 - \frac{x}{x + \frac{c}{2} + 2t_m}\right) - \sigma_{\max} \leq 0$$

where the search is across

$$(t_m, D_h, h_b, x, d) \in \mathbb{R}^5 \qquad (16.39)$$

with

$$(c, \tau, t, L, W, H, \sigma_{\max}, K_{\text{cycles}}) \in \mathbb{R}^8 \qquad (16.40)$$

as constants.

Applying the Solver function in Excel, a solution to this product model is shown in Figure 16.18. Thus, a finger force of approximately 2 lbs is required in a $2.5'' \times .5'' \times .688''$ clipper, with a plastic material that has a 15,000 psi tensile strength. This solution does not satisfy the target values for a fingernail clipper (a 1 lb finger force, with a material that has a lower tensile strength). This difficulty in the concept arises due to the mechanical disadvantage produced by the finger force being closer to the fulcrum than the force to cut the nail. In the next example, let's consider an alternative concept that overcomes this issue.

Example: Fingernail Clipper Product—Shear Blades on Side

In Chapter 13, a product model is developed for a fingernail clipper concept that uses sharp side blades to shear a nail. The resulting model is given by

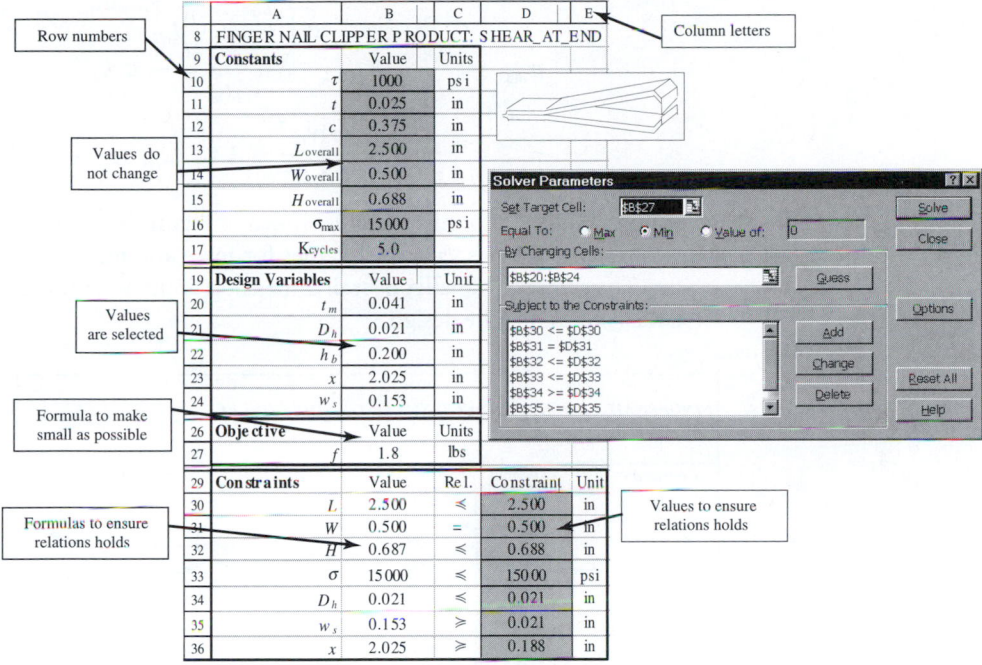

▼ Figure 16.18.
Excel solution to a fingernail product model.

minimize

$$f = \frac{L_F}{L_f}\tau\frac{W_b}{h_o}t^2$$

subject to

$$W_b \geq 5D_h$$

$$W_b \geq W_f$$

$$L_F \geq W_b$$

$$L_{overall} = L_s + L_f + L_g$$

$$L_s \geq 5t_m$$

$$L_f \geq L_F + \tfrac{1}{2}d_c$$

$$L_g = \tfrac{1}{2}d_c$$

$$H_{overall} = 2h_b + 2t_m \Big|_{\substack{\text{assumes stowage in} \\ \text{closed-locked position}}}$$

$$h_b \geq h_o + 3D_h$$

$$h_n, h_o \geq 1.1t$$

$$t_m \geq 2D_h$$

$$W_{overall} \geq W_f$$

(16.41)

$$\sigma_{\text{max}} = \frac{My}{I} = K_{\text{cycles}} \frac{6f\left(\dfrac{L_f}{L_F} - 1\right)L_F}{W_b t_m^2} \le S_y$$

$$\tau_{\text{rivet}} = \frac{1}{2} \frac{4F}{\pi D_h^2} = \frac{2\tau W_b t^2}{\pi h_o D_h^2} \le 1.5 S_{\text{sy,rivet}}$$

An Excel solution to this model is shown in Figure 16.19. Notice that the solution satisfies the target values for force and material properties. The choices of smaller overall dimension limits and different materials may refine the design.

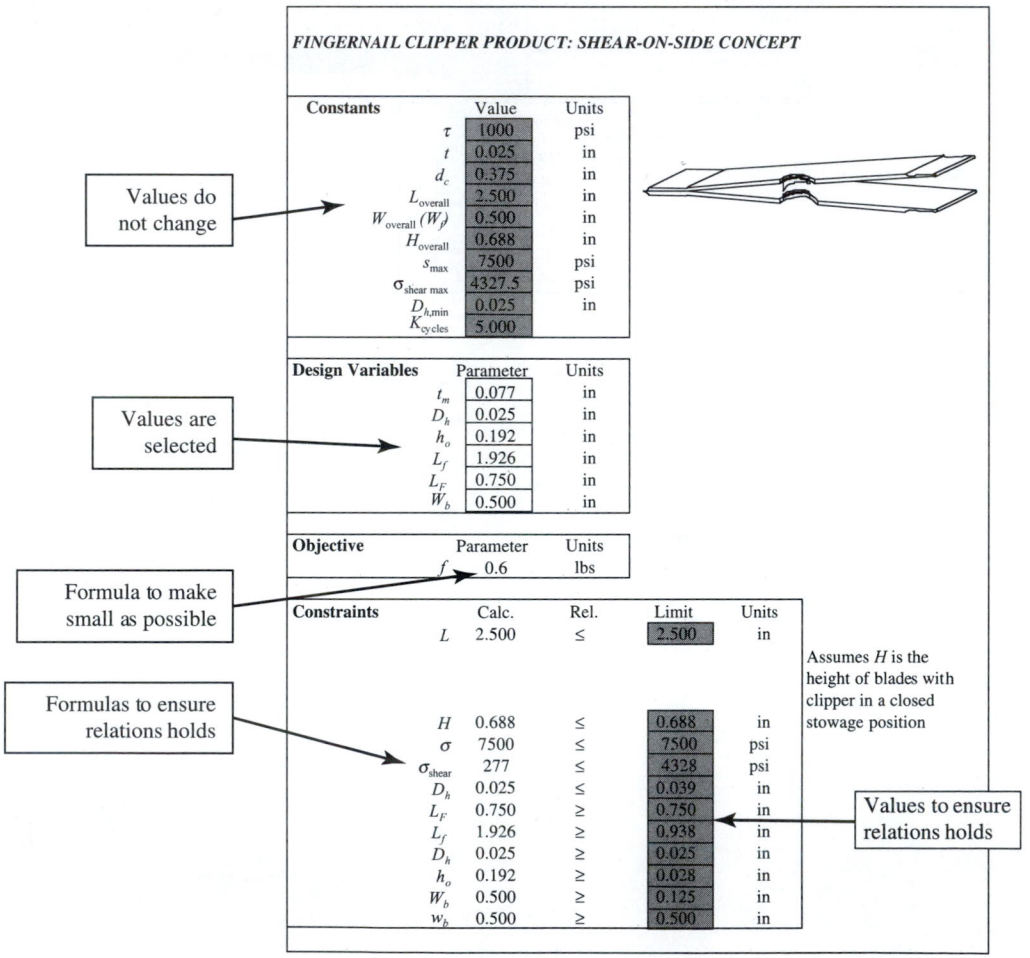

FINGERNAIL CLIPPER PRODUCT: SHEAR-ON-SIDE CONCEPT

Constants		Value	Units
	τ	1000	psi
	t	0.025	in
	d_c	0.375	in
	L_{overall}	2.500	in
	$W_{\text{overall}}\ (W_f)$	0.500	in
	H_{overall}	0.688	in
	s_{max}	7500	psi
	$\sigma_{\text{shear max}}$	4327.5	psi
	$D_{h,\text{min}}$	0.025	in
	K_{cycles}	5.000	

Values do not change

Design Variables		Parameter	Units
	t_m	0.077	in
	D_h	0.025	in
	h_o	0.192	in
	L_f	1.926	in
	L_F	0.750	in
	W_b	0.500	in

Values are selected

Objective		Parameter	Units
	f	0.6	lbs

Formula to make small as possible

Constraints		Calc.	Rel.	Limit	Units
	L	2.500	$\le$	2.500	in
	H	0.688	$\le$	0.688	in
	σ	7500	$\le$	7500	psi
	σ_{shear}	277	$\le$	4328	psi
	D_h	0.025	$\le$	0.039	in
	L_F	0.750	$\ge$	0.750	in
	L_f	1.926	$\ge$	0.938	in
	D_h	0.025	$\ge$	0.025	in
	h_o	0.192	$\ge$	0.028	in
	W_b	0.500	$\ge$	0.125	in
	w_b	0.500	$\ge$	0.500	in

Formulas to ensure relations holds

Assumes H is the height of blades with clipper in a closed stowage position

Values to ensure relations holds

▼ **Figure 16.19.**
Excel solution to a side-blade fingernail clipper concept.

To evaluate such choices, we need to understand the solutions charac-teristics, including binding constraints, sensitivities, and slack variables. Figure 16.19 shows these output data for the fingernail concept. The binding constraints identify the limits in the solution where a bound-ary is active. The "Not Binding" constraints do not define the solution or the design variable values. If improvements are desired in the solu-tion (about the identified optimum point), the binding constraint lim-its will need to be relaxed (or tightened if a target is desired for the objective). The Lagrange multipliers in the output identify the most sensitive constraints for the solution. For example, the constraint on D_h is critical for the fingernail clipper. In addition, the slack variables de-fine the "distance" that the solution is from constraint boundaries. Taken together, the binding constraints, Lagrange multipliers, and slack variables provide insights for modifying the optimization result. In the case of the clipper, it appears that the active material stress constraint can be modified without greatly affecting the objective. Because of the height dimensions' dependency on D_h, however, creating a more com-pact clipper in height will greatly affect the objective function.

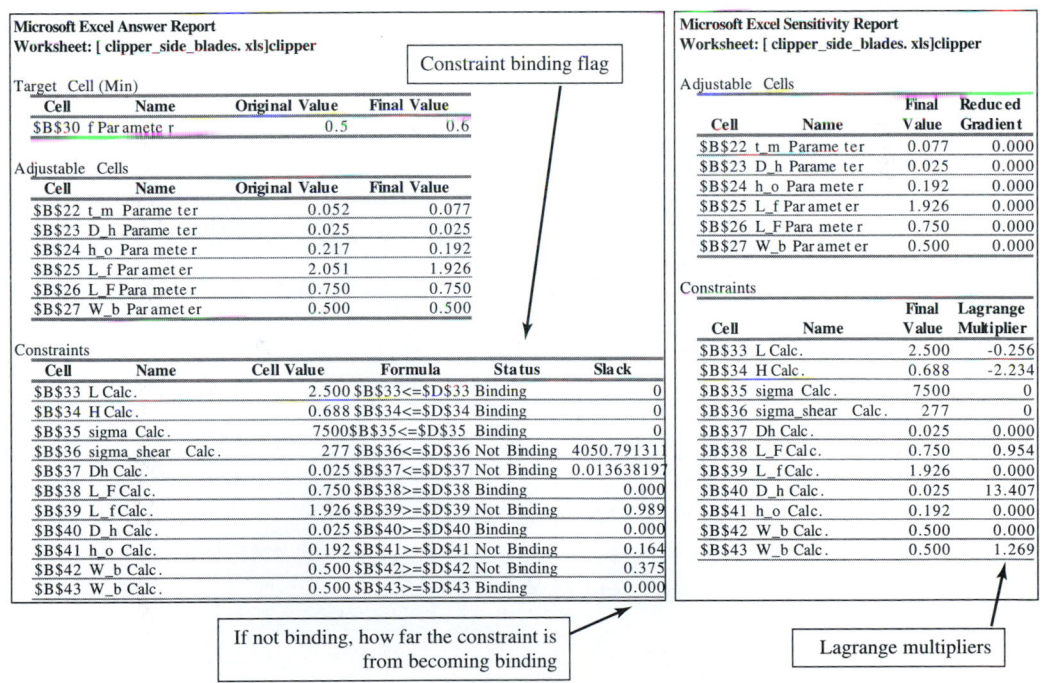

▼ **Figure 16.20.**

Optimization outputs for the Excel solution of the side-blade clipper concept.

VII. PRODUCT APPLICATIONS

In this section, two product applications of optimization methods are described. These applications provide additional insights into the use of product modeling during product development.

Application: Redesign of a TOMY™ Push 'n Go™ Train

The mission is to redesign the TOMY™ Push 'n Go™ Train (Patterson, et al., 1994). Based on functional modeling, customer needs assessment, product dissection, and initial modeling, two engineering specifications are critical to the evolution of the toy train: "inexpensive (minimize cost)" and "durability." An engineering model of the toy vehicle (shown schematically in Figures 16.21 and 16.22) is developed below and solved, focusing on these specifications.

A cost breakdown analysis of the train reveals that the compression spring (helical) is the most expensive single component/subassembly. Thus, as a first step to reducing cost, we investigate a spring model. A first question concerns the performance impacts of modifying the spring variables. The basic scaling law is the energy stored in the spring,

▼ *Figure 16.21.*
The TOMY Push 'n Go vehicle.

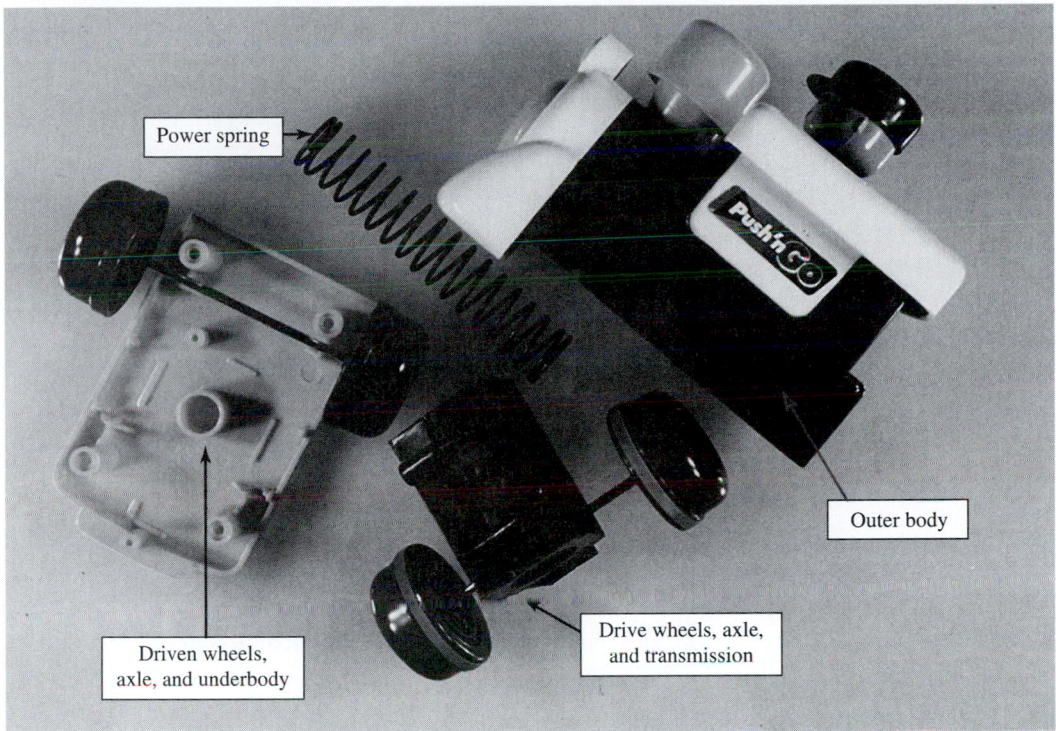

▼ **Figure 16.22.**
Push 'n Go vehicle spring energy system.

given by $E = \dfrac{1}{2}kx^2$, where k is the spring constant, and x is the spring deflection relative to the undeformed state. Force-deflection measurements show that the current spring constant is $k = 390$ N/m. Using this spring, and modifying the spring deflection with plastic collars of various heights, a plot may be generated of distance traveled versus energy storage, as shown in Figure 16.23.

It may be inferred from this plot that a large change in stored energy (50%) will only reduce the distance traveled by 17%. A 5-meter travel is also acceptable to the interviewed customers, if the product cost may be reduced.

Based on this experimental study, lower and upper bounds on the spring stiffness may be stated with respect to maintaining the performance requirements:

$$\left(\frac{2E}{x_{\text{max}}^2}\right)\frac{N}{m} \leq k \leq 390\frac{N}{m} \tag{16.42}$$

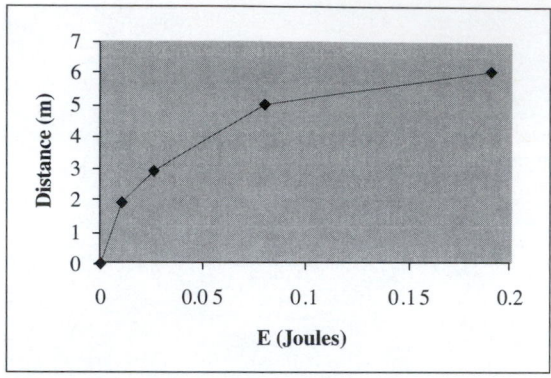

▼ *Figure 16.23.*
Plot of distance traveled versus stored energy.

where the stiffness is given by (Juvinall and Marshek, 1991)

$$k = \frac{(d^4 G)}{(8D^3 N)} \qquad (16.43)$$

d = wire diameter (0.0014 m current design)

D = coil diameter (0.019 m currently)

G = modulus of rigidity (79 GPa for steel)

x_{max} = maximum spring deflection (0.031 m currently)

L_f = free length of spring (0.106 m currently)

N = number of active coils (16, currently, of 18 total coils)

E = required energy storage = 0.1 J

The above ranges and relationships provide a statement of what may be chosen with respect to the spring design. In addition, we require relationships for the geometric bounds, cost, and spring durability. The current train geometry limits the overall spring size. Likewise, a buckling constraint provides a lower bound (Juvinall and Marshek, 1991). These limits may be stated as

$$0.018 \, m \le \left(D + \frac{d}{2} \right) \le 0.023 \, m \qquad (16.44)$$

Besides geometry, the durability of the spring may be represented by the spring's endurance limit

$$S_n = 0.145 S_u \qquad (16.45)$$

where S_u is the ultimate strength of the spring material. The endurance limit for a compression spring is governed by the shear stress such that, assuming a constant-life fatigue model,

$$\tau_a \equiv \frac{(0.0065dG)}{(D^2N)} + \frac{(0.0096d^2G)}{(D^3N)}(Pa) \le 0.145S_u(Pa) \qquad (16.46)$$

where τ_a is the average shear stress.

Since our goal is to minimize cost, it is assumed that a reduction in material volume will model spring cost. A cost function may then be stated as

$$C \propto DNd^2R \qquad (16.47)$$

where R is a cost factor related to the type of material, and the terms (Dd^2N) are proportional to the material volume. Three typical and available spring materials are listed in Table 16.1.

The overall model for redesigning the train spring becomes:

minimize

$$C = \frac{DNd^2R}{D_cN_cd_c^2R_c} \qquad (16.48)$$

subject to

$$\left(\frac{2E}{x_{max}^2}\right)\frac{N}{m} \le \frac{d^4G}{8D^3N} \le 390\frac{N}{m}$$

$$0.018\,m \le D + \frac{d}{2} \le 0.023\,m$$

$$\frac{0.065dG}{D^2N} + \frac{0.0096d^2G}{D^3N}\,Pa \le 0.145S_u\,Pa$$

where the "c" subscript denotes the current design. A validity analysis of the model reveals that the relative cost is a monotonic variable in the objective function. It does not satisfy the first principle of monotonicity, implying that it is not properly bounded from below with a

TABLE 16.1.   SPRING WIRE MATERIALS (CARLSON, 1992)

Material	R (relative $/volume)	S_u (Mpa)
Cold Drawn Steel (ASTM 227)	1.0	1750
Oil-Tempered Steel (ASTM 229)	1.3	1850
Music Steel (ASTM 228)	2.0	2200

formal constraint. This analysis further implies that the cheapest material listed in Table 16.1 will lead to the cheapest design, assuming relatively equal manufacturing costs. Thus, $R = 1.3$, and the material choice is ASTM 227.

It is likewise apparent that the number of turns of the spring is not bounded by commercial standards. Thus, $N \geq 12$ (integer) (Spring, 1990), and the new model becomes

minimize

$$C = \frac{DNd^2R}{D_cN_cd_c^2R_c}$$

subject to

$$\left(\frac{2E}{x_{max}^2}\right)\frac{N}{m} \leq \frac{d^4G}{8D^3N} \leq 390\frac{N}{m} \tag{16.49}$$

$$0.018 \, m \leq D + \frac{d}{2} \leq 0.023 \, m$$

$$\frac{0.065dG}{D^2N} + \frac{0.0096d^2G}{D^3N} \, Pa \leq 0.145S_u \, Pa$$

$$N \geq 12$$

Numerical optimization (e.g., Microsoft Excel) of this design model results in the following variable choices:

Material	d (m)	D(m)	N	Relative Cost
ASTM 227	0.0011	0.0175	12	0.426

These results predict a cheaper design that satisfies energy storage and durability requirements. The new spring design can also be used with the current toy train geometry.

Tolerance Analysis

While the results are promising, a tolerance analysis is a good idea to verify the design choices and determine the potential impacts of the uncertainties (noise). More details are presented on this topic in Chapter 19. Manufacturing tolerances on D, N, and d are given by 0.00035 m, 0.125 coils (*Spring Design Manual*, 1990), and 0.000025 m (SAE,

1992), respectively. Likewise, the uncertainty in the modulus of rigidity for steel is approximately 10% (Callister, 1991). Assuming the tolerances are normally distributed and equal to three standard deviations, a tolerance model for the spring constant is thus

$$t_k = \left[\left(\frac{\partial k}{\partial d} \right)^2 t_d^2 + \left(\frac{\partial k}{\partial D} \right)^2 t_D^2 + \left(\frac{\partial k}{\partial N} \right)^2 t_N^2 + \left(\frac{\partial k}{\partial G} \right)^2 t_G^2 \right]^{0.5} \qquad (16.50)$$

$$= \left[\left(\frac{4d^3 G}{8D^3 N} \right)^2 t_d^2 + \left(\frac{-3d^4 G}{8D^4 N} \right)^2 t_D^2 + \left(\frac{-d^4 G}{8D^3 N^2} \right)^2 t_N^2 + \left(\frac{d^4}{8D^3 N} \right)^2 t_G^2 \right]^{0.5}.$$

Substituting the design variable values and tolerances,

$$k = 220 \pm 33 \tfrac{N}{m}. \qquad (16.51)$$

This result suggests that k will violate the lower bound, set by the required energy storage. In addition, the tolerance on D will violate the lower bound on the geometry $(D + \tfrac{d}{2})$, determined by buckling conditions. To account for these uncertainties, the lower limits on k and D are modified to

$$k \geq \left(\frac{2E}{x_{max}^2} \right) \tfrac{N}{m}; \quad \text{where } E \equiv 0.12 \text{ J} \qquad (16.52)$$

$$D \geq 0.018 \ m$$

The final results, including tolerance analysis, are given by

Material	d (m)	D(m)	N	Relative Cost
ASTM 227	0.0012	0.018	12	0.520

These results satisfy the durability, performance, and geometry constraints, while minimizing cost. The designer is also confident in the results at 99% of the expected values of the variables.

Verification

A spring manufacturer from New York was contacted to bid on the new spring design. A quote of $1.30 was given for the new design (quantities of 1,000), compared to $1.70 for the original design, where the offset in relative costs is due to manufacturing. These relative costs demonstrate a significant improvement in the design −$0.40 savings per toy train.

Application: Electric Wok Product

In Chapter 13, a calibrated model is developed for an electric wok. Having this model, we can develop a spreadsheet to optimize the design model. An objective may be stated as the norm of the steady state temperature variations

$$\|\Delta \vec{T}\| = \sqrt{(T_m - T_c)^2 + (T_m - T_o)^2}, \qquad (16.53)$$

which we want to minimize, subject to the other product metrics being acceptable. The optimization formulation of the model then becomes

minimize	$\|\Delta \vec{T}\|$	(temp variation)
subject to	$t_r \leq t_\tau$	(temp rise time)
	$\|T(t_r) - T_{ss}\| \leq \varepsilon$	(temp at SS)
	$\dot{T}(t) + f(t) = 0$	(transcients)
	$W \leq W_\tau$	(weight)
	$T_c \geq T_{\tau c}$	(min center temp)
	$T_h \leq T_{\tau h}$	(max handle temp)

$$(16.54)$$

This formulation is the model of the wok concept. The search space is across the geometric variables, such as thickness and radii.

With exploration of the model, one finds that the middle ring width should be maximized to minimize the temperature errors. There is a model accuracy limit, though, with an excessive large middle ring width because of view factors. Given this limit, a reasonable approximation is to limit the center disk at about 5 cm, with the heater dimension at 16 mm. This result then leaves the wok thickness as the primary remaining variable. We can therefore tabulate performance metric values for different values of the thickness. Figure 16.24 shows the initial model, with constants shaded and design variables highlighted. This is the same model as was constructed and calibrated in Chapter 13, Figure 13.30.

Several exploratory optimization solutions can be pursued. For example, Figure 16.25 shows numerical optimal solutions for different materials, and the different material thicknesses required. This shows that changes to a high performing high-cost materials such as copper are not particularly effective, as compared to design changes such as bowl thickness. For one effective material, aluminum, one can solve the model at multiple thickness values to determine how the performance metrics vary (Figure 16.25). In particular, the temperature uniformity and the temperature rise time

ELECTRIC WOK REDESIGN

Constants		Value	Units
Boltzmann Constant	σ	5.7E-08	W/m^{2}K^4
Temp Environ	T_∞	25	.C
Bowl Conductivity	k	180	W/m.C
Convection Coeff	h	4.0	W/m^2
Wok Radius	X_{max}	0.18	m
Wok Height	Y_{max}	0.09	m
	R_{max}	0.20	m
Density	ρ	2800	kg/m^3
Emissivity	ε	0.80	
Heat Capacity	c_p	900	J/kg C
Wattage	q_{in}	1000	W
Duty Cycle		12%	
Handle Conductivity	k_h	1.0	W/m.C
Handle Conv Coef	h_h	4.0	W/m^2

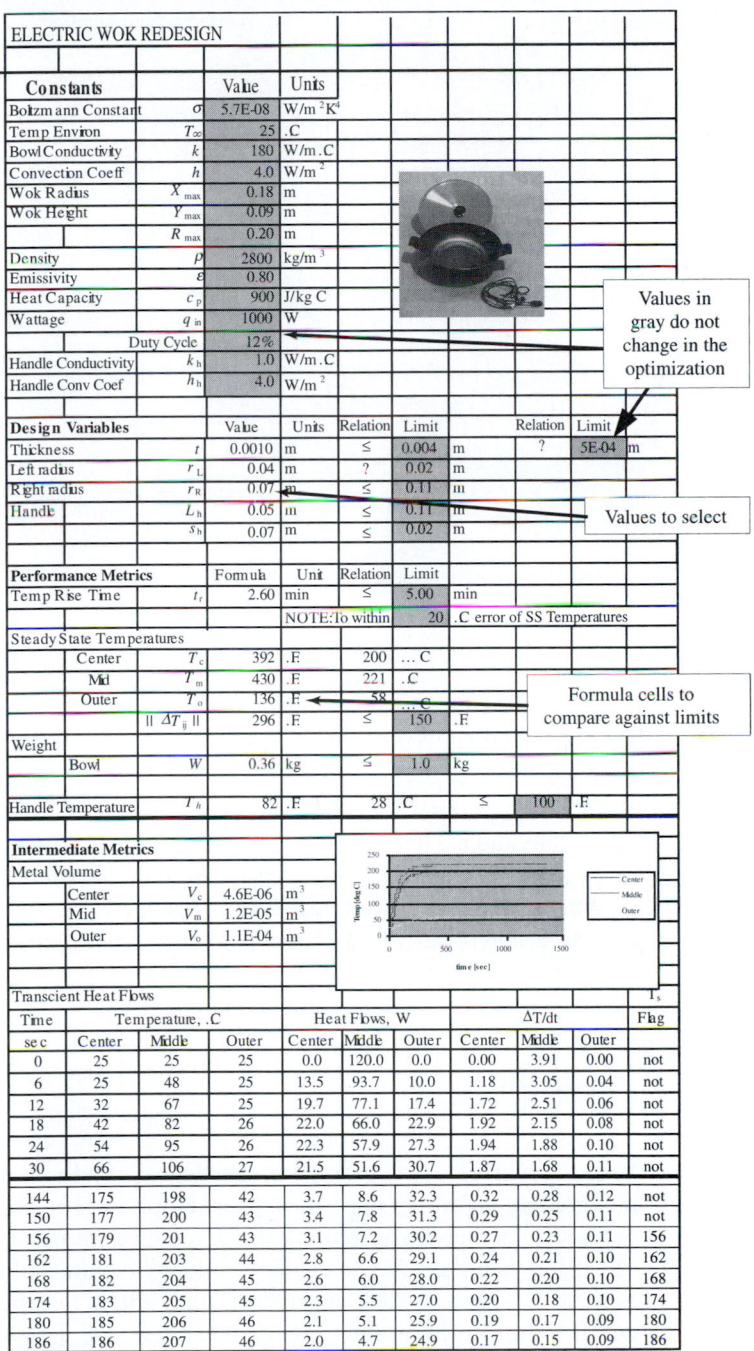

Values in gray do not change in the optimization

Design Variables		Value	Units	Relation	Limit		Relation	Limit
Thickness	t	0.0010	m	$\leq$	0.004	m	?	5E-04	m
Left radius	r_L	0.04	m	?	0.02	m		
Right radius	r_R	0.07	m	$\leq$	0.11	m		
Handle	L_h	0.05	m	$\leq$	0.11	m		
	s_h	0.07	m	$\leq$	0.02	m		

Values to select

Performance Metrics		Formula	Unit	Relation	Limit	
Temp Rise Time	t_r	2.60	min	$\leq$	5.00	min
			NOTE: To within		20	.C error of SS Temperatures

Steady State Temperatures

Center	T_c	392	.F	200	... C	
Mid	T_m	430	.F	221	.C	
Outer	T_o	136	.F	58	... C	
	$\|\Delta T_{ij}\|$	296	.F	$\leq$	150	.F

Formula cells to compare against limits

Weight

Bowl	W	0.36	kg	$\leq$	1.0	kg

Handle Temperature	T_h	82	.F	28	.C	$\leq$	100	.F

Intermediate Metrics

Metal Volume

Center	V_c	4.6E-06	m^3
Mid	V_m	1.2E-05	m^3
Outer	V_o	1.1E-04	m^3

Transcient Heat Flows

Time	Temperature, .C			Heat Flows, W			ΔT/dt			Flag
sec	Center	Middle	Outer	Center	Middle	Outer	Center	Middle	Outer	
0	25	25	25	0.0	120.0	0.0	0.00	3.91	0.00	not
6	25	48	25	13.5	93.7	10.0	1.18	3.05	0.04	not
12	32	67	25	19.7	77.1	17.4	1.72	2.51	0.06	not
18	42	82	26	22.0	66.0	22.9	1.92	2.15	0.08	not
24	54	95	26	22.3	57.9	27.3	1.94	1.88	0.10	not
30	66	106	27	21.5	51.6	30.7	1.87	1.68	0.11	not
144	175	198	42	3.7	8.6	32.3	0.32	0.28	0.12	not
150	177	200	43	3.4	7.8	31.3	0.29	0.25	0.11	not
156	179	201	43	3.1	7.2	30.2	0.27	0.23	0.11	156
162	181	203	44	2.8	6.6	29.1	0.24	0.21	0.10	162
168	182	204	45	2.6	6.0	28.0	0.22	0.20	0.10	168
174	183	205	45	2.3	5.5	27.0	0.20	0.18	0.10	174
180	185	206	46	2.1	5.1	25.9	0.19	0.17	0.09	180
186	186	207	46	2.0	4.7	24.9	0.17	0.15	0.09	186

▼ **Figure 16.24.**
Calibrated electric wok optimization model.

Different configurations

Design Variables			Steel	Copper	Aluminum	Steel	Aluminum	Aluminum
Material								
Thickness (m)		t	0.0010	0.0010	0.0010	0.0017	0.0017	0.0033
Left radius (m)		r_L	0.04	0.04	0.04	0.04	0.04	0.04
Right radius (m)		r_R	0.07	0.07	0.07	0.07	0.07	0.07
Handle	Length (m)	L_h	0.05	0.05	0.05	0.05	0.05	0.05
	Width (m)	s_h	0.02	0.02	0.02	0.02	0.02	0.02
Performance Metrics								
Temp Rise Time		t_r	4.40	3.10	2.60	6.90	4.10	7.30
Steady State Temp.		$\| \Delta T_{ij} \|$	482	171	296	394	209	124
Bowl Weight		W	0.99	1.13	0.36	1.69	0.61	1.18
Handle Temperature		T_h	71	79	78	77	79	79

Compare the performance

▼ **Figure 16.25.**
Pareto optimal solutions for the electric wok concept.

are Pareto related. Increasing the temperature uniformity by increasing the thickness results in drastically increased times to raise the temperature.

Based on the results, 1.0 mm to 3.5 mm thickness aluminum bowls are acceptable, and a 2.0 mm is recommended. This thickness represents a design that is reasonable in material cost, yet performing. The temperature distribution is not adversely affected, the time to heat up is improved, and the handle design will not be too hot to the touch.

VIII. SUMMARY AND "GOLDEN NUGGETS"

The discussion of numerical models includes basic and advanced methods, presenting process steps, theory, and examples. The process steps provide guidelines for systematically developing numerical models. The theory helps us to understand the underlying limits of the methods, and the examples illustrate practical applications. In either case, the goal is to transform product models into choices of design variables that satisfy the engineering specifications and target benchmarks.

Golden nuggets that may be gleaned from this chapter include:

▶ As a basic method, spreadsheets are a numerical tool for solving product models. They are prevalently used due to their flexibility and relatively benign learning curve.

▶ The addition of optimization solvers to spreadsheets greatly enhances the abilities of the designer. Search may be performed more efficiently and with greater confidence in the results.

▶ Engineering product models can lead to very good predictions of real world devices, as demonstrated in this chapter by the fingernail clipper, electric wok, and toy rocket products.

References

Arora, J. *Introduction to Optimum Design*. McGraw-Hill Book Company, New York, 1989.

Brinkman, L. *Mathematical Introduction to Linear Programming and Game Theory*. Springer-Verlag, New York, 1989.

Callister, *Materials Science and Engineering*, John Wiley and Sons, New York, 1991.

Carlson, *Spring Manufacturing Handbook*, Marcel Dekker, New York, 1982.

Dixon, J. R., *Design Engineering: Inventiveness, Analysis, and Decision Making*, McGraw-Hill, NY, 1966.

Eschenauer, H., Koski, J. and A. Osyczka, editors. *Multicriteria Design Optimization*. Springer-Verlag, Berlin, 1990.

Gill, P., Murray, W. and M. Wright. *Practical Optimization*. Academic Press, New York, 1981.

Juvinall and Marshek, *Fundamentals of Machine Component Design*, John Wiley and Sons, New York, 1991.

Osycska, A. *Multi-Criterion Optimization in Engineering with Fortran Examples*. Halstad Press, New York, 1984.

Papalambros, P. and D. Wilde. *Principles of Optimal Design*. Cambridge University Press, New York, 1988.

Patterson, J., B. Spears, S. Vega, and P. Yu , Reverse Engineering Report, The University of Texas at Austin, Dept. of Mechanical Engineering, 1994.

Raiffa, H. and R. Keeney. *Decisions with Multiple Objectives: Preferences and Value Tradeoffs*. Wiley, New York, 1976.

Ramaswamy, R. and K. Ulrich, "A Designer's Spreadsheet," *ASME Design Theory and Methodology Conference*, DE-Vol. 53, 1993, pp. 104–113.

Serrano, D., *Constraint Management in Conceptual Design*, PhD Thesis, Dept. of Mechanical Engineering, Massachusetts Institute of Technology, 1987.

SAE Handbook Vol. 1: Materials, SAE, Warrendale, PA, 1992.

Spring Design Manual, "Design and Application of Helical and Spiral Springs," No. SAE HS 795, SAE, 1990.

Steuer, R. *Multiple Criteria Optimization: Theory, Computation, and Application*. J. Wiley, New York, 1986.

17 Physical Prototypes

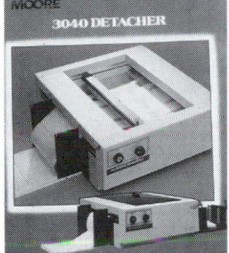

Original Moore Detacher

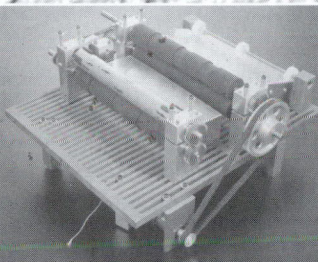

Mechanical breadboard proof
of concept prototype

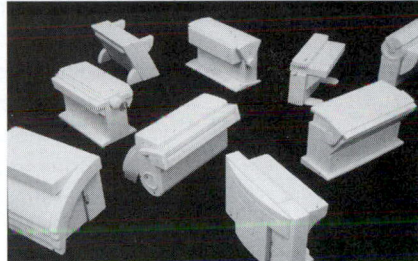

Machined foam industrial
design prototypes

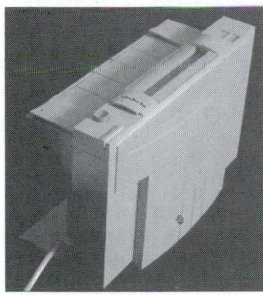

Non-functioning industrial
design prototype conveys
look and feel

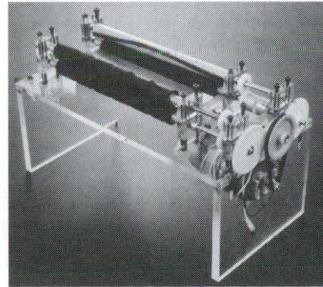

Experimental prototype

Alpha prototype, with fully functional hardware

Final product

Photos Product Genesis, Inc., by permission.

Product Genesis Inc. is a technology-oriented product development firm with expertise in both engineering and industrial design. They perform complete product development services for mechanical and electromechanical products on a contract basis. A key aspect of the rapid, high-technology Product Genesis Product Development Process is effective prototyping. A *prototype* is a physical instantiation of a product, meant to be used to help resolve one or more issues during the product development. Prototyping communicates the visual layout and a product's look and feel; furthermore, experimental prototyping enables the exploration, optimization, and validation of mechanical hardware. Final hardware prototyping helps communicate fabrication and assembly issues. As captured in the pictures, the activity of creating physical models transmutes product development from concept to form. Product Genesis wanted to design a small desktop "burster-trimmer" product for smaller billing agencies that could separate roll paper along perforations. Foam prototypes of different styles were created to provide a representation that the customer could hold. For the style selected, a completely painted and finished prototype was made to convey the look and feel. The concept behind the burster-trimmer is to pull the paper apart by rollers working at different speeds. A simple experimental prototype was mechanically *breadboarded* (assembled with stock mechanical parts on a board) to test this concept. Once the physics were understood, adjustable experimental prototypes were made and used to help select a layout and components. The first combination of the internal functional components as a system with the industrial design shell created the alpha prototype.

Prototypes help the design process in that they may be visually inspected, tactilely experienced, tested, modeled, varied, and simply observed as a 3-D entity. In this sense, physical models provide a leap toward product realization, where ultimately the machinery of manufacturing processes create suitable batch sizes of the product in its consumer form. In this context, this chapter investigates the applications of physical models in the product development process. Topics of investigation include the range of common models, techniques for creating a physical prototype, and the materials/processes for constructing effective models. A straightforward product example is shown to illustrate the factors discovered in these topics.

I. CHAPTER ROADMAP

Figure 17.1 illustrates the layout of this physical modeling chapter. A preview begins the chapter, discussing example objectives of physical models and the expected outcomes. The chapter then continues with prevalent

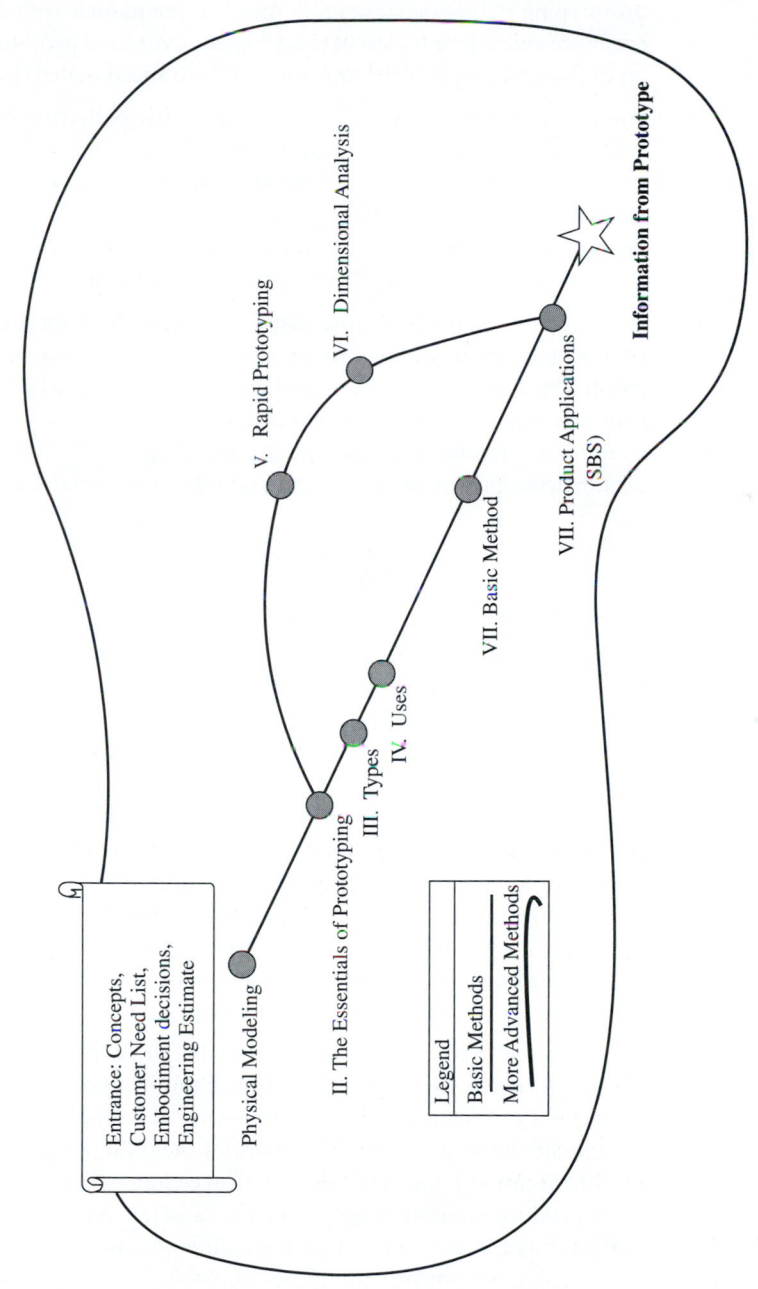

Entrance: Concepts,
Customer Need List,
Embodiment decisions,
Engineering Estimate

Physical Modeling

II. The Essentials of Prototyping

III. Types

IV. Uses

VII. Basic Method

V. Rapid Prototyping

VI. Dimensional Analysis

VII. Product Applications
(SBS)

Information from Prototype

Legend
Basic Methods
More Advanced Methods

▼ **Figure 17.1.**
Chapter roadmap.

prototyping technologies, ranging from simple models with common hardware and simple materials to rapid prototypes and precision archetypes that utilize specialized processes and advanced materials.

Following these introductory remarks, alternative paths may be used to explore the techniques discussed in the chapter. A basic path to navigate physical modeling is to explore an example prototyping effort and learn a simplified systematic approach for testing a prototype. This approach provides the necessary foundation for understanding how to model product performance with physical realizations.

An alternative path through the chapter begins with the basic approach of modeling products, but also discusses dimensional analysis and the construction of prototypes in sizes different than the actual product, for products where doing so is prohibitive in time or cost. With these advanced concepts, the limits of physical modeling may be understood in greater detail, providing more depth to the designer's modeling toolbox.

II. PROTOTYPING ESSENTIALS

As discussed in Chapter 13, two choices exist for modeling and simulating a product's performance: analytical models and physical models. A number of reasons clarify and distinguish these choices: model accuracy versus prototype expense (Table 13.4), comprehensive versus a focused study of the model, cycle time required for modeling versus project schedule, and the intent of the model. Clearly, all models for a product, during the preproduction phases of development, cannot be solely analytical in nature. Simulation of system models, whether based on back-of-the-envelope estimates, analytical closed-form models, or numerical approximations, fall short when verifying a product against its specifications. Unexpected phenomena and effects are always discovered in the physical reality of a product.

Many reasons exist for this realistic shortfall of analytical models, including accuracy, development time, and model intent. For example, the accuracy of such models, due to theoretical limitations, may poorly estimate the actual product behavior. If the underlying physics of a product is not well understood, analytical estimates cannot be expected to produce accurate results. Likewise, a detailed numerical model (such as finite element analysis or finite difference models) may require months of development. The efficacy of physical models is pronounced for this situation. Finally, an extreme case of incomplete analytical modeling relates to the intent of a prototype. The simple intent may be to satisfy a milestone or demonstrate ergonomic effects that do not have analytical representations. In this circumstance, physical prototype construction and analysis is, and will continue to be, a critical aspect of product realization.

Thus, on the one hand, product development demands the effective use of prototypes throughout the development cycle. Paper and virtual models hide many mysteries of how a product will actually perform. On the other hand, it is not practical to build a large number of physical prototypes at every miniscule step of a product's creation. Product development drivers inhibit the fabrication of even a fraction of the prototypes that would be needed.

To understand the relationship between virtual and physical models, let's consider the drivers of physical models. Two primary drivers in the development of physical models are cost and cycle time. During the decades of the 1970s and early 1980s, physical prototyping was demanded for virtually every stage of the development process. Physical models were scheduled for construction whether or not clear questions were being answered from prototype tests. This practice lead to higher cost and cycle time for product development. With increasing pressure from international competition, a dramatic change in philosophy was demanded to revise this practice. The rule became one of entering the market with new products at a fast pace, at the same time achieving and maintaining a high level of quality.

The emphasis on cycle time and quality led to a number of advancements in the product development process. For example, concurrent engineering, with product-function teams (Chapter 2), helped streamline the historical "toss the design over the fence" approaches. These teams are composed of multidisciplinary members so that design for manufacturing and other decisions can be made early in the process and with less risk. Likewise, more emphasis is placed on utilizing stable technologies and virtual modeling. In doing so, analytical models can be used to make product configuration and variable decisions, thus replacing more costly and time-expensive physical prototypes.

An exemplifying case of this change of philosophy is the Xerox product development process. Classically, Xerox created at least three major prototypes during a product's creation or evolution (Bebb 1990). These prototypes were primarily full scale, required up to months of planning and construction and incrementally approached a preproduction model of the product. Due to increased competition, Xerox lost significant market share in the copier industry (Bebb 1990). To regain their status as the market leader, Xerox had to implement concurrent engineering processes and reduce their prototyping efforts to a single full scale prototype prior to production. Through these efforts, Xerox greatly streamlined product development and again became a world leader in copier technologies (Bebb 1990; Xerox 1996).

The moral of this case is simply stated as "minimize the product development cycle while delivering a quality product that satisfies the customer needs over varying conditions of use." One step for satisfying this goal, as implemented by Xerox and many other companies, was to reduce the number of physical prototypes and their associated resources in time and cost. But herein lies the dilemma. We have also argued for the necessity of creating physical prototypes; they uncover physical effects and phenomena that are not predicted by the analytical models.

The solution to this conflict lies in technological developments. Computational modeling can become more complex, but at less expense in time due to advances in software tools and hardware performance. Further, physical prototyping itself is becoming faster. Stereolithography, selective laser sintering, and other rapid prototyping technologies permit physical prototyping with limited impact on schedule. With such recent advances in prototyping processes, many product development teams are evaluating and adjusting the balance between virtual and physical analysis. Instead of assuming that physical prototypes always increase cycle time and cost, teams can exploit recent advances to both reduce time-to-market and positively impact product quality. Physical and virtual models are seen now as competing technologies with clear decision criteria. Depending on the questions to be answered, studies to be conducted, and decisions to be made, the valid development choice may be a physical model, a virtual model, or some combination.

In this section, we explore further the contemporary uses of physical models. This exploration will help us to choose when physical models should be constructed and tested. It will also provide basic insights on the creation process of prototypes, including material and fabrication choices.

What Are Physical Models/Prototypes?

A *physical model* is an object or set of objects that is fabricated from a variety of materials to approximate an aspect(s) of how a product concept will perform. Synonymously, we refer to physical models as *physical prototypes*. Generally, a physical prototype is a simplification of a product concept. It is tested under a certain range of conditions to approximate the performance, constructed to control possible variability in the tests, and is ultimately used to communicate empirical data about the product so that development decisions may be made with high confidence and reduced risk.

Based on this view, a number of important questions must be asked when considering physical prototypes:

- ▶ Should the product development team build a prototype(s) at a certain time?
- ▶ What is the purpose(s) of the prototyping efforts?
- ▶ What are the possible forms of the prototype?
- ▶ What simplifications can be made that are independent of the prototype's purpose?
- ▶ What types of tests will be applied to the prototypes?
- ▶ What is the risk of constructing prototypes or continuing without them?

These questions must be considered carefully, especially in the context of prototyping efforts in industry. Concrete answers to these questions will increase the probability of success from the prototyping efforts. It will also reduce the risk of wasted resources and missed milestones. Without a careful assessment of these questions, the common tendency is to say "let's just build." This gut reaction can lead to poor results. Instead of creating effective prototypes that answer important product questions, the development team can become bogged down in iterative hardware construction and debugging. Significant slowdown will then occur, wasting time on activities that have very little to do with the final product configuration or manufacturing.

III. TYPES OF PROTOTYPES

One approach for aiding the prototype decision process is to consider the current trends of prototype use in industry. This section studies these trends. As a context to visualize and understand the different classes of prototypes, we consider a product development project throughout this section. Keurig Inc. of Boston, Mass. is the manufacturer of coffee makers that quickly brew fresh individual servings of coffee. The product was developed in association with Product Genesis Inc. of Cambridge, Mass. The original product that Keurig developed is shown in Figure 17.2, and they wanted to redesign it to perform better. As a part of the process of developing a new machine, several prototypes were used.

For contemporary product development processes (Chapter 1), six general classes of prototypes are typically used:

1. proof-of-concept models
2. industrial design prototypes
3. DOE experimental prototypes

▼ Figure 17.2.
Keurig individual serving coffee brewer. (Photos Product Genesis, Inc. and Keurig, Inc., by permission.)

4. alpha (same material & geometry; different manufacturing) prototype
5. beta (final part production; special assembly) prototype
6. preproduction (pilot production, limited capacity) prototype

Proof-of-concept models are used to answer specific questions of feasibility about a product. They are usually fabricated from simple, readily available materials, they focus on a component or subsystem of the product, and they are constructed post-concept generation, usually during concept selection and product embodiment. The general question proof-of-concepts answer is whether the imagined physics of the concept on paper indeed actually happen, and what any unforeseen physics might be. As an example, one idea that Product Genesis had for the burster-trimmer project mentioned in the chapter introduction was to have two rollers running at different speeds to tear the paper at the perforations. Would it work? A proof of concept mock-up was developed using mechanical breadboard components as shown in the cover figure. The Keurig coffee brewer project did not require proof of concept design, since it was a redesign effort of a previous Keurig model.

Industrial design prototypes demonstrate the look and feel of the product. Generally, they are initially constructed out of simple materials such as foam or foam core and seek to demonstrate many options quickly. For the Keurig coffee brewer, the first industrial design concepts were shown as renderings, or drawings, that show the look and feel of the product. Four concept renderings are shown in Figure 17.3. With these, Keurig and Product Genesis selected a concept layout. At that point, an industrial design mock-up was constructed as shown in Figure 17.4. The mock-up looks exactly like the final product but in fact is made of plastic blocks with no working internal components.

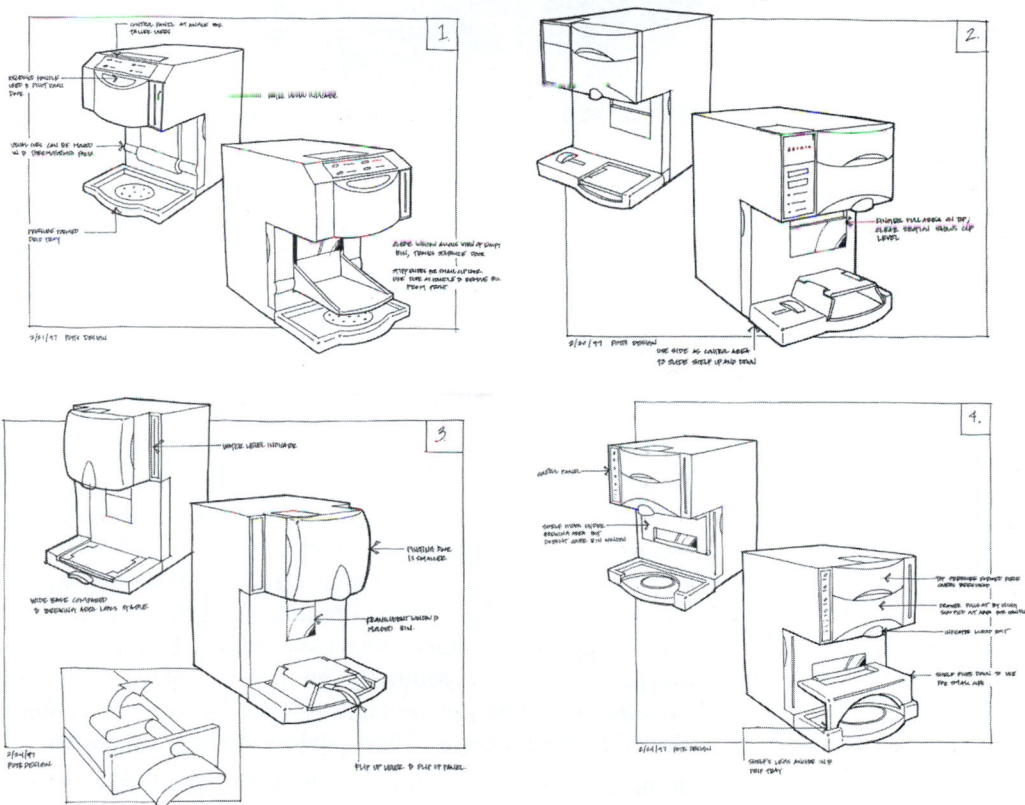

▼ *Figure 17.3.*
Keurig coffee maker industrial design renderings.

▼ Figure 17.4.
Keurig coffee brewer industrial design
prototype. The parts are completely
nonfunctional solid blocks, but the prototype
conveys the look and feel. (Photos Product
Genesis, Inc. and Keurig, Inc., by permission.)

Design of experiments (DOE) experimental prototypes are focused physical models where empirical data is sought to parameterize, lay out, or shape aspects of the product. The focus with DOE prototypes is usually to model a subsystem of a product while converging to a target performance of the subsystem. This class of prototypes is fabricated from similar materials and geometry (perhaps scaled) as the actual product, with the DOE prototype being just similar enough to replicate the real product's physics, but otherwise made as simply, cheaply, and as quickly as possible. DOE in product design is developed in Chapter 18. For the coffee brewer, an experimental prototype is shown in Figure 17.5, where different heating and flow experiments were conducted.

To answer questions regarding overall layout, *alpha prototypes* are constructed. Alpha prototypes are fabricated using materials, geometry, and layout that the design team believes will be used for the actual product. The alpha is the first system construction of the subsystems

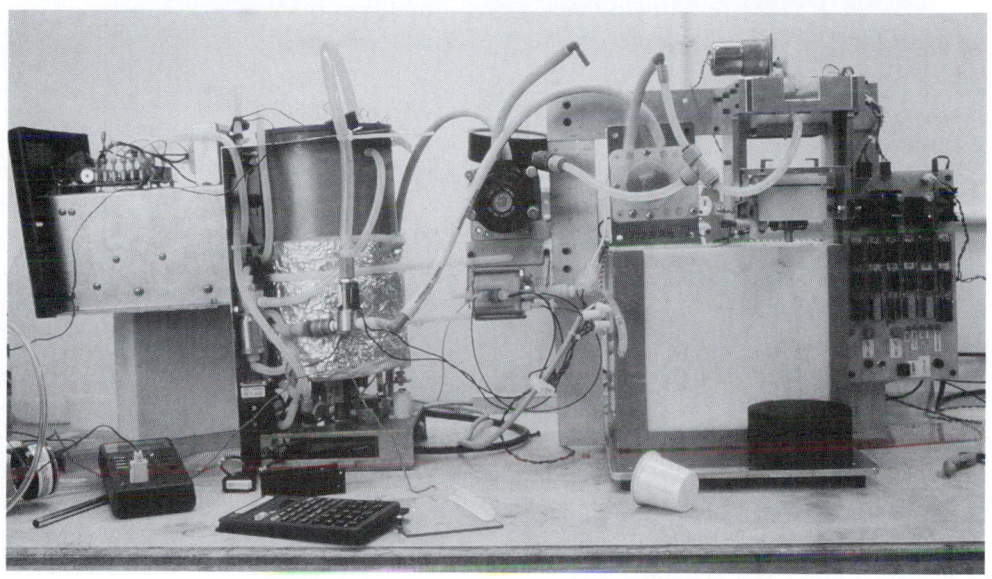

▼ *Figure 17.5.*
Keurig coffee brewer experimental prototype.(Photos Product Genesis, Inc. and Keurig, Inc., by permission.)

that are individually proven in the subsystem DOE prototyping and/or design. Alphas also usually include some functional features for testing and measurement of the product as a system. The alpha prototype of the internals of the Keurig coffee brewer is shown in Figure 17.6. Generally, the subsystems do not work together exactly as perceived by the various subsystem engineers, so alphas usually include only about 80–90% of the actual geometry. They contain limited numbers of the parts manufactured as in final product, though original equipment manufacturers (OEM) components are exact.

Beta prototypes, on the other hand, are the first full-scale functional prototypes of a product, constructed from the actual materials as the final product. They may not necessarily be fabricated using the same production processes as the final product though. Plastic parts on beta prototypes are typically CNC machined rather than individually injection molded. The beta prototype of the Keurig coffee brewer is shown in Figure 17.7. Test plans are applied to the functional beta prototype over a range of product functions and operating conditions. Failure modes and effects (Chapter 12) are typically analyzed by considering extremes in the operational range of the product.

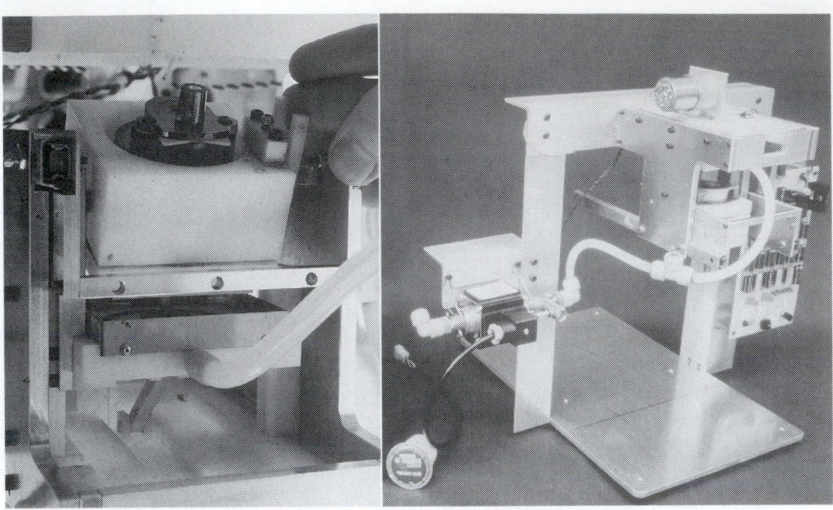

▼ *Figure 17.6.*
Keurig coffee brewer alpha prototype.(Photos Product Genesis, Inc. and Keurig, Inc., by permission.)

▼ *Figure 17.7.*
Keurig coffee brewer beta prototype. (Photos Product Genesis, Inc. and Keurig, Inc., by permission.)

Preproduction prototypes are the final class of physical models. These prototypes are used to perform a final part production and assembly assessment using the actual production tooling. Many design firms strive to make the beta prototype be both the preproduction unit and the actual unit; that is, to have no corrections needed in the beta prototype.

In any case, small batches of the product are produced for the production prototypes to verify product performance for predicted full-scale capacities. For example, final part tolerances on some parts may have to be loosened and tolerances on other parts or assembly operations tightened. The preproduction prototype, produced in a statistically valid quantity (dozens to hundreds, depending on the cost and complexity), answers these questions.

Prototype Goals

For these classes of prototypes, the product development team may seek a number of outcomes. Important questions to pose include the purpose of the prototypes and example outcomes. When answering these questions, we should consider the following typical goals of prototypes:

- a "rough version" to answer a single or set of binary (yes/no) questions (using materials such as cardboard, foam, and soft wood)
- a concept model with no working features to obtain early feedback from customers
- a rough version to visualize and brainstorm possible improvements
- a study of the product features and models to refine difficult features
- functional and semifunctional models
- a simulated walk-through of product activities
- the creation of a photographic quality model to create a demonstration video for marketing and evaluating the product in use
- a study of appearance and semantic "feel" of a product
- a sample batch to resolve possible manufacturing problems and process variables
- a small batch to obtain customer feedback in focus group settings

The goals of prototypes may also be considered from the viewpoint of different subgroups of a product development team (Chapter 2). Group roles and the corresponding prototypes are shown in Table 17.1. Generally, industrial designers use prototypes to develop the look and feel of the product, electrical engineers use prototypes to validate the variety of states that systems can achieve and change, and mechanical engineers use prototypes to develop the physical behavior of a product.

TABLE 17.1.   USE OF PROTOTYPES BY DISCIPLINE

Industrial design	• Alternatives for testing aesthetics and artistic impression (feel), usually embodied in early foam models • Studies of semantic product statement • Arrangement of internal components and its effect on shape • New product concepts • Ergonomic studies
Electrical engineering	• Layout and physical models of printed circuit boards • Test fixtures for electronic function and control • Electronic function (breadboarding) • Power supply, modulation, and control • Assessment of UL ratings • Standard component studies and integration
Mechanical engineering	• Functional proof of concepts • Product component layout and interconnects • Verification of virtual modeling (acoustics, vibration, heat transfer, stress levels, kinematics, etc.) • Machine design (elements and mechanisms) • Fabrication and testing of packaging • Studies of manufacturing processes • Material selection • Tool design (sheet metal, plastic moldings, castings, etc.) • Assemblability analysis and time motion studies • Housing studies for mechanical deformations, stress/strain, heat transfer, vibration, ESD protection, and mechanical interfaces • Validation of mechanical assembly and component drawings

IV. USES OF PROTOTYPES

Many types and classes of physical prototypes exist, but their planned uses may be categorized rather simply. The foundational category is to minimize risk during the product development process, but specific categories, and their intentions, may be listed as follows:

▶ **Communication:** to obtain feedback from customers, suppliers, vendors, and management.

▶ **Demonstration:** to show the accomplishment of project goals and milestones, especially to customers, clients, and investors.

▶ **Scheduling/milestones:** to force preliminary decisions and avoid terminal cycles in concept development and embodiment design.

▶ **Feasibility:** to focus on specific ideas, to determine if they will work, to determine if they satisfy customer needs, and to uncover unpredicted phenomena. Measurements are typically recorded and analyzed.

- ▶ **Parametric modeling:** to systematically develop a phenomenological model of the product or subsystem. Experimentation is required, and the results lead to decisions for design variables to optimize performance.

- ▶ **Architectural interfacing:** to determine if module interfaces perform properly, to test the compatibility and assembly of components, and to explore manufacturing processes and standard part issues.

With proper planning and contemporary prototyping processes, these uses should lead to a number of benefits for the product design team. Such benefits include:

- ▶ *Greater freedom and care in allocating resources.* Prototypes uncover priorities that must be solved to realize a product. Implementing Pareto's 20-80 rule, the top priority items must be solved first with intense focus, resulting in many others problems being solved simultaneously and, quite often, serendipitously.

- ▶ *Reduction of costly iterations.* Early prototypes provide insights into manufacturability and assemblability. These insights reduce the problems that must be solved later in the process. They also aid in early decisions that might otherwise be tabled for later consideration. For example, many companies report a 50–60% reduction in turnaround time when showing suppliers prototypes as opposed to just assembly drawings.

- ▶ *Acceleration of parallel activities.* By communicating to customers, vendors, and investors early product versions, team dynamics are facilitated. Improvements that are identified by these parties may be addressed and integrated early in the process.

- ▶ *More flexible product choices.* Through modeling, a greater number of options may be explored, resulting in clearer choices for shape, dimensions, material properties, processes, and so forth.

While these benefits exist, a number of pitfalls of prototypes must be avoided. Such pitfalls include delaying time-to-market because of unnecessary prototype iterations or poor prototype fabrication plans; fabricating unnecessarily detailed or complex prototypes when simplifications are possible; creating schedules that do not allow for prototype test results to be integrated into the final product; and constructing prototypes (time + money!) that do not contribute to the product goals. The basic method at the end of this chapter helps avoid these pitfalls and create physical models that are productive and have intrinsic value.

Mock-up Materials and Processes

Types of Materials and Processes

Assuming that a prototype type is selected for a particular use, the product development team must then consider alternative designs for the prototype in conjunction with possible model-making processes (Figure 17.8). Within these considerations, it is important to evaluate various technologies available for prototyping. Overall issues in this evaluation include the number of prototype units needed, the type of units needed and their intended use, the prototyping materials, and the choice of particular measurements and test plans. With respect to

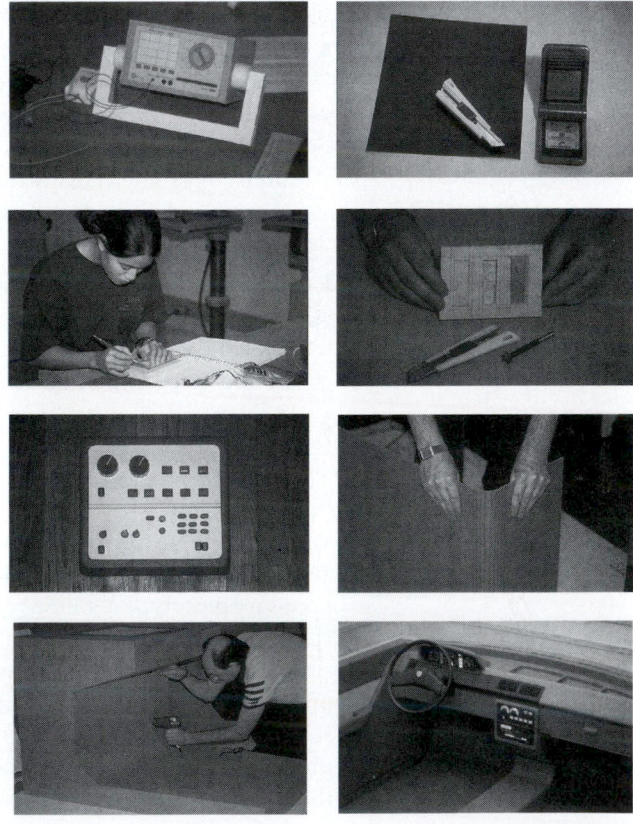

▼ *Figure 17.8.*
Example model-making scenarios (Wallace, 2000).

the materials issue, a number of criteria should be applied when making a decision:

1. *Cost.* Should be minimized without sacrificing the prototype goals and uses. What is the risk of using a material(s) for the prototype? What advantages do certain materials provide for prototype testing?

2. *Availability.* Readily available materials (raw, stock, and off-the-shelf components) should be chosen to avoid delays in prototype construction.

3. *Ability to accept changes.* It is highly probable that a prototype will be changed as it is prepared. The choice of materials should thus accommodate modifications in dimensions, surface requirements, and so forth.

4. *Ease of use and forming capability.* Special tools, safety equipment, fixtures, and setup should be avoided. The material should also be relatively easy to form. If these issues are carefully contemplated, time will be conserved.

5. *Scalable geometry.* Simplified and convenient prototype shapes will be needed to eliminate unnecessary forming operations.

6. *Scalable properties.* Is the material choice scalable to the actual product's function? Valid material properties must be chosen to satisfy the prototype use and level of approximation. Example properties to consider are mechanical (such as strength, endurance limit, durability, wear, and moduli), thermal (such as conductivity, mass, and expansion coefficient) electrical (such as inductance and resistance), and chemical (such as corrosion resistance).

Based on these criteria, let's explore a range of materials that are available for prototyping. Each of these materials is not suitable for all types of physical models, but collectively they cover a large spectrum from proof-of-concept or appearance prototypes to fully functional models.

Wood and Wood Products

Overall, wood materials are very useful when constructing early prototypes. Wood materials are readily available, easy to work, using inexpensive hand and power tools, and available in a large variety of sizes (Table 17.2).

Plastics

A very large number of polymer materials are readily available as prototyping materials. Table 17.3 contains a summary of those most commonly used.

TABLE 17.2.   PROTOTYPING PROPERTIES OF WOOD

Pine	• Good overall modeling material • Inexpensive • Easy to tool
Hardwoods (oak, cherry, birch, mahogany, etc.)	• Generally higher quality than pine in terms of appearance, absence of knots, etc. • More expensive than pine • Makes more precise joints • Holds fasteners better • Higher load capability
Plywood	• High strength • Available in large sheets • Both hardwood and softwood varieties • Choice of grades
Balsa	• Light weight • Low strength (in general, but good static strength along grains) • Versatile and easy to form (hand tools)
Wood products (particle board, pressboard, hardboard, etc.)	• Wood products impregnated with resins • Very dense • Available in sheets

Metals

The use of metals may be appropriate for prototype construction. In certain industries, such as the aerospace and automotive industries, prototyping includes fabrication of full-scale functional components in target materials. Such prototyping may require production of tooling as well as the prototypes and require the use of metals.

For smaller scale consumer products, off-the-shelf components are frequently used when metal subsystems are needed in prototypes. In some cases, it is necessary to fabricate metal components from raw materials. A general categorization of metals is listed Table 17.4 to aid in decisions when raw stock is needed.

Adhesives

Adhesives provide a useful means of fastening items within a prototype. Although adhesives reduce the flexibility in adjusting and reworking a prototype, they save time and construction operations as a fastening technology. To determine the choice of adhesive, one must consider the types of materials being joined in addition to the characteristics listed in Table 17.5.

TABLE 17.3.   PROTOTYPING PROPERTIES OF PLASTICS

ABS	• Good for structural applications • Thermoformable • Bonds well • High impact strength, good fracture resistance • Expensive • Available in sheets and rods
Polymethyl methacrylate (Plexiglas™)	• Excellent optical clarity • Good impact strength and durability • Easy to tool • Bonds with cyanoacrylate (super glue)
Polycarbonate (Lexan®)	• High impact strength • High tensile strength • Transparent in thickness up to 1/2″ • Available in colors • Available in sheets, rods, tubes, and film
Polystyrene	• Relatively brittle • Easy to mold • Useful in sheets: 1/64″, 1/32″, 1/16″, 1/8″, and –″ thickness • Tubing, strips, sheets, and extruded shapes are available
Acrylic	• Useful in sheets • Transparent applications • Fairly brittle
Nylon	• Low coefficient of friction • High strength • FDA approved for medical and food-processing applications • Available in sheets, bars, rods, and tubes (bushings)
Delrin	• Machines well • Durable, excellent toughness • Resists many solvents • Does not bond well • Available in sheets, rods, bars, and strips
Teflon™	• Low coefficient of friction • High impact resistance • FDA approved for medical and food-processing applications • Available in sheets and rods
Ultra high molecular weight polyethylene (UHMW)	• Low coefficient of friction • "Poor person's Teflon™" • Available in sheets, rods, and tubes
Renwood	• Wood-like material • Easy to work with
Hard-urethane foam (Last-A-Foam)	• Very easy to work with • Good for dimensional mockups and layout planning • Available from General Plastics, Tacoma, Wash.
Bondo	• Epoxy-like material used in automotive bodywork • Good for filling gaps and providing smooth contours over most materials • Widely available
Phenolic (Garolite)	• Inexpensive, durable • Nice finish • Higher temperature capability

TABLE 17.4. **PROTOTYPING PROPERTIES OF METALS**

Aluminum	• Machines well
	• High strength-to-weight ratio
	• Corrosion resistant
	• Difficult to weld
	• Available in variety of alloys, heat treatments, and shapes
Brass	• Machines well (self-lubricating)
	• Can be brazed
	• Corrosion resistant
	• Relatively expensive
	• Available in sheets, rods, bars, and extrusions
Steel	• High strength
	• Welds easily
	• Available in extrusions (angles, etc.) and bar stock
Lead	• Very dense
	• Low melting temperature
	• Easy to mold
	• Toxic—care must be taken to dispose of properly
	• Available in sheets, bricks, shot, discs, and ingots

Other Materials

Beyond these basic classes of materials, there are also others used for structural and mechanical functions. Table 17.6 lists some different materials that one should consider using in prototype construction.

Prototyping Processes

Numerical models enable the "experimentation" of products without trail-and-error fabrication. Likewise, computer solid modeling and rendering have enabled the visualization and error checking of product geometry with the ability to make quick changes. Nevertheless, the production of a "hands-on-product" is essential. Prototyping fabrication processes make this step possible.

A number of fabrication processes exist for producing prototypes. They range from handworking techniques (hand tools, knives, sanding blocks, and small power tools) to advanced precision equipment. Examples of advanced fabrication processes include:

▶ epoxy molds

▶ CNC machining

▶ cast metal molds

▶ machined aluminum molds

▶ injection molding

TABLE 17.5.   PROTOTYPING PROPERTIES OF ADHESIVES

Cyanoacrylate (super glue)	• Quick drying • Joins almost anything, although will not work on polypropylene • Discolors clear plastics • Activators (spray or applicator brush) will result in immediate setup • Contains harmful volatile organic compounds
Wood/white glue	• Versatile • Strong joint when impregnates materials and clamped • Requires long setup time • Good as a secondary fastening method
Hot-melt glue	• Convenient and versatile • Low strength • Fast setup
Epoxy	• Base material and a catalyst, mixed before use • Very strong joint • Sets relatively quickly • Contains harmful volatile organic compounds
Methyl chloride	• Solvent to join plastics • Works well with ABS and polystyrene
Trichloroethane (contact cement)	• Convenient and versatile, joins most anything • 30 minutes drying time • Contains harmful volatile organic compounds
Silicone (Shoe Goop™, RTV)	• Translucent clear or colored • Convenient and versatile • Weak joints • Will bond with fabric and polypropylene • Contains harmful volatile organic compounds
SprayMount	• Adhesive in a spray can • Useful for adhering paper or plastic labels, lettering • Contains harmful volatile organic compounds

- ▶ vacuum forming
- ▶ silicon rubber (RTV) molds
- ▶ electronic breadboarding
- ▶ mechanical breadboarding
- ▶ rapid prototyping

These processes have been born from traditional manufacturing techniques or model-making technologies. In addition to these processes, a new field of prototyping processes has been applied, with increasing fervor, since the mid-1980s. This field is known as *rapid prototyping*, also referred to as *solid freeform fabrication*.

TABLE 17.6. PROTOTYPING PROPERTIES OF MISCELLANEOUS MATERIALS

Modeling clay	• Easy to work with hands and hand tools • Useful for visualization and air flow studies • Always remains soft • Available in hobby and craft shops
Machining wax	• Machines well • Useful for patterns for prototyping tooling
Foam, foam board	• Easily carved • Good finish • Useful for painting (aesthetic/appearance models) • Machinable • Pressurized cans of insulating foam available that harden quickly and may be cut and formed with a knife and sanding board
Foam core	• Sheets of hard paper with internal foam • Useful for mockups and layout of square objects • Can be used with bondo/clay for more complex shapes • More durable and rigid than cardboard
Rubber, elastomers	• Useful in energy absorption applications or seals • Can be used as a removable molds for castings of other materials • Can be carved
Fiberglas™	• Good strength • Low weight • Variety of shapes • Requires lay-up process and drying • Solarez™ as a filler compound dries in minutes of direct sunlight • Easily finished
Cardboard, paper, cloth	• Use foam core instead of cardboard • Cloth and paper can serve as joints in mockups • Very cheap

V. RAPID PROTOTYPING TECHNIQUES

Solid freeform fabrication (SFF) is a categorization of processes that produce 3-D shapes by forming the geometry in successively additive steps. Virtually no part-specific tooling is required for SFF processes. Instead, a part is produced from a 3-D computer model (solid model) of the part. This model is sliced into 2-D layers and consecutively fabricated from the first layer to the last, using control schemes to direct the shaping of each layer. Figure 17.9 illustrates this general process, from computer model to final part.

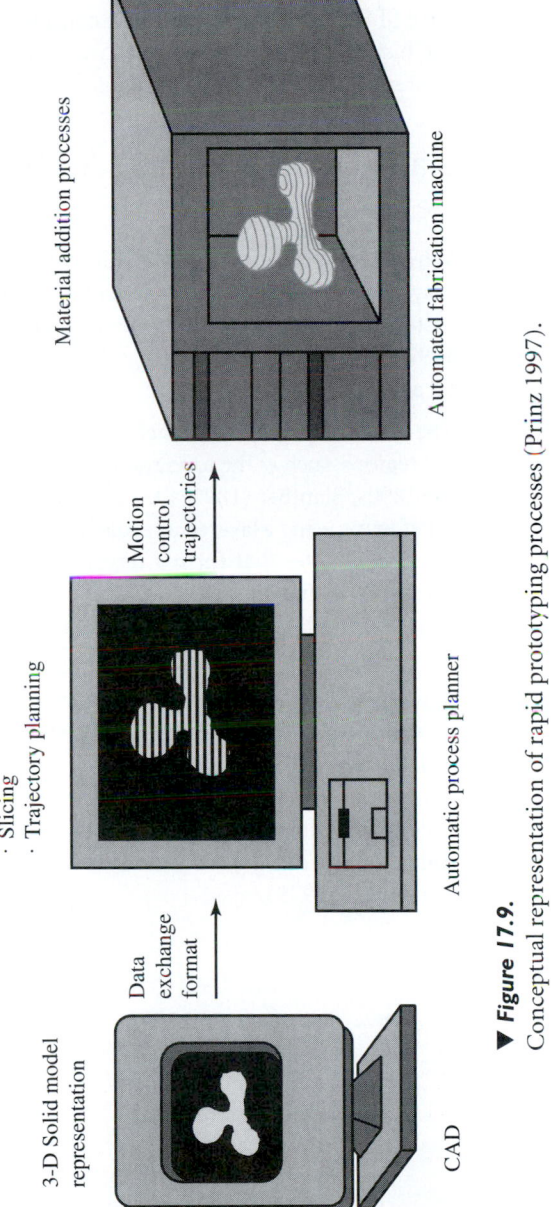

▼ *Figure 17.9.*
Conceptual representation of rapid prototyping processes (Prinz 1997).

A number of commercial processes exist for fabricating prototypes with SFF. Before discussing these commercial processes, let's consider the history of this technological area, especially as it has grown into a contemporary product domain.

Rapid Prototyping: A Historical Perspective

While rapid prototyping is a relatively new area of technology, its roots may be traced to the nineteenth century. Beaman has closely studied the historical development from its roots to today's applications (Prinz 1997). According to his findings, SFF can be traced to two fundamental areas of technology: topography and photosculpture.

The field of *topography* is concerned with creating contour maps of spatial regions, such as the topography of landmasses on Earth's surface. In the 1890s, Blanther (1892) filed a patent to make molds for topological relief maps using a layered approach. The essence of the method was to stack wax plates that form a terrain corresponding to contour lines. These plates then form a mold from which raised contour maps could be produced. Figure 17.10 illustrates the original patent by Blanther.

This original concept from topography resulted in similar ideas, both in map producing methods and visual aid devices (Perera 1940; Zang 1964; Gaskin 1973). It also led (directly or indirectly) to a number of innovations in manufacturing technology. Examples ranged from casting molds (Beaman 1996) to surface fabrication methods utilized for producing complex shapes, such as propellers, airfoils, 3-D cams, and forming of dies for punch presses (DiMatteo 1976; Nakagawa et al. 1979).

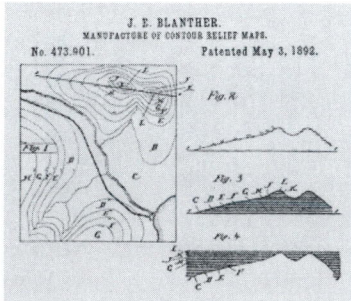

▼ *Figure 17.10.*
Layered map concept developed by Blanther (1892; Prinz 1997).

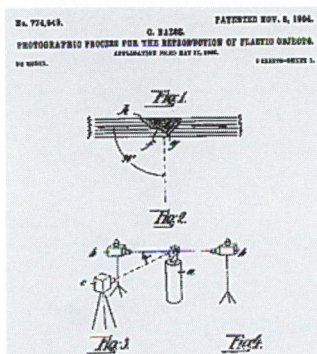

▼ *Figure 17.11.*
Photographic process for creating 3-D sculptures (Baese 1904).

Another field that influenced contemporary rapid prototyping is known as *photosculpture.* Photosculpture actually arose back in the nineteenth century in an attempt to create exact 3-D replicas of objects, including human forms (Bogart 1979; Prinz 1997). Innovative photographic techniques were used to first create hand carvings of subjects (Bogart 1979), followed by less labor-intensive approaches based on exposing photosensitive gelatin to graduated light sources (Baese 1904). Figure 17.11 illustrates one of these early concepts.

These concepts led to a number of improvements since their origination in the late nineteenth and early twentieth centuries (Monteah 1924). They also led to photographic methods for creating stacked shapes or original carvings from a variety of materials (Morioka 1944) and ultimately led to methods that mirror rapid prototyping processes currently in production today. One such mirror process was developed by Munz and is shown in Figure 17.12 (Beaman 1996). This process used scanned cross sections of an object and photo emulsion to form a layered version of the object. Photochemical etching was then used to remove sacrificial material from the layering process.

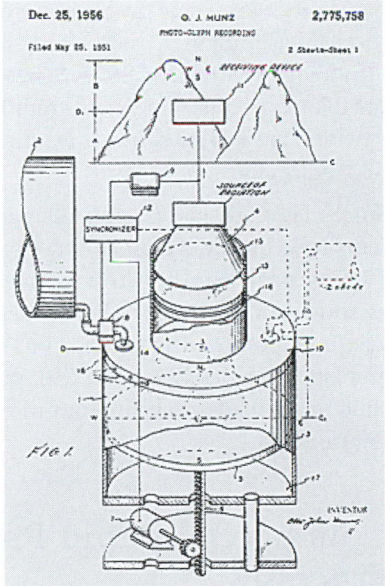

▼ *Figure 17.12.*
Process for forming a 3-D object from photo emulsion and subsequent photochemical etching (Munz 1956).

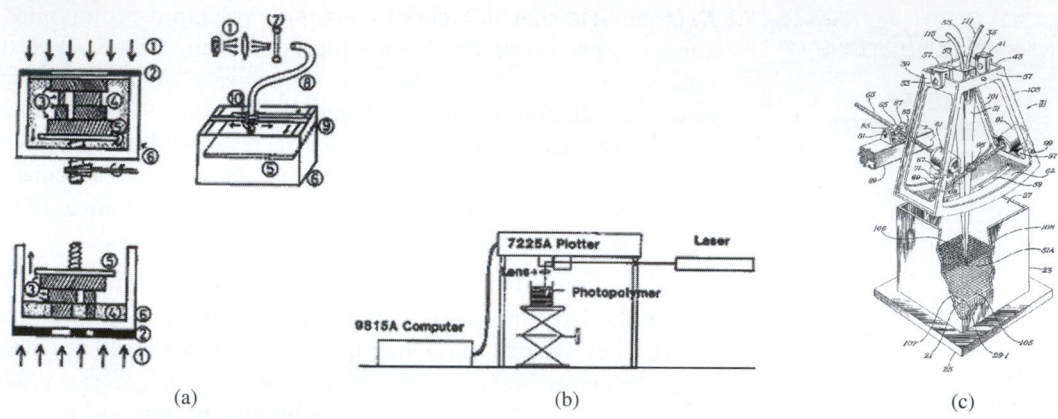

(a)　　　　　　　　　　(b)　　　　　　　　　　(c)

▼ Figure 17.13.
Early SFF concepts. (a) Kodma's (1981) three photopolymer systems. (b) Herbert's (1982) photopolymer process. (c) Householder's (1979) grid mold process.

Since the early developments of topography and photosculpture, a number of concepts were created for fabricating 3-D objects from layer-based processes. These processes were founded on photochemical, photosensitive polymer, and laser-melted particle or solid-layer technologies (Swainson 1977; Schwerzel 1984; Ciraud 1972). Following these concepts, more viable rapid prototyping techniques began to emerge. Figure 17.13 shows three such techniques (Kodama 1981; Herbert 1982; Housholder 1979).

Since these early viable concepts, a number of rapid prototyping systems have been developed and commercialized, beginning around 1986. Beaman lists the active patents covering these existing processes as shown in Table 17.7 (Prinz 1997). He also shows the U.S. development of existing processes as well as parallel efforts in Europe and Japan (Tables 17.8 and 17.9). Within a complete historical snapshot, this provides a chronology of rapid prototyping as illustrated in Figure 17.14.

Commercial Rapid Prototyping Processes

In the product development process, it is vital to become proficient at rapidly prototyping product concepts. Commercialized rapid prototyping processes, developed in the last decade, have helped accelerate development cycle times and productivity. In fact, rapid prototyping has

**TABLE 17.7.   CURRENT PATENTS IN THE RAPID PROTOTYPING FIELD
(BEAMAN 1996, PRINZ 1997)**

Name	Title	Filed	Country
Householder	Molding process	December 1979	U.S.
Murutani	Optical molding method	May 1984	Japan
Masters	Computer automated manufacturing process and system	July 1984	U.S.
André et al.	Apparatus for making a model of an industrial part	July 1984	France
Hull	Apparatus for making three-dimensional objects by sterolithography	August 1984	U.S.
Pomerantz et al.	Three-dimensional mapping and modeling apparatus	June 1986	Israel
Feygin	Apparatus and method for forming an integral object from laminations	June 1986	U.S.
Deckard	Method and apparatus for producing parts by selective sintering	October 1986	U.S.
Fudim	Method and apparatus for producing three-dimensional objects by photo-solidification, radiating an uncured photo polymer	February 1987	U.S.
Arcella et al.	Casting shapes	March 1987	U.S.
Crump	Apparatus and method for creating three dimensional objects	October 1989	U.S.
Helinski	Method and means for constructing three dimensional articles by particle desposition.	November 1989	U.S.
Marcus	Gas phase selective beam deposition; three-dimensional, computer controlled	December 1989	U.S.
Sachs et al.	Three-dimensional printing	December 1989	U.S.
Levant et al.	Method and apparatus for fabricating three-dimensional articles by thermal spray deposition	December 1990	U.S.
Penn	System, method, and process for making three-dimensional objects	June 1992	U.S.

**TABLE 17.8.   RAPID PROTOTYPE DEVELOPMENTS IN THE UNITED STATES
(BEAMAN 1997; PRINZ 1997)**

Company	Process	Venture start	Shipment	Notes
Aaroflex	Stereo lithography	1995	n/a	License from DuPont
BPM	Ink jet	1989	1995	
DTM	Selective laser sintering	1987	1992	Operated a service bureau from 1990–93
DuPont Somos	Stereo lithography	1987	n/a	Licensed to Teijin Seiki 1991, Aaroflex 1995
Helisys	Laminated object	1985	1991	Founded as Hydronetics
Light Sculpting	Photomasking	1986	n/a	Operates as a service bureau
Quadrax	Stereo lithography	1990	1990	Technology acquired by 3D in 1992
Sanders Prototyping	Ink jet	1994	1994	Partially developed at E-Systems
Soligen	3D printing	1991	1993	Operates as a service bureau
Sratasys	Fused deposition	1988	1991	
3D Systems	Stereo lithography	1986	1988	First commercial shipment equipment

**TABLE 17.9. RAPID PROTOTYPE DEVELOPMENTS IN EUROPE AND JAPAN
(BEAMAN 1997; PRINZ 1997)**

Company	Process	Venture start	Shipment	Notes
CMET	Stereolithography	1988, Japan	1990	
Cubital	Photomasking	1987, Japan	1991	
Denken	Stereolithography	1985, Japan	1993	
DMEC	Stereolithography	1990, Japan	1990	
EOS	Stereolithography, Selective Laser Sintering	1989, Germany	1990	
Fockele & Swarze	Stereolithography	1991, Germany	1994	Service bureau since 1992
Kira	Laminated Object	1992, Japan	1994	
Meiko	Stereolithography	1991, Japan	1994	
Mitsui	Stereolithography	1991, Japan	1991	
Sparx	Laminated Object	Sweden	1994	Foam machine
Teijin Seiki	Stereolithography	1991, Japan	1992	License from DuPont
Ushio	Stereolithography	Japan	1994	

become a significant base of manufacturing technology in many nations, particularly the United States. According to Wohlers (1996), rapid prototyping is growing quickly as a technological sector, accounting for approximately one-half billion dollars of revenue.

In this context, rapid prototyping doesn't replace other, more traditional prototyping processes, but it does add flexibility in the types of

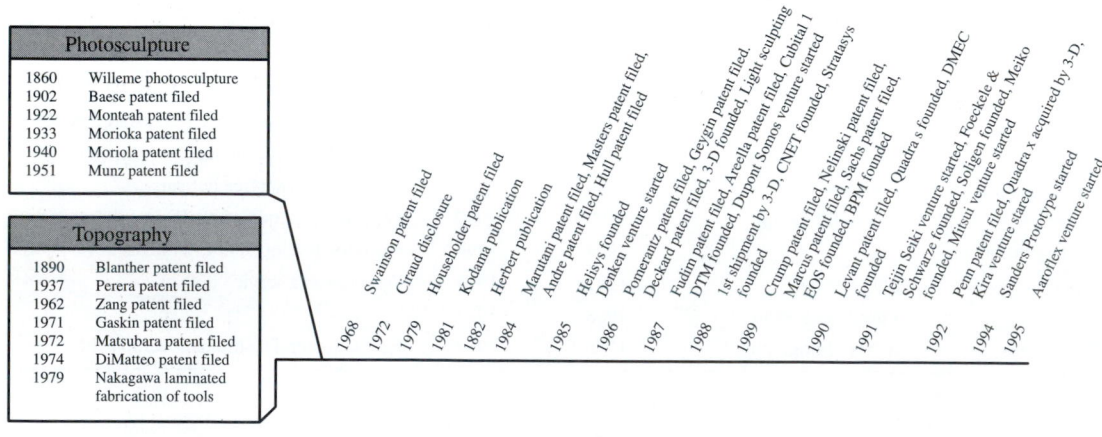

▼ *Figure 17.14.*
Chronology of rapid prototype development from its root technologies (Prinz 1997).

prototypes and provides alternative processes for fulfilling the known uses of prototypes, including:

▶ visualizing concepts as physical entities

▶ market research for ergonomic use and aesthetics

▶ prototypes for functional testing

▶ assembly and manufacturing feasibility

▶ molds for castings

▶ verification of design changes

▶ cost analysis

▶ early marketing promotions

Provided that material properties, surface finish, and other prototype needs are satisfied, rapid prototyping offers these uses with increased speed and lower costs.

A variety of commercial rapid prototyping processes currently exist to add increased productivity in product development. These processes may be categorized as laser fusion, stereolithography, lamination, extrusion, and inkjet printing (Prinz 1997). Figure 17.15 lists the commercialized processes from the United States, Europe, and Japan (Prinz 1997).

Laser fusion processes use lasers to fuse powder layers, creating the final part cross section in each layer. A well-known process is *selective laser sintering,* created at the University of Texas and commercialized by Desktop Manufacturing (DTM) Corporation. Selective laser sintering starts with a thin, evenly distributed layer of powder. A laser is then used to sinter only the powder that is inside a cross section of the part. The energy added by the laser heats the powder into a melted state, and individual particles coalesce into a solid. Once the laser has scanned the entire cross section, another layer of powder is spread on top of the part, and the entire process is repeated.

Based on the process description shown in Figure 17.16, laser fusion processes are able to create prototypes in polymer (plastics), wax, and polymer-coated metal materials. Example parts produced from the DTM process are shown in Figure 17.17.

Alternatively, stereolithography processes use light or laser sources to solidify photosensitive resins into solid shapes. A popular stereolithography process, the first to enter the market as a rapid prototyping technology, is SLA, manufactured by 3-D systems. In this process, a laser selectively scans a vat of photochemical polymer, solidifying the scanned regions into a solid structure. The structure is then lowered into the vat and the next layer scanned. Figure 17.18 depicts the SLA process.

Manufacturer	Process Name	Process Type	Materials
United States			
3-D Systems	Stereolithography Apparatus (SLA)	Laser Photolithography	acrylate, epoxy
Helisys	Laminated Object Manufacturing (LOM)	Lamination, Laser Cut	paper, tape castings
Stratasys	Fused Deposition Modeling (FDM)	extrusion	U.S.
DTM	Selective Laser Sintering (SLS)	power based, laser fusion	Nylon, wax, polycarbonate, Polymer-coated metal
Sanders Prototype	Model Maker	liquid jetting	low-melt plastic
Soligen	Direct Shell Production Casting (DSPC)	Powder-based, 3-D printing of binder	ceramics
BPM	Ballistic Particle Manufacturing (BPM)	liquid jetting	low-melt plastic
3D Systems	Multi-Jet Modeling	liquid jetting	wax
Japan			
CMET (NTT Data Communications)	Solid Object Ultraviolet Plotter (SOUP)	Laser Photolithography	epoxy
D-MEC (JSR/Sony)	Sony's Solid Creation System (SCS)	Laser Photolithography	urethane acrylate
Kira Corp.	Solid Center	lamination, knife-cut	paper
Teijin Seiki	Solid Forming System (Soliforming)	Laser Photolithography	urethane acrylate, glass-filled resin
Denken Engineering	Solid Laser Plotter (SLP)	Laser Photolithography	acrylate
Meiko Corp.	Meiko	Laser Photolithography	acrylate
Mitsui Zosen	COLAMM	Laser Photolithography	
Ushio, Inc.	Uni-Rapid	Laser Photolithography	
Europe			
EOS (Germany)	STEREOS	Laser Photolithography	acrylate, epoxy
EOS (Germany)	EOSINT	power based, laser fusion	Nylon, wax, polycarbonate, Polymer-coated metal
Cubital (Germay/Israel)	Solid Ground Curing (SGC)	photomasking	acrylate, wax
Fockele & Schwarze (Germany)	LMS	Laser Photolithography	

▼ *Figure 17.15.*
Commercialized rapid prototyping processes (Prinz 1997).

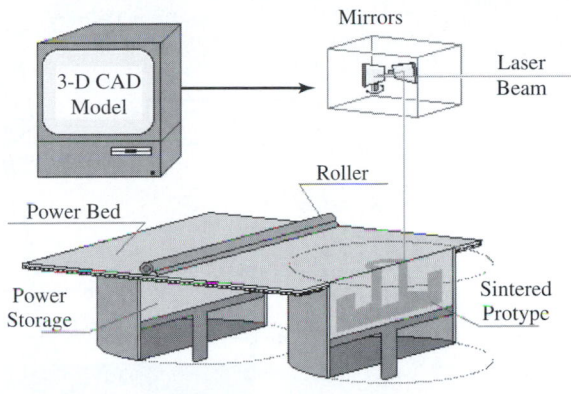

UT Austin, 1995

▼ *Figure 17.16.*
Example laser fusion process diagram (selective laser
sintering) (Beaman et al. 1997).

Besides laser fusion and stereolithography, rapid prototyping processes may be categorized as lamination, extrusion, and inkjet printing
(Prinz 1997). Lamination processes build shapes with paper or plastic
by "attaching" layers of the material with an adhesive through a heating process. A laser or knife is then used to crosshatch the part and remove material outside the cross section. Extrusion processes, such as the
Fused Deposition Process developed by Stratasys, Inc., deposit materials from a heated extrusion head. The deposited material forms layers
of the part being prototyped. Finally, a number of rapid prototyping
processes implement technology similar to inkjet printing. In these

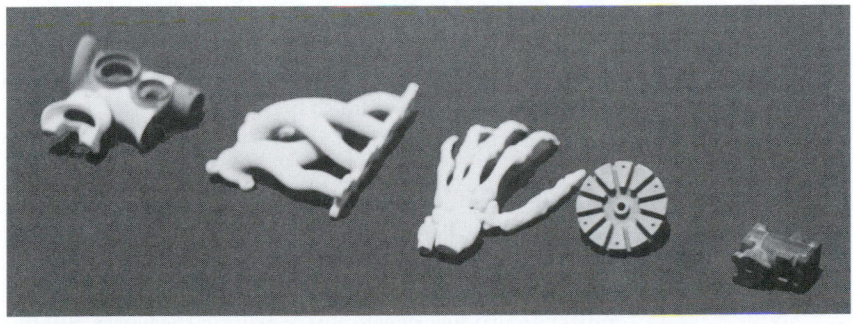

▼ *Figure 17.17.*
Example parts created using the DTM selective laser sintering process.

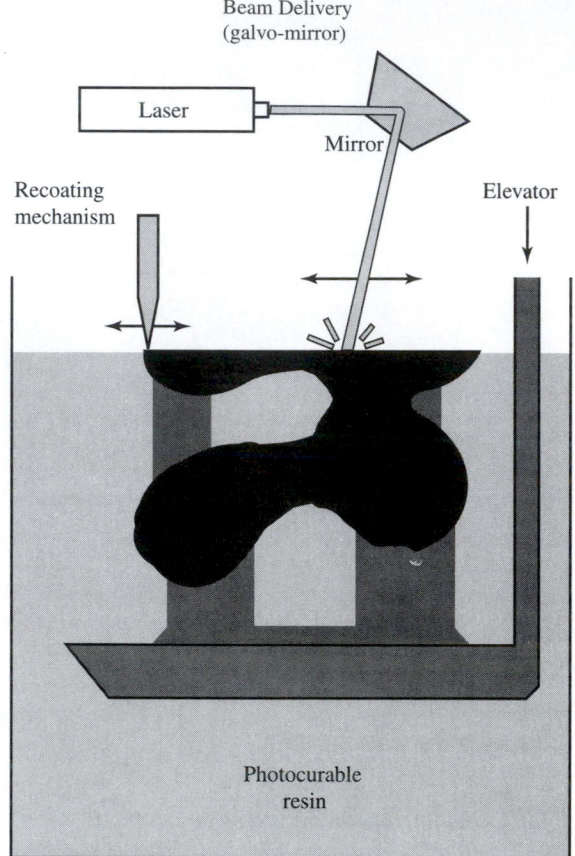

▼ *Figure 17.18.*
Stereolithography process (Prinz 1997; Beaman et al 1997).

processes, an inkjet head prints layers of material in a bin or platform. An example is 3-D printing developed at MIT (Sachs et al. 1995). In this process, a powder of material is selectively injected with a binder to fuse the material into solid shapes.

Choosing Rapid Prototyping for a Product?

Commercialized rapid prototyping processes have had a significant impact on product development around the world. But two common-sense questions must be pondered when considering rapid prototyping. First, should a rapid prototyping process be chosen over other prototyping technologies for a given type of prototype? Second, which rapid prototyping process should be chosen?

A number of criteria must be evaluated when pondering the first question. These criteria depend on the product application and may be listed as:

- cost trade-off
- cycle time (and speed of fabrication)
- accuracy of the prototyped parts, including tolerances and surface finish
- material properties (type of materials)
- part size
- part strength

These criteria must be carefully judged, predicted, and traded off between competing technologies. Business cases (Chapter 3) and decision processes (Chapter 11) should be implemented to make sound choices for each prototype produced.

Values for the criteria will be a function of the particular product being developed. But they will also be a function of in-house model making versus out-sourcing prototype builds. If in-house model making is not available, a number of service bureaus exist for prototyping parts with traditional and rapid prototyping processes (lists of bureaus are available at, for example, http://www.geocities.com/). Production data may be obtained from these service bureaus as a function of their expertise and available processes.

The second question, which rapid prototyping process is preferred, also depends on the criteria listed above. When considering the purchase of a rapid prototyping process, a number of other factors must be added to this list. These factors include the complexity of the system; size, weight, and portability of the system; operating environment, office versus shop floors; operating paraphernalia; material waste; production of CAD data; and personnel requirements (Wohlers 1992).

Overall, rapid prototyping processes will continue to impact product development. It is the industrial trend to verify and adjust product performance and quality through rapid physical testing. Rapid prototyping usually requires the least cost and time for the fabrication of a single part, but rapid prototyping materials are currently too limited to perform diverse functional tests. Many additional processes are currently in research and promise to appear as new technologies in the near future. The variety of materials and process capabilities are also evolving at an alarming rate. Complex heterogeneous part structures, implementing multiple materials for function testing, are only a matter of time.

VI. SCALE, DIMENSIONAL ANALYSIS, AND SIMILITUDE

While prototype materials and processes are important factors in creating physical prototypes, an equally important factor is the scale of the model, such as the scale of size, shape, materials, loading conditions, and operating conditions. What scale should we choose to make our prototype? In terms of size, three direct options exist: actual scale; larger scale (e.g., the kinematics and operation of small mechanisms may be more easily tested with larger scale models [at reduced cost] and smaller scale (e.g., space, cost, or processing capabilities may prohibit the testing of an actual scale mode). After choosing the scale of a prototype, similarity methods provide techniques for relating the results of scaled models to the actual product scale.

The objective of a *similarity method* is to experimentally predict the behavior of a target system (the product) through an indirect scaled testing, alleviating complex system construction and testing effort. Traditionally, a scaled physical model that can show similar behavior is designed on the basis of *dimensional analysis,* and the behavior of the target system is estimated from mathematically derived scaling laws (Kline 1975; Baker et al. 1992). In some special cases, more expensive scaled-up models are prepared. However, it is typical to geometrically scale down, change materials, and/or simplify models (Fay 1993) to reduce the physical modeling effort. In the context of product design, scale models and similarity methods may accelerate design processes, if they can yield a reliable prediction.

The underlying reasons to apply scaled models are to provide a means of significantly reducing the effort to build empirical models by reducing the number of required experiments (dimensional analysis), to design a scaled system, and to predict the behavior of the system of interest by testing this scaled system. This application is known as similarity, the similitude method, or comparative dimensional analysis (emphasizing two systems). The fundamentals of the Buckingham Π theorem are mathematically described here by highlighting comparative dimensional analysis.

Buckingham Π Theorem and Scaled Testing

The Buckingham Π theorem-based similarity method (Kline 1965; Baker et al 1991) provides a method to design scaled physical models and to predict the behavior of target systems (full-scale systems) by providing scaling laws. To properly predict the full-scale system behavior with this traditional similarity method, it is very important to be aware of the implicit assumptions in the Π theorem. For this purpose, the fundamental mathematical description of the theorem is provided prior to the details of the method.

In general, the state of interest x (dependent variable) can be represented as:

$$x = f(p_1, p_2, \cdots, p_n) \tag{17.1}$$

where the details of the function f are not well known, and p_i are system parameters, viz. material constants, geometric parameters, boundary conditions, and independent time and spatial variables. Consider two systems Σ_A and Σ_B, which can be represented as:

$$x^A = f(p_1^A, p_2^A, \cdots, p_n^A),$$
$$x^B = f(p_1^B, p_2^B, \cdots, p_n^B) \tag{17.2}$$

where the superscripts A and B represent the system Σ_A and Σ_B, respectively. These systems can also be equivalently represented as:

$$\pi_x^A = \phi(\pi_1^A, \pi_2^A, \cdots, \pi_N^A),$$
$$\pi_x^B = \phi(\pi_1^B, \pi_2^B, \cdots, \pi_N^B) \tag{17.3}$$

where $\pi_i (i = x, \text{ and } 1, 2, \ldots, N)$ are dimensionless parameters that are related to the $p_k (k = 1, 2, \ldots, N)$. As $n > N$, Eq. (17.3) can be considered as a compact representation of the original system equation (Eq. 17.2). If $\pi_i^A = \pi_i^B$ for any $i = 1, 2, \ldots, N$, then $\pi_x^A = \pi_x^B$ from Eq. (17.3). As a result, one can predict x^A from the following *prediction equation*:

$$x^A = x^B \cdot \prod_{k=1}^{n} \left(\frac{p_k^B}{p_k^A} \right)^{r_k} \tag{17.4}$$

where r_k is a rational number that can be decided from π^A (or π^B).

In summary, one should notice that the traditional similarity method implicitly assumes that full-scaled and scaled-system states are related to system parameters through an identical functional form as shown in Eq. (17.2). In addition to the identity of the governing equation, the scaled models should be designed to make all parameters dimensionless (excluding the one variable that includes the state variable). The overall procedure of the method is shown in Figure 17.19, and the details are described through a product subsystem example.

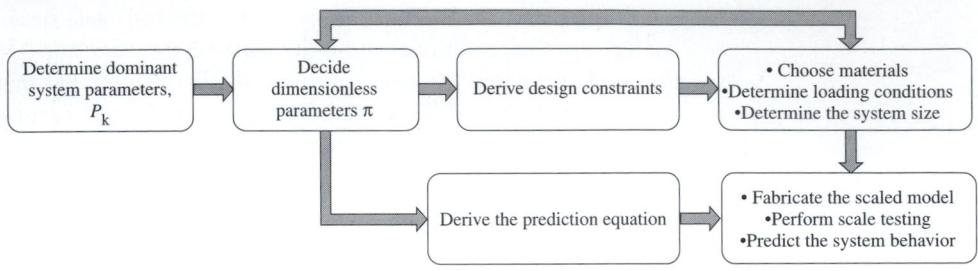

▼ Figure 17.19.
Overall procedure of the similarity method.

Buckle Design Example

Let's consider the design of a quick-connect buckle as part of a product. The stress level at critical points and the reaction force to release the buckle need to be modeled in designing the buckle. The procedure to experimentally derive this information through the traditional similarity method is explained in detail.

The buckle of interest is illustrated in Figure 17.20. The horizontal force F_x to release the buckle and the stress levels of several critical points (A, B, C, and D) under maximum deformation are the states of interest. The buckle should be designed so that the force F_x is within the customer's comfort zone and the maximum stress level within the elastic range. In this example, we illustrate the procedure to correlate an aluminum buckle and a nylon one (material parameters); however, one can follow the same procedure for other scaling (e.g., geometrical scaling) problems. The detailed steps are as followed.

Step 1: Select dominant system parameters: The dominant system parameters of the buckle problem are the Young's modulus $E[\mathrm{FL}^{-2}]$, characteristic length $L_c[\mathrm{L}]$, resulting stress $\sigma[\mathrm{FL}^{-2}]$, deflection $d[\mathrm{L}]$, Poisson's ratio $\nu[\ ]$, and the force $F_x[\mathrm{F}]$ required to deform the buckle tabs.

Step 2: Find dimensionless Π groups. The Π groups can be efficiently determined using the echelon matrix procedure (Barr 1984). The echelon matrix procedure transforms an initial dimensional matrix into a simpler form by changing the fundamental (base) dimensions. The dimensional matrix of the buckle design problem can be represented in the following form. In this procedure, the matrix elements are the power terms of fundamental units of a system parameter:

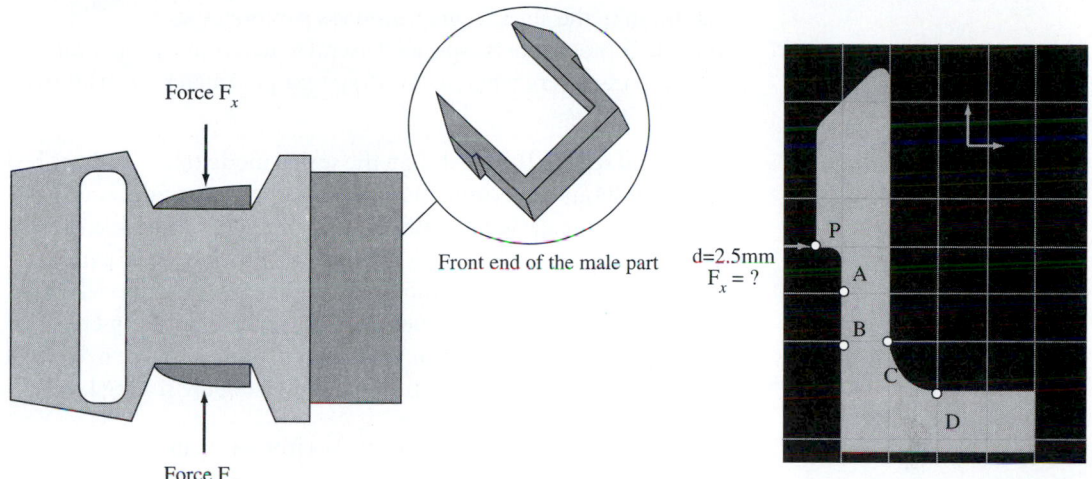

Grid size: 5 by 5 mm

▼ **Figure 17.20.**
Schematic and details of the side buckle release.

	$E[\text{FL}^{-2}]$	$d[\text{L}]$	$L_c[\text{L}]$	$\sigma[\text{FL}^{-2}]$	$F_x[\text{F}]$
F	1	0	0	1	1
L	−2	1	1	−2	0

If one takes the dimensions of E and d as the base dimensions instead of the fundamental dimensions [F] and [L], the dimension of the parameters can be represented as follows in terms of the new base dimension:

	$E[\text{FL}^{-2}]$	$d[\text{L}]$	$L_c[\text{L}]$	$\sigma[\text{FL}^{-2}]$	$F_x[\text{F}]$
$(E)[\text{FL}^{-2}]$	1	0	0	1	1
$(d)[\text{L}]$	0	1	1	0	2

Just as $[E] = [\text{FL}^{-2}]$, $[d] = [\text{L}]$, $[L_c] = [\text{L}]$, $[\sigma] = [\text{FL}^{-2}]$, and $[F_x] = [\text{F}]$ in the first matrix, we obtain the following reduced dimensional relations from the second matrix: $[\sigma] = [E]$, $[F_x] = [E\, d^2]$, and $[L_c] = [d]$ (where $[E] = [E]$ and $[d] = [d]$ are automatically satisfied). These three reduced relations form the following dimensionless parameters:

$$\pi_1 = \frac{L_c}{d}, \pi_2 = \frac{\sigma}{E}, \pi_3 = \frac{F_x}{E \cdot d^2},$$

in addition to the already dimensionless parameter $\pi_4 = \nu$. This echelon matrix procedure is especially useful to determine proper dimensionless parameters when a set of system parameters is arbitrarily selected.

Step 3: Find scaling laws to design the scaled model(s).   As described above, one should determine the system parameters of the scaled system, maintaining the identity of corresponding dimensionless parameters, π_i. Using the scale factor λ_p, defined as the ratio of the parameter p of the full-scale system (aluminum buckle) to the scaled model (nylon buckle), the constraints on scaled models (scaling laws) can be represented with a simple expression. As π_1, π_2, π_3, and π_4 of the nylon and aluminum buckles should be the same, following the scaling laws,

$$\lambda_d = \lambda_{L_c} \text{ (from } \pi_1\text{)}, \ \lambda_\sigma = \lambda_E \text{ (from } \pi_2\text{), and } \lambda_{F_x}$$
$$= \lambda_E \cdot \lambda_d^2 \text{ (from } \pi_3\text{)}, \ \lambda_\nu = 1 \text{ (from } \pi_4\text{)}$$

can be derived.

Step 4: Design a scaled model so that all of the above scaling laws are satisfied.   From the material properties given in Table 17.10, the scale factors are given by $\lambda_E = 73/0.5 = 146$, and $\lambda_\nu \approx 1$. The last scaling law is already satisfied, and the only restriction on the model is $\lambda_d = \lambda_{L_c}$. In this example, we consider the model with same geometrical scale, i.e. $\lambda_d = \lambda_{L_c} = 1$.

Step 5: Calculate the state of the full-scale system.   From the scaling law $\lambda_\sigma = \lambda_E$, the stress of the aluminum buckle is a multiple of 146 of that of the nylon. Similairly, the force scale is also a multiple of 146 from $\lambda_{F_x} = \lambda_E \cdot \lambda_d^2$.

Simulation Results

The y-directional normal stress plots (simulated with ANSYS™) for the nylon (scaled model) and aluminum buckles are shown in Figures 17.21 (a), (b). As expected, the contour plots are similar, and the ratio of the stress at any corresponding point is identical. The y-directional

TABLE 17.10.   MATERIAL PROPERTIES OF THE ALUMINUM AND NYLON

	Young's Modulus E (GN/m²)	Poisson's Ratio
Aluminum	73.0	0.33
Nylon	0.5	0.30

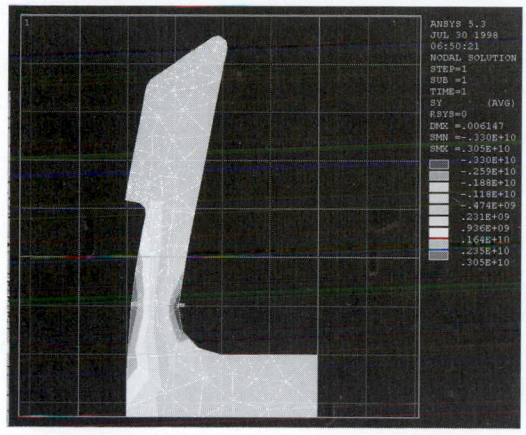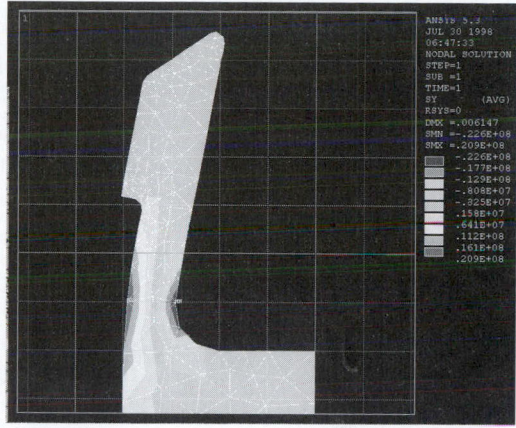

▼ *Figure 17.21.*
Normal stress of the nylon and aluminum buckle front end.

stress at the four critical points is listed in Table 17.11, and the derived stress scale matches well to the stress scale factor, $\lambda_\sigma = 146$. The simulated reaction force to fully deform the nylon and aluminum buckle is 8.1 N and 1.19 kN, respectively, and the force ratio also matches well to the derived force scale factor.

VII. BASIC METHOD: PHYSICAL PROTOTYPE DESIGN AND PLANNING

We have explored the basics of physical modeling, from model types and uses to prototype materials, processes, and scaling effects. With this information in hand, a basic method is needed to design and plan a physical prototype. The following eight-step procedure provides a skeleton checklist for systematic prototype creation, while not subsuming the creativity and innovation required by the product development team in prototype design.

TABLE 17.11. SIMULATED STRESS AT FOUR POINTS OF INTEREST (GN/M²)

	A	B	C	D
Nylon	1.55	3.0	−3.29	−1.44
Aluminum	0.0114	0.0212	0.0225	0.00986

Prototype design procedure:

1. Identify the purpose(s) of the prototype in the context of customer needs.

2. Document the functionality (function chains) for these customer needs (Chapter 5). Identify module interfaces (if applicable).

3. Answer the question, What physical principles are needed to understand possible experiments to be performed on a physical model?

4. Answer the questions, How would the physical model(s) be "measured?" Does the measurement(s) (metric[s]) directly relate to the customer needs and correspond to the engineering specifications (QFD, Chapter 7)?

5. Decide if the prototype will be focused or comprehensive, scaled or actual geometry, and produced from actual materials or not.

6. Answer the questions, For a physical prototype, could rapid prototyping technology be used? If so, which technology is appropriate? If not, what other fabrication methods and materials could be used?

7. Sketch alternative prototype concepts (based on the product concept); determine cost, appropriate scale, and alternative build plans; choose a preferred concept (Chapter 11); and develop a fabrication process plan.

8. Answer the questions, How will the prototype be tested? What factors will be controlled to minimize experimental error? What response(s) will be measured and with what sensors? How many tests will be conducted/replicated? Will the tests be destructive or nondestructive? What is the desired accuracy of the measurements (Chapter 6)?

Table 17.12 provides an example template for recording the prototype-planning information. This template should be augmented with sketches (or solid models) of the prototype design, in addition to a schedule for procurement, fabrication, and testing. In some cases, prototype fabrication will require a long lead time. For example, during the early 1990s, Ford Motor Company reported a 14-week average time to obtain prototype injection-molded dunnage (packaging and carrying crates) for certain automobiles.

Guidelines for Prototype Design

In conjunction with the prototype design procedure, a number of guidelines should be considered during prototype development, as shown in Table 17.13. These guidelines simply augment the checklist contained in the procedure. If followed (as appropriate), they can save time and prevent wasted resources.

TABLE 17.12.   TEMPLATE FOR DESIGNING AND PLANNING A PHYSICAL PROTOTYPE

Prototype Name. Define the Purpose and Type of Prototype	Virtual Prototype or Physical Prototype or combination? Focused or comprehensive?	Relationship to Customer Needs, Metrics, Physical Principles, and Functional Relationships	Prototype Materials, Scale, Accuracy, and Fabrication Processes?	Rapid Prototyping Technology?	Cost and Build Time?	Testing Approach? Control Factors, Noise Factors, and Response?

Sample Prototype Application

Shaken Baby Syndrome (SBS) is a disorder that is afflicted on young children. Typical ages of the children range from infancy to 2 years old. Injuries associated with SBS include cerebral and retinal hemorrhage, coma, and permanent brain damage. SBS often occurs when parents or caregivers are incapable of handling a persistently crying infant. In an effort to quiet the child, the baby is shaken, resulting in great harm, even death.

TABLE 17.13.   GUIDELINES DURING PROTOTYPE DEVELOPMENT

Judge the risk of not creating a prototype vs. cost and build time. If there exists minimal risk, proceed forward without the physical prototype.

If a prototype is needed, don't be tenuous in its fabrication, but move ahead with enthusiasm. Each prototype simply adds to a strong foundation for creating a successful product.

During proof-of-concept, prototype the difficult, trickiest components first. These undoubtedly will be the weakest links in the product concept.

Draw, sketch, or create a solid model of the prototype at an appropriate level of detail.

For electronic prototypes or subsystems, wire-wrapping and breadboarding work well.

Stick to basic machines.

Use off-the-shelf components when possible.

Remove features from the prototype that are not necessary for the tests.

Know the catalogs: Thomas Registry (http://www.thomasregister.com/), McMaster-Carr (http://www.mcmaster.com/), Edmund Scientific (http://www.edsci.com/), etc.

Hardware stores, novelty stores, and product catalogs are great sources for ideas and parts.

Scavenge parts from similar or analogous products when developing early prototypes.

Prototyping with molds, at low cost, is possible.

Rapid prototyping is not always rapid and is not always cheaper. Comparisons must be made with other technologies.

Dual prototypes at a given stage of the product development process are common if a singular prototype cannot be fabricated to satisfy multiple purposes. Be willing to create two or more simple prototypes instead of one that is overly complex.

In an effort to address these concerns, a Texas county health office created a program to educate people of the dangers associated with shaking a child. Skits, where young parents attempt to cope with a crying child, are a large part of the educational effort. To enhance the program, a training doll, the product, was desired. This product originally called for a number of global needs to be satisfied (as originally developed by Jeff Norrell, Department of Mechanical Engineering, The University of Texas at Austin), including approximating the weight of a 2–4-month-old infant, demonstrating physiological effects contributing to SBS, namely a proportionally weak neck and large head, and demonstrating damage typically seen in SBS, such as subdural hematomas.

During the development of this product, the design team developed a budget and basic business case, interviewed customers, created a functional model of the product, and developed engineering specifications. Using the results of these steps, the design team members then generated concepts, fabricated proof-of-concept models, chose a final concept, embodied the design, and fabricated an alpha prototype to be reviewed by the customers. Based on customer feedback, requested revisions were documented, and improvements were made to the product.

At this stage in the process, a beta prototype was needed to demonstrate the full functionality of the product. Applying the prototype development procedure, Table 17.14 shows the major decisions regarding the prototype layout and fabrication. Figure 17.22 then illustrates the concept sketches for the beta prototype.

Based on the prototype sketch and plan shown in Table 17.14 and Figure 17.22, a detailed fabrication plan is developed. An abbreviated plan is shown below, originally developed by Jeff Norrell at The University of Texas at Austin, to illustrate this step.

Neck and Weight Installation

The first step in construction is to procure an off-the-shelf, life like doll. In this case, the chosen doll is a La Baby® Softbody baby doll, as shown in Figure 17.23.

Next, we remove the head and all of the fiber stuffing from the cloth doll's body until we are left with the empty body with arms and legs attached, as shown in Figure 17.24.

Next, the long end of the blue hose is inserted into the doll's cloth body. Placing the original fiber batting around the tube, the cloth body is filled. Once the batting is firmly in place, the short end of the tube is sticking out of the cloth body's opening, as shown in Figure 17.25. This provides a hose attachment as an elastic attachment point.

TABLE 17.14.   SBS BETA PROTOTYPE PLAN

Prototype Name. Define the Purpose and Type of Prototype	Virtual Prototype or Physical Prototype or combination? Focused or comprehensive?	Relationship to Customer Needs, Metrics, Physical Principles and Functional Relationships	Prototype Materials, Scale, Accuracy and Fabrication Processes?
SBS Beta I; Purpose: Communication and Architectural Interfacing (feedback on fully functional product; testing of manufacturing and assembly)	Physical Prototype. Comprehensive, including all functional and aesthetic features.	Full set of customer needs. Primary function carriers: display injury, hide injury, distribute mass, allow dof, regulate rotation, support loads, display baby feel, etc. Metrics: weight, center of mass, distance to observe injury, durability, steps to set up, negative reaction by majority of sample size, etc.	Materials: Off-the-shelf, life-like doll; stuffing fabric and metal weights; wooden neck system; Lexan, transparent skull; fabric covering; molded polymer brain; painted clay hematomas; plastic connectors; etc. Scale: Actual. Accuracy: tolerances within 0.1-0.2". Processes: band saw, drill, hand tools (exacto knife, etc.), sand castings/thermal mold, hand clay forming/painting, etc.
Rapid Prototyping Technology?	*Cost and Build Time?*	*Testing Approach? Control Factors, Noise Factors, and Response?*	
No, unless brian mold or skull cover is intended for mass production.	Total: $XXX Breakdown: Materials: $YY Labor: $ZZZ Materials Procurement: 2 days. Build Time: 2 days.	Responses: Measurement of weight, center of mass with scale. Measure reaction to baby appearance, feel, shaking reaction, durability, and exposed injury with customer sample size. Control factors: weight, placement of mass, color and size of hematomas, etc. Approach: test with clients through all product activities. Test with multiple teenage parent groups.	

Molding of Clear Dome

Once the doll is appropriately weighted and the neck has been constructed, creation of the clear dome begins. This starts with the fabrication of the necessary simple molds. First, we fabricate the male mold by placing the rear portion of the doll's head into sand, as shown in Figure 17.26.

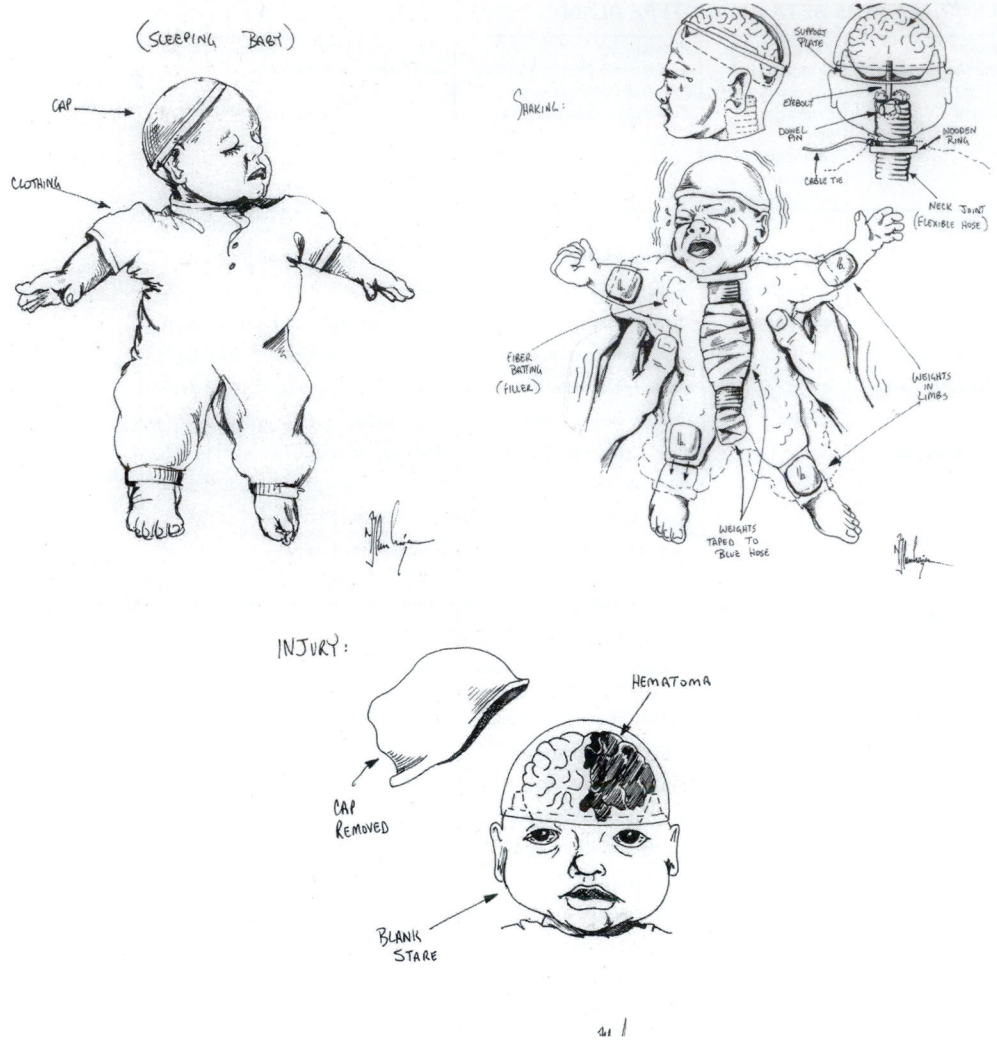

(SLEEPING BABY)

CAP

CLOTHING

SHAKING:

SUPPORT PLATE

EYEBOLT

DOWEL PIN

CABLE TIE

WOODEN RING

NECK JOINT (FLEXIBLE HOSE)

FIBER BATTING (FILLER)

WEIGHTS IN LIMBS

WEIGHTS TAPED TO BLUE HOSE

INJURY:

HEMATOMA

CAP REMOVED

BLANK STARE

▼ *Figure 17.22.*
SBS beta prototype sketches.

After carefully removing the head, a cavity is left in the sand, as shown in Figure 17.27.

Care is required not to push the doll's head into the sand too forcefully. This action can result in deformation of the rubber head, leaving a misshapen cavity. Once the cavity is formed, the cavity is filled with Plaster of Paris as shown in Figure 17.28.

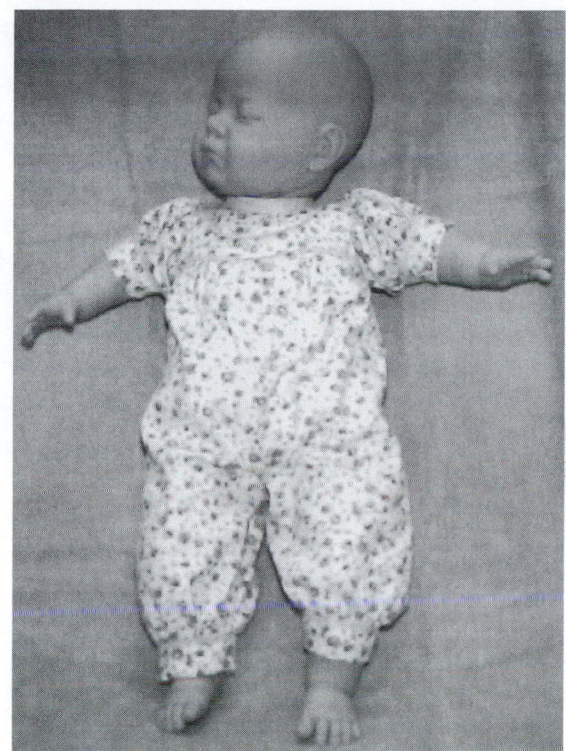

▼ *Figure 17.23.*
Doll just out of the box.

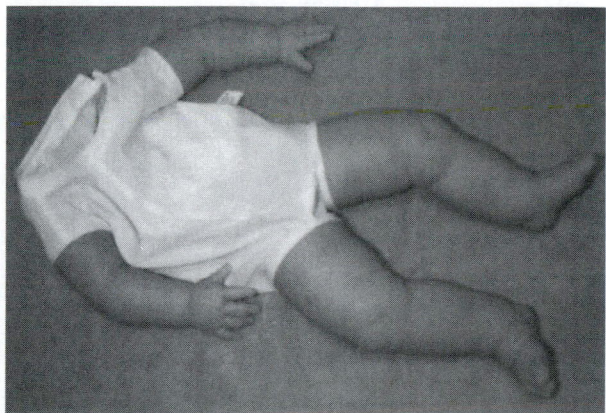

▼ *Figure 17.24.*
Doll's body after stuffing has been removed.

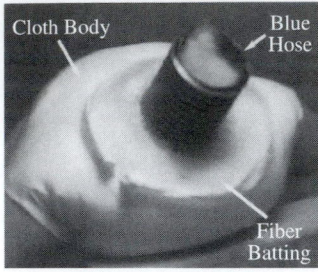

▼ *Figure 17.25.*
Neck assembly placed inside
doll's cloth body.

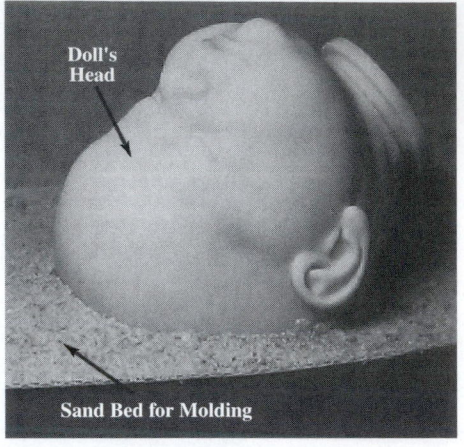

▼ *Figure 17.26.*
Doll's head placed in sand.

▼ *Figure 17.27.*
Cavity left behind after doll's head is
removed from the sand bed.

▼ *Figure 17.28.*
Male mold cavity filled with Plaster of Paris.

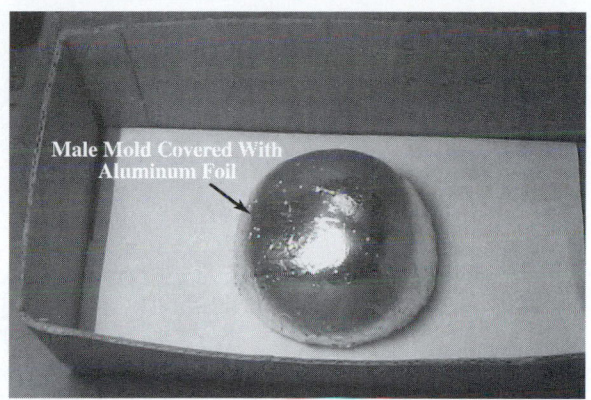

▼ *Figure 17.29.*
Male mold covered with aluminum foil and placed in the
bottom of a box.

Next, we need to fabricate the female mold. We start by covering the
male mold with aluminum foil and placing it in the bottom of a box,
as shown in Figure 17.29.

Then we pour enough Plaster of Paris into the box to cover the male
mold to a depth of at least 1″, as shown in Figure 17.30.

Once the plaster has cured, we remove the female mold from the box.
After removing the foil-covered male mold, we have a female mold, as
shown in Figure 17.31.

We next sand the inside of the mold cavity smooth, and the result is the
full set of molds shown in Figure 17.32, necessary to create a clear dome.

To continue the dome molding, we next need Lexan® disks that will become the transparent portion of the skull showing the hematoma on the
brain. We first create a template consisting of a cardboard circle 8″ in diameter. We draw an 8″ circle on the Lexan® and then cut out the disk.

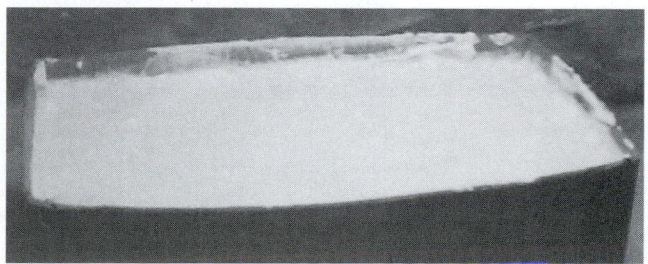

▼ *Figure 17.30.*
Box filled with Plaster of Paris to create the female mold.

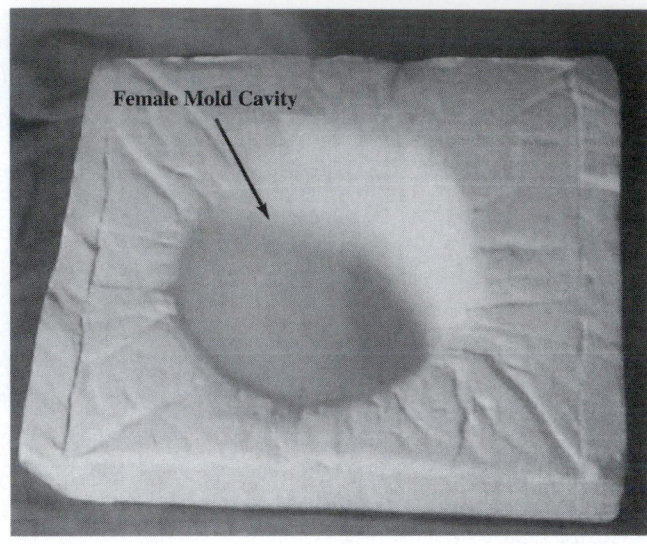

▼ *Figure 17.31.*
Female mold showing cavity left after fabrication.

Lexan® has a fractional content of dissolved water that had to first be removed by drying with heat. If the water is not removed, bubbles will occur in the plastic when exposed to temperatures needed to soften Lexan®. Thus, we need to preheat the molds and dry the disks. In a cold oven, we place the disks as well as both molds as shown in Figure 17.33.

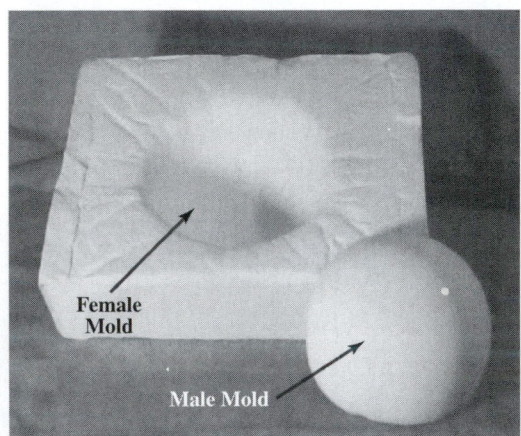

▼ *Figure 17.32.*
The female and male molds necessary for thermoforming.

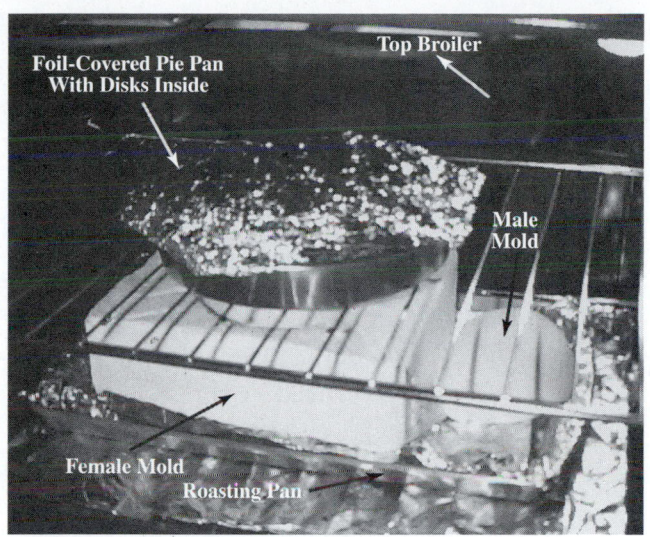

▼ *Figure 17.33.*
Disks and molds placed inside oven for heating and drying.

When everything is inside the oven, we bake the molds and Lexan® at a temperature of approximately 260°F for five hours. This heating period ensures that any moisture dissolved in the Lexan® disks will be driven off, preventing bubbles from forming during the molding process.

After removing the moisture, the plastic is heated to a softening temperature. The male mold and female mold are then used to form the dome (Figures 17.34 and 17.35).

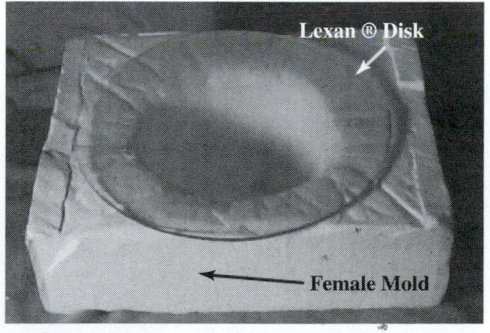

▼ *Figure 17.34.*
Lexan® disk placed on the female mold in preparation for the forming process.

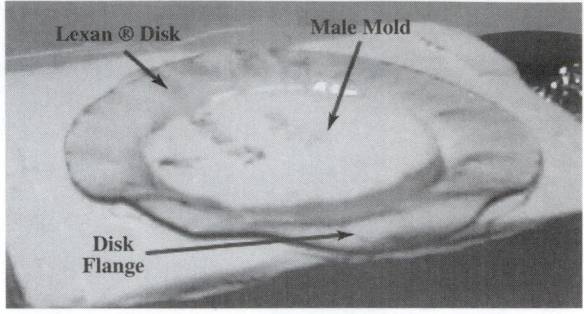

▼ *Figure 17.35.*
Male mold pressing Lexan® disk into female mold.

Very little force is needed to compress the male mold into the female cavity. After several minutes pass and the plastic disk is no longer soft, we remove the male mold, followed by the disk. To quickly cool the disk, we plunge it into a container of room-temperature water. With the disk removed, we have the clear dome as shown in Figure 17.36.

With the dome molded, we must separate it from the flange. We do this by carefully cutting the flange off. Once we finish the cut and remove any tailings from the cut edge, we are left with a clear dome as shown in Figure 17.37.

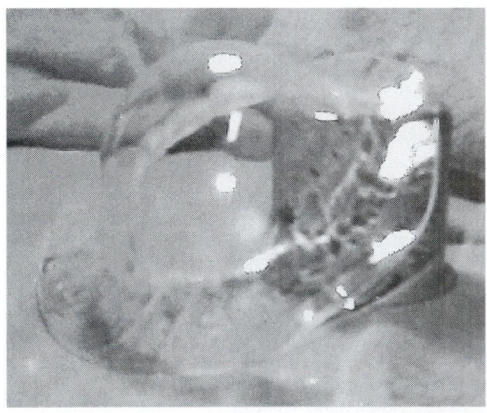

▼ *Figure 17.36.*
Clear dome after molding.

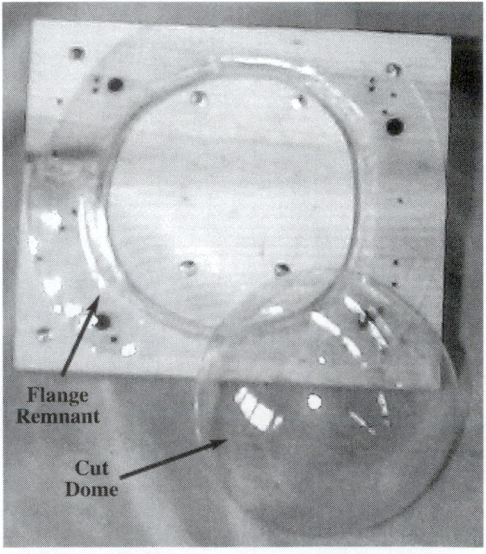

▼ *Figure 17.37.*
Dome after being separated from flange.

Modification to Doll's Head

Once the dome is prepared, we can begin the modification of the doll's head. First, we press the cut dome onto the doll's head ensuring that the dome is properly aligned with the head as shown in Figure 17.38. We then scribe a line on the head along the bottom edge of the dome.

We subsequently cut the portion of the doll head off along the line (Figure 17.39). The doll's head is now ready for attachment to the hose neck.

Brain Support Plate Fabrication

A plate is needed in the hollow head of the doll for supporting the brain. The support plate fabrication begins by creating a cardboard template using the plate outline shown in Figure 17.40.

After creating the template, the pattern is transferred to both a piece of 1/4″ oak plywood veneer and a piece of 2″ × 6″ pine. Using a band saw, we cut out the wood pieces as shown in Figure 17.41.

We then push the support plate into the head until approximately 1-1/4″ of an eyebolt protrudes through the support plate hole. We use a nut and washer to secure the support plate to the eyebolt (Figure 17.42). This step completes the head construction, leaving only the brain modification to be fabricated.

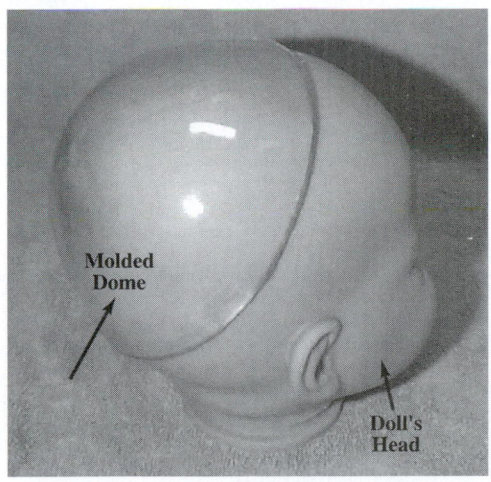

▼ **Figure 17.38.**
Molded dome after pressing onto doll's head.

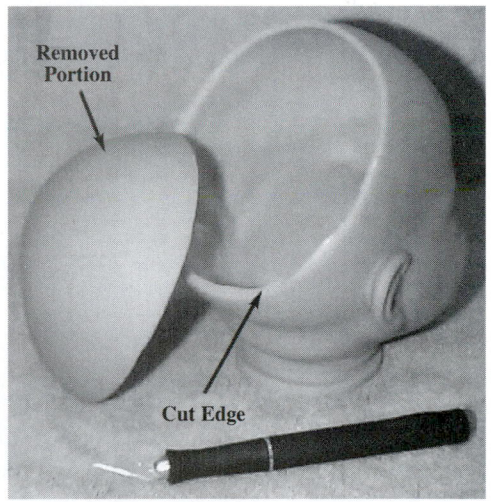

▼ **Figure 17.39.**
Doll's head with a portion removed.

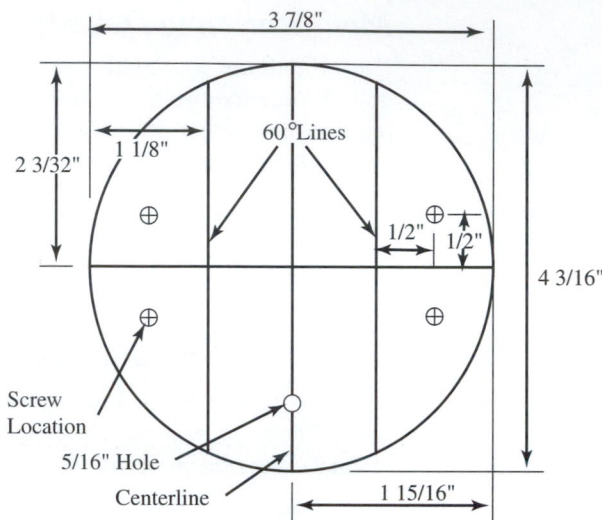

▼ *Figure 17.40.*
Template needed to create the brain support plate.

Hematoma Fabrication and Brain Modification

The final steps in the assembly of the doll include the fabrication of cranial hematomas and modification of the standard brain, shown in Figure 17.43.

We place a segment of Creative Paperclay® about the size of a large marble onto the brain at the first desired hematoma location and then slowly spread the clay out and smooth the surface as shown in Figure 17.44.

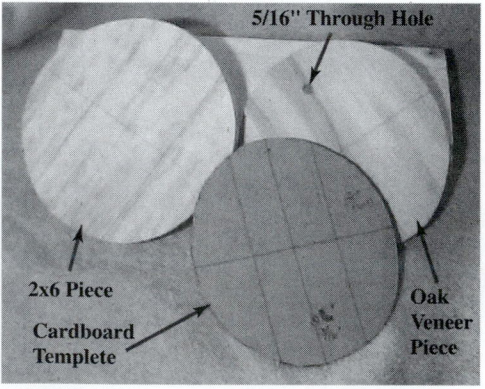

▼ *Figure 17.41.*
Wood pieces after cutting.

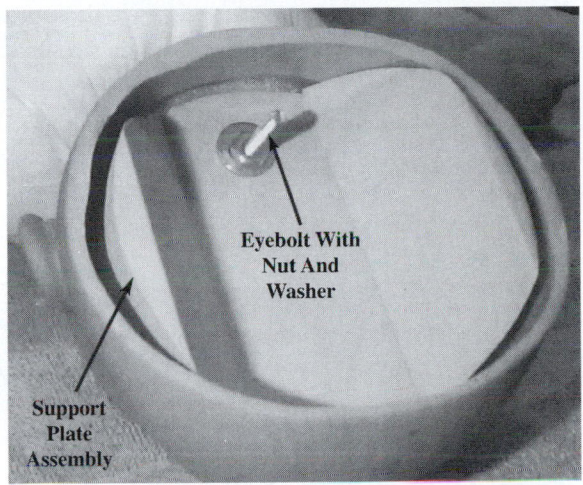

▼ *Figure 17.42.*
Support plate assembly placed into head with a nut and
washer on the eyebolt.

We then repeat this process for a second hematoma. Then we paint the
hematomas a realistic dark red color as shown in Figure 17.45. After the
paint dries, the brain is ready for assembly.

Final Assembly

The first step in final assembly is to place the brain into the doll's head.
We press the slit over the exposed eyebolt shaft and press the molded

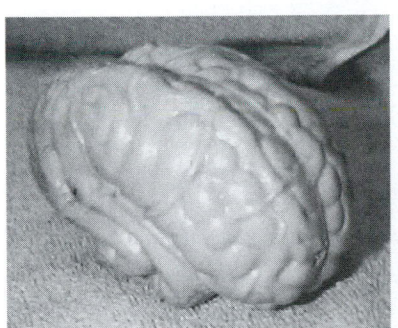

▼ *Figure 17.43.*
Vinyl brain.

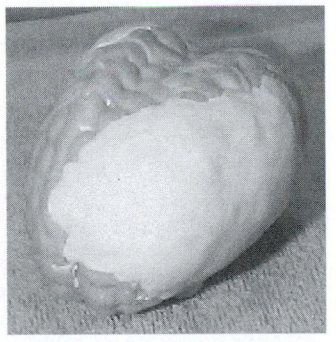

▼ *Figure 17.44.*
Clay hematoma on the brain's
frontal lobe.

▼ *Figure 17.45.*
Brain with painted hematomas
attached.

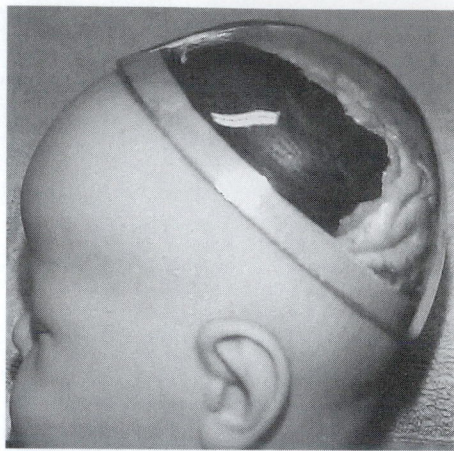

▼ *Figure 17.46.*
Modified brain after being placed into head
and covered by molded dome.

dome onto the doll's head, aligning the holes in the dome and in the
head. We attach the dome to the head by inserting cut wire nails into
the aligned holes. This last step results in a brain showing hematomas
underneath a clear, protective dome as shown in Figure 17.46.

The final step is to redress the doll with the included jumper and a
stocking cap, as shown in Figure 17.47. We then have a completed SBS
beta prototype, ready for customer testing. The stocking cap serves to

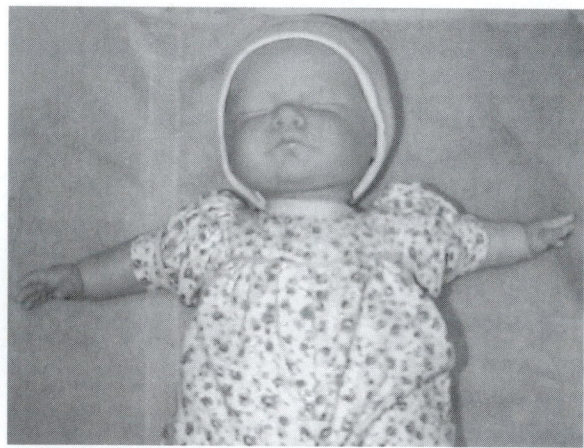

▼ *Figure 17.47.*
Doll fully dressed in jumper and stocking cap.

conceal the brain until we wish to show the potential cranial damage Shaken Baby Syndrome can cause.

The prototype was provided for testing to hospitals and care facilities with good results, leading to the production of the product. As a product, differences in fabrication would be required. Plastic molds would be developed for parts such as the brain, complete with attachment features. Nonetheless, this beta prototype of Figure 17.47 was very effective, completely functional, and true to size, weight, and strength.

VII. SUMMARY AND "GOLDEN NUGGETS"

This chapter has taken us on a journey through physical modeling and prototyping essentials. The discussion should provide a basic understanding on how we use, plan, and fabricate physical prototypes. In the next chapter, we focus on the complementary concept of how to test and statistically model prototype performance measurements to make design decisions. For the purpose of physical prototypes, however, the golden nuggets may be summarized as:

▶ There exists a paradox between the necessity of constructing physical prototypes for successful product development and the desire to minimize physical prototypes to manage cycle time and cost. Recent advances in prototyping technology have helped to solve this conflict by creating competitive processes that reduce cycle time and cost.

▶ Strategic choices of prototype types and uses greatly enhance the likelihood of product success.

▶ A variety of materials and manufacturing processes exist for creating physical prototypes. A product development team must understand the range of materials and processes to be the most efficient and effective with prototyping.

▶ Rapid prototyping processes offer novel opportunities for prototype fabrication and testing.

▶ Similitude and empirical models of prototypes are needed to explore the design choices for a product concept.

References

Baese, C. 1904. U.S. Patent 774,549.

Baker, W., Westine, P., and Dodge, F., 1992, Similarity Methods and Engineering Dynamics: Theory and Practice of Scale Modeling, Elsevier, New York.

Bearman, J., Marcus, H., Bourell, D., Barlow, J., Crawford, R., McAlea, K., 1996, Solid Freeform Fabrication: A New Direction in Manufacturing, Kluuser Academic Publishers, Massachusetts: Boston.

Bebb, H. B. 1990. Quality design engineering: The missing link in U.S. competitiveness. Byron Short Lecture Series, Dept. of Mechanical Engineering, The University of Texas, Austin, 23 April.

Blanther, J. E. 1892. U.S. Patent 473,901.

Bogart, M. 1979. In art the ends don't always justify the means. *Smithsonian* 104–110.

Ciraud, P. A. 1972. *FRG Disclosure Publication 2263777*.

DiMatteo, P. L. 1976. U.S. Patent 3,932,923.

Fay, R., 1993, "Scale Model Tests of Vehicle Motion," SAE Publication No. 795, SAE, pp. 51–54, Warrendale, PA.

Gaskin, T. A. 1973. U.S. Patent 3,751,827.

Herbert, A. J. 1982. Solid object generation. *Journal of Applied Photographic Engineering* 8 (4): 185–188.

Housholder, R. F. 1979. U.S. Patent 4,247,508.

Kodama, H. 1981. Automatic method for fabricating a three-dimensional plastic model with photo-hardening polymer. *Review of Scientific Instrumentation* 1770–1773.

Kline, S. J., 1975. Similitude and Approximation Theory, McGraw-Hill, New York.

Kunieda, M., and Nakagawa, T. 1984. Development of laminated drawing dies by laser cutting. *Bulletin, of JSPE* 353–354.

Matsubara, K. 1974. Japanese Kokai Patent Application, Sho 51[1976]-10813.

ME/RPA. 1996. Proceedings of the Rapid Prototyping and Manufacturing '96 Conference, sponsored by the Society of Manufacturing Engineers (SME) and the Rapid Prototyping Association (RPA), Dearborn, MI, 23–25 April.

Monteah, F. H. 1924. U.S. Patent 1,516,199.

Morioka, I. 1944. U.S. Patent 2,350,796.

Nakagawa, T. 1979. Blanking tool by stacked bainite steel plates. *Press Technique:* 93–101.

Nakagawa, T., Kunieda, M., and Liu, S. 1985. Laser cut sheet laminated forming dies by diffusion bonding. Proceedings of the 25th International MTDR Conference, pp. 505–510.

Perera, B. V. 1940. U.S. Patent 2,189,592.

Prinz, F. B., ed. 1997. *Site reports*. Vols. I and II, *JTEC/WTEC panel report on rapid prototyping in Europe and Japan.* Baltimore, MD: Loyola College. NTIS Rep. #PB96-199583.

Sachs, E. et al. 1995. Production of injection molding tooling with conformal cooling channels using the three dimensional printing process. Proceedings of Solid Freeform Fabrication Symposium, pp. 448–467.

Schwerzel, R. E. et al. 1984. Three-dimensional photochemical machining with lasers. *Applications of Lasers to Industrial Chemistry* 90–97.

Swainson, W. K. 1977. U.S. Patent 4,041,476.

Ulrich, K., and Eppinger, S. 1995. *Product design and development.* New York: McGraw-Hill.

Wallace, D. 2000. Sketch Modeling. Lecture in 2.744 Product Design, http://me.mit.edu/lectures/sketch-modelling/.

Web sites: http://www.3dsystems.com/library/usrfocus/mattel.htm, http://conceptual-reality.com/, http://www.itr.loyola.edu.

Wohlers, T. 1996. Rapid prototyping state of the industry: 1995–96 worldwide progress report. Society of Manufacturing Engineers Symposium, Dearborn, MI, April.

Wohlers, T. T. 1992. Chrysler compares rapid prototyping systems, *Computer-Aided Engineering,* Oct.

Xerox. 1996. http://www.xerox.com/factbook/1996/brief.html.

Zang, E. E. 1964. U.S. Patent 3,137,080.

18 Physical Models and Experimentation

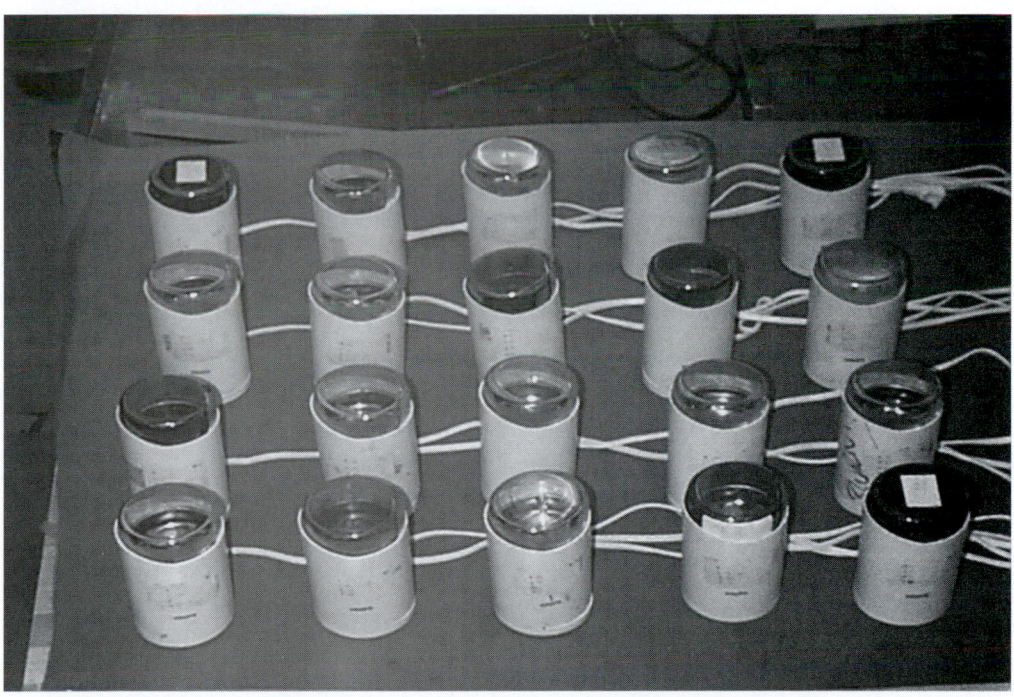

Experimental configurations used to model noise in a coffee mill.

The last chapter discusses fundamentals of prototype construction and mockups. One key goal of prototyping is to improve the observed, physical behavior of a product by making effective design changes. We can build a product prototype in one configuration, build it again in a different configuration, and select the better performing experiment as the configuration to use for the actual product.

A more systematic approach is to interpolate among the experiments. Consider the redesign of a coffee mill. Most mills are notoriously loud, generating over 80 dB of noise, even though they are intended for use during the early morning hours. To explore noise absorption and reduction, an analytical model could be developed to representation how vibrations are generated, transmitted, and emitted to the acoustical environment. The resulting model could then be used to change the mill's dimensions for noise reduction. To be sufficiently accurate, the modeling effort would require significant time resources, in addition to an in-depth knowledge of the field.

An alternative approach is to visualize how the vibrations are generated, transmitted, and emitted, we can generate alternative concepts to absorb or block these vibrations, and then prototype these potential solutions. This approach, while pragmatic, can be very laborious in prototype construction. It would also require significant resources in both time and materials.

A better method is to combine the two approaches. For example, one might consider a very simple and general relationship (equation) to model noise. A selection of experimental trials may then be fit to this relationship. The results may then be interpolated to predict the response of alternative configurations that are not explicitly tested during experimentation.

As an example of this approach, let's consider the coffee mill, where experiments were performed by inserting vibration absorbers at different locations in the mill using different materials. A simple linear equation was fit over these experiments, to provide a model of noise, measured in decibels. This model was expressed in terms of the design variables, that is, materials and locations. Preferred materials and locations were then chosen based on the model. From this approach, the noise produced by the mill was reduced by over 50%.

I. CHAPTER ROADMAP

The material in this chapter is diagrammed in Figure 18.1. Section II provides a basic working understanding of design of experiments. It describes the two-level-factorial approach and graphical analysis. Blender

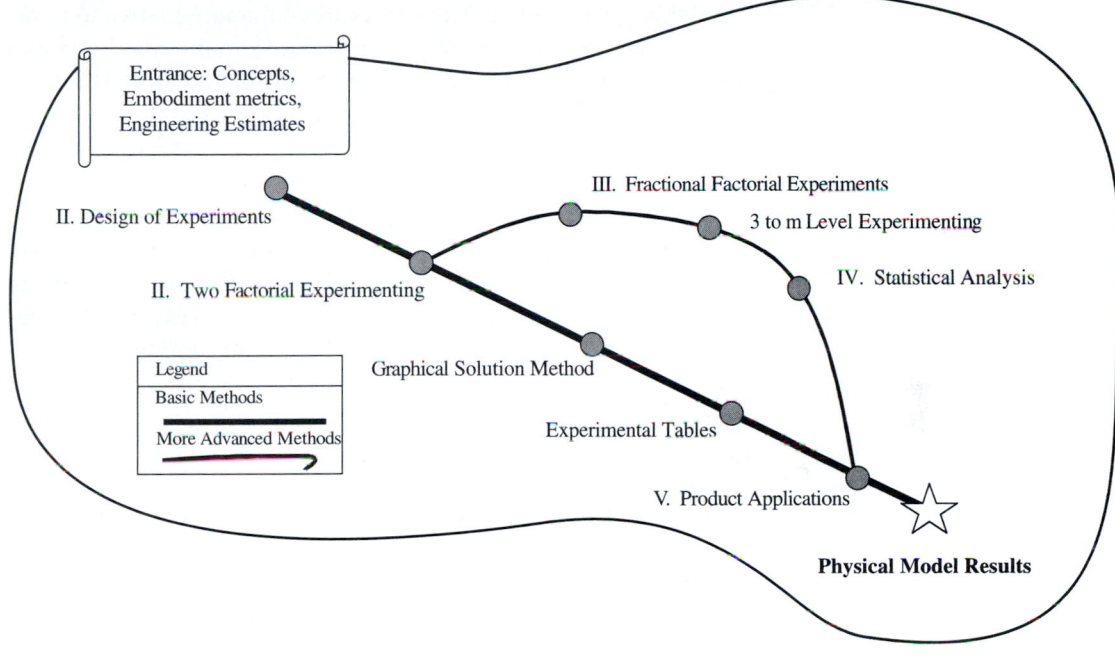

Entrance: Concepts,
Embodiment metrics,
Engineering Estimates

II. Design of Experiments

III. Fractional Factorial Experiments

3 to m Level Experimenting

II. Two Factorial Experimenting

IV. Statistical Analysis

Legend
Basic Methods

More Advanced Methods

Graphical Solution Method

Experimental Tables

V. Product Applications

Physical Model Results

▼ **Figure 18.1.**
Chapter roadmap for physical models and experimentation.

and solar-car product applications in Section II provide examples. A more advanced analysis is given in the subsequent sections, including a derivation for determining the number of experiments needed and development of relevant statistical tests.

II. DESIGN OF EXPERIMENTS

The development of a physical model requires that we define the type of prototype, its use, and so forth, as in the previous chapter. Often it is also important that we decide how to implement a prototype as a *model* of the final product. Generally speaking, a model is a representation that can be used to answer a question or set of questions. This model may be anything from a yes/no feasibility experiment to a fully developed performance expression over a feasible range of product design variables.

We have studied two types of design models in Chapters 13 and 16: analytical and numerical. These types are commonly referred to as *virtual models,* that is, they are not physical, we cannot reach out and touch them; rather, they are ostensibly mathematical formulations.

An alternative type is an *empirical model,* one developed not based on analysis but instead physical experiments—measurements of physical systems that we can see, feel, and smell. If virtual models of performance are too difficult to develop, then prototypes can be fabricated and tested to determine performance. As discussed in the previous section, key issues in prototype fabrication and testing are how many to build, what variables to vary, how to control noise and experimental uncertainties, and how to formalize the results of the testing.

These issues give rise to the subject of *design of experiments* (DOE). DOE is an experimental theory and methodology. It is used to determine the minimum number of experiments that will be required for adequately predicting a physical phenomenon. It also provides a statistical basis for monitoring and analyzing the inherent noise in an experiment. This basis can be used to determine the significance and relevance of the experimental results.

In the following sections, the basics of DOE are presented in the context of motivating examples and practical use. There is no attempt to present the entire theoretical foundation. Instead, generalizations are discussed to obtain a working knowledge of the field.

It is also important to note that DOE is also applied in product development to ensure the quality of products and processes as they are designed. We discuss this topic, known as design-for-quality and Taguchi methods, that is, *robust design,* in the next chapter.

Basics of Designed Experiments

As design experimenters, we are *experimentally optimizing* our identified design using physical models. Out of the possible design choices for a concept, we experimentally determine which combinations are the most preferred. Our goal is *not* to fit a model *to* the experiments for the sake of observation. Instead, we are answering the questions of, "What are the preferred choices for the product?" and "What are the most sensitive choices for maximizing the performance of the product?"

We start our discussions on design of experiments here with the concept of a formal engineering model, as introduced in Chapter 13. A design space $\mathbb{D}$ and a performance metric(s) p are identified for modeling a product. The problem addressed by DOE concerns the existence of no formal analytical or numerical relationship for representing p as a function of the de-

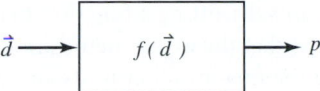

▼ **Figure 18.2.**
Design equation for empirical
model building.

sign variables $d \in \mathbb{D}$. We have no expression f in the equation $p = f(\vec{d})$. Figure 18.2 illustrates the abstract relationship between the product performance metric and the design variable choices. The inputs are known, and the metric for measuring the output is chosen. We need the machinery in the box to relate d to p, resulting in a means to estimate the performance.

More generally, as introduced in Chapter 13, there are other factors beyond design variables that can affect the performance p, as shown in Figure 18.3. Manufacturing variations and fluctuations in the product environment can lead to differences in customer experienced performance, represented as noise $n \in \mathbb{N}$. These other factors are discussed in the next chapter; for now we restrict our study to the model of Figure 18.2.

As depicted in Figure 18.2, we can estimate performance by experimentally determining a phenomenological model. The basic idea is to "fit" a relationship f over $\mathbb{D}$ using statistical concepts of measured performance states. Based on this function fit, we can then experimentally maximize f over $\mathbb{D}$ and determine the most sensitive design variables. In Chapter 19, this approach is augmented by simultaneously considering the effect of noise variables on product performance (Figure 18.3).

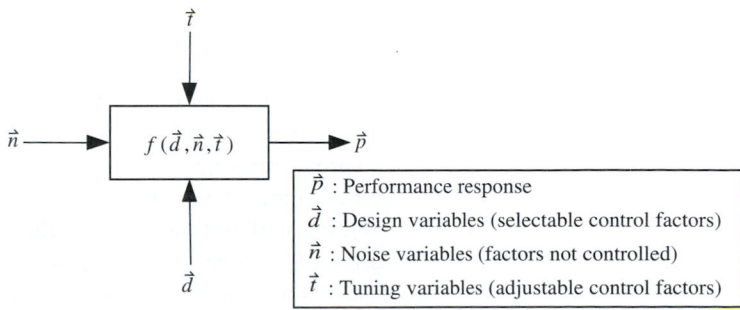

▼ **Figure 18.3.**
Black-box input-output concept of empirical model building.

As we consider fitting a function to experimental data, it must be understood that the experiments are not the "real" product. We are not testing the final product in its complete environment and operating conditions. Experiments thus involve a model:

- We assume a design space $\mathbb{D}$. This assumption implies a restriction on the possible design choices selected by the product development team, implying a simplification or model.

- We assume a performance metric p. How does our metric really measure activities experienced by a consumer of our product? In a dynamic environment?

- What effects will variations in the surface, material wear, environment, or any other uncontrollable factor have on our objective? Such noise variables are ignored for now.

It must also be realized that an experimental approach determines coefficients for fitting an assumed model of p over $\mathbb{D}$. This model must be validated with further measurement data for accuracy and robustness.

Problem Representation

After constructing a high-level model of a product concept, the design variables are known as is a measurable performance metric. The problem now is that $\mathbb{D}$ usually contains an infinite set of $\vec{d}$. For example, if $d \in [1,5] \subset \mathbb{R}$, an infinite number of length choices exist from 1 to 5 centimeters.

We therefore must choose a finite set of experimental arrangements, denoted $\vec{d}_h$, to approximate the range of $\mathbb{D}$. Only a finite number of experiments are possible, since only finite time and resources exist for each project deadline. In the length example, we might choose to test values of 1 and 5 cm. In general, the number of values of each d_i chosen for the experimental trials is dependent, in a large part, on how many experiments can be run within the given resources.

Overall, the first step in organizing tests of a physical prototype is to determine how many experiments N are affordable, in terms of cost, time, patience, and so on. If it is desired to test 10 variables for a product prototype, with 10 values for each variable, 10^{10} trials are needed to test all combinations/permutations of the variables choices. Experimenting at all permutations of the variables levels is known as a *full-factorial experiment*. For the case of 10 levels of 10 variables, this level of experimentation is obviously not possible. In fact, for most prototype tests, the number of affordable experiments is less than 100, usually on the order of 64 or less. A strategy is thus needed to design an experiment that balances engineering judgment, resources, and desired results.

Experimental Design

This section provides a synopsis of procedures for determining the minimum number of required experiments. The basis of this synopsis is the simultaneous testing and adjustment of multiple variables, followed by the interpolation and limited extrapolation of the results to estimate performance results for values not explicitly tested. Contrast this approach to traditional experimental prototyping. Intuitively, we may first try changing one variable while keeping all other variables at predetermined values. The results may demonstrate performance improvement or not. We might then focus on another variable, followed by yet another one, until the variable list is exhausted. This *one-variable-at-a-time* approach is not efficient, because it requires a large number of tests, and it may miss many combination effects. Instead, the approach discussed here applies a factorial technique of planned experimentation.

In this context, a number of questions should begin the experimental design:

▶ Are we sure that the proper variables have been chosen? Do we need to contemplate high-level physical principles to obtain additional insights? Does there exist a more effective choice of variables for experimentation?

▶ Are we sure that the range of the variables and their number is not too broad, unnecessarily complicating the experiments?

▶ What is the trade-off point, for the development team, between the cost of varying more variables versus expected payoff in performance increase?

Overall, these questions boil down to validating the choice of the design space, $\mathbb{D}$. Is it valid and complete, yet judicious so resources are not wasted?

Given this background, a general approach is needed for experimenting with physical prototypes. Again, the goal is to create an empirical model, relating the performance metric to the design variable choices (Figure 18.2). A five-step process is listed below to address this need.

Procedure for Experimental Design:

Step 1: *Model variables.* Identify performance metrics, noise variables (uncontrolled factors), and control factors (design and tuning variables). In addition to these variables, list high-level physical principles that will provide insights into the experimental design.

Step 2: *Variable targets and boundaries.* For each performance metric to be evaluated, specify a target value (QFD, Chapter 7) and determine boundaries (ranges) for each of the tested variables. Determine values to fix the noise variables. These fixed values must be controlled during the experimentation.

Step 3: *Experimental plan and matrix.* Design the experiment, including the number of trials, levels of the design variables per trial, and number of replicates (repeated tests). The results are captured in an experimental matrix. In addition, appropriate sensors and equipment should be chosen for measuring the experimental response. Issues for these choices include the magnitude, accuracy, and resolution of the measurements (Chapter 6).

Step 4: *Testing.* Perform the tests in random order, adhering to the design matrix. Record the data, including the needed replicates to check the experimental error. Ensure that the noise variables remain fixed.

Step 5: *Analysis.* Analyze the results using statistical concepts, and select new test conditions if needed.

This procedure begins with the identification of performance variables, design variables, and possible sources of noise. The noise variables must be clearly identified and eliminated so that they will not dominate the tests as they are executed. Noise variables are eliminated by ensuring they remain fixed over all the experimental trials. If noise does dominate, the test results will be indistinguishable from the experimental error, and no choices will be possible. In Chapter 19, this procedure is extended to include noise proactively so that the product design will be robust, or insensitive, to uncontrolled factors during its use.

After identifying the required variable sets, physical principles should be listed to give insight into the performance measurements. By listing and carefully considering physical principles, the choices of design variables and identification of possible noise can be verified and modified, if necessary.

Subsequent to Step 1, target values for the performance responses and boundaries for the design variables must be chosen. The target values provide a means of normalizing and evaluating the experimental results. The design variable boundaries, on the other hand, define the range of each design choice that needs to be tested. These boundaries should never be too broad due to the possibility of nonlinearities affecting the results.

The requisite information is now available for designing the experiment. The experimental design focuses on number of issues: estimating a structure for the empirical model (linear, quadratic, or even higher order), choosing the appropriate number of trials to accommodate the design variable boundaries, determining the level (value) of each design choice per trial, and choosing the number of repeated trials that are needed to estimate the magnitude of experimental error. In addition to these issues, the experimental design also considers a number of practical concerns, including measurement equipment (Chapter 6), support equipment, the means for recording data, and a means for preventing noise factors from influencing the experimental results.

With the experiment planned, the trials and their replicates should be run in a random order. This process reduces the likelihood of biasing the results by a predetermined trial structure. Data must be carefully measured and recorded for each trail, and a careful eye must be trained on possible environmental noise sources that could distort the results.

The final step in the process is to analyze the results of the experiments. Statistical tools are implemented at this point for three purposes. The first is to predict the magnitude of experimental error (based on the replicates tested). This experimental-error estimate may be compared with the change in response values across trails to determine if the results are significant or, alternatively, if the results have been dominated by experimental error. The second purpose is to validate the modeling assumptions. For example, if a certain order of the model (such as linearity) is assumed, statistical tools may be used to determine if the experimental data match this assumption or not. If the assumption is not valid, more test conditions will be needed to expand the model. Finally, a purpose of the experimental step is to convert the measured experimental data into an analytical model that estimates response values for the full range of design choices. The sections below expound further on this step with details of the statistical tools.

Experimental Formulation

Let's now consider, more closely, the third step of the experimental design procedure. First we restrict the design variable choices to a number of levels (values), usually two or three levels. This restriction approximates $\mathbb{D}$ with, say, 2^n possible (factorial) experimental arrangements for two levels on each variable. It also presumes a model fit between the levels (linearity for two levels and quadratic for three levels). An experiment using two levels for all n design variables is termed a 2^n *factorial design*: 2^n tests are needed, in this two-level experimental approach, to study all combinations of design variables.

For most physical tests, two or three levels are sufficient to estimate the performance relationship as a function of the design choices. Very few factorial designs use more than three levels to empirically model products or processes. Practically, it is difficult to experiment with more levels since the number of experimental trails increases exponentially, to the nth power of the number of levels.

Statistically, the analysis is easier when we use normalized variables. That is, after approximating the range of a variable with a finite number of points, we denote the lowest level by d_{low} and the highest level by d_{high}. These levels are the boundaries for each design variable and are then mapped to a normalized domain with span $[-1, +1]$. This -1 to $+1$ range is a normalized representation of each design variable.

For two-level experiments, we denote the higher value of d_i with a $(+1)$ and the smaller value with a (-1), where a value of (0) represents the midpoint of the range. If $d_i \in \mathbb{R}$, we think of $\hat{d}_i$ as a normalized scale for d_i, from a transformation:

$$\hat{d}_i = \frac{2d_i - d_{i\text{high}} - d_{i\text{low}}}{d_{i\text{high}} - d_{i\text{low}}} \tag{18.1}$$

and the inverse:

$$d_i = \frac{(\hat{d}_i - 1)(d_{i\text{high}} - d_{i\text{low}})}{2} + d_{i\text{low}} \tag{18.2}$$

If d_i simply represents a binary decision, such as "use vibration isolation, do not use vibration isolation," then there is no implied scale associated to the range between numbers $(+1)$ and (-1). The extremes are the only points of interest. For the remainder of this chapter, we will shorten the notation of $\hat{d}_i$ to d_i and consider all design variables as normalized.

Experimental Matrix

Assume we have reduced each variable range to a finite set of points, say two or three levels. We now need to assign the levels of each d_i to experimental arrangements $\bar{d}_h$. A methodical assignment is to start with the first variable and create a repeated list of its levels. For two-level experiments, this list is given by $(-1, +1, -1, +1, \ldots)$. Then for the second variable, we assign the lowest level to the first sequence of the first variable, the next level to the second sequence, and so on. For two-level experiments, this approach results in the structure $(-1, -1, +1, +1, -1, -1, +1, +1, \ldots)$ for the second variable, and so on, until the last variable, which has its levels split evenly between low and high states, $(-1, \ldots, -1, +1, \ldots, +1)$.

TABLE 18.1. EXPERIMENTAL MATRIX OF A 2^3 FACTORIAL DESIGN

Trial	Trial Vector Label	d_1	d_2	d_3
1	$\vec{d}_1$	-1	-1	-1
2	$\vec{d}_2$	$+1$	-1	-1
3	$\vec{d}_3$	-1	$+1$	-1
4	$\vec{d}_4$	$+1$	$+1$	-1
5	$\vec{d}_5$	-1	-1	$+1$
6	$\vec{d}_6$	$+1$	-1	$+1$
7	$\vec{d}_7$	-1	$+1$	$+1$
8	$\vec{d}_8$	$+1$	$+1$	$+1$

For a three-variable, two-level formulation, the 2^3 (read "two-to-the-third") experimental matrix is shown in Table 18.1. Each row defines an experimental arrangement, called an *experiment, trial, test,* or *configuration*. The number of trials is calculated using the relationship $k = 2^n = 2^3 = 8$.

Basic Method: Two Factorial Experiments

A two-level design is the simplest set of experiments for testing a product prototype. Such experimental designs provide the most direct interpolation of the performance responses between design variable values (Box 1978; John 1971; John 1990; Phadke 1989). They can capture *interactions* (compounding effects only brought out by multiple variables) as well as *main effects* (the general trend of a single variable). They also require tests only at the boundary values of d_i.

Regression: Creating a Phenomenological Model from Prototypes

Consider formulating a 2^3 experimental matrix as in Table 18.2. We seek to conduct experiments at each configuration. This process provides a finite set of values that we can use to fit a functional expression for a product's performance metric. This expression then provides our mapping between $\mathbb{D}$ and the single metric range, where the expression may be interpolated between design variable boundaries. To perform this interpolation, we need two basic results: a finite, representative set

TABLE 18.2. EXPERIMENTAL MATRIX OF A 2^3 FACTORIAL DESIGN, LISTING RESPONSE TERMS

Trial	Trial Vector	d_1	d_2	d_3	$y(\vec{d})$
1	$\vec{d}_1$	-1	-1	-1	$y(\vec{d}_1)$
2	$\vec{d}_2$	$+1$	-1	-1	$y(\vec{d}_2)$
3	$\vec{d}_3$	-1	$+1$	-1	$y(\vec{d}_3)$
4	$\vec{d}_4$	$+1$	$+1$	-1	$y(\vec{d}_4)$
5	$\vec{d}_5$	-1	-1	$+1$	$y(\vec{d}_5)$
6	$\vec{d}_6$	$+1$	-1	$+1$	$y(\vec{d}_6)$
7	$\vec{d}_7$	-1	$+1$	$+1$	$y(\vec{d}_7)$
8	$\vec{d}_8$	$+1$	$+1$	$+1$	$y(\vec{d}_8)$

of experimental points (including both configuration points and metric values (responses) as pairs) and a functional expression to fit.

A functional expression is a relationship that we believe represents how the metric values relate to the design variables. If this expression is unknown, a linear relationship can be used to provide a first-order indication of behavior, such as:

$$y = f(\vec{d}) = \beta_0 + \beta_1 d_1 + \cdots + \beta_n d_n \qquad (18.3)$$

If more information is known, a more complex expression of behavior can be used.

Assuming a relationship given by Eq. (18.3), the problem is to determine the β coefficients that provide the best fit among the known experimental values. For example, consider a two-factorial experiment with three variables and the experimental matrix given by Table 18.2. Now suppose we seek to fit a "best" linear function:

$$y = f(\vec{d}) = \beta_0 + \beta_1 d_1 + \beta_2 d_2 + \beta_3 d_3. \qquad (18.4)$$

We need to determine the four β coefficients that fit the $N = 2^3 = 8$ experimental data points "best." To do this, we need to establish what "best" means. Consider the difference between the prediction of Eq. (18.4) and the measured values:

$$y_{\text{measured}} = y_{\text{predicted}} + \varepsilon = \beta_0 + \beta_1 d_1 + \beta_2 d_2 + \beta_3 d_3 + \varepsilon \qquad (18.5)$$

where ε is an error term for the model. This error term will equal zero if the linear function passes through all the measured responses. As the line deviates from the measurements, more error will occur.

One way to consider the total error of the expression Eq. (18.5) is to consider the RMS sum of the error terms:

$$
\begin{aligned}
\varepsilon^2 &= \sum_h \varepsilon_h^2 \\
&= \sum (y_{\text{measured}} - y_{\text{predicted}})^2 \qquad (18.6) \\
&= \sum_h (y_h - \beta_0 - \beta_1 d_{1h} - \beta_2 d_{2h} - \beta_3 d_{3h})^2
\end{aligned}
$$

To determine the values β that provide the best fit, we need to minimize Eq. (18.6) by differentiating Eq. (18.6) with respect to each β_i:

$$
\frac{\partial \varepsilon^2}{\partial \beta_i} = 0 \qquad (18.7)
$$

Applying this minimization to the three-variable example, we derive:

$$
\frac{\partial \varepsilon^2}{\partial \beta_0} = 0 = -2 \sum_h (y_h - \beta_0 - \beta_1 d_{1h} - \beta_2 d_{2h} - \beta_3 d_{3h}) \qquad (18.8)
$$

$$
\frac{\partial \varepsilon^2}{\partial \beta_i} = 0 = -2 \sum_h (y_h - \beta_0 - \beta_1 d_{1h} - \beta_2 d_{2h} - \beta_3 d_{3h}) d_{ih} \qquad (18.9)
$$

This result defines $n + 1$ linear equations for $n + 1$ variables. For our three-variable problem, this expression corresponds to four linear equations for the four coefficients β:

$$
\begin{bmatrix}
N & \sum d_{1h} & \sum d_{2h} & \sum d_{3h} \\
\sum d_{1h} & \sum d_{1h}^2 & \sum d_{1h}d_{2h} & \sum d_{1h}d_{3h} \\
\sum d_{2h} & \sum d_{1h}d_{2h} & \sum d_{2h}^2 & \sum d_{2h}d_{3h} \\
\sum d_{3h} & \sum d_{1h}d_{3h} & \sum d_{2h}d_{3h} & \sum d_{3h}^2
\end{bmatrix}
\begin{pmatrix}
\beta_0 \\ \beta_1 \\ \beta_2 \\ \beta_3
\end{pmatrix}
=
\begin{pmatrix}
\sum y_h \\ \sum y_h d_{1h} \\ \sum y_h d_{2h} \\ \sum y_h d_{3h}
\end{pmatrix}
\qquad (18.10)
$$

or

$$
[A]\vec{\beta} = \vec{B} \qquad (18.11)
$$

which can be inverted to obtain:

$$
\vec{\beta} = [A]^{-1}\vec{B} \qquad (18.12)
$$

where this result is the least squares regression fit for β. With this approach, the best set of coefficients β are the ones that minimize the least square error.

Linear Algebra Derivation

We can also formulate this result another way. Consider that the experimental matrix, augmented with a column of 1's, as the following mathematical relationship:

$$X = \begin{bmatrix} 1 & 1 & 1 & 1 \\ 1 & 1 & 1 & -1 \\ 1 & 1 & -1 & 1 \\ 1 & 1 & -1 & -1 \\ 1 & -1 & 1 & 1 \\ 1 & -1 & 1 & -1 \\ 1 & -1 & -1 & 1 \\ 1 & -1 & -1 & -1 \end{bmatrix} \tag{18.13}$$

We can use this relationship to express:

$$\begin{bmatrix} 1 & 1 & 1 & 1 \\ 1 & 1 & 1 & -1 \\ 1 & 1 & -1 & 1 \\ 1 & 1 & -1 & -1 \\ 1 & -1 & 1 & 1 \\ 1 & -1 & 1 & -1 \\ 1 & -1 & -1 & 1 \\ 1 & -1 & -1 & -1 \end{bmatrix} \begin{pmatrix} \beta_0 \\ \beta_1 \\ \beta_2 \\ \beta_3 \end{pmatrix} = \begin{pmatrix} y_1 \\ y_2 \\ y_3 \\ y_4 \\ y_5 \\ y_6 \\ y_7 \\ y_8 \end{pmatrix} + \begin{pmatrix} \varepsilon_1 \\ \varepsilon_2 \\ \varepsilon_3 \\ \varepsilon_4 \\ \varepsilon_5 \\ \varepsilon_6 \\ \varepsilon_7 \\ \varepsilon_8 \end{pmatrix} \tag{18.14}$$

or

$$X\vec{\beta} = \vec{y} + \vec{\varepsilon}. \tag{18.15}$$

If X is a full $N \times N$ matrix of full rank, it could be inverted for β, ignoring ε. It is not $N \times N$, so inverting X is not possible. So we first premultiply by X^T:

$$X^T X \vec{\beta} = X^T \vec{y} + X^T \vec{\varepsilon} \tag{18.16}$$

This expression reduces the number of equations from N (number of experiments) to $n + 1$ (number of variables + 1). In our three-variable example, it reduces the number of equations from eight to four. Now equating:

$$\begin{aligned} A &= X^T X \\ \vec{B} &= X^T \vec{y} \end{aligned}, \tag{18.17}$$

we have the same basic expression as before, derived using calculus. Thus, solving the n simultaneous equations through:

$$\vec{\beta} = [X^T X]^{-1} X^T \vec{y} \tag{18.18}$$

provides the coefficients β that give the best fit for factorial product test, where "best" is defined as the one that minimizes the RMS error of the empirical equation compared to the measured data points (responses).

The regression formulation represents the basic idea behind design of experiments. By performing a minimum number of experiments at two or three levels, a relationship (phenomenological model) may be derived that estimates the full range of responses for the bounded design variables. The remainder of the theory of DOE involves methods to choose experimental points to use, methods of fitting equations, methods to minimize the effect of noise (uncontrolled factors), and methods to determine confidence in the results (Box 1978; John 1971; Montgomery 1991).

Model Solution: Main Effects

If we have performed a full 2^n set of experiments, the maximum and minimum of the performance response will appear at a tested point. This result is not true for *partial factorial* designs involving only a fraction of the possible 2^n experiments. In this case, we need a way to choose which combination of variables we should use as an optimal solution, since it may be one that is different from the experimental trials.

The next expansion of the model is to predict the nominal impact caused by changing any one variable. The *main effect* of a variable represents the contribution or sensitivity of a given design variable, d_i, on the performance response. By calculating a main effect, we can determine if a design variable is significant in changing the response (implying that it has a greater effect than the experimental error). We can also determine if the design variable has a relatively small or large effect on the response (its sensitivity). The more sensitive the variable, the more important it will be to the design team.

If we believe the main effects are the primary contributors to the response, we can determine which $\bar{d}_h$ is the optimal solution for the problem. That is, for each design variable d_i, we select d_{i*} such that the following expression is minimized:

$$\text{minimize} \qquad f(\bar{d}) = \beta_0 + \sum_{i=1}^{n} \beta_i d_i \qquad (18.19)$$

$$\text{subject to} \qquad -1 \leq d_i \leq 1$$

One can show the solution to this equation can also be determined another way. Since the equation is linear, the solution lies at one of the vertices of the hypercube in $\mathbb{D}$. The effect that each variable has on the response is independent of the others, and so we can inspect each variable independently. On each variable, we can choose the level $(+1)$ or (-1) that provides the lower average response. That is, pick the value of d_i by

$$d_i^- \text{ if } \frac{1}{2^{n-1}} \sum_{\bar{d}_h \text{ using } d_i^-} f(\bar{d}_h) \leq \frac{1}{2^{n-1}} \sum_{\bar{d}_h \text{ using } d_i^+} f(\bar{d}_h), \text{ else } d_i^+ \quad (18.20)$$

We now present a graphical method to solve this equation.

Graphical Solution Method—The Response Diagram

Equation 18.20 permits us to develop a simple graphical solution method, rather than directly applying the mathematical definition. Along a graph of each variable d_i, we can plot all of the data points y. For example, in Table 18.2, we can simply plot the d_1 column data along an x-axis, and the y response along a y-axis, and study the graphed data to understand how d_1 impacts the response y. We will call this plot the *response diagram*.

Generally, the data stacks up as two data sets, one at the (-1) x-axis value and one at the $(+1)$ x-axis value. If the difference between the averages of these two data sets is large, then one should go ahead and use the x-axis value that has the lower average. This example is shown in Figure 18.4(a). On the other hand, if the difference in average is small, then the design variable is not explaining the difference in measured

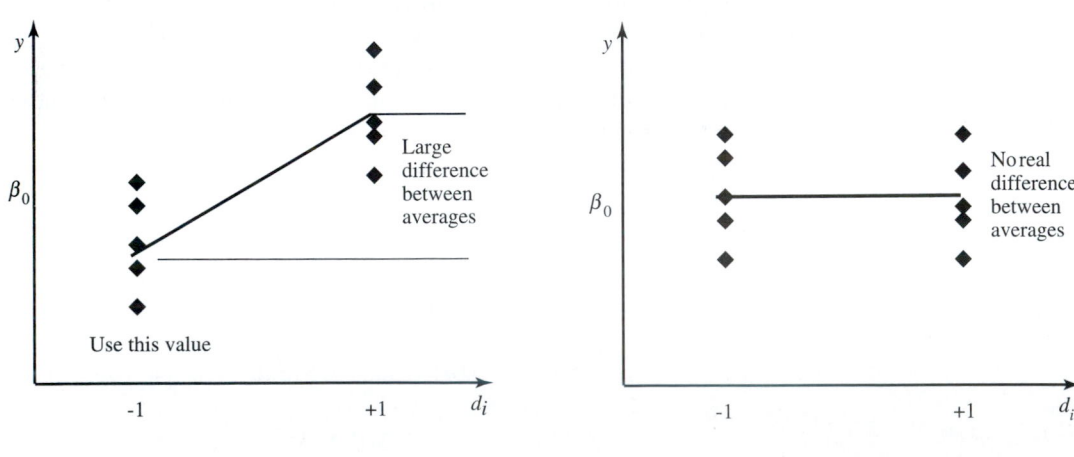

(a) Data indicates a value to use (-1).   (B) Data indicates d_i does not affect y.

▼ **Figure 18.4.**
Graphical method of main effect model. (a) Data indicates a value to use (-1). (b) Data indicates d_i does not affect y.

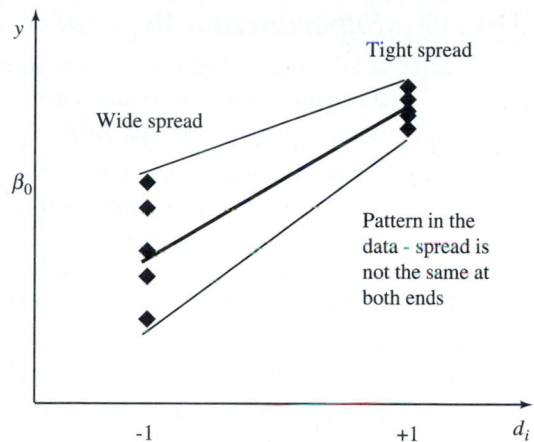

▼ *Figure 18.5.*
Graphical method indicating the main effect model
is not sufficient.

values, as shown in Figure 18.4(b). It does not matter what value along
the x-axis is used, the response cannot be predicted by it.

Note that the spread about the average line of the graph is not important.
Other variables such as d_2 or d_3 may be causing the response y to vary in
a controlled way, where this situation will plot as a large spread at each
value of d_1. On the other hand, patterns in the spread, such as shown in
Figure 18.5, are an indication of a weakness in the linear model. Log
transformations ($y_{\log} = \ln(y)$) or interaction models should be tried.

The response graph is precisely related to the regression equation. In
particular, the average of all the data points is exactly β_0. The best fit line
through the graph has slope β_1 and passes through β_0 at the value
$d = 0$. Thus, the graphical technique should not be considered a separate technique from regression fitting, but rather a graphical solution
to the regression equation.

Other good graphical methods include plotting the line of the average
of all experiments with a different design variable fixed at its high value
and then again at its low value. This plot will produce two average lines,
each with different slopes. This *two-way diagram* can visually indicate
any interactions between the variables, depending on magnitude of
difference in the slopes.

While the arguments above are not proven rigorously here, the approach is sound. Later sections discuss actual statistical measures that
compare these two spreads to provide a numerical indicator of whether
any design variable has any impact on the response.

Experimental Replicates

When conducting a set of experiments, it is important to determine the significance of the results. Do the measured responses (performance measurements) have meaning? Are the data valid and useful?

An important consideration in answering these questions concerns the inherent experimental error in running tests. Error will occur in physical measurements, in changing environmental conditions, and in changing experimental conditions, due to contamination or other sources. It is impossible to control and remove all the possible experimental error. The goal then exists to control noise sources so that the error is minimized, at least well below the expected magnitude of the measured responses.

Even if we identify and control potential error sources, we must check an experiment to verify that the results are significant. The concept of experimental replicates helps us to perform this check. A *replicate* is simply a repeat of an experimental trial, measuring the test response. If the replicate is identical to the original response measurement, the experimental error is zero. If it is not identical, the variation represents a sampled estimate of the experimental error. With this meaning, replicates should be randomly distributed within an original experimental plan to obtain the best estimate possible for the experimental error.

In general, replicates will not identically equal the original response measurements. At each experimental configuration, the variation data provides us the ability to check the significance of the experimental results. A basic method is to calculate the variance for each experimental configuration, h, of an experiment:

$$s_h^2 = \frac{(y_{h1} - \bar{y}_h)^2 + \cdots + (y_{hr} - \bar{y}_h)^2}{r - 1}, \qquad (18.21)$$

where r is the number of replicates per experimental configuration, and the average response at each experimental configuration is given by:

$$\bar{y}_h = \frac{\sum_{j=1}^{r} y_{hj}}{r}. \qquad (18.22)$$

An estimate of the nominal measurement variance in the total set of experiments may then be calculated as:

$$s_T^2 = \frac{s_1^2 + \cdots + s_N^2}{N}, \qquad (18.23)$$

where N is the number of experimental configurations. Taking the square root of this variance gives the estimated standard deviation of the experimental error for the entire designed experiment.

The measurement error s_h or s_T is different from the fit error ε_h discussed earlier (18.19). The fit error exists even with perfect measurements; it is the difference between the exact measurements and the approximate model equation used. The measurement error would exist even with a perfect model; that is, with a model that exactly predicted the replicate-average values (Eq. 18.22).

The measurement standard deviation s_T is important to consider when making design inferences with the experimental results. Any conclusion can only be drawn to within the error of s_T. If a variable's effect range in the graphical analysis of Figure 18.4 is on the order of s_T, then one should not expect that variable to have much impact. Subsequent sections present more advanced approaches for significance tests, such as the *t*-test and *F*-test statistical analyses.

Product Application (Lee et al. 1997)

Let's consider the redesign of a household blender product. The purpose of this product is to mix food products, both liquids and solids. Figure 18.6 shows an example household blender.

Important customer needs for this product include uniform blending of food products; the ability to blend thick food items, including vegetables

▼ Figure 18.6.
Common household blender.

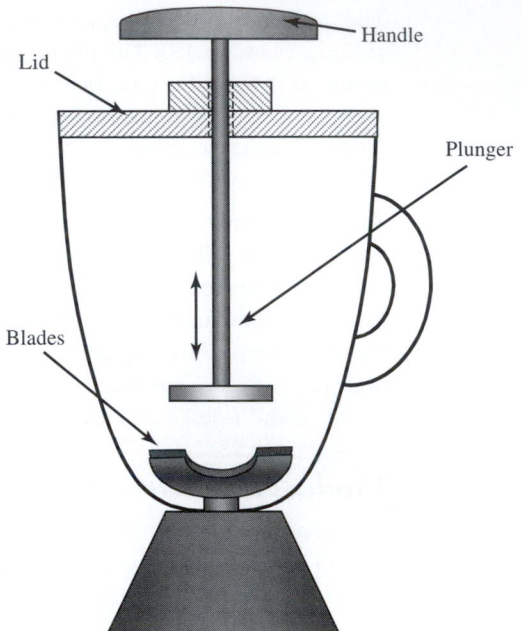

▼ Figure 18.7.
Plunger concept for household blender.

and ice; a large capacity container; a sturdy assembly; and small storage volume. For this application, let's focus on the important customer need of mixing thick food items, without interrupting the blending process. After implementing the product development process, through concept development and selection (Chapters 1–11), the general concept of adding a hand-actuated plunger to the blender is chosen to address the need. The current blender configuration produces sufficient cutting force to shear ice and other solids. However, the blender performs poorly in overall blending because there is insufficient driving force to feed the solid into the rotating blades. A novel plunger module, adapted through the lid of the blender, is feasible for providing this feeding force.

Figure 18.7 shows the general configuration of the plunger adaptive design. This concept must now be embodied as part of the blender, in conjunction with the redesigns of other subsystems. The general embodiment process (Chapter 12) is implemented to perform this task. As part of this process, a model of the blender and plunger system must be developed. This model will be used to select appropriate dimensions, materials, and processes. It can be based on analytics, empirical data, or a combination.

The primary performance metric for the model is blending efficiency, the quantity of blended solid per unit time. The embodiment checklist and FMEA of Chapter 12 identify additional metrics of buckling resistance, torsional strength, wear resistance, and so forth. These latter metrics may be modeled analytically with follow-up verification through experimentation. The first metric, however, is difficult to model analytically. It is a function of the overall shape of the plunger head. The power input from the rotating blades would need to be modeled as it interacts with the plunger head and is transformed into a shearing forces and dynamic motion of the blended substances.

Such a model would be very difficult to create and calibrate for a given blender configuration. Thus, as an alternative, DOE is applied. DOE will result in a phenomenological model of the blending process.

Step 1: Model variables.   The model includes many possible performance, design, and noise variables. For this embodiment, the performance variable is blending efficiency. Design variables include plunger-head shape, plunger-head size, type of plunger surface, type of plunger force, and rotation of the plunger. The noise variables include room temperature, blender setting, size and hardness of blended solid, and transients of the blender motor.

Step 2: Variable targets and boundaries.   Table 18.3 shows the target values for the performance and design variables. The bounds on the design variables define a number of plunger configurations. These bounds are developed based on the blender geometry and physics of process. Figure 18.8 illustrate the various configurations for testing.

TABLE 18.3.   TARGET VALUES AND BOUNDARIES FOR THE BLENDER DOE

Variable	Target	Lower bound ($-$)	Upper bound ($+$)
p, Blending efficiency	75% of 200 grams in 10 seconds		
d_1, plunger shape		Rectangular (aspect ratio, 2:1)	Circular
d_2, plunger size (length or radius)		6.4 cm	9.7 cm
d_3, type of surface		Concentrated (no prongs)	Distributed (prongs of 1.3 cm in length)
d_4, plunger force		No force	Force (1 Hz, oscillating, 40 N)
d_5, rotation		No rotation	Rotation

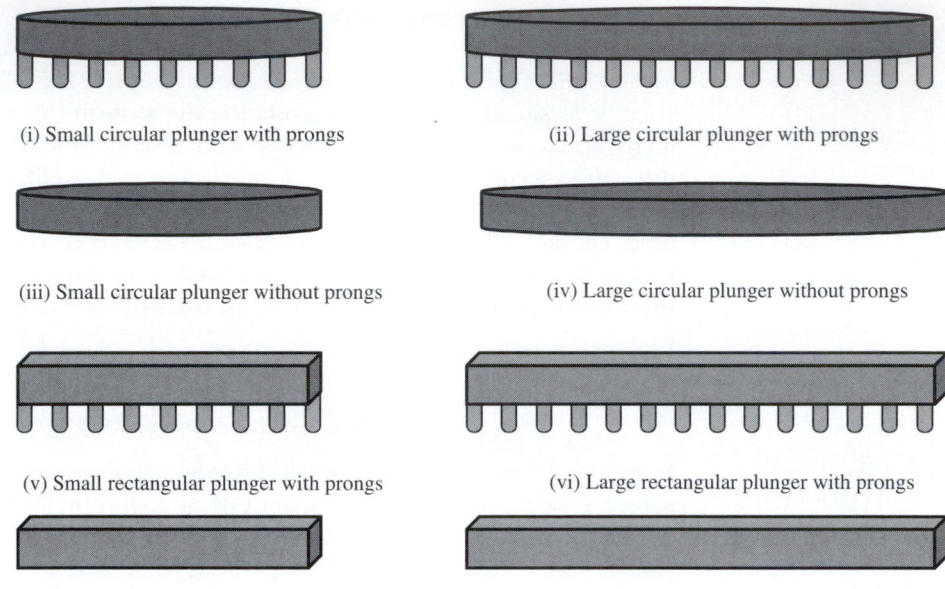

(i) Small circular plunger with prongs

(ii) Large circular plunger with prongs

(iii) Small circular plunger without prongs

(iv) Large circular plunger without prongs

(v) Small rectangular plunger with prongs

(vi) Large rectangular plunger with prongs

(vii) Small rectangular plunger without prongs

(viii) Large rectangular plunger without prongs

▼ *Figure 18.8.*
Plunger variants used for blender experimentation.

The identified noise variables will be controlled during the experiment. For example, the solid to be blended is ice. This choice represents the worst case of blended substances, as identified from customer needs.

Step 3: Experimental plan and matrix. Based on the model variables, $2^5 = 32$ experiments are needed for all possible variable combinations. Because the resources to construct and test the plunger variants are minimal, all tests are run, resulting in a full-factorial experiment. These experiments will include the introduction of 200 grams of ice into the blender. This quantity of ice corresponds to a single blended frozen drink.

After introduction of the ice, it is blended for 10 seconds, implementing the plunger configuration for a randomly chosen trial. Each trail corresponds to one of the 32 experimental combinations. The response of the blending experiment is then measured by pouring the contents of the blender through a mesh screen with 1-cm squares. The ice that passes through the mesh is collected and weighted. Dividing the result by the original 2000 g gives a blending efficiency.

Table 18.4 shows the experimental matrix. The 32 experimental configurations are randomized to remove bias from the trials. In addition,

TABLE 18.4. EXPERIMENTAL MATRIX AND RECORDED RESPONSES FOR THE BLENDER REDESIGN

Trial (h)	d_1	d_2	d_3	d_4	d_5	$y_{h,1}$ (%)	$y_{h,2}$ (%)	$\bar{y}_h$ (%)	s_h^2
1	−	−	−	−	−	33.3	30.0	31.7	5.4
2	+	−	−	−	−	13.3	33.3	23.3	200.0
3	−	+	−	−	−	25.0	16.7	20.9	34.4
4	+	+	−	−	−	16.7	16.7	16.7	0.0
5	−	−	+	−	−	26.7	25.0	25.9	1.4
6	+	−	+	−	−	26.7	21.7	24.2	12.5
7	−	+	+	−	−	43.3	22.3	32.8	220.5
8	+	+	+	−	−	26.7	23.3	25.0	5.8
9	−	−	−	+	−	83.3	93.3	88.3	50.0
10	+	−	−	+	−	58.3	60.0	59.2	1.4
11	−	+	−	+	−	83.3	75.0	79.2	34.4
12	+	+	−	+	−	25.0	28.3	26.7	5.4
13	−	−	+	+	−	48.3	51.7	50.0	5.8
14	+	−	+	+	−	58.3	60.0	59.2	1.4
15	−	+	+	+	−	51.7	41.7	46.7	50.0
16	+	+	+	+	−	41.7	45.0	43.4	5.4
17	−	−	−	−	+	41.7	30.0	35.9	68.4
18	+	−	−	−	+	30.0	33.3	31.7	5.4
19	−	+	−	−	+	33.3	41.7	37.5	35.3
20	+	+	−	−	+	20.0	30.0	25.0	50.0
21	−	−	+	−	+	40.0	41.7	40.9	1.4
22	+	−	+	−	+	48.3	53.3	50.8	12.5
23	−	+	+	−	+	48.3	46.7	47.5	1.3
24	+	+	+	−	+	41.7	38.3	40.0	5.8
25	−	−	−	+	+	68.3	65.0	66.7	5.4
26	+	−	−	+	+	83.3	71.7	77.5	67.3
27	−	+	−	+	+	91.7	83.3	87.5	35.3
28	+	+	−	+	+	15.0	33.3	24.2	167.4
29	−	−	+	+	+	81.7	83.3	82.5	1.3
30	+	−	+	+	+	95.0	83.0	89.0	72.0
31	−	+	+	+	+	66.7	75.0	70.9	34.4
32	+	+	+	+	+	58.3	66.7	62.5	35.3

$N = 32$						$\Sigma/N = 47.6$	$\Sigma s_h^2 = 1232$

$$s_T = \sqrt{\frac{\sum\limits_h s_h^2}{N}}$$

$$s_T = 6.2$$

the entire experiment of all 32 configurations was run twice: a full replicate. Calculating standard deviations at each experimental configuration allows us to check the significance of the results. If a large
experimental error exists, the measured responses will be a function
of the inherent variability in the experiments, not the design variables.

Step 4: Testing.   Each of the trials is tested in random order. The replicate experiments are integrated randomly into this testing. After each
test, the blender is dried, a new plunger variant is inserted through the
blender lid, a new quantity of ice is introduced, and the blender is cy-
cled for 10 seconds.

After measuring the blender efficiency, the results are entered into the
design matrix, as shown in Table 18.4. The average response, $\bar{y}_h$, and
variance, s_h^2, are calculated for each trial.

Step 5: Analysis.   Let's consider analyzing the results with a graphical approach. First, response diagrams will let us focus on the main effects of the experiments. Figure 18.9 shows an example response
diagram for d_4, the plunger force. Two-way diagrams may then be created to determine interaction effects.

Based on this analysis, plunger force is shown to be critical, it causes a
range of output of about 40% (the difference between the high and
low point of the line in Figure 18.9), and the experimental error was

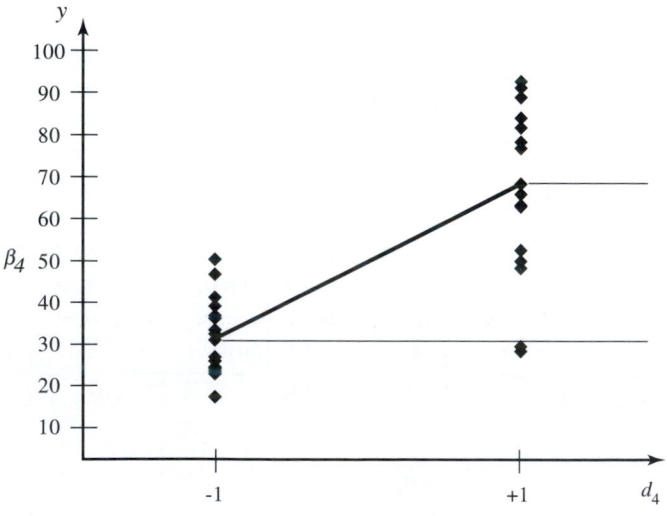

▼ *Figure 18.9.*
Response diagram for the plunger-force design variable.

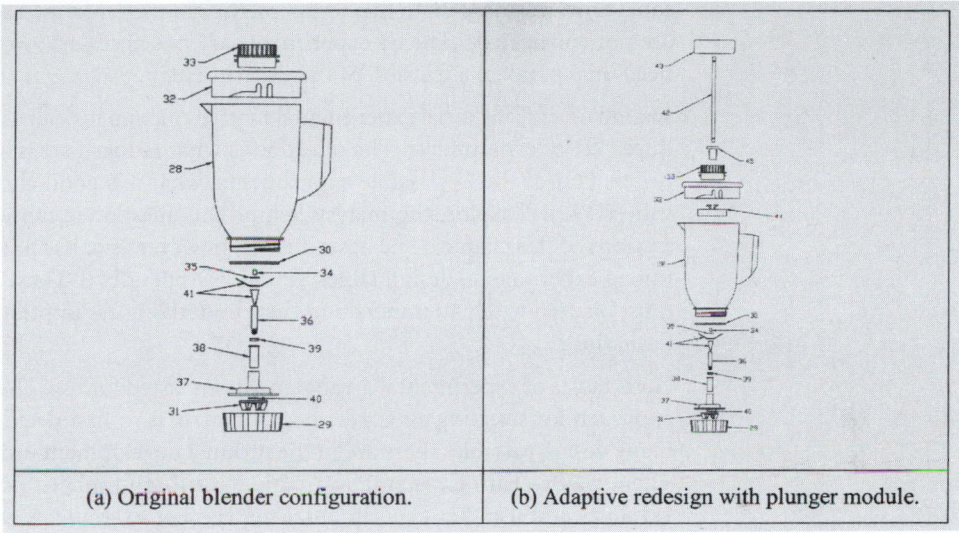

| (a) Original blender configuration. | (b) Adaptive redesign with plunger module. |

▼ *Figure 18.10.*
Original and adaptive-plunger blender configurations. (a) Original blender
configuration. (b) Adaptive redesign with plunger module.

about 6%. The high level of plunger force is needed to approach the target value of the performance metric. The remaining variables are significant to the blender performance; in fact, they interact to produce varying results across the experiments.

The preferred choices of variable combinations are highlighted with boldface in Table 18.4. Each of these combinations is significant compared to the other trials (based on the overall standard deviations). The question arises as to which combination should be chosen. Considering cost, mold complexity, and operations, trial 9 is chosen from the table. This combination of variables results in a small rectangular plunger, without prongs, and without rotation.

This embodiment choice must now be detailed. Specific materials, fasteners, tolerances, and so forth must be chosen to complete the adaptive design of the plunger module. Figure 18.10 shows example results of this process.

Fractional Factorial Designs

For problems with six or more variables, experimenting with a full-factorial set of experiments is not usually practical. For six variables and experimenting at two levels, 64 experiments are needed, but, at the same time, we only seek to fit a seven-coefficient model (if interactions are not considered). The 64 experiments are not necessary; we really need only perform a fraction of the experiments.

Instead of a full-factorial experimental matrix, one can instead use a reduced set of experiments. The question is what reduced set is appropriate. Out of the 2^n possible experiments, what is a good choice of values? We will explore the analysis behind this question in subsequent sections of this chapter. The basic answer, however, is to use predetermined experimental design tables, found in Appendix B. These tables must be used wisely to understand their underlying assumptions and limitations.

Here, tables of experimental arrangements are listed for use. The basic approach for selecting an experimental matrix is to first decide how many design variables there are in the product development problem. Then, we determine the smallest table with that number of design variables and use that experimental matrix. For example, with four design variables, Table B.3 can be used, with eight experimental trials. For 10 design variables, Table B.4 can be used, with 32 experimental trials. This selection defines the experimental matrix to use. The responses should be determined at each experimental arrangement, and then the graphical technique to solving regression equations can be used to determine better performing configuration values. Alternatively, a regression equation may be explicitly calculated for the experimental results.

Extended Method: Interactions

The basic discussion above presents a linear coefficient model. Higher order terms might also need to be included in the regression, for more complex phenomena. To present this material, we first review main effect modeling from a statistical point of view.

Main Effects

The previous section derives the experimental response using regression fitting of a linear model. Let's now relax the linearity assumption and construct a model of the response. First, the *average response* over all experiments is defined as the average of all response values:

$$\beta_0 = \bar{y} = \frac{1}{2^n} \sum_{h=1}^{2^n} f(\vec{d}_h) \qquad (18.24)$$

Given no other statements, the average response is a good estimate as to what the result will be.

The next expansion of the model is to predict the nominal impact of changing any one variable. The *main effect* of a variable represents the contribution or sensitivity of a given design variable, d_i, on the performance response. By calculating a main effect, we can determine if a design variable is significant in changing the response (implying that it has a greater effect than the experimental error). We can also determine if the design variable has a relatively small or large effect on the response (its sensitivity). The more sensitive the variable, the more important it will be to the design team.

Mathematically, we define the main effect of a d_i as the difference between the average of f with d_i at each of its extremes, a $(+1)$ or a (-1), and the overall average of f across all experiments. That is, we first define:

$$\beta_i^+ = \text{Main Effect of } d_+^i = \frac{1}{2^{n-1}} \sum_{\vec{d}_h : d^i = +} f(\vec{d}_h) - \beta_0 \quad (18.25)$$

A similar expression exists for d_i^- but with d_i at the (-1) value. This expression defines the function β_i^+ at both its high and low levels, that is, $(+1)$ and (-1) values. For example, for a 2^3 design, the main effect of d_2^+ is:

$$\beta_2^+ = \frac{y_b + y_{ab} + y_{bc} + y_{abc}}{4} - \qquad (18.26)$$

$$\frac{y_1 + y_a + y_b + y_{ab} + y_c + y_{ac} + y_{bc} + y_{abc}}{8}$$

Then, the *main effect of d_i*, considering both levels of d_i, is defined as the difference between these high and low levels:

$$\text{Main Effect of } d_i = \beta_i^+ - \beta_i^- . \qquad (18.27)$$

That is, the main effect of d_i is the difference in the response when d_i is changed, determined by averaging over any other variable changes. With some algebra, one can show that:

$$\beta_i^+ - \beta_i^- = \beta_i(d_i^+ - d_i^-) \quad \text{or} \quad \beta_i = \frac{\beta_i^+ - \beta_i^-}{(d_i^+ - d_i^-)} \quad (18.28)$$

where β_i is determined in a linear regression analysis as before. So for normalized variables, one can simply inspect the regression coefficient β_i to determine the main effect of d_i, since $d_i^+ - d_i^- = 2$ for all variables.

Eq. (18.27) represents the effect of a design variable on the response as the design variable is changed from its low level to its high level. The reason this main effect value is important is that, conceptually, if it is near zero, then varying d_i has no effect on f, implying that the design

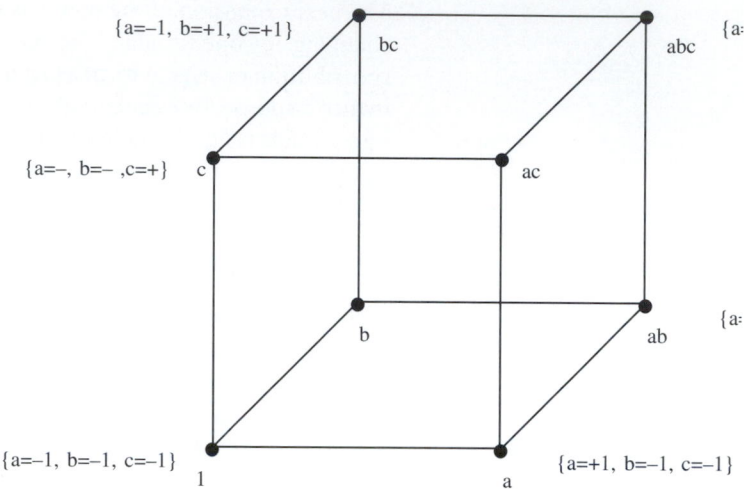

▼ Figure 18.11.
Graphical representation of a 2^3-factorial design. Each orthogonal face of the cube represents an axis of one design variable.

variable is irrelevant. On the other hand, if the main effect is relatively large, then d_i has a large effect on the performance.

A graphical interpretation can be given to the main effects if we represent the experiments as a hypercube. Consider a 2^3 experiment. Then, each experimental trail (and its response value) can be considered a vertex of the cube (Figure 18.11).

The main effect of each variable can be considered as the average of the values of the cube face with a high or low value of the variable. For example, the average of the high and low face of d_1 is shown in Figure 18.12.

The β_i coefficients themselves can also be graphically interpreted, equivalently to the coefficients as shown now as Figure 18.13. Each β_i times the distance between levels is one half the difference between the two averages of the faces with d_i at its high and low value (Eq. 18.25).

Interaction Effects

If we only consider the main effects, we are only considering the average performance deviation, by examining one variable at a time. The resulting solution assumes independence of the variables, that is, f will improve by changing the value of d_i as specified no matter what the values of d_j. This solution hides any possible performance increase achievable by simultaneously moving the design variable values away

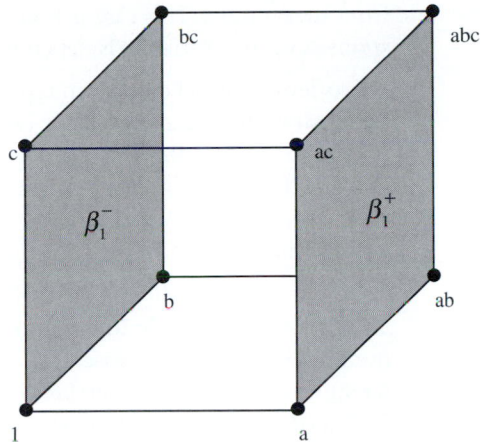

▼ **Figure 18.12.**
Graphical representation of a 2^3-factorial design, where the high and low planes of d_1 are highlighted. The values shown are the averages over the gray areas.

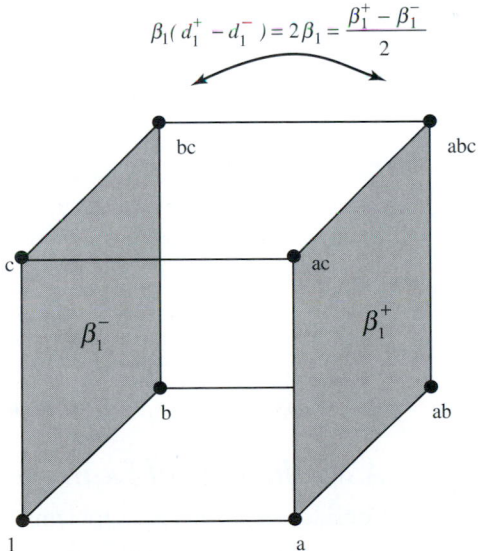

▼ **Figure 18.13.**
Graphical representation of a 2^3-factorial design, where the high and low planes of d_1 are highlighted for the regression model coefficient.

from their means. This case is known as an *interaction*, when the response over two variables is dependent on their combined effect.

A two-level factorial model can represent more than just main effects. To illustrate this idea, consider a two-variable design: $2^2 = 4$ experimental points. A main effect model is linear only: It defines a planar surface over $\mathbb{D}$. We need three points to define a plane, but we have four experimental values. We can use the fourth point to either obtain a better fit of the plane or to obtain a higher order fit other than a plane. This argument extends to higher dimensions.

To observe the possible interaction effects, we expand our additive model to include effects caused by combinations of the variables at different levels. However, understanding the interaction effects is greatly facilitated by considering the case of $d_i \in \mathbb{R}$, and $f:\mathbb{R}^n \to \mathbb{R}$. Then we can interpret the β_{ij} as coefficients in a Taylor series approximation of f and skip the additivity discussion. That is, if we additionally assume a coefficient model at the outset and approximate f by:

$$f(\vec{d}) = \beta_0 + \sum_{i=1}^{n} \beta_i d_i + \sum_{i=1}^{n} \sum_{j=i+1}^{n} \beta_{ij} d_i d_j \tag{18.29}$$

then the coefficients β can again be solved for using a regression fit similar to the previous development; the only difference is that the matrix X is expanded to include $d_i d_j$ terms. When the number of variables is greater than two, this expansion can be extended to include $d_i d_j d_k$ for three-variable models, $d_i d_j d_k d_l$ for four-variable models, etc. The expanded model, given by Eq. (18.29) with or without further terms, is known as a multilinear model.

To determine $\vec{d}*$, we now have more information than a simple main-effect model, we have Eq. (18.29). We want to know where it has a minimum over our domain $\mathbb{D}$. Generally, this can be determined through a constrained optimization:

minimize $\quad f(\vec{d}) = \beta_0 + \sum_{i=1}^{n} \beta_i d_i + \sum_{i=1}^{n} \sum_{j=i+1}^{n} \beta_{ij} d_i d_j$

subject to $\quad\quad\quad\quad\quad -1 \le d_i \le 1$

$$\tag{18.30}$$

A Mathematical Example

Consider a mathematical problem involving three variables. The experimental data are shown in Table 18.5.

We can fit a model:

$$f(\vec{d}) = \beta_0 + \beta_1 d_1 + \beta_2 d_2 + \beta_3 d_3 + \beta_{12} d_1 d_2 + \beta_{13} d_1 d_3 + \beta_{23} d_2 d_3$$
$$+ \beta_{123} d_1 d_2 d_3$$

TABLE 18.5.   EXAMPLE DATA FOR A 2^3-FACTORIAL EXPERIMENT

Trial	Yates order	d_1	d_2	d_3	y
1	1	-1	-1	-1	30
2	a	$+1$	-1	-1	34
3	b	-1	$+1$	-1	37
4	ab	$+1$	$+1$	-1	38
5	c	-1	-1	$+1$	32
6	ac	$+1$	-1	$+1$	52
7	bc	-1	$+1$	$+1$	31
8	abc	$+1$	$+1$	$+1$	56

Notice that this model has eight terms, that is, a mean followed by three main effects, two second-order interaction effects, and one third-order interaction effect. Solving, the model for the regression terms (coefficients) gives:

$$\beta_0 = 38.75$$
$$\beta_1 = 6.25$$
$$\beta_2 = 1.75$$
$$\beta_3 = 4.0$$
$$\beta_{12} = 0.25 \tag{18.31}$$
$$\beta_{13} = 5.0$$
$$\beta_{23} = -1.0$$
$$\beta_{123} = 1.0$$

Both the main effect and the interaction effect models can be solved that minimize the function representing the response, linear or multilinear. Comparing solution predictions at the low levels of all design variables:

Main effect	$\vec{d}* = (-1,-1,-1)$	$y_{predicted} = 26.5$
Interaction effect (up to 2nd-order terms only)	$\vec{d}* = (-1,-1,-1)$	$y_{predicted} = 31.0$
Interaction effect (including all terms)	$\vec{d}* = (-1,-1,-1)$	$y_{predicted} = 30.0$
Experiments	$\vec{d}* = (-1,-1,-1)$	$y_{measured} = 30.0$

Note the results. The main effect, 2nd order, full interaction, and experimental data all result in the same solution point $(-1,-1,-1)$. They vary in their accuracy of predicted y.

A Basic Product Application: Redesign of a Toy Solar Car

Let's apply this factorial design approach to a product design. In this case, the product is a toy solar car, sold in hobby and toy stores. We wish to evolve the product to improve its performance for the customers.

The product design statement is "develop an evolved a toy solar car product that is competitive, robust, and satisfies the customer needs" (Figure 18.14). Using this problem statement, possible improvements of the solar car, from customer need analysis and product teardown, include solar panel/collection areas, power efficiency, manufacturing cost and assemblability, aesthetics (model an actual race car instead of a pinewood derby car), speed, and maneuverability. For the purposes of this example, increasing car speed is our focus.

Let's now consider a clarification of the product development to understand what parameters affect car speed. A high-level physical principle of the car speed may be stated as:

$$v_{car} = \left(\frac{2\pi}{60}\right) \frac{\omega \cdot r_g \cdot r_{rw}}{r_{rg}}$$

where ω is the angular velocity of the motor output (rpm), r_g is the radius of the pinion (motor) gear, r_{rw} is the radius of the rear wheel, and r_{rg} is the radius of the rear internal gear. Likewise, general physical principles for the motor speed and torque are given by:

$$\omega = f(T) \text{ and } T = g(r_g, r_{rw}, r_{fw}, r_{rg}, \mu_r, \mu_f, W_r, W_f, P, \dots)$$

where T is the motor torque, μ_r is the coefficient of friction in the rear axle, μ_f is the coefficient of friction in the front axle, W_r is the weight

▼ **Figure 18.14.**
Pinewood derby solar car product.

on the rear axle, W_f is the weight on the front axle, and P is the available motor power. These variables should be considered as possible factors in our physical model.

Let's consider when the motor speed and torque relationships (functions f and g) are not readily available. For this situation, the motor may also be modeled with a dc torque speed relationship, estimated by

$$\omega = mT + b \quad \text{or} \quad T = m'\omega + b'$$

This linear relationship of torque-speed for a dc motor may be determined a variety of ways. For example, if we consider the average speed of the current solar car, $v = D/t$, where D = distance traveled and t = time, we can determine the torque speed curve of the motor by placing different weights on the car, measuring the time to travel a given distance (holding solar power constant), and backing out torque through the wheels and drive train of the car.

These qualitative models provide us with a number of insights into the engineering problem at hand, that is, development of an empirical model for improving the speed of the solar car (our customer need focus). Let's now apply the experimental design procedure for developing an empirical model.

Step 1: Model the Variables

For this solar car design problem, we need to determine the response (metric) to measure, the possible noise factors to be minimized, and the control factors (design variables) to vary. These model variables are chosen as follows:

1. response (performance variable)—car speed (measured by timing the car over a fixed distance)

2. noise factors (noise variables)—friction surface, car steering position, temperature and humidity, power source proximity (intensity) to solar panels, solar collector position, wind, and so forth

3. control factors (design variables)—we desire to change/choose the solar panel characteristics, motor parameters, gear/wheel radii, bearing friction, car weight, aerodynamics, and so on. A variable power source will emulate varying motor parameters and solar panel characteristics.

In our case, for the purpose of this product redesign, we consider two control factors, while attempting to minimize the effects of the noise factors. The power source (equivalent to increased solar panel array) and the car weight are chosen, since they are expected to affect the car speed to the greatest degree. Car weight, in this case, emulates a larger motor range if more power is input to the car. We also wish to minimize car weight if it greatly affects the car speed (in a short distance).

Step 2: Determine the Boundaries for the Two Control Factors (Design Variables)

The choice of boundaries for the physical model are given by:

$d_1 \equiv$ power source (50W, 100W light bulb)

$d_2 \equiv$ car weight (low: current weight, high: double current weight).

where we wish to maximize car speed. The boundaries of the power variable represent a modest increase in solar panel size. The car weight range is chosen based on double the motor size.

For these boundaries, the current solar car will be used as the physical prototype. Weight will be added but not removed from the current car. By changing the power source (equivalent to increasing the area of the solar panel), no modifications are needed to the current car.

Step 3: Design the Experiment Matrix

Based on our experimental plan to this point, we will need to run 2^2 experiments to test all combinations of the design variables, assuming a multilinear model. We will also run a replicate, repeat experiment, for each case to check the quantity of experimental error that exists. Thus, the experimental matrix will include four trials, with each trial tested twice (in random order). In this matrix, the response of the car will be recorded for each trial and each replicate.

Table 18.6 shows the resulting experimental matrix. For this experiment, we are assuming a multilinear response of the solar car speed to changes in the control factors of power and car weight. The relationships used to formulate and complete the design matrix are as follows:

$$d_1^- = 55 \text{ W bulb}$$
$$d_1^+ = 100 \text{ W bulb}$$

TABLE 18.6. EXPERIMENTAL MATRIX FOR THE SOLAR CAR REDESIGN

Trials	d_1	d_2	d_{12}	y_{h1}	y_{h2}	$\bar{y}_h$	s_h^2	s_h
1	−1	−1	+1	5.60	5.80	5.70	0.0200	0.140
2	+1	−1	−1	4.70	4.64	4.67	0.0018	0.042
3	−1	+1	−1	7.20	7.00	7.10	0.0200	0.140
4	+1	+1	+1	5.40	5.50	5.45	0.0050	0.071

d_2^- = original car weight

d_2^+ = added fishing weights

y_{hj} = speed of car (actually the time of car in seconds, for a distance of 10 ft or 3 m)

r = number of replicates (repeated experiments) to determine experiment error ($r = 2$)

$h = 1, \ldots, N, j = 1, \ldots, r$

$\bar{y}_h$ = response avg. at a configuration = $\dfrac{\displaystyle\sum_{j=1}^{r} y_{hj}}{r}$

s_h^2 = variance = $\dfrac{(y_{h1} - \bar{y}_h)^2 + \cdots + (y_{hr} - \bar{y}_h)^2}{r}$

s_h = the standard deviation at an experimental configuration

Step 4: Perform the Experiment

The car performance is varied by changing the light source and adding weight at the center of gravity. Eight total trials are run in random order (four configurations multiplied by two replicates); the results are recorded in Table 18.6. The light source moves with the car and is held a fixed distance above the solar panel. Times are recorded from when the light source is turned on to when the car crosses a distance of 3 m. "Time" is thus the metric response for measuring the speed of the car. Time and speed are inversely related.

Step 5: Analyze, Verify, and Test Results

We now wish to analyze the experimental results to create an empirical model of the solar car product. To analyze the data, we calculate the mean for each experimental configuration, in addition to the standard deviation. We also calculate the experimental error s_T, here ± 0.10 sec. If a multiple of the standard deviation (e.g., $3s$) is greater than the response coefficient, then the controlled change of design variables cannot be distinguished from experimental error. Therefore, a factor is significant if its response coefficient is larger than 0.30 sec.

In our case, we calculate a simple regression model of the empirical data and determine the significance of the terms in the model by considering the standard deviation. The regression model (bilinear) is:

$$y = \bar{y} + \beta_1 d_1 + \beta_2 d_2 + \beta_{12} d_1 d_2$$

where y-bar is the average response for all trials, and the beta terms are the sensitivity coefficients of the main and interaction (coupled)

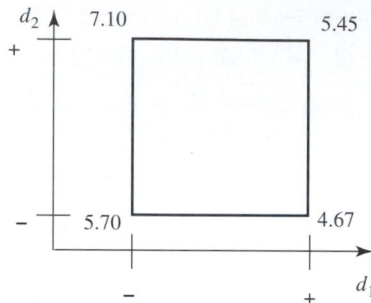

▼ Figure 18.15.
Design space and response values for
the solar car empirical model.

effects of the control factors. The response diagram is then given by
Figure 18.15.

Consider the "planes" of the control factors in Figure 18.15, as they
vary from a "+1" state to a "−1" state. What insights do we have by
inspection? For example, the weight affects the car speed less than the
change in power source (for the ranges chosen). Next we calculate the
model coefficients. For the main effects (average of the high state minus
avg. of low state):

$$\beta_1 = \frac{(4.67 + 5.45)}{2} - \frac{(5.70 + 7.10)}{2} = -1.34 \text{ sec}$$

This result implies that as we increase the power from its "+1" state to
its "−1" state (two units of change), the time will decrease by 1.34/2 sec.
Likewise, for car weight:

$$\beta_2 = \frac{(7.10 + 5.45)}{2} - \frac{(5.70 + +4.67)}{2} = 1.09 \text{ sec}$$

In turn, this result implies that as we increase the weight of the car, the
time will increase, but not as much as the power effect! When comparing the main effects to the experimental noise, both effects are significant, since they are greater than three times the average standard
deviation of the trials.

The interaction between the power input and the design weight of the
car may now be calculated. This interaction effect is given by:

$$\beta_{12} = (\text{diff. in } x_2 \text{ high}) - (\text{diff. in } x_2 \text{ low})$$

$$= \frac{(5.45 - 7.10)}{2} - \frac{(4.67 - 5.70)}{2} = -0.31 \text{ sec}$$

This result is close to the experimental noise; however, the interaction shows that weight and power are coupled. This means that the speed of the car does not change independently with weight and power. As we increase weight, the power will also increase and compensate for the weight increase, thus causing a decrease in coupled time.

Step 6: Evaluate the Results (Conclusions)

A number of conclusions may be extracted from the experimental results and empirical model. First, a 20–30% performance increase can be achieved by increasing power or decreasing weight (assuming that a new, more powerful motor would add double the weight, which is very conservative). Second, weight and power interact, implying that we need to be concerned about which motor torque-speed curve we are on for efficiency. Third, because weight and power interact, we can actually achieve better speed than predicted by the main effects of power and weight. To understand this interaction further, let's consider the two-way diagram for the experiments (Figure 18.16).

This diagram confirms the results of the regression models. Because the slopes of the weight design variable for its low state versus its high state are not parallel, an interaction exists between weight and power. This interaction implies that we must carefully choose a motor for the solar car.

Based on these findings, we conclude that we can obtain significantly increased speeds, even if we add a larger motor. Applying this conclusion, Figure 18.17 shows a possible evolved concept for the solar car product. The new solar car's feel and its aesthetics have been evolved to a modern race car concept; the motor is larger, the solar panel is larger, weight is reduced in the body panels, and multiple gears have been

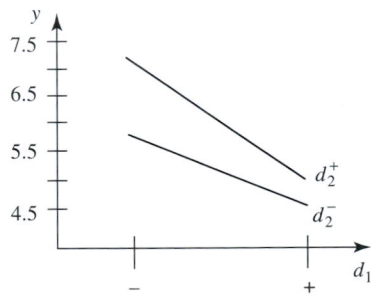

▼ **Figure 18.16.**
Two-way diagram.

▼ *Figure 18.17.*
A possible evolved solar panel product, replacing the pinewood derby concept.

added to adjust torque speed characteristics. The empirical tests lead to the motor, solar panel, and weight modifications.

Summary: Basic DOE Method for Product Testing

Building on the solar car application, we can summarize the basic approach to design of experiments with full-factorial configurations. This approach utilizes the mathematical methods discussed in this section.

Step 1:   Identify the performance variables (responses), design variables (control factors), and noise variables (uncontrolled factors). State all high-level physical principles to clarify and understand the product testing.

Step 2:   Define the target values for the responses and boundaries for the design variables.

Step 3:   Plan the prototype testing by developing an experimental matrix, choosing the number of trails, levels for each design variable, number of replicates, and how the response will be measured. In addition, determine the experimental apparatus and methods for eliminating noise sources.

Step 4:   Execute experiments by randomizing the trails and replicates; record the data to the accuracy needed for analysis.

Step 5: Analyze the results by constructing a regression model or response diagram(s). Calculate and average the variances of the replicates; determine the overall standard deviation from the average variance and compare the standard deviation to coefficients of the regression model. After determining the significant results, use the regression model and/or graphs to make design changes to the product.

III. DESIGN OF EXPERIMENTS: REDUCED TESTS AND FRACTIONAL EXPERIMENTS

Full-Factorial Inefficiencies

Next let's discuss experimenting with two levels at each design variable, but not with a full set of $N = 2^n$ experimental arrangements (Box 1978; John 1971). A full-factorial analysis is inefficient in the economy of experimental trials. Usually we only want to model zero-, main-, and second-order interaction effects, which really only requires

$$1 + n + \frac{n(n-1)}{2}$$ variables to fill the model. On the other hand, a

full-factorial analysis performs 2^n experiments. For example, consider 10 design variables. $N = 2^{10} = 1024$ experiments. In the expression:

$$y = \bar{y} + \beta_i d_i + \beta_{ij} d_i d_j + \beta_{ijk} d_i d_j d_k + \cdots, \qquad (18.36)$$

the distribution of effect terms shown in Table 18.7 could be determined.

TABLE 18.7.   10-FACTOR DISTRIBUTION OF COEFFICIENTS

# of effects	Effect type	Term
1	Average response	β_0
10	Main effects	β_i
45	2-factor interactions	β_{ij}
120	3-factor interactions	β_{ijk}
210	4-factor interactions	β_{ijkl}
252	5-factor interactions	β_{ijklm}
210	6-factor interactions	$\beta_{ij\ldots}$
120	7-factor interactions	$\beta_{ij\ldots}$
45	8-factor interactions	$\beta_{ij\ldots}$
10	9-factor interactions	$\beta_{ij\ldots}$
1	10-factor interaction	$\beta_{ij\ldots n}$

If we only wish to consider up to 2-factor interactions, we really only need to be running, at most, around 60 experiments (disregarding replicates) to determine the 56 terms, not 1024 experiments. This analysis outlines a method for determining how many experiments must be carried out.

Yates Standard Order

Instead of labeling the experiments by $\vec{d}_h$, a more informative reference scheme is to use what is known as the *Yate's Standard Order*. The Yates order applies only for two-level designs. With the Yate's order, we label each design variable with letters α rather than numbers i:

$$a \approx d_1 \quad b \approx d_2 \qquad (18.32)$$

We then label each experimental arrangement $\vec{d}_h$ by whether or not α is a $(+1)$. If so, we include α in the label. If $\vec{d}_h$ has α at a (-1) level, we do not include α in the label. The trial $\vec{d}_1$, where all α are at low $(-)$ levels, is a special case. We identify $\vec{d}_1$ as 1. The remaining trials are defined as above. Thus, $\vec{d}_2$ is identified as a. $\vec{d}_4$ is identified as ab. The Yates order is shown for the three-variable experimental matrix in Table 18.8. Note that we should beware of the possibility for slight confusion in this nomenclature: a can mean d_1 or it can mean $\vec{d}_2$ depending on the context.

We also define y_α as:

$$y_\alpha = f(\vec{d}_\alpha)$$

where α is the Yates order label, and $\vec{d}_\alpha$ is the experimental arrangement corresponding to α. Thus, $y_{ab} = f(+1, +1, -1)$ in a three-variable design.

TABLE 18.8. YATES ORDER FOR A 2^3-FACTORIAL DESIGN

Trial	Trial vector label	Yates order	d_1	d_2	d_3
1	$\vec{d}_1$	1	-1	-1	-1
2	$\vec{d}_2$	a	$+1$	-1	-1
3	$\vec{d}_3$	b	-1	$+1$	-1
4	$\vec{d}_4$	ab	$+1$	$+1$	-1
5	$\vec{d}_5$	c	-1	-1	$+1$
6	$\vec{d}_6$	ac	$+1$	-1	$+1$
7	$\vec{d}_7$	bc	-1	$+1$	$+1$
8	$\vec{d}_8$	abc	$+1$	$+1$	$+1$

Yates Interaction Labels

The Yates labeling can also be used to describe interaction effects. To define the Yates labels for an interaction effect, we make the same associations as in Eq. (18.32). That is:

$$\text{interactions with } d_1 d_2 \approx ab, \, d_2 d_3 \approx bc, \text{ etc.} \qquad (18.33)$$

We can arrange this ordering as an experimental matrix, similar to the full-factorial table (Table 18.9). Each row represents an experimental arrangement, and the Yates label is shown. We extend the matrix horizontally to describe the possible interactions in each experimental arrangement. Each possible interaction is shown along the columns, again as a Yates label. The "1" column is a special *identity interaction,* used to represent the average of the entire matrix. Its effect is defined as always positive.

For the remainder of the nontrivial effects, we need to determine whether an interaction effect requires a $(+1)$ or a (-1) in the matrix. This operation is completed by simply "multiplying" the $(+1)$'s and (-1)'s together in each column.

Determining the Number of Experiments

Let m_i be the number of levels for each d_i. For two-level factorial designs, $m_i = 2$ for all i. In the general case, to fit an interaction coefficient model the number of experiments to execute is:

$$N = 1 + \sum_{1}^{n} (m_i - 1) + \sum_{i=1}^{n} \sum_{j=i+1}^{n} (m_i - 1)(m_j - 1) \qquad (18.34)$$

TABLE 18.9.   YATES INTERACTION LABELS

	Yates order	d_1 a	d_2 b	d_3 c	Interactions				
					ab	ac	bc	abc	1
$\bar{d}_1$	1	-1	-1	-1	$+1$	$+1$	$+1$	-1	$+1$
$\bar{d}_2$	a	$+1$	-1	-1	-1	-1	$+1$	$+1$	$+1$
$\bar{d}_3$	b	-1	$+1$	-1	-1	$+1$	-1	$+1$	$+1$
$\bar{d}_4$	ab	$+1$	$+1$	-1	$+1$	-1	-1	-1	$+1$
$\bar{d}_5$	c	-1	-1	$+1$	$+1$	-1	-1	$+1$	$+1$
$\bar{d}_6$	ac	$+1$	-1	$+1$	-1	$+1$	-1	-1	$+1$
$\bar{d}_7$	bc	-1	$+1$	$+1$	-1	-1	$+1$	-1	$+1$
$\bar{d}_8$	abc	$+1$	$+1$	$+1$	$+1$	$+1$	$+1$	$+1$	$+1$

where the first term (1) is the average response, the first sum corresponds to the main effects, and the last sum counts all two-level interactions. If not all two-level interactions are deemed important, the last term in Eq. (18.34) interaction can be reduced appropriately. Likewise, if higher order interactions are desired, they can be added to the expression. For two-factorial experiments, Eq. (18.34) reduces to:

$$N = 1 + n + \frac{n(n-1)}{2}. \qquad (18.35)$$

This analysis provides the minimum number of experiments that must be performed for the accuracy of the model proposed. The next question is what experimental arrangements $\vec{d}$ out of the total should we use and which can we ignore? For example, in the above example, out of the 1024 total combinations, which 60 should we test? Intuitively, some combinations of 60 are better than others.

The overall goal is to reduce the number of trials from 2^n to a "reasonable" number. This number of trails is equivalent to the number of terms to use in the additive (regression) model. For the minimum number of experiments, each experiment provides another coefficient. The model we are using is basically a coefficient model:

$$y = \bar{y} + \beta_i d_i + \beta_{ij} d_i d_j + \beta_{ijk} d_i d_j d_k + \cdots \qquad (18.36)$$

To reduce the number of experiments, consider each unnecessary higher order term one at a time, and split the total number of experiments into groups, one group for each value of coefficient $\beta_{ij...}$. Each of these equal sized subsets we will call a *block*.

Using only one of the blocks for experimentation eliminates our ability to resolve that term. It also cuts the number of experiments to be completed from the total possible. For example, with two-level experiments, $m_i = 2$ for all variables i, and so a block will reduce the number of experiments by $\frac{1}{2}$. If this block is executed for the higher order, unnecessary terms, the number of experiments is reduced.

As an example, consider a six-variable design. Let Block I be defined as all experiments with *abcdef* = $(+1)$, and let Block II be defined as all experiments with *abcdef* = (-1). the resulting design matrix is shown in Table 18.10.

There are 32 trials in each block. If we only experiment with Block I or II exclusively, we *confound* the *abcdef* interaction. We cannot determine the β_{abcdef} term's values, since in each block *abcdef* is either all $(+1)$ or all (-1). On the other hand, the number of experiments is also reduced from 64 to 32. This approach provides a basis for fractionalizing and is the reason for developing the Yates ordering scheme for 2-factorial experiments.

TABLE 18.10. DESIGN MATRIX FOR A 2^6-FACTORIAL DESIGN

Experiment	$d_1 = a$	$d_2 = b$	$d_3 = c$	$d_4 = d$	$d_5 = e$	$d_6 = f$	$abcdef$	Test block
			Design variable (Yates order)					
1	−1	−1	−1	−1	−1	−1	+1	I
2	+1	−1	−1	−1	−1	−1	−1	II
3	−1	+1	−1	−1	−1	−1	−1	II
4	+1	+1	−1	−1	−1	−1	+1	I
5	−1	−1	+1	−1	−1	−1	−1	II
6	+1	−1	+1	−1	−1	−1	+1	I
7	−1	+1	+1	−1	−1	−1	+1	I
8	+1	+1	+1	−1	−1	−1	−1	II
9	−1	−1	−1	+1	−1	−1	−1	II
10	+1	−1	−1	+1	−1	−1	+1	I
.								
.								
.								
63	−1	+1	+1	+1	+1	+1	−1	II
64	+1	+1	+1	+1	+1	+1	+1	I

Orthogonality

With a reduction from the 2^n experiments, there is still a capability to determine some, but not all, of the total 2^n possible coefficients in the additive model. Care must be taken, however, not to bias their values during the model reductions in terms of the design variable values. The same number of (+1) and (−1) values must occur for the remaining interactions. This condition is an *orthogonal* requirement in the experiments.

Mathematically, two experiments $\vec{d}_1$, $\vec{d}_2$ are *orthogonal* if the sum of their multiplied elements ($[−1,1]$ normalized values) equals zero. That is, if:

$$\sum_{h=1}^{N} d_{ih} d_{jh} = 0, \qquad (18.37)$$

where $\vec{d}_{ih} \in [−1,1]$. An experimental matrix is called orthogonal if all of its experimental arrangements are mutually orthogonal. As an example, a full-factorial matrix is orthogonal. It is a unique two-level factorial matrix with 2^n experiments. The experiments are called orthogonal, because if one considers a 2^n dimensional space, such as $\mathbb{R}^{2^n}$, the experimental arrangement vectors are orthogonal in this space; for a Euclidean vector space, the angle between the vectors is 90°.

The orthogonality property is the requirement that must be maintained to ensure the experimental results are not biased when determining the performance response. This requirement forces a constraint on the process of determining which arrangements of variable values to choose for experimentation. The arrangements must be mutually orthogonal.

Base Design Method

A method to determine an orthogonal allocation, referred to as the *base design method,* is discussed below. This method helps us to choose partial factorial designs when experimenting with physical product prototypes.

The method involves several steps:

▶ Determine the number of experiments to execute (to the nearest 2^b), where b is a chosen exponent.

▶ Form a 2^b base design.

▶ Add n-b columns to the matrix, each new column a higher order column for the 2^b base design, but that will be used for the remaining b to n design variables in the full n variable design.

Thus, a *base design* is formed, and it is expanded out for the remaining variables, but in a way such that the additional variable is only confounded with a higher order term. This process ensures orthogonality in the matrix.

Overall, we first determine how many experiments to perform. In this method, this choice must be a number among $2^b = 2, 4, 8, 16, 32, 64, \ldots$ for a 2-factorial design. We then complete a full-factorial matrix, 2^b rows of experimental arrangements, and b columns of design variables.

At this stage, b of the n design variables are allocated. This allocation is the "base design." There are n-b design variables to still allocate for experimental values. To do this, we add n-b columns to the matrix. Each new column will correspond to the levels tested for the remaining variables to complete the full n variable allocation. In each new column, we select levels $(+1, -1)$ for the additional design variable, which correspond to the levels of a higher order interaction of the first b variables. The levels selected will then be orthogonal to the base design levels.

As an example, consider five design variables and a capability to execute only around 10 experiments. If we want a full second-order interaction model, we need $1 + 5 + 10 = 16$ experiments. We are only

TABLE 18.11.   FOUR-VARIABLE DESIGN OF EXPERIMENT TABLE

Exp.	d_1 a	d_2 b	d_3 c	d_4 $d = abc$	ab	ac	ad	bc	bd	cd	abc	abd	acd	bcd	$abcd$	1
1	−1	−1	−1	−1	+1	+1	+1	+1	+1	+1	−1	−1	−1	−1	+1	+1
2	+1	−1	−1	+1	−1	−1	+1	+1	−1	−1	+1	−1	−1	+1	+1	+1
3	−1	+1	−1	+1	−1	+1	−1	−1	+1	−1	+1	−1	+1	−1	+1	+1
4	+1	+1	−1	−1	+1	−1	−1	−1	−1	+1	−1	−1	+1	+1	+1	+1
5	−1	−1	+1	+1	+1	−1	−1	−1	−1	+1	+1	+1	−1	−1	+1	+1
6	+1	−1	+1	−1	−1	+1	−1	−1	+1	−1	−1	+1	−1	+1	+1	+1
7	−1	+1	+1	−1	−1	−1	+1	+1	−1	−1	−1	+1	+1	−1	+1	+1
8	+1	+1	+1	+1	+1	+1	+1	+1	+1	+1	+1	+1	+1	+1	+1	+1

carrying out around 10, so we must confound the second-order inter-action effects.

In this situation, the closest 2^b to 10 is $2^3 = 8$, so $m = 3$. We form a 2^3 full-factorial matrix, and add two columns for d_4 and d_5. Then, we choose values for d_4 that exactly correspond to the highest order effect in the d_m matrix, here abc. d_4 gets $(+1)$ and (-1) whenever abc does. Next, we choose d_5 from the next lower equivalence class of interactions, here ab, ac, or bc. And so on. Without performing a full factorial analysis, there will *always* be confounding effects; we as designers must choose what we want confounded.

To determine which effects are confounded, we denote d_4 by d and d_5 by e in the Yates order and fill out the columns. Some interactions of d and e will have their columns appear exactly like the original variables. These are confounded. A four design variable example is shown in Table 18.11, using eight experiments. Notice which interactions are confounded and that confounding is unavoidable. In this allocation:

$d_1 = a$ and bcd,	ab, and cd
$d_2 = b$ and acd,	ac, and bd
$d_3 = c$ and abd,	bc, and ad
$d_4 = d$ and abc,	1, and $abcd$

Each of these pairs are confounded. If a coefficient value is determined for each effect independently, the exact same number will be determined for both. Stated another way, if an experimental matrix X is constructed with both terms in it and using the experimental trials as indicated, then their columns in X are identical and $X^T X$ will not be invertable. We cannot solve for both effects' coefficients. The effects of these interactions are combined together into the one number determined; their effects cannot be isolated from each other.

Thus, in this experiment, if the $O(3)$ terms are ignored, the main effects are not confounded. Pairs of the interaction effects are confounded. If one of these pairs is shown to have a large effect on the performance response (relative to the main effects), this result is an indication to the design team that more experiments should be performed to unconfound these variables.

The generalization of the base design method to a general 2^{n-b} fractional factorial design is as follows:

1. Identify n design variables, and the two levels for each variable.
2. Determine how many experiments N will be performed, where $N = 2^b > n$.
3. Complete a 2^b full-factorial base design.
4. Expand the matrix with n-b columns.
5. Fill in the new n-b columns with higher order interaction effect levels of the base design.

This technique is often called the *2^{n-k} fractional factorial method,* where $b = n - k$ defines k.

Higher Dimensions Fractional Factorial Designs

With three or more values per variable, the number of experiments that must be executed rises drastically. The growth rate is m^n, where m is the number of levels and n is the number of variables. Clearly the number of experiments becomes impractical very quickly.

Therefore, for most experimentation, levels higher than two are not usually used, especially during the initial set of experiments. This result may seem counterintuitive, but nonetheless true. In product development, we are usually very resource constrained. Further, we usually start with something that will work at some level. Typically, we are just finding a better performing, feasible configuration. In this situation, two-level experimentation offers the most information per experiment.

At other times, a multilevel set of experiments is desired, such as might be the case for

▶ objectives that rapidly change with the experimental variables
▶ discrete option variables that have more than two possibilities

In these situations, multilevel experiments are used on some or all of the design variables, since a linear interpolation is not accurate or has no meaning. More terms are thus needed in our additive model. In a three-level factorial design, for example:

$$y = \beta_0 + \beta_i d_i + \beta_{ij} d_i d_j + \beta_{ijk} d_i d_j d_k + \cdots \qquad (18.38)$$

is not a complete model. Thus, terms such as $a^2 b$ can be considered in addition to ab and abc.

The basic approach, however, remains the same as for 2-factorial experiments.

▶ Orthogonality must be ensured.

▶ Confounding of effects must be considered carefully.

For three-level experimentation, we will use the normalized variable values $\{-1, 0, +1\}$, similar to using $\{-1, +1\}$ for the two-level experimentation. For higher levels, one can use fractions.

Experimental Arrangements

Selecting a reduced set of experimental arrangements is more complicated than with two levels and involves the use of *contrasts* (Montgomery 1991). The reason for using more than two levels is that additional interactions can be considered. These interactions also make the partial factorial experimental design more difficult. Generally, for three and more level designs,

▶ Main effects $a, b, \ldots$ can be considered.

▶ Two-way interactions $ab, ac, \ldots$ can be considered.

▶ Three-way interactions $ab^2, abc, ac^2, \ldots$ can be considered.

Interactions beyond three-way are not typically considered; one is usually better off forming a nonlinear expression analytically and fitting physically derived coefficients.

The Yates ordering scheme and the base design approach do not work with higher order designs; they are for 2-factorial experiments. Instead, one determines what interactions are not important, and, from these blocks, the m^n possible experiments are reduced to a reasonable number. For the most part, others who have faced the same experimental design problem have already completed such blocking. One can find tables of experimental arrangements in Appendix B that are orthogonal and valid for use.

IV. STATISTICAL ANALYSIS OF EXPERIMENTS

When testing physical product prototypes, the regression fit (empirical model) must be statistically analyzed for confidence in the fit and significance of each design variable. There are at least five parameters that are useful:

1. standard error of the replicates
2. correlation coefficient of the fit
3. standard error of the residuals of the fit
4. t-ratio (*t*-statistic or *t*-test) of each variable of the fit or replicates
5. Analysis of Variance (ANOVA): *F*-ratio of the fit and replicates

Beyond calculating these values analytically, they may also be graphically determined, such as the concept of *plotting the residuals*. One can then visually examine plots for trends in the random data. We have previously discussed the standard error of the replicates; now we explore the remaining parameters. To define these parameters, however, we must first define statistical degrees of freedom.

Degrees of Freedom

A *degree of freedom* is the number of independent variables that contribute to a statistical quantity. Degrees of freedom are used to properly weight component variances into the summary quantities.

- A set of nine experiments has nine degrees of freedom.
- The overall average, on the other hand, has one degree of freedom.
- The number of degrees of freedom associated with a variable d_i is equal to one less than the number of levels M_i for that variable (the one being subtracted for the average).
- The degree of freedom in a sum of squares is $M - 1$, where M is the number of elements in the sum. The -1 comes from the constraint that $\sum(y_i - \bar{y})^2 = 0$. Since the mean is subtracted, the value of y_i is subtracted from itself, resulting in a loss of degree of freedom.
- The degrees of freedom associated with interaction effects are given by the products of the degrees of freedom for each of the two factors. That is, the degrees of freedom for the interaction effect between two variables a and b is $(m_a - 1)(m_b - 1)$.

Correlation Coefficient

An indication of how well experimental data reflect a linear fit is the *correlation coefficient*. It ranges from -1 to $+1$ inclusive, where 0 indicates a purely random set of data with no correlation, and ± 1 indicates a positive or negative relationship. Table 18.12 depicts several values of the correlation coefficient for one-dimensional data. Note that the information in Table 18.12 is much different than the response diagrams of Figure 18.4. The response diagram has spread at each level due to other design variables. The plots in Table 18.12 show spread due to unexplained phenomena—the model for the plots of Table 18.12 is a straight line.

The correlation coefficient indicates the overall linearity of the measurements with the predicted model. The basic definition is

$$r^2 = \frac{1}{dof_{\text{error}}} \frac{\sum y_h^2 - \dfrac{(\sum y_h)^2}{N}}{\sum y_h^2 - \sum f(\vec{d}_h)^2} \qquad (18.39)$$

TABLE 18.12.   CORRELATION COEFFICIENT FOR VARIOUS ONE-DIMENSIONAL DATA

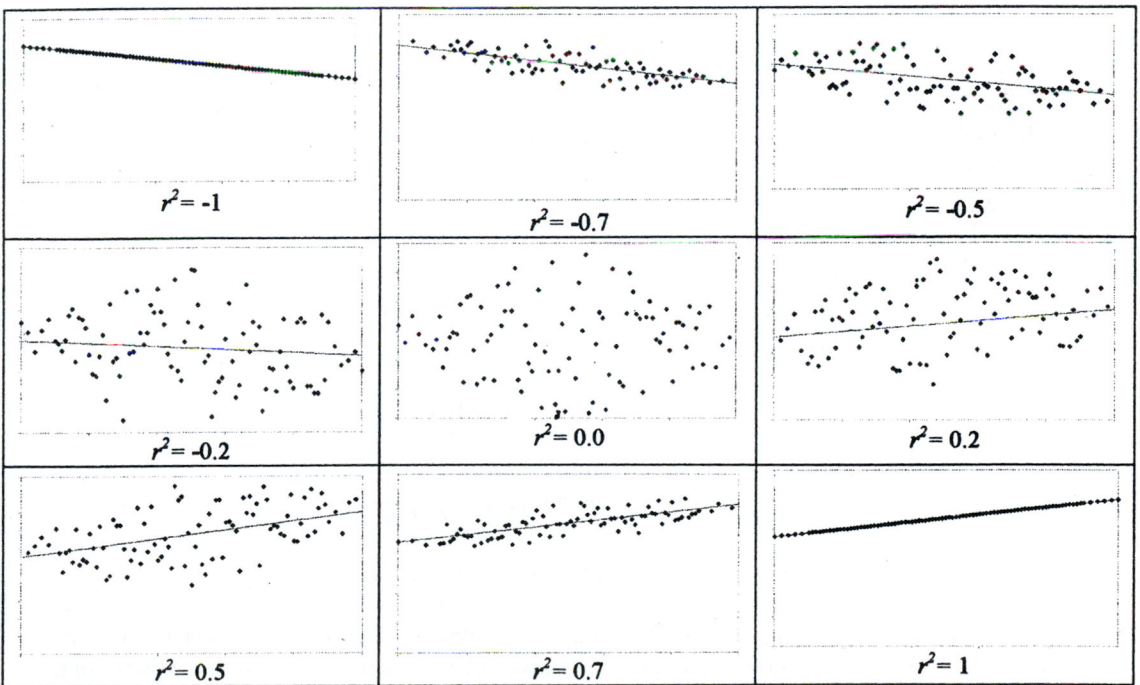

The basic r^2 is a poor indicator of fit due to the fact that it is nonlinear in the number of variables. More regression variables generally then have larger correlation coefficients. This can be accounted for in part through "adjustment" to:

$$r^2_{\text{adjusted}} = 1 - \frac{N-1}{N-n-1}(1-r^2) \qquad (18.40)$$

Most DOE software and spreadsheet programs include this coefficient as an output with any regression analysis. While always easy to provide, useful interpretation of correlation is difficult as it has no units. If the correlation coefficient is 0.9, then what will the product reliability be? How often will it be in specification and how often out? You cannot tell from the correlation coefficient.

Standard Error of the Residual

To overcome the difficulty with the correlation coefficient, one can examine the standard error term of an experiment. Recall ε_h is the difference between the actual measurement at $\vec{d}_h$ and the predicted value $f(\vec{d}_h)$ using the β coefficients. One can assess the standard deviation of the error terms ε_h, and interpret its size.

$$\sigma_{\text{error}} = \sqrt{\frac{1}{N-n}\sum_{h=1}^{N}\varepsilon_h^2} \qquad (18.41)$$

σ_{error} is called the *standard error of the residuals* or the *residual error.*

The good thing about the standard error of the residuals is that it is easily interpreted, since it is expressed in the units of the response y. Basically, any predicted value $y_{\text{predicted}} = f(\vec{d})$ is accurate to $\pm\sigma_{\text{error}}$ at 66.7% confidence, and accurate to $\pm 3\sigma_{\text{error}}$ at 99.7% confidence (ignoring any experimental measurement error as reflected by s_T). Also, again, most DOE software and spreadsheet programs include this coefficient as an output with any regression analysis.

If replicates were used in the experiment, then the error of the residual σ_{error} can also be compared against the experimental error s_T. If they are of the same size, then the discrepancy between the model and the data is experimental.

t Test

A statistical test of use in regression is the t *test,* used to determine significance of experimental variables. It is a method to compare two populations and predict if they have essentially the same mean, using a measure over the two distributions called the t-*statistic*, or t-*test*. As an example, we can use a t test to determine whether the average of a part dimension is different for sample measurements taken before and after a treatment. In the DOE regression context, the t test can be used to compare each variable's variance $(\beta_i d_{ih})$ with the residual error (σ_{error}) or the experimental error (s_T). Thus, we can predict whether each variable is useful in the prediction.

The t-statistic compares the values of the average response at different levels of a variable to the residual error. If this difference is large, then the variable has an effect on the response that is much larger than the residual error, and so the variable is important. Mathematically:

$$t = \frac{\beta_i}{\sigma_{\beta_i}} \tag{18.42}$$

where σ_{β_i} is the standard error of the β_i coefficient, as defined below.

The t-statistic is equivalent in interpretation to the standard deviation multiplier needed to make a normal distribution with mean zero and standard deviation of one become the normal distribution of the data. The interpretation corresponds to the number of standard deviations that the design variable's β is from being zero. A high t-statistic indicates that we are highly confident that the β_i is not zero; a low t-statistic indicates that we are not confident if β_i is zero or not.

The t-statistic can also be directly converted into a probability α, using the cumulative probability function for a normal distribution of mean zero and standard deviation one. This number is useful in determining our confidence level. It is the probability that the β_i coefficient is actually zero, and so the variable has no effect. It is a confidence factor, on a probability scale of 0–1.

Again, most DOE software and spreadsheet programs include both the t-statistic and the confidence factor as an output with any regression analysis. Be aware, though, that some systems indicate the probability that β_i is zero; others indicate the probability that β_i is not zero.

Main effect variables that have low β_i coefficients will also have low slopes on the response diagrams. The probability calculation of the t test provides a confidence factor α on this decision. Variables that have low confidence α should be dropped from the regression analysis. They are indistinguishable from the experimental error.

Another indicator often provided in statistical software during regression analysis is the *standard error of the coefficient* for each β_i. That is, an indicator σ_{β_i} is shown for each β_i. One might think that these error values can be used in the predictor equation:

$$y = \beta_0 + \sum \beta_i d_i \tag{18.43}$$

and propagate onto the variances of y, considering β_i as really being $\beta_i \pm \sigma_{\beta_i}$. This is *not* correct, as these σ_{β_i} terms are *not* independent error terms. They are simply the standard error of the residual as projected onto the d_i variable. Explicitly, the $\sigma_{\beta_i}^2$ terms are given by the diagonal terms of the coefficient covariance matrix, where

$$\operatorname{cov}(\bar{\beta}) = \sigma^2 [X^T X]^{-1} \tag{18.44}$$

where σ is σ_{error} or s_T, depending on if we are considering residual error or experimental error. Note the σ_{β_i} terms are the diagonal terms of $\operatorname{cov}(\bar{\beta})$. In the special case of two factorial experiments with normalized design variables, then Eq. (18.44) reduces to:

$$\sigma_{\beta_i} = \sigma \sqrt{\frac{4}{Nr}} \tag{18.45}$$

where, again, σ is σ_{error} or s_T. This becomes clearly evident when one uses normalized variables. The standard error of each coefficient β_i is identical for the same number of levels. It can be confusing, however, when using nonnormalized data, as then each σ_{β_i} value is different for each design variable d_i. They are not independent, however; they are just different scaled versions of the same number, the standard error of the residual σ_{error} or s_T. On the other hand, the standard error and the t-statistic are related by Eq. (18.42).

ANOVA: F-Ratio Test

A second indicator of how well the experimental data fits the regression model is an *Analysis of Variance* (ANOVA). Having determined an empirical model from prototype tests, the problem exists to determine whether the solution is believable and worth using. To consider this question, a sensitivity analysis is in order.

ANOVA is a technique for decomposing the total variation observed across experiments into sources of variation. The sources of variation are the important variables to consider; the variables that have no effect on the variation do not affect the product performance.

ANOVA also has another important use: The total variability of the experiment can be quantified. Based on this quantification, the error variation in the model can be compared to the controlled variation of product design variables. Thus, the quality of an experimental model can be quantified.

Theory

ANOVA is a method of analyzing the variation in response across an entire data set by decomposing the variation into the possible sources: the design variable changes and the experimental error. Thus ANOVA can be used to judge the importance of the different effects in a model and the quality of the model.

First we define some terminology. The *overall average response* (sometimes called the grand mean) for the entire set of experiments is denoted by:

$$\bar{\bar{y}} = \frac{1}{Nr} \sum_{h=1}^{N} \sum_{j=1}^{r} y_{hj} \qquad (18.46)$$

Note this result is equivalent to β_0.

Now, if the same test is repeated with the same design variable arrangement $\vec{d}_h$, in general the same value will not result, due to measuring error or uncontrolled noise. Assume for this analysis that we repeat r measurements at each $\vec{d}_h$. The average response at $\vec{d}_h$ is:

$$\bar{y}_h = \frac{1}{r} \sum_{j=1}^{r} y_{hj} \qquad (18.47)$$

where y_{hj} is the *jth replicate* experiment using $\vec{d}_h$.

Recall our terminology, where we denote

▶ r as the number of replicates

▶ N as the number of different experimental arrangements $\vec{d}_h$.

Note that these variables are related by:

$$\bar{\bar{y}} = \frac{1}{N} \sum_{h=1}^{N} \sum_{j=1}^{r} y_{hj}. \qquad (18.48)$$

Derivation

Consider a decomposition of the total variation *SS (sum of squares)* into sources due to

▶   the mean

▶ $\vec{d}$ to $\vec{d}$ variation

▶   variation at a fixed $\vec{d}$ (replication error)

We will ignore interaction effects explicitly in this derivation, but its extension is similar. With this assumption, we can model the response as:

$$y_{hj} = \bar{\bar{y}} + (\bar{y}_h - \bar{\bar{y}}) + (y_{hj} - \bar{y}_h) \qquad (18.49)$$

or in words: average $+ \vec{d}$ to $\vec{d}$ variations $+$ fixed $\vec{d}$ variations. The first term is the grand mean of the experiments, the second term is the effects due to changes in the design variables, and the last term is the inner error, due to experimental replication noise. If we square both sides of this expression and sum over all experiments, we will produce a result referred to as the *total sum of squares variance*, SS_{total}:

$$\sum_{h=1}^{N} \sum_{j=1}^{r} y_{hj}^2 = \sum_{h=1}^{N} \sum_{j=1}^{r} (\bar{\bar{y}} + (\bar{y}_h - \bar{\bar{y}}) + (y_{hj} - \bar{y}_h))^2 \quad (18.50)$$

or

$$\sum_{h=1}^{N} \sum_{j=1}^{r} y_{hj}^2 = \sum_{h=1}^{N} \sum_{j=1}^{r} \bar{\bar{y}}^2 + \sum_{h=1}^{N} \sum_{j=1}^{r} (\bar{y}_h - \bar{\bar{y}})^2 + \sum_{h=1}^{N} \sum_{j=1}^{r} (y_{hj} - \bar{y}_h)^2$$

$$+ 2 \sum_{h=1}^{N} \sum_{j=1}^{r} \bar{\bar{y}}(\bar{y}_h - \bar{\bar{y}}) + 2 \sum_{h=1}^{N} \sum_{j=1}^{r} \bar{\bar{y}}(y_{hj} - \bar{y}_h) \quad (18.51)$$

$$+ 2 \sum_{h=1}^{N} \sum_{j=1}^{r} (\bar{y}_h - \bar{\bar{y}})(y_{hj} - \bar{y}_h)$$

But note that by substituting Eqs. (18.47–49) and by manipulating the result, all cross-product terms vanish from (18.51):

$$\sum_{h=1}^{N} \sum_{j=1}^{r} \bar{\bar{y}}(\bar{y}_h - \bar{\bar{y}}) = \bar{\bar{y}} \sum_{h=1}^{N} (r\bar{y}_h - r\bar{\bar{y}}) = \bar{\bar{y}}(N\bar{\bar{y}} - N\bar{\bar{y}}) = 0$$

$$\sum_{h=1}^{N} \sum_{j=1}^{r} \bar{\bar{y}}(y_{kj} - \bar{y}_h) = \bar{\bar{y}} \sum_{h=1}^{N} \left(\sum_{j=1}^{r} r y_{hj} - r\bar{y}_h \right)$$

$$= \bar{\bar{y}} \sum_{h=1}^{N} (r\bar{y}_h - r\bar{y}_h) = 0 \quad (18.52)$$

$$\sum_{h=1}^{N} \sum_{j=1}^{r} (\bar{y}_h - \bar{\bar{y}})(y_{hj} - \bar{y}_h) = \sum_{h=1}^{N} (\bar{y}_h - \bar{\bar{y}}) \left(\sum_{k=1}^{r} y_{hj} - r\bar{y}_h \right)$$

$$= \sum_{h=1}^{N} (\bar{y}_h - \bar{\bar{y}})(r\bar{y}_h - r\bar{y}_h) = 0$$

This result thus reduces to

$$\sum_{h=1}^{N} \sum_{j=1}^{r} y_{hj}^2 = \sum_{h=1}^{N} \sum_{j=1}^{r} \bar{\bar{y}}^2 + \sum_{h=1}^{N} \sum_{j=1}^{r} (\bar{y}_h - \bar{\bar{y}})^2$$

$$+ \sum_{h=1}^{N} \sum_{j=1}^{r} (y_{hj} - \bar{y}_h)^2 \quad (18.53)$$

In other words, the overall experimental results may be decomposed into primary sources:

$$SS_{total} = SS_{average} + SS_{d\,to\,d} + SS_{fixed\,d} \qquad (18.54)$$

In general, we will use the sum of squares to decompose the total response variation into the mean effect, main effects, interaction effects, residual error, and experimental error contributions. We can then use these values to compare relative variance effects and therefore determine the importance of each.

Mean Squares

The *mean square* is defined as the sum of squares over the degrees of freedom:

$$MS = \frac{SS}{dof} \qquad (18.55)$$

This relationship can be used for variables and interactions, for example. The mean square indicates the average squared response for a variable (analogous to the variance).

The mean square provides an indication of the relative importance of each variable or interaction relative to the error, for example. It is the primary quantity of interest of the ANOVA analysis for any particular input factor.

F *Test*

An important statistical test used in ANOVA is the F *test.* It is a method to compare two population variances. For example, we can use an F test to determine whether the dimensional variances of a part vary for samples from two different plants.

For design of experiments, we are concerned about whether the variations across the changes in one variable are different from the random error variations. If so, the variable results are significant. If the variances are the same, then the performance changes, observed when an input variable is modified, have no more effect than randomness. Thus, design variable effects cannot be distinguished from the experimental error.

The F-*statistic* of a variable or interaction x is defined as the mean square of the variable or interaction over the error mean square.

$$F(x) = \frac{MS(x)}{MS_{error}} \qquad (18.56)$$

The larger the F-statistic, generally the more important the variable or interaction is to the response.

Similar to the t-statistic, the F-statistic is equivalent in interpretation to a multiplier of a standard deviation on a normal distribution with mean zero and standard deviation of one. It represents the number of standard deviations that a design variable's standard deviation is from being the same as the error's standard deviation over random data. A high F-statistic indicates that we are confident that the variation is not due to randomness; a low F-statistic indicates a low confidence level.

The F-statistic can also be directly converted into a probability α, using the cumulative probability function for a normal distribution of mean zero and standard deviation one. This number is useful in quantifying the confidence level. It is the probability that the factor has no additive effect. It is a confidence factor, on a probability scale of 0–1.

Again, most DOE software and spreadsheet programs include both the F-statistic and the confidence factor as an output with any regression analysis. It does this in two groupings, the sum of squares for the experiment (the prediction) and the sum of squares for the residual. One should compare these two values. If the predicted fraction of the sum of squares is much larger than the sum of squares for the residual, the regression factors capture the variation. If the sum of squares of the residual is very high, the changes in the chosen variables do not explain the experiments. The F-statistic for the prediction is a useful measure for this judgment.

ANOVA Tables

The ANOVA analysis can be conveniently summarized into an ANOVA table. We break down the effects into

▶ mean
▶ main effects
▶ interaction effects
▶ error

and then examine the sum of squares, degrees of freedom, mean square, and F-statistic.

Mean: The mean's effect on the sum of squares is the grand mean squared:

$$SS_{\text{mean}} = Nr\left(\frac{1}{Nr}\sum_{h=1}^{N}\sum_{j=1}^{r}y_{hj}\right)^2 = Nr\,\overline{\overline{y}}^2. \qquad (18.57)$$

The mean includes a single degree of freedom:

$$dof_{\text{mean}} = 1. \tag{18.58}$$

The mean square of the mean is the mean:

$$MS_{\text{mean}} = \frac{MS_{\text{mean}}}{dof_{\text{mean}}}. \tag{18.59}$$

Main effects: A main effect's variance is the sum of the variance over the experiments at each level. We denote $\bar{y}_{d_i=\alpha}$ as the average response over all experiments with d_i fixed at a specific value α. There are M_i levels experimented on variable d_i, and $|\alpha|$ experiments (including replicates) with $d_i = \alpha$,

$$SS_i = \sum |\alpha| \bar{y}^2_{d_i=\alpha} - SS_{\text{mean}}. \tag{18.60}$$

The degree of freedom of each main effect is the number of levels, minus one for the average:

$$dof_i - M_i - 1. \tag{18.61}$$

The mean square of a main effect is as defined:

$$MS_i = \frac{SS_i}{dof_i}. \tag{18.62}$$

Interaction effects: An interaction effect's variance is the sum of the variance over the experiments with both variables at the specified level. So we denote $\bar{y}_{d_i=\alpha,d_j=\beta}$ as the average response over all experiments with d_i and d_j fixed at the specified values α and β, respectively:

$$SS_{ij} = \sum \sum |\alpha\beta| \bar{y}^2_{d_i=\alpha,d_j=\beta} - SS_i - SS_j - SS_{\text{mean}}. \tag{18.63}$$

The degree of freedom of each interaction is the number of values each interaction can take on:

$$dof_{ij} = (M_i - 1)(M_j - 1). \tag{18.64}$$

The mean square of an interaction effect is as defined:

$$MS_{ij} = \frac{SS_{ij}}{dof_{ij}}. \tag{18.65}$$

Total: The total sum of squares is:

$$SS_{\text{total}} = \sum_{h=1}^{N} \sum_{j=1}^{r} y^2_{hj}. \tag{18.66}$$

The total degree of freedom is the total number of experiments, including replicates:

$$dof_{\text{total}} = Nr. \qquad (18.67)$$

The mean square of the total is:

$$MS_{\text{total}} = \frac{SS_{\text{total}}}{dof_{\text{total}}}. \qquad (18.68)$$

Experimental error: The experimental error is:

$$SS_{\text{experimental}} = \sum_{h=1}^{N} \sum_{j=1}^{r} (y_{hj} - \bar{y}_h)^2 \qquad (18.69)$$

The degree of freedom is:

$$dof_{\text{experimental}} = N(r - 1). \qquad (18.70)$$

The mean square of the experimental error is:

$$MS_{\text{experimental}} = \frac{SS_{\text{experimental}}}{dof_{\text{experimental}}}. \qquad (18.71)$$

Residual error: The residual error degree of freedom is the remaining degrees of freedom:

$$SS_{\text{residual}} = SS_{\text{total}} - SS_{\text{mean}}$$
$$- \underbrace{\sum SS_i}_{\text{main effects}} - \underbrace{\sum SS_{ij}}_{\text{interactions}} - SS_{\text{experimental}} \qquad (18.72)$$

The error degree of freedom is the remaining degrees of freedom:

$$dof_{\text{residual}} = dof_{\text{total}} - 1 - \underbrace{\sum dof_i}_{\text{main effects}} - \underbrace{\sum dof_{ij}}_{\text{interactions}} - dof_{\text{experimental}}. \qquad (18.73)$$

The mean square of the error is:

$$MS_{\text{error}} = \frac{SS_{\text{error}}}{dof_{\text{error}}}. \qquad (18.74)$$

The table: All of this data summarizes the percentage effects for our model of the experimental results. This summary is presented in what is called the *ANOVA table*, shown in Table 18.13. The confidence is the associated probability of the *F*-statistic. It can be interpreted as the confidence that the factor is contributing to the observed variation. A low confidence means that a factor is ostensibly not contributing in an additive way toward the response.

TABLE 18.13.   ANOVA EFFECTS TABLE

Variation source	Sum of squares	Degrees of freedom	Mean square	Replicate F test	Replicate Confidence	Residual F-test	Residual Confidence
Mean	$Nr\bar{\bar{y}}^2$	1	$Nr\bar{\bar{y}}^2$	$\dfrac{MS_{mean}}{MS_{exp}}$	1-Fdist(F_{mean}, 1, dof_{exp})	$\dfrac{MS_{mean}}{MS_{res}}$	1-Fdist(F_{mean}, 1, dof_{res})
Main effect i	SS_i	$M_i - 1$	$\dfrac{SS_i}{dof_i}$	$\dfrac{MS_i}{MS_{exp}}$	1-Fdist(F_i, 1, dof_{exp})	$\dfrac{MS_i}{MS_{res}}$	1-Fdist(F_i, 1, dof_{res})
Interaction ij	SS_{ij}	$(M_i - 1)*(M_j - 1)$	$\dfrac{SS_{ij}}{dof_{ij}}$	$\dfrac{MS_{ij}}{MS_{exp}}$	1-Fdist(F_{ij}, 1, dof_{exp})	$\dfrac{MS_{if}}{MS_{res}}$	1-Fdist(F_{ij}, 1, dof_{res})
Experimental	SS_{exp}	$N(r - 1)$	$\dfrac{SS_{exp}}{dof_{exp}}$				
Residual	SS_{res}	dof_{res}	$\dfrac{SS_{res}}{dof_{res}}$				
Total	SS_{total}	Nr	$\dfrac{SS_{total}}{dof_{total}}$				

The confidence factor itself is a function of the F-statistic F and the two degrees of freedom. It can be calculated using the F-distribution integral:

$$\alpha(F, dof_1, dof_2) = \frac{\Gamma((dof_1 + dof_2)/2)}{\Gamma(dof_1/2)\Gamma(dof_2/2)}$$

$$\left(\frac{dof_1}{dof_2}\right)^{dof_1/2} \int_F^\infty f^{(dof_1/2)-1}\left(1 + \frac{dof_1}{dof_2}f\right)^{-(dof_1+dof_2)/2} df \quad (18.75)$$

Typically, this calculation is directly implemented as a function in most DOE software and spreadsheet programs.

Other Indicators: Residual Plots

Other than these derived statistical variables, there are less rigorous approaches to ensuring confidence in a regression fit. One good approach is to plot the residual errors ε_h versus each variable. If the regression analysis is valid, the residual plots should appear as a random distribution with expected value zero. Any trend in the residual data indicates unmodeled behavior.

Summary: Advanced DOE Method for Product Testing

Considering the topics of fractional experiments and statistical analysis, a more advanced method for DOE is presented below. There exists redundancy between this approach and the basic method. The primary differences arise in the design of the experimental matrix, significance analysis of the results, and validation of the regression model's structure.

Step 1: Identify the high-level physical principles, followed by the performance variables (responses), the design variables (control factors), and the noise variables (uncontrolled factors).

Step 2: Define the target values for the responses, the boundaries for the design variables, and the nominal values for the noise variables, where they will remain fixed during the experiments (Chapter 19 discusses experiments with noise variables).

Step 3: Plan the prototype testing by developing an experimental matrix, choosing the number of levels for each design variable, the number of experiments to execute, and the base design. Add columns to the matrix to determine confounding variables, the number of replicates, measurements for the response, experimental apparatus and procedure, and methods for eliminating noise sources.

Step 4: Execute prototype testing by randomizing the trials and replicates, and record the data to the accuracy needed for analysis.

Step 5: Analyze the test results by constructing a regression model, applying ANOVA and regression tests to the main and interaction effects of the model. Reduce the model by removing coefficients that are not significant, check the model's structure with residual analysis, and apply the reduced regression model to the product design.

V. PRODUCT APPLICATIONS OF PHYSICAL MODELING AND DOE

Product Application I: Nerf™ Missilestorm™ (Norrell 1995)

In this application, we consider the redesign of a Nerf™ Missilestorm™ product. This product is a children's toy that shoots soft projectiles with human actuation. Three primary customer needs are to minimize pull force, to increase projectile range, and to increase accuracy. Figure 18.18 shows a photograph of the current product.

Preliminary Experimentation: Preliminary experiments allow the determination of various product component properties. Several items must be investigated within the Missilestorm™. First, for the barrel rotation spring, the spring constant, k, must be determined. Figure 18.19 shows the input force as a function of spring displacement.

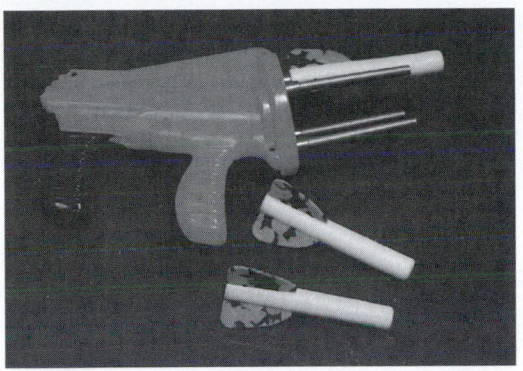

▼ *Figure 18.18.*
Nerf™ Missilestorm™ product.

Concurrent with these spring tests is the measurement of the maximum force encountered during a pull action on the blue handle, or launch base rotation. Ten different pull readings are taken using a spring-based fishing scale. The average value of the readings is approximately 19.6 N. Given the coarse value of the spring scale, an accuracy of approximately ±2 N is assumed. It is quite apparent that something aside from spring force is dominating the maximum force needed to withdraw the blue handle.

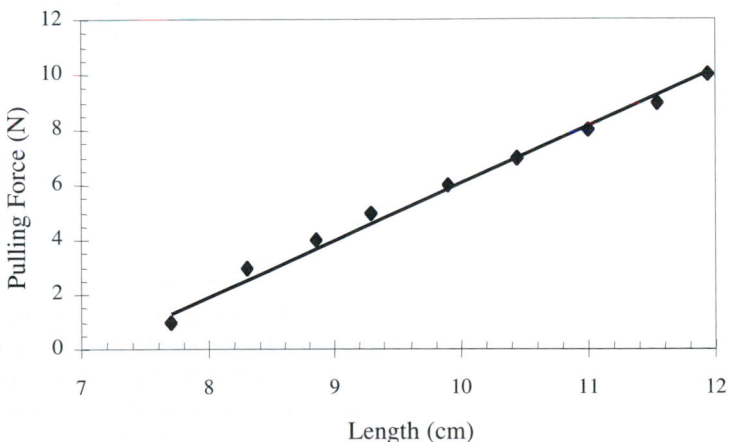

▼ *Figure 18.19.*
Determination of spring constant (k = 1.88 N/cm).

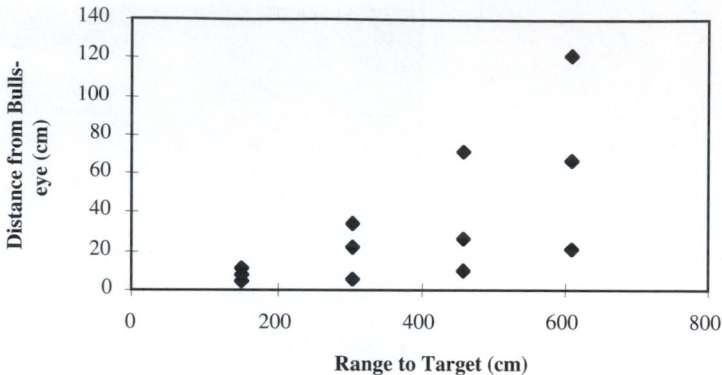

▼ *Figure 18.20.*
Accuracy as a function of distance from target, using *stock* missiles.

The last preliminary test completed on the original product considers accuracy of the device. A base point is needed to develop an appropriate target value. The test consists of several launches over a range of distances, attempting to hit a 7.5 × 7.5-cm-square target. The results are shown in Figure 18.20.

The middle data points are the averages for the four shots performed at the given range. The upper and lower points represent the maximum and minimum distance from the target respectively and are included to provide an indication of the Missilestorm™ scatter. The plot shows that accuracy falls off nonlinearly with distance. A log transform of the data would give a reasonable means of predicting the current accuracy capability. Instead, though, we focus on means to redesign the product.

Morphology: As a step in the product development process, a morphological matrix (Chapter 10), shown in Table 18.14, is created to tabulate the solution principles used by the original Missilestorm™ to solve some of the crucial functions. During preliminary analysis, some alternative solution principles are generated and presented here.

Adaptive Redesign: By considering the missiles of the Missilestorm™, we realize an immediate improvement in the product. Of the four missiles received with the toy, none of them had fins placed in the same location about the circumference of the missile body. More consistent and accurate flight will be achieved with tighter control over fin placement.

In addition to fin placement, tighter control needs to be exacted over placement of missile mounting holes (i.e., the holes the launch tubes are slipped into). On all of the four missiles, the missile-mounting hole is off

TABLE 18.14.   MORPHOLOGICAL MATRIX

Function	Current device	Possible solutions					
Stabilize missile during flight	Fins on missile (Irregularly placed)	Precise location of fins	Angled fins for missile spin	Stiffer fins	More (>3) fins	Increase angular momentum	Active fin control
Enable missile location	Bright colors	Scent-emitting missiles	Impact noise	Beeper	Whistle (during flight)	Smoke trail	Trailing string	Light flash
		Larger missiles	Glow-in-the-dark				
Store missile	None	Reception holes in body	Velcro attachments	Separate carrying case	Bandolier worn around chest		
Transform energy (For launch)	Piston	Combustion	Spring	Hydrolysis			
Aim missile	Handles, tilted tubes	Integrated aiming reticule	Laser sight	Tripod mount			
Absorb missile impact energy	Foam missiles, blunt noses	Crushable zone on missile	Spring in missile body	Telescoping body (collapses on impact)			
Release missile	Static friction between tube/missile	Latch on missile, connected to trigger	Magnetic coupling	Adhesive bond to overcome			
Input energy	Human (kinetic)	Chemical (pot.)	Battery	Compressed air			
Resist damage by animals	None	Tough (chew-proof) materials	Flavored with tastes bad to animals	Scented to scare off animals			

center. Additionally, the four stock Missilestorm™ missiles do not have the same cavity length, the length of space in the missile past the launch tube tip. Values of cavity length, without modification, range from 0 to 3 cm.

In Table 18.14, several different solution principles are presented to stabilize the missile during flight. Offering the most promise are a combination of solution principles, joining precise location of fins, angling of fins, and an increase of angular momentum. Whereas fin location and an increase in angular momentum fall under the domain of parametric design, the angling of fins will add some entirely new functionality to the missile. Not only should angled fins increase the accuracy and stabilize the missile, it will cause the missile to spin! At first, this change simply seems to be a means to an end, until we remember we are considering a toy during this redesign effort. One primary purpose of a toy is to entertain and a spinning missile flying through the air will increase children's enjoyment of the Missilestorm™.

Parametric Design: As mentioned above, there are two areas for parametric redesign of the Missilestorm™ missile, precise location of fins and an increase in angular momentum. There are two principal ways of increasing angular momentum: Increase rotational speed or move mass from the center of the rotating body to the outer edges. Adding mass to the missile raises several more issues including where to place the mass. The other concern is the fact that the mass of the missile and its minimization relates to a safety customer need of minimizing the projectile mass. Therefore, efforts are directed at increasing the angular spin rate of the missile, a goal obtainable with canted, or angled, fins. These canted fins will be used to increase the accuracy and precision of the missile.

In conjunction with increasing the angular momentum, let's consider the device creating the energy transfer from blue piston handle to missile. A high-level model predicts that an increase in missile cavity length should result in a greater launch velocity. Additionally, the model predicts that decreasing friction between the missile and the launch tube should result in greater launch velocity and subsequently larger flight distance. However, as the old saying goes, "The proof is in the pudding," so experiments must be designed and performed to validate the model.

Physical Modeling and Experimentation: Due to the inherently complex nature of analytically modeling a spinning missile, experimentation is undertaken to determine the effects of changes to the standard, stock Nerf™ missile. Three factors are chosen for variation throughout the experiment (Table 18.15).

TABLE 18.15.   CONTROL FACTORS AND THEIR LEVELS

Factor	+, High level	−, Low level
d_1: Fin arrangement	Canted fins (angled)	Straight fins
d_2: Lubrication	With graphite powder	Without graphite powder
d_3: Cavity size	0 cm cavity size	Full (2 cm or 3 cm) cavity size

The goals of the tests are to examine the effects of the above factors on the accuracy of the Missilestorm™ as well as the maximum distance achievable. Before the tests begin, planning and construction must be undertaken.

Planning of accuracy tests: To assess accuracy changes of the Missilestorm™, a 3″ × 3″ square target is affixed to a wall, approximately 5 ft from the ground. The current tester then steps to the firing line (either 10′ or 20′ from the target) and takes a shot with the given missile (that which is being tested). The distance between missile impact and target center is then measured and recorded. A standard desk ruler is used for measurements.

Planning of maximum distance tests: To reliably assess the maximum distance capability of the Nerf™ Missilestorm™, a stable shooting platform must be constructed. As expected, the greater the initial height of a missile, the farther it will travel. To avoid data inconsistencies due to differing launch heights, a platform 1.0 m from the ground is constructed. The platform allows the user to take the Missilestorm™ through its full range of motion while maintaining the position of the handle bottoms on the platform.

This approach to the maximum distance test allows all maximum distance tests to take place from a launch height of 1.0 m (floor to base of Missilestorm™) and a launch angle of approximately 16° (the Nerf™ Missilestorm™ launch tube angle). In addition, the 1.0-m launch height precludes any missile impacts with the ceiling, which would invalidate a particular trial.

Construction: The stock Nerf™ Missilestorm™ missiles come with three soft, foam fins that are often placed irregularly around the missile circumference. In an attempt to reduce noise from manufacturing inconsistencies in these tests, both straight and canted fins are fabricated. Standard blue styrofoam housing insulation (approximately 1″ thick) is cut into 1″-wide strips, approximately 1/8″ thick, using a razor blade and craft knife.

A cardboard template is then made from a stock missile fin. Using the template, fins are made from the previously cut styrofoam. Stock fins are removed from the missile bodies using a razor blade. It is important to note that only two missiles are modified for these tests.

The new fins are attached to the missile bodies using a thin bead of glue from a low-temperature hot glue gun. Regardless of where the stock fins are located on the missile body, the three replacement fins are placed at 120° intervals around the missile body.

In addition to fin construction, blue styrofoam housing insulation is used to create foam "plugs." The plugs are used to fill the missile cavity when needed. To fabricate the plugs, rough parallelepiped blanks are cut with a craft knife and then sanded to appropriate size with a fine-grit sandpaper. The plugs are inserted into the missiles using a small wooden dowel to force the plug into position.

To remove the plugs, a small dab of hot glue is placed on the wooden dowel. The dowel is then inserted into the missile and the glue is allowed to contact the plug. When the glue cures, the dowel/plug could easily be removed. Given the nature of hot glue, the plug is easily separated from the wooden dowel once removed from the missile.

To lubricate the missiles, graphite powder in a plastic squeeze bottle is used. Three to four "squirts" of graphite are placed inside a missile. A tissue is placed over the end and the missile is then shaken for several minutes, ensuring good graphite dispersion within the missile body. Any remaining loose graphite is dumped out before the missile is tested.

To remove the graphite, a small piece of paper towel is hot glued to a wooden dowel. The paper towel is then moistened and slowly slid into the missile. After several repetitions, the paper towel would come out of the missile clean. Similarly, a damp cloth is used to wipe down the Missilestorm™ to remove graphite powder.

Noise factors and blocking: There are many different potential noise factors for tests such as those presented here. One of the primary noise sources is user-skill variation. Ideally, we want the Missilestorm™ to be accurate for skilled and nonskilled users alike. In an attempt to capture this effect, two different users tested the Missilestorm™. Neither user had any extensive pretest experience firing the Missilestorm™.

In addition to user skill, another factor outside of the Missilestorm™ designers' control is the distance from user to target. Not every shot with the Missilestorm™ can be expected to take place from 1–2 m away. To address this effect, ranges to target of 3 and 6 m are considered.

Another (unintentional) noise source is the length of cavity in the missile. It is discovered prior to testing that cavity lengths in the missiles ranged from 0 to 3 cm. The two missiles to be used during the testing have a 2-cm gap (straight-finned missile) and a 3-cm gap (curve-finned missile).

Environmental effects, such as wind, are minimized by performing all tests in one day, inside a building hallway. Weight differences between the stock missiles and the modified missiles are another possible source

TABLE 18.16.   MASSES OF STOCK AND MODIFIED MISSILES

Item weighed	Features (cm gap)	Mass (g)
Stock missile (stock fins)	0	2.307
Stock missile (stock fins)	3	2.011
Modified missile (curved fins)	3	2.353
Modified missile (straight fins)	2	2.562
2-cm plug	n/a	0.035
3-cm plug	n/a	0.049

of noise in these experiments. A digital scale calibrated to $+/-0.005$ g is used, and the results are shown in Table 18.16.

As can be shown from the first two entries in the table, there is a significant amount of variation among *stock missiles*. In addition, the modified missiles are approximately the same mass as the unmodified ones. Lastly, the mass of the plugs added to the missiles does not significantly increase the total mass.

One very important and likely source of noise is contamination between tests, in particular where graphite is concerned. The graphite is difficult to remove. To block out this effect and force it into the higher order terms, the experimental trials are ran in a random blocked order. The tests are ran with a conscious effort put toward delaying the addition of graphite to the system until all other tests are complete.

TABLE 18.17.   EXPERIMENTAL MATRIX

Trial	Run #: accur.	d_1	d_2	d_3	3 m Accuracy[a] Operator 1	Operator 2	6 m Accuracy[a] Operator 1	Operator 2	Run #: max. dist.	Max. distance (m)	
1	2	−	−	−	0.23	0.32	0.36	0.50	6	9.30	8.31
2	1	+	−	−	0.08	0.05	0.64	0.33	5	9.60	8.08
3	7	−	+	−	0.22	0.09	0.35	0.53	3	10.36	8.84
4	8	+	+	−	0.37	0.20	0.49	0.47	4	9.83	10.11
5	4	−	−	+	0.20	0.43	0.32	1.05	7	9.65	7.98
6	3	+	−	+	0.15	0.08	0.18	0.46	8	8.26	6.99
7	5	−	+	+	0.30	0.32	0.81	0.37	2	8.23	9.19
8	6	+	+	+	0.24	0.08	0.21	0.80	1	8.53	8.36

[a]Meters from target.

TABLE 18.18. ANOVA TABLE FOR 3-M ACCURACY TEST

	Mean	Cavity d_3	Graphite d_2	Fins d_1	d_2d_3	d_1d_3	d_1d_2	$d_1d_2d_3$	meters y_1	y_2	$\bar{y}$ (m)	Error
# Trials	1	−1	−1	−1	1	1	1	−1	0.23	0.32	0.273	0.004
8	1	−1	−1	1	1	−1	−1	1	0.08	0.05	0.064	0.0003
	1	−1	1	−1	−1	1	−1	1	0.22	0.09	0.155	0.0074
# Repl.	1	−1	1	1	−1	−1	1	−1	0.37	0.20	0.283	0.0145
2	1	1	−1	−1	−1	−1	1	1	0.20	0.43	0.315	0.0273
	1	1	−1	1	−1	1	−1	−1	0.15	0.08	0.114	0.0029
	1	1	1	−1	1	−1	−1	−1	0.30	0.32	0.311	8E-05
	1	1	1	1	1	1	1	1	0.24	0.08	0.159	0.0136
β_i	0.209	0.031	0.036	−0.109	−0.015	−0.068	0.097	−0.072			SS_{exp}	0.0701
SS_i	0.701	0.004	0.005	0.047	0.001	0.019	0.037	0.021			MS_{exp}	0.009
MS_i	0.701	0.004	0.005	0.047	0.001	0.019	0.037	0.021				
F_i	79.9	0.44	0.58	5.38	0.11	2.11	4.25	2.39				
Pr_i	1.00	0.46	0.53	0.95	0.24	0.82	0.93	0.84				

Experimental results: Given the three control (design) factors, each at two levels, a full 2^3-factorial experiment is to be run, with one replicate (for each particular test). The tests and the trial order are given in Table 18.17. Data for the experiments is collected in one day, and the trials were run in a random order for each of the measured responses, as shown by the "Run #" columns. The results of the experiments are shown in the table.

The first step in analysis of the physical tests is to perform an ANOVA on the data. A template for this analysis is shown in Table 18.18. As an example analysis for the Nerf™ product, the results for the 3-m accuracy tests are included in the table.

Based on a 95% confidence test, we obtain a reduced regression model of:

$$y_{acc}^{6\,m} = 0.21 - 0.05d_1 \quad \text{(m)}. \qquad (18.76)$$

Keeping in mind that the reduced model indicates distance from the intended target, we want to *minimize y*. To this end, we choose d_1 +, in this instance, curved fins, which gives an output of 0.16 m. This result makes sense, since curved fins increase the rotational inertia of the missiles, thereby increasing the accuracy. The trials also indicated, however, that the rotation did not adversely affect throwing distance. Similar results occurred from the distance and 20′ accuracy tests.

▼ *Figure 18.21.*
Common household blender.

Blender Panel Display Evaluation

The previous example used design of experiments to construct a simple expression of physical performance in terms of design variables. In a sense, it is a localized experimental construction of an equation that, with thought, could also have been developed from first principles. The choice is one between ease and labor of physical experimenting versus analytic modeling and the accuracy of the results in either approach.

As a different problem, consider again the design of the display panel on a blender, as introduced in Chapter 4. Here, there is no available underlying physics model.

Figure 18.21 shows a common blender used for making mixed drinks and sauces. How many speed choices should be offered? What size buttons? What should be used as speed indicators? The customer needs list provides general needs such as "intuitive control panel," but the list does not have enough detail to indicate how to engineer the display to make it intuitive.

One approach to answering this question is to build many prototypes of the display and ask many customers. Then analyze these results sta-

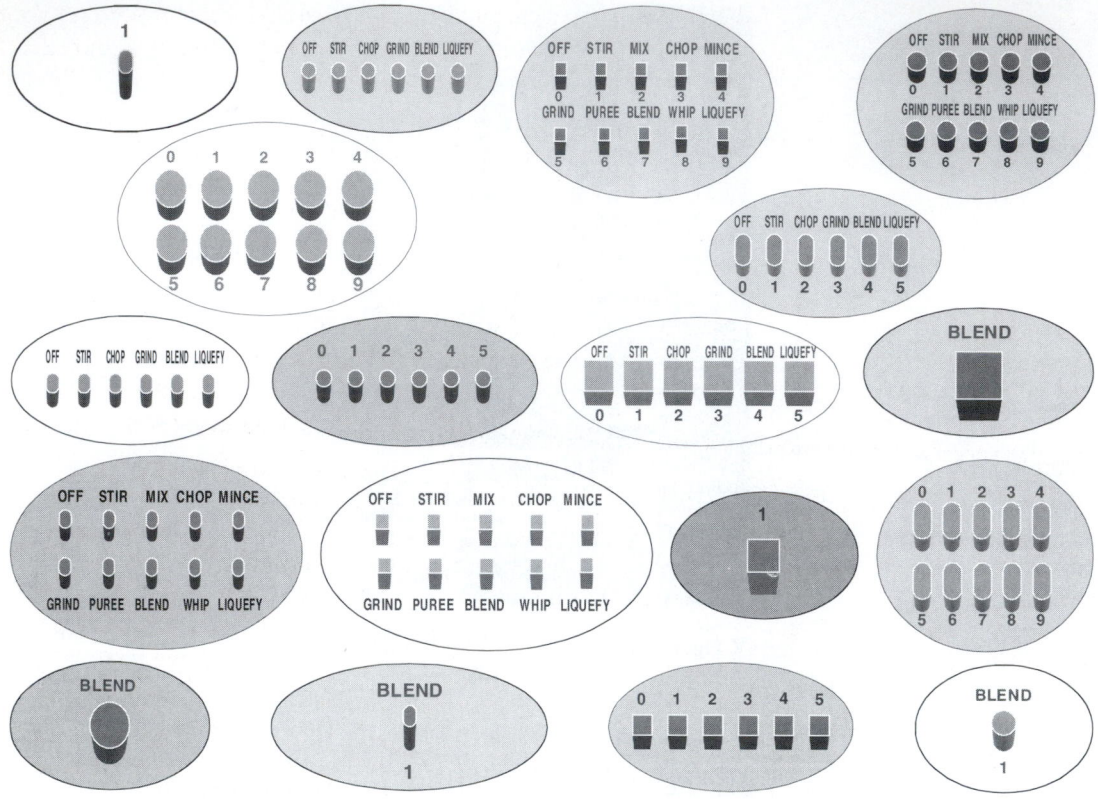

▼ *Figure 18.22.*
Experimental panels.

tistically and select a good display. This approach is often called *conjoint analysis.*

For the blender, several design variables were available to the engineers, including letter size, button size, button color, letter color, background color, button shape, number of speeds, and whether typical usage labels should be provided at the different speeds (e.g., blend and liquify).

A 3-factorial designed experiment was completed with eight design variables and 18 experimental arrangements (Table B.7 in Appendix B); each of the 18 experimental panels is shown in Figure 18.22. Next, some form of customer preference measurement is needed for a performance metric. One way to do this is to show the panels to customers and have them rank order the panels in terms of how well they like them. This ranks each trial into a numbering scheme from 1 to 18, best to worst. This can be replicated with many customers, and so a replicated set of data is developed.

The data in Figure 4.16 now forms a typical design of experiments' data set, ready for analysis. First, an ANOVA analysis was completed on the main effects as shown in Table 18.19. This result was not very promising, indicating only two variables were of any use.

Not to be daunted, the response diagrams were next plotted for the data, as shown in Figure 18.23, complete with one sigma error curves. Doing this, it became clear there was predictive power in the variables; it was just that there were quadratic trends in the data. This seemed to make sense, as people prefer buttons to be not too small and not too large. A linear or even an additive model would not provide a good fit.

The next step was to try and fit various quadratic and two-way interactions. The most effective approach to this is to apply principle component analysis among the main effect and interaction variables, but this is beyond the scope of the book. Rather, various quadratic and two-way interactions models were tested by repeated trials. This result is shown in Table 18.20, where the main effects and each interaction effect is run as an individual model, and that is repeated for each interaction effect. While this 18-level experimental matrix is designed only for main and quadratic effects, the interactions were explored to see if, even when partially confounded with main effects, an interaction term might be important, since any final chosen new panel will be later validated in customer trials anyway. Note that terms such as (Letter size)^2 could not be determined, as it is confounded in this set of 18 experiments. Examining the others variables, it is clear that others do not explain the variation well. These factors should be dropped.

The final regression factors kept in the model are shown in Table 18.21. The final correlation coefficient for the model was 0.971, and the standard error of the residual was ± 0.674. The resulting equation is then:

$$\mu(\vec{d}) = 12.8 + 3.3d_G + 1.5d_H - 0.7d_B^2 - 1.2d_C^2$$
$$+ 1.5d_D^2 - 4.5d_G^2 - 0.5d_D d_E, \qquad (18.74)$$

Note the quadratic and interaction terms. This equation can be used as a performance expression of customer satisfaction over the button panel configuration choices as equipped with an ordinal scale. The interaction term $d_D d_E$ might be suspect, since this matrix has that effect partially confounded with main effects, but it is worth keeping in mind.

TABLE 18.19. ANOVA ANALYSIS OF THE MAIN EFFECTS FOR THE BLENDER PANEL

Source	dof	Sum of squares	F-statistic	$Pr(MS_i = 0)$
Letter size	1	0.154	0.171	0.720
Button size	1	1.260	1.40	0.359
Button color	1	0.001	0.001	0.976
Letter color	1	0.544	0.603	0.519
Background color	1	1.120	1.24	0.381
Shape	1	0.202	0.223	0.683
Number	1	122.	136.	0.007
Labels	1	29.7	32.9	0.029

At this point, a model is in hand for predicting the average response of the market. One might ask if the average carries meaning. To explore this question, we can look at the replicate variance. We have responses from a sample of nine individuals representing the market. How does the person-to-person variance impact the result?

The first analysis to this question is to determine the overall replicate variance s_T. Using Eq. (18.23), we find $s_T = \pm 3.72$, a very large number. Therefore, generally the resulting model Eq. (18.77) fits the market average very well but may not represent any particular person well. There is substantial person-to-person variation from market average.

Exploring this further, we can propagate the replicate variance onto each term in Eq. (18.77) to determine s_{β_i} using Eq. (18.44) as shown in Table 18.22. Here we see that there is substantial replicate variance on four variables and less replicate variance on three variables (as indicated by the Pr column in Table 18.22). Interpreting, we can see that no single market-average value of button size, button color, letter color, and letter/background color is satisfactory. However, average values of button number and labeling are satisfactory to the whole market. Therefore, the replicate analysis helps to determine how to offer product variety, similar to methods in Chapter 8, presuming platform portfolio architecture.

This example indicates the sophistication to which market analysis can become. Given stratified responses as above, statistical techniques in market segmentation can be applied to determine clusters of similarly responding persons. To each of these market clusters, similar conjoint analyses can be performed to determine particular configurations for each segment. This presumes, of course, that a company is willing to let each design variable target be different for each cluster.

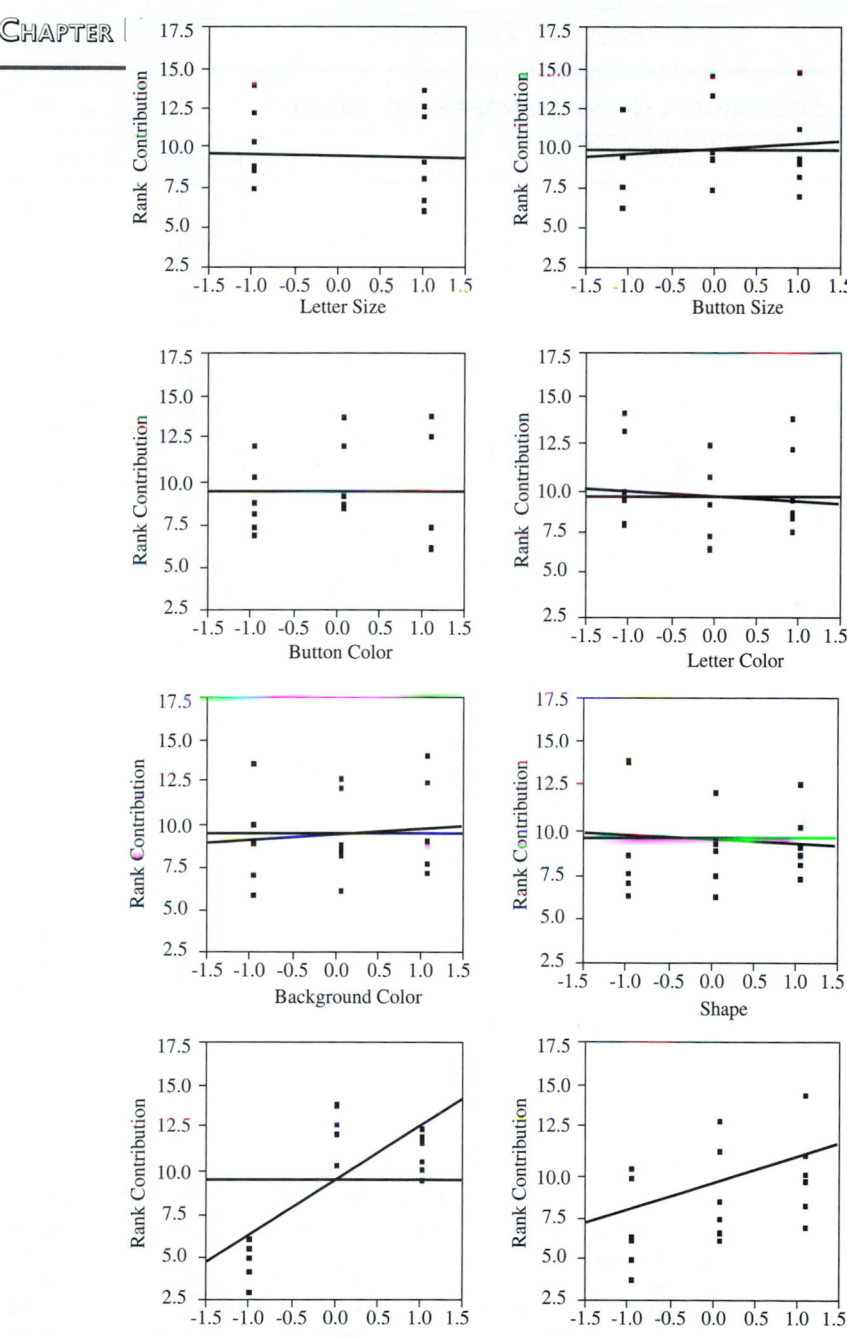

▼ *Figure 18.23.*
Response diagrams for the design variables.

TABLE 18.20. INTERACTION MODEL OF THE BLENDER PANEL DESIGN

Term	Variable	β	σ_{β_i}	t-statistic	$\Pr(\beta = 0)$
Intercept	1	13.3	± 0.868	15.33	0.004
Letter size	A	-0.093	± 0.224	-0.41	0.720
Button size	B	0.324	± 0.274	1.18	0.359
Button color	C	-0.009	± 0.274	-0.03	0.976
Letter color	D	-0.213	± 0.274	-0.78	0.519
Background color	E	0.306	± 0.274	1.11	0.381
Shape	F	-0.130	± 0.274	-0.47	0.683
Number	G	3.19	± 0.274	11.64	0.007
Labels	H	1.57	± 0.274	5.74	0.029
(Letter size)^2	$A^2 = I$				
(Button size)^2	B^2	-0.972	± 0.475	-2.05	0.177
(Button color)^2	C^2	-1.306	± 0.475	-2.75	0.111
(Letter color)^2	D^2	1.472	± 0.475	3.10	0.090
(Background color)^2	E^2	-0.028	± 0.475	-0.06	0.959
Shape^2	F^2	0.111	± 0.475	0.23	0.837
Number^2	G^2	-4.639	± 0.475	-9.76	0.010
Labels^2	H^2	-0.333	± 0.475	-0.70	0.556
(Letter size)(Button size)	AB	0.120	± 1.058	0.11	0.912
(Letter size)(Button color)	AC	-0.704	± 1.477	-0.48	0.646
(Letter size)(Letter color)	AD	-2.056	± 0.926	1.70	0.128
(Letter size)(Background color)	AE	2.694	± 1.156	2.33	0.048
(Letter size)(Shape)	AF	1.102	± 1.446	0.76	0.468
(Letter size)(Number)	AG	-0.398	± 1.491	-0.27	0.796
(Letter size)(Labels)	AH	-0.639	± 1.480	-0.43	0.678
(Button size)(Button color)	BC	0.310	± 1.958	0.16	0.878
(Button size)(Letter color)	BD	-2.103	± 1.814	-1.16	0.280
(Button size)(Background color)	BE	0.484	± 1.953	-0.25	0.811
(Button size)(Shape)	BF	-1.952	± 1.835	-1.06	0.319
(Button size)(Number)	BG	-1.119	± 1.920	-0.58	0.576
(Button size)(Labels)	BH	4.381	± 1.202	3.64	0.007
(Button color)(Letter color)	CD	-1.297	± 2.402	-0.54	0.604
(Button color)(Background color)	CE	-2.204	± 1.969	-1.12	0.296
(Button color)(Shape)	CF	-0.967	± 1.746	-0.55	0.595
(Button color)(Number)	CG	0.000	± 1.640	0.00	1.000
(Button color)(Labels)	CH	-2.514	± 2.437	-1.03	0.332
(Letter color)(Background color)	DE	-4.056	± 2.162	-1.88	0.098
(Letter color)(Shape)	DF	-1.506	± 1.552	-0.97	0.360
(Letter color)(Number)	DG	-1.592	± 2.041	-0.78	0.458
(Letter color)(Labels)	DH	-1.516	± 1.697	-0.89	0.398
(Background color)(Shape)	EF	-1.259	± 2.404	-0.52	0.615
(Background color)(Number)	EG	-0.150	± 1.779	-0.08	0.935
(Background color)(Labels)	EH	-1.117	± 1.592	-0.70	0.503
(Shape)(Number)	FG	-1.069	± 2.567	-0.42	0.688
(Shape)(Labels)	FH	-2.574	± 1.912	-1.35	0.215
(Number)(Labels)	GH	0.988	± 2.420	0.41	0.694

TABLE 18.21. FINAL INTERACTION MODEL OF THE BLENDER PANEL DESIGN

Term	Variable	β	σ_{β_i}	t Ratio	$\Pr(\beta = 0)$
Intercept	1	12.79	±0.504	25.40	<.0001
Number	G	3.32	±0.203	16.34	0.114
Labels	H	1.45	±0.203	7.13	<.0001
(Button size)^2	B^2	−0.72	±0.359	−2.00	0.073
(Button color)^2	C^2	−1.18	±0.336	−3.51	0.006
(Letter color)^2	D^2	1.47	±0.328	4.49	0.001
Number^2	G^2	−4.51	±0.336	−13.42	<.0001
(Letter color)(Background color)	DE	−0.51	±0.294	−1.72	0.116

Sometimes this can create excessive production variety. Analytically solving the problem of providing maximum market variety (all variables have many different levels) with minimum production complexity (some variables have fixed single average levels) effectively for a variety of industries is at the forefront of product development research.

Coffee Mill Experimental Optimization

As a final example, consider the redesign of a coffee mill to reduce the noise generated during use (Hirschi, 1997). In survey responses, customers indicated that they found the sound produced by the Krups coffee chopper to be too loud, especially during the initial chopping action. Although some customers keyed off a reduction in grinding noise volume to determine when the grinding process was nearing completion, all customers also desired a quieter product. The basic need was for a coffee grinder that would be less likely to wake someone when used in the early morning. Because of this clear trend in customer preference, it was decided to investigate a low noise redesign for the Krups coffee chopper .

This was done by performing a set of designed experiments that would allow changes in design variables to be linked to measured changes in an output variable, namely the audible noise produced by grinder vibrations during operation. A first step was to investigate which parts in the mill were vibrating and then subsequently determine possible

TABLE 18.22. PERSON-TO-PERSON REPLICATE VARIANCE FOR THE BLENDER PANEL DESIGN

Term	Variable	β	s_{β_i}	t Ratio	Pr
Intercept	1	12.79	±2.85	4.48	0.00
Number	G	3.32	±1.15	2.88	0.02
Labels	H	1.45	±1.15	1.26	0.24
(Button size)^2	B^2	−0.72	±2.04	−0.35	0.73
(Button color)^2	C^2	−1.18	±1.91	−0.62	0.55
(Letter color)^2	D^2	1.47	±1.86	0.79	0.45
Number^2	G^2	−4.51	±1.91	−2.37	0.04
(Letter color)(Background color)	DE	−0.51	±1.66	−0.30	0.77

changes to these parts to reduce generated noise. Using an accelerometer, vibration data was recorded from various positions on the parts of the running coffee grinder as shown in Figure 18.24. The accelerometer signal was recorded and Fourier transformed into a spectrogram, a representation intensity of each vibration component, graphed with respect to time and frequency. A sample spectrogram from accelerometer readings when attached to the outer shell is shown in Figure 18.25.

The accelerometer data provided a voltage signal on a scale of ±1 V. Therefore, the amplitude data was normalized with respect to the peak amplitude that occurred at the fundamental frequency of 390 Hz. Next, the acoustic noise frequencies from Figure 18.25 were listed and then compared with the frequencies of vibration of each part within the mill. These data are shown in Table 18.23, where, for each acoustic noise frequency recorded, parts that were vibrating at that frequency are listed. This provided an indication of which parts to attack to minimize the transmission and broadcasting of the acoustic noise.

Next, a set of experimental variables was devised, basically just modifications to the structure and construction of the original mill. Then, after building prototypes, audible noise recordings were made and analyzed for each experimental configuration. In this manner, the quantity and quality of noise produced by each experimental prototype could be determined and a statistical model developed.

The best way to reduce the grinder's audible output was to incorporate sound-and-vibration-absorbing materials into the redesigned product. This included placing elastomeric damping layers between the stainless steel grinding chamber and the plastic shell and also between the base of the motor and the bottom plastic base plate. Elastomeric motor

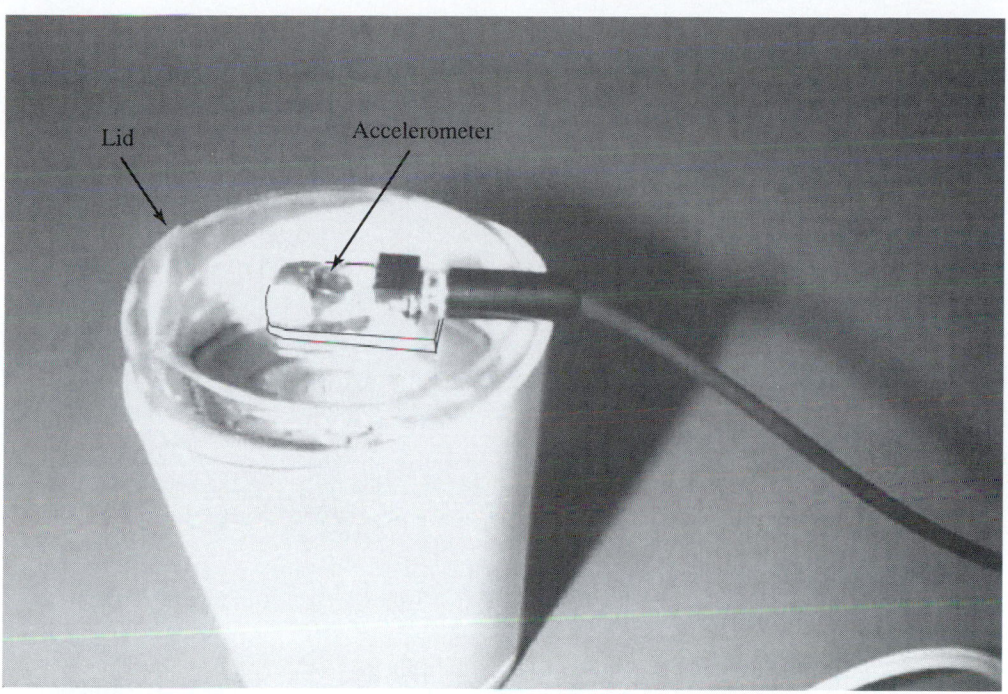

▼ *Figure 18.24.*
Accelerometer reading of vibrations on the grinder lid.

mounts replaced the plastic motor mounts used. Elastomeric pads were tried to decouple the grinder from the surface on which it was operated. Finally, a shear damping layer, consisting of an inner elastomer and two outer constraining layers of thin polypropylene or aluminum, was explored as a part of the grinder body. In theory, the elastic shear damping layer would create a viscous coupling between the polypropylene outer layer of the grinder's body and lid and thereby resist vibration deflections. The resulting design variables are shown in Table 18.24, along with the materials and dimensions used.

Two different types of rubber, neoprene and Buna-N, were used because of their differing hardness and the appreciable difference in their vibration-damping characteristics. The neoprene was rated Shore A 35–45, while the Buna-N rubber was rated 45–55. Motor mounts, constructed of neoprene in all cases, were rated either Shore A 5–10 for the "soft" design or 65–75 for the more rigid design. Rather than just trying two different thickness' of plastic for the shear damping layer inside the shell, two different materials, aluminum and polypropylene, were selected to explore how the material's stiffness would factor into its shear damping

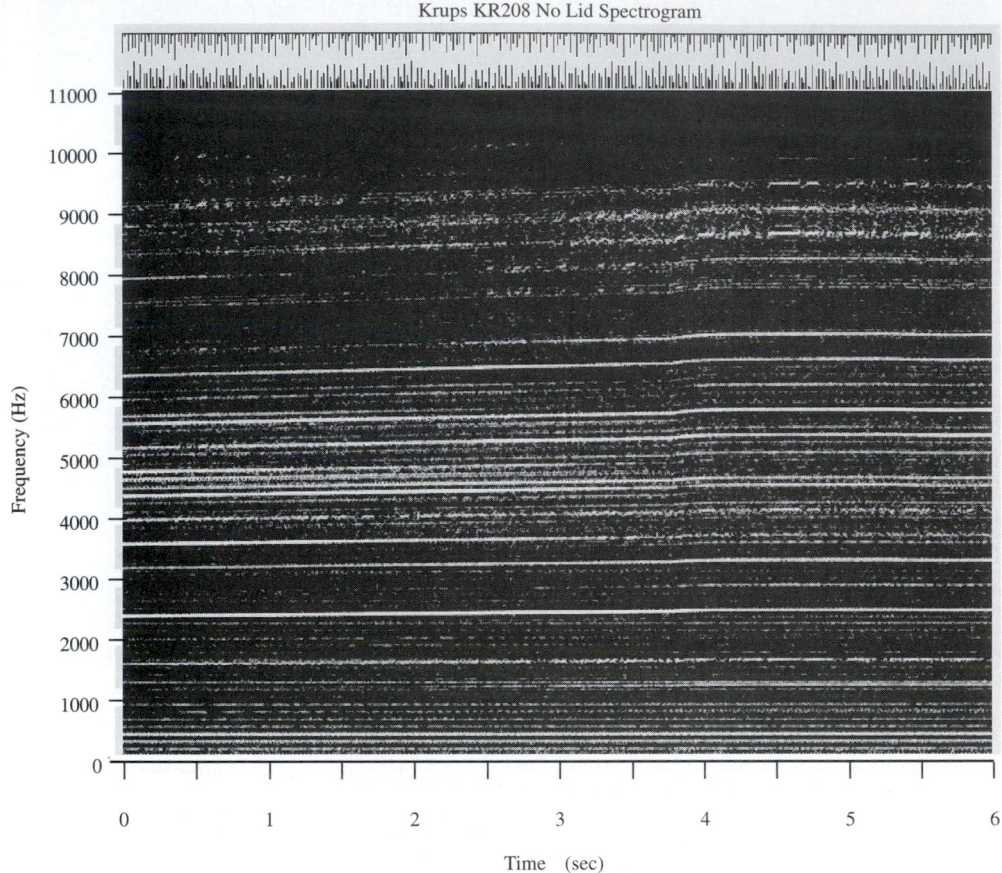

▼ Figure 18.25.
Spectrogram of typical accelerometer reading.

characteristics. All prototyping materials and grinder parts were adhered together with a clear silicon adhesive, selected because of its ability to bond both plastic and metal and also because it formed a firm yet flexible bond that would allow the shear-damping effect to occur.

Once the important design variables had been selected, an experimental matrix was developed. With 12 variables, each having two possible values, it was obvious that a full-factorial set of 2^{12} experiments including all possible permutations would be excessive and wasteful. Time and resources permitted 16 grinders to be purchased. This made 16 grinders available for experiments, thereby making exactly 16 ways to combine all the damping and constraining layer possibilities for the body of the grinder. Also, however, the swapping of various easily interchangeable parts such as the lid further increased the number of experiments that could be conducted with 16 grinders. The final experimental matrix, derived from Table B.4 in Appendix B, is shown in Table 18.26, with 32 different experimental configurations.

The partial factorial design forced confounded variables. In order to ensure this did not pose a problem, variables whose interactions were confounded were chosen such that they physically could not di-

TABLE 18.23. FREQUENCIES OF MEASURED ACOUSTIC NOISE AND COMPONENTS WITH MEASURED MATCHING VIBRATING FREQUENCIES

SIDE Amplitude (%)	Frequency Hz	Models that Correlate				BOTTOM Amplitude (%)	Frequency Hz	Models that Correlate		
0.4	20–60	Line Voltage				1.0	398	Motor	Anul	
0.1	253	Sensor	Misalign			0.1	522	Misalign	Anul	
1.0	393	Motor				0.1	646	Motor	Anul	
0.5	511	Sensor	Misalign			0.9	802	Motor		
0.7	775	Motor				0.1	921	Misalign	Anul	
0.1	894	Motor				0.1	1206	Motor		
0.1	1168	Motor	Shell RA	Plate AX	S5	0.1	1604	Motor		
0.2	1550	Motor				0.8	2003	Motor		
0.1	1938	Motor	Shell RA	S5		0.4	2401	Motor		
0.7	2236	Motor	Shell RA			0.1	2541	Motor	Shell RA	
0.5	2713	Motor	S5			0.4	2842	Motor		
0.7	3110	Motor	Shell RA	S5	Lid	0.4	3208	Motor		
0.1	3252	Motor				0.2	3618	Motor		
0.5	3510	Motor	S5			0.4	4050	Motor		
0.5	4285	Motor				0.1	4479	Motor		
0.5	4673	Motor				0.1	4565	Motor		
0.4	5050	Motor	Shell RA			0.3	5254	Motor		
0.1	5512	Motor	Shell RA	Lid		0.2	5556	Motor	Shell RA	Cage
						0.2	8010	Motor		

rectly interact. For example, the lid-damping material would have little interaction with the rubber feet, and so the lid material was selected as variable A and foot pad material as variable B. Since their interaction effect would be negligible, the fact that their interaction effect AB could not be isolated from variable G in this experimental design is irrelevant.

After developing the experimental array and variables, experimental prototypes were constructed as shown in Figure 18.26.

Each experimental configuration was recorded for a full operational cycle both with and without beans in the grinding chamber. Replicates were deemed not necessary, since it was clear the sound recordings were identical trial to trial (s_T is very small). The operational cycle consisted of turning the grinder on and letting it operate until the beans were completely ground, which ordinarily took about 12 sec. Ordinary store-bought beans were used for these experiments. During operation, the lid of the mill was gently depressed in order to keep it from flying off and also to keep the torque generated by the motor from moving the grinder around on the lab bench. However, care was taken not to allow anything to touch the side walls of the grinder body or to get between the grinder and the sound level meter and microphone, so as not to effect the experimental results.

This process was repeated for each of the 32 prototype models, yielding a total of 64 grinder noise recordings. In addition, recordings were also made of the unmodified mill, both with and without beans.

TABLE 18.24. KRUPS GRINDER NOISE REDUCTION DESIGN VARIABLES

Factor	Variable	Level +1	Level −1
Lid damping material	d_1	0.03″ Buna-N	0.03″ Neoprene
Body damping material	d_2	Buna-N	Neoprene
Body damping material thickness	d_3	0.03″	0.06″
Body constraining layer material	d_4	Aluminum	Polypropylene
Body constraining layer thickness	d_5	0.016″ Al, 0.02″ PP	0.012″ Al, 0.06″ PP
Lid constraining layer	d_6	0.03″ PP	0.06″ PP
Pan damping material	d_7	0.03″ Buna-N	0.03″ Neoprene
Pan damping material area	d_8	Small (Ring)	Large (Full)
Base outside material	d_9	0.06″ Buna-N	0.06″ Neoprene
Base outside area	d_{10}	Small (feet)	Large (Full)
Motor mounts	d_{11}	Hard Neoprene	Soft Neoprene
Base inside material thickness	d_{12}	Thin	Thick

After the experimental data had been collected, the analysis process began. Each noise signal recording was split into transient and steady-state segments and Fourier transformed into frequency components. Figure 18.25 shows a typical spectrogram. Using the spectrogram as a guide, the transient segment of the grinders noise output was judged to occur from the onset of grinding to the point where the motor frequency reached about 90% of its steady-state value of approximately 390 Hz. Generally, this consisted of 1.5–2 sec of recorded material, which was then clipped and separately analyzed as a transient response. Steady-state material was considered to be any sound produced by the grinder from the point at which the motor reached 98% of its full operational speed to just before the conclusion of grinding. This too was clipped and analyzed.

An important step in the analysis was achieving proper weighting by frequency. That is, the human ear has a nonlinear frequency response. If a test tone of constant amplitude were swept up from the lowest humanly perceivable frequency, about 20 Hz, to the highest, near 20 kHz, its perceived loudness would change with respect to frequency, even though the actual volume of the tone remained constant during the test. The human ear is insensitive to low frequencies, exhibiting a roll-off in response of about 20 dB/decade starting at about 1 kHz. On the other hand, the human ear is sensitive to frequencies in the 2–10-kHz range, which correspond to the components of speech that are most critical for speech comprehension, such as sibilants and the overtones that differentiate one vowel from another. To solve this frequency per-

TABLE 18.25.   COFFEE MILL EXPERIMENTAL MATRIX

Experimental configuration	d_1 A	d_2 B	d_3 C	d_4 D	d_5 E	Design variable d_6 F	d_7 G	d_8 H	d_9 I	d_{10} J	d_{11} K	d_{12} L
1		−1	−1	−1	−1	−1	1	1	1	1	1	1
2	−1	−1	−1	−1	1	1	1	1	1	−1	−1	1
3	−1	−1	−1	1	−1	1	1	1	−1	1	−1	1
4	−1	−1	−1	1	1	−1	1	1	−1	−1	1	1
5	−1	−1	1	−1	−1	1	1	−1	1	1	−1	−1
6	−1	−1	1	−1	1	−1	1	−1	1	−1	1	−1
7	−1	−1	1	1	−1	−1	1	−1	−1	1	1	−1
8	−1	−1	1	1	1	1	1	−1	−1	−1	−1	−1
9	−1	1	−1	−1	−1	1	−1	1	1	1	−1	−1
10	−1	1	−1	−1	1	−1	−1	1	1	−1	1	−1
11	−1	1	−1	1	−1	−1	−1	1	−1	1	1	−1
12	−1	1	−1	1	1	1	−1	1	−1	−1	−1	−1
13	−1	1	1	−1	−1	−1	−1	−1	1	1	1	1
14	−1	1	1	−1	1	1	−1	−1	1	−1	−1	1
15	−1	1	1	1	−1	1	−1	−1	−1	1	−1	1
16	−1	1	1	1	1	−1	−1	−1	−1	−1	1	1
17	1	−1	−1	−1	−1	1	−1	−1	−1	−1	1	1
18	1	−1	−1	−1	1	−1	−1	−1	−1	1	−1	1
19	1	−1	−1	1	−1	−1	−1	−1	1	−1	−1	1
20	1	−1	−1	1	1	1	−1	−1	1	1	1	1
21	1	−1	1	−1	−1	−1	−1	1	−1	−1	−1	−1
22	1	−1	1	−1	1	1	−1	1	−1	1	1	−1
23	1	−1	1	1	−1	1	−1	1	1	−1	1	−1
24	1	−1	1	1	1	−1	−1	1	1	1	−1	−1
25	1	1	−1	−1	−1	−1	1	−1	−1	−1	−1	−1
26	1	1	−1	−1	1	1	1	−1	−1	1	1	−1
27	1	1	−1	1	−1	1	1	−1	1	−1	1	−1
28	1	1	−1	1	1	−1	1	−1	1	1	−1	−1
29	1	1	1	−1	−1	1	1	1	−1	−1	1	1
30	1	1	1	−1	1	−1	1	1	−1	1	−1	1
31	1	1	1	1	−1	−1	1	1	1	−1	−1	1
32	1	1	1	1	1	1	1	1	1	1	1	1

ception problem, engineers have developed weighting filters that simulate the sensitivity of the human ear, among other things.

Various filters have been developed and given letter designations, as shown in Figure 18.27. The A-weighting filter approximates the fre-

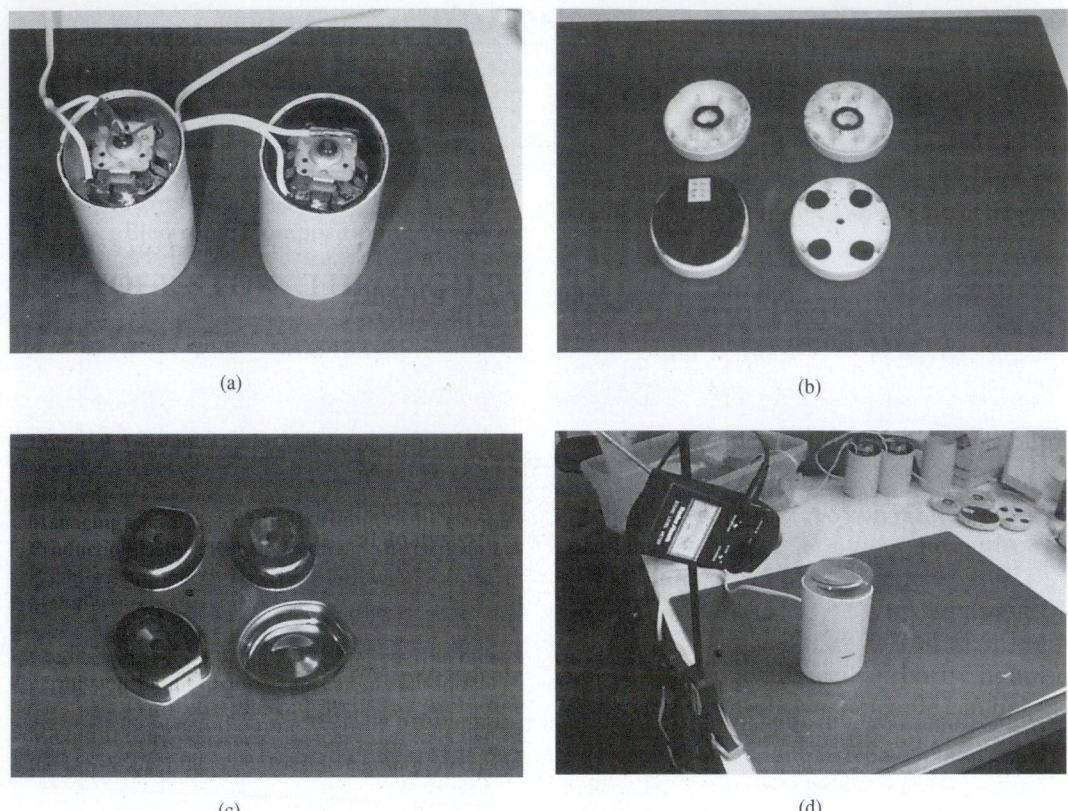

▼ *Figure 18.26.*
Experimental prototype construction. (a) Modified grinder shells, with aluminum constraining layer on the left and polyproylene on the right. (b) Bases showing full pad and small feet, as well as the elastomer added directly below the motor. (c) Pans showing elastomeric damping materials. (d) The experimental setup.

quency response of the human ear, and its sound pressure level response has been shown to correlate well with human-perceived differences in loudness.

Since the A-weighted response was an important factor for gauging the human perception of the grinder's loudness, all recordings of the grinder were modulated with the A-weighting filter. A typical A-weighted noise spectrum of the mill from an experimental trial is shown in Figure 18.28.

The next step was to convert each of the power spectrums into a single noise rating, by integrating the A-weighted noise amplitudes over the frequencies:

$$P_{\text{acoustic noise}} = \frac{\int A(\omega)\,d\omega}{\int d\omega}, \tag{18.77}$$

where $A(\omega)$ is the A-weighted amplitude at frequency ω. The result is average amplitude on a decibel scale. This basically involves integrating the area under the curve in Figure 18.28, which provides a single number that can be statistically fit to the experimental arrangements as shown in Table 18.26. There are four sets of data shown, discriminated by the noise generated with and without beans in the grinder and the noise generated during the initial transient stage versus the latter stages of grinding when the motor no longer accelerates. The initial stage is generally louder due to the beans being greatly reduced in size.

The unmodified coffee mill had A-weighted decibel readings of 82 dB with beans and 71 dB when operated without beans, at steady state. The experimental data ranged from about 65 to 75 dB. Recall every 10 dB increase represents a doubling of acoustic noise, so a 10 dB decrease in acoustic noise is a substantial and noticeable 50% noise reduction.

Experimental analysis was made on the experimental data of Table 18.26. The correlation coefficient of the main effect model for each output is shown in Table 18.27 and are not particularly great for any of the outputs.

Given these weak results with the main effect model, next higher order models were explored. Interaction terms were expected, since some of the variables represented stiffeners and some of the design variables represented absorbers, and these should work in concert. Trying fits of different interaction effects, a reasonable model was developed of the transient noise with beans as shown in Table 18.28. The regression equation was:

$$P_{\text{acoustic noise}} = 71 + 0.74d_2 + 0.71d_3 - 0.12d_4 + 0.16d_5$$
$$- 0.16d_8 - 0.15d_{12} - 0.18d_2d_4 - 0.20d_2d_5$$
$$+ 0.54d_3d_4 + 0.35d_3d_5 + 0.15d_4d_5 \tag{18.78}$$

which fit to a correlation coefficient of 0.75, and the standard error of the residual was ± 0.72 dB. This was not the greatest fit, but it was not that bad either.

Using this model, a configuration that minimized noise was determined and constructed. This redesigned mill had a generated acoustic noise of 68 dB on an equivalent A-weighted scale, compared with 82 dB of the original design. This represents 14 dB or well over a 50% reduction in acoustic noise.

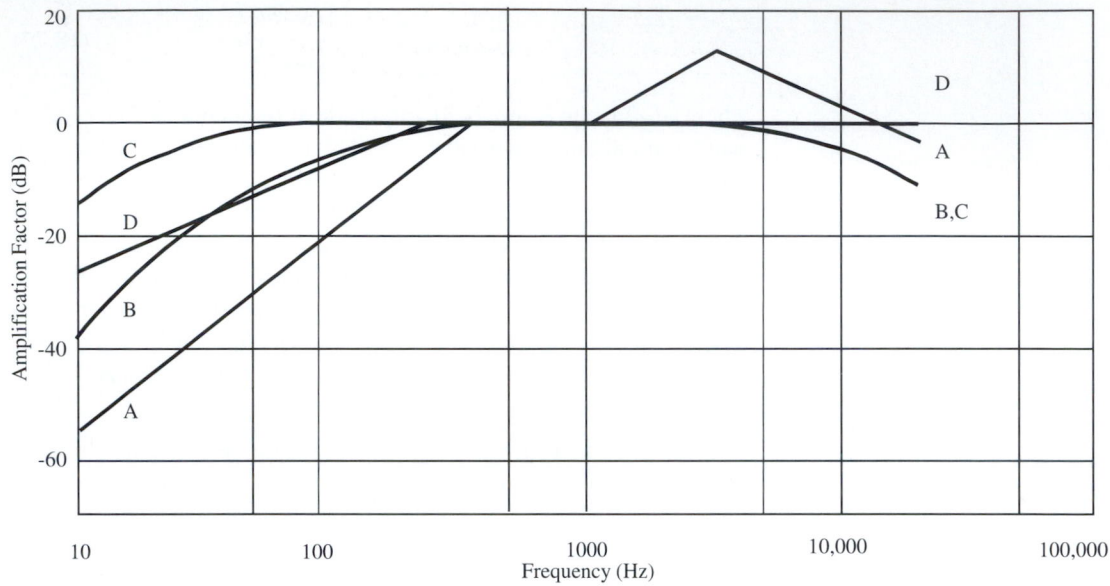

▼ *Figure 18.27.*
Noise importance weightings by frequency.

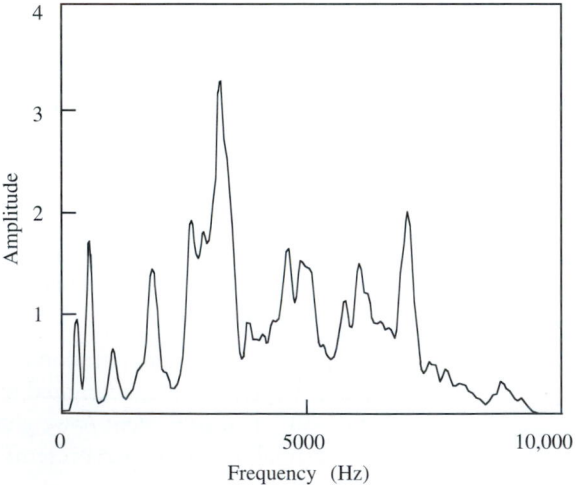

▼ *Figure 18.28.*
Typical A-weighted noise spectrum experimental data.

TABLE 18.26.   COFFEE MILL EXPERIMENTAL A-WEIGHTED ACOUSTIC NOISE PERFORMANCE METRIC DATA (DB)

Experiment	Without beans		With beans	
	Transient	Steady state	Initial chop	Steady state
1	70.53	72.33	68.83	70.19
2	66.01	66.09	69.29	64.40
3	71.52	69.71	69.27	65.36
4	66.30	66.61	69.57	65.61
5	68.13	68.99	70.79	68.01
6	68.54	68.04	71.69	67.15
7	63.69	63.85	70.56	64.72
8	71.99	72.26	72.87	69.28
9	68.06	66.93	73.53	66.83
10	68.99	69.60	72.21	65.68
11	67.56	66.20	70.93	63.99
12	69.43	70.54	69.56	70.36
13	70.01	68.36	72.41	66.90
14	69.12	68.97	72.12	67.62
15	69.32	68.83	73.05	66.70
16	64.69	65.83	73.69	67.95
17	74.11	74.83	70.76	68.84
18	65.12	69.45	70.01	69.01
19	68.81	70.78	68.87	64.65
20	68.26	69.62	69.99	66.22
21	67.68	70.37	70.77	66.62
22	69.82	70.25	70.83	65.91
23	64.65	63.21	70.76	65.40
24	70.34	71.41	72.07	68.47
25	67.74	65.61	72.42	66.36
26	68.77	71.45	71.92	66.38
27	68.32	65.38	70.58	64.53
28	71.35	70.79	69.63	68.76
29	67.71	68.61	70.74	66.63
30	69.60	69.27	72.34	68.85
31	68.96	69.88	71.96	66.28
32	65.31	66.24	73.58	67.56

TABLE 18.27. CORRELATION COEFFICIENTS OF THE MAIN EFFECT MODELS

Scenario	r^2
Transient, no beans	+0.36
Transient with initial chopping	−0.42
Steady state, no beans	−0.21
Steady state, with chopping	−0.19

TABLE 18.28. REGRESSION COEFFICIENTS FOR THE COFFEE MILL INITIAL CHOPPING

	Variable	β_i	σ_{β_i}	t-statistic	$\Pr(\beta = 0)$
Intercept	1	71	±0.126	563	0.00
B—Body damping material	d_2	0.74	±0.126	5.87	0.00
C—Body damping material thickness	d_3	0.71	±0.126	5.65	0.00
D—Body constraining layer material	d_4	−0.12	±0.126	−0.920	0.37
E—Body constraining layer thickness	d_5	0.16	±0.126	1.27	0.22
AC—Pan damping material area	d_8	−0.16	±0.126	−1.27	0.22
BC—Base inside material thickness	d_{12}	−0.15	±0.126	−1.15	0.27
Interaction BD	$d_2 d_4$	−0.18	±0.126	−1.41	0.17
Interaction BE	$d_2 d_5$	−0.20	±0.126	−1.55	0.14
Interaction CD	$d_3 d_4$	0.54	±0.126	4.31	0.00
Interaction CE	$d_3 d_5$	0.35	±0.126	2.76	0.01
Interaction DE	$d_4 d_5$	0.15	±0.126	1.19	0.25

VI. SUMMARY AND "GOLDEN NUGGETS"

Chapter 17 presents the basics of physical modeling and prototyping. This chapter builds on these concepts through systematic methods of experimenting with prototypes. Empirical models may be derived from these methods, creating basic relationships for specifying design configurations.

▶ Design-of-experiments (DOE) is a systematic method for determining a sufficient and comprehensive set of experimental trials for product development.

▶ Basic DOE methods can provide rich insights into a product's performance, even if only main effects are analyzed.

▶ Interaction effects are a direct extension to DOE main effect models. An understanding of design-variable interactions can greatly aid in bounding a product's configuration.

▶ All experiments must be verified through the analysis of replicates and experimental error.

▶ Fractional-factorial designs help us to balance resources and the level of information obtained from experimentation.

▶ A number of statistical tools exist to verify and understand the results of experimentation.

References

Box, G. 1978. *Statistics for experimenters.* New York: Wiley.

Hirschi, N. 1997. *Optimal redesign of a Krups coffee grinder,* S. B. thesis, Massachusetts Institute of Technology, Cambridge.

John, P. 1971. *Statistical design and analysis of experiments.* New York: Macmillan.

John, P. 1990. *Statistical methods in engineering and quality assurance.* New York: Wiley.

Lee, K., 1997. Redesign of a Hamilton Beach 10-Speed Blendmaster Blender, Model #50120, Dept. of Mechanical Engineering, The University of Texas, Austin.

Montgomery, D. 1991. *Design and analysis of experiments.* New York: Wiley.

Norrell, J. 1995. Redesign of a Nerf™ Misslestorm™ product, Dept. of Mechanical Engineering, The University of Texas, Austin.

Phadke, M. 1989. *Quality engineering using robust design.* Englewood Cliffs, NJ: Prentice Hall.

19
Design for
Robustness

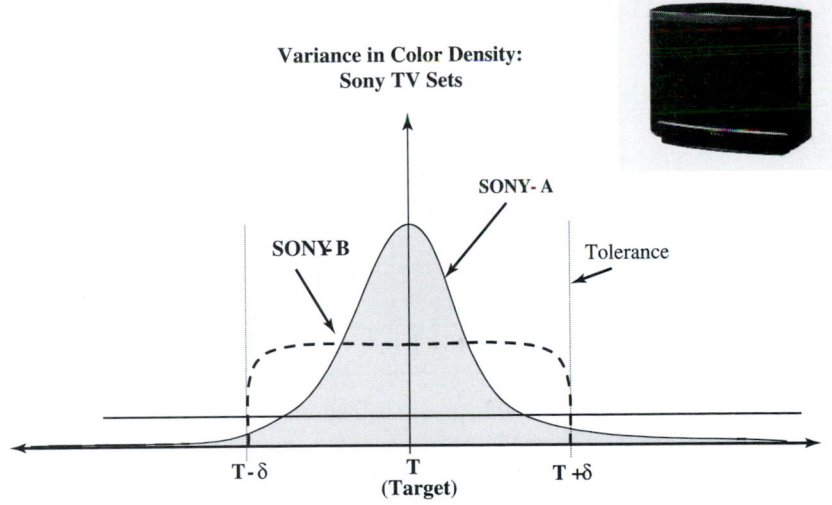

**Variance in Color Density:
Sony TV Sets**

SONY- A

SONY B

Tolerance

T-δ

T
(Target)

T +δ

The objective of product development engineers is to develop a working product that most satisfies the customer: a product design that works every time (all that are manufactured) and all the time (for the life of the product). Design is, in part, forecasting and dealing with future scenarios that the product will be subject to, and ensuring the product functions well. This process includes forecasting the future production system that will make the product, and forecasting the different uses the customer will make of the product. These tasks are difficult and inherently risky.

In the early 1980's, Sony Corporation produced television sets at two plants. One plant applied a robust design and manufacturing philosophy, working to ensure minimum variation in its activities. The other plant worked to meet tolerance specifications. The difference in delivered television sets as measured by color density (related to picture quality) is represented in the figure above. The plant that focused on minimizing variation is still making televisions, the other was shut down.

I. CHAPTER ROADMAP

In this chapter, we first develop the underlying objectives and modeling theory for robust design (Figure 19.1). Then in Section III we present a basic method, the experimental approach of Taguchi. In Section IV we present an advanced mathematical treatment of robust design, and a standard null form expression of the robust design configuration selection problem.

II. QUALITY DESIGN THEORY

With an effective specification process, as in Chapter 7, developing a product that works well becomes one of achieving the target specifications most consistently in the physical system. To do this, we need to apply the modeling approach of Chapter 13. That is, we can consider the performance specifications of the product design, but then explore causes of the product to deviate from the performance specifications. We can also think about what configuration specifications we have to impact these deviations. The task of *robust design* is then to select a best set of nominal configuration parameters that satisfies the performance specifications with minimum deviation due to manufacture, material, or use variations. This viewpoint of robust design conforms to the ideas of customer and engineering quality, as discussed in Chapter 12.

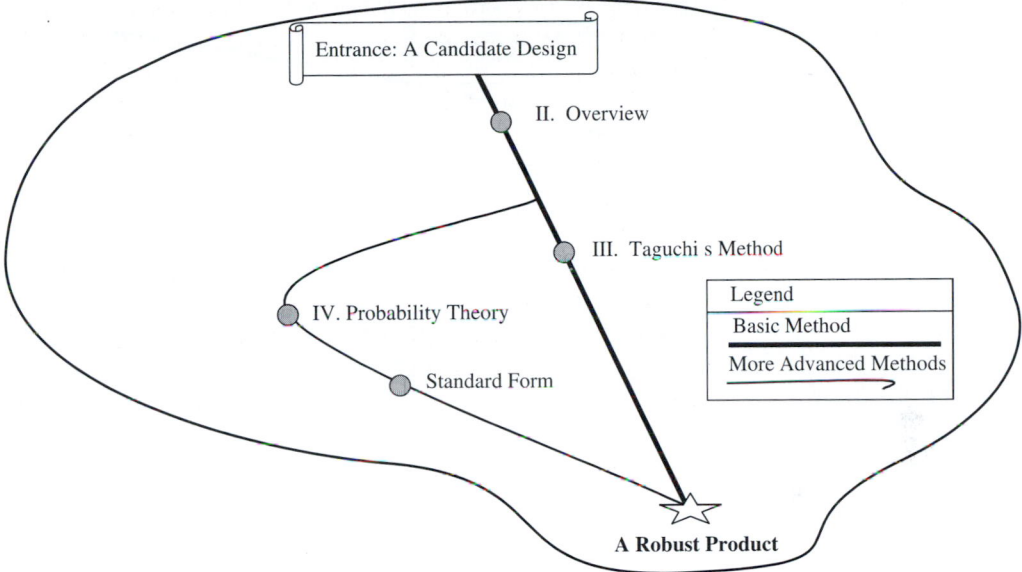

▼ *Figure 19.1.*
Chapter roadmap.

General Robust Design Model

Consider a product as a black box input-output model, the simplest form of the functional model of Chapter 5. The product uses materials, information, and energy to produce output. A designer can apply a performance specification to the difference between the inputs and outputs, the *performance variables*. These are the outputs to be ensured by the designer.

To do this, the product designer must make configuration and parametric choices to specify the product configuration, a complete dimensional, material, and manufacturing description. These choices are the *design variables*. It takes a skilled engineer to determine what makes effective design variables—what variables can be specified that impact the variation of the performance variables?

The variation in the performance variables is caused by something. Materials into the manufacturing process, deviations in the operations of production, and differences in the working or user environment all cause the performance of a product to not be on target. These causes of variation are the *noise variables*.

The reason selecting a different design configuration can improve robustness is that the sensitivity of the performance variables changes.

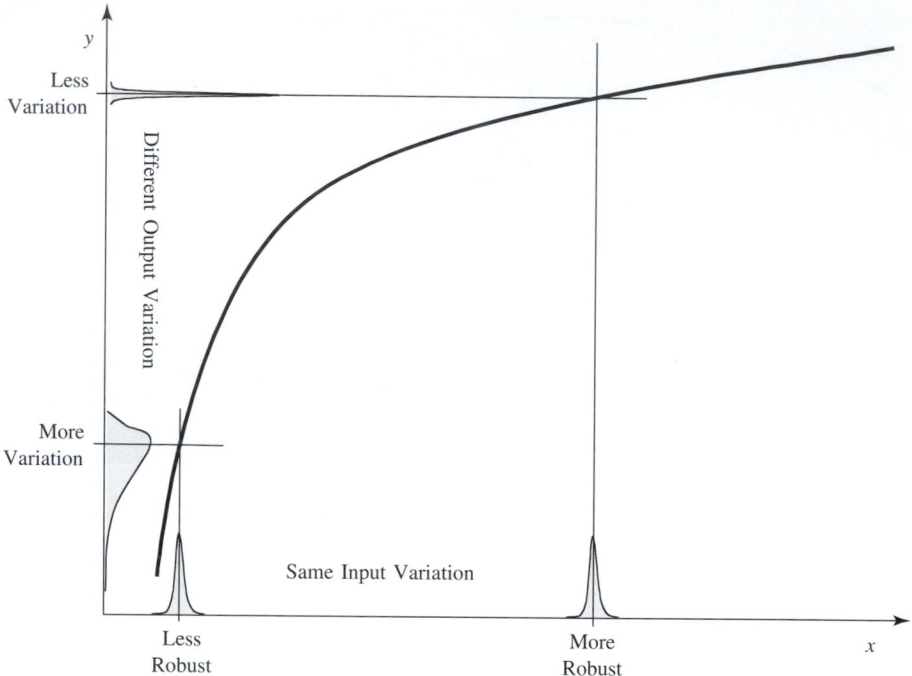

▼ **Figure 19.2.**
Sensitivity of performance varies with configuration.

This idea is depicted in Figure 19.2. Here, a performance variable y is graphed against a design variable x that has super-imposed on it an additional uniform noise variation δx about any nominal design variable value. At higher values of x, the sensitivity of y is less, and so the variation about x results in less variation in the performance y. In this case, higher values of x are more robust.

A way to think about this circumstance in terms of product design is to consider the various factors that can change the value of performance. As shown in Figure 19.3, three different types of inputs determine the performance that any product will exhibit at any point in time. There is a nominal design that is selected; altering it changes the performance. This nominal design is the choice of the design team.

There are also other factors, however; manufacturing variations and variations in material or environment can all change the performance level. These are noise variations. There are also other variations that are unique forms of "noise," such as differences in how the operator uses the product. This type of noise is discussed in greater depth, later in the chapter.

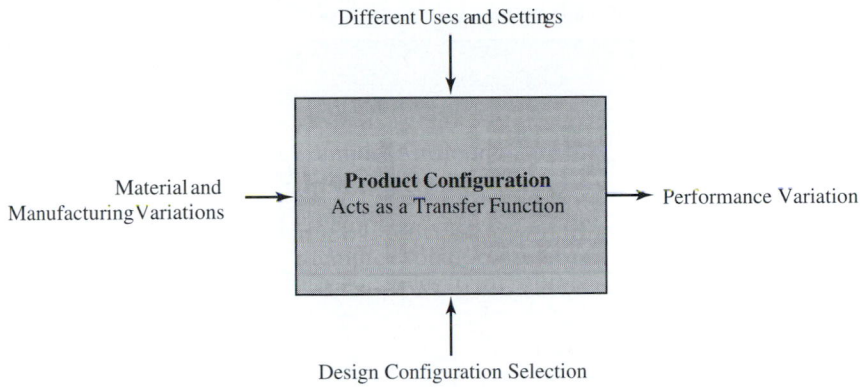

▼ **Figure 19.3.**
Factors that can change delivered performance.

The task of robust design is then to construct a product model including performance, design, and noise variables, and to then use the model to improve the design by selecting a design configuration that provides low performance deviations when the noise variables are free to vary. This robust selection process is depicted in Figure 19.4. There are basically three phases to the process. First, the model must be constructed. The performance, noise, and design variables must be identified. Next, the performance variation must be measured as the noise varies. This step must be repeated at different design configurations, to determine which design configuration has less performance variation, all for the same input noise variation. At each product configuration, this task involves taking a set of performance measurements and reducing them to a single rating for the design configuration. Lastly, the most robust configuration must be selected, based upon this robustness rating.

Robust Design Model Construction

The first step in any robust design activity is identification: determining what the variation problem is, what might cause it, how to measure its impact, and what might be done about it.

What to Measure—Performance Variables

To have effective results in making a product robust (i.e., a high quality product), it is important for a design team to consider the right performance criteria. Improving and ensuring quantities the customer does

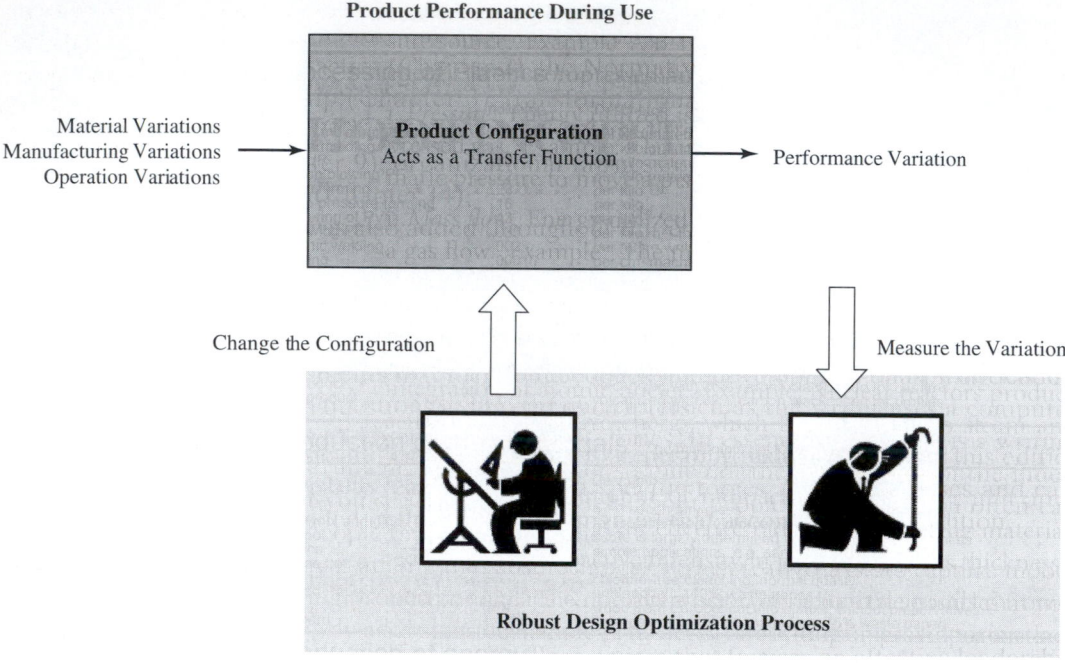

▼ Figure 19.4.
Overview of the Robust Design Selection.

not care about is a waste of effort. This measurement selection is a problem in generating effective specifications, as discussed in Chapter 7.

Even with effective performance specifications, however, there remains a step to specifying effective performance variables for robust design. For example, a performance variable often considered is the product yield—a measurement of the percentage of products that are within the output tolerance specifications. This choice of metrics is not a good idea for at least three reasons. First, to measure yield, many measurements are required—yield is a function of the entire noise space, it is not a function of a single noise variable value. A second reason is that yield is notorious for varying non-linearly over design variables, since it incorporates failure from a variety of failure modes. Third, yield is an indirect and aggregate metric. It is not tied directly to the physics underlying a product's performance, and it also averages across all possible processes causing failures. This averaging masks the fingerprint of why a product does not perform at its target specification.

A better approach is to consider measures that are not purely statistical in nature (metrics that are averages, standard deviations or other statistical parameters). Rather, one should think as a physicist and operate with continuously varying performance metrics of both the design and noise variables. As discussed in Chapter 7, a performance metric is a function complete with dimensional units that represents an objective in a one-to-one way. Here, this description means that a performance metric is a measurement of a failure mode. One should consider identifying the failure mode (for example with FMEA, Chapter 12), and then what can be measured, complete with units, to represent this failure in a one-to-one manner. An increase in the metric means an increase in the extent of failure on any one product instance.

Besides failure modes, it is important to consider the system performance of a product. Systems models, such as function structures (Chapter 5) and analytical representations (Chapter 13), estimate the underlying physics of a product. Metrics for a product may be chosen as any intermediate state of these models. For example, let's consider the design of a CNC mill as a product. An important customer metric for this product is the creation of small magnitude surface roughness on processed materials. Surface roughness may be chosen as a performance metric to evaluate robustness; however, surface roughness does not capture the product's physical sources causing variation. A more useful metric is to measure the energy into and out-of the machine. The more energy variation and energy use, the greater variation that will be caused by the product (Taguchi, 1998).

Other guidelines on performance metrics include selecting variables that are easy to measure. For example, hardness tests may provide adequate information on what the customer means by strength, rather than tensile tests. A good heuristic is that a quality characteristic should also reflect an input/output transfer of the product, such as energy transfer through a mechanism.

Product example: As an example of robust design, let's refer back to the Nerf™ Missilestorm™ product introduced in Chapters 5 and 18. In Chapter 18, a physical model is created for the toy product, based on design of experiments. A number of noise factors are encountered in the use of the product. These factors include user skill, wind, temperature, target elevation, humidity, etc. Two metrics are used for the model: accuracy and distance in firing a missile. For this redesign, the chosen metric (response) for robustness is firing distance. Firing distance is a direct function of the Nerf™ gun physics and the mechanics of missile flight. It directly represents the highly rated customer need of "shoot far."

In the sections below, this example is explored further. The goal is ultimately to consider what nominal design variables for the missiles should be chosen to create a robust toy product. The design variables include fin arrangement, lubrication, and cavity size (Chapter 18).

Causes of Variation—Noise Variables

The next related step in robust model construction is to identify variables that represent the underlying cause of the variation observed in the performance variables. It is of use to consider three types of noise that can be expected in product development.

▶ *Manufacturing noise.* These are manufacturing variations on product variables. In other words, an internal noise is a random δ on a design variable d_i due to manufacturing.

▶ *Internal noise.* These are product variations caused by degradation, and may not behave randomly, for example, wear.

▶ *External noise.* These are variations external to a product, such as temperature variations, user interaction, and other environmental variations.

When forming a model with random errors, these basic categories should be considered as a checklist, to ensure model completeness.

Product example: Continuing the Nerf™ Missilestorm™ application, all three types of noise exist. Focusing on the missile redesign, variational noise include cavity tolerances, effective coefficient of friction between the cavity and the launch pegs, fin shape from manufacturing, and the missile's aggregate mass. Likewise, internal noise consist of the cavity-peg coefficient of friction, size of the cavity due to wear, shape of the missile nose due to repeated use, etc. Finally, the external (or customer environment) noise are those listed in the previous section.

What to Specify—Design Variables

The final step in robust model construction is to identify variables that are available to specify, and, when changed, can reduce the sensitivity of the output performance variables to the input noise variations. Typically, the more complex a product is, the more design variables there are.

The same heuristics that apply in selecting performance metrics apply in selecting design variables. Namely, one should look for input/output transfer mechanisms, and examine them for variables. Energy transfers are a first step. What causes the output metric to be powered?

Factors that can be selected around that mechanism make good design variables.

Product example: A number of design variables may be considered to affect the robustness of shooting a missile a long distance. These include missile materials, missile cross-sectional area, fin size, fin arrangement, number of fins, cavity size, peg material, piston size and volume, lubrication properties, etc. For this application, fin arrangement, lubrication, and cavity size are considered as the focus design variables. They potentially provide the most impact on performance, while minimizing costly changes to the missile design.

III. BASIC METHOD: TAGUCHI'S METHOD

This section discusses an experimental robust design technique, *Taguchi's Method*. In the late 1950's, Genichi Taguchi was a manager in charge of developing telecommunications products at the Electrical Communications Laboratories of Nippon Telephone and Telegraph Company in Japan. In this role he developed the foundations of the method and applied it to many products.

The basic approach is experimental. For each design and noise variable, different values are chosen for experimentation—high/low, high/medium/low, etc. At different combinations of these values, a performance variable is measured. Then an equation is fitted to the data, and the most robust design configuration selected.

The first step is to determine values for experimentation. The next section focuses on this step.

Noise Variable Matrix

Consider a noise variable that behaves according to a normal distribution. To characterize this variable, two statistical parameters are needed: an average and a standard deviation. The question arises: how do we choose experimental values, assuming a normal distribution for the noise variables?

For two-level experimentation, one thought is to experiment at the points $(\mu - \sigma, \mu + \sigma)$. On the other hand, why not $\pm 2\sigma$? $\pm 4\sigma$? Or perhaps the $\pm \sigma$ is a good idea, to stay close to the average. To consider this decision, the philosophy of robustness must be applied.

TABLE 19.1. NOISE VARIABLE MATRIX

n_1	n_1^1	n_2^1	$\cdots$	n_M^1
n_2	n_1^2	n_2^2	$\cdots$	n_M^2
	$\vdots$	$\vdots$		$\vdots$
n_m	n_1^m	n_2^m	$\cdots$	n_M^m
	$\bar{n}_1$	$\bar{n}_2$	$\cdots$	$\bar{n}_M$

A design team wants the product to be robust—to work in all circumstances. Given this desire, the next observation is that the product will be subject to the extreme tails of the noise distribution at some point in the product life cycle. These conditions are usually difficult to satisfy; however, if the product does work at these points, then it works very well at more typical operating points.

These observations and philosophy then answer the question of this section. For a product to be robust, then one should always experiment at the limits of the distribution. One should *attack the limits*—experiment at points $(\mu - 3\sigma, \mu + 3\sigma)$, since if the products works at these points, it will work virtually always. It will be a robust product.

If there are m noise variables, the limits define $M = 2^m$ possible values for experimentation. This approach assumes the measured response will behave linearly for the range defined by the limits. This approach is also excessive from a resource viewpoint. M experiments will be quite large, even for a small number of variables, m.

Thus, the number of experimental cases must be pared down to something that is physically possible, given finite resources. To carry out this pruning operation, we apply orthogonality conditions on the experiments, and use the experimental matrices of Appendix B. This amended approach reduces to a small reasonable number for M. The allocation results in a *noise matrix*, shown in Table 19.1, which is an allocation of noise variable values to noise experimental arrangements.

It should be noted that the experimental matrices provided in Appendix B must be used thoughtfully. A number of assumptions are made regarding the variables in the matrix. First, only a finite set of noise variables will impact the product performance. Second, the higher-order effects of the noise variables are negligible, except where explicitly chosen for a given experimental matrix. Third, the product performance may be represented between the chosen noise variable limits, but not necessarily extrapolated outside these limits.

Design Variable Matrix

Similar to the noise variable matrix, a design variable matrix must be constructed from the identified design variables (Chapter 18). Here, for each design variable, a nominal design value is chosen, in addition to a range over which design choices may be made.

In this case, the attack-the-limits robust design philosophy does not apply, since the design variables are not causes of variation. Rather, the design variables can be thought of as defining a vector space. What is now needed is a good description of how the performance varies over that space—the surface plot of performance over the space. If one is only willing to experiment at two points, then selecting the outer limits of the range does not make sense. This approach provides good information locally about each experimental point, and less information in between (via interpolation). Yet at the limit points, only one side of the information can be used, toward the nominal. Therefore, it is better to back the experimental points away from the limits, and experiment at points equidistant from the nominal and range limit.

If there are n design variables, this defines $N = 2^n$ possible values for experimentation (factorial design, Chapter 18). This number of experiments is excessive, and must be pruned to something that is physically possible. To execute this pruning operation, we apply orthogonality, and use the matrices of Appendix B. This approach reduces our experiments to a small and reasonable number N. The allocation results in a *design matrix*, shown in Table 19.2, which is an allocation of design variable values to design experimental arrangements.

Experimental Matrix

The noise matrix is combined with the design matrix to produce the overall *experimental matrix*, as shown in Table 19.3. This matrix is also referred to as an *inner-outer array*. The inner portion corresponds to the design variables. The values of these variables define a product's con-

TABLE 19.2. DESIGN VARIABLE MATRIX

d_1	d_2	$\cdots$	d_n	
d_1^1	d_2^1	$\cdots$	d_N^1	$\vec{d}_1$
d_1^2	d_2^2	$\cdots$	d_N	$\vec{d}_2$
$\vdots$	$\vdots$		$\vdots$	$\vdots$
d_1^n	d_2^p	$\cdots$	d_N^n	$\vec{d}_N$

TABLE 19.3.   EXPERIMENTAL MATRIX FOR TAGUCHI'S METHOD

					n_1	n_1^1	n_2^1	$\cdots$	n_M^1
					n_2	n_1^2	n_2^2	$\cdots$	n_M^2
						$\vdots$	$\vdots$		$\vdots$
					n_m	n_1^m	n_2^m	$\cdots$	n_M^m
d_1	d_2	$\cdots$	d_n			$\bar{n}_1$	$\bar{n}_2$	$\cdots$	$\bar{n}_M$
d_1^1	d_2^1	$\cdots$	d_N^1	$\bar{d}_1$					
d_1^2	d_2^2	$\cdots$	d_N^2	$\bar{d}_2$					
$\vdots$	$\vdots$		$\vdots$	$\vdots$					
d_1^n	d_2^n	$\cdots$	d_N^n	$\bar{d}_N$					

figuration and are ultimately chosen by the design team. On the other hand, the outer portion represents the noise states that will be tested. These states are controlled during the experiments, but can not be controlled when the product is introduced into the market. Thus, there are NM experiments to perform. The response (performance metric) y must be evaluated at each $(\bar{d}_i, \bar{n}_k)$.

Determining the solution to this experimental system involves finding the best $\bar{d}$ given the experimental performance. The problem is that at each configuration $\bar{d}$, there is not just a single value of y, but many, one for each $\bar{n}$.

To resolve this issue, the M values of y at each design configuration $\bar{d}$ must be reduced to a single number. Then an interpolation over the performed experiments must be done, using the *expected* performance surface. We next discuss the initial reduction of the noise space.

Product Example

Returning to the Nerf™ Missilestorm™ application, an experimental matrix must consider the simultaneous issues of design choices and expected noise. As presented above, three design variables are considered for this product analysis: fin arrangement, cavity size, and lubrication. The inner array thus requires $N = 2^3$ experiments. There also exists a number of noise variables. For the sake of this application, the most dramatic noise if the user skill. All other noise variables will be controlled at constant levels by running the experiments in an environmentally controlled building. Thus, the outer array requires $M = 2^1$ experiments, resulting in a total of $NM = 16$ experiments. The experimental matrix for this design is shown in Table 19.4, not considering replicates.

TABLE 19.4.   EXPERIMENTAL MATRIX FOR THE NERF™ MISSILESTORM™ PRODUCT

d_1	d_2	d_3	n_1	$n_1^1 = -$ $\overline{n}_1$	$n_2^1 = +$ $\overline{n}_2$
$-$	$-$	$-$	$\overrightarrow{d}_1$	y_{11}	y_{12}
$+$	$-$	$-$	$\overrightarrow{d}_2$	y_{21}	y_{22}
$\vdots$	$\vdots$	$\vdots$	$\vdots$	$\vdots$	$\vdots$
$+$	$+$	$+$	$\overrightarrow{d}_8$	y_{81}	y_{82}

Signal to Noise Ratios

To reduce the M values of y at each design configuration $\overrightarrow{d}$ to a single number, Taguchi developed the concept of a statistical *quality loss function*. A loss function $L(p)$ is any function over a performance variable p that represents quality lost in terms of the performance deviation $(p - \tau)$, where τ is a desired nominal output specification value on p. Such deviations on output p are due to noise variations.

The next question is what is $L(p)$? What is a function that can represent the quality lost by a product when it is not designed and manufactured and performing exactly on specification? To answer this, consider a design performance $p = f(\overrightarrow{d})$. The quality lost is a function of the performance p, call it $L(p)$. Suppose our target value of performance p is a value τ. Expanding the loss L in a Taylor series:

$$L(p) = L(\tau) + \frac{\partial L}{\partial p}\bigg|_\tau (p - \tau) + \frac{1}{2}\frac{\partial^2 L}{\partial p^2}\bigg|_\tau (p - \tau)^2 + \cdots \quad (19.1)$$

Now there is no quality lost, by definition, when the design produced is on target. Thus

$$L(\tau) = 0. \quad (19.2)$$

Further, the quality lost increases when moving away from target. That is, τ is a local minimum for L. Therefore,

$$\frac{\partial L}{\partial p}\bigg|_\tau = 0 \quad (19.3)$$

These results lead to the following quality loss function

$$L(p) \approx K(p - \tau)^2. \quad (19.4)$$

Thus, the loss is proportional to the square of the deviation from specification target. Quality loss L is effectively represented as a quadratic loss function.

Before one readily accepts this argument, we should consider some examples. Fast food chains make hamburgers with very little variance across the entire planet, but most people do not consider them to be a fancy gourmet meal. This observation is an issue over whether the target specification represents quality.

Another issue is over whether a company really should maximize quality. Shouldn't they maximize the magnitude of quality per unit cost, or "value"? To argue against this philosophy, Taguchi claims that companies should take a more global view. If one maximizes the quality per company cost, then someone else's cost will increase by the reduced quality that they must accommodate. So applying this method will reduce cost to "society as a whole."

Even with the loss function of Equation 19.4, we do not yet have a means of transforming M experimental points, for a design configuration $\vec{d}$, reduced to a single number. To do this, we will make a thought extension to the quality loss function. Taguchi roughly considered quality (as observed on the output p) as the difference between the effects of the intended output as compared to the undesired variations. That is, quality is a measure of the intended output—the *signal*—relative to the *noise*. Thus, Taguchi calls a measure of quality incorporating the noise variations as a *signal-to-noise ratio*, or *S/N* ratio. An *S/N* ratio is, generally speaking, any function that takes, for any design $\vec{d}$, the values of p across the M noise experiments and converts them into a single number. Of importance are the particular functions that also are quality loss functions.

There are many such *S/N* ratios. We develop here forms suitable for different types of problems, depending upon the nature of the specifications available on the performance variable p.

Target Problems

The first *S/N* ratio to consider is when we have a target value τ for performance p. Then we define our loss as above, evaluated across our noise variations. Also, Taguchi always suggests transforming variations through a logarithmic transformation. The logarithmic transformation converts errors that are multiplicative into being additive, which Taguchi argues is appropriate for manufacturing. Thus, our first *S/N* ratio is

$$S/N(\vec{d}) = -10 \log\left(\frac{1}{M-1} \sum_{j=1}^{M} (f(\vec{d}, \vec{n}_j) - \tau)^2 \right) \quad (19.5)$$

The $M-1$ normalizes the sum, and the form is a quadratic loss function.

Continuous (or Multiple) Target Problems

Another *S/N* ratio used is for problems where we do not have a single, static target value τ for our performance, but we want to minimize the variation. This type of *S/N* ratio is appropriate for designs that have an offset adjustment. For example, we may produce a radio tuning circuit. We need to make radio circuits that do not vary off frequency with noise variations, such as heat, humidity, *etc.* The consumer can twist the dial to select the station; thus a bias does not matter. Once it is set, though, we want no variation. An *S/N* ratio appropriate for these type problems is

$$S/N(\vec{d}) = -10\log\left(\frac{1}{M-1}\sum_{j=1}^{M}(f(\vec{d}, \vec{n}_j) - \bar{p}(\vec{d}))^2\right) \quad (19.6)$$

where $\bar{p}$ is the average across the noise,

$$\bar{p}(\vec{d}) = \frac{1}{M}\sum_{j=1}^{M}f(\vec{d}, \vec{n}_j). \quad (19.7)$$

This problem is very important in product development: we can usually move a bias with a change in $\vec{d}$, but reducing variation is more difficult. This type of minimization is also known as a "two-model problem" in the statistics literature.

Minimization Problems

For problems which need p to be minimized, an *S/N* ratio of

$$S/N(\vec{d}) = -10\log\left(\frac{1}{M-1}\sum_{j=1}^{M}(f(\vec{d}, \vec{n}_j))^2\right) \quad (19.8)$$

is suggested.

Maximization Problems

For problems that need p to be maximized, an *S/N* ratio of

$$S/N(\vec{d}) = -10\log\left(\frac{1}{M-1}\sum_{j=1}^{M}\left(\frac{1}{f(\vec{d}, \vec{n}_j)}\right)^2\right) \quad (19.9)$$

is suggested.

Selection of a Target Design

Having defined these S/N ratios, we select a value for each design variable d_i by using the value that maximizes the S/N ratio. That is, for each d_{ih}, evaluate:

$$\sum_{\bar{d}_h \in S(d_{ih})} S/N(\bar{d}_h) \frac{1}{|S(d_{ih})|} \qquad (19.10)$$

where

$$S(d_{ih}) = \{\text{experiments with } d_i \text{ fixed at } d_{ih}\}$$

$$|S(d_{ih})| = \text{number of experiments with } d_i \text{ fixed at } d_{ih}$$

For each d_i we evaluate this sum for each d_{ih}, and pick the value of d_{ih} that maximizes this sum. This approach applies a main-effect model, no interaction effects are considered. As discussed in Chapter 18, often it is effective to plot the values of (19.10), to help understand the relative first order sensitivity of each variable. Two-way plots can also help visualize interactions.

Product Example

For the Nerf™ Missilestorm™ application, the chosen performance metric is travel distance of the missiles. The goal of the product is to maximize launch distance; thus, Equation 19.9 is selected as the appropriate signal-to-noise ratio.

Parameter Design and the Taguchi Philosophy

The underlying principle behind Taguchi's method is the idea of *maximizing quality*. Taguchi advocates that the quality of a product must not only be ensured to within specified tolerances, but variations from the specified target should be further minimized.

To accomplish this task, Taguchi introduced the idea of what he called *parameter design*, which has been extended into what is presented here as robust design. According to Taguchi, we as mechanical engineers typically skip a step in our design process. We perform *system design* to develop a product. System design entails the material covered in Chapters 1–11 of this book: at the end of system design, we have a complete set of drawings specifying our product design.

Next, Taguchi claims we proceed straight to *tolerance design*. Here, should our manufacturing process fail to provide the specified product, then we increase quality by specifying tighter tolerances on our manufacturing processes. We inspect to achieve tight tolerances. This inspection process involves a significant expenditure of resources, in addition to numerous design changes over the life of production.

Taguchi claims that, between these steps, we must introduce *parameter design*. In parameter design, we consider changes to the nominal configuration that we can easily accommodate. Then experiments are performed across these possible designs, as well as over the noise variables of the product. Then the changed design variables that maximizes "quality" are taken as the new target design, rather than the original design determined in the system design stage.

Parameter design is a powerful concept for increasing satisfaction in a product. The product will work at least as well as the original nominal design, and typically will make the product work much better. Today, the concept is extended into a pervasive activity in the product development process, as discussed in Chapter 12. The idea of considering noise and understanding its impact on performance is used at every phase of product design, from concept development, through embodiment and parameter design, and into detailed design. This idea forms the spirit of robust design.

Product Example of Parameter Design (Norrell, 1995)

Returning to the Nerf™ Missilestorm™ application, the design variables, noise variables, and performance metric have all been chosen. In addition, an experimental matrix has been configured. The experiment must now be conducted for each combination of design variables and noise variables. The trails in the experimental matrix, Table 19.4, are randomized to avoid experimental bias. Blocking of the effects is also considered; however, the experiments will not require a change in environmental conditions or multiple days.

Table 19.5 shows the results of the experiments, where the "run #" corresponds to the randomized order for the trails during testing. As a first step in analyzing the data, ANOVA is applied to the results, averaging across the noise variable. Table 19.6 lists the ANOVA results.

Based on the ANOVA analysis, a modest confidence level gives the following phenomenological model:

$$y_{\text{max. dist.}} = 8.85 - 0.13d_1 + 0.33d_2 - 0.45d_3 + 0.16d_1d_2$$
$$- 0.23d_1d_3 - 0.15d_2d_3.$$

Assuming that this model reflects maximum distance, we want to maximize y, and chose $d_1 +$, $d_2 +$, $d_3 -$ (curved fins with graphite and full

TABLE 19.5. EXPERIMENTAL MATRIX AND RESULTS FOR THE NERF™ MISSILESTORM™

Trial	Run #: Max. Dist.	d_1	d_2	d_3	p = Max. Distance (meters) $n_1=-$	$n_1=+$
1	6	−	−	−	9.30	8.31
2	5	+	−	−	9.60	8.08
3	3	−	+	−	10.36	8.84
4	4	+	+	−	9.83	10.11
5	7	−	−	+	9.65	7.98
6	8	+	−	+	8.26	6.99
7	2	−	+	+	8.23	9.19
8	1	+	+	+	8.53	8.36

Factor	+, High Level	−, Low Level
d_1: Fin arrangement	Canted fins (angled)	Straight fins
d_2: Lubrication	With graphite powder	Without graphite powder
d_3: Cavity size	0 cm cavity size	Full (2cm or 3cm) cavity size
n_1: User skill	Unskilled Operator	Skilled Operator

TABLE 19.6. ANOVA FOR MAXIMUM DISTANCE DATA

	Mean	Cavity d_3	Graphite d_2	Fins d_1	d_2d_3	d_1d_3	d_1d_2	$d_1d_2d_3$	y_1 (m)	y_2(m)	y_{ave} (m)	Exp. Error
# Trials	1	-1	-1	-1	1	1	1	-1	9.3	8.31	8.805	0.49005
8	1	-1	-1	1	1	-1	-1	1	9.6	8.08	8.84	1.1552
	1	-1	1	-1	-1	1	-1	1	10.36	8.84	9.6	1.1552
# Repl.	1	-1	1	1	-1	-1	1	-1	9.83	10.11	9.97	0.0392
2	1	1	-1	-1	-1	-1	1	1	9.65	7.98	8.815	1.39445
	1	1	-1	1	-1	1	-1	-1	8.26	6.99	7.625	0.80645
	1	1	1	-1	1	-1	-1	-1	8.23	9.19	8.71	0.4608
	1	1	1	1	1	1	1	1	8.53	8.36	8.445	0.01445
β_i	8.8503	-0.905	0.6636	-0.264	-0.302	-0.467	0.3143	0.1492				
SS_i	1253.5	3.28	1.74	0.28	0.37	0.87	0.40	0.09			SS_{exp}	5.52
MS_i	1253.5	3.28	1.74	0.28	0.37	0.87	0.40	0.09			MS_{exp}	0.69
F_i	1818.1	4.75	2.53	0.40	0.53	1.25	0.58	0.13				
Pr_i	1.00	0.94	0.85	0.46	0.51	0.70	0.53	0.27				

Trial #	S/N_{max}
1	18.85
2	18.83
3	19.56
4	19.97
5	18.79
6	17.55
7	18.76
8	18.53

cavity), which gives an output of 10.0 meters. These results are a bit counterintuitive in that a spinning missile's drag (due to a curved fin arrangement) would lead one to think it would decrease maximum distance (part of the energy is used in rotation instead of linear translation). However, interactions between the three design variables have a greater benefit than the loss seen due to spinning-missile drag.

How does this result compare with the stock missile maximum distance? From tests with the missiles as purchased, the mean stock maximum distance is 8.78 meters. At 10.0 meters, we have produced approximately a 15% improvement in maximum distance.

But is this design robust? Let's consider an analysis of the results as an inner-outer array. Equation 19.9 may then be applied to calculate the signal-to-noise ratios across the design variables. Table 19.7 lists the results of this extended analysis.

As shown in the table, largest signal-to-noise ratio is seen with d_1+, d_2+, d_3-. However, all of the S/N ratios are relatively high. When this case occurs, we have several options. The first option is to add other constraints to the problem and determine if these constraints aid us in choosing the final design variable values. For instance, one variable set may offer much cheaper manufacturing with only a slight loss in signal-to-noise ratio. Another option is to select the variable set which maximizes our model if it doesn't cause a drop in the S/N ratio. In this instance, d_1+, d_2+, d_3- gives both the highest S/N ratio and maximizes the output of maximum distance. If we consider variance in the data, reflected by the error column in the ANOVA, we see that this factor combination also has one of the lower magnitudes of experimental errors. Thus, the variable set of d_1+, d_2+, d_3- not only maximizes flight distance, but it also gives the highest signal-to-noise ratio and the smallest variance. It is chosen as the preferred design configuration for a robust product.

Application II: Automotive Connector Design (Engine Fluid Line Connector) (Byrne and Taguchi, 1986)

An automotive connector assembly consists of an elastomeric tubing that is inserted into a nylon housing, as shown in Figure 19.5. We are asked to design the connector (parametric, embodiment design) for maximum pull-off force and robustness (insensitive to noise/variance). From preliminary analysis, differences in pull-off force of greater than or equal to 0.25 lb are significant. Four design variables (control factors)

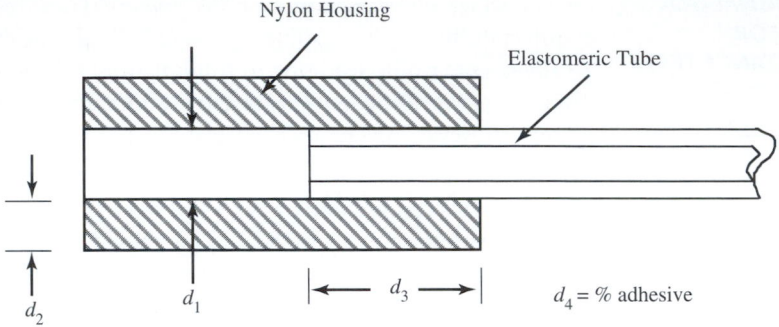

▼ Figure 19.5.
Automotive tubing connector.

are identified for this problem: (1) interference; (2) wall thickness; (3) insertion depth; and (4) percent adhesive in tubing pre-dip.

Experimental Design Strategy: A 3^2 fractional-factorial design experiment is conducted to compare the design variables (inner array). Three noise factors (uncontrolled when installed/used) are also considered in a 2^3 factorial design (outer array) (Tables 19.8 and 19.9). The factors for the experiment and the experimental results are shown in the tables below.

Analysis: ANOVA may be applied to determine the significant effects and the "optimal" choice of control factors (design variables). This analysis shows that the d_1 and d_3 are the most significant design variables, and n_3 is the most significant noise factor. Based on these results, we can apply a simple signal-to-noise analysis to determine the choices of design variables that maximize pull force, while being insensitive to noise.

TABLE 19.8.   CONNECTOR PRODUCT: FACTORS/VARIABLES AND LEVELS

Factors/Variables	Type	Symbol	Low level (−)	Medium (0)	High level (+)
Interference	Control	d_1	Low	Medium	High
Wall Thickness	Control	d_2	Thin	Nominal	Thick
Insertion Depth	Control	d_3	Shallow	Nominal	Deep
% Adhesive	Control	d_4	Low	Medium	High
Conditioning Time	Noise	n_1	24 hrs	N/A	120 hrs
Conditioning Temp.	Noise	n_2	72 °F	N/A	150 °F
Relative Humidity	Noise	n_3	25%	N/A	75%

TABLE 19.9.　EXPERIMENTAL RESULTS FROM 3^2 INNER ARRAY AND 2^3 OUTER ARRAY

				8	7	6	5	4	3	2	1	Test
				+	+	+	+	−	−	−	−	n_1
				+	+	−	−	+	+	−	−	n_2
				+	−	+	−	+	−	+	−	n_3
Test	d_1	d_2	d_3	d_4								
1	−	−	−	−	19.1	20.0	19.6	19.6	19.9	16.9	9.5	15.6	
2	−	0	0	0	21.9	24.2	19.8	19.7	19.6	19.4	16.2	15.0	
3	−	+	+	+	20.4	23.3	18.2	22.6	15.6	19.1	16.7	16.3	
4	0	−	0	+	24.7	23.2	18.9	21.0	18.6	18.9	17.4	18.3	
5	0	0	+	−	25.3	27.5	21.4	25.6	25.1	19.4	18.6	19.7	
6	0	+	−	0	24.7	22.5	19.6	14.7	19.8	20.0	16.3	16.2	
7	+	−	+	0	21.6	24.3	18.6	16.8	23.6	18.4	19.1	16.4	
8	+	0	−	+	24.4	23.2	19.6	17.8	16.8	15.1	15.6	14.2	
9	+	+	0	−	28.6	22.6	22.7	23.1	17.3	19.3	19.9	16.1	

Analysis Approach:

(i) First, we average the experimental results over the noise factors (outer array):

Test	Average force (lb)	d_1	d_3
1	17.5	−	−
2	19.5	−	0
3	19.0	−	+
4	20.0	0	0
5	22.8	0	+
6	19.2	0	−
7	19.9	+	+
8	18.3	+	−
9	21.2	+	0

(ii) Next, we determine the main effects for each design parameter: average x_1 over x_3 and vice versa. We then plot the results (force versus design parameter).

Avg. force (lb)	d_1
18.7	$d_1\,(−)$
20.7	$d_1\,(0)$
19.8	$d_1\,(+)$

Avg. force (lb)	d_3
18.3	$d_3\,(−)$
20.2	$d_3\,(0)$
20.6	$d_3\,(+)$

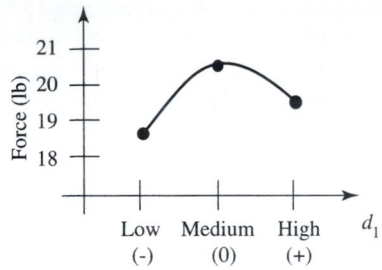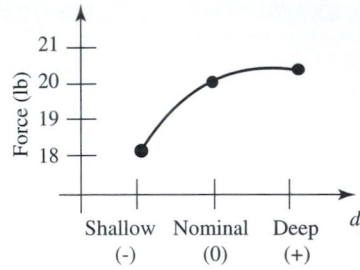

▼ **Figure 19.6.**
Main effects.

(iii) Next, we plot the interaction between d_1 and n_3 (averaging across all other design/noise parameters):

d_1	n_3	Avg. force (lb)
−	−	19.3
−	+	18.0
0	−	20.6
0	+	20.9
+	−	18.9
+	+	20.7

Conclusions:

1. Figure 19.6 shows that the medium value of d_1 is clearly the most feasible setting based on average pull-off force across all other parameters.

2. The medium and high levels of d_3 show little force difference (same figure).

3. Figure 19.7 demonstrates that a strong (significant) interaction exists between d_1 and n_3. Clearly the medium level of d_1 is the least sensitive to noise.

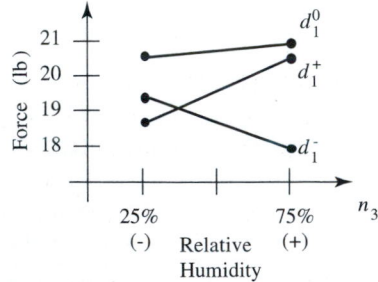

▼ **Figure 19.7.**
Interaction effect.

4. Overall, the medium value of d_1 and the high (deep) level of d_3 appear to be "optimal" embodiment choices shown by test trial no. 5. The medium (nominal) level of d_3 is not appropriate due to the large *relative* decrease in average force (test trial no. 4).

IV. ADVANCED ANALYSIS: PROBABILITY THEORY

The previous discussion on Taguchi's method provides one approach to robustness; however, we should examine its mathematical underpinnings. Consider a slightly rearranged version of Equation (19.5)

$$S/N(\vec{d}) = -10 \log\left(\sum_{j=1}^{M} (f(\vec{d}, \vec{n}_j) - \tau)^2 \frac{1}{M-1} \right) \quad (19.11)$$

The sum in (19.11) is over the noise space, approximated as M experimental points. When any product is manufactured, it has some probability of its noise being more similar to any one of the $\vec{n}_j$ rather than any of the others. Each trial point $\vec{n}_j$, though, has equal weighting in the sum of (19.11), namely $\frac{1}{M-1}$. We could theoretically improve the accuracy of (19.11) if we weighted each $\vec{n}_j$ not uniformly, but by the probability that any given product made will look more like the experimental point $\vec{n}_j$ than any of the others. That is,

$$S/N(\vec{d}) = -10 \log\left(\sum_{j=1}^{M} (f(\vec{d}, \vec{n}_j) - \tau)^2 \Pr(\vec{n}_j) \right) \quad (19.12)$$

Now let us suppose we do not conduct just M experiments, but rather many, to make the approximate of the noise space by M experiments more accurate. In fact, let's experiment at every possible point in the noise space, possibly an infinite set. Then,

$$S/N(\vec{d}) = -10 \log\left(\int_N (f(\vec{d}, \vec{n}_j) - \tau)^2 \, d\Pr(\vec{n}) \right) \quad (19.13)$$

where N is the noise space, the set of all possible noise variable values. For real-valued spaces, this expression is typically evaluated using probability density functions and integrated over the noise space as

$$S/N(\vec{d}) = -10 \log\left(\int_N (f(\vec{d}, \vec{n}_j) - \tau)^2 \, pdf(\vec{n}) \, d\vec{n} \right) \quad (19.14)$$

Examining Equation (19.14), we can understand two things. First, Taguchi's method is an experimental approximation to integration over a noise space. If we do not wish to conduct experiments, for example, we can apply analytical methods instead using Equation (19.14). A second observation is that Taguchi's method considers the squared deviation from target nominal τ. A reasonable quality loss to society argument is made for why this should be done, but on the other hand one can come up with other approaches. We now explore these concepts from first principles to derive methods in robust design.

Sizing the Variation

In the robust design formulation, at any configuration $\vec{d}$, we have M values of performance deviations that must be considered. The question becomes how to weight each deviation. Generally, this weighting is a question in "sizing" the differences in performance. Mathematically, we can generalize the concept to vector norms. A *norm* on a vector space V is a map

$$\|\cdot\| : V \to \mathbb{R} \qquad (19.20)$$

such that

1) $\forall \vec{\nu} \in V, \|\vec{\nu}\| \geq 0$

   $\|\vec{\nu}\| = 0$ iff $\vec{\nu} = \vec{0}$ $\qquad$ (boundary conditions)

2) $\forall \lambda \in \mathrm{R}, \vec{\nu} \in V, \|\lambda \vec{\nu}\| = |\lambda| \|\vec{\nu}\|$ $\qquad$ (linear)

3) $\|\vec{\nu}_1 \oplus \vec{\nu}_2\| \leq \|\vec{\nu}_1\| + \|\vec{\nu}_2\|$ $\qquad$ (triangle inequality)

Common Norms

Consider $\mathbb{R}^n$. The *2-norm* is a measure of length along a vector, also called the *Euclidean norm,* and is given by

$$\|\cdot\|_2 : \mathbb{R}^n \to \mathbb{R}$$

$$\vec{n} \to \sqrt{\frac{1}{m} \sum_{i=1}^{m} n_i^2} \qquad (19.21)$$

This result is the approach used by Taguchi. It is not the only norm, though. The *1-norm*, also known as the *average,* is

$$\|\cdot\|_1 : \mathbb{R}^n \to \mathbb{R}$$

$$\vec{n} \to \frac{1}{m} \sum_{i=1}^{m} |n_i| \qquad (19.22)$$

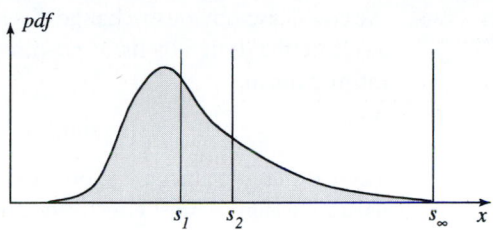

▼ *Figure 19.8.*
Rating a deviation from target.

This result is often used in more traditional design-of-experiments' literature, that are not focused on robust design. More generally, the *p-norm* is

$$\| \cdot \|_p : \mathbb{R}^n \to \mathbb{R}$$
$$\vec{n} \to \sqrt[p]{\frac{1}{m} \sum_{i=1}^{m} |n_i|^p} \qquad (19.23)$$

And taking the limit as $p \to \infty$, the ∞-*norm* is

$$\| \cdot \|_\infty : \mathbb{R}^n \to \mathbb{R}$$
$$\vec{n} \to \max\{n_i\}, \qquad (19.24)$$

which is useful when looking over 3σ spaces that must be ensured. These are all norms, and each has its use in design.

To demonstrate this result, consider a manufacturing error on a nominal target, as shown in Figure 19.8. In performing calculations, how should we rate the effects of this disturbance? We need a single number that represents this distribution. We need to apply a norm over the distribution.

We might rate the signal by its *expected value*. That is, we might take the one norm,

$$s_1 = \int x \, pdf(x) \, dx. \qquad (19.25)$$

Alternatively, we might rate the signal by its *RMS value*. That is, we might take the two norm,

$$s_2 = \left(\int x^2 \, pdf(x) \, dx \right)^{\frac{1}{2}}. \qquad (19.26)$$

We could also obviously change this result to any arbitrary p. We could also rate the design by its *worst-case value*. That is, we might take the infinity norm,

$$s_\infty = \sup\{x \in \text{supp}(pdf)\}. \tag{19.27}$$

These different values are shown in Figure 19.8. Clearly, norms can be used in design to most effectively consider noise.

General Robust Design Problem Formulation

Given this analysis, we can now state the general problem formulation for robust design. This formulation is not limited to a single output variable as is Taguchi's method, and can be applied to a variety of implementation approaches, from designed experiments to mathematical analysis.

We develop here a standard null form expression of the problem. This form means that we select one performance variable out of the set to use as an objective function, and place constraint limits on the others (Chapter 16).

The problem in robust design is that a suite of performances must be evaluated at every design configuration, due to the noise space. For the objective function, it makes sense to use a squared deviation norm to evaluate the objective function over the noise space, since we will then minimize the performance deviations, on average.

$$\overline{p}(\vec{d}) = \int_{\mathbb{N}} (f(\vec{d}, \vec{n}) - \tau)^2 \, pdf(\vec{n}) \, d\vec{n} \tag{19.28}$$

This reasoning is equivalent to that given by Taguchi on quality loss. This result should please the likely customer. An average performance can also easily be considered.

On the other hand, the constraints must be ensured. This extension becomes difficult with random variations. To accomplish this task, we should apply an infinity norm across the constraint variations. However, many have infinite tails, and we would be going to extremes to ensure unlikely events. Therefore, we can instead consider ensuring the constraints are satisfied for a percentage of the trials.

This approach is traditionally executed using a *probability constraint level* α. Consider a random constraint

$$g(\vec{d}, \vec{n}) \leq 0. \tag{19.29}$$

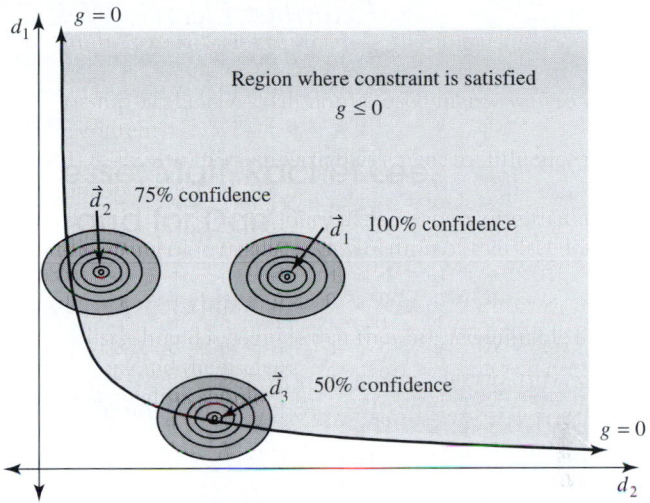

▼ *Figure 19.9.*
Different confidence levels in satisfying a constraint.

For any given $\vec{d}$, is this constraint satisfied? Generally, this question can only be answered to a given level of probability α. For example, in Figure 19.9, the different nominal points $\vec{d}$ have different probabilities of the resulting produced products violating the constraint g.

Thus, we specify a confidence level α, and convert the random constraint equations above into

$$\Pr(\vec{g}(\vec{d}, \vec{n}) \leq \vec{0}) \geq \alpha, \tag{19.30}$$

where the Pr can be evaluated through a Monte-Carlo simulation across $\mathbb{N}$, or across a pre-determined factorial approximation of $\mathbb{N}$.

In summary, to include random errors into a standard null form problem, two different norms can be used. An average can be used for the objective function, and a worst case model can be used with the constraints. The standard null form expression is

minimize $\qquad \int_{\mathbb{N}} (f(\vec{d}, \vec{n}) - \tau)^2 \, pdf(\vec{n}) \, d\vec{n}$

$\hfill (19.31)$

subject to $\qquad \Pr(\vec{g}(\vec{d}, \vec{n}) \leq 0) \geq \alpha$

Example: Electric Wok

As an example of this formulation, consider the electric wok variant re-design problem introduced in Chapter 13, and formulated as an optimization problem in Chapter 16. In Chapter 16, one form of the equations we solve is

$$
\begin{array}{lll}
\text{minimize} & t_r & \text{(temp rise time)} \\
\text{subject to} & \|\Delta \vec{T}\| \le \|\Delta \vec{T}\|_\tau & \text{(temp variation)} \\
& |T(t_r) - T_{ss}| \le \varepsilon & \text{(temp at SS)} \\
& \dot{T}(t) + f(t) = 0 & \text{(transients)}
\end{array}
\tag{19.32}
$$

This produces a solution of

$$
t = 1.3 \text{ mm}
$$
$$
r_R = 2 \text{ cm}
$$
$$
r_L = 11 \text{ cm}
\tag{19.33}
$$
$$
t_r = 4.1 \text{ min}
$$
$$
\|\Delta T_{ij}\| = 150 \text{ °F}
$$

This result basically ensures that the heat flux enters over as wide an areas as possible to make it uniform (make r_R small and r_L large). We then choose a bowl thickness that is a trade-off between heat rise time (thin t) and temperature uniformity (thick t).

Here, we might wonder how this solution from Chapter 16 might change when robustness to variations are considered. Suppose we consider different heat transfer surface characteristics as noise. This noise is relatively straightforward to handle using our current mathematical model; we can vary the emissivity ε and the convective heat transfer coefficient h.

Suppose these constants are now permitted to vary as noise variables. We might consider how they vary as normal distributions about their nominal value with a 10% variance. Doing this, we set up the following robust design problem, using Equation (19.31)

$$
\begin{array}{lll}
\text{minimize} & \displaystyle\int_\varepsilon \int_h t_r \, pdf(\varepsilon) \, pdf(h) \, d\varepsilon \, dh & \text{(temp rise time)} \\
\text{subject to} & \Pr(\|\Delta \vec{T}\| \le \|\Delta \vec{T}\|_\tau) \ge \alpha & \text{(temp variation)} \\
& |T(t_r) - T_{ss}| \le \varepsilon & \text{(temp at SS)} \\
& \dot{T}(t) + f(t) = 0 & \text{(transients)}
\end{array}
\tag{19.34}
$$

where α is set to 10%, or the constraints are satisfied 10% of the time in use over that noise distribution.

Practically, we can solve this problem with, for example, Matlab™ using the optimization and statistical toolboxes, or with Microsoft Excel™ using the solver and a third party Monte-Carlo simulation toolbox for Excel™, Crystal Ball™. Taking the latter approach, we determine that a solution to Equation (19.34) is

$$t = 1.4 \text{ mm}$$

$$r_R = 2 \text{ cm}$$

$$r_L = 11 \text{ cm} \qquad (19.35)$$

$$\bar{t}_r = 4.3 \text{ min}$$

$$\overline{\|\Delta T_{ij}\|} = 146 \text{ °F}$$

Note that the bowl geometry (r_L and r_R) does not change at all, and the thickness decreases by 0.1 mm. The distributions on $\|\Delta T_{ij}\|$ and t_r due to the noise variation is shown in Figure 19.10, at the optimal solution point of Equation (19.35). Note the temperature rise time decreases the minimum rise time by 0.2 minutes, and the average uniformity (the active constraint) improves. Peak determined t_r performance is thus sacrificed for more assuredness on the temperature uniformity.

The solution of Equation (19.35) is not entirely surprising. The thickness impacts $\|\Delta T_{ij}\|$ in conjunction with the noise variables ε and h, and so its value has changed. The r_L and r_R variables, however, have monotonic impact on $\|\Delta T_{ij}\|$, and so their values are unchanged from the traditional formulation. No matter what value of noise, small r_R is best and large r_L is best. Such observations are one reason why analytical solutions can be more effective than experimental approaches.

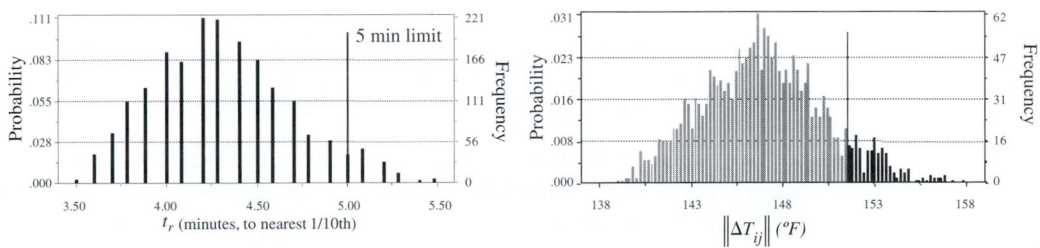

▼ **Figure 19.10.**
Wok performance equation distributions when t_r is minimized. 2000 trials are made over the noise variables ε and h.

V. CHAPTER SUMMARY AND "GOLDEN NUGGETS"

Robust Design as a Design Philosophy

We can think more broadly about the philosophy of robust design, as compared to a traditional approach. Generally, variation exists in products, both due to the user's environment and due to manufacturing. The issue is over how to consider the variation. The standard approach is to reduce the noise space to a set of fixed values, and then select the design configuration using these fixed numbers. The noise variable space is reduced to a single point—a single value for each variable, called a *constant*. This approach applies even in physical prototyping, where one might use one typical set of operating and use conditions. They are held constant. Of course, the constants are never really constant. To compensate for this reality, a *factor of safety* is applied to the constant, in the hopes that one can ensure the risk is acceptable.

A robust design approach is different. Rather than first reducing the noise space to a constant and then making design choices, the noise space is considered, and a suite of performances are evaluated. The results are subsequently reduced to a single evaluation, but only after the range of output is understood.

The differences in philosophies are diagrammed in Figure 19.11. Notice the use of factors of safety versus norms and multiple performance evaluations.

Given these observations, it should become clear that one misses the point in debates over topics such as what norm ought to used, whether a main effect model is appropriate, or what Taguchi's logarithmic transformation really does. The primary concern that makes robust design effective is that the noise space should be considered and reduced only on the performance space. Reducing the noise space variations before any evaluations are made is only safe when the mapping is well understood—we believe we know what will happen with the introduction of the noise.

The issue, of course, reduces to one of comparing the cost of repeated performance evaluations versus the confidence one has in not doing so. Performing the many more evaluations required by robust design will generally be more costly during product development: physical experiments, finite element analysis, or computer calculations are required to evaluate any performance variable value, and these generally are not trivial in time or expense.

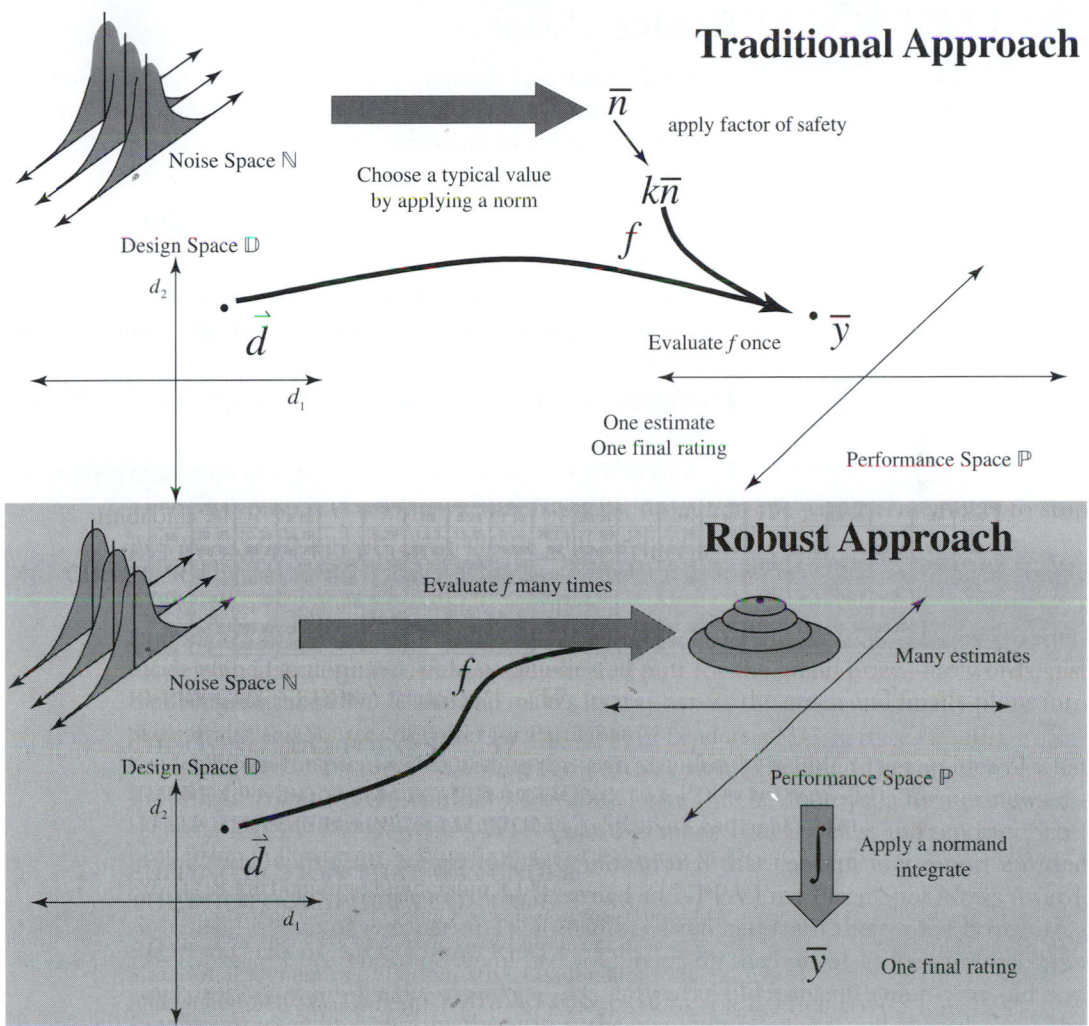

▼ *Figure 19.11.*
The difference between robust and non-robust design.

Making the decision over when to take a robust design approach is one of engineering and business judgement. Generally speaking, though, it is almost always the case that a design team is better off in the robust design approach, even in the early conceptual phases of design. It is better to spend the extra resources during product development. This investment will ensure that there is less opportunity for failure at later stages due to unforeseen performance responses due to noise.

Golden Nuggets

Robust design is a philosophy in design as well as a set of methods. When implemented, it will slow the design process down, but will increase the effectiveness and improve the result. Some key ideas in robust design include:

▶ Always consider the impact that input variation will have upon the delivered performance.

▶ Repeated evaluations over the noise variations are a much more accurate approach than choosing a nominal value of noise and estimating with a factor of safety.

▶ Taguchi's method is a simple experimental approach to robust design that is very effective.

▶ Variations are usually the study of random noise, and expectation integration forms the basis of modeling variation.

▶ Worst case norms are better suited for constraint type performance variables that must be ensured. Euclidean norms are better suited for objective function type performance variables that are always better when smaller.

References

Antonsson, E. and K. Otto. Imprecision in Engineering Design. Special combined issue of the *Journal of Mechanical Design* and the *Journal of Vibration and Acoustics*, 117(B):11–16, 1995, invited paper.

Byrne, D. and Taguchi, S., "The Taguchi Approach to Parameter Design," *ASQC Quality Congress Transaction*, Anaheim, CA, 1986, pp. 168–177.

Clausing, D. *Total Quality Development*. ASME Press, New York, 1994.

Dudewicz, E. "Basic Statistical Methods," in Juran, J. et al. *Quality Control Handbook*. McGraw-Hill, New York, third edition, 1979.

French, S. *Decision Theory*, Ellis Horwood, New York, 1988.

Juran, J. et al. *Quality Control Handbook*. McGraw-Hill, New York, third edition, 1979.

Norrell, J., "Redesign of a Nerf™ Misslestorm™ Product," The University of Texas, Dept. of Mechanical Engineering, ME 392M Report, Austin, TX, 1995.

Montgomery, D. *Design and Analysis of Experiments*, Wiley, 1991.

Otto, K., and E. Antonsson. Extensions to the Taguchi Method of Product Design. *Journal of Mechanical Design*, 115(1):5–13, 1993.

Phadke, M. *Quality Engineering Using Robust Design*. Prentice Hall, Englewood Cliffs, NJ, 1989.

Taguchi, G. *Introduction to Quality Engineering*. Asian Productivity Organization, Unipub, White Plains, NY, 1986.

Taguchi, G., Plenary Talk, *The Quality Symposium*, Texas A&M, Dept. of Mechanical Engineering, College Station, TX, November 4, 1998.

Appendix A
Function Structure Definition

I. FLOW DEFINITIONS

As pointed out in Stone (1997) and adapted here, matter and energy receive equal stature in the theory of relativity. In addition, humans comprehend these concepts only with reference to time. Thus, we understand matter and energy based on how it changes with time, or, how it flows.

Energy, matter, and information are considered basic concepts in a mechanical design problem (Pahl and Beitz, 1988). It is the flow of these three concepts that concerns design engineers. Matter is better represented as material. Information is more concretely expressed as a signal. Signals in actuality are either flows of material or energy, but receive a special classification because their function is to carry information.

All design problems deal with these three basic flows, but it seldom advances the design solution to deal with flows at this highest level. How do we specify these flows more accurately? The flow categories in Table 5.2 of Chapter 5 provide a greater specification of flows; however, ambiguity can still exist in their use. Definitions of flows are provided here to eliminate this ambiguity and make function structure decomposition a repeatable exercise. For materials, basic physics provides a suitable starting point for definitions. The energy flow concept can be specified further by a bond graph approach that considers efforts and flows (Karnopp et al., 1990). The effort and flow analogies used do not all have power as their product. However, they may all be scaled to produce power as their product, with the exception of the thermal energy basic category. There, a pseudo bond graph approach can be used to provide a more practical set of flows for designers to work with. Signals can defined from a human factors standpoint (Federal Aviation Administration, 1996).

The set of flows that follows is part of the functional basis described in Chapter 5. A flow from this list is selected to fill the object position of the verb-object functional description. Flows in the functional basis set are a more abstract representation of the actual problemís flows. The given definitions make the transformation from actual flow to basis set flow more methodical and repeatable. An example of the flow usage follows each definition.

Material

a) *Solid.* Any object having a definite, firm shape. Example: The flow of sand paper into a hand sander is transformed into a solid entering the sander.

b) *Liquid.* A readily flowing fluid, specifically having its molecules moving freely with respect to each other, but because of cohesive forces, not expanding indefinitely. Example: The flow of water through a coffee maker is a liquid.

c) *Gas.* Any collection of molecules which are characterized by random motion and the absence of bonds between the molecules. Example: An oscillating fan moves air by rotating blades. The air is transformed as gas flow.

d) *Human.* All or part of a person who crosses the device boundary. Example: Most coffee makers require the flow of a human hand to actuate (or start) the electricity and thus heat the water.

Energy

a) *Human.* Work performed by a person on the device. Example: An automobile requires the flow of human energy to steer and accelerate the vehicle.

i. *Force.* The activity which requires effort into the system while not requiring significant motion. Example: Human force is needed to actuate the trigger of a toy gun.

ii. *Motion.* The activity of moving all or part of the body through a prescribed path. Example: The trackpad on a PowerBook receives the flow of human motion to control the cursor.

b) *Acoustic.* Work performed in the production and transmission of sound. Example: The motor of a power drill generates a flow of acoustic energy in addition to the torque.

i. *Pressure.* Energy utilized is the pressure field of the sound waves. Example: A condenser microphone has a diaphragm which vibrates in response to acoustic pressure. This vibration changes the capacitance of the diaphragm, thus superimposing an alternating voltage on the direct voltage applied to the circuit.

ii. *Particle velocity.* The speed at which sound waves travel through a conducting medium. Example: Sonar devices rely on the flow of acoustic particle velocity to determine the range of an object.

c) *Biological.* Work produced by or connected with plants or animals. Example: In poultry houses, the biological energy produced by thousands of chickens (in the form of heat) is an important flow in determining cooling requirements for the house.

i. *Pressure.* Energy utilized is the pressure field exerted by a compressed biological fluid. Example: The high concentration of sugars and salts inside a cell causes the entry, via osmosis, of water into the vacuole, which in turn expands the vacuole and generates a hydrostatic biological pressure, called turgor, that presses the cell membrane against the cell wall. Turgor is the cause of rigidity in living plant tissue.

ii. *Volumetric flow.* Energy utilized is the kinetic energy of molecules in a biological fluid flow. Example: Increased metabolic activity of tissues such as muscles or the intestine automatically induces increased volumetric flow of blood through the dilated vessels.

d) *Chemical.* Work resulting from the reactions by which substances are produced from or converted into other substances. Example: A battery converts the flow of chemical energy into electrical energy.

i. *Pressure.* Energy utilized is the pressure field exerted by reactants. Example: The chemical pressure from combustion of a gas is the flow that moves the pistons in an engine.

ii. *Volumetric flow.* Energy utilized is the kinetic energy of molecules or ions.

e) *Electrical.* Work resulting from the flow of electrons from a negative to a positive source. Example: A power belt sander imports a flow of electricity from a wall outlet and transforms it into a rotation.

i. *Electromotive force.* Potential difference across the positive and negative sources. Example: Household electrical receptacles provide a flow of voltage of 110 V.

ii. *Current.* The flow or rate of flow of electric charge in a conductor or medium between two points having a difference in potential.

f) *Electromagnetic.* Energy that is propagated through free space or through a material medium in the form of electromagnetic waves (Britannica Online, 1997). It has both wave and particle-like properties. Example: Solar panels convert the flow electromagnetic energy into electricity.

i. *Optical.* Work associated with the nature and properties of light and vision. Example: A car visor refines the flow of optical energy that its passengers receive.

a) *Intensity.* The amount of optical energy per unit area. Example: Tinted windows reduce the optical intensity of the light entering.

b) *Velocity.* The speed of light in its conducting medium.

ii. *Solar.* Work produced by or coming from the sun. Example: Solar panels collect the flow of solar energy and transform it into electricity.

a) *Intensity.* The amount of solar energy per unit area. Example: A cloudy day reduces the solar intensity available to solar panels for conversion to electricity.

b) *Velocity.* The speed of light in free space.

g) *Hydraulic.* Work that results from the movement and force of a liquid, including hydrostatic forces. Example: Hydroelectric dams generate electricity by harnessing the hydraulic energy in the water that passes through the turbines.

i. *Pressure.* Energy utilized is the pressure field exerted by a compressed liquid. Example: A hydraulic jack uses the flow hydraulic pressure to lift heavy objects.

ii. *Volumetric flow.* Energy utilized is the kinetic energy of molecules in a fluid flow. Example: A water meter measures the volumetric flow of water without a significant pressure drop in the line.

h) *Magnetic.* Work resulting from materials that have the property of attracting metals, whether that quality is naturally occurring or electrically induced. Example: The magnetic energy of a magnetic lock is the flow that keeps it secured to the iron based structure.

i) *Magnetomotive force.* The driving force which sets up the magnetic flux inside of an iron core. Example: Magnetomotive force is directly proportional to the current in the coil surrounding the iron core.

ii. Time rate of change of magnetic flux. Flux is the magnetic displacement variable in an iron core induced by the flow of current through a coil. The magnetic flow variable is the time rate of change of the flux. Example: The voltage across a magnetic coil is directly proportional to the time rate of change of magnetic flux.

i) *Mechanical.* Work provided by moving parts of a machine. Example: An elevator often converts electrical or hydraulic energy into mechanical energy.

i. *Rotational energy.* Energy that results from a rotation or an attempted rotation. Example:

a) *Torque.* Pertaining to the moment that produces or tends to produce rotation. Example: In a power screwdriver, electricity is converted into the flow torque. The relevant flow is torque since the primary customer need is to insert screws easily, not quickly.

b) *Angular velocity.* Pertaining to the orientation or the magnitude of the time rate of change of angular position about a specified axis. Example: A centrifuge is used to separate out liquids of different densities from a mixture. The primary flow it produces is that of angular velocity, since the rate of rotation about an axis is the main concern.

ii. *Translational energy.* Energy that results from a translation or an attempted translation. Example:

a) *Force.* The action that produces or attempts to produce a translation. Example: In a tensile testing machine, the primary flow is that of a force which produces a stress in the test specimen. Although there is translation associated with the machine, it is a result of the force.

b) *Linear velocity.* Motion that can be described by three component directions. Example: An elevator car receives a flow of linear velocity to move between floors.

iii. *Vibrational energy.* Oscillating translational or rotational energy that is characterized by an amplitude and frequency. In the rotational case, motion does not complete a 360 cycle. Example: In many block sanders, the sanding surface receives a flow of vibration to remove the wood surface. Vibration is produced by an off-center mass on the motor shaft.

a) *Amplitude.* Energy is characterized by the magnitude of the generalized force or displacement. Example: In fatigue testing, the vibrational amplitude of the tensile stress is more important than the speed of each loading cycle.

b) *Frequency.* Energy is characterized by the speed of the oscillation. Example. Human exposure to vibration can induce sickness at certain vibrational frequencies.

j) *Pneumatic.* Work resulting from a compressed gas flow or pressure source. Example: A B-B gun relies on the flow of pneumatic energy (from compressed air) to propel the projectile (B-B).

i. *Pressure.* Energy utilized is the pressure field exerted by a compressed gas. Example: Certain cylinders rely on the flow of pneumatic pressure to move a piston or support a force.

ii.*Mass flow.* Energy utilized is the kinetic energy of molecules in a gas flow. Example: The mass flow of air is the flow that transmits the thermal energy of a hair dryer to damp hair.

k) *Radioactive.* Work resulting from or produced by particles or rays, such as alpha, beta and gamma rays, by the spontaneous disintegration of atomic nuclei. Example: Nuclear reactors produce a flow of radioactive energy which heats water into steam and then drives electricity generating turbines.

i. *Intensity.* The amount of radioactive particles per unit area. Example: Concrete is an effective radioactive shielding material, reducing the radioactive intensity in proportion to its thickness.

ii. *Decay rate.* The rate of emission of radioactive particles from a substance. Example: The decay rate of carbon provides a method to date pre-historic objects.

l) *Thermal.* A form of energy that is transferred between bodies as a result of their temperature difference. Example: A coffee maker converts the flow of electricity into the flow of thermal energy which it transmits to the water.

Note: A pseudo bond graph approach is used here. The true effort and flow variables are temperature and the time rate of change of entropy. However, a more practical pseudo-flow of heat flow is chosen here.

i. *Temperature.* The degree of heat of a body. Example: A coffee maker brings the temperature of the water to boiling in order to siphon the water from the holding tank to the filter basket.

ii. *Heat flow.* (Note: this is a pseudo-flow) The time rate of change of heat energy of a body. Example: Fins on a motor casing take heat flow away from the motor by conduction (through the fin), convection (to the air) and radiation (to the environment).

Signal

a) *Status.* A condition of some system, as in information about the state of the system. Example: Automobiles often measure the engine water temperature and send a status signal to the driver via a temperature gage.

i. *Auditory*. A condition of some system as displayed by a sound. Example: Pilots receive an auditory status, often the words "pull up," when their aircraft reaches a dangerously low altitude.

ii. *Olfactory*. A condition of some system as related by the sense of smell or particulate count. Example: Carbon monoxide detectors receive an olfactory status signal from the environment and monitor it for high levels of CO.

iii. *Tactile*. A condition of some system as perceived by touch or direct contact. Example: Temperature, pressure and roughness measurements are examples of tactile status signals.

iv. *Taste*. A condition of some dissolved substance as perceived by the sense of taste. Example: In an electric wok, the taste status signal from the human chef is used to determine when to turn off the wok.

v. *Visual*. A condition of some system as displayed by some image. Example: A power screwdriver provides a visual status of its direction through the display of arrows on the switch.

Control. A command sent to an instrument or apparatus to regulate a mechanism. Example: An airplane pilot sends a control signal to the elevators through movement of the yoke. The yoke movement is transformed into an electrical signal, sent through wiring to the elevator, and then transformed back into a physical elevator deflection.

II. FUNCTIONS DEFINITIONS

The function classes are introduced in Chapter 5. Definitions for each class and basic function are presented below. Examples are given for the basic functions. These definitions first appeared in work by Little et al. (1997), following on work by Pahl and Beitz (1988) who defined a more restricted set of 5 "golden functions." Used with the flow definitions of Appendix A.1, the function definitions here complete the functional basis set which improves repeatability of function structure development and provides a standard level of detail at which the decomposition process stops.

Channel

a) *Import*. To bring in an energy or material from outside the system boundary. Example: A physical opening at the top of a blender pitcher imports a solid (food) into the system. Also, a handle on the blender pitcher imports a human hand. The blender system imports electricity via an electric plug.

b) *Export.* To send an energy or material outside the system boundary. Example: Pouring blended food out of a standard blender pitcher is exporting liquid from the system. The opening at the top of the blender is a solution to the export sub-function.

c) *Transfer.* To shift, or convey, a flow from one place to another.

i. *Transport.* To move a material from one place to another. Example: A coffee maker transports liquid (water) from its reservoir through its heating chamber and then to the filter basket.

ii. *Transmit.* To move an energy from one place to another. Example: In a hand held power sander, human force is transmitted from the human to the object being sanded through the housing of the sander.

d) *Guide.* To direct the course of an energy or material along a specific path. Example: In a domestic HVAC system, gas (air) is guided around the house to the correct locations via a set of ducts.

i. *Translate.* To fix the movement of a material into one linear direction. Example: In an assembly line, partially completed products are translated straight from one assembly station to another by a conveyor belt.

ii. *Rotate.* To fix the movement of a material around one axis. Example: In a computer disk drive, the magnetic disks rotate around an axis so that data can be read by the head.

iii. *Allow degree of freedom.* To control the movement of a material into one or more directions. Example: To provide easy trunk access and close appropriately, trunk lids need to move along a specific degree of freedom. A four bar linkage might give the trunk lid this degree of freedom.

Support

To firmly fix a material into a defined location, or secure an energy into a specific course.

a) *Stop.* To cease, or prevent, the transfer of a material or energy. Example: The transmission of UV radiation through a window is stopped by applying a reflective coating to the window.

b) *Stabilize.* To prevent a material or energy from changing course or location. Example: Auto shock absorbers stabilize the vehicle by preventing the axle from changing its course from that of the car.

c) *Secure.* To firmly fix a material or energy path. Example: On a bicycling glove, a velcro strap is used to secure the human hand in the correct place.

d) *Position.* To place a material or energy into a specific location or orientation. Example: The coin slot on a soda machine is used to position the coin to begin the coin evaluation and transportation procedure.

Connect

To bring two or more energies or materials together.

a) *Couple.* To join or bring together energies or materials such that the members are still distinguishable from each other. Example: On a standard pencil, an eraser is coupled to the shaft. The coupling is performed using a metal sleeve that is crimped to the eraser and the shaft.

b) *Mix.* To combine two materials into a single, uniform homogeneous mass. Paint is mixed before application. Its base and dyes are mixed to form an homogeneous liquid.

Branch

To cause an material or energy to no longer be joined or mixed.

a) *Separate.* To isolate a material or energy into distinct components. The separated components are distinct from the flow before separation, as well as each other. Example: Light is separated into different wavelength components to produce a rainbow. A glass prism performs this separation.

   i. *Remove.* To take away a part of a material from its prefixed place. Example: Small pieces of the surface of wood are removed by a sander to smooth the wood.

b) *Refine.* To reduce a material or energy such that only the desired elements remain. Example: Coffee grounds are refined by passing a hot liquid (water) through the grounds. A filter retains the refined coffee grounds and allows the new liquid (coffee) to pass through.

c) *Distribute.* To cause a material or energy to break up. The individual bits are similar to each other and the undistributed flow. Example: Hair-styling liquids are distributed over the head to hold the hair in the desired style. An atomizer is used to distribute, or spray, the liquid.

d) *Dissipate.* To break up and drive away or dispel. Example: The steel safety cage of an automobile dissipates impact energy to prevent injury to its passengers. The energy is dissipated by the steel members bending and twisting during impact.

Provide

To accumulate or provide material or energy.

a) *Store.* To accumulate material or energy. Example: Energy is stored in a flashlight. A DC electrical battery is used to store the energy.

b) *Supply.* To provide material or energy from storage. Example: In a flashlight, energy is supplied to the bulb by the battery.

c) *Extract.* To draw, or forcibly pull out, a material or energy. Example: Mechanical kinetic energy is extracted from the wind in a windmill. The extraction is performed by the airfoil (windmill blades).

Control Magnitude

To alter or govern the size or amplitude of material or energy.

a) *Actuate.* To commence the flow of energy or material in response to an imported control signal. Example: Actuating the flow of electrical energy turns on a light bulb. The actuation is performed using a circuit switch.

b) *Regulate.* To adjust the flow of energy or material in response to a control signal, such as a characteristic of a flow. Example: The liquid flowing from a faucet is regulated to allow different flow rates.

c) *Change.* To adjust the flow of energy or material in a predetermined and fixed manner. In a hand held drill, the electrical energy flow to the motor is changed thus changing the speed the drill turns. This change is accomplished by a variable resistor.

i. *Form.* To mold or shape a material. Example: In the auto industry, sheet metal is formed into contoured surfaces that become fenders, hoods and trunks. These parts are formed by large presses.

ii. *Condition.* To render an energy appropriate for the desired use. Example: To prevent damage to electrical equipment, electrical energy is conditioned by excluding spikes and noise from the energy path.

Convert

To change from one form of energy or material to another. For completeness, any type of flow conversion is valid. In practice, conversions such as convert electricity to torque will be more common than convert solid to optical energy. Example: An electrical motor converts electricity to rotational energy.

Signal

To provide information.

a) *Sense.* To perceive, or become aware, of a signal. Example: An audio cassette machine senses if the end of the tape has been reached. A spring senses an increase in tension when the tape ends.

b) *Indicate.* To make something known to the user. Example: A coffee maker indicates the level of water in the machine. A small window into the water container indicates the level to the user.

c) *Display.* To show a visual effect. Example: The face and needle of an air pressure gage display the status of the pressure vessel.

d) *Measure.* To determine the magnitude of a material or energy flow. Example: A thermostat measures temperature. A bimetallic strip is used to measure the temperature.

III. FUNCTION STRUCTURES FOR EXAMPLE PRODUCTS

Function structures are presented here for example products. The structures themselves include sufficient detail to cover the customer needs elicited for each product. This may lead to apparent discrepancies when comparing similar products. For example, the function structure for the VersaPak charger has explicit detail on the ability to connect to a wall outlet, yet many of the other products that also plug into a wall simply have an "Import Electricity" sub-function. There is no correctness problem here, only a matter of detail. Every product that uses a power cord and has the "Import Electricity" sub-function could as well be expanded into the set shown in the VersaPak charger.

Another related discrepancy across similar products is the difference in flows. Some structures separate the flows of a material and its kinetic energy, for example, while other structures combine both aspects into a single flow. Again, this is simply reflective of resolution needed by the customer needs. Generally, two often-conflicting rules apply: material and energy flows must be conserved (per unit time), and one should use as little detail as is necessary. Given that, when a type of flow disappears or appears, it is because of one of two reasons. First, a flow of a different type actually brought it in or carries it out (such as a material flow bringing in kinetic energy that is extracted into a useful force flow as the material flow continues). Second, the flow can disappear or appear because where it went or came from is irrelevant to the customer needs (such as with human forces causing actuation or regulation, and the relatively trivial force being dissipated somehow).

As a second point, one should take particular notice of when sub-functions become isolated from others in a network. This indicates functional independence. For example, many products must be positioned for use, and the functions associated with this, such as a "position tooling" sub-function, are isolated. This should be no surprise, since the function can be completed irrespective of the remaining functions: the power need not be on, the tooling need not be attached, etc., since once can still point the product. The functions are independent.

Similarly, the tooling used in the operation of a product, such as a drill bit with drilling holes, should never be flowing through the sub-function that "does" the operation. The drill bit does not flow through a "Drill Hole" sub-function, though the bit does the drilling. This is similar to the fact that no function structure has a flow of the product it represents: you always represent what flows through the product (be the product...what flows to/from/through you?). With the drill, the bit physically does the drilling, and objects that physically "do" a function should not flow through the function. Instead, the bit, when attached, generates an information signal indicating that it is attached, which a human processes to tell the hand to input the "on" signal once the drill is positioned. These external human functions are not product functions (though they are needed to functionally complete the operation that the product is used for).

As an exercise, for any product with separated sub-functions, the student should add the user functions into the product function structure, to understand how the product/human pair make a complete functions structure network for the particular operation.

Figure A.1

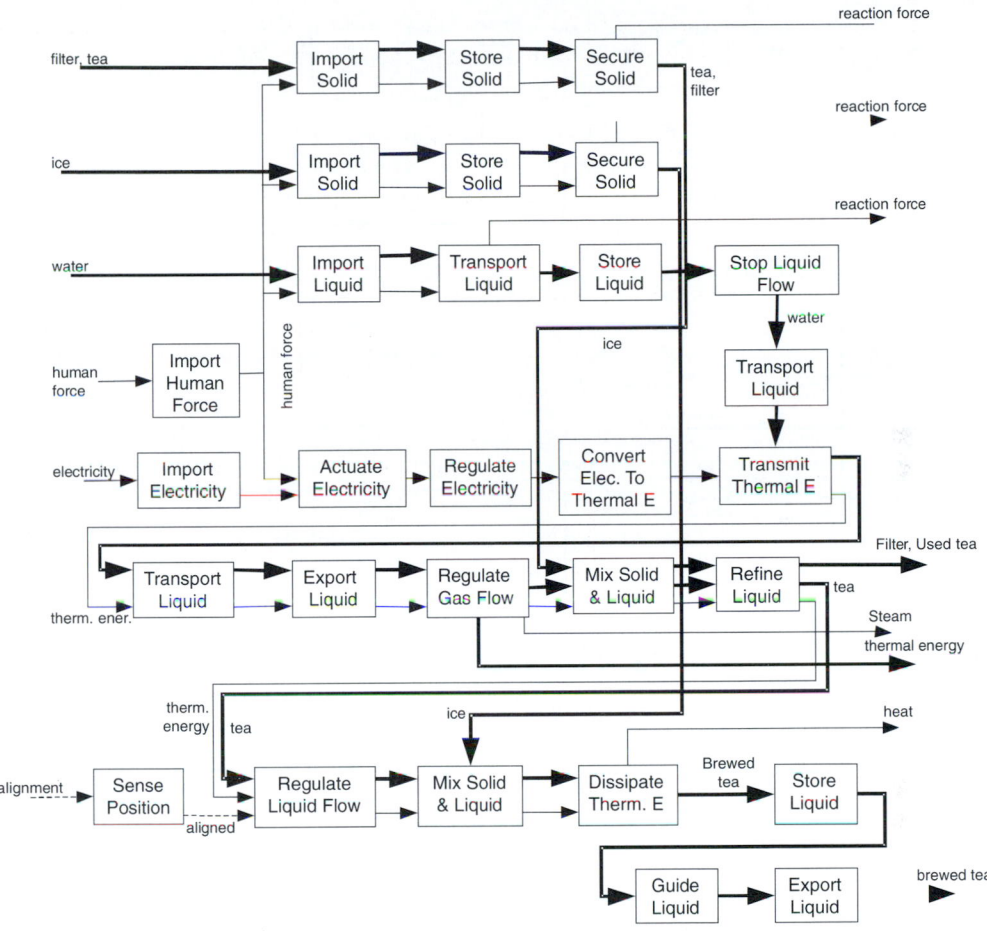

Function structure for a Mr. Coffee and West Bend iced tea/coffee brewer.

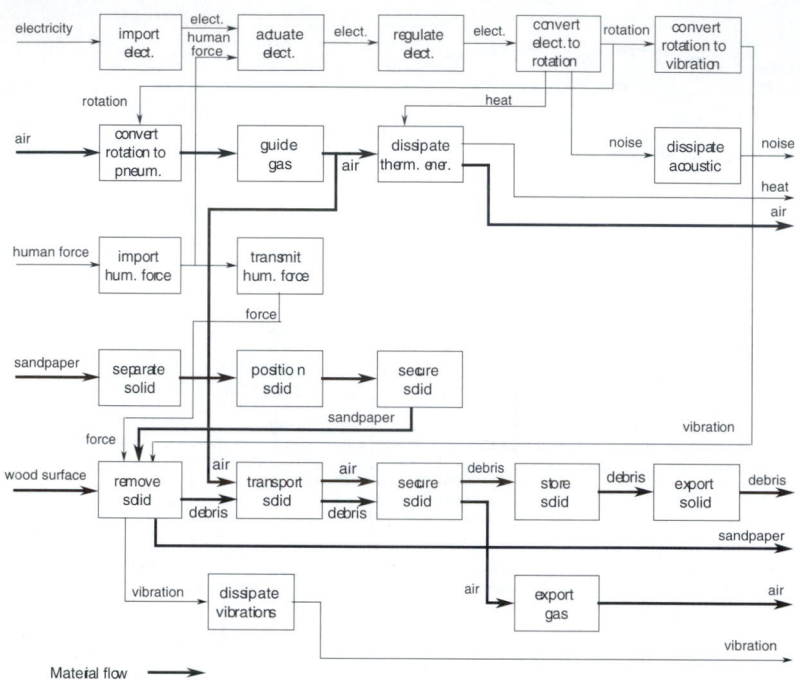

Figure A.2 Function structure for a DeWalt palm grip sander with debris bag.

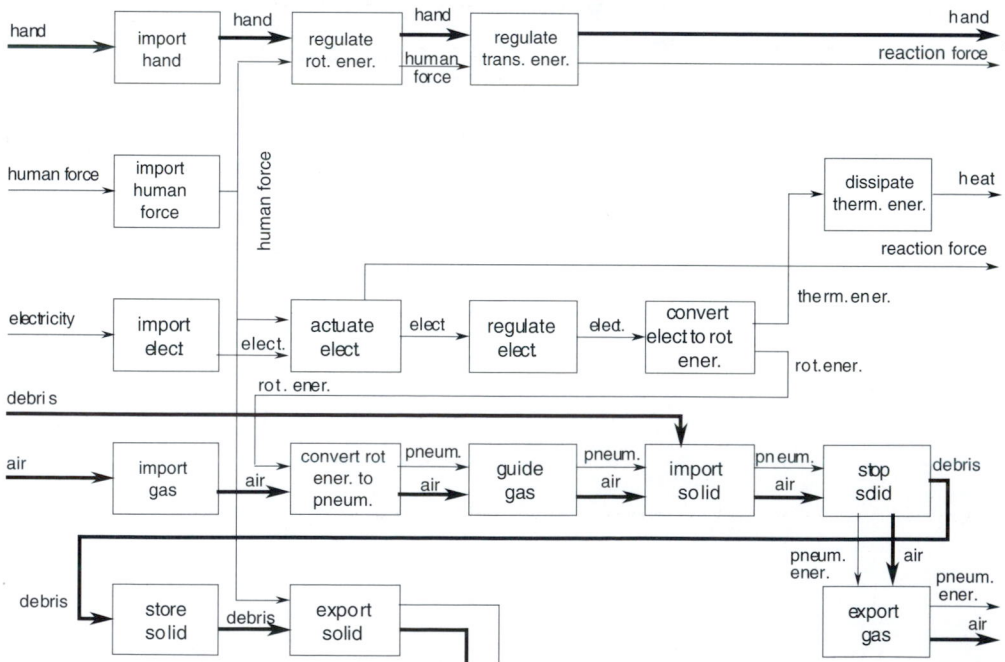

Figure A.3 Function structure for a Bissel hand vacuum.

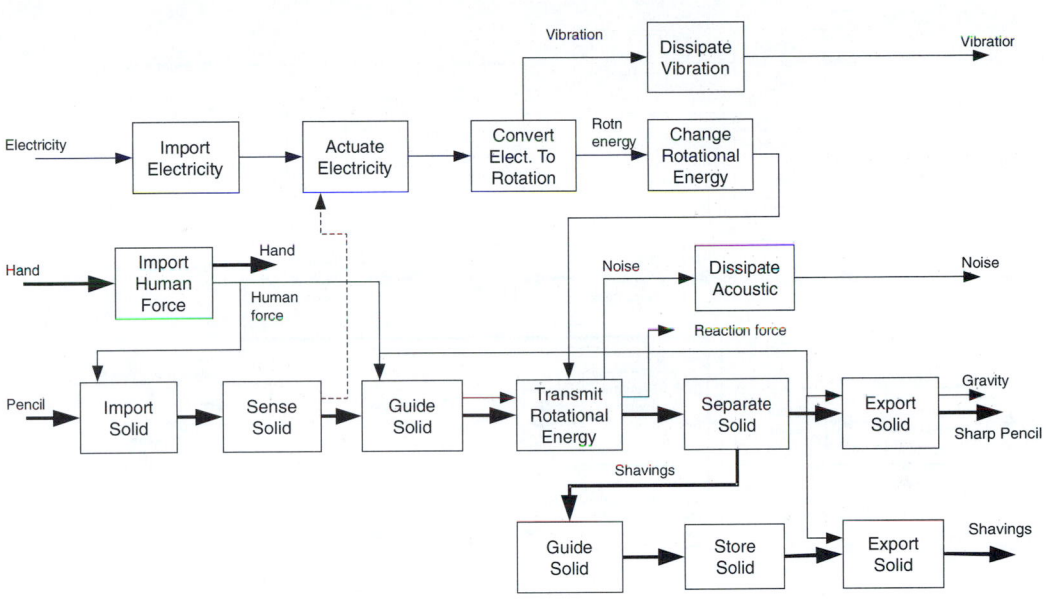

Figure A.4 Function structure for an electric pencil sharpener.

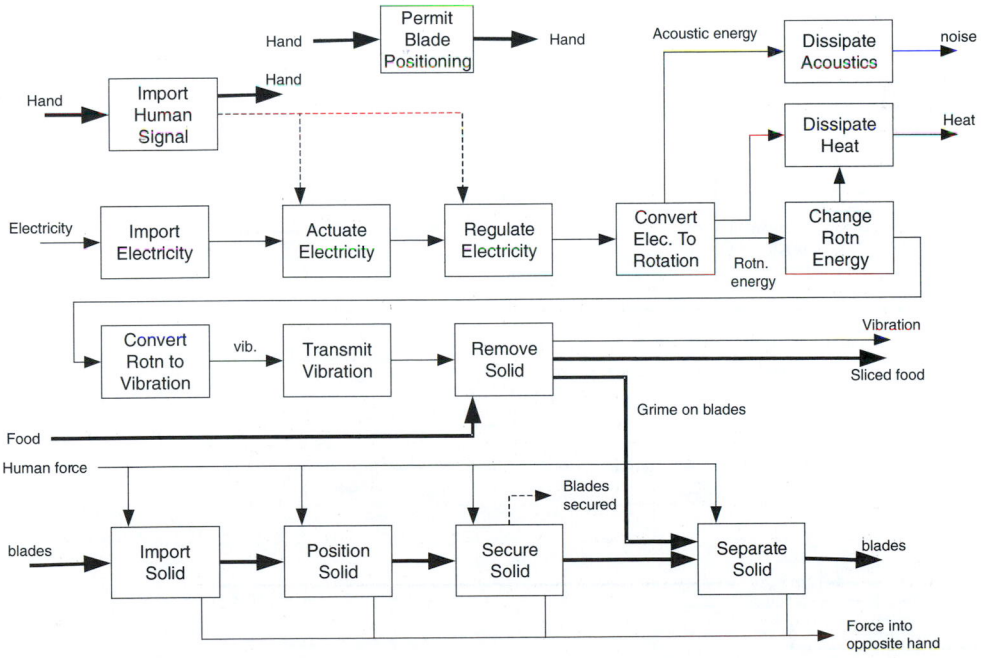

Figure A.5 Function structure for a Black and Decker electric knife.

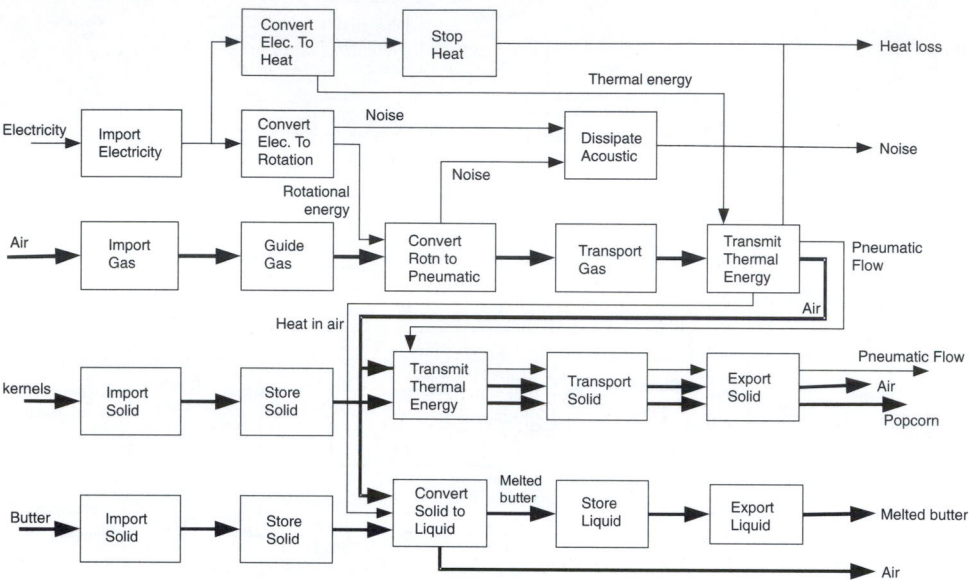

Figure A.6 Function structure for a Presto hot air popcorn popper.

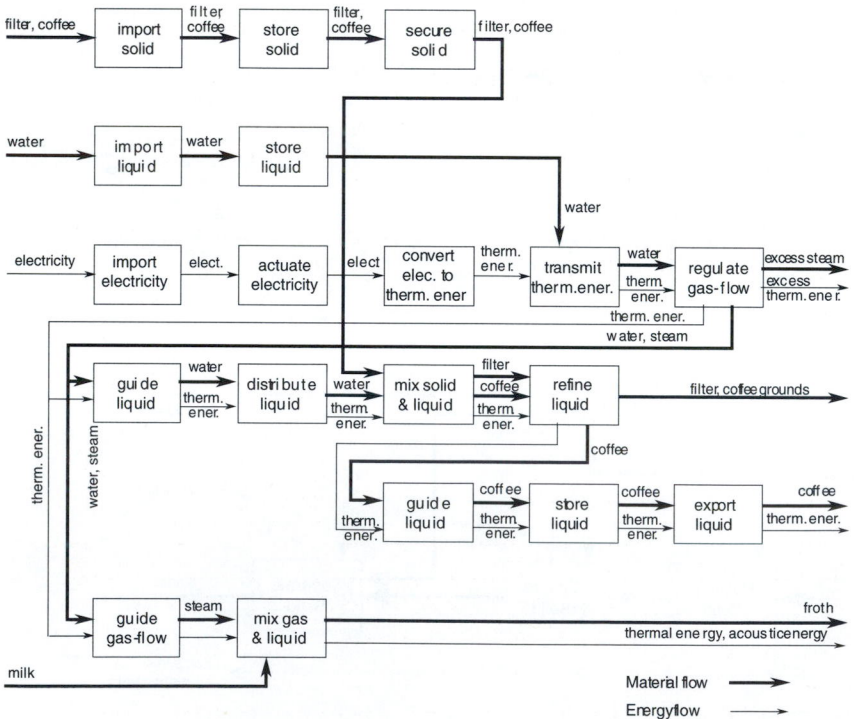

Figure A.7 Function structure for a Krups cafe trio (coffee/expresso maker).

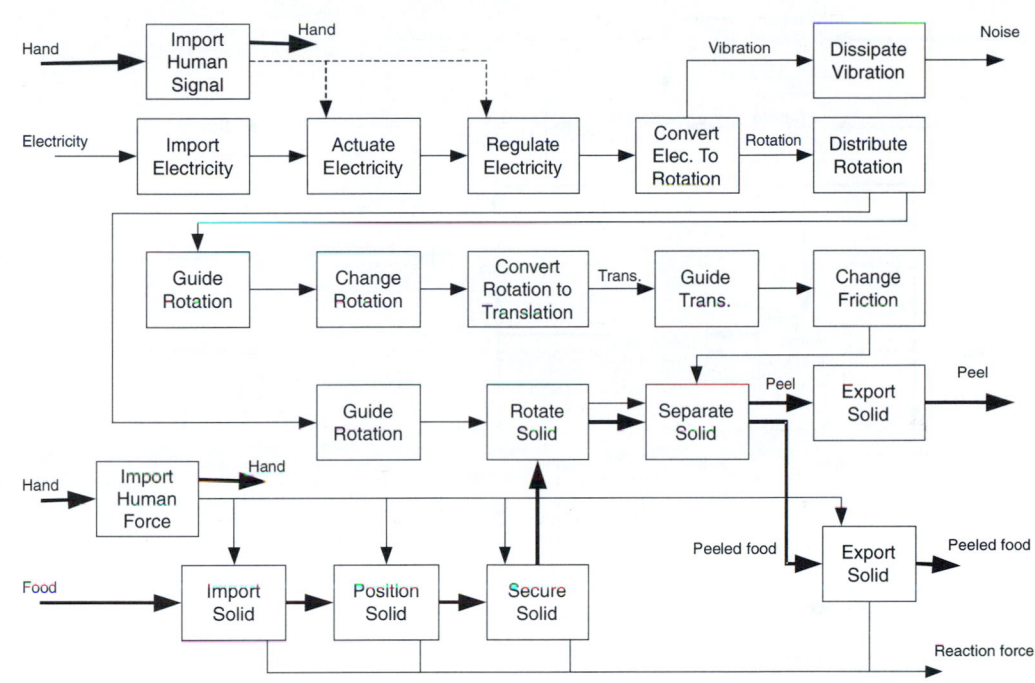

Figure A.8 Function structure for a Dazey fruit/veggie peeler.

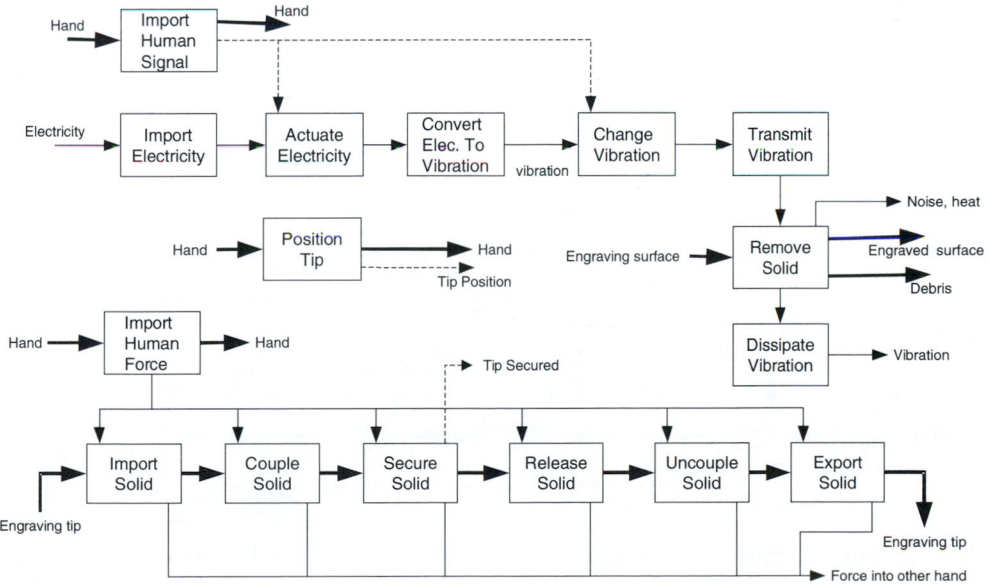

Figure A.9 Function structure for a Dremel engraver.

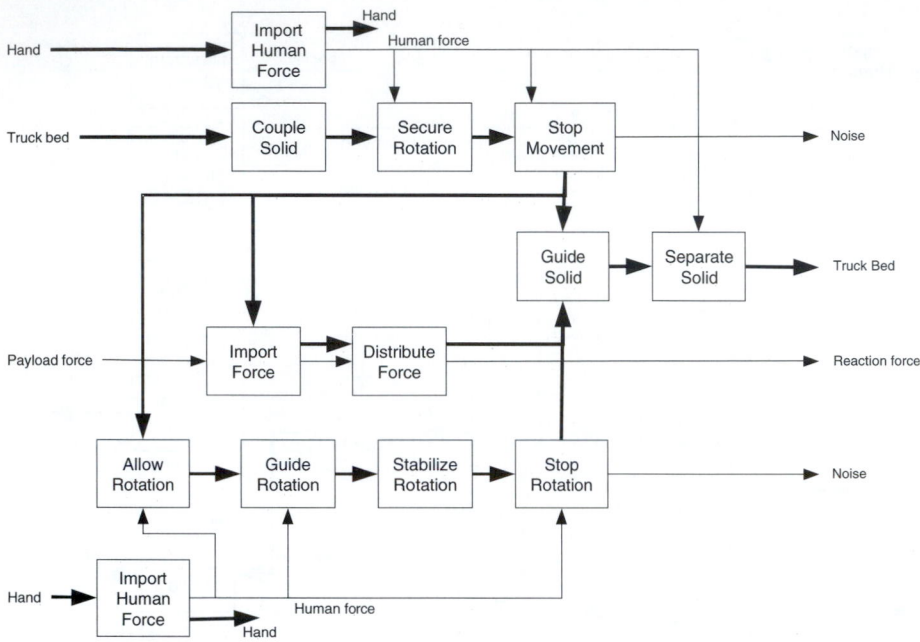

Figure A.10 Function structure for a 1974 Chevrolet Pickup Truck tailgate.

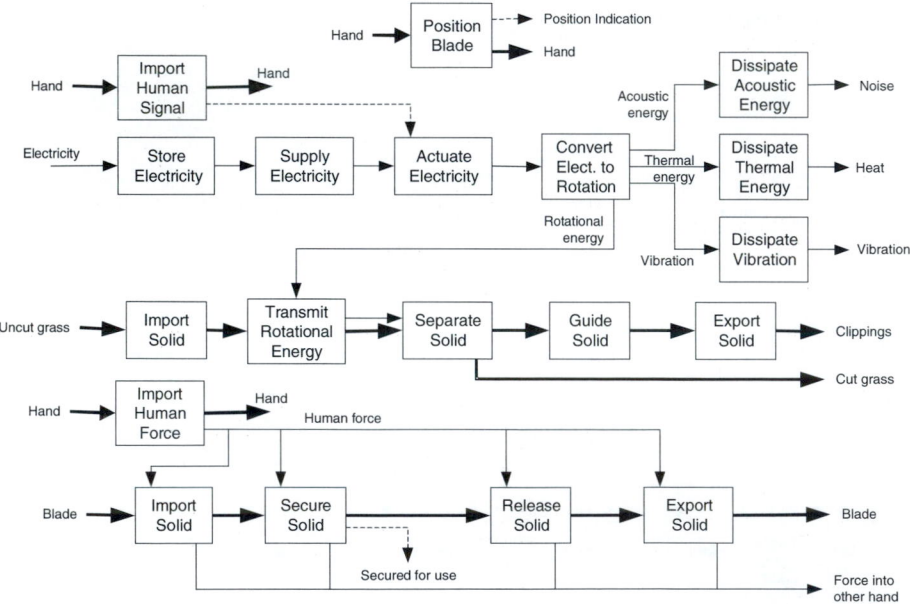

Figure A.11 Function structure for a Black and Decker (non-VersaPak) weed trimmer.

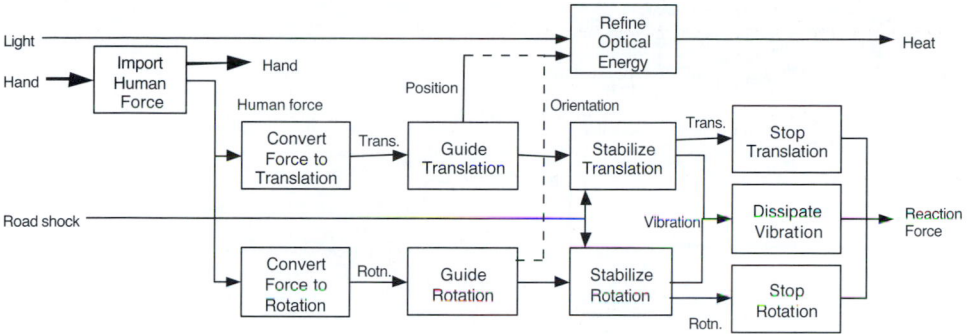

Figure A.12 Function structure for a Cadillac auxiliary visor.

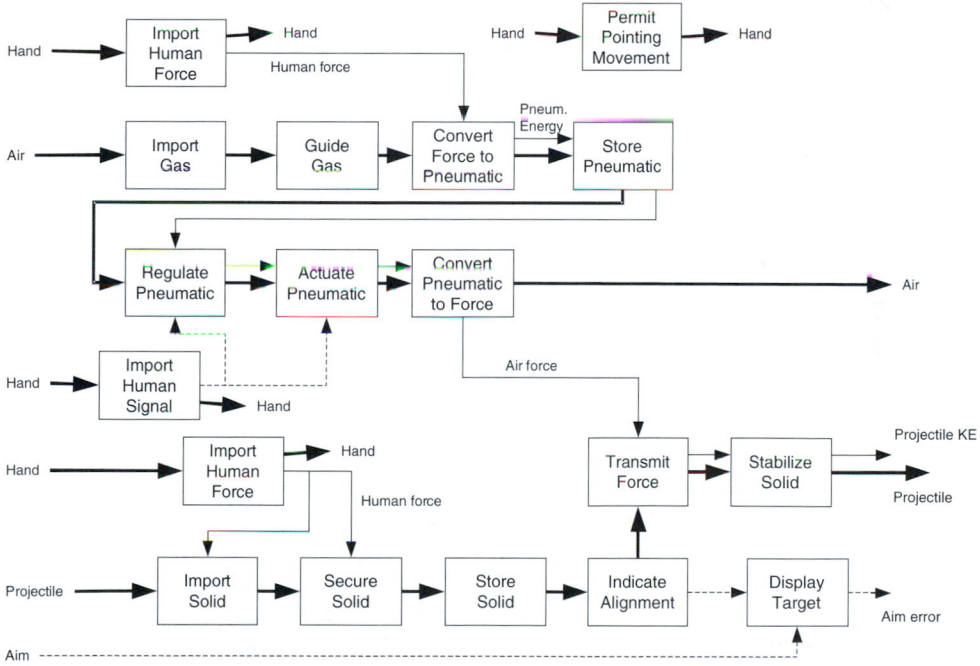

Figure A.13 Function structure for a Super Maxx ball shooter.

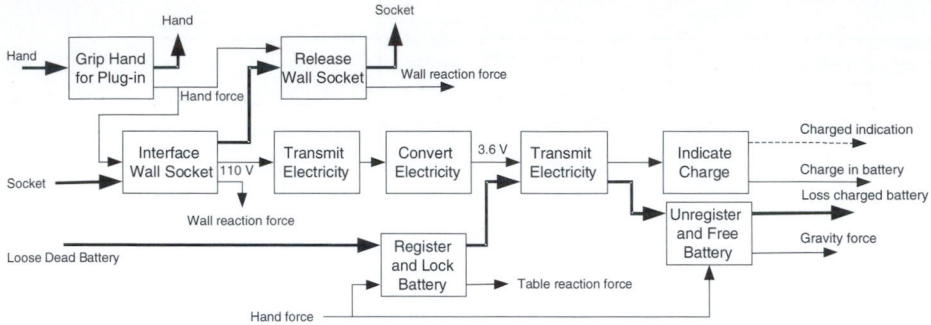

Figure A.14 Function structure for a VersaPak battery charger.

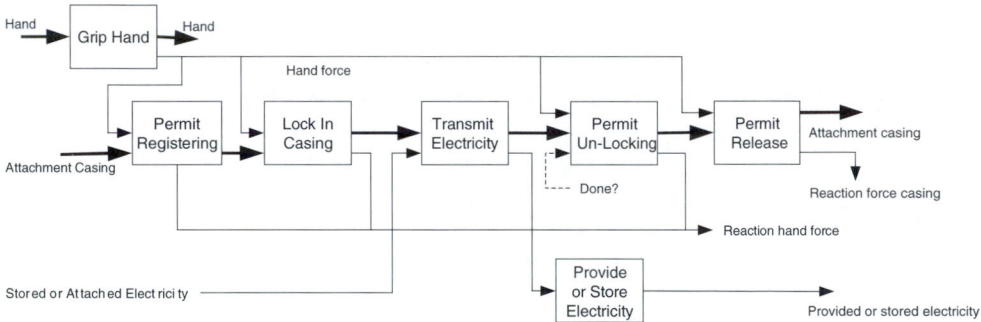

Figure A.15 Function structure for a VersaPak battery.

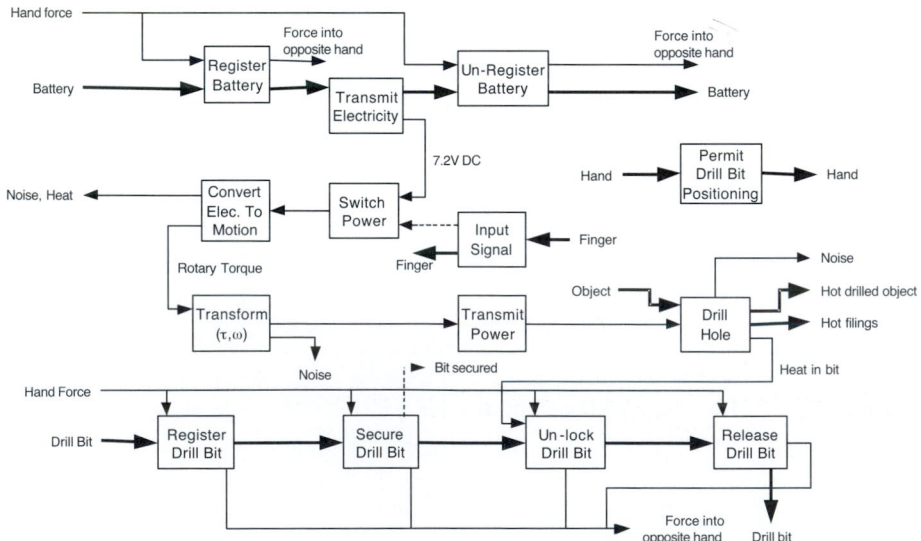

Figure A.16 Function structure for a VersaPak cordless drill.

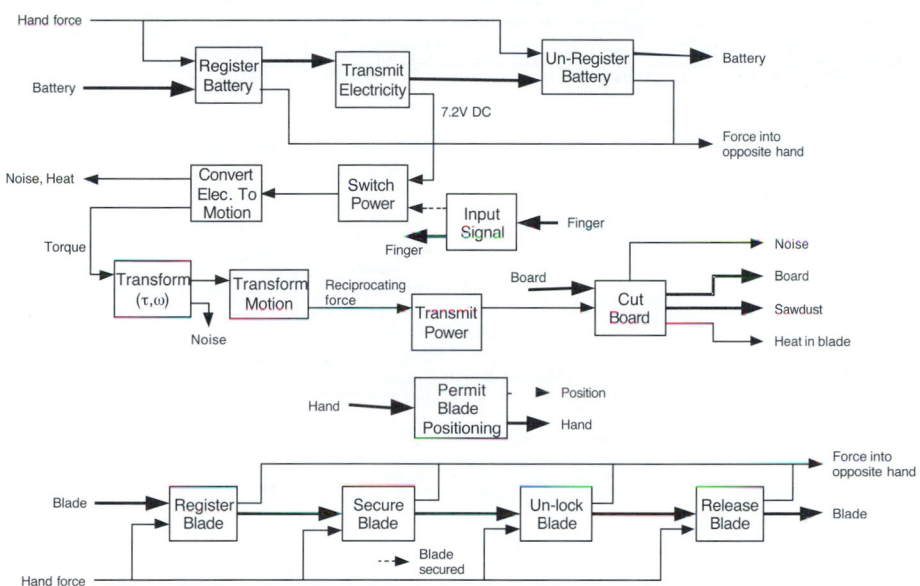

Figure A.17 Function structure for a VersaPak cordless saber saw.

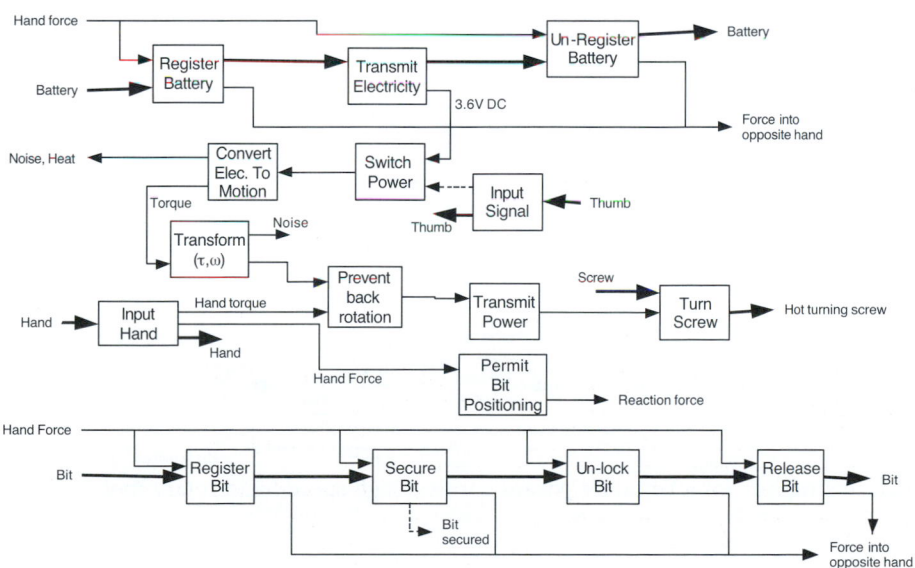

Figure A.18 Function structure for a VersaPak cordless screwdriver.

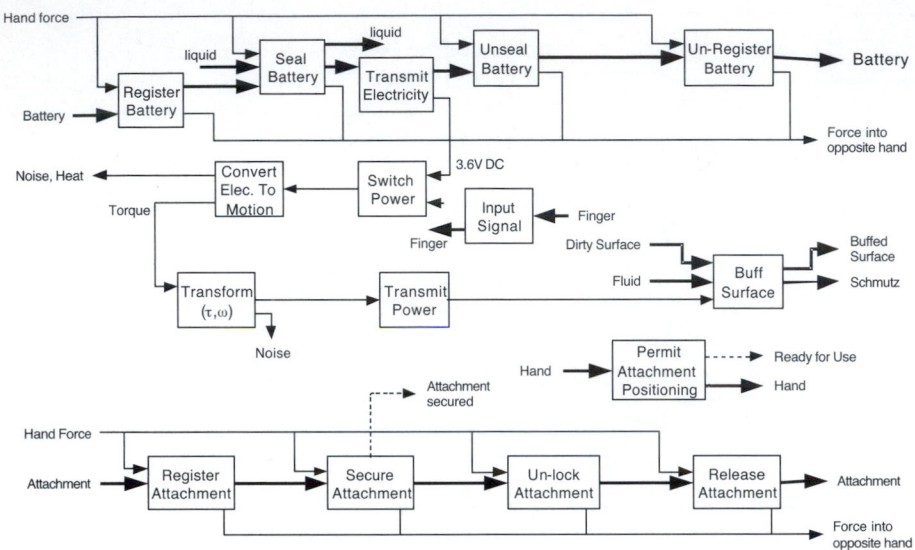

Figure A.19 Function structure for a VersaPak ScumBuster cordless power brush.

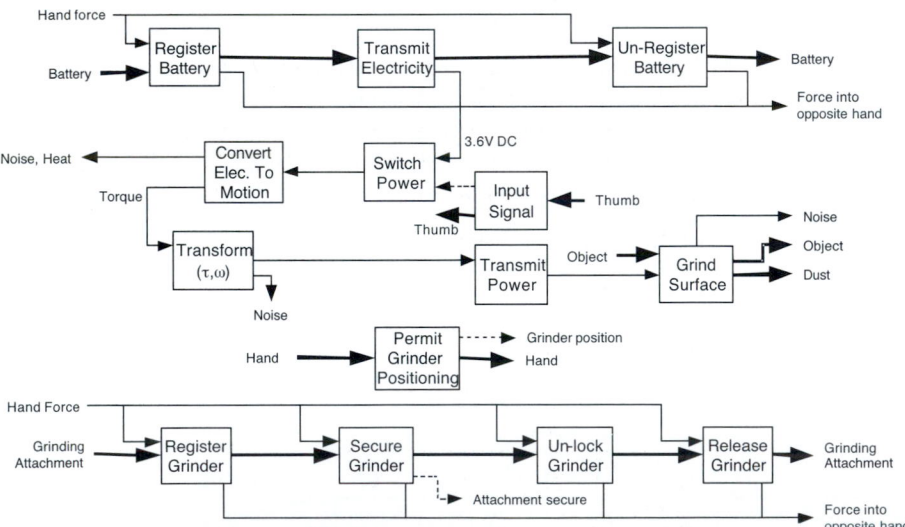

Figure A.20 Function structure for a VersaPak Wizard cordless high speed hand rotary tool.

Appendix B

I. TWO FACTORIAL EXPERIMENTAL TABLES

These experimental matrices are derived using the base design method as described in Chapter 18. Using these matrices, main effects will not be confounded with any second order interaction effects. Some second order interaction effects will be confounded with other second order interaction effects. All are orthogonal.

Table B.1 Two or three variables

| | x_1 | x_2 | x_3 |
	A	B	C
$\vec{x}_1$	-1	-1	1
$\vec{x}_2$	1	-1	-1
$\vec{x}_3$	-1	1	-1
$\vec{x}_4$	1	1	1
Main Effect Confounded with Interaction	BC	AC	AB

Table B.2 Three to five variables

	x_1	x_2	x_3	x_4	x_5	x_6	x_7
	A	B	C	D	E	F	G
$\vec{x}_1$	-1	-1	-1	-1	1	1	1
$\vec{x}_2$	1	-1	-1	1	-1	-1	1
$\vec{x}_3$	-1	1	-1	1	-1	1	-1
$\vec{x}_4$	1	1	-1	-1	1	-1	-1
$\vec{x}_5$	-1	-1	1	1	1	-1	-1
$\vec{x}_6$	1	-1	1	-1	-1	1	-1
$\vec{x}_7$	-1	1	1	-1	-1	-1	1
$\vec{x}_8$	1	1	1	1	1	1	1
Main Effect Confounded with Interaction	BD	AD	AF	AG	AB	AC	AD
	CF	CG	BG	BF	CE	BD	BC
	EG	EF	DE	CE	FG	EG	EF

Table B.3 Four to eleven variables.

	x_1	x_2	x_3	x_4	x_5	x_6	x_7	x_8	x_9	x_{10}	x_{11}	x_{12}	x_{13}	x_{14}	x_{15}
	A	B	C	D	E	F	G	H	I	J	K	L	M	N	O
$\vec{x}_1$	-1	-1	-1	-1	1	-1	-1	-1	-1	1	1	1	1	1	1
$\vec{x}_2$	1	-1	-1	-1	-1	1	1	1	-1	-1	-1	-1	1	1	1
$\vec{x}_3$	-1	1	-1	-1	-1	1	1	-1	1	-1	1	1	-1	-1	1
$\vec{x}_4$	1	1	-1	-1	1	-1	-1	1	1	1	-1	-1	-1	-1	1
$\vec{x}_5$	-1	-1	1	-1	1	1	-1	1	1	1	1	-1	-1	1	-1
$\vec{x}_6$	1	-1	1	-1	1	-1	1	-1	1	-1	1	-1	-1	1	-1
$\vec{x}_7$	-1	1	1	-1	1	-1	1	1	-1	-1	-1	1	1	-1	-1
$\vec{x}_8$	1	1	1	-1	-1	1	-1	-1	-1	1	1	-1	1	-1	-1
$\vec{x}_9$	-1	-1	-1	1	-1	-1	1	1	1	1	1	-1	1	-1	-1
$\vec{x}_{10}$	1	-1	-1	1	1	1	-1	-1	1	-1	-1	1	1	-1	-1
$\vec{x}_{11}$	-1	1	-1	1	1	1	-1	1	-1	-1	1	-1	-1	1	-1
$\vec{x}_{12}$	1	1	-1	1	-1	-1	1	-1	-1	-1	1	-1	1	1	-1
$\vec{x}_{13}$	-1	-1	1	1	1	1	1	1	-1	1	-1	-1	-1	-1	1
$\vec{x}_{14}$	1	-1	1	1	-1	-1	-1	1	-1	-1	1	1	-1	-1	1
$\vec{x}_{15}$	-1	1	1	1	-1	-1	-1	-1	1	-1	-1	-1	1	1	1
$\vec{x}_{16}$	1	1	1	1	1	1	1	1	1	1	1	1	1	1	1
Main Effect Confounded with Interaction	BJ	AJ	AK	AL	AI	AM	AN	AO	AE	AB	AC	AD	AF	AG	AH
	CK	CM	BM	BN	BH	BK	BL	BE	BO	CF	BF	BG	BC	BD	BI
	DL	DN	DO	CO	CG	CJ	CE	CL	CN	DG	DH	CH	DI	CI	CD
	EI	EH	EG	EF	DF	DE	DJ	DK	DM	EO	EN	EM	EL	EK	EJ
	FM	FK	FJ	GJ	JO	GO	FO	FN	FL	HI	GI	FI	GH	HF	FG
	GN	GL	HL	HK	KN	HN	HM	GM	GK	KM	JM	JN	JK	JL	KL
	HO	IO	IN	IM	LM	IL	IK	JI	HJ	LN	LO	OK	ON	OM	MN

Table B.4 Five to twenty five variables

	x_1	x_2	x_3	x_4	x_5	x_6	x_7	x_8	x_9	x_{10}	x_{11}	x_{12}	x_{13}	x_{14}	x_{15}	x_{16}	x_{17}	x_{18}	x_{19}	x_{20}	x_{21}	x_{22}	x_{23}	x_{24}	x_{25}	x_{26}	x_{27}	x_{28}	x_{29}	x_{30}	x_{31}
	A	B	C	D	E	F	G	H	I	J	K	L	M	N	O	P	Q	R	S	T	U	V	W	X	Y	Z	a	b	c	d	e
$\vec{x}_1$	-1	-1	-1	-1	-1	-1	1	1	1	1	1	1	1	-1	-1	-1	-1	-1	-1	-1	1	1	1	1	1	1	1	1	1	1	1
$\vec{x}_2$	1	-1	-1	-1	-1	1	1	1	-1	-1	-1	-1	-1	-1	-1	1	1	1	1	1	1	-1	-1	-1	-1	1	1	1	1	1	1
$\vec{x}_3$	-1	1	-1	-1	-1	1	-1	1	-1	-1	-1	-1	1	1	1	-1	-1	-1	1	1	1	-1	1	1	1	-1	-1	-1	1	1	1
$\vec{x}_4$	1	1	-1	-1	-1	-1	-1	-1	1	1	1	1	1	1	1	1	1	-1	-1	-1	1	-1	-1	-1	-1	-1	-1	-1	1	1	1
$\vec{x}_5$	-1	-1	1	-1	-1	1	1	-1	1	1	-1	-1	1	-1	1	1	-1	1	1	-1	1	1	-1	1	1	-1	1	-1	-1	-1	1
$\vec{x}_6$	1	-1	1	-1	-1	-1	1	-1	1	1	-1	-1	1	-1	1	-1	1	1	-1	1	1	-1	1	-1	1	1	-1	1	-1	1	1
$\vec{x}_7$	-1	1	1	-1	-1	-1	1	-1	1	1	1	1	-1	-1	-1	1	1	1	1	-1	1	1	-1	1	-1	1	1	1	-1	-1	1
$\vec{x}_8$	1	1	1	-1	-1	1	1	1	-1	-1	1	1	-1	-1	-1	1	1	1	1	1	1	-1	1	1	-1	1	1	-1	-1	-1	1
$\vec{x}_9$	-1	-1	-1	1	-1	1	1	-1	-1	1	1	1	1	1	-1	1	1	-1	-1	1	1	-1	1	1	-1	1	1	-1	1	1	-1
$\vec{x}_{10}$	1	-1	-1	1	-1	-1	-1	1	1	1	-1	1	1	1	-1	1	-1	-1	1	1	1	-1	1	1	1	-1	1	-1	1	1	-1
$\vec{x}_{11}$	-1	1	-1	1	-1	-1	1	-1	1	-1	1	1	-1	-1	1	1	-1	1	1	1	1	-1	1	1	-1	1	1	-1	-1	1	-1
$\vec{x}_{12}$	1	1	-1	1	-1	1	1	1	1	1	-1	1	-1	-1	1	-1	1	-1	-1	-1	1	1	-1	1	-1	1	1	-1	1	1	-1
$\vec{x}_{13}$	-1	-1	1	1	-1	1	1	1	-1	1	1	-1	1	1	1	1	1	1	1	-1	1	1	-1	-1	-1	1	1	-1	-1	-1	-1
$\vec{x}_{14}$	1	-1	1	1	-1	-1	1	1	1	1	1	-1	-1	-1	1	1	1	-1	-1	-1	1	1	-1	1	1	1	1	-1	1	-1	-1
$\vec{x}_{15}$	-1	1	1	1	-1	-1	1	-1	1	1	-1	1	1	-1	1	1	-1	-1	1	1	1	1	-1	1	-1	1	1	-1	1	-1	-1
$\vec{x}_{16}$	1	1	1	1	-1	1	1	-1	-1	-1	-1	-1	1	1	1	1	-1	1	1	1	1	1	-1	1	1	1	1	-1	-1	1	-1
$\vec{x}_{17}$	-1	-1	-1	-1	1	1	1	-1	1	-1	1	1	1	-1	1	1	-1	1	1	1	1	1	-1	1	-1	1	1	-1	-1	1	-1
$\vec{x}_{18}$	1	-1	-1	-1	1	-1	1	1	1	1	-1	1	-1	1	1	-1	1	-1	-1	1	1	-1	-1	1	1	-1	1	-1	-1	1	-1
$\vec{x}_{19}$	-1	1	-1	-1	1	-1	1	1	1	1	1	-1	-1	-1	1	1	1	1	-1	-1	1	-1	-1	1	1	1	1	-1	-1	1	-1
$\vec{x}_{20}$	1	1	-1	-1	1	1	1	-1	-1	-1	1	1	1	1	1	1	1	1	1	1	1	-1	-1	1	-1	1	1	-1	-1	1	-1
$\vec{x}_{21}$	-1	-1	1	-1	1	1	1	1	-1	1	-1	1	-1	1	-1	1	1	-1	1	1	1	1	-1	1	-1	1	1	-1	-1	-1	1
$\vec{x}_{22}$	1	-1	1	-1	1	1	1	-1	1	1	-1	1	1	-1	1	1	-1	-1	1	-1	1	1	-1	1	-1	1	-1	-1	1	1	-1
$\vec{x}_{23}$	-1	1	1	-1	1	1	1	-1	-1	1	1	-1	1	-1	-1	1	1	-1	1	1	1	-1	-1	1	1	1	1	-1	1	1	-1
$\vec{x}_{24}$	1	1	1	-1	1	-1	-1	1	1	-1	1	-1	1	-1	1	1	1	-1	1	-1	1	1	-1	1	1	1	1	-1	1	1	-1
$\vec{x}_{25}$	-1	-1	-1	1	1	-1	1	-1	-1	1	-1	1	-1	1	1	1	1	-1	1	1	1	-1	-1	1	1	-1	-1	-1	-1	-1	1
$\vec{x}_{26}$	1	-1	-1	1	1	-1	1	1	-1	1	-1	1	-1	1	1	-1	1	-1	1	1	1	-1	-1	1	1	-1	-1	1	-1	-1	1
$\vec{x}_{27}$	-1	1	-1	1	1	1	1	1	1	1	1	-1	1	1	-1	1	1	-1	-1	1	1	-1	-1	1	-1	1	1	-1	1	1	1
$\vec{x}_{28}$	1	1	-1	1	1	-1	-1	-1	1	-1	-1	1	-1	-1	1	-1	1	-1	1	1	1	-1	-1	1	1	-1	1	-1	1	1	1
$\vec{x}_{29}$	-1	-1	1	1	1	1	1	-1	1	1	1	1	1	-1	1	1	-1	1	1	1	-1	-1	1	-1	-1	1	-1	-1	1	1	1
$\vec{x}_{30}$	1	-1	1	1	1	-1	1	-1	-1	-1	-1	-1	1	1	1	-1	1	1	1	-1	-1	-1	1	1	1	-1	-1	1	1	1	1
$\vec{x}_{31}$	-1	1	1	1	1	-1	1	-1	-1	-1	-1	1	1	1	1	-1	-1	-1	-1	-1	-1	-1	1	-1	-1	-1	1	1	1	1	1
$\vec{x}_{32}$	1	1	1	1	1	1	1	1	1	1	1	1	1	1	1	1	1	1	1	1	1	1	1	1	1	1	1	1	1	1	1

Main effect confounded with interaction																															
	BV	AV	AW	AX	AY	AG	AF	AL	AM	AN	AO	AH	AI	AJ	AK	Ae	Ad	Ac	Ab	Aa	AZ	AB	AC	AD	AE	AU	AT	Ae	AR	AQ	AP
	CW	CZ	BZ	Ba	Bb	BH	Ad	BF	BP	BQ	BR	BG	Be	Bd	Bc	BI	BJ	BK	BY	BX	BW	CU	BU	BT	BS	BC	BD	AS	BO	BN	BM
	DX	Da	Dc	Cc	Cd	CI	Bc	CP	CF	CS	CT	Ce	CG	Cb	Ca	CH	CY	CX	CJ	CK	CV	DT	DR	CR	CQ	DO	CO	CN	CD	CE	CL
	EY	Eb	Ed	Ee	De	DJ	CM	DQ	DS	DF	DU	Dd	Db	DG	DZ	DY	DH	DW	DI	DV	DK	ES	EQ	EP	DP	EN	EM	DM	EL	Ce	DE
	FG	FH	FI	FJ	FK	EK	DN	ER	ET	EU	EF	Ec	Ea	EZ	EG	EX	EW	EH	EV	EI	EJ	FL	FM	FN	FO	FP	FQ	EB	FS	DL	FU
	IM	GL	GM	GN	GO	LV	EO	GV	GW	GX	GY	FV	FW	FX	FY	FZ	Fa	Fb	Fc	Fd	Fe	GH	GI	GJ	GK	Ge	Gd	FR	Gb	FT	GZ
	JN	JQ	HP	HQ	HR	MW	HV	IZ	HZ	Ha	Hb	IU	HU	HY	HS	GU	GT	GS	GR	GQ	GP	Ie	He	Hd	Hc	HI	HJ	Gc	HY	Ga	HW
	KO	KR	JS	KU	IT	NX	IW	Ja	Jc	Ic	Id	JT	JR	IR	IQ	JO	JO	IO	IN	HO	HN	HM	Jd	Jb	Ib	Ia	JY	IY	HK	IJ	HX	JK
	LR	Me	KT	Ld	JU	OY	JX	Kb	Kd	Ke	Je	KS	KQ	KP	JP	KN	KM	JM	KL	JL	IL	Kc	Ka	KZ	JZ	KX	KW	IX	KV	IK	NO
	Pe	Nd	Le	Mb	Lc	PZ	KY	MU	LU	LT	LS	MZ	LZ	La	Lb	LW	LX	LY	MX	NY	MP	LP	LQ	LR	LM	LN	JW	MN	JV	RQ
	Qd	Oc	Nb	OZ	Ma	Qa	LB	NT	NR	MR	MQ	Na	Nc	Mc	Md	MV	NV	OV	NW	OW	OX	NQ	NS	MS	MT	QS	PS	LO	QP	MO	ST
	Rc	Pl	Oa	PY	NZ	Rb	PU	OS	OQ	OP	NP	Ob	Od	Oe	Ne	Qc	Pc	Pd	Pa	Pb	Qb	OR	OT	OU	NU	TR	RU	PT	TU	PR	VI
	Sb	SY	QY	RW	PX	Sc	QT	We	Ve	Vd	Vc	PW	PV	QV	RV	Rd	Re	Qe	QZ	RZ	Ra	WZ	VZ	Va	Vb	VW	VX	QU	WX	SU	XY
	Ta	TX	RX	TV	QW	Td	RS	Xd	Xb	Wb	Wa	QX	SX	SW	TW	Sa	SZ	TZ	Te	Se	Sd	Xa	Xc	Wc	Wd	ac	Zc	VY	Za	WY	ab
	UZ	UW	UV	VI	SV	Ue	Ze	Yc	Ya	YZ	XZ	RY	TY	UY	UX	Tb	Ub	Ua	Ud	Uc	Tc	Yb	Yd	Ye	Xe	bd	be	Zd	de	Zb	cd

II. THREE FACTORIAL EXPERIMENTAL TABLES

These experimental matrices are derived from the Taguchi L-arrays, see Taguchi (1986). When using these matrices, main effect and second order interaction terms are not confounded. All are orthogonal.

Table B.5 Two variables

	x_1 A	x_2 B
$\vec{x}_1$	-1	-1
$\vec{x}_2$	0	1
$\vec{x}_3$	1	0

Table B.6 Two to four variables

	x_1 A	x_2 B	x_3 C	x_4 D
$\vec{x}_1$	-1	-1	-1	-1
$\vec{x}_2$	-1	0	0	0
$\vec{x}_3$	-1	1	1	1
$\vec{x}_4$	1	-1	0	1
$\vec{x}_5$	1	0	1	-1
$\vec{x}_6$	1	1	-1	0
$\vec{x}_7$	0	-1	1	0
$\vec{x}_8$	0	0	-1	1
$\vec{x}_9$	0	1	0	-1
Main effect confounded with interaction	BC BD CD	AC AD CD	AB AD BD	AB AC BC

Table B.7 Three to eight variables, with one 2 factorial

	x_1 A	x_2 B	x_3 C	x_4 D	x_5 E	x_6 F	x_7 G	x_8 H
$\vec{x}_1$	-1	-1	-1	-1	-1	-1	-1	-1
$\vec{x}_2$	-1	-1	0	0	0	0	0	0
$\vec{x}_3$	-1	-1	1	1	1	1	1	1
$\vec{x}_4$	-1	0	-1	-1	0	0	1	1
$\vec{x}_5$	-1	0	0	0	1	1	-1	-1
$\vec{x}_6$	-1	0	1	1	-1	-1	0	0
$\vec{x}_7$	-1	1	-1	0	-1	1	0	1
$\vec{x}_8$	-1	1	0	1	0	-1	1	-1
$\vec{x}_9$	-1	1	1	-1	1	0	-1	0
$\vec{x}_{10}$	1	-1	-1	1	1	0	0	-1
$\vec{x}_{11}$	1	-1	0	-1	-1	1	1	0
$\vec{x}_{12}$	1	-1	1	0	0	-1	-1	1
$\vec{x}_{13}$	1	0	-1	0	1	-1	1	0
$\vec{x}_{14}$	1	0	0	1	-1	0	-1	1
$\vec{x}_{15}$	1	0	1	-1	0	1	0	-1
$\vec{x}_{16}$	1	1	-1	1	0	1	-1	0
$\vec{x}_{17}$	1	1	0	-1	1	-1	0	1
$\vec{x}_{18}$	1	1	1	0	-1	0	1	-1

Only main effects, quadratic effects (A2, B2, etc.) and the AB interaction can be studied with this array. All interactions are all confounded with the main effects.

Table B.8 Five to thirteen variables.

	x_1 / A	x_2 / B	x_3 / C	x_4 / D	x_5 / E	x_6 / F	x_7 / G	x_8 / H	x_9 / I	x_{10} / J	x_{11} / K	x_{12} / L	x_{13} / M
$\vec{x}_1$	-1	-1	-1	-1	-1	-1	-1	-1	-1	-1	-1	-1	-1
$\vec{x}_2$	-1	-1	-1	-1	0	0	0	0	0	0	0	0	0
$\vec{x}_3$	-1	-1	-1	-1	1	1	1	1	1	1		1	1
$\vec{x}_4$	-1	0	0	0	-1	-1	-1	0	0	0	1	1	1
$\vec{x}_5$	-1	0	0	0	0	0	0	1	1	1	-1	-1	-1
$\vec{x}_6$	-1	0	0	0	1	1	1	-1	-1	-1	0	0	0
$\vec{x}_7$	-1	1	1	1	-1	-1	-1	1	1	1	0	0	0
$\vec{x}_8$	-1	1	1	1	0	0	0	-1	-1	-1	1	1	1
$\vec{x}_9$	-1	1	1	1	1	1	1	0	0	0	-1	-1	-1
$\vec{x}_{10}$	0	-1	0	1	-1	0	1	-1	0	1	-1	0	1
$\vec{x}_{11}$	0	-1	0	1	0-	1	-1	0	1	-1	0	1	-1
$\vec{x}_{12}$	0	-1	0	1	1	-1	0	1	-1	0	1	-1	0
$\vec{x}_{13}$	0	0	1	-1	-1	0	1	0	1	-1	1	-1	0
$\vec{x}_{14}$	0	0	1	-1	0	1	-1	1	-1	0	-1	0	1
$\vec{x}_{15}$	0	0	1	-1	1	-1	0	-1	0	1	0	1	-1
$\vec{x}_{16}$	0	1	-1	0	-1	0	1	1	-1	0	0	1	-1
$\vec{x}_{17}$	0	1	-1	0	0	1	-1	-1	0	1	1	-1	0
$\vec{x}_{18}$	0	1	-1	0	1	-1	0	0	1	-1	-1	0	1
$\vec{x}_{19}$	1	-1	1	0	-1	1	0	-1	1	0	-1	1	0
$\vec{x}_{20}$	1	-1	1	0	0	-1	1	0	-1	1	0	-1	1
$\vec{x}_{21}$	1	-1	1	0	1	0	-1	1	0	-1	1	0	-1
$\vec{x}_{22}$	1	0	-1	1	-1	1	0	0	-1	1	1	0	-1
$\vec{x}_{23}$	1	0	-1	1	0	-1	1	1	0	-1	-1	1	0
$\vec{x}_{24}$	1	0	-1	1	1	0	-1	-1	1	0	0	-1	1
$\vec{x}_{25}$	1	1	0	-1	-1	1	0	1	0	-1	0	-1	1
$\vec{x}_{26}$	1	1	0	-1	0	-1	1	-1	1	0	1	0	-1
$\vec{x}_{27}$	1	1	0	-1	1	0	-1	0	-1	1	-1	1	0

Main effect Confounded with Interaction	A	B	C	D	E	F	G	H	I	J	K	L	M
	BC	AC	AB	AB	AF	AE	AE	AI	AH	AH	AL	AK	AK
	BD	AD	AD	AC	AG	AE	AF	AJ	AJ	AI	AM	AL	AL
	CD	CD	BD	BC	BH	BI	BJ	BE	BF	BG	BG	BF	BG
	EF	EH	EI	EJ	BK	BL	BM	BK	BL	BM	BH	BI	BJ
	EG	EK	EM	EL	CI	CJ	CH	CG	CE	CF	CF	CG	CE
	FG	FI	FJ	FH	CM	CK	CL	CL	CM	CK	CJ	CH	CI
	HI	FL	FK	FM	DJ	DH	DI	DF	DJ	DE	DG	DE	DF
	HJ	GJ	GH	GI	DL	DM	DK	DM	DL	DL	DI	DJ	DH
	IJ	GM	GL	GK	FG	EG	EF	EK	EH	EH	EJ	EI	EI
	KL	HK	HL	HM	HK	HM	HL	FM	FL	FK	FJ	FI	FH
	KM	IL	IM	IK	IM	IL	IK	GL	GK	GM	GI	GH	GJ
	LM	JM	JK	JL	JL	JK	JM	IJ	HJ	HI	LM	KM	KL

Appendix C
TRIZ Relationship Matrix

The TRIZ relation matrix relates generalized performance parameters (Table 10.7) to generalized solution principles (Tables 10.8 and 10.9). Each column and row represent the generalized performance parameters, and the cell contents represent the suggested generalized solution principles. The rows represent "What should be improved" and the columns represent "What deteriorates due to the conflict."

	1	2	3	4	5	6	7	8	9	10	11	12	13	14	15	16	17	18	19	20
1			15 8 / 29 34		29 17 / 38 34		29 2 / 40 28		2 8 / 15 38	8 10 / 18 37	10 36 / 37 40	10 14 / 35 40	1 35 / 19 39	28 27 / 18 40	5 34 / 31 35		6 29 / 4 38	19 1 / 32	35 12 / 34 31	
2			10 1 / 29 35		35 30 / 13 7			5 35 / 14 2		8 10 / 19 35	13 29 / 10 18	13 10 / 29 14	26 39 / 1 40	28 2 / 10 27		2 27 / 19 6	28 19 / 32 22	35 19 / 32		18 19 / 28 1
3	15 8 / 29 34				15 17 / 4		7 17 / 4 35		13 4 / 8	17 10 / 4	1 8 / 35	1 8 / 10 29	1 8 / 15 34	8 35 / 29 34	19		10 15 / 19	32	8 35 / 24	
4		35 28 / 40 29					17 7 / 10 40		35 8 / 2 14		28 10	1 14 / 35	13 14 / 15 7	39 37 / 35	15 14 / 28 26		1 40 / 35	3 35 / 38 18	3 25	
5	2 17 / 29 4		14 15 / 18 4				7 14 / 17 4		29 30 / 4 34	19 30 / 35 2	10 15 / 36 28	5 34 / 29 4	11 2 / 13 39	3 15 / 40 14	6 3		2 15 / 16	15 32 / 19 13	19 32	
6		30 2 / 14 18		26 7 / 9 39							1 18 / 35 36	10 15 / 36 37	2 38	40		2 10 / 19 30	35 39 / 38			
7	2 26 / 29 40		1 7 / 35 4		1 7 / 4 17				29 4 / 38 34	15 35 / 36 37	6 35 / 36 37	1 15 / 29 4	28 10 / 1 39	9 14 / 15 7	6 35 / 4		34 39 / 10 18	10 13 / 2	35	
8	35 10 / 19 14		19 14 / 8	35 8 / 2 14							2 18 / 37	24 35 / 35	7 2 / 35	34 28 / 35 40	9 14 / 17 15	35 34 / 38	35 6 / 4			
9	2 28 / 13 38		13 14 / 8		29 30 / 34		7 29 / 34			13 28 / 15 19	6 18 / 38 40	35 15 / 18 34	28 33 / 1 18	8 3 / 26 14	3 19 / 35 5		28 30 / 36 2	10 13 / 19	8 15 / 35 38	
10	8 1 / 37 18	18 13 / 1 28	17 19 / 9 36	28 10	19 10 / 15	1 18 / 36 37	15 9 / 12 37	2 36 / 18 37	13 28 / 15 12		18 21 / 11	10 35 / 40 34	35 10 / 21	35 10 / 14 27	19 2		35 10 / 21		19 17 / 10	1 16 / 36 37
11	10 36 / 37 40	13 29 / 10 18	35 10 / 36	35 1 / 14 16	10 15 / 36 28	10 15 / 36 37	6 35 / 10	35 24	6 35 / 36	36 35 / 21			35 4 / 15 10	35 33 / 2 40	9 18 / 3 40		35 39 / 19 2	14 24	10 37	
12	8 10 / 29 40	15 10 / 26 3	29 34 / 5 4	13 14 / 10 7	5 34 / 4 10		14 4 / 15 22	7 2 / 35	35 15 / 34 18	35 10 / 37 40	34 15 / 10 14		33 1 / 18 4	30 14 / 10 40	14 26 / 9 25		22 14 / 19 32	13 15 / 32	2 6 / 34 14	
13	21 35 / 2 39	26 39 / 1 40	13 15 / 1 28	37	2 11 / 13	39	28 10 / 19 39	34 28 / 35 40	33 15 / 28 18	10 35 / 21 16	2 35 / 40	22 1 / 18 4		17 9 / 15	13 27 / 10 35	39 3 / 35 23	35 1 / 32	32 3 / 27 15	13 19	27 4 / 29 18
14	1 8 / 40 15	40 26 / 27 1	1 15 / 8 35	15 14 / 28 26	3 34 / 40 29	9 40 / 28	10 15 / 14 7	9 14 / 17 15	8 13 / 26 14	10 18 / 3 14	10 3 / 18 40	10 30 / 35 40	13 17 / 35		27 3 / 26		30 10 / 40	35 19	19 35 / 10	35
15	19 5 / 34 31		2 19 / 9		3 17 / 19		10 2 / 19 30		3 35 / 5	19 2 / 16	19 3 / 27	14 26 / 28 25	13 3 / 35	27 3 / 10			19 35 / 39	2 19 / 4 35	28 6 / 35 18	
16	6 27 / 19 16			1 40 / 35					35 34 / 38				39 3 / 35 23				19 18 / 36 40			
17	36 22 / 6 38	22 35 / 32	15 19 / 9	15 19 / 9	3 35 / 39 18	35 38	34 39 / 40 18	35 6 / 4	2 28 / 36 30	35 10 / 3 21	35 39 / 19 2	14 22 / 19 32	1 35 / 32	10 30 / 22 40	19 13 / 39	19 18 / 36 40		32 30 / 21 16	19 15 / 3 17	
18	19 1 / 32	2 35 / 32	19 32 / 16		19 32 / 26		2 13 / 10		10 13 / 19	26 19 / 6		32 30	32 3 / 27	35 19	2 19 / 6		32 35 / 19		32 1 / 19	32 35 / 1 15
19	12 18 / 28 31		12 28		15 19 / 25		35 13 / 18		8 35	16 26 / 21 2	23 14 / 25	12 2 / 29	19 13 / 17 24	5 19 / 9 35	28 35 / 6 18		19 24 / 3 14	2 15 / 19		
20		19 9 / 6 27									36 37		27 4 / 29 18	35				19 2 / 35 32		

	21	22	23	24	25	26	27	28	29	30	31	32	33	34	35	36	37	38	39
1	12 36 18 31	6 2 34 19	5 35 3 31	10 24 35	10 35 20 28	3 26 18 31	3 11 1 27	28 27 35 26	28 35 26 18	22 21 18 27	22 35 31 39	27 28 1 36	35 3 2 24	2 27 28 11	29 5 15 8	26 30 36 34	28 29 26 32	26 35 18 19	35 3 24 37
2	15 19 18 22	18 19 28 15	5 8 13 30	10 15 35	2 19 35 26	19 6 18 26	10 28 8 3	18 26 28	10 1 35 17	2 19 22 37	35 22 1 39	28 1 9	6 13 1 32	2 27 28 11	19 15 29	1 10 26 39	25 28 17 15	2 26 35	1 28 15 35
3	1 35	7 2 35 39	4 29 23 10	1 24	15 2 29	29 35	10 14 29 40	28 32 4	10 28 29 37	1 15 17 24	17 15	1 29 17	15 29 35 4	1 28 10	14 15 1 16	1 19 26 24	35 1 16	17 24 26 16	14 4 28 29
4	12 8	6 28	10 28 24 35	24 26	30 29 14		15 29 28 3	32 28 3	2 32 10	1 18		15 17 27	2 25	3	1 35	1 26	26		30 14 7 26
5	19 10 32 18	15 17 30 26	10 35 2 39	30 26	26 4	29 30 6 13	29 9	26 28 32 3	2 32	22 33 28 1	17 2 18 39	13 1 26 24	15 17 13 16	15 13 10 1	15 30	14 1 13	2 36 26 18	14 30 28 23	10 26 34 2
6	17 32	17 7 30	10 14 18 39	30 16	10 35 4 18	2 18 40 4	32 35 40 4	26 28 32 3	2 29 18 36	27 2 39 35	22 1 40	40 16	16 4	16	15 16	1 18 36	2 35 30 18	23	10 15 17 7
7	35 6 13 18	7 15 13 16	36 39 34 10	2 22	2 6 34 10	29 30 7	14 1 40 11	25 26 28	25 28 2 16	22 21 27 35	17 2 40 1	29 1 40	15 13 30 12	10	15 29	26 1	29 26 4	35 34 16 24	10 6 2 34
8	30 6		10 39 35 34		35 16 32 18	35 3	2 35 16		35 10 25	34 39 19 27	30 18 35 4	35		1		1 31	2 17 26		35 37 10 2
9	19 35 38 2	14 20 19 35	10 13 28 38	13 26		10 19 29 38	11 35 27 28	28 32 1 24	10 28 32 25	1 28 35 23	2 24 32 21	35 13 8 1	32 28 13 12	34 2 28 27	15 10 26	10 28 4 34	3 34 27 16	10 18	
10	19 35 18 37	14 15	8 35 40 5		10 37 36	14 29 18 36	3 35 13 21	35 10 23 24	28 29 37 36	1 35 40 18	13 3 36 24	15 37 18 1	1 28 3 25	15 1 11	15 17 18 20	26 35 10 18	36 37 10 19	2 35	3 28 35 37
11	10 35 14	2 36 25	10 36 37		37 36 4	10 14 36	10 13 19 35	6 28 25	3 35	22 2 37	2 33 27 18	1 35 16	11	2	35	19 1 35	2 36 37	35 24	10 14 35 37
12	4 6 2	14	35 29 3 5		14 10 34 17	36 22	10 40 16	28 32 1	32 30 40	22 1 2 35	35 1	1 32 17 28	32 15 26	2 13 1	1 15 29	16 29 1 28	15 13 39	15 1 32	17 26 34 10
13	32 35 27 31	14 2 39 6	2 14 30 40		35 27	15 32 35		13	18	35 23	35 40 27 39	35 19	32 35 30	2 35 10 16	35 30 34 2	2 35 22 26	35 22 39 23	1 8 35	23 35 40 3
14	10 26 35 28	35	35 28 31 40		29 3 28 10	29 10 27	11 3	3 27 16	3 27	18 35 37 1	15 35 22 2	11 3 10 32	32 40 28 2	27 11 3	15 3 32	2 13 28	27 3 15 40	15	29 35 10 14
15	19 10 35 38		28 27 3 18	10	20 10 28 18	3 35 10 40	11 2 13	3	3 27 16 40	22 15 33 28	21 39 16 22	27 1 4	12 27	29 10 27	1 35 13	10 4 29 15	19 29 39 35	6 10	35 17 14 19
16	16		27 16 18 38	10	28 20 10 16	3 35 31	34 27 6 40	10 26 24		17 1 40 33	22	35 10	1	1	2		25 34 6 35	1	20 10 16 38
17	2 14 17 25	21 17 35 38	21 36 29 31		35 28 21 18	19 35 30 39	32 19 3 10	32 19 24	24	22 33 35 2	22 35 2 24	26 27	26 27	4 10 16	2 18 27	2 17 16	3 27 35 31	23 2 19 16	15 28 35
18	32	19 16 1 6	13 1	1 6	19 1 26 17	1 19		11 15 32	3 32	15 19	35 19 32 39	19 35 28 26	28 26 19	15 17 13 16	15 1 19	6 32 13	32 15	2 26 10	2 25 16
19	6 19 37 18	12 22 15 24	35 24 18 5		35 38 19 18	34 23 16 18	19 21 11 27	3 1 32		1 35 6 27	2 35 6	28 26 30	19 35	1 15 17 28	15 17 13 16	2 29 27 28	35 38	32 2	12 28 35
20			28 27 18 31			3 35 31	10 36 23			10 2 22 37	19 22 18	1 4					19 35 16 25	1 6	

	1	2	3	4	5	6	7	8	9	10	11	12	13	14	15	16	17	18	19	20
21	8 36 38 31	19 26 17 27	1 10 35 37		19 38 13 38	17 32 38	35 6 25	30 6 25	15 35 2	26 2 36 35	22 10 35	29 14 2 40	35 32 15 31	26 10 28	19 35 10 38	16	2 14 17 25	16 6 19	16 6 19 37	
22	15 6 19 28	19 6 18 9	7 2 6 13	6 38 7	15 26 17 30	17 7 30 18	7 18 23	7	16 35 38	36 38			14 2 39 6	26			19 38 7	1 13 32 15		
23	35 6 23 40	35 6 22 32	14 29 10 39	10 28 24	35 2 10 31	10 18 39 31	1 29 30 36	3 39 18 31	10 13 28 38	14 15 18 40	3 36 37 10	29 35 3 5	2 14 30 40	35 28 31 40	28 27 3 18	27 16 18 38	21 36 39 31	1 6 13	35 18 24 5	28 27 12 31
24	10 24 35	10 35 5	1 26	26	30 26	30 16		2 22	26 32						10	10		19		
25	10 20 37 35	10 20 26 5	15 2 29	30 24 14 5	26 4 5 16	10 35 17 4	2 5 34 10	35 16 32 18		10 37 36 5	37 36 4	4 10 34 17	35 3 22 5	29 3 28 18	20 10 28 18	28 20 10 16	35 29 21 18	1 19 26 17	35 38 19 18	1
26	35 6 18 31	27 26 18 35	29 14 35 18		15 14 29	2 18 40 4	15 20 29		35 29 34 28	35 14 3	10 36 14 3	35 14	15 2 17 40	14 35 34 10	3 35 10 40	3 35 31	3 17 30 39		34 29 16 18	3 35 31
27	3 8 10 40	3 10 8 28	15 9 14 4	15 29 28 11	17 10 14 16	32 35 40 4	3 10 14 24	2 35 24	21 35 11 28	8 28 10 3	10 24 35 19	35 1 16 11		11 28	2 35 3 25	34 27 6 40	3 35 10	11 32 13	21 11 27 19	36 23
28	32 35 26 28	28 35 25 26	28 26 5 16	32 28 3 16	26 28 32 3	26 28 32 3	32 13 6		28 13 32 24	32 2	6 28 32	6 28 32	32 35 13	28 6 32	28 6 32	10 26 24	6 19 28 24	6 1 32	3 6 32	
29	28 32 13 18	28 35 27 9	10 28 29 37	2 32 10	28 33 29 32	2 29 18 36	32 28 2	25 10 35	10 28 32	28 19 34 36	3 35	32 30 40	30 18	3 27	3 27 40		19 26	3 32	32 2	
30	22 21 27 39	2 22 13 24	17 1 39 4	1 18	22 1 33 28	27 2 39 35	22 23 37 35	34 39 19 27	21 22 35 28	13 35 39 18	22 2 37	35 1	35 24 30 18	18 35 37 1	22 15 33 28	17 1 40 33	22 33 35 2	1 19 32 13	1 24 6 27	10 2 22 37
31	19 22 15 39	35 22 1 39	17 15 16 22		17 2 18 39	22 1 40	17 2 40	30 18 35 4	35 28 3 23	35 28 1 40	2 33 27 18	35 1	35 40 27 39	15 35 22 2	21 39 16 22	22	22 35 2 24	19 24 39 32	2 35 6	19 22 18
32	28 29 15 16	1 27 36 13	1 29 13 17	15 17 27	13 1 26 12	16 40	13 29 1 40	35	35 13 8 1	35 12	35 19 1 37	1 28 13 27	11 13 1	1 3 10 32	27 1 4	35 16	27 26 18	28 24 27 1	28 26 27 1	1 4
33	25 2 13 15	6 13 1 25	1 17 13 12		1 17 13 16	18 16 15 39	1 16 35 15	4 18 39 31	18 13 34	28 13 35	2 32 12	15 34 29 28	32 35 30	32 40 3 28	29 3 8 25	1 16 25	26 27 13	13 17 1 24	1 13 24	
34	2 27 35 11	2 27 35 11	1 28 10 25	3 18 31	15 13 32	16 25	25 2 35 11	1	34 9	1 11 10	13	1 13 2 4	2 35	11 1 2 9	11 29 28 27	1	4 10	15 1 13	15 1 28 16	
35	1 6 15 8	19 15 29 16	35 1 29 2	1 35 16	35 30 29 7	15 16	15 35 29		35 10 14	15 17 20	35 16	15 37 1 8	35 30 14	35 3 32 6	13 1 35	2 16	27 2 3 35	6 22 26 1	19 35 29 13	
36	26 30 34 36	2 26 35 39	1 19 26 24	26	14 1 13 16	6 36	34 26 6	1 16	34 10 28	26 16	19 1 35	29 13 28 15	2 22 17 19	2 13 28	10 4 28 15		2 17 13	24 17 13	27 2 29 28	
37	27 26 28 13	6 13 28 1	16 17 26 24	26	2 13 18 17	2 39 30 16	29 1 4 16	2 18 26 31	3 4 16 35	36 28 40 19	35 36 37 32	27 13 1 39	11 22 39 30	27 3 15 28	19 29 39 25	25 34 6 35	3 27 35 16	2 24 26	35 38	19 35 16
38	28 26 18 35	28 26 35 10	14 13 28 17	23	17 14 13		35 13 16		28 10	2 35	13 35	15 32 1 13	18 1	25 13	6 9		26 2 19	8 32 19	2 32 13	
39	35 26 24 37	28 27 15 3	18 4 28 38	30 7 14 26	10 26 34 31	10 35 17 7	2 6 34 10	35 37 10 2		28 15 10 36	10 37 14	14 10 34 40	35 3 22 39	29 28 10 18	35 10 2 18	20 10 16 38	35 21 28 10	26 17 19 1	35 10 38 19	1

	21	22	23	24	25	26	27	28	29	30	31	32	33	34	35	36	37	38	39
21		10 35 / 38	28 27 / 18 38	10 19	35 20 / 10 6	4 34 / 19	19 24 / 26 31	32 15 / 2	32 2	19 22 / 31 2	2 35 / 18	26 10 / 34	26 35 / 10	35 2 / 10 34	19 17 / 34	20 19 / 30 34	19 35 / 16	28 2 / 17	28 35 / 34
22	3 38		35 27 / 2 37	19 10	10 18 / 32 7	7 18 / 25	11 10 / 35	32		21 22 / 35 2	21 35 / 2 22		35 32 / 1	2 19		7 23	35 3 / 15 23	2	28 10 / 29 35
23	28 27 / 18 38	35 27 / 2 31			15 18 / 35 10	6 3 10 / 24	10 29 / 39 35	16 34 / 31 28	35 10 / 24 31	33 22 / 30 40	10 1 / 34 29	15 34 / 33	32 28 / 2 24	2 35 / 34 27	15 10 / 2	35 10 / 28 24	35 18 / 10 13	35 10 / 18	28 35 / 10 23
24	10 19	19 10			24 26 / 28 32	24 28 / 35	10 28 / 23			22 10 / 1	10 21 / 22	32	27 22				35 33	35	13 23 / 15
25	35 20 / 10 6	10 5 / 18 32	35 18 / 10 39	24 26 / 28 32		35 38 / 18 16	10 30 / 4	24 34 / 28 32	24 26 / 28 18	35 18 / 34	35 22 / 18 39	35 28 / 34 4	4 28 / 10 34	32 1	35 28	6 29	18 28 / 32 10	24 28 / 35 30	
26	35	7 18 / 25	6 3 / 10 24	24 28 / 35	35 38 / 18 16		18 3 / 28 40	3 2 / 28	33 30	35 33 / 29 31	3 35 / 40 39	29 1 / 35 27	35 29 / 10 25	2 32 / 10 25	15 3 / 29	3 13 / 27 10	3 27 / 29 18	8 35	13 29 / 3 27
27	21 11 / 26 31	10 11 / 35	10 35 / 29 39	10 28	10 30 / 4	21 28 / 40 3		32 3 / 11 23	11 32 / 1	27 35 / 2 40	35 2 / 40 26		27 17 / 40	1 11	13 35 / 8 24	13 35 / 1	27 40 / 28	11 13 / 27	1 35 / 29 38
28	3 6 / 32	26 32 / 27	10 16 / 31 28		24 34 / 28 32	2 6 / 32	5 11 / 1 23			28 24 / 22 26	3 33 / 39 10	6 35 / 25 18	1 13 / 17 34	1 32 / 13 11	13 35 / 2	27 35 / 10 34	26 24 / 32 28	28 2 / 10 34	10 34 / 28 32
29	32 2	13 32 / 2	35 31 / 10 24		32 26 / 28 18	32 30	11 32 / 1			26 28 / 10 36	4 17 / 34 26		1 32 / 35 23	25 10		26 2 / 18	22 19	26 28 / 18 23	10 18 / 32 39
30	19 22 / 31 2	21 22 / 35 2	33 22 / 19 40	22 10 / 2	35 18 / 34	35 33 / 29 31	27 24 / 2 40	28 33 / 23 26	26 28 / 10 18			24 35 / 2	2 25 / 28 39	35 10 / 2	35 11 / 22 31	22 19 / 29 40	22 19 / 29 40	33 3 / 34	22 35 / 13 24
31	2 35 / 18	21 35 / 22 2	10 1 / 34	10 21 / 29	1 22	3 24 / 39 1	24 2 / 40 39	3 33 / 26	4 17 / 34 26							19 1 / 31	2 21 / 27 1	2	22 35 / 18 39
32	27 1 / 12 24	19 35 / 33	15 34 / 18 16	32 24	35 28 / 34 4	35 23 / 1 24		1 35 / 12 18		24 2			2 5 / 13 16	35 1 / 11 9	2 13 / 15	27 26 / 1	6 28 / 11 1	8 28 1	35 1 / 10 28
33	35 34 / 2 10	2 19 / 13	28 32 / 2 24	4 10 / 27 22	4 28 / 10 34	12 35	17 27 / 8 40	25 13 / 2 34	1 32 / 35 23	2 25 / 28 39		2 5 / 12		12 26 / 1 32	15 34 / 1 16	32 25 / 12 17		1 34 / 12 3	15 1 / 28
34	15 10 / 32 2	15 1 / 32 19	2 35 / 34 27		32 1 / 10 25	2 28 / 10 25	11 10 / 1 16	10 2 / 13	25 10	35 10 / 2 16			1 35 / 11 10	1 12 / 26 15	7 1 / 4 16	35 1 / 13 11		34 35 / 7 13	1 32 / 10
35	19 1 / 29	18 15 / 1	15 10 / 2 13		35 28	3 35 / 15	35 13 / 8 24	35 5 / 1 10		35 11 / 32 31			1 13 / 31	15 34 / 1 16	1 16 / 7 4	15 29 / 37 28	1	27 34 / 35	35 28 / 6 37
36	20 19 / 30 34	10 35 / 13 2	35 10 / 28 29		6 29	13 3 / 27 10	13 35 / 1	2 26 / 10 34	26 24 / 32	22 19 / 29 40	19 1	27 26 / 1 13	27 9 / 26 24	1 13	29 15 / 28 37		15 10 / 37 28	15 1 / 24	12 17 / 28
37	19 1 / 16 10	35 3 / 15 19	1 18 / 10 24	35 33 / 27 22	18 28 / 32 9	3 27 / 29 18	27 40 / 28 8	26 24 / 32 28		22 19 / 29 28	2 21	5 28 / 11 29	2 5	12 26	1 15	15 10 / 37 28		34 21	35 18
38	28 2 / 27	23 28	35 33 / 18 5	35 33	24 28 / 35 30	35 13	11 27 / 32	28 26 / 10 34	28 26 / 18 23	2 33	2	1 26 / 13	1 12 / 34 3	1 35 / 13	1 35	27 4 / 1 35	15 24 / 10	34 27 / 25	5 12 / 35 26
39	35 20 / 10	28 10 / 29 35	28 10 / 35 23	13 15 / 23		35 38	1 35 / 10 38	1 10 / 34 28	32 1 / 18 10	22 35 / 13 24	35 22 / 18 39	35 28 / 2 24	1 28 / 7 19	1 32 / 10 25	1 35 / 28 37	12 17 / 28 24	35 18 / 27 2	5 12 / 35 26	

Appendix D
Eco-Indicator-95 Environment Assessment

The Eco-Indicator environmental assessment is a evaluation method developed by PrÈ Associates on the initiative of the Ministry of Housing, Spatial Planning and the Environment in the Netherlands, called the Eco-indicator 95 (Goedkoop et al., 1996). The Eco-indicator system provides weightings by mass for materials, treatment processes, transport processes, energy generation processes, and disposal scenarios. As shown in Figure D.1, the weightings themselves are based upon a valuation of damage to public health and to the ecosystem through contribution to several effects, such as ozone layer depletion, smog, etc. The actual effects considered, and from what impact sources, are shown in Table D.1.

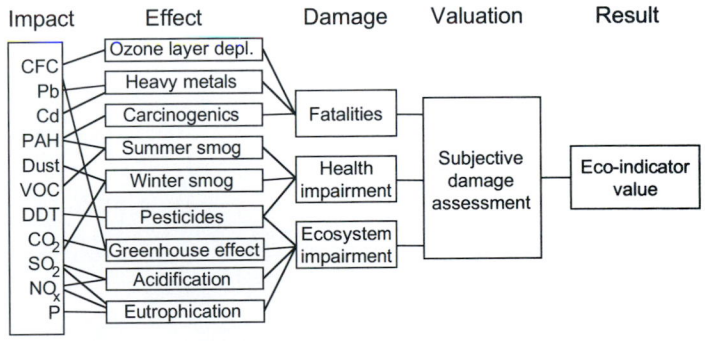

Figure D.1 Development of weightings for the EcoIndicator95.

Table D.1 Development of weightings for the EcoIndicator95.

Environmental effect	Weighting Factor	Criteria
Greenhouse effect	2.5	0.1YC rise every 10 years, 5% ecosystem degredation
Ozone Layer Depletion	100	Probability of 1 fatality per year per million inhabitants
Acidification	10	5% ecosystem degredation
Eutrophication	5	Rivers and lakes, degradation of an unknown number of aquatic ecosystems (5% degradation)
Summer smog	2.5	Occurrence of smog periods, health complaints, particularly amongst asthma patients and the elderly, prevention of agricultural damage
Winter smog	5	Occurrence of smog periods, health complaints, particularly amongst asthma patients and the elderly
Pesticides	25	5% ecosystem degredation
Airborne heavy metals	5	Lead content in children's blood, reduced life expectancy and learning performance in an unknown number of people
Waterborne heavy metals	5	Cadmium content in rivers, ultimately also impacts on people (see airborne)
Carcinogenic substances	10	Probability of 1 fatality per year per million people

Table D.2 Eco-Indicator 95 worksheet, page 1

Production			
Materials, processing, transport and extra energy			
Material or Process	Amount	Indicator	Result
		Total	

Production			
Materials, processing, transport and extra energy			
Material or Process	Amount	Indicator	Result
		Total	

Use			
Transport, energy and any auxiliary materials			
Process	Amount	Indicator	Result
		Total	

Use			
Transport, energy and any auxiliary materials			
Process	Amount	Indicator	Result
		Total	

Disposal			
Disposal processes per type of material			
Material and Type of Processing	Amount	Indicator	Result
		Total	
Total of All Phases			

Disposal			
Disposal processes per type of material			
Material and Type of Processing	Amount	Indicator	Result
		Total	
Total of All Phases			

Table D.2, contd: Eco-Indicator 95 worksheet, page 2

Product or component	Project
Date	Author

Notes and Conclusions

Production
Materials, processing, transport and extra energy

Material or Process	Amount	Indicator	Result
			Total

Use
Transport, energy and any auxiliary materials

Process	Amount	Indicator	Result
			Total

Disposal
Disposal processes per type of material

Material and Type of Processing	Amount	Indicator	Result
			Total

Table D.3 Eco-indicator 95 Impact Weightings

Production of Metals (in millipoints / kg)

	Indicator	Descripton
Secondary Aluminum	1.8	made completely from secondary material (more difficult to obtain)
Aluminum	18	containing an average of 20% recycled material
Copper, primary	85	primary electrolytic copper from relatively modern American factories
Copper, 60% primary	60	normal proportion secondary and primary copper
Secondary copper	23	100% secondary copper (more difficult to obtain)
Other non-ferrous metals	50-200	estimate for zinc, brass, chromium, nickel etc.; lackof data
Stainless steel	17	sheet mateial, grade 18-8, 1mm thickness
Secondary steel	1.3	block material with average 100% scrap
Steel	4.1	block material with average 20% scrap
Sheet steel	4.3	cold-rolled sheet with average 20% scrap

Production of Steel (in millipoints)

	Indicator	Descripton
Bending steel	0.0021	one sheet of 1mm over width of 1 meter; straight angle
Bending stainless steel	0.0029	one sheet of 1mm over width of 1 meter; straight angle
Cutting steel	0.0015	one sheet of 1mm over width of 1 meter
Cuttin stainless steel	0.0022	one sheet of 1mm over width of 1 meter
Pressing and deep-drawing	0.58	per kilo deformed steel (do not include non-deformed parts)
Rolling (cold)	0.46	per pass, per m^2
Spot welding	0.0074	per weld of 7mm diameter, sheet thickness 2mm
Machining	0.42	per kilo machined materiaI! (turning, milling, boring)
Machining	0.0033	per cm^3 machined materiaI! (turning, milling, boring)
Hot-galvanizing	17	per m^2, 10 micrometres, double-sided; data fairly unreliable
Electrolytic galvanizing	22	per m^2, 2.5 micrometres, double-sided; data fairly unreliable
Electroplating (chrome)	70	per m^2, 1 micrometres thick, double-sided; data fairly unreliable

Processing of aluminum (in millipoints)

	Indicator	Descripton
Blanking and cutting	0.00092	one sheet of 1mm over width of 1 meter
Bending	0.0012	one sheet of 1mm over width of 1 meter
Rolling (cold)	0.28	per pass, per m^2
Spot - welding	0.068	per weld of 7mm diameter, sheet thickness 2mm
Machining	0.12	per kilo machined mateial! (turning, milling, boring)
Machining	0.00033	per cm^3 machined materiaI! (turning, milling, boring)
Extrusion	2.0	per kilogram

Production of plastic granulate (in millipoints / kg)

	Indicator	Descripton and explanation score
ABS	9.3	high energy input for production, therefore high emission output
HDPE	2.9	relatively simple production process
LDPE	3.8	score possibly flattered by lack of CFC emission
PA	13	high energy input for production, therefore high emission output
PC	13	high energy input for production, therefore high emission output
PET amorphous	7.1	used for fibers and foil
PET bottle grade	7.4	used for bottles
PP	3.3	relatively simple production process
PPE / PS	5.8	a commonly used blend, identical to PPO/PS
PS rigid foam	13	block of foam with pentane as blowing agent (causes smog)
PS high impact (HIPS)	8.3	High-impact polystyrene
PUR flexible block foam	5.9	for furniture, bedding, clothing, leisure goods (water blown)
PUR rigid foam	8.4	In white goods, insulation, construction materials (pentane blown)
PUR semi rigid foam	6.9	used in dashboards (pentane blown)
PUR energy absorbing	8.7	bumpers (pentane blown)
PVC	4.2	calculated as pure PVC, without adding stabilizers or plasticizers
PVDC	9.1	for thin coatings, calculated without stabilizers and other additives

Table D.3, contd: Eco-indicator 95 Impact Weightings

Processing of Plastics (in millipoints)

	Indicator	Descripton
Injection molding general	.53	per kilo material, may also be used as estimate forextrusion
Injection molding PVC& PC	1.1	per kilo material, may also be used as estimate forextrusion
RIM, PUR	.30	per kilo material
Extrusion blowing PE	.72	per kilo material, for bottles
Vacuum forming	.23	per kilo
Vacuum pressure forming	.16	per kilo
Calandering of PVC	.43	per kilo
Foil blowing PE	.030	per kilo m², thin foil (for bags)
Ultrasonic welding	.0025	per metre weld length
Machining	.00016	per cm³ machined material

Production of rubbers and elastomers (in millipoints in kg)

	Indicator	Descripton
Raw natural rubber	1.5	dried and baled natural from latex, for vulcanization
Natural rubber product	4.3	vulcanized with 28% carbon black; used for truck tires
SBR product	5.6	vulcanized with 26% carbon black; used for car tires
EPDM product	4.1	vulcanized with 32% carbon black; used for profiles

Production of other materials (in millipoints per kg)

	Indicator	Descripton
Glass	2.1	57% secondary glass
Glass wool and glass fibre	2.1	for isolation and reinforcement
Rockwool	4.3	score is largely determined by carcinogenic substances
Ceramics	0.47	simple applications, e.g. sanitary fittings etc.
Cellulose board	3.4	this material is particularly used in dashboards
Paper	3.3	chlorine-free bleaching, normal quality
Recycled Paper	1.5	unbleached, 100% waste paper
Wood	0.74	wood from Europe, cut into planks, without preservatives
Cardboard	1.4	corrugated cardboard made from 75% waste paper

Production of energy (in millipoints)

	Indicator	Descripton
Electricity high voltage	.57	per kWh, for industrial use
Electricity low voltage	.67	per kWh, for industrial use
Heat from gas (MJ)	.063	per MJ heat
Heat from oil (MJ)	.15	per MJ heat
Mechanical (Diesel, MJ)	.17	per MJ mechanical energy from a diesel engine

Transport (in millipoints)

	Indicator	Descripton
Truck(28 ton)	.34	per ton kilometer, 60% loading, European average
Truck (75m³)	.13	per m³ km, 60% loading, European average
Train	.043	per ton kilometer, European average for diesel and electric traction
Container ship	.056	per ton kilometer, fast ship, with relatively high fuel consumption
Aircraft (continental)	1.7	per kg!, with continental flights the distance is not relevant
Aircraft (intercontinental)	.81	per ton kilometer

Table D.3, contd: Eco-indicator 95 Impact Weightings

Self-made Indicators for Components (in millipoints)

	Indicator	Descripton

Waste Processing and Recycling (in millipoints per kg)

Fraction	Indicator	Notes

Incineration (in a modern waste incinerator with heat recovery and flue-gas treatment)

Fraction	Indicator	Notes
Glass	0.89	almost inert material on incineration
Ceramics	0.020	almost inert material on incineration
Plastics and Rubber	1.8	plastics contain heavy metals, but also have a high energy yield
PVC	6.9	PVC contains heavy metals and has a relatively low energy yield
Paper and Cardboard	0.56	heavy metals (ink) are dominant, energy yield is relatively high
Steel and Iron	1.8	70% is recovered from slag, particularly larger pieces
Aluminium*	-7	30% is recovered from slag, not valid for very thin materials
Copper*	-16	30% is recovered from slag, not valid for very thin materials

Landfill (in a modern landfill site with percolation water treatment and dense base)

Fraction	Indicator	Notes
Glass	0	almost inert material on a landfill
Ceramics	0.027	almost inert material on a landfill
Plastics and Rubber	0.035	0.1% of all heavy metals released
PVC	0.077	0.1% of all heavy metals released
Paper and Cardboard	0.16	10% of all heavy metals (mainly in ink) released
Steel and Iron	0.80	small proportion (ca. 1%) of heavy metals released
Aluminium*	0.003	mainly due to contaminants
Copper*	4.6	0.1% of copper released

Recycling (these values cannot be used for recycling of secondary material)

Fraction	Total:	process	avoided emissions*	Total score is split into a score for the recycling process and avoided product
Glass	-1.5	0.5	-2	recycling avoids glass production
Ceramics				no usefull recycling possible
Plastics (PE ad PP)	-0.46	2.2	-2.66	if not mixed with other plastics
Engineering plastics*	-3 - 9.5	2.2	-5.7 - -11.7	avoided emission is 90% of production
PVC	-1.6	2.2	-3.8	if not mixed with other plastics
Paper and Cardboard	-1.8	0.2	-2	recycling avoids pulp production
Steel and Iron	-2.9	0.8	-3.7	recycling avoids pig iron production
Aluminium*	-13	2	-15	85% aluminium recovery
Copper*	-35	22	-58	96% copper recovery

Municipal waste (processing of waste by average Dutch municipality)

Fraction	Indicator	Notes
Glass	0.35	63% landfilled, 37% incinerated
Ceramics	0.041	63% landfilled, 37% incinerated
Plastics and Rubber	0.69	63% landfilled, 37% incinerated
PVC	2.6	63% landfilled, 37% incinerated
Paper and Cardboard	0.33	63% landfilled, 37% incinerated
Steel and Iron	1.2	63% landfilled, 37% incinerated, from which 70% is recovered
Aluminium*	-3	63% landfilled, 37% incinerated (30% recovery)
Copper*	-2.6	63% landfilled, 37% incinerated (30% recovery)

Household Waste (same, but with average separation by consumer, example: paper and glass)

Fraction	Indicator	Notes
Glass	-0.80	61% separated and recycled, the rest is municipal waste (above)
Ceramics	0.041	almost all processed as municipal waste
Plastics and Rubber	.66	2% separated and recycled, the rest is municipal waste
PVC	2.5	2% separated and recycled, the rest is municipal waste
Paper and Cardboard	-0.43	35% separated and recycled, the rest is municipal waste
Steel and Iron	-0.28	36% separated and recycled, the rest is municipal waste

Index

D

H

I